COMPUTER GRAPHICS THROUGH OPENGL®

2nd EDITION

From Theory to Experiments

COMPUTER GRAPHICS THROUGH OPENGL®

2nd EDITION

From Theory to Experiments

Sumanta Guha
Asian Institute of Technology, Thailand

CRC Press
Taylor & Francis Group
Boca Raton London New York

CRC Press is an imprint of the
Taylor & Francis Group, an **informa** business

AN A K PETERS BOOK

Cover Designed by Somying Pongpimol.

CRC Press
Taylor & Francis Group
6000 Broken Sound Parkway NW, Suite 300
Boca Raton, FL 33487-2742

Printed on acid-free paper
Version Date: 20140505

Printed and bound in India by Replika Press Pvt. Ltd.

International Standard Book Number-13: 978-1-4822-5839-4 (Hardback)

Library of Congress Cataloging-in-Publication Data

Guha, Sumanta.
Computer graphics through openGL : from theory to experiments / author, Sumanta Guha. -- Second edition.
pages cm
Includes bibliographical references and index.
ISBN 978-1-4822-5839-4 (hardback)
1. Computer graphics. 2. Microcomputers--Programming. 3. OpenGL. I. Title.

T385.G85 2014
006.6'633--dc23
2014016287

Visit the Taylor & Francis Web site at
http://www.taylorandfrancis.com

and the CRC Press Web site at
http://www.crcpress.com

**To my late parents Santa and Utpal Chandra
and to Kamaladi**

Contents

Preface

Welcome to the second edition of *Computer Graphics Through OpenGL: From Theory to Experiments*! The first edition appeared in late 2010. In the nearly four years since, I have been fortunate enough to have received much thoughtful and, mostly, positive feedback. Happily, too, there was a fair bit of reassurance that my way of doing things, somewhat different from my peers', was on the right track. And, of course, the field of computer graphics as always has been evolving rapidly, of a particular impact being the maturing of the fourth generation of OpenGL. The upshot was that about a year and a half ago I began working on a new edition and am glad now that the finished text is in your hands. Let's get to the facts.

About the Book

This is an introductory textbook on computer graphics with equal emphasis on theory and practice. The programming language used is C++, with OpenGL as the graphics API, which means calls are made to the OpenGL library from C++ programs. OpenGL is taught from scratch.

After Chapters 1-14 – the undergraduate core of the book – the reader will have a good grasp of the concepts underpinning 3D computer graphics, as well as an ability to code fairly sophisticated 3D scenes and animation, including games and movies. With, additionally, Chapters 20-21, which can, in fact, be read following Chapter 13, she will have command over fourth-generation OpenGL, particularly version 4.3. Chapters 15-19, though advanced, but still mainstream, could be selected topics for an undergraduate course or part of a second course.

The book has been written to be used as a textbook for a first college course, as well as for self-study.

Specs

This book, comprising 21 chapters, comes with approximately 170 programs, 250 experiments based on these programs, 650 exercises, including theory and programming exercises and programming projects, 100 worked examples, and 600 four-color illustrations. The book was typeset using LaTeX and figures drawn in Adobe Illustrator.

New in the Second Edition

- 30 more programs, 50 more experiments, 50 more exercises
- Vertex buffer objects
- Vertex array objects
- Occlusion culling
- Occlusion queries and conditional rendering
- Texture matrices
- Multitexturing and texture combining
- Multisampling
- Point sprites
- Image and pixel manipulation
- Pixel buffer objects
- Shadow mapping
- OpenGL 4.3, shaders and the programmable pipeline:
  - Complete coverage over two chapters
  - OpenGL Shading Language (GLSL)
  - Vertex, fragment, tessellation and geometry shaders
  - From basic methods, such as animation, lighting and textures, to advanced topics, including instanced rendering, shader subroutines, transform feedback, texture buffer objects, several others
  - 19 example programs

Pedagogical Approach

Code and theory have been intertwined as far as possible in what may be called a discuss-experiment-repeat loop: often, following a theoretical discussion, the reader is asked to perform validating experiments (run code, that is); sometimes, too, the other way around, an experiment is followed by an explanation of what is observed. It's kind of like discovering physics.

Why use an API?
Needless to say, I am not a believer in an API-agnostic approach to teaching CG, where focus is on principles only, with no programming practice.

Undergrads, typically, love to code and make things happen, so there is little justification to denying the new student the joy of creating scenes, movies and games, not to mention the pride of achievement. And, why not leverage the way code and theory reinforce one another when teaching the subject, or learning on one's own, when one can? Would you want Physics 101 without a lab section?

Moreover, OpenGL is very well-designed and the learning curve short enough to fully integrate into a first CG course. And, it is supported on every OS platform with drivers for almost every graphics card on the market; so, in fact, OpenGL is there to use for anyone who cares to.

Note to student: Our pedagogical style means that for most parts of the book you want a computer handy to run experiments. So, if you are going to snuggle up with it at night, make it a threesome with a notebook.

Note to instructor: Lectures on most topics – both of the theory and programming practice – are best based around the book's experiments, as well as those you develop yourself. The *Experimenter* teaching resource makes this convenient. Slides, otherwise, are rarely necessary.

How to teach modern shader-based OpenGL?

Our point of view needs careful explanation as it is different from some of our peers'. Firstly, to push the physics analogy one more time, even though relativistic mechanics seems to rule the universe, in the classroom one might prefer doing classical physics before relativity theory.

Shaders, which are the programmable parts of the modern OpenGL pipeline, add great flexibility and power. But, so too, do they add a fair bit of complexity – even a cursory comparison of our very first program `square.cpp` from Chapter 2 with its equivalent in OpenGL 4.3, `squareShaderized.cpp` complemented with a vertex and a fragment shader in Chapter 20, should convince the reader of this.

Consider more carefully, say, a vertex shader. It must compute the position coordinates of a vertex, taking into account all transformations, both modelview – such as translation, rotation, scaling and viewing – and projection. In the classical fixed-function pipeline, the user can simply issue commands such as `glTranslatef()`, `glRotatef()`, etc., leaving to OpenGL actual computation of the transformed coordinates; not so for the programmable pipeline, where the reader must write herself all the needed matrix operations in the vertex shader.

We firmly believe that the new student is best served learning first how to transform objects according to an understanding of simply how a scene comes together *physically* (e.g., a ball falls to the ground, a robot arm bends at the elbow, etc.) with the help of ready-to-use commands like `glTranslatef()`, and, only later, the actual mathematics behind them.

Such consideration applies as well to other automatic services of the fixed-function pipeline which allow the student to focus on phenomena, disregarding *initially* implementation. For example, as an instructor, I would much prefer to teach first how diffuse light lends three-dimensionality, specular light highlights, and so on, gently motivating Phong's lighting equation, leaving OpenGL to grapple with its actual implementation, which is exactly what we do in Chapter 11.

In fact, we find an understanding of the fixed-function pipeline makes the subsequent learning of the programmable one easier because it's then clear exactly what the shaders should try to accomplish. For example, following the fixed-function groundwork in Chapter 11, writing shaders to implement Phong lighting, as we do in Chapter 20, is near trivial.

We take a similarly laissez-faire attitude to classical OpenGL syntax. So long as it eases the learning curve we'll put up with it. Take for example the following snippet from our very first program `square.cpp`:

```
glBegin(GL_POLYGON);
   glVertex3f(20.0, 20.0, 0.0);
   glVertex3f(80.0, 20.0, 0.0);
   glVertex3f(80.0, 80.0, 0.0);
   glVertex3f(20.0, 80.0, 0.0);
glEnd();
```

Does it not scream square – even though it's immediate mode and uses the discarded polygon primitive? So, we prefer this for our first lesson, avoiding thus the distraction of a vertex array and the call `glDrawArrays(GL_TRIANGLE_STRIP, 0, 4)`, as in the 4.3-program `squareShaderized.cpp`, our goal on Day 1 being a simple introduction of the synthetic camera model.

Of course, as we move along, we introduce each modern construct in its logical place, but with an eye always toward the overall learning process. For example, we introduce vertex arrays and their drawing commands in Chapter 3 on OpenGL gadgets, from then on making a point of using them, except for objects with few vertices when the overhead seems more distraction than convenience. Vertex buffer objects (VBOs) and vertex array objects (VAOs) are introduced in Chapter 3, as well, following logically vertex arrays; however, we counsel the reader against using them, until she gets to OpenGL 4.3, where they are mandatory, because they add a layer of coding complexity one can very well do without when learning fundamental concepts.

Does this kind of staggered introduction to modern OpenGL, with the old still around, not lead to bad practice? Not at all from our experience. When push comes to shove, how hard is to replace polygons with triangle strips? Or, for that matter, use VBOs and VAOs to store data? In fact, as we remarked earlier, grasp of the old motivates the step up to the new (there's virtue it seems then in retracing the path of the graybeards!).

So, practically, our code is backward-compatible OpenGL 4.3, which allows use of legacy syntax, for the first nineteen chapters. Then, Chapters 20-21, which together give a comprehensive coverage of OpenGL 4.3, use forward-compatible core OpenGL 4.3 (the strictest form).

The reader might note, as well, that OpenGL ES (Embedded Systems) 3.0, the latest OpenGL version for mobile devices, and WebGL, the emerging

3D standard supported by almost all the newer browsers, are syntax-wise very close to OpenGL 4.3, so assimilation of the latter means ability to code 3D graphics on multiple platforms.

On the other hand, there are millions of currently live applications written in legacy OpenGL, which are not going to be discarded or rewritten any time soon – the reason, in fact, for the Khronos Group to retain the compatibility version of the API – so familiarity with older syntax might well be useful for the intending professional.

Does our approach cost timewise? If the goal is OpenGL 4.3, then, yes, it does take a bit more time, but not much. Chapters 20-21 can be read after Chapter 13; in fact, they can be taught in parallel with Chapters 11-13. So, a one-semester course can perfectly well cover OpenGL 4.3. We discuss various possible learning sequences through the book later on in the preface.

Target Audience

- Students in a first university CG course, typically offered by a CS department at a junior/senior level (though, often, graduate students can take it for credit). This is the primary audience for which the book was written.
- Students in a second or advanced CG course, who may use the book as preparation or reference, depending on the goals. For example, the book would be a useful reference for a study of 3D design – particularly, Bézier, B-spline and NURBS theory – and of projective transformations and their applications to CG.
- Students in a non-traditional setting, e.g., studying alone or in a short course or an on-line program. The author has tried to be especially considerate of the reader on her own.
- Professional programmers, to use the book as a reference.

Prerequisites

Zero knowledge of computer graphics is presumed. However, the student is expected to know the following:

(1) Basic C++ programming. There is no need to be an expert programmer. The C++ program serves mainly as an environment for the OpenGL calls, so there's rarely need for fancy footwork in the C++ part itself.

(2) Basic math. This includes coordinate geometry, trigonometry and linear algebra, all at college first-course level (or, even strong high school in some cases). For intended readers of the book who may be unsure of their math preparation, we have a self-test in Appendix B, with solutions in Appendix C. The test should tell exactly how ready you are and where the weaknesses are.

Resources

The following are available through the book's website `www.sumantaguha.com`:

- Program source code – which runs on Windows, Mac OS and Linux platforms. The programs are arranged chapter-wise in the top-level folder `ExperimenterSource`.
- Guide to installing OpenGL and running the programs.
- Multiplatform *Experimenter* software to help run the experiments – whose interface is a pdf file containing all the experiments from the book, each being clickable to bring up the related program and, in a Windows environment, the workspace as well. *Experimenter* is only an aid and not mandatory – each program is stand-alone. However, it is the most convenient way to run the book's code, and instructors are strongly encouraged to use it.
- Book figures in `jpg` format arranged in sequence as one PowerPoint presentation per chapter.
- Instructor's manual with solutions to 100 problems (only for instructors who have adopted this textbook).
- Contributory resource bank with homework and examination questions, experiments and other teaching and learning aids.
- Other resources as they are developed (suggestions welcome).

Capsule Chapter Descriptions

Part I: Hello World

Chapter 1: An Invitation to Computer Graphics
A non-technical introduction to the field of computer graphics.

Chapter 2: On to OpenGL and 3D Computer Graphics
Begins the technical part of the book. It introduces OpenGL and fundamental principles of 3D CG.

Part II: Tricks of the Trade

Chapter 3: An OpenGL Toolbox
Describes a collection of OpenGL programming devices, including vertex arrays, vertex buffer and array objects, mouse and key interaction, pop-up menus, and several more.

Part III: Movers and Shapers

Chapter 4: Transformation, Animation and Viewing

Introduces the theory and programming of animation and the virtual camera. Explains user interactivity via object selection. Foundational chapter for game and movie programming.

Chapter 5: Inside Animation: The Theory of Transformations

Presents the mathematical theory behind animation, particularly linear and affine transformations in 3D.

Chapter 6: Advanced Animation Techniques

Describes frustum culling, occlusion culling as well as orienting animation using both Euler angles and quaternions, techniques essential to programming games and busy scenes.

Part IV: Geometry for the Home Office

Chapter 7: Convexity and Interpolation

Explains the theory of convexity and the role it plays in interpolation, which is the procedure of spreading material properties from the vertices of a primitive to its interior.

Chapter 8: Triangulation

Describes how and why complex objects should be split into triangles for efficient rendering.

Chapter 9: Orientation

Describes how the orientation of a primitive is used to determine the side of it that the camera sees, and the importance of consistently orienting a collection of primitives making up a single object.

Part V: Making Things Up

Chapter 10: Modeling in 3D Space

Systematizes the principles of modeling both curves and surfaces, including Bézier and fractal. Foundational chapter for object design.

Part VI: Lights, Camera, Equation

Chapter 11: Color and Light

Explains the theory of light and material color, the interaction between the two, and describes how to program light and color in 3D scenes. Foundational chapter for scene design.

Chapter 12: Textures

Explains the theory of texturing and how to apply textures to objects.

Chapter 13: Special Visual Techniques

Describes a set of special techniques to enhance the visual quality of a scene, including, amongst others, blending, billboarding, aliasing and multisampling, stencil buffer methods, and image and pixel manipulation.

Part VII: Pixels, Pixels, Everywhere

Chapter 14: Raster Algorithms
Describes low-level rendering algorithms to determine the set of pixels on the screen corresponding to a line or a polygon.

Part VIII: Anatomy of Curves and Surfaces

Chapter 15: Bézier
Describes the theory and programming of Bézier primitives, including curves and surfaces.

Chapter 16: B-Spline
Describes the theory and programming of (polynomial) B-spline primitives, including curves and surfaces.

Chapter 17: Hermite
Introduces the basics of Hermite curves and surfaces.

Part IX: Well Projected

Chapter 18: Applications of Projective Spaces
Applies the theory of projective spaces to deduce the projection transformation in the graphics pipeline, following up with shadow mapping as a case study. Introduces rational Bézier and B-spline, particularly NURBS, theory and practice.

Part X: The Time is Pipe

Chapter 19: Fixed-Functionality Pipelines
Gives a detailed view of the synthetic-camera and ray-tracing pipelines and introduces radiosity.

Part XI: Rendering Pipe Dreams

Chapter 20: OpenGL 4.3, Shaders and the Programmable Pipeline: Liftoff
Introduces OpenGL 4.3, GLSL (OpenGL Shading Language) 4.3, and writing vertex and fragments shaders to program the pipeline, particularly to animate, light and apply textures.

Chapter 21: OpenGL 4.3, Shaders and the Programmable Pipeline: Escape Velocity
Continuing the previous chapter onto advanced OpenGL 4.3 topics, including, amongst others, instanced rendering, shader subroutines and transform feedback, as well as tessellation and geometry shaders.

Appendix A: Projective Spaces and Transformations

A CG-oriented introduction to the mathematics of projective spaces and transformations. Provides a complete theoretical background for Chapter 18 on applications of projective spaces.

Appendix B: Math Self-Test
A self-test to assess math readiness for intended readers.

Appendix C: Math Self-Test Solutions
Solutions for the preceding self-test.

Suggested Course Outlines

See the chapter dependencies in Figure 1.

(1) Undergraduate first CG course:

This course should be based on Chapters 1-14 + Chapters 20-21, though full coverage might be ambitious for one semester. Instructors may pick topics to emphasize or skip, depending on their goals for the course and the chapter dependence chart.

For example, for more practice and less theory, a possible sequence would be 1 → 2 → 3 → 4 → 6 (only frustum culling) → 7 → 8 → 9 → 10 (skip curve/surface theory) → 11 → 12 → 13 (→ 14, low-level raster algorithms are independent of the higher-level topics of the preceding chapters, and may be taught depending on time) → 20 → 21 (20-21, on OpenGL 4.3, can be taught in parallel with 11-13, with discussion of a topic using the fourth-generation pipeline following its discussion using the classical one, e.g., Section 20.5 on shader-based lighting following Section 11.7 deducing Phong's lighting equation).

Time permitting, selected topics may come from Chapter 5 (theory of transformations), Chapters 15-16 (Bézier and B-spline modeling, respectively, which should be taught in sequence), Chapter 17 (Hermite curves and surfaces), Chapter 18 (rational Bézier and NURBS modeling), and Chapter 19 (graphical pipelines, including the synthetic-camera and ray-tracing), which may be read independently of each other.

Note to instructor: The most effective teaching method with this book is to base discussion around experiments – both from the book and those you develop yourself. Our *Experimenter* software makes this especially convenient. Students should be involved in the experiments, running code simultaneously on their own machines in class. Minimize use of slides except, possibly, for the book figures; for your convenience these are available to download, arranged as one PowerPoint presentation per chapter.

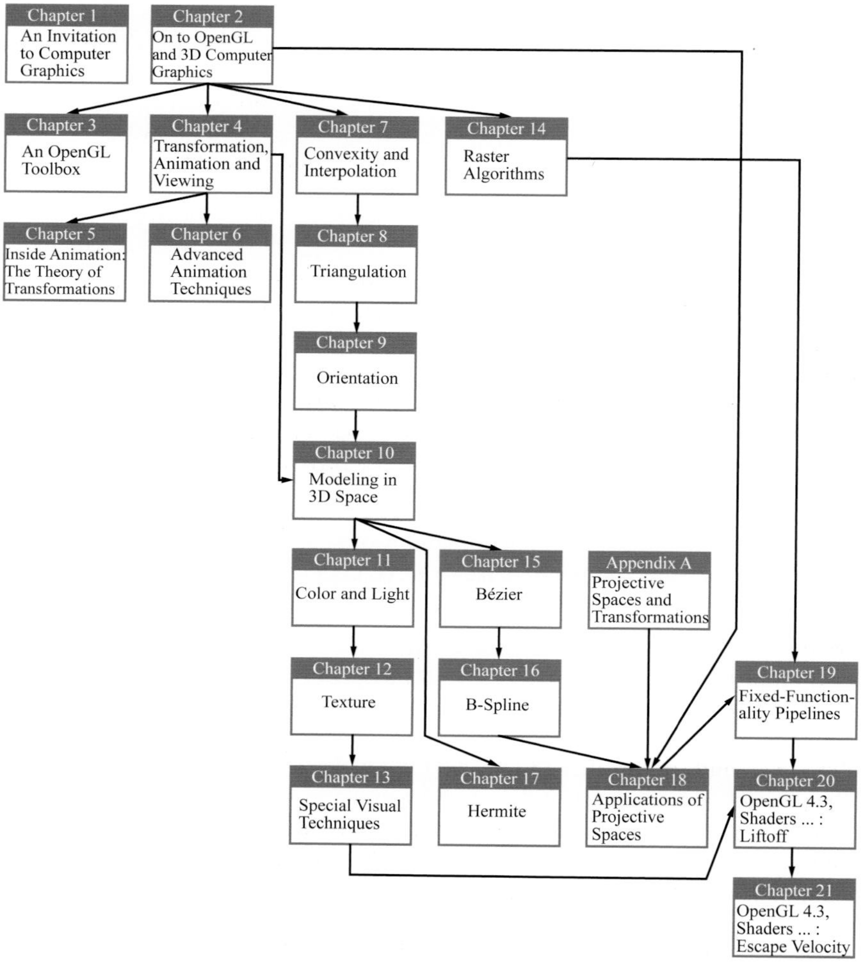

Figure 1: Chapter dependence chart: dashed arrows represent weak dependencies.

(2) Advanced CG courses:

This book could serve as a reference for a study of 3D design – particularly, Bézier (Chapter 15), B-spline (Chapter 16) and rational Bézier and NURBS theory (Chapter 18) – and of projective transformations and their applications (Appendix A and Chapter 18). From a practical point of view, Chapters 20-21 go fairly deep into the fourth generation of OpenGL and the GLSL, useful for students who may be familiar with only the classical pipeline.

(3) Self-study:

A recommended first pass would be $1 \rightarrow 2 \rightarrow 3 \rightarrow 4 \rightarrow 7 \rightarrow 8 \rightarrow$

9 (go light on 7-9 if your math is rusty) → 10 (skip theory) → 11 → 12 → 13 → 20 → 21.

Following this the student should take up a fair-sized programming project, returning to the book as needed. For the theoretically-inclined student there's a lot to keep her busy in Chapters 5 and 15-19.

Code

All the book's programs, written in C++ with OpenGL, were developed in a Microsoft Visual Studio 2010 IDE running on Windows 7. However, they can run as well on Linux and Mac OS platforms with possibly some modification depending on the exact environment. The programs can be downloaded from `www.sumantaguha.com`, where they are arranged chapter-wise in the top-level folder `ExperimenterSource`. The reader will find there, as well, a guide to installing OpenGL and running the programs on various platforms.

Acknowledgments

I owe a lot to many people, most of all students whom I have had the privilege of teaching in my CG classes over the years at UW-Milwaukee and the Asian Institute of Technology.

I thank Tarun Mukherjee at Jadavpur University for being a constant source of inspiration, not to mention help with various technical questions.

I thank KV, Ichiro Suzuki, Glenn Wardius, Mahesh Kumar, Le Phu Binh, Maria Sell and, especially, Paul McNally, for their support at UWM, where I began to teach CG and learn OpenGL.

I am grateful to my colleagues and the staff and students at AIT for such a pleasant environment, which allowed me to combine teaching and research commitments with the writing of a book.

Particular thanks at AIT to Vu Dinh Van, Nguyen Duc Cong Song, Ahmed Waliullah Kazi, Hameedullah Kazi, Long Hoang, Songphon Klabwong, Robin Chanda, Sutipong Kiatpanichgij, Samitha Kumara, Somchok Sakjiraphong, Pyae Phyo Myint Soe, Adbulrahman Otman, Sushanta Paudyal (a summer's worth of help revising code), Akila de Silva, Nitchanun Saksinchai, Thee Thet Zun, Suwanna Xanthavanij, and our ever-helpful secretaries K. Siriporn and K. Tong.

I am grateful to Kumpee Teeravech, Kanit Tangkathach, Thanapoom Veeranitinun and Pongpon Nilaphruek, students in my CG course at AIT, for allowing me to use programs they wrote.

I thank Somying Pongpimol for her Illustrator drawings. She drew most of the figures for the first edition based on my original sketches done rather amateurishly in Xfig, and then revised several and created new ones for the current edition. Somying also designed the cover for both editions.

I would like to thank Olivier Nicole for revising the book's website for the new edition.

My special thanks to reader Denis Dalpé for an extensive list of typos.

I am especially grateful to Brian Barsky for encouraging me to persevere after seeing an early and awkward draft of the first edition, and subsequently inviting the book to his series. I want to acknowledge the production team at Taylor & Francis who went out of their way for this book. Particularly, I want to thank my editor Randi Cohen who has been simply tremendous to work with. I really appreciate her making so stress-free the "business" of publishing. I am grateful to Mimi Williams for her careful and expeditious proofreading.

I am grateful to the numerous reviewers and readers of the first edition whose comments helped immeasurably improve the current one.

I acknowledge the many persons and businesses who were kind enough to allow me to include images to which they own copyrights.

On a personal note, I express my deep gratitude to Dr. Anupam De for keeping Kamaladi healthy enough that I could concentrate on the first edition thorough the few years that I spent writing it.

By far my biggest debt of gratitude is to my student and friend Chansophea Chuon. Helping me with the first edition, Chansophea developed the LaTeX style sheet, supervised the drawings while doing several of the Illustrator figures himself, laid out the manuscript, developed the multi-platform program template, designed the *Experimenter* software to help run the book experiments and created the book's initial website, all the while putting out countless fires as they happened. Chansophea's layout, in particular, transformed a rather dowdy set of notes into a handsome four-color textbook. There is no doubt that, without Chansophea working shoulder to shoulder with me, I would not have finished even the first edition and this book would never have happened.

Finally, I must say that had I not had the opportunity to study computer science in the United States and teach there, I would never have reached a position where I could even contemplate writing a textbook. It's true, too, that had I not moved to Thailand, this book would never have begun to be written. This is an enchanting country with a strangely liberating and lightening effect – to which thousands of expats can attest – that encourages one to express oneself.

Website and Contact Information

The book's website is at `www.sumantaguha.com`. Users of the book will find there various resources, including downloads, and a few links specially for instructors. The author welcomes feedback, corrections and suggestions for improvement emailed to him at `sg@sumantaguha.com`.

About the Author

Sumanta Guha obtained a Ph.D. in mathematics from the Indian Statistical Institute, Kolkata, in 1987. From 1984 to 1987 he taught mathematics at Jadavpur University in Kolkata. He left in 1987 to study computer science at the University of Michigan in Ann Arbor, where he obtained a Ph.D. in 1991. On graduating from Michigan he joined the computer science faculty of the University of Wisconsin-Milwaukee where he taught from 1991 to 2002. In 2002 he moved to the information management and computer science program of the Asian Institute of Technology in Thailand, where he is currently a professor. His research interests include computational geometry, computer graphics, computational topology, robotics and data mining.

Part I

Hello World

CHAPTER 1

An Invitation to Computer Graphics

Computer graphics, or CG as it is often simply called, is the use of computers to generate images. This is as opposed to the capture of images of real-world or imagined objects which would be, for example, photography or the work of an artist with pencil and paper.

To not see the end product of CG, that being computer-generated imagery (CGI), throughout your day, you would have to be on a deserted island. Images on the screen of the cell phone you probably check first thing on waking are digitally synthesized by a processor. Almost every frame on the TV showing the morning news has CGI in some part. If you commute, then the vehicle which carries you to school or work likely communicates with its operator through multiple computer-managed console panels, displaying information ranging from fuel level to geographical location.

Figure 1.1: A cell phone, news opening graphics, car dashboard.

At work, if at all you use a computer, then, of course, there you are sitting right at a fountainhead of computer graphics. And, CGI probably plays an even more important role in your recreational life. Even the most casual video games amusing commuters heading home nowadays have sophisticated interactive 3D graphics. The web on which we spend so many hours a day is increasingly becoming a multimedia smorgasbord synthesizing animation, movie clips, CGI and sound.

Figure 1.2: A computer at work, handheld game player, AIT home page (used with permission of the Asian Institute of Technology).

When you watch a movie you are seeing a product from an industry, which together with the gaming industry, has the biggest relationship with CG of any other, not only as a consumer of the latest and greatest in technique, but also as promoter, with hundreds of millions of dollars in investment, of cutting-edge research. A little blue elephant which grows into a mighty warrior, an eccentric mouse with a ribald sense of humor, and a massive dinosaur looking so hungrily for food that you would think its species had never really become extinct more than fifty million years ago – to contemplate such achievements is to be in awe of the human imagination, as well as the ingenuity of the engineers and programmers who materialize these fantastical conceptions as palpable and believable digital presences.

Figure 1.3: Khan Kluay, the first 3D animated Thai movie (courtesy Kantana Animation), an anthropomorphic mouse, a massive (fortunately herbivorous) dinosaur.

Then there's the quiet CG impacting our lives some would say even more profoundly than its more flamboyant manifestations. Doctors and surgeons practice their craft in simulated environments detailed to the tiniest capillaries. Commercial pilots put in hundreds of hours on a flight simulator before entering a real cockpit. (Flight simulators are a sentimental favorite because they were the first killer CG app, drawing attention and investment dollars to the then nascent field in the sixties.)

Automobiles, airplanes and almost any fairly complex manufactured object we see around us are designed, fabricated and even put through regulatory tests as virtual entities – which exist entirely as a collection of bits perceptible only as an image on a monitor – gestating often for years before the first physical prototype is ever built. Supercomputers implement extremely complex mathematical models of the weather, but their predictions have to be visualized – again CGI – in order to be meaningful to humans.

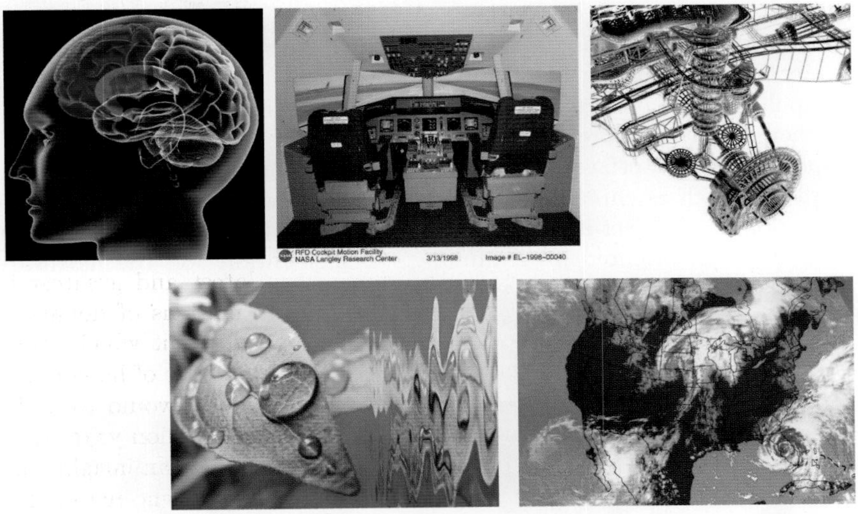

Figure 1.4: Clockwise from top left: Image of the human brain, flight simulator cockpit (from NASA), engine design, hurricane over Florida, water drop on a leaf.

Because its business is the creation of pictures, computer graphics has an immediate allure. But, it is a science as well, with intellectual challenges ranging from the routine to about as deep and hard as you please. Think of modeling a drop of water rolling off a leaf. There would be a fair amount of physics and, probably, a differential equation or two to solve on the way to getting just the mechanics of the rolling drop right, not to mention texturing the leaf, creating a translucent (and changing) shape for the drop, and determining illumination.

The field of computer graphics brings particular pleasure to students and practitioners alike because it's always about making something – just like sculpting or painting. Part by part you watch your creation come together, and alive even, if it is animated. Aside from the aesthetic, there are more tangible rewards to be had too. One would be hard pressed to name a sphere of social or scientific or industrial activity where CGI does not have a role. Wherever it is that ultimately you want to be, medicine or fashion, rocket science or banking, weapons development or teaching yoga, sales and marketing or environmental modeling, CG skills not only can make a difference, but also make you a career.

1.1 Brief History of Computer Graphics

Although the term "computer graphics" itself was coined in 1960 by William Fetter, a designer at Boeing, to describe his own job, the field can be said to have first arrived with the publication in 1963 of Ivan Sutherland's Sketchpad program, as part of his Ph.D. thesis at MIT.

Sketchpad, as its name suggests, was a drawing program. Beyond the interactive drawing of primitives such as lines and circles and their manipulation – in particular, copying, moving and constraining – with use of the then recently invented light pen, Sketchpad had the first fully-functional graphical user interface (GUI) and the first algorithms for geometric operations such as clip and zoom. Interesting, as well, is that Sketchpad's innovation of an object-instance model to store data for geometric primitives foretold object-oriented programming. Coincidentally, on the hardware side, the year 1963 saw the invention by Douglas Engelbart at the Stanford Research Institute of the mouse, the humble device even today carrying so much of GUI on its thin shoulders.

Figure 1.5: Ivan Sutherland operating Sketchpad on a TX-2 (courtesy of Ivan Sutherland), Douglas Engelbart's original mouse (courtesy of John Chuang).

Although Sketchpad ran on a clunky Lincoln TX-2 computer with only 64KB in memory and a bulky monochrome CRT monitor as its front-end, nevertheless, it thrust CG to the attention of the early researchers by showing what was possible. Subsequent advances through the sixties came thick and fast: raster algorithms, the implementation of parametric surfaces, hidden-surface algorithms and the representation of points by homogeneous coordinates, the latter crucially presaging the foundational role of projective geometry in 3D graphics, to name a few. Flight simulators were the killer app of the day and companies such as General Electric and Evans & Sutherland, co-founded by Douglas Evans and Ivan Sutherland, wrote simulators with real-time graphics.

Interestingly, the advent of flight simulators actually predated that of CG – at least Sutherland and his Sketchpad – by nearly two decades, when the US Navy began the funding of Project Whirlwind at MIT during the Second World War for the purpose of creating simulators to train bomber crews. Those early devices had actually little graphics and consisted essentially of a simulated instrument panel reacting in real-time to control input from the pilots, but Project Whirlwind helped fund the talent and research environment at MIT which enabled Sutherland to create Sketchpad, launch computer graphics and, finally, complete the circle by establishing a company to make flight simulators.

The next decade, the seventies, brought the z-buffer for hidden surface removal, texture mapping, Phong's lighting model – all crucial components of the OpenGL API (Application Programming Interface) we'll be using soon – as well as keyframe-based animation. Photorealistic rendering of animated movie keyframes almost invariably deploys ray tracers, which were born in the seventies too. Emblematic of the advances in 3D design was Martin Newell's 1975 Utah teapot, composed entirely of bicubic Bézier patches, which became the testbed of choice for CG algorithms. The latter half of the decade saw, too, the Apple I and II personal computers make their debut, bringing CG for the first time to the mass market.

Figure 1.6: Utah teapot (from Wikimedia), Apple II Plus (courtesy of Steven Stengel), SIGGRAPH 2006 expo floor in Boston (courtesy of Jason Della Rocca).

From the academic point of view, particularly important were the establishment in 1969 of the SIGGRAPH (Special Interest Group in Graphics) by the ACM (Association for Computing Machinery, the premier academic society for computers and computing) and, subsequently, the first annual SIGGRAPH conference in 1973. These two developments signaled the emergence of computer graphics as a major subdiscipline of computer science. The SIGGRAPH conference has since then become the foremost annual event in the CG world. In addition to being the most prestigious forum for research papers, it hosts a giant exhibition which attracts hundreds of companies, from software developers to book publishers, who set up booths to promote their wares and recruit talent.

Since the early eighties, CG, both software and hardware, began rapidly to assume the form we see today. The IBM PC, the Mac and the x86 chipsets all arrived, sparking off the race to become faster (processor), smaller (size), bigger (memory) and cheaper (particularly important if you are going to school). As computers became consumer goods, the market for software spilled over from academia to individuals and businesses.

Nintendo released Donkey Kong in 1981, the wildly successful arcade video game which revolutionized the genre, and soon after Wavefront Technologies released its Preview software, used then to create opening graphics for television programs. Now, of course, Nintendo is a star of the video games industry producing the Wii and it successors, while Wavefront has morphed into Alias (owned by Autodesk) whose 3D graphics modeling package Maya is ubiquitous in the design world.

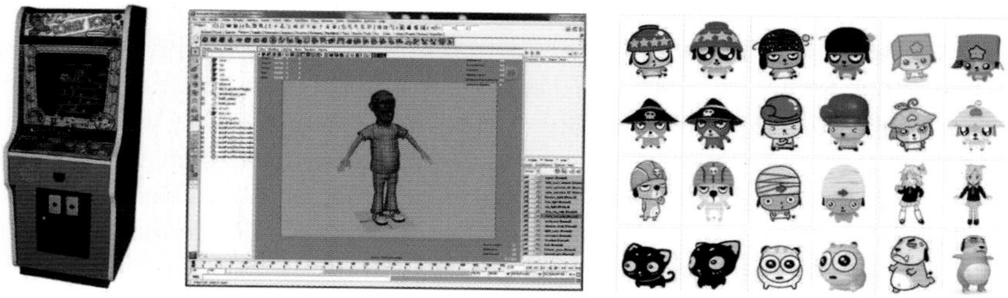

Figure 1.7: Donkey Kong arcade game (from Wikimedia), Maya screenshot of Scary Boris (courtesy of Sateesh Malla at `www.sateeshmalla.com`), 2D characters on the left versus 3D on the right (© Mediafreaks Cartoon Pte. Ltd., 2006. All rights reserved.).

3D graphics began to displace its plainer 2D sister through the nineties as hardware increasingly became capable of supporting the rendering needs of 3D models, even in real-time, thus allowing interaction and its myriad consequences (such as gaming). The difference between 2D and 3D graphics is that models in the latter are created in a (virtual) 3D world, geometrically the same as the real world, and then projected onto the viewing screen, while all drawings in 2D graphics are on a flat plane.

Models drawn in 3D are more realistic because they have all the three dimensions we humans can perceive, but they are more complex as well; moreover, the projection step, non-existent for 2D graphics, is computation-intensive too. Graphics cards, manufactured by companies such as ATI and Nvidia, which not only manage the image output to the display unit, but have, as well, additional hardware support for rendering of 3D primitives, are now inexpensive enough that desktops and even notebooks can run high-end 3D applications. How well they run 3D games often, in fact, is used to benchmark personal computers.

Through the nineties, as well, the use of 3D effects in movies became pervasive. The Terminator and Star Wars series, and Jurassic Park, were among the early movies to set the standard for CGI. Toy Story from Pixar, released in 1995, has special importance in the history of 3D CGI as the

Figure 1.8: T. Rex: heartthrob of the Jurassic Park movies, Quake 1 game (courtesy of Quake ® © 1996 id Software LLC, a ZeniMax Media Company, All Rights Reserved).

first movie to be entirely computer-generated – no scene was ever pondered through a glass lens, nor any recorded on a photographic reel! It was cinema without film. Quake, released in 1996, the first of the hugely popular Quake series of games, was the first fully 3D game.

Another landmark from the nineties of particular relevance to us was the release in 1992 of OpenGL, the open-standard cross-platform and cross-language 3D graphics API, by Silicon Graphics. OpenGL is actually a library of calls to perform 3D tasks, which can be accessed from programs written in various languages and running over various operating systems. That OpenGL was high-level (in that it frees the applications programmer from having to care about such low-level tasks as representing primitives like lines and triangles in the raster, or rendering them to the window) and easy to use (much more so than its predecessor 3D graphics API, PHIGS, standing for Programmer's Hierarchical Interactive Graphics System) first brought 3D graphics programming to the "masses". What till then had been the realm of a specialist was now open to a casual programmer following a fairly amicable learning curve.

Since its release OpenGL has been rapidly adopted throughout academia and industry. It's only among game developers that Microsoft's proprietary 3D API, Direct3D, which came soon after OpenGL bearing an odd similarity to it but optimized for Windows, is more popular.

Figure 1.9: OpenGL and OpenGL ES logos (used with permission of Khronos).

The story of the past decade has been one of steady progress, rather than spectacular innovations in CG. Hardware continues to get faster, better, smaller and cheaper, continually pushing erstwhile high-end software downmarket, and raising the bar for new products. The almost complete displacement of CRT monitors by LCD and the emergence of high-definition television are familiar consequences of recent hardware evolution.

Of likely even greater economic impact is the migration of sophisticated software applications – ranging from web browsers to 3D games – to handheld devices like smartphones, on the back of small yet powerful processors. CG has now been untethered from large immobile devices and placed into the hands and pockets of consumers. In fact, a lightweight subset of OpenGL called OpenGL ES – ES abbreviating Embedded Systems – released by the Khronos Group in 2003, is now the most popular API for programming 3D graphics on small devices.

1.2 Overview of a Graphics System

The operation of a typical graphics system can be split into a three-part sequence:

$$\text{Input} \longrightarrow \text{Processing} \longrightarrow \text{Output}$$

The simplest example of this is when you click on a thumbnail image in, say, YouTube, and a video clip pops up and begins to play. The click is the input. Your computer then reacts to this input by processing, which involves downloading the movie file and running it through the Adobe Flash Player, which in turn outputs video frames to your monitor.

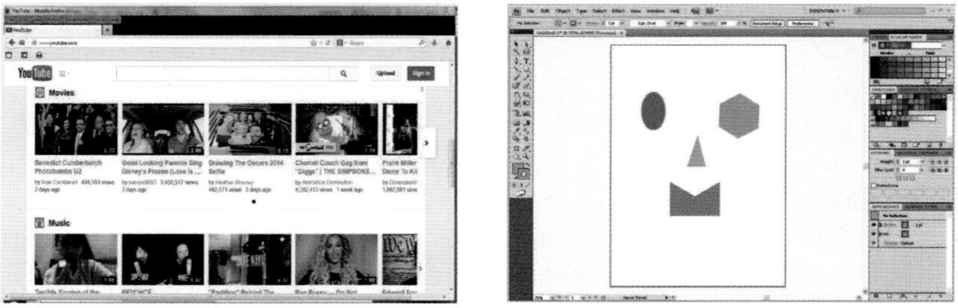

Figure 1.10: YouTube and Adobe Illustrator screenshots.

Graphics systems can be of two types, non-interactive and interactive. The playing of a YouTube clip is an example of a non-interactive one: beyond the first click to get the movie started you have little further say over the output process, other than maybe to stop it or manipulate the window. On the other hand, if, say, you are using a package like Adobe Illustrator, then the output – what you have drawn – changes in real-time in response to input you provide by pressing keys and moving and clicking the mouse; e.g., you can create shapes, color and move them, and so on. In an interactive system output continuously reacts to input via the processor.

Input/output devices (or I/O devices, or peripheral devices, as they are also called) are of particular importance in interactive systems because they determine the scope of the interaction. For example, an input device that functions like a steering wheel would be essential to a video game to race cars; simulating flight through a virtual 3D environment, on the other hand, needs something akin to a joystick used to maneuver an aircraft.

Because it is, in fact, interactive computer graphics – theory and programming – which we'll be studying the next nineteen chapters, let's quickly survey first the most common I/O devices found in graphics systems nowadays. As for the processors that may come between the I and the O, from the point of view of CG, essentially, these are just boxes to be coded in order to obtain the desired input-to-output mapping. For the sake of completeness, though, here's a list of the important ones, all somewhat different one from the other in the context of CG (Figure 1.11 pictures them):

Computer: As far as we are concerned, this category includes PC's, workstations, servers and the like.

Portable computer: This, of course, is simply a small and light computer with a built-in display, keyboard and pointing device. Because of the size constraint, limited power supply and also the lack of space for a large cooling fan, CPU's and graphics cards in portable computers tend to underperform their desktop counterparts. Software writers need to take this into account, especially for graphics-intensive applications.

Figure 1.11: Processing devices clockwise from top: laptop, smartphone, game console (used with permission from Microsoft), computer box.

Handheld device: The size-weight constraint on this class of devices – of which the mobile phone is the most visible example – is even more severe than for portable computers. Handhelds are expected to travel in bags and pockets. Low-end handhelds often have no peripheral other than a limited keypad, while higher-end ones may come equipped with a full QWERTY keypad and touchscreen. In addition to the possibly small RAM and anemic CPU, another consideration to keep in mind for graphics developers for handhelds is the limited real estate of the display: busy scenes tend to become "chaotic" on a handheld.

Game consoles: All stops are off for programming these devices. Running graphics-intensive applications at blinding speeds is what these machines were born for.

1.2.1 Input Devices

The following is by no means a complete list of input devices, but it does cover the ones we are most likely to encounter in everyday use. The devices are all pictured in Figure 1.12, ringing the processing devices in the middle, and our list goes clockwise starting from the top rightmost.

Keyboard: This device is a mandatory peripheral for any computer. Its alphanumeric keys, evidently derived from the traditional typewriter, are used to enter text strings, while additional keys, such as the arrow and function keys, perform special actions.

Mouse: This is an example of a *pointing device* which inputs spatial data to the computer. As the mouse is moved by the user's hand on a flat surface, a mechanical ball or optical sensor at its base signals the amount of movement to the computer, which correspondingly moves a cursor on the screen. Effectively, then, the user determines the location of the cursor. Strictly speaking, a mouse is more than just a pointing device if it has buttons, as most do, each of which can be clicked to give binary input.

Touchpad: Another 2D pointing device, particularly common on portable computers, the touchpad is a small rectangular area embedded with electronic sensors to determine the position of a touching finger or stylus. Movement of the finger or stylus is echoed by movement of the cursor.

Pointing stick: Yet another 2D pointing device common on portable computers, the pointing stick is, typically, a rubber peg located between the 'G', 'H' and 'B' keys, which moves the cursor in response to pressure applied with a finger.

Trackball: This is essentially an upside-down mouse, with a socket containing a ball which the user manipulates with her hand to make the cursor move.

Figure 1.12: Input devices clockwise from top right (surrounding processing devices in the middle): keyboard, mouse, touchpad, pointing stick (courtesy of Long Zheng), trackball, spaceball (courtesy of Logitech), tablet, haptic device (© SensAble Technologies, Inc.), joystick, wheel, gamepad, webcam, touchscreen, data gloves (courtesy of `www.5dt.com`).

Spaceball: This is a pointing device with six degrees of freedom versus the two of an ordinary mouse. It is used in special applications such as manipulating a camera in a 3D scene: not only is the camera moved, but also rotated, affording it multiple degrees of freedom, each of which the user controls. The spaceball itself consists of a pressure-sensitive ball which can distinguish different kinds of forces, including forward/backward, lateral and twist, responding by moving and orienting the selected object.

Tablet: This is a *digitizing device* which has a surface embedded with sensors to pick up the successive coordinates of a stylus head or fingertip as it travels over the surface (in effect converting physical motion into digital data). The user can write or draw on a tablet, just as on paper with pen, the output being displayed on the monitor. The monitor is usually separate, though, on devices like a tablet PC, the display and the sensing surface are the same.

Haptic device: This is a pointing device which gives physical feedback to the user based on the location of the cursor or, possibly, that of an object being moved along with the cursor. The easiest way to understand the functioning of a haptic device, if you have never used one, is to imagine a mouse with a mechanical ball which is (somehow) programmed to lock and stop rolling when the cursor reaches the side of the screen. The reaction the user then has is of that of the cursor running into a physical obstacle at the edge of the screen, though evidently it is moving in virtual space. The device depicted in Figure 1.12 is not a haptic mouse, of course, but one commonly seen in HCI (human-computer interaction) labs. The three-link arm swiveling on a ball gives it six degrees of freedom.

Haptics has numerous applications, a couple of noteworthy ones being the teleoperation of robots (where the operator gets haptic feedback as she manipulates a robot in either a virtual or a remote real environment) and simulated surgery training in medicine (which is similar to training pilots on a flight simulator, except that surgery has the added component of tactile feedback, mostly absent in flying).

Joystick: This is an input device popular in video games and applications such as flight simulators. It originated from its namesake found in real aircraft cockpits. A joystick pivots around a fixed base, gaining thus two degrees of freedom, and usually has buttons which can be depressed to provide additional input. In a game or simulator setting a joystick is typically used to control an object traveling through space. Nowadays, high-end joysticks have embedded motors to provide haptic feedback to user motion, e.g., resistance as a plane is banked.

Wheel: This again is a specialized input device for games and simulators, obviously derived from the car steering wheel, and provides rotational input in an exactly similar manner, most often to a virtual automobile. Again, haptic feedback to give the user a sense of the vehicle's response, and even of the terrain over which it is traveling, is becoming increasingly popular.

Gamepad: This device is the standard controller for many modern game consoles. Usual features include action buttons operated usually with the right thumb and a cross-shaped directional controller with the left.

Camera: Although this input device needs no introduction, it's worth noting the increasingly sophisticated uses a peripheral camera is being put to with the help, e.g., of software to recognize faces, gestures and expressions.

Touchscreen: Increasingly popular as the interface of handheld devices such as smartphones, a touchscreen is a display which can accept input via touch. It is similar to touchpads and tablets in that it senses the location of a finger or stylus – one or the other is usually preferable based on the particular technology used to make the screen – on the display area. A common application of touchscreens is to eliminate the need for a physical keyboard by displaying a virtual one responding to taps on the screen.

Touchscreens often respond not only to the location of the touch, but also the motion of the touching object. For example, a flicking motion with a finger may cause a window to scroll. Multi-touch capability, now increasingly common, makes possible for the device to respond to gestures with more than one finger, e.g., pinching and spreading with two fingers.

Data gloves: This device is used particularly in virtual reality environments which are programmed to react to the position of the gloves, the direction in which fingers are pointing, as well as to hand motion and gestures. The gloves themselves are wired to transmit not only their location, but also their configuration and orientation to the processor, so that the latter can display the environment accordingly. For example, an index finger pointing at a particular atom in a virtual-reality display of a molecule may cause this atom to zoom up to the viewer.

1.2.2 Output Devices

Again, the following list is not meant to be comprehensive, but, rather, representative of the most common output devices. We go clockwise around the outer ring of devices pictured in Figure 1.13 beginning with the rightmost.

CRT (cathode-ray tube) monitor: A CRT monitor has phosphors of the three primary colors – R(ed), G(reen) and B(lue) – located at each one of a rectangular array of pixels, called the raster. Additionally, it has three electron guns inside, causing its infamous bulk, that each fires a beam at phosphors of one color. A mechanism to aim and control their intensities causes the beams to travel together, striking one pixel after another, row after row, exciting the RGB phosphors at each pixel to the values specified for it in the color buffer. Figure 1.14(a) shows the electron beams striking one pixel on a dog.

From the point of view of OpenGL and, indeed, most CG theory, what matters is that the pixels in a monitor are, in fact, arranged in a rectangular raster (as depicted in Figure 1.14(b)). For, this layout is the basis of the

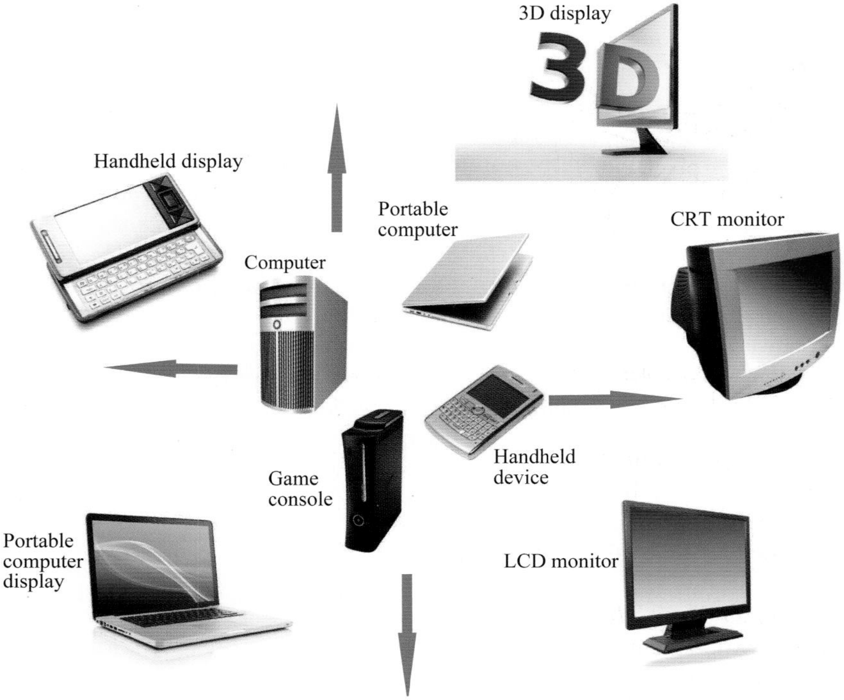

Figure 1.13: Output devices clockwise from the rightmost (surrounding processing devices in middle): CRT monitor, LCD monitor, notebook, mobile phone, 3D LCD monitor.

lowest-level CG algorithms, the so-called raster algorithms, which actually select and color the pixels to represent user-specified shapes such as lines and triangles on the monitor. Figure 1.14(b), for example, shows the rasterization of a right-angled triangle (with terrible jaggies because of the low resolution).

The number of rows and columns of pixels in the raster determines the monitor's resolution. Typical for a CRT monitor is a resolution in the range of 1024×768 (which means 1024 columns and 768 rows). High-definition monitors (as needed, say, for high-definition TV, or HDTV as it's acronymed) have higher resolution, e.g., 1920×1080 is common.

Moreover, a memory location called the color buffer, either in the CPU or graphics card, contains, typically, 32 bits of data per raster pixel – 8 bits for each of RGB, and 8 for the alpha value (used in blending images). It is the RGB values in the color buffer which determine the corresponding raster pixel's color intensities. The values in the color buffer are read by the raster – in other words, the raster is refreshed – at a rate called the monitor's refresh rate. Beyond this, the technology underlying the particular display device, no matter how primitive or how fancy, really matters little

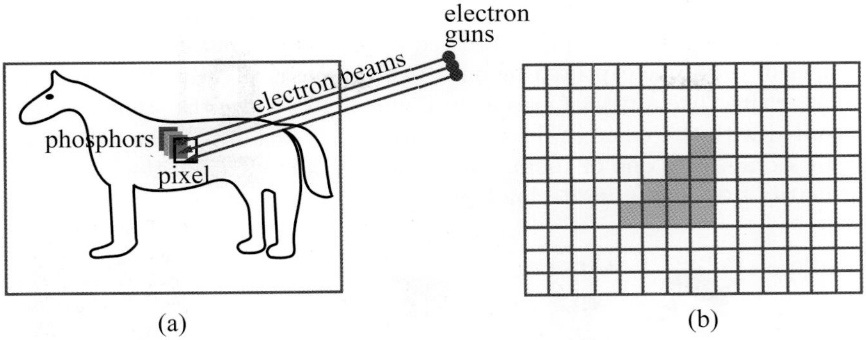

Figure 1.14: (a) Color CRT monitor with electron beams aimed at a pixel with phosphors of the 3 primaries (b) A raster of pixels showing a rasterized triangle.

to the CG programmer.

For decades a bulky CRT monitor, or two, was a fixture atop work tables. Now, of course, they have been nearly totally supplanted by a sleeker successor which we discuss next.

LCD (liquid crystal display) monitor: Pixels in an LCD monitor each consist of three subpixels made of liquid crystal molecules, which separately filter lights of the primary colors. The amount of light emerging through a subpixel is controlled by an electric charge whose intensity is determined by subpixel's corresponding value in the color buffer. The absence of electron guns allows LCD monitors to be made flat and thin – unlike CRT monitors – so they are one of the class of flat panel displays.

Technologies other than LCD, e.g., plasma and OLED (organic light emitting diode), are used as well in flat panel displays, though LCD is by far the most common one found with computers.

Again, for all practical purposes, the view to keep in mind of the LCD monitor, as of other flat panel displays, is of a rectangular raster of pixels whose RGB intensities are individually set by values in the computer's color buffer.

Portable computer display: This display is again a raster of pixels whose RGB values are read from a color buffer. The technology employed, typically, is TFT-LCD, a variant of LCD which uses thin film transistors to improve image quality.

Handheld display: Handheld displays, such as those on devices like mobile phones, commonly use the same TFT-LCD technology as portable computers. The resolution, though, is necessarily smaller, e.g., 480×640 would be in the ballpark for low-end mobiles.

3D display: Almost all 3D displays are based on the principle of stereoscopy, in which an illusion of depth is created by showing either eye of the viewer images of the scene captured by one of two cameras slightly offset from one another (just like a pair of eyes as in Figure 1.15). Once the scene has been recorded with two cameras, it is in ensuring that each eye of the viewer sees frames only from one of them, called stereoscopic viewing, that there are primarily two competing technologies.

Figure 1.15: Dual cameras filming a motorbike for subsequent 3D viewing with a pair of polarized glasses.

In the first, frames alternately from either camera are displayed on the monitor, a process called alternate frame sequencing. Simultaneously, the viewer wears LCD shutter glasses embedded with a polarizing filter which can be darkened with an electrical signal. The glasses are synchronized with the monitor's refresh rate via a link such as Bluetooth, either lens being alternately darkened with successive frames. Consequently, each eye sees images from only one of the two cameras, resulting in a stereoscopic effect. Typically, the frame rate is increased to 48 per second as well, so that both eyes experience a smooth-seeming 24 frames each second. The great advantage of LCD shutter glasses is that they can be used with any computer which has a monitor with a refresh rate fast enough to support alternate frame sequencing, as well as a graphics card with enough buffer space for two video streams. So with these glasses even a high-end home system would qualify to play 3D movies and games.

In the second, polarized 3D glasses are used to view two images, from either camera, projected simultaneously on the same screen through orthogonal polarizing filters. The lenses too contain orthogonal polarizing filters, each allowing through only light of like polarization. Consequently, either lens sees images from only one or other camera, engendering a stereoscopic view. Polarized 3D glasses are significantly less expensive than LCD shutter glasses and, moreover, require no synchronization with the monitor. However, the projection system is complicated and expensive and primarily used to equip theaters for 3D viewing.

OpenGL, the API we'll be using, is well-suited to making scenes and movies for 3D viewing because it allows one or more (virtual) cameras to be positioned arbitrarily.

1.3 Quick Preview of the Adventures Ahead

To round out this invitation to CG we want to show you three programs written by students in their first college 3D CG course, taught using a draft of this book. They were written in C++ with calls to OpenGL.

But, first, what exactly is OpenGL? You may have been wondering this awhile. We said earlier in the section on CG history that OpenGL is a cross-platform 3D graphics API. It consists of a library of over 500 commands to perform 3D tasks, which can be accessed from programs written in various languages. Well, here's a glimpse of something concrete – an example snippet from a C++ environment to draw 10 red points:

```
glColor3f(1.0, 0.0, 0.0);
glBegin(GL_POINTS);
   for(int i = 0; i < 10; i++)
   {
      glVertex3i(i, 2*i, 0);
   }
glEnd();
```

The first function call `glColor3f(1.0, 0.0, 0.0)` declares the red drawing color, while the loop bracketed between the `glBegin(GL_POINTS)` and `glEnd()` calls draws a point at $(i, 2i, 0)$ in each of ten iterations. There are many more calls in the OpenGL library, for example, to draw straight lines, triangles, create light sources, apply textures, move and rotate objects, maneuver the camera, and so on – in fact, not surprisingly, pretty much all one needs to create and animate realistic (or fantastic) 3D scenes.

Isn't that old OpenGL, though, you show above? Yes, it is. Precisely, it's pre-shader OpenGL.

But, the fact you asked this question probably means you are not familiar yet with our pedagogical approach, which is described in the book's preface. We explain there why we believe in setting the reader's foundations in the classical version of OpenGL before proceeding to the new (fourth generation, version 4.3, to be precise) of which there is complete coverage later in the book. We urge you to read at least that part of the preface in order to be comfortable with how we plan on doing things.

Getting back to the student programs, the code itself is not of importance and would actually be a distraction at this time. Instead, just running the programs and viewing the output will give an idea of what can be accomplished even in a fairly short time (ranging from 3 weeks to 3 months for the different programs) by persons coming to CG with little more than a good grasp of C++ and some basic math. Of course, we'll get a feel as well for what goes into making 3D scenes.

Experiment 1.1. Open `ExperimenterSource/Chapter1/Ellipsoid` and, hopefully, you'll be able to run at least one of the two executables there for the `Ellipsoid` program – one for Windows and one for the Mac. The program draws an ellipsoid (an egg shape). The left of Figure 1.16 shows the initial screen. There's plenty of interactivity to try as well. Press any of the four arrow keys, as well as the page up and down keys, to change the shape of the ellipsoid, and 'x', 'X', 'y', 'Y', 'z' and 'Z' to turn it.

It's a simple object, but the three-dimensionality of it comes across rather nicely does it not? As with almost all surfaces that we'll be drawing ourselves, the ellipsoid is made up of triangles. To see these press the space bar to enter wireframe mode. Pressing space again restores the filled mode. Wireframe reveals the ellipsoid to be a mesh of triangles decorated with large points. A color gradient has apparently been applied toward the poles as well.

Drawing an ellipsoid with many triangles may seem a hard way to do things. Interestingly, and often surprisingly for the beginner, OpenGL offers the programmer only a tiny set of low-level geometric primitives with which to make objects – in fact, points, lines and triangles are, basically, it. So, a curved 3D object like an ellipsoid has to be made, or, more accurately, *approximated*, using triangles. But, as we shall see as we go along, the process really is not all that difficult.

That's it. There's really not much more to this program: no lighting or blending or other effects you may have heard of as possible using OpenGL (understandably, as the program was written just a few weeks into the semester). It's just a bunch of colored triangles and points laid out in 3D space. The magic is in those last two words: *3D space*. 3D modeling is all about making things in 3D – not a flat plane – to create an illusion of depth, even when viewing on a flat plane (the screen). **End**

(a) (b) (c)

Figure 1.16: Screenshots of (a) `Ellipsoid` (b) `AnimatedGarden` (c) `Dominos`.

Experiment 1.2. Our next program is animated. It creates a garden which grows and grows and grows. You will find executables in `ExperimenterSource/Chapter1/AnimatedGarden`. Press enter to start the animation; enter again to stop it. The delete key restarts the animation, while the period key toggles between the camera rotating and not. Again, the space key toggles between wireframe and filled. The middle of Figure 1.16 is a screenshot a few seconds into the animation.

As you can see from the wireframe, there's again a lot of triangles (in fact, the flowers might remind you of the ellipsoid from the previous program). The plant stems are thick lines and, if you look carefully, you'll spot points as well. The one special effect this program has that `Ellipsoid` did not is blending, as is not hard to see. **End**

Experiment 1.3. Our final program is a movie which shows a Rube Goldberg domino effect with "real" dominos. The executables are in `ExperimenterSource/Chapter1/Dominos`. Simply press enter to start and stop the movie. The screenshot on the right of Figure 1.16 is from part way through.

This program has a bit of everything – textures, lighting, camera movement and, of course, a nicely choreographed animation sequence, among others. Neat, is it not? **End**

Hopefully, these three programs have got you all fired up and ready to rumble on into OpenGL. Great! Get yourself a coffee and flip the page.

Acknowledgments: Kumpee Teeravech wrote `Ellipsoid` and `AnimatedGarden`, while Kanit Tangkathach and Thanapoom Veeranitinunt wrote `Dominos`.

CHAPTER 2

On to OpenGL and 3D Computer Graphics

The primary goal for this chapter is to get acquainted with OpenGL and begin our journey into computer graphics using OpenGL as our API (Application Programming Interface) of choice. We shall apply an experiment-discuss-repeat approach where we run code and ask questions of what is seen, acquiring thereby an understanding not only of the way the API functions, but underlying CG concepts as well. Particularly, we want to gain insight into:

(a) The synthetic-camera model to record 3D scenes, which OpenGL implements.

(b) The approach of approximating curved objects, such as circles and spheres, with the help of straight and flat geometric primitives, such as line segments and triangles, which is fundamental to object design in computer graphics.

We begin in Section 2.1 with our first OpenGL program to draw a square, the computer graphics equivalent of "Hello World". Simple though it is, with a few careful experiments and their analysis, `square.cpp` yields a surprising amount of information through Sections 2.1-2.3 about orthographic projection, the fixed world coordinate system OpenGL sets up and how the so-called viewing box in which the programmer draws is specified in this system. We gain insight as well into the 3D-to-2D rendering process.

Adding code to `square.cpp` we see in Section 2.4 how parts of objects outside the viewing box are clipped off. Section 2.5 discusses OpenGL as a state machine. We have in this section as well our first glimpse of property values, such as color, initially specified at the vertices of a primitive, being interpolated throughout its interior.

Next is the very important Section 2.6 where all the drawing primitives of OpenGL are introduced. These are the parts at the application programmer's disposal with which to assemble objects from thumbtacks to spacecrafts.

The first use of straight primitives to approximate a curved object comes in Section 2.7: a curve (a circle) is drawn using straight line segments. To create more interesting and complex objects one must invoke OpenGL's famous three-dimensionality. This involves learning first in Section 2.8 about perspective projection and also hidden surface removal using the depth buffer.

After a bunch of drawing exercises in Section 2.9 for the reader to practice her newly-acquired skills, the topic of approximating curved objects is broached again in Section 2.10, this time to approximate a surface with triangles, rather than a curve with straight segments as in Section 2.7. Section 2.11 is a review of all the syntax that goes into making a complete OpenGL program.

We conclude with a summary, brief notes and suggestions for further reading in Section 2.12.

2.1 First Program

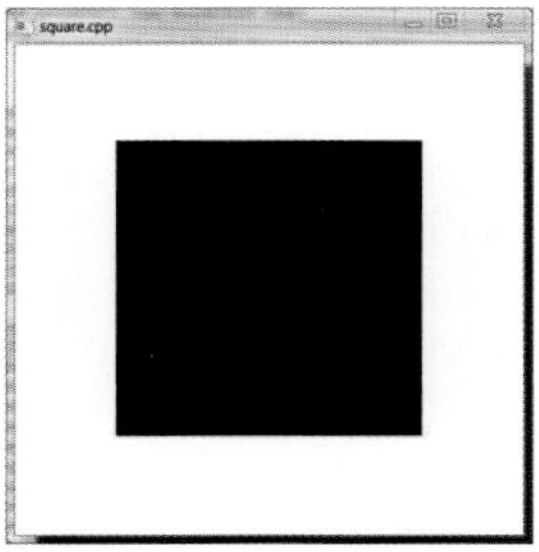

Figure 2.1: Screenshot of `square.cpp`.

Figure 2.2: OpenGL window of `square.cpp` (bluish green pretending to be white).

Experiment 2.1. Run `square.cpp`.

Note: Visit the book's website `www.sumantaguha.com` for a guide on how to install OpenGL and run our programs on various platforms.*

In the OpenGL window appears a black square over a white background. Figure 2.1 is an actual screenshot, but we'll draw it as in Figure 2.2, bluish green standing in for white in order to distinguish it from the paper. We are going to understand next how the square is drawn, and gain some insight as well into the workings behind the scene. **End**

The following six statements in `square.cpp` create the square:

```
glBegin(GL_POLYGON);
   glVertex3f(20.0, 20.0, 0.0);
   glVertex3f(80.0, 20.0, 0.0);
   glVertex3f(80.0, 80.0, 0.0);
   glVertex3f(20.0, 80.0, 0.0);
glEnd();
```

Important! If, from what you might have seen elsewhere, you have the notion that `glBegin()-glEnd()`, and even `GL_POLYGON`, specifications are

*If your program compiles but doesn't run, this might be because your graphics card doesn't support OpenGL 4.3. See the note following item 2 in Section 2.11 for what to do in this case.

classical and don't belong in the newest version of OpenGL, then you are right insofar as they are not in the core profile of the latter. They are, though, accessible via the compatibility profile which allows for backward compatibility. Moreover, we explain carefully in the book's preface why we don't subscribe to the "shaders from the opening bell and toss everything classical" school of thought as far as *teaching* OpenGL is concerned. Of course, we shall cover thoroughly the most modern – in fact, fourth generation – OpenGL later in the book. If you have not done so yet, we urge you to read about our pedagogical approach in the preface.

The corners of the square are specified by the four vertex declaration statements between `glBegin(GL_POLYGON)` and `glEnd()`. Let's determine how the `glVertex3f()` statements correspond to corners of the square.

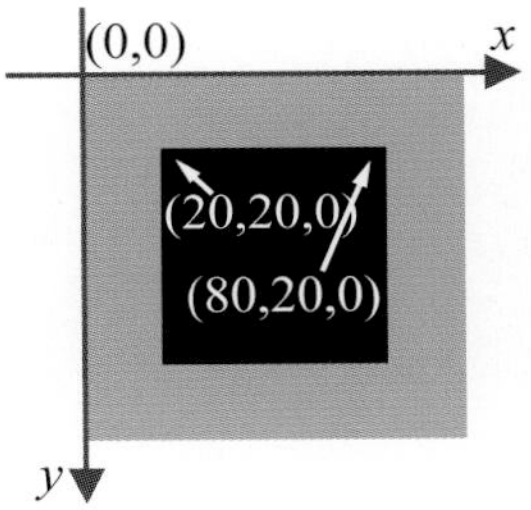

Figure 2.3: The coordinate axes on the OpenGL window of `square.cpp`? *No.*

If, suppose, the vertices are specified in some coordinate system that is embedded in the OpenGL window – which certainly is plausible – and if we knew the axes of this system, the matter would be simple. For example, *if* the x-axis increased horizontally rightwards and the y-axis vertically downwards, as in Figure 2.3, then `glVertex3f(20.0, 20.0, 0.0)` would correspond to the upper-left corner of the square, `glVertex3f(80.0, 20.0, 0.0)` to the upper-right corner and so on.

However, even assuming that there do exist these invisible axes attached to the OpenGL window, how do we find out where they are or how they are oriented? One way is to "wiggle" the corners of the square! For example, change the first vertex declaration from `glVertex3f(20.0, 20.0, 0.0)` to `glVertex3f(30.0, 20.0, 0.0)` and observe which corner moves. Having determined in this way the correspondence of the corners with the vertex statements, we ask the reader to deduce the orientation of the hypothetical coordinate axes. Decide where the origin is located too.

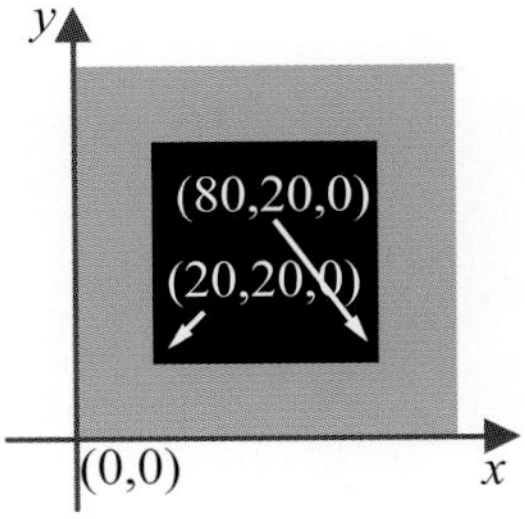

Figure 2.4: The coordinate axes on the OpenGL window of `square.cpp`? *Almost there ....*

Well, it seems then that `square.cpp` sets up coordinates in the OpenGL window so that the increasing direction of the x-axis is horizontally rightwards, that of the y-axis vertically upwards and, moreover, the origin seems to correspond to the lower-left corner of the window, as in Figure 2.4. We're making progress but there's more to the story, so read on!

The last of the three parameters of a `glVertex3f(*, *, *)` declaration is evidently the z coordinate. Vertices are specified in *3-dimensional* space (simply called 3-space or, mathematically, $\mathbb{R}^3$). Indeed, OpenGL allows us to draw in 3-space and create truly 3D scenes, which is its major claim to fame. However, we *perceive* the 3-dimensional scene as a picture *rendered* to a 2-dimensional part of the computer's screen, the rectangular OpenGL window. Shortly we'll see how OpenGL converts a 3D scene to its 2D rendering.

2.2 Orthographic Projection, Viewing Box and World Coordinates

What exactly do the vertex coordinate values mean? For example, is the vertex at (20.0, 20.0, 0.0) of `square.cpp` 20 mm., 20 cm. or 20 pixels away from the origin along both the x-axis and y-axis, or is there some other absolute unit of distance native to OpenGL?

Experiment 2.2. Change the `glutInitWindowSize()` parameter values of `square.cpp`* – first to `glutInitWindowSize(300, 300)` and then `glutInitWindowSize(500, 250)`. The square changes in size, and even shape, with the OpenGL window. Therefore, coordinate values appear not to be in any kind of absolute units on the screen. **End**

Remark 2.1. Of course, you could have reshaped the OpenGL window directly by dragging one of its corners with the mouse, rather than resetting `glutInitWindowSize()` in the program.

Understanding what the coordinates actually represent involves understanding first OpenGL's rendering mechanism, which itself begins with the program's *projection statement*. In the case of `square.cpp` the projection statement is

```
glOrtho(0.0, 100.0, 0.0, 100.0, -1.0, 1.0)
```

in the `resize()` routine, which determines an imaginary *viewing box* inside which the programmer draws. Generally,

`glOrtho(`*left*`,` *right*`,` *bottom*`,` *top*`,` *near*`,` *far*`)`

sets up a viewing box, as in Figure 2.5, with corners at the 8 points:

(*left*, *bottom*, −*near*), (*right*, *bottom*, −*near*), (*left*, *top*, −*near*), (*right*, *top*, −*near*), (*left*, *bottom*, −*far*), (*right*, *bottom*, −*far*), (*left*, *top*, −*far*), (*right*, *top*, −*far*)

It's a box with sides aligned along the axes, whose span along the x-axis is from *left* to *right*, along the y-axis from *bottom* to *top*, and along the z-axis from −*far* to −*near*. Note the little quirk of OpenGL that the *near* and *far* values are flipped in sign.

The viewing box corresponding to the projection statement `glOrtho(0.0, 100.0, 0.0, 100.0, -1.0, 1.0)` of `square.cpp` is shown in Figure 2.6(a). The reader may wonder at this time how the initial coordinate axes are *themselves* calibrated – e.g., is a unit along an axis one inch, one centimeter or something else – as the size of the viewing box and that of the objects drawn inside it depend on this. The answer will be evident once the rendering process is explained momentarily.

*When we refer to `square.cpp`, or any `program.cpp`, it's always to the original version in a folder in the `ExperimenterSource` directory, so if you've modified the code for an earlier experiment you'll need to copy back the original.

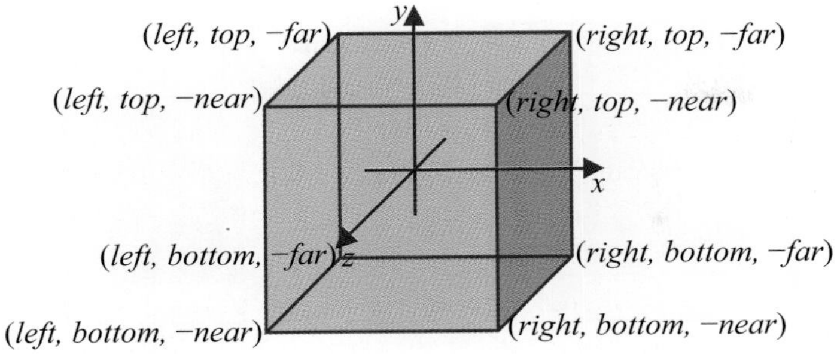

Figure 2.5: Viewing box of `glOrtho(`*left*, *right*, *bottom*, *top*, *near*, *far*`)`.

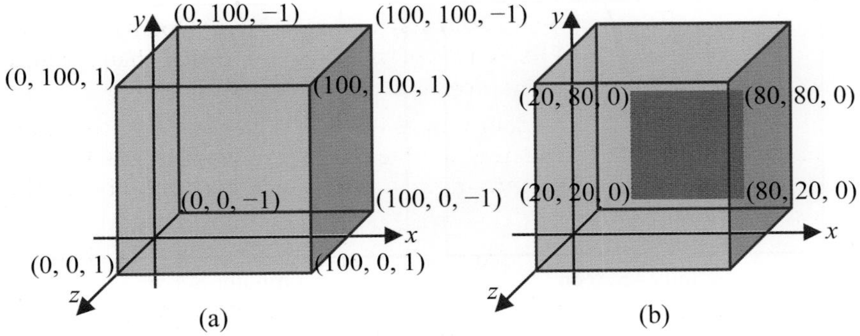

Figure 2.6: (a) Viewing box of `square.cpp` (b) With the square drawn inside.

As for drawing now, the vertex declaration `glVertex3f(x, y, z)` corresponds to the point (x, y, z). For example, the corner of the square declared by `glVertex3f(20.0, 20.0, 0.0)` is at (20.0, 20.0, 0.0). The square of `square.cpp`, then, is as depicted in Figure 2.6(b).

Once the programmer has drawn the entire scene, if the projection statement is `glOrtho()` as in `square.cpp`, then the rendering process is two-step:

1. *Shoot*: First, objects are *projected perpendicularly* onto the front face of the viewing box, i.e., the face on the $z = -near$ plane. For example, the square in Figure 2.7(a) (same as Figure 2.6(b)) is projected as in Figure 2.7(b). The front face of the viewing box is called the *viewing face* and the plane on which it lies the *viewing plane*.

   This step is like shooting the scene on film. In fact, one can think of the viewing box as a giant version of those archaic *box cameras* where the photographer ducks behind the film – the viewing face – and covers her head with a black cloth; so big, in fact, that the whole scene is actually inside the box! Moreover, mind that with this analogy there's

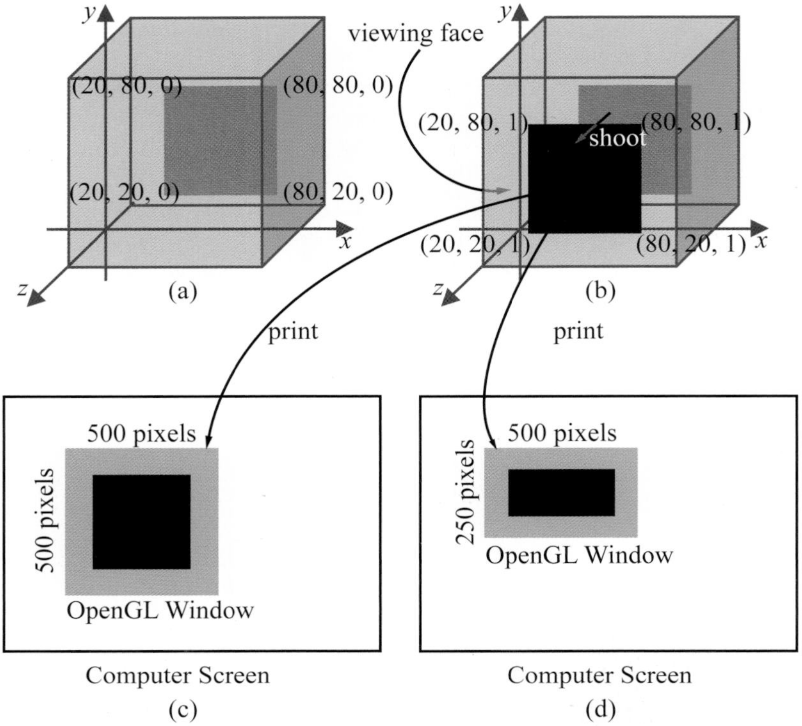

Figure 2.7: Rendering with `glOrtho()`.

no lens, only the film.

2. *Print*: Next, the viewing face is *proportionately scaled* to fit the rectangular OpenGL window. This step is like printing the film on paper. In the case of `square.cpp`, printing takes us from Figure 2.7(b) to (c).

   If, say, the window size of `square.cpp` were changed to one of *aspect ratio* (= width/height) of 2, by replacing `glutInitWindowSize(500, 500)` with `glutInitWindowSize(500, 250)`, printing would take us from Figure 2.7(b) to (d) (which actually distorts the square into a rectangle).

The answer to the earlier question of how to calibrate the coordinate axes of the space in which the viewing box is created should be clear now: the 2D rendering finally displayed is the same *no matter* how they are calibrated, because of the proportionate scaling of the viewing face of the box to fit the OpenGL window. So it does not matter what unit we use, be it an inch, millimeter, mile, ...! Here's a partly-solved exercise to drive home the point.

Exercise 2.1.

(a) Suppose the viewing box of `square.cpp` is set up in a coordinate system where one unit along each axis is 1 cm. Assuming pixels to be 0.2 mm. × 0.2 mm. squares, compute the size and location of the square rendered by shoot-and-print to a 500 pixel × 500 pixel OpenGL window.

(b) Suppose next that the coordinate system is re-calibrated so that a unit along each axis is 1 meter instead of 1 cm., everything else remaining same. What then are the size and location of the rendered square in the OpenGL window?

(c) What is rendered if, additionally, the size of the OpenGL window is changed to 500 pixel × 250 pixel?

Part answer:

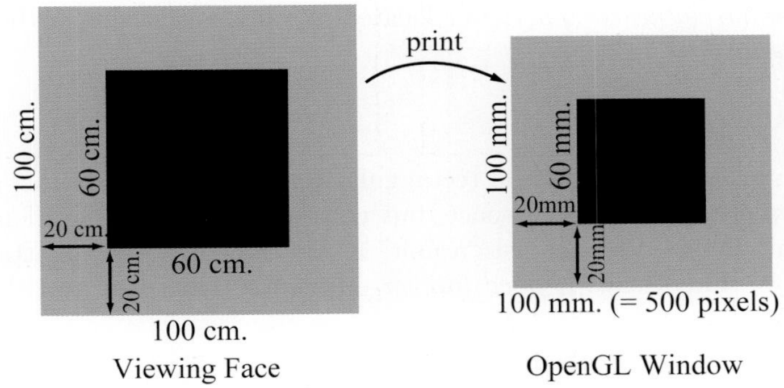

Figure 2.8: The viewing face for `square.cpp`, given that one unit along each coordinate axis is 1 cm., scaled to a 500 pixel × 500 pixel OpenGL window.

(a) Figure 2.8 on the left shows the square projected to the viewing face, which is 100 cm. square. The viewing face is then scaled to the OpenGL window on the right, which is a square of sides 500 pixels = 500 × 0.2 mm. = 100 mm. The scaling from face to the window, therefore, is a factor of 1/10 in both dimensions. It follows that the rendered square is 60 mm. × 60 mm., with its lower-left corner located both 20 mm. above and to the right of the lower-left corner of the window.

(b) Exactly the same as in part (a) because, while the viewing box and viewing face are now 10 times larger, the scaling from face to window is now a factor of 1/100, rather than 1/10.

We conclude that the size and location of the rendering in each coordinate direction are independent of how the axes are calibrated, but determined rather by the *ratio* of the original object's size to that of the viewing box in that direction.

Although the calibration of the axes doesn't matter, nevertheless, we'll make the sensible assumption that all three are calibrated *identically*, i.e., one unit along each axis is of equal length (yes, oddly enough, we could make them different and still the rendering would not change, which you can verify yourself by re-doing Exercise 2.1(a), after assuming that one unit along the x-axis is 1 cm. and along the other two 1 meter). The only other assumptions about the initial coordinate system that we make are conventional ones:

(a) It is *rectangular*, i.e., the three axes are mutually perpendicular.

(b) The x-, y- and z-axes in that order form a *right-handed* system in the following sense: the rotation of the x-axis 90° about the origin so that its positive direction matches with that of the y-axis appears *counter-clockwise* to a viewer located on the positive side of the z-axis (Figure 2.9).

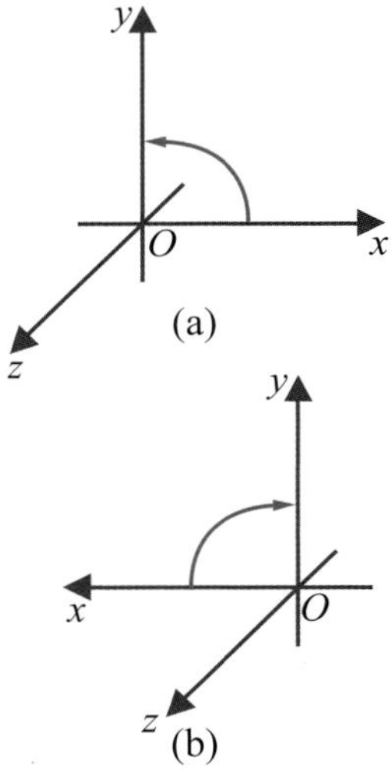

Figure 2.9: The x-, y- and z-axes are rectangular and form a (a) right-handed system (b) left-handed system.

Fixed World System

To summarize, set up an initial rectangular right-handed coordinate system located wherever you like in space, but with axes all calibrated identically. Call a unit along each axis just "a unit" as we know it doesn't matter what the unit is. Then leave it *fixed forever* – imagine it screwed to the top of your desk!

Figure 2.10: A dedicated 3D graphics programmer in a world all her own.

See Figure 2.10. This system coordinatizes *world space* and, in fact, we shall refer to it as the *world coordinate system*. All subsequent objects, including the viewing box and those that we create ourselves, inhabit world space and are specified in world coordinates. These are all virtual objects, of course!

Remark 2.2. Because it's occupied by user-defined objects, world space is sometimes called *object space*.

Experiment 2.3. Change only the viewing box of `square.cpp` by replacing `glOrtho(0.0, 100.0, 0.0, 100.0, -1.0, 1.0)` with `glOrtho(-100, 100.0, -100.0, 100.0, -1.0, 1.0)`. The location of the square in the new viewing box is different and, so as well, the result of shoot-and-print. Figure 2.11 explains how. **End**

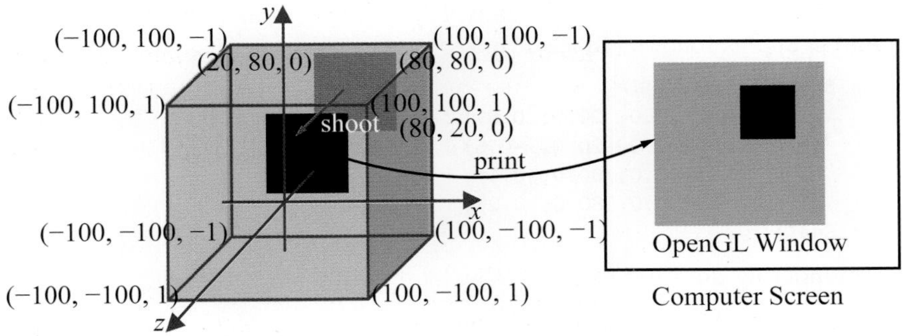

Figure 2.11: The viewing box of `square.cpp` defined by `glOrtho(-100, 100.0, -100.0, 100.0, -1.0, 1.0)`.

Exercise 2.2. (Programming) Change the viewing box of `square.cpp` by replacing `glOrtho(0.0, 100.0, 0.0, 100.0, -1.0, 1.0)` successively with the following, in each case trying to predict the output before running:

(a) `glOrtho(0.0, 200.0, 0.0, 200.0, -1.0, 1.0)`

(b) `glOrtho(20.0, 80.0, 20.0, 80.0, -1.0, 1.0)`

(c) `glOrtho(0.0, 100.0, 0.0, 100.0, -2.0, 5.0)`

Exercise 2.3. If the viewing box of square.cpp is changed by replacing `glOrtho(0.0, 100.0, 0.0, 100.0, -1.0, 1.0)` with `glOrtho(-100.0, 100.0, -100.0, 100.0, -1.0, 1.0)`, and the OpenGL window size changed replacing `glutInitWindowSize(500, 500)` with `glutInitWindowSize(500, 250)`, then calculate the area (in *number of pixels*) of the image of the square.

Exercise 2.4. (Programming) We saw earlier that, as a result of the print step, replacing `glutInitWindowSize(500, 500)` with `glutInitWindowSize(500, 250)` in `square.cpp` causes the square to be distorted into a rectangle. By changing *only one* numerical parameter elsewhere in the program, eliminate the distortion to make it appear square again.

Incidentally, it's clear now that our working hypothesis after the first experiment in Section 2.1, that the OpenGL window comes with axes fixed to it, though not unreasonable, was not accurate either. The OpenGL window it turns out is simply an empty target rectangle on which the front face of the viewing box is printed. This rectangle is called *screen space*.

So there are two spaces we'll be interacting with: world and screen. The former is a virtual 3D space in which we create our scenes, while the latter is a real 2D space where images concocted from our scenes by shoot-and-print are rendered for viewing.

Exercise 2.5. (Programming) Alter the z coordinates of each vertex of the "square" – we should really call it a polygon if we do this – of `square.cpp` as follows (Block 1*):

```
glBegin(GL_POLYGON);
   glVertex3f(20.0, 20.0, 0.5);
   glVertex3f(80.0, 20.0, -0.5);
   glVertex3f(80.0, 80.0, 0.1);
   glVertex3f(20.0, 80.0, 0.2);
glEnd();
```

The rendering does not change. Why?

Remark 2.3. Always set the parameters of `glOrtho(`*left*`,` *right*`,` *bottom*`,` *top*`,` *near*`,` *far*`)` so that *left* < *right*, *bottom* < *top*, and *near* < *far*.

Remark 2.4. The aspect ratio (= width/height) of the viewing box should be set same as that of the OpenGL window or the scene will be distorted by the print step.

Remark 2.5. The perpendicular projection onto the viewing plane corresponding to a `glOrtho()` call is also called *orthographic projection* or *orthogonal projection* (hence the name of the call). Yet another term is *parallel projection* as the lines of projection from points in the viewing box to the viewing plane are all parallel.

2.3 The OpenGL Window and Screen Coordinates

We've already had occasion to use the `glutInitWindowSize(`w`,` h`)` command which sets the size of the OpenGL window to width w and height h measured in pixels. A companion command is `glutInitWindowPosition(`x`,` y`)` to specify the location (x, y) of the upper-left corner of the OpenGL window on the computer screen.

*To cut-and-paste you can find the block in text format in the file `chap2codeModifications.txt` in the directory `ExperimenterSource/CodeModifications`.

Experiment 2.4. Change the parameters of `glutInitWindowPosition(`x`, `y`)` in `square.cpp` from the current (100, 100) to a few different values to determine the location of the origin (0, 0) of the computer screen, as well as the orientation of the screen's own x-axis and y-axis. **End**

The origin $(0, 0)$ of the screen it turns out is at its upper-left corner, while the increasing direction of its x-axis is horizontally rightwards and that of its y-axis vertically downwards; moreover, one unit along either axis is *absolute* and represents a pixel. See Figure 2.12, which shows as well the coordinates of the corners of the OpenGL window initialized by `square.cpp`.

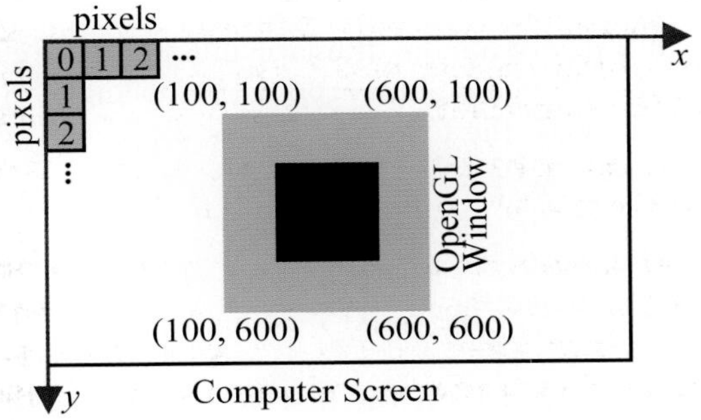

Figure 2.12: The screen's coordinate system: a unit along either axis is the pitch of a pixel.

Note the inconsistency between the orientation of the screen's y-axis and the y-axis of the world coordinate system, the latter being directed *up* the OpenGL window (after being projected there). One needs to be careful about this, especially when coding programs where data is read from one system and used in the other.

2.4 Clipping

A question may have come to the reader's mind about objects which happen to be drawn outside the viewing box. Here are a few experiments to clarify how they are processed.

Experiment 2.5. Add another square by inserting the following right after the code for the original square in `square.cpp` (Block 2):

```
glBegin(GL_POLYGON);
   glVertex3f(120.0, 120.0, 0.0);
   glVertex3f(180.0, 120.0, 0.0);
   glVertex3f(180.0, 180.0, 0.0);
   glVertex3f(120.0, 180.0, 0.0);
glEnd();
```

From the value of its vertex coordinates the second square evidently lies entirely outside the viewing box.

If you run now there's no sign of the second square in the OpenGL window! This is because OpenGL *clips* the scene to within the viewing box before rendering, so that objects or parts of objects drawn outside are not rendered. Clipping is a stage in the graphics pipeline. We'll not worry about its implementation at this time, only the effect it has. **End**

Exercise 2.6. (**Programming**) In the preceding experiment can you redefine the viewing box by changing the parameters of `glOrtho()` so that both squares are visible?

Figure 2.13: Screenshot of a triangle.

Experiment 2.6. For a more dramatic illustration of clipping, first replace the square in the original `square.cpp` with a triangle; in particular, replace the polygon code with the following (Block 3):

```
glBegin(GL_POLYGON);
   glVertex3f(20.0, 20.0, 0.0);
   glVertex3f(80.0, 20.0, 0.0);
   glVertex3f(80.0, 80.0, 0.0);
glEnd();
```

See Figure 2.13. Next, lift the first vertex up the z-axis by changing it to `glVertex3f(20.0, 20.0, 0.5)`; lift it further by changing its z-value to 1.5 (Figure 2.14 is a screenshot), then 2.5 and, finally, 10.0. Make sure you believe that what you see in the last three cases is indeed a triangle clipped to within the viewing box – Figure 2.15 may be helpful. **End**

Figure 2.14: Screenshot of the triangle clipped to a quadrilateral.

Exercise 2.7. (**Programming**) A triangle was clipped to a quadrilateral in the viewing box in the preceding experiment. What is the maximum number of sides of a figure to which you can clip a triangle in the box (quadrilateral, pentagon, hexagon, ...)? Code and show.

Exercise 2.8. Use pencil and paper to guess the output if the polygon declaration part of `square.cpp` is replaced with the following (Block 4):

```
glBegin(GL_POLYGON);
   glVertex3f(-20.0, -20.0, 0.0);
   glVertex3f(80.0, 20.0, 0.0);
   glVertex3f(120.0, 120.0, 0.0);
   glVertex3f(20.0, 80.0, 0.0);
glEnd();
```

The viewing box has six faces that lie on different six planes and, effectively, OpenGL clips *off* the scene on one side of each of these six planes, accordingly called *clipping planes*. Imagine a knife slicing down each plane as in Figure 2.15. Specifically, in the case of the viewing box set up by `glOrtho(`*left*`,` *right*`,` *bottom*`,` *top*`,` *near*`,` *far*`)`, clipped off is to the left of the plane $x = left$, to the right of the plane $x = right$ and so on.

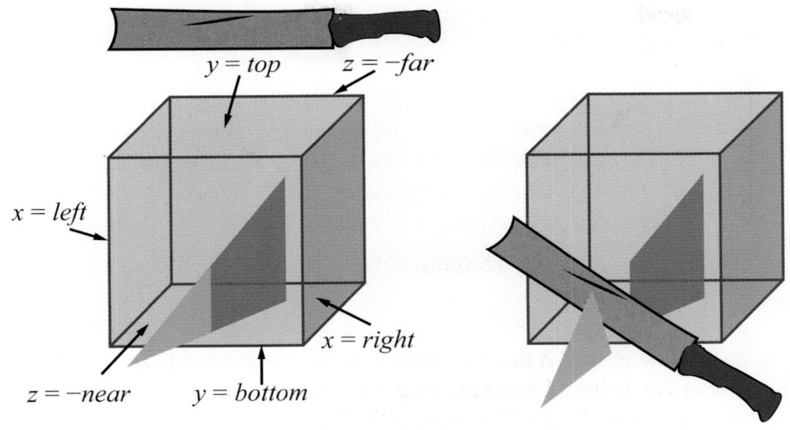

Figure 2.15: Six clipping planes of the `glOrtho(`*left*, *right*, *bottom*, *top*, *near*, *far*`)` viewing box. The lightly shaded part of the triangle sticking out of the box is clipped by a "clipping knife".

Remark 2.6. As we shall see in Chapter 3, the programmer can define clipping planes in addition to the six that bound the viewing box.

We'll leave this section with a rather curious phenomenon for the reader to explain.

Exercise 2.9. (**Programming**) Raising the first vertex of (the original) `square.cpp` from `glVertex3f(20.0, 20.0, 0.0)` to `glVertex3f(20.0, 20.0, 1.5)` causes the square – actually, the new figure which is not a square any more – to be clipped. If, instead, the second vertex is raised from `glVertex3f(80.0, 20.0, 0.0)` to `glVertex3f(80.0, 20.0, 1.5)`, then the figure is clipped too, *but* very differently from when the first vertex is raised. Why? Should not the results be similar by symmetry?
Hint: OpenGL draws polygons after triangulating them as so-called *triangle fans* with the first vertex of the polygon the center of the fan. For example, the fan in Figure 2.16 consists of three triangles around vertex v_0.

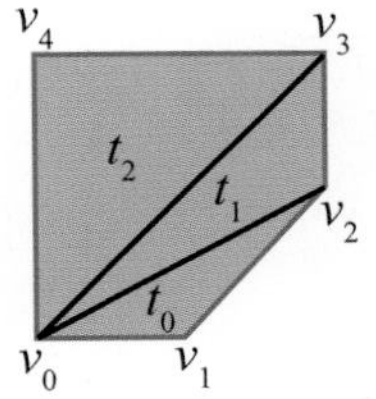

Figure 2.16: A triangle fan.

2.5 Color, OpenGL State Machine and Interpolation

Experiment 2.7. The color of the square in `square.cpp` is specified by the three parameters of the `glColor3f(0.0, 0.0, 0.0)` statement in the `drawScene()` routine, each of which gives the value of one of the three primary components, *blue*, *green* and *red*.

Determine which of the three parameters of `glColor3f()` specifies the blue, green and red components by setting in turn each to 1.0 and the others to 0.0. In fact, verify the following table:

Call	Color
`glColor3f(0.0, 0.0, 0.0)`	Black
`glColor3f(1.0, 0.0, 0.0)`	Red
`glColor3f(0.0, 1.0, 0.0)`	Green
`glColor3f(0.0, 0.0, 1.0)`	Blue
`glColor3f(1.0, 1.0, 0.0)`	Yellow
`glColor3f(1.0, 0.0, 1.0)`	Magenta
`glColor3f(0.0, 1.0, 1.0)`	Cyan
`glColor3f(1.0, 1.0, 1.0)`	White

End

Generally, the `glColor3f(`*red*`, `*green*`, `*blue*`)` call specifies the *foreground color*, or *drawing color*, which is the color applied to objects being drawn. The value of each color component, which ought to be a number between 0.0 and 1.0, determines its *intensity*. For example, `glColor3f(1.0, 1.0, 0.0)` is the brightest yellow while `glColor3f(0.5, 0.5, 0.0)` is a weaker yellow

Remark 2.7. The color values are each *clamped* to the range $[0, 1]$. This means that, if a value happens to be set greater than 1, then it's taken to be 1; if less than 0, it's taken to be 0.

Exercise 2.10. (Programming) Both `glColor3f(0.2, 0.2, 0.2)` and `glColor3f(0.8, 0.8, 0.8)` should be grays, having equal red, green and blue intensities. Guess which is the darker of the two. Verify by changing the foreground color of `square.cpp`.

The call `glClearColor(1.0, 1.0, 1.0, 0.0)` in the `setup()` routine specifies the *background color*, or *clearing color*. Ignore for now the fourth parameter, which is the *alpha* value. The statement `glClear(GL_COLOR_BUFFER_BIT)` in `drawScene()` actually clears the window to the specified background color, which means that every pixel in the color buffer is set to that color.

Experiment 2.8. Add the additional color declaration statement `glColor3f(1.0, 0.0, 0.0)` just after the existing one `glColor3f(0.0, 0.0, 0.0)` in the drawing routine of `square.cpp` so that the foreground color block becomes

```
// Set foreground (or drawing) color.
glColor3f(0.0, 0.0, 0.0);
glColor3f(1.0, 0.0, 0.0);
```

The square is drawn red because the *current* value of the foreground color is red when each of its vertices is specified. **End**

Foreground color is one of a collection of variables, called *state variables*, which determine the state of OpenGL. Among other state variables are point size, line width, line stipple, material properties, etc. We'll meet several as we go along or you can refer to the red book* for a full list. OpenGL remains and functions in its current state until a declaration is made changing a state variable. For this reason, OpenGL is often called a *state machine*. The following simple experiment illustrates a couple of important points about how state variables control rendering.

Experiment 2.9. Replace the polygon declaration part of `square.cpp` with the following to draw two squares (Block 5):

```
glColor3f(1.0, 0.0, 0.0);
glBegin(GL_POLYGON);
   glVertex3f(20.0, 20.0, 0.0);
   glVertex3f(80.0, 20.0, 0.0);
   glVertex3f(80.0, 80.0, 0.0);
   glVertex3f(20.0, 80.0, 0.0);
glEnd();

glColor3f(0.0, 1.0, 0.0);
glBegin(GL_POLYGON);
   glVertex3f(40.0, 40.0, 0.0);
   glVertex3f(60.0, 40.0, 0.0);
   glVertex3f(60.0, 60.0, 0.0);
   glVertex3f(40.0, 60.0, 0.0);
glEnd();
```

A small green square appears inside a larger red one (Figure 2.17). Obviously, this is because the foreground color is red for the first square, but green for the second. One says that the color red *binds* to the first square – or, more precisely, to each of its four specified vertices – and green to the second square. These bound values specify the color *attribute* of either square. Generally, the values of those state variables which determine how it is rendered collectively form a primitive's attribute set.

Flip the order in which the two squares appear in the code by cutting the seven statements that specify the red square and pasting them after those to do with the green one. The green square is overwritten by the red one and no longer visible because OpenGL draws in *code order*.

End

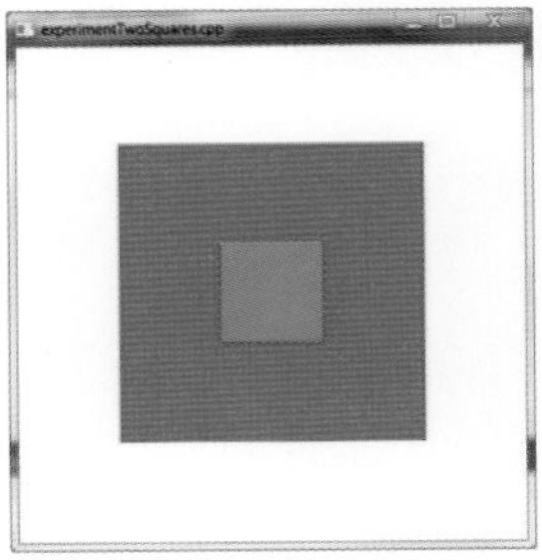

Figure 2.17: Screenshot of a green square drawn in the code after a red square.

Experiment 2.10. Replace the polygon declaration part of `square.cpp` with (Block 6):

```
glBegin(GL_POLYGON);
   glColor3f(1.0, 0.0, 0.0);
```

*The *OpenGL Programming Guide* [104] and its companion volume, the *OpenGL Reference Manual* [105], are the canonical references for the OpenGL API and affectionately referred to as the red book and blue book, respectively. Note that the on-line reference docs at `www.opengl.org` pretty much cover all that is in the blue book.

```
      glVertex3f(20.0, 20.0, 0.0);
      glColor3f(0.0, 1.0, 0.0);
      glVertex3f(80.0, 20.0, 0.0);
      glColor3f(0.0, 0.0, 1.0);
      glVertex3f(80.0, 80.0, 0.0);
      glColor3f(1.0, 1.0, 0.0);
      glVertex3f(20.0, 80.0, 0.0);
   glEnd();
```

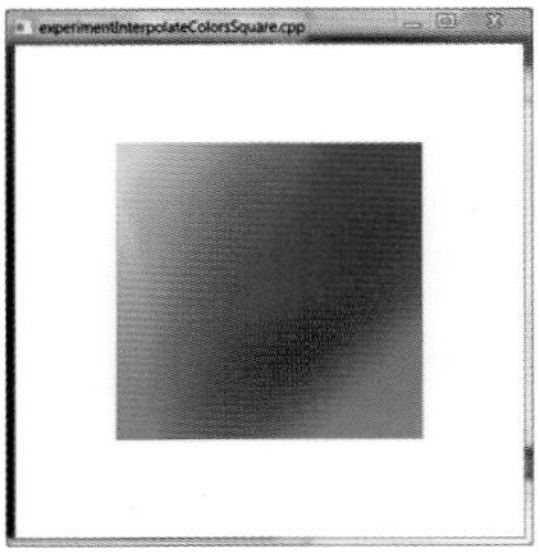

Figure 2.18: Screenshot of a square with differently colored vertices.

The different color values bound to the four vertices of the square are evidently *interpolated* over the rest of the square as you can see in Figure 2.18. In fact, this is most often the case with OpenGL: numerical attribute values specified at the vertices of a primitive are interpolated throughout its interior. In a later chapter we'll see exactly what it means to interpolate and how OpenGL goes about the task. End

2.6 OpenGL Geometric Primitives

The geometric primitives – also called drawing primitives or, simply, primitives – of OpenGL are the parts that programmers use in Lego-like manner to create mundane objects like balls and boxes, as well as elaborate spacecrafts, the worlds to which they travel, and pretty much everything in between. The only one we've seen so far is the polygon. It's time to get acquainted with the whole family.

Experiment 2.11. Replace `glBegin(GL_POLYGON)` with `glBegin(GL_POINTS)` in `square.cpp` and make the point size bigger with a call to `glPointSize(5.0)`, so that the part drawing the polygon is now

```
   glPointSize(5.0); // Set point size.
   glBegin(GL_POINTS);
      glVertex3f(20.0, 20.0, 0.0);
      glVertex3f(80.0, 20.0, 0.0);
      glVertex3f(80.0, 80.0, 0.0);
      glVertex3f(20.0, 80.0, 0.0);
   glEnd();
```

End

Experiment 2.12. Continue, replacing `GL_POINTS` with `GL_LINES`, `GL_LINE_STRIP` and, finally, `GL_LINE_LOOP`. End

In the explanation that follows of how OpenGL draws each primitive, assume that the n vertices declared in the code between `glBegin(`*primitive*`)` and `glEnd()` are $v_0, v_1, \ldots, v_{n-1}$ in that order, i.e., the declaration of the primitive is of the form:

```
glBegin(primitive);
   glVertex3f(*, *, *); // v0
   glVertex3f(*, *, *); // v1
   ...
   glVertex3f(*, *, *); // vn-1
glEnd();
```

Refer to Figure 2.19 as you read.

GL_POINTS draws a point at each vertex

$$v_0, v_1, \ldots, v_{n-1}$$

GL_LINES draws a *disconnected* sequence of straight line segments (henceforth, we'll simply use the term "segment") between the vertices, taken two at a time. In particular, it draws the segments

$$v_0v_1,\ v_2v_3,\ \ldots,\ v_{n-2}v_{n-1}$$

if n is even. If n is not even then the last vertex v_{n-1} is simply ignored.

GL_LINE_STRIP draws the *connected* sequence of segments

$$v_0v_1,\ v_1v_2,\ \ldots,\ v_{n-2}v_{n-1}$$

Such a sequence is called a *polygonal line* or *polyline*.

GL_LINE_LOOP is the same as `GL_LINE_STRIP`, *except* that an additional segment $v_{n-1}v_0$ is drawn to complete a loop:

$$v_0v_1,\ v_1v_2,\ \ldots,\ v_{n-2}v_{n-1},\ v_{n-1}v_0$$

Such a segment sequence is called a *polygonal line loop*.

The thickness of lines can be set by a `glLineWidth(width)` call.

Remark 2.8. In world space points have zero dimension and lines zero width; values specified by `glPointSize()` and `glLineWidth()` are used only for rendering. Otherwise, it would be rather hard to see a point actually of zero dimension or a line of zero width!

Why does OpenGL provide separate primitives to draw polygonal lines and line loops when both can be viewed as a collection of segments and drawn using `GL_LINES`? For example,

```
glBegin(GL_LINE_STRIP);
   v0;
   v1;
   v2;
   ...
glEnd();
```

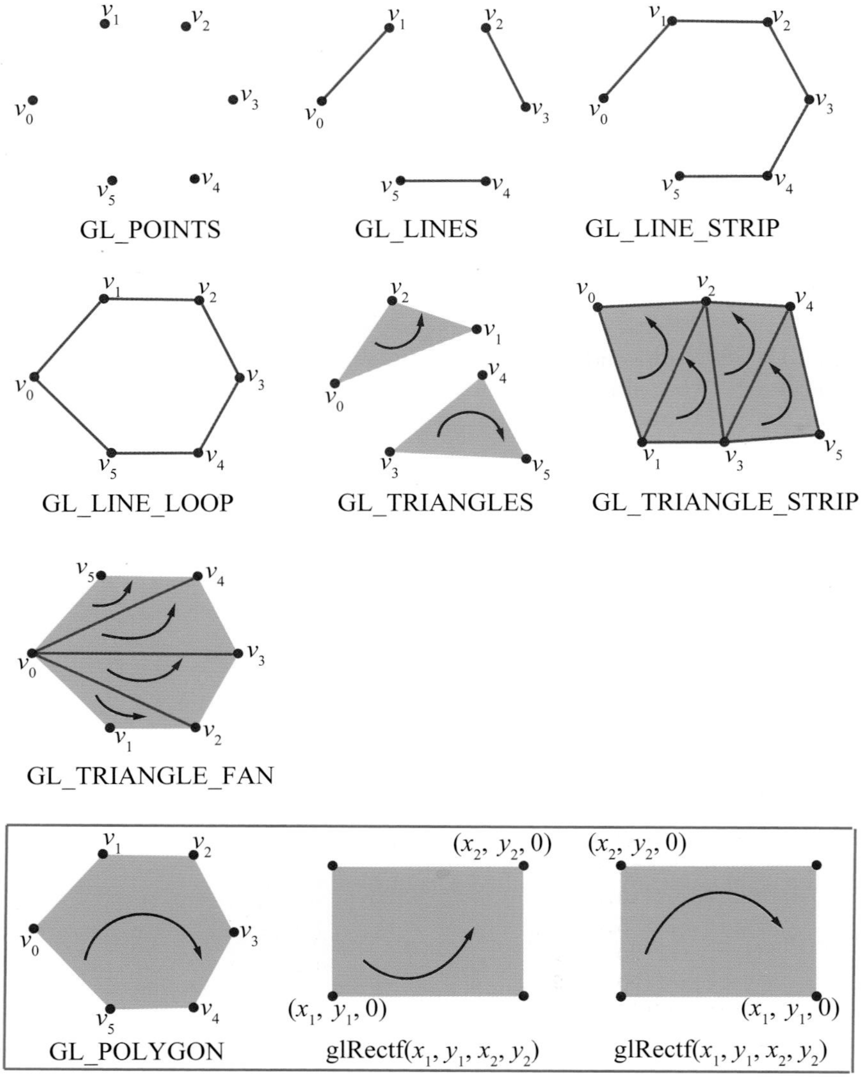

Figure 2.19: OpenGL's geometric primitives. Vertex order is indicated by a curved arrow. Primitives inside the red rectangle have been discarded from the core profile of later versions of OpenGL, e.g., 4.3; however, they are accessible via the compatibility profile.

is equivalent to

```
glBegin(GL_LINES);
   v0;
   v1
```

```
    v1;
    v2
    v2;
    ...
glEnd();
```

The answer is first to avoid redundancy in vertex data. Secondly, possible rendering error is avoided as well because OpenGL does not know that the two v1s in the GL_LINES specification above are supposed to represent the same vertex, and may render them at slightly different locations because of differences in floating point round-offs.

Exercise 2.11. (Programming) This relates to the brief discussion on interpolation at the end of Section 2.5. Replace the polygon declaration part of square.cpp with (Block 7):

```
glLineWidth(5.0);
glBegin(GL_LINES);
    glColor3f(1.0, 0.0, 0.0);
    glVertex3f(20.0, 20.0, 0.0);
    glColor3f(0.0, 1.0, 0.0);
    glVertex3f(80.0, 20.0, 0.0);
glEnd();
```

Can you say what the color values should be at the midpoint $(50.0, 20.0, 0.0)$ of the segment drawn? Check your answer by drawing a point with those color values just above the midpoint, say at $(50.0, 22.0, 0.0)$, and comparing.

Experiment 2.13. Replace the polygon declaration part of square.cpp with (Block 8):

```
glBegin(GL_TRIANGLES);
    glVertex3f(10.0, 90.0, 0.0);
    glVertex3f(10.0, 10.0, 0.0);
    glVertex3f(35.0, 75.0, 0.0);
    glVertex3f(30.0, 20.0, 0.0);
    glVertex3f(90.0, 90.0, 0.0);
    glVertex3f(80.0, 40.0, 0.0);
glEnd();
```

End

GL_TRIANGLES draws a sequence of triangles using the vertices three at a time. In particular, the triangles are

$$v_0v_1v_2,\ v_3v_4v_5,\ \ldots,\ v_{n-3}v_{n-2}v_{n-1}$$

if n is a multiple of 3; if it isn't, the last one, or two, vertices are ignored.

The given order of the vertices for each triangle, in particular, v_0, v_1, v_2 for the first, v_3, v_4, v_5 for the second and so on, determines its *orientation* as perceived by a viewer. Figure 2.19 indicates orientation with curved arrows.

Orientation is important because it enables OpenGL to decide which side of a primitive, front or back, the viewer sees. We'll deal with this topic separately in Chapter 9. Till then disregard orientation when drawing, listing the vertices of a primitive in any order you like.

`GL_TRIANGLES` is a *2-dimensional* primitive and, by default, triangles are drawn filled. However, one can choose a different drawing mode by applying the `glPolygonMode(`*face*`,` *mode*`)` command where *face* may be one of `GL_FRONT`, `GL_BACK` or `GL_FRONT_AND_BACK`, and *mode* one of `GL_FILL`, `GL_LINE` or `GL_POINT`. Whether a primitive is front-facing or back-facing depends, as said above, on its orientation. To keep matters simple for now, though, we'll use only `GL_FRONT_AND_BACK` in a `glPolygonMode()` call, which applies the given drawing mode to a primitive regardless of which face is visible. The `GL_FILL` option is, of course, the default filled option for 2D primitives, while `GL_LINE` draws the primitive in *outline* (or *wireframe* as it's also called), and `GL_POINT` only the vertices.

Experiment 2.14. In fact, it's often easier to decipher a 2D primitive by viewing it in outline. Accordingly, continue the preceding experiment by inserting the call `glPolygonMode(GL_FRONT_AND_BACK, GL_LINE)` in the drawing routine and, further, replacing `GL_TRIANGLES` with `GL_TRIANGLE_STRIP`. The relevant part of the display routine then is as below:

```
// Set polygon mode.
glPolygonMode(GL_FRONT_AND_BACK, GL_LINE);

// Draw a triangle strip.
glBegin(GL_TRIANGLE_STRIP);
   glVertex3f(10.0, 90.0, 0.0);
   glVertex3f(10.0, 10.0, 0.0);
   glVertex3f(35.0, 75.0, 0.0);
   glVertex3f(30.0, 20.0, 0.0);
   glVertex3f(90.0, 90.0, 0.0);
   glVertex3f(80.0, 40.0, 0.0);
glEnd();
```

End

GL_TRIANGLE_STRIP draws a sequence of triangles – called a *triangle strip* – as follows: the first triangle is $v_0v_1v_2$, the second $v_1v_3v_2$ (v_0 is dropped and v_3 brought in), the third $v_2v_3v_4$ (v_1 dropped and v_4 brought in), and so on. Formally, the triangles in the strip are

$$v_0v_1v_2,\ v_1v_3v_2,\ v_2v_3v_4,\ \ldots,\ v_{n-3}v_{n-2}v_{n-1} \qquad \text{(if } n \text{ is odd)}$$

or

$$v_0v_1v_2,\ v_1v_3v_2,\ v_2v_3v_4,\ \ldots,\ v_{n-3}v_{n-1}v_{n-2} \qquad \text{(if } n \text{ is even)}$$

Exercise 2.12. (Programming) Create a square annulus as in Figure 2.20(a) using a *single* triangle strip. You may first want to sketch the

annulus on graph paper to determine the coordinates of its eight corners. The figure depicts one possible *triangulation* – division into triangles – of the annulus.
Hint: A solution is available in `squareAnnulus1.cpp` of Chapter 3.

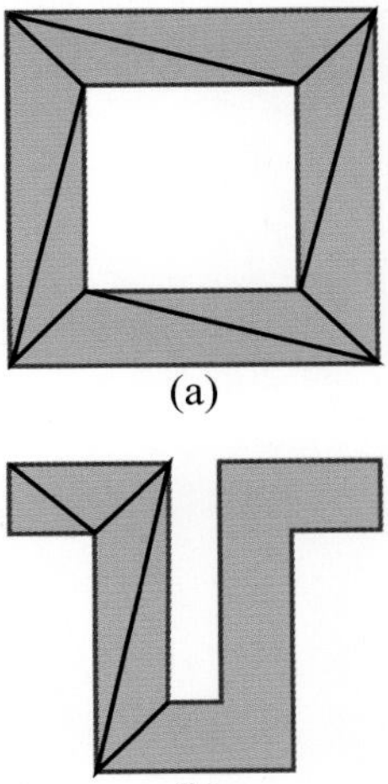

Figure 2.20: (a) Square annulus – the region between two bounding squares – and a possible triangulation (b) A partially triangulated shape.

Exercise 2.13. (**Programming**) Create the shape of Figure 2.20(b) using a single triangle strip. A partial triangulation is indicated.

Experiment 2.15. Replace the polygon declaration part of `square.cpp` with (Block 9):

```
glBegin(GL_TRIANGLE_FAN);
   glVertex3f(10.0, 10.0, 0.0);
   glVertex3f(15.0, 90.0, 0.0);
   glVertex3f(55.0, 75.0, 0.0);
   glVertex3f(80.0, 30.0, 0.0);
   glVertex3f(90.0, 10.0, 0.0);
glEnd();
```

Apply both the filled and outlined drawing modes. **End**

GL_TRIANGLE_FAN draws a sequence of triangles – called a *triangle fan* – around the first vertex as follows: the first triangle is $v_0v_1v_2$, the second $v_0v_2v_3$ and so on. The full sequence is

$$v_0v_1v_2,\ v_0v_2v_3,\ \ldots,\ v_0v_{n-2}v_{n-1}$$

Exercise 2.14. (**Programming**) Create a square annulus using *two* triangle fans. First sketch a triangulation different from that in Figure 2.20(a).

GL_POLYGON draws a polygon with the vertex sequence

$$v_0\, v_1\, \ldots\, v_{n-1}$$

(n must be at least 3 for anything to be drawn).

Finally:

glRectf(*x1*, *y1*, *x2*, *y2*) draws a rectangle lying on the $z = 0$ plane with sides parallel to the x- and y-axes. In particular, the rectangle has diagonally opposite corners at (*x1*, *y1*, 0) and (*x2*, *y2*, 0). The full list of four vertices is (*x1*, *y1*, 0), (*x2*, *y1*, 0), (*x2*, *y2*, 0) and (*x1*, *y2*, 0). The rectangle created is 2-dimensional and its vertex order depends on the situation of the two vertices (*x1*, *y1*, 0) and (*x2*, *y2*, 0) with respect to each other, as indicated by the two drawings at the lower right of Figure 2.19.

Note that `glRectf()` is a stand-alone call; it is not a parameter to `glBegin()` like the other primitives.

Important: The preceding two, `GL_POLYGON` and `glRectf()`, have both been discarded from the core profile of later versions of OpenGL, e.g., the one we are going to study ourselves later in the book, namely, fourth generation 4.3; however, they are accessible via the compatibility profile.

The reason that polygons and rectangles have been discarded is not hard to understand: both can be made from triangles, so are really redundant. The reason we do use them in the first part of this book is because they afford an easily understood way to make objects – e.g., a polygon is certainly more intuitive for a beginner than a triangle strip.

However, when drawing a polygon one must be careful in ensuring that it is a *plane convex* figure, i.e., it lies on one plane and has no "bays" or "inlets" (see Figure 2.21); otherwise, rendering is unpredictable. So, even though we draw them occasionally for convenience, we recommend that the reader, in order to avoid rendering issues and to prepare for the fourth generation of OpenGL, *avoid polygons and rectangles altogether* in her own projects, and, instead, *use exclusively triangles*.

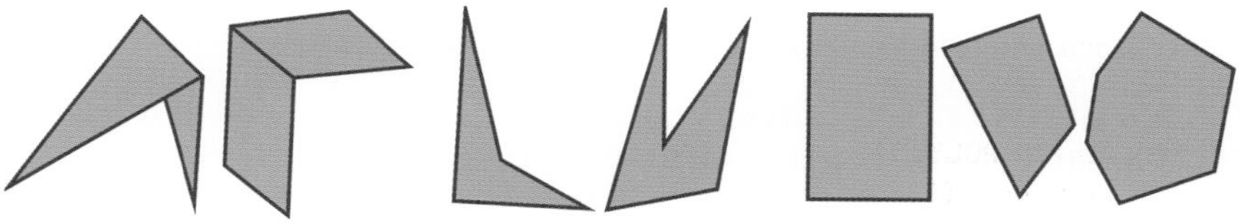

Figure 2.21: OpenGL polygons should be planar and convex.

In fact, following are a couple of experiments, the second one showing how polygon rendering can behave oddly indeed if one is not careful.

Experiment 2.16. Replace the polygon declaration of `square.cpp` with (Block 10):

```
glBegin(GL_POLYGON);
   glVertex3f(20.0, 20.0, 0.0);
   glVertex3f(50.0, 20.0, 0.0);
   glVertex3f(80.0, 50.0, 0.0);
   glVertex3f(80.0, 80.0, 0.0);
   glVertex3f(20.0, 80.0, 0.0);
glEnd();
```

You see a convex 5-sided polygon (Figure 2.22(a)). **End**

Experiment 2.17. Replace the polygon declaration of `square.cpp` with (Block 11):

```
glBegin(GL_POLYGON);
```

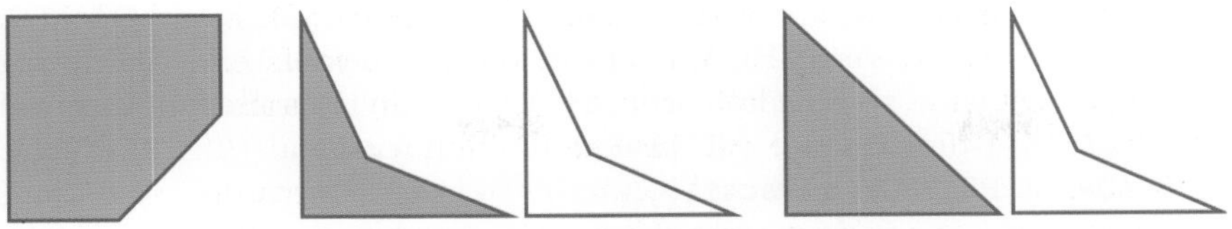

(a) (b) (c)

Figure 2.22: Outputs: (a) Experiment 2.16 (b) Experiment 2.17 (c) Experiment 2.17, vertices cycled.

```
    glVertex3f(20.0, 20.0, 0.0);
    glVertex3f(80.0, 20.0, 0.0);
    glVertex3f(40.0, 40.0, 0.0);
    glVertex3f(20.0, 80.0, 0.0);
glEnd();
```

Display it *both* filled and outlined using appropriate `glPolygonMode()` calls. A non-convex quadrilateral is drawn in either case (Figure 2.22(b)). Next, keeping the same *cycle* of vertices as above, list them starting with `glVertex3f(80.0, 20.0, 0.0)` instead (Block 12):

```
glBegin(GL_POLYGON);
    glVertex3f(80.0, 20.0, 0.0);
    glVertex3f(40.0, 40.0, 0.0);
    glVertex3f(20.0, 80.0, 0.0);
    glVertex3f(20.0, 20.0, 0.0);
glEnd();
```

Make sure to display it both filled and outlined. When filled it's a triangle, while outlined it's a non-convex quadrilateral (Figure 2.22(c)) identical to the one output earlier! Because the cyclic order of the vertices is unchanged, shouldn't it be as in Figure 2.22(b) both filled and outlined? **End**

We'll leave the apparent anomaly* of this experiment as a mystery to be resolved in Chapter 8 on triangulation. But, if you are impatient then refer to the hint provided with Exercise 2.9.

Exercise 2.15. (Programming) Verify, by cycling the vertices, that no such anomaly arises in the case of the convex polygon of Experiment 2.16.

Exercise 2.16. (Programming) Draw the double annulus (a figure '8') shown in Figure 2.23 using as few triangle strips as possible. Introduce additional vertices on the three boundary components if you need to (in addition to the original twelve).

Note: Such additional vertices are called *Steiner vertices*.

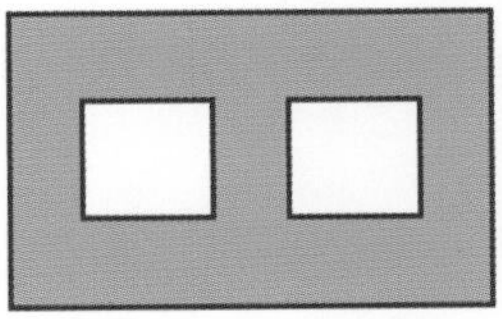

Figure 2.23: Double annulus.

*The rendering depends on the particular OpenGL implementation. However, all implementations that we are aware of show identical behavior.

Remark 2.9. Here's an interesting semi-philosophical question. OpenGL claims to be a 3D drawing API. Yet, why does it not have a single 3D drawing primitive, e.g., cube, tetrahedron or such? All its primitives are 0-dimensional (`GL_POINTS`), 1-dimensional (`GL_LINE*`) or 2-dimensional (`GL_TRIANGLE*`, `GL_QUAD*`, `GL_POLYGON`, `glRectf()`).

The answer lies in how we humans (the regular ones that is and not supers with X-ray vision) perceive 3D objects such as cubes, tetrahedrons, chairs and spacecraft: *we see only the surface, which is two-dimensional*. It makes sense for a 3D API, therefore, to draw only as much as can be seen.

2.7 Approximating Curved Objects

Looking back at Figure 2.19 we see that the OpenGL geometric primitives are composed of points, straight line segments and flat pieces, the latter being triangles, rectangles and polygons. How, then, to draw curved objects such as discs, ellipses, spirals, beer cans and flying saucers? The answer is to *approximate* them with straight and flat OpenGL primitives well enough that the viewer cannot tell the difference. As a wag once put it, "Sincerity is a very important human quality. If you don't have it, you *gotta* fake it!" In the next experiment we fake a circle.

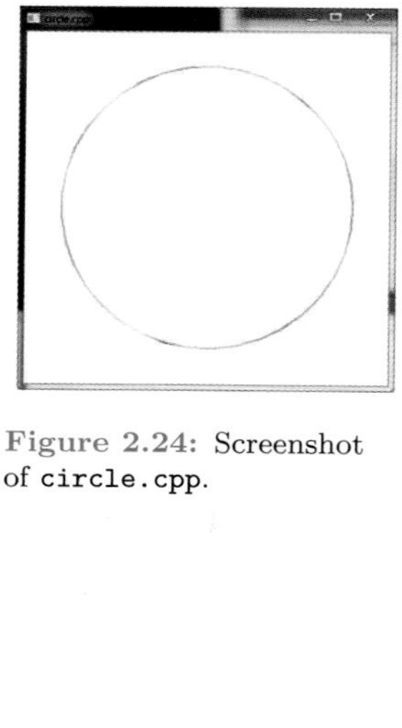

Figure 2.24: Screenshot of `circle.cpp`.

Experiment 2.18. Run `circle.cpp`. Increase the number of vertices in the line loop

```
glBegin(GL_LINE_LOOP);
   for(i = 0; i < numVertices; ++i)
   {
      glColor3f((float)rand()/(float)RAND_MAX,
                (float)rand()/(float)RAND_MAX,
                (float)rand()/(float)RAND_MAX);
      glVertex3f(X + R * cos(t), Y + R * sin(t), 0.0);
      t += 2 * PI / numVertices;
   }
glEnd();
```

by pressing '+' till it "becomes" a circle, as in the screenshot of Figure 2.24. Press '-' to decrease the number of vertices. The randomized colors are just a bit of eye candy. **End**

The vertices of the loop of `circle.cpp`, which lie evenly spaced on the circle, are collectively called a *sample of points* or, simply, *sample* from the circle. See Figure 2.25(a). The denser the sample evidently the better the approximation.

The parametric equations of the circle implemented are

$$x = X + R\cos t,\ y = Y + R\sin t,\ z = 0, \quad 0 \le t \le 2\pi \tag{2.1}$$

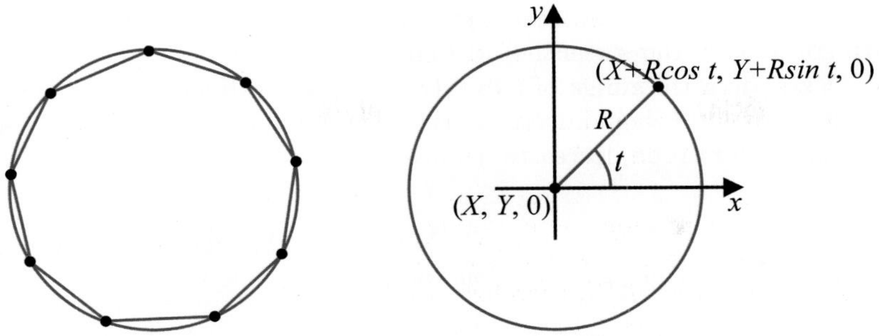

(a) (b)

Figure 2.25: (a) A line loop joining a sample of points from a circle (b) Parametric equations for a circle.

where $(X, Y, 0)$ is the center and R the radius of the circle. See Figure 2.25(b). A `numVertices` number of sample points equally spaced apart is generated by starting with the angle $t = 0$ and then incrementing it successively by $2\pi/$`numVertices`.

Observe that the vertex specifications occur within a loop construct, which is pretty much mandatory if there is a large number of vertices.

Incidentally, the program `circle.cpp` also demonstrates output to the command window, as well as non-trivial user interaction via the keyboard. The routine `keyInput()` is registered as the key handling routine in `main()` by the `glutKeyboardFunc(keyInput)` statement. Note the calls to `glutPostRedisplay()` in `keyInput()` asking the display to be redrawn after each update of `numVertices`.

Follow these conventions when writing OpenGL code:

1. Program the "Esc" key to exit the program.
2. Describe user interaction at two places:
   (a) The command window using `cout()`.
   (b) Comments at the *top* of the source code.

Here's a parabola.

Experiment 2.19. Run `parabola.cpp`. Press '+/-' to increase/decrease the number of vertices of the approximating line strip. Figure 2.26 is a screenshot with enough vertices to make a smooth-looking parabola.

The vertices are equally spaced along the x-direction. The parametric equations implemented are

$$x = 50 + 50t,\ y = 100t^2,\ z = 0, \quad -1 \le t \le 1$$

the constants being chosen so that the parabola is centered in the OpenGL window.

End

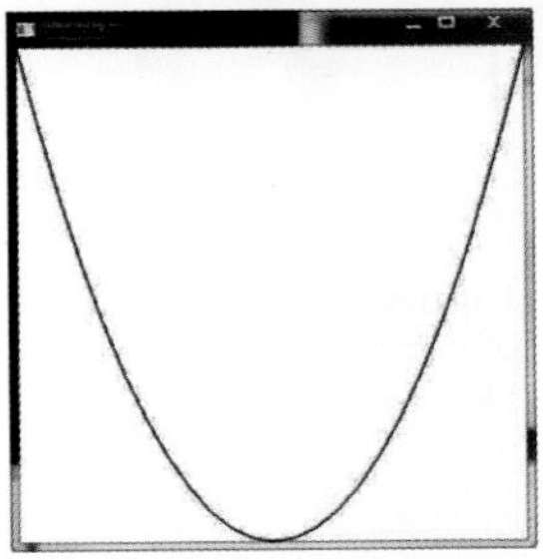

Figure 2.26: Screenshot of `parabola.cpp`.

Exercise 2.17. (**Programming**) Modify `circle.cpp` to draw a flat 3-turn spiral as in the screenshot of Figure 2.27.
Hint: Extending the range of t to 0 to 6π gives 3 full turns. At the same time the "radius" R should decrease with each iteration of the `for` loop, e.g., `R-=20.0/numVertices` decreases it linearly from 40 to 20.

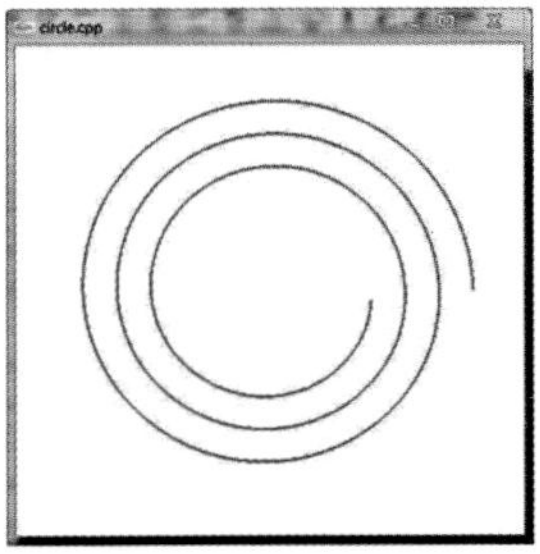

Figure 2.27: Flat spiral.

Exercise 2.18. (**Programming**) Modify `circle.cpp` to draw a disc (a filled circle) by way of (a) a polygon and (b) a triangle fan.

Exercise 2.19. (**Programming**) Draw a flat leaf like the one in Figure 2.28.

Exercise 2.20. (**Programming**) Modify `circle.cpp` to draw a circular annulus, like one of those shown in Figure 2.29, using a triangle strip. Don't look at the program `circularAnnuluses.cpp`!

We'll be returning shortly to the topic of approximating curved objects, but it's on to 3D next.

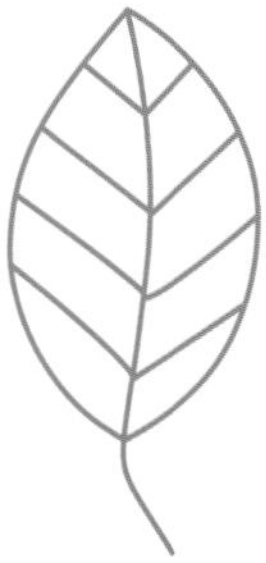

Figure 2.28: Flat leaf.

2.8 Three Dimensions, the Depth Buffer and Perspective Projection

The reader by now may be getting impatient to move on from the plane (pun intended) and simple to full 3D. Okay then, let's get off to an easy start in 3-space by making use of the third dimension to fake a circular annulus. Don't worry, we'll be doing fancier stuff soon enough!

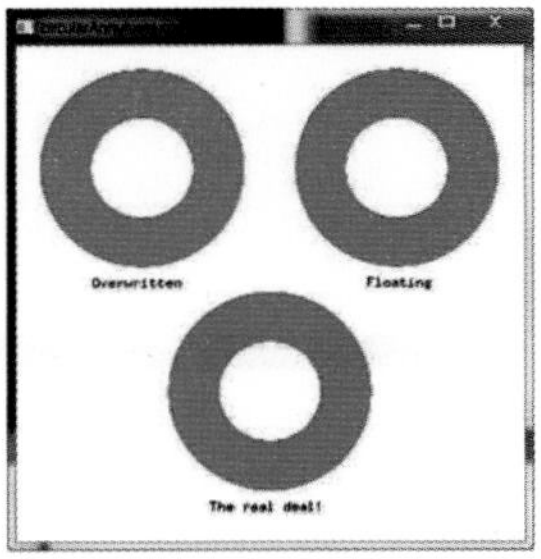

Figure 2.29: Screenshot of `circularAnnuluses.cpp`.

Experiment 2.20. Run `circularAnnuluses.cpp`. Three identical-looking red circular annuluses (Figure 2.29) are drawn in three *different* ways:

i) Upper-left: There is not a real hole. The white disc *overwrites* the red disc as it appears later in the code.

```
glColor3f(1.0, 0.0, 0.0);
drawDisc(20.0, 25.0, 75.0, 0.0);
glColor3f(1.0, 1.0, 1.0);
drawDisc(10.0, 25.0, 75.0, 0.0);
```

Note: The first parameter of `drawDisc()` is the radius and the remaining three the coordinates of the center.

ii) Upper-right: There is not a real hole either. A white disc is *drawn closer* to the viewer than the red disc thus blocking it out.

```
glEnable(GL_DEPTH_TEST);
glColor3f(1.0, 0.0, 0.0);
drawDisc(20.0, 75.0, 75.0, 0.0);
glColor3f(1.0, 1.0, 1.0);
drawDisc(10.0, 75.0, 75.0, 0.5);
glDisable(GL_DEPTH_TEST);
```

Observe that the z-value of the white disc's center is greater than the red disc's. We'll discuss the mechanics of one primitive blocking out another momentarily.

iii) Lower: A true circular annulus with a real hole.

```
if (isWire) glPolygonMode(GL_FRONT, GL_LINE);
else glPolygonMode(GL_FRONT, GL_FILL);
glColor3f(1.0, 0.0, 0.0);
glBegin(GL_TRIANGLE_STRIP);
...
glEnd();
```

Press the space bar to see the wireframe of a triangle strip. **End**

Exercise 2.21. (Programming) Interchange in `circularAnnuluses.cpp` the drawing orders of the red and white discs – i.e., the order in which they appear in the code – in either of the top two annuluses. Which one is affected? (*Only the first!*) Why?

Remark 2.10. Note the use of a text-drawing routine in `circularAnnuluses.cpp`. OpenGL offers only rudimentary text-drawing capability but it often comes in handy, especially for annotation. We'll discuss text-drawing in fair detail in Chapter 3.

By far the most important aspect of `circularAnnuluses.cpp` is its use of the *depth buffer* to draw the upper-right annulus. Following is an introduction to this critical utility which enables realistic rendering of 3D scenes.

2.8.1 A Vital 3D Utility: The Depth Buffer

Enabling the depth buffer, also called the *z-buffer*, causes OpenGL to eliminate, prior to rendering, parts of objects that are *obscured* (or, *occluded*) by others.

Precisely, a point of an object is not drawn if its projection – think of a ray from that point – toward the viewing face is obstructed by another object. See Figure 2.30(a) for the making of the upper-right annulus of `circularAnnuluses.cpp`: the white disc obscures the part of the red disc behind it (because the projection is orthogonal, the obscured part is exactly

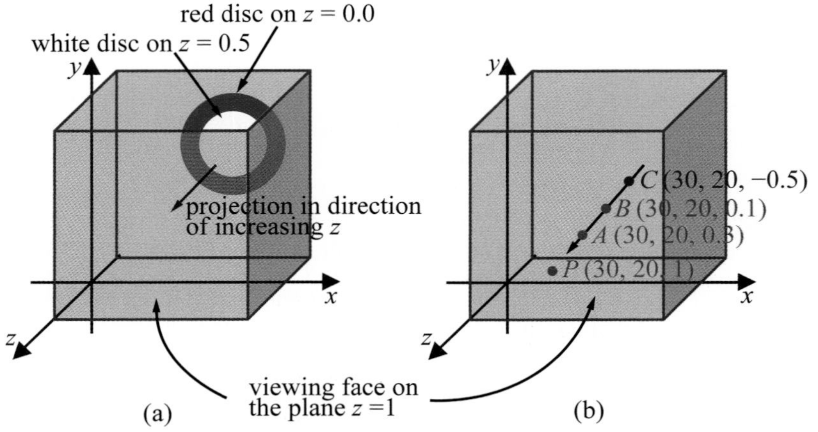

Figure 2.30: (a) The front white disc obscures part of the red one (b) The point A with largest z-value is projected onto the viewing plane so P is red.

the same shape and size as the white disc). This process is called *hidden surface removal* or *depth testing* or *visibility determination.*

Stated mathematically, the result of hidden surface removal in case of orthographic projection is as follows. Suppose that the set of points belonging to drawn objects in the viewing box, with their first two coordinate values particularly equal to X and Y, respectively, is $S = \{(X, Y, z)\}$, where z varies. In other words, S is the set of drawn points lying on the straight line through $(X, Y, 0)$ parallel to the z-axis.

Then only the point (X, Y, Z) of S, with the *largest* z-value, say, Z, lends its color attributes to their shared projection $(X, Y, -near)$ on the viewing face. The implication is that only (X, Y, Z) is drawn of the points in S, the rest obscured.

For example, in Figure 2.30(b), the three points A, B and C, colored red, green and blue, respectively, share the same first two coordinate values, namely, $x = 30$ and $y = 20$. So, all three project to the same point P on the viewing face. As A has the largest z coordinate of the three, it obscures the other two and P, therefore, is drawn red.

The z-buffer itself is a block of memory containing z-values, one per pixel. If depth testing is enabled, then, as a primitive is processed for rendering, the z-value of each of its points – or, more accurately, each of its pixels – is compared with that of the one with the same (x, y)-values currently resident in the z-buffer. If an incoming pixel's z-value is greater, then its RGB attributes and z-value replace those of the current one; if not, the incoming pixel's data is discarded. For example, if the order in which the points of Figure 2.30(b) happen to appear in the code is C, A and B, here's how the color and z-buffer values at the pixel corresponding to P change:

```
draw C; // Pixel corresponding to P gets color blue
```

```
         // and z-value -0.5.
draw A; // Pixel corresponding to P gets color red
         // and z-value 0.3: A's values overwrite C's.
draw B; // Pixel corresponding to P retains color red
         // and z-value 0.3: B is discarded.
```

Next, note in `circularAnnuluses.cpp` the enabling syntax of hidden surface removal so that you can begin to use it:

1. The `GL_DEPTH_BUFFER_BIT` parameter of `glClear(GL_COLOR_BUFFER_BIT | GL_DEPTH_BUFFER_BIT)` in the `drawScene()` routine causes the depth buffer to be cleared.

2. The command `glEnable(GL_DEPTH_TEST)` in the `drawScene()` routine turns hidden surface removal on. The complementary command is `glDisable(GL_DEPTH_TEST)`.

3. The `GLUT_DEPTH` parameter of `glutInitDisplayMode(GLUT_SINGLE | GLUT_RGB | GLUT_DEPTH)` in `main()` causes the depth buffer to be initialized.

Figure 2.31: Bull's eye target.

Exercise 2.22. (**Programming**) Draw a bull's eye target as in Figure 2.31 by means of five discs of different colors, sizes and depths.

2.8.2 A Helix and Perspective Projection

We get more seriously 3D next by drawing a spiral or, more scientifically, a helix. A helix, though itself 1-dimensional – drawn as a line strip actually – can be made authentically only in 3-space.

Open `helix.cpp` but don't run it as yet! The parametric equations implemented are

$$x = R\cos t,\ y = R\sin t,\ z = t - 60.0, \quad -10\pi \le t \le 10\pi \qquad (2.2)$$

See Figure 2.32. Compare these with Equation (2.1) for a circle centered at (0, 0, 0), putting $X = 0$ and $Y = 0$ in that earlier equation. The difference is that the helix climbs up the z-axis *simultaneously* as it rotates circularly with increasing t (so, effectively, it coils around the z-axis). Typically, one writes simply $z = t$ for the last coordinate; however, we tack on "-60.0" to push the helix far enough down the z-axis so that it's contained entirely in the viewing box.

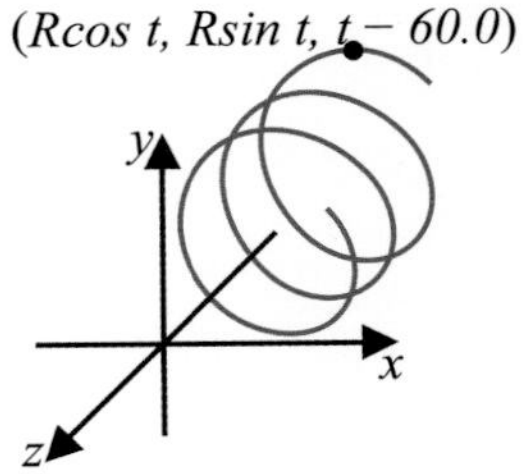

Figure 2.32: Parametric equations for a helix.

Exercise 2.23. Even before viewing the helix, can you say from Equation (2.2) how many times it is supposed to coil around the z-axis, i.e., how many full turns it is supposed to make?
Hint: One full turn corresponds to an interval of 2π along t.

Experiment 2.21. Okay, run `helix.cpp` now. All we see is a circle as in Figure 2.33(a)! There's no sign of any coiling up or down. The reason, of course, is that the orthographic projection onto the viewing face flattens the helix. Let's see if it makes a difference to turn the helix upright, in particular, so that it coils around the y-axis. Accordingly, replace the statement

```
glVertex3f(R * cos(t), R * sin(t), t - 60.0);
```

in the drawing routine with

```
glVertex3f(R * cos(t), t, R * sin(t) - 60.0);
```

Hmm, not a lot better (Figure 2.33(b))! **End**

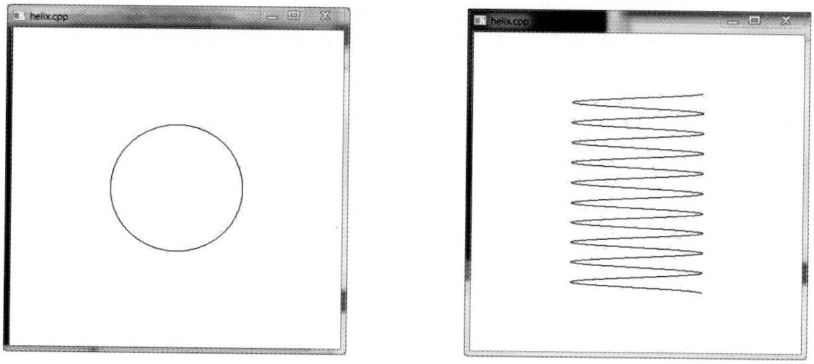

(a) (b)

Figure 2.33: Screenshots of `helix.cpp` using orthographic projection with the helix coiling around the: (a) z-axis (b) y-axis.

Because it squashes a dimension, typically, orthographic projection is not suitable for 3D scenes. OpenGL, in fact, provides another kind of projection, called *perspective projection*, more appropriate for most 3D applications. Perspective projection is implemented with a `glFrustum()` call.

Instead of a viewing box, a `glFrustum(`*left*`,` *right*`,` *bottom*`,` *top*`,` *near*`,` *far*`)` call sets up a *viewing frustum* – a frustum is a *truncated pyramid* whose top has been cut off by a plane parallel to its base – in the following manner (see Figure 2.34):

The apex of the pyramid is at the origin. The front face, called the *viewing face*, of the frustum is the rectangle, lying on the plane $z = -near$, whose corners are $(left, bottom, -near)$, $(right, bottom, -near)$, $(left, top, -near)$, and $(right, top, -near)$. The plane $z = -near$ is called the *viewing plane*. The four edges of the pyramid emanating from the apex pass through the four corners of the viewing face. The base of the frustum, which is also the base of the pyramid, is the rectangle whose vertices are precisely where the pyramid's four edges intersect the $z = -far$ plane. By proportionality with the front vertices, the coordinates of the base vertices are:

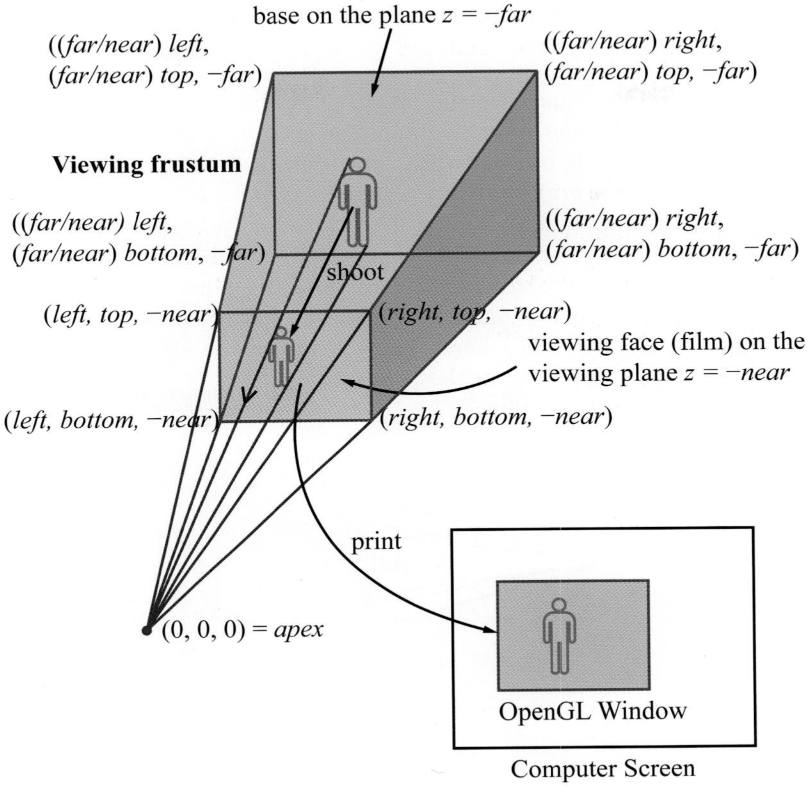

Figure 2.34: Rendering with `glFrustum()`.

```
((far/near) left,(far/near) bottom, −far),
((far/near) right,(far/near) bottom, −far),
((far/near) left, (far/near) top, −far),
((far/near) right, (far/near) top, −far)
```

Values of the `glFrustum()` parameters are typically set so that the frustum lies symmetrically about the z-axis; in particular, *right* and *top* are chosen to be positive, and *left* and *bottom* their respective negatives. The parameters *near* and *far* should both be positive and $near < far$.

Example 2.1. Determine the corners of the viewing frustum created by the call `glFrustum(-15.0, 15.0, -10.0, 10.0, 5.0, 50.0)`.

Answer: By definition, the corners of the front face are $(-15.0, -10.0, -5.0)$, $(15.0, -10.0, -5.0)$, $(-15.0, 10.0, -5.0)$ and $(15.0, 10.0, -5.0)$. The x- and y-values of the vertices of the base (or back face) are scaled from those on the front by a factor of 10 (because $far/near = 50/5 = 10$). The base vertices are, therefore, $(-150.0, -100.0, -50.0)$, $(150.0, -100.0, -50.0)$, $(-150.0, 100.0, -50.0)$ and $(150.0, 100.0, -50.0)$.

Exercise 2.24. Determine the corners of the viewing frustum created by the call `glFrustum(-5.0, 5.0, -5.0, 5.0, 5.0, 100.0)`.

The rendering sequence in the case of perspective projection is a two-step shoot-and-print, similar to orthographic projection. The shooting step again consists of projecting objects within the viewing frustum onto the viewing face, *except that the projection is no longer perpendicular*. Instead, each point is projected along the line joining it to the apex, as depicted by the black dashed lines from the bottom and top of the man in Figure 2.34. Perspective projection causes *foreshortening* because objects farther away from the apex appear smaller (a phenomenon also called *perspective transformation*). For example, see Figure 2.35 where A and B are of the same height, but the projection pA is shorter than the projection pB.

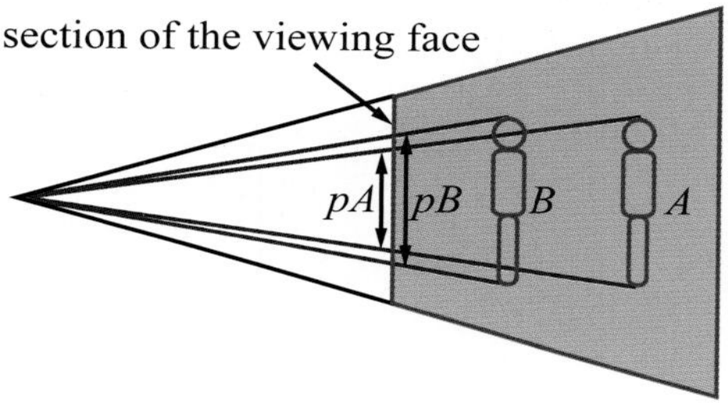

Figure 2.35: Section of the viewing frustum showing foreshortening.

Time now to see perspective projection turn on its magic!

Experiment 2.22. Fire up the original `helix.cpp` program. Replace orthographic projection with perspective projection; in particular, replace the projection statement

```
glOrtho(-50.0, 50.0, -50.0, 50.0, 0.0, 100.0);
```

with

```
glFrustum(-5.0, 5.0, -5.0, 5.0, 5.0, 100.0);
```

You can see a real spiral now (Figure 2.36(a)). View the upright version as well (Figure 2.36(b)), replacing

```
glVertex3f(R * cos(t), R * sin(t), t - 60.0);
```

with

```
glVertex3f(R * cos(t), t, R * sin(t) - 60.0);
```

A lot better than the orthographic version is it not?! **End**

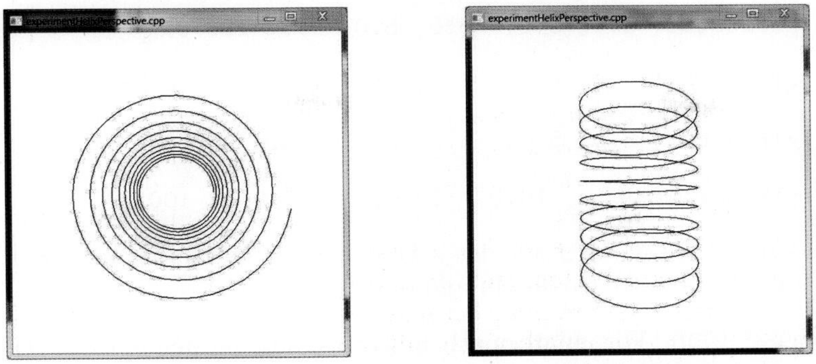

(a) (b)

Figure 2.36: Screenshots of `helix.cpp` using perspective projection with the helix coiling up the (a) z-axis (b) y-axis.

Perspective projection is more realistic than orthographic projection as it mimics how images are formed on the retina of the eye by light rays traveling toward a fixed point. And, in fact, it's precisely foreshortening that cues us humans to the distance of an object.

Remark 2.11. One can think of the apex of the frustum as the location of a *point camera* and the viewing face as its film.

The second rendering step where the viewing face is proportionately scaled to fit onto the OpenGL window is exactly as for orthographic projection. Similar to orthographic projection as well, the scene is clipped to within the viewing frustum by the 6 planes that bound the latter.

Remark 2.12. One might think of orthographic and perspective projections *both* as being along lines of projection convergent to a single point, the center of projection (COP). In the case of perspective projection, this is a regular point with finite coordinates; however, for orthographic projection the COP is a "point at infinity" – i.e., infinitely far away – so that lines toward it are parallel.

Remark 2.13. There do exist 3D applications, e.g., in architectural design, where foreshortening amounts to distortion, so, in fact, orthographic projection is preferred.

Remark 2.14. It's because it captures the image of an object by intersecting rays projected from the object – either orthographically or perspectively – with a plane, which is similar to how a real camera works, that OpenGL is said to implement the *synthetic-camera* model.

Exercise 2.25. (Programming) Continuing from where we were at the end of the preceding experiment, successively replace the `glFrustum()` call as follows, trying in each case to predict the change in the display before running the code:

(a) `glFrustum(-5.0, 5.0, -5.0, 5.0, 5.0, 120.0)`

(b) `glFrustum(-5.0, 5.0, -5.0, 5.0, 10.0, 100.0)`

(c) `glFrustum(-5.0, 5.0, -5.0, 5.0, 2.5, 100.0)`

(d) `glFrustum(-10.0, 10.0, -10.0, 10.0, 5.0, 100.0)`

Parts (b) and (c) show, particularly, how moving the film forward and back causes the camera to "zoom" in and out, respectively.

Exercise 2.26. Formulate mathematically how hidden surface removal should work in the case of perspective projection, as we did in Section 2.8.1 for orthographic projection.

Figure 2.37: Screenshot of `moveSphere.cpp`.

Experiment 2.23. Run `moveSphere.cpp`, which simply draws a movable sphere in the OpenGL window. Press the left, right, up and down arrow keys to move the sphere, the space bar to rotate it and 'r' to reset.

The sphere appears distorted as it nears the periphery of the window, as you can see from the screenshot in Figure 2.37. Can you guess why? Ignore the code, especially unfamiliar commands such as `glTranslatef()` and `glRotatef()`, except for the fact that the projection is perspective.

This kind of *peripheral distortion* of a 3D object is unavoidable in any viewing system which implements the synthetic-camera model. It happens with a real camera as well, but we don't notice it as much because the field of view when snapping pictures is usually quite large and objects of interest tend to be centered. **End**

2.9 Drawing Projects

Here are a few exercises to stretch your drawing muscles. The objects may look rather different from what we have drawn so far, but as programming projects aren't really. In fact, you can probably cannibalize a fair amount of code from earlier programs.

Exercise 2.27. (Programming) Draw a sine curve between $x = -\pi$ and $x = \pi$ (Figure 2.38(a)). Follow the strategy of `circle.cpp` to draw a polyline through a sample from the sine curve.

Exercise 2.28. (Programming) Draw an ellipse. Recall the parametric equations for an ellipse on the xy-plane, centered at (X, Y), with semi-major axis of length A and semi-minor axis of length B (Figure 2.38(b)):

$$x = X + A\cos t,\ y = Y + B\sin t,\ z = 0, \quad 0 \le t \le 2\pi$$

Again, `circle.cpp` is the template to use.

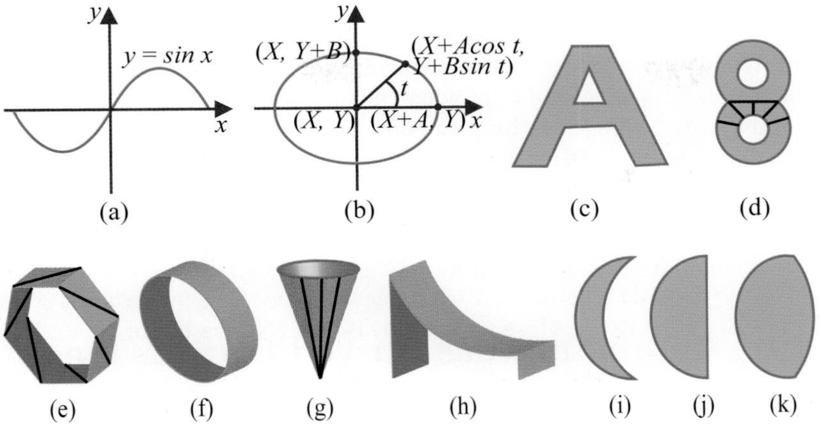

Figure 2.38: Draw these!

Exercise 2.29. (Programming) Draw the letter 'A' as a *two-dimensional* figure like the shaded region in Figure 2.38(c). It might be helpful to triangulate it first on graph paper.

Allow the user to toggle between filled and wireframe *a la* the bottom annulus of `circularAnnuluses.cpp`.

Exercise 2.30. (Programming) Draw the number '8' as the 2D object in Figure 2.38(d). Do this in two different ways: (i) drawing 4 discs and using the z-buffer and (ii) as a true triangulation, allowing the user to toggle between filled and wireframe. For (ii), a method of dividing the '8' into two triangle strips is suggested in Figure 2.38(d).

Exercise 2.31. (Programming) Draw a ring with cross-section a regular (equal-sided) polygon as in Figure 2.38(e), where a scheme to triangulate the ring in one triangle strip is indicated. Allow the user to change the number of sides of the cross-section. Increasing the number of sides sufficiently should make the ring appear cylindrical as in Figure 2.38(f). Use perspective projection and draw in wireframe.

Exercise 2.32. (Programming) Draw a cone as in Figure 2.38(g) where a possible triangulation is indicated. Draw in wireframe and use perspective projection.

Exercise 2.33. (Programming) Draw a children's slide as in Figure 2.38(h). Choose an appropriate equation for the cross-section of the curved surface – part of a parabola, maybe – and then "extrude" it as a triangle strip. (If you did Exercise 2.31 then you've already extruded a polygon.) Draw in wireframe and use perspective projection.

Remark 2.15. Your output from Exercises 2.31-2.33 may look a bit "funny", especially viewed from certain angles. For example, the ring viewed head-on down its axis may appear as two concentric circles on a single plane. This problem can be alleviated by drawing the object with a different alignment or, equivalently, changing the viewpoint. In Experiment 2.24, coming up shortly, we'll learn code for the user to change her viewpoint.

Exercise 2.34. (Programming) Draw in a single scene a crescent moon, a half-moon and a three-quarter moon (Figures 2.38(i)-(k)). Each should be a true triangulation. Label each as well using text-drawing.

2.10 Approximating Curved Objects Once More

Our next 3-space drawing project is a bit more challenging: a hemisphere, which is a 2-dimensional object. We'll get in place, as well, certain design principles which will be expanded in Chapter 10 dedicated to drawing (no harm starting early).

Remark 2.16. A hemisphere is a 2-dimensional object because it is a surface. Recall that a helix is 1-dimensional because it's line-like. Now, both hemisphere and helix need 3-space to "sit in"; they cannot do with less. For example, you could sketch either on a piece of paper (2-space) but it would not be the real thing. On the other hand, a circle – another 1D object – does sit happily in 2-space.

Consider a hemisphere of radius R, centered at the origin O, with its circular base lying on the xz-plane. Suppose the spherical coordinates of a point P on this hemisphere are a longitude of θ (measured clockwise from the x-axis when looking from the plus side of the y-axis) and a latitude of ϕ (measured from the xz-plane toward the plus side of the y-axis). See Figure 2.39(a). The Cartesian coordinates of P are by elementary trigonometry

$$(R\cos\phi\cos\theta,\ R\sin\phi,\ R\cos\phi\sin\theta)$$

The range of θ is $0 \le \theta \le 2\pi$ and of ϕ is $0 \le \phi \le \pi/2$.

Exercise 2.35. Verify that the Cartesian coordinates of P are as claimed. *Suggested approach*: From the right-angled triangle OPP' one has $|PP'| = R\sin\phi$ and $|OP'| = R\cos\phi$. $|PP'|$ is the y-value of P. Next, from right-angled triangle $OP'P''$ find $|OP''|$ and $|P'P''|$, the x- and z-values of P, respectively, in terms of $|OP'|$ and θ.

Sample the hemisphere at a mesh of $(p+1)(q+1)$ points P_{ij}, $0 \le i \le p$, $0 \le j \le q$, where the longitude of P_{ij} is $(i/p) * 2\pi$ and its latitude $(j/q) * \pi/2$. In other words, $p+1$ longitudinally equally-spaced points are chosen along each of $q+1$ equally-spaced latitudes. See Figure 2.39(b), where $p = 10$ and

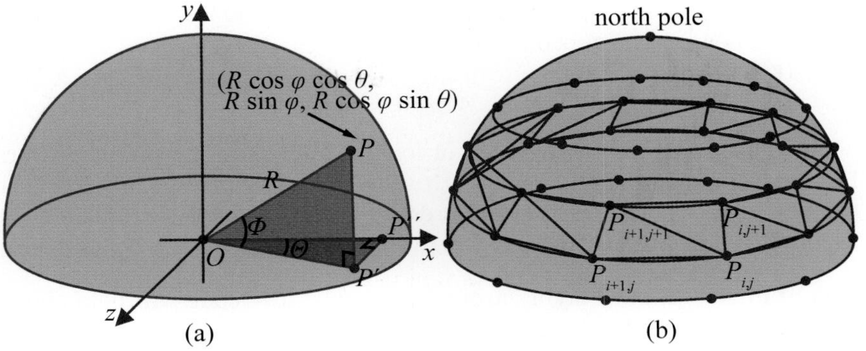

Figure 2.39: (a) Spherical and Cartesian coordinates on a hemisphere (b) Approximating a hemisphere with latitudinal triangle strips.

$q = 4$. The sample points P_{ij} are not all distinct. In fact, $P_{0j} = P_{pj}$, for all j, as the same point has longitude both 0 and 2π; and, the point P_{iq}, for all i, is identical to the north pole, which has latitude $\pi/2$ and arbitrary longitude.

The plan now is to draw one triangle strip with vertices at

$$P_{0,j+1},\ P_{0j},\ P_{1,j+1},\ P_{1j}, \ldots,\ P_{p,j+1},\ P_{pj}$$

for each j, $0 \le j \le q-1$, for a total of q triangle strips. In other words, each triangle strip takes its vertices alternately from a pair of adjacent latitudes and, therefore, approximates the circular band between them. Figure 2.39(b) shows one such strip. The stack of all q triangle strips approximates the hemisphere itself.

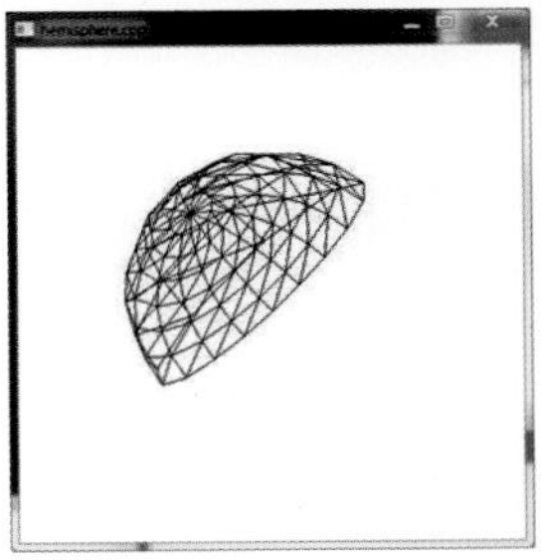

Figure 2.40: Screenshot of `hemisphere.cpp`.

Experiment 2.24. Run `hemisphere.cpp`, which implements exactly the strategy just described. You can verify this from the snippet that draws the hemisphere:

```
for(j = 0; j < q; j++)
{
   // One latitudinal triangle strip.
   glBegin(GL_TRIANGLE_STRIP);
      for(i = 0; i <= p; i++)
      {
         glVertex3f(R * cos((float)(j+1)/q * PI/2.0) *
                        cos(2.0 * (float)i/p * PI),
                    R * sin((float)(j+1)/q * PI/2.0),
                    R * cos((float)(j+1)/q * PI/2.0) *
                        sin(2.0 * (float)i/p * PI));
         glVertex3f(R * cos((float)j/q * PI/2.0) *
                        cos(2.0 * (float)i/p * PI),
                    R * sin((float)j/q * PI/2.0),
                    R * cos((float)j/q * PI/2.0) *
```

```
                                        sin(2.0 *(float)i/p * PI));
      }
   glEnd();
}
```

Increase/decrease the number of longitudinal slices by pressing 'P/p'. Increase/decrease the number of latitudinal slices by pressing 'Q/q'. Turn the hemisphere about the axes by pressing 'x', 'X', 'y', 'Y', 'z' and 'Z'. See Figure 2.40 for a screenshot. End

Experiment 2.25. Playing around a bit with the code will help clarify the construction of the hemisphere:

(a) Change the range of the hemisphere's outer loop from

```
for(j = 0; j < q; j++)
```

to

```
for(j = 0; j < 1; j++)
```

Only the bottom strip is drawn. The keys 'P/p' and 'Q/q' still work.

(b) Change it again to

```
for(j = 0; j < 2; j++)
```

Now, the bottom two strips are drawn.

(c) Reduce the range of both loops:

```
for(j = 0; j < 1; j++)
...
      for(i = 0; i <= 1; i++)
      ...
```

The first two triangles of the bottom strip are drawn.

(d) Increase the range of the inner loop by 1:

```
for(j = 0; j < 1; j++)
...
      for(i = 0; i <= 2; i++)
      ...
```

The first four triangles of the bottom strip are drawn. End

There's syntax in **hemisphere.cpp** – none to do with the actual making of the hemisphere – which you may be seeing for the first time. The command `glTranslatef(0.0, 0.0, -10.0)` is used to move the hemisphere, drawn initially centered at the origin, into the viewing frustum, while the `glRotatef()` commands turn it. We'll explain these so-called *modeling transformations* in Chapter 4 but you are encouraged to experiment with them even now as the syntax is fairly intuitive. The set of three `glRotatef()`s, particularly, comes in handy to re-align a scene.

Exercise 2.36. (Programming) Modify **hemisphere.cpp** to draw:

(a) the bottom half of a hemisphere (Figure 2.41(a)).

(b) a 30° slice of a hemispherical cake (Figure 2.41(b)). Note that simply reducing the range of the inner loop of **hemisphere.cpp** produces a slice of cake without two sides and bottom, so these have to be added in separately to close up the slice.

Make sure the 'P/p/Q/q' keys still work.

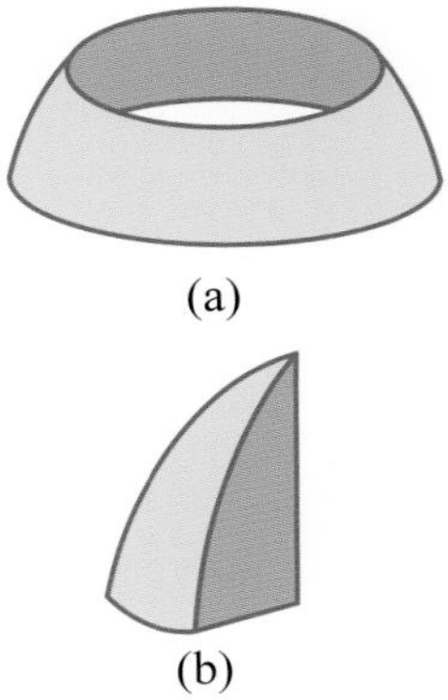

Figure 2.41: (a) Half a hemisphere (b) Slice of a hemisphere.

Exercise 2.37. (Programming) Just to get you thinking about animation, which we'll be studying in depth soon enough, guess the effect of replacing `glTranslatef(0.0, 0.0, -10.0)` with `glTranslatef(0.0, 0.0, -20.0)` in **hemisphere.cpp**. Verify.

And, here are some more things to draw.

Exercise 2.38. (Programming) Draw the objects shown in Figure 2.42. Give the user an option to toggle between filled and wireframe renderings.

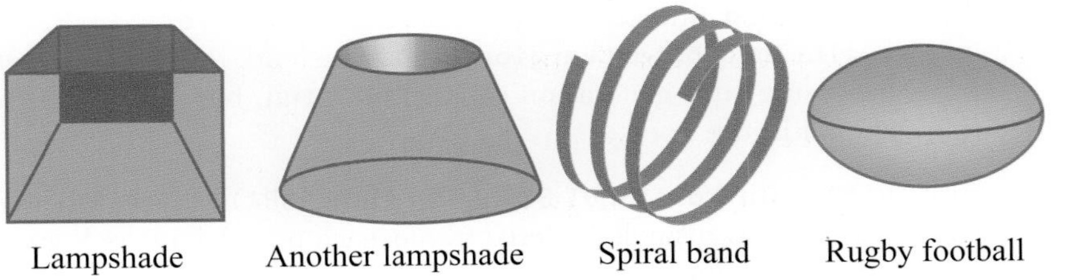

Figure 2.42: More things to draw.

A suggestion for the football, or ellipsoid, is to modify **hemisphere.cpp** to make half of an ellipsoid (a hemi-ellipsoid?). Two hemi-ellipsoids back to back would then give a whole ellipsoid.

Remark 2.17. Filled renderings of 3D scenes, even with color, rarely look pleasant in the absence of lighting. See for yourself by applying color to 3D objects you have drawn so far (remember to invoke a `glPolygonMode(*, GL_FILL)` call). For this reason, we'll draw mostly wireframe till Chapter 11, which is all about lighting. You'll have to bear with this. Wireframe, however, fully exposes the geometry of an object, which is not a bad thing when one is learning object design.

2.11 An OpenGL Program End to End

Of `square.cpp`, in particular, we have touched on almost every command which is functional from a graphics points of view. However, let's run over the whole program to see all that goes into making OpenGL code tick.

We start with `main()`:

1. `glutInit(&argc, argv)` initializes the FreeGLUT library. FreeGLUT, successor to GLUT (OpenGL Utility Toolkit), is a library of calls to manage a window holding OpenGL contexts (the reason such a separate library is needed is that OpenGL itself is only a library of graphics calls).

2. ```
   glutInitContextVersion(4, 3)
   glutInitContextProfile(GLUT_COMPATIBILITY_PROFILE)
   ```
   ask FreeGLUT to provide an OpenGL 4.3 context which is backward-compatible in that legacy commands are implemented. This, for example, allows us to draw with the `glBegin()-glEnd()` operations from OpenGL 2.1, which do not belong in the core profile of OpenGL 4.3.

   Important: If your graphics card doesn't support OpenGL 4.3 then the program may compile but not run as the system is unable to provide the context asked. What you might do in this case is thin the context by replacing the first line above with `glutInitContextVersion(3, 3)`, or even `glutInitContextVersion(2, 1)`, instead. Of course, then, programs using later-generation calls will not run, but you should be fine early on in the book.

3. `glutInitDisplayMode(GLUT_SINGLE | GLUT_RGBA)` asks the OpenGL context to support a single-buffered frame, each pixel having red, green, blue and alpha values.

4. ```
   glutInitWindowSize(500, 500)
   glutInitWindowPosition(100, 100)
   ```
   as we have already seen, set the size of the OpenGL window and the location of its top left corner on the computer screen.

5. `glutCreateWindow("square.cpp")` creates the window (precisely, the rendering context) with the specified string parameter as title.

6. ```
   glutDisplayFunc(drawScene)
   glutReshapeFunc(resize)
   ```
   ```

```
glutKeyboardFunc(keyInput)
```

register the routines to call – so-called *callback routines* – when the OpenGL window is to be drawn, when it is resized (and first created), and when keyboard input is received, respectively.

7. `glewExperimental = GL_TRUE`
   `glewInit()`

   initializes GLEW (the OpenGL Extension Wrangler Library) which handles the loading of OpenGL extensions, with the switch set so that extensions implemented in even pre-release drivers are exposed.

8. `setup()` invokes the initialization routine.

9. `glutMainLoop` begins the event-processing loop, calling registered callback routines as needed.

We have already seen that the only command in the initialization routine `setup()`, namely, `glClearColor(1.0, 1.0, 1.0, 0.0)`, specifies the clearing color of the OpenGL window.

The callback routine to draw the OpenGL window is:

```
void drawScene(void)
{
   glClear(GL_COLOR_BUFFER_BIT);

   glColor3f(0.0, 0.0, 0.0);

   // Draw a polygon with specified vertices.
   glBegin(GL_POLYGON);
      ...
   glEnd();

   glFlush();
}
```

The first command clears the OpenGL window to the specified clearing color, in other words, paints in the background color. The next command `glColor3f()` sets the foreground, or drawing, color, which is used to draw the polygon specified within the `glBegin()-glEnd()` pair (we have already examined this polygon carefully). Finally, `glFlush()` forces the prior draw calls to actually execute, which, in this case, means the polygon is drawn.

The callback routine when the OpenGL window is resized, and first created, is `void resize(int w, int h)`. The window manager supplies the width `w` and height `h` of the resized OpenGL window (or, initial window, when it is first created) as parameters to the resize routine.

The first command

```
glViewport(0, 0, w, h);
```

of `square.cpp`'s resize routine specifies the rectangular part of the OpenGL window in which actual drawing is to take place; with the given parameters it is the entire window. We'll be looking more carefully into `glViewPort()` and its applications in the next chapter.

The next three commands

```
glMatrixMode(GL_PROJECTION);
glLoadIdentity();
glOrtho(0.0, 100.0, 0.0, 100.0, -1.0, 1.0);
```

activate the projection matrix stack, place the identity matrix at the top of this stack, and then multiply the identity matrix by the matrix corresponding to the final `glOrtho()` command, effectively setting up the viewing box of `square.cpp` described in Section 2.2.

The final two commands

```
glMatrixMode(GL_MODELVIEW);
glLoadIdentity();
```

of the resize routine activate the modelview matrix stack and place the identity matrix at the top in readiness for modelview transformation commands in the drawing routine, of which there happen to be none in `square.cpp`. We'll be learning much more about OpenGL's matrix stacks and how they are at the heart of its operation in Chapters 4 and 5.

The callback routine to handle ASCII keys is `keyInput(unsigned char key, int x, int y)`. When an ASCII key is pressed it is passed in the parameter `char` to this callback routine, as is the location of the mouse in the parameters `x` and `y`. As is easy to see, all that `keyInput` of `square.cpp` does is terminate the program when the escape key is pressed. In the next chapter we'll see callback routines to handle non-ASCII keys, as well as interaction via the mouse.

As the reader might well guess, the guts of an OpenGL program are in its drawing routine. Interestingly, the initialization routine often pulls a fair load, too, because one would want to locate there tasks that need to be done once at start-up, e.g., setting up data structures. In fact, it's a common beginner's mistake to place initialization chores in the drawing routine, as the latter is invoked repeatedly if there is animation, leading to inefficiency.

The other routines, such as `main()` and the interactivity and reshape callbacks, are either standard and pretty much transferable from program to program (e.g., `main()`), or easy to code.

2.12 Summary, Notes and More Reading

In this chapter we began the study of 3D CG, looking at it through the "eyes" of OpenGL. OpenGL itself was presented to the extent that the reader acquires functional literacy in this particular API. The drawing primitives were probably the most important part of the API's vernacular.

We discovered as well how OpenGL functions as a state machine, attributes such as color defining the current state. Moreover, we learned that quantifiable attribute values, e.g., RGB color, are typically interpolated from the vertices of a primitive throughout its interior. We saw that OpenGL clips whatever the programmer draws to within a viewing volume, either a box or frustum.

Beyond acquaintance with the language, we were introduced as well to the synthetic-camera model of 3D graphics, which OpenGL implements via two kinds of projection: orthographic and perspective. This included insights into the world coordinate system, the viewing volume – box or frustum – which is the stage in which all drawings are made, the shoot-and-print rendering process to map a 3D scene to a 2D window, as well as hidden surface removal. We made a first acquaintance as well with another cornerstone of 3D graphics: the technique of simulating curved objects using straight and flat primitives like line segments and triangles.

Historically, OpenGL evolved from SGI's IRIS GL API, which popularized the approach to creating 3D scenes by drawing objects in actual 3-space and then rendering them to a 2D window by means of a synthetic camera. IRIS GL's efficiently pipelined architecture enabled high-speed rendering of animated 3D graphics and, consequently, made possible as well real-time interactive 3D. The ensuing demand from application developers for an open and portable (therefore, platform-independent) version of their API spurred SGI to create the first OpenGL specification in 1992, as well as a sample implementation. Soon after, the OpenGL ARB (Architecture Review Board), a consortium composed of a few leading companies in the graphics industry, was established to oversee the development of the API. Stewardship of the OpenGL specification passed in 2006 to the Khronos Group, a member-funded industry consortium dedicated to the creation of open-standard royalty-free API's. (That no *one* owns OpenGL is a good thing.) The canonical, and very useful, source for information about OpenGL is its own home page [103].

Microsoft has a non-open Windows-specific 3D API – Direct3D [90, 141] – which is popular among game programmers as it allows optimization for the pervasive Windows platform. However, outside of the games industry, where it nonetheless competes with Direct3D, and leaving aside particular application domains with such high-quality rendering requirements that ray tracers are preferred, by far the dominant graphics API is OpenGL.

It's safe to say that OpenGL is the de facto standard 3D graphics API. A primary reason for this, other than the extremely well-thought-out design which it had from inception – initial versions of Direct3D in contrast were notoriously buggy and hard to use – is OpenGL's portability. With their recent versions, though, OpenGL and Direct3D seem to be converging, at least in functionality (read an interesting comparison in Wikipedia [27]). It's worth knowing as well that, despite its intended portability, OpenGL can take advantage of platform-specific and card-specific capabilities via so-called extensions, at the cost of clumsier code.

An unofficial clone of OpenGL, Mesa 3D [93], which uses the same syntax, was originally developed by Brian Paul for the Unix/X11 platform, but there are ports now to other platforms as well.

Perhaps the best reason for OpenGL to be *the* API of choice for students of 3D computer graphics is – and this is a consequence of its almost universal adoption by the academic, engineering and scientific communities – the sheer volume of learning resources available. Not least among these is the number of textbooks that teach computer graphics with the help of OpenGL. Search amazon.com with the keywords "computer graphics opengl" and you'll see what we mean. Angel [2], Buss [22], Govil-Pai [60], Hearn & Baker [69], Hill & Kelley [72] and McReynolds & Blythe [92] are some introductions to computer graphics via OpenGL that the author has learned much from.

In case the reader prefers not to be distracted by code, here are a few API-independent introductions: Akenine-Möller, Haines & Hoffman [1], Foley et al. [47, 48] (the latter being an abridgment of the former), Shirley & Marschner [130], Watt [147] and Xiang & Plastock [154]. Keeping different books handy in the beginning is a good idea as, often, when you happen to be confused by one author's presentation of a topic, simply turning to another for help on just that may clear the way.

With regard to the code which comes with this book, we don't make much use of OpenGL-defined data types, which are prefixed with `GL`, e.g., `GLsizei`, `GLint`, etc., though the red book advocates doing so in order to avoid type mismatch issues when porting. Fortunately, we have not yet encountered a problem in any implementation of OpenGL that we've tried.

In addition to the code that comes with this book, the reader should try to acquire OpenGL programs from as many other sources as possible, as an efficient way to learn the language – any language as a matter of fact – is by modifying live code. Among the numerous sources on the web – there are pointers to several coding tutorials at the OpenGL site [103] – special mention must be made of Jeff Molofee's excellent tutorials at NeHe Productions [102], covering a broad spectrum of OpenGL topics, each illustrated with a well-documented program. The book by Wright, Lipchak & Haemel [129] is specifically about programming OpenGL and has numerous example programs. The red book comes with example code as well. Incidentally, in addition to the somewhat bulky red and blue books, a handy reference manual for OpenGL is Angel's succinct primer [3].

Hard-earned wisdom: Write experiments of your own to clarify ideas. Even if you are sure in your mind that you understand something, do write a few lines of code in verification. As the author has repeatedly been, you too might be surprised!

Part II

Tricks of the Trade

CHAPTER 3

An OpenGL Toolbox

Before getting to animation and other fun stuff in the next chapter, here are a few practical skills worth acquiring first. Our goal this chapter is to learn the following frequently-used OpenGL programming devices:

1. Vertex arrays and their drawing commands: storing geometric data in a single location for efficient access.

2. Vertex buffer objects: storage for vertex-related data on the graphics server to save client-to-server transfer time.

3. Vertex array objects: encapsulating the set of calls defining an object's vertex arrays.

4. Display lists: "macros" to store frequently-invoked pieces of code.

5. Drawing of text.

6. Programming the mouse – for button clicks, turning the wheel and mouse motion.

7. Programming non-ASCII keys.

8. Programming pop-up menus.

9. Line stipples: applying patterns to lines.

10. FreeGLUT objects: ready-made library objects.

11. Clipping planes: planes to clip a scene in addition to the automatic six that bound the viewing box or frustum.

12. `gluPerspective()`: a more intuitive version of the `glFrustum()` projection statement with fewer parameters.

13. Viewports: specifiable parts of the OpenGL window to which a drawing is rendered.

14. Multiple windows: multiple top-level OpenGL windows.

None is particularly challenging or deep and the reader may choose to flip quickly through the pages to just see what each is about in order to be able to return later for how to implement when the need arises.

However, the exceptions we would make to this approach are the first three sections. The reader should master vertex arrays in Section 3.1 and begin using them right away. As for vertex buffer objects and vertex array objects in Sections 3.2-3.3, though we don't expect the user to code much of them at first, they are indispensable in the newer versions of OpenGL, e.g., 4.3 of the fourth generation, which we shall cover in depth down the road, so the reader should at least make their acquaintance at this time.

The next fourteen sections follow the order of the list.

3.1 Vertex Arrays and Their Drawing Commands

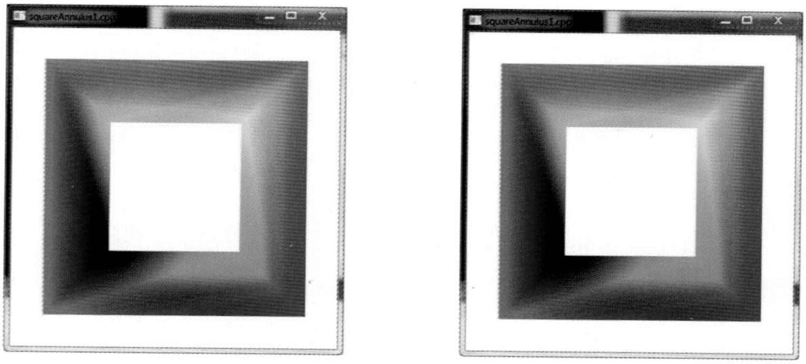

(a) (b)

Figure 3.1: Screenshots of `squareAnnulus1.cpp`.

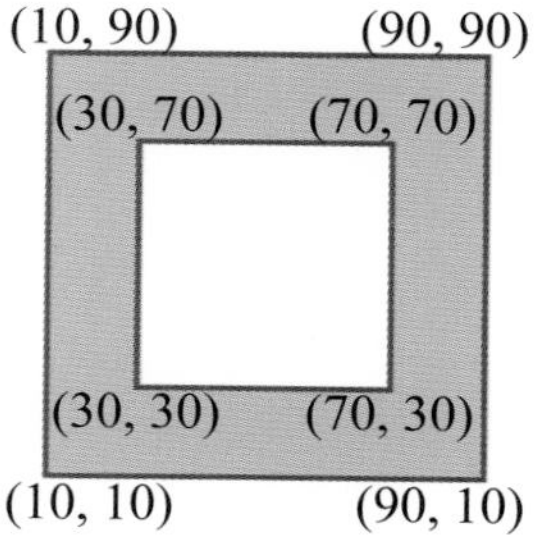

Figure 3.2: Square annulus (z coordinates all 0).

Experiment 3.1. Run `squareAnnulus1.cpp`. A screenshot is seen in Figure 3.1(a). Press the space bar to see the wireframe in Figure 3.1(b).

It is a plain-vanilla program which draws the square annulus diagrammed in Figure 3.2 using a single triangle strip defined by a giant `glBegin(GL_TRIANGLE_STRIP)-glEnd()` block containing 10 vertices and their color attributes (the last two vertices being identical to the first two in order to close the strip).

End

Experiment 3.2. Run `squareAnnulus2.cpp`.

It draws the same annulus as `squareAnnulus1.cpp`, except that the vertex coordinates and color data are now separately stored in two-dimensional global arrays, `vertices` and `colors`, respectively. Moreover, in each iteration, the loop

```
glBegin(GL_TRIANGLE_STRIP);
   for(int i = 0; i < 10; ++i)
   {
      glColor3fv(colors[i%8]);
      glVertex3fv(vertices[i%8]);
   }
glEnd();
```

retrieves a vector of coordinate values by the *pointer form* (also called *vector* form) of vertex declaration, namely, `glVertex3fv(`**pointer*`)`, and as well a vector of color values with the pointer form `glColor3fv(`**pointer*`)`. **End**

Compared with `squareAnnulus1.cpp`, an obvious efficiency gained in `squareAnnulus2.cpp` is in placing vertex and color data at one place in the code to be able simply to point to them from elsewhere. This allows the triangle strip block, though still containing 10 vertices and their colors, to be coded as a short loop.

It's always good practice, as in the last program, to collect and place data for a program at a single location separate from the procedures which access the data. Redundancy and consequent errors tend to be eliminated, memory usage is more efficient, and it's easier to modularize and debug those procedures which access data.

OpenGL offers specific devices – the *vertex array* data structures – which make it easy and efficient for the user to centralize and share data. Let's learn them from live code.

Experiment 3.3. Run `squareAnnulus3.cpp`.

It again draws the same colorful annulus as before. The coordinates and color data of the vertices are stored in one-dimensional global vertex arrays, `vertices` and `colors`, respectively, as in `squareAnnulus2.cpp`, except, now, the arrays are flat and not 2D (because of the way C++ stores array data, we could, in fact, have specified `vertices` and `colors` as 2D arrays exactly as in `squareAnnulus2.cpp` if we had so wanted). **End**

Now, to the magic: within the triangle strip loop

```
glBegin(GL_TRIANGLE_STRIP);
   for(int i = 0; i < 10; ++i) glArrayElement(i%8);
glEnd();
```

the ith vector of values from both coordinates *and* color arrays are retrieved *simultaneously* with a single `glArrayElement(`*i*`)` call.

Note the steps in setting up the vertex arrays in the initialization routine:

1. Two vertex arrays are enabled by calling `glEnableClientState(`*array*`)`, where *array* is, successively, `GL_VERTEX_ARRAY` and `GL_COLOR_ARRAY`, for vertex coordinate and color values, respectively. There are other possible values for the parameter *array* to store additional kinds of vertex data, e.g., normal values and texture coordinates.

2. The data for the two vertex arrays is specified with a call to `glVertexPointer(`*size*`, `*type*`, `*stride*`, `**pointer*`)` and a call to `glColorPointer(`*size*`, `*type*`, `*stride*`, `**pointer*`)`. The parameter *pointer* is the address of the start of the data array, *type* declares the data type, *size* is the number of values per vertex (both coordinate and color arrays store 3 values for each vertex) and *stride* is the byte offset between the start of the values for successive vertices (0 indicates that values for successive vertices are not separated).

Experiment 3.4. Run `squareAnnulus4.cpp`.

The code is even more concise with the single call

```
glDrawElements(GL_TRIANGLE_STRIP, 10, GL_UNSIGNED_INT, stripIndices)
```

replacing the entire `glBegin(GL_TRIANGLE_STRIP)-glEnd()` block. **End**

The general form of this call is `glDrawElements(`*primitive*`, `*count*`, `*type*`, `**indices*`)` where parameter *primitive* is a geometric primitive, *indices* is the address of the start of an array of indices, *type* is the data type of the *indices* array and *count* is the number of indices to use. What this call does is pick *count* number of vertices for *primitive* from the enabled vertex arrays in the sequence specified by *indices*, equivalent, therefore, to the loop

```
glBegin(primitive);
   for(i = 0; i < count; i++) glArrayElement(indices[i]);
glEnd();
```

Exercise 3.1. **(Programming)** Rewrite `hemisphere.cpp` to use vertex arrays and a loop of `glDrawElements()` commands.

When there are multiple objects in a scene it's convenient to keep their data separately in different vertex arrays, as in the following program.

Experiment 3.5. Run `squareAnnulusAndTriangle.cpp`, which adds a triangle inside the annulus of the `squareAnnulus*.cpp` programs. See Figure 3.3 for a screenshot. **End**

Figure 3.3: Screenshot of `squareAnnulusAndTriangle.cpp`.

This program demonstrates the use of multiple vertex arrays. The vertex arrays `vertices1` and `colors1` contain the coordinate and color data, respectively, for the annulus, exactly as in `squareAnnulus4.cpp`.

The single vertex array `vertices2AndColors2Intertwined` for the triangle, on the other hand, is *intertwined* in that it contains both coordinate and color data together. When pointing to data for the triangle, the *stride* parameter of both the `glVertexPointer()` and `glColorPointer()` calls is set to 6 times the number of bytes in a `float` data item, as there are 6 such items between the start of successive coordinate or color vectors in the intertwined array.

The statement

```
glDrawArrays(GL_TRIANGLES, 0, 3)
```

drawing the triangle introduces a new drawing command to use with vertex arrays. Generally, `glDrawArrays(`*primitive*`, `*first*`, `*count*`)` draws the geometric primitive *primitive*, using *count* elements from the vertex array, starting with the element at position *first*. This is the command of choice when the drawing needs simply to process elements in a vertex array linearly, in particular, without needing to bounce around with something like an *indices* array supplying an intermediate level of indirection.

Exercise 3.2. (Programming) Rewrite `circle.cpp` to use vertex arrays and a single `glDrawArrays()` command.

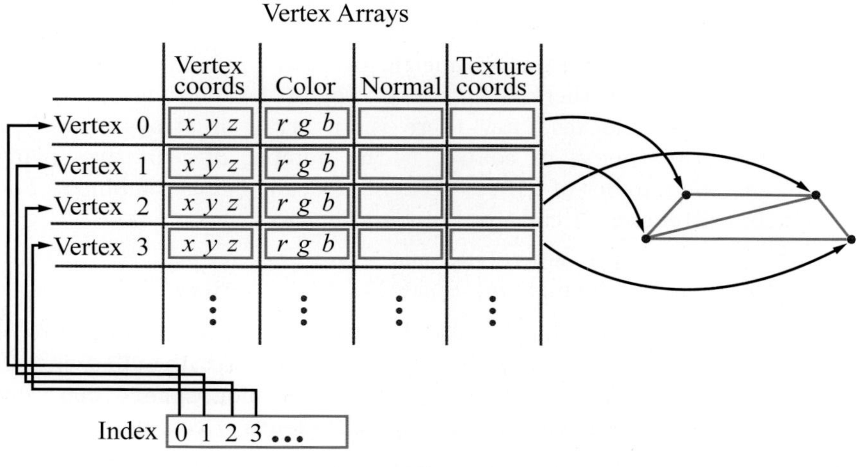

Figure 3.4: Logical representation of data in vertex arrays.

Vertex arrays make for efficient, logical and conceptually clean OpenGL code. Figure 3.4 illustrates this (it shows additional vertex attributes which are also stored in vertex arrays – we'll be discussing these later). Moreover, they are *mandatory* in the latest versions of OpenGL, e.g., 4.3, which we'll be covering later on. *Make a habit of using vertex arrays*!

Caveat: Henceforth, we'll be implementing vertex arrays consistently ourselves except, possibly, for programs with few vertices where the overhead might be a distraction.

Remark 3.1. Commands like `glDrawElements()` and `glDrawArrays()` are for *retained mode* rendering vs. `glBegin()-glEnd()`-type commands for *immediate mode* rendering. In immediate mode, the client (the machine running the program) forces rendering by the server (the GPU), while, in retained mode, the client provides the server only with instructions to perform and the data to use, allowing the latter to optimize prior to rendering.

Remark 3.2. Keep in mind that the display routine is called repeatedly if there is animation. It's particularly inefficient, therefore, and, unfortunately, a common beginner's mistake to store static data in this routine, or perform computations there which actually can be done once initially and the results saved. The rule is to store vertex attributes in vertex arrays, while the initialization routine is the place for one-time computation.

Before closing this section we'll introduce a couple more drawing commands to use with vertex arrays. The first,

`glMultiDrawElements(`*primitive*, **count*, *type*, ***indices*, *primitivesCount*)

is a powerlifter with the capacity of multiple `glDrawElements()`. In fact, the parameters of `glMultiDrawElements()` are much as you would expect in order to combine many

`glDrawElements(` *primitive*, *count*, *type*, **indices*)

calls, each drawing the same geometric primitive *primitive*: instead of one *count* value, now there is an array **count* of values; instead of an array **indices* of indexes, now there is an array of arrays ***indices*; finally, *primitivesCount*, of course, is the number of *primitive*s being drawn, i.e., the number of `glDrawElements()` calls being combined. The `glMultiDrawElements()` call itself is equivalent to

```
for (int i = 0; i < primitivesCount; i++)
   glDrawElements(primitive, count[i], type, indices[i]);
```

It's more efficient to draw an object composed of multiple instances of the same geometric primitive using one (or few) `glMultiDrawElements()` calls versus several `glDrawElements()`. Let's redo `hemisphere.cpp` from the last chapter using a single `glMultiDrawElements()` command.

Experiment 3.6. Run `hemisphereMultidraw.cpp`, whose sole purpose is to draw the loop

```
for(j = 0; j < q; j++)
{
   // One latitudinal triangle strip.
   glBegin(GL_TRIANGLE_STRIP);
   ...
}
```

of triangle strips in `hemisphere.cpp` using the single

```
glMultiDrawElements(GL_TRIANGLE_STRIP, count, GL_UNSIGNED_BYTE,
                    (const void **)indices, q)
```

command instead. **End**

We'll leave the reader to deconstruct the code of `hemisphereMulti-draw.cpp`, which is fairly straightforward and mostly dedicated to setting up arrays for the `glMultiDrawElements()` call. Particularly, the reader should convince herself that `fillIndices()` does its job of filling the 2D *indices* array correctly (this is the part where one usually needs to be most careful in transitioning from `glDrawElements()` to `glMultiDrawElements()`).

Exercise 3.3. **(Programming)** If you drew any of the objects from Exercise 2.38 of the last chapter, redo the code to use `glDrawElements()`, or, if possible, `glMultiDrawElements()`.

Finally,

`glMultiDrawArrays(` *primitive*, **first*, **count*, *primitivesCount*`)`

is related to `glDrawArrays()` exactly as `glMultiDrawElements()` is related to `glDrawElements()`: a single `glMultiDrawArrays()` command can encapsulate multiple `glDrawArrays()` commands.

Exercise 3.4. **(Programming)** Draw the bull'e eye of of Exercise 2.22 of the last chapter using a single `glMultiDrawArrays()` call.

3.2 Vertex Buffer Objects

OpenGL's client-server model means that each time the server requires vertex data – e.g., coordinates, color or such to execute, say, a `glDrawElements()` call – it must be fetched from the client. On a PC, for example, this translates to a transfer across the bus connecting the CPU (the client holding the application and data) to the GPU (graphics processing unit, being the server which does the drawing). Now, accessing data across a bus is, typically, many times slower than accessing it locally. Moreover, the access might even be redundant if the same data had been retrieved for an earlier command and, subsequently, not changed. To save such inefficiency, *buffer objects* allow the programmer to explicitly ask that some particular set of data, possibly vertex or pixel, be stored server-side, e.g., in the GPU memory in case of a PC.

We will focus now on buffer objects that store vertex data, such being called *vertex buffer objects*, or *VBOs*. Let's get straight to code showing how to create, initialize and update a VBO.

Experiment 3.7. Fire up `squareAnnulusVBO.cpp`, which modifies `squareAnnulus4.cpp` to store vertex-related data in VBOs. There is a simple animation, too, through periodically changing color values in a VBO. Figure 3.5 is a screenshot, colors having already changed. **End**

Figure 3.5: Screenshot of `squareAnnulusVBO.cpp`.

Let's understand how `squareAnnulusVBO.cpp` works. The `setup()` routine is the one to look at first. The call

```
glGenBuffers(2, buffer)
```

returns two available buffer IDs which we'll use to identify two VBOs, in the array `buffer`. Generally, a call of the form `glGenBuffers(`*n*`,` *buffer*`)` returns *n* such IDs. Two vertex arrays, one for coordinate data and the other for color data, are enabled next by

```
glEnableClientState(GL_VERTEX_ARRAY);
glEnableClientState(GL_COLOR_ARRAY);
```

The binding command

```
glBindBuffer(GL_ARRAY_BUFFER, buffer[VERTICES])
```

activates the first VBO `buffer[VERTICES]`, the parameter `GL_ARRAY_BUFFER` declaring it to be for vertex data. Next,

```
glBufferData(GL_ARRAY_BUFFER, sizeof(vertices) + sizeof(colors),
             NULL, GL_STATIC_DRAW)
```

reserves `sizeof(vertices) + sizeof(colors)` bytes of space for the VBO currently bound to `GL_ARRAY_BUFFER`, that being `buffer[VERTICES]`, of course. The parameter `NULL` indicates that the buffer is not at this time initialized with data. The last parameter `GL_STATIC_DRAW` is a usage hint to the OpenGL system that the data will be specified once and used multiple times as a source for drawing commands.

The general form of the command `glBufferData(`*target*`,` *size*`,` **data*`,` *usage*`)` allocates *size* bytes of storage to the buffer object currently bound to *target*, filling it with application memory data pointed to by **data*, provided this pointer is not `NULL`, supplying, as well, the usage hint *usage*. The usage hint allows the system to optimize the data for performance.

The next two commands

```
glBufferSubData(GL_ARRAY_BUFFER, 0, sizeof(vertices), vertices);
glBufferSubData(GL_ARRAY_BUFFER, sizeof(vertices), sizeof(colors),
                colors);
```

are update commands. In particular, we use them to update the VBO `buffer[VERTICES]` with coordinate and color values. What the command `glBufferSubData(`*target*`,` *offset*`,` *size*`,` **data*`)` does is copy *size* bytes of application data pointed to by **data* into the buffer object currently bound to *target*, starting at an offset of *offset* bytes from the start of the buffer. So, the two commands above evidently fill the first half of buffer[VERTICES] with vertex coordinate values and the latter half with color values.

Next,

```
glBindBuffer(GL_ELEMENT_ARRAY_BUFFER, buffer[INDICES]);
glBufferData(GL_ELEMENT_ARRAY_BUFFER, sizeof(stripIndices),
             stripIndices, GL_STATIC_DRAW);
```

activate the second VBO buffer[INDICES], the parameter GL_ELEMENT_-ARRAY_BUFFER declaring it to be for index data, and initialize it with data from the stripIndices array.

Finally,

```
glVertexPointer(3, GL_FLOAT, 0, 0);
glColorPointer(3, GL_FLOAT, 0, (GLvoid*)(sizeof(vertices)));
```

specify vertex pointers as in the discussion following Experiment 3.3; *however*, the final parameter, instead of being a pointer to application memory as in squareAnnulus4.cpp, now is an offset relative to the start of the currently-bound VBO.

To the much simpler drawing routine next. The interesting thing to notice here is another update method, different from glBufferSubData(). Firstly,

```
float* bufferData = (float*)glMapBuffer(GL_ARRAY_BUFFER, GL_WRITE_ONLY);
```

retrieves into the variable bufferData a pointer to the data store for the VBO currently bound to GL_ARRAY_BUFFER, that being buffer[VERTICES]. The second parameter GL_WRITE_ONLY says access will only be to write into the VBO. The general form of this command is glMapBuffer(*target*, *access*), where *target* is the target buffer object and *access* is one of GL_READ_ONLY, GL_WRITE_ONLY and GL_READ_WRITE.

The loop

```
for (int i = 0; i < sizeof(colors)/sizeof(float); i++)
   bufferData[sizeof(vertices)/sizeof(float) + i]
   = (float)rand()/(float)RAND_MAX;
```

randomly updates color values in buffer[VERTICES], keeping in mind that the type of bufferData means that we'll be offsetting into the buffer storage in units of float size (not byte as earlier). Once updating is done,

```
glUnmapBuffer(GL_ARRAY_BUFFER)
```

releases the VBO, following which

```
glDrawElements(GL_TRIANGLE_STRIP, 10, GL_UNSIGNED_INT, 0);
```

draws the square annulus. Note, again, that the final parameter, instead of being a pointer to the start of the index array in application memory, is now an offset relative to the start of the (index data) VBO.

The reader can at this time safely ignore the functioning of the little gadget `animate()`, which we include to periodically update color values and actually display the changed annulus.

Our next project is to buffer the vertex and index data of `hemisphereMultidraw.cpp`.

Experiment 3.8. Run `hemisphereMultidrawVBO.cpp`. The code, which buffers the vertex and index data of `hemisphereMultidraw.cpp` along the lines of `vbo.cpp`, should be intelligible to one who followed our analysis of the latter program. **End**

The current program, however, dispenses with the interactivity of `hemisphereMultidraw.cpp`, which allowed the user to alter the number of the hemisphere's latitudinal and longitudinal slices. The reason is that a different number of slices requires vertex and index arrays of different sizes and contents so, therefore, corresponding new VBOs which need to be filled anew with data from the application, which goes against the "ship data once and use many times" principle underlying the utility of VBOs.

The problem of multiple data shipments might be circumvented by reserving space for one giant VBO, but this solution seems not particularly elegant. We shall learn an efficient way to interactively refine the hemisphere when we come to the tessellation shaders of fourth generation OpenGL.

The one command of `hemisphereMultidrawVBO.cpp` we would like to draw particular attention to, though, is the program's only drawing command:

```
glMultiDrawElements(GL_TRIANGLE_STRIP, count, GL_UNSIGNED_INT,
                    (const void **)offsets, LAT_SLICES);
```

In particular, note that the fourth parameter is now a pointer to the array *offsets* of offsets in the index VBO to the start of 1D index subarrays, one subarray per triangle strip. The somewhat ugly `(const void **)` casting of *offsets* in the `glMultiDrawElements()` command is necessary owing to the type specification of this command's parameters.

Exercise 3.5. Modify `circle.cpp` of the last chapter to use VBOs to hold vertex data.

Exercise 3.6. **(Programming)** Continue with Exercise 3.4 to redo the bull's eye yet again, this time using VBOs to hold vertex data.

3.3 Vertex Array Objects

A busy scene with many objects, each coded up with its own vertex arrays, possibly buffered in VBOs as well, will likely require switching multiple times between these sets of arrays and buffers, leading to a proliferation of calls like, for example, `glBindBuffer()` and `glVertexPointer()`.

Since version 3.0, OpenGL provides a mechanism to deal with this problem: a *vertex array object*, or *VAO*, is a container to hold all the calls specifying one or more vertex arrays. So, once all the calls specifying a particular object's vertex arrays have been associated with a VAO, one need only activate that VAO prior to drawing the object; in other words, the VAO can be thought of as encapsulating the storage states associated with the object. Let's get to code.

Experiment 3.9. Run `squareAnnulusAndTriangleVAO.cpp`. This program builds on `squareAnnulusVBO.cpp`. We add to it the triangle from `squareAnnulusAndTriangle.cpp` in order to have two VAOs. Note, however, that we separate out the vertex coordinates and color arrays of the triangle, as intertwined they are difficult to manage in a VBO. The outputs of the current program and `squareAnnulusAndTriangle.cpp` are identical. **End**

VAOs are simple to code. The call

```
glGenVertexArrays(2, vao)
```

in the initialization routine of `squareAnnulusAndTriangleVAO.cpp` returns two available IDs for VAOs in the array `vao`. Generally, a call of the form `glGenVertexArrays(`*n*`, `*vao*`)` returns n such IDs.

Next, the first block of statements bracketed by a `// BEGIN...-// END...` comment pair, namely,

```
glBindVertexArray(vao[ANNULUS]);
glGenBuffers(2, buffer); // Generate buffer ids.
...
glVertexPointer(3, GL_FLOAT, 0, 0);
glColorPointer(3, GL_FLOAT, 0, (GLvoid*)(sizeof(vertices1)));
```

begins with the binding command `glBindVertexArray(vao[ANNULUS])`, which activates the first VAO `vao[ANNULUS]`. Then, associated to this VAO are the rest of the calls in the above block, which are copied line for line from `squareAnnulusVBO.cpp`, in particular, from the block dedicated to setting up the vertex arrays for the square annulus in the latter program.

The next block of statements bracketed by a `// BEGIN...-// END...` comment pair, in particular,

```
glBindVertexArray(vao[TRIANGLE]);
glGenBuffers(1, buffer); // Generate buffer ids.
...
glVertexPointer(3, GL_FLOAT, 0, 0);
glColorPointer(3, GL_FLOAT, 0, (GLvoid*)(sizeof(vertices2)));
```

likewise associates to the VAO `vao[TRIANGLE]` all the calls after the first, setting up vertex arrays for the triangle.

Let's see now the drawing routine, much simplified courtesy the VAOs. The two blocks of statement pairs in

```
glBindVertexArray(vao[ANNULUS]);
glDrawElements(GL_TRIANGLE_STRIP, 10, GL_UNSIGNED_INT, 0);

glBindVertexArray(vao[TRIANGLE]);
glDrawArrays(GL_TRIANGLES, 0, 3);
```

successively activate the VAOs corresponding to the annulus and the triangle, drawing these objects. See again the code between the two `// BEGIN...-// END...` comment pairs in the initialization routine to appreciate the savings made in the drawing routine (which would have been even greater had there been multiple occurrences of the annulus or triangle).

Note that `glBindVertexArray(`*vaoID*`)` creates a new VAO if *vaoID* has been freshly returned by a `glGenVertexArrays()` call, and associates with it subsequent vertex array specifications. On the other hand, if *vaoID* is the ID of an already-created VAO, then `glBindVertexArray(`*vaoID*`)` activates that VAO.

Exercise 3.7. (**Programming**) Put the VBOs and related data of the hemisphere of Experiment 3.8 into a VAO and draw a sphere as one hemisphere on top of another.

Exercise 3.8. (**Programming**) Continue with Exercise 3.5 putting all the data for the circle into a VAO.

Remark 3.3. *Important*! Having thus introduced VBOs and VAOs we are, oddly enough, going to counsel the reader against using them (*for now*)! The reason is that the simple (low poly count) programs she, presumably, is going to be writing for a while will hardly benefit from their usage, not really justifying, therefore, the added layer of complexity. It's best to focus on the fundamentals at this stage. Just keep the resource in mind for when your data sets get big.

So that you know, though, VBOs and VAOs will truly come into their own when we study shader-based OpenGL, where using them is *mandatory*! But, not to worry as they are nothing conceptually difficult as we have seen.

3.4 Display Lists

A set of commands, e.g., to define an object such as a wheel or robot arm, which is invoked repeatedly can be cached in a so-called *display list*. The display list is stored on the machine which runs the display unit and, often, pre-compiled and optimized. When the set of commands needs to be invoked, the program simply calls the display list rather than reissue them.

Display lists are particularly efficient in a client-server environment where the two communicate over a network and a goal is to minimize traffic. Once a display list has been saved by the server (the machine running the display unit), it can be invoked on a single command from the client (the machine running the program).

Another evident advantage of display lists is that they provide a logical way to encapsulate objects (think classes of objects, say, in C++).

Experiment 3.10. Run `helixList.cpp`, which shows many copies of the same helix, variously transformed and colored. Figure 3.6 is a screenshot.
End

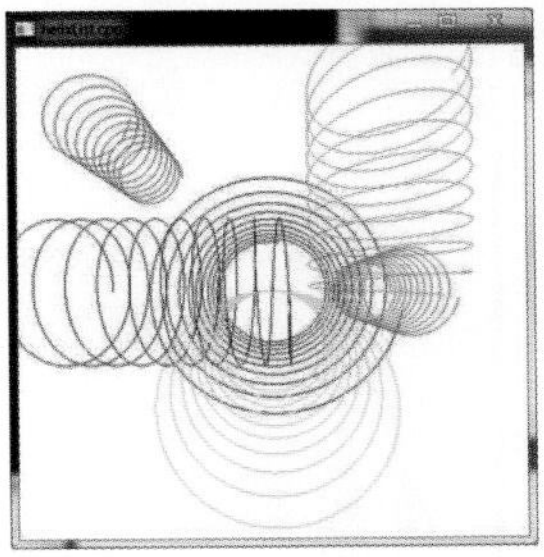

Figure 3.6: Screenshot of `helixList.cpp`.

Here's the snippet from the initialization routine of `helixList.cpp` which creates the display list to draw the helix:

```
aHelix = glGenLists(1);
glNewList(aHelix, GL_COMPILE);
glBegin(GL_LINE_STRIP);
for(t = -10 * PI; t <= 10 * PI; t += PI/20.0)
   glVertex3f(20 * cos(t), 20 * sin(t), t);
glEnd();
glEndList();
```

The call `glGenLists(`*range*`)` returns an integer which starts a block of size *range* of available display list indices. If a block of size *range* is not available, 0 is returned.

The set of commands to be cached in a display list – a helix-drawing routine in the case of `helixList.cpp` – is grouped between a `glNewList(`*listName*`,` *mode*`)` and a `glEndList()` statement. The parameter *listName* – `aHelix` in `helixList.cpp` – is the index which identifies the list. The parameter *mode* may be `GL_COMPILE` (only store, as in the program) or `GL_COMPILE_AND_EXECUTE` (store and execute immediately).

Finally, the drawing routine of `helixList.cpp` invokes `glCallList-(aHelix)` six times to execute the display list. The `glPushMatrix()-glPopMatrix()` statement pairs, as also the modeling transformations (viz. `glTranslatef()`, `glRotatef()`, `glScalef()`) within these pairs, are used to position and scale copies of the helix. Ignore them if they don't make sense at present.

Remark 3.4. Display lists should be created (i.e., the `glNewList()-glEnd-List()` piece of code should be located) in the initialization routine if one wants the efficiency of optimization. If the code is in the drawing routine instead, then, actually, a different display list will be created every frame at run-time! There is little optimization possible in this case. However, the benefit of encapsulation from a programming practice point of view stays.

Exercise 3.9. (Programming) Put the hemisphere-drawing routine of `hemisphere.cpp` into a display list and call the list twice to make a sphere – apply the scaling transformation `glScalef(1.0, -1.0, 1.0)` to one of the hemispheres to flip it over.

Exercise 3.10. (**Programming**) Make a ring of concentric circles of multiple colors on the xy-plane by repeatedly calling a display list containing a circle-drawing routine based on `circle.cpp`. Scale each invocation of the circle by a factor of u with a call to `glScalef(`u`, `u`, 1.0)`.

There is a special mechanism in OpenGL to execute several display lists together.

Figure 3.7: Screenshot of `multipleLists.cpp`.

Experiment 3.11. Run `multipleLists.cpp`. See Figure 3.7 for a screenshot. Three display lists are defined in the program: to draw a red triangle, a green rectangle and a blue pentagon, respectively. **End**

The call `glCallLists(`*n*`, `*type*`, `**lists*`)` causes n display list executions (n is 6 in the program). The indices of the lists to be executed are obtained by adding the current display list base – this base is specified by `glListBase(`*base*`)` – to the successive offset values of type *type* in the array pointed by *lists*.

Exercise 3.11. (**Programming**) Modify `multipleLists.cpp` to draw a vertical black line between each object and the next. The line itself should be in a display list.

3.5 Drawing Text

Graphical text can be of two types: *bitmapped* (also called *raster*) and *stroke* (also called *vector*). Characters of bitmapped text are defined as a pattern of on and off bits in a rectangular block, while characters of stroke text are created using line primitives. For example, in Figure 3.8, the letter 'E' is represented as a bitmap consisting of 10 on bits and 5 off in a 3×5 raster, as well as in stroke form as a union of four straight segments.

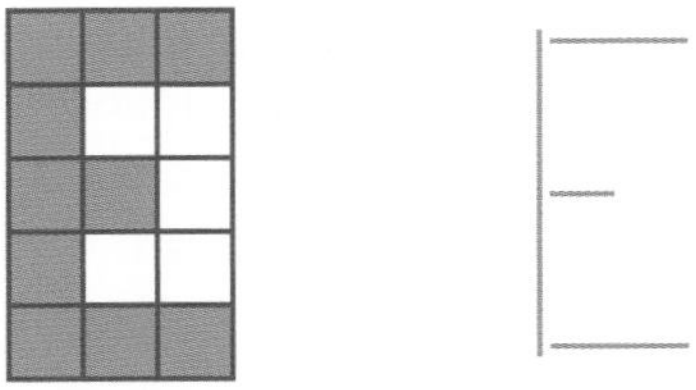

Figure 3.8: Bitmapped versus stroke text.

The FreeGLUT library offers both bitmapped and stroke characters. The calls `glutBitmapCharacter(`**font*`, `*character*`)` and `glutStrokeCharacter-(`**font*`, `*character*`)` render *character* in the specified *font*.

Fonts available for bitmapped characters include:

GLUT_BITMAP_8_BY_13
GLUT_BITMAP_9_BY_15
GLUT_BITMAP_TIMES_ROMAN_10
GLUT_BITMAP_TIMES_ROMAN_24
GLUT_BITMAP_HELVETICA_10
GLUT_BITMAP_HELVETICA_12
GLUT_BITMAP_HELVETICA_18

Fonts available for stroke characters include:
GLUT_STROKE_ROMAN
GLUT_STROKE_MONO_ROMAN
Stroke characters offer an advantage over bitmapped ones in that they can be scaled in size and rotated, because line segments can be so transformed, whereas bitmapped characters, being fixed patterns, are always aligned with the axes.

Experiment 3.12. Run `fonts.cpp`. Displayed are the various fonts available through the FreeGLUT library. See Figure 3.9. **End**

Figure 3.9: Screenshot of `fonts.cpp`.

The canonical routine we use to draw bitmapped text is the following:

```
void writeBitmapString(void *font, char *string)
{
   char *c;
   for (c = string; *c != '\0'; c++) glutBitmapCharacter(font, *c);
}
```

Accordingly, a subsequent call block

```
glRasterPos3f(p, q, r);
writeBitmapString(font, string);
```

renders *string* in bitmapped *font* starting from position (p, q, r) in world coordinates. Keep in mind that these coordinates are transformed by prior modelview transformations, e.g., `glTranslatef()`, `glRotatef()` and such, though, as we said earlier, the bitmapped text itself is always drawn axis-aligned.
Our canonical routine to draw stroke text is

```
void writeStrokeString(void *font, char *string)
{
   char *c;
   for (c = string; *c != '\0'; c++) glutStrokeCharacter(font, *c);
}
```

which renders the text starting from $(0, 0, 0)$ in world coordinates. Note that in addition to scaling and rotation, one can apply a `glLineWidth()` call to alter the thickness of stroke characters as well, as FreeGLUT uses `GL_LINE*` primitives to draw them.

Exercise 3.12. (**Programming**) Locate the labels of `circularAnnuluses.cpp` in the white center of each annulus (you may have to split the labels into more than one line to fit them).

Exercise 3.13. (**Programming**) Modify `fonts.cpp` to be able to cycle through stroke fonts of different line widths by pressing the space bar.

3.6 Programming the Mouse

The mouse can be programmed to respond to button clicks, motion and the wheel turning.

Clicks

Figure 3.10: Screenshot of `mouse.cpp`.

Experiment 3.13. Run `mouse.cpp`. Click the left mouse button to draw points on the canvas and the right one to exit. Figure 3.10 is a screenshot of "OpenGL" scrawled in points. **End**

A mouse callback routine *mouse_callback_func*() is registered to handle mouse events by the FreeGLUT statement `glutMouseFunc(`*mouse_callback_func*`)` in the main routine. In the case of `mouse.cpp`, the callback routine is `mouseControl()`:

```
void mouseControl(int button, int state, int x, int y)
{
   if (button == GLUT_LEFT_BUTTON && state == GLUT_DOWN)
      points.push_back( Point(x, height - y) );
   if (button == GLUT_RIGHT_BUTTON && state == GLUT_DOWN) exit(0);
   glutPostRedisplay();
}
```

The callback routine itself has the form *mouse_callback_func*(*button*, *state*, x, y), where *button* is one of:

```
GLUT_LEFT_BUTTON,   GLUT_RIGHT_BUTTON,   GLUT_MIDDLE_BUTTON
```

and *state* is one of:

```
GLUT_UP,   GLUT_DOWN
```

The coordinates (x, y) return the location in the OpenGL window where the mouse event occurs. They are measured similarly as for screen coordinates – recall from Section 2.3 that screen coordinates are measured in pixels starting from the origin at the upper-left corner of the screen with the x-axis heading right and the y-axis down. The only difference in the case of a mouse click is that the origin is at the upper-left corner of the OpenGL window, rather than screen. Units are still pixels and the x-axis still heads right and the y-axis down. See Figure 3.11. This necessitates care when using the coordinates of

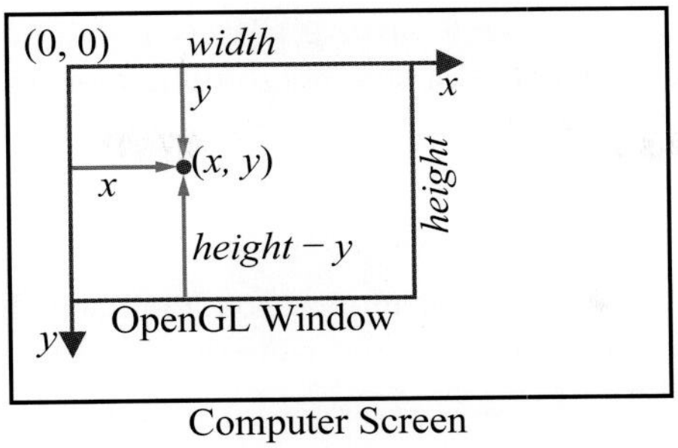

Figure 3.11: Mouse event coordinates (x, y).

a mouse event in the OpenGL program itself, because there is no *a priori* connection between the former and the world coordinates used by the latter.

In particular, note the following two steps in `mouse.cpp`:

1. The call

   ```
   glOrtho(0.0, w, 0.0, h, -1.0, 1.0);
   ```

   in the reshape routine `resize(w, h)` ties screen coordinates to world coordinates by making the dimensions of the viewing face *equal* the actual physical dimensions of the OpenGL window, the latter being passed to the reshape routine via the parameters `w` and `h`. Because viewing face and OpenGL window are now the same size in their respective coordinate systems (world and screen), effectively, one unit along the viewing face along either the x- or y-axis is a pixel.

   The only correction remaining to be made is owing to the y-axis being "upside down" going from one coordinate system to the other. This is done next.

2. The statement

   ```
   points.push_back( Point(x, height - y) );
   ```

   in the mouse callback routine to store points in the `points` vector when the mouse is clicked makes the final correction from the event's screen coordinates to world coordinates, as (x, y) on the screen corresponds to $(x, height - y, 0)$ in the world.

Remark 3.5. A point of note in `mouse.cpp` is the use of an STL `vector` to store `Point` objects. STL stands for the *Standard Template Library*, a C++ library of container classes, e.g., vector, list, set and so forth, together with routines to manipulate container objects. It is a part of the current ANSI C++ standard. The STL is extremely useful and saves a lot of repetitive programming. Readers not already familiar with the STL are well-advised to pick it up. There's no need to devote time separately for this purpose. Keeping a book like Schildt [124] handy while you code should be enough.

Exercise 3.14. (**Programming**) Write a program to draw a circle on a canvas after two left clicks of the mouse. The first click picks the center and the second a point on the circle.

Motion

Experiment 3.14. Run `mouseMotion.cpp`, which enhances `mouse.cpp` by allowing the user to drag the just-created point using the mouse with the left button still pressed. **End**

The additional capability of `mouseMotion.cpp` is obtained as follows. First, when the left mouse button is clicked, the mouse callback routine `mouseControl()` stores a point at the clicked position in the variable `currentPoint` of type `Point`. Only when the button is released is the new point added to the `points` vector by the same routine.

In the interim, between the press and release of the left mouse button, if the mouse moves, then its motion is tracked by the mouse motion callback routine `mouseMotion()`:

```
void mouseMotion(int x, int y)
{
   currentPoint.setCoords(x, height - y);
   glutPostRedisplay();
}
```

This routine simply keeps updating the coordinates of `currentPoint` with the current location of the mouse as the latter moves with the button pressed. The result is that this point, which is drawn separately in the `drawScene()` routine, travels with the mouse. Note that, just as the mouse callback routine is registered in `main()`, so is the motion callback routine, the latter by `glutMotionFunc(mouseMotion)`.

One can also track so-called *passive* motion of the mouse – when it moves with no button pressed – via a passive motion callback function, which is registered in main with a `glutPassiveMotionFunc()` call.

Exercise 3.15. (**Programming**) Enhance the previous circle-drawing exercise by allowing the user to view the changing circle as the mouse is dragged with the second click.

Exercise 3.16. (**Programming**) Modify `mouseMotion.cpp` to make a program which allows the user to sketch on a canvas.

Turning the Wheel

Experiment 3.15. Run `mouseWheel.cpp`, which further enhances `mouse-Motion.cpp` with the capability to change the size of the points drawn by turning the mouse wheel. **End**

The wheel callback routine

```
void mouseWheel(int wheel, int direction, int x, int y)
{
   (direction > 0) ? pointSize++ : pointSize--;
   glutPostRedisplay();
}
```

updates the *wheel* number, which is 0 if there is a single wheel; *direction* of rotation, which is either $+1$ or -1; and the location (x, y) of the mouse in screen coordinates.

To change the size of all drawn points with the wheel, the current program sets the point size in the drawing routine to be the value of the global `pointSize`, rather than the value of the `Point` member variable `size` as in `mouseMotion.cpp`. And, of course, the wheel callback routine is registered in `main()` by `glutMouseWheelFunc(mouseWheel)`.

Exercise 3.17. (**Programming**) Further enhance the circle-drawing program by allowing the user to change the circle's size by scrolling the mouse wheel.

3.7 Programming Non-ASCII Keys

In various programs to date, we've already interacted with the OpenGL window through keyboard entry by registering a handling function *keyboard_handling_func*() in the main routine via a call to `glutKeyboardFunc(`*keyboard_handling_func*`)`. To interact with non-ASCII keys such as the arrow, F, and page up and down keys, one needs likewise to register a handling function *special_key_handling_func*() with a call to `glutSpecialFunc(`*special_key_handling_func*`)`.

Experiment 3.16. Run `moveSphere.cpp`, a program we saw earlier in Experiment 2.23, where you can see a screenshot as well. Press the left, right, up and down arrow keys to move the sphere, the space bar to rotate it and 'r' to reset.

Note how the `specialKeyInput()` routine is written to enable the arrow keys to change the location of the sphere. Subsequently, this routine is registered in `main()` as the handling routine for non-ASCII entry. **End**

Exercise 3.18. (**Programming**) Write a program to cycle through the FreeGLUT fonts applied to the string "I am having so much fun with OpenGL it can't be legal!" by pressing the left and right arrow keys.

3.8 Menus

The FreeGLUT library provides pop-up menus.

Experiment 3.17. Run `menus.cpp`. Press the right mouse button for menu options which allow you to change the color of the initially red square or exit. Figure 3.12 is a screenshot. **End**

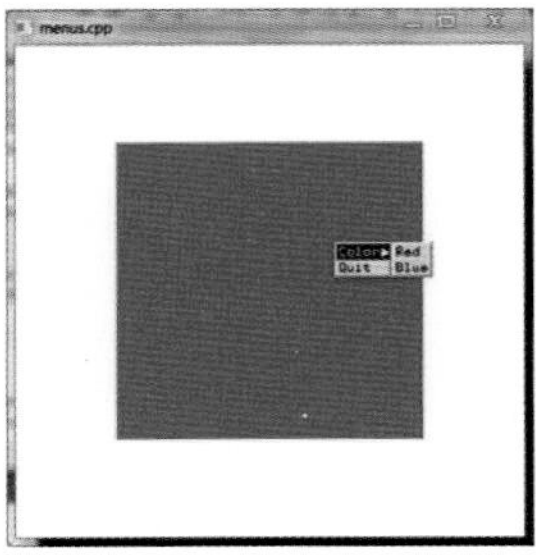

Figure 3.12: Screenshot of `menus.cpp`.

A `glutCreateMenu(`*menu_function*`)` declaration in the `makeMenu()` routine creates a menu, registers *menu_function*() as its callback function and returns a unique integer identifying the menu – to be used by any higher-level menu which may call the current one.

`glutAddMenuEntry(`*tag*`, `*returned_value*`)` creates a menu entry titled *tag* which, when clicked, returns *returned_value* to the callback function *menu_function*(). The latter, therefore, must be of the form *menu_function*(*type_of_returned_value*).

`glutAddSubMenu(`*tag*`, `*sub_menu*`)` is similar to `glutAddMenuEntry()`, except that when *tag* is clicked a sub-menu pops up whose ID is *sub_menu*. Evidently, the statement creating a sub-menu must precede that for a higher-level menu which calls it, as the former's ID *sub_menu* is needed in order to create the latter. So, menus have to be created "bottom-up".

`glutAttachMenu(`*button*`)` attaches the menu to a mouse button.

Exercise 3.19. (Programming) Enhance `menus.cpp` to add two more items to the top-level pop-up menu:

(a) A "Mode" option allowing the rectangle to be shown either "Outlined" or "Filled".

(b) A "Size" option which leads to two sub-menu options "Width" and "Height", either of which has options "Small", "Medium" and "Large".

3.9 Line Stipples

One can create *stippled*, i.e., broken, lines in OpenGL by specifying and applying a stipple pattern.

Experiment 3.18. Run `lineStipple.cpp`. Press the space bar to cycle through stipples. A screenshot is shown in Figure 3.13. **End**

Figure 3.13: Screenshot of `lineStipple.cpp`.

Stippling is enabled with a call to `glEnable(GL_LINE_STIPPLE)` and disabled by calling `glDisable(GL_LINE_STIPPLE)`. The stipple pattern itself is specified by the call `glLineStipple(`*factor*`, `*pattern*`)`.

Parameter *pattern* is a hex string of the form $0\text{x}X_3X_2X_1X_0$ where each X_i is a hexadecimal symbol (equivalent to 4 bits). Thus $X_3X_2X_1X_0$ represents a 16 bit string, say, $a_{15}a_{14}\ldots a_0$. Parameter *factor* is a positive integer.

The stipple pattern is applied as follows: if a_0 is 1, then the first *factor* pixels starting from the first vertex of the line primitive are set on; if a_0 is 0, the first *factor* pixels are off. If a_1 is 1, the next *factor* pixels of the line are on; if a_1 is 0, they are off. And so on .... Note that the lower bits of the stipple pattern come first and that *factor* simply scales the pattern.

For example, suppose the stipple is specified by `glLineStipple(1, 0x5555)`. Since 0x5555 = 0101010101010101, alternate pixels of the line are on and off with the first one being on. See Figure 3.14(a).

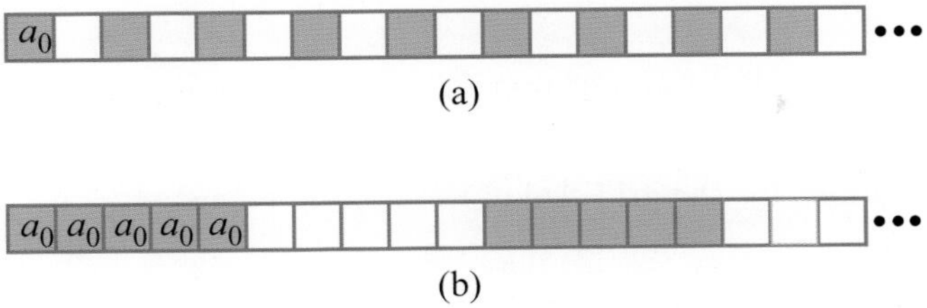

Figure 3.14: (a) Line stipple specified by `glLineStipple(1, 0x5555)` (b) Line stipple specified by `glLineStipple(5, 0x5555)`.

If the stipple is specified by `glLineStipple(5, 0x5555)` then alternate groups of five pixels on the line are on and off. See Figure 3.14(b).

Exercise 3.20. (Programming) Apply the different line stipples of `lineStipple.cpp` to the circle in `circle.cpp`.

FreeGLUT stroke characters can be stippled to interesting effect as well, as `GL_LINE*` primitives are used to draw them.

Exercise 3.21. (Programming) Display the text "I am having so much fun with OpenGL it can't be legal!" using variously stippled stroke fonts.

A 2D Drawing Program

Experiment 3.19. Run `canvas.cpp`, a simple program to draw on a flat canvas with menu and mouse functionality.

Left click on an icon to select it. Then left click on the drawing area to draw – click once to draw a point, twice to draw a line or rectangle. Right click for a pop-up menu. Figure 3.15 is a screenshot. **End**

Figure 3.15: Screenshot of `canvas.cpp`.

Exercise 3.22. (Programming) Enhance `canvas.cpp`:

(a) Add a polyline (multi-segment line) drawing capability. Create a suitable icon. Left clicking this icon picks the polyline option. Subsequent left clicks pick successive segment endpoints until a middle click completes the polyline.

(b) Add a circle drawing capability. After left clicking the circle icon, the next two left clicks pick the center and a point on the circle, respectively, following which the circle is drawn.

(c) Add a regular (equal-sided) hexagon drawing capability. After left clicking the hexagon icon, the next two left clicks pick the center and a vertex, respectively, following which the hexagon is drawn.

(d) For the existing line segment and rectangle options, as well as for the new polyline, circle and hexagon options, use mouse motion tracking to allow the user to see the newly-created primitive change in real-time as the mouse moves, before it is saved with a final click.

(e) Add functionality to input text from the keyboard.

(f) Give options for the grid size in the pop-up menu.

(g) Add color options through the pop-up menu.

(h) Add an outlined/filled option through the pop-up menu.

(i) For the preceding three parts, make the pop-up menu depend on where the mouse is right-clicked. In particular, the color option should be available when any of the primitive icons on the left is clicked; the filled option, on the other hand, should appear only upon clicking icons of the 2D primitives, namely, the rectangle, circle and hexagon; finally, if the click is on the drawing area, then the options are, simply, grid-clear-quit (grid having a size sub-menu as well). For the color and filled options, the icon should change as well to represent the choice made.

A way to make the pop-up menu location-sensitive is by having the mouse callback routine, rather than the main routine, call the menu-making routine, so that the latter can access mouse event coordinates.

(j) Here's something not to do with drawing per se but to get you to revisit vertex arrays in the first section: the `drawGrid()` routine draws lines over a `for` loop of vertices; use vertex array commands instead.

3.10 FreeGLUT Objects

The FreeGLUT library offers a collection of standard objects. Each object is available in two flavors: solid and wireframe. The respective calls are shown in the table below. The objects are all drawn centered at the origin. The parameters, if any, determine the object's size and the fineness of its triangulation. All the FreeGLUT objects are depicted in wireframe in Figure 3.16.

Solid	Wireframe
glutSolidSphere(*radius*, *slices*, *stacks*)	glutWireSphere(*radius*, *slices*, *stacks*)
glutSolidCube(*size*)	glutWireCube(*size*)
glutSolidCone(*base*, *height*, *slices*, *stacks*)	glutWireCone(*base*, *height*, *slices*, *stacks*)
glutSolidTorus(*inRadius*, *outRadius*, *sides*, *rings*)	glutWireTorus(*inRadius*, *outRadius*, *sides*, *rings*)
glutSolidDodecahedron(*void*)	glutWireDodecahedron(*void*)
glutSolidOctahedron(*void*)	glutWireOctahedron(*void*)
glutSolidTetrahedron(*void*)	glutWireTetrahedron(*void*)
glutSolidIcosahedron(*void*)	glutWireIcosahedron(*void*)
glutSolidTeapot(*size*)	glutWireTeapot(*size*)

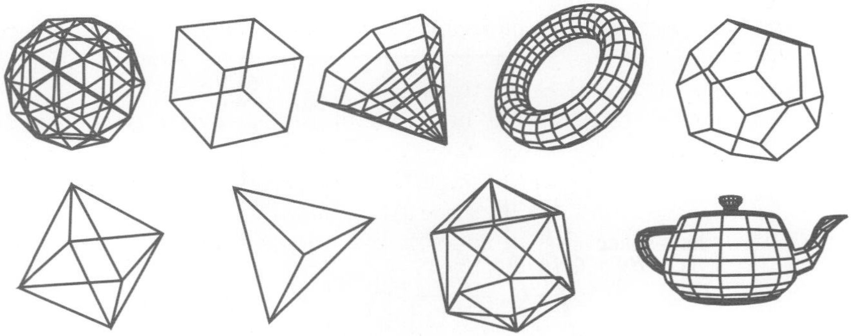

Figure 3.16: Wireframe FreeGLUT objects.

Experiment 3.20. Run `glutObjects.cpp`. Press the arrow keys to cycle through the various FreeGLUT objects and 'x/X', 'y/Y' and 'z/Z' to turn them. **End**

3.11 Clipping Planes

We saw in Section 2.4 that OpenGL clips a scene to within a viewing volume (box or frustum), a process which can be thought of as clipping the scene off on one side of each of the six planes which bound the volume. These six planes, called *clipping planes*, are automatically implied by the projection statement, such as `glOrtho()` or `glFrustum()`, which defines the box or frustum. However, the programmer can specify additional clipping planes.

The call

`glClipPlane(GL_CLIP_PLANE`*i*`, `**equation*`);`

specifies an ith additional clipping plane, where *equation* points to an array giving the four coefficients of the equation

$$Ax + By + Cz + D = 0$$

of the new clipping plane. If this plane is enabled with the call `glEnable(GL_CLIP_PLANE`i`)`, then the points (x, y, z) of objects which lie in the open half-space

$$Ax + By + Cz + D < 0$$

are clipped off; equivalently, only those points (x, y, z) of objects lying in the closed half-space

$$Ax + By + Cz + D \geq 0$$

are rendered. The ith additional clipping plane is disabled with a call to `glDisable(GL_CLIP_PLANE`i`)`.

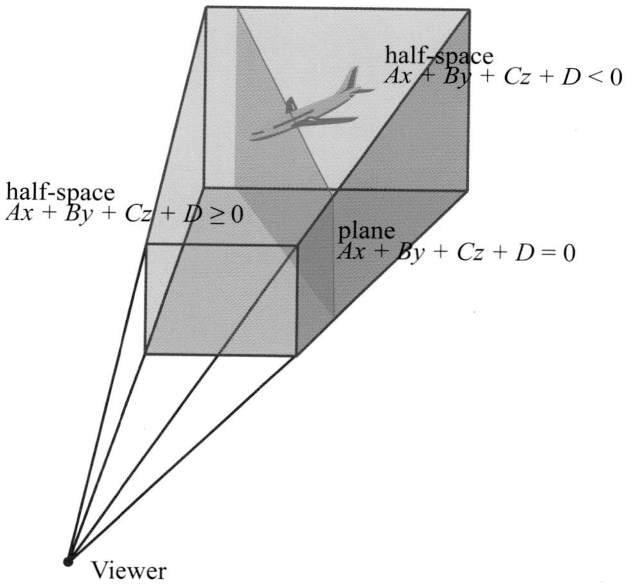

Figure 3.17: A clipping plane clipping a plane in half.

In Figure 3.17, for example, only the front part of the aircraft will be visible. Although we depict it in the figure, the clipping plane itself is not drawn by OpenGL.

Experiment 3.21. Run `clippingPlanes.cpp`, which augments `circular-Annuluses.cpp` with two additional clipping planes which can be toggled on and off by pressing '0' and '1', respectively.

The first plane clips off the half-space $-z + 0.25 < 0$, i.e., $z > 0.25$, removing the floating white disc of the annulus on the upper-right. The second one clips off the half-space $x + 0.5y < 60.0$, which is the space below an angled plane parallel to the z-axis. Figure 3.18 is a screenshot of both clipping planes activated. **End**

Figure 3.18: Screenshot of `clippingPlanes.cpp`.

Exercise 3.23. (**Programming**) Change the equations of the two clipping planes of `clippingPlanes.cpp` so that enabling both leaves only the red disc of the upper-right annulus visible.

Example 3.1. Replace the data

```
double eqn0[4] = 0.0, 0.0, -1.0, 0.25;
```

for the first clipping plane of `clippingPlanes.cpp` with

```
double eqn0[4] = 0.0, 0.0, 1.0, -0.25;
```

Apparently, we are replacing $-z + 0.25 = 0$ with $z - 0.25 = 0$, which are both equations of the same plane. Why, then, is the result of clipping different for the two?

Answer: The half-space clipped, given the equation $-z + 0.25 = 0$, is $-z + 0.25 < 0$, i.e., $z > 0.25$. On the other hand, the half-space clipped, given the equation $z - 0.25 = 0$, is $z - 0.25 < 0$, i.e., $z < 0.25$.

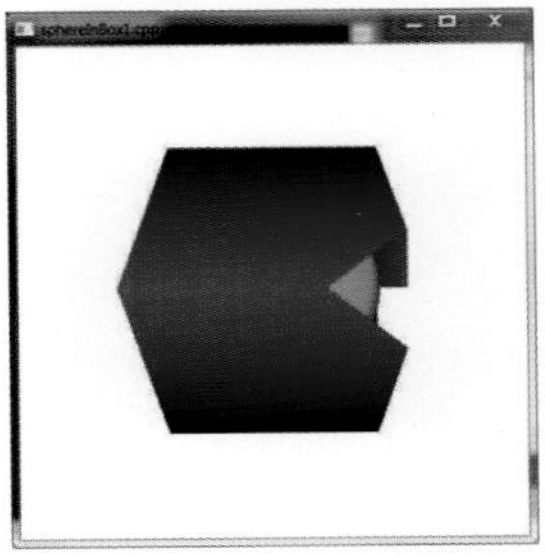

Figure 3.19: Screenshot of `sphereInBox1.cpp` with a corner clipped off.

Exercise 3.24. (**Programming**) Add a clipping plane to `sphereIn-Box1.cpp` (see Chapter 11, ignore lighting) to clip off a corner of the box, revealing the sphere inside. Your output should look like Figure 3.19.

Exercise 3.25. (**Programming**) Add a clipping plane to `moveSphere.cpp` to turn the movable sphere into a movable hemisphere.

Clipping planes cause OpenGL not to display parts of an object which are otherwise computed. For example, if one draws a hemisphere by clipping off half a FreeGLUT sphere, then OpenGL first computes geometric data (vertices, etc.) for the entire sphere and then suppresses the part on one side of the clipping plane before rendering. See Figure 3.20. Clearly, this is doubly inefficient for the suppressed part, as OpenGL computes the location of each of its vertices, *and then* computes again to decide that they are actually invisible!

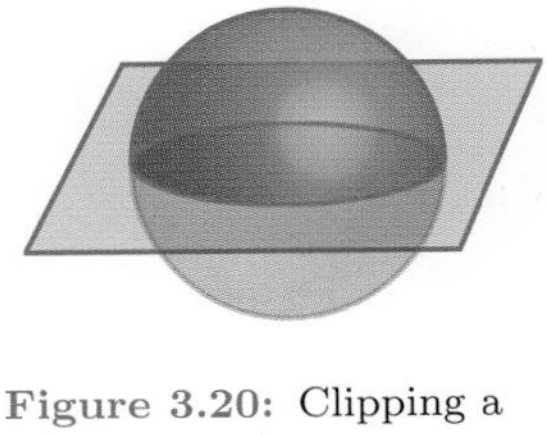

Figure 3.20: Clipping a sphere to make a hemisphere: the clipped half is computed *and* suppressed.

Keep in mind as well that clipping planes don't act just on any one object, but across the whole scene. So, a careless programmer could very well end up unintentionally slicing a remote object. Clipping planes though are ideal for the purpose of displaying a *cut-away view* of an object, as in Figure 3.19.

Bottom line: Use clipping planes as a viewing and not a drawing device.

3.12 gluPerspective()

The statement

`gluPerspective(`*fovy*`, `*aspect*`, `*near*`, `*far*`);`

calls a utility library routine built on top of `glFrustum()`, the perspective projection command introduced in Section 2.8. It creates a viewing frustum as does `glFrustum()`. However, the frustum is specified differently:

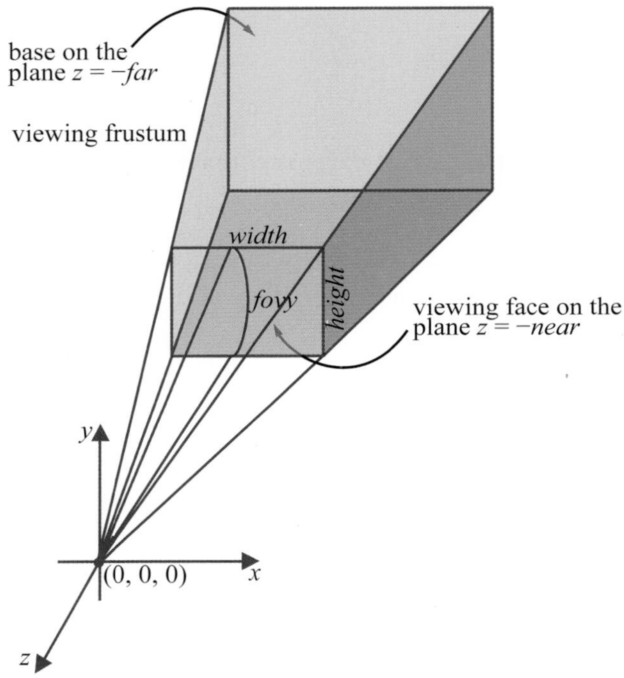

Figure 3.21: Viewing frustum created by `gluPerspective(`*fovy*`, `*aspect*`, `*near*`, `*far*`)`.

The parameter *fovy*, called the *field of view angle*, is the angle along the *yz*-plane at the apex of the pyramid (of which the frustum is a truncation); *aspect* is the *aspect ratio* = *width*/*height* of the front face of the frustum; and *near* and *far* remain as for `glFrustum()`. See Figure 3.21. These four parameters it turns out are, in fact, enough for OpenGL to determine the eight vertices of a frustum which is *symmetric* about the *z*-axis, in other words, a frustum corresponding to a

`glFrustum(`*left*`, `*right*`, `*bottom*`, `*top*`, `*near*`, `*far*`);`

call where $left = -right$ and $bottom = -top$. Such frustums are, in fact, most typical in applications and rarely does one have occasion to create one not symmetric about the *z*-axis.

Example 3.2. The projection statement of `hemisphere.cpp` is the symmetric

```
glFrustum(-5.0, 5.0, -5.0, 5.0, 5.0, 100.0);
```

Determine the equivalent `gluPerspective()` call.

Answer: The aspect ratio of the front face of the frustum created by `glFrustum(-5.0, 5.0, -5.0, 5.0, 5.0, 100.0)` is 1 as both its width ($= right - left$) and height ($= top - bottom$) are 10.0. To determine *fovy*, see Figure 3.22, which shows the section of the viewing frustum by the yz-plane. By elementary trigonometry the half-angle at the apex is 45°, so that $fovy = 90.0$. Therefore, the equivalent call is

```
gluPerspective(90.0, 1.0, 5.0, 100.0);
```

Check it out! Replace the current `glFrustum()` projection statement of `hemisphere.cpp` with the `gluPerspective()` statement just computed.

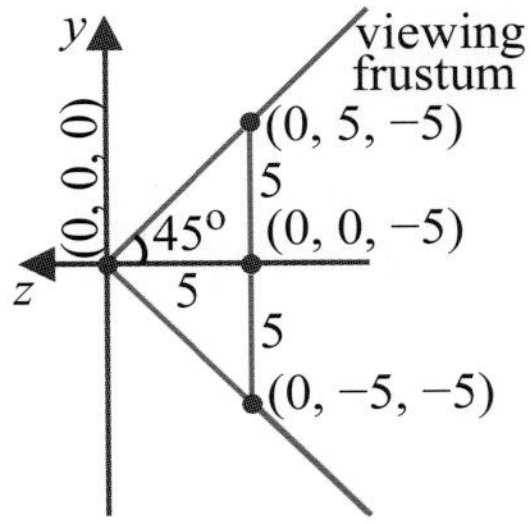

Figure 3.22: Section by the yz-plane (i.e., $x = 0$ plane) of the viewing frustum (bold) created by `glFrustum(-5.0, 5.0, -5.0, 5.0, 5.0, 100.0)`.

Exercise 3.26. Change the *near* value of the projection statement of `hemisphere.cpp` as follows:

```
glFrustum(-5.0, 5.0, -5.0, 5.0, 10.0, 100.0);
```

Determine the equivalent `gluPerspective()` call.

Exercise 3.27. Determine the equivalent `glFrustum()` call of the following projection statement:

```
gluPerspective(60.0, 2.0, 10.0, 100.0);
```

Hint: Use trigonometry in the yz-section to determine first the *top* and *bottom* values and then the aspect ratio to determine *left* and *right*.

Whether to define a perspective projection by a `glFrustum()` or a `gluPerspective()` call is a matter of personal preference. As we've seen, they are equivalent provided one is interested only in frustums symmetric about the z-axis.

However, a convenience of `gluPerspective()` in certain applications arises from the fact that the aspect ratio of the viewing face is an explicit parameter, making it easy to bind it to the aspect ratio of the OpenGL window itself. This comes in handy if you recall the final step of the rendering process when the viewing face is scaled to fit onto the OpenGL window – resulting in distortion if the aspect ratios of the two differ. Let's see this in code.

Experiment 3.22. Run `hemisphere.cpp`.

The initial OpenGL window is a square 500 × 500 pixels. Drag a corner to change its shape, making it tall and thin. The hemisphere is distorted to become ellipsoidal (Figure 3.23(a)). Replace the perspective projection statement

```
glFrustum(-5.0, 5.0, -5.0, 5.0, 5.0, 100.0);
```

with

```
gluPerspective(90.0, 1.0, 5.0, 100.0);
```

As this is equivalent to the original `glFrustum()` call, there is still distortion if the window's shape is changed. Next, replace the projection statement with

```
gluPerspective(90.0, (float)w/(float)h, 5.0, 100.0);
```

which sets the aspect ratio of the viewing frustum equal to that of the OpenGL window. Resize the window – the hemisphere is no longer distorted (Figure 3.23(b))! **End**

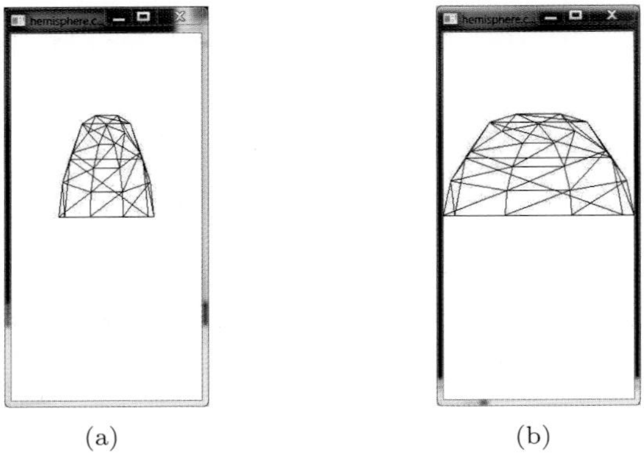

Figure 3.23: Screenshots of `hemisphere.cpp` with the window squished and the projection statement (a) `glFrustum(-5.0, 5.0, -5.0, 5.0, 5.0, 100.0)` and (b) gluPerspective(90.0, (float)w/(float)h, 5.0, 100.0).

3.13 Viewports

The *viewport* of a scene is that region of the OpenGL window in which it is drawn. By default, it is the entire window. However, a `glViewPort()` call may be used to draw to a smaller rectangular subregion.

The call `glViewport(`x, y, w, h`)` specifies the viewport as the rectangular subregion of the OpenGL window which has its lower-left corner at the point (x, y), and is of width w and height h. Units are pixels and the coordinates in the OpenGL window are such that the origin is located at the lower-left corner, the increasing direction of the x-axis is rightwards, and that of the y-axis upwards. See Figure 3.24.

Multiple viewports can be created in a single OpenGL window by invoking more than one `glViewport()` call in the drawing routine. The contents of

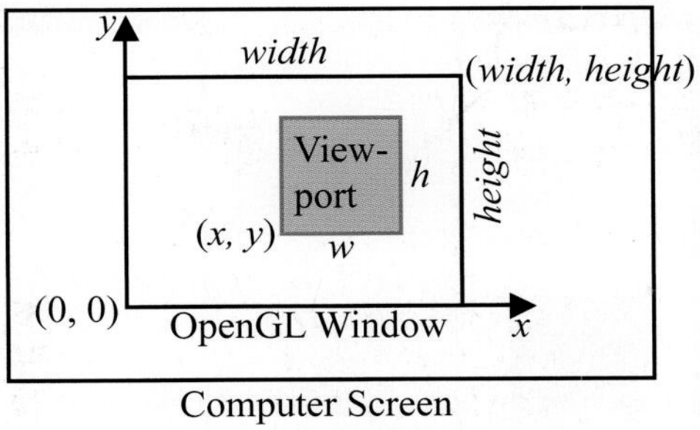

Figure 3.24: Viewport specified by `glViewport(x, y, w, h)`.

a particular viewport are defined by the statements following its defining `glViewport()` call and before the next one (if any).

Experiment 3.23. Run `viewports.cpp` where the screen is split into two viewports with contents a square and a circle, respectively. Figure 3.25 is a screenshot.

A vertical black line is drawn (in the program) at the left end of the second viewport to separate the two. As the aspect ratio of both viewports differs from that of the viewing face, the square and circle are squashed laterally. **End**

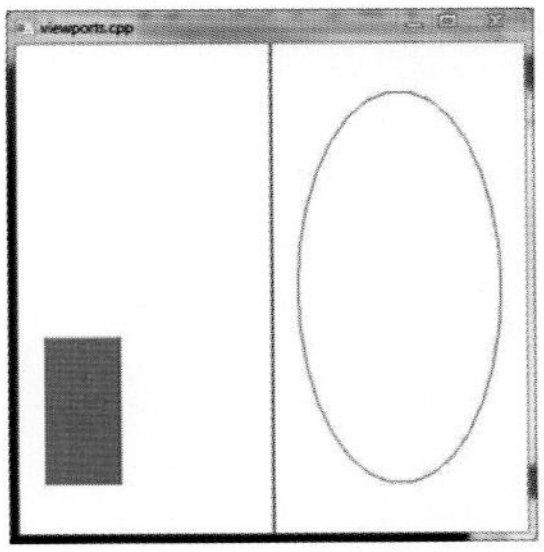

Figure 3.25: Screenshot of `viewports.cpp`.

Viewports are particularly useful in games to show split-screen views of different scenes, or perhaps the same scene from different cameras. We'll see a nice application of this in the program `spaceTravel.cpp` coming up in the next chapter, where the user maneuvers a spacecraft through an asteroid field with a split-screen view in the OpenGL window – one from a global fixed camera and the other from the craft itself.

Exercise 3.28. (Programming) Change the orthographic projection statement of `viewports.cpp` so that the square and circle are no longer distorted.

Exercise 3.29. (Programming) Create a 2×2 grid of four equally-sized viewports with one of the words "This", "is", "so" and "easy" in each. Add lines to separate the viewports.

3.14 Multiple Windows

The `glutCreateWindow()` call of the FreeGLUT library may be invoked more than once in the main routine to create multiple top-level OpenGL windows. Each call to `glutCreateWindow()` returns an integer *id* which is then passed to a `glutSetWindow(`*id*`)` call in the display routine to determine the current drawing window. Properties such as the display routine, resize routine, etc., of each top-level window may be specified independently.

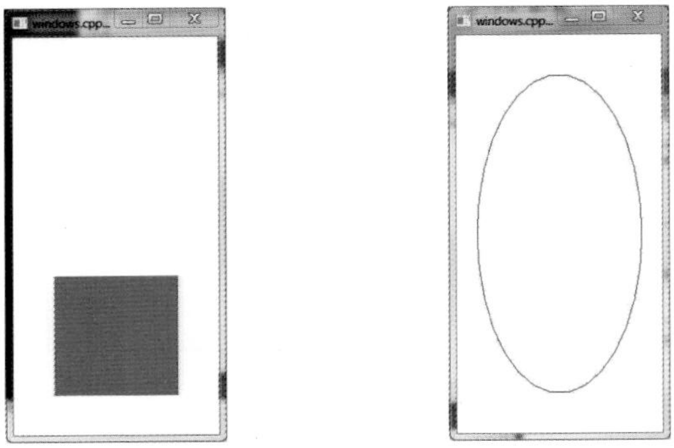

Figure 3.26: Screenshot of `windows.cpp`.

Experiment 3.24. Run `windows.cpp`, which creates two top-level windows (Figure 3.26). **End**

Exercise 3.30. **(Programming)** Create three top-level windows with red, green and blue backgrounds, and containing the words "Red", "Green" and "Blue", respectively.

3.15 Summary, Notes and More Reading

In this chapter we learned a number of different coding utilities, none difficult, but all useful. Vertex arrays and their access commands are particularly important for efficient OpenGL code and the reader should make a practice of using them. The coverage of syntax was by no means complete, nor was it meant to be.

For OpenGL utilities the reader should refer to the red and blue books for a full description. FreeGLUT's home page [49] doesn't seem to have much by way of documentation but Lighthouse 3D [87] has a tutorial on FreeGLUT's predecessor GLUT which applies to FreeGLUT. NeHe Productions [102] has, among many, tutorials on fonts and display lists.

Keep in mind that the purpose of GLUT originally, as of FreeGLUT now, is to provide a few easy-to-use platform-independent utilities to build a simple GUI, not provide a full-featured suite.

Programmers who do require sophisticated interfaces should employ platform-specific utilities, e.g., the MFC Library for Windows. Readers may also find helpful Paul Rademacher's GLUI User Interface Library [57], which provides a collection of GLUT-based utilities such as buttons and checkboxes. Trolltech's Qt [142] may be of interest to those planning commercial-grade GUI's.

Part III

Movers and Shapers

CHAPTER 4

Transformation, Animation and Viewing

The goal for this chapter is to understand how to move and manipulate objects, and maneuver the camera, skills essential to making movies and games. OpenGL provides the laboratory for us to explore. The modeling transformations of OpenGL – including translation, scaling and rotation – control object motion, while the viewing transformation manages the camera. We'll examine the syntax of the transformation commands and how they are composed and applied to achieve animation. To efficiently and creatively animate it's essential to have some grasp of its implementation, so we'll examine parts of OpenGL's animation engine as well. An experiment-discuss-repeat approach is used throughout, each new idea introduced and illustrated with the help of live code.

When objects move, especially in an interactive and unscripted environment like that of a game, they can collide. We'll discuss collision detection, therefore, in the context of animation. Related to animation as well is the notion of the orientation of an object, which we'll see how to quantify.

Section 4.1 introduces the three modeling transformations – translation, rotation and scaling. Sections 4.2 and 4.3 discuss composing transformations and how such composition places multiple objects relative to one another. The modelview matrix stack facilitates the application of transformations to multiple objects, as we see in Section 4.4. Section 4.5 analyzes a few instructional animation programs and concludes with a bunch of exercises.

The viewing transformation is introduced in Section 4.6. After grasping its functioning we find that a viewing transformation is actually a bit of a "fake", being simulated by OpenGL with the help of modeling transformations. An understanding of the viewing transformation leads to a preliminary discussion in Section 4.6, as well, of orientation and how it is specified by Euler angles. We present as well an application of the viewing transformation to animate a camera, together with rudimentary collision detection, in a space-travel program.

More animation code, including programs to develop key-frame animation sequences for a man-like articulated figure, as well as simple shadow animation, is presented in Section 4.7.

Section 4.8 describes methods to enable a user to choose an object on the screen with a mouse-like device, a facility critical in interactive programs like games. Section 4.9 concludes the chapter with a summary, notes and suggestions for more reading.

This chapter is a longish slog but it gets you well on the way to designing realistic 3D applications.

4.1 Modeling Transformations

Translation, scaling and rotation, the so-called *modeling transformations* of OpenGL, are applied to objects to change their location and shape.

4.1.1 Translation

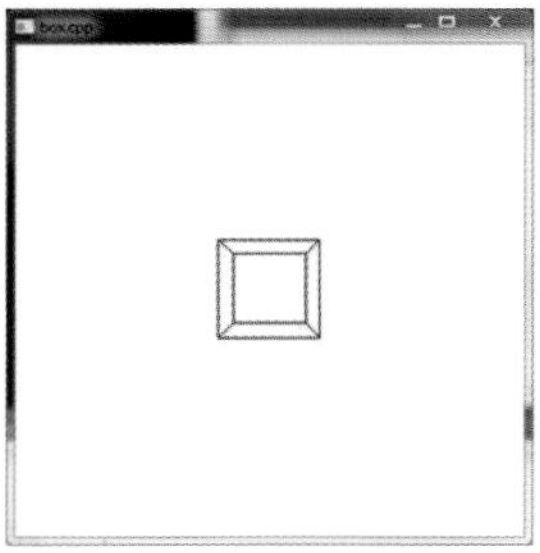

Figure 4.1: Screenshot of `box.cpp`.

Experiment 4.1. Run `box.cpp`, which shows an axis-aligned – i.e., with sides parallel to the coordinate axes – FreeGLUT wireframe box of dimensions $5 \times 5 \times 5$. Figure 4.1 is a screenshot. Note the foreshortening – the back of the box appears smaller than the front – because of perspective projection in the viewing frustum specified by the `glFrustum()` statement.

Comment out the statement

```
glTranslatef(0.0, 0.0, -15.0);
```

What do you see now? *Nothing*! We'll explain why momentarily. **End**

The *translation* command `glTranslatef(`p`, `q`, `r`)` translates an object p units in the x-direction, q units in the y-direction and r units in the z-direction. Precisely, each point (x, y, z) of the object is mapped to the point $(x + p,\ y + q,\ z + r)$. See Figure 4.2, which also shows a whole box translated by `glTranslatef(`p`, `q`, `r`)`.

Returning to `box.cpp`, the command `glutWireCube(5.0)` itself creates a box of side length 5 centered at the origin, with vertices, therefore, at $(\pm 2.5, \pm 2.5, \pm 2.5)$, each vertex corresponding to one of the eight possible combinations of signs. The box clearly lies entirely outside the viewing frustum specified by `glFrustum(-5.0, 5.0, -5.0, 5.0, 5.0, 100.0)` – in fact, entirely on the clipped side of the viewing plane $z = -5$. However, `glTranslatef(0.0, 0.0, -15.0)` pushes the box 15 units in the $-z$ direction, to place it inside the viewing frustum and make it visible (see

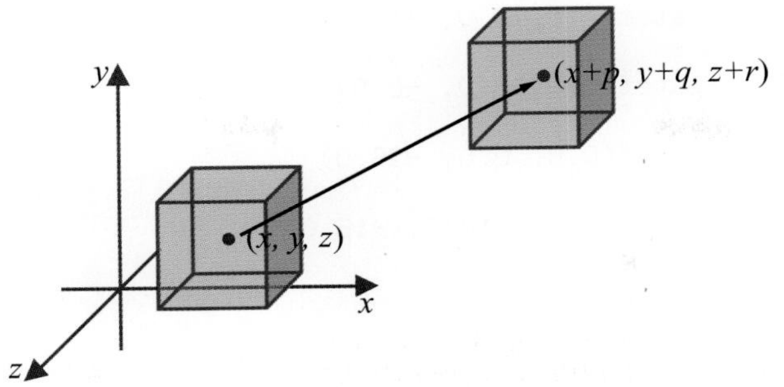

Figure 4.2: Translation: `glTranslatef(p, q, r)`.

Figure 4.3). That is why commenting out this statement results in a blank window.

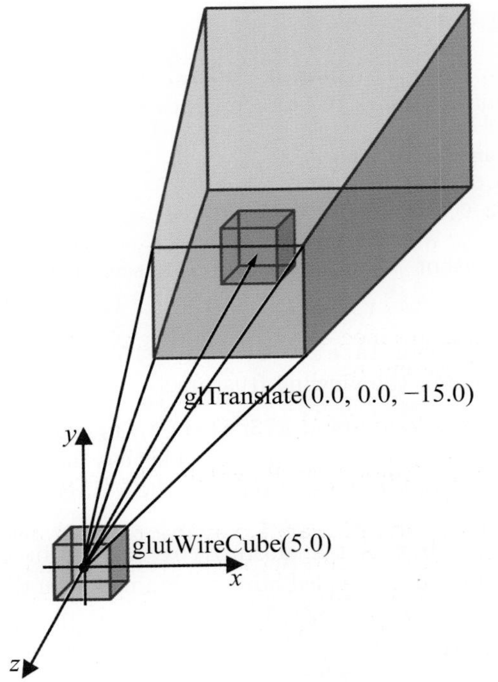

Figure 4.3: Translating into the viewing frustum.

Experiment 4.2. Successively replace the translation command of `box.cpp` with the following, making sure that what you see matches your understanding of where the command places the box. Keep in mind foreshortening, as well as clipping to within the viewing frustum.

1. `glTranslatef(0.0, 0.0, -10.0)`

2. `glTranslatef(0.0, 0.0, -5.0)`

3. `glTranslatef(0.0, 0.0, -25.0)`

4. `glTranslatef(10.0, 10.0, -15.0)`

End

Exercise 4.1. To what point is $(-2.0, 3.0, 9.0)$ transformed by `glTranslatef(3.0, 1.0, -8.0)`?

Exercise 4.2. What is the OpenGL translation that takes $(3.0, -1.0, 2.0)$ to $(3.0, 5.0, 9.0)$?

4.1.2 Scaling

Experiment 4.3. Add a scaling command, in particular, replace the modeling transformation block of `box.cpp` with (Block 1*):

```
// Modeling transformations.
glTranslatef(0.0, 0.0, -15.0);
glScalef(2.0, 3.0, 1.0);
```

Figure 4.4 is a screenshot – compare with the unscaled box of Figure 4.1.

End

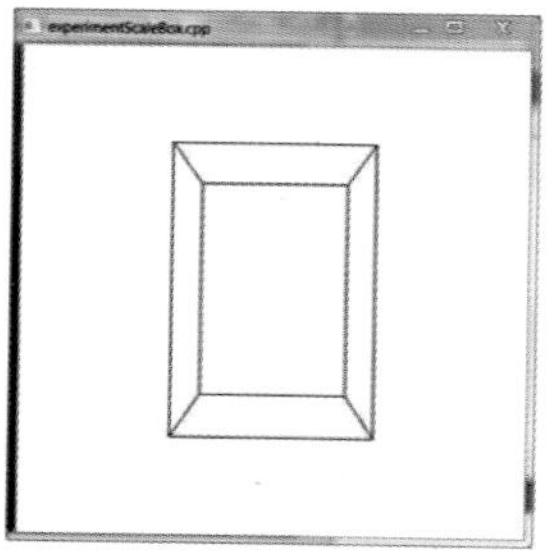

Figure 4.4: Screenshot of Experiment 4.3.

Remark 4.1. The `glTranslatef(0.0, 0.0, -15.0)` call is retained to "kick" the scaled box into the viewing frustum.

Precisely, the *scaling* command `glScalef(`u`, `v`, `w`)` maps each point (x, y, z) of an object to the point (ux, vy, wz). This has the effect of *stretching* objects by a factor of u in the x-direction, v in the y-direction, and w in the z-direction. See Figure 4.5.

Let's see how the box is transformed by scaling in the preceding experiment. The vertices of the scaled box are obtained from the original ones by the transformation $(x, y, z) \mapsto (2x, 3y, 1z)$. For example, $(2.5, 2.5, 2.5) \mapsto (5.0, 7.5, 2.5)$, $(-2.5, 2.5, 2.5) \mapsto (-5.0, 7.5, 2.5)$, and so on. So, the new vertices are $(\pm 5.0, \pm 7.5, \pm 2.5)$, which gives a $10 \times 15 \times 5$ box as one would expect from applying `glScalef(2.0, 3.0, 1.0)` to a $5 \times 5 \times 5$ box.

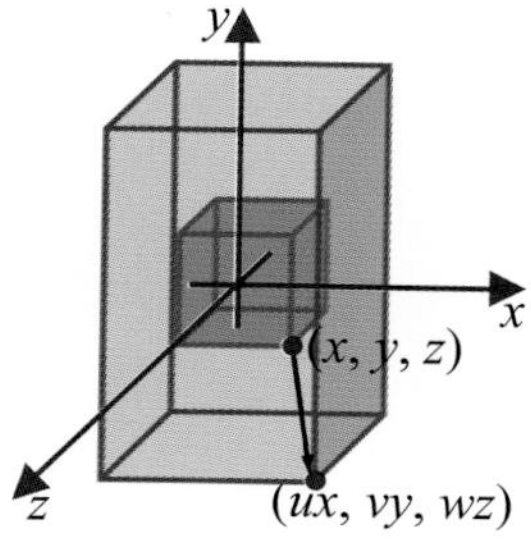

Figure 4.5: Scaling: glScalef(u, v, w).

Experiment 4.4. An object less symmetric than a box is more interesting to work with. Care for a teapot? Accordingly, change the modeling transformation and object definition part of `box.cpp` to (Block 2):

*To cut-and-paste you can find the block in text format in the file `chap4codeModifications.txt` in the directory `ExperimenterSource/CodeModifications`.

```
// Modeling transformations.
glTranslatef(0.0, 0.0, -15.0);
glScalef(1.0, 1.0, 1.0);

glutWireTeapot(5.0); // Teapot.
```

Of course, `glScalef(1.0, 1.0, 1.0)` does nothing and we see the original unscaled teapot (Figure 4.6).

Next, successively change the scaling parameters by replacing the scaling command with the ones below. In each case, make sure your understanding of the command matches the change that you see in the shape of the teapot.

Figure 4.6: Screenshot of initial configuration of Experiment 4.4.

1. `glScalef(2.0, 1.0, 1.0)`
2. `glScalef(1.0, 2.0, 1.0)`
3. `glScalef(1.0, 1.0, 2.0)`

End

Exercise 4.3. (Programming) Continuing with the preceding experiment, try to guess first, for the scalings below, each of which has at least one negative parameter, the difference you will see from the initial configuration shown in Figure 4.6.
Hint: The transformation $(x, y, z) \mapsto (-x, y, z)$, for instance, is a mirror-like reflection about the yz-plane. See Figure 4.7.

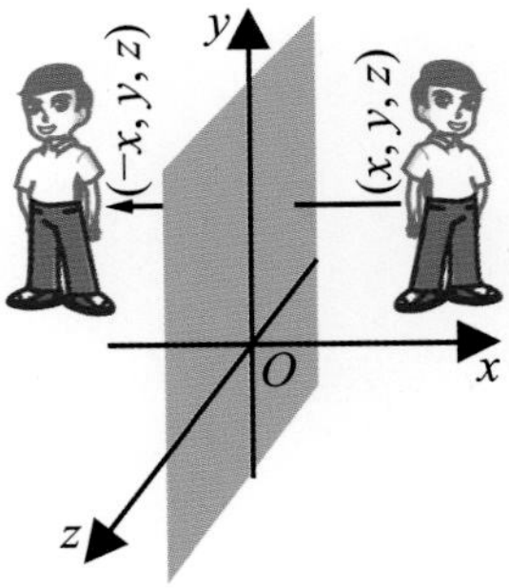

Figure 4.7: Reflection in the yz-plane.

4. `glScalef(-1.0, 1.0, 1.0)`
5. `glScalef(1.0, -1.0, 1.0)`
6. `glScalef(1.0, 1.0, -1.0)`
7. `glScalef(-1.0, -1.0, 1.0)`

Exercise 4.4. (Programming) Continue with the preceding exercise and replace the scaling command with the following, each of which has a zero parameter:

8. `glScalef(1.0, 1.0, 0.0)`

Hint: The transformation

$$(x, y, z) \mapsto (1x,\ 1y,\ 0z) = (x,\ y,\ 0)$$

"collapses" all z-values to 0.0.

9. `glScalef(1.0, 0.0, 1.0)`
10. `glScalef(0.0, 1.0, 1.0)`

Not very interesting the last two! A scaling transformation where one or more of the scaling factors is zero is said to be *degenerate*. Although not common, there is the occasional application where a degenerate scaling transformation comes in handy. We'll see one such in drawing a shadow later in this chapter in Experiment 4.35.

Exercise 4.5. To what point is $(-2.0, 3.0, 9.0)$ transformed by `glScalef(3.0, 1.0, -8.0)`?

Exercise 4.6. What is the OpenGL scaling that transforms $(3.0, -1.0, 2.0)$ to $(3.0, 5.0, 9.0)$?

We have so far scaled only FreeGLUT wire cubes and teapots, whose own axes are aligned with the coordinate axes, so that, effectively, they are only stretched and not skewed. Let's try one that's not so aligned.

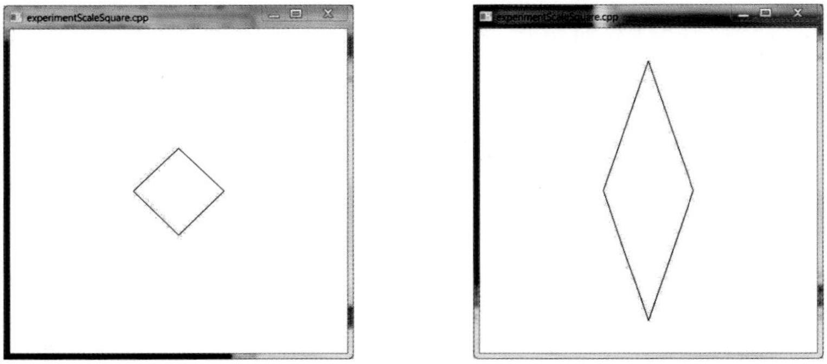

(a) (b)

Figure 4.8: Screenshots of Experiment 4.5: (a) before scaling (b) after.

Experiment 4.5. Replace the cube of `box.cpp` with a square whose sides are not parallel to the coordinate axes. In particular, replace the modeling transformation and object definition part of that program with (Block 3):

```
// Modeling transformations.
glTranslatef(0.0, 0.0, -15.0);
// glScalef(1.0, 3.0, 1.0);

glBegin(GL_LINE_LOOP);
   glVertex3f(4.0, 0.0, 0.0);
   glVertex3f(0.0, 4.0, 0.0);
   glVertex3f(-4.0, 0.0, 0.0);
   glVertex3f(0.0, -4.0, 0.0);
glEnd();
```

See Figure 4.8(a). Verify by elementary geometry that the line loop forms a square with sides of length $4\sqrt{2}$ angled at 45° to the axes.

Uncomment the scaling. See Figure 4.8(b). The square now seems skewed to a non-rectangular parallelogram. Mathematically verify that the new vertices after the transformation $(x, y, z) \mapsto (x, 3y, z)$ is applied to the square's vertices are indeed those of a parallelogram. **End**

4.1.3 Rotation

Experiment 4.6. Add a rotation command by replacing the modeling transformation and object definition part – we prefer a teapot – of `box.cpp` with (Block 4):

```
// Modeling transformations.
glTranslatef(0.0, 0.0, -15.0);
glRotatef(60.0, 0.0, 0.0, 1.0);

glutWireTeapot(5.0);
```

Figure 4.9: Screenshot for Experiment 4.6.

Figure 4.9 is a screenshot.

The *rotation* command `glRotatef(`A`, `p`, `q`, `r`)` rotates each point of an object about an axis from the origin $O = (0, 0, 0)$ to the point (p, q, r). The amount of rotation is $A°$, measured counter-clockwise looking *from* (p, q, r) to the origin. In this experiment, then, the rotation is 60° CCW (counter-clockwise) looking down the z-axis. **End**

If the intuitive idea you have of rotating a point P about an axis is of the point turning along a cylinder on that axis as in Figure 4.10, then, well, you are perfectly correct. What we'll do next, though, is describe the rotation `glRotatef(`A`, `p`, `q`, `r`)` as a physical process for which we, hopefully, can find a formula.

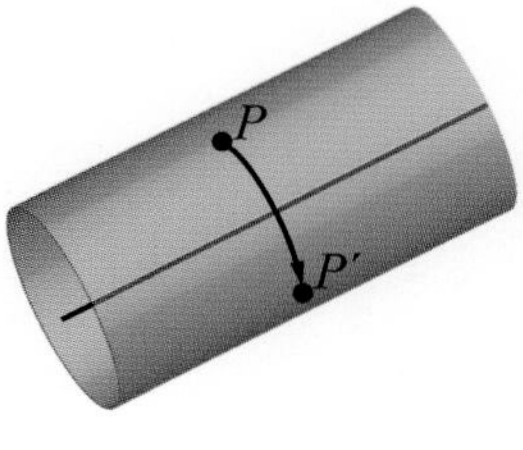

Figure 4.10: Turning along a cylinder.

Refer to Figure 4.11 as you read on. Assume $(p, q, r) \neq O$ so that the axis l through (p, q, r) and the origin O can indeed be drawn; in fact, if $(p, q, r) = O$ then `glRotatef(`A`, `p`, `q`, `r`)` is not a valid operation. Now, first, if a given point P lies on l itself then the situation is simple – the rotation does not move it. Suppose, then, that P does not lie on l. Here's how it's mapped by the rotation:

1. Drop the perpendicular from P to the point Q on l. Denote as L the segment PQ. L lies on the plane h perpendicular to l through Q.

2. Locate a viewer at V far enough along l, on the side of (p, q, r), as to be able to see h when looking toward the origin.

3. Rotate the segment L about Q (on the plane h) an angle $A°$ counter-clockwise, as measured by the viewer.

4. If L' is the new position of L after rotation, then P is mapped to the corresponding endpoint P' of L'.

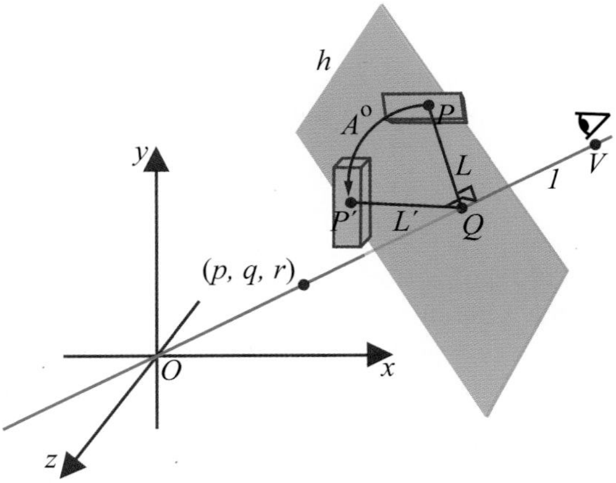

Figure 4.11: Rotation: `glRotatef(`A`, `p`, `q`, `r`)`. The point P is rotated according to the 4-step process in the text. The rotation of a box is also shown.

Note: In Experiment 4.6, the axis of rotation, the z-axis, happens to intersect the object rotated, which is the teapot.

Experiment 4.7. Continuing with Experiment 4.6, successively replace the rotation command with the ones below, in each case trying to match what you see with your understanding of how the command should turn the teapot. (It can occasionally be a bit confusing because of the perspective projection.)

1. `glRotatef(60.0, 0.0, 0.0, -1.0)`
2. `glRotatef(-60.0, 0.0, 0.0, 1.0)`
3. `glRotatef(60.0, 1.0, 0.0, 0.0)`
4. `glRotatef(60.0, 0.0, 1.0, 0.0)`
5. `glRotatef(60.0, 1.0, 0.0, 1.0)` **End**

The alert reader probably noticed in the 4-step definition of rotation earlier, that the purpose of the point (p, q, r), apart from specifying the axis l joining it to the origin, is to specify the *side* of the origin on l that the viewer is located. If (p, q, r) were replaced by another point (p', q', r') on l on the same side of O as (p, q, r), then the rotation would be *exactly same*. This is illustrated in the next experiment.

Experiment 4.8. Appropriately modify `box.cpp` to compare the effects of each of the following pairs of rotation commands:

1. `glRotatef(60.0, 0.0, 0.0, 1.0)` and `glRotatef(60.0, 0.0, 0.0, 5.0)`

2. `glRotatef(60.0, 0.0, 2.0, 2.0)` and `glRotatef(60.0, 0.0, 3.5, 3.5)`

3. `glRotatef(60.0, 0.0, 0.0, -1.0)` and `glRotatef(60.0, 0.0, 0.0, -7.5)`

There is no difference in each case. One concludes that the rotation command `glRotatef(`A, p, q, r`)` is equivalent to `glRotatef(`$A, \alpha p, \alpha q, \alpha r$`)`, where α is any *positive* scalar. **End**

Exercise 4.7. Relate the three commands `glRotatef(`A, p, q, r`)`, `glRotatef(`$-A, p, q, r$`)` and `glRotatef(`$A, \beta p, \beta q, \beta r$`)`, where β is a *negative* scalar.

Now, the general formula for how a point $P = (x, y, z)$ is mapped by the rotation `glRotatef(`A, p, q, r`)` is complicated – in fact, significantly more so than the corresponding formulae in case of translation and scaling. Nevertheless, we'll ask the reader to derive the formula in the three simple cases where the rotation is about a coordinate axis. We'll defer the general formula to the next chapter.

Exercise 4.8. Deduce the formula for how $P = (x, y, z)$ is mapped by each of the rotations:

(a) `glRotatef(`A`, 1.0, 0.0, 0.0)`

(b) `glRotatef(`A`, 0.0, 1.0, 0.0)`

(c) `glRotatef(`A`, 0.0, 0.0, 1.0)`

Part answer: See Figure 4.12 for (c). The axis of rotation is the z-axis. The point $P = (x, y, z)$ is mapped to $P' = (x', y', z')$. We'll find expression for x', y' and z' in terms x, y, z and the angle parameter A.

Draw $L = PQ$, the perpendicular from P to the z-axis. Further, draw the line k through Q parallel to the x-axis and the perpendicular PR from P to k. If $\angle PQR = \alpha$, then

$$\begin{aligned} x &= |RQ| = |L|\cos\alpha \\ y &= |RP| = |L|\sin\alpha \end{aligned}$$

Now, the rotated segment $L' = P'Q$ makes an angle of $\alpha + A$ with k, so that $\angle P'QR' = \alpha + A$, where R' is the foot of the perpendicular from P' to k. Therefore,

$$\begin{aligned} x' &= |L'|\cos(\alpha + A) = |L|\cos(\alpha + A) \\ y' &= |L'|\sin(\alpha + A) = |L|\sin(\alpha + A) \end{aligned}$$

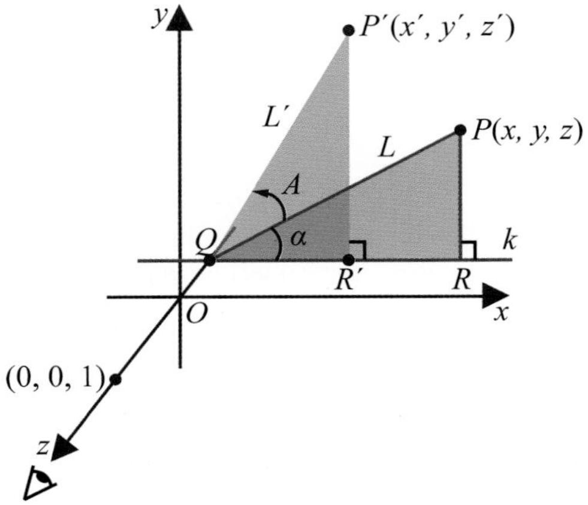

Figure 4.12: `glRotatef(A, 0.0, 0.0, 1.0)`.

as $|L'| = |L|$ because rotation does not change length. Apply trigonometric formulae to expand the rightmost sides of the two equations above:

$$\begin{aligned} x' &= |L|\cos\alpha\cos A - |L|\sin\alpha\sin A = x\cos A - y\sin A \\ y' &= |L|\cos\alpha\sin A + |L|\sin\alpha\cos A = x\sin A + y\cos A \end{aligned}$$

using the expressions for x and y derived earlier. And, of course,

$$z' = z$$

because rotation about the z-axis does not change the z-value.

Exercise 4.9. To what point is $(2.0, 3.0, 9.0)$ transformed by

(a) `glRotatef(90.0, 0.0, 0.0, 1.0)`?

(b) `glRotatef(90.0, 0.0, 0.0, 5.0)`?

(c) `glRotatef(90.0, 0.0, 0.0, -5.0)`?

(d) `glRotatef(60.0, 0.0, 0.0, 1.0)`?

(e) `glRotatef(180.0, 0.0, 1.0, 0.0)`?

(f) `glRotatef(45.0, 1.0, 0.0, 0.0)`?

4.2 Composing Modeling Transformations

In most of the previous experiments we successively applied more than one modeling transformation to an object – a translation plus one other – but never explained exactly how it is that OpenGL goes about *composing* multiple transformations. There is magic to this as we'll see, but first a couple of motivating experiments.

Experiment 4.9. Apply three modeling transformations by replacing the modeling transformations block of `box.cpp` with (Block 5):

```
// Modeling transformations.
glTranslatef(0.0, 0.0, -15.0);
glTranslatef(10.0, 0.0, 0.0);
glRotatef(45.0, 0.0, 0.0, 1.0);
```

It seems the box is *first* rotated 45° about the z-axis and *then* translated right 10 units. See Figure 4.13(a). The first translation `glTranslatef(0.0, 0.0, -15.0)`, of course, serves to "kick" the box down the z-axis into the viewing frustum.

Next, interchange the last two transformations, namely, the rightward translation and the rotation, by replacing the modeling transformations block with (Block 6):

```
// Modeling transformations.
glTranslatef(0.0, 0.0, -15.0);
glRotatef(45.0, 0.0, 0.0, 1.0);
glTranslatef(10.0, 0.0, 0.0);
```

It seems that the box is now *first* translated right and *then* rotated about the z-axis causing it to "rise". See Figure 4.13(b). **End**

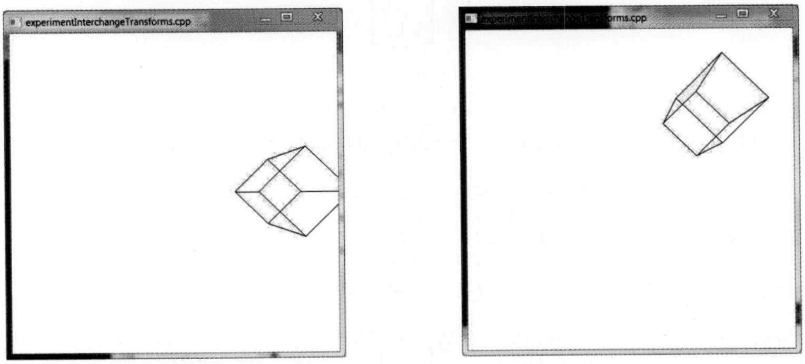

(a) (b)

Figure 4.13: Screenshots from Experiment 4.9.

Exercise 4.10. (Programming) Again, apply three modeling transformations, this time by replacing the modeling transformations block of `box.cpp` with (Block 7):

```
// Modeling transformations.
glTranslatef(0.0, 0.0, -15.0);
glRotatef(45.0, 0.0, 0.0, 1.0);
glScalef(1.0, 3.0, 1.0);
```

Interchange the rotation and scaling by replacing the modeling transformation block with (Block 8):

```
// Modeling transformations.
glTranslatef(0.0, 0.0, -15.0);
glScalef(1.0, 3.0, 1.0);
glRotatef(45.0, 0.0, 0.0, 1.0);
```

Keeping the conclusions of the preceding experiment in mind, can you explain what you see?

Apparently transformations are applied to an object in *backward* order through the code from where the object is created! This is correct and, hopefully, once it's explained how the OpenGL engine composes transformations, will not seem as idiosyncratic as it might at first.

We need, though, a quick acquaintance first with a concept which will be discussed in depth in the next chapter – that transformations correspond to matrices. We'll present here just enough that the reader can follow along. Our goal is a conceptual understanding of how transformations are composed.

A vertex $V = (x, y, z)$ is represented in OpenGL as a 4×1 column matrix

$$\begin{bmatrix} x \\ y \\ z \\ 1 \end{bmatrix}$$

Take that extra 1 in the 4th row for granted for now – it's to do with so-called homogeneous coordinates. We'll use V to denote this column matrix as well.

A modeling transformation t is represented by a 4×4 matrix of the form

$$M = \begin{bmatrix} a_{11} & a_{12} & a_{13} & a_{14} \\ a_{21} & a_{22} & a_{23} & a_{24} \\ a_{31} & a_{32} & a_{33} & a_{24} \\ a_{41} & a_{42} & a_{43} & a_{44} \end{bmatrix}$$

Applying this transformation to the vertex V consists of multiplying V from the left by the transformation matrix. In particular, V is transformed by t to the vertex $t(V)$ where

$$\begin{aligned} t(V) = MV &= \begin{bmatrix} a_{11} & a_{12} & a_{13} & a_{14} \\ a_{21} & a_{22} & a_{23} & a_{24} \\ a_{31} & a_{32} & a_{33} & a_{24} \\ a_{41} & a_{42} & a_{43} & a_{44} \end{bmatrix} \begin{bmatrix} x \\ y \\ z \\ 1 \end{bmatrix} \\ &= \begin{bmatrix} a_{11}x + a_{12}y + a_{13}z + a_{14} \\ a_{21}x + a_{22}y + a_{23}z + a_{24} \\ a_{31}x + a_{32}y + a_{33}z + a_{24} \\ a_{41}x + a_{42}y + a_{43}z + a_{44} \end{bmatrix} \end{aligned}$$

Here's an example.

Example 4.1. The transformation t_1 given by the translation command `glTranslatef(5.0, 0.0, 0.0)` corresponds to the matrix

$$M_1 = \begin{bmatrix} 1.0 & 0.0 & 0.0 & 5.0 \\ 0.0 & 1.0 & 0.0 & 0.0 \\ 0.0 & 0.0 & 1.0 & 0.0 \\ 0.0 & 0.0 & 0.0 & 1.0 \end{bmatrix}$$

This is verified by the multiplication

$$\begin{bmatrix} 1.0 & 0.0 & 0.0 & 5.0 \\ 0.0 & 1.0 & 0.0 & 0.0 \\ 0.0 & 0.0 & 1.0 & 0.0 \\ 0.0 & 0.0 & 0.0 & 1.0 \end{bmatrix} \begin{bmatrix} x \\ y \\ z \\ 1 \end{bmatrix} = \begin{bmatrix} x+5.0 \\ y \\ z \\ 1 \end{bmatrix}$$

Similarly, verify that the transformation t_2 given by the translation command `glTranslatef(0.0, 10.0, 0.0)` corresponds to the matrix

$$M_2 = \begin{bmatrix} 1.0 & 0.0 & 0.0 & 0.0 \\ 0.0 & 1.0 & 0.0 & 10.0 \\ 0.0 & 0.0 & 1.0 & 0.0 \\ 0.0 & 0.0 & 0.0 & 1.0 \end{bmatrix}$$

Now, if one applies t_2 followed by t_1 to a vertex V, then V is mapped as follows:

$$V \mapsto t_1(t_2(V)) = M_1(M_2V) = (M_1M_2)V$$

(the associativity of matrix multiplication was applied in the second equality). The skeptical reader may multiply matrices as below to verify that

$$\begin{aligned}(M_1M_2)V &= \left(\begin{bmatrix} 1.0 & 0.0 & 0.0 & 5.0 \\ 0.0 & 1.0 & 0.0 & 0.0 \\ 0.0 & 0.0 & 1.0 & 0.0 \\ 0.0 & 0.0 & 0.0 & 1.0 \end{bmatrix} \begin{bmatrix} 1.0 & 0.0 & 0.0 & 0.0 \\ 0.0 & 1.0 & 0.0 & 10.0 \\ 0.0 & 0.0 & 1.0 & 0.0 \\ 0.0 & 0.0 & 0.0 & 1.0 \end{bmatrix} \right) \begin{bmatrix} x \\ y \\ z \\ 1 \end{bmatrix} \\ &= \begin{bmatrix} 1.0 & 0.0 & 0.0 & 5.0 \\ 0.0 & 1.0 & 0.0 & 10.0 \\ 0.0 & 0.0 & 1.0 & 0.0 \\ 0.0 & 0.0 & 0.0 & 1.0 \end{bmatrix} \begin{bmatrix} x \\ y \\ z \\ 1 \end{bmatrix} \\ &= \begin{bmatrix} x+5.0 \\ y+10.0 \\ z \\ 1 \end{bmatrix}\end{aligned}$$

which indeed corresponds to how (x, y, z) is transformed by the code sequence

```
glTranslatef(5.0, 0.0, 0.0);
glTranslatef(0.0, 10.0, 0.0);
```

Put simply, the matrix of the composition of two transformations is the product of their matrices. This generalizes. If one applies successively the transformations $t_n, t_{n-1}, \ldots, t_1$ (in that order, t_n being first) to a vertex V, then it is mapped to

$$t_1(t_2(\ldots t_n(V)\ldots)) = M_1(M_2(\ldots(M_nV)\ldots)) = (M_1M_2\ldots M_n)V \quad (4.1)$$

again with the help of associativity of matrix multiplication, where matrix M_i corresponds to transformation t_i, $1 \leq i \leq n$. *One sees that the matrix of the composition of transformations is precisely the product of the matrices corresponding to the individual transformations.*

We now have enough to explain exactly how OpenGL itself goes about composing transformations. Consider the code sequence:

```
modelingTransformation 1;   // t1
modelingTransformation 2;   // t2
...
modelingTransformation n-1; // tn-1
modelingTransformation n;   // tn
object;
```

where the transformation t_i corresponds to the statement `modelingTransformation i`.

Now, OpenGL maintains a 4×4 *modelview matrix*, call it M, which is initialized to the identity

$$I = \begin{bmatrix} 1.0 & 0.0 & 0.0 & 0.0 \\ 0.0 & 1.0 & 0.0 & 0.0 \\ 0.0 & 0.0 & 1.0 & 0.0 \\ 0.0 & 0.0 & 0.0 & 1.0 \end{bmatrix}$$

As the drawing routine is processed during run-time, the matrix of each successive modeling transformation encountered is multiplied from the *left* by the current modelview matrix, the product becoming the new modelview matrix. For example, assuming the matrix of t_i is M_i and that there were no earlier transformations, the successive values of the modelview matrix M for the code sequence above are indicated in the comments below:

```
                            // M = I, initially
modelingTransformation 1;   // M = IM1 = M1
modelingTransformation 2;   // M = M1M2
...
modelingTransformation n-1; // M = M1M2... Mn-1
modelingTransformation n;   // M = M1M2... Mn-1Mn
object;
```

Moreover, an object drawing statement is processed by multiplying the object's vertices from the left by the current modelview matrix, e.g., for the code sequence above, each vertex V of `object` is transformed as follows:

$$V \mapsto MV = (M_1M_2\ldots M_{n-1}M_n)V$$

However, by associativity

$$\begin{aligned}(M_1 M_2 \ldots M_{n-1} M_n)V &= M_1(M_2(\ldots M_{n-1}(M_n V)\ldots)) \\ &= t_1(t_2(\ldots t_{n-1}(t_n(V)\ldots))\end{aligned}$$

We see, from the last line of the preceding equation, that transformation t_n is applied to V, then t_{n-1} and so on, until, finally, t_1, *indeed backward in code order*!

The conclusion, then, is that the backward order in which OpenGL applies transformations is simply a consequence of the particular way it processes their matrices. It does take a little getting used to, but (trust us) by the end of this chapter you will be quite comfortable applying multiple transformations.

Remark 4.2. That the matrices corresponding to successive modeling transformations are being multiplied into one matrix, the current modelview matrix, means that OpenGL is effectively *composing multiple modeling transformations into one transformation*, which is extremely important from the point of view of run-time efficiency.

Remark 4.3. Here's another more informal way to understand how multiple transformations are applied. Transformation t_n, given by the statement `modelingTransformation n` which is *closest* in the code to `object`, is applied first, then, t_{n-1}, given by the next closest statement `modelingTransformation n-1`, etc. Indeed, transformations are applied to the `object` as one works *away* from it, which is not unfamiliar if one recalls evaluating mathematical expressions such as $\cos(\exp(\sin x))$.

Remark 4.4. There is one other kind of transformation, in addition to the three modeling transformations, which can modify the modelview matrix – the *viewing transformation* `gluLookAt()`. We'll discuss viewing transformations in Section 4.6. Modelview matrices, in fact, get their name from these two kinds of transformations.

Exercise 4.11. For each of the following, give (x, y, z) coordinates of the point where the center of the sphere is transformed by the given piece of code in the display routine.

(a)
```
glTranslatef(0.0, 2.0, 2.0);
glTranslatef(4.0, 0.0, 2.0);
glutWireSphere(2.0, 10, 8);
```

(b)
```
glRotatef(90.0, 0.0, 0.0, 1.0);
glTranslatef(4.0, 0.0, 0.0);
glutWireSphere(2.0, 10, 8);
```

(c)
```
glRotatef(90.0, 1.0, 0.0, 0.0);
glTranslatef(4.0, 0.0, 0.0);
glRotatef(90.0, 0.0, 0.0, 1.0);
glutWireSphere(2.0, 10, 8);
```

(d)
```
glRotatef(90.0, 1.0, 0.0, 0.0);
glRotatef(90.0, 0.0, 1.0, 0.0);
glTranslatef(4.0, 0.0, 0.0);
glutWireSphere(2.0, 10, 8);
```

(e)
```
glScalef(1.0, 2.0, 3.0);
glRotatef(45.0, 1.0, 0.0, 0.0);
glRotatef(90.0, 0.0, 0.0, 1.0);
glTranslatef(4.0, 0.0, 0.0);
glutWireSphere(2.0, 10, 8);
```

(f)
```
glRotatef(90.0, 0.0, 0.0, 1.0);
glRotatef(45.0, 1.0, 0.0, 0.0);
glRotatef(90.0, 0.0, 1.0, 0.0);
glTranslatef(4.0, 0.0, 0.0);
glutWireSphere(2.0, 10, 8);
```

Example 4.2. Replace the object definition statement

```
glutWireCube(5.0); // Box.
```

of `box.cpp` with

```
glRectf(5.0, 5.0, 10.0, 10.0); // Square
```

to draw, instead of a box centered at the origin, an axis-aligned square some ways north-east of it, centered at $(7.5, 7.5, 0.0)$.

Now, add transformation(s) to rotate the square 45° counter-clockwise about its *own center*, as indicated in Figure 4.14(a).

Answer: Inserting the command `glRotatef(45.0, 0.0, 0.0, 1.0)` just before `glRectf()` will not do as it rotates the square about the origin, and not its own center, as shown in Figure 4.14(b). What one must do instead (see Figure 4.14(c)) is first (i) translate the square so that its center is at the origin, then (ii) rotate it about the origin and, finally, (iii) translate it back. This is equivalent to the following modeling transformation block (Block 9):

```
// Modeling transformations.
glTranslatef(0.0, 0.0, -15.0);

glTranslatef(7.5, 7.5, 0.0); // Translate back.
glRotatef(45.0, 0.0, 0.0, 1.0); // Rotate about origin.
glTranslatef(-7.5, -7.5, 0.0); // Translate to origin.
```

Such a maneuver is unavoidable as OpenGL's own rotations, as we know from Section 4.1.3, are each about an axis through the origin, such being called a *radial* axis. Therefore, a non-radial axis needs to be translated to the origin and back again in order to be rotated about. This "tricky" maneuver and its variants come up so often that we'll give them a collective name: the Trick.

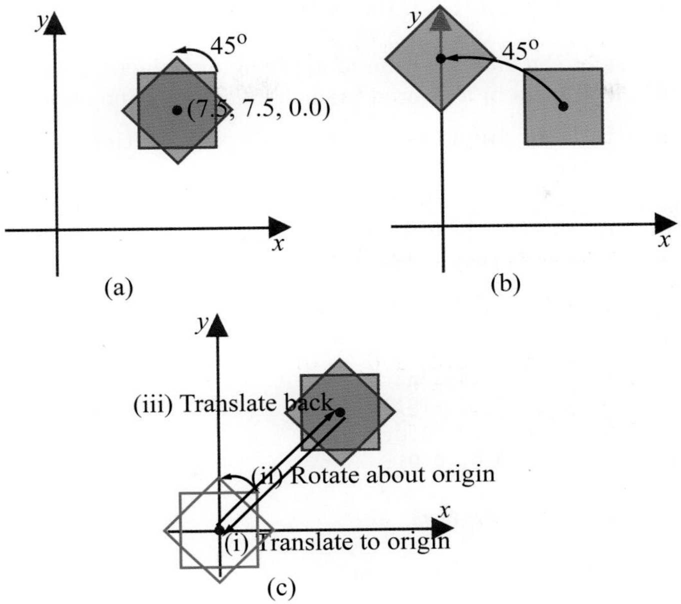

Figure 4.14: Rotating a square about its own center.

Exercise 4.12. (Programming) As in the preceding example, replace the object definition statement

```
glutWireCube(5.0); // Box.
```

of `box.cpp` with

```
glRectf(5.0, 5.0, 10.0, 10.0); // Square
```

Now, scale the square so that its center is unchanged, but its shape changes to a rectangle of aspect ratio 2. Use the Trick.

Exercise 4.13. Prove that a composition of multiple translations is a single translation and that a composition of multiple scalings is a single scaling.

Remark 4.5. A composition of multiple rotations is a single rotation as well, but this is much harder to prove generally and we'll leave it to Chapter 5.

Exercise 4.14. What is the *inverse* of a translation? Specifically, what modeling transformation composed with a translation `glTranslatef(`p`, `q`, `r`)` "undoes" its effect, so that all points remain stationary?

How about scalings and rotations? What are their inverses?

4.3 Placing Multiple Objects

We next consider the vital problem of applying modeling transformations to place multiple objects in a desired manner *relative* to one another.

Experiment 4.10. Replace the entire display routine of the original `box.cpp` with (Block 10):

```
void drawScene(void)
{
   glClear(GL_COLOR_BUFFER_BIT);
   glColor3f(0.0, 0.0, 0.0);
   glLoadIdentity();

   // Modeling transformations.
   glTranslatef(0.0, 0.0, -15.0);
   // glRotatef(45.0, 0.0, 0.0, 1.0);
   glTranslatef(5.0, 0.0, 0.0);

   glutWireCube(5.0); // Box.

   //More modeling transformations.
   glTranslatef(0.0, 10.0, 0.0);

   glutWireSphere(2.0, 10, 8); // Sphere.

   glFlush();
}
```

See Figure 4.15(a) for a screenshot. The objects are a box and a sphere.

End

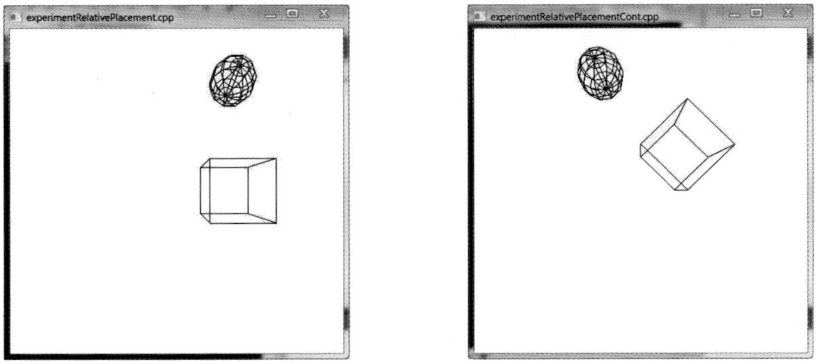

(a) (b)

Figure 4.15: Screenshots: (a) Experiments 4.10 and (b) 4.11.

Let's understand the placement of the box and sphere in the preceding experiment individually first and then with respect to each other.

It's actually fairly straightforward to understand the placements individually. For example, to place the sphere, work backwards from where it's created in the code, applying to it the successive modeling transformations (all translations in this case) encountered, and ignoring the one non-transformation statement `glutWireCube()` on the way. The result is that the sphere is centered at $(5.0, 10.0, -15.0)$. Likewise, the box is seen to be centered at $(5.0, 0.0, -15.0)$.

The relative placement in this case is not difficult either. Clearly, the sphere is transformed by `glTranslatef(0.0, 10.0, 0.0)` – which is the transformation "between them" – with respect to the box. The result is that the sphere's center is 10 units vertically above the box's.

Experiment 4.11. Continuing with the previous experiment, uncomment the `glRotatef()` statement. Figure 4.15(b) is a screenshot.

Again, the individual placements are fairly straightforward. Working backwards from where it is created we see that, after being translated to (5.0, 10.0, 0.0), the sphere is rotated 45° counter-clockwise about the z-axis and, of course, finally pushed 15 units in the $-z$ direction. We'll not compute the exact final coordinates of its center. The individual placement of the box is simple to parse as well and left to the reader.

It's the relative placement which is particularly interesting in this case. The sphere is no longer vertically above the box, though the transformation between them is still `glTranslatef(0.0, 10.0, 0.0)`! Before trying to explain what's going on, let's return to the basics for a moment. **End**

Consider the code sequence below which draws two objects:

```
modelingTransformation 1;   // t1
modelingTransformation 2;   // t2
...
modelingTransformation n-1; // tn-1
modelingTransformation n;   // tn
object 1;
modelingTransformation n+1; // tn+1
...
modelingTransformation m;   // tm
object 2;
```

Assuming that the transformation t_i specified by `modelingTransformation i` corresponds to the matrix M_i, for $1 \leq i \leq m$, the successive values of the modelview matrix M are indicated below:

```
                            // M = I, initially
modelingTransformation 1;   // M = IM1 = M1
modelingTransformation 2;   // M = M1M2
...
modelingTransformation n-1; // M = M1M2... Mn-1
modelingTransformation n;   // M = M1M2... Mn-1Mn
```

```
object 1;                          // M does not change
modelingTransformation n+1; // M  =  M1M2 ...  Mn-1MnMn+1
...
modelingTransformation m;     // M  =  M1M2 ...  Mn-1MnMn+1 ...  Mm
object 2;
```

Accordingly, each vertex V of the final `object 2` call is transformed according to:

$$V \mapsto (M_1 \dots M_{m-1} M_m)V = t_1(\dots t_{m-1}(t_m(V)) \dots) \tag{4.2}$$

exactly as we would expect by working backwards in the code from `object 2`. Now, how about the placement of `object 2` *with respect to* `object 1`?

Let's repeat the transformation for a vertex V of `object 2` by stepping backward through the right side of Equation (4.2): first transform V by t_m to $t_m(V)$, then by t_{m-1} to $t_{m-1}(t_m(V))$, ..., then by t_{n+1} to $t_{n+1}(\dots t_{m-1}(t_m(V)) \dots)$. *Stop*!

At this time `object 1` is drawn. Imagine that a part of `object 1` is a set of three directed line segments (drawn, say, using `GL_LINES`), aligned with the three world coordinate axes and calibrated identically. These lines are said to represent the *local coordinate system* of `object 1`. See Figure 4.16. So, at the time of its creation the local coordinate system of `object 1` coincides with the world coordinate system.

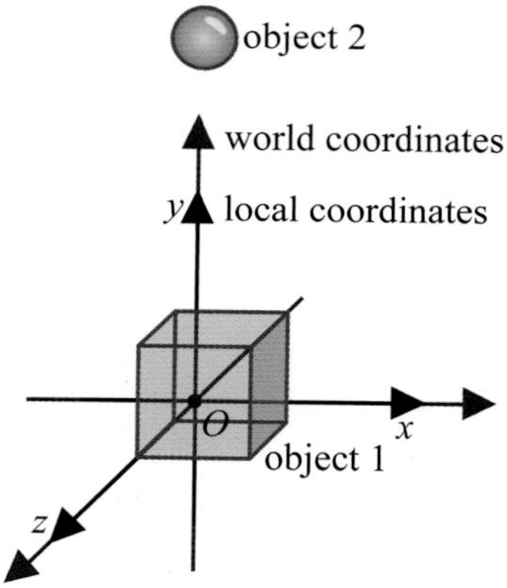

Figure 4.16: Local system (bold) coincides with the global initially. The global system is fixed.

Further, suppose at the time of `object 1`'s creation that the so-far transformed V, i.e., $t_{n+1}(\dots t_{m-1}(t_m(V)) \dots)$, is located at (a, b, c) in the local coordinates of `object 1` (same as world coordinates, of course, at that moment).

Let's get back now to applying transformations backwards from where we had stopped. Next was t_n. Now, t_n applies to *both* V and `object 1`. Three cases arise according to the type of t_n.

1. t_n is a translation specified by `glTranslatef(`p`, `q`, `r`)`:

   This translation applies to V and `object 1` and, so, to the local coordinate system of the latter as well. That is, they all "move together". Therefore, the location (a, b, c) of V with respect to the local coordinate system of `object 1` does not change. The location of V in world coordinates, of course, changes to $(a + p, b + q, c + r)$.

2. t_n is a rotation specified by `glRotatef(`A`, `p`, `q`, `r`)`:

   Same argument as for translation. Again, the location (a, b, c) of V with respect to the local coordinate system of `object 1` does not change.

3. t_n is a scaling specified by `glScalef(`u`, `v`, `w`)`:

   The location of V in world coordinates is changed by the scaling to (ua, vb, wc). However, as the same scaling applies to the axes of the local coordinate system of `object 1` – particularly the units calibrating them – the location (a, b, c) of V with respect to this system again does not change.

   Sci-fi analogy: Prior to an experiment in a lab you measure yourself with a tape to be 6 feet tall. The experiment goes horribly wrong and radiation causes you to shrink by a factor of 12, leaving you at a Lilliputian 6 inches. However, if everything around you including the tape shrank by exactly the same factor, you would still believe yourself to be 6 feet.

Continue, applying transformations $t_{n-1}, t_{n-2}, \ldots t_1$, successively, and reason as above for each. The conclusion is that the location of V at the point (a, b, c) of the local coordinate system of `object 1` at the time of the latter's creation is not altered by any subsequent transformation, i.e., those in the code prior to `object 1`. Neither, obviously, is it changed by transformations in the code after `object 2`, because their corresponding matrices multiply into the modelview matrix only after both `object 1` and `object 2` have already been drawn. We have, therefore, the following:

Proposition 4.1. *If* `object 1` *precedes* `object 2` *in code, then the location of* `object 2` *in the local coordinate system of* `object 1` *is determined by the transformation statements between the two and nothing else.* □

What the proposition says is that, if `object 1` precedes `object 2` in code, then the latter is *frozen* in the former's coordinate system at a position determined solely by the transformation statements between the two. Accordingly, moving `object 2` *with respect to* `object 1` requires changing transformations between them. The practical importance of this, as we'll see, cannot be over-emphasized.

Let's try and understand now the relative position of the sphere with respect to the box in Experiment 4.11 in light of the preceding proposition. We'll do this by the oft-useful technique of deconstructing code by incrementally adding back transformations after stripping them all off.

Experiment 4.12. Repeat Experiment 4.11. The modeling transformation and object definition part are as below (Block 11):

```
// Modeling transformations.
glTranslatef(0.0, 0.0, -15.0);
glRotatef(45.0, 0.0, 0.0, 1.0);
glTranslatef(5.0, 0.0, 0.0);

glutWireCube(5.0); // Box.

//More modeling transformations.
glTranslatef (0.0, 10.0, 0.0);

glutWireSphere (2.0, 10, 8); // Sphere.
```

First, comment out the last two statements of the first modeling transformations block as below (the first translation is always needed to place the entire scene in the viewing frustum):

```
// Modeling transformations.
glTranslatef(0.0, 0.0, -15.0);
// glRotatef(45.0, 0.0, 0.0, 1.0);
// glTranslatef(5.0, 0.0, 0.0);

glutWireCube(5.0); // Box.

//More modeling transformations.
glTranslatef (0.0, 10.0, 0.0);

glutWireSphere (2.0, 10, 8); // Sphere.
```

The output is as depicted in Figure 4.17(a).

Next, uncomment `glTranslatef(5.0, 0.0, 0.0)` as below:

```
// Modeling transformations.
glTranslatef(0.0, 0.0, -15.0);
// glRotatef(45.0, 0.0, 0.0, 1.0);
glTranslatef(5.0, 0.0, 0.0);

glutWireCube(5.0); // Box.

//More modeling transformations.
glTranslatef (0.0, 10.0, 0.0);

glutWireSphere (2.0, 10, 8); // Sphere.
```

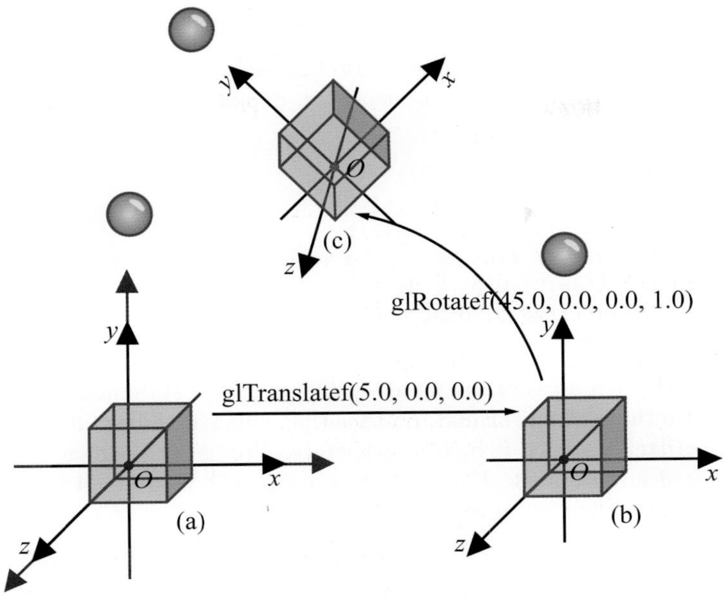

Figure 4.17: Transitions of the box, the box's local coordinates system (bold) and the sphere. The world coordinate system, which never changes, coincides with the box's initial local.

The output is as in Figure 4.17(b). Finally, uncomment `glRotatef(45.0, 0.0, 0.0, 1.0)` as follows:

```
// Modeling transformations.
glTranslatef(0.0, 0.0, -15.0);
glRotatef(45.0, 0.0, 0.0, 1.0);
glTranslatef(5.0, 0.0, 0.0);

glutWireCube(5.0); // Box.

//More modeling transformations.
glTranslatef (0.0, 10.0, 0.0);

glutWireSphere (2.0, 10, 8); // Sphere.

glFlush();
```

The result is seen in Figure 4.17(c). Figure 4.17 shows the box's local coordinate system as well after each transition. Observe that the sphere is *always* 10 units vertically above the box in the latter's coordinate system, as one would expect from the `glTranslatef (0.0, 10.0, 0.0)` call between the two.
End

The following program should solidify the reader's understanding of how

Proposition 4.1 governs relative placement.

Experiment 4.13. Run `relativePlacement.cpp`. Pressing the up arrow key once causes the last statement, viz. `drawBlueMan`, of the following piece of code to be executed:

```
glScalef(1.5, 0.75, 1.0);
glRotatef(30.0, 0.0, 0.0, 1.0);
glTranslatef(10.0, 0.0, 0.0);
drawRedMan; // Also draw grid in his local coordinate system.
glRotatef(45.0, 0.0, 0.0, 1.0);
glTranslatef(20.0, 0.0, 0.0);
drawBlueMan;
```

With each press of the up arrow we go back a statement and successively execute that statement *and* the ones that follow it. The statements executed are written in black text, the rest white. Pressing the down arrow key goes forward a statement. Figure 4.18 is a screenshot after all transformations from the scaling on have been executed. **End**

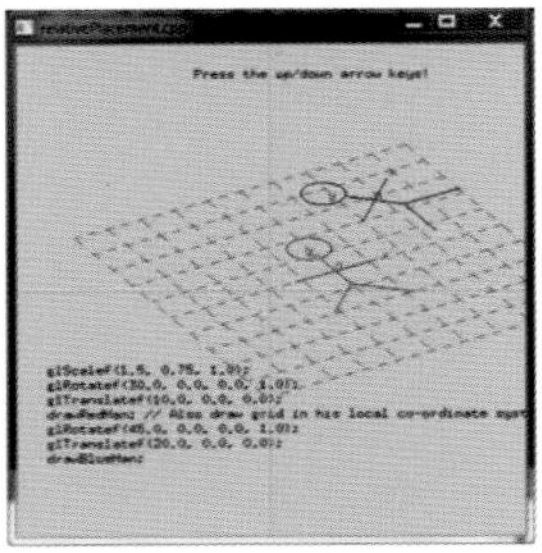

Figure 4.18: Screenshot of `relativePlacement.cpp` after all transformations from the scaling down have been executed.

The torso and arms of both men are aligned along their respective local coordinate axes (only the x and y we care about, z being 0 always). The world coordinate axes which never change are drawn in cyan. At the time of the red man's creation also drawn is a 10×10 grid of boxes in his local coordinate system, the sides of each box being 5 units long.

With each transformation going back from the red man's creation, observe – focus on a point like the blue man's origin and trust your eyes – how the blue man stays static in the red man's local coordinate system. A simple calculation shows that the blue man's origin is actually at $(20/\sqrt{2},\ 20/\sqrt{2}) \simeq (14.14, 14.14)$ in the red man's system. Even when scaling skews the red man's system so that it's not rectangular any more, the blue man skews the same way as well, staying put in the red system. Proposition 4.1 is not to be denied!

Exercise 4.15. For the following two pieces of code in the drawing routine give (x, y, z) coordinates of the point to which the center of the sphere is transformed. Explain as well the relative positions of the sphere and box.

(a)
```
glRotatef(90.0, 1.0, 0.0, 0.0);
glutWireCube(1.0);
glRotatef(90.0, 0.0, 0.0, 1.0);
glTranslatef(4.0, 0.0, 0.0);
glutWireSphere(2.0, 10, 8);
```

(b)
```
glTranslatef(2.0, 0.0, 0.0);
glScalef(2.0, 2.0, 2.0);
glutWireCube(1.0);
glRotatef(90.0, 0.0, 1.0, 0.0);
glTranslatef(0.0, 0.0, 4.0);
glutWireSphere(2.0, 10, 8);
```

If you're impatient to get to animation hang on – there's one final piece to get in place!

4.4 Modelview Matrix Stack and Isolating Transformations

The modelview matrix, which we have described as being modified by modeling transformations by multiplication on the right, is actually the topmost one of a *modelview matrix stack*. This particular matrix is called the *current modelview matrix*. In fact, OpenGL maintains three different matrix stacks: modelview, projection and texture. A `glMatrixMode(`*mode*`)` command, where *mode* is `GL_MODELVIEW`, `GL_PROJECTION` or `GL_TEXTURE`, determines which stack is currently active.

Here's an experiment to motivate use of the modelview matrix stack:

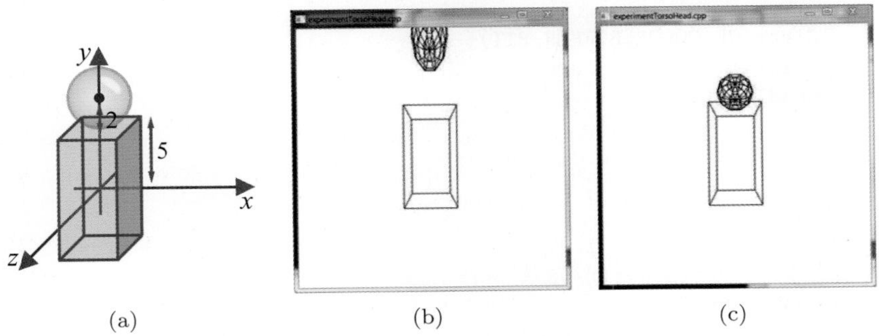

Figure 4.19: Planning a head on a torso: (a) The plan (b) Drawn without isolating the scaling (c) After isolating the scaling.

Experiment 4.14. We want to create a human-like character. Our plan is to start by drawing the torso as an elongated cube and placing a round sphere as its head directly on top of the cube (no neck for now). To this end replace the drawing routine of `box.cpp` with (Block 12):

```
void drawScene(void)
{
   glClear(GL_COLOR_BUFFER_BIT);
   glColor3f(0.0, 0.0, 0.0);
   glLoadIdentity();

   glTranslatef(0.0, 0.0, -15.0);

   glScalef(1.0, 2.0, 1.0);
   glutWireCube(5.0); // Box torso.
```

```
    glTranslatef(0.0, 7.0, 0.0);
    glutWireSphere(2.0, 10, 8); // Spherical head.

    glFlush();
}
```

Our calculations are as follows: (a) the scaled box is $5 \times 10 \times 5$ and, being centered at the origin, is 5 units long in the $+y$ direction; (b) the sphere is of radius 2; (c) therefore, if the sphere is translated $5 + 2 = 7$ in the $+y$ direction, then it should sit exactly on top of the box (see Figure 4.19(a)).

It doesn't work: the sphere is no longer round and is, moreover, some ways above the box (Figure 4.19(b)). Of course, because the sphere is transformed by `glScalef(1.0, 2.0, 1.0)` as well! So, what to do? A solution is to *isolate* the scaling by placing it within a *push-pop pair* as below (Block 13):

```
void drawScene(void)
{
    glClear(GL_COLOR_BUFFER_BIT);
    glColor3f(0.0, 0.0, 0.0);
    glLoadIdentity();

    glTranslatef(0.0, 0.0, -15.0);

    glPushMatrix();
    glScalef(1.0, 2.0, 1.0);
    glutWireCube(5.0); // Box torso.
    glPopMatrix();

    glTranslatef(0.0, 7.0, 0.0);
    glutWireSphere(2.0, 10, 8); // Spherical head.

    glFlush();
}
```

The resulting screenshot is Figure 4.19(c), which shows a round head on a neckless torso as desired. **End**

What the `glPushMatrix()` command does is make a *copy* of the current (i.e., current topmost) matrix in the modelview matrix stack and place it on top of the stack; consequently, upon execution of a `glPushMatrix()`, the two top matrices of the stack are identical. The `glPopMatrix()` statement, on the other hand, deletes the topmost matrix of the modelview matrix stack so the one underneath becomes the current one.

Let's follow the modelview matrix stack through the code above. Assume that the matrix corresponding to the translation `glTranslatef(0.0, 0.0, -15.0)` is M_1, to `glScalef(1.0, 2.0, 1.0)` is M_2, and to `glTranslatef(0.0, 7.0, 0.0)` is M_3. The transitions of the stack are shown in Figure 4.20, starting from the top.

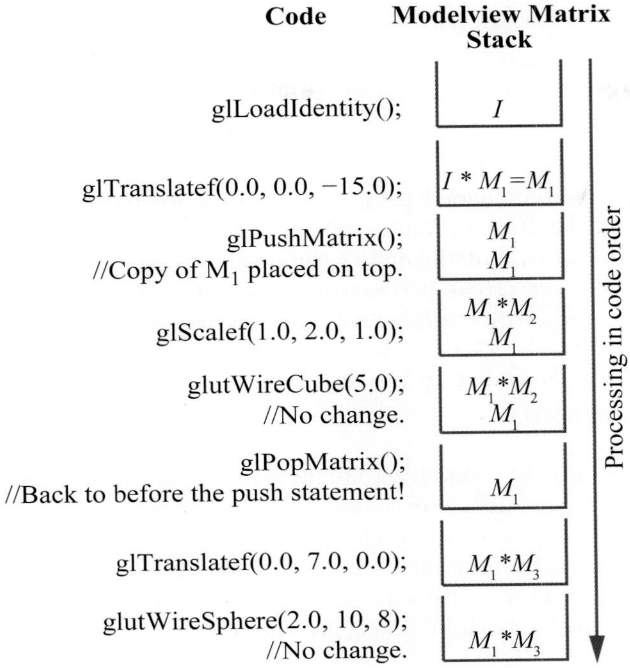

Figure 4.20: Transitions of the modelview matrix stack.

As you see, the push-pop pair *stores* the current modelview matrix *prior* to the scaling transformation and then *restores* it once the cube has been drawn, effectively localizing the effect of the scaling to only the cube.

Exercise 4.16. Give (x, y, z) coordinates of the points where the centers of the four spheres in it are located by the drawing routine below, assuming no prior transformations.

```
glPushMatrix();

glTranslatef(2.0, 0.0, 0.0);
glutWireSphere(2.0, 10, 8); // Sphere A

glPushMatrix();
glScalef(2.0, 2.0, 2.0);
glutWireSphere(2.0, 10, 8); // Sphere B
glPopMatrix();

glPushMatrix();
glRotatef(90.0, 1.0, 0.0, 0.0);
glTranslatef(0.0, 0.0, 4.0);
glutWireSphere(2.0, 10, 8); // Sphere C
```

```
glPopMatrix();

glTranslatef(0.0, 4.0, 0.0);
glutWireSphere(2.0, 10, 8); // Sphere D

glPopMatrix();
```

Remark 4.6. It's recommended programming practice to enclose all the transformations in the drawing routine in one giant push-pop pair, as in the preceding exercise, so that at the end of the routine the modelview matrix stack is guaranteed to revert to its initial state of containing a single identity matrix (though, admittedly, we don't follow this ourselves).

4.5 Animation

We're there! Animation in computer graphics is really just a sequence of still frames, just like those in a movie reel, smoothness being achieved by drawing frames rapidly one after another, each a little different from the previous. Successive frames in an animation are created by a "transform-draw" loop: the scene is redrawn after transformations in the drawing routine change the location or shape, or both, of objects in the scene.

4.5.1 Animation Technicals

Before analyzing animated programs we need first to explain a couple of animation-related technicalities.

Controlling Animation

OpenGL provides essentially three different methods to control animation:

1. Interactively, via keyboard or mouse input, with the help of their callback routines to invoke transformations.

   Experiment 4.15. Run `rotatingHelix1.cpp` where each press of space calls the `increaseAngle()` routine to turn the helix. Note the `glutPostRedisplay()` command in `increaseAngle()` which asks the screen to be redrawn. Keeping the space bar pressed turns the helix continuously. Figure 4.21 is a screenshot. **End**

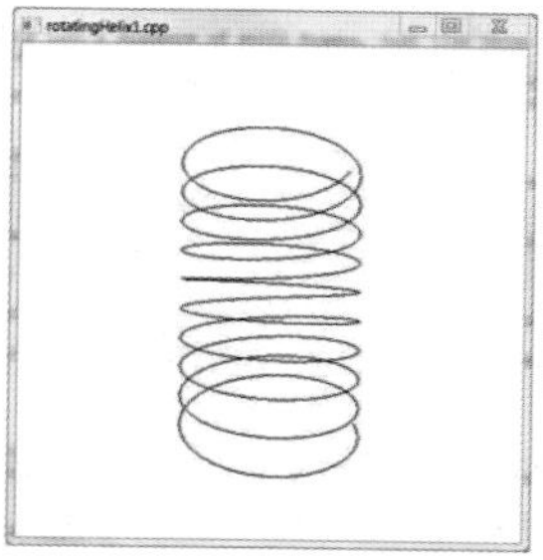

Figure 4.21: Screenshot of `rotatingHelix1.cpp`.

2. Automatically, by specifying a function `idle_function`, called the *idle function*, with the statement `glutIdleFunc(idle_function)`. The idle function is called whenever no OpenGL event is otherwise pending.

   Experiment 4.16. Run `rotatingHelix2.cpp`, a slight modification of `rotatingHelix1.cpp`, where pressing space causes the routines `increaseAngle()` and `NULL` (do nothing) to be alternately specified as idle functions.

The speed of animation is determined by the processor speed – in particular, the speed at which frames can be redrawn – and the user cannot influence it. **End**

3. Automatically, by specifying a routine *timer_function*, called the *timer function*, with a call to `glutTimerFunc(`*period*, *timer_function*, *value*`)`. The timer function is called *period* milliseconds after the `glutTimerFunc()` statement is executed and with the parameter *value* being passed to it.

Experiment 4.17. Run `rotatingHelix3.cpp`, another modification of `rotatingHelix1.cpp`, where the particular timer function `animate()` calls itself recursively after `animationPeriod` number of msecs., by means of its own `glutTimerFunc(animationPeriod, animate, 1)` statement. The parameter value 1 passed to `animate()` is not used in this program. The routine `increaseAngle()` called by `animate()` turns the helix as before. Figure 4.22 shows the animation scheme.

The user can vary the speed of animation by changing the value of `animationPeriod` by pressing the up and down arrow keys. **End**

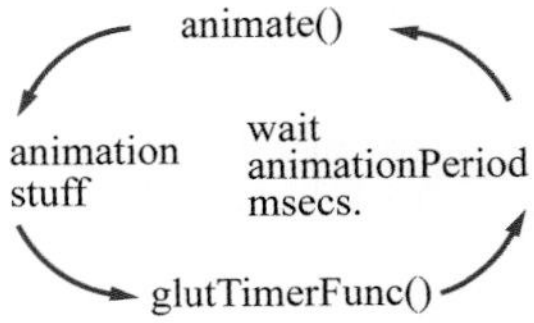

Figure 4.22: Animation control in `rotatingHelix3.cpp`.

The speed of animation or, equivalently, *frame rate* – the rate at which the screen is redrawn – cannot be increased arbitrarily by lowering the value of `animationPeriod` because redrawing the scene takes some minimum amount of time, depending on its complexity and the speed of the processor (or graphics card).

Moreover, the frame rate can never exceed the monitor's installed refresh rate, particularly, if the latter is n Hz then the maximum achievable fps (frames per second) is n. The next program shows how to count the fps with help of OpenGL.

Experiment 4.18. Run `rotatingHelixFPS.cpp`, which enhances `rotatingHelix2.cpp` adding the routine `frameCounter()` to count the number of times the `drawScene()` routine is called, equivalently, the number of frames drawn, per second. The fps is output every second to the debug window.

The way this works is by `drawScene()` incrementing the global `frameCount` every time it is called, and `frameCounter()` outputting the value of `frameCount` each second – because of its final `glutTimerFunc(1000, frameCounter, 1)` statement, `frameCounter()` calls itself after a second – as well as resetting the value of `frameCount` to 0. The `if` conditional in `frameCounter()` is so that no fps is output when it is first called from `main()` with the value passed being 0. **End**

Double Buffering

The second technicality critical to smooth animation is *double buffering*.

Space for two color buffers is provided in a double-buffered system in such a manner that one buffer, the *viewable buffer*, displays the current frame while the next frame is being drawn in the second buffer, the *drawable buffer*. When the drawing of the frame in the drawable buffer is complete, the buffers are swapped, so that the next frame now becomes viewable and, at the same time, the one following it begins to be drawn. This *draw-and-swap* loop repeats through the animation. Figure 4.23 illustrates the process.

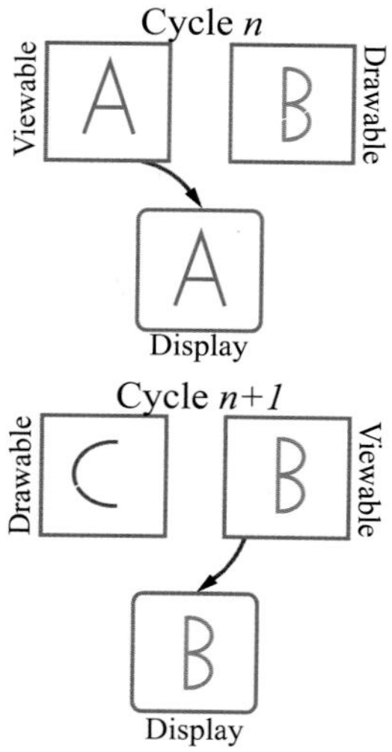

Figure 4.23: Successive cycles in double buffering.

Terminology: The viewable buffer is often called the *front buffer* or *main buffer*, while the drawable buffer is called the *back buffer* or *swap buffer*. Either buffer is also called a *refresh buffer*.

Remark 4.7. There is a subtle difference between the "draw" in the transform-draw animation loop described earlier as how animation is implemented and the "draw" in the draw-and-swap loop just described as how double buffering operates. The first is a programmer-instigated operation – typically, with a `glutPostRedisplay()` call – in which the world space is projected and scaled (recall shoot-and-print from Chapter 2) *and* rasterized into the color buffer. The second actually draws the screen, in particular, the OpenGL window, with the contents of the color buffer.

Remark 4.8. There are two ways double buffering can be implemented. In a software implementation, the back buffer is in non-video RAM, while the front buffer is in video RAM (VRAM), and the swap is done by copying the contents of the back buffer into the front one. In a hardware implementation, often called *ping-pong buffering*, both buffers are in VRAM, the swap switching the buffer being displayed by simply modifying a pointer. Not surprisingly, ping-pong buffering runs much faster.

Double buffering greatly improves the quality of animation by hiding transition between successive frames from the viewer. With single buffering, on the other hand, the viewer "sees" the next frame being drawn in the same buffer that contains the current one. The result can be unpleasant *ghosting*, so called because a prior image persists while the next is being created.

The double buffering display mode is enabled by calling `glutInitDisplayMode()` in `main` with `GLUT_DOUBLE` as one of the arguments (instead of `GLUT_SINGLE` *and* inserting a call to `glutSwapBuffers()` at the end of the drawing routine (*instead of* `glFlush()`). The `rotatingHelix*.cpp` programs are all double buffered.

Experiment 4.19. Disable double buffering in `rotatingHelix2.cpp` by replacing `GLUT_DOUBLE` with `GLUT_SINGLE` in the `glutInitDisplayMode()` call in `main`, and replacing `glutSwapBuffers()` in the drawing routine with `glFlush()`. Ghostly is it not?! **End**

4.5.2 Animation Code

Ball Flying About a Torus

Experiment 4.20. Run `ballAndTorus.cpp`. Press space to start the ball both flying around (longitudinal rotation) and in and out (latitudinal rotation) of the torus. Press the up and down arrow keys to change the speed of the animation. Press 'x/X', 'y/Y' and 'z/Z' to change the viewpoint. Figure 4.24 is a screenshot.

The animation of the ball is interesting and we'll deconstruct it. Comment out all the modeling transformations in the ball's block, except the last translation, as follows:

Figure 4.24: Screenshot of `ballAndTorus.cpp`.

```
// Begin revolving ball.
// glRotatef(longAngle, 0.0, 0.0, 1.0);

// glTranslatef(12.0, 0.0, 0.0);
// glRotatef(latAngle, 0.0, 1.0, 0.0);
// glTranslatef(-12.0, 0.0, 0.0);

glTranslatef(20.0, 0.0, 0.0);

glColor3f(0.0, 0.0, 1.0);
glutWireSphere(2.0, 10, 10);
// End revolving ball.
```

The ball is centered at $(20, 0, 0)$, its start position, by `glTranslatef(20.0, 0.0, 0.0)`. See Figure 4.25. There is no animation.

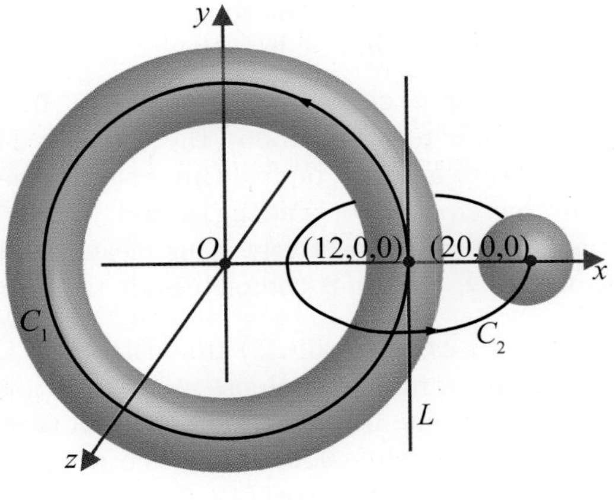

Figure 4.25: The ball's axis of latitudinal rotation from its start position is L.

The ball's intended latitudinal rotation is in and out of the circle C_1 through the middle of the torus. C_1's radius, called the *outer radius* of the torus, is 12.0, as specified by the second parameter of `glutWireTorus(2.0,`

`12.0, 20, 20)`. Moreover, C_1 is centered at the origin and lies on the xy-plane. Therefore, ignoring longitudinal motion for now, the latitudinal rotation of the ball *from its start position* is about the line L through $(12, 0, 0)$ parallel to the y-axis (L is tangent to C_1). This rotation will cause the ball's center to travel along the circle C_2 centered at $(12, 0, 0)$, lying on the xz-plane, of radius 8.

As `glRotatef()` always rotates about a radial axis, how does one obtain the desired rotation about L, a non-radial line? Employ the Trick (see Example 4.2 if you don't remember). First, translate left so that L is aligned along the y-axis, then rotate about the y-axis and, finally, reverse the first translation to bring L back to where it was. This means uncommenting the corresponding three modeling transformations as below:

```
// Begin revolving ball.
// glRotatef(longAngle, 0.0, 0.0, 1.0);

glTranslatef(12.0, 0.0, 0.0);
glRotatef(latAngle, 0.0, 1.0, 0.0);
glTranslatef(-12.0, 0.0, 0.0);

glTranslatef(20.0, 0.0, 0.0);

glColor3f(0.0, 0.0, 1.0);
glutWireSphere(2.0, 10, 10);
// End revolving ball.
```

Press space to view only latitudinal rotation.

Note: The two consecutive translation statements could be combined into one, but then the code would be less easy to parse.

Finally, uncomment `glRotatef(longAngle, 0.0, 0.0, 1.0)` to implement longitudinal rotation about the z-axis. The angular speed of longitudinal rotation is set to be five times slower than that of latitudinal rotation – the increments to `longAngle` and `latAngle` in the `animate()` routine being 1° and 5°, respectively. This means the ball winds in and out of the torus five times before it completes one trip around it. **End**

Exercise 4.17. (Programming) It's instructive as well to uncomment the three modeling transformations used to apply the Trick in the preceding experiment one by one, rather than all together. So uncomment `glTranslatef(-12.0, 0.0, 0.0)` first, then `glRotatef(latAngle, 0.0, 1.0, 0.0)` and, last, `glTranslatef(12.0, 0.0, 0.0)`. Check if animation can be activated and explain the output at each step.

Exercise 4.18. (Programming) Now here's something rather funny. Actually, what we'll show is not an uncommon accidental error. Cut the `glLoadIdentity()` call from the drawing routine of `ballAndTorus.cpp` and paste it as the last line of the window reshape routine (as, say, in `square.cpp`).

Oops! The ball and torus speed away together and are out of sight pretty quickly. Explain.
Hint: The current modelview matrix is not automatically cleared to identity between successive calls to the drawing routine.

Exercise 4.19. (Programming) Add a red and a green ball to the existing blue ball so that the three are always 120° from each other and follow a similar rotate-revolve path one after the other.
Hint: Copy and paste the revolving ball code a couple of times, making sure to isolate each instance with a `glPushMatrix()-glPopMatrix()` pair, and add in appropriate `glRotatef(*, 0.0, 0.0, 1.0)` calls.

Experiment 4.21. We want to add a satellite that tags along with the ball of `ballAndTorus.cpp`. The following piece of code added to the end of the drawing routine – just before `glutSwapBuffers()` – does the job (Block 14):

```
glTranslatef(4.0, 0.0, 0.0);

// Satellite
glColor3f(1.0, 0.0, 0.0);
glutWireSphere(0.5, 5, 5);
```

Figure 4.26: Screenshot from Experiment 4.21.

See Figure 4.26 for a screenshot. For a revolving satellite add the following instead (Block 15):

```
glRotatef(10*latAngle, 0.0, 1.0, 0.0);
glTranslatef(4.0, 0.0, 0.0);

// Satellite
glColor3f(1.0, 0.0, 0.0);
glutWireSphere(0.5, 5, 5);
```

Observe how Proposition 4.1 is being applied in both cases to determine the motion of the satellite *relative to the* ball by means of transformation statements between the two. End

Exercise 4.20. (Programming) Thinking that the Trick should be invoked to revolve the satellite about the ball, exactly as was done to obtain the latitudinal rotation of the ball itself, suppose we code the satellite as below (Block 16):

```
// Trick code block.
glTranslatef(4.0, 0.0, 0.0);
glRotatef(10*latAngle, 0.0, 1.0, 0.0);
glTranslatef(-4.0, 0.0, 0.0);
```

```
glTranslatef(4.0, 0.0, 0.0);

// Satellite.
glColor3f(1.0, 0.0, 0.0);
glutWireSphere(0.5, 5, 5);
```

The satellite still follows the ball, but does *not* revolve about it. Why? *Hint*: A good way to verify your answer is to stop the ball from moving by commenting out both `glRotatef()`'s in its definition block and observing only the satellite.

Exercise 4.21. (**Programming**) Continuing with Experiment 4.21, add a second satellite. Both should revolve around the ball, but in different orbits.

Throwing a Ball

Figure 4.27: Screenshot of `throwBall.cpp`.

Experiment 4.22. Run `throwBall.cpp`, which simulates the motion of a ball thrown with a specified initial velocity subject to the force of gravity. Figure 4.27 is a screenshot.

Press space to toggle between animation on and off. Press the right/left arrow keys to increase/decrease the horizontal component of the initial velocity, up/down arrow keys to increase/decrease the vertical component of the initial velocity and the page up/down keys to increase/decrease gravitational acceleration. Press 'r' to reset. The values of the initial velocity components and of gravitational acceleration are displayed on the screen.

End

The equation determining the horizontal motion of the ball in `throwBall.cpp`, in terms of time t, is

$$x(t) = ht$$

where h is the horizontal component of the initial velocity; that determining vertical motion is

$$y(t) = vt - \frac{g}{2}t^2$$

where v is the vertical component of the initial velocity and g is gravitational acceleration (a basic physics or calculus book should have a derivation of these standard equations).

Motion is simulated by repeatedly redrawing the ball at the new location it's mapped to by `glTranslatef`$(x(t), y(t), 0)$, incrementing t by 1 each time.

Remark 4.9. The techniques to animate the spheres in `ballAndTorus.cpp` and `throwBall.cpp` are interesting to compare. One could say that the first is "physical" while the latter "equational".

Exercise 4.22. (Programming) Animate a ball thrown toward and bouncing off a wall. See Figure 4.28. The initial force on the ball is horizontal – allow the user to change the amount of this force. Also allow the user to adjust gravitational acceleration and the "springiness" of the ball. Animation can end when the ball hits the floor.

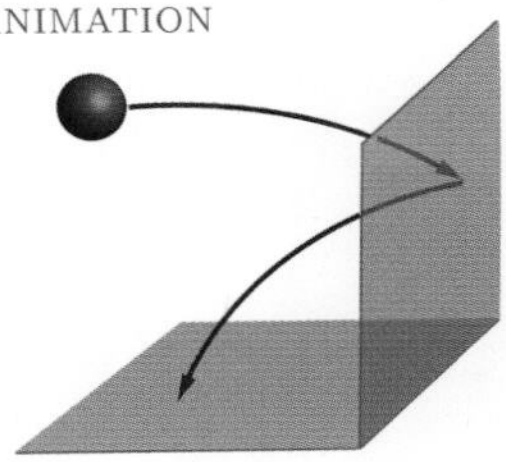

Figure 4.28: Ball bouncing off wall.

Ball Facing Friction

Experiment 4.23. Run `ballAndTorusWithFriction.cpp`, which modifies `ballAndTorus.cpp` to simulate an invisible viscous medium through which the ball travels.

Press space to apply force to the ball. It has to be kept pressed in order to continue applying force. The ball comes to a gradual halt after the key is released. Increase or decrease the level of applied force by using the up and down arrow keys. Increase or decrease the viscosity of the medium using the page up and down keys. Press 'x/X', 'y/Y' and 'z/Z' to rotate the scene.

End

The equation of motion implemented takes the frictional drag (or, equivalently, deceleration) on the ball of `ballAndTorusWithFriction.cpp` to be proportional to its velocity, a valid assumption from physics [52]. So, the equation is

$$drag\_deceleration = drag * velocity$$

where the *drag* (in real life) is a constant depending on the medium through which the object moves, as well as its shape. When space is pressed, the external acceleration applied is `applied_acceleration`, resulting in the net acceleration calculated in the line of code

```
acceleration = applied_acceleration - drag*velocity;
```

When space is not pressed, there is no applied acceleration but only frictional drag, so the equation is instead

```
acceleration = -drag*velocity;
```

At every time step we find the change in velocity from the equation

$$\frac{\Delta(velocity)}{\Delta(time)} = acceleration$$

which is certainly true in the limit as $\Delta(time) \to 0$. However, we approximate change through a unit time step by setting $\Delta(time) = 1$ to get

$$\Delta(velocity) = acceleration$$

which is implemented by the program statement

```
velocity += acceleration;
```

Finally, change per time step in the `latAngle` and `longAngle` variables is taken proportional to the current value `velocity`.

Exercise 4.23. **(Programming)** Simulate a ball falling through air, landing upon and continuing on through a viscous medium such as water. You need not simulate splashing. To differentiate air and water simply use color, e.g., the upper half of your window may be white, and the lower blue.

Remark 4.10. The last two programs, `throwBall.cpp` and `ballAndTorusWithFriction.cpp`, demonstrated simple applications of *physics in graphics*. This is a fascinating field – also known as *physically-based modeling* and *game physics* – of great importance in realistic animation. Plausible simulation of such phenomena as a wall of bricks crashing down, clothes and hair blowing in the wind, a drop of water rolling off a leaf, and smoke, fire and explosions, to mention a few, all require the programmer to take into account the real-world physics of the setting.

Special effects in a Hollywood production are almost always physics in graphics in action. There are two overarching and competing considerations in this discipline – realism versus computational efficiency.

A couple of books for the interested reader include Bourg & Bywalec [20] and Eberly [40]. A comprehensive list of pointers to ongoing research in the field is maintained by Simon Clavet [110].

Clown Head

Our next project is a program, which we'll develop incrementally, to draw a clown's head.

(a) (b) (c)

Figure 4.29: Screenshot of (a) `clown1.cpp` (b) `clown2.cpp` (c) `clown3.cpp`.

Experiment 4.24. We start with simply a blue sphere for the head. See `clown1.cpp` which has the following drawing routine (note that `clown1.cpp` and `clown2.cpp` are not separate programs but incremental stages of `clown3.cpp` which is in `ExperimenterSource/Chapter4`):

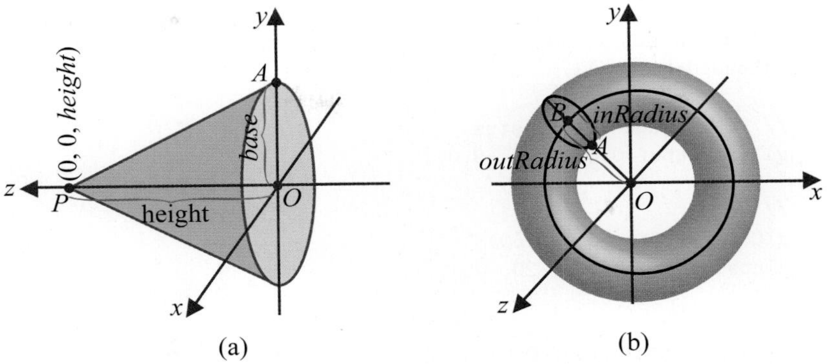

Figure 4.30: (a) Cone drawn by `glutWireCone(`*base*`,` *height*`,` *slices*`,` *stacks*`)` (b) Torus drawn by `glutWireTorus(`*inRadius*`,` *outRadius*`,` *sides*`,` *rings*`)`. Note that the axes are depicted differently in each diagram.

```
void drawScene(void)
{
   glClear(GL_COLOR_BUFFER_BIT);
   glLoadIdentity();

  // Place scene in frustum.
   glTranslatef(0.0, 0.0, -9.0);

   // Head.
   glColor3f(0.0, 0.0, 1.0);
   glutWireSphere(2.0, 20, 20);

   glutSwapBuffers();
}
```

Figure 4.29(a) is a screenshot.

Next, we want a green conical hat. The command `glutWireCone(`*base*`,` *height*`,` *slices*`,` *stacks*`)` draws a wireframe cone of base radius *base* and height *height*. The base of the cone lies on the xy-plane with its axis along the z-axis and its apex pointing in the positive direction of the z-axis. See Figure 4.30(a). The parameters *slices* and *stacks* determine the fineness of the mesh (not shown in the figure).

Accordingly, insert the lines

```
   // Hat.
   glColor3f(0.0, 1.0, 0.0);
   glutWireCone(2.0, 4.0, 20, 20);
```

in `clown1.cpp` after the call that draws the sphere, so that the drawing routine becomes (Block 17):

```
void drawScene(void)
```

```
{
   glClear(GL_COLOR_BUFFER_BIT);
   glLoadIdentity();

   // Place scene in frustum.
   glTranslatef(0.0, 0.0, -9.0);

   // Head.
   glColor3f(0.0, 0.0, 1.0);
   glutWireSphere(2.0, 20, 20);

   // Hat.
   glColor3f(0.0, 1.0, 0.0);
   glutWireCone(2.0, 5.0, 20, 20);

   glutSwapBuffers();
}
```

Not good! Because of the way `glutWireCone()` aligns, the hat covers the clown's face. This is easily fixed. Translate the hat 2 units up the z-axis and rotate it $-90°$ about the x-axis to arrange it on top of the head. Finally, rotate it a rakish $30°$ about the z-axis! Here's the modified drawing routine of `clown1.cpp` at this point (Block 18):

```
void drawScene(void)
{
   glClear(GL_COLOR_BUFFER_BIT);
   glLoadIdentity();

   // Place scene in frustum.
   glTranslatef(0.0, 0.0, -9.0);

   // Head.
   glColor3f(0.0, 0.0, 1.0);
   glutWireSphere(2.0, 20, 20);

   // Transformations of the hat.
   glRotatef(30.0, 0.0, 0.0, 1.0);
   glRotatef(-90.0, 1.0, 0.0, 0.0);
   glTranslatef(0.0, 0.0, 2.0);

   // Hat.
   glColor3f(0.0, 1.0, 0.0);
   glutWireCone(2.0, 5.0, 20, 20);

   glutSwapBuffers();
}
```

Let's add a brim to the hat by attaching a torus to its base. The command `glutWireTorus(`*inRadius*`, `*outRadius*`, `*sides*`, `*rings*`)` draws a wireframe torus of inner radius *inRadius* (the radius of a circular section of the torus), and outer radius *outRadius* (the radius of the circle through the middle of the torus). The axis of the torus is along the z-axis and centered at the origin. See Figure 4.30(b). Insert the call `glutWireTorus(0.2, 2.2, 10, 25)` right after the call that draws the cone, so the drawing routine becomes (Block 19):

```
void drawScene(void)
{
   glClear(GL_COLOR_BUFFER_BIT);
   glLoadIdentity();

   // Place scene in frustum.
   glTranslatef(0.0, 0.0, -9.0);

   // Head.
   glColor3f(0.0, 0.0, 1.0);
   glutWireSphere(2.0, 20, 20);

   // Transformations of the hat and brim.
   glRotatef(30.0, 0.0, 0.0, 1.0);
   glRotatef(-90.0, 1.0, 0.0, 0.0);
   glTranslatef(0.0, 0.0, 2.0);

   // Hat.
   glColor3f(0.0, 1.0, 0.0);
   glutWireCone(2.0, 5.0, 20, 20);

   // Brim.
   glutWireTorus(0.2, 2.2, 10, 25);

   glutSwapBuffers();
}
```

Observe that the brim is drawn suitably at the bottom of the hat and stays there despite modeling transformations between head and hat – a consequence of Proposition 4.1.

To animate, let's spin the hat about the clown's head by rotating it around the y-axis. We rig the space bar to toggle between animation on and off and the up/down arrow keys to change speed. All updates so far are included in `clown2.cpp`. Figure 4.29(b) is a screenshot.

What's a clown without little red ears that pop in and out?! Spheres will do for ears. An easy way to bring about oscillatory motion is to make use of the function $\sin(angle)$ which varies between -1 and 1. Begin by translating either ear a unit distance from the head, and then repeatedly translate each a distance of $\sin(angle)$, incrementing *angle* each time.

Note: A technicality one needs to be aware of in such applications is that angle is measured in *degrees* in OpenGL syntax, e.g., in `glRotatef(`*angle*`,`

$p,\ q,\ r$), while the C++ math library assumes angles to be given in *radians*. Multiplying by $\pi/180$ converts degrees to radians.

The ears and head are physically separate, though. Let's connect them with springs! Helixes are springs. We borrow code from `helix.cpp`, but modify it to make the length of the helix 1, its axis along the x-axis and its radius 0.25. As the ears move, either helix is scaled along the x-axis so that it spans the gap between the head and an ear. The completed program is `clown3.cpp`, of which a screenshot is seen in Figure 4.29(c). **End**

Exercise 4.24. (Programming) Comment out the push-pop pair isolating the hat and brim in `clown3.cpp`. Explain the new situation of the ears.

Exercise 4.25. Proposition 4.1 came before our discussion of push-pop pairs, so the assumption there is that there are none. Do we have to revise the proposition to take into account possible push-pop pairs?

Blooming Flower

Figure 4.31: Screenshot of `floweringPlant.cpp` in mid-bloom.

Experiment 4.25. Run `floweringPlant.cpp`, an animation of a flower blooming. Press space to start and stop animation, delete to reset, and 'x/X', 'y/Y' and 'z/Z' to change the viewpoint. Figure 4.31 is a screenshot. **End**

The stem of the plant of `floweringPlant.cpp` consists of four straight segments, the sepal (base of the flower) is modeled as a hemisphere, while the six petals are circles. Both the hemisphere and circle are reshaped by scaling during animation. The code to draw the two is modified from `circle.cpp` and `hemisphere.cpp`.

As calls to display lists cannot be parametrized at run-time, those defining a sepal and a petal have to be placed, unfortunately, in the drawing routine to allow them access to the changing global variable `t` via the variables – `angleFirstSegment`, ..., `petalOpenAngle` – at the top of `drawScene()`. (See Remark 3.4 of Section 3.4 for the evils of this practice.) Another option for modular code would have been to write these parts of the plant as C++ objects.

The parameters involved in configuring the stem, sepal and petal all change from a start value to an end one via linear interpolation using the animation parameter t. For example,

$$\texttt{hemisphereScaleFactor} = (1-t)*0.1 \;+\; t*0.75$$

linearly changes `hemisphereScaleFactor` from 0.1 to 0.75 as t goes from 0 to 1.

4.5.3 Animation Projects

Exercise 4.26. (Programming) Starting from `clown3.cpp`, add to the clown's head a conical nose which changes in length and color, as well as eyes that rotate and change in size and color.

Exercise 4.27. (Programming) Animate a ball *rolling* down a fixed flat inclined plane. See Figure 4.32(a). The ball should not slip or slide. Make the plane a wireframe mesh of triangles and the ball a wireframe sphere, as well, so that relative motion is apparent.

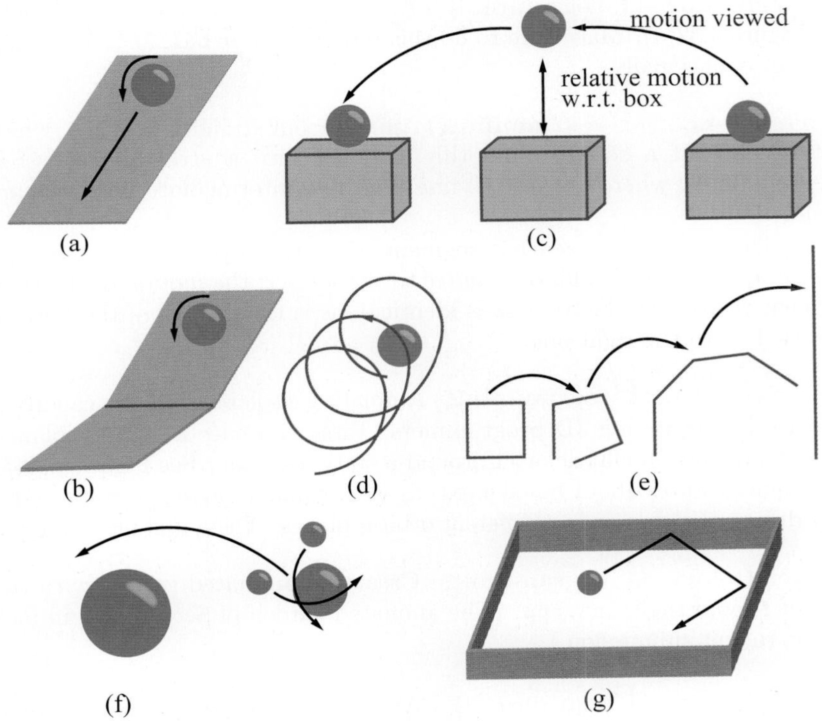

Figure 4.32: (a) Ball rolling down one plane (b) Ball rolling down two planes (c) Ball bouncing on a box (d) Ball traveling along a helix (e) Four segments opening from a square into a straight line (f) Solar system with a sun, one planet and two moons (g) Pool table with one ball.

Exercise 4.28. (Programming) Add physics to the preceding exercise by allowing the incline of the plane to be changed even as the ball rolls down, the latter's speed depending obviously on the angle of inclination.

Exercise 4.29. (Programming) Yet another extension of Exercise 4.27: add another plane at the bottom so that the ball rolls from the first onto the second. See Figure 4.32(b).

Exercise 4.30. (**Programming**) Roll a ball down the curved children's slide of Exercise 2.33 of Chapter 2, if you did that particular exercise.

Exercise 4.31. (**Programming**) Animate a ball bouncing up and down a box which itself moves in a straight line. See Figure 4.32(c).

First, code the straight-line motion of the box and, then, that of the ball relative to the box, which is straight, too. The resultant motion of the ball as viewed in the OpenGL window, which is, of course, as that seen by a stationary external observer, is parabolic.

Exercise 4.32. (**Programming**) Animate a ball traveling a helical path. See Figure 4.32(d). Make sure to do this physically *a la* `ballAndTorus.cpp`, and not equationally.

Exercise 4.33. (**Programming**) Animate four straight segments, which initially bound a square, smoothly opening into a straight line. See Figure 4.32(e), where the initial, final and two intermediate positions are depicted.
Hint: Draw first the fixed left segment, then the top segment as a copy. Determine the latter's motion *relative* to the former; the motion of the right segment relative to the top one is identical, as is the motion of the bottom one relative to the right one.

Exercise 4.34. (**Programming**) Animating a solar system is a canonical exercise for beginning 3D programmers. First, animate a solitary planet, with two moons, in elliptic orbit around a stationary sun. See Figure 4.32(f). The planet rotates about its own axis as well, while its moons revolve about it at different speeds and on different orbital planes. Then, add more planets.

Exercise 4.35. (**Programming**) Create an animated garden with the help of `floweringPlant.cpp`. (The animated garden of Section 1.3, in fact, was a student submission.)

Exercise 4.36. (**Programming**) Animate a lone cue ball moving on a pool table. The table should simply be a rectangle enclosed by four low walls – no need to make pockets. See Figure 4.32(g).

The ball should initially be stationary at a fixed position on the table. Then allow the user, with the help of a simple visual interface, to choose a direction and speed to get the ball moving – you don't need to draw a cue stick.

Animate the subsequent motion of the ball as it *rolls* along the surface of the table and *bounces* off its sides. You can either choose not to program in any deceleration, so that the ball keeps moving at uniform speed, or to incorporate frictional resistance to ultimately bring the ball to rest.

4.6 Viewing Transformation

We begin our discussion of the viewing transformation `gluLookAt()`, whose function is to arrange OpenGL's camera, by systematically deciphering its somewhat non-trivial syntax.

4.6.1 Understanding the Viewing Transformation

Think of the OpenGL camera as located at the origin with its lens pointing down the $-z$ direction (the *line of sight*) and with its top aligned along the $+y$ direction (the *up direction*). This, in fact, is the *default pose* of the OpenGL camera. See Figure 4.33(a).

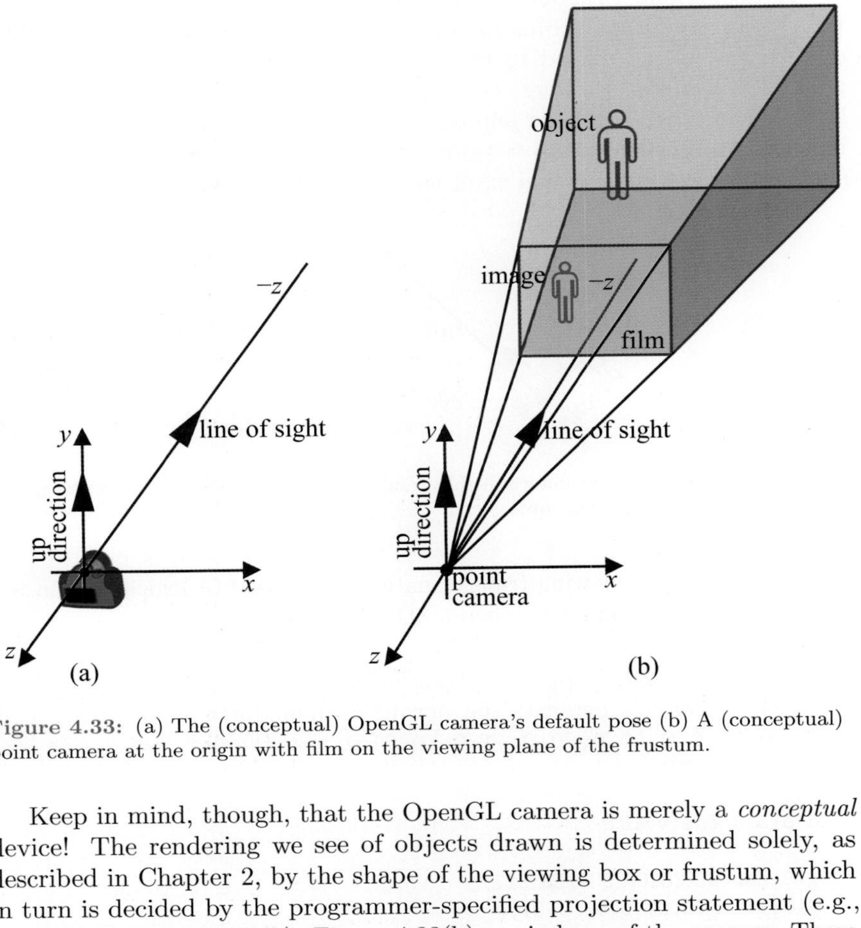

Figure 4.33: (a) The (conceptual) OpenGL camera's default pose (b) A (conceptual) point camera at the origin with film on the viewing plane of the frustum.

Keep in mind, though, that the OpenGL camera is merely a *conceptual* device! The rendering we see of objects drawn is determined solely, as described in Chapter 2, by the shape of the viewing box or frustum, which in turn is decided by the programmer-specified projection statement (e.g., `glOrtho()`, `glFrustum()`). Figure 4.33(b) reminds us of the process. There is *no* camera as such!

Nevertheless, it appeals to the intuition to imagine that what we're viewing is through a camera. In the case of a viewing frustum, particularly, one can imagine a point camera at the origin with the film lying in front of it on the viewing face, as indicated in Figure 4.33(b). It's intuitive as well to think of changing the view by moving and turning the camera. This is exactly where the viewing transformation `gluLookAt()` comes in.

Note: For now, we ask the reader to assume that we have a viewing frustum defined by a `glFrustum()` statement, rather than a viewing box by `glOrtho()`, as the point camera is logically placed at the origin in the case of the former, but it's not evident where to place it for the latter. However, this apparent problem will be sorted out as soon as the working of `gluLookAt()` becomes clear.

The command `gluLookAt(`*eyex*`, `*eyey*`, `*eyez*`, `*centerx*`, `*centery*`, `*centerz*`, `*upx*`, `*upy*`, `*upz*`)` *simulates* – mark the word *simulates* – OpenGL's camera first being moved to the location *eye* = (*eyex*, *eyey*, *eyez*) and pointed at *center* = (*centerx*, *centery*, *centerz*); next, it is rotated about its line of sight (*los*) – the line joining *eye* to *center* – so that its up direction is one determined from *up* = (*upx*, *upy*, *upz*). See Figure 4.34. We'll see shortly how the up direction is, in fact, determined from *up*.

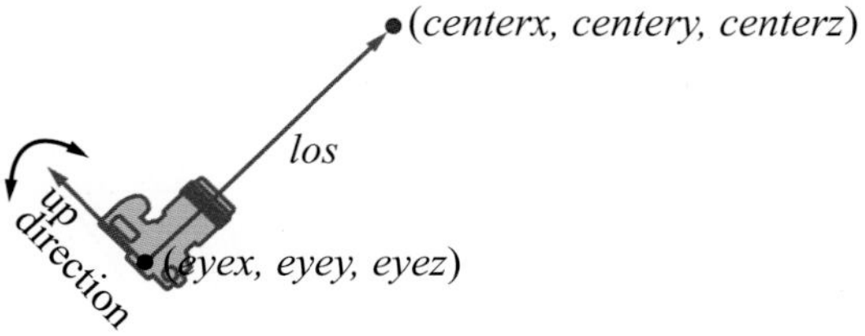

Figure 4.34: Camera pose determined by `gluLookAt(`*eyex*`, `*eyey*`, `*eyez*`, `*centerx*`, `*centery*`, `*centerz*`, `*upx*`, `*upy*`, `*upz*`)`.

Remark 4.11. The viewing transformation `gluLookAt()` logically then is a function of *three* parameters, each a 3D point vector.

Experiment 4.26. Replace the translation command `glTranslatef(0.0, 0.0, -15.0)` of `box.cpp` with the viewing command `gluLookAt(0.0, 0.0, 15.0, 0.0, 0.0, 0.0, 0.0, 1.0, 0.0)` so that the drawing routine is as below (Block 20):

```
void drawScene(void)
{
   glClear(GL_COLOR_BUFFER_BIT);
   glColor3f(0.0, 0.0, 0.0);
   glLoadIdentity();
```

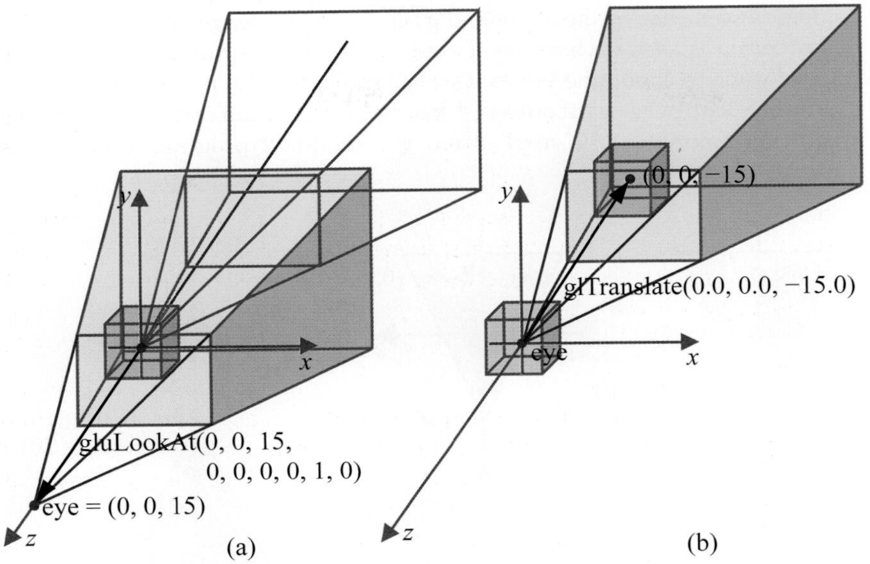

Figure 4.35: (a) `gluLookAt()`: the broken frustum is the original viewing frustum, the unbroken one is where it's translated by the `gluLookAt()` call, the box doesn't move. (b) `glTranslatef()`: the viewing frustum doesn't move, rather the box is translated by the `glTranslatef()` call.

```
   // Viewing transformation.
   gluLookAt(0.0, 0.0, 15.0, 0.0, 0.0, 0.0, 0.0, 1.0, 0.0);

   glutWireCube(5.0); // Box.

   glFlush();
}
```

There is no change in what is viewed. The commands `glTranslatef(0.0, 0.0, -15.0)` and `gluLookAt(0.0, 0.0, 15.0, 0.0, 0.0, 0.0, 0.0, 1.0, 0.0)` are *exactly equivalent*.

Note: A convenient way to compare side-by-side the output of two pieces of code is to put them into the two drawing routines of `windows.cpp` of Chapter 3.

End

To understand why the two statements are equivalent, note that `gluLookAt(0.0, 0.0, 15.0, 0.0, 0.0, 0.0, 0.0, 1.0, 0.0)` takes the *eye* to $(0, 0, 15)$ looking down the z-axis toward the *center* at $(0, 0, 0)$, the frustum traveling with it. Now, compare Figures 4.35(a) and (b): should the box appear different in the first (the frustum translated back) than the second (the box translated forward)? *No*, because its position relative to the frustum is the same in both.

The convenience of the command gluLookAt() over glTranslatef() in this program is that we have been able to arrange the camera according to how we want to shoot the box, rather than moving the box itself.

As box.cpp with gluLookAt() instead of glTranslatef(), as in the preceding experiment, is used often, the modified program is stored as boxWithLookAt.cpp.

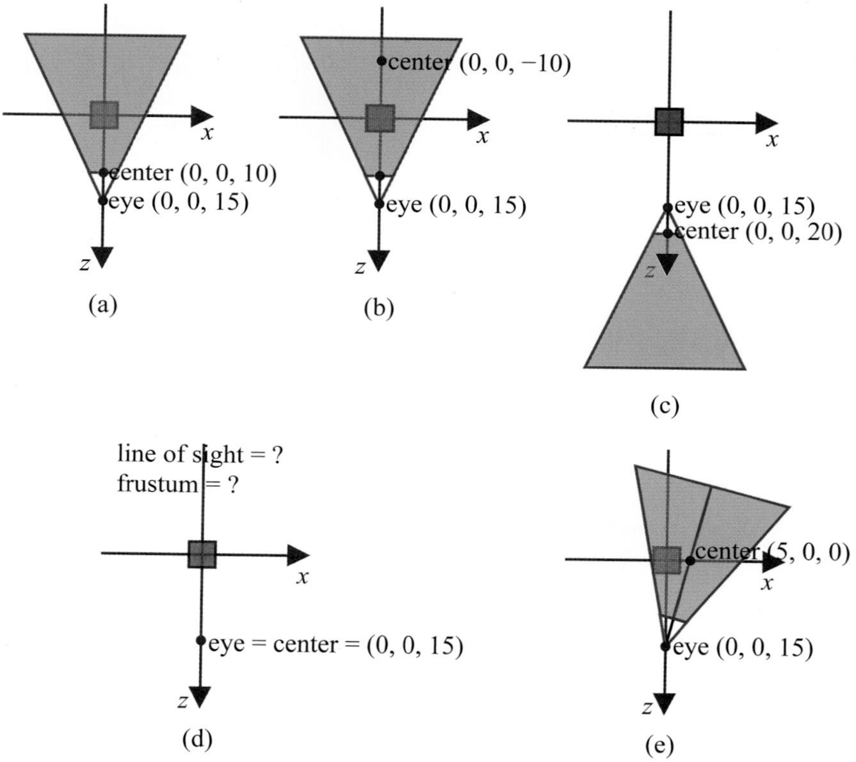

Figure 4.36: Sectional diagrams of the (simulated) configuration of the eye and frustum for various gluLookAt() calls in boxWithLookAt.cpp.

Experiment 4.27. Continue the previous experiment, or run boxWithLookAt.cpp, successively changing only the parameters *centerx*, *centery*, *centerz* – the middle three – of the gluLookAt() call to the following:

1. $0.0, 0.0, 10.0$
2. $0.0, 0.0, -10.0$
3. $0.0, 0.0, 20.0$
4. $0.0, 0.0, 15.0$

End

The view does not change with the first two parameter sets of the experiment as the viewer's line of sight from eye to center does not change. Figures 4.36(a) and (b) show the respective configurations. The third set (Figure 4.36(c)) produces a blank screen because the eye is looking the "wrong way". The last set (Figure 4.36(d)) confuses OpenGL because eye and center coincide, making it impossible to decide a line of sight. Again a blank screen appears. Note in all cases that the shape of the frustum is not changed by `gluLookAt()`, only its placement and alignment.

Here are a few more *center* sets for you to try.

Exercise 4.37. (**Programming**) Restore the original `boxWithLookAt.cpp` program with its `gluLookAt(0.0, 0.0, 15.0, 0.0, 0.0, 0.0, 0.0, 1.0, 0.0)`. Next, successively change only the parameters *centerx*, *centery*, *centerz* – the middle three of `gluLookAt()` – to the following, drawing diagrams as in Figure 4.36 to explain what is seen in each case:

1. $5.0, 0.0, 0.0$ (*Answer*: See Figure 4.36(e))

2. $-5.0, 0.0, 0.0$

3. $0.0, 5.0, 0.0$

4. $0.0, -5.0, 0.0$

5. $5.0, 5.0, 0.0$

Let's change the *eye* next. It's still pretty much the same game as changing *center*, though keep in mind that the frustum's apex moves with the eye.

Exercise 4.38. (**Programming**) Restore the original `boxWithLookAt.cpp` program with its `gluLookAt(0.0, 0.0, 15.0, 0.0, 0.0, 0.0, 0.0, 1.0, 0.0)` call. First, replace the box with a `glutWireTeapot(5.0)`, a non-symmetric object. Next, successively change only the parameters *eyex*, *eyey*, *eyez* – the first three parameters of `gluLookAt()` – to the following, drawing diagrams as in Figure 4.36 to explain what is seen in each case:

1. $0.0, 0.0, 10.0$

2. $0.0, 0.0, 25.0$

3. $0.0, 0.0, -15.0$

4. $15.0, 0.0, 15.0$

5. $15.0, 0.0, 0.0$

6. $15.0, 15.0, 15.0$

Let's get a feel now for how the up vector $up = (upx,\ upy,\ upz)$ works.

Experiment 4.28. Restore the original `boxWithLookAt.cpp` program with its `gluLookAt(0.0, 0.0, 15.0, 0.0, 0.0, 0.0, 0.0, 1.0, 0.0)` call and, again, first replace the box with a `glutWireTeapot(5.0)`. Run: a screenshot is shown in Figure 4.37(a). Next, successively change the parameters *upx*, *upy*, *upz* – the last three parameters of `gluLookAt()` – to the following:

1. $1.0, 0.0, 0.0$ (Figure 4.37(b))
2. $0.0, -1.0, 0.0$ (Figure 4.37(c))
3. $1.0, 1.0, 0.0$ (Figure 4.37(d))

Screenshots of the successive cases are shown in Figures 4.37(b)-(d). The camera indeed appears to rotate *about* its line of sight, the z-axis, so that its up direction points along the *up* vector (*upx*, *upy*, *upz*) each time. **End**

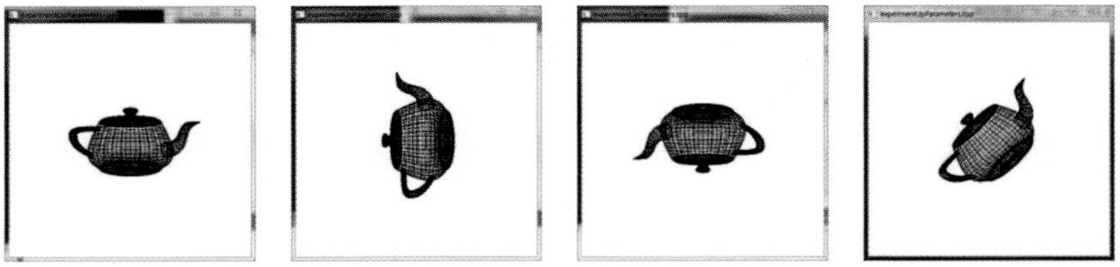

(a) (b) (c) (d)

Figure 4.37: Screenshots from Experiment 4.28.

Before we can state the rule for how the *up* vector determines the camera's up direction generally, here are some facts about the dot product of vectors which we'll need. Skip this part if you already have dot product basics.

Sidebar on Dot Products

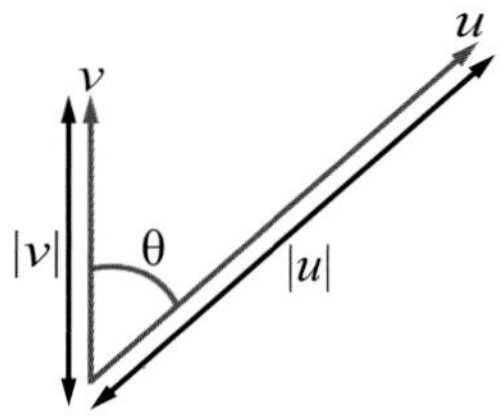

Figure 4.38: Taking the dot product: $u \cdot v = |u||v| \cos\theta$.

The *dot product* (also called *scalar product*) of two vectors u and v in $\mathbb{R}^3$ is a scalar, denoted $u \cdot v$, defined as follows:

(a) if either of u or v is zero, then $u \cdot v$ is zero;

(b) if not, then the value of $u \cdot v$ is $|u||v| \cos\theta$, where θ is the angle between u and v.

See Figure 4.38.

It turns out that $u \cdot v$ is given by the following simple formula, where $u = (u_x, u_y, u_z)$ and $v = (v_x, v_y, v_z)$:

$$u \cdot v = u_x v_x + u_y v_y + u_z v_z \quad (4.3)$$

This makes the dot product useful in calculating angles between pairs of vectors.

Example 4.3. Determine the angle θ between the two vectors $u = (1, 0, 2)$ and $v = (-2, 3, 4)$.

Answer:

$$|u||v| \cos\theta = u \cdot v = u_x v_x + u_y v_y + u_z v_z = 1 * -2 \; + \; 0 * 3 \; + \; 2 * 4 = 6$$

Therefore,

$$\cos\theta = \frac{6}{|u||v|} = \frac{6}{\sqrt{1^2 + 0^2 + 2^2}\sqrt{(-2)^2 + 3^2 + 4^2}} = \frac{6}{\sqrt{5}\sqrt{29}} \simeq 0.49827$$

which gives $\theta \simeq 60.11439°$.

Exercise 4.39. Determine the angle between each pair from the three vectors $(1, 0, 0)$, $(\frac{1}{\sqrt{2}}, \frac{1}{\sqrt{2}}, 0)$ and $(\frac{1}{\sqrt{3}}, \frac{1}{\sqrt{3}}, \frac{1}{\sqrt{3}})$.

Exercise 4.40. Prove the following about dot products where u, v and w are any three vectors and c any scalar:

(a) Assuming that they are both non-zero, u and v are perpendicular if and only if $u \cdot v = 0$ (*perpendicularity test*)

(b) $u \cdot u = |u|^2$

(c) $u \cdot v = v \cdot u$ (*dot product is commutative*)

(d) $(cu) \cdot v = u \cdot (cv) = c(u \cdot v)$

(e) $u \cdot (v + w) = u \cdot v + u \cdot w$ (*dot product distributes over a sum*)

(f) $|u \cdot v| \leq |u||v|$

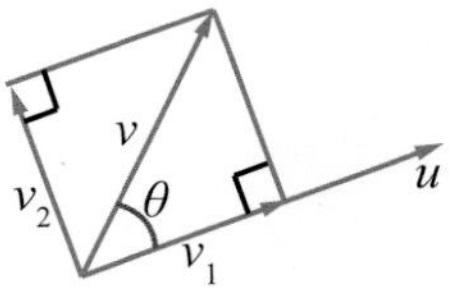

(a)

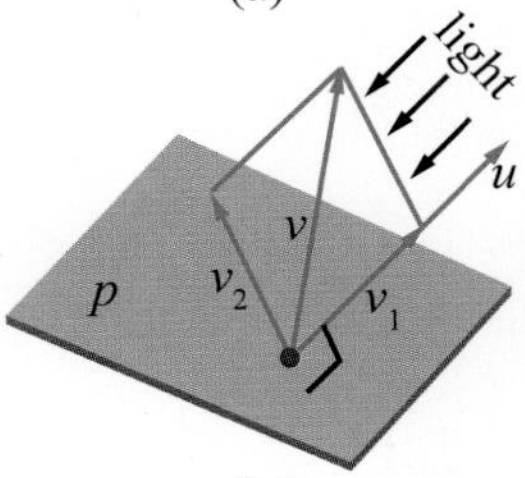

(b)

Figure 4.39: (a) Splitting v into components v_1 and v_2, parallel and perpendicular to u, respectively (b) v_2 as the "shadow" of v on a plane p perpendicular to u.

A particularly useful application of the dot product is when one wants to split a given vector v as $v = v_1 + v_2$, where the components v_1 and v_2 are, respectively, parallel and perpendicular to a given non-zero vector u. See Figure 4.39(a). An intuitive way to think of v_2 is as the shadow of v cast on a plane p perpendicular to u by a light shining from the direction of u, as depicted in Figure 4.39(b).

The component v_1 is the perpendicular projection of v onto the line of u, so its *signed length* is

$$|v| \cos\theta = \frac{|u||v| \cos\theta}{|u|} = \frac{u \cdot v}{|u|}$$

where θ is the angle between u and v. Multiplying the value of the signed length by the unit vector in the direction of u, which is $u/|u|$, one obtains the formula for v_1:

$$v_1 = \frac{u \cdot v}{|u|^2} \, u \qquad (4.4)$$

The formula for the component v_2 of v that is perpendicular to u follows, as the sum of v_1 and v_2 is v:

$$v_2 = v - v_1 = v - \frac{u \cdot v}{|u|^2}\, u \tag{4.5}$$

The preceding formulae have particularly simple forms if u is a unit vector, as we ask the reader to show next.

Exercise 4.41. If u is a unit vector and v arbitrary, prove the following:

(a) The component of v parallel to u is $v_1 = (u \cdot v)\, u$.

(b) The component of v perpendicular to u is $v_2 = v - (u \cdot v)\, u$.

Example 4.4. Split $v = (-2, 3, 4)$ into components parallel and perpendicular to $u = (1, 0, 2)$.

Answer: The component parallel to u is

$$v_1 = \frac{u \cdot v}{|u|^2}\, u = \frac{6}{5}\,(1, 0, 2) = \left(\frac{6}{5},\ 0,\ \frac{12}{5}\right)$$

and that perpendicular to u is

$$v_2 = v - v_1 = (-2, 3, 4) - \left(\frac{6}{5},\ 0,\ \frac{12}{5}\right) = \left(-\frac{16}{5},\ 3,\ \frac{8}{5}\right)$$

The following worked example shows a neat matrix expression for the component of one vector parallel to another. The vectors themselves are written as column matrices.

Example 4.5. Show that if $u = [u_x\ u_y\ u_z]^T$ and $v = [v_x\ v_y\ v_z]^T$ are two vectors in $\mathbb{R}^3$, such that u is not zero, then the component v_1 of v parallel to u is given by

$$v_1 = \frac{1}{|u|^2} \begin{bmatrix} u_x^2 & u_x u_y & u_x u_z \\ u_x u_y & u_y^2 & u_y u_z \\ u_x u_z & u_y u_z & u_z^2 \end{bmatrix} v$$

Answer: We have that the component of v parallel to u is

$$\begin{aligned}
v_1 &= \frac{u \cdot v}{|u|^2}\, u \\
&= \frac{1}{|u|^2}\,(u^T v)u \quad && (u^T v \text{ denotes its lone scalar entry, viz. } u \cdot v, \\
& && \text{as a } 1 \times 1 \text{ matrix product}) \\
&= \frac{1}{|u|^2}\, u(u^T v) \quad && (\text{as } (u^T v)u = u(u^T v), \text{ where } u^T v \text{ denotes a scalar} \\
& && \text{on the LHS and a } 1 \times 1 \text{ matrix on the RHS}) \\
&= \frac{1}{|u|^2}\,(uu^T)v \quad && (\text{by associativity}) \\
&= \frac{1}{|u|^2}\begin{bmatrix} u_x^2 & u_x u_y & u_x u_z \\ u_x u_y & u_y^2 & u_y u_z \\ u_x u_z & u_y u_z & u_z^2 \end{bmatrix} v \ (\text{multiplying as matrices } u \text{ and } u^T)
\end{aligned}$$

For a more thorough discussion of dot products refer to any book on linear algebra, e.g., Banchoff and Wermer [8].

Back to OpenGL

It's simple now to explain how OpenGL uses the $up = (upx, upy, upz)$ vector to align the top of its camera – in other words, determine its up direction – upon the call `gluLookAt(`*eyex*`, `*eyey*`, `*eyez*`, `*centerx*`, `*centery*`, `*centerz*`, `*upx*`, `*upy*`, `*upz*`)`.

Denote the camera's line of sight vector $(centerx, centery, centerz) - (eyex, eyey, eyez)$ by *los*. What OpenGL does is split *up* into components up_1 and up_2 parallel and perpendicular, respectively, to *los*. *The up direction is then taken to be* up_2. In particular, think of the camera, which is located at $eye = (eyex, eyey, eyez)$ and pointing down *los*, as being rotated about *los* till its top points in the direction parallel to up_2.

For an example, see Figure 4.40. Imagine the camera lying with its back on this page (call it the plane p) facing up, so that the line of sight *los* emerges perpendicularly from p (toward the reader). The specified *up* vector is drawn in the figure starting from the camera, as also its components up_1 and up_2, the latter lying on the page. The camera, then, is rotated about *los* with its back always on the page till its top points along up_2.

The magnitude of *up* or of up_2 is of no consequence as long as it's not zero, because it's only the direction that matters in aligning the top; if either is zero, then OpenGL is unable to determine the alignment and renders a blank screen.

Exercise 4.42. Of course, if *up* is zero then its component up_2 is zero. Can it happen that *up* is non-zero and yet up_2 is zero?

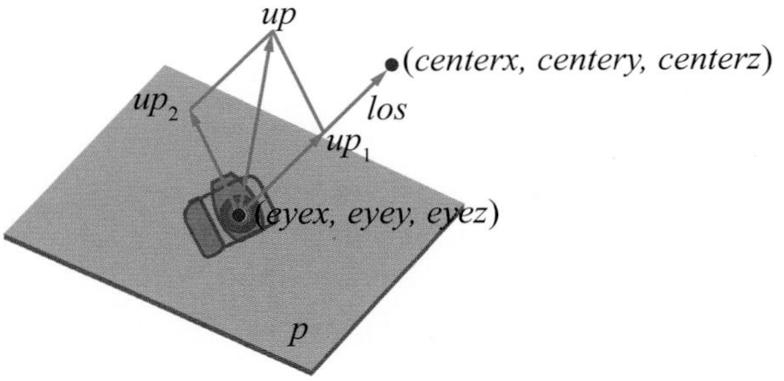

Figure 4.40: The camera is seen face-forward so that its back-plane p lies on the page. The line of sight los comes perpendicularly up from the page toward the reader. The components of the vector up, parallel and perpendicular to los, respectively, are up_1 and up_2 (the latter lying on the page).

Experiment 4.29. Replace the wire cube of `boxWithLookAt.cpp` with a `glutWireTeapot(5.0)` and replace the `gluLookAt()` call with:

```
gluLookAt(0.0, 0.0, 15.0, 0.0, 0.0, 0.0, 1.0, 1.0, 1.0)
```

The vector $los = (0.0, 0.0, 0.0) - (15.0, 0.0, 0.0) = (-15.0, 0.0, 0.0)$, which is down the z-axis. The component of $up = (1.0, 1.0, 1.0)$, perpendicular to the z-axis, is (1.0, 1.0, 0.0), which then is the up direction. Is what you see the same as Figure 4.37(d), which, in fact, is a screenshot for `gluLookAt(0.0, 0.0, 15.0, 0.0, 0.0, 0.0, 1.0, 1.0, 0.0)`? **End**

Exercise 4.43. **(Programming)** Change (upx, upy, upz) of `gluLookAt()` in `boxWithLookAt.cpp` to (0.0, 0.0, 1.0). What do you see? *Nothing*! Why?

Exercise 4.44. Compute the direction of the top of the camera for each of the following viewing transformations:

(a) `gluLookAt(0.0, 0.0, 5.0, 5.0, 0.0, 0.0, 0.0, 1.0, 1.0)`

(b) `gluLookAt(0.0, 5.0, 5.0, 0.0, 0.0, 0.0, 1.0, 1.0, 1.0)`

(c) `gluLookAt(10.0, 5.0, 5.0, 0.0, 5.0, 0.0, 5.0, 1.0, 1.0)`

(d) `gluLookAt(10.0, 10.0, 5.0, 0.0, 5.0, 0.0, 1.0, 2.0, 3.0)`

Part answer:

(a) The line of sight vector

$$\begin{aligned} los &= (centerx, centery, centerz) - (eyex, eyey, eyez) \\ &= (5, 0, 0) - (0, 0, 5) \;=\; (5, 0, -5) \end{aligned}$$

The component of $up = (0, 1, 1)$ perpendicular to los is

$$\begin{aligned} up_2 &= up - \frac{los \cdot up}{|los|^2}\, los \\ &= (0,1,1) - \frac{(5,0,-5)\cdot(0,1,1)}{50}\,(5,0,-5) \\ &= (0,1,1) + \frac{1}{10}\,(5,0,-5) \\ &= (0.5, 1, 0.5) \end{aligned}$$

which, therefore, is the direction of the top of the camera. It is perpendicular, of course, to the line of sight and, as easily verified, tilted about 35.3° from the direction of the y-axis. See Figure 4.41.

The answer can be verified by alternately plugging in `gluLookAt(0.0, 0.0, 5.0, 5.0, 0.0, 0.0, 0.0, 1.0, 1.0)` and `gluLookAt(0.0, 0.0, 5.0, 5.0, 0.0, 0.0, 0.5, 1.0, 0.5)` into `boxWithLookAt.-cpp` to see the same (clipped) box (screenshot in Figure 4.42).

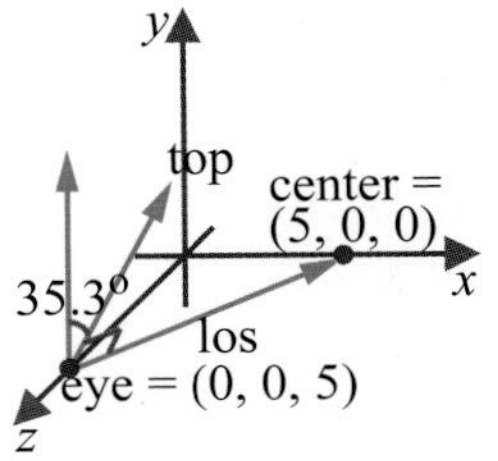

Figure 4.41: Solution to Exercise 4.44(a).

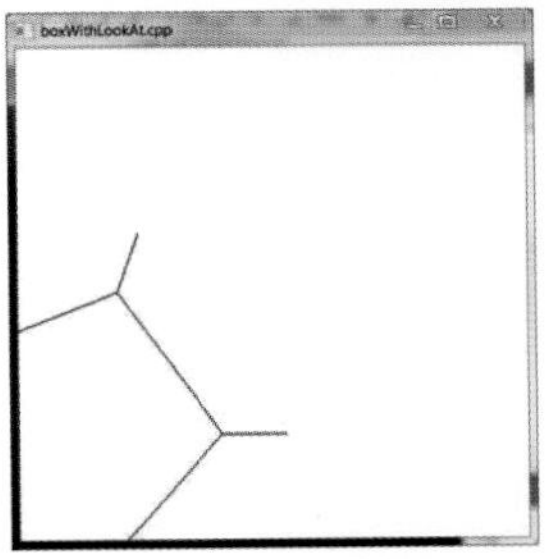

Figure 4.42: Checking the solution to Exercise 4.44(a).

Remark 4.12. Collectively, the modeling transformations `glTranslatef()`, `glScalef()` and `glRotatef()` and the viewing transformations `gluLookAt()` are called *modelview transformations*.

Exercise 4.45. (Programming) Program a camera flying at a height of 3 units over a sequence of balls arranged along the x-axis, looking ahead and down at the balls. See Figure 4.43.
Hint: Coordinates for the eye and center are suggested in the figure.

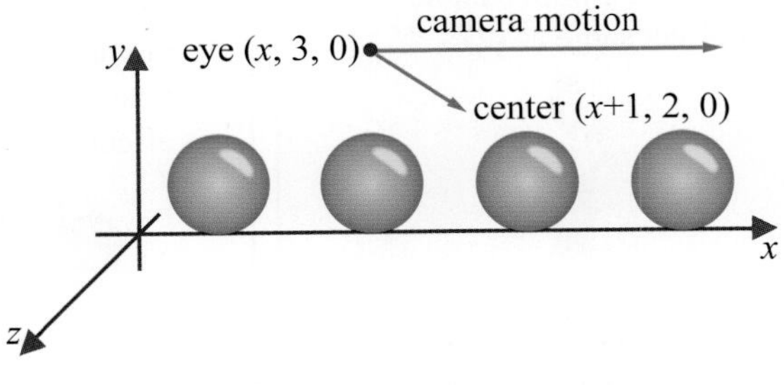

Figure 4.43: Camera flying over balls.

Figure 4.44: Camera rotated on an imaginary sphere enclosing a teapot.

Exercise 4.46. (Programming) Place a wire teapot centered at the origin. Program a camera which can be moved by the user anywhere on an imaginary sphere enclosing the teapot, the direction of the camera being always toward the origin. Appropriately program keys to move the camera. See Figure 4.44, where a couple of positions of the camera are indicated.

4.6.2 Simulating a Viewing Transformation with Modeling Transformations

You can skip this section on a first reading.

When we introduced `gluLookAt()` we said that it *simulates* OpenGL camera movement. This is exactly right. The OpenGL camera *never* leaves its default pose at the origin with its lens pointing down the $-z$ direction and with its top aligned along the $+y$ direction. In fact, the viewing transformation is simulated by replacing it with an equivalent sequence of modeling transformations. We actually saw a simple example of this earlier in Section 4.6.1 where the commands `glTranslatef(0.0, 0.0, -15.0)` and `gluLookAt(0.0, 0.0, 15.0, 0.0, 0.0, 0.0, 0.0, 1.0, 0.0)` were found to be equivalent.

Here's a motivating thought experiment for the general case:

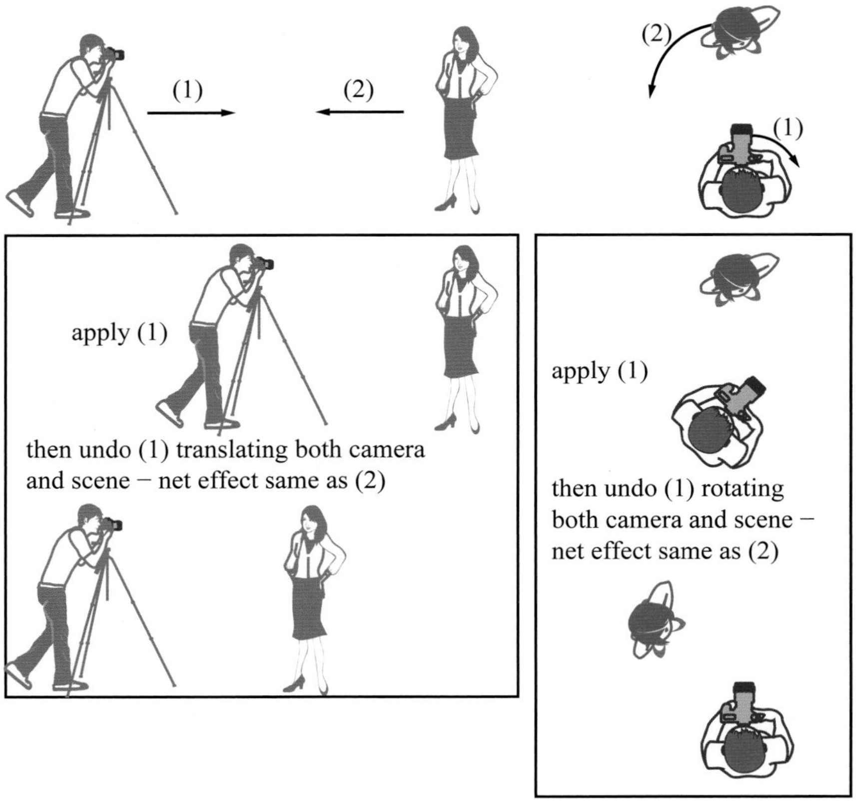

Figure 4.45: Relative movement of the camera and scene.

You are out on an open field with a friend and a camera. She stands 10 meters in front of you, but looking through the viewfinder you think she should be, say, 3 meters closer. There are two options: (1) you, i.e., the camera, translate (walk) 3 meters toward her, or (2) she, i.e., the scene, translates 3 meters toward you. See the top left of Figure 4.45. The picture is same in either case. Ignore the backdrop, as it's a homogeneous open field!

Here's another way to arrive at the equivalence of the two options. Say you had already applied (1) when the guy you had borrowed the camera from starts yelling that it's really expensive and would you mind not moving it around but just keep it where it was first set up. In other words, you have to manage by rearranging the scene instead. So, to undo the effect of (1) and bring the camera back to its original position, you apply the reverse of (1) to *both* camera and scene (so as not to alter the picture). The result, of course, is the same as applying just (2) in the first place. See the two diagrams in the big box on the lower left of Figure 4.45.

Looking through the viewfinder again, you feel it'll be a nicer composition if your friend stands not at the center of the frame but to a side. Again, (1) you can rotate the camera, say, 45° *clockwise*, or (2) your friend can sidle 45° *counter-clockwise* along a circle centered where you are, as in the top right of Figure 4.45. The picture is exactly the same in either case. And, again, one can imagine arriving at (2) by first applying (1), and then undoing it by applying the reverse of (1) to both camera and scene, as the two diagrams in the big box on the lower right of Figure 4.45 indicate.

It should now be fairly straightforward understanding the equivalence of a viewing transformation to a sequence of modeling transformations. The transformation

`gluLookAt(`*eyex*, *eyey*, *eyez*, *centerx*, *centery*, *centerz*, *upx*, *upy*, *upz*`)`

asks that the camera be (i) first translated to the position (*eyex*, *eyey*, *eyez*), then, (ii) rotated at that position till it's pointing at (*centerx*, *centery*, *centerz*) and, finally, (iii) rotated about its line of sight till its up direction is parallel to the vector up_2, the component of (*upx*, *upy*, *upz*) perpendicular to the line of sight.

Let's move the camera as asked. Figure 4.46(a) shows the resulting configuration. Next – it's the owner yelling again – we'll restore the camera to its default pose by incrementally undoing its movements, moving instead the scene as in the preceding thought experiment. The sum total, then, of these reverse movements to bring the camera back to default will be equivalent to the viewing transform.

The first translation is undone by applying `glTranslatef(`$-eyex$, $-eyey$, $-eyez$`)`. The camera is then at the origin, but still pointing parallel to the line of sight vector

$$los = (centerx, centery, centerz) - (eyex, eyey, eyez)$$

and with its top still parallel to up_2. See Figure 4.46(b).

Suppose that p is a plane that contains both los and the z-axis – shaded in the figure. If los does not lie along the z-axis then p is unique; if it

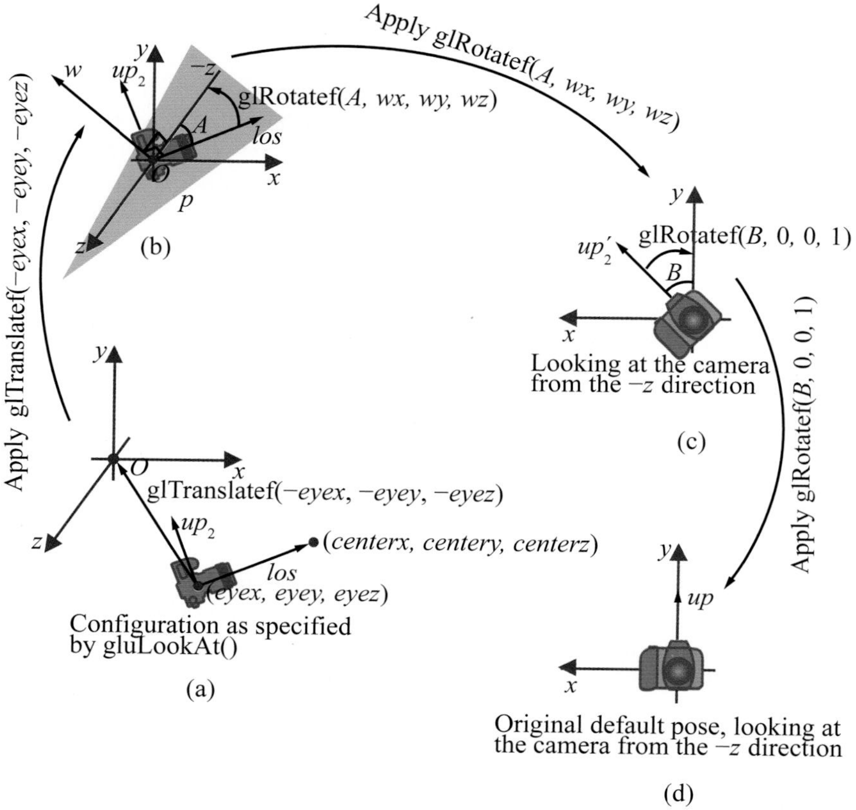

Figure 4.46: Restoring the camera to its default pose: broken arrows indicate movements which applied take the camera to the next configuration in the sequence (a)-(d).

does then p can be any plane that contains the common line. Choose a non-zero vector $w = (wx, wy, wz)$ perpendicular to p, i.e., w is perpendicular to both los and the z-axis. Let A be the angle from los to $-z$ on the plane p measured counter-clockwise when looking down from w.

Applying `glRotatef(`A, wx, wy, wz`)` then rotates the camera till its line of sight matches the $-z$ direction. Moreover, its top then is parallel to the vector, call it up_2', which is the result of `glRotatef(`A, wx, wy, wz`)` applied to up_2. Now, up_2', the new top direction, is perpendicular to the new line of sight down $-z$, because both were obtained by the same rotation `glRotatef(`A, wx, wy, wz`)` applied to perpendicular vectors up_2 and los. Therefore, up_2' lies on the xy-plane. See Figure 4.46(c) where the camera is seen from the negative side of the z-axis.

Finally, all that remains to restore the camera to its default position is a rotation `glRotatef(`$B, 0.0, 0.0, 1.0$`)`, of angle B about the z-axis, to align its top along $+y$. See Figure 4.46(d).

We conclude that the viewing transformation

gluLookAt($eyex$, $eyey$, $eyez$, $centerx$, $centery$, $centerz$, upx, upy, upz)

is, indeed, equivalent to a sequence of modeling transformations, in particular, a translation followed by two rotations:

glRotatef(B, 0.0, 0.0, 1.0);
glRotatef(A, wx, wy, wz);
glTranslatef($-eyex$, $-eyey$, $-eyez$);

We do not attempt to express the parameters A, B, wx, wy and wz in terms of the parameters $eyex$, $eyey$, ..., upz of the `gluLookAt()` command. Generally, this would be a tedious computation, but for simple settings of the camera it is not, as the experiment next shows.

Experiment 4.30. Replace the display routine of `box.cpp` with (Block 21):

```
void drawScene(void)
{
   glClear(GL_COLOR_BUFFER_BIT);
   glColor3f(0.0, 0.0, 0.0);

   glLoadIdentity();

   // Viewing transformation.
   gluLookAt(0.0, 0.0, 15.0, 15.0, 0.0, 0.0, 0.0, 1.0, 0.0);

   // Modeling transformation block equivalent
   // to the preceding viewing transformation.
   // glRotatef(45.0, 0.0, 1.0, 0.0);
   // glTranslatef(0.0, 0.0, -15.0);

   glutWireCube(5.0);

   glFlush();
}
```

Run. Next, both comment out the viewing transformation and uncomment the modeling transformation block following it. Run again. The displayed output, shown in Figure 4.47, is the same in both cases. The reason, as Figures 4.48(a)-(c) explain, is that the viewing transformation is equivalent to the modeling transformation block. In particular, the former is undone by the latter. **End**

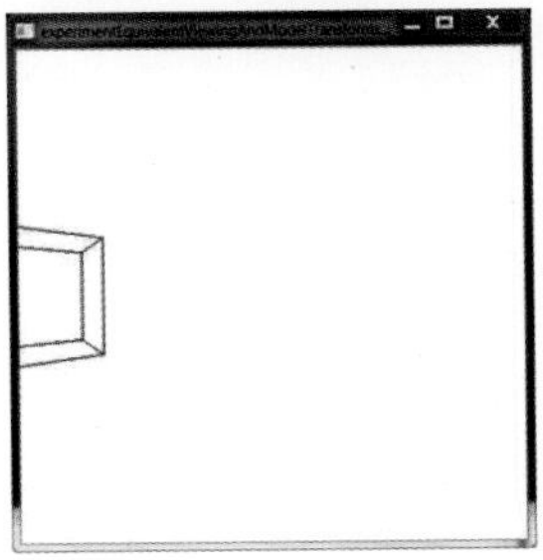

Figure 4.47: Screenshot from Experiment 4.30.

Exercise 4.47. (Programming) Replace the display routine of `box.cpp` with (Block 22):

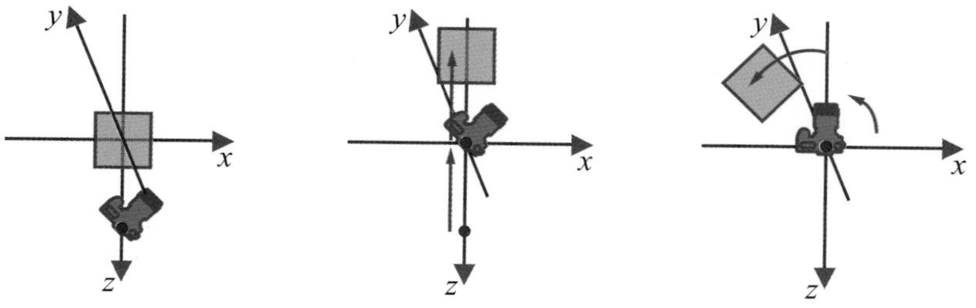

gluLookAt(0.0, 0.0, 15.0, 15.0, 0.0, 0.0, 0.0, 1.0, 0.0);
(a)

glTranslatef(0.0, 0.0, −15.0);
(b)

glRotatef(45.0, 0.0, 1.0, 0.0);
glTranslatef(0.0, 0.0, −15.0);
(c)

Figure 4.48: Viewing transformation equivalent to a sequence of modeling transformations.

```
void drawScene(void)
{
   glClear(GL_COLOR_BUFFER_BIT);
   glColor3f(0.0, 0.0, 0.0);

   glLoadIdentity();

   // Viewing transformation.
   gluLookAt(-30.0, 0.0, 30.0, 0.0, 0.0, 0.0, 0.0, 1.0, 0.0);

   // Modeling transformation block equivalent
   // to the preceding viewing transformation.
   // glRotatef(45.0, 0.0, 1.0, 0.0);
   // glTranslatef(30.0, 0.0, -30.0);

   glutWireCone(3.0, 10.0, 20, 20);

   glFlush();
}
```

Draw diagrams as in Figure 4.48 to show the equivalence of the viewing transformation and the modeling transformation block following it.

Exercise 4.48. Show that the viewing transformation

```
gluLookAt(30.0, 0.0, 30.0, 0.0, 0.0, 0.0, 1.0, 0.0, -1.0);
```

is equivalent to the sequence

```
glRotatef(90, 0.0, 0.0, 1.0);
glRotatef(45, 0.0, -1.0, 0.0);
glTranslatef(-30.0, 0.0, -30.0);
```

of modeling transformations. Pay particular attention to the alignment of the top of the camera.

Exercise 4.49. The sequence of modeling transformations equivalent to a given viewing transformation is not unique. In fact, for the preceding exercise find a sequence of modeling transformations, different from the one given, yet equivalent to the viewing transformation there.

Exercise 4.50. What sequence of modeling transformations is equivalent to each of the following viewing transformations:

(a) `gluLookAt(0.0, 0.0, 0.0, -15.0, 0.0, 15.0, 0.0, 1.0, 0.0)`

(b) `gluLookAt(0.0, 0.0, 15.0, 0.0, 0.0, 0.0, 1.0, 0.0, 0.0)`

(c) `gluLookAt(0.0, 0.0, 15.0, -15.0, 0.0, 0.0, 1.0, 0.0, -1.0)`

(d) `gluLookAt(0.0, 0.0, 15.0, 0.0, 1.0, 14.0, 0.0, 1.0, 0.0)`

(e) `gluLookAt(0.0, 0.0, 15.0, 0.0, 1.0, 14.0, 1.0, 0.0, 0.0)`

Part answer:

(c) One solution:

```
glRotatef(90, 0.0, 0.0, 1.0);
glRotatef(-45, 0.0, 1.0, 0.0);
glTranslatef(0.0, 0.0, -15.0);
```

Exercise 4.51. What is the viewing transformation equivalent to the following sequence of modeling transformations:

```
glRotatef(45.0, 0.0, 1.0, 0.0);
glTranslatef(0.0, 0.0, -5.0);
```

Remark 4.13. It's invariably good programming practice for there to be at most a single viewing transformation in a program, which comes in the code before all modeling transformations; in other words, the viewing transformation is applied last. Logically this means that objects are drawn first and placed as desired with respect to each other using modeling transformations and, then, a `gluLookAt()` is applied *finally* to transport the *entire* scene together.

Exercise 4.52. A programmer writes the following near the top of his drawing routine:

```
gluLookAt(0.0, 10.0, 10.0, 0.0, 0.0, 0.0, 0.0, 1.0, 0.0);
gluLookAt(0.0, 0.0, 15.0, 0.0, 0.0, 0.0, 0.0, 1.0, 0.0);
```

As it is bad practice to have two `gluLookAt()` statements like this, replace them with one `gluLookAt()` having the same effect.

Remark 4.14. Toward the beginning of this section on the viewing transformation we asked the reader to assume that we had a viewing frustum defined by a `glFrustum()` statement, rather than a `glOrtho()`-defined viewing box, because a point camera is logically placed at the origin in case of the former, but it's not clear where to place it for the latter.

So what happens when one applies `gluLookAt()` in the drawing routine when the projection statement, in fact, is a `glOrtho()`? The answer, as the reader has probably already guessed, is that OpenGL simply replaces the viewing transformation with its corresponding sequence of modeling transformations *whatever* may be the projection statement. The imaginary camera we thought of as being manipulated by a viewing transformation is simply an intuitive gadget for us programmers; it has no place inside OpenGL!

Remark 4.15. We have been insistent that the viewing transformation `gluLookAt()`'s purported manipulation of the camera is simulated entirely by modeling transformations. Indeed, we showed in this section how this can be done. But, is this *really* what OpenGL does? For, it's plausible that OpenGL actually does move the viewing frustum, with the camera at its apex, as directed by a `gluLookAt()` call, rather than apply any modeling transformations. For example, is Figure 4.35(a) or (b) in Section 4.6.1 the "truth"?

Here's how to decide. Modeling transformations change the current modelview matrix at the top of the modelview matrix stack. The viewing frustum, on the other hand, is determined by the current projection matrix at the top of the projection matrix stack, which is altered, among others, by projection statements such as `glFrustum()`. So, a way to find out what really happens inside the OpenGL engine is to read both the current modelview and projection matrices, both before and after issuing a `gluLookAt()`, and see which changes.

So what does happen? Only the current modelview matrix changes! The current projection matrix remains at the value it had prior to the `gluLookAt()` call. Take this in good faith now – you'll be able to verify the claim when we learn to access the modelview and projection matrix stacks in Chapter 5. In fact, we'll see then that the modelview matrix changes exactly as if multiplied on the right by the matrices corresponding to a sequence of modeling transformations equivalent to the given viewing transformation.

Remark 4.16. An interesting upshot of all this is that viewing transformations are not really needed, as any such transformation can always be manufactured from modeling transformations! Later versions of OpenGL, as we shall see, take this to heart.

4.6.3 Orientation and Euler Angles

This section may be skipped on a first reading. You will need it, though, before Section 6.3 about animating orientation with the help of Euler angles.

The viewing transformation leads nicely to a method of specifying the orientation of a camera. Recall the conclusion in Section 4.6.2 that the viewing transformation

```
gluLookAt(eyex, eyey, eyez, centerx, centery, centerz, upx, upy, upz)
```

is equivalent to a translation followed by two rotations:

```
glRotatef(B, 0.0, 0.0, 1.0);
glRotatef(A, wx, wy, wz);
glTranslatef(-eyex, -eyey, -eyez);
```

The axis of the particular rotation `glRotatef(`A, wx, wy, wz`)` is variable and depends on the line of sight. It was chosen, in fact, perpendicular to both line of sight and the z-axis. It's possible, however, to find a translation followed by a sequence of rotations, each about a *fixed* axis, equivalent to the given viewing transformation. In particular, one can show that

```
gluLookAt(eyex, eyey, eyez, centerx, centery, centerz, upx, upy, upz)
```

is equivalent to:

```
glRotatef(-γ, 0.0, 0.0, 1.0);
glRotatef(-β, 0.0, 1.0, 0.0);
glRotatef(-α, 1.0, 0.0, 0.0);
glTranslatef(-eyex, -eyey, -eyez);
```

where rotations are each about a coordinate axis, for suitable angles α, β and γ (the minus signs are for simpler notation later on). Here's how.

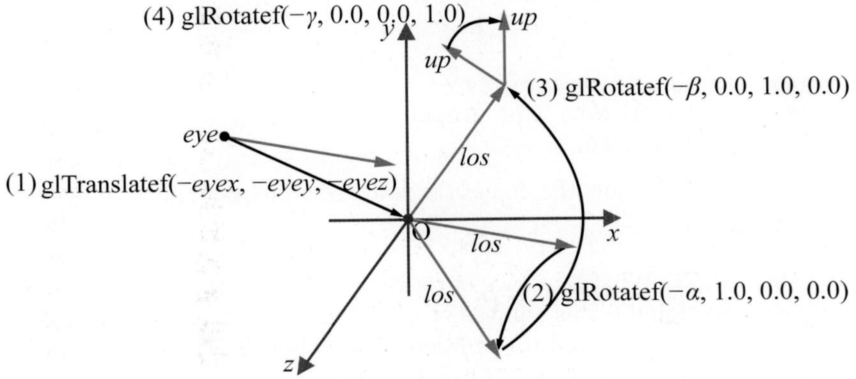

Figure 4.49: Applying a translation (1) and rotations (2)-(4) about the three coordinate axes to bring the camera back to its default pose. The original line of sight is bold. The up direction is shown only at the end.

Figure 4.49 – an all-in-one version of Figure 4.46 – shows that the sequence of four transformations below restores the camera to its default pose from the one specified by gluLookAt(*eyex*, *eyey*, *eyez*, *centerx*, *centery*, *centerz*, *upx*, *upy*, *upz*)

so, indeed, they are equivalent.
(*Note*: Figure 4.49 looks busy but it's not hard to read. The best way is to start with the bold vector indicating the camera's initial configuration and follow the sequence (1)-(4) of transformations one by one.)

(1) glTranslatef($-eyex$, $-eyey$, $-eyez$), to bring the eye to the origin.

(2) glRotatef($-\alpha, 1.0, 0.0, 0.0$), $-\alpha$ chosen to rotate the *los* about the x-axis till it lies on the xz-plane.

(3) glRotatef($-\beta, 0.0, 1.0, 0.0$), $-\beta$ chosen to rotate the *los* about the y-axis till it points down the $-z$ direction.

(4) glRotatef($-\gamma, 0.0, 0.0, 1.0$), $-\gamma$ chosen to rotate the camera about its *los* (pointing down the z-axis) till its top is aligned in the $+y$ direction.

Remark 4.17. Evidently, from the above, we can reduce the number of parameters required to specify camera movement from the nine of gluLookAt() to only six: α, β, γ, *eyex*, *eyey* and *eyez*. This indicates redundancy in the construction of gluLookAt(), but it has the virtue of being intuitive to use.

Example 4.6. Express the viewing transformation

```
gluLookAt(0.0, 0.0, 0.0, 1.0, 1.0, 0.0, -1.0, 1.0, 0.0);
```

as a sequence of rotations about the coordinate axes (no translation is needed as the eye is already at the origin).

Answer:

```
glRotatef(90.0, 0.0, 0.0, 1.0);
glRotatef(135.0, 0.0, 1.0, 0.0);
glRotatef(90.0, 1.0, 0.0, 0.0);
```

See Figure 4.50 for how the camera is restored to its default pose by these three rotations.

Exercise 4.53. What sequence of rotations would have been found by the method of Section 4.6.2 as equivalent to the viewing transformation of the preceding example? Would they all have been about the coordinate axes?

If its first three parameters (*eyex*, *eyey*, *eyez*) = (0, 0, 0), then a gluLookAt()'s translational component is zero, so that it alters only a camera's orientation, or pose. From the preceding discussion, such a gluLookAt() call is equivalent to a sequence

```
glRotatef(−γ, 0.0, 0.0, 1.0);
glRotatef(−β, 0.0, 1.0, 0.0);
glRotatef(−α, 1.0, 0.0, 0.0);
```

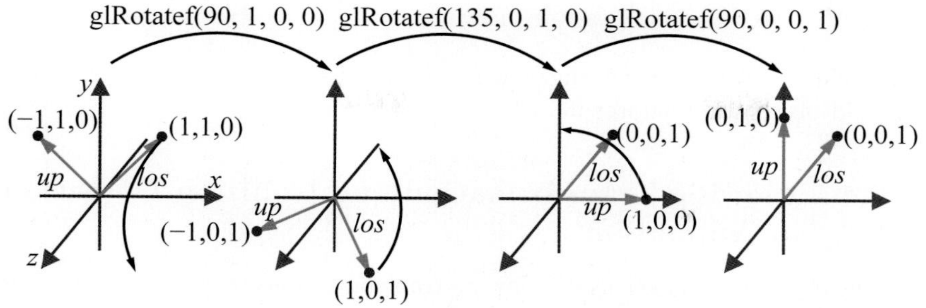

Figure 4.50: Solution to Example 4.6: the configuration of the camera given by `gluLookAt(0.0, 0.0, 0.0, 1.0, 1.0, 0.0, -1.0, 1.0, 0.0)` is at left; the line of sight and up vectors are indicated by blue arrows; rotations are both annotated at the top and indicated in the figures themselves by broken arrows, the result of each rotation being the *next* configuration.

because this sequence restores the camera *to* its default pose. In the opposite direction, therefore, the orientation of the camera resulting from this particular `gluLookAt()` call is obtained *from* its default pose by applying the inverse of the above sequence, viz.

```
glRotatef(α, 1.0, 0.0, 0.0);
glRotatef(β, 0.0, 1.0, 0.0);
glRotatef(γ, 0.0, 0.0, 1.0);
```

The angles α, β and γ are called the camera's *Euler angles*.

So, we see that Euler angles determine the camera's orientation: specifically, if they are α, β and γ, then the camera's orientation is obtained by applying first `glRotatef(`$\gamma, 0.0, 0.0, 1.0$`)`, then `glRotatef(`$\beta, 0.0, 1.0, 0.0$`)` and, finally, `glRotatef(`$\alpha, 1.0, 0.0, 0.0$`)` to its default pose.

Euler angles are not unique. For example, it's clear in Figure 4.49 that `glRotatef(`$-\alpha \pm 180°, 1.0, 0.0, 0.0$`)` could have been applied in Step (2), instead of `glRotatef(`$-\alpha, 1.0, 0.0, 0.0$`)`, to still place the camera's *los* on the xz-plane. So the orientation given by Euler angles α, β and γ is the same as the ones given by $\alpha \pm 180°$, β' and γ', for some, possibly, new β' and γ'.

Exercise 4.54. What are the Euler angles of a camera

(a) at the origin pointing at $(1, 0, 0)$?

(b) at the origin pointing at $(1, 1, 1)$?

(c) at $(1, 0, 0)$ pointing at $(1, 1, 1)$?

Note: In each case assume the *up* vector to be $(0, 1, 0)$. To determine the Euler angles of a camera not at the origin simply translate it first to the origin.

Part answer:

(a) $0°$, $-90°$, $0°$ (one possible answer)

We'll see more of Euler angles when we discuss animating the orientation of rigid objects in Chapter 6.

4.6.4 Viewing Transformation and Collision Detection in Animation

Our next program makes use of viewing transformations to simulate a moving camera in an animated environment. It also has another aspect of interest, particularly to those programming interactive applications such as games, namely, collision detection.

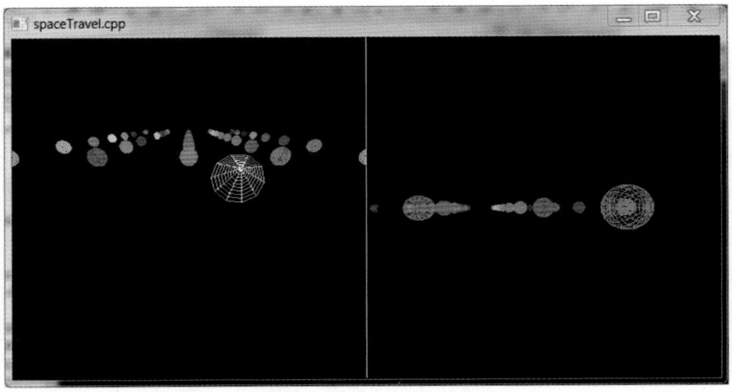

Figure 4.51: Screenshot of `spaceTravel.cpp`.

Experiment 4.31. Run `spaceTravel.cpp`. The left viewport shows a global view from a fixed camera of a conical spacecraft and 40 stationary spherical asteroids arranged in a 5×8 grid. The right viewport shows the view from a front-facing camera attached to the tip of the craft. See Figure 4.51 for a screenshot of the program.

Press the up and down arrow keys to move the craft forward and backward and the left and right arrow keys to turn it. Approximate collision detection is implemented to prevent the craft from crashing into an asteroid.

The asteroid grid can be changed in size by redefining `ROWS` and `COLUMNS`. The probability that a particular row-column slot is filled is specified as a percentage by `FILL_PROBABILITY` – a value less than 100 leads to a non-uniform distribution of asteroids. **End**

We'll discuss next the two most interesting aspects of `spaceTravel.cpp`: (a) the viewing transformation that defines the scene in the right viewport and (b) collision detection.

Viewing Transformation

The shape of the craft is defined by the `glutWireCone(5.0, 10.0, 10, 10)` statement; precisely, it is a cone of base radius 5 and height 10. The configuration of the spacecraft is specified by the values of $xVal$, $zVal$ and $angle$, all three global variables of `spaceTravel.cpp`. Figure 4.52(a) is a generic configuration in section along the xz-plane. The coordinates of the center of the craft's base are $(xVal, 0, zVal)$, while the angle its axis makes with the negative z-direction is $angle$. The middle A of the craft's axis will be of use in collision detection.

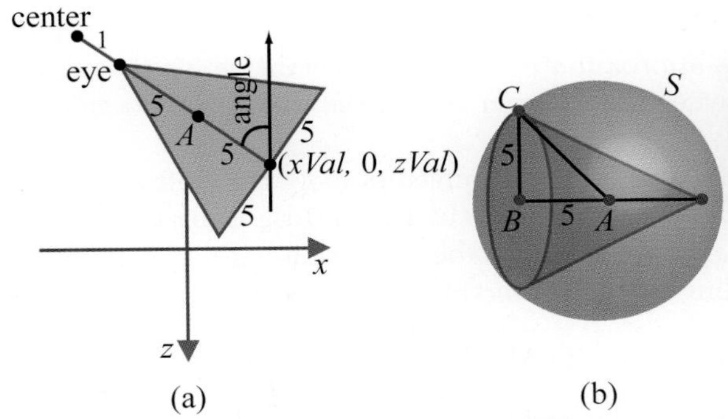

Figure 4.52: Spacecraft diagrams.

The camera for the right viewport is situated at the tip of the craft pointing straight ahead. It's straightforward trigonometry, now, to calculate the coordinates of *eye*, i.e., the tip of the craft, and of an imaginary point *center* to which it points, located 1 unit ahead of the tip:

$$\begin{aligned} eye &= (\, xVal - 10\sin(angle),\ 0,\ zVal - 10\cos(angle)\,) \\ center &= (\, xVal - 11\sin(angle),\ 0,\ zVal - 11\cos(angle)\,) \end{aligned}$$

These equations for *eye* and *center* explain the parameters of the `gluLookAt()` command for the right viewport.

Collision Detection

Collision detection as implemented in `spaceTravel.cpp` is simple though approximate. The spacecraft is enclosed in an imaginary bounding sphere S centered at the middle A of the cone's axis, with radius equal to the distance $|AC|$ from A to a point C on the boundary of its base. See Figure 4.52(b).

If B is the center of the base, then it follows from the dimensions of the cone that $|AB| = |BC| = 5$; therefore,

$$|AC| = \sqrt{|AB|^2 + |BC|^2} = \sqrt{50} = 7.071\ldots$$

Accordingly, we specify the radius of S to be 7.072 (slightly larger, in fact, than $|AC|$). The coordinates of the center A of S are obtained by trigonometry from Figure 4.52(a):

$$A = (\,xVal - 5\sin(angle),\, 0,\, zVal - 5\cos(angle)\,)$$

To detect collision between the spacecraft and an asteroid T, we detect instead collision between the craft's bounding sphere S and T. It's easy to determine if there is a collision between the two spheres S and T: compare the distance d between their centers with the sum $r_1 + r_2$ of their radii; there is collision if $d \le r_1 + r_2$ (e.g., as in Figure 4.52(c)), and not otherwise. This check is implemented in the routine `checkSphereCollision()`. This collision-detection test is approximate, in fact, conservative, as the craft's bounding sphere may intersect an asteroid even if the craft itself doesn't (as shown in Figure 4.52(c)).

The up and down arrow keys are programmed to move the craft a distance of 1 in either direction along its axis, and the left and right arrows to turn the craft an angle of 5°, *only if* there'll not be a collision with an asteroid in the new position (according to the conservative test above).

Exercise 4.55. (Programming) Modify `spaceTravel.cpp` as follows:

(a) Make the left viewport the view from the front of the spacecraft (currently, it is the right viewport).

(b) Make one of the asteroids the "big golden asteroid" by drawing it larger than the others and painting it suitably. Make it glow as well by oscillating the intensity of its color.

(c) Place a camera on the golden asteroid whose location is fixed but which rotates to track the spacecraft, i.e., its direction is pointed always toward the craft. Attach a tall antenna to the craft so that, even if it's obscured by other asteroids, at least the antenna will be visible from the big golden asteroid. Show the view from the golden asteroid's camera in the right viewport.

(d) When the spacecraft reaches the big golden asteroid, flash the text "You have found gold!".

Exercise 4.56. (Programming) Modify `spaceTravel.cpp` as follows:

(a) All the asteroids are currently colored spheres. Make them more interesting by using a few different FreeGLUT objects, e.g., cube, tetrahedron, octahedron, etc. You can also combine more than one object, e.g., one sphere on top of another, or design your own.

(b) Currently, the spacecraft moves interactively. Change this to program an automated tour which takes a fixed but zig-zag path through the asteroids and returns to the start position. Plan a path so that the craft comes close to a few interesting asteroids, visible in the right viewport. Pressing space should start/stop the movement.

(c) Currently, the camera on the craft always points straight ahead. Program occasional rotation of the camera, e.g., when the craft passes a strange asteroid, pan the camera to keep it in view.

Exercise 4.57. (Programming) Place a camera on top of the rolling ball of Exercise 4.27, pointing always down the plane. This camera does *not* rotate with the ball, but stays always at the top, so its motion is entirely linear. (How would you even install such a camera in real life? Well, that is a great thing about CG: you are entirely free from real-life constraints!)

Place a box just beyond the bottom of the plane so that the ball's camera sees an approaching object. Place an additional fixed camera on the box pointing at the plane to observe the ball. See Figure 4.53. Give a split-screen view as in `spaceTravel.cpp`.

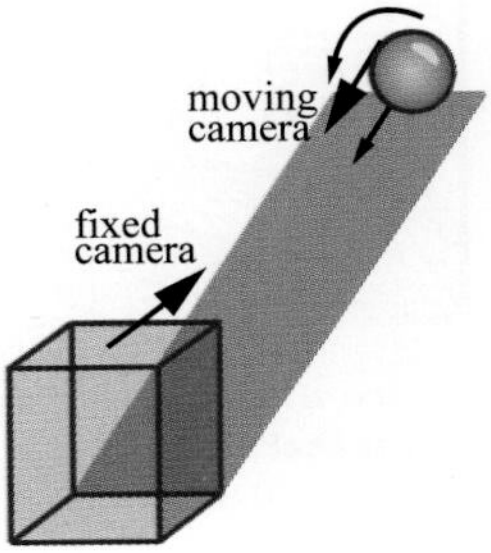

Figure 4.53: Ball rolling toward a box.

The following experiment is to whet your appetite for the topic of *frustum culling*, critical to the efficient rendering of complex scenes with large numbers of objects.

Experiment 4.32. Run `spaceTravel.cpp` with `ROWS` and `COLUMNS` both increased to 100. The spacecraft now begins to respond so slowly to key input that its movement seems clunky, unless, of course, you have a super-fast computer (in which case, increase the values of `ROWS` and `COLUMNS` even more). **End**

The reason for the degradation in the preceding experiment is that, every time an arrow key is pressed, OpenGL processes 10,000 asteroids, which is an enormous amount of computing. However, of these 10,000 only a few (about 100, or 1%) are ultimately rendered, as you can roughly count on the screen! The rest, of course, are outside the viewing frustum and clipped.

Unfortunately, by the time the decision to clip is made in the graphics pipeline, a large amount of computation has already been invested. Frustum culling is a technique to reduce this burden on OpenGL, whereby the programmer leverages her knowledge of the scene to pre-filter objects lying beyond the viewing frustum, not letting them into the pipeline at all.

We'll discuss frustum culling in detail in Section 6.1. There's really not much more by way of prerequisites needed to read that particular section though, so if you're anxious to learn this technique, which is so important in coding busy games and movies, feel free to jump right there.

We are not done yet with animation, though, and have a bunch more fun code for you.

4.7 More Animation Code

4.7.1 Animating an Articulated Figure

Our next project is a "studio" to develop animation sequences for an articulated figure.

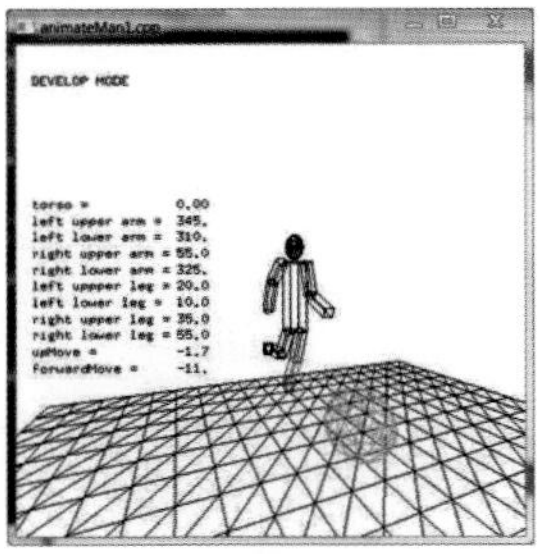

Figure 4.54: Screenshot of `animateMan1.cpp`.

Experiment 4.33. Run `animateMan1.cpp`. This is a fairly complex program to develop a sequence of key frames for a man-like figure, which can subsequently be animated. In addition to its spherical head, the figure consists of nine box-like body parts which can rotate about their joints. See Figure 4.54. All parts are wireframe. We'll explain the program next. **End**

It's advisable to learn to use the program before studying the code. There are two modes, develop and animate, and the program starts in the develop mode with the man facing you with his currently highlighted part, the torso, colored red. The rest of the body is black. Press the space bar to cycle through the man's movable parts, successively highlighting each. There are nine movable parts, all OpenGL wire cubes: the torso, the upper and lower arms on either side, and the upper and lower legs on either side.

Rotate the currently highlighted part by pressing the page-up and page-down keys. To move the man as a whole press the left/right and up/down arrow keys. The angles at which the 9 movable parts are currently rotated, as well as the vertical and horizontal translational components of the man as a whole, are shown as text data in the window in develop mode.

While arranging the man into a desired configuration, you can rotate your own viewpoint by pressing 'r/R', or zoom in and out pressing 'z/Z'.

Once the first configuration is completed to your satisfaction, press 'n'. This creates a new configuration which cannot be seen immediately as it's a copy of the previous one. Press, say, the right arrow key to separate the new configuration from the previous one. The (current) new configuration is bright, while the other(s) are ghosted. Again, use the space key to select a part, the page-up and page-down keys to rotate that part, and the arrow keys to move the entire configuration until it is arranged suitably.

Figure 4.55: Screenshot of `animateMan1.cpp` in develop mode.

Press 'n' to create new configurations until the key frames sequence is complete. Figure 4.55 shows a screenshot part way through the develop mode. You can edit the sequence at any time as follows.

Press the tab key to cycle through the sequence of configurations – the currently selected configuration is bright, while the rest ghosted. Press backspace to reset the currently selected configuration, delete to remove it altogether, or you can rearrange it using keys as already described.

When the key frames sequence is complete, pressing 'a' begins an animation which cycles through the programmer-created configurations. Pressing the up or down arrow keys speeds up or slows down the animation. Pressing 'a' again returns the program to develop mode.

Switching to animation mode also causes the program to write out to the file `animateManDataOut.txt` successive configurations of the animation sequence, stored currently in the vector manVector. Configuration are stored in successive lines of animateManDataOut.txt, each consisting of 11 floating point values – partAngles[0]-[8], upMove and forwardMove – the same as are displayed on the screen in develop mode.

Experiment 4.34. Run animateMan2.cpp. This is simply a pared-down version of animateMan1.cpp, whose purpose is to animate the sequence of configurations listed in the file animateManDataIn.txt, likely generated from the develop mode of animateMan1.cpp. Press 'a' to toggle between animation on/off. As in animateMan1.cpp, pressing the up or down arrow key speeds up or slows down the animation. The camera functionalities via the keys 'r/R' and 'z/Z' remain as well. Think of animateMan1.cpp as the studio and animateMan2.cpp as the movie.

The current contents of animateManDataIn.txt cause the man to do a handspring over the ball. Figure 4.56 is a screenshot. **End**

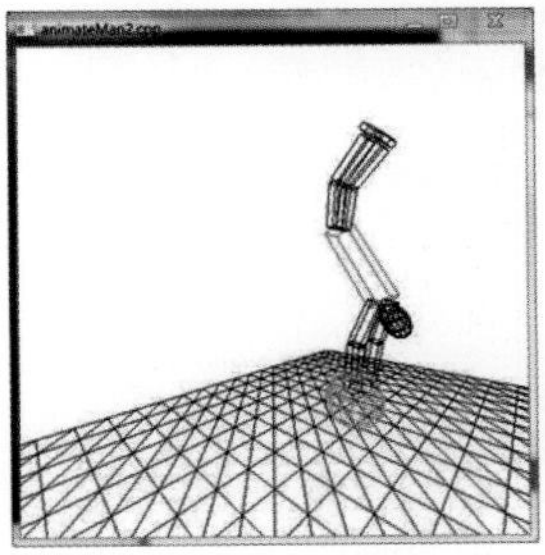

Figure 4.56: Screenshot of animateMan2.cpp.

Now let's look at the code of animateMan1.cpp. From an OpenGL point of view, most interesting possibly is the drawing of a configuration by the function Man::draw(). The best way to understand it is to analyze the successive placement of parts. We'll do this our usual way of deconstructing a program by first commenting out most of it and then restoring code piece by piece.

Accordingly, first comment out all parts except the torso as below:

```
// Function to draw man.
void Man::draw()
{
   if (highlight||animateMode) glColor3fv(highlightColor);
   else glColor3fv(lowlightColor);

   glPushMatrix();

   // Up and forward translations.
   glTranslatef(0.0, upMove, forwardMove);

   // Torso begin.
   if (highlight && !animateMode) if (selectedPart == 0)
      glColor3fv(partSelectColor);

   glRotatef(partAngles[0], 1.0, 0.0, 0.0);

   glPushMatrix();
   glScalef(4.0, 16.0, 4.0);
   glutWireCube(1.0);
   glPopMatrix();
   if (highlight && !animateMode) glColor3fv(highlightColor);
   // Torso end.
```

```
   /*
   // Head begin.
   .
   .
   .
   // Right upper and lower leg with foot end.
   */

   glPopMatrix();
}
```

Next, uncomment the head:

```
// Function to draw man.
void Man::draw()
{
   if (highlight||animateMode) glColor3fv(highlightColor);
   else glColor3fv(lowlightColor);

   glPushMatrix();

   // Up and forward translations.
   glTranslatef(0.0, upMove, forwardMove);

   // Torso begin.
   if (highlight && !animateMode) if (selectedPart == 0)
      glColor3fv(partSelectColor);

   glRotatef(partAngles[0], 1.0, 0.0, 0.0);

   glPushMatrix();
   glScalef(4.0, 16.0, 4.0);
   glutWireCube(1.0);
   glPopMatrix();
   if (highlight && !animateMode) glColor3fv(highlightColor);
   // Torso end.

   // Head begin.
   glPushMatrix();

   glTranslatef(0.0, 11.5, 0.0);
   glPushMatrix();
   glScalef(2.0, 3.0, 2.0);
   glutWireSphere(1.0, 10, 8);
   glPopMatrix();

   glPopMatrix();
   // Head end.

   /*
```

```
   // Left upper and lower arm begin.
   .
   .
   .
   // Right upper and lower leg with foot end.
   */

   glPopMatrix();
}
```

Continue – as you successively uncomment each body part, it'll be clear how it's being placed with respect to existing ones.

The creation of the `camera` as an object of the `Camera` class may be of interest as well and we'll leave the reader to relate the parameter values of the `gluLookAt()` command to the member variables `viewDirection` and `zoomDistance` of the `Camera` class.

Much of the rest of the code consists simply of managing and using `manVector`, which stores the sequence of configurations.

Remark 4.18. Even though he himself is 3D, the man moves and rotates his parts always parallel to the yz-plane, so he's not really capable of 3D motion!

Exercise 4.58. (Programming) Use `animateMan*.cpp` to animate a character kicking a football.

Exercise 4.59. (Programming) Enhance `animateMan*.cpp`:

(a) The character's body parts, except for the head, are currently all cubes. Make them more realistically rounded using cylinders.

(b) Add movement to the character's feet, which are currently fixed with respect to his lower legs. Give him movable hands as well.

(c) As remarked earlier, all the character's movements are currently parallel to a single plane. Enhance to true 3D.

Exercise 4.60. (Programming) Stick a camera to the front of the man's head and give a split-screen view of what he sees as he advances through an animation sequence and what is seen from a separate fixed camera focused on him.

Exercise 4.61. (Programming) By scaling individual body parts, create a second character who looks different from the first, though with identical functionality. Make a simple movie with the two.

It would be particularly effective in such a sequence to occasionally switch to a camera located in front of either one of their heads, to record how one sees the other.

Exercise 4.62. (Programming) Smoothly animating even a short movie requires several key frames (approximately 20 per second). However, the "important" ones are likely far fewer in number. For example, if a man kicks a ball, these are probably his wound-up pose ready to kick, the pose when his foot makes contact with the ball, a follow-through pose having kicked and, possibly, a few more in between to guide the sequence; certainly, far fewer than the 40 or so key frames needed for even a 2-second kicking sequence.

A movie-maker, therefore, saves a lot of tedious labor by simply drawing the important key frames, leaving an interpolating routine to fill in enough frames to make the animation smooth, a process called *tweening*.

Write a simple tweening routine based on `animateMan*.cpp`. In particular, use linear interpolation to fill configurations – each being an 11-vector of floats – in between successive programmer-created ones.

4.7.2 Simple Shadow Animation

When the scaling transformation was introduced at the beginning of this chapter we said that degenerate scalings have the occasional application. Here's one to create and animate a simple shadow.

Figure 4.57: Screenshot of `ballAndTorusShadowed.cpp`.

Experiment 4.35. Run `ballAndTorusShadowed.cpp`, based on `ballAndTorus.cpp`, but with additional shadows drawn on a checkered floor. Press space to start the ball traveling around the torus and the up and down arrow keys to change its speed. Figure 4.57 is a screenshot. **End**

There are parts of the program to make the picture look nice, e.g., lighting and material properties, which may not make sense currently, but neither are they relevant to drawing shadows, so ignore them for now.

Note, first, that the routine `drawFlyingBallAndTorus()` repositions the ball and torus from `ballAndTorus.cpp` horizontally so that their shadow, thrown supposedly by a distant overhead light source, falls on the floor. That the (imaginary) light source is vertically far above is important, as it justifies drawing the shadows as if cast by rays parallel to the y-axis. The actual drawing itself is quite simple – the following few lines in the drawing routine do the trick:

```
glPushMatrix();
glScalef(1.0, 0.0, 1.0);
drawFlyingBallAndTorus(1);
glPopMatrix();
```

The argument value 1 to `drawFlyingBallAndTorus()` causes both ball and torus to be drawn black, while the degenerate scaling command `glScalef(1.0, 0.0, 1.0)` collapses the y-values of all their vertices to 0, creating a flat black object which is precisely their shadow on the xz-plane from light rays parallel to the y-axis.

Remark 4.19. Since `ballAndTorusShadowed.cpp` evidently contains code to light the scene, you might think that OpenGL can compute shadows automatically. This is not the case: *OpenGL does not automatically compute secondary consequences of lighting such as shadows and reflection.* These have to be implemented separately by the programmer.

Remark 4.20. Note that the shadow of the ball on the torus is missing, even when it flies directly above. Our simple blacken-and-flatten method cannot draw shadows on curved surfaces. We'll learn a way to do this, however, later on in Section 18.2 on shadow mapping.

4.8 Selection and Picking

Strictly speaking, this section does not fit in a chapter about animation and viewing. However, countless animated applications ask the user to pick and move an object on the screen with a mouse or mouse-like device (shoot-em-up games come to mind). We thought it important, therefore, to explain how to implement such interactivity.

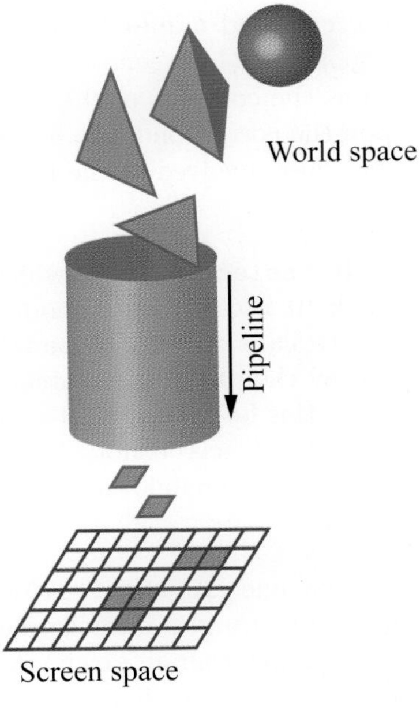

Figure 4.58: OpenGL's synthetic-camera pipeline (highly simplified!).

Unfortunately, picking an object on the screen – which, effectively, means deciding to which object a picked pixel belongs – is not a simple operation given how the synthetic-camera pipeline functions. Particularly, objects enter the pipeline, are processed and emerge each as a set of fragments (fragment = pixel + color values), which are then rendered to the screen. Figure 4.58 is a conceptual diagram.

The pipeline is not designed to be reversible, so there's no easy way to "climb back up" from screen space to world space. How then does one go about picking? Fortunately, OpenGL provides support for picking as well as a process it calls selection, which, in fact, enables picking. Let's begin with selection.

4.8.1 Selection

The idea underlying selection is simple. In a nutshell, it is to allow the user to specify a viewing volume and to then find the objects that intersect, or *hit*, this volume. To this end the user must first enter a rendering mode, called *selection mode*, by invoking `glRenderMode(GL_SELECT)`. In selection mode nothing is drawn to the frame buffer; rather, primitives are processed simply to determine their intersections with the specified viewing volume and generate so-called *hit records*.

To help determine from a hit record the primitive, or primitives, which produced it, OpenGL provides a so-called *name stack* which the user manipulates. The user can load names onto the name stack in a manner that establishes correspondence between primitives and names.

A hit record contains the contents of the name stack at the time of its creation so, based upon the correspondence between primitives and names, the user can determine those involved in the hit. Let's get to specifics with the help of live code.

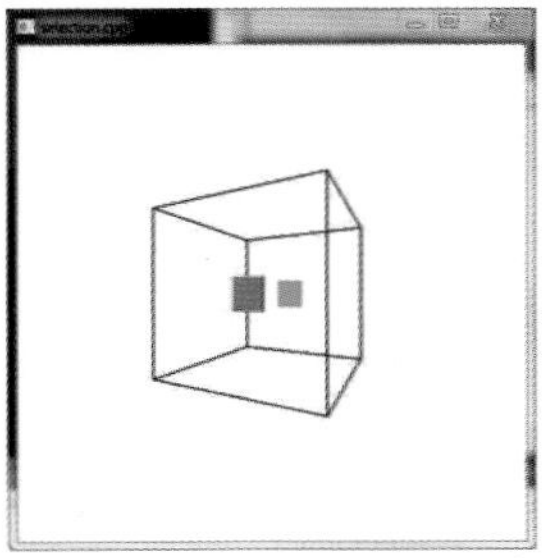

Figure 4.59: Screenshot from `selection.cpp`.

Experiment 4.36. Run `selection.cpp`, which is inspired by a similar program in the red book. It uses selection mode to determine the identity of rectangles, drawn with calls to `drawRectangle()`, which intersect the viewing volume created by the projection statement `glOrtho (-5.0, 5.0, -5.0, 5.0, -5.0, 5.0)`, this being a $10 \times 10 \times 10$ axis-aligned box centered at the origin. Figure 4.59 is a screenshot. Hit records are output to the command window. In the discussion following, we parse the program carefully. **End**

We'll call the viewing volume `glOrtho (-5.0, 5.0, -5.0, 5.0, -5.0, 5.0)`, used to "select" the rectangles intersecting it, the *selection volume*. Note that it is different from the program's own viewing volume defined by the `glFrustum(-5.0, 5.0, -5.0, 5.0, 5.0, 100.0)` call in the `resize()` routine.

Displayed by the `drawConfiguration` routine is the outline of the selection volume and two rectangles, one red and one green, both inside it. If you don't trust the perspective view of the scene in Figure 4.59, as probably you shouldn't, verify from the parameters of the `drawRectangle()` call that the two rectangles indeed lie inside the selection volume.

The `selectHits()` routine, which comes next in the code, is where all the action is. Let's step through it carefully. The first statement

```
glSelectBuffer(1024, buffer);
```

specifies the array, called the *hit buffer*, to store hit records, as well as its size. The next statement

```
glRenderMode(GL_SELECT);
```

makes OpenGL enter selection mode. The next block of statements

```
glMatrixMode(GL_PROJECTION);
glPushMatrix();
glLoadIdentity();
glOrtho(-5.0, 5.0, -5.0, 5.0, -5.0, 5.0);
glMatrixMode(GL_MODELVIEW);
glLoadIdentity();
```

causes the matrix mode to change to projection, the current projection matrix (i.e., the one defined in the `resize()` routine) to be saved, that corresponding to the selection volume for hit testing to be placed on top of the projection matrix stack and, finally, modelview matrix mode to be re-entered and the current modelview matrix set to identity.

The statement pair next, viz.

```
glInitNames();
glPushName(0);
```

initializes an empty name stack and pushes the name 0 on it (names are always non-negative integers). We'll not be using 0 to name any primitive, but push it on so that we have something to replace with "real" names when using `glLoadName()`. The initial configuration is depicted in Figure 4.60(a).

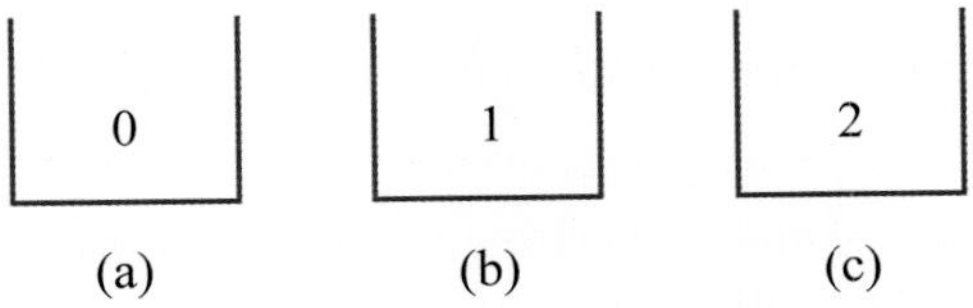

Figure 4.60: Name stack configurations: (a) Initial (b) When the red rectangle is drawn (c) When the green rectangle is drawn.

The following set of commands both manipulates the name stack and correspondingly "draws" primitives. Keep in mind that in selection mode nothing is *actually* drawn to the frame buffer, in other words, nothing is seen to happen.

```
glLoadName(1);
drawRectangle(0.0, 0.0, 3.0, 1.0, 0.0, 0.0); // Rect 1 (red).
```

```
glLoadName(2);
drawRectangle(0.0, 0.0, -3.0, 0.0, 1.0, 0.0); // Rect 2 (green).
```

Figures 4.60(b) and (c) depict the name stack as it is at the time of drawing of the first and second rectangles, respectively. The next statement

```
hits = glRenderMode(GL_RENDER);
```

takes OpenGL back to the default rendering mode where objects are indeed drawn to the frame buffer, at the same time returning the number of hit records currently in the hit buffer. Note that the return value of `glRenderMode()` has meaning only when transiting out of selection mode or another mode called feedback, which we'll not use, and not when leaving rendering mode. Finally,

```
glMatrixMode(GL_PROJECTION);
glPopMatrix();
glMatrixMode(GL_MODELVIEW);
```

restore the projection matrix from the `resize()` routine and return OpenGL to modelview matrix mode.

As `selectHits()` was being executed, hit records were written into the hit buffer following rules we'll describe next. A hit record is written into the hit buffer when *both* of the following conditions hold:

(a) A name stack manipulation or `glRenderMode()` command is encountered, *and*

(b) a hit has occurred (i.e., a primitive drawn that intersects the selection volume) since the previous instance of such a command.

Each hit record contains four fields in the following order:

1. The number of names in the name stack at the time of writing the record.

2. The minimum z-value of vertices belonging to primitives which have hit the selection volume since the last hit record was written. This value is normalized by dividing by the depth of the selection volume to a number in the range $[0, 1]$, which is then multiplied by $2^{32} - 1$, rounded, and stored in the hit record as a 32-bit unsigned integer.

3. The maximum z-value of vertices belonging to primitives which have hit the selection volume since the last hit record was written, stored likewise.

4. The sequence of the names in the name stack at the time of writing the record with the bottom one first. (This sequence may be empty.)

It is the `processHitBuffer()` routine, called by `drawScene()`, which steps through the hit buffer, outputting its contents to the command window. Items 2 and 3 above, the minimum and maximum z-values of vertices, are normalized back to between 0 and 1 by dividing by $2^{32} - 1$.

There are two hit records, as you can see in the command window. The first one $(1, 0.2, 0.2, 1)$ was generated on processing the `glLoadName(2)` call because a hit (the red rectangle) occurred after the previous name stack manipulation command (`glLoadName(1)`). The contents of this record are easy to understand if one observes that the configuration of the name stack at the time of the record's creation was as in Figure 4.60(b); moreover, the depth of *all* vertices of the red rectangle from the front face of the viewing box is 2, which becomes $2/10 = 0.2$ when normalized by division by the box's depth.

The second hit record $(1, 0.8, 0.8, 2)$ is generated on processing the

```
hits = glRenderMode(GL_RENDER);
```

statement, and we leave the reader to parse its contents.

The following exercises should, hopefully, completely clarify how hit records are generated.

Exercise 4.63. (Programming) Add one more rectangle, but in two different ways. First, insert the statement

```
drawRectangle(0.0, 0.0, 0.0, 0.0, 0.0, 1.0);
```

in the `selectHits()` routine (a) *just before* the `glPushName(0)` call (and after `glInitNames()`), and (b) *just after* `glPushName(0)`. What are the hit records generated in each case? When is each of these hit records generated?

Note: To *see* the new rectangle, make sure to add an identical drawing statement in the `drawConfiguration()` routine!

Part answer: In either case a new hit record comes before the two from the original program. When the statement is before `glPushName(0)`, the new record is $(0, 0.5, 0.5, \ )$ with an empty name list.

Exercise 4.64. (Programming) Restore the original `selection.cpp` program and insert the rectangle-drawing statement

```
drawRectangle(0.0, 0.0, 0.0, 0.0, 0.0, 1.0);
```

of the preceding exercise in the `selectHits()` routine, just after the statement that draws the first (red) rectangle. Explain the hit records, particularly, the z-values of the first one.

Exercise 4.65. (Programming) Restore the original `selection.cpp` program, but then change the command to draw the red rectangle to

```
drawRectangle(5.5, 0.0, 3.0, 1.0, 0.0, 0.0);
```

so that it intersects the selection volume without being entirely inside it. Observe that the hit records are still the same.

Exercise 4.66. (**Programming**) Restore the original `selection.cpp` program and insert the pair of name stack manipulation commands

```
glLoadName(3);
glLoadName(4);
```

right after the statement that draws the second (green) rectangle. The output is the same as for the unmodified `selection.cpp`. Why?

Exercise 4.67. (**Programming**) Restore the original `selection.cpp` program and add a new name stack manipulation command

```
glPushName(3);
```

between the `glLoadName(2)` call and the statement that draws the green rectangle. Predict the output before running.

Remark 4.21. The previous exercise shows a way of tagging an object with multiple names (in this case the green rectangle with 2 and 3) which is particularly useful in a scene where there is a hierarchy of objects. For example, we may want to tag the tail fin of the fourth aircraft with the names 4 and 7, if 7 is the part number of a tail fin.

Exercise 4.68. (**Programming**) The one remaining name stack manipulation command, which we have not used yet, is `glPopName()`, whose action the user can easily guess.

Insert a `glPopName()` statement in the `selectHits()` routine of `selection.cpp` in such a manner that the second hit record generated is $(0, 0.8, 0.8, \ )$.

4.8.2 Picking

Now that we have an understanding of the selection process, let's move on to picking, which is really selection plus a little help from OpenGL in setting up a selection volume to track a user-specified point on the screen.

Figure 4.61 illustrates the idea. V is the viewing frustum defined by the projection statement of a program. Objects are, therefore, drawn to the OpenGL window following perspective projection to the viewing face of V (we'll identify V's viewing face with the OpenGL window without harm, because going from one to the other is a simple scaling).

Accordingly, one can find objects picked from choosing a point P in the OpenGL window by determining those that intersect a long thin frustum like V' whose base is centered at P, because it's precisely these objects whose projections intersect P. Of course, there's some error depending on the size of V', the thinner being V' the more accurate the picking. And, as you might guess, it's in detecting intersection with V' that selection comes in.

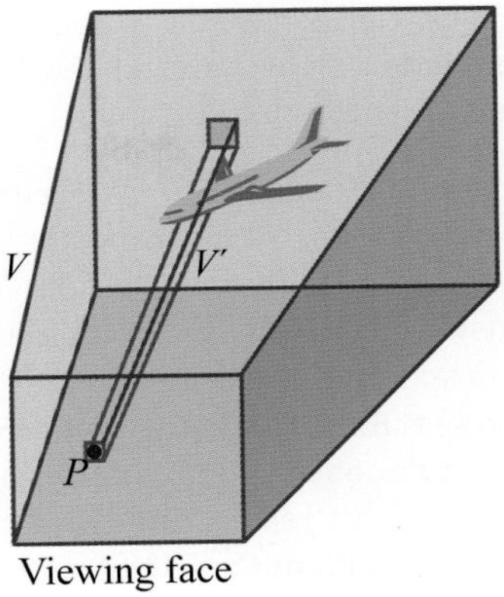

Figure 4.61: Clicking P "hits" the aircraft because the latter intersects V'.

In addition to the selection mechanism, there's even more help to be had from OpenGL: the GLU routine `gluPickMatrix()` defines a selection volume for use in picking, which is a frustum of user-specified size centered at a user-specified point. Here's how it works. The sequence of commands

```
glLoadIdentity();
gluPickMatrix(pickX, pickY, width, height, viewport[4]);
glFrustum(); or gluPerspective(); or glOrtho(); // Copied from the
                                                // reshape routine.
```

causes the top matrix of the projection matrix stack to be replaced by one corresponding to a selection volume whose front face is a *width* × *height* rectangle centered at the point of the OpenGL window with x and y world coordinate values equal to *pickX* and *pickY*, respectively. The *viewport[]* array supplies the current viewport boundaries and may be set by calling `glGetIntegerv(GL_VIEWPORT, ` *viewport*`)`. Functionally, the `gluPickMatrix()` command actually generates a matrix, called the *pick matrix*.

Let's get to work using the pick mechanism in a simple game-like application.

Experiment 4.37. Run `ballAndTorusPicking.cpp`, which preserves all the functionality of `ballAndTorus.cpp` upon which it is based and adds the capability of picking the ball or torus with a left click of the mouse. The picked object blushes. See Figure 4.62 for a screenshot. **End**

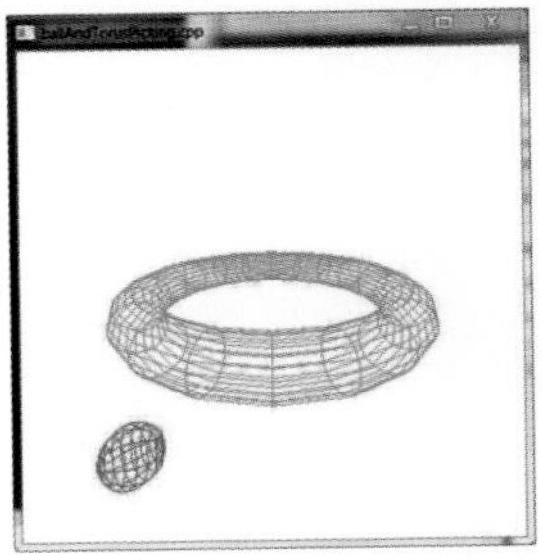

Figure 4.62: Screenshot of `ballAndTorus-Picking.cpp` moments after the ball has been clicked.

The `drawBallAndTorus()` routine of `ballAndTorusPicking.cpp` is pretty much the whole `drawScene()` routine of `ballAndTorus.cpp`, except with two main differences:

(a) In selection mode, `glLoadName()` is invoked to tag the torus with the name 1 and the ball with name 2.

(b) If one of the torus or ball is picked – the name being contained in the global `closestName` – it is painted red for as long as the global `highlightFrames` is greater than 0.

The mouse callback `pickFunction()` is written along the lines of `selectHits()` of `selection.cpp`. The important difference is that the selection volume for hits is specified with help of a `gluPickMatrix()` call. And, of course, instead of drawing rectangles as in `selection.cpp`, `drawBallAndTorus()` is executed in selection mode.

The routine `findClosestHit()` called by `pickFunction()` is an interesting modification of the `processHitBuffer()` routine of `selection.cpp`. In case there is more than one hit record, implying that both ball and torus fell under the mouse click, `findClosestHit()` compares their min-z fields to determine the one closer to the viewer.

Note: Sometimes an object doesn't light up on what seems like a definite click or the farther object lights up when both fall under the same mouse click. That's because the click fell between mesh wires! Possible solutions include making the meshes finer or the picking less sensitive by increasing the *width* and *height* parameters of `gluPickMatrix()` from the current values of 3 for both.

Picking plus dragging with mouse motion (see Section 3.6 for the latter) make a potent duo. Give it a go in the next exercise.

Exercise 4.69. (**Programming**) Enhance `canvas.cpp`, from the previous chapter, so that figures in the drawing area can be picked and moved.

Exercise 4.70. (**Programming**) Referring again to Exercise 4.19, where you added two more balls to `ballAndTorus.cpp`, now add the functionality of being able to pick any of the four objects *a la* `ballAndTorusPicking.cpp`.

Exercise 4.71. (**Programming**) Create a game, be it a shoot-em-up or drag-em-down or .... Use your imagination.

Picking by Color Coding

Yet another method to pick objects in OpenGL is by means of so-called *color coding*. We'll describe the idea briefly, but not go into detail, nor use it in a program. Picking by color coding requires use of the back buffer, so the program must run in double-buffered mode.

Here's how it works. When the user picks a pixel the entire scene is *redrawn* to the back buffer, but with objects of interest drawn in *different* colors. In other words, objects are color coded there. Next, data from the picked pixel is read from the back buffer with help of a `glReadPixels()` call and its color decoded to determine the picked object.

Figure 4.63 illustrates the idea – the disc and the rectangle are distinguishable in the back buffer by means of their color. So, e.g., if

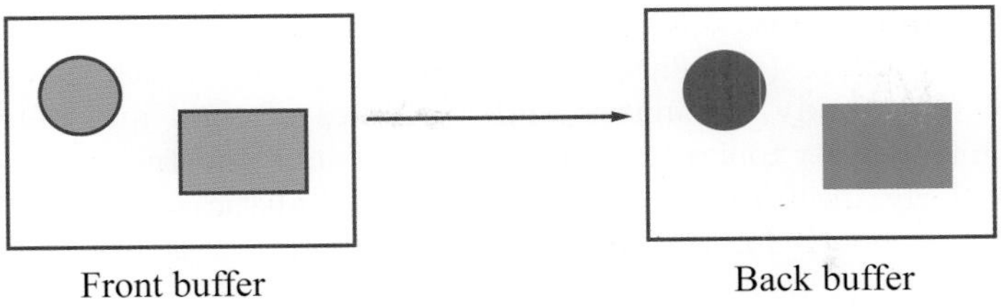

Figure 4.63: Color coding.

the picked pixel is red in the back buffer then the primitive picked is the disc.

4.9 Summary, Notes and More Reading

If animation is like a car, then we've just gotten our driver's license. In this chapter we learned the basics of the modeling and viewing transformations and how to use them to move objects and change their shape, as well as to manipulate the camera. We also peeked under the hood at OpenGL's engine, particularly in order to understand how transformations are composed and how they are used to place objects relative to one another. Collision detection, which is often crucial in interactive game programming, was discussed in the context of animation. We also began a discussion of orientation and Euler angles which will be continued in a more advanced chapter on animation. We learned as well the techniques of selection and picking, essential in interactive environments. And we saw plenty of live code along the way.

The topics covered in this chapter are at the very heart of computer graphics. Every introduction to the subject will have some coverage – see, for example, any of the introductory references, both OpenGL-based and API-independent, listed in Section 2.12 – differing perhaps in style and extent. It can only reinforce understanding to get more than one point of view, so the reader is encouraged to pick up other CG books which may be handy and turn to the relevant chapters. She will also find useful several of the on-line tutorials listed at the OpenGL site [103]. Particularly noteworthy is Nate Robins' finely-designed suite [100].

Collision detection is also to a greater or lesser extent covered in most introductory CG books, often in the context of ray tracing, which is a technique of rendering where light rays are followed from source to collision with an object (and possibly reflection again). For further reading about collision detection, however, the reader is well-advised to consult books on game programming, where it is especially important. See, e.g., Lengyel [85] and van Verth & Bishop [145]. Specialized books on collision detection include Ericson [43] and van den Bergen [144]. An extensive repository of resources on collision detection, including research papers and code, is at the UNC Gamma Research Group website [143].

This chapter is also a lead-in to the extremely important discipline of physics in graphics (popularly called game physics), which includes the study of multi-body kinematics and dynamics of rigid and deformable bodies, among other topics from real-world physics. Real-time game physics is particularly important in the creation of realistic interactive games. Introductory books for the interested reader include Bourg & Bywalec [20] and Eberly [40].

Undoubtedly, the best way for the reader to build on this chapter is to write lots and lots of animation code. In fact, this is a good time for her to begin coding, if she hasn't already done so, a significant project, e.g., a game or movie. She can get the essentials in place now and then embellish her project as we go along – with more complex objects, color, light and texture.

Chapter 5

Inside Animation: The Theory of Transformations

We studied transformations and their application to animation in Chapter 4. The goal for this chapter is to understand the underlying theory. We want to look under the hood of the graphics engine and understand exactly how transformations are implemented. What we'll encounter is the mathematics of geometric transformations.

We begin our discussion of geometric transformations in Section 5.1 in the simple surroundings of Flatland (2-dimensional space), the objective being to get concepts in place and prove results that will extend fairly easily to the real world. This program starts in Sections 5.1.1-5.1.4 with the expression of familiar geometric transformations, in particular, translations, scalings, rotations and reflections, by means of matrices.

Next, we briefly interrupt our pursuit of 2D geometric transforms to digress in Section 5.2 into linear algebra, particularly for an understanding of affine transformations. Affine transformations will provide a unifying perspective of all the geometric transformations that we encounter. In Sections 5.2.1-5.2.3 we define affine transformations as a generalization of linear transformations, prove that they are particularly pleasant in their geometric behavior, understand the central role they play in the design of a graphics API such as OpenGL and, finally, learn the use of homogeneous coordinates to facilitate the application of affine transformations.

We resume our study of 2D geometric transforms in Section 5.3 with our newfound knowledge of affine transformations. We begin in 5.3.1 by placing the transformations of Section 5.1 in context as affine geometric transformations. In 5.3.2 comes the notion of Euclidean transformations, and their subclass of rigid transformations, neither of which distorts the shape of an object. Consequently, these are the transformations to use to animate rigid objects. The exploration of 2D transformations concludes with a discussion of shears, a commonly occurring shape-distorting transformation, in 5.3.3.

Geometric transformations of the real world or 3-space – transformations that OpenGL actually implements – is the topic of Section 5.4. The development parallels that of the previous section on 2D transformations. Matrix expressions for translations, scalings and reflections generalize easily from their 2D counterparts in Sections 5.4.1-5.4.2 and 5.4.4. 3D rotations, however, require considerably more work in the longish 5.4.3.

Observing in 5.4.5 that translations, scalings and rotations about radial axes are fundamental affine transformations, in the sense that they can be used to generate all other affine transformations, lends insight into the design of a CG animation engine. We realize that, however exciting the game or movie is that we happen to be enjoying, most of what is going on inside the machine is the distinctly unglamorous activity of multiplying 4×4 matrices – lots and lots of them and very, very fast! We learn to access and manipulate the OpenGL modelview matrix stack in 5.4.6. Euclidean and rigid 3D transformations are discussed next in Section 5.4.7. The ability to access the modelview matrix stack comes in handy in 5.4.8 when we learn about 3D shears and how to manually code and insert their matrices into the stack.

We conclude in Section 5.5 with a summary, notes and suggestions for further reading.

Important: We assume for this chapter familiarity with basic linear algebra, as, say, would be found in a first college course or in a multitude of introductory texts, e.g., [8, 50, 73, 80, 84, 138] and others. Section 5.2 needs familiarity as well with the basics of convex sets, again from an introductory geometry text.*

In fact, you can safely defer this somewhat theoretical chapter to a second pass through the book.

Note to the reader about projection transformations: This chapter is devoted to the theory of modelview transformations. You will find a thorough coverage of projection transformations and their matrices in Chapter 18.

5.1 Geometric Transformations in 2-Space

We begin discussion of geometric transformations in 2D space, rather than real-life 3D, in order to develop our intuition in a simpler setting. Much of what we say and prove, though, will generalize fairly easily to 3D.

*The needed material on convexity can also be found in our own Chapter 7, which logically belongs to the group Chapters 7-9. Nevertheless, it's easily readable at this time and, in fact, you are strongly recommended to flip through at least Sections 7.1-7.2.

5.1.1 Translation

A translation is specified by a *displacement vector* $D = [d_x\ d_y]^T$, which is added to the location vector of each point. Precisely, the image of the point $P = [x\ y]^T$ by this translation is $P' = [x'\ y']^T$, where

$$\begin{bmatrix} x' \\ y' \end{bmatrix} = \begin{bmatrix} x \\ y \end{bmatrix} + \begin{bmatrix} d_x \\ d_y \end{bmatrix} = \begin{bmatrix} x + d_x \\ y + d_y \end{bmatrix} = \begin{bmatrix} 1 & 0 \\ 0 & 1 \end{bmatrix} \begin{bmatrix} x \\ y \end{bmatrix} + \begin{bmatrix} d_x \\ d_y \end{bmatrix}$$

See Figure 5.1. Concisely:

$$P' = \begin{bmatrix} 1 & 0 \\ 0 & 1 \end{bmatrix} P + D \qquad (5.1)$$

The matrix multiplication may seem redundant, but it serves to put the RHS in a general form which will soon prove useful.

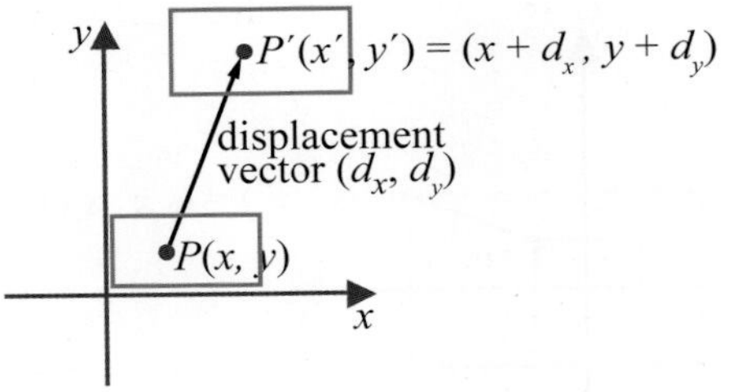

Figure 5.1: Translation.

Terminology: We'll use the coordinate notation (x, y) and the matrix notation $[x\ y]^T$ for a point interchangeably, particularly preferring the latter when we want to treat the point's location as a vector.

Exercise 5.1. Prove that the composition of translations is a translation and that the inverse of a translation is another translation.

Part answer: We'll prove that the composition of translations is again a translation the long way, using the matrix form of the translation equation, but it's good practice. So suppose the translations t_1 and t_2 are specified by the displacement vectors D_1 and D_2, respectively. Then

$$\begin{aligned} (t_1 \circ t_2)(P) &= t_1(t_2(P)) = t_1\left(\begin{bmatrix} 1 & 0 \\ 0 & 1 \end{bmatrix} P + D_2\right) \\ &= \begin{bmatrix} 1 & 0 \\ 0 & 1 \end{bmatrix}\left(\begin{bmatrix} 1 & 0 \\ 0 & 1 \end{bmatrix} P + D_2\right) + D_1 \\ &= \left(\begin{bmatrix} 1 & 0 \\ 0 & 1 \end{bmatrix} P + D_2\right) + D_1 = \begin{bmatrix} 1 & 0 \\ 0 & 1 \end{bmatrix} P + (D_2 + D_1) \end{aligned}$$

proving that $t_1 \circ t_2$ is indeed a translation, specified by the displacement $D_2 + D_1$.

5.1.2 Scaling

A scaling is specified by a scaling factor s_x along the x-axis and a scaling factor s_y along the y-axis. The image of a point P by this scaling is the one whose x coordinate value is s_x times that of P and y coordinate value s_y times that of P (see Figure 5.2). Precisely, the image of $P = [x\ y]^T$ is $P' = [x'\ y']^T$, where

$$\begin{bmatrix} x' \\ y' \end{bmatrix} = \begin{bmatrix} s_x x \\ s_y y \end{bmatrix} = \begin{bmatrix} s_x & 0 \\ 0 & s_y \end{bmatrix} \begin{bmatrix} x \\ y \end{bmatrix}$$

Concisely:

$$P' = \begin{bmatrix} s_x & 0 \\ 0 & s_y \end{bmatrix} P \tag{5.2}$$

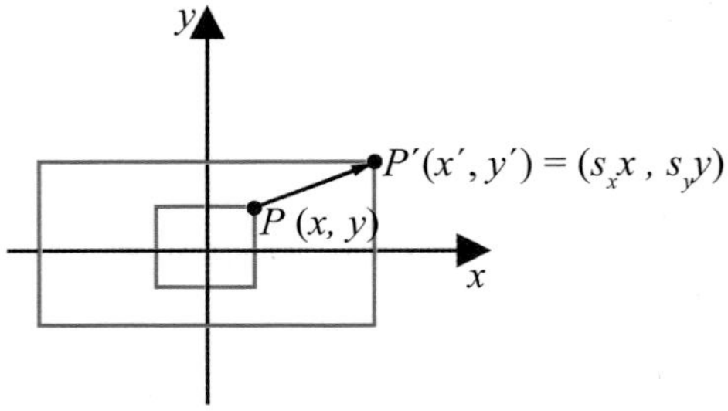

Figure 5.2: Scaling.

If either, or both, of the scaling factors s_x and s_y is zero, then the scaling is said to be *degenerate*; if neither is zero, it is *non-degenerate*. By a scaling we shall always mean a non-degenerate one, unless specifically stated otherwise.

Exercise 5.2. Show that the scaling given by Equation (5.2) is non-degenerate if and only if its matrix is non-singular (i.e., has non-zero determinant).

Exercise 5.3. Use Equation (5.2) to prove that the composition of scalings is a scaling and that the inverse of a non-degenerate scaling is another non-degenerate scaling. Are degenerate scalings invertible?

5.1.3 Rotation

A rotation about the origin is specified by an angle θ measured counter-clockwise as seen by a viewer V located on the positive side of the z-axis (in a hypothetical right-handed 3D coordinate system made by adding a z-axis to the x- and y-axes of the given 2D plane). See Figure 5.3(a).

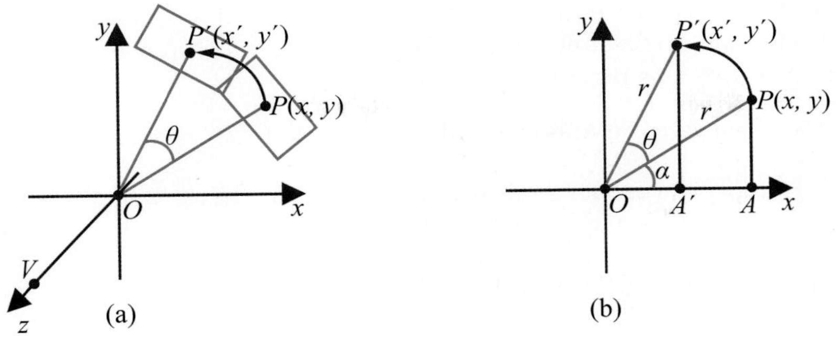

Figure 5.3: Rotation.

Note: We had to add the z-axis and place a viewer on a particular side of it because it's not enough to simply say that a rotation on the xy-plane is counter-clockwise: the same rotation appears counter-clockwise from one side and clockwise from the other.

In the future, to avoid tedious language, we'll always assume that a viewer is located at a point such as V, on the positive side of the z-axis.

We want to calculate the image $P' = [x' \ y']^T$ of the point $P = [x \ y]^T$ by this rotation. The method is exactly the same as the solution given for Exercise 4.8(c), in the case of a 3D rotation about the z-axis. The reader can refer again to that exercise or deduce herself the following equations from Figure 5.3(b):

$$\begin{aligned} x &= OA = r\cos\alpha \\ y &= PA = r\sin\alpha \end{aligned}$$

which are used in

$$\begin{aligned} x' &= OA' = r\cos(\alpha+\theta) = r\cos\alpha\cos\theta - r\sin\alpha\sin\theta \\ &= x\cos\theta - y\sin\theta \\ y' &= P'A' = r\sin(\alpha+\theta) = r\cos\alpha\sin\theta + r\sin\alpha\cos\theta \\ &= x\sin\theta + y\cos\theta \end{aligned}$$

Therefore, the image of $P = [x \ y]^T$ by a rotation of angle θ counter-clockwise about the origin is $P' = [x' \ y']^T$, where

$$\begin{bmatrix} x' \\ y' \end{bmatrix} = \begin{bmatrix} x\cos\theta - y\sin\theta \\ x\sin\theta + y\cos\theta \end{bmatrix} = \begin{bmatrix} \cos\theta & -\sin\theta \\ \sin\theta & \cos\theta \end{bmatrix} \begin{bmatrix} x \\ y \end{bmatrix} \tag{5.3}$$

or, concisely,

$$P' = \begin{bmatrix} \cos\theta & -\sin\theta \\ \sin\theta & \cos\theta \end{bmatrix} P \tag{5.4}$$

The matrix in the preceding equation is often called a *rotation matrix*.

Example 5.1. Write the matrix form as in (5.4) of a counter-clockwise rotation by an angle of 60° about the origin. To which point is $[1 \ -2]^T$ transformed by this particular rotation?

Answer: The given rotation will take $P = [x \ y]^T$ to $P' = [x' \ y']^T$, where

$$P' = \begin{bmatrix} \cos 60^\circ & -\sin 60^\circ \\ \sin 60^\circ & \cos 60^\circ \end{bmatrix} P = \begin{bmatrix} 1/2 & -\sqrt{3}/2 \\ \sqrt{3}/2 & 1/2 \end{bmatrix} P$$

Therefore, $[1 \ -2]^T$ is transformed to

$$\begin{bmatrix} 1/2 & -\sqrt{3}/2 \\ \sqrt{3}/2 & 1/2 \end{bmatrix} \begin{bmatrix} 1 \\ -2 \end{bmatrix} = \begin{bmatrix} 1/2 + \sqrt{3} \\ -1 + \sqrt{3}/2 \end{bmatrix} \simeq \begin{bmatrix} 2.23 \\ -0.13 \end{bmatrix}$$

Exercise 5.4. Is the matrix of a rotation about the origin always non-singular?

Exercise 5.5. Determine the matrix expression for a counter-clockwise rotation by an angle θ about an arbitrary point $O' = [a \ b]^T$, not necessarily the origin.

Suggested approach: Use the Trick of Example 4.2 to express this rotation as a composition of three successive transformations:

1. A translation by the displacement vector $[-a \ -b]^T$ taking O' to the origin.

2. A counter-clockwise rotation by θ about the origin.

3. A translation by the displacement vector $[a \ b]^T$ restoring O' to its original position.

Next, compose the expressions corresponding to these three transformations (make sure to do this in the right order). The result will be a transformation of the form $P \mapsto MP + D$.

Exercise 5.6. Determine the matrix expression for a rotation of 45° counter-clockwise about the point $[2 \ 3]^T$.

Exercise 5.7. Use Equation (5.4) to prove that the composition of rotations about the origin is another such and that so is the inverse of a rotation about the origin.

Exercise 5.8. How about the composition of rotations about some fixed point other than the origin? Is this again a rotation about that point?

Example 5.2. Is the composition of rotations about *different* points necessarily equivalent to a *single* rotation about some one point?

Answer: Consider rotations r_1 and r_2, both of 180°, about the two points $O_1 = [0\ 0]^T$ and $O_2 = [1\ 0]^T$, respectively. We'll show that $r_2 \circ r_1$ is not a rotation about any point, answering the question asked in the negative.

It's easy to check that $(r_2 \circ r_1)(O_1) = r_2(r_1(O_1)) = r_2(O_1) = [2\ 0]^T$, while $(r_2 \circ r_1)(O_2) = r_2(r_1(O_2)) = r_2([-1\ 0]^T) = [3\ 0]^T$. See Figure 5.4(a).

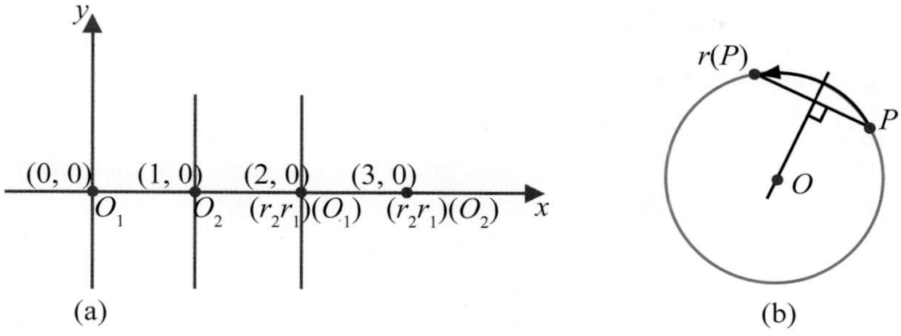

Figure 5.4: Illustrations for Example 5.2.

Next, observe that for any (non-identity) rotation r, and any point P which is not itself the center O of the rotation, O lies on the perpendicular bisector of the segment joining P and $r(P)$. Figure 5.4(b) indicates why.

Now, *if* $r_2 \circ r_1$ were indeed a rotation, its center, first, is not either O_1 or O_2, as both points are moved by $r_2 \circ r_1$. Therefore, its center must lie on the perpendicular bisector of the segment joining O_1 and $(r_2 \circ r_1)(O_1)$, as also on the perpendicular bisector of the segment joining O_2 and $(r_2 \circ r_1)(O_2)$. But this is not possible as the two bisectors are straight lines through $[1\ 0]^T$ and $[2\ 0]^T$, respectively, both parallel to the y-axis and, therefore, non-intersecting. One concludes that $r_2 \circ r_1$ is not a rotation about any point.

Exercise 5.9. The composition $r_2 \circ r_1$ of the preceding example, though not a rotation, is, nevertheless, a familiar kind of transformation. Can you identify it?
Hint: It's a translation!

Exercise 5.10. Can you give an example where the composition of two (non-trivial) rotations about different points *is* equivalent to a single rotation about some point?
Hint: Consider rotating 90° counter-clockwise around $[-1\ 1]^T$ and then around $[1\ 1]^T$ the same amount. Show that this composition is equivalent to a rotation of 180° around the origin.

Remark 5.1. From the preceding two exercises we see one case where the composition of two rotations is a translation and one where it is again a rotation. It turns out that these are the only two possibilities in general. We'll prove this fact in Section 5.3.2, in particular, when we classify rigid transformations, and see an easy rule as well to decide the nature of a composition of rotations.

5.1.4 Reflection

The image of the point $P = [x\ y]^T$ by a reflection about a straight line l, called the *mirror*, is $P' = [x'\ y']^T$ such that:

(a) if P lies on l, then $P' = P$;

(b) if P does not lie on l, then P' is the point on the other side of l such that PP' is perpendicular to l, and P' is the same distance from l as P. See Figure 5.5.

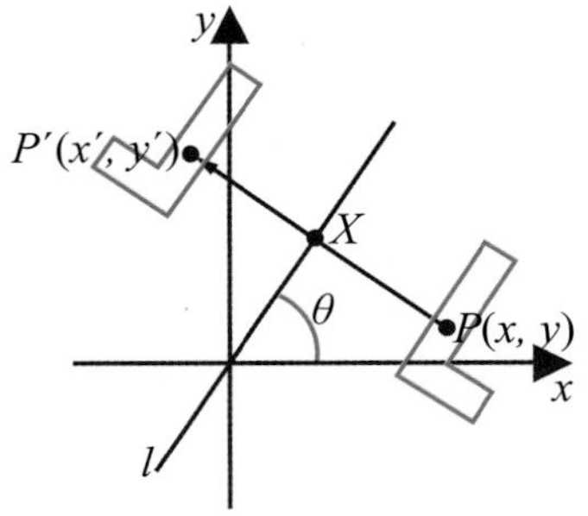

Figure 5.5: Reflection ($|XP| = |XP'|$).

A reflection is, therefore, specified by the mirror about which it occurs. Let's analyze first the reflection about a radial mirror l at an angle θ counter-clockwise from the positive direction of the x-axis (as depicted in Figure 5.5). We claim that reflection about l maps the point P to P' where

$$P' = \begin{bmatrix} \cos 2\theta & \sin 2\theta \\ \sin 2\theta & -\cos 2\theta \end{bmatrix} P \tag{5.5}$$

and leave the proof to the reader in the next exercise.

Note: A *radial* line or plane is one which passes through the origin.

Exercise 5.11. Verify Equation (5.5).

Suggested approach: Use the Trick to express this reflection as the composition of three successive transformations:

1. A rotation of $-\theta$ about the origin to align l along the x-axis.
2. A reflection about the x-axis. This is given by $[x\ y]^T \mapsto [x\ \ -y]^T$, which is simply scaling by a factor of 1 along the x-axis and -1 along the y-axis.

3. A rotation of θ about the origin to restore l to its original alignment.

Exercise 5.12. Write the matrix form, as in (5.5), of a reflection about the radial mirror at an angle of 30° to the positive x-axis. To which point is $[1\ 1]^T$ transformed by this reflection?

Exercise 5.13. What is the determinant of the matrix of a reflection about a radial mirror? Is the matrix always non-singular?

Exercise 5.14. Use the Trick to prove that a reflection about an arbitrary mirror, not necessarily radial, is a composition of two translations, two rotations about the origin and one scaling.

A consequence of the preceding exercise is that one of those highly-paid Flatland programmers developing a graphics API has only to implement translations, rotations about the origin and scalings to get reflections for free.

Exercise 5.15. What is the inverse of a reflection?

Exercise 5.16. Show that any non-identity translation can be obtained by composing reflections about two parallel mirrors. Show that any non-identity rotation can be obtained by composing reflections about two intersecting mirrors. (The identity transformation itself can be obtained obviously by composing reflections about the same mirror twice.)

Exercise 5.17. A reflection about a mirror l, followed by translation by a displacement vector (d_x, d_y) which is either zero or parallel to l, is called a *glide reflection*. See Figure 5.6.

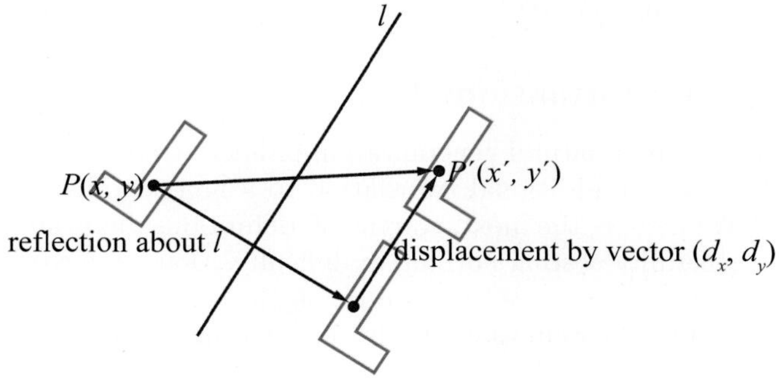

Figure 5.6: Glide reflection.

Determine the matrix expression for the glide reflection which uses the mirror l and displacement vector (d_x, d_y). Assume l goes through the point (a, b) and makes an angle θ measured counter-clockwise from the positive direction of the x-axis.

Exercise 5.18. Two transformations f_1 and f_2 are said to *commute* (or be *commutative*) if $f_1 \circ f_2 = f_2 \circ f_1$, in other words, if applying f_1 followed by f_2 is the same as applying f_2 followed by f_1.

(a) Do translations commute with each other?

(b) Do scalings commute with each other?

(c) Do rotations about the same point commute with each other?

(d) Does a rotation about one point commute with another about a different point?

Hint: A counter-example to show that, generally, rotations about different points don't commute can be obtained, in fact, from the configuration given in the answer to Example 5.2: consider if $(r_2 \circ r_1)(O_1)$ and $(r_1 \circ r_2)(O_1)$ are the same. If they are not, then, of course, $r_2 \circ r_1$ and $r_1 \circ r_2$ are not the same either.

(e) Do translations and rotations commute?

(f) Do reflections about two different mirrors *ever* commute?

Hint: Keep in mind the special case of perpendicular mirrors.

5.2 Affine Transformations

Before resuming our pursuit of geometric transformations in 2-space, we change pace a bit to learn about affine transformations because they will provide a unifying framework in which to locate the seemingly disparate geometric transformations that we have encountered (and will encounter).

5.2.1 Affine Transformations Defined

Affine transformations are a natural generalization of linear transformations, obtained by tacking on an additional translation to a non-singular linear transformation. We'll give the next couple of definitions in arbitrary dimensions – this generality costing nothing in difficulty. Down the road, of course, we can specialize to $\mathbb{R}^2$ or $\mathbb{R}^3$ depending on the setting.

For the record, here's the definition of a linear transformation, which is our starting point.

Definition 5.1. The *linear transformation* f^M of $\mathbb{R}^m$, with the $m \times m$ *defining matrix*

$$M = \begin{bmatrix} a_{11} & a_{12} & \dots & a_{1m} \\ a_{21} & a_{22} & \dots & a_{2m} \\ & \dots & \dots & \\ a_{m1} & a_{m2} & \dots & a_{mm} \end{bmatrix}$$

is the transformation $f^M : \mathbb{R}^m \to \mathbb{R}^m$ specified by the equation

$$f^M(P) = MP, \qquad P \in \mathbb{R}^m \tag{5.6}$$

In other words, f^M maps $P = [x_1\ x_2\ \ldots\ x_m]^T$ to $f^M(P) = [x'_1\ x'_2\ \ldots\ x'_m]^T$, where

$$\begin{aligned} x'_1 &= a_{11}x_1 + a_{12}x_2 + \ldots + a_{1m}x_m \\ x'_2 &= a_{21}x_1 + a_{22}x_2 + \ldots + a_{2m}x_m \\ &\ldots \quad \ldots \\ x'_m &= a_{m1}x_1 + a_{m2}x_2 + \ldots + a_{mm}x_m \end{aligned} \tag{5.7}$$

Note: It's called a linear transformation because the power on each x_i on the right of (5.7) is 1, i.e., each x'_i is a linear combination of the x_is.

As promised, an additional translation next gives affine transformations.

Definition 5.2. An *affine transformation* of $\mathbb{R}^m$ is a transformation $g : \mathbb{R}^m \to \mathbb{R}^m$ specified by an equation of the form

$$g(P) = f^M(P) + D = MP + D \tag{5.8}$$

for $P \in \mathbb{R}^m$, where f^M is a non-singular linear transformation of $\mathbb{R}^m$ and D is an m-vector. The matrix M, which is non-singular as f^M is non-singular, is called the *defining matrix* of g. The vector D is called the *translational component* of g.

Accordingly, if

$$M = \begin{bmatrix} a_{11} & a_{12} & \ldots & a_{1m} \\ a_{21} & a_{22} & \ldots & a_{2m} \\ & \ldots & \ldots & \\ a_{m1} & a_{m2} & \ldots & a_{mm} \end{bmatrix} \text{ is non-singular and } D = \begin{bmatrix} d_1 \\ d_2 \\ \ldots \\ d_m \end{bmatrix} \text{ arbitrary,}$$

then the affine transformation g defined by $g(P) = MP + D$ maps $P = [x_1\ x_2\ \ldots\ x_m]^T$ to $g(P) = [x'_1\ x'_2\ \ldots\ x'_m]^T$, where

$$\begin{aligned} x'_1 &= a_{11}x_1 + a_{12}x_2 + \ldots + a_{1m}x_m + d_1 \\ x'_2 &= a_{21}x_1 + a_{22}x_2 + \ldots + a_{2m}x_m + d_2 \\ &\ldots \quad \ldots \\ x'_m &= a_{m1}x_1 + a_{m2}x_2 + \ldots + a_{mm}x_m + d_m \end{aligned} \tag{5.9}$$

If its translational component is zero, then an affine transformation evidently reduces to a non-singular linear transformation. Conversely, a non-singular linear transformation is an affine transformation with zero translational component.

Example 5.3. $g : \mathbb{R}^2 \to \mathbb{R}^2$ given by

$$g([x\ y]^T) = \begin{bmatrix} 2 & 1 \\ 0 & 4 \end{bmatrix} [x\ y]^T + [4\ 6]^T$$

is affine. Writing out the formula for g we have

$$g([x\ y]^T) = \begin{bmatrix} 2x + y + 4 \\ 4y + 6 \end{bmatrix}$$

So, e.g.,

$$g([-1\ 2]^T = [4\ 14]^T \qquad \text{and} \qquad g([0\ 3]^T = [7\ 18]^T$$

Exercise 5.19. What are the images of the points $[0\ 0\ 0]^T$ and $[1\ -1\ 1]^T$ by the affine transformation $g : \mathbb{R}^3 \to \mathbb{R}^3$ given by

$$g([x\ y\ z]^T) = \begin{bmatrix} -1 & -2 & 3 \\ 4 & 0 & 2 \\ 0 & -3 & 1 \end{bmatrix} [x\ y\ z]^T + [-1\ 6\ 3]^T\ ?$$

The next example says that an affine transformation is respectful of convex combinations.

Example 5.4. Show that an affine transformation g of $\mathbb{R}^m$ preserves convex combinations and barycentric coordinates in that

$$g(c_1P_1 + c_2P_2 + \ldots + c_kP_k) = c_1g(P_1) + c_2g(P_2) + \ldots + c_kg(P_k)$$

for any m-vectors P_i and scalars c_i, $1 \le i \le k$, such that $0 \le c_i \le 1$ and $c_1 + c_2 + \ldots + c_k = 1$.

Answer: Suppose that $g(P) = MP + D$, where M is the defining matrix and D the translational component of g. Then

$$\begin{aligned} g(c_1P_1 + c_2P_2 + \ldots + c_kP_k) &= M(c_1P_1 + c_2P_2 + \ldots + c_kP_k) + D \\ &= M(c_1P_1 + c_2P_2 + \ldots + c_kP_k) + \\ &\quad (c_1 + c_2 + \ldots + c_k)D \\ &\quad (\text{because } c_1 + c_2 + \ldots + c_k = 1) \\ &= c_1MP_1 + c_2MP_2 + \ldots + c_kMP_k + \\ &\quad c_1D + c_2D + \ldots + c_kD \\ &= c_1(MP_1 + D) + c_2(MP_2 + D) + \ldots + \\ &\quad c_k(MP_k + D) \\ &= c_1g(P_1) + c_2g(P_2) + \ldots + c_kg(P_k) \end{aligned}$$

Exercise 5.20. Prove that an affine transformation which fixes the origin (i.e., maps the origin to itself) is a non-singular linear transformation.

Exercise 5.21. Prove that the composition of affine transformations is again an affine transformation.

Exercise 5.22. Determine the affine transformation $g_1 \circ g_2$, where

$$g_1(P) = \begin{bmatrix} 2 & 1 \\ 0 & 4 \end{bmatrix} P + [4\ 6]^T \quad \text{and} \quad g_2(P) = \begin{bmatrix} -1 & 3 \\ 1 & -2 \end{bmatrix} P + [-1\ 0]^T$$

Example 5.5. An affine transformation g is always invertible. In fact, if g is defined by $g(P) = MP + D$, then show that its inverse, also affine, is given by

$$g^{-1}(Q) = M^{-1}Q - M^{-1}D$$

Answer: For any $P \in \mathbb{R}^m$, we have

$$(g^{-1} \circ g)(P) = g^{-1}(g(P)) = M^{-1}g(P) - M^{-1}D = M^{-1}(MP + D) - M^{-1}D = P$$

proving that $g^{-1} \circ g$ is the identity on $\mathbb{R}^m$. Likewise, it can be seen that $g \circ g^{-1}$ is the identity, proving that g^{-1}, as defined above, indeed is the inverse of g.

Exercise 5.23. Determine the inverse of the affine transformation g of $\mathbb{R}^2$ given by

$$g([x\ y]^T) = \begin{bmatrix} 2 & 1 \\ 0 & 4 \end{bmatrix} [x\ y]^T + [4\ 6]^T$$

The following important proposition says that affine transformations are particularly well-behaved from a geometric point of view, in particular, that they preserve straightness, planarity, parallelism and convexity.

Proposition 5.1.

(a) An affine transformation g of $\mathbb{R}^2$ maps straight lines to straight lines. Moreover, it maps parallel straight lines to parallel straight lines and intersecting straight lines to intersecting straight lines.

(b) An affine transformation g of $\mathbb{R}^2$ maps convex sets to convex sets. Moreover, it maps the convex hull of $\{P_1, P_2, \ldots, P_k\}$ to the convex hull of $\{f^M(P_1), f^M(P_2), \ldots, f^M(P_k)\}$.

(c) An affine transformation g of $\mathbb{R}^3$ maps straight lines to straight lines and planes to planes. Moreover, it maps parallel straight lines to parallel straight lines, intersecting straight lines to intersecting straight lines, parallel planes to parallel planes and intersecting planes to intersecting planes.

(d) An affine transformation g of $\mathbb{R}^3$ maps a convex set lying on one plane of $\mathbb{R}^3$ to a convex set on another plane. Moreover, it transforms the convex hull of a set of points $\{P_1, P_2, \ldots, P_k\}$ on one plane to the convex hull of $\{g(P_1), g(P_2), \ldots, g(P_k)\}$ lying on another plane.

Figure 5.7 illustrates the actions of an affine transformation in $\mathbb{R}^3$.

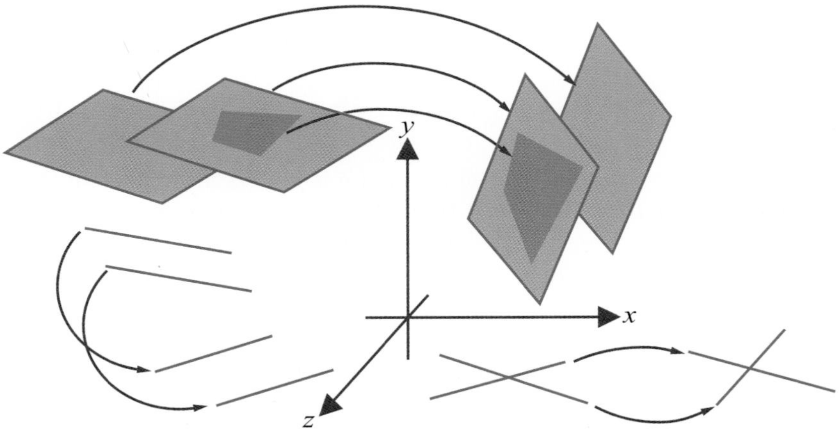

Figure 5.7: Affine transformation in $\mathbb{R}^3$.

Proof. (a) Say $g(P) = MP + D$, for $P \in \mathbb{R}^3$, where M is a non-singular 2×2 matrix and D a column 2-vector.

Suppose, first, that the equation of a given straight line l in $\mathbb{R}^2$ is $ax + by = c$, which can be written as $[a\ b][x\ y]^T = c$. If the point $[x\ y]^T$ is on l, let's see where its image $g([x\ y]^T) = M\,[x\ y]^T + D$ lies. Now

$$([a\ b]\,M^{-1})\,(M\,[x\ y]^T + D) = [a\ b][x\ y]^T + [a\ b]M^{-1}D = c + [a\ b]M^{-1}D$$

Writing $[a'\ b'] = [a\ b]M^{-1}$ and the scalar $c' = c + [a\ b]M^{-1}D$, the preceding equation gives

$$[a'\ b']\,(M\,[x\ y]^T + D) = c'$$

which shows that if $[x\ y]^T$ lies on l, then $g([x\ y]^T)$ lies on the straight line $[a'\ b'][x\ y]^T = c'$, proving that the image $g(l)$ of l is indeed a straight line.

Say next that l and l' are two parallel straight lines in $\mathbb{R}^2$, whose equations can then be written as $[a\ b][x\ y]^T = c$ and $[a\ b][x\ y]^T = d$, respectively, where $c \neq d$.

From the first part of the proof it's seen that $g(l)$ is the straight line whose equation is

$$[a'\ b'][x\ y]^T = c', \quad \text{where} \quad [a'\ b'] = [a\ b]M^{-1} \quad \text{and} \quad c' = c + [a\ b]M^{-1}D$$

Likewise, $g(l')$ is the straight line whose equation is

$$[a'\ b'][x\ y]^T = d', \quad \text{where} \quad [a'\ b'] \ \text{ is as before and } \ d' = d + [a\ b]M^{-1}D$$

Moreover, it follows from $c \neq d$, that $c' \neq d'$. We conclude that $g(l)$ and $g(l')$ are indeed parallel straight lines.

Finally, it's easy to see that two straight lines l and l' which intersect at P are mapped by g to straight lines which intersect at $g(P)$.

(b) Again, say, $g(P) = MP + D$, for $P \in \mathbb{R}^3$, where M is a non-singular 2×2 matrix and D a column 2-vector.

Suppose, first, that S is a convex subset of $\mathbb{R}^2$. To prove that $g(S)$ is convex as well, it is sufficient to show that $cP + (1-c)Q \in g(S)$, given two points P and Q in $g(S)$ and c in $0 \le c \le 1$.

Since $P, Q \in g(S)$, there exist $P', Q' \in S$ such that $g(P') = P$ and $g(Q') = Q$. As S is convex $cP' + (1-c)Q' \in S$. Applying g to both sides of the preceding inclusion we have that $g(cP' + (1-c)Q') \in g(S)$, but

$$\begin{aligned}
g(cP' + (1-c)Q') &= M(cP' + (1-c)Q') + D \\
&= M(cP' + (1-c)Q') + cD + (1-c)D \\
&= cMP' + cD + (1-c)MQ' + (1-c)D \\
&= c(MP' + D) + (1-c)(MQ' + D) \\
&= cg(P') + (1-c)g(Q') \\
&= cP + (1-c)Q
\end{aligned}$$

proving that indeed $cP + (1-c)Q \in g(S)$, so that the latter is a convex set.

We leave the proof of the second part of (b) as well as those of (c) and (d) to the reader. □

Exercise 5.24. Does an affine transformation of $\mathbb{R}^2$ or $\mathbb{R}^3$ necessarily map radial lines to radial lines or radial planes to radial planes? How about a linear transformation?

5.2.2 Affine Transformations and OpenGL

Proposition 5.1 says that affine transformations preserve straightness and flatness, among other properties, because they keep straight lines straight and planes plane. It's not hard to see, if one works through the proof, that this is a consequence of the fact that their defining Equations (5.9) (shown again below)

$$\begin{aligned}
x'_1 &= a_{11}x_1 + a_{12}x_2 + \ldots + a_{1m}x_m + d_1 \\
x'_2 &= a_{21}x_1 + a_{22}x_2 + \ldots + a_{2m}x_m + d_2 \\
&\ldots \quad \ldots \\
x'_m &= a_{m1}x_1 + a_{m2}x_2 + \ldots + a_{mm}x_m + d_m
\end{aligned}$$

are of degree one, in particular, that the maximum degree of a variable x_i on the right side of each of these equations is one.

Here's an example of what happens if this were not the case, and if even we go to degree two.

Example 5.6. The quadratic transformation h of $\mathbb{R}^2$ defined by $h([x\ y]^T) = [x\ y^2]^T$ doesn't necessarily keep straight lines straight. In fact, we'll show that it maps at least one straight segment into an arc of a parabola.

Write the transformation as

$$\begin{aligned} x' &= x \\ y' &= y^2 \end{aligned}$$

Now consider how the straight line

$$y = x$$

is mapped. We have, using the preceding 3 equations

$$y' = y^2 \implies y' = x^2 \implies y' = x'^2$$

which is the equation of a parabola. It follows that h maps the straight segment between $(0,0)$ and $(1,1)$ to the arc of the parabola $y = x^2$ joining the same two points, as shown in Figure 5.8.

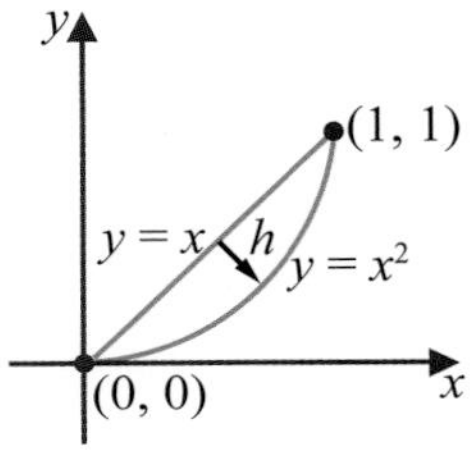

Figure 5.8: Quadratic transform h of $\mathbb{R}^2$ takes a straight segment to a parabolic arc.

One notices, further, that affine transformations are the *most general* class of transformations of degree one, because the right side of each one of the Equations (5.9) has its full complement of terms possible up to degree one – specifically, *every* x_i is present with degree one and there is, as well, the constant term d_i of degree zero.

One concludes that not only do affine transformations of $\mathbb{R}^2$ or $\mathbb{R}^3$ (or $\mathbb{R}^m$ in general) preserve straightness and flatness, they are the most general class of transformations to do so. Put another way, one cannot hope to go beyond affine transformations if straightness and flatness are not to be broken.

What has all this to do with OpenGL? Because they preserve straightness and flatness, affine transformations preserve as well the primitives of OpenGL, in particular, they map primitives of one type to another of the same type. The following exercise asks the reader to prove the specifics of this claim.

Exercise 5.25. Given an affine transformation g of $\mathbb{R}^2$ or $\mathbb{R}^3$, prove that

(a) g maps the straight segment joining two points P and Q to the straight segment joining $g(P)$ and $g(Q)$.

(b) g maps the triangle with vertices at P, Q and R to the triangle with vertices at $g(P)$, $g(Q)$ and $g(R)$.

(c) g maps the n-sided polygon with vertices at $P_1, P_2, \ldots, P_n$ to the n-sided polygon with vertices at $g(P_1), g(P_2), \ldots, g(P_n)$. If the original polygon is planar, then so is the transformed one. If the original polygon is convex, then so is the transformed one.

Hint: Use Proposition 5.1.

Non-affine transformations may not treat OpenGL primitives with quite as much respect, as the following exercise shows.

Exercise 5.26. We already saw in Example 5.6 the non-affine quadratic transformation h of $\mathbb{R}^2$, given by $h([x\ y]^T) = [x\ y^2]^T$, take a straight segment to a parabolic arc. How does h transform the triangle with corners at $(0,0)$, $(1,0)$ and $(1,1)$? Figure 5.9 is a gentle hint.

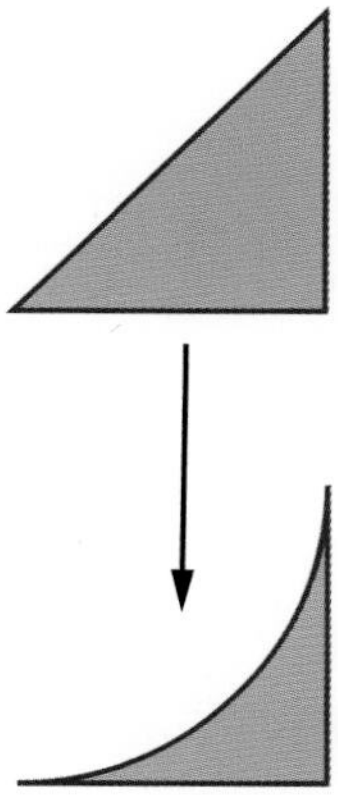

Figure 5.9: Hint for Exercise 5.26.

Now, it's desirable for a graphics API such as OpenGL to implement only modeling transformations which preserve its drawing primitives – specifically, mapping each one to another of the same type. Why? Consider OpenGL in particular. At the rendering end of its pipeline are evidently modules to render points, segments and triangles (mind that even a general polygon is triangulated prior to rendering). Suppose, then, that a particular scene is specified by the programmer as a list of n primitives:

$$primitive1,\ primitive2,\ \ldots,\ primitiveN$$

where each $primitiveI$, $1 \leq I \leq N$, is a point, segment or triangle. The scene is rendered essentially in a simple loop:

`for` $(I = 1;\ \ I\ \leq\ N;\ \ I{+}{+})$ `render` $primitiveI$

where each iteration invokes the appropriate primitive rendering module.

Suppose, next, that a modeling transformation g is applied to the scene. The transformed scene is given by the list:

$$g(primitive1),\ g(primitive2),\ \ldots,\ g(primitiveN)$$

If g preserves primitives, then $g(primitiveI)$ is of the same class as $primitiveI$, for $1 \leq I \leq N$, and the transformed scene is rendered in the loop

`for` $(I = 1;\ \ I\ \leq\ N;\ \ I{+}{+})$ `render` $g(primitiveI)$

invoking the same modules as before.

On the other hand, if g doesn't map primitives of one class to another of the same, e.g., if a triangle can change to something that is no longer one, as in Figure 5.9, then the situation becomes significantly more complicated. In this case, either there have to be modules to render all possible target objects of all the drawing primitives, or modules to approximate them using existing primitives, or, maybe, a combination of both. See Figure 5.10 for an illustration of both situations.

If the API designer is understandably reluctant to open this particular can of worms, then she should restrict herself to modeling transformations which do keep primitives within their class. In the case of OpenGL, this calls for transformations preserving straightness and flatness. However, even given

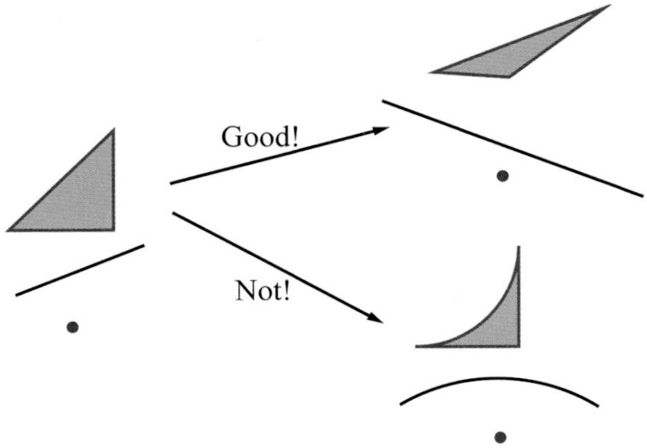

Figure 5.10: Transformations that are good from the API programmer's point of view, and not so good.

this constraint, the designer would reasonably want as many as possible at her disposal. Transformations of degree one preserve straightness and flatness, so the designer would want them all if possible; in other words, *all affine transformations*.

And, in fact, we shall see that the designers of OpenGL have implemented, barring a few degenerate calls, exactly the class of affine transformations as their modeling transformations.

5.2.3 Affine Transformations and Homogeneous Coordinates

Despite their virtues listed in the previous two sections, there is potentially a serious computational problem with applying affine transformations rather than linear ones. The source lies in the difference in how the two are defined. A linear transformation is given by an equation of the form

$$f^M(P) = MP$$

while an affine transformation by one of the form

$$g(P) = MP + D$$

The former is expressed as a single matrix-vector multiplication, while the latter by a matrix-vector multiplication *followed* by a vector-vector sum. It is the additional sum step which cascades when composing affine transformations.

For example, if $f^{M_1}, f^{M_2}, f^{M_3}, \ldots$ are linear transformations of $\mathbb{R}^m$, then for an m-vector P,

$$\begin{aligned} f^{M_1}(P) &= M_1P \\ (f^{M_2} \circ f^{M_1})(P) &= (M_2M_1)P \\ (f^{M_3} \circ f^{M_2} \circ f^{M_1})(P) &= (M_3M_2M_1)P \end{aligned} \tag{5.10}$$

and so on. On the other hand, if $g_1, g_2, g_3, \ldots$ are affine transformations of $\mathbb{R}^m$ given by $g_1(P) = M_1P + D_1$, $g_2(P) = M_2P + D_2$, $g_3(P) = M_3P + D_3, \ldots$, respectively, then for an m-vector P,

$$\begin{aligned} g_1(P) &= M_1P + D_1 \\ (g_2 \circ g_1)(P) &= (M_2M_1)P + M_2D_1 + D_2 \\ (g_3 \circ g_2 \circ g_1)(P) &= (M_3M_2M_1)P + M_3M_2D_1 + M_3D_2 + D_3 \end{aligned} \tag{5.11}$$

It's not hard to see that the number of matrix operations grows quadratically with the number n of affine transformations $g_n \circ \ldots \circ g_2 \circ g_1$ being composed, versus linearly in the case $f^{M_n} \circ \ldots \circ f^{M_2} \circ f^{M_1}$ of linear transformations. Composing affine transformations, at least by means of equations as above, therefore, is highly inefficient. There is an elegant way, however, to rectify the problem. It is with the help of so-called homogeneous coordinates.

Definition 5.3. A point

$$P = [x_1\ x_2\ \ldots\ x_m]^T$$

belonging to $\mathbb{R}^m$ is represented in *homogeneous coordinates* by *any* $m+1$-tuple of the form

$$[cx_1\ cx_2\ \ldots\ cx_m\ c]^T$$

where c is a *non-zero* scalar. Homogeneous coordinates, therefore, are not unique. And, note they live one dimension higher.

Example 5.7. Possible homogeneous coordinates of the point $P = [3\ 7]^T \in \mathbb{R}^2$ include $[3\ 7\ 1]^T$, $[16.5\ 38.5\ 5.5]^T$, $[-6\ -14\ -2]^T$, etc.

For our current purposes, though, it's good enough to fix the scalar c in Definition 5.3 to be 1. We'll have use for the general c later when studying projective spaces. So, for the present, assume that the point

$$P = [x_1\ x_2\ \ldots\ x_m]^T$$

is represented in homogeneous coordinates by

$$[x_1\ x_2\ \ldots\ x_m\ 1]^T$$

For example, $[3\ 7]^T$ would be homogenized to $[3\ 7\ 1]^T$. To save space we'll often write $[x_1\ x_2\ \ldots\ x_m\ 1]^T$ as $[P\ 1]^T$.

Observe now that Equations (5.9)

$$\begin{aligned} x_1' &= a_{11}x_1 + a_{12}x_2 + \ldots + a_{1m}x_m + d_1 \\ x_2' &= a_{21}x_1 + a_{22}x_2 + \ldots + a_{2m}x_m + d_2 \\ &\ldots \ \ldots \\ x_m' &= a_{m1}x_1 + a_{m2}x_2 + \ldots + a_{mm}x_m + d_m \end{aligned}$$

defining an affine transformation g are equivalent to the *single* matrix equation

$$\begin{bmatrix} x_1' \\ x_2' \\ \ldots \\ x_m' \\ 1 \end{bmatrix} = \begin{bmatrix} a_{11} & a_{12} & \ldots & a_{1m} & d_1 \\ a_{21} & a_{22} & \ldots & a_{2m} & d_2 \\ & \ldots & \ldots & & \\ a_{m1} & a_{m2} & \ldots & a_{mm} & d_m \\ 0 & 0 & 0 & 0 & 1 \end{bmatrix} \begin{bmatrix} x_1 \\ x_2 \\ \ldots \\ x_m \\ 1 \end{bmatrix}$$

(as is easily verified by multiplying the two matrices on the right). Concisely:

$$\begin{bmatrix} g(P) \\ 1 \end{bmatrix} = \begin{bmatrix} P' \\ 1 \end{bmatrix} = \begin{bmatrix} M & D \\ 0 & 1 \end{bmatrix} \begin{bmatrix} P \\ 1 \end{bmatrix} \qquad (5.12)$$

Presto! Computation of the affine transformation g, which earlier required a matrix-vector multiplication followed by a vector-vector addition, has now become a *single* matrix-vector multiplication with the use of homogeneous coordinates and, albeit, a bigger matrix. The translational component has, evidently, been subsumed into the extra dimension of the larger matrix.

Example 5.8. The affine transformation $g : \mathbb{R}^2 \rightarrow \mathbb{R}^2$ given by

$$g\left(\begin{bmatrix} x \\ y \end{bmatrix}\right) = \begin{bmatrix} x' \\ y' \end{bmatrix} = \begin{bmatrix} 2 & 1 \\ 0 & 4 \end{bmatrix} \begin{bmatrix} x \\ y \end{bmatrix} + \begin{bmatrix} 4 \\ 6 \end{bmatrix}$$

can be written using homogeneous coordinates as

$$\begin{bmatrix} x' \\ y' \\ 1 \end{bmatrix} = \begin{bmatrix} 2 & 1 & 4 \\ 0 & 4 & 6 \\ 0 & 0 & 1 \end{bmatrix} \begin{bmatrix} x \\ y \\ 1 \end{bmatrix}$$

Let's give it a check for, say, the point $[1\ 1]^T$. Now,

$$g\left(\begin{bmatrix} 1 \\ 1 \end{bmatrix}\right) = \begin{bmatrix} 2 & 1 \\ 0 & 4 \end{bmatrix} \begin{bmatrix} 1 \\ 1 \end{bmatrix} + \begin{bmatrix} 4 \\ 6 \end{bmatrix} = \begin{bmatrix} 7 \\ 10 \end{bmatrix}$$

and with homogeneous coordinates

$$\begin{bmatrix} 2 & 1 & 4 \\ 0 & 4 & 6 \\ 0 & 0 & 1 \end{bmatrix} \begin{bmatrix} 1 \\ 1 \\ 1 \end{bmatrix} = \begin{bmatrix} 7 \\ 10 \\ 1 \end{bmatrix}$$

the RHS of the preceding equation indeed being the homogenization of the RHS of the one before it.

Exercise 5.27. Express the affine transformation g of $\mathbb{R}^2$ given by

$$g\left(\begin{bmatrix} x \\ y \end{bmatrix}\right) = \begin{bmatrix} -1 & 2 \\ -3 & 0 \end{bmatrix}\begin{bmatrix} x \\ y \end{bmatrix} + \begin{bmatrix} 2 \\ 1 \end{bmatrix}$$

as a single matrix vector multiplication using homogeneous coordinates.

The composition of affine transformations is drastically simplified with use of homogeneous coordinates. For example, the last equation of (5.11) becomes

$$\begin{bmatrix} (g_3 \circ g_2 \circ g_1)(P) \\ 1 \end{bmatrix} = \left(\begin{bmatrix} M_3 & D_3 \\ 0 & 1 \end{bmatrix}\begin{bmatrix} M_2 & D_2 \\ 0 & 1 \end{bmatrix}\begin{bmatrix} M_1 & D_1 \\ 0 & 1 \end{bmatrix}\right)\begin{bmatrix} P \\ 1 \end{bmatrix}$$

the number of matrix operations now growing linearly with the number of affine transformations being composed, instead of quadratically.

Exercise 5.28. If you did Exercise 5.22, then you have determined the affine transformation $f \circ g$, where

$$f(P) = \begin{bmatrix} 2 & 1 \\ 0 & 4 \end{bmatrix} P + [4\ 6]^T \quad \text{and} \quad g(P) = \begin{bmatrix} -1 & 3 \\ 1 & -2 \end{bmatrix} P + [-1\ 0]^T$$

Now, verify your answer by multiplying the 3×3 matrices corresponding to f and g, and checking if the result corresponds to the composed transformation already computed.

5.3 Geometric Transformations in 2-Space Continued

We resume our study of 2D geometric transformations, equipped now with a newfound grasp of affine transformations. Keep in mind that, as in the first section, by default we are in 2D space.

5.3.1 Affine Geometric Transformations

Are translations, scalings, rotations about the origin and reflections about radial mirrors, which we studied in the opening section, affine transformations? Of course, they all are! This follows easily from the non-singularity of the matrix on the RHS of each of the Equations (5.1), (5.2), (5.4) and (5.5).

Exercise 5.29. Prove that rotations about arbitrary points (not necessarily the origin) and reflections about arbitrary mirrors (not necessarily radial) are affine as well.

That translations, scalings, rotations and reflections are affine means they are geometrically well-behaved, preserving straightness, parallelism and convexity, as well. We record this fact as a proposition.

Proposition 5.2. *Let g be either a translation, a scaling, a rotation (about an arbitrary point) or a reflection (about an arbitrary mirror). Then:*

(a) g maps straight lines to straight lines. Moreover, it maps parallel straight lines to parallel straight lines and intersecting straight lines to intersecting straight lines.

(b) g maps convex sets to convex sets. Moreover, it maps the convex hull of $\{P_1, P_2, \ldots, P_k\}$ to the convex hull of $\{f^M(P_1), f^M(P_2), \ldots, f^M(P_k)\}$.

Proof. Follows from Proposition 5.1 for 2D affine transformations in general. □

Geometric Transformation Equations Using Homogeneous Coordinates

In Section 5.2.3 we learned how to express an affine transformation as a single matrix-vector multiplication, after writing points in homogeneous coordinates. In the case of $\mathbb{R}^2$ this means writing $P = [x\ y]^T$ as $[x\ y\ 1]^T$ or $[P\ 1]^T$ for short. Let's now rewrite Equations (5.1), (5.2), (5.4) and (5.5) of the basic 2D transformations using homogeneous coordinates:

Translation by displacement vector $[d_x\ d_y]^T$:

$$\begin{bmatrix} P' \\ 1 \end{bmatrix} = \begin{bmatrix} 1 & 0 & d_x \\ 0 & 1 & d_y \\ 0 & 0 & 1 \end{bmatrix} \begin{bmatrix} P \\ 1 \end{bmatrix} \tag{5.13}$$

Scaling by a factor of s_x along the x-axis and s_y along the y-axis:

$$\begin{bmatrix} P' \\ 1 \end{bmatrix} = \begin{bmatrix} s_x & 0 & 0 \\ 0 & s_y & 0 \\ 0 & 0 & 1 \end{bmatrix} \begin{bmatrix} P \\ 1 \end{bmatrix} \tag{5.14}$$

Rotation by an angle θ counter-clockwise about the origin:

$$\begin{bmatrix} P' \\ 1 \end{bmatrix} = \begin{bmatrix} \cos\theta & -\sin\theta & 0 \\ \sin\theta & \cos\theta & 0 \\ 0 & 0 & 1 \end{bmatrix} \begin{bmatrix} P \\ 1 \end{bmatrix} \tag{5.15}$$

Reflection about a radial mirror l at an angle of θ counter-clockwise from the positive x-axis:

$$\begin{bmatrix} P' \\ 1 \end{bmatrix} = \begin{bmatrix} \cos 2\theta & \sin 2\theta & 0 \\ \sin 2\theta & -\cos 2\theta & 0 \\ 0 & 0 & 1 \end{bmatrix} \begin{bmatrix} P \\ 1 \end{bmatrix} \tag{5.16}$$

Exercise 5.30. Write the 3×3 matrix corresponding to each of the following affine transformations:

(a) Translation by the displacement vector $[-2\ 3]^T$.

(b) Scaling by a factor of 2 in the x-direction and 4 in the y.

(c) Counter-clockwise rotation by an angle of $-45°$ about the origin.

(d) Reflection about the radial mirror making an angle of $30°$ measured counter-clockwise from the positive direction of the x-axis.

Factoring Affine Transformations

We know then that affine transformations include translations, scalings and rotations. But are they more than just these three special kinds of transformations? It's extremely important that the answer is *no*! In fact, *any* affine transformation can be "made from" translations, scalings and rotations. Precisely, any affine transformation can be expressed as a composition of transformations of just these three kinds. Here is the formal statement:

Proposition 5.3. *Any affine transformation of $\mathbb{R}^2$ is the composition in some order of translations, scalings and rotations about the origin.*

In particular, any affine transformation $g : \mathbb{R}^2 \to \mathbb{R}^2$ can be factored into a composition $g = g_4 \circ g_3 \circ g_2 \circ g_1$, where g_1 is a rotation about the origin, g_2 a scaling, g_3 another rotation about the origin and g_4 a translation.

Proof. Let

$$g(P) = MP + D$$

where $M = \begin{bmatrix} a_{11} & a_{12} \\ a_{21} & a_{22} \end{bmatrix}$ is g's non-singular 2×2 defining matrix and the 2-vector $D = \begin{bmatrix} d_x \\ d_y \end{bmatrix}$ is its translational component.

We claim first that it is possible to find 2×2 matrices M_1, M_2 and M_3, corresponding, respectively, to a rotation about the origin, a scaling and another rotation about the origin, such that

$$M = M_3 M_2 M_1$$

Say M_1 corresponds to a rotation by angle θ, M_2 to scaling by a factor of s_x along the x-axis and s_y along the y-axis and M_3 to a rotation by angle ϕ. The preceding equation gives, therefore, that

$$\begin{bmatrix} a_{11} & a_{12} \\ a_{21} & a_{22} \end{bmatrix} = \begin{bmatrix} \cos\phi & -\sin\phi \\ \sin\phi & \cos\phi \end{bmatrix} \begin{bmatrix} s_x & 0 \\ 0 & s_y \end{bmatrix} \begin{bmatrix} \cos\theta & -\sin\theta \\ \sin\theta & \cos\theta \end{bmatrix} \quad (5.17)$$

which we'll show next can be solved to find ϕ, θ, s_x and s_y.

Multiply the three matrices on the RHS of the preceding equation and then equate terms of the resulting matrix with the corresponding ones on the LHS to see that:

$$\begin{aligned}
a_{11} &= s_x \cos\phi \cos\theta - s_y \sin\phi \sin\theta \\
a_{12} &= -s_x \cos\phi \sin\theta - s_y \sin\phi \cos\theta \\
a_{21} &= s_x \sin\phi \cos\theta + s_y \cos\phi \sin\theta \\
a_{22} &= -s_x \sin\phi \sin\theta + s_y \cos\phi \cos\theta
\end{aligned} \tag{5.18}$$

Four equations in four unknowns seems right. Check that:

$$\begin{aligned}
a_{21} - a_{12} &= (s_x + s_y) \sin(\phi + \theta) \\
a_{11} + a_{22} &= (s_x + s_y) \cos(\phi + \theta) \\
a_{21} + a_{12} &= (s_x - s_y) \sin(\phi - \theta) \\
a_{11} - a_{22} &= (s_x - s_y) \cos(\phi - \theta)
\end{aligned} \tag{5.19}$$

Assuming for the moment that neither $a_{11} + a_{22}$ nor $a_{11} - a_{22}$ is zero, divide the first equation above by the second and the third by the fourth to get:

$$\begin{aligned}
\tan(\phi + \theta) &= \frac{a_{21} - a_{12}}{a_{11} + a_{22}} \\
\tan(\phi - \theta) &= \frac{a_{21} + a_{12}}{a_{11} - a_{22}}
\end{aligned} \tag{5.20}$$

which implies:

$$\begin{aligned}
\phi + \theta &= \tan^{-1}\left(\frac{a_{21} - a_{12}}{a_{11} + a_{22}}\right) \\
\phi - \theta &= \tan^{-1}\left(\frac{a_{21} + a_{12}}{a_{11} - a_{22}}\right)
\end{aligned} \tag{5.21}$$

These two equations can be solved to determine ϕ and θ. Furthermore, the values of $\phi+\theta$ and $\phi-\theta$ can then be substituted back into equation set (5.19) to determine equations for $s_x + s_y$ and $s_x - s_y$, which can then be solved to find s_x and s_y.

The earlier claim that (5.17) can be solved to find ϕ, θ, s_x and s_y is proved and, therefore, $M = M_3 M_2 M_1$, in the manner claimed at the start of the proof as well – *except* when either or both of $a_{11} + a_{22}$ and $a_{11} - a_{22}$ is 0, in which case Exercise 5.32 below verifies the claim.

As $g(P) = MP + D = (M_3 M_2 M_1)P + D$ one concludes, finally, that indeed $g = g_4 \circ g_3 \circ g_2 \circ g_1$, where g_1 is the counter-clockwise rotation about the origin by an angle of θ, g_2 the scaling by a factor of s_x along the x-axis and s_y along the y-axis, g_3 the counter-clockwise rotation about the origin by an angle of ϕ and g_4 translation by the displacement vector D. □

Example 5.9. Factor the affine transformation

$$g(P) = \begin{bmatrix} 1 & \frac{\sqrt{3}}{2} \\ \frac{\sqrt{3}}{2} & 0 \end{bmatrix} P + \begin{bmatrix} 2 \\ 1 \end{bmatrix}$$

according to the proposition.

Answer: From Equations (5.21) we have

$$\phi + \theta = \tan^{-1} 0 = 0^\circ \quad \text{and} \quad \phi - \theta = \tan^{-1} \sqrt{3} = 60^\circ$$

which solve to

$$\phi = 30^\circ \quad \text{and} \quad \theta = -30^\circ$$

Plugging the values of $\phi + \theta$ and $\phi - \theta$ into the second and fourth equations of (5.19) we have

$$1 = (s_x + s_y)\cos 0^\circ = s_x + s_y \quad \text{and} \quad 1 = (s_x - s_y)\cos 60^\circ = \frac{1}{2}(s_x - s_y)$$

which solve to

$$s_x = \frac{3}{2} \quad \text{and} \quad s_y = -\frac{1}{2}$$

(If the reader is wondering about the other two equations in (5.19) – the first and third – she may check that these are satisfied as well by the values found above for ϕ, θ, s_x and s_y.)

Therefore, $g = g_4 \circ g_3 \circ g_2 \circ g_1$, where g_1 is the clockwise rotation about the origin by an angle of 30°, g_2 the scaling by a factor of $\frac{3}{2}$ along the x-axis and $-\frac{1}{2}$ along the y-axis, g_3 the counter-clockwise rotation about the origin by an angle of 30° and g_4 translation by the displacement vector $[2\ 1]^T$.

Exercise 5.31. Factor the affine transformation

$$g(P) = \begin{bmatrix} 1 & \frac{1}{2} \\ \frac{1}{2} & 0 \end{bmatrix} P + \begin{bmatrix} -1 \\ 1 \end{bmatrix}$$

according to the proposition.

Exercise 5.32. Fill in the gap in the proof of the preceding proposition, where it was assumed (just after Equations (5.19)) that neither $a_{11} + a_{22}$ nor $a_{11} - a_{22}$ is zero. In particular, even if one or both of these quantities is zero, show how to proceed again from (5.19) to solve for ϕ, θ, s_x and s_y.

Exercise 5.33. Factor the affine transformation

$$g(P) = \begin{bmatrix} \frac{1}{2} & \sqrt{3} \\ 0 & \frac{1}{2} \end{bmatrix} P$$

according to the proposition.

Exercise 5.34. Give an example of an affine transformation which itself is neither a translation, nor scaling nor rotation about the origin, so one *must* compose in order to obtain it.

Remark 5.2. The reader may have noticed that we never used the non-singularity of M in the proof of Proposition 5.3. As a matter of fact, even if M is singular, it can be written as $M = M_3 M_2 M_1$ as in the proposition, *except* that the scaling M_2 turns out to be degenerate.

Proposition 5.3 suggests that translations, scalings and rotations about the origin are fundamental in the sense that they can be used to generate *all* affine transformations, a particularly useful insight for anyone in Flatland trying to implement a graphics API. For, all such a programmer has to code is an implementation of each of those three special kinds of affine transformations, to get the rest automatically.

Since a non-singular linear transformation of $\mathbb{R}^2$ is simply an affine transformation with null translational component, we have also proved the following on the way to proving Proposition 5.3:

Proposition 5.4. *Any non-singular linear transformation of $\mathbb{R}^2$ is the composition successively of a rotation about the origin, a scaling and another rotation about the origin.* □

5.3.2 Euclidean and Rigid Transformations

Proposition 5.2 tells us that transformations such as translations, scalings, rotations and reflections are respectful of a bunch of geometric attributes, from straightness to convexity. How about that most important geometric attribute of all, though, namely, distance? We would say a transformation g *preserves distance* if it were true that, for any pair of points P and Q, the distance between $f(P)$ and $f(Q)$ is the same as that between P and Q.

It's clear, if one thinks of scalings, that distance is not preserved by transformations in general. However, there certainly are transformations that seem to preserve distance. Translations come to mind, as points are "carried together" by a translation, so neither pulled apart nor drawn closer together. Similar thoughts apply to rotations. We'll see soon that translations and rotations do indeed preserve distance.

Distance-preserving transformations are important in animation because they preserve *shape* as well. In fact, an object's shape is not changed precisely when the distance between *every* pair of points belonging to it is not changed. See Figure 5.11. Comparing the pre-hit and post-hit heads, one observes that the distance between at least two pairs of points is different from those between the transformed pairs: the eyeballs, and P and Q. On the other hand, the distance between any pair of points of the book remains unchanged.

Transformations preserving distances are the ones, therefore, to use when animating *rigid* objects such as balls, bats (not the flying kind) and houses.

Figure 5.11: Square-headed student struck by a CG book: the shape of the head is distorted, but not that of the book.

They are important enough, in fact, to have been honored with the name of the great ancient geometer Euclid. Here's a formal definition.

Definition 5.4. A *Euclidean transformation* (also called *isometry*) of $\mathbb{R}^2$ is one that preserves distance. Precisely, $f : \mathbb{R}^2 \to \mathbb{R}^2$ is Euclidean if $|f(P)f(Q)| = |PQ|$ for any two points $P, Q \in \mathbb{R}^2$.

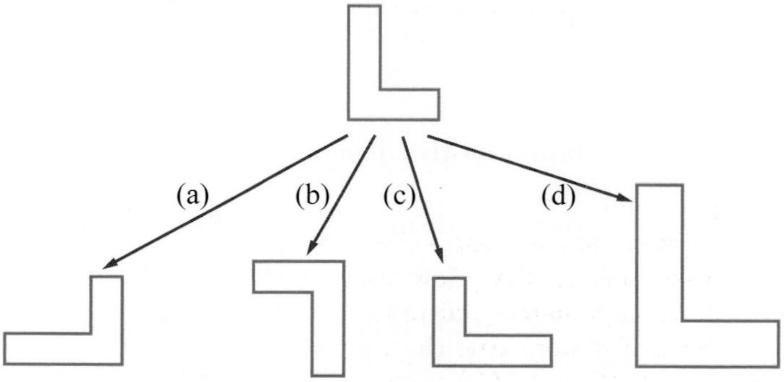

Figure 5.12: Transformations (a)-(c) are Euclidean, (d) is not.

See Figure 5.12 for three simple examples of Euclidean and one of non-Euclidean transformation. It may seem, as it cannot alter shape, that all a Euclidean transformation can do is "slide" an object around the plane, which, if true, would imply that it is merely a composition of translations and rotations. However, compare the Euclidean transformations in cases (a), (b) and (c) of Figure 5.12. The first two can certainly be obtained by sliding the top L around the page. However, it's not hard to convince oneself that (c) cannot and, therefore, is not a combination of translations and rotations.

Let's examine (c). As indicated in Figure 5.13, it can, in fact, be obtained by applying a reflection about a vertical mirror l, followed by translation and rotation. A reflection is required because (c) is a so-called orientation-reversing transformation. Here's the relevant definition:

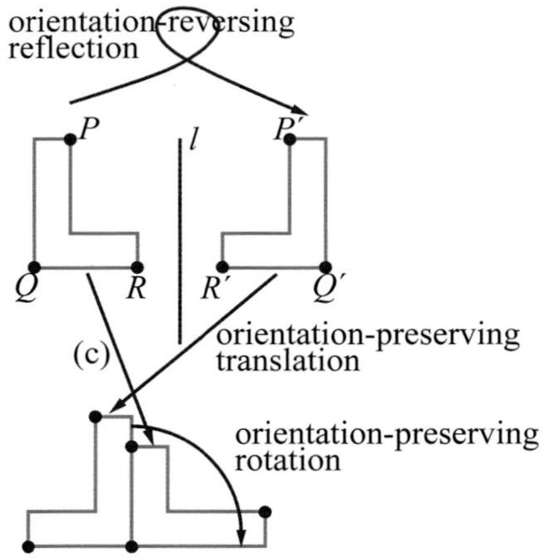

Figure 5.13: Executing (c) of Figure 5.12 by a reflection about the mirror l followed by translation and rotation.

Definition 5.5. A Euclidean transformation f of $\mathbb{R}^2$ is said to be *orientation-reversing* if there exist three non-collinear points P, Q and R in $\mathbb{R}^2$ such that, looking at $\mathbb{R}^2$ from a fixed side, one of the two sequences PQR and $f(P)f(Q)f(R)$ appears clockwise (CW) and the other counter-clockwise (CCW).

A Euclidean transformation that is not orientation-reversing is said to be *orientation-preserving*.

Orientation-preserving Euclidean transformations are also called *rigid transformations*.

Remark 5.3. The property of the transformation f described in the first paragraph of the preceding definition does not depend on the choice of the non-collinear points P, Q and R. In fact, as we'll see, if for *some* three non-collinear points P, Q and R it is true that PQR and $f(P)f(Q)f(R)$ appear oriented differently, then this is true for *any* three non-collinear points.

Rigid transformations are so called because they model the physical motion of a rigid object restricted always to a plane – such motion can never reverse orientation. Conceptually, reversing orientation requires the object to be "lifted off" the plane, "flipped" and "placed back" again.

The sequence PQR in Figure 5.13 appears CCW to the reader, while that of their images $P'Q'R'$ by reflection about l appears CW, proving that the reflection is indeed orientation-reversing and, therefore, not rigid.

Exercise 5.35. Show that a Euclidean transformation f preserves angles, i.e., $\angle ABC = \angle f(A)f(B)f(C)$, where A, B, C are any three points on the plane.

We'll see next how to determine algorithmically if PQR appears CW or CCW to a given viewer, which will in turn help decide if a transformation is orientation-preserving or not.

Lemma 5.1. *Let $P = [x_1\ y_1]^T$, $Q = [x_2\ y_2]^T$ and $R = [x_3\ y_3]^T$ be three points on the plane. Define the scalar D by*

$$D = x_1y_2 - x_2y_1 + x_2y_3 - x_3y_2 + x_3y_1 - x_1y_3 = \begin{vmatrix} x_1 & x_2 & x_3 \\ y_1 & y_2 & y_3 \\ 1 & 1 & 1 \end{vmatrix}$$

the rightmost term being called the discriminant determinant.

Let V be a viewer on the positive side of the z-axis of a hypothetical right-handed system. We have then the following:

1. *If $D = 0$, then P, Q and R are collinear.*

2. *If $D < 0$, then V perceives the order PQR as CW.*

3. *If $D > 0$, then V perceives the order PQR as CCW.*

Note: The column vectors of the discriminant determinant are the coordinates of P, Q and R, respectively, homogenized; so, it can be written

$$D = \begin{vmatrix} P & Q & R \\ 1 & 1 & 1 \end{vmatrix}$$

Proof. We'll first prove the lemma assuming that $R = O$, the origin, in which case

$$D = \begin{vmatrix} x_1 & x_2 & 0 \\ y_1 & y_2 & 0 \\ 1 & 1 & 1 \end{vmatrix} = x_1y_2 - x_2y_1$$

If $P = O$ as well, then P, Q and R are trivially collinear, and it's easily seen that the determinant $D = 0$, too, which falls into case 1 of the lemma. Accordingly, suppose that $P \neq O$ as in Figure 5.14. The straight line l through P and R has the equation

$$x_1y - y_1x = 0$$

If Q does not lie on l, then whether PQR appears CW or CCW to V depends on which half-plane of l contains Q. In particular, if Q lies in the half-plane $x_1y - y_1x > 0$ – the case depicted in the figure – then PQR appears CCW to V; if in the half-plane $x_1y - y_1x < 0$, then CW. Plugging in Q's coordinates means that PQR appears CCW to V if $x_1y_2 - y_1x_2 > 0$ and CW if $x_1y_2 - y_1x_2 < 0$. Of course, $x_1y_2 - y_1x_2 = 0$ if Q lies on l, in which case P, Q and R are collinear. Therefore, we've proved the lemma assuming $R = O$.

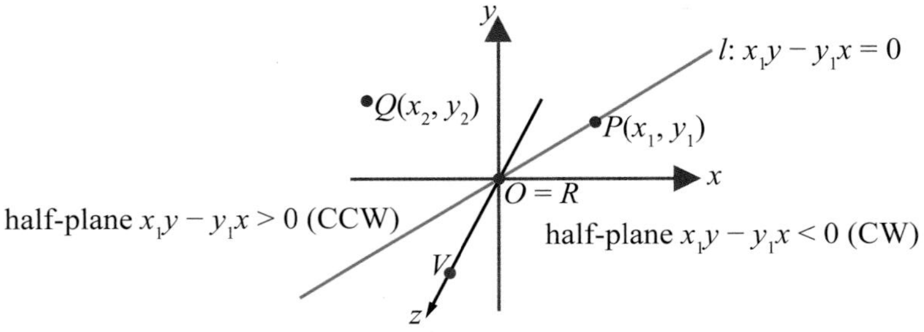

Figure 5.14: The orientation of PQR perceived by V depends on the half-plane of l containing Q (Q is depicted here in the half-plane $x_1y - y_1x > 0$).

The case of arbitrary R can be reduced to that of $R = O$ by applying the translation $-R$ to all three points, because the relative dispositions of P, Q and R as they appear to V are the same as those of $P - R$, $Q - R$ and $R - R$ ($= O$). Now, $P - R = [x_1 - x_3 \;\; y_1 - y_3]^T$ and $Q - R = [x_2 - x_3 \;\; y_2 - y_3]^T$, and we leave it to the reader to use the case $R = O$ to finish up the proof. □

Exercise 5.36. Verify the lemma for the following triples by plotting the points on graph paper:

(a) $P = [1\ 0]^T$, $Q = [0\ 1]^T$, $R = [0\ 0]^T$

(b) $P = [-1\ \ -1]^T$, $Q = [-2\ 1]^T$, $R = [3\ 4]^T$

The following proposition is intuitively fairly clear but, nevertheless, has to be proved formally.

Proposition 5.5. *A translation or a rotation about an arbitrary point is a rigid transformation of 2-space. A reflection about an arbitrary mirror is an orientation-reversing Euclidean transformation of 2-space.*

Proof. We'll prove first that a translation t by the displacement vector $D = [d_x\ d_y]^T$ preserves both distance and orientation.

Let $P = [x_1\ y_1]^T$ and $Q = [x_2\ y_2]^T$ be two points in $\mathbb{R}^2$. The images of P and Q by t are, respectively, $P' = P + D = [x_1 + d_x \;\; y_1 + d_y]^T$ and $Q' = Q + D = [x_2 + d_x \;\; y_2 + d_y]^T$. Now,

$$|PQ| = \sqrt{(x_1 - x_2)^2 + (y_1 - y_2)^2}$$

and

$$\begin{aligned} |P'Q'| &= \sqrt{((x_1 + d_x) - (x_2 + d_x))^2 + ((y_1 + d_y) - (y_2 + d_y))^2} \\ &= \sqrt{(x_1 - x_2)^2 + (y_1 - y_2)^2} \end{aligned}$$

proving t indeed preserves distance.

Let $P = [x_1\ y_1]^T$, $Q = [x_2\ y_2]^T$ and $R = [x_3\ y_3]^T$ be three points in $\mathbb{R}^2$, and V a viewer on the positive side of the z-axis. Lemma 5.1 says that PQR are collinear, appear CW to V, or CCW to V, according as the determinant

$$D = \begin{vmatrix} x_1 & x_2 & x_3 \\ y_1 & y_2 & y_3 \\ 1 & 1 & 1 \end{vmatrix}$$

is equal to, less than, or greater than 0.

The images of P, Q and R by t are $P' = P + D = [x_1 + d_x\ \ y_1 + d_y]^T$, $Q' = Q + D = [x_2 + d_x\ \ y_2 + d_y]^T$ and $R' = R + D = [x_3 + d_x\ \ y_3 + d_y]^T$, respectively. By another application of Lemma 5.1, $P'Q'R'$ are collinear, appear CW to V, or CCW to V, according as the determinant

$$D' = \begin{vmatrix} x_1 + d_x & x_2 + d_x & x_3 + d_x \\ y_1 + d_y & y_2 + d_y & y_3 + d_y \\ 1 & 1 & 1 \end{vmatrix}$$

is equal to, less than, or greater than 0.

However, subtracting d_x times the third row of D' from its first and d_y times the third row from its second, we see that, in fact, $D = D'$. It follows that the relative dispositions of PQR and of $P'Q'R'$ (either CCW or CW) with respect to V are identical, giving the conclusion that t indeed preserves orientation.

The proofs for rotations and reflections are left to the reader. □

Exercise 5.37. Scalings in general are not Euclidean transformations, but for certain choices of scaling factors they are. List these choices and for each say if it preserves or reverses orientation.

Exercise 5.38. Show that the composition of two Euclidean transformations is Euclidean and that of two rigid transformations is rigid.

Exercise 5.39. Show that the composition of two orientation-reversing Euclidean transformations is an orientation-preserving Euclidean transformation (in other words, rigid). Show that the composition of an orientation-preserving and an orientation-reversing Euclidean transformation is orientation-reversing.

We saw in Proposition 5.3 that an affine transformation can be factored as a composition of translations, scalings and rotations about the origin. The following proposition shows how Euclidean and rigid transformations can be factored. The first part verifies our intuition that a rigid transformation slides an object around the plane by translation and rotation, while the second says that a Euclidean transformation is at most one reflection away from being rigid.

Proposition 5.6. *A rigid transformation of $\mathbb{R}^2$ keeping the origin fixed is a rotation about the origin, while an arbitrary rigid transformation is a composition of a rotation about the origin followed by a translation.*

A Euclidean transformation of $\mathbb{R}^2$ is a composition of a rotation about the origin, followed by a translation, possibly followed again by a reflection.

Proof. Consider, first, a rigid transformation $f : \mathbb{R}^2 \to \mathbb{R}^2$ keeping the origin fixed, i.e., $f(O) = O$. Let $P \in \mathbb{R}^2$ be different from the origin. By the distance-preserving property

$$|OP| = |f(O)f(P)| = |Of(P)|$$

so both P and $f(P)$ lie on a circle c centered at O. See Figure 5.15(a) or (b). Say the angle from OP to $Of(P)$ is θ measured counter-clockwise. We'll show for any point $Q \in \mathbb{R}^2$ that its image $f(Q)$ is obtained by rotating Q counter-clockwise by an angle of θ about the origin as well, proving the claim that f is a rotation about the origin.

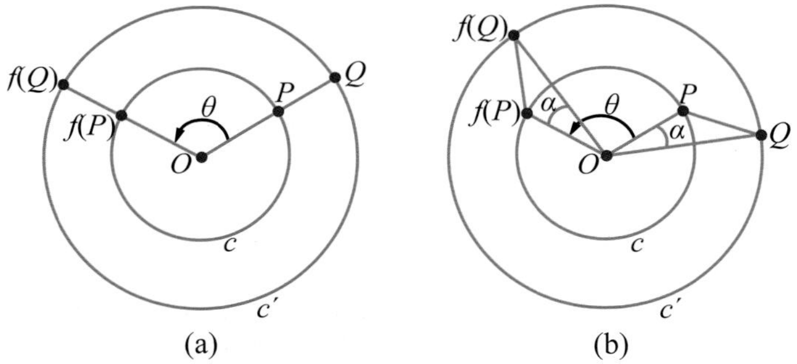

Figure 5.15: Illustrations for the proof of Proposition 5.6.

Let Q be an arbitrary point on the plane. Without loss of generality assume $Q \neq O$. Reasoning as before, then, both Q and $f(Q)$ lie on some circle c' centered at O. By the distance-preserving property

$$|PQ| = |f(P)f(Q)|$$

Consider first the case that Q lies on the straight line through O and P (Figure 5.15(a)). Because $|PQ| = |f(P)f(Q)|$, it's seen that $f(Q)$ lies at the intersection with c' of the straight line through O and $f(P)$, as all other points on c' are at a distance more than $|PQ|$ from $f(P)$. In this case, $f(Q)$ is indeed obtained by rotating Q counter-clockwise by an angle of θ about the origin.

Next, consider the case that Q does not lie on the straight line through O and P. First, suppose that the vertex order POQ of triangle POQ appears CCW to the viewer (Figure 5.15(b)). Let the angle POQ be α.

The congruence of the triangles POQ and $f(P)Of(Q)$, a consequence of the distance-preserving property, implies that angle $f(P)Of(Q)$ is α as well. Furthermore, f being rigid preserves orientation, so $f(P)Of(Q)$ appears CCW to the viewer as well. It follows from simple angular arithmetic that $f(Q)$ is θ counter-clockwise about the origin from Q.

If the vertex order POQ appears CW instead, a similar conclusion can still be reached. This completes the proof that, if f is a rigid transformation keeping the origin fixed, then it is a rotation about the origin.

Suppose, next, that f is an arbitrary rigid transformation, not necessarily fixing the origin. Let $f(O) = O'$ and t be translation by the displacement vector $O'O$. Then the transformation $f' = t \circ f$ is a rigid transformation such that $f'(O) = O$, i.e., fixing the origin. Therefore, as proved earlier, f' is a rotation about the origin. Consequently, $f = t^{-1} \circ f'$ is a rotation about the origin followed by a translation, proving the statement of the proposition about arbitrary rigid transformations and completing the proof of the first paragraph.

Note: If f is itself a translation then, of course, the rotation f' about the origin is the identity, i.e., zero rotation.

For the second paragraph of the proposition, suppose that f is an orientation-reversing Euclidean transformation, because if f is orientation-preserving, then it is rigid, and there is nothing to prove after the first paragraph.

Let w be a reflection about any mirror, an orientation-reversing Euclidean transformation by Proposition 5.5. Then $f' = w \circ f$, being a composition of two orientation-reversing Euclidean transformations, is rigid by Exercise 5.39. By the first part of the proposition, f' is a rotation about the origin followed by a translation, implying that $f = w^{-1} \circ f'$ is the composition of a rotation about the origin, followed by a translation and, then, a reflection. This completes the proof of the second paragraph. □

Exercise 5.40. Apply the proposition to show that a rigid transformation which keeps

(a) no point fixed is a translation.

(b) exactly one point fixed is a rotation (about the fixed point as center).

(c) more than one point fixed is the identity (which, therefore, keeps every point fixed, so also is a zero translation and a zero rotation).

Exercise 5.41. Use Exercises 5.16 and 5.40 to prove that any Euclidean transformation can be obtained by composing reflections about at most three mirrors.

Exercise 5.42. At the end of Section 5.1.3 we saw one case that the composition of two rotations is a translation and one where it is again a rotation. Use Exercise 5.40 to prove now that these are the only two possibilities in general for the composition of two rotations.

Moreover, show how to decide which case arises by proving:

1. The composition of two rotations, either both counter-clockwise or both clockwise, one of angle θ_1 and one of angle θ_2, about arbitrary centers, is a translation if either $\theta_1 = \theta_2 = 0$ or $\theta_1 + \theta_2 = 2\pi$ (assume $0 \le \theta_1, \theta_2 < 2\pi$); otherwise, it is a rotation.

2. The composition of two rotations, one counter-clockwise of angle θ_1 and the other clockwise of angle θ_2, about arbitrary centers, is a translation if $\theta_1 = \theta_2$ (assume $0 \le \theta_1, \theta_2 < 2\pi$); otherwise, it is a rotation.

Proposition 5.7. *Affine, Euclidean and rigid transformations of 2-space are related by the following inclusions, which are each proper:*

$$\textit{rigid transforms} \subset \textit{Euclidean transforms} \subset \textit{affine transforms}$$

Proof. The first inclusion follows from the definitions. It is proper because a reflection about any mirror is Euclidean but not rigid.

From Proposition 5.6 it follows that a Euclidean transformation is a composition of affine transformations (because translations, rotations and reflections are all affine) and, therefore, itself affine, proving the second inclusion. The inclusion is proper because a scaling by factors not all of unit magnitude is affine but not Euclidean. □

Remark 5.4. An interesting perspective on the proposition is to think of affine transformations as being made from translations, rotations, reflections and scalings; Euclidean transformations from translations, rotations and reflections; and rigid transformations from translations and rotations.

The worked example next says that Definition 5.5 about whether a Euclidean transformation reverses or preserves orientation is, in fact, independent of the choice of the three non-collinear points P, Q and R.

Example 5.10. Suppose that an affine transformation g of $\mathbb{R}^2$ maps *some* three non-collinear points P, Q and R in a manner that, looking at $\mathbb{R}^2$ from a fixed side, one of the sequences PQR and $g(P)g(Q)g(R)$ appears CW and the other CCW.

Show, then, that for *any* three non-collinear points X, Y and Z, one of the sequences XYZ and $g(X)g(Y)g(Z)$ appears CW and the other CCW, looking at $\mathbb{R}^2$ from the same side.

Answer: Use homogeneous coordinates to write $[g(W)\ 1]^T = M[W\ 1]^T$, where M is a fixed non-singular 3×3 matrix, and $W = [x\ y]^T$ is an arbitrary point of the plane.

Suppose that $P = [x_1\ y_1]^T$, $Q = [x_2\ y_2]^T$ and $R = [x_3\ y_3]^T$. Consider the equation

$$M \begin{bmatrix} x_1 & x_2 & x_3 \\ y_1 & y_2 & y_3 \\ 1 & 1 & 1 \end{bmatrix} = \begin{bmatrix} x_1' & x_2' & x_3' \\ y_1' & y_2' & y_3' \\ 1 & 1 & 1 \end{bmatrix}$$

in matrices, which gives the following

$$det(M) * \begin{vmatrix} x_1 & x_2 & x_3 \\ y_1 & y_2 & y_3 \\ 1 & 1 & 1 \end{vmatrix} = \begin{vmatrix} x'_1 & x'_2 & x'_3 \\ y'_1 & y'_2 & y'_3 \\ 1 & 1 & 1 \end{vmatrix}$$

relating determinants. Now, $[x'_1\ y'_1\ 1]^T = M[x_1\ y_1\ 1]^T = M[P\ 1]^T = [g(P)\ 1]^T$. Likewise, $[x'_2\ y'_2\ 1]^T = [g(Q)\ 1]^T$ and $[x'_2\ y'_2\ 1]^T = [g(R)\ 1]^T$. Therefore, the preceding equation can be written

$$det(M) * \begin{vmatrix} P & Q & R \\ 1 & 1 & 1 \end{vmatrix} = \begin{vmatrix} g(P) & g(Q) & g(R) \\ 1 & 1 & 1 \end{vmatrix}$$

Considering the signs of the three determinants above, and applying Lemma 5.1, one sees that PQR and $g(P)g(Q)g(R)$ appear differently oriented, from a fixed side of the plane, if and only if $det(M)$ is negative. But, then, by similar calculations, exactly the same would be true of XYZ and $g(X)g(Y)g(Z)$, for any points X, Y and Z.

5.3.3 Shear

With translations, rotations and scalings, and their compositions, we know that we "cover" all affine transformations. Shears, though, are a particularly distinctive kind of affine transformation that arise naturally from physical processes. For this reason they merit separate discussion. Roughly, a shear is the kind of distortion caused by placing a lump of putty between a pair of palms and then moving one palm parallel to the other.

A 2D shear s is uniquely determined by two parameters:

1. A directed line l called the *line of shear*.

2. An angle α called the *angle of shear*.

Here's how a point $P \in \mathbb{R}^2$ is mapped to the point P' by s (see Figure 5.16(a)):

(a) If P lies on l, then it is unchanged.

(b) If P lies a distance of h left of l, then it moves parallel to l in the positive direction of l a distance of $h \tan \alpha$.

(c) If P lies a distance of h right of l, then it moves parallel to l in the negative direction of l a distance of $h \tan \alpha$.

Note: Left or right is according to a viewer standing upright on the plane at a point of l, head pointing toward the positive z-axis (of a hypothetical right-handed coordinate system) and facing toward the direction of l.

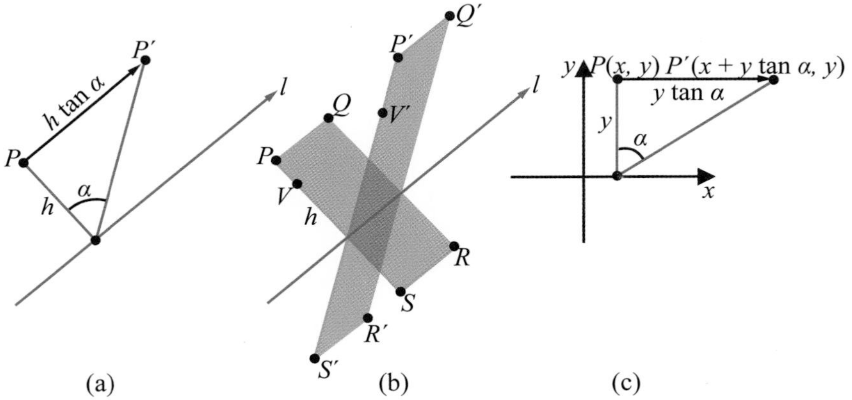

Figure 5.16: 2D shears: l is a directed line, α the angle of shear.

Another way to think of the shear is as a force parallel to l which "bends" each perpendicular to it a fixed angle α. The rectangle $PQRS$ in Figure 5.16(b) is sheared into the parallelogram $P'Q'R'S'$. The farther points are from l, the proportionately more they travel under the shear; e.g., compare $V \mapsto V'$ and $P \mapsto P'$ in Figure 5.16(b). Figure 5.17 shows a sheared sheep (pictorial pun).

Figure 5.17: Sheared sheep.

If the directed line l of the shear is the x-axis, then it is particularly simple to determine its transformation equation. See Figure 5.16(c). The shear maps the point $P = [x\ y]^T$ to the point

$$P' = [x + y\tan\alpha \quad y]^T = \begin{bmatrix} 1 & \tan\alpha \\ 0 & 1 \end{bmatrix} P \tag{5.22}$$

A shear along the x-axis, then, is a non-singular linear transformation given by Equation (5.22). Therefore, by Proposition 5.4, it is equivalent to a rotation about the origin, followed by a scaling, followed by another rotation about the origin. In fact, write

$$\begin{bmatrix} 1 & \tan\alpha \\ 0 & 1 \end{bmatrix} = \begin{bmatrix} \cos\phi & -\sin\phi \\ \sin\phi & \cos\phi \end{bmatrix} \begin{bmatrix} s_x & 0 \\ 0 & s_y \end{bmatrix} \begin{bmatrix} \cos\theta & -\sin\theta \\ \sin\theta & \cos\theta \end{bmatrix} \tag{5.23}$$

It turns out that solving this equation is simpler than solving the more general (5.17). Indeed, it may be verified that the following four equations derive from (5.23):

$$\tan^2\theta + \tan\alpha\tan\theta - 1 = 0 \tag{5.24}$$

$$\phi = \theta - 90^\circ \tag{5.25}$$

$$s_x = \tan\theta \tag{5.26}$$

$$s_y = \frac{1}{\tan\theta} \tag{5.27}$$

The value of θ can then be calculated from (5.24) and those of ϕ, s_x and s_y subsequently from (5.25)-(5.27). In fact, it's interesting to visualize a shear along the x-axis as a rotation-scaling-rotation as in Figure 5.18.

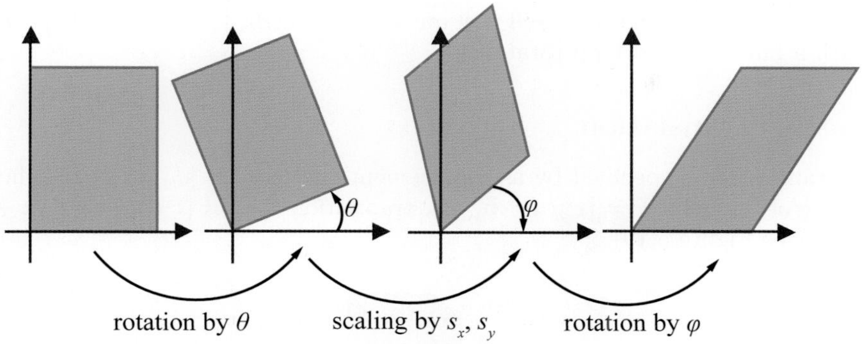

Figure 5.18: A shear as a rotation-scaling-rotation.

Exercise 5.43. Verify that the Equations (5.24)-(5.27) indeed follow from (5.23).

Example 5.11. Let s be the shear of angle $\alpha = 45°$ along the x-axis. Then $\tan \alpha = 1$ and Equation (5.24) in this case becomes

$$\tan^2 \theta + \tan \theta - 1 = 0$$

solving to (ignoring the negative root)

$$\tan \theta = \frac{-1 + \sqrt{5}}{2} \simeq 0.618034$$

so that

$$\theta \simeq 31.72°$$

Equations (5.25)-(5.27) give next:

$$\begin{aligned} \phi &\simeq -58.28° \\ s_x &\simeq 0.62 \\ s_y &\simeq 1.62 \end{aligned}$$

Therefore, s is equivalent to a rotation of 31.72° counter-clockwise about the origin, followed by scaling by factors 0.62, 1.62 along the x- and y-axes, respectively, followed by a rotation of 58.28° clockwise about the origin.

Exercise 5.44. Express the shear of angle 30° along the y-axis as a rotation-scaling-rotation.

Remark 5.5. We'll code shears in Section 5.4.8 following a discussion of their 3D version.

5.4 Geometric Transformations in 3-Space

Finally, the real world. Our discussions will mirror those of the previous section. In fact, extending translations, scalings and reflections from 2D to 3D is almost automatic. We'll pay our dues, though, for entering 3-space with a fair bit of work on rotations.

5.4.1 Translation

A translation is specified by a displacement vector $D = [d_x \ d_y \ d_z]^T$. The image of the point $P = [x \ y \ z]^T$ by this translation is $P' = [x + d_x \ y + d_y \ z + d_z]^T$ (see Figure 5.19).

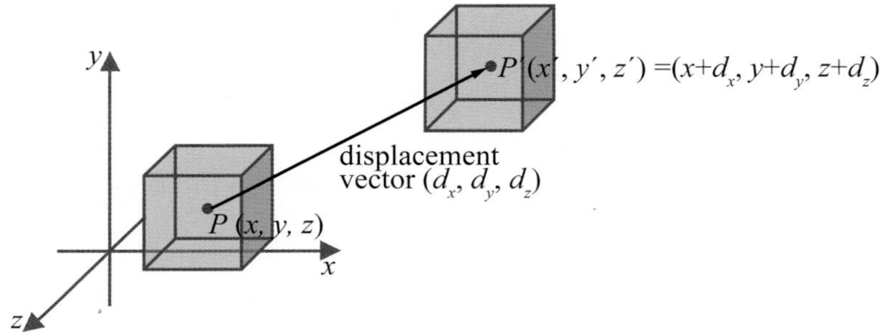

Figure 5.19: Translation.

Equivalently,

$$P' = P + D = \begin{bmatrix} 1 & 0 & 0 \\ 0 & 1 & 0 \\ 0 & 0 & 1 \end{bmatrix} P + D$$

which in homogeneous form, analogous to the 2D version (5.13), is

$$\begin{bmatrix} P' \\ 1 \end{bmatrix} = \begin{bmatrix} 1 & 0 & 0 & d_x \\ 0 & 1 & 0 & d_y \\ 0 & 0 & 1 & d_z \\ 0 & 0 & 0 & 1 \end{bmatrix} \begin{bmatrix} P \\ 1 \end{bmatrix} \tag{5.28}$$

For the record, the 4×4 matrix corresponding to translation by the displacement vector $[d_x \ d_y \ d_z]^T$ is denoted $T(d_x, d_y, d_z)$ and given by

$$T(d_x, d_y, d_z) = \begin{bmatrix} 1 & 0 & 0 & d_x \\ 0 & 1 & 0 & d_y \\ 0 & 0 & 1 & d_z \\ 0 & 0 & 0 & 1 \end{bmatrix} \tag{5.29}$$

Exercise 5.45. Write the 4×4 matrix corresponding to translation by the displacement vector $[3\ 0\ -1]^T$.

Exercise 5.46. Use Equation (5.28) to prove that the composition of translations is a translation and that the inverse of a translation is a translation as well.

Note: By default we're in 3-space from now on and all exercises and examples are in 3D.

5.4.2 Scaling

A scaling is specified by scaling factors s_x, s_y and s_z along the x-, y- and z-axis, respectively. The image of the point $P = [x\ y\ z]^T$ by this scaling is $P' = [s_x x\ \ s_y y\ \ s_z z]^T$ (see Figure 5.20).

Without further ado we write the 4×4 matrix corresponding to this scaling as (compare the 2D Equation (5.14))

$$S(s_x, s_y, s_z) = \begin{bmatrix} s_x & 0 & 0 & 0 \\ 0 & s_y & 0 & 0 \\ 0 & 0 & s_z & 0 \\ 0 & 0 & 0 & 1 \end{bmatrix} \tag{5.30}$$

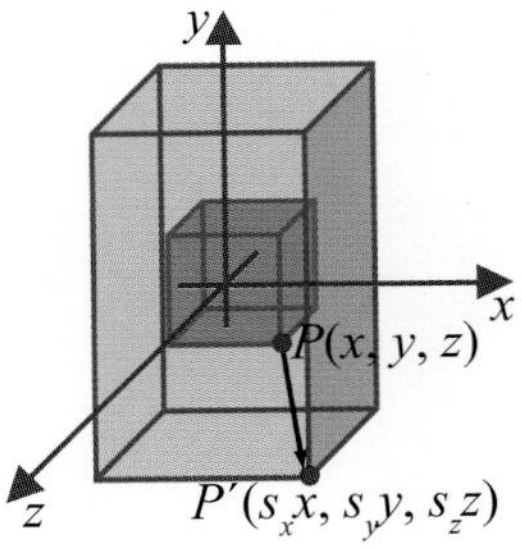

Figure 5.20: Scaling.

If any one or more of the scaling factors s_x, s_y and s_z is zero, the scaling is said to be *degenerate*; otherwise, it is *non-degenerate*. Clearly, a scaling is non-degenerate if and only if its matrix is non-singular. By a scaling we shall always mean a non-degenerate one, unless stated otherwise.

Exercise 5.47. Write the 4×4 matrix corresponding to scaling by the factors -1, 3 and 4 along the x-, y- and z-axis, respectively.

Exercise 5.48. Use (5.30) to prove that the composition of scalings is a scaling and that the inverse of a non-degenerate scaling is a non-degenerate scaling.

5.4.3 Rotation

Warning upfront: This section is much longer than the corresponding Section 5.1.3 on 2D rotations as there's much more magic going around in 3D!

A rotation about a radial axis is specified by (a) a directed line l through the origin, which is the axis of rotation, and (b) the angle θ of the rotation.

We'll describe as a physical process how such a rotation maps a point P. First, if P lies on the axis l itself, then it does not move. Suppose, then, that P does not lie on l. Here's how it's mapped by the rotation (see Figure 5.21):

1. Drop the perpendicular from P to the point Q on l. Denote as L the segment PQ. L lies on the plane h perpendicular to l through Q.

   Note that Figure 5.21 has Q and h on the positive side of l, but they could very well be on the other side, or even touching the origin, depending on where P is.

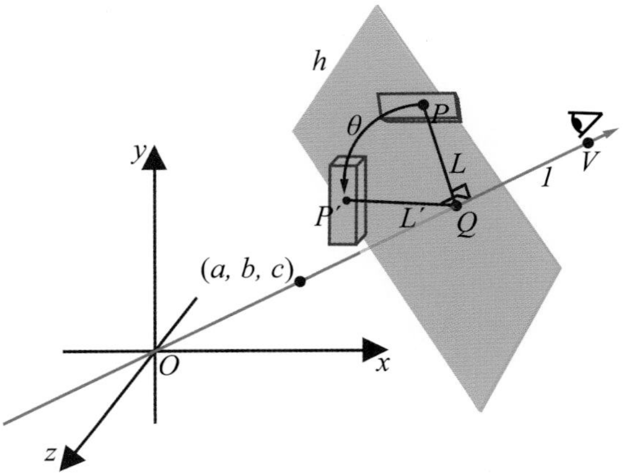

Figure 5.21: Rotation.

2. Locate a viewer at V far enough in the positive direction of l as to be able to see h when looking toward the origin.

3. Rotate the segment L about Q (on the plane h) an angle θ counter-clockwise, as measured by the viewer.

4. If L' is the new position of L after rotation, then P is mapped to the corresponding endpoint P' of L'.

Remark 5.6. Giving a single point (a, b, c), not equal to the origin, is enough to specify the directed radial line l through it, as indicated in Figure 5.21. Therefore, all that remains to specify a rotation about l is the angle θ. This, of course, is exactly how the OpenGL command `glRotatef(θ, a, b, c)` works, as described earlier in Section 4.1.3.

Rotation about the Coordinate Axes

The matrices corresponding to rotations in 3D about the coordinate axes are straightforwardly deduced from the 2D equation (5.3), reproduced below

$$\begin{bmatrix} x' \\ y' \end{bmatrix} = \begin{bmatrix} \cos\theta & -\sin\theta \\ \sin\theta & \cos\theta \end{bmatrix} \begin{bmatrix} x \\ y \end{bmatrix} \tag{5.31}$$

where $[x' \; y']^T$ is the image of $[x \; y]^T$ by a rotation on the xy-plane by an angle of θ about the origin, measured counter-clockwise by a viewer V on the positive side of the z-axis (Figure 5.22(a)).

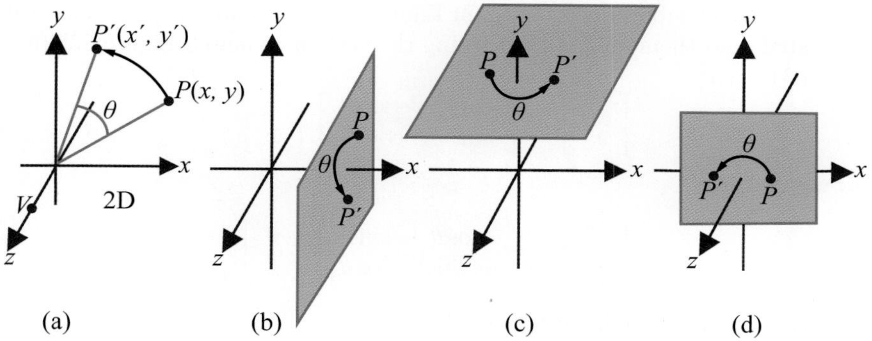

Figure 5.22: (a) 2D rotation on the xy-plane (b)-(d) 3D rotations about the coordinate axes.

In 3D, rotation about the x-axis by an angle θ (Figure 5.22(b)) maps a point $P = [x \; y \; z]^T$ to the point $P' = [x' \; y' \; z']^T$, where

(a) $x' = x$, because P travels parallel to the yz-plane, so its x value never changes.

(b) $[y \; z]^T \mapsto [y' \; z']^T$ is precisely as for a *2D rotation* by an angle θ CCW about the origin on the yz-plane, looking from the positive side of the x-axis.

Therefore, replacing x with y and y with z in (5.31), we have

$$\begin{bmatrix} y' \\ z' \end{bmatrix} = \begin{bmatrix} \cos\theta & -\sin\theta \\ \sin\theta & \cos\theta \end{bmatrix} \begin{bmatrix} y \\ z \end{bmatrix}$$

Therefore, the 4×4 matrix of 3D rotation about the x-axis is

$$R_x(\theta) = \begin{bmatrix} 1 & 0 & 0 & 0 \\ 0 & \cos\theta & -\sin\theta & 0 \\ 0 & \sin\theta & \cos\theta & 0 \\ 0 & 0 & 0 & 1 \end{bmatrix} \tag{5.32}$$

(the first row serving to keep x unchanged).

Rotation about the y-axis by an angle θ (Figure 5.22(c)) maps a point $P = [x \; y \; z]^T$ to the point $P' = [x' \; y' \; z']^T$, where:

(a) $y' = y$

(b) $[x \; z]^T \mapsto [x' \; z']^T$ is as for a 2D rotation by an angle θ CCW about the origin on the xz-plane, looking from the positive side of the y-axis.

We have to be careful, though, in applying (5.31). For, compare Figure 5.22(a) with Figure 5.22(c) to observe that the role of x in the 2D figure is played by z in the 3D one, that of the 2D y by the 3D x, and, of course, that of the 2D z (the viewer's axis) by the 3D y. (You can verify this by scratching out the current labels on the axes in Figure 5.22(a), relabeling them as just suggested, and then "mentally" turning the system to match Figure 5.22(c).) So, (5.31) gives

$$\begin{bmatrix} z' \\ x' \end{bmatrix} = \begin{bmatrix} \cos\theta & -\sin\theta \\ \sin\theta & \cos\theta \end{bmatrix} \begin{bmatrix} z \\ x \end{bmatrix}$$

or, equivalently,

$$\begin{bmatrix} x' \\ z' \end{bmatrix} = \begin{bmatrix} \cos\theta & \sin\theta \\ -\sin\theta & \cos\theta \end{bmatrix} \begin{bmatrix} x \\ z \end{bmatrix}$$

Finally, since y is fixed, we have

$$R_y(\theta) = \begin{bmatrix} \cos\theta & 0 & \sin\theta & 0 \\ 0 & 1 & 0 & 0 \\ -\sin\theta & 0 & \cos\theta & 0 \\ 0 & 0 & 0 & 1 \end{bmatrix} \tag{5.33}$$

We ask the reader to verify that the matrix of rotation about the z-axis by an angle θ (Figure 5.22(d)) is

$$R_z(\theta) = \begin{bmatrix} \cos\theta & -\sin\theta & 0 & 0 \\ \sin\theta & \cos\theta & 0 & 0 \\ 0 & 0 & 1 & 0 \\ 0 & 0 & 0 & 1 \end{bmatrix} \tag{5.34}$$

Exercise 5.49. Write the 4×4 matrix corresponding to rotation by an angle of 30° about the y-axis.

Rotation about an Arbitrary Radial Axis

It is a bit of work to find the matrix corresponding to rotation about an arbitrary axis through the origin. But it's important enough that we'll do it in two different ways. The first is mainly geometric and fairly intuitive. The second involves a bit of algebraic legerdemain, so it is a little less intuitive, but the final form it yields is more compact than that of the first.

Let the axis of rotation be specified as the directed line l through the origin O toward a point $P = (a, b, c)$ ($\neq O$), and the angle of rotation as θ. See Figure 5.23. To simplify computation we'll assume that P is a unit vector, i.e., $|P| = \sqrt{a^2 + b^2 + c^2} = 1$. There is no loss in generality because, if P is not of unit length, we can always divide it by $|P|$ to obtain a unit vector specifying the same rotation. Our goal is to compute the matrix, denote it $R_{a,\,b,\,c}(\theta)$, corresponding to this rotation.

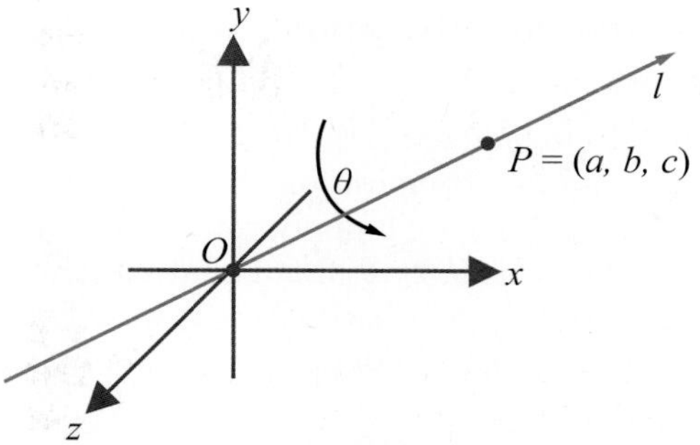

Figure 5.23: Rotating about an arbitrary radial axis.

Remark 5.7. To be honest, at this point we don't even know if a rotation about an arbitrary radial axis has a matrix representation at all, in other words, if it is a linear transformation!

Before we proceed, here's a possible temptation, and, then, an exercise to nip it in the bud.

Can't we simply "add rotational axes" like vectors? For example, isn't it true, say, that `glRotatef(90.0, 0.0, 1.0, 1.0)` is the same as `glRotatef(90.0, 0.0, 1.0, 0.0)` followed by `glRotatef(90.0, 0.0, 0.0, 1.0)` or, maybe, the other way around? Certainly, translations do work this way: `glTranslatef(0.0, 1.0, 1.0)` is, indeed, `glTranslatef(0.0, 1.0, 0.0)` followed by `glTranslatef(0.0, 0.0, 1.0)`, or vice versa.

If rotational axes could be so added, then writing the matrix corresponding to `glRotatef(90.0, 0.0, 1.0, 1.0)` would be simple: it would be the product of the matrices corresponding to `glRotatef(90.0, 0.0, 1.0, 0.0)` and `glRotatef(90.0, 0.0, 0.0, 1.0)` in some order, both matrices easily written from what we know already about rotating about the coordinate axes themselves.

Exercise 5.50. Prove that we *cannot*, in general, add rotational axes. In fact, show that `glRotatef(90.0, 0.0, 1.0, 1.0)` is neither `glRotatef(90.0, 0.0, 1.0, 0.0)` followed by `glRotatef(90.0, 0.0, 0.0, 1.0)`, nor the other way around.

Hint: If, say, `glRotatef(90.0, 0.0, 1.0, 1.0)` were equal to the transformation `glRotatef(90.0, 0.0, 1.0, 0.0)` followed by `glRotatef(90.0, 0.0, 0.0, 1.0)`, then the two would move all points identically. Consider the point $(0, 1, 1)$. How is it moved by `glRotatef(90.0, 0.0, 1.0, 1.0)`? By `glRotatef(90.0, 0.0, 1.0, 0.0)` followed by `glRotatef(90.0, 0.0, 0.0, 1.0)`?

A Method to Compute the Rotation Matrix Which Is Mainly Geometric

Even though we can't quite add axes, the plan is still to express the rotation of θ about the radial axis l as a composition of rotations about the coordinate axes. We'll use the Trick. First we'll apply rotations to align l along one of the coordinate axes, then rotate by θ about that coordinate axis and, last, undo the initial rotations to bring l back where it was. For our plan to work, of course, the rotations to align l along a coordinate axis must themselves be about coordinate axes!

Here's a simple motivating experiment.

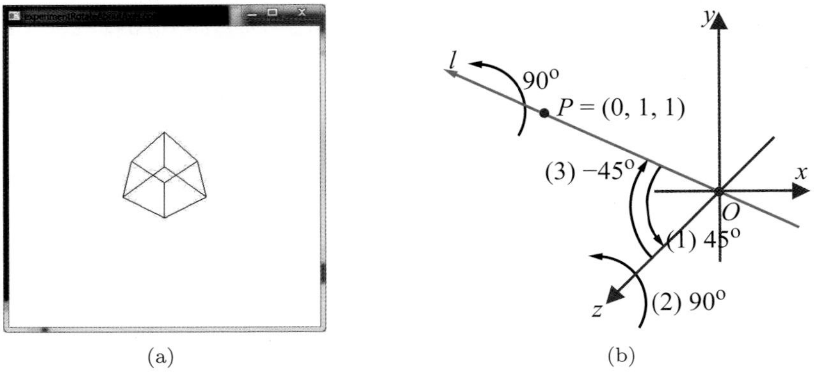

Figure 5.24: Experiment 5.1: (a) Screenshot of output (b) Trick-based rotation scheme.

Experiment 5.1. Fire up `box.cpp` and insert a rotation command – in fact, the same one as in the previous exercise – just before the box definition so that the transformation and object definition part of the drawing routine becomes:

```
// Modeling transformations.
glTranslatef(0.0, 0.0, -15.0);

glRotatef(90.0, 0.0, 1.0, 1.0);
glutWireCube(5.0); // Box.
```

The rotation command asks to rotate 90° about the line l from the origin through $(0, 1, 1)$. See Figure 5.24(a) for the displayed output.

Let's try now, instead, to use the strategy suggested above to express the given rotation in terms of rotations about the coordinate axes. Figure 5.24(b) illustrates the following simple scheme. Align l along the z-axis by rotating it 45° about the x-axis. Therefore, the given rotation should be equivalent to (1) a rotation of 45° about the x-axis, followed by (2) a rotation of 90° about the z-axis followed, finally, by a (3) rotation of −45° about the x-axis.

Give it a whirl. Replace the single rotation command glRotatef(90.0, 0.0, 1.0, 1.0) with a block of three as follows:

```
// Modeling transformations.
glTranslatef(0.0, 0.0, -15.0);

glRotatef(-45.0, 1.0, 0.0, 0.0);
glRotatef(90.0, 0.0, 0.0, 1.0);
glRotatef(45.0, 1.0, 0.0, 0.0);
glutWireCube(5.0); // Box.
```

Seeing is believing, is it not?! **End**

Returning to the general problem, let's plan to rotate to align l along the z-axis in a manner that $P = (a, b, c)$ maps to the point $P'' = (0, 0, 1)$ on the positive side of the z-axis. We accomplish this by applying two successive rotations (see Figure 5.25):

(1) Rotate l an angle α about the x-axis onto a line l' on the xz-plane, taking P to P'.

(2) Rotate l' an angle $-\beta$ about the y-axis till it's aligned along the z-axis, taking P' to P'' (the minus sign in front of β is because the rotation is CW).

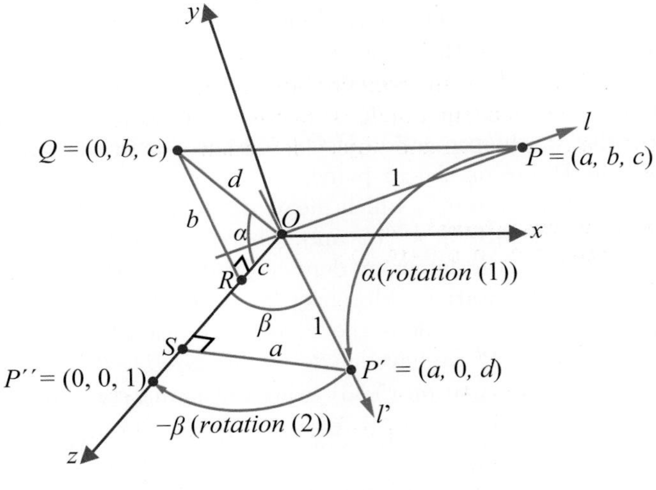

Figure 5.25: Aligning l along the z-axis.

Note: The choice of the z-axis as l's final alignment was arbitrary – it could have been any of the three coordinate axes.

We must determine α and β. In fact, we'll simply determine the sine and cosine of both, which is sufficient to write the matrices $R_x(\alpha)$ and $R_y(-\beta)$ corresponding to the rotations (1) and (2), respectively.

Observe that the angle α that OP turns by rotation (1) about the x-axis is the same as the angle between its projection OQ on the yz-plane and the positive direction of the z-axis. In fact, imagine OQ as the "shadow" of OP cast on the yz-plane by a light shining down the x-axis – as OP turns so does its shadow, and by the same amount.

The coordinates of Q are $[0\ b\ c]^T$ as Q is the projection of $[a\ b\ c]^T$ on the yz-plane. Drop the perpendicular from Q to the point R on the z-axis. The angle QOR then is equal to α. We see from the coordinates of Q that $|OR| = c$ and $|RQ| = b$. Denoting $|OQ|$ by d, it follows from the right-angled triangle ORQ that $d = \sqrt{b^2 + c^2}$. Therefore, *assuming that* $d \neq 0$, we have $\sin\alpha = \frac{b}{d}$ and $\cos\alpha = \frac{c}{d}$.

Note: We don't lose any generality in assuming $d \neq 0$, because $d = 0$ means $Q = O$, which in turn means P is on the x-axis, implying that l lies along the x-axis as well, in which case we already know the matrix for rotation about l.

We can now use Equation (5.32) to write the matrix of rotation (1) as

$$R_x(\alpha) = \begin{bmatrix} 1 & 0 & 0 & 0 \\ 0 & \cos\alpha & -\sin\alpha & 0 \\ 0 & \sin\alpha & \cos\alpha & 0 \\ 0 & 0 & 0 & 1 \end{bmatrix} = \begin{bmatrix} 1 & 0 & 0 & 0 \\ 0 & c/d & -b/d & 0 \\ 0 & b/d & c/d & 0 \\ 0 & 0 & 0 & 1 \end{bmatrix}$$

After rotation (1), l coincides with l' on the xz-plane, and P with P'. The x coordinate of P' is a, the same as that of P, because rotation about the x-axis leaves this value unchanged; the z coordinate is d because rotation by α about the x-axis causes OQ, the shadow of OP whose length is d, to coincide with OS, the projection of OP' on the z-axis; the y coordinate, of course, is 0 as P' lies on the xz-plane.

Therefore, $P' = [a\ 0\ d]^T$, which means that in the right-angled triangle OSP', $|OS| = d$, $|SP'| = a$, and, therefore, $|OP'| = \sqrt{a^2 + d^2} = \sqrt{a^2 + b^2 + c^2} = 1$ (the latter evident as well from the fact that OP' is the unit vector OP rotated). Moreover, angle $P'OS = \beta$, where $-\beta$ is the angle turned by l' to align along the z-axis in rotation (2) above. Therefore, from the triangle OSP' we have $\sin\beta = a$ and $\cos\beta = d$.

We can now use Equation (5.33) to write the matrix of rotation (2) as

$$R_y(-\beta) = \begin{bmatrix} \cos(-\beta) & 0 & \sin(-\beta) & 0 \\ 0 & 1 & 0 & 0 \\ -\sin(-\beta) & 0 & \cos(-\beta) & 0 \\ 0 & 0 & 0 & 1 \end{bmatrix} = \begin{bmatrix} d & 0 & -a & 0 \\ 0 & 1 & 0 & 0 \\ a & 0 & d & 0 \\ 0 & 0 & 0 & 1 \end{bmatrix}$$

Returning to our original Trick-based plan, the first step of aligning l along the z-axis is accomplished, then, by the composition $R_y(-\beta)\ R_x(\alpha)$.

The next, of rotating by θ about the z-axis, is simply a matter of applying $R_z(\theta)$. Finally, the initial rotations aligning l along the z-axis are undone by the inverse transformation $(R_y(-\beta)\ R_x(\alpha))^{-1} = R_x(\alpha)^{-1}\ R_y(-\beta)^{-1} = R_x(-\alpha)\ R_y(\beta)$.

Putting everything together we have, finally,

$$\begin{aligned} R_{a,\,b,\,c}(\theta) &= R_x(-\alpha)\ R_y(\beta)\ R_z(\theta)\ R_y(-\beta)\ R_x(\alpha) \\ &= \begin{bmatrix} 1 & 0 & 0 & 0 \\ 0 & c/d & b/d & 0 \\ 0 & -b/d & c/d & 0 \\ 0 & 0 & 0 & 1 \end{bmatrix} \begin{bmatrix} d & 0 & a & 0 \\ 0 & 1 & 0 & 0 \\ -a & 0 & d & 0 \\ 0 & 0 & 0 & 1 \end{bmatrix} \\ &\quad \begin{bmatrix} \cos\theta & -\sin\theta & 0 & 0 \\ \sin\theta & \cos\theta & 0 & 0 \\ 0 & 0 & 1 & 0 \\ 0 & 0 & 0 & 1 \end{bmatrix} \begin{bmatrix} d & 0 & -a & 0 \\ 0 & 1 & 0 & 0 \\ a & 0 & d & 0 \\ 0 & 0 & 0 & 1 \end{bmatrix} \\ &\quad \begin{bmatrix} 1 & 0 & 0 & 0 \\ 0 & c/d & -b/d & 0 \\ 0 & b/d & c/d & 0 \\ 0 & 0 & 0 & 1 \end{bmatrix} \end{aligned} \tag{5.35}$$

The five matrices on the right side of the formula can be multiplied to give a single matrix, which would then be the value of $R_{a,\,b,\,c}(\theta)$. However, we'll not do so as the next method to calculate $R_{a,\,b,\,c}(\theta)$ gives a more concise form directly.

Exercise 5.51. Is rotation about an arbitrary radial axis a linear transformation? If so, is it always non-singular, or can it be singular?

Exercise 5.52. Use the Trick to write a rotation about an arbitrary axis l, not necessarily radial, as a seven-matrix product.

Example 5.12. (a) Determine the 4×4 matrix corresponding to a $90°$ rotation about the radial axis l directed toward the point $[1\ 1\ 1]^T$, which corresponds to the OpenGL command `glRotatef(90.0, 1.0, 1.0, 1.0)`.

(b) Express `glRotatef(90.0, 1.0, 1.0, 1.0)` as a composition of five successive rotations about the coordinate axes and experimentally verify.

Answer:

(a) The unit vector along l in the direction of $[1\ 1\ 1]^T$ is $P = [\frac{1}{\sqrt{3}}\ \frac{1}{\sqrt{3}}\ \frac{1}{\sqrt{3}}]^T$. Accordingly, keeping the notation used above,

$$a = b = c = \frac{1}{\sqrt{3}}\,, \quad d = \sqrt{b^2 + c^2} = \frac{\sqrt{2}}{\sqrt{3}} \quad \text{and, of course,} \quad \theta = \pi/2.$$

Plugging these values into (5.35) we get the required matrix as

$$R_{\overline{P}}(\theta) = \begin{bmatrix} 1 & 0 & 0 & 0 \\ 0 & \frac{1}{\sqrt{2}} & \frac{1}{\sqrt{2}} & 0 \\ 0 & -\frac{1}{\sqrt{2}} & \frac{1}{\sqrt{2}} & 0 \\ 0 & 0 & 0 & 1 \end{bmatrix} \begin{bmatrix} \frac{\sqrt{2}}{\sqrt{3}} & 0 & \frac{1}{\sqrt{3}} & 0 \\ 0 & 1 & 0 & 0 \\ -\frac{1}{\sqrt{3}} & 0 & \frac{\sqrt{2}}{\sqrt{3}} & 0 \\ 0 & 0 & 0 & 1 \end{bmatrix}$$

$$\begin{bmatrix} 0 & -1 & 0 & 0 \\ 1 & 0 & 0 & 0 \\ 0 & 0 & 1 & 0 \\ 0 & 0 & 0 & 1 \end{bmatrix}$$

$$\begin{bmatrix} \frac{\sqrt{2}}{\sqrt{3}} & 0 & -\frac{1}{\sqrt{3}} & 0 \\ 0 & 1 & 0 & 0 \\ \frac{1}{\sqrt{3}} & 0 & \frac{\sqrt{2}}{\sqrt{3}} & 0 \\ 0 & 0 & 0 & 1 \end{bmatrix} \begin{bmatrix} 1 & 0 & 0 & 0 \\ 0 & \frac{1}{\sqrt{2}} & -\frac{1}{\sqrt{2}} & 0 \\ 0 & \frac{1}{\sqrt{2}} & \frac{1}{\sqrt{2}} & 0 \\ 0 & 0 & 0 & 1 \end{bmatrix}$$

$$= \begin{bmatrix} \frac{1}{3} & \frac{1}{3} - \frac{1}{\sqrt{3}} & \frac{1}{3} + \frac{1}{\sqrt{3}} & 0 \\ \frac{1}{3} + \frac{1}{\sqrt{3}} & \frac{1}{3} & \frac{1}{3} - \frac{1}{\sqrt{3}} & 0 \\ \frac{1}{3} - \frac{1}{\sqrt{3}} & \frac{1}{3} + \frac{1}{\sqrt{3}} & \frac{1}{3} & 0 \\ 0 & 0 & 0 & 1 \end{bmatrix}$$

after some tedious computation.

(b) Now, (5.35) says that

$$R_{\overline{P}}(\theta) = R_x(-\alpha)\ R_y(\beta)\ R_z(\theta)\ R_y(-\beta)\ R_x(\alpha)$$

where $\alpha = \sin^{-1}\frac{1}{\sqrt{2}} = 45°$ and $\beta = \sin^{-1}\frac{1}{\sqrt{2}} = 35.26°$. So,

$$R_{\overline{P}}(\theta) = R_x(-45°)\ R_y(35.26°)\ R_z(90°)\ R_y(-35.26°)\ R_x(45°)$$

which means `glRotatef(90.0, 1.0, 1.0, 1.0)` is equivalent to the sequence

```
glRotatef(-45.0, 1.0, 0.0, 0.0);
glRotatef(35.26, 0.0, 1.0, 0.0);
glRotatef(90.0, 0.0, 0.0, 1.0);
glRotatef(-35.26, 0.0, 1.0, 0.0);
glRotatef(45.0, 1.0, 0.0, 0.0);
```

We'll leave verification along the lines of Experiment 5.1 to the reader.

Exercise 5.53. Determine the 4×4 matrix corresponding to a 90° rotation about the radial axis l directed toward the point $[0\ 1\ 1]^T$, which corresponds to the OpenGL command `glRotatef(90.0, 0.0, 1.0, 1.0)`. (Recall that in Experiment 5.1 we had already written `glRotatef(90.0, 0.0, 1.0, 1.0)` as a composition of rotations about the coordinate axes.)

Exercise 5.54. (a) Determine the 4×4 matrix corresponding to a $45°$ rotation about the radial axis l directed toward the point $[1 \ -1 \ 1]^T$, which corresponds to the OpenGL command `glRotatef(45.0, 1.0, -1.0, 1.0)`.

(b) Express `glRotatef(45.0, 1.0, -1.0, 1.0)` as a composition of five successive rotations about the coordinate axes and experimentally verify.

Before discussing the second method to compute the matrix corresponding to rotation about an arbitrary axis, here are some facts about cross-products that we'll need. Skip this part if you are already familiar with cross-products of vectors.

Sidebar on Cross-Products

The *cross-product* (also called *vector product*) of two vectors u and v in $\mathbb{R}^3$ is another vector, denoted $u \times v$, defined as follows:

(a) If u and v are collinear, then $u \times v$ is the zero vector.

Note: Two vectors are collinear if and only if any one is a scalar (positive, zero or negative) multiple of the other. Therefore, if at least one of the vectors is zero, the two are trivially collinear. (Figure 5.26(a) shows an example of three non-zero vectors, each pair being collinear.)

(b) If u and v are not collinear, then $u \times v$ is the vector whose (a) magnitude is $|u||v||\sin\theta|$, where θ is the angle between u and v, and (b) direction is perpendicular to the plane spanned by u and v, such that u, v and $u \times v$ form a right-handed system. See Figure 5.26(b).

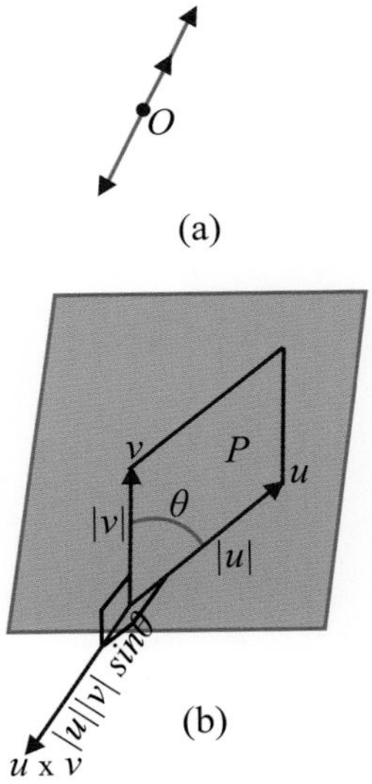

Figure 5.26: (a) Non-zero collinear vectors drawn from the origin (b) Taking the cross-product.

Here's another way to think of the cross-product. The magnitude $|u||v||\sin\theta|$ of the cross-product is nothing but the area of the parallelogram P with u and v as adjacent sides. The area of this parallelogram is, in fact, zero if and only if u and v are collinear. Consequently, the following is an alternate definition:

$u \times v$ is the vector whose magnitude is the area of the parallelogram with u and v as adjacent sides; if the magnitude is non-zero, then the direction of $u \times v$ is perpendicular to the plane spanned by u and v, such that u, v and $u \times v$ form a right-handed system.

If $u = [u_x \ u_y \ u_z]^T$ and $v = [v_x \ v_y \ v_z]^T$, a formula for the cross-product is the following:

$$u \times v = [u_y v_z - v_y u_z \quad v_x u_z - u_x v_z \quad u_x v_y - v_x u_y]^T \tag{5.36}$$

A convenient way to remember this formula is with the help of a determinant, as you are asked to show in the following exercise.

Exercise 5.55. If

$$u = u_x\mathbf{i} + u_y\mathbf{j} + u_z\mathbf{k} \quad \text{and} \quad v = v_x\mathbf{i} + v_y\mathbf{j} + v_z\mathbf{k}$$

where **i**, **j** and **k** are the unit vectors in the directions of the positive x-, y- and z-axes, show that

$$u \times v = \begin{vmatrix} \mathbf{i} & u_x & v_x \\ \mathbf{j} & u_y & v_y \\ \mathbf{k} & u_z & v_z \end{vmatrix} \tag{5.37}$$

Example 5.13. Determine the cross-product $[2\ 1\ 0]^T \times [1\ -2\ 4]^T$.

Answer:

$$[2\ 1\ 0]^T \times [1\ -2\ 4]^T = \begin{vmatrix} \mathbf{i} & 2 & 1 \\ \mathbf{j} & 1 & 2 \\ \mathbf{k} & 0 & 4 \end{vmatrix} = 4\mathbf{i} - 8\mathbf{j} - 5\mathbf{k} = [4\ -8\ -5]^T$$

Exercise 5.56. Determine the cross-product $[3\ -1\ 2]^T \times [-1\ 0\ 3]^T$.

Exercise 5.57. Write the result of the cross-product of every ordered pair from the three vectors **i**, **j** and **k** (there are 9 such products if you include products of vectors with themselves).

Remark 5.8. An easy way to remember the answer to the preceding exercise is the following:

The cross-product of any of **i**, **j** and **k** with itself is the zero vector. For the product of two different ones from **i**, **j** and **k**, keep in mind the cyclic order $\mathbf{i} \to \mathbf{j} \to \mathbf{k} \to \mathbf{i}$. Then, if two successive elements in this order are multiplied, the result is the next; if two successive elements are multiplied in reverse order, then the result is the negative of the next element. For example, $\mathbf{j} \times \mathbf{k} = \mathbf{i}$ and $\mathbf{k} \times \mathbf{j} = -\mathbf{i}$.

Exercise 5.58. Prove the following about cross-products, where u, v and w are any three vectors, and c an arbitrary scalar:

(a) u and v are collinear if and only if $u \times v = \mathbf{0}$ (*collinearity test*)

Note: The "only if" direction follows from the definition of cross-product; "if" needs to be proved.

(b) $u \times u = \mathbf{0}$

(c) $u \times v = -(v \times u)$ (*cross-product is anti-commutative*)

(d) It may not be true that $(u \times v) \times w = u \times (v \times w)$ (*cross-product is not associative*). Give an example of u, v and w where it isn't true.

(e) $(cu) \times v = u \times (cv) = c\,(u \times v)$

(f) $u \times (v + w) = u \times v + u \times w$ and $(v + w) \times u = v \times u + w \times u$
(*cross-product distributes over a sum*)

Exercise 5.59. Prove that if u is a unit vector and v arbitrary, then the vector $(u \times v) \times u$ is the component of v perpendicular to u.

Interestingly, it turns out, as the next example shows, that a cross-product with one fixed vector is a linear transformation.

Example 5.14. Show for a fixed vector $u = u_x\mathbf{i} + u_y\mathbf{j} + u_z\mathbf{k}$ that the transformation of $\mathbb{R}^3$ defined by $v \mapsto u \times v$ is linear.

Answer: Check from the formula (5.36) for $u \times v$ that

$$u \times v = Mv$$

where

$$M = \begin{bmatrix} 0 & -u_z & u_y \\ u_z & 0 & -u_x \\ -u_y & u_x & 0 \end{bmatrix}$$

which proves that $v \mapsto u \times v$ is indeed linear, with defining matrix M.

A Method to Compute the Rotation Matrix Which Is Part Geometry and Part Algebra

The problem statement again:

The axis of rotation is the directed line l through the origin toward a point $P = [a\ b\ c]^T$ and the angle of rotation is θ. The goal is to compute the matrix $R_{a,\,b,\,c}(\theta)$ corresponding to this rotation. As before, we assume without loss that $|P| = 1$.

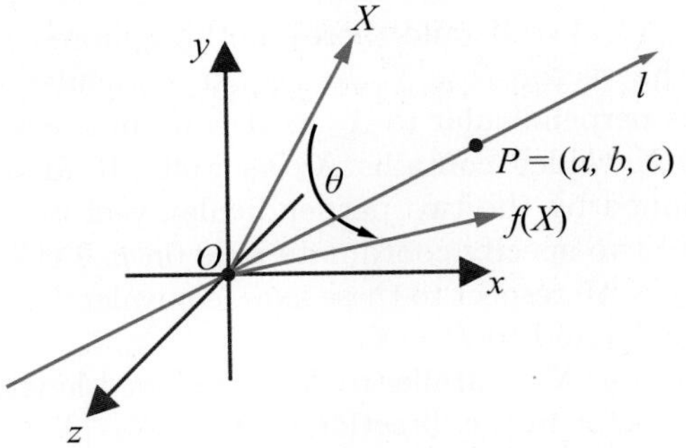

Figure 5.27: The vector $f(X)$ is obtained by rotating X an angle of θ about the radial line l.

Let the image of a vector X by the given rotation be $f(X)$. See Figure 5.27. First, split X as

$$X = X_1 + X_2$$

into components X_1 and X_2 parallel and perpendicular, respectively, to l. See Figure 5.28.

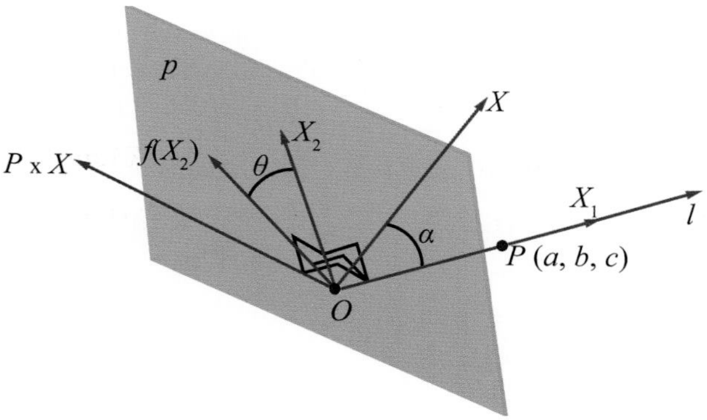

Figure 5.28: X_1 and X_2 are components of X parallel and perpendicular, respectively, to l; X_2, $f(X_2)$ and $P \times X$ all lie on the plane p through O perpendicular to l. X_2 and $P \times X$ are mutually perpendicular as well.

Since a rotation is a linear transformation, we have

$$f(X) = f(X_1) + f(X_2) \tag{5.38}$$

As X_1 lies on l, rotation about l leaves it unchanged. So

$$f(X_1) = X_1 \tag{5.39}$$

Now, X_2 lies on the plane p through O perpendicular to l and rotates by an angle θ about l, to $f(X_2)$. Therefore, $f(X_2)$ lies on p as well. We'll assume for now that X_2 is non-zero, meaning that X is not parallel to l, for, otherwise, $f(X) = f(X_1) = X_1$ and there's nothing more to do.

Observe that the vector $P \times X$, being perpendicular to l, lies on p; moreover, $P \times X$ is perpendicular to X_2 as it is perpendicular to the plane containing P and X, which contains X_2 as well. It follows, then, that the plane p is spanned by the two perpendicular vectors X_2 and $P \times X$. Consequently, these two specify coordinate axes on p. Let's determine the coordinates of $f(X_2)$ with respect to these axes, equivalently, the components of $f(X_2)$ parallel to X_2 and to $P \times X$.

The component of $f(X_2)$ parallel to X_2 has signed length $|f(X_2)| \cos \theta$. Moreover, the unit vector in the direction of X_2 is $X_2/|X_2|$. Therefore, the component of $f(X_2)$ parallel to X_2 is

$$|f(X_2)| \cos \theta \times X_2/|X_2| = X_2 \cos \theta \tag{5.40}$$

using the fact that $|f(X_2)| = |X_2|$, because $f(X_2)$ is X_2 rotated.

The component of $f(X_2)$ parallel to $P \times X$ has signed length

$$\begin{aligned} |f(X_2)| \sin\theta &= |X_2| \sin\theta \quad \text{(using again } |f(X_2)| = |X_2|\text{)} \\ &= |X \sin\alpha| \sin\theta \quad \text{(where } \alpha \text{ is the angle between } P \text{ and } X\text{)} \\ &= |P||X|| \sin\alpha| \sin\theta \quad \text{(as } |P| = 1\text{)} \\ &= |P \times X| \sin\theta \quad \text{(by definition of the cross-product)} \end{aligned}$$

The unit vector in the direction of $P \times X$ is $(P \times X)/|P \times X|$. It follows that the component of $f(X_2)$ parallel to $P \times X$ is

$$|P \times X| \sin\theta \, (P \times X)/|P \times X| = (P \times X) \sin\theta \tag{5.41}$$

Adding its components parallel to X_2 and $P \times X$ with help of (5.40) and (5.41), we conclude that

$$f(X_2) = X_2 \cos\theta + (P \times X) \sin\theta \tag{5.42}$$

Plugging the values from (5.39) and (5.42) into (5.38) we see that

$$\begin{aligned} f(X) &= X_1 + X_2 \cos\theta + (P \times X) \sin\theta \\ &= X_1 + (X - X_1) \cos\theta + (P \times X) \sin\theta \\ &= X \cos\theta + X_1(1 - \cos\theta) + (P \times X) \sin\theta \end{aligned} \tag{5.43}$$

Note: The preceding equation is valid even if X_2 is the zero vector, for, then, $X = X_1$ and $P \times X = 0$, so that the equation says $f(X) = X$, which is correct. So we're completely general from here on.

Use the results of Example 4.5 and Example 5.14 to replace X_1 and $P \times X$, respectively, with their equivalent matrix products:

$$\begin{aligned} f(X) &= \cos\theta \begin{bmatrix} 1 & 0 & 0 \\ 0 & 1 & 0 \\ 0 & 0 & 1 \end{bmatrix} X + (1 - \cos\theta) \begin{bmatrix} a^2 & ab & ac \\ ab & b^2 & bc \\ ac & bc & c^2 \end{bmatrix} X + \\ & \sin\theta \begin{bmatrix} 0 & -c & b \\ c & 0 & -a \\ -b & a & 0 \end{bmatrix} X \\ &= \left(\begin{bmatrix} a^2 & ab & ac \\ ab & b^2 & bc \\ ac & bc & c^2 \end{bmatrix} + \cos\theta \begin{bmatrix} 1-a^2 & -ab & -ac \\ -ab & 1-b^2 & -bc \\ -ac & -bc & 1-c^2 \end{bmatrix} + \right. \\ & \left. \sin\theta \begin{bmatrix} 0 & -c & b \\ c & 0 & -a \\ -b & a & 0 \end{bmatrix} \right) X \end{aligned} \tag{5.44}$$

Finally, adding the matrices in the parentheses and writing the result in homogeneous coordinates, we get a second form for the rotation matrix, different from the earlier geometrically-derived (5.35):

$$R_{a,\,b,\,c}(\theta) =$$

$$\begin{bmatrix} a^2(1-\cos\theta)+\cos\theta & ab(1-\cos\theta)-c\sin\theta & ac(1-\cos\theta)+b\sin\theta & 0 \\ ab(1-\cos\theta)+c\sin\theta & b^2(1-\cos\theta)+\cos\theta & bc(1-\cos\theta)-a\sin\theta & 0 \\ ac(1-\cos\theta)-b\sin\theta & bc(1-\cos\theta)+a\sin\theta & c^2(1-\cos\theta)+\cos\theta & 0 \\ 0 & 0 & 0 & 1 \end{bmatrix} \tag{5.45}$$

Remark 5.9. One can replace X_1 in (5.43) by $(X \cdot P)P$, as X_1 is the component of X parallel to the vector P of unit length (see Exercise 4.41(a)). The resulting equation

$$f(X) = X\cos\theta + (X \cdot P)P(1-\cos\theta) + (P \times X)\sin\theta \tag{5.46}$$

is called *Rodrigues' rotation formula.*

Exercise 5.60. Verify the result of Example 5.12 by computing the rotation matrix using Equation (5.45) instead of (5.35).

Exercise 5.61. Verify your answer to Exercise 5.53 by computing the rotation matrix using Equation (5.45) instead of (5.35).

Remark 5.10. An exercise which may have been conspicuous by its absence so far is to show that the composition of two rotations about radial axes is also a rotation about a radial axis. Unfortunately, though true, this is not easy to prove.

In fact, unlike its 2D counterpart, it's not even obvious that it is true. For example, is it evident that, say, a rotation of 45° about the x-axis followed by another, say, of 30° about the y-axis is a rotation about some axis in the first place? We'll prove that rotations do, in fact, compose to rotations using properties of rigid transformations in Section 5.4.5.

Whew, we told you this section was going to be long! It turned out to be fairly technical, too. We'll be coasting downhill the rest of the way and, believe it or not, get to see some code before long.

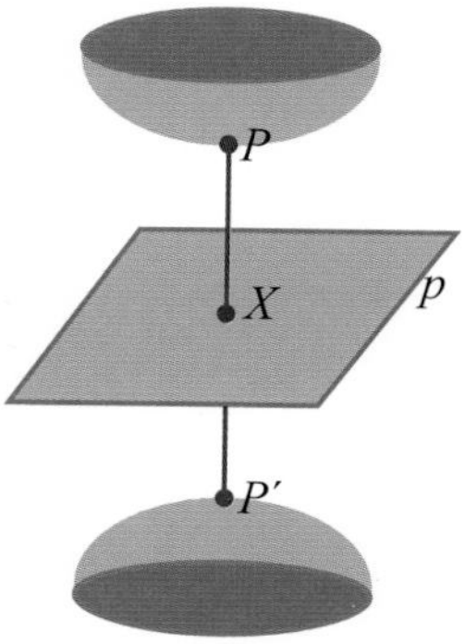

Figure 5.29: Reflection about plane p ($|XP| = |XP'|$).

5.4.4 Reflection

The image of the point $P = [x\ y\ z]^T$ by reflection about a plane p, called the *mirror*, is $P' = [x'\ y'\ z']^T$ such that:

(a) if P lies on p, then $P' = P$;

(b) if P does not lie on p, then P' is the point on the other side of p such that PP' is perpendicular to p, and P' is the same distance from p as P. See Figure 5.29.

Reflection about the xy-plane is simply scaling by the factors $s_x = 1$, $s_y = 1$ and $s_z = -1$. Its matrix, therefore, is

$$M = S(1, 1, -1) = \begin{bmatrix} 1 & 0 & 0 & 0 \\ 0 & 1 & 0 & 0 \\ 0 & 0 & -1 & 0 \\ 0 & 0 & 0 & 1 \end{bmatrix} \tag{5.47}$$

Exercise 5.62. Let p be an arbitrary plane mirror. Use the Trick to show that the matrix corresponding to reflection about p is of the form

$$M_1^{-1} M_2^{-1} S(1, 1, -1) M_2 M_1$$

where M_1 corresponds to a translation and M_2 to a rotation about a radial axis. You don't need to find exact values for M_1 and M_2.

Exercise 5.63. Write the 4×4 matrix corresponding to reflection about the plane $x - z = 0$.

Exercise 5.64. What is the result of composing reflections about the same mirror? What transformation is the result of composing reflections about two parallel mirrors? About two perpendicular mirrors? What is the inverse of a reflection?

Exercise 5.65. (Commutativity of transformations of 3-space)

(a) Do translations commute with each other?

(b) Do scalings commute with each other?

(c) Do rotations about the same radial axis commute with each other?

(d) Does a rotation about one radial axis commute with another about a different radial axis?

(e) Do translations and rotations commute?

(f) Do reflections about two different mirrors ever commute?

5.4.5 Affine Geometric Transformations

From Equations (5.29), (5.30), (5.45) and (5.47) and judicious applications of the Trick one sees that translations, rotations about arbitrary axes, scalings and reflections about arbitrary mirrors are all affine transformations of 3-space. Consequently, one has the following 3D analogue of Proposition 5.2 about their geometric niceness.

Proposition 5.8. *Let g be either a translation, a scaling, a rotation (about an arbitrary radial axis) or a reflection (about an arbitrary mirror). Then:*

(a) *g maps straight lines to straight lines and planes to planes. Moreover, it maps parallel straight lines to parallel straight lines, intersecting straight lines to intersecting straight lines, parallel planes to parallel planes and intersecting planes to intersecting planes.*

(b) *g maps a convex set on one plane to a convex set on another plane. Moreover, it transforms the convex hull of a set of points* $\{P_1, P_2, \ldots, P_k\}$ *on one plane to the convex hull of* $\{g(P_1), g(P_2), \ldots, g(P_k)\}$ *lying on another plane.*

Proof. Follows from Proposition 5.1 about affine transformations in general. □

Translations, rotations, scalings and reflections are all affine. In the opposite direction, the following analogues of the 2D Propositions 5.3 and 5.4 are true as well, though there seem to be no "low-level" proofs similar to the 2D ones. Fairly sophisticated linear algebra appears unavoidable. So at this time we'll only state Propositions 5.9 and 5.10, deferring the proof of the latter to later in this chapter as optional reading for the mathematically inclined. Mind that Proposition 5.9 follows straightforwardly from Proposition 5.10.

Proposition 5.9. *Any affine transformation of* $\mathbb{R}^3$ *is the composition in some order of translations, scalings and rotations about radial axes.*

In particular, an affine transformation $g : \mathbb{R}^3 \to \mathbb{R}^2$ *can be factored into a composition* $g = g_4 \circ g_3 \circ g_2 \circ g_1$*, where* g_1 *is a rotation about a radial axis,* g_2 *a scaling,* g_3 *another rotation about a radial axis and* g_4 *a translation.* □

Proposition 5.10. *Any non-singular linear transformation of* $\mathbb{R}^3$ *is the composition successively of a rotation about a radial axis, a scaling and another rotation about a radial axis.* □

OpenGL and Affine Transformations

The importance of Proposition 5.9, particularly to the design of an API like OpenGL, cannot be overstated. The modeling transformations one creates using OpenGL are compositions of translations (`glTranslatef()`), scalings (`glScalef()`, excluding for the moment degenerate calls) and rotations about a radial axis (`glRotatef()`). The whole collection, therefore, is affine as a composition of affine transformations is affine. Proposition 5.9 tells us that, conversely, any affine transformation of 3-space is a composition of translations, scalings and rotations about a radial axis and, so, may be implemented in OpenGL.

Conclusion: barring degenerate scalings, *the modeling transformations one can create in OpenGL are precisely the affine transformations of 3-space and nothing else*. And, as we argued in Section 5.2.2, this is a welcome situation both from the API designer's point of view of being able to implement a simple rendering engine, and the application programmer's point of view of having a comprehensive set of transformations at her disposal.

Example 5.15. Express the affine transformation

$$f([x \ \ y \ \ z]^T) = [-y \ \ x \ \ z+2]^T$$

as a composition of OpenGL transformations.

Answer: The mapping by f is the composition

$$[x \ \ y \ \ z]^T \mapsto [-y \ \ x \ \ z]^T \mapsto [-y \ \ x \ \ z+2]^T$$

Now, the first map is easily seen to be a rotation of $\pi/2$ about the z-axis, while the second is a translation of 2 in the z-direction. We have, therefore, the required block of OpenGL transformations:

```
glTranslatef(0.0, 0.0, 2.0);
glRotatef(90.0, 0.0, 0.0, 1.0);
```

Exercise 5.66. Express each of the following affine transformations as a composition of OpenGL transformations:

(a) $f([x \; y \; z]^T) = [y \; x \; z]^T$

(b) $f([x \; y \; z]^T) = [y \; z \; x]^T$

(c) $f([x \; y \; z]^T) = [x-y \ \ x+y \ \ -z]^T$

Verifying the Matrices Generated by OpenGL

Appendix E of the red book lists the matrices which OpenGL generates for the modeling transformations.

The translation and scaling matrices are simple and seen to agree with Equations (5.29) and (5.30), respectively. We leave it to the reader to verify that the rotation matrix R which OpenGL generates for the rotation transformation `glRotate{fd}(`*a*`, `*x*`, `*y*`, `*z*`)` is equivalent to that of Equation (5.45); in fact, the red book expresses it in the form of the prior more expansive Equation (5.44).

Incidentally, it's clear now and worth emphasizing that the composition of modelview transformations is implemented in the OpenGL engine as 4×4 matrix multiplication. Almost all of the "action" in animation is, therefore, matrix multiplication. In fact, it's not an oversimplification to say that a graphics card animates as fast as it multiplies.

Projection Transformations

There is another set of transformations which OpenGL implements as 4×4 matrix multiplication as well: these are the projection transformations used to transform the viewing box (respectively, frustum) created by a `glOrtho()`

(respectively, `glFrustum()`) call into a cubical so-called canonical viewing box.

However, we defer consideration of projection transformations to Chapter 18 because transformation at least of a frustum into a box cannot be done affinely, but requires an understanding of projective spaces (if the transformation matrix is not to be pulled out of a hat). Nevertheless, a reader with some math preparation who is eager to see all of OpenGL's matrices can proceed to Chapter 18 right away as it may be read independently. Moreover, it is written in a manner to be at least accessible to the reader even without much projective math under her belt.

5.4.6 Accessing and Manipulating the Current Modelview Matrix

Finally, we surface from deep theory to see – *code*!

There are four commonly-used methods to access the current modelview matrix, i.e., the matrix at the top of the OpenGL modelview matrix stack, three to change its value and one to read it. After setting the matrix mode to `GL_MODELVIEW` with the command `glMatrixMode(GL_MODELVIEW)` if need be, the call

1. `glLoadIdentity()` sets the current modelview matrix to the identity matrix I_4.

2. `glLoadMatrix(`**matrixData*`)` sets the current modelview matrix to the 4×4 matrix whose elements are listed in the one-dimensional array pointed by **matrixData* in column-major order (which means elements of the first column are listed first in order of increasing row, then those of the second column and so on).

3. `glMultMatrix(`**matrixData*`)` multiplies the current modelview matrix on the right by the 4×4 matrix whose elements are listed in column-major order in the one-dimensional array pointed by **matrixData*.

4. `glGetFloatv(GL_MODELVIEW_MATRIX,` **modelviewMatrixData*`)` stores the 16 elements of the current modelview matrix in column-major order in the one-dimensional array pointed by **modelviewMatrixData*.

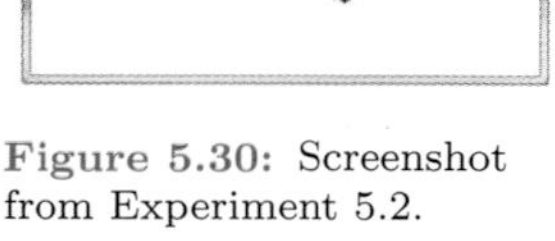

Figure 5.30: Screenshot from Experiment 5.2.

Experiment 5.2. Run `manipulateModelviewMatrix.cpp`. Figure 5.30 is a screenshot, although in this case we are really more interested in the transformations in the program rather than its visual output.

The `gluLookAt(0.0, 0.0, 10.0, 0.0, 0.0, 0.0, 0.0, 1.0, 0.0)` statement we understand to multiply the current modelview matrix on the right by the matrix of its equivalent modeling transformation. The current modelview matrix is changed again by the `glMultMatrixf(matrixData)` call, which multiplies it on the right by the matrix corresponding to a rotation of 45° about the z-axis, equivalent to a `glRotatef(45.0, 0.0, 0.0, 1.0)` call. It's changed one last time by `glScalef(1.0, 2.0, 1.0)`.

The current modelview matrix is output to the command window initially and then after each of the three modelview transformations. We'll see next if the four output values match our understanding of the theory. **End**

As expected, the first matrix output by the program is the identity

$$I_4 = \begin{bmatrix} 1 & 0 & 0 & 0 \\ 0 & 1 & 0 & 0 \\ 0 & 0 & 1 & 0 \\ 0 & 0 & 0 & 1 \end{bmatrix}$$

The `gluLookAt(0.0, 0.0, 10.0, 0.0, 0.0, 0.0, 0.0, 1.0, 0.0)` command is equivalent to `glTranslatef(0.0, 0.0, -10.0)`, whose matrix, from Equation (5.29), is

$$T(0, 0, -10) = \begin{bmatrix} 1 & 0 & 0 & 0 \\ 0 & 1 & 0 & 0 \\ 0 & 0 & 1 & -10 \\ 0 & 0 & 0 & 1 \end{bmatrix}$$

Therefore, after the `gluLookAt()` call the current modelview matrix should equal

$$\begin{bmatrix} 1 & 0 & 0 & 0 \\ 0 & 1 & 0 & 0 \\ 0 & 0 & 1 & 0 \\ 0 & 0 & 0 & 1 \end{bmatrix} \begin{bmatrix} 1 & 0 & 0 & 0 \\ 0 & 1 & 0 & 0 \\ 0 & 0 & 1 & -10 \\ 0 & 0 & 0 & 1 \end{bmatrix} = \begin{bmatrix} 1 & 0 & 0 & 0 \\ 0 & 1 & 0 & 0 \\ 0 & 0 & 1 & -10 \\ 0 & 0 & 0 & 1 \end{bmatrix}$$

which indeed is the second one output.

Next, the current modelview matrix is multiplied on the right by the matrix

$$M = \begin{bmatrix} X & -X & 0 & 0 \\ X & X & 0 & 0 \\ 0 & 0 & 1 & 0 \\ 0 & 0 & 0 & 1 \end{bmatrix}$$

where $X = 0.70710678 \simeq 1/\sqrt{2}$, so M corresponds to a rotation of 45° about the z-axis. The third matrix output then is, as expected,

$$\begin{bmatrix} 1 & 0 & 0 & 0 \\ 0 & 1 & 0 & 0 \\ 0 & 0 & 1 & -10 \\ 0 & 0 & 0 & 1 \end{bmatrix} \begin{bmatrix} X & -X & 0 & 0 \\ X & X & 0 & 0 \\ 0 & 0 & 1 & 0 \\ 0 & 0 & 0 & 1 \end{bmatrix} = \begin{bmatrix} X & -X & 0 & 0 \\ X & X & 0 & 0 \\ 0 & 0 & 1 & -10 \\ 0 & 0 & 0 & 1 \end{bmatrix}$$

The final matrix output is after the call `glScalef(1.0, 2.0, 1.0)`, a scaling by a factor of 2 along the y-axis. From Equation (5.30)

$$S(1,2,1) = \begin{bmatrix} 1 & 0 & 0 & 0 \\ 0 & 2 & 0 & 0 \\ 0 & 0 & 1 & 0 \\ 0 & 0 & 0 & 1 \end{bmatrix}$$

which multiplies the third matrix on the right to indeed give the final output matrix as

$$\begin{bmatrix} X & -X & 0 & 0 \\ X & X & 0 & 0 \\ 0 & 0 & 1 & -10 \\ 0 & 0 & 0 & 1 \end{bmatrix} \begin{bmatrix} 1 & 0 & 0 & 0 \\ 0 & 2 & 0 & 0 \\ 0 & 0 & 1 & 0 \\ 0 & 0 & 0 & 1 \end{bmatrix} = \begin{bmatrix} X & -2X & 0 & 0 \\ X & 2X & 0 & 0 \\ 0 & 0 & 1 & -10 \\ 0 & 0 & 0 & 1 \end{bmatrix}$$

Exercise 5.67. (Programming) Replace the `gluLookAt()` statement in `manipulateModelviewMatrix.cpp` with the following

```
gluLookAt(0.0, 10.0, 10.0, 0.0, 0.0, 0.0, 0.0, 1.0, 0.0);
```

Theoretically verify the correctness of the modelview matrix output by the program after the new `gluLookAt()` statement.
Hint: The new `gluLookAt()` statement is simulated as a translation by displacement vector $[0\ -10\ -10]^T$, followed by a rotation of 45° about the x-axis.

Exercise 5.68. (Programming) Verify your answer to Exercise 5.53 by comparing it with the output from an appropriately modified `manipulate-ModelviewMatrix.cpp`.

Exercise 5.69. (Programming) What is the current modelview matrix after the following piece of code in the drawing routine:

```
glMatrixMode(GL_MODELVIEW);
glLoadIdentity();
glScalef(1.0, 2.0, 2.0);
glTranslatef(2.0, 1.0, 0.0);
glRotatef(90.0, 1.0, 0.0, 0.0);
```

Find the answer theoretically by multiplying appropriate 4×4 matrices and then verify with the help of `manipulateModelviewMatrix.cpp`.

Exercise 5.70. (Programming) Verify your answers to Exercise 4.50(a)-(e) by using `manipulateModelviewMatrix.cpp` to find the matrix corresponding to the given `gluLookAt()` call, as well as that corresponding to the composed sequence of modeling transformations which you gave as being equivalent.

Exercise 5.71. (**Programming**) In Remark 4.15 about the viewing transformation being simulated by modeling transformations, we claimed that, following a `gluLookAt()` call, the current modelview matrix changes, but not the current projection matrix.

The part about the current modelview matrix is clearly true from what we have just seen in Experiment 5.2.

We ask the reader to verify the claim about the current projection matrix, particularly for the program `manipulateModelviewMatrix.cpp`, by reading the current projection matrix, both before and after the `gluLookAt()` statement, with the help of `glGetFloatv(GL_PROJECTION_MATRIX,` **projectionMatrixData*`)` calls.

5.4.7 Euclidean and Rigid Transformations

We have definitions analogous to the 2D case for Euclidean and rigid transformations of 3-space. Euclidean transformations are important because they preserve distance and, therefore, shape as well.

Definition 5.6. A *Euclidean transformation* (also called *isometry*) f of $\mathbb{R}^3$ is such that $|f(P)f(Q)| = |PQ|$ for any two points $P, Q \in \mathbb{R}^3$.

For the discussion of orientation coming up next, we need first to know when triples of vectors in 3-space are right-handed or left-handed.

Definition 5.7. An ordered triple of non-coplanar vectors $\{u, v, w\}$, each assumed originating from the same point O, is said to form a *right-handed system* (or, simply, be *right-handed*) if the rotation of u about O toward v, along the plane containing u and v, and along the smaller of the angles between the two, appears counter-clockwise to a viewer watching from the endpoint of w; otherwise, it is said to be *left-handed*. See Figures 5.31(a) and (b) for examples.

Remark 5.11. We actually first discussed handedness in the context of coordinate systems way back in Section 2.2.

Remark 5.12. Another term for handedness – more scientific-sounding, but less used – is *chirality*.

Definition 5.8. A Euclidean transformation f of $\mathbb{R}^3$ is said to be *orientation-reversing* if there exist four non-coplanar points P, Q, R and S in $\mathbb{R}^3$ such that one of the two ordered triples of vectors $\{PQ, PR, PS\}$ and $\{f(P)f(Q), f(P)f(R), f(P)f(S)\}$ is right-handed, while the other is left-handed.

A Euclidean transformation that is not orientation-reversing is said to be *orientation-preserving*.

Orientation-preserving Euclidean transformations are called *rigid transformations*.

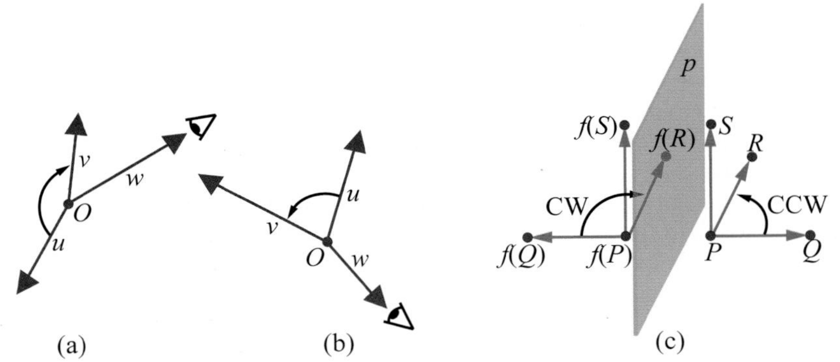

Figure 5.31: (a) $\{u, v, w\}$ is left-handed (b) $\{u, v, w\}$ is right-handed (c) The reflection f about the plane p is orientation-reversing, because the triple $\{PQ, PR, PS\}$ is right-handed, while the triple of images $\{f(P)f(Q), f(P)(R), f(P)(S)\}$ is left-handed.

Remark 5.13. The property of the transformation f described in the first paragraph of the definition does not depend on the choice of the non-coplanar points: we'll see that if it is true for *some* four non-coplanar points P, Q, R and S in $\mathbb{R}^3$ that one of $\{PQ, PR, PS\}$ and $\{f(P)f(Q), f(P)f(R), f(P)f(S)\}$ is right-handed, while the other left-handed, then it is true for *any* four non-coplanar points.

Example 5.16. The reflection f of Figure 5.31(c) about the plane p is an orientation-reversing Euclidean transformation.

Exercise 5.72. Scalings in general are not Euclidean transformations, but with certain choices of scaling factors they are. List these choices and for each say if the scaling preserves or reverses orientation.

Exercise 5.73. Show that the composition of two Euclidean transformations is Euclidean and that of two rigid transformations is rigid.

Exercise 5.74. Show that the composition of two orientation-reversing Euclidean transformations is an orientation-preserving Euclidean transformation (in other words, rigid). Show that the composition of an orientation-preserving and an orientation-reversing Euclidean transformation is orientation-reversing.

The following result gives a way to decide if an ordered triple of vectors is right-handed or left-handed.

Lemma 5.2. *Assuming that the coordinate axes themselves form a right-handed system, then an ordered triple $\{u, v, w\}$ of non-coplanar vectors, where $u = [u_x\ u_y\ u_z]^T$, $v = [v_x\ v_y\ v_z]^T$ and $w = [w_x\ w_y\ w_z]^T$, is right-handed or left-handed according as the determinant*

$$\begin{vmatrix} u_x & v_x & w_x \\ u_y & v_y & w_y \\ u_z & v_z & w_z \end{vmatrix}$$

is greater or less than zero (it cannot be zero as $\{u, v, w\}$ is non-coplanar).

Proof. The proof is not difficult, but uses more linear algebra than we would like to assume at this time, so we ask the reader to refer to a text such as [8]. □

The next exercise, analogue of the 2D Example 5.10, says that Definition 5.8 about a Euclidean transformation reversing or preserving orientation is independent of the choice of the four non-coplanar points P, Q, R and S.

Exercise 5.75. Suppose that an affine transformation f of $\mathbb{R}^3$ maps some four non-coplanar points P, Q, R and S in $\mathbb{R}^3$ such that one of the two ordered triples of vectors $\{PQ, PR, PS\}$ and $\{f(P)f(Q), f(P)f(R), f(P)f(S)\}$ is right-handed, while the other is left-handed.

Show, then, that for any four non-coplanar points, P', Q', R' and S', one of the two ordered triples of vectors $\{P'Q', P'R', P'S'\}$ and $\{f(P')f(Q'), f(P')f(R'), f(P')f(S')\}$ is right-handed, while the other is left-handed.

Hint: Use the same approach as for Example 5.10. You will need as well to apply the preceding lemma.

Next is the 3D analogue of the 2D Proposition 5.5.

Proposition 5.11. *A translation or a rotation about an arbitrary axis is a rigid transformation of 3-space. A reflection about an arbitrary mirror is an orientation-reversing Euclidean transformation of 3-space.*

Proof. The proof that a 3D translation t is distance-preserving is exactly similar to the 2D version in the proof of Proposition 5.5. That t is orientation-preserving is even simpler, because the two ordered triples of vectors $\{PQ, PR, PS\}$ and $\{t(P)t(Q), t(P)t(R), t(P)t(S)\}$ are identical for any four points P, Q, R and S in $\mathbb{R}^3$.

We'll leave the reader to prove the claims for rotations and reflections. □

The following is the 3D version of the 2D Proposition 5.6.

Proposition 5.12. *A rigid transformation of $\mathbb{R}^3$ keeping the origin fixed is a rotation about a radial axis, while an arbitrary rigid transformation of $\mathbb{R}^3$ is a composition of a rotation about a radial axis followed by a translation.*

A Euclidean transformation of $\mathbb{R}^3$ is a composition of a rotation about a radial axis followed by a translation, possibly followed again by a reflection.

Proof. The first statement of the proposition can be proved using linear algebra, but more interesting is to apply elementary arguments along the lines of Proposition 5.6. In fact, we ask the reader who doesn't mind wallowing in a bit of geometry to follow the approach suggested below to make a proof herself. If you are not so inclined it won't hurt to skip the proof altogether.

Suggested approach: Show first that a rigid transformation f of 3-space fixing the origin O is a rotation about a radial axis as follows:

If f is the identity, then it is trivially a rotation.

If f is not the identity, then suppose that P is a point such that $f(P) \neq P$. There are three possibilities:

(a) $f(f(P)) \neq P$.

Argue that the three points P, $f(P)$ and $f(f(P))$ cannot be collinear. Therefore, they belong to a unique plane p. Show that the line l through O perpendicular to p is the axis of f; further, the angle of rotation θ is the angle between the perpendiculars from P and $f(P)$ to l. See Figure 5.32(a).

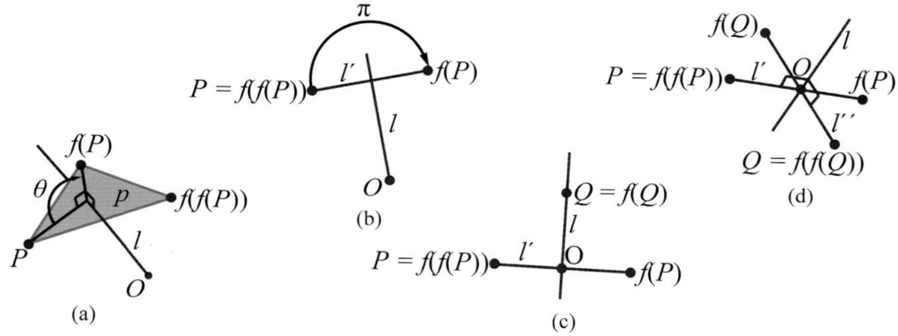

Figure 5.32: Finding the axis of a rigid transformation that fixes the origin.

(b) $f(f(P)) = P$ and the line l' joining P and $f(P)$ does not contain O.

Show in this case that the line l through O perpendicular to l' is the axis of f; furthermore, the angle of rotation is π. See Figure 5.32(b).

(c) $f(f(P)) = P$ and the line l' joining P and $f(P)$ does contain O.

In this case, let Q be a point not lying on l'.

If $f(Q) = Q$, then show that the line l through O and Q is the axis of f and the angle of rotation is π. See Figure 5.32(c).

If $f(Q) \neq Q$, then assume that $f(f(Q)) = Q$ and that the line l'' joining Q and $f(Q)$ contains O, for, if not, this case is equivalent to one of (a) or (b). Then show that the line l through O perpendicular to both l and l' is the axis and the angle of rotation is π. See Figure 5.32(d).

The rest of the proposition follows easily from the first part. □

Remark 5.14. The first part of Proposition 5.12, that a rigid transformation of $\mathbb{R}^3$ which keeps the origin fixed is a rotation about a radial axis, is often called Euler's Theorem. It is actually one of several theorems, proved by the great eighteenth century Swiss mathematician Leonhard Euler, bearing his name.

Finally, here's the 3D analogue of the 2D Proposition 5.7.

Proposition 5.13. *Affine, Euclidean and rigid transformations of 3-space are related by the following inclusions, which are each proper:*

$$\textit{rigid transforms} \subset \textit{Euclidean transforms} \subset \textit{affine transforms}$$

Proof. We leave this to the reader. □

A proposition whose proof we kept putting off, because of its apparent difficulty, is now all of a sudden simple to prove:

Proposition 5.14. *The composition of two rotations about radial axes in 3-space is another such.*

Proof. Let f_1 and f_2 be rotations about radial axes in 3-space. By Proposition 5.11 they are rigid transformations and, moreover, both fix the origin. By Exercise 5.73 the composition $f_1 \circ f_2$ is rigid, and it obviously fixes the origin because both f_1 and f_2 do so. Proposition 5.12 then completes the proof. □

Exercise 5.76. Consider reflections through *points*. For example, the reflection through the origin corresponds to the transformation $[x\ y\ z]^T \mapsto [-x\ \ -y\ \ -z]^T$. This transformation is clearly affine, in fact, linear. Is it Euclidean? Rigid? How about reflections through arbitrary points?

Sketch how the boy of Figure 4.7 of the last chapter would be transformed by reflection through the origin.

Proof of Proposition 5.10

This section is only for those with a good knowledge of linear algebra and may be safely skipped by others with no consequences to their learning of CG.

Lemma 5.3. *A special orthogonal transformation of $\mathbb{R}^3$, i.e., one whose matrix is orthogonal with determinant 1, is a rotation about a radial axis.*

Proof. It's easy to verify that a linear transformation f of $\mathbb{R}^3$ defined by an orthogonal matrix of determinant 1 preserves both distance and orientation. Therefore, it is rigid and the lemma follows from Proposition 5.12. □

The author learned the proof of the following lemma from T. K. Mukherjee [97].

Lemma 5.4. *For any non-singular real $n \times n$ matrix M, there exist special orthogonal matrices P and Q and a real diagonal matrix D with all non-zero entries such that*

$$M = PDQ$$

Proof. Consider the product MM^T. As it is symmetric, by a standard result of linear algebra there exists a real orthogonal matrix P such that

$$P^{-1}(MM^T)P = D'$$

where D' is a diagonal matrix.

Moreover, MM^T is positive definite, which implies that D' is as well. Therefore, each element of the diagonal of D' is positive. Let $D = \sqrt{D'}$. In particular, if $D' = [d'_i]$, then $D = [d_i]$, where d_i is the positive square root of d'_i.

Accordingly,

$$P^{-1}MM^TP = D^2$$

It follows that

$$\begin{aligned} I &= D^{-1}(P^{-1}MM^TP)D^{-1} = (D^{-1}P^{-1}M)(M^TPD^{-1}) \\ &= (D^{-1}P^{-1}M)(D^{-1}P^{-1}M)^T \end{aligned}$$

as $(D^{-1})^T = D^{-1}$ and $(P^{-1})^T = P$.

Writing

$$Q = D^{-1}P^{-1}M$$

we have from the above that $QQ^T = I$, so Q is orthogonal

Now

$$M = PDQ$$

and we can assume that both P and Q are, in fact, special orthogonal, for, if either is not, then it can be multiplied by a diagonal matrix with determinant -1, viz.

$$R = \begin{bmatrix} -1 & 0 & 0 & \dots & 0 \\ 0 & 1 & 0 & \dots & 0 \\ 0 & 0 & 1 & \dots & 0 \\ & \dots & \dots & \dots & \\ 0 & 0 & 0 & \dots & 1 \end{bmatrix}$$

and, correspondingly, D multiplied by R^{-1} (=R). □

The two lemmas combine to establish Proposition 5.10. □

Remark 5.15. Proposition 5.10 and its 2D sibling Proposition 5.4 are both deducible as cases of the Singular Value Decomposition Theorem. For an account of this fundamental theorem of linear algebra see, for example, Roman [119].

5.4.8 Shear

Thinking back to the analogy made in Section 5.3.3, imagine again placing a lump of putty between your palms and then moving one palm parallel to the other. In proper 3D there are three choices to make: (1) how to initially align your palms in space with putty between them, (2) which direction to move, say, the upper palm, but keeping it parallel to the fixed lower one and, finally, (3) how far to move the upper palm. Accordingly, a 3D shear s is uniquely determined by three parameters:

1. A plane p called the *plane of shear*.
2. A directed line l called the *line of shear*, which is parallel to p.
3. An angle α called the *angle of shear*.

The action of s is equivalent to that of "infinitely many" 2D shears applied to parallel planes. Here's how (see Figure 5.33)(a)):

Given a point $P \in \mathbb{R}^3$, let q be the unique plane containing P which is both perpendicular to the plane p of shear s *and* parallel to the line l of shear s. Therefore, q intersects p in a directed line, denote it l', parallel to l. P, then, is transformed by s to the point P' *exactly* as it would by the particular 2D shear s', on the plane q, specified by the directed line l' and the angle α.

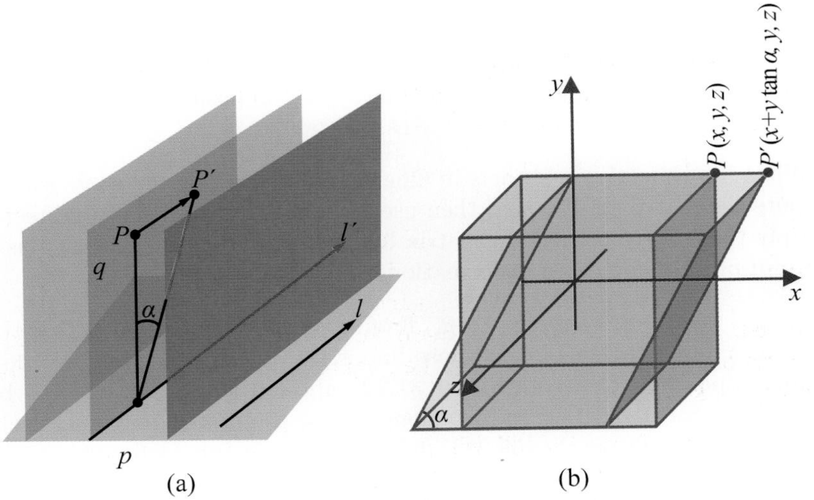

Figure 5.33: (a) A 2D slice of a 3D shear on the plane q and two more "copies" of q (b) A shear along the xz-plane whose line is the x-axis.

In other words, imagine 3D space "sliced" into infinitely many parallel planes, each perpendicular to p and parallel to l, and s as the "union of identical 2D shears" on each of these slices (Figure 5.33)(a) depicts two more slices parallel to q).

Example 5.17. Figure 5.33(b) depicts a 3D shear along the xz-plane whose line is the x-axis and angle α, shearing a cube into a parallelepiped. As the "slices" are all parallel to the xy-plane and the line of the shear in each parallel to the x-axis, one can straightforwardly apply the 2D Equation (5.22) to write the equation of this shear as follows:

$$P' = \begin{bmatrix} 1 & \tan\alpha & 0 & 0 \\ 0 & 1 & 0 & 0 \\ 0 & 0 & 1 & 0 \\ 0 & 0 & 0 & 1 \end{bmatrix} P$$

Exercise 5.77. Verify that the 4×4 matrix of a 3D shear along the xz-plane, whose line is the z-axis and angle α, is:

$$\begin{bmatrix} 1 & 0 & 0 & 0 \\ 0 & 1 & 0 & 0 \\ 0 & \tan\alpha & 1 & 0 \\ 0 & 0 & 0 & 1 \end{bmatrix}$$

Exercise 5.78. What is the 4×4 transformation matrix of a 3D shear along the xy-plane, whose line is the x-axis and angle α?

Exercise 5.79. Prove that a shear along a radial plane is a non-singular linear transformation of 3-space, while an arbitrary 3D shear is an affine transformation. Conclude that any 3D shear is a composition of translations, scalings and rotations about radial axes.

When implementing a shear s in OpenGL, it's typically more efficient to compute the matrix M of s and then use a `glMultMatrix()` call to directly multiply the current modelview matrix by M, rather than expressing s as a composition of modeling transformations.

Exercise 5.80. (Programming) From Example 5.11 it follows that the 3D shear s along the xz-plane, whose line is the x-axis and angle $\alpha = 45°$, is the composition of a rotation of $31.72°$ about the z-axis, followed by scaling by factors 0.62, 1.62 and 1.0 along the x-, y- and z-axes, respectively, followed by a rotation of $-58.28°$ about the z-axis (up to floating point error).

Verify this by modifying `manipulateModelviewMatrix.cpp` to output the current modelview matrix after applying the preceding modeling transformations and separately computing the value of the shear matrix.

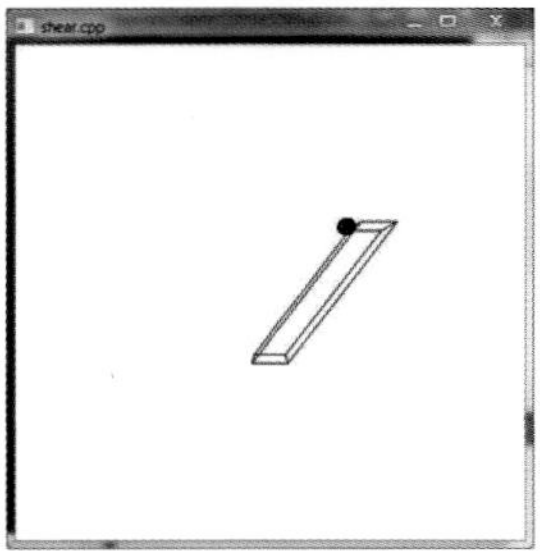

Figure 5.34: Screenshot of `shear.cpp`.

Experiment 5.3. Run `shear.cpp`. Press the left and right arrow keys to move the ball. The box begins to shear when struck by the ball. See Figure 5.34.

The shear matrix is explicitly computed in the code and multiplied from the right into the current modelview matrix via a `glMultMatrix()` call.
End

Exercise 5.81. (**Programming**) Recall the program `ballAndTorus-Shadowed.cpp` from Experiment 4.35. Modify the shadow-drawing part of the program to cast shadows by the sun at 45° in the sky. Figure 5.35 indicates that the transformation to be implemented in the xy-plane is

$$(x, y) \mapsto (x + y, 0)$$

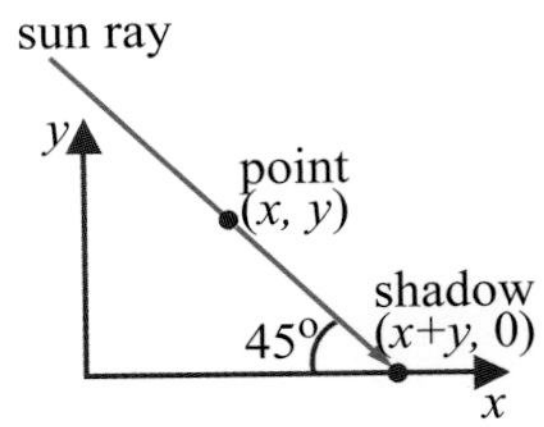

Figure 5.35: Shadow of a point cast by the sun at 45° in the sky.

5.5 Summary, Notes and More Reading

In this chapter we opened up the graphics animation engine to find out what makes it tick. The short answer is 4×4 matrix multiplication and we learned why.

Each modeling transformation corresponds to an affine transformation of 3-space, which is represented using homogeneous coordinates by a 4×4 matrix. The composition of modeling transformations corresponds to multiplication of their matrices. Viewing transformations, being compositions of modeling transformations, each corresponds to a 4×4 matrix as well. We learned the particular matrix representations of basic geometric transformations such as translation, scaling, rotation and reflection.

We learned as well that the modeling transformations of OpenGL – translation, scaling and rotation – were chosen for good reasons by the designers of the API. Not only are these transformations affine, but they combine to generate all affine transformations.

We studied certain particularly useful subclasses of affine transformations. One was that of shape-preserving transformations, the Euclidean transformations, which in turn include rigid transformations that model the motion of rigid objects in space. Another special class of affine transformations which we studied was that of the shears.

We learned to access and manipulate OpenGL's modelview matrix stack as well, allowing us to program transformations directly into the stack if need be, rather than through calls to OpenGL's own modeling transformations.

Till Chapter 4 we were primarily interested in *using* OpenGL. Now we have an understanding of the *working* of this API as well. True, familiarity, say, with the functioning of an internal combustion engine does not necessarily make one a better driver; however, it certainly does help one better understand technical issues, more confidently deal with them and, generally, be a more informed consumer, which has its value.

One topic the knowledgeable reader might think missing from this chapter on transformations is discussion of the so-called projection transformation in the graphics pipeline – which is critical to the shoot part of the shoot-and-print rendering paradigm from Chapter 2 – and how it is implemented by means of mathematical projective transformations. However, we thought it best to introduce the projection transformation as an application of projective spaces and their transformations later in the book in Chapter 18.

The reason is that the choice of the particular transformations applied in the graphics pipeline is hard to motivate without an understanding of projective spaces, and we were reluctant to pull it out of a hat.

For further reading about geometric transformations the reader is recommended to see Mortenson [95] and Yaglom's series [155, 156, 157]. Articles about transformations and their role in computer graphics, written in recreational style and yet very informative, can be found in the books by Blinn [17, 18] and Glassner [54, 55].

CHAPTER 6

Advanced Animation Techniques

The goal for this chapter is to learn techniques to cope with two issues that arise often in animation projects. The first is managing large worlds where the polygon count may painfully slow down the rendering pipeline. The programmer can help ease the logjam by prefiltering objects lying outside the camera's field of view. This process is called frustum culling and we describe how to do it by means of space partitioning in Section 6.1. The related process of occlusion culling, where objects blocked from the camera's view by other objects are filtered, is the topic of Section 6.2. This section discusses, as well, occlusion queries and conditional rendering.

The second issue is that of animating the orientation of an object. In Chapter 4 we learned all about animating motion, coding balls and boxes that flew, fell, spun and revolved around one another. But how about animating orientation or pose? For example, an aircraft maneuvering in a dogfight or a camera tracking a scene. Changing orientation involves modeling transformations as well, particularly rotation, but first one must develop a method to *quantify* orientation, just as (x, y, z) quantify position. Only then comes the question of moving between two orientations.

In Section 6.3 we learn how to use Euler angles – which we first encountered in Section 4.6.3 – to quantify orientation in 3D. Animating between a pair of orientations given by their Euler angle tuples is possible but, as we shall see, potentially problematic. A more sophisticated approach is with the use of quaternions. This is the topic of Section 6.4, which begins with an introduction to the mathematics of quaternions, and then goes on to describe their application to representing and animating orientation.

There is a fair amount of theory in this chapter but it has important practical applications and we back it all the way with code.

6.1 Frustum Culling by Space Partitioning

Frustum culling is *de rigueur* for game programmers or, for that matter, anyone creating scenes with large polygon counts. We'll motivate the proceedings by rerunning Experiment 4.32.

Experiment 6.1. First, run the program `spaceTravel.cpp` with its current values of `ROWS` and `COLUMNS`, being the size of the asteroid grid, as 8 and 5, respectively. Move the spacecraft with the arrow keys. Next, increase both `ROWS` and `COLUMNS` to 100. Figure 6.1 is a screenshot. The spacecraft now begins to respond sluggishly to the arrow keys, at least on a typical desktop. You may have to pump up even more the values of `ROWS` and `COLUMNS` if yours is exceptionally fast. **End**

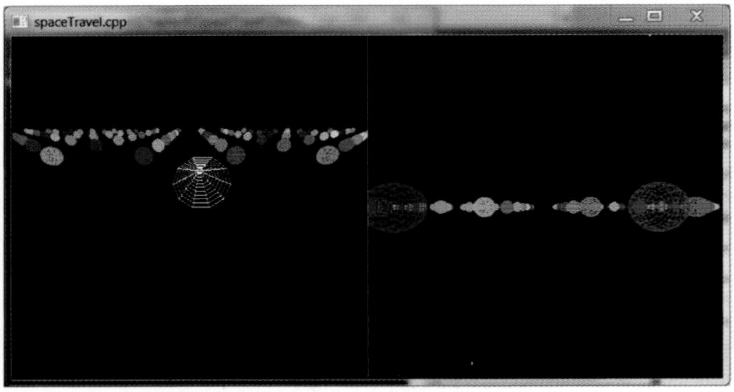

Figure 6.1: Screenshot of `spaceTravel.cpp` with a 100×100 array of asteroids.

Let's first do a back-of-the-envelope calculation to understand what's happening. Assuming the viewable space of `spaceTravel.cpp` to be a box of sides 250, significantly larger, in fact, than the viewing frustum defined by the `glFrustum(-5.0, 5.0, -5.0, 5.0, 5.0, 250.0)` statement of the program, and noting that asteroids are 30 units apart in both the x and z directions, one deduces that at most 9*9=81 asteroids are viewable in either viewport at any given time. That's not a lot for OpenGL to draw. In fact, set ROWS to 9 and COLUMNS to 9 to find no perceptible slow-down! So what's going on? Why the slow-down in simply *creating* a larger asteroid field?

A moment's consideration of the rendering pipeline reveals the answer. At every redisplay, in other words, every arrow press, the `ROWS` $\times$ `COLUMNS` number of asteroids in `arrayAsteroids[ROWS][COLUMNS]` are *all* – in fact, their *collective polygons* are *all* – zapped first with the modelview matrix, then with the projection matrix, and *then* those that fall outside the viewing volume are clipped, and the rest projected to the viewing face and rendered. Specifically, if `ROWS` and `COLUMNS` are both 100, then the polygons of all 10,000 asteroids, several hundred thousand in total, enter the rendering pipeline before only those belonging to approximately 80 are drawn. That's more than 99% of the polygons, each incurring computational cost in the pipeline, ultimately not being drawn. Talk about waste!

It's not OpenGL's fault though. OpenGL finds out which polygons belong on-screen and off only *after* transforming and clipping, operations well into the pipeline. However, the programmer can help by pre-identifying as many objects as possible that do not intersect the viewing frustum, which means that they will end up being clipped, and not letting them into the pipeline in the first place. This is a process called *frustum culling*, which consists, then, of adding to the program routines to check if a polygon (or object) intersects the frustum and filtering through to the drawing routine only those which do or, equivalently, culling those which do not.

6.1.1 Space Partitioning

The most straightforward way to frustum cull is to test each object individually if it intersects the frustum. This may, in fact, give decent speed-up if the objects are simple enough that the combined cost of testing them all is still cheap.

However, frustum culling is typically more efficient if space is first partitioned in a hierarchical manner into cells which each contain only a few objects. Cost-effective partitioning must be driven by the distribution of the objects – subdividing into smaller ones only cells containing many objects. Once space is partitioned, frustum culling can be accomplished by hierarchically checking cells to determine if they intersect the frustum and passing to the drawing routine objects belonging only to those which do. This approach is based on a few premises:

(a) That partitioning space and determining the distribution of the objects in individual cells is primarily a one-time pre-processing cost, which is true if most objects in the scene are static. The few moving objects, in this case, can be passed mandatorily to the drawing routine.

(b) That the cells are of a shape easy to check for intersection with the frustum.

(c) That the hierarchical nature of the partition leads to efficiency because, if a cell is found not to intersect the frustum, then its sub-cells and the objects which they contain can all be eliminated from further consideration, a process called *pruning*.

(d) That the partitioning process is efficient in that, if a cell contains only a few objects, then it is not subdivided, so that the final partition reflects the distribution of the objects in space.

There's more than one way to hierarchically partition space, but intuitively simplest is the *octree* for 3-space and its analogue, the *quadtree*, for

2-space. We'll explain quadtree-based space partitioning using the scenario of spaceTravel.cpp as a running example because we have, in fact, an implementation in the program spaceTravelFrustumCulled.cpp.

6.1.2 Quadtrees

Note, first, that spaceTravel.cpp is essentially a 2D problem as the spacecraft, asteroids and frustum can all be projected onto the xz-plane and intersection testing done on that plane – note that the spacecraft has no motion in the y-direction. Therefore, a quadtree is appropriate for spaceTravel.cpp even though, nominally, the scene is 3D.

A quadtree partitions 2-space into axis-aligned squares, the collection having the hierarchy of a tree. In frustum culling applications the root node corresponds to a square large enough to contain all objects that might potentially be culled.

Terminology: We'll use the terms "node" and "square" interchangeably in the context of quadtrees.

Figure 6.2(a) illustrates a hypothetical projected scenario of spaceTravel.cpp (an irregular distribution of asteroids is obtained by setting FILL_PROBABILITY to a value less than 100). The craft itself is ignored because it moves; therefore, it's always passed into the pipeline.

To begin with, the root square of the quadtree is chosen big enough to bound the entire asteroid field – see the setting of the initialSize variable of the setup() routine of spaceTravelFrustumCulled.cpp. (Note that we'll be running spaceTravelFrustumCulled.cpp soon enough, but, for now, let's see how the code is developed through incorporating frustum culling into spaceTravel.cpp.)

Subsequently, at each level, each square may be subdivided into four equal sub-squares (quadrants). The criterion to subdivide is generally determined by the programmer. In the particular case of spaceTravelFrustumCulled.cpp we subdivide a square if it intersects more than one asteroid. If a square is subdivided, then its four quadrants become its children in the tree hierarchy and are denoted SW, NW, NE and SE according to their location in the parent square. Given our condition for when to subdivide, evidently leaf squares – which are not further subdivided – intersect either one asteroid or none.

The squares of the quadtree corresponding to the arrangement of asteroids in Figure 6.2(a) are shown in Figure 6.2(b) and the underlying tree structure in Figure 6.2(c).

Once the quadtree is built, culling is straightforward: check, starting at the root, for squares that intersect the frustum; if a non-leaf square intersects the frustum, then recursively process its children; if a leaf square intersects the frustum, then pass its asteroid (if any) to the drawing routine.

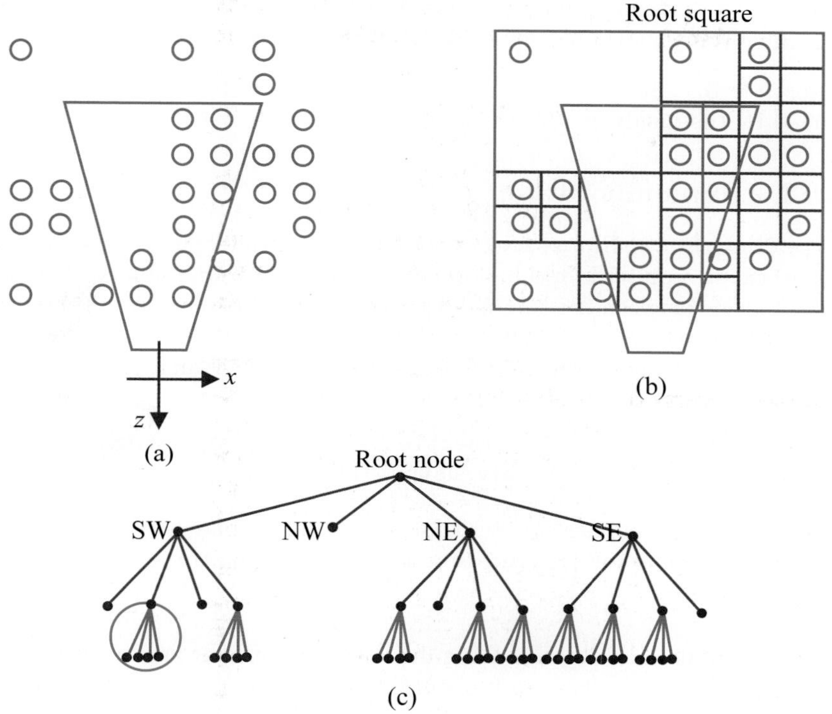

Figure 6.2: (a) Projection of the asteroids and the frustum of `spaceTravel.cpp` onto the xz-plane. (b) Corresponding quadtree squares (the root square is bold) (c) The tree structure with children at each node drawn SW, NW, NE, SE from left to right; the nodes in the red circle are *some* of those pruned.

Let's verify the premises (a)-(d) for space partitioning, mentioned earlier, in the case of the `spaceTravelFrustumCulled.cpp` quadtree:

(a) The asteroids are all static while the spacecraft is the only object which is not, so a one-time quadtree is built for the asteroids, while the craft itself is always passed to the drawing routine.

In the second viewport the camera moves, which, as we know, is implemented by actually transforming the scene. However, in order not to have to update the quadtree structure, it's preferable to imagine the viewing frustum itself moving, attached to the front of the spacecraft, as in Figure 6.3.

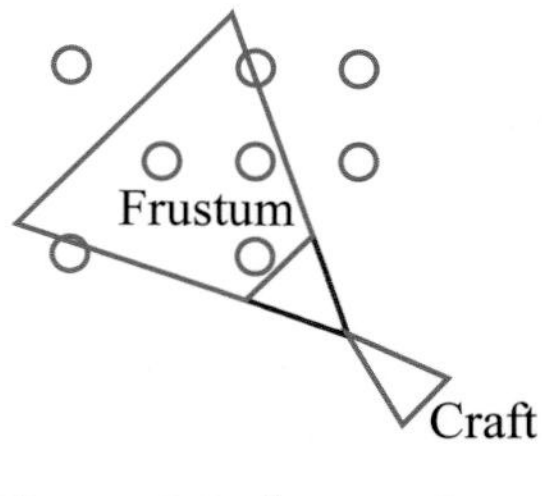

Figure 6.3: Spacecraft carrying a viewing frustum "attached" to its front.

(b) The cells are each a square, a shape easy to test for intersection with the trapezoidal projection of a frustum.

(c) Several of the nodes, even in the simple example of Figure 6.2, are pruned, e.g., the ones inside the red circle (and others).

(d) It can be seen from Figure 6.2 that the spatial distribution of the quadtree squares indeed tracks that of the asteroids.

Exercise 6.1. Indicate *all* the nodes of the tree of Figure 6.2 which are pruned by the quadtree-based frustum culling.

6.1.3 Implementation

Experiment 6.2. Run `spaceTravelFrustumCulled.cpp`, which enhances `spaceTravel.cpp` with optional quadtree-based frustum culling. Pressing space toggles between frustum culling enabled and disabled. As before, the arrow keys maneuver the craft.

The current size of the asteroid field is 100×100. Dramatic isn't it, the speed-up from frustum culling?!

Note: Make sure the file `intersectionDetectionRoutines.cpp` is in the same directory as `spaceTravelFrustumCulled.cpp`.

Note: When the number of asteroids is large, the display may take a while to come up because of pre-processing to build the quadtree structure.

End

We have already described the development of `spaceTravelFrustumCulled.cpp`, which follows pretty much word for word the quadtree-based strategy described at the start of the section. Here are specifics.

The quadtree `asteroidsQuadtree` is an object of the `Quadtree` class containing nodes belonging to the `QuadtreeNode` class. The member function `numberAsteroidsIntersected()` of the class `QuadtreeNode` helps decide for each quadtree square if it is to be subdivided, while the member list `asteroidList` stores for each leaf square the asteroid (if any) intersecting it.

In addition to `checkSpheresIntersection()` from the original `spaceTravel.cpp`, used in `asteroidCraftCollisiond()` to detect (approximately) intersection between the spacecraft and an asteroid, routines from the program `intersectionDetectionRoutines.cpp` are invoked for other intersection tests. In particular, `checkQuadrilateralsIntersection()` determines if the frustum in either viewport intersects a quadtree square, while `checkDiscRectangleIntersection()` if an asteroid intersects a quadtree square.

In Figure 6.2, asteroids lie either entirely inside or outside a square. This need not be the case in general, of course, and in our code an asteroid straddling the boundary of a quadtree square is associated with it.

With large numbers of asteroids, the speed-up through frustum-culling is clearly enormous. Even so, our implementation `spaceTravelFrustumCulled.cpp` is minimal and there are further optimizations to be made. We ask the reader to explore a couple next.

Exercise 6.2. (**Programming**) A large quadtree costs both in RAM space and pre-processing time. Try the following two options in `spaceTravelFrustumCulled.cpp` to control its size:

(a) The size of the quadtree tends to grow exponentially with its height. Accordingly, set a *cut-off depth* beyond which nodes cannot be partitioned.

(b) The criterion for subdividing a square, currently if it intersects more than one asteroid, can be made stricter by setting a larger threshold for the number to be intersected, again reducing the size of the quadtree.

6.1.4 More about Space Partitioning

Octrees are a straightforward generalization of quadtrees to 3-space – space is partitioned into a tree-like hierarchy of axis-aligned cubes. Each cube in an octree can be partitioned into 8 child octants (see Figure 6.4).

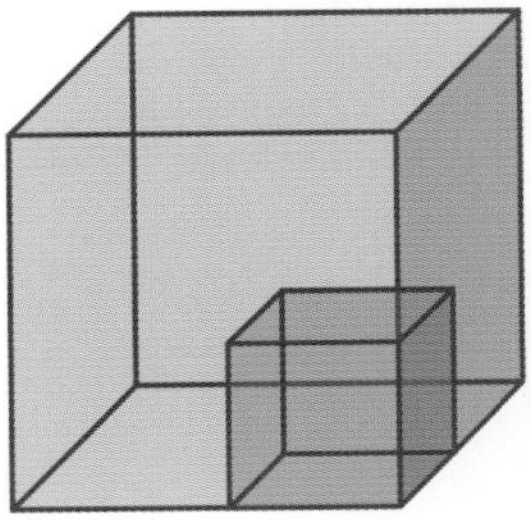

Figure 6.4: An octree cube and one of its 8 octants.

Quadtrees and octrees are not the only ways to partition space. There are more sophisticated data structures such as kd-trees, range trees and BSP (Binary Space Partitioning) trees, which can be applied in two and higher dimensions.

Moreover, applications of space partitioning are not limited to frustum culling either. Another important one is collision detection. The principle is that two objects can intersect only if they belong to the same or adjacent cells; accordingly, one can pre-process and pass only "nearby" pairs to the, typically, costly intersection-checking routines.

Dynamic scenes with multiple mobile objects are a challenge for any space partitioning application. Options include predictively locating moving objects in cells if there is prior knowledge of their trajectories, followed, possibly, by a repartitioning of space or a redistribution of objects to cells.

Bottom line: There is overhead both in code and pre-processing in setting up a space-partitioning structure, but if the application is appropriate, e.g., frustum culling a fairly static scene, the bang for the buck is enormous.

Exercise 6.3. (**Programming**) Currently, `spaceTravelFrustumCulled.cpp` checks for intersection between the spacecraft and *every* asteroid. Use the quadtree to improve on this. In particular, check only for collision between the craft and each asteroid associated with a leaf square that the craft intersects.

Exercise 6.4. (**Programming**) Design a busy scene, maybe part of a game, and draw it using frustum culling.

6.2 Occlusion Culling

Often, a programmer's global view of a scene allows her to conclude for a given position of the camera that certain objects are blocked from view and, therefore, need not be sent into the rendering pipeline. For example, in the scene of Figure 6.5 of two rooms opening to a hall, if the camera is inside one room, then objects in the other are always occluded by the wall in between.

Therefore, part of the drawing routine could be as the pseudo-code below:

```
if (camera is in Room1)
{
   draw Obj1;
   draw Obj2;
}
if (camera is in Room2)
{
   draw Obj3;
   draw Obj4;
   draw Obj5;
}
```

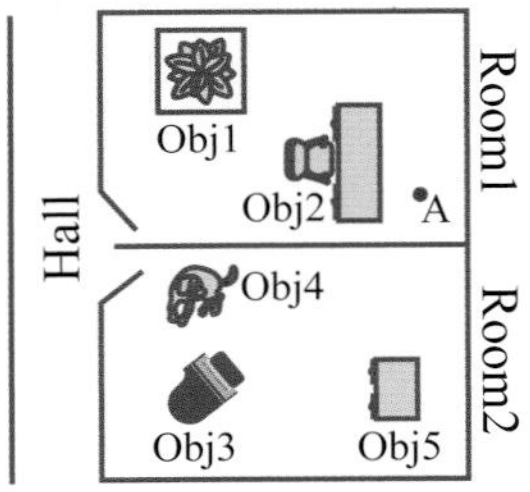

Figure 6.5: Two rooms off a hall. A dashed bounding box is shown containing the first object.

This technique of *occlusion culling* can improve immensely efficiency of rendering, as can be imagined if, say, the objects in Figure 6.5 are complex. In this example, occlusion culling is coded in by the programmer herself from a fairly coarse global understanding of the scene. Evidently, she could dig even deeper and implement more refined tests, e.g., one which determines that a camera at location `A` will find `Obj1` occluded, applying, possibly, geometric intersection tests of the kind used in `spaceTravelFrustumCulled.cpp`.

OpenGL itself provides a powerful device to help with occlusion culling: one can set up a so-called *occlusion query* to determine if an object is visible after depth testing. Before we get to an actual example, here's the general idea. In a situation as depicted in Figure 6.5, the camera being at `A`, before attempting to render (the complex) `Obj1`, one could query if its bounding box, the (much simpler) dashed rectangles, is visible. If the box is visible, then render `Obj1`; if not, don't. Hopefully, in more configurations than not, the investment in querying occlusion for a simple object is more than offset by being able, subsequently, to toss a complex object from the pipeline. Let's get to code.

Figure 6.6: Screenshot of `occlusion.cpp`.

Experiment 6.3. Run `occlusion.cpp`, which initially shows only a green rectangle. Use the arrow keys to move a solid red cube into view from behind the rectangle (see Figure 6.6). Press the space bar, which is a toggle, to reveal a red wire sphere contained in the box. The sphere is our "complex" object, while the cube, of course, is its bounding box. We'll analyze this program next. **End**

The call `glGenQueries(1, &query)` in the setup routine generates a query object, storing its id in the variable `query`.

The drawing routine begins with a piece of code to draw the obscuring green rectangle. Next comes the block to draw the bounding box enclosed in a query:

```
glBeginQuery(GL_SAMPLES_PASSED, query);
if (!boxVisible)
{
   glColorMask(GL_FALSE, GL_FALSE, GL_FALSE, GL_FALSE);
   glDepthMask(GL_FALSE);
}
glutSolidCube(1.0);
if (!boxVisible)
{
   glDepthMask(GL_TRUE);
   glColorMask(GL_TRUE, GL_TRUE, GL_TRUE, GL_TRUE);
}
glEndQuery(GL_SAMPLES_PASSED);
```

The command `glBeginQuery(GL_SAMPLES_PASSED, query)` in the block above tells OpenGL to start counting samples – as we're not multisampling, a sample is simply a fragment – which pass the depth test. The complementary call `glEndQuery(GL_SAMPLES_PASSED)` stops the counting. For now, ignore the four masking statements.

Next,

```
while (!resultAvailable) glGetQueryObjectuiv(query,
                    GL_QUERY_RESULT_AVAILABLE, &resultAvailable);
```

places a Boolean in `resultAvailable` saying if the query had indeed completed counting. For, observe that, especially if multiple queries are being run, it is quite possible that all the drawing commands before a `glEndQuery()` had not completed before the fragment total was retrieved into `result`, in which case the latter value may not be accurate. So, the `while` loop spins until the query has, in fact, completed.

The next statement in the drawing routine

```
glGetQueryObjectuiv(query, GL_QUERY_RESULT, &result);
```

retrieves the total number of fragments counted by the query and places it in `result`. Obviously, if this number is more than zero, then the box, the only object for which depth-test passed fragments were counted, is visible. The following statement

```
if (result) glutWireSphere(0.5, 16, 10);
```

draws the sphere only if the box is visible.

Messages at the top left of the display indicate the current values of `result` (the box being visible if it's more than zero) and `resultAvailable`.

Finally, we deal with the technical issue that, while we dropped the bounding box into the pipeline in order to query it, we may not want to see it actually displayed in a real scene. This is where the four masking statements, conditional on making the box invisible, in the `glBeginQuery()`-`glEndQuery()` block above, come into play.

First, we disable writes to the color buffer prior to rendering the box so that indeed it cannot display. Moreover, we obviously want the box to compete in depth tests with the green rectangle in order to resolve our query; however, we do *not* want it to update the depth buffer (when it wins) because, at time of drawing the actual scene, the sphere should compete only with the green rectangle, certainly not its own bounding box. For this reason, we further emasculate the box by setting the depth mask so that it cannot write the depth buffer. The masks are obviously reset once the box is drawn.

Is the program really doing what it's supposed to when the box is behind the green rectangle, which is suppress rendering of the sphere? A neat way to check this is to make the box smaller by replacing `glutSolidCube(1.0)` with `glutSolidCube(0.5)`. Of course, the box no longer bounds the sphere, but moving it around we can see now that the sphere indeed disappears as soon as the box does.

Conditional Rendering

OpenGL versions 3.0 and on support *conditional rendering* which makes for even more efficient handling of the query outcome. Conditional rendering allows the programmer to specify that the results of a particular block of rendering statements are to be discarded if the query result is zero. This obviates the need for explicit `glGetQueryObject*` calls on the programmer's part to fetch its result or determine if a query had completed. Code then is simpler, and, more importantly, the query outcome can be handled *entirely within* the GPU – e.g., without having to ship values of variables like `result` or `resultAvailable` in `occlusion.cpp` back to the application for further instruction.

The following reworking of `occlusion.cpp` invokes conditional rendering, but, otherwise, is functionally equivalent to the original.

Experiment 6.4. Run `occlusionConditionalRendering.cpp`. The major change from `occlusion.cpp` is in the following code block in the drawing routine, where the rendering of the sphere is made conditional upon the outcome of the query with id `query`. The parameter `GL_QUERY_WAIT` says to wait for the query to complete.

```
glBeginConditionalRender(query, GL_QUERY_WAIT);
glutWireSphere(0.5, 16, 10);
glEndConditionalRender();
```

Since OpenGL knows itself what not to render if the query outcome is zero, there is no need for user-instigated determination of results, so the variables `result` and `resultAvailable` are now gone. **End**

Exercise 6.5. **(Programming)** If you move the box in `occlusion.cpp` out of the display, i.e., out of the viewing frustum, then the occlusion query indicates it to be invisible – of course, because no fragment even made it to the depth test, let alone past. So, it seems occlusion queries might be applied to frustum culling as well.

Rewrite `spaceTravelFrustumCulled.cpp` to apply occlusion queries instead of a quadtree. Is the new program faster?

6.3 Animating Orientation Using Euler Angles

The lead-in to this section is the discussion of viewing transformations in Section 4.6, particularly orientation and Euler angles in 4.6.3, so you might want to review this earlier material.

6.3.1 Euler Angles and the Orientation of a Rigid Body

Consider this for a second: it's no different to locate and orient a camera in 3-space than it is to locate and orient an aircraft, spacecraft or any other freely-movable rigid object. The `gluLookAt()` command simply happens to provide intuitive syntax to use, in particular, with a camera, namely, translate to (*eyex*, *eyey*, *eyez*), point at (*centerx*, *centery*, *centerz*) and turn about the line of sight according to the (*upx*, *upy*, *upz*) value. As you can see from Figure 6.7, the captain of a spacecraft could use similar syntax to steer her ship.

Figure 6.7: Video camera and spacecraft.

Accordingly, replace the camera with an arbitrary rigid object B. Assume that the location of B is fixed or, more precisely, that the location of a point P belonging to B is fixed, say, at the origin – see Figure 6.8(a), where the point P at the end of the long leg of an L-shaped object B is fixed at the origin.

Now we are exactly at the point in Section 4.6.3 that we considered a `gluLookAt()` command with zero translational component, i.e., (*eyex*, *eyey*, *eyez*) = (0, 0, 0). From discussions in that section, the orientation of B is specified by *Euler angles* α, β and γ such that it can be obtained from a fixed reference orientation – in the case of the OpenGL camera this is its default pose – by applying the following rotation sequence:

```
glRotatef(α, 1.0, 0.0, 0.0);
glRotatef(β, 0.0, 1.0, 0.0);
glRotatef(γ, 0.0, 0.0, 1.0);
```

Of the two orientations shown of B in Figure 6.8(a), the reference one is bold.

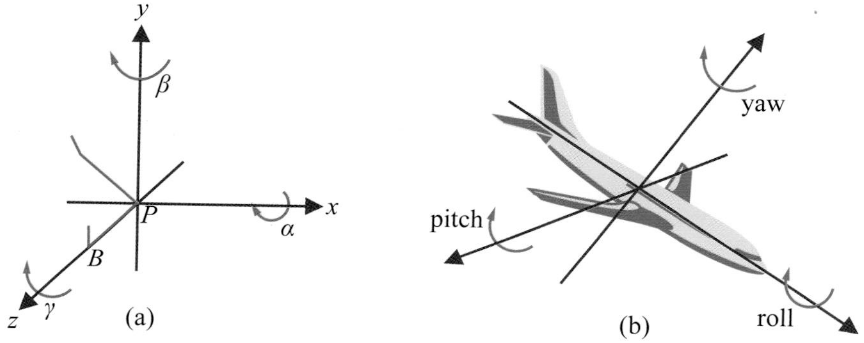

Figure 6.8: (a) Orienting an object in space with respect to fixed axes – the fixed reference orientation of the L is shown in bold (b) Orientation of an aircraft with respect to local axes "carried" by it.

Remark 6.1. The yaw, pitch and roll of an aircraft to which pilots refer are nothing but Euler angles, the difference being that the axis system is carried by the aircraft itself (e.g., the roll axis is the line through the middle of the craft from tail to nose). See Figure 6.8(b).

6.3.2 Animating Orientation

It seems, then, that animating orientation is a matter simply of changing Euler angles. This is true.

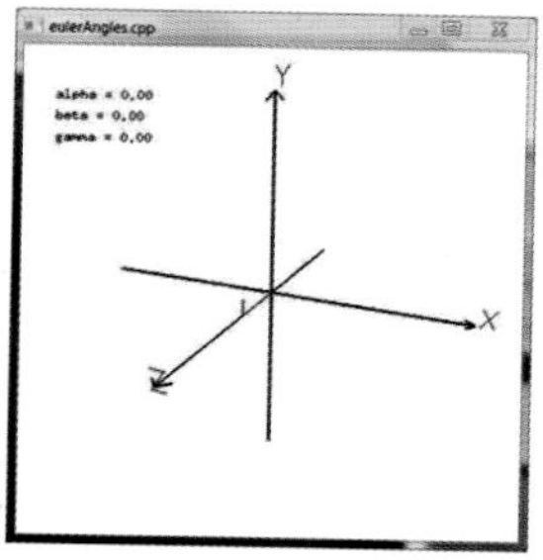

Figure 6.9: Screenshot of `eulerAngles.cpp`.

Experiment 6.5. Run `eulerAngles.cpp`, which shows an L, similar to the one in Figure 6.8(a), whose orientation can be interactively changed.

The original orientation of the L has its long leg lying along the z-axis and its short leg pointing up parallel to the y-axis. Pressing 'x/X', 'y/Y' and 'z/Z' changes the L's Euler angles and delete resets. The Euler angle values are displayed on-screen. Figure 6.9 is a screenshot of the initial configuration. **End**

Remark 6.2. If the commands

```
glRotatef(Xangle, 1.0, 0.0, 0.0);
glRotatef(Yangle, 0.0, 1.0, 0.0);
glRotatef(Zangle, 0.0, 0.0, 1.0);
```

in `eulerAngles.cpp` to change the Euler angles look familiar, well, we've been using similar ones since the second chapter to rotate scenes.

It all seems easy enough so far. However, things can get rather strange with Euler angles. Run `eulerAngles.cpp`, or use paper and pencil, to determine the two orientations of the L specified by the following distinct tuples of Euler angles (all angles are in degrees in this section):

(a) $\alpha = 0$, $\beta = 90$, $\gamma = 0$

(b) $\alpha = -90$, $\beta = 90$, $\gamma = 90$

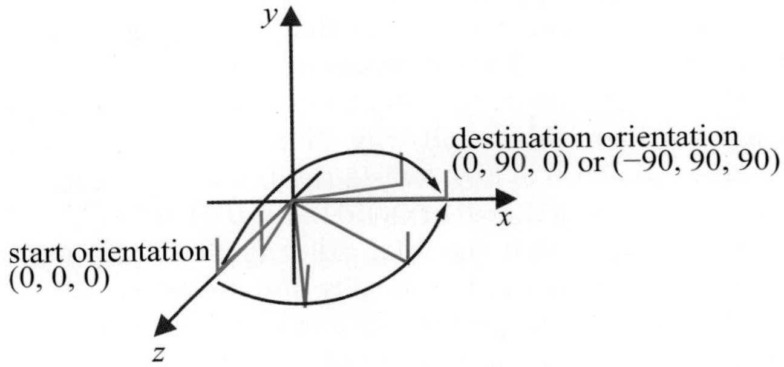

Figure 6.10: The bold blue start orientation is given by the Euler angle tuple $(0, 0, 0)$ and the bold blue destination one by either $(0, 90, 0)$ or $(-90, 90, 90)$. Intermediate orientations (green) in the linear interpolation between $(0, 0, 0)$ and $(0, 90, 0)$ all lie on the xz-plane, while those (red) between $(0, 0, 0)$ and $(-90, 90, 90)$ arc above it.

The two orientations are identical, both equal to the destination orientation of Figure 6.10, with the L's long leg along the x-axis and short leg pointing up parallel to the y-axis. Imagine now that the L is a spacecraft you want to take from its start orientation specified by the Euler angle tuple $(0, 0, 0)$ to the destination one specified by either of the tuples (a) or (b).

What comes to mind naturally is a linear interpolation between the start and destination orientations, exactly as if one had to translate the spacecraft from a start to a destination *location*, except that, instead of intermediate positions, one has now intermediate orientations. However, the ambiguity in representing the destination leads to a surprise we see next.

6.3.3 Problems with Euler Angles: Gimbal Lock and Ambiguity

Experiment 6.6. Run `interpolateEulerAngles.cpp`, which is based on `eulerAngles.cpp`. It simultaneously interpolates between the tuples $(0, 0, 0)$

and (0, 90, 0) and between (0, 0, 0) and (−90, 90, 90). Press the left and right arrow keys to step through the interpolations (delete resets). For the first interpolation (green L) the successive tuples are $(0, angle, 0)$ and for the second (red L) they are $(-angle, angle, angle)$, *angle* changing by 5 at each step in both.

The paths are different! The green L seems to follow the intuitively straighter path by keeping its long leg always on the xz-plane as it rotates about the y-axis, while the red L arcs above the xz-plane, as diagrammed in Figure 6.10. Figure 6.11 is a screenshot of `interpolateEulerAngles.cpp` part way through the interpolation. **End**

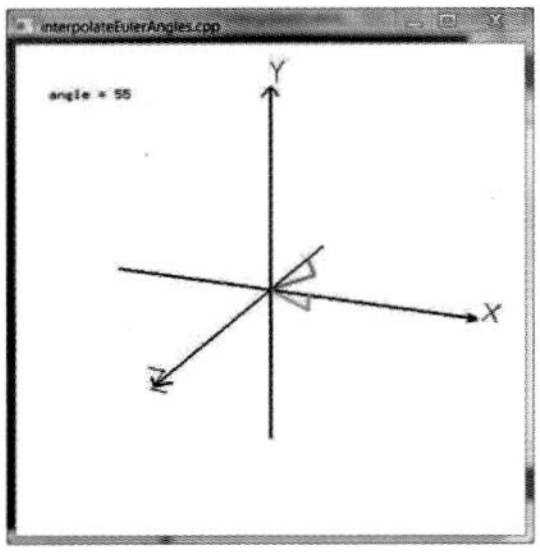

Figure 6.11: Screenshot of `interpolateEulerAngles.cpp`.

Two different paths arise in the program, of course, because of the non-unique representation of the destination orientation by Euler angles. Are (0, 90, 0) and (−90, 90, 90) the only tuples of Euler angles that represent this particular destination? Emphatically no, as we see next!

Use `eulerAngles.cpp` to see that any tuple of the form $(-A, 90, A)$ does as well: the $-A$ rotation about the x-axis seems always to cancel the A rotation about the z-axis to make an equivalence to (0, 90, 0)! It's not hard to understand why. The first rotation that is applied, viz. `glRotatef(`A`, 0.0, 0.0, 1.0)`, twists the L about its long leg lying along the z-axis, the second `glRotatef(90, 0.0, 1.0, 0.0)` rotates the long leg to line it up with x-axis, so that the final `glRotatef(`$-A$`, 1.0, 0.0, 0.0)` again twists the L about its long leg, but equally and oppositely to the first rotation.

We've run into the problem of *gimbal lock* which afflicts the Euler angle representation of orientation. Let's get a better understanding of this phenomenon.

Experiment 6.7. Run `eulerAngles.cpp` again.

Press 'x' and 'X' a few times each – the L turns longitudinally. Reset by pressing delete. Press 'y' and 'Y' a few times each – the L turns latitudinally. Reset. Press 'z' and 'Z' a few times each – the L twists. There appear to be three physical *degrees of freedom* of the L derived from rotation about the three coordinate axes which we might descriptively term, respectively, longitudinal, latitudinal and "twisting".

Now, from the initial configuration of `eulerAngles.cpp` press 'y' till $\beta = 90$. Next, press 'z' or 'Z' – the L twists. Then press 'x' or 'X' – the L still twists! **End**

Even with β fixed in `eulerAngles.cpp` one would expect two degrees of freedom to remain, viz. twisting *and* longitudinal. However, because of the particular value of β, namely, 90, which takes the L from the z-axis to the x-axis, we seem to have lost the longitudinal one.

Well, physical space hasn't changed and all three degrees of freedom are obviously still there for the taking. It's simply because of the particular value of β that the other two Euler angles α and γ both "act on the same degree of freedom", disallowing access to the remaining one.

This *apparent* loss of a degree of freedom is precisely gimbal lock, and it's the reason as well for the multiple representations of the destination orientation of the L in `interpolateEulerAngles.cpp`: rotations about the x- and z-axes both acting on the same degree of freedom when $\beta = 90$, one can be used to cancel the other.

Exercise 6.6. Show that gimbal lock also arises when $\beta = -90$.

Even though infinitely many different representations of an orientation occur only at the two gimbal lock values of $\beta = \pm 90$, there's actually instability near these values as well. In fact, two paths beginning and ending at nearby orientations which happen to be close to gimbal lock values can differ widely.

Exercise 6.7. (Programming) Verify the preceding remark by modifying `interpolateEulerAngles.cpp` to simultaneously interpolate between the Euler angle tuples $(0, 0, 0)$ and $(5, 95, 0)$ and the tuples $(0, 0, 0)$ and $(-95, 85, 95)$.

Moreover, there's another pesky problem with Euler angles: that of *non-unique*, in fact, *dual* representation of *every* orientation. To begin with, mentally visualize, or run `eulerAngles.cpp`, to see that the Euler angle tuple $(180, 180, 180)$ is equivalent to the Euler angle tuple $(0, 0, 0)$, both representing the initial orientation of the L.

This generalizes:

Proposition 6.1. *The Euler angle tuples* (α, β, γ) *and* $(\alpha + 180, -\beta + 180, \gamma + 180)$ *are equivalent in that they both represent the same orientation.* (Note the minus sign in front of the β in the second tuple.)

Proof. We'll not prove this formally, but ask the reader to "visually" verify it in the next exercise. □

Exercise 6.8. (Programming) Verify the claim of the proposition by comparing the orientation of the L of `eulerAngles.cpp`, as specified by the Euler angle tuples (α, β, γ) and $(\alpha + 180, -\beta + 180, \gamma + 180)$, for a few different values of α, β and γ.

Note: If you happen to go outside the range of $-180°$ to $180°$ with α or β or γ, then you can get back in again by adding or subtracting $360°$.

Of course, an Euler angle tuple (α, β, γ) is always equivalent to the Euler angle tuple $(\alpha \pm 360, \beta \pm 360, \gamma \pm 360)$ for any choice of the pluses and minuses, simply because angular arithmetic is modulo 360. However, the equivalence of the Euler angle tuples (α, β, γ) and $(\alpha + 180, -\beta + 180, \gamma + 180)$ is *not* a consequence of angular arithmetic, but a true geometric *duality* intrinsic to Euler angles.

Exercise 6.9. (Programming) Modify `interpolateEulerAngles.cpp` to simultaneously interpolate between the initial orientation $(0, 90, 0)$ and the destination orientation represented dually by the distinct Euler angle tuples $(0, 0, 0)$ and $(180, 180, 180)$.

Again, the two paths are different.

All this is not the best news if one is in the business of animating the orientation of a camera or rigid body, as one then wants to be able to *unambiguously* choose representations of the start and destination orientations in order to interpolate between the two. It turns out that there is, in fact, a more efficient representation by means of mathematical thingies called quaternions, in which case there's never gimbal lock and, though there's still an issue with ambiguous representation, it's one that can be elegantly resolved prior to interpolation. Quaternions and their application to orientation are the topic of the next section.

6.4 Quaternions

We are going to continue our discussion of animating orientation from the previous section. However, we'll now invoke quaternions for this particular application.

The air of mathematical mystique surrounding them has caused a fair bit of (mis)apprehension of quaternions among game programmers, which is particularly unfortunate as it is interactive applications like games that stand to benefit most from their use. Hopefully, this section will convince at least the reader that not only are quaternions fairly benign – about as hard as complex numbers as a matter of fact – they are not hard to apply either.

6.4.1 Quaternion Math 101

Quaternions were invented by the Irish mathematician William Hamilton in the mid-1900s as part of his investigation into 3D mechanics. Think of them as complex numbers on steroids. Whereas complex numbers extend the reals with one imaginary i, a square root of -1, quaternions add three such numbers i, j and k, all square roots of -1. Formally:

Definition 6.1. A *quaternion* q is a number of the form

$$q = w + xi + yj + zk$$

where w, x, y and z are real numbers

It's often written as $q = w + \mathbf{v}$, where w is the *real* or *scalar* part, while $\mathbf{v} = xi + yj + zk$ is the *vector* or *pure quaternion* part. A quaternion of the form $q = xi + yj + zk$, with a zero scalar part, is also called a pure quaternion.

Terminology: In this section vector parts will be denoted in bold to distinguish them from the scalar (except for i, j and k themselves, as there's little risk of ambiguity with the three).

The set of quaternions is commonly denoted $\mathbb{H}$ in honor of its inventor.

The real numbers are a subset of the quaternions, the real w being identified with the quaternion w ($= w + 0i + 0j + 0k$) with a zero vector part.

Example 6.1. Some quaternions:

$$2 - 3.4i + 4.8j + 2k, \quad -i + \sqrt{2}k, \quad 10.9, \quad -6.3j, \quad 0$$

Quaternions are added component-wise, just as complex numbers:

Definition 6.2. The *sum* of two quaternions $q_1 = w_1 + x_1 i + y_1 j + z_1 k$ and $q_2 = w_2 + x_2 i + y_2 j + z_2 k$ is the quaternion

$$q_1 + q_2 = (w_1 + w_2) + (x_1 + x_2)i + (y_1 + y_2)j + (z_1 + z_2)k$$

Exercise 6.10. Add the quaternions

(a) $2 - 3.4i + 4.8j + 2k$ and $-6.3j$

(b) $-i + \frac{1}{\sqrt{2}}k$ and $\frac{1}{\sqrt{2}}k$

The "square root of -1" property kicks in when multiplying quaternions. Here are the rules to keep in mind:

$$i^2 = j^2 = k^2 = -1$$

$$ij = k \qquad ji = -k$$

$$jk = i \qquad kj = -i$$

$$ki = j \qquad ik = -j$$

They're not hard to remember. The square of i, j and k each is -1. For the rest, simply keep in mind the cyclic order $i \rightarrow j \rightarrow k \rightarrow i$. If two successive elements in this order are multiplied, the result is the next element; if two successive elements are multiplied in reverse order, the result is the negative of the next element. This second rule is a replica of that of taking the cross-product of two different ones from the three unit vectors **i**, **j** and **k** (see Remark 5.8).

When multiplying two quaternions, it's a matter of applying distributivity and the preceding rules. Here's an example:

Example 6.2.

$$\begin{aligned}(2-3i+2j)(3+i-k) &= 2(3+i-k)-3i(3+i-k)+2j(3+i-k)\\ &= 2*3\ +\ 2*i\ +\ 2*-k\ \ -3i*3\ \ -3i*i\\ &\quad -3i*-k\ +\ 2j*3\ +\ 2j*i\ +\ 2j*-k\\ &= 6+2i-2k-9i+3-3j+6j-2k-2i\\ &= 9-9i+3j-4k\end{aligned}$$

Exercise 6.11. Multiply

$$(4+2i+2j-k)(1-k)$$

Exercise 6.12. Prove the following:

(a) The addition of quaternions is commutative and associative. In particular,

$$q_1+q_2=q_2+q_1 \quad \text{and} \quad (q_1+q_2)+q_3=q_1+(q_2+q_3)$$

for any three quaternions q_1, q_2 and q_3.

(b) The formula for the product of two quaternions $q_1 = w_1+x_1i+y_1j+z_1k$ and $q_2 = w_2+x_2i+y_2j+z_2k$ is

$$\begin{aligned}q_1q_2 = &(w_1w_2-x_1x_2-y_1y_2-z_1z_2)+(w_1x_2+x_1w_2+y_1z_2-z_1y_2)\,i\\ &(w_1y_2+y_1w_2-x_1z_2+z_1x_2)j+(w_1z_2+z_1w_2+x_1y_2-y_1x_2)k\end{aligned} \tag{6.1}$$

(c) The multiplication of quaternions is associative and, moreover, distributes both ways over addition. In particular,

$$(q_1q_2)q_3=q_1(q_2q_3),\ \ q_1(q_2+q_3)=q_1q_2+q_1q_3,\ \ (q_2+q_3)q_1=q_2q_1+q_3q_1$$

for any three quaternions q_1, q_2 and q_3. *Alert*: Not hard, but tedious!

(d) The multiplication of quaternions is *not* commutative. In particular, it need not be true that $q_1q_2 = q_2q_1$ for two quaternions q_1 and q_2. Give an example.

(e) The additive identity is 0 and the multiplicative identity 1. In particular,

$$q+0=0+q=q=q1=1q$$

for any quaternion q.

There is a useful shorter expression for the product of two quaternions in terms of vector operations:

Exercise 6.13. Prove that if $q_1 = w_1 + \mathbf{v_1}$ and $q_2 = w_2 + \mathbf{v_2}$, then

$$q_1 q_2 = w_1 w_2 - \mathbf{v_1} \cdot \mathbf{v_2} + w_1 \mathbf{v_2} + w_2 \mathbf{v_1} + \mathbf{v_1} \times \mathbf{v_2} \tag{6.2}$$

where $\cdot$ and $\times$ represent the vector dot and cross-product, respectively.

Note: Here we treat the pure quaternion (vector) part $\mathbf{v} = xi + yj + zk$ of the quaternion $q + \mathbf{v}$ as the geometric vector $x\mathbf{i} + y\mathbf{j} + z\mathbf{k}$.

Similar to complex numbers, quaternions each have a magnitude and a conjugate.

Definition 6.3. The *magnitude* of the quaternion $q = w + xi + yj + zk$, denoted $|q|$, is the non-negative value of the square root

$$\sqrt{w^2 + x^2 + y^2 + z^2}$$

Therefore, if we write $q = w + \mathbf{v}$, then its magnitude $|q| = \sqrt{w^2 + |\mathbf{v}|^2}$, where $|\mathbf{v}|$ denotes as usual the magnitude of the vector $\mathbf{v}$. A *unit quaternion* is one with magnitude 1.

Definition 6.4. The *conjugate* of the quaternion $q = w + xi + yj + zk$ is the quaternion

$$\overline{q} = w - xi - yj - zk$$

In other words, if $q = w + \mathbf{v}$ then its conjugate $\overline{q} = w - \mathbf{v}$.

And, as with complex numbers, a quaternion and its conjugate and magnitude are related:

Exercise 6.14. Prove that $q\overline{q} = \overline{q}q = |q|^2$.

Definition 6.5. The *inverse* of a quaternion q, if it exists, is the quaternion q^{-1} such that

$$qq^{-1} = q^{-1}q = 1$$

Exercise 6.15. Use Exercise 6.14 to prove that a quaternion q has an inverse if and only if it is non-zero, in which case

$$q^{-1} = \frac{\overline{q}}{|q|^2}$$

Therefore, $q^{-1} = \overline{q}$ for a unit quaternion q.

Example 6.3. Determine the inverse of the quaternion $q = 1 + i + j + k$.

Answer:

$$q^{-1} = \frac{\overline{q}}{|q|^2} = \frac{1 - i - j - k}{1^2 + 1^2 + 1^2 + 1^2} = \frac{1}{4}(1 - i - j - k)$$

Exercise 6.16. Determine the inverses of the quaternions

(a) j

(b) $\sqrt{3}i + \sqrt{2}k$

(c) $-1 + i + 2j - k$

Exercise 6.17. Prove the following if q_1 and q_2 are quaternions and c a scalar:

(a) $\overline{(q_1 q_2)} = \overline{q_2}\,\overline{q_1}$

(b) $(cq_1)^{-1} = c^{-1}q_1^{-1}$, provided both c and q_1 are non-zero.

(c) $(q_1 q_2)^{-1} = q_2^{-1}q_1^{-1}$, provided both q_1 and q_2 are non-zero.

(d) $|q_1 q_2| = |q_1||q_2|$

6.4.2 Quaternions and Orientation

So what do quaternions have to do with orienting a rigid body? The answer comes by way of the rotation transformation. Recall from the last section that the orientation of a rigid object B is specified by Euler angles α, β and γ, so that the specified orientation can be obtained from a fixed reference orientation by applying the following sequence of rotations:

```
glRotatef(α, 1.0, 0.0, 0.0);
glRotatef(β, 0.0, 1.0, 0.0);
glRotatef(γ, 0.0, 0.0, 1.0);
```

Now, we know from Proposition 5.14 that the composition of rotations about radial axes is another such. Therefore, the three above can be combined into a *single* rotation

```
glRotatef(θ, x, y, z)
```

for some angle θ and some values of x, y and z.

It follows that, instead of Euler angles, one can represent an orientation in 3D by means of a single 3D rotation about a radial axis. Moreover, it turns out that each quaternion, as we shall see, represents a 3D rotation about a radial axis. The conclusion, then, is that each quaternion represents a 3D orientation. The next proposition says how a quaternion determines a 3D rotation.

Note: In what follows we'll often identify the quaternions i, j and k with their vector counterparts **i**, **j** and **k**, respectively. It should be clear from the context whether we mean a quaternion or vector.

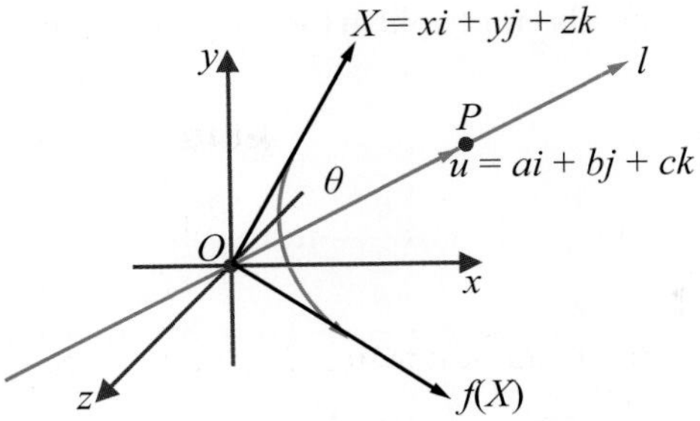

Figure 6.12: The vector $f(\mathbf{X})$ is obtained by rotating $\mathbf{X}$ about the line l.

Proposition 6.2. *Suppose the axis of a 3D rotation is specified by the directed line l through the origin O toward a point $P = [a\ b\ c]^T$. Assume that $|P| = 1$. Denote the unit vector $OP = a\mathbf{i} + b\mathbf{j} + c\mathbf{k}$ by $\mathbf{u}$.* (See Figure 6.12.)

If $\mathbf{X} = xi + yj + zk$ is an arbitrary vector in $\mathbb{R}^3$ then the image, call it $f(\mathbf{X})$, of $\mathbf{X}$ by a rotation of angle θ about l, is given by

$$f(\mathbf{X}) = q\,\mathbf{X}\,q^{-1} \tag{6.3}$$

where q is the unit quaternion

$$q = \cos\frac{\theta}{2} + \mathbf{u}\sin\frac{\theta}{2} \tag{6.4}$$

Let's put the proposition into context first. It gives one more way to determine the image $f(\mathbf{X})$ of a vector $\mathbf{X}$ by rotation of an angle θ about a radial axis l. The first, which we derived in Section 5.4.3, expressed $f(\mathbf{X})$ as the matrix product $R_{a,\,b,\,c}(\theta)\,\mathbf{X}$, where $R_{a,\,b,\,c}(\theta)$ was, in fact, given in a couple of different ways by the equations (5.35) and (5.45). Now, instead of a matrix, we manufacture a unit quaternion q such that the rotated vector $f(\mathbf{X})$ is $q\,\mathbf{X}\,q^{-1}$ (this operation of pre-multiplying by an element and then post-multiplying by its inverse is called an *inner automorphism* by that element).

The proof itself is a straight slog.

Proof. As q is a unit quaternion, its inverse is

$$q^{-1} = \overline{q} = \cos\frac{\theta}{2} - \mathbf{u}\sin\frac{\theta}{2}$$

Repeatedly applying the multiplication formula (6.2) we get the following equations:

$$
\begin{aligned}
& q\,\mathbf{X}\,q^{-1} \\
= \;& (\cos\frac{\theta}{2} + \mathbf{u}\sin\frac{\theta}{2})\,\mathbf{X}\,(\cos\frac{\theta}{2} - \mathbf{u}\sin\frac{\theta}{2}) \\
= \;& (-(\mathbf{u}\cdot\mathbf{X})\sin\frac{\theta}{2} + \mathbf{X}\cos\frac{\theta}{2} + (\mathbf{u}\times\mathbf{X})\sin\frac{\theta}{2})\;(\cos\frac{\theta}{2} - \mathbf{u}\sin\frac{\theta}{2}) \\
= \;& -(\mathbf{u}\cdot\mathbf{X})\sin\frac{\theta}{2}\cos\frac{\theta}{2} + (\mathbf{u}\cdot\mathbf{X})\sin\frac{\theta}{2}\cos\frac{\theta}{2} + (\mathbf{u}\times\mathbf{X})\cdot\mathbf{u}\sin^2\frac{\theta}{2} \\
& + (\mathbf{u}\cdot\mathbf{X})\,\mathbf{u}\sin^2\frac{\theta}{2} + \mathbf{X}\cos^2\frac{\theta}{2} + (\mathbf{u}\times\mathbf{X})\sin\frac{\theta}{2}\cos\frac{\theta}{2} \\
& + (\mathbf{u}\times\mathbf{X})\sin\frac{\theta}{2}\cos\frac{\theta}{2} - ((\mathbf{u}\times\mathbf{X})\times\mathbf{u})\sin^2\frac{\theta}{2}
\end{aligned}
$$

Of the eight terms in the final expression, the first two cancel, while the third is 0 because $\mathbf{u}\times\mathbf{X}$ is perpendicular to $\mathbf{u}$. Let $\mathbf{X} = \mathbf{X_1} + \mathbf{X_2}$, where $\mathbf{X_1}$ and $\mathbf{X_2}$ are the components of $\mathbf{X}$ parallel and perpendicular, respectively, to $\mathbf{u}$. Use the facts that $\mathbf{X_1} = (\mathbf{u}\cdot\mathbf{X})\,\mathbf{u}$ and $\mathbf{X_2} = \mathbf{X} - \mathbf{X_1} = (\mathbf{u}\times\mathbf{X})\times\mathbf{u}$, which we know from Exercises 4.41 and 5.59, respectively, to further simplify the final expression:

$$
\begin{aligned}
q\,\mathbf{X}\,q^{-1} &= \mathbf{X_1}\sin^2\frac{\theta}{2} + \mathbf{X}\cos^2\frac{\theta}{2} + (\mathbf{u}\times\mathbf{X})\sin\frac{\theta}{2}\cos\frac{\theta}{2} \\
&\quad + (\mathbf{u}\times\mathbf{X})\sin\frac{\theta}{2}\cos\frac{\theta}{2} - (\mathbf{X} - \mathbf{X_1})\sin^2\frac{\theta}{2} \\
&= \mathbf{X}(\cos^2\frac{\theta}{2} - \sin^2\frac{\theta}{2}) + 2\mathbf{X_1}\sin^2\frac{\theta}{2} + 2(\mathbf{u}\times\mathbf{X})\sin\frac{\theta}{2}\cos\frac{\theta}{2} \\
&= \mathbf{X}\cos\theta + \mathbf{X_1}(1-\cos\theta) + (\mathbf{u}\times\mathbf{X})\sin\theta \qquad (6.5)
\end{aligned}
$$

Comparing (6.5) with the formula (5.43) derived for $f(\mathbf{X})$ in Section 5.4.3 completes the proof. □

So, indeed, we have the correspondence:

orientation → radial rotation → quaternion

Example 6.4. Let's verify the preceding proposition in the case of rotating the vector i by an angle of $\pi/2$ about the axis k.

It's easily checked by hand that this particular rotation takes i to j.

Next, to use the proposition, write first $\mathbf{u} = k$, $\mathbf{X} = i$ and $\theta = \pi/2$. Then the quaternion

$$q = \cos\frac{\theta}{2} + \mathbf{u}\sin\frac{\theta}{2} = \frac{1}{\sqrt{2}} + \frac{1}{\sqrt{2}}k$$

Therefore, by the proposition the image of i by the rotation of $\pi/2$ about axis k is

$$\begin{aligned} q\,\mathbf{X}\,q^{-1} &= (\frac{1}{\sqrt{2}} + \frac{1}{\sqrt{2}}k)\,i\,(\frac{1}{\sqrt{2}} - \frac{1}{\sqrt{2}}k) \\ &= (\frac{1}{\sqrt{2}}i + \frac{1}{\sqrt{2}}j)\,(\frac{1}{\sqrt{2}} - \frac{1}{\sqrt{2}}k) \\ &= \frac{1}{2}i + \frac{1}{2}j + \frac{1}{2}j - \frac{1}{2}i \\ &= j \end{aligned}$$

which, indeed, matches what we checked.

Exercise 6.18. Verify the proposition in the case of rotating the vector $i + k$ by an angle of π about j.

Proposition 6.2 says that every radial rotation corresponds to a unit quaternion. How about the other way around: does every unit quaternion correspond to a radial rotation in the sense that it gives that rotation by an inner automorphism? The answer is yes:

Proposition 6.3. *Let $q = w + xi + yj + zk$ be a unit quaternion. If $q = \pm 1$, then $\mathbf{X} \mapsto q\,\mathbf{X}\,q^{-1}$ is the identity transformation, i.e., a zero rotation about an arbitrary axis. If $q \neq \pm 1$, then there exists a unique pair $(\mathbf{u}, \theta)$, such that $\mathbf{u}$ is a unit vector and $\theta \in (0, 2\pi)$, and such that*

$$q = \cos\frac{\theta}{2} + \mathbf{u}\sin\frac{\theta}{2}\ ,$$

which implies that $\mathbf{X} \mapsto q\,\mathbf{X}\,q^{-1}$ is a rotation by angle θ about the radial axis directed along $\mathbf{u}$.

Proof. If $q = 1$ or $q = -1$ then it's obvious that $\mathbf{X} \mapsto q\,\mathbf{X}\,q^{-1}$ is the identity transformation.

Suppose, then, that $q \neq \pm 1$. As $w^2 + x^2 + y^2 + z^2 = |q|^2 = 1$, we must have $-1 \leq w \leq 1$. However, if $w = \pm 1$, then we would have $x = y = z = 0$, so that $q = \pm 1$, contradicting our assumption. One deduces, therefore, that $-1 < w < 1$, which implies that there is a unique $\theta/2 \in (0, \pi)$ – equivalently, a unique $\theta \in (0, 2\pi)$ – such that $\cos\frac{\theta}{2} = w$. Accordingly, $\sin\frac{\theta}{2} = \sqrt{1 - w^2}$, where the RHS is the positive square root. It follows that

$$\begin{aligned} q &= w + xi + yj + zk \\ &= w + (\frac{x}{\sqrt{1-w^2}}\,i + \frac{y}{\sqrt{1-w^2}}\,j + \frac{z}{\sqrt{1-w^2}}\,k)\sqrt{1-w^2} \\ &= \cos\frac{\theta}{2} + \mathbf{u}\sin\frac{\theta}{2} \end{aligned}$$

where

$$\mathbf{u} = \frac{x}{\sqrt{1-w^2}}\,i + \frac{y}{\sqrt{1-w^2}}\,j + \frac{z}{\sqrt{1-w^2}}\,k$$

is a unit vector because $x^2 + y^2 + z^2 = 1 - w^2$. The conclusion in the last line of the proposition now follows from an application of Proposition 6.2. □

Example 6.5. Determine the rotation corresponding to the unit quaternion

$$\frac{1}{\sqrt{3}}i + \frac{1}{\sqrt{3}}j + \frac{1}{\sqrt{3}}k$$

and write it in OpenGL form.

Answer: We want to express

$$q = \frac{1}{\sqrt{3}}i + \frac{1}{\sqrt{3}}j + \frac{1}{\sqrt{3}}k$$

in the form

$$\cos\frac{\theta}{2} + \mathbf{u}\sin\frac{\theta}{2}$$

Following the preceding proposition, write

$$q = w + xi + yj + zk \quad \text{where} \quad w = 0 \quad \text{and} \quad x = y = z = \frac{1}{\sqrt{3}}$$

Next, set

$$\cos\theta/2 = w = 0 \implies \theta/2 = \frac{\pi}{2} \implies \theta = \pi$$

and

$$\mathbf{u} = \frac{x}{\sqrt{1-w^2}}i + \frac{y}{\sqrt{1-w^2}}j + \frac{z}{\sqrt{1-w^2}}k = \frac{1}{\sqrt{3}}i + \frac{1}{\sqrt{3}}j + \frac{1}{\sqrt{3}}k$$

It follows that the given quaternion corresponds to the OpenGL rotation (up to round-off error)

```
glRotatef(180.0, 0.58, 0.58, 0.58)
```

Exercise 6.19. Determine the rotation corresponding to the unit quaternion

$$\frac{1}{\sqrt{2}} + \frac{1}{2}i + \frac{1}{2}k$$

and write it in OpenGL form.

We prove next a couple of useful facts related to Proposition 6.2:

Proposition 6.4. *(a) If the rotation f_1 corresponds to the quaternion q_1 and the rotation f_2 to q_2, then the composed rotation $f_1 \circ f_2$ corresponds to the product $q_1 q_2$.*

In other words, rotations can be composed by multiplying their corresponding quaternions.

(b) If q is a unit quaternion and c an arbitrary non-zero scalar, then the transformation

$$\mathbf{X} \mapsto (cq)\,\mathbf{X}\,(cq)^{-1}$$

is equivalent to the rotation

$$\mathbf{X} \mapsto q\,\mathbf{X}\,q^{-1}$$

In other words, q and any non-zero scalar multiple cq give the same rotation by inner automorphism.

Proof. (a) For a vector $\mathbf{X}$,

$$\begin{aligned}(f_1 \circ f_2)(\mathbf{X}) &= f_1(f_2(\mathbf{X})) = q_1(q_2\,\mathbf{X}\,q_2^{-1})q_1^{-1} \\ &= (q_1q_2)\,\mathbf{X}\,(q_2^{-1}q_1^{-1}) = (q_1q_2)\,\mathbf{X}\,(q_1q_2)^{-1}\end{aligned}$$

completing the proof.

(b) Since $(cq)^{-1} = c^{-1}q^{-1}$, the equalities

$$(cq)\,\mathbf{X}\,(cq)^{-1} = (cq)\,\mathbf{X}\,(c^{-1}q^{-1}) = (cc^{-1})q\,\mathbf{X}\,q^{-1} = q\,\mathbf{X}\,q^{-1}$$

complete the proof (note, for the second equality, that a scalar can be moved in and out of a product with impunity). □

Exercise 6.20. Let f_1 be a rotation of $60°$ about the x-axis and f_2 a rotation of $90°$ about the y-axis. Determine the composed rotations $f_1 \circ f_2$ and $f_2 \circ f_1$ (by giving their respective axis and amount of rotation).

Part answer: The plan is to go from rotation space to quaternion space, multiply and return to rotation space.

Proposition 6.2 tells us that rotation f_1 corresponds to the quaternion

$$q = \cos\frac{\theta}{2} + \mathbf{u}\sin\frac{\theta}{2}$$

where $u = i$ and $\theta = \pi/3$. In other words, f_1 corresponds to

$$\frac{\sqrt{3}}{2} + \frac{1}{2}i$$

Likewise, f_2 corresponds to

$$\frac{1}{\sqrt{2}} + \frac{1}{\sqrt{2}}j$$

By Proposition 6.4(a), $f_1 \circ f_2$ corresponds to

$$\left(\frac{\sqrt{3}}{2} + \frac{1}{2}i\right)\left(\frac{1}{\sqrt{2}} + \frac{1}{\sqrt{2}}j\right) = \frac{\sqrt{3}}{2\sqrt{2}} + \frac{1}{2\sqrt{2}}i + \frac{\sqrt{3}}{2\sqrt{2}}j + \frac{1}{2\sqrt{2}}k$$

Applying Proposition 6.3 to determine the rotation corresponding to the above quaternion, one finds after some calculation $f_1 \circ f_2$ to be `glRotatef(104.48, 0.45, 0.77, 0.45)` written as an OpenGL rotation up to round-off error.

We leave the reader to apply the same method to $f_2 \circ f_1$.

Proposition 6.4 has a couple of interesting consequences.

Firstly, part (a) has an implication for computational efficiency. If a rotation is represented by a matrix, as in Equation (5.45), then the complexity of composing two rotations, equivalent to multiplying the corresponding matrices, is dominated by 27 scalar multiplications – treating the matrix of (5.45) as 3×3 because the fourth row and column don't add complexity and observing that each of the 9 entries of the product involves 3 multiplications.

On the other hand, if a rotation is represented by a quaternion, then composing two rotations, equivalent to multiplying the corresponding quaternions by Proposition 6.4(a), requires 16 scalar multiplications. The conclusion is that an efficient way to compose multiple rotations is via quaternion representation.

Secondly, a consequence of part (b) of the proposition is a mathematically useful, though somewhat abstract, representation of 3D rotations. Note, first, that the set $\mathbb{H}$ of quaternions is in one-to-one correspondence with 4D space $\mathbb{R}^4$ via the association $w + xi + yj + zk \leftrightarrow [w\ x\ y\ z]^T$. We can, therefore, identify $\mathbb{H}$ with $\mathbb{R}^4$ and refer to points of the latter as quaternions.

Now, Proposition 6.4(b) says that all non-zero quaternions on a given radial line in $\mathbb{R}^4$ correspond to the same rotation, as they differ one from another by a scalar multiple. It can also be verified that non-zero quaternions on distinct radial lines correspond to different rotations.

So here's a summary of the situation. Quaternions are in one-to-one correspondence with points of $\mathbb{R}^4$. Each non-zero quaternion also corresponds *uniquely* to a rotation of 3-space, which it gives by inner automorphism. Rotations, on the other hand, are not as "faithful" because each corresponds to infinitely many quaternions, in fact, a whole radial line's worth (except that the origin O does not correspond to a rotation). See Figure 6.13. Rotations of 3-space, therefore, are in one-to-one correspondence with the set of radial lines in $\mathbb{R}^4$.

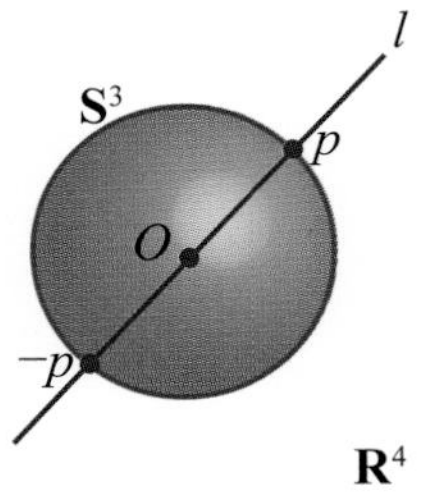

Figure 6.13: The unit sphere $\mathbb{S}^3$ in $\mathbb{R}^4$ with a radial line l, representing a rotation of $\mathbb{R}^3$, passing through a pair of antipodal points.

If one is uncomfortable with identifying rotations of $\mathbb{R}^3$ with lines in $\mathbb{R}^4$, then here's a way to identify them with points instead: since a radial line of $\mathbb{R}^4$ intersects $\mathbb{S}^3$, the unit sphere of $\mathbb{R}^4$, in two antipodal points, rotations are in 1-1 correspondence with the points of $\mathbb{S}^3$, *provided* one is willing to undertake the mental trick of identifying each antipodal pair as a single point. $\mathbb{S}^3$ with its antipodal points identified is actually the so-called projective 3-space $\mathbb{P}^3$, so, finally, one identifies the space of 3D rotations with $\mathbb{P}^3$.

Incidentally, note that since the space $\mathbb{H}$ of all quaternions is identified with $\mathbb{R}^4$, the space of unit quaternions identifies with the unit sphere $\mathbb{S}^3$ of $\mathbb{R}^4$.

Quaternion to Rotation Matrix

Proposition 6.3 enables us to find the unique rotation corresponding to a unit quaternion in terms of its axis and rotation angle. However, in various applications, e.g., when using OpenGL, the rotation matrix itself is more useful. We ask the reader to find the 4×4 rotation matrix corresponding to a given quaternion by completing the solution to the following exercise.

Exercise 6.21. Show that the 4×4 matrix representing the rotation corresponding to the unit quaternion

$$q = w + xi + yj + zk$$

is

$$\begin{bmatrix} w^2+x^2-y^2-z^2 & 2xy-2wz & 2xz+2wy & 0 \\ 2xy+2wz & w^2-x^2+y^2-z^2 & 2yz-2wx & 0 \\ 2xz-2wy & 2yz+2wx & w^2-x^2-y^2+z^2 & 0 \\ 0 & 0 & 0 & 1 \end{bmatrix} \quad (6.6)$$

Part answer: Verify first the case when $q = \pm 1$ (keep in mind that this implies that $w = \pm 1$ and that $x = y = z = 0$). Suppose, next, that $q \neq \pm 1$. Proposition 6.3 says, then, that q gives, by inner automorphism, the rotation of angle θ about the unit vector $\mathbf{u}$ where

$$\cos\frac{\theta}{2} = w \quad \text{and} \quad \mathbf{u} = (x/\sqrt{1-w^2})\,i + (y/\sqrt{1-w^2})\,j + (y/\sqrt{1-w^2})\,k$$

In Section 5.4.3 – see Equation (5.45) – we derived the following matrix corresponding to a rotation of angle θ about the radial axis l toward the point $P = [a\ b\ c]^T$, where $|P| = 1$.

$R_{a,\,b,\,c}(\theta) =$

$$\begin{bmatrix} a^2(1-\cos\theta)+\cos\theta & ab(1-\cos\theta)-c\sin\theta & ac(1-\cos\theta)+b\sin\theta & 0 \\ ab(1-\cos\theta)+c\sin\theta & b^2(1-\cos\theta)+\cos\theta & bc(1-\cos\theta)-a\sin\theta & 0 \\ ac(1-\cos\theta)-b\sin\theta & bc(1-\cos\theta)+a\sin\theta & c^2(1-\cos\theta)+\cos\theta & 0 \\ 0 & 0 & 0 & 1 \end{bmatrix}$$

We ask the reader to finish the exercise by deriving the matrix (6.6) after plugging the following into the matrix expression above for $R_{a,\,b,\,c}(\theta)$:

$$\begin{aligned} a &= x/\sqrt{1-w^2} \\ b &= y/\sqrt{1-w^2} \\ c &= z/\sqrt{1-w^2} \\ 1-\cos\theta &= 2\sin^2\frac{\theta}{2} = 2(1-w^2) \\ \cos\theta &= 2\cos^2\frac{\theta}{2} - 1 = 2w^2 - 1 \\ \sin\theta &= 2\sin\frac{\theta}{2}\cos\frac{\theta}{2} = 2w\sqrt{1-w^2} \end{aligned}$$

Spherical Linear Interpolation

Let's return to our original objective of applying quaternions to the animation of orientation. We have a 3-step strategy: (1) represent the start and end orientations corresponding, say, to the rotations f_1 and f_2 by unit quaternions, q_1 and q_2, respectively; (2) interpolate between the two in quaternion space, traveling from q_1 to q_2 along a path of unit quaternions as well; (3) finally, map this path back to a path in the space of rotations from f_1 to f_2, which, of course, is equivalent to a path in the space of orientations between the original start and end ones. Figure 6.14(a) is a conceptual diagram (numbers in parentheses indicate steps).

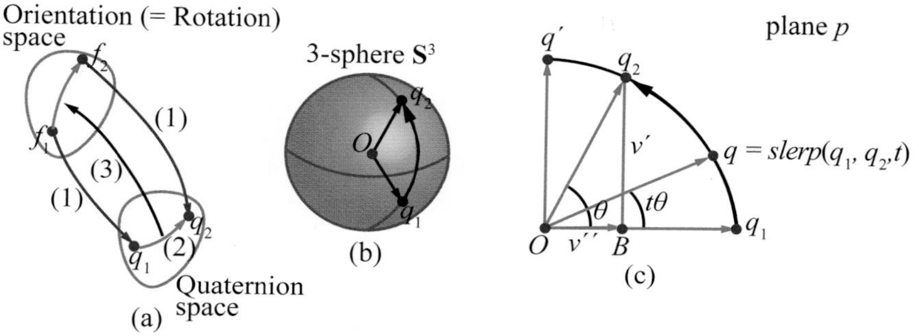

Figure 6.14: (a) Conceptual plan to use quaternion space to interpolate in orientation space (numbers in parentheses indicate steps) (b) The geodesic path from q_1 to q_2 on $\mathbb{S}^3$ (c) Slerping from q_1 to q_2.

Figure 6.14(b) shows the representing unit quaternions q_1 and q_2 as points on the unit sphere $\mathbb{S}^3$ of $\mathbb{R}^4$, the latter being identified with the set $\mathbb{H}$ of all quaternions. A shortest path on $\mathbb{S}^3$ between q_1 and q_2 is along a great circle – a *great circle* is the intersection of a radial plane with $\mathbb{S}^3$. A shortest path itself is called a *geodesic* path. If q_1 and q_2 are not antipodal, then there is a unique geodesic path joining them; otherwise, there are infinitely many (each half a great circle).

Remark 6.3. It helps the intuition to think of shortest paths on the surface of the Earth, where the situation is exactly the same, only one dimension lower.

Suppose, first, that q_1 and q_2 are not antipodal. We want to interpolate at a constant angular rate from q_1 to q_2 along the unique geodesic joining them. This is called *spherical linear interpolation*, or *slerp* for short. Even though q_1 and q_2 are points of $\mathbb{R}^4$, the slerp between them takes place entirely on the unique 2D plane p containing the two and the origin. See Figure 6.14(c).

Suppose that θ is the smaller of the angles between q_1 and q_2 on the great circle containing them. The point q, an angle of $t\theta$ from q_1 toward q_2 on that same circle, is denoted $slerp(q_1, q_2, t)$. As t varies from 0 to 1, $slerp(q_1, q_2, t)$ travels from q_1 to q_2. We seek, therefore, a formula for $slerp(q_1, q_2, t)$.

Let q' be the unit quaternion on the plane p on the same side of q_1 as q_2 and such that Oq' is perpendicular to Oq_1 (this assumes, additionally, that $q_2 \neq q_1$, for slerping between them is trivial otherwise). Drop the perpendicular from q_2 to the point B on Oq_1 and denote the 4-vector Bq_2 by v' and OB by v''.

Observe, first, that

$$|v'| = |q_2| \sin\theta = \sin\theta \quad \text{and} \quad |v''| = |q_2| \cos\theta = \cos\theta$$

because $|q_2| = 1$. Moreover, $v'' = |v''|q_1$, because q_1 is the unit vector parallel to v'', which means $v'' = q_1 \cos\theta$. Therefore, we have

$$v' \;=\; q_2 - v'' \;=\; q_2 - q_1 \cos\theta$$

Since q' is of unit length and parallel to v', we have as well

$$q' = \frac{v'}{|v'|} = \frac{v'}{\sin\theta} = \frac{q_2 - q_1\cos\theta}{\sin\theta}$$

(that q_1 and q_2 are neither equal nor antipodal ensures that θ and $\sin\theta$ are both non-zero). Therefore,

$$\begin{aligned}
q \;=\; slerp(q_1, q_2, t) \;&=\; q_1 \cos(t\theta) + q' \sin(t\theta) \\
&\qquad \text{(adding the components of } q \text{ along } q_1 \text{ and } q'\text{)} \\
&=\; q_1 \cos(t\theta) + \frac{q_2 - q_1\cos\theta}{\sin\theta}\sin(t\theta) \\
&=\; q_1\left(\cos(t\theta) - \frac{\cos\theta}{\sin\theta}\sin(t\theta)\right) + q_2\frac{\sin(t\theta)}{\sin\theta} \\
&=\; q_1\frac{\sin((1-t)\theta)}{\sin\theta} + q_2\frac{\sin(t\theta)}{\sin\theta} \qquad (6.7)
\end{aligned}$$

which gives the formula sought for $slerp(q_1, q_2, t)$ (note that θ, the angle between them, is easily determined from q_1 and q_2).

Remark 6.4. Expectedly, linear interpolation has not been able to escape the ugly diminutive *lerp* in the CG literature, with the obvious formula

$$lerp\,(q_1, q_2, t) = (1 - t)q_1 + tq_2$$

Interpolating Orientations via Quaternions

We now have all the pieces in place to implement the following scheme to interpolate between two orientations corresponding to the rotations f_1 and f_2:

1. Go from rotation space to quaternion space by finding unit quaternions q_1 and q_2 corresponding to f_1 and f_2, respectively.

2. Observe that both q_2 and $-q_2$ represent the same rotation but one of the two makes an angle of at most $\pi/2$ with q_1 and the other an angle at least $\pi/2$. We want to interpolate to the one closer to q_1, in case the two are at different angular distances.

   Accordingly, determine $q_1 \cdot q_2$. If its value is negative, then the angle between q_1 and q_2 is greater than $\pi/2$, in which case set $q_2 = -q_2$; otherwise, leave it as it is.

   Note: It's in this last step that we resolve the problem from potentially ambiguous representation of a rotation. In fact, it is only when q_1 and q_2 are orthogonal that both q_2 and $-q_2$ are at the same angle of $\pi/2$ from q_1, and we *could* choose either. The last step above finesses this choice by leaving q_2 unchanged in this case. However, see Exercise 6.22 below.

   Recall in this connection how much more troublesome were ambiguous Euler angle representations in the last section.

3. Compute $slerp(q_1, q_2, t)$, t varying from 0 to 1.

4. Return to rotation space by computing the rotation $f(t)$ corresponding to $slerp(q_1, q_2, t)$. Then $f(t)$ interpolates between the rotations f_1 and f_2 (equivalently, the corresponding orientations).

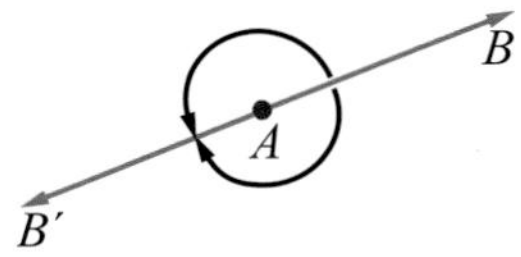

Figure 6.15: Changing the orientation from AB to AB' is inherently ambiguous.

Exercise 6.22. Show that if q_1 and q_2 are orthogonal then ambiguity is *inherent*, in that the orientation corresponding to one is a rotation of 180° about some axis of the orientation corresponding to the other. In this case, rotating one way about that axis, to change one orientation to the other, is exactly symmetric to going the other way, and there is no procedure to prefer one over the other. For example, consider the two ways of going from the orientation AB to AB' of the solid straight arrow in Figure 6.15.

Time for code.

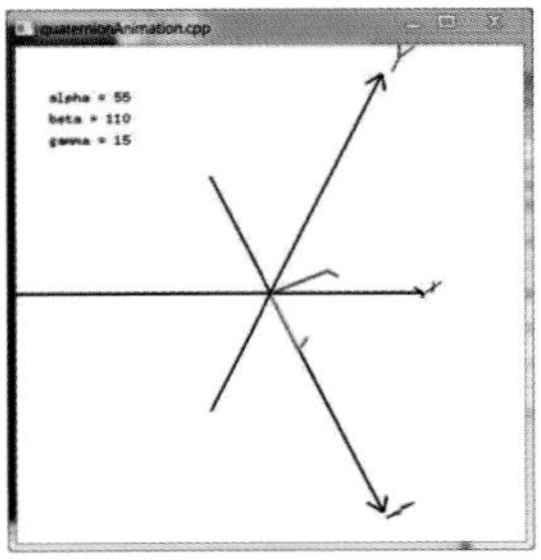

Figure 6.16: Screenshot of `quaternionAnimation.cpp`.

Experiment 6.8. Run `quaternionAnimation.cpp`, which applies the preceding ideas to animate the orientation of our favorite rigid body, an L, with the help of quaternions. Press 'x/X', 'y/Y' and 'z/Z' to change the orientation of the blue L, whose current Euler angles are shown on the display. Its start orientation is the currently fixed red L. See Figure 6.16 for a screenshot.

Pressing enter at any time begins an animation of the red L from the start to the blue's current orientation. Press the up and down arrow keys to change the speed and delete to reset. End

The routine `eulerAnglesToQuaternion()` of the program determines the unit quaternion corresponding to an orientation specified by three Euler angles, by computing first the unit quaternion corresponding to the rotation about a coordinate axis connoted by each Euler angle, and then multiplying the three.

The routine `slerp()` implements formula (6.7), except for the following "hack" to avoid problems with division by zero, or near-zero numbers, when θ is small. Applying the approximation $\sin\alpha \simeq \alpha$ for small α, formula (6.7) is modified as follows if $\theta \leq 0.000001$:

$$\begin{aligned} slerp(q_1, q_2, t) &= q_1 \frac{\sin((1-t)\theta)}{\sin\theta} + q_2 \frac{\sin(t\theta)}{\sin\theta} \\ &= q_2 \frac{(1-t)\theta}{\theta} + q_2 \frac{t\theta}{\theta} \quad = \quad (1-t)q_1 + tq_2 \end{aligned}$$

(the value 0.000001 having been chosen arbitrarily).

The routine `quaternionToRotationMatrix()` gets us back to rotation space with help of (6.6) to find the 4×4 rotation matrix corresponding to a given quaternion.

Exercise 6.23. (Programming) Extend `quaternionAnimation.cpp` to animate the motion of a rigid body, say a spacecraft, from a start disposition to a user-specified target disposition which can differ *both* in location and orientation from the start. Transformation should be simultaneous in both position and orientation.

The utility of quaternions in interactive animation cannot be overemphasized and they should be in every game programmer's tool kit. For instance, try to do what `quaternionAnimation.cpp` does *without* using quaternions (good luck!).

6.5 Summary, Notes and More Reading

In this chapter we learned a few practically useful techniques. Particularly indispensable to programmers of heavily-populated environments are the methods of frustum culling and occlusion culling. For further reading, books on game programming, e.g., Lengyel [85] and Eberly [39], will, typically, contain descriptions of structures such as quadtrees, octrees, kd-trees and BSP trees in the context of frustum culling, occlusion culling and collision detection. An excellent computational geometry reference for space partitioning data structures, including range trees and kd-trees, is the one by de Berg et al. [11]. Samet [121] is a must for anyone seeking to learn about spatial data structures in depth. See Slater et al. [135] and the paper by Kumar et al. [83] for literature on the important problem of partitioning a space encompassing a dynamic scene.

We learned how to animate orientation with the use of both Euler angles and quaternions. This will come in handy in camera control and rigid body animation. The Euler angle representation, as we saw, suffers from certain problems, surmounted subsequently by the slick mathematics of quaternions. The books by Buss [22], Lengyel [85] and Watt [147], among others, contain discussions of Euler angles and quaternions and their relation to rigid-body kinematics. The ones by Hanson [63] and Kuipers [82] are all about quaternions and their applications.

Part IV

Geometry for the Home Office

CHAPTER 7

Convexity and Interpolation

It's time now to get some of the geometric concepts underlying 3D modeling and lighting in place before we reach those particular chapters. We've seen programs where colors, defined at the vertices of a primitive, are mixed and spread throughout the primitive's interior. This is done by means of interpolation. In this chapter we'll study the exact mechanics of the interpolation process.

Section 7.1 motivates the process of interpolation with simple examples. Section 7.2 gets to the heart of the matter by showing first that line segment and triangle primitives are particularly suited for interpolation because of the property they share that any point of such a primitive can be uniquely represented as a so-called convex combination of its vertices. Section 7.3 shows precisely how this property is used by OpenGL to interpolate values such as color.

The geometric property which some objects have of being convex is closely related to interpolation. Section 7.4 defines convexity and the notion of the convex hull of a set of points and applies them to understanding if objects more complicated than line segments and triangles, e.g., polygons in general, can be equally easily interpolated. We see that the answer is no and that line segments and triangles are indeed special. We conclude with Section 7.5.

This chapter and the next two on triangulation and orientation, respectively, are intimately related and should be read one after another. The material is somewhat mathematical. However, the math involved is geometric, which means that it can be "seen to work", and not particularly abstract. The importance of these three chapters at the conceptual foundation of 3D computer graphics cannot be overemphasized. Having said this, it's true that this particular chapter will be fairly light reading for someone already familiar with linear interpolation and convexity, possibly from an earlier math class. If this is the case, then flip quickly through the pages – make sure the parts to do with OpenGL make sense – and move on.

7.1 Motivation

OpenGL has three favorites among its several drawing primitives: points, segments (by segment we'll always mean a straight line segment) and triangles. This is not owing to some idiosyncrasy of its specification as an API, but for a deeper reason. We've already seen that material values such as color, specified at a primitive's vertices, are apparently interpolated throughout its interior.

So here, briefly, is why points, segments and triangles are favored: they have the property that every point in each can be *unambiguously* (or, *uniquely*, same thing) represented in terms of its vertices. This makes it possible for values defined at the vertices to be *unambiguously* – therefore, *automatically*, by means of a program – interpolated throughout the primitive. We'll clarify all this soon, but to begin with here is a simple example.

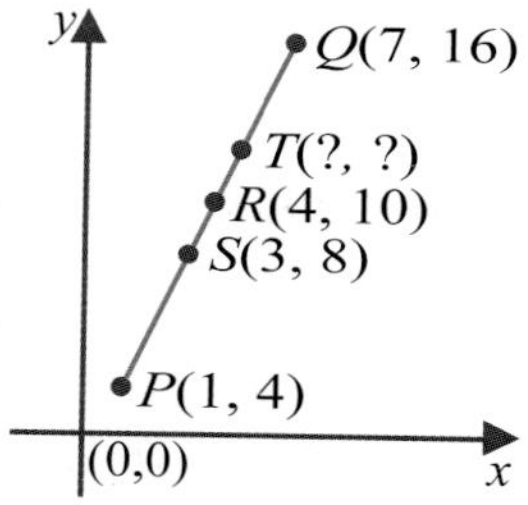

Figure 7.1: Points that split segment PQ.

Example 7.1. Using graph paper if you like, draw the segment PQ joining the point P, with coordinates $(1, 4)$, to the point Q, with coordinates $(7, 16)$. See Figure 7.1. Measure off the midpoint R of the segment. Verify that its coordinates are as indicated in the figure. Since the midpoint is halfway from either endpoint it does make sense that the coordinates of R are an exact average of those of P and Q, viz.

$$\frac{1}{2} * (1,4) + \frac{1}{2} * (7,16) = \left(\frac{1}{2} * 1 + \frac{1}{2} * 7,\ \frac{1}{2} * 4 + \frac{1}{2} * 16\right) = (4,10)$$

How about the point S a third of the way from P to Q? Again, measure it off and see if the coordinates shown in the figure are correct. S splits PQ in the ratio $\frac{2}{3} : \frac{1}{3}$, where it's $\frac{2}{3}$ toward P from Q and $\frac{1}{3}$ toward Q from P. Ergo, S's coordinates are

$$\frac{2}{3} * (1,4) + \frac{1}{3} * (7,16) = \left(\frac{2}{3} * 1 + \frac{1}{3} * 7,\ \frac{2}{3} * 4 + \frac{1}{3} * 16\right) = (3,8)$$

Exercise 7.1. We ask you to calculate the coordinates of T, which is two-thirds of the way from P to Q, and verify by actual measurement.

Interestingly, P itself, which splits PQ in the ratio $1 : 0$, has coordinates

$$1 * (1,4) + 0 * (7,16) = (1,4)$$

while Q, which splits it in the ratio $0 : 1$, has coordinates

$$0 * (1,4) + 1 * (7,16) = (7,16)$$

It seems then that in the expression

$$X = c * (1, 4) + (1 - c) * (7, 16)$$

the variable c acts as a "dial" which can be turned from 1 to 0 to move the point X from P to Q.

Here's an exercise to get you thinking about using an expression like the last one to interpolate material properties.

Exercise 7.2. If the end vertex P of the segment in the preceding example is specified red (RGB = (1, 0, 0)) and Q green (RGB = (0, 1, 0)), then what should the colors be at the midpoint R? At the point S?

7.2 Convex Combinations

We said at the beginning of the last section that points, segments and triangles are favored by OpenGL. In fact, they are *the* building blocks of OpenGL. Even quadrilaterals and polygons in general, as we'll see, are first sub-divided into triangles before being processed. We informally described a property shared by the three primitives – that each point belonging to one has a unique expression in terms of its vertices – which makes unambiguous interpolation possible.

Segments

Here is the formal statement of the aforementioned property in the case of a segment:

Proposition 7.1. *If P and Q are two points in $\mathbb{R}^3$, then a point V lies on the segment PQ if and only if it can be expressed as*

$$V = c_1 P + c_2 Q$$

where $0 \leq c_i \leq 1$, for both $i = 1$ and $i = 2$, and where $c_1 + c_2 = 1$.

Further, if P and Q are distinct – so that PQ does not degenerate to a point – then this expression for V is unique.

Before the proof, take a second to match the proposition with the example of the previous section: the points P, Q, R, S and T on the segment PQ could each indeed be expressed in the form $c_1 P + c_2 Q$, where c_1 and c_2 lie between 0 and 1, and add up to 1.

Proof. *Skip this proof if you start to get bogged down in the math. Just make sure to understand what the proposition says.* For the first part of the proposition, suppose initially that $P \neq Q$ so that PQ is a non-degenerate segment. Consider a point V on this segment. The vector $V - P$ clearly is parallel to the vector $Q - P$ (see Figure 7.2). Therefore, one is obtained from the other by multiplying by the ratio of their lengths:

$$V - P = \frac{|V - P|}{|Q - P|}(Q - P) \quad \Longrightarrow \quad V - P = c(Q - P)$$
$$\Longrightarrow \quad V = (1 - c)P + cQ$$

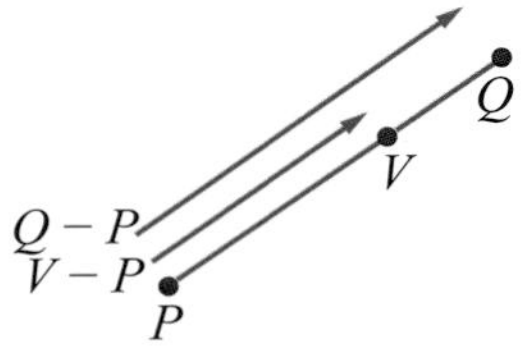

Figure 7.2: Illustration for the proof of Proposition 7.1.

where c denotes $\frac{|V-P|}{|Q-P|}$. Note then that $0 \leq c \leq 1$, as $|V - P| \leq |Q - P|$. Writing $c_1 = 1 - c$ and $c_2 = c$ we have, indeed, that

$$V = c_1 P + c_2 Q$$

where $0 \leq c_i \leq 1$, for both $i = 1$ and $i = 2$, and, moreover, $c_1 + c_2 = 1$. This proves the "only if" direction of the first part, provided $P \neq Q$.

Conversely, for the "if" direction, assuming again $P \neq Q$, suppose that

$$V = c_1 P + c_2 Q$$

where $0 \leq c_i \leq 1$, for both $i = 1$ and $i = 2$, and $c_1 + c_2 = 1$. Then writing $c = c_2$ we have

$$V = (1 - c)P + cQ = P + c(Q - P) \tag{7.1}$$

V is seen to be the point at a distance of $c|Q - P|$ from P in the direction of Q. As $0 \leq c \leq 1$, V indeed lies on the segment joining P and Q. This completes the proof of the first part of the proposition when $P \neq Q$. If $P = Q$, the first part is actually trivial to prove because the segment PQ degenerates to a point (so any point on it is P itself – we leave the rest to the reader).

For the second part regarding uniqueness, suppose that the point V on PQ can be expressed as both $V = c_1 P + c_2 Q$ and $V = d_1 P + d_2 Q$, where $c_1 + d_1 = d_1 + d_2 = 1$. Then

$$V = c_1 P + c_2 Q = d_1 P + d_2 Q \quad \Longrightarrow \quad (c_1 - d_1)P = (d_2 - c_2)Q \tag{7.2}$$

From $c_1 + c_2 = d_1 + d_2 = 1$ we have that $c_1 - d_1 = d_2 - c_2$. Therefore, if these two equal quantities are not 0 we could multiply Equation (7.2) by $\frac{1}{c_1 - d_1}(= \frac{1}{d_2 - c_2})$ to deduce that $P = Q$, contradicting the hypothesis of the second part. We are led to conclude that $c_1 - d_1 = d_2 - c_2 = 0$, so that $c_1 = d_1$ and $c_2 = d_2$, proving that the expression for V in the proposition is indeed unique. □

Remark 7.1. To minimize notation we wrote the endpoints of the segment in the proposition as single variables P and Q. Of course, one can write out their coordinates as, say, $P = (p_x, p_y, p_z)$ and $Q = (q_x, q_y, q_z)$ and, correspondingly, the equation for V in the statement of the proposition as

$$(v_x, v_y, v_z) = c_1(p_x, p_y, p_z) + c_2(q_x, q_y, q_z) = (c_1 p_x + c_2 q_x, c_1 p_y + c_2 q_y, c_1 p_z + c_2 q_z)$$

For example, if $P = (1, 4, 3)$ and $Q = (2, 5, 2)$, then the proposition says that points on the segment PQ are of the form

$$(c_1 + 2c_2, 4c_1 + 5c_2, 3c_1 + 2c_2), \text{ where } 0 \le c_1, c_2 \le 1 \text{ and } c_1 + c_2 = 1.$$

Remark 7.2. We could have saved ourselves a variable and written $V = cP + (1-c)Q$, instead of $V = c_1P + c_2Q$, because $c_1 + c_2 = 1$, but chose not to in order to have separate variables for the coefficients of P and Q. This keeps our notation consistent with the proposition for triangles coming up.

Definition 7.1. A point V of the form

$$V = c_1P + c_2Q$$

where $0 \le c_i \le 1$, for both $i = 1$ and $i = 2$, and where $c_1 + c_2 = 1$, is said to be a *convex combination* – or *barycentric combination* – of P and Q. The scalars c_1 and c_2 are called the *barycentric coordinates* of V.

The following corollary then is just a rewrite of the first part of Proposition 7.1.

Corollary 7.1. *The segment joining two points P and Q consists of all their convex combinations.* □

A point $V = c_1P + c_2Q$ on the segment PQ can be usefully thought of as a *weighted sum* of P and Q, where the barycentric coordinates c_1 and c_2 are the weights – or influence, or ownership, if you will – of P and Q, respectively, on the location of V. The next exercise follows up on this idea.

Exercise 7.3. Suppose that $V = c_1P + c_2Q$ lies on the segment PQ.

(a) If $c_1 = c_2 = \frac{1}{2}$, *prove* that V is the midpoint of PQ; in other words, if the weights of P and Q on V are equal, then it's in the middle of the two. (We saw an illustration, though not proof, of this in Example 7.1.)

Suggested approach:

Without using vectors: Say $P = (p_x, p_y, p_z)$ and $Q = (q_x, q_y, q_z)$. Then

$$V = \frac{1}{2}(p_x, p_y, p_z) + \frac{1}{2}(q_x, q_y, q_z) = \left(\frac{p_x + q_x}{2}, \frac{p_x + q_x}{2}, \frac{p_x + q_x}{2}\right)$$

Determine the distance between P and V and between Q and V using the formula for distance between a pair of points, and find that the two are equal.

Note: The distance between the points (x, y, z) and (x', y', z') is $\sqrt{(x-x')^2 + (y-y')^2 + (z-z')^2}$.

Using vectors: $V - P = (\frac{1}{2}P + \frac{1}{2}Q) - P = \frac{1}{2}Q - \frac{1}{2}P$. Similarly, determine $Q - V$ and find that it equals $V - P$.

(b) Prove generally that $PV : VQ$ equals $c_2 : c_1$.

(c) If $c_1 > c_2$, prove that V is closer to P than Q, and vice versa.

Exercise 7.4. Say $P \neq Q$ and that $V = c_1P + c_2Q$, where c_1 and c_2 are *any* real numbers such that $c_1 + c_2 = 1$ (the condition that the c_i's must lie between 0 and 1 is dropped). Show then that V may be any point on the (infinite) straight line through P and Q.
Hint: $V = (1 - c_2)P + c_2Q = P + c_2(Q - P)$. How does this point change as c_2 varies?

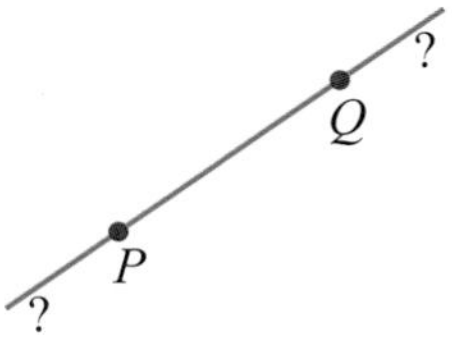

Figure 7.3: What are the conditions on c_1 and c_2 for $V = c_1P + c_2Q$ to lie on either side of PQ?

Exercise 7.5. The previous exercise says that points on the whole straight line through P and Q (given $P \neq Q$) are of the form $c_1P + c_2Q$, where $c_1 + c_2 = 1$. We already know that points on this line *between* P and Q additionally satisfy $0 \leq c_1, c_2 \leq 1$. How about those on either side of PQ – what are the conditions on c_i? See Figure 7.3.

Triangles

Statements analogous to Proposition 7.1 can be proved for triangles as well:

Proposition 7.2. *If P, Q and R are three points in $\mathbb{R}^3$, then a point V lies on the triangle PQR if and only if it can be expressed as*

$$V = c_1P + c_2Q + c_3R$$

where $0 \leq c_i \leq 1$, for $1 \leq i \leq 3$, and where $c_1 + c_2 + c_3 = 1$.

Further, if P, Q and R are not collinear – so that PQR does not degenerate to a segment or a point – then this expression for V is unique.

Proof. *Skip this proof if you start to get bogged down in the math. Just make sure to understand what the proposition says.*

For the first part of the proposition, suppose initially that the triangle PQR is non-degenerate, in other words, that P, Q and R are not collinear. Consider a point V on this triangle.

If $V = P$, then, of course, $V = 1P + 0Q + 0R$, which is an expression of the form required.

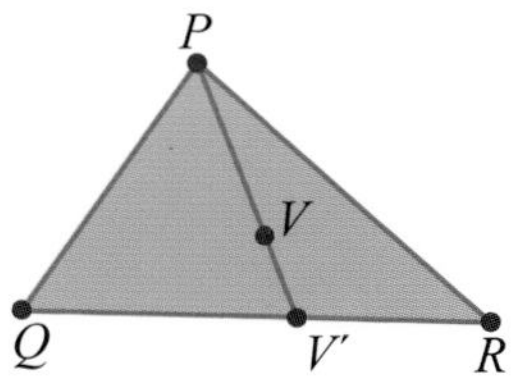

Figure 7.4: Illustration for the proof of Proposition 7.2.

If $V \neq P$, suppose that the straight line from P through V intersects the edge QR at V' (see Figure 7.4). As V lies on the segment joining P and V' the previous proposition gives an expression

$$V = c_1'P + c_2'V' \tag{7.3}$$

where $0 \leq c_i' \leq 1$, for $i = 1$ and $i = 2$, and $c_1' + c_2' = 1$.

Again, by the previous proposition, since V' lies on the segment joining Q and R,

$$V' = c_1''Q + c_2''R \tag{7.4}$$

where $0 \le c''_i \le 1$, for $i = 1$ and $i = 2$, and $c''_1 + c''_2 = 1$.

We have by using both (7.3) and (7.4) that

$$\begin{aligned} V &= c'_1 P + c'_2(c''_1 Q + c''_2 R) \\ &= c'_1 P + c'_2 c''_1 Q + c'_2 c''_2 R \end{aligned}$$

Writing $c_1 = c'_1$, $c_2 = c'_2 c''_1$ and $c_3 = c'_2 c''_2$, we see that

$$V = c_1 P + c_2 Q + c_3 R \tag{7.5}$$

where it may be verified that $0 \le c_i \le 1$, for $1 \le i \le 3$, and, moreover, $c_1 + c_2 + c_3 = c'_1 + c'_2 c''_1 + c'_2 c''_2 = c'_1 + c'_2(c''_1 + c''_2) = c'_1 + c'_2 * 1 = 1$. This proves the "only if" direction of the first part, provided PQR is not degenerate; we leave the proof in case PQR is degenerate (so, either a point or segment) to the reader.

We leave the proof of the "if" direction and of the uniqueness of the expression in the case that PQR is non-degenerate, to the reader as well. □

Definition 7.2. A point V of the form

$$V = c_1 P + c_2 Q + c_3 R$$

where $0 \le c_i \le 1$, for $1 \le i \le 3$, and where $c_1 + c_2 + c_3 = 1$, is said to be a *convex combination* – or *barycentric combination* – of P, Q and R. The scalars c_1, c_2 and c_3 are called the *barycentric coordinates* of V.

The following corollary is a rewrite of the first part of Proposition 7.2.

Corollary 7.2. *The triangle with vertices at P, Q and R consists of all their convex combinations.* □

Similarly as for a segment, a point $V = c_1 P + c_2 Q + c_3 R$ on the triangle PQR can be thought of as a weighted sum of P, Q and R, where the barycentric coordinates c_1, c_2 and c_3 are the respective weights.

Exercise 7.6. Suppose that $V = c_1 P + c_2 Q + c_3 R$ lies on the triangle PQR. Where is V when

(a) one of the c_i's is 1 and the other two are 0?

(b) one of the c_i's is 0 and the other two equal to $\frac{1}{2}$?

(c) all the c_i's are equal to $\frac{1}{3}$?

Example 7.2. If $P = (0, 0, 0)$, $Q = (20, 0, 0)$ and $R = (20, 30, 0)$, does the point $V = (10, 20, 0)$ lie on the triangle PQR? If so, express V as a convex combination of the three vertices.

Answer: Let's try to solve the equations

$$\begin{aligned} V &= c_1P + c_2Q + c_3R \\ c_1 + c_2 + c_3 &= 1 \end{aligned}$$

The first gives

$$(10, 20, 0) = c_1(0, 0, 0) + c_2(20, 0, 0) + c_3(20, 30, 0)$$

Equating the values in each coordinate on either side of the above equation we get the following (the equation in the z coordinate is trivial and not written):

$$\begin{aligned} 20c_2 + 20c_3 &= 10 \\ 30c_3 &= 20 \end{aligned}$$

With the additional

$$c_1 + c_2 + c_3 = 1$$

one solves to find that

$$c_1 = \frac{1}{2}, \qquad c_2 = -\frac{1}{6}, \qquad c_3 = \frac{2}{3}$$

As the c_i's do not all lie between 0 and 1 we conclude that V is not a convex combination of P, Q and R and, therefore, does not lie on the triangle PQR.

Exercise 7.7. If $P = (0, 0, 0)$, $Q = (20, 0, 0)$ and $R = (20, 30, 0)$, does the point $V = (15, 15, 0)$ lie on the triangle PQR? If so, express V as a convex combination of the three vertices.

Exercise 7.8. What if $V = c_1P + c_2Q + c_3R$, where c_i, $1 \leq i \leq 3$, are *any* real numbers such that $c_1 + c_2 + c_3 = 1$ (the condition that the c_i's must lie between 0 and 1 is dropped)? Where does V lie?

Exercise 7.9. If $P = (30, 50, 45)$, $Q = (40, 20, 5)$ and $R = (30, 20, 0)$, which of the points $V = (35, 25, 20)$, $V' = (35, 25, 15)$ and $V'' = (28, 80, 89)$ lie on the triangle PQR? Which of them lie on the plane containing P, Q and R, but not on the triangle PQR?

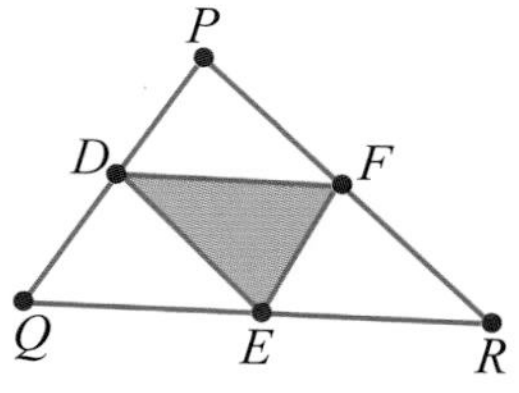

Figure 7.5: A triangle DEF with vertices at the midpoints of the edges of a larger triangle.

Exercise 7.10. In Figure 7.5, points D, E and F are midpoints of the edges PQ, QR and RP, respectively. Are points in the triangle DEF a special kind of convex combination of P, Q and R? Precisely, if $V \in DEF$ is expressed as $V = c_1P + c_2Q + c_3R$, what restrictions, if any, are there on the values that the c_i can have?

7.3 Interpolation

It is straightforward now to explain how OpenGL interpolates property values over its favorite primitives. Let's begin with a non-degenerate triangle $P_1P_2P_3$.

Suppose that the RGB color tuples at the vertices P_1, P_2 and P_3 are specified by the programmer to be (R_1, G_1, B_1), (R_2, G_2, B_2) and (R_3, G_3, B_3), respectively. Let V be any point of $P_1P_2P_3$. See Figure 7.6.

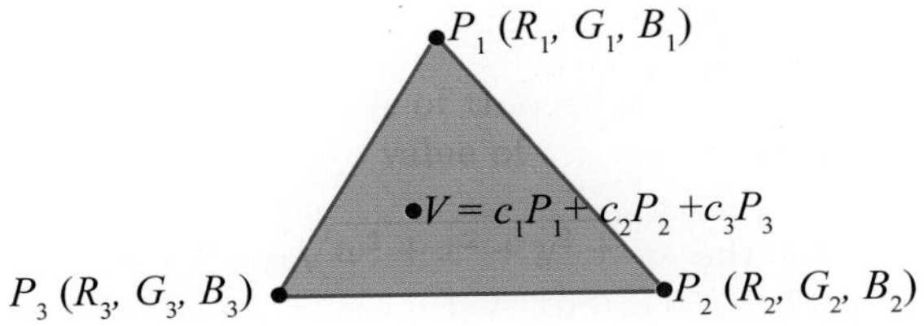

Figure 7.6: Color values specified at the vertices V_1, V_2 and V_3 of a triangle are interpolated at V.

By Proposition 7.2, there is a unique expression

$$V = c_1P_1 + c_2P_2 + c_3P_3 \tag{7.6}$$

of V as a convex combination of the vertices P_1, P_2 and P_3. OpenGL, in fact, determines this expression, particularly, the values of c_1, c_2 and c_3. The color at V is then set to

$$\begin{aligned} & c_1(R_1, G_1, B_1) + c_2(R_2, G_2, B_2) + c_3(R_3, G_3, B_3) \qquad (7.7) \\ = \quad & (c_1R_1 + c_2R_2 + c_3R_3,\ c_1G_1 + c_2G_2 + c_3G_3,\ c_1B_1 + c_2B_2 + c_3B_3) \end{aligned}$$

Simply put, OpenGL uses the weight of each vertex on the location of V as its weight on the color of V as well. As the weights c_1, c_2 and c_3 are unique, the interpolation process is *unambiguous* (and, therefore, *programmable*).

OpenGL performs an exactly similar computation to interpolate color values specified at the end vertices of a non-degenerate segment. In the case of the third of OpenGL's favorite primitives, the point, there is obviously nothing to interpolate.

Remark 7.3. What we call interpolation is often (more accurately) referred to as *linear interpolation* because the interpolation parameters enter as linear variables – i.e., with power one – in the expression for the interpolated value; e.g., the c_i's in the expression for V in Equation (7.6), viz. $V = c_1P_1 + c_2P_2 + c_3P_3$.

Remark 7.4. Color values aren't the only ones to be interpolated. In fact, any attribute specified *numerically* at a triangle's vertices can be interpolated through its interior. For example, in Phong's shading model, normal vector values defined at a triangle's vertices are interpolated.

Figure 7.7: Screenshot of Experiment 7.1.

Experiment 7.1. Replace the polygon declaration part of our old favorite `square.cpp` with the following (Block 1*):

```
glBegin(GL_TRIANGLES);
   glColor3f(1.0, 0.0, 0.0);
   glVertex3f(20.0, 20.0, 0.0);
   glColor3f(0.0, 1.0, 0.0);
   glVertex3f(80.0, 20.0, 0.0);
   glColor3f(0.0, 0.0, 1.0);
   glVertex3f(80.0, 80.0, 0.0);
glEnd();
```

Observe how OpenGL interpolates vertex color values throughout the triangle. Figure 7.7 is a screenshot. **End**

Exercise 7.11. For the triangle of Experiment 7.1 calculate the RGB colors at the point $(70.0, 50.0, 0.0)$. Verify your answer by drawing a point with those colors at $(70.0, 50.0, 0.0)$. You should not be able to see this point but, then, of course, you do see the point (why it should be invisible)!

Exercise 7.12. Show by computation that the interpolated color value at the *centroid* (whose barycentric coordinates are all equal) of the triangle of Experiment 7.1 is a darkish gray. Again, verify your answer by drawing a point of that color at the centroid.

Exercise 7.13. If the vertices of a triangle at $P = (0, 0, 0)$, $Q = (20, 0, 0)$ and $R = (20, 30, 0)$ are colored red, cyan and magenta, respectively, what is the color at the point $(15, 15, 0)$?

Exercise 7.14. If the vertices of a triangle at $P = (10, 10, 0)$, $Q = (40, 10, 0)$ and $R = (30, 30, 0)$ are colored white, black and white, respectively, what is the color at the point $(25, 20, 0)$?

Remark 7.5. It's clear now from the per-triangle interpolation process that the computation involved in rendering a scene is proportional to its triangle count (or polygon count). In animated applications, in particular, where the scene is repeatedly re-rendered, an important objective then is to minimize this count without compromising visual quality.

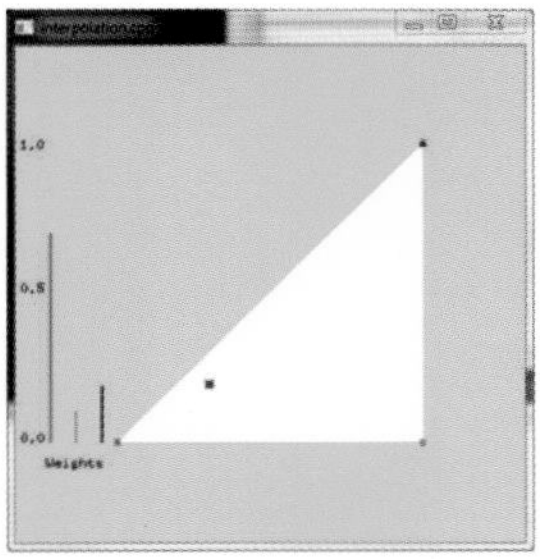

Figure 7.8: Screenshot of `interpolation.cpp`.

Experiment 7.2. Run `interpolation.cpp`, which shows the interpolated colors of a movable point inside a triangle with red, green and blue vertices. The triangle itself is drawn white. See Figure 7.8 for a screenshot.

As the arrow keys are used to move the large point, the height of each of the three vertical bars on the left indicates the weight of the respective triangle vertex on the point's location. The color of the large point itself is interpolated (by the program) from those of the vertices. **End**

*To cut-and-paste you can find the block in text format in the file `chap7codeModifications.txt` in the directory `ExperimenterSource/CodeModifications`.

Exercise 7.15. (Programming) Replace the triangle declaration of `interpolation.cpp` with (Block 2):

```
glBegin(GL_TRIANGLES);
   glColor3f(1.0, 0.0, 0.0);
   glVertex3f(20.0, 20.0, 0.0);
   glColor3f(0.0, 1.0, 0.0);
   glVertex3f(80.0, 20.0, 0.0);
   glColor3f(0.0, 0.0, 1.0);
   glVertex3f(80.0, 80.0, 0.0);
glEnd();
```

The movable large point is no longer visible except when it pokes out of the triangle. Why?

Exercise 7.16. (Programming) The interpolation procedure described above requires the triangle or segment to be non-degenerate. Find out by writing code how OpenGL draws a degenerate segment or triangle.
Hint: It doesn't!

Exercise 7.17. It is easy to test a segment with vertices at $P = (x_1, y_1, z_1)$ and $Q = (x_2, y_2, z_2)$ for degeneracy: simply check if the end vertices are identical, i.e., if $x_1 = x_2$ and $y_1 = y_2$ and $z_1 = z_2$.

How about a triangle? How does one test if the triangle with vertices at $P = (x_1, y_1, z_1)$, $Q = (x_2, y_2, z_2)$ and $R = (x_3, y_3, z_3)$ is degenerate?
Hint: PQR is degenerate if and only if at least one of the two vectors $Q - P$ and $R - P$ is zero, *or*, if they are parallel. It's easy to test if one of them is zero; if not, their being parallel implies that each is a multiple of the other.

Exercise 7.18. If you know about determinants, then write a succinct condition for the degeneracy of a triangle lying on the xy-plane with vertices at $P = (x_1, y_1)$, $Q = (x_2, y_2)$ and $R = (x_3, y_3)$, using a single determinant.

Remark 7.6. A practical application of interpolation to rendering must take into account the fact that screen space is not actually a 2D continuum but, in practice, a rectangular array, called a *raster*, of finitely many pixels. Each pixel is not a point either but a square of non-zero size.

We know – see the discussion of shoot-and-print in Section 2.2 – that a primitive object, such as a triangle t, drawn in the viewing volume is projected to the volume's front and, then, scaled to its image t' on the OpenGL window. See the left and middle diagrams of Figure 7.9. The scaled t' is actually *rendered* by a set of pixels in the OpenGL window – the shaded ones in the raster on the right of Figure 7.9 (admittedly at a rather lousy resolution). A part of the print process, called *rasterization* or *scan conversion*, in fact, consists of choosing and coloring the pixels to render t'.

We'll be studying rasterization algorithms in depth later on, but it's worth noting a couple of issues at this time in relation to interpolation. Since a pixel is a square that contains not one point, but infinitely many, OpenGL

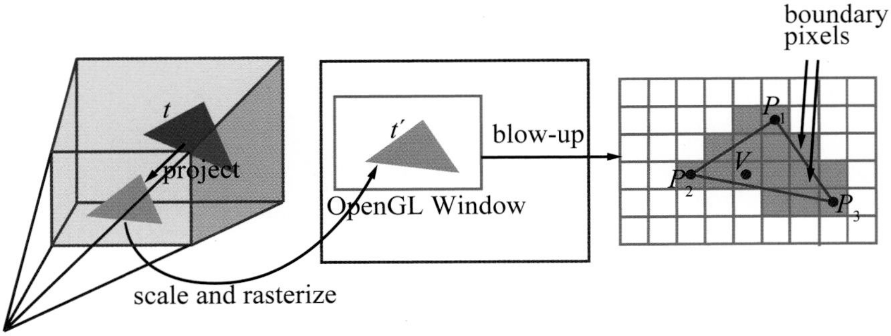

Figure 7.9: Project, scale and rasterize.

must pick a representative one at which to interpolate the color values from the vertices and then set the entire pixel RGB to those particular values.

For a pixel in the interior of the primitive, a valid choice is its center point, e.g., V on the right of Figure 7.9 for the pixel to which it belongs. The color of that pixel is then determined by formula (7.7), viz.

$$c_1(R_1, G_1, B_1) + c_2(R_2, G_2, B_2) + c_3(R_3, G_3, B_3)$$

where $V = c_1P_1 + c_2P_2 + c_3P_3$, and the programmer-specified color at P_i is (R_i, G_i, B_i), for $1 \leq i \leq 3$.

Coloring boundary pixels is more complicated, as both foreground and background colors need to be taken into account. In practice, depending on if effects such as antialiasing are in force, weights may be decided by the area of the pixel occupied by the primitive and the background, respectively. For example, compare the two boundary pixels pointed at by arrows in the right of Figure 7.9: the lower one should give greater weight to the foreground color (that of the triangle) than the background, while the opposite is the case for the upper pixel.

The reader may have been wondering the following for a while now. Points, segments and triangles seem to be programmably interpolatable, but how about quadrilaterals and, in general, polygons with four or more vertices? Isn't it true that points belonging to such figures have unique expressions in terms of their vertices as well, in which case they are programmably interpolatable too? The answer is no, as we'll see. First, though, let's learn about convexity and the convex hull, which will lead to the answer and more.

7.4 Convexity and the Convex Hull

There's nothing about convex combinations in Definitions 7.1 and 7.2 that's specific to two or three points. They can be defined for arbitrary numbers of points:

Definition 7.3. If $F = \{P_1, P_2, \ldots, P_k\}$ is a set of k points in $\mathbb{R}^2$, then a point V of the form

$$V = c_1 P_1 + c_2 P_2 + \ldots + c_k P_k$$

where $0 \le c_i \le 1$, for $1 \le i \le k$, and where $c_1 + c_2 + \ldots + c_k = 1$, is said to be a *convex combination* – or *barycentric combination* – of F. The scalars c_i, $1 \le i \le k$, are the *barycentric coordinates* of V.

Remark 7.7. We restrict to $\mathbb{R}^2$ as we'll be computing convex combinations only on a plane. Definitions and results can all be generalized to $\mathbb{R}^3$ and higher.

Corollaries 7.1 and 7.2 tell us that the convex combinations of two points form the segment joining them and those of three points the triangle with corners at these points. How about an arbitrary set $F = \{P_1, P_2, \ldots, P_k\}$ of points on the plane? What object is formed by its convex combinations? To answer the question we need first to define convexity.

Definition 7.4. A non-empty set S of points in $\mathbb{R}^2$ is said to be *convex* if, for any two points $P, Q \in S$, it is true that the segment $PQ \subset S$; in other words, if it is true that, if the endpoints of a segment are in S, then the segment itself is contained in S.

See Figure 7.10 for examples of both convex and non-convex subsets of $\mathbb{R}^2$. Intuitively, convexity ensures that the object has neither holes nor depressions.

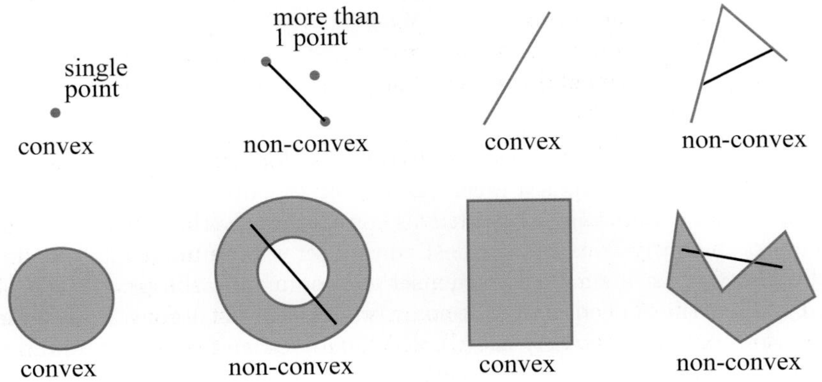

Figure 7.10: Convex and non-convex subsets of $\mathbb{R}^2$ – non-convexity is indicated by an example of a black line with endpoints in the object but that itself is not contained in it.

Exercise 7.19. Prove that points, segments and triangles are always convex.

Exercise 7.20. Prove that the entire plane $\mathbb{R}^2$ is itself convex. What about a *half-plane*, i.e., the part of the plane to only one side of a straight line? Assume it is a *closed* half-plane (i.e., one which includes its border straight line).

Exercise 7.21. Prove that the intersection of (any number of) convex sets is again convex. Is the union of two convex sets necessarily convex?

A polygon which happens to be a convex set is, of course, a *convex polygon*. Convex polygons are particularly important in OpenGL. As we noted even in the second chapter, a programmer should ensure that polygons he draws with `GL_POLYGON` calls are convex; otherwise, rendering is unpredictable (we'll see why in the next chapter).

Exercise 7.22. Show that it's possible to tell if a (plane) polygon is convex by measuring the internal angle at each vertex.
Hint: Compare the two polygons at the right of Figure 7.10, one of which is convex and the other not.

Exercise 7.23. For the four non-convex figures shown in Figure 7.10, fill them out *minimally* to make them convex; in particular, for each, shade in on the page itself an additional area as small as possible which, together with the original, forms a convex figure.

Part answer: See Figure 7.11.

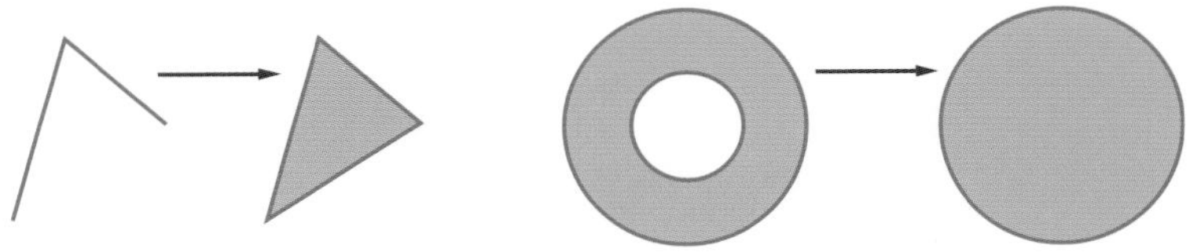

Figure 7.11: Part answer to Exercise 7.23.

The previous exercise leads to the consideration of the *smallest* possible convex set which contains a given planar set F – obtained by "filling out the holes and depressions" of F. If F is convex already then there's nothing to do as, obviously, F is the smallest convex set containing itself. But does there always exist a smallest convex set containing an arbitrary F?

Consider the collection of *all* convex sets containing a given set F. This collection includes certainly the whole plane itself and, possibly, infinitely many other sets. Now, the intersection X of this collection surely contains F as each member does. And Exercise 7.21 tells us that X is convex as well. Moreover, X is no bigger than any convex set C containing F, because C was one of the collection that was intersected to derive X in the first place. So the answer is yes, there always exists a smallest convex set containing a given planar set F: it is simply the intersection of all convex sets containing F.

Definition 7.5. The smallest convex set containing a set F of points on the plane is called its *convex hull*, denoted $ch(F)$.

Remark 7.8. Again, the restriction to the plane is for our purposes only. It can all be made to work in higher dimensions as well.

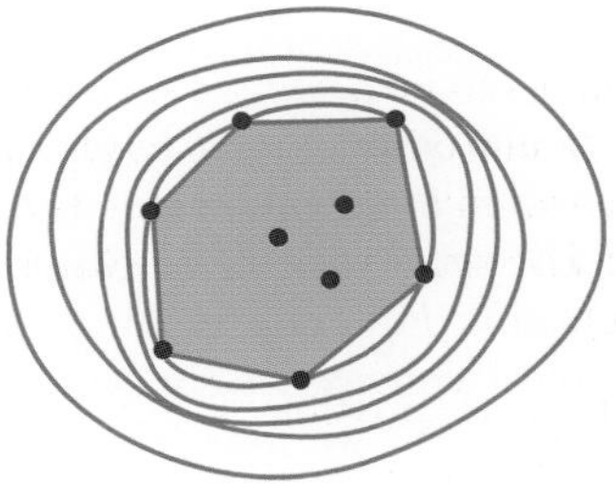

Figure 7.12: A rubber band snapping to bound the convex hull of the nails.

The intersection of infinitely many convex sets is a rather abstract notion. It's more intuitive to think of $ch(F)$ as the "limit" of a shrinking sequence of convex sets containing F. The process is equivalent to sticking a nail at each point of F, stretching a rubber band around all the nails, then releasing it. See Figure 7.12. When the rubber band becomes taut it bounds the convex hull of F. Figure 7.13 shows the convex hulls of a few small sets of points.

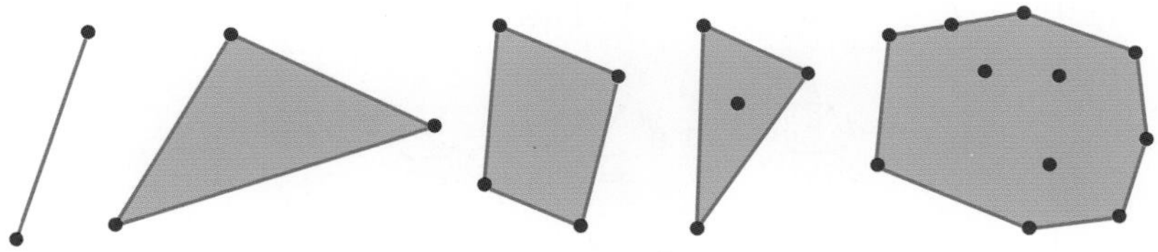

Figure 7.13: Convex hulls.

Experiment 7.3. Run `convexHull.cpp`, which shows the convex hull of 8 points on a plane. Use the space bar to select a point and the arrow keys to move it. Figure 7.14 is a screenshot.

Note: The program implements a very inefficient (but easily coded) algorithm to compute the convex hull of a set F as the union of all triangles with vertices in F.

End

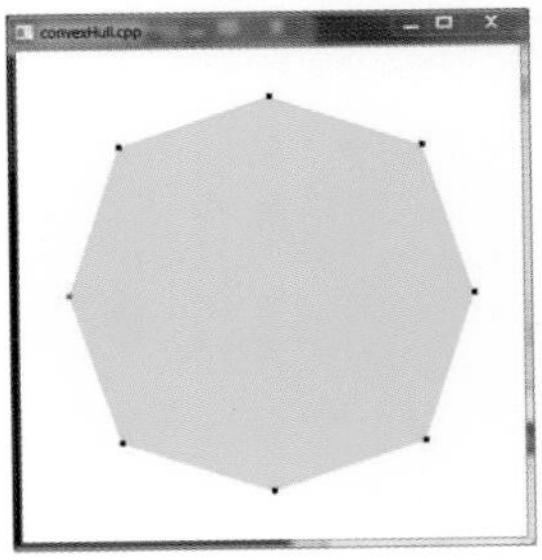

Figure 7.14: Screenshot of `convexHull.cpp`.

Example 7.3. What is the convex hull of the set consisting of two opposite edges of a parallelogram? How about that consisting of two adjacent edges?

Answer: The whole (filled) parallelogram. The triangle on the two edges.

Exercise 7.24. What are the convex hulls of the figures $+$ and $\times$?

Exercise 7.25. Earlier, in Exercise 7.20, you showed that a closed half-plane H is convex. Can you find two straight lines S and T, not necessarily finite, such that $H = ch(S \cup T)$?

We relate next convex hulls to convex combinations.

Proposition 7.3. *Given a set $F = \{P_1, P_2, \ldots, P_k\}$ of k points in $\mathbb{R}^2$, $ch(F)$ is exactly the set of convex combinations of F.*

Proof. *Skip this proof if you start to get bogged down in the math. Just make sure to understand what the proposition says.*

We'll prove first that the set X of convex combinations of F is a convex set. As X obviously contains F, a consequence will be that $ch(F) \subset X$, because $ch(F)$ is the smallest convex set containing F.

Accordingly, given $V, V' \in X$, we have to show that the segment VV' lies in X. As V and V' are convex combinations of F, they can be written as

$$V = \sum_{i=1}^{k} c_i P_i \quad \text{and} \quad V' = \sum_{i=1}^{k} c'_i P_i$$

where $0 \leq c_i, c'_i \leq 1$, for each i, and $\sum_{i=1}^{k} c_i = \sum_{i=1}^{k} c'_i = 1$. A point W on the segment VV' is of the form

$$W = dV + d'V'$$

where $0 \leq d, d' \leq 1$ and $d + d' = 1$. Therefore,

$$W = d\sum_{i=1}^{k} c_i P_i + d'\sum_{i=1}^{k} c'_i P_i = \sum_{i=1}^{k}(dc_i + d'c'_i)P_i$$

It's easily verified that $0 \leq dc_i + d'c'_i \leq 1$, for each i. Moreover, $\sum_{i=1}^{k}(dc_i + d'c'_i) = d\sum_{i=1}^{k} c_i + d'\sum_{i=1}^{k} c'_i = d.1 + d'.1 = 1$. One concludes that W is a convex combination of F as well and, therefore, belongs to X. As W was an arbitrary point of VV', we have proved that VV' indeed lies in X and, therefore, X is convex.

Next, we'll prove that any convex set Y containing F contains X as well. This will imply that $ch(F) \supset X$, which, together with the proof above that $ch(F) \subset X$, will complete the proof of the proposition.

In fact, we'll prove by induction that Y contains all convex combinations of the sets $\{P_1, P_2, \ldots, P_r\}$, for $r = 1, 2, \ldots, k$ (the case $r = k$ will, of course, prove that Y contains X). Starting the induction at $r = 1$ is trivial as the only convex combination of $\{P_1\}$ is P_1, which belongs to F and, therefore, to Y.

Suppose, inductively, that Y contains all convex combinations of the set $\{P_1, P_2, \ldots, P_r\}$, for some r, $1 \leq r \leq k-1$. Let $V = \sum_{i=1}^{r+1} c_i P_i$ be a convex combination of $\{P_1, P_2, \ldots, P_r, P_{r+1}\}$. We'll prove that $V \in Y$, completing the induction. We can assume that $c_{r+1} < 1$, because, if $c_{r+1} = 1$, then $V = P_{r+1}$ which trivially belongs to Y. Now,

$$\begin{aligned} V &= \sum_{i=1}^{r} c_i P_i + c_{r+1} P_{r+1} \\ &= c\left(\sum_{i=1}^{r} \frac{c_i}{c} P_i\right) + c_{r+1} P_{r+1} \end{aligned}$$

writing $c = \sum_{i=1}^{r} c_i$, which is not 0 as $c_{r+1} < 1$.

Since $\sum_{i=1}^{r} \frac{c_i}{c} = \frac{\sum_{i=1}^{r} c_i}{c} = 1$, apply the inductive hypothesis to conclude that the point

$$V' = \sum_{i=1}^{r} \frac{c_i}{c} P_i$$

is in Y. Next, use the preceding expression for V' to rewrite the earlier equation for V as

$$V = cV' + c_{r+1} P_{r+1}$$

which is a convex combination of V' and P_{r+1}, because $c + c_{r+1} = 1$, and so lies on the segment $V'P_{r+1}$. Since Y is a convex set containing both V' and P_{r+1}, it contains the segment $V'P_{r+1}$ and, therefore, V as well.

The proof of the proposition is complete. □

Extreme Points

The reader, contemplating again Figure 7.13, will note that some of the points of the two sets on the right lie at the corners of their respective hulls, while others are inside or elsewhere on the boundary. Here's a definition that classifies points accordingly.

Definition 7.6. If a point P of a set F of points on the plane is such that the convex hulls of F and $F - \{P\}$ (i.e., F with P deleted) are the same, then P is said to be a *non-extreme* point of F; otherwise, it is an *extreme point* of F.

Remark 7.9. Colloquially, non-extreme points are "expendable" in that the elimination of any one doesn't affect the convex hull. However, removing an extreme point will change the hull; it will, in fact, become smaller.

Exercise 7.26. What are the extreme points of a set of 10 points chosen arbitrarily from a given circle?

Interpolation and Convexity

We've covered a fair amount of the theory of convexity so far. Let's pause to take stock of how it impacts our understanding of the interpolation process.

We asked in the last paragraph of Section 7.3 if it's true that points in a polygon with four or more vertices have unique expressions in terms of these vertices. Since the expressions that we seek are convex combinations and since the convex hull of the vertex set consists precisely of its convex combinations, a fair question to ask first is if each point in the convex hull of a set with at least four points has a unique expression as a convex combination of these points.

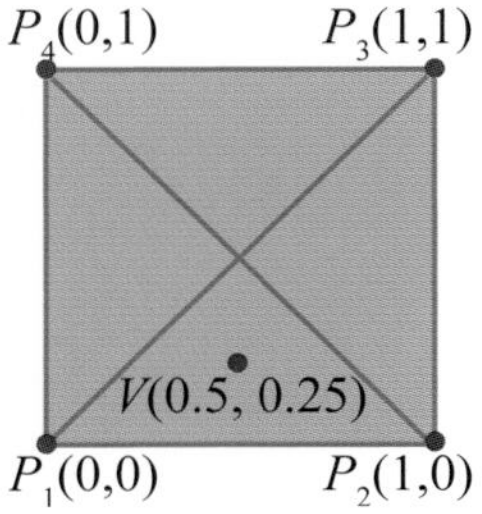

Figure 7.15: V has at least two expressions as a convex combination of P_1, P_2, P_3 and P_4.

The answer to this question is *without exception* no. Figure 7.15 indicates a counter-example in the case of a four-point set $\{P_1, P_2, P_3, P_4\}$ whose members are at the corners of a square. The convex hull of the set, of course, is the square itself. Observe now that the point V has at least two distinct expressions as a convex combination of P_1, P_2, P_3 and P_4:

(i) $V = 0.25P_1 + 0.5P_2 + 0.0P_3 + 0.25P_4$, obtained by computing for V as belonging to the triangle $P_1P_2P_4$ (so the coefficient of P_3 is 0).

(ii) $V = 0.5P_1 + 0.25P_2 + 0.25P_3 + 0.0P_4$, obtained by computing for V as belonging to the triangle $P_1P_2P_3$ (so the coefficient of P_4 is 0).

In such a situation, there is no unambiguous way in which to interpolate values, such as color, which are defined at the vertices. For example, which set of weights would one use to interpolate at V: $(0.25, 0.5, 0.0, 0.25)$ or $(0.5, 0.25, 0.25, 0.0)$? This problem is actually general and arises in the case of any convex polygon with at least four vertices, as we ask the reader to show.

Exercise 7.27. Show that in any convex polygon with four or more vertices a point can be found which has more than one expression as a convex combination of its vertices.
Hint: If the polygon is $P_1P_2 \ldots P_n$, a point V in the intersection of the triangles $P_1P_2P_3$ and $P_1P_2P_4$ will do.

Exercise 7.28. Could it be that some strange plane *non-convex* polygon has the property that all its points are uniquely expressible as convex combinations of its vertices?

Suggested approach: If you did Exercise 7.22 you know that a plane non-convex polygon $P_1P_2 \ldots P_n$ has at least one vertex, say P_r, where the internal angle is greater than 180°. Show, in fact, that P_r is a non-extreme member of the set of vertices $\{P_1, P_2, \ldots, P_n\}$ and, moreover, contained in a triangle with corners at some three other vertices, say P_{i_1}, P_{i_2} and P_{i_3} (can you find such a triangle for the lone non-extreme vertex of the non-convex polygon on the right of Figure 7.10?). Then P_r itself has more than one expression as a convex combination of the vertices $P_1, P_2, \ldots, P_n$.

We conclude that any polygon with four or more vertices, be it convex or not, contains points with ambiguous representation as a convex combination of its vertices. It is with good reason, therefore, that OpenGL has chosen points, segments and triangles, and *none other*, as its fundamental primitives.

Planarity

One final note is that we have been making the tacit assumption that polygons under consideration are all planar, as evident from the fact that our definitions of convex combination and convexity have been for a plane. This assumption is valid because a polygon is *not even properly defined* if its vertices do not lie on one plane! The following example illustrates the problem.

Example 7.4. A quadrilateral $q = P_1P_2P_3P_4$ is made starting from a square of sides 2 units on the xy-plane and then lifting one vertex a unit in the z-direction. Specifically, the vertices of q are $P_1 = (1, 1, 0)$, $P_2 = (3, 1, 0)$, $P_3 = (3, 3, 1)$ and $P_4 = (1, 3, 0)$. P_1, P_2 and P_4 lie on the xy-plane, while P_3 is one unit above.

Now, which is q: is it the union of the two triangles $P_1P_2P_3$ and $P_1P_4P_3$ (Figure 7.16(a)) or is it the union of the two triangles $P_2P_1P_4$ and $P_2P_3P_4$ (Figure 7.16(b))? These two shapes are completely disjoint except for the shared boundary (appropriately bending a couple of sheets of paper will convince you of this)!

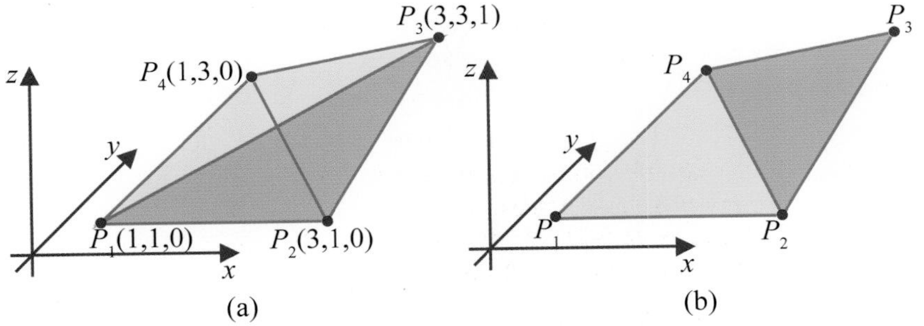

Figure 7.16: A non-planar quad q drawn in two different ways.

As a set of at most three vertices is always planar, no such problem as in the preceding example can arise for a point, a segment or a triangle – yet another reason for OpenGL to favor these three. In fact, as we said in Section 2.6, general polygons, drawn using `GL_POLYGON`, have been discarded from the core of the newer versions of OpenGL, and even readers using the compatibility profile are urged to use triangles always instead of polygons.

7.5 Summary, Notes and More Reading

In this chapter we learned that points, segments and triangles are preferred by a drawing API such as OpenGL for the reason that each of their points can be unambiguously interpolated from their vertices. This led to exploring the definition of a convex combination and the notions of convexity and a convex hull.

To learn more about convexity and related algorithms the place to look is the computational geometry literature. The introductory computational geometry texts by de Berg et al. [11] and O'Rourke [107] are good starting points. The mathematical introduction to computer graphics by Buss [22] has a chapter on averaging and interpolation and an interesting discussion of convexity as well.

CHAPTER 8

Triangulation

To create a 3D object in CG one assembles its surface, or a good likeness, with help of 2D primitives. For example, the (seemingly 3D) solid ball and torus in Figure 8.1 are depicted by the *2D* surface sphere and surface torus which bound them, respectively (note that the calls `glutSolidSphere()` and `glutSolidTorus()` in `ballAndTorusShadowed.cpp` refer to *filled* triangles, not solid 3D objects). Boris's head in Figure 8.2 is a mesh of quads.

Figure 8.1: Screenshot of `ballAndTorus-Shadowed.cpp`.

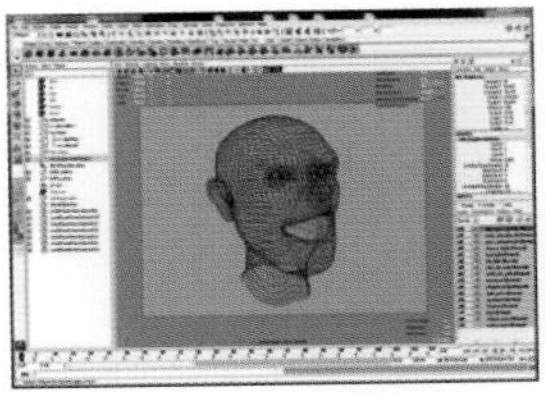

Figure 8.2: Mesh of Boris's head (courtesy of Sateesh Malla at `www.sateeshmalla.com`).

The triangle, as we learned in the previous chapter, is the preferred of the 2D primitives. It turns out, though, that simply cobbling together a collection of triangles that resembles the desired object may not be good enough. In order to avoid problems at the time of rendering the collection must follow certain rules. The goal of this chapter is to formulate these rules and understand their importance, particularly in the context of OpenGL.

Section 8.1 begins by listing the rules that make a collection of triangles a so-called triangulation. After several examples of triangulations and non-triangulations it then explains the logic behind these rules, in particular why they make a collection behave predictably at the time of rendering. The next section is a brief discussion of so-called Steiner vertices and how they can be included to improve the quality of a triangulation. Section 8.3 explains OpenGL's own somewhat simple-minded triangulation mechanism and the consequent importance of making sure that all polygons specified in a program are convex. Section 8.4 concludes.

This short chapter, together with its sisters Chapter 7 on convexity and interpolation and Chapter 9 on orientation, goes to the geometric core of CG.

8.1 Definition and Justification

Definition 8.1. Suppose $\mathcal{T}$ is a collection of triangles whose union is an object X. Then $\mathcal{T}$ is said to be a *triangulation* of X if, given any two triangles t_1 and t_2 from $\mathcal{T}$, exactly one of the following three is true:

1. t_1 and t_2 are disjoint, i.e., do not intersect at all.
2. t_1 and t_2 intersect in a vertex of both.
3. t_1 and t_2 intersect in an edge of both.

Informally, triangles in a triangulation are asked to intersect "nicely" or not at all. If $\mathcal{T}$ is a triangulation of X then X is said to be *triangulated* by $\mathcal{T}$.

A collection of triangles may be such that its union is X, but without being a triangulation of X according to the preceding rules – such a collection is called an *invalid triangulation* of X.

Remark 8.1. We should probably call triangulations "valid triangulations" to make the contrast with invalid ones clear, but that would be cumbersome.

Figure 8.3(a) shows the triangulation of a simple non-convex polygon. In Figure 8.3(b), $\{ABC, ADC\}$ is a triangulation of the rectangle $ABCD$.

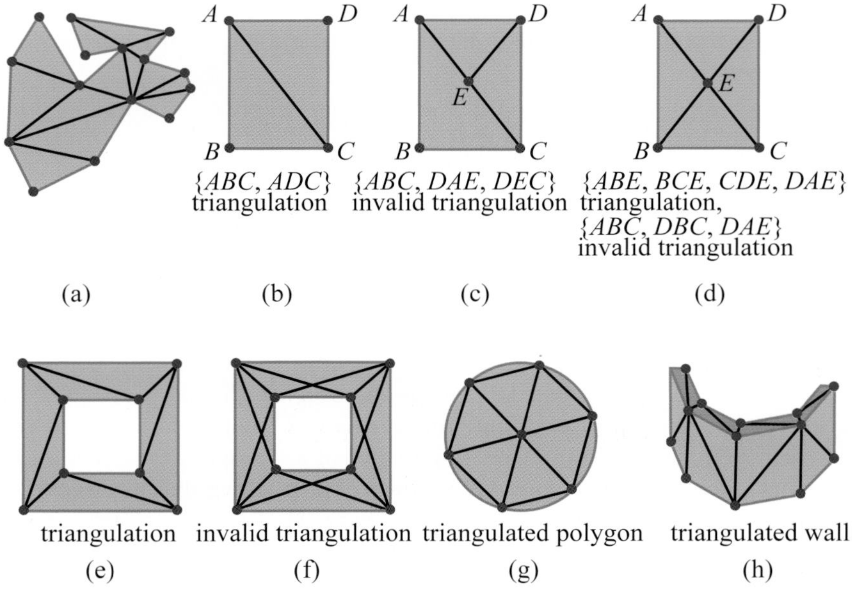

Figure 8.3: Triangulations.

In Figure 8.3(c), $\{ABC, DAE, DEC\}$ is an invalid triangulation of $ABCD$ as ABC intersects DAE in AE, which is an edge of DAE, but *not* of ABC (it is only *part* of the edge AC). Generally, if the vertex of one triangle lies on another, then it should be a vertex of the second triangle as well, or there is a violation of the triangulation rules.

In Figure 8.3(d), $\{ABE, BCE, CDE, DAE\}$ is a triangulation of $ABCD$, while $\{ABC, DBC, DAE\}$ is an invalid triangulation because ABC and DBC intersect in a triangle EBC, which, of course, is neither an edge nor vertex of either.

Figure 8.3(e) is a familiar triangulation of a square annulus, while Figure 8.3(f) is an often drawn invalid triangulation. Figure 8.3(g) is a triangulated polygon approximating a disc. The approximation may be made closer by increasing the number and decreasing the size of the sides of the polygon. Likewise, the wall of Figure 8.3(h) may be made to appear smoothly curved by increasing the number of flat panels and making them narrower.

Triangulation is not a unique process and the same object may have multiple triangulations.

Exercise 8.1. Draw a different triangulation of the polygon of Figure 8.3(a). Does the number of triangles change?

Exercise 8.2. Can you give a formula for the number of triangles that any triangulation of a simple polygon with n vertices must have? Assume that there is no triangle vertex other than those from the original polygon (e.g., like Figure 8.3(a), but unlike Figure 8.3(c) where vertex E doesn't belong to the input rectangle).

Why the rules for triangulation? As long as a collection of triangles looks like the desired object why should one care if it happens to be a triangulation according to Definition 8.1 or not?

To answer this question consider the invalid triangulation $\{ABC, DBC, DAE\}$ of the rectangle in Figure 8.3(d). Say the programmer has specified color values at the vertices A-E. These are interpolated separately through the three triangles ABC, DBC and DAE. What happens though in the region of overlap EBC of the two triangles ABC and DBC? The color of EBC is determined by the one of triangles ABC and DBC which appears *later* in the code, as its color values overwrite those of the earlier one. This exact situation is implemented in the following experiment.

Experiment 8.1. Run `invalidTriangulation.cpp`, which implements exactly the invalid triangulation $\{ABC, DBC, DAE\}$ of the rectangle in Figure 8.3(d). Colors have been arbitrarily fixed for the five vertices A-E. Press space to interchange the order that ABC and DBC appear in the code. Figure 8.4 shows the difference. **End**

Exercise 8.3. (Programming) Theoretically, in Figure 8.3(c) a similar ambiguity arises in the coloring of the segment AC, depending on the order

(a)

(b)

Figure 8.4: Screenshots of `invalidTriangulation.cpp` with two different drawing orders: (a) ABC drawn first in code, DBC next (b) DBC drawn first, ABC next.

of the triangles ABC, DAE and DEC in the code. Try to write code like `invalidTriangulation.cpp` to demonstrate this.

Interestingly, you will find that the expected ambiguity does not arise in practice. The reason is that, at the time of rasterization, pixels on the edge shared between two abutting polygons are assigned uniquely to one or the other, independent of code order, by the polygon rasterizing algorithm (we'll see how in Section 14.4). Accordingly, color values of these border pixels are obtained each from its "owning" polygon and there is no ambiguity.

Generally, though, it is not desirable that the image of an object be sensitive to the order in which the collection of triangles composing it *happens* to appear in the code. From a designer's perspective it should be enough to simply specify (a) a set of vertices, possibly, with color and other data at each, and (b) a collection of triangles with corners among these vertices, in other words, vertex adjacency data. It should *not* be necessary to additionally specify a particular order on the collection. We have now the following proposition which the reader is asked next to prove.

Proposition 8.1. *If the collection of triangles composing an object satisfies the properties of a triangulation, then its image is independent of the order in which the triangles are rendered, i.e., independent of their code order.*

Exercise 8.4. Prove the proposition just stated.
Suggested approach: Let t_1 and t_2 be two triangles of a triangulation. If they don't intersect then, of course, they cannot "conflict" in coloring a region and code order between them does not matter. If they intersect in a vertex v of both then there cannot be a conflict to be resolved by code order either, because colors are specified per-vertex. If they intersect in an edge ....

8.2 Steiner Vertices and the Quality of a Triangulation

A vertex used in the triangulation of an input object which is not a vertex of the object itself is called a *Steiner vertex*. For example, E is a Steiner vertex of the triangulation $\{ABE, BCE, CDE, DAE\}$ of the rectangle $ABCD$ of Figure 8.3(d). Even though they may not be necessary per se in order to triangulate, Steiner vertices are often inserted to improve the "quality" of a triangulation.

Roughly, a good triangulation is one where there are few long and thin triangles, called *slivers*, and where most triangles are of nearly equal size and relatively small with respect to the entire object. We'll not try to formalize any further the notion of the quality of a triangulation, but here's an example to explain how it can impact rendering.

Consider the long rectangular floor $ABCD$, triangulated as in the top of Figure 8.5. Located at either end are two lamps emitting white light – one can indeed create light sources in OpenGL, as we'll see. The vertices A, B, C and D are clearly well-lit; say the color tuple computed at each is (0.9, 0.9, 0.9). Recalling that OpenGL interpolates over a triangle the values evaluated at its vertices, the color at *every* point of $ABCD$ turns out to be (0.9, 0.9, 0.9), even though, realistically, a point in its interior, such as P, should appear darker.

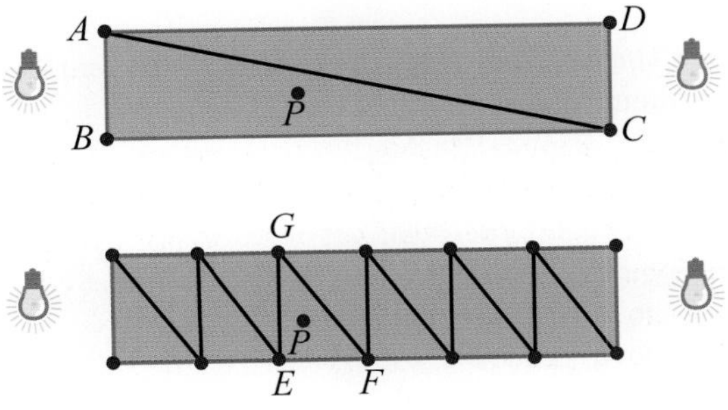

Figure 8.5: Triangulations of different quality.

Unfortunately, the sliver ABC causes what should be a local brightness to propagate globally. Such a problem can often be alleviated by improving the triangulation – e.g., the lower one of Figure 8.5 which uses Steiner vertices. There, the color intensities computed at E, F and G will be less owing to the increased distance from the light sources, so the interpolated values at P less too.

Exercise 8.5. Typical triangulations of a disc that come to mind are shown in Figures 8.6(a) and (b), but in either the problem with slivers becomes severe with an increasing number of edges. Can you suggest a better quality of triangulation? Maybe so that most of the disc is covered with "good" triangles and only a small area with slivers.

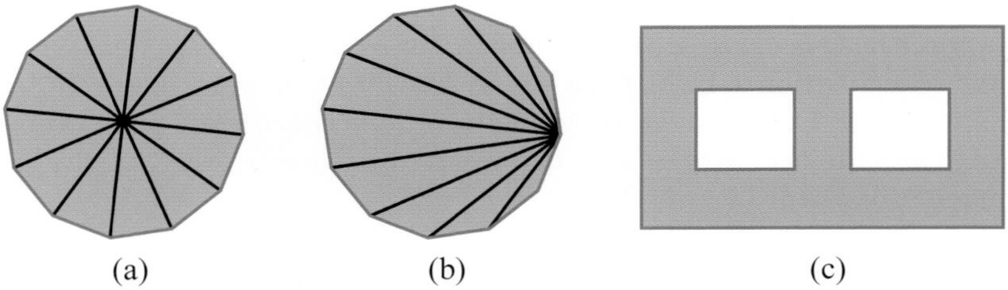

Figure 8.6: (a) and (b) Triangulations of a disc (c) Double annulus.

Exercise 8.6. Triangulate the double annulus depicted in Figure 8.6(c) using exactly one triangle strip, with the help of Steiner vertices.

Exercise 8.7. Can the invalid triangulation of Figure 8.3(f) be made valid by declaring Steiner vertices?

8.3 Triangulation in OpenGL and the Trouble with Non-Convexity

We'll now resolve the mystery that arose in Experiment 2.17 of Chapter 2. Here's the experiment again.

Experiment 8.2. Replace the polygon declaration of `square.cpp` with (Block 1*):

```
glBegin(GL_POLYGON);
   glVertex3f(20.0, 20.0, 0.0);
   glVertex3f(80.0, 20.0, 0.0);
   glVertex3f(40.0, 40.0, 0.0);
   glVertex3f(20.0, 80.0, 0.0);
glEnd();
```

Display it *both* filled and outlined using appropriate `glPolygonMode` calls. A non-convex quadrilateral is drawn in either case (Figure 8.7(a)).

Next, keeping the *same* cycle of vertices as above, list them starting with `glVertex3f(80.0, 20.0, 0.0)` instead (Block 2):

```
glBegin(GL_POLYGON);
   glVertex3f(80.0, 20.0, 0.0);
   glVertex3f(40.0, 40.0, 0.0);
```

*To cut-and-paste you can find the block in text format in the file `chap8codeModifications.txt` in the directory `ExperimenterSource/CodeModifications`.

```
    glVertex3f(20.0, 80.0, 0.0);
    glVertex3f(20.0, 20.0, 0.0);
glEnd();
```

Make sure to display it both filled and outlined. When filled it's a triangle, while outlined it's a non-convex quadrilateral identical to the one output earlier (see Figure 8.7(b))! Because the cyclic order of the vertices is unchanged, shouldn't it be as in Figure 8.7(a) both filled and outlined? **End**

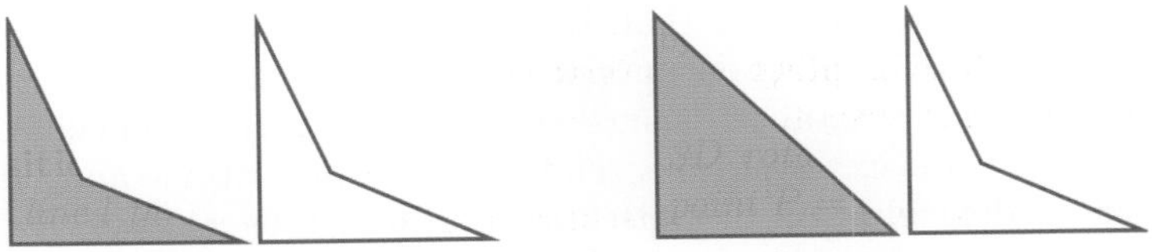

(a) (b)

Figure 8.7: Outputs: (a) Experiment 8.2 (b) Experiment 8.2, vertices cycled.

Here is what's happening. When OpenGL is asked to draw a *filled* polygon P with n vertices $v_0, v_1, \dots, v_{n-1}$, it renders a fan of $n-2$ triangles around the first vertex, exactly as though the call was made using the primitive `GL_TRIANGLE_FAN` instead of `GL_POLYGON`; in particular, the triangles of the fan are $v_0v_1v_2$, $v_0v_2v_3$, …, $v_0v_{n-2}v_{n-1}$.*

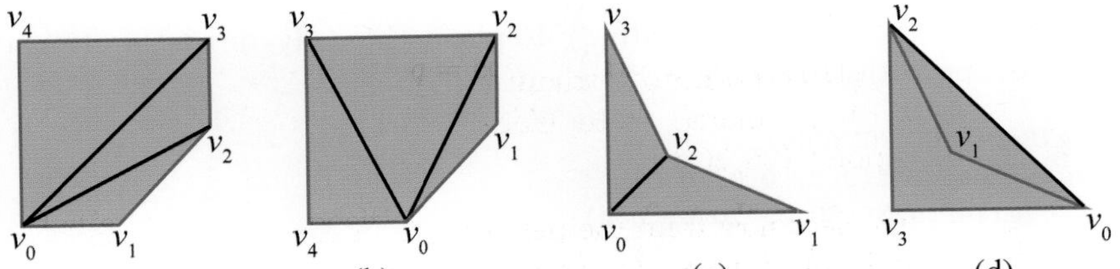

(a) (b) (c) (d)

Figure 8.8: Triangle fans.

Now, if the polygon $P = v_0v_1 \dots v_{n-1}$ is convex then the fan around v_0 is a triangulation of P. In fact, if P is convex then it does not matter how the cycle of vertices is listed, i.e., at which vertex it starts: the fan around the first vertex, or any vertex for that matter, is *always* a triangulation of P. For example, Figures 8.8(a) and (b) show the triangulation fans corresponding to cyclically rotated listings of the vertices of a convex pentagon.

Remark 8.2. That OpenGL seeks to triangulate the polygon before rendering is understandable given, from what we learned in Chapter 7, that it can

*This triangulation is implementation-dependent but all implementations that we are aware of behave as described.

then unambiguously interpolate property values from the vertices through individual triangles.

However, if the polygon $P = v_0v_1 \ldots v_{n-1}$ is not convex then there is no guarantee that the fan $v_0v_1v_2$, $v_0v_2v_3$, $\ldots$, $v_0v_{n-2}v_{n-1}$ is a triangulation of P. For example, the listing $v_0v_1v_2v_3$ as in Figure 8.8(c) of the vertices of the non-convex quadrilateral of the preceding experiment does, in fact, give the fan triangulation $\{v_0v_1v_2, v_0v_2v_3\}$. However, the fan from the listing as in Figure 8.8(d) not only does not give a triangulation, it does not even give an invalid triangulation, as the union of the triangles $v_0v_1v_2$ and $v_0v_2v_3$ is $v_0v_2v_3$ itself, which is larger than the input quadrilateral $v_0v_1v_2v_3$! This explains the differing filled outputs of Experiment 8.2.

It's accordingly vital to ensure that a filled polygon specified using `GL_POLYGON` is convex; otherwise, rendering is unpredictable and may even be incorrect depending on the vertex listing. In fact, as we have said before: avoid polygons, using triangles instead; indeed, newer versions of OpenGL (like 4.3, which we'll study extensively) have discarded polygons altogether.

When asked to draw a polygon in outline, OpenGL draws a line loop using the given vertex sequence, which is always valid however the vertices are listed. This explains the correctness of the outlined outputs in both cases in Experiment 8.2.

Important: As the reader may easily verify, a triangle strip specified with `GL_TRIANGLE_STRIP` and a triangle fan with `GL_TRIANGLE_FAN` are automatically valid triangulations provided they don't self-intersect, so use these constructs as much as possible.

Exercise 8.8. (Programming) Replace the polygon declaration of `square.cpp` with (Block 3):

```
glBegin(GL_POLYGON);
   glColor3f(0.0, 0.0, 0.0);
   glVertex3f(80.0, 20.0, 0.0);
   glColor3f(1.0, 0.0, 0.0);
   glVertex3f(40.0, 40.0, 0.0);
   glColor3f(0.0, 0.0, 0.0);
   glVertex3f(20.0, 80.0, 0.0);
   glVertex3f(20.0, 20.0, 0.0);
glEnd();
```

It's actually the second listing of the polygon of Experiment 8.2 with all vertices, except the second, colored black, while the second vertex itself is colored red. The rendered figure is all black with no sign of red at all. Why?

Exercise 8.9. (Programming) Replace the polygon declaration of `square.cpp` with the following piece of code specifying a non-convex pentagon (Block 4):


```
glBegin(GL_POLYGON);
```

```
   glVertex3f(50, 10, 0);
   glVertex3f(40, 50, 0);
   glVertex3f(10, 60, 0);
   glVertex3f(90, 60, 0);
   glVertex3f(60, 50, 0);
glEnd();
```

Sketch the pentagon on graph paper first and then predict the filled output each time as you rotate the vertices cyclically.

Even in the case of a convex polygon, different triangulations may lead to different renderings as the following exercise illustrates.

Exercise 8.10. (Programming) Replace the polygon declaration of `square.cpp` with the following to make a colored pentagon (Block 5):

```
glBegin(GL_POLYGON);
   glColor3f(0.0, 0.0, 0.0);
   glVertex3f(20.0, 20.0, 0.0);
   glVertex3f(50.0, 20.0, 0.0);
   glVertex3f(80.0, 50.0, 0.0);
   glVertex3f(80.0, 80.0, 0.0);
   glColor3f(1.0, 0.0, 0.0);
   glVertex3f(20.0, 80.0, 0.0);
glEnd();
```

All the vertices are black except the last one listed, which is red. Next, cyclically rotate the vertices, *preserving* their colors (Block 6):

```
glBegin(GL_POLYGON);
   glColor3f(0.0, 0.0, 0.0);
   glVertex3f(50.0, 20.0, 0.0);
   glVertex3f(80.0, 50.0, 0.0);
   glVertex3f(80.0, 80.0, 0.0);
   glColor3f(1.0, 0.0, 0.0);
   glVertex3f(20.0, 80.0, 0.0);
   glColor3f(0.0, 0.0, 0.0);
   glVertex3f(20.0, 20.0, 0.0);
glEnd();
```

Explain the difference in rendering. Verify your understanding by calculating the color of the point $(50.0, 70.0, 0.0)$ in either pentagon and actually drawing a point of that color at $(50.0, 70.0, 0.0)$ (which should then be invisible). *Hint*: See Figures 8.8(a) and (b).

Exercise 8.11. A polygon with more than three vertices might be worse than non-convex – it might even be non-planar. Explain what might happen with different fan-triangulations of such a "polygon", particularly using Example 7.4 of the last chapter.

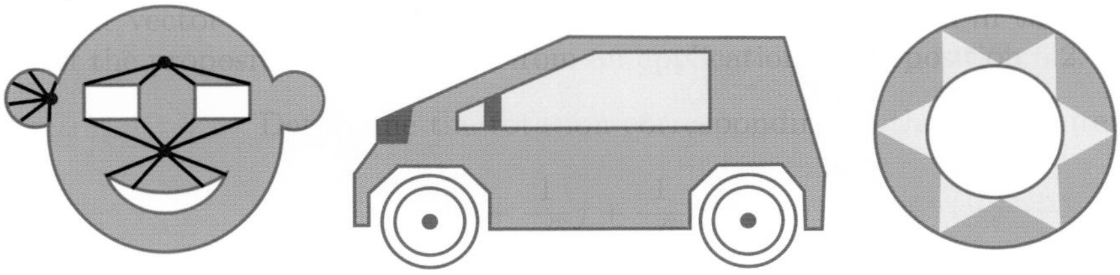

Figure 8.9: Objects to draw.

Exercise 8.12. (Programming) Draw the objects in Figure 8.9 after first triangulating them. Allow rendering both filled and wireframe.

Make true holes for the eyes and mouth of the mask, which is a flat object. In addition to the vertices you'll need on the circular arcs in order to approximate them, it may be useful to situate Steiner vertices in the interior of the mask as well (possible strategic locations are indicated in the figure).

Make the car simple and boxy. Keep in mind that it is 3D and depicted in Figure 8.9 is just a side view. In fact, copy code from `hemisphere.cpp` to rotate an object so that the car may be viewed from all angles.

For the flat decorated annulus make sure that the triangles in the darkly shaded part are separate from those in the lightly shaded part.

8.4 Summary, Notes and More Reading

In this chapter we learned the importance of triangulation and the potential problems arising from an invalid one. We learned as well of OpenGL's somewhat cavalier default attitude toward triangulation and the actually good reasons for it.

For more on the topic, computational geometry literature, in particular, deals extensively with theory of triangulation – and tetrahedralization, its 3D equivalent – as well as practical algorithms for both. There are practical algorithms to triangulate a simple plane polygon with n vertices in $O(n \log n)$ time. You will find them described in de Berg et al. [11] and O'Rourke [107], both good introductions to computational geometry in general. The CGAL [28] library is a marvelous source of ready-to-use algorithms for various geometric applications, including triangulation.

Mesh generation, as triangulation is often called, is obviously key to object creation in computer graphics. We'll be seeing much more of this process as we go along and Chapter 10 on design is mostly devoted to the topic.

An advanced text on mesh generation is by Edelsbrunner [41]. The proceedings of the annual Meshing Roundtable Conference organized by Sandia National Laboratories [74] is a source for the latest developments in the field.

CHAPTER 9

Orientation

The notion of orientation is vitally important in CG when drawing 3D scenes but, unfortunately, often confusing for the beginner. OpenGL itself makes critical use of orientation to determine the visible side of a surface. Note that the word "orientation" in the current context relates to handedness, e.g., clockwise or counter-clockwise, as we shall see, and has nothing to do with the orientation of a camera as discussed in Section 4.6.3, where the word meant pose or arrangement. The goal for this chapter is an understanding of orientation and its utility in CG.

The first section motivates the concept of orientation with a benign thought experiment. Section 9.2 describes how OpenGL applies orientation to determine the particular side of a 2D primitive which the viewer sees and then renders it with that side's specified material properties. If an object is specified as a collection of triangles, as in a triangulation, the question then arises of consistently orienting the collection. This is the topic of Section 9.3. Section 9.4 describes how OpenGL can make use of orientation to improve the efficiency of its rendering pipeline by culling certain triangles belonging to a closed surface, a procedure called back-face culling. In Section 9.5 we see how geometric transformations affect the perceived orientation of a primitive. We conclude in Section 9.6.

Although the three are conceptual in nature without a lot of excitement by way of programming, this chapter and the two preceding ones form a good part of the geometric core of CG.

9.1 Motivation

A thought experiment:

You and your friend, environmentally-conscious types both, are headed separately toward a meeting of the Tree Huggers' Union. The meeting is out in the open in a field with, well, lots of trees and no other landmarks. There is, though, a triangle of long helium-filled balloons with the letters T, H and U at the corners floating high above the meeting site. See Figure 9.1 (ignore superman with a spray can and the sheet in the middle for now).

Figure 9.1: Meeting of the Tree Huggers' Union.

Now, you want to meet up with your friend before running into the crowd. So while walking you call him on his cell phone to try to figure out how he is currently situated with respect to you. How do you do this?

As both can see the balloons, a start is to determine if you are on the same side or not. Unfortunately, the letters at the corners (carefully chosen, of course!) are of no help as they each look the same from either side.

What you can do, though, is ask your friend, "Does the vertex sequence THU – that's T→H→U – appear CW (clockwise) or CCW (counter-clockwise) from where you are?" If the orientation appears the same for both, then you are on the same side of the balloons; if not, you are on opposite sides.

OpenGL, as well, must determine for each triangle if the viewer currently sees one side or the other. And, as we'll see, it does so in an exactly similar manner. We'll understand as well the reason for this (seemingly) roundabout method. Why does OpenGL need to distinguish sides in the first place? Because they may have properties (e.g., outlined/filled, color, etc.) specified differently by the programmer and OpenGL is obliged to display accordingly.

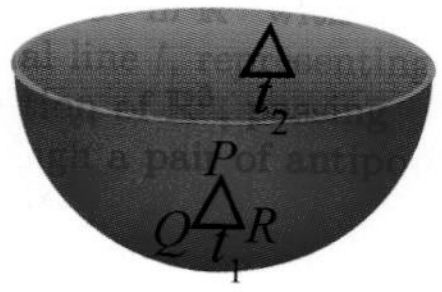

Figure 9.2: A bowl of two colors.

For example, if the inside of a triangulated bowl is green and the outside red, then the two sides of every triangle composing it are colored differently as well. Given a viewpoint, OpenGL must determine the visible side of each triangle and render it with the appropriate color. See Figure 9.2. From the current (reader's) viewpoint the red side of triangle t_1 and the green side of triangle t_2 are visible. If the viewpoint travels 180° around the bowl, then the visible side is reversed for both.

9.2 OpenGL Procedure to Determine Front and Back Faces

Here then is the procedure that OpenGL follows.

(1) First, it obtains the vertex orders of each 2D primitive *from the code*. For example, the declaration

```
glBegin(GL_TRIANGLES);
   v0; v1; v2; v3; v4; v5;
glEnd();
```

specifies the order of the vertices of the first triangle as `v0, v1, v2` and that of the second as `v3, v4, v5` (these orders are part of the `GL_TRIANGLES` definition; see Section 2.6). The declaration

```
glBegin(GL_TRIANGLE_STRIP);
   v0; v1; v2; v3; v4; v5;
glEnd();
```

specifies the vertex orders of the four successive triangles in the strip as `v0, v1, v2` and `v1, v3, v2` and `v2, v3, v4` and `v3, v5, v4`. And, similarly, for the other 2D primitives.

(2) Second, OpenGL determines for each component primitive if the order of its vertices as determined in Step (1) is *perceived* as CW or CCW by the viewer. This is said to be the *orientation* of the primitive with respect to the viewer (keep in mind that orientation as just defined has nothing to do with the identical word used to describe the pose of a camera in Section 4.6.3).

OpenGL can make this determination because it knows both the location of the viewer – at the origin in case of perspective projection and at some point on the viewing face (it doesn't matter which) in case of orthographic projection – and those of the primitive's vertices. For example, in Figure 9.3, if the vertex order of the triangle is P, Q, R, then it is perceived as CCW by the viewer. We'll see later in this section a specific algorithm to output the orientation given these respective inputs.

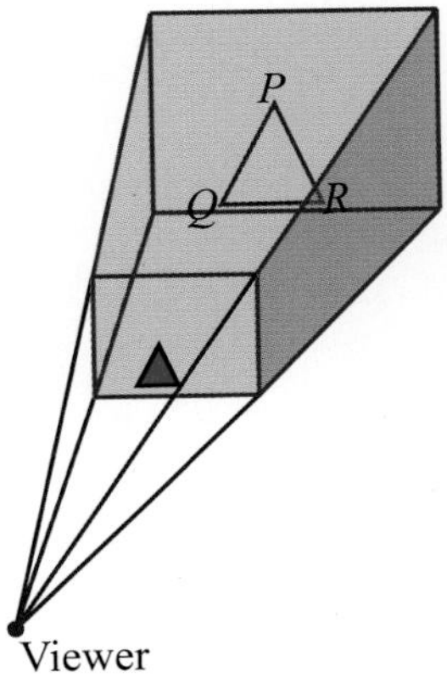

Figure 9.3: *PQR* is oriented CCW to the viewer, so it's rendered (red triangle on the viewing face) according to properties for its front face.

(3) Finally, those component primitives whose orientation the viewer perceives as CCW are presumed to be *front-facing*, i.e., the viewer is presumed by OpenGL to see their front faces, while those whose

orientation is perceived as CW are *back-facing*. This is actually the default, which can be flipped with a `glFrontFace(GL_CW)` call. Front-facing components are rendered with properties specified for their front faces, and back-facing ones with those for their back faces.

For example, if the vertex order of triangle t_1 in Figure 9.2 happens to be P, Q, R and the viewer is the reader, then OpenGL determines that this triangle is oriented CCW with respect to the viewer, who sees, therefore, the front face. Accordingly, t_1 is rendered red, assuming that the code indeed specifies that front-facing triangles are red. In Figure 9.3 we show the red rendering on the viewing face itself, pretending that it is the OpenGL window.

The reader may wonder at this point why one needs to invoke a *particular* viewpoint to distinguish sides. In real life the inside of the bowl (which is *absolute* and does not depend on the location of any viewer) is painted green and the outside (absolute as well) red. Subsequently, a viewer's perception is determined simply by the laws of nature, in particular, how light from the bowl travels to her eyes.

Why doesn't OpenGL try and simulate this phenomenon? The answer is that, yes, it is true that the inside and outside of the bowl are absolute irrespective of the viewer, but *only after the entire bowl has been created*! If one breaks off a tiny piece of the bowl – a tiny flat triangle, if you will – and shows it to someone who has never seen the whole, then it is not possible for that person to decide which side of the piece originally lay on the bowl's inside and which the outside (Figure 9.4). OpenGL has no global notion of objects either as it simply draws them triangle by triangle, and, therefore, requires direction from the programmer as to which side of each triangle is which.

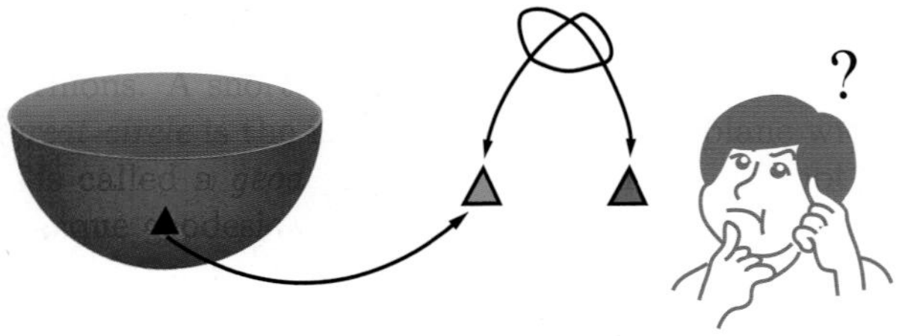

Figure 9.4: Was the inside of the bowl red or green?

The three-step procedure described above provides exactly a mechanism for such direction. Let's return to the thought experiment at the start of the section and assume that there is a giant triangular sheet of paper attached to the balloons (as in Figure 9.1) which you know is colored differently on either side. Then, of course, you could ask your friend, "What color do you see up there?" instead of "Does the vertex sequence THU – that's

T→H→U – appear CW or CCW from where you are?" The point is that the two questions are exactly equivalent in that those who perceive a particular orientation see a particular side and vice versa.

Continuing with this line of thought, suppose as you are walking that you notice a man high up about to spray-paint the sides of the triangular paper and he has a cell phone which you can call. You could then either ask him to arbitrarily paint one side green and the other red, which would at least serve the purpose of locating your friend, or you could ask him to paint your side green and the other red, which allows you (the programmer) to dictate that "CCW-seers" see red and "CW-seers" green.

There are three points worth emphasizing:

(a) A real-life 2D object (like a piece of paper) actually has two physical sides regardless of which an observer sees. This is not true of OpenGL, whose objects are all, of course, virtual. An OpenGL 2D primitive such as a triangle consists simply of data, e.g., vertex coordinates, color values, etc., residing inside the computer.

When asked to draw, OpenGL determines *if* the viewer is *supposed* to see the front or the back face according to the procedure described earlier and then *displays* the primitive with properties specified for that face. And, what it displays, of course, is simply a set of colored pixels in the OpenGL window (which has only one side!).

(b) The terms "front-facing" and "back-facing" are simply used to indicate one side and the other. There is no *intrinsic* front or back of an OpenGL triangle or other 2D primitive. If we didn't use these terms, we would have to say things like "the side which the viewer sees when the order `v0v1v2` appears clockwise from the origin".

(c) OpenGL draws primitives *one by one* as they occur in the code. It has no global understanding of the objects formed by these primitives *together*.

Exercise 9.1. If a triangle t is specified by

`glBegin(GL_TRIANGLES);` v_0; v_1; v_2; `glEnd();`

where the vertices are as below, in each case determine which side of t, front or back, a viewer at the origin sees, assuming the default of `glFrontFace(GL_CCW)`:

(a) $v_0 = (1, 0, 0)$, $v_1 = (0, 1, 0)$, $v_2 = (0, 0, 1)$

Answer:

The back face because $v_0v_1v_2$ appears CW from O. See Figure 9.5.

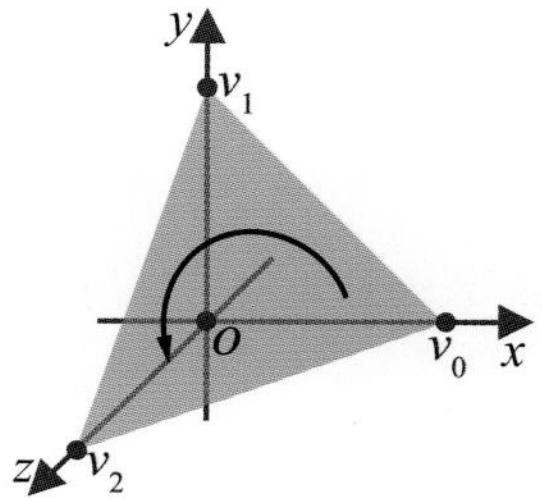

Figure 9.5: $v_0v_1v_2$ appears CW from O.

(b) $v_0 = (0, 1, 0)$, $v_1 = (1, 0, 0)$, $v_2 = (0, 0, 1)$

(c) $v_0 = (-1, 0, 0)$, $v_1 = (0, -1, 0)$, $v_2 = (0, 0, -1)$

(d) $v_0 = (1, 1, 1)$, $v_1 = (1, 1, -2)$, $v_2 = (-1, 1, -2)$

Exercise 9.2. A tacit assumption in all of the preceding discussion is that a viewer at a particular location sees, in fact, *only one* side – front or back – of a 2D primitive. For example, if a viewer could see both sides of a triangle, then is it front or back facing (or both)? Moreover, how then would one reconcile the situation with the three-step procedure at the start of the section, which purports to determine a unique orientation for the primitive?

So, is the assumption that only one side is visible a valid one?

Hint: A triangle is always *flat* (planar), while a polygon should be specified to be so.

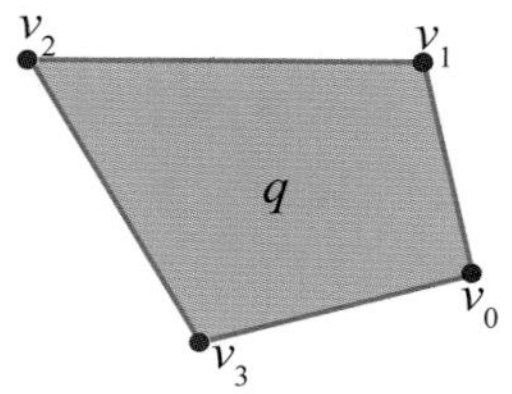

Figure 9.6: A quadrilateral.

Definition 9.1. Two orders of the vertices of a polygon are said to be *equivalent* if one can be *cyclically rotated* into the other.

It follows that the sequence of vertices around any given polygon can be written in exactly *two* inequivalent orders. For example, the sequence of vertices around the quadrilateral q of Figure 9.6 can be written in eight different ways:

$$\begin{array}{cccc} v_0v_1v_2v_3 & v_1v_2v_3v_0 & v_2v_3v_0v_1 & v_3v_0v_1v_2 \\ v_0v_3v_2v_1 & v_3v_2v_1v_0 & v_2v_1v_0v_3 & v_1v_0v_3v_2 \end{array}$$

The orders on the top line are all equivalent to each other, while those on the second all to each other as well, and none on the first equivalent to any on the second. The notion of equivalence is important precisely because of the fact that a viewer on one side of a polygon perceives equivalent orders of vertices as either all CW or all CCW.

Experiment 9.1. Replace the polygon declaration part of `square.cpp` with (Block 1*):

```
glPolygonMode(GL_FRONT, GL_LINE);
glPolygonMode(GL_BACK, GL_FILL);
glBegin(GL_POLYGON);
   glVertex3f(20.0, 20.0, 0.0);
   glVertex3f(80.0, 20.0, 0.0);
   glVertex3f(80.0, 80.0, 0.0);
   glVertex3f(20.0, 80.0, 0.0);
glEnd();
```

This simply adds the two `glPolygonMode()` statements to the original `square.cpp`. In particular, they specify that front-facing polygons are to be drawn in outline and back-facing ones filled. Now, the order of the vertices is (20.0, 20.0, 0.0), (80.0, 20.0, 0.0), (80.0, 80.0, 0.0), (20.0, 80.0, 0.0), which

*To cut-and-paste you can find the block in text format in the file `chap9codeModifications.txt` in the directory `ExperimenterSource/CodeModifications`.

appears CCW from the viewing face. Therefore, the square is drawn in outline.

Next, rotate the vertices cyclically so that the declaration becomes (Block 2):

```
glPolygonMode(GL_FRONT, GL_LINE);
glPolygonMode(GL_BACK, GL_FILL);
glBegin(GL_POLYGON);
   glVertex3f(20.0, 80.0, 0.0);
   glVertex3f(20.0, 20.0, 0.0);
   glVertex3f(80.0, 20.0, 0.0);
   glVertex3f(80.0, 80.0, 0.0);
glEnd();
```

As the vertex order remains equivalent to the previous one, the square is still outlined.

Reverse the listing next (Block 3):

```
glPolygonMode(GL_FRONT, GL_LINE);
glPolygonMode(GL_BACK, GL_FILL);
glBegin(GL_POLYGON);
   glVertex3f(80.0, 80.0, 0.0);
   glVertex3f(80.0, 20.0, 0.0);
   glVertex3f(20.0, 20.0, 0.0);
   glVertex3f(20.0, 80.0, 0.0);
glEnd();
```

The square is drawn filled as the vertex order now appears CW from the front of the viewing box. **End**

Exercise 9.3. (Programming) Return to the first part of the preceding experiment, where you replaced the polygon declaration part of `square.cpp` with Block 1 from `chap9codeModifications.txt`. Next, add the following statements just before the polygon declaration.

```
glTranslatef(50.0, 0.0, 0.0);
glRotatef(180.0, 0.0, 1.0, 0.0);
glTranslatef(-50.0, 0.0, 0.0);
```

Explain what you observe.

Exercise 9.4. (Programming) If the polygon declaration part of `square.cpp` is replaced with the following piece of code (Block 4), then is an outlined or filled triangle seen? Try to answer first without running the program.

```
glFrontFace(GL_CW);
glPolygonMode(GL_FRONT, GL_LINE);
glPolygonMode(GL_BACK, GL_FILL);
glBegin(GL_TRIANGLES);
```

```
      glVertex3f(80.0, 10.0, -1.0);
      glVertex3f(90.0, 75.0, 1.0);
      glVertex3f(15.0, 10.0, 0.5);
   glEnd();
```

Remark 9.1. Before we get to Chapter 11 and learn about material properties and how to color the sides of an object differently, we'll have to do with distinguishing them by the unglamorous means of drawing one in outline and the other filled.

Algorithm to Decide the Orientation Perceived by a Viewer

An algorithmic question, that we did not address then, arose earlier in this section in Step (2) of OpenGL's procedure to determine the side of a primitive a viewer sees: given a viewpoint and a primitive with its vertices ordered, how to decide if the given order appears CW or CCW? We invite the reader to answer this for herself in the following exercise, with a fair amount of input from our end.

Exercise 9.5. Assume that the viewpoint is at the origin O and that the vertices of a triangle are $P = (x_1, y_1, z_1)$, $Q = (x_2, y_2 z_2)$ and $R = (x_3, y_3, z_3)$. See Figure 9.7. Determine if the viewer at O perceives the order PQR as CCW or CW.

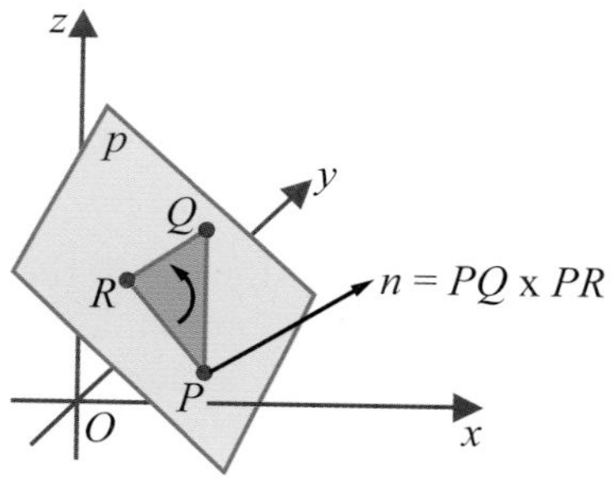

Figure 9.7: The plane p contains the triangle PQR: the orientation of PQR depends on which side of p the viewer is located.

If you don't do the exercise do at least read the conclusion below in terms of the determinant D.

Suggested approach: Supposing, first, that P, Q and R are not collinear, i.e., PQR is a non-degenerate triangle, determine the equation $ax+by+cz+d = 0$ of the unique plane p containing P, Q and R.

A point (x, y, z) lies on p if $ax + by + cz + d = 0$. A point lies in one *half-space* of p, i.e., on one side of p or the other, depending on whether $ax + by + cz + d < 0$ or $ax + by + cz + d > 0$.

Observe, next, that a viewer located *on* p sees triangle PQR "edge-on", in other words, as a line and not a triangle, so the question of orientation does not arise. A viewer not on p, on the other hand, perceives the orientation of PQR depending on the half-space she is in: particularly, all viewers inside one half-space perceive CCW, while those in the other CW.

Therefore, the perception at viewpoint O, in particular, depends on whether O lies on p or, if not, on which side.

Finally, conclude the following:

Let D be the determinant

$$\begin{vmatrix} x_1 & x_2 & x_3 \\ y_1 & y_2 & y_3 \\ z_1 & z_2 & z_3 \end{vmatrix}$$

1. If $D = 0$, then either (a) P, Q and R are collinear, in which case PQR is a degenerate triangle and the question of an orientation of PQR does not arise, or (b) O lies on the plane p containing P, Q and R, so that the viewer at O sees triangle PQR edge-on and, again, the question of orientation does not arise.

2. If $D > 0$, then the viewer at O perceives the order PQR as CW.

3. If $D < 0$, then the viewer at O perceives the order PQR as CCW.

Another approach is with the use of cross-products, by observing that $n = PQ \times PR$ is normal to the plane p and, in fact, points to the half-space where observers perceive PQR as CCW. Therefore, if the eye direction vector PO makes an angle of less than 90° with n – placing it in the same half-space as n – then the viewer at O perceives the order PQR as CCW as well; if greater, then as CW (in the configuration depicted in the figure the angle is, in fact, greater than 90°). Whether the angle between the two vectors n and PO is greater or less than 90° can be decided from the sign of the dot product $n \cdot PO$.

Note: If you're not familiar with the dot or cross-product of vectors, we have short sidebars in Sections 4.6.1 and 5.4.3, respectively.

Exercise 9.6. Does a viewer at the origin perceive the order PQR of the points $P = (-1, 2, 0)$, $Q = (3, 2, 2)$ and $R = (-3, -8, 6)$ as CW or CCW?

Exercise 9.7. Does a viewer at the point $O' = (1, 3, 2)$ perceive the order PQR of the points $P = (3, 7, 5)$, $Q = (4, 1, 2)$ and $R = (0, 1, 2)$ as CW or CCW?
Hint: Translate all points by $(-1, -3, -2)$ to bring O' to the origin and then apply the result of Exercise 9.5.

Exercise 9.8. Relate Lemma 5.1 to the answer to Exercise 9.5.

9.3 Consistently Oriented Triangulation

The notion of orientation gets even more interesting when one considers a collection of triangles, as in a triangulation. The issue arises then of *consistency*. We have the following definition:

Definition 9.2. Suppose an order is given of the vertices of each triangle belonging to some triangulation $\mathcal{T}$ of an object X. $\mathcal{T}$ is said to be *consistently oriented* if any two triangles of $\mathcal{T}$ which share an edge order the shared edge oppositely; otherwise, $\mathcal{T}$ is *inconsistently oriented*.

Figure 9.8(a) shows a consistently oriented triangulation. For example, the edge shared by the two triangles $v_0v_1v_2$ and $v_1v_3v_2$ is ordered v_1v_2 by the first and v_2v_1 by the second. The triangulation of Figure 9.8(b) is not consistently oriented as the edge shared by the two leftmost triangles is ordered v_1v_2 by both.

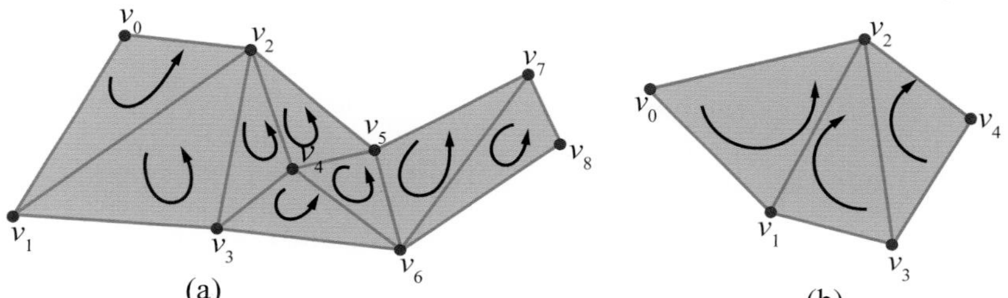

Figure 9.8: (a) Consistently oriented triangulation (b) Inconsistently oriented triangulation.

Intuitively, triangles in a consistently oriented triangulation of X appear oriented either all CW or all CCW "looking at one side of X". What exactly does this mean?

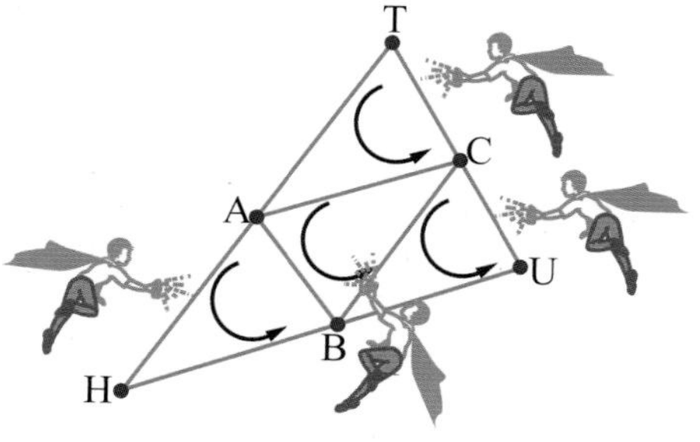

Figure 9.9: OpenGL spray-painting bots.

Let's return again to the earlier thought experiment at the point when you were about to call the painter. Looking up again you make out that the large triangular sheet is actually composed of four smaller ones and that there's a painter for each, so you'll have to call them separately (Figure 9.9). Moreover, all that you are allowed to specify to each is the order of his triangle's vertices – e.g., you can specify to the painter at the top his vertex order as either C→A→T or T→A→C – for these painters are nothing but OpenGL bots that have been programmed to do the following:

Determine if you perceive the order that you just called in as CCW or CW; if CCW then paint your side red, if not green.

Clearly, the onus then is on you to call in the four orders so that the small triangles are consistently oriented or else your side of the large triangle will be colored disparately.

Are we saying that an observer at a given position can see only one side of a consistently oriented surface? Not at all. For example, the man in Figure 9.10 can see parts of both sides of the consistently oriented triangulated wall. However, he sees a change in side, according to the CW/CCW rule, only across boundary edges, never across an internal edge – which is physically authentic. If the wall were not consistently oriented, though, then this would not be the case. For example, the reader using the CW/CCW rule would believe herself to be seeing two different sides of the polygon of Figure 9.8(b) along the edge v_1v_2.

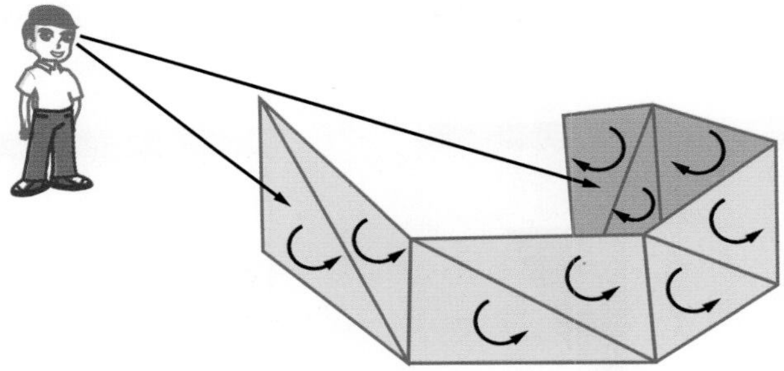

Figure 9.10: Man looking at both sides of a consistently oriented wall.

Recall again the bowl of Figure 9.2 with its inside green and outside red. If it's created in OpenGL as a triangulation, the programmer should then (a) specify that all front faces are of one color and back faces of the other, *and* (b) ensure consistent orientation of the triangulation so that the entire inside and entire outside appear of the desired colors, respectively.

In fact, the preceding rule should apply to all surfaces that we create. Here's what can happen if it doesn't.

Experiment 9.2. Replace the polygon declaration part of `square.cpp` with (Block 5):

```
glPolygonMode(GL_FRONT, GL_LINE);
glPolygonMode(GL_BACK, GL_FILL);
```

```
glBegin(GL_TRIANGLES);
   // CCW
   glVertex3f(20.0, 80.0, 0.0);
   glVertex3f(20.0, 20.0, 0.0);
   glVertex3f(50.0, 80.0, 0.0);

   //CCW
   glVertex3f(50.0, 80.0, 0.0);
   glVertex3f(20.0, 20.0, 0.0);
   glVertex3f(50.0, 20.0, 0.0);

   // CW
   glVertex3f(50.0, 20.0, 0.0);
   glVertex3f(50.0, 80.0, 0.0);
   glVertex3f(80.0, 80.0, 0.0);

   // CCW
   glVertex3f(80.0, 80.0, 0.0);
   glVertex3f(50.0, 20.0, 0.0);
   glVertex3f(80.0, 20.0, 0.0);
glEnd();
```

The specification is for front faces to be outlined and back faces filled, but, as the four triangles are not consistently oriented, we see both outlined and filled triangles (Figure 9.11(a)). **End**

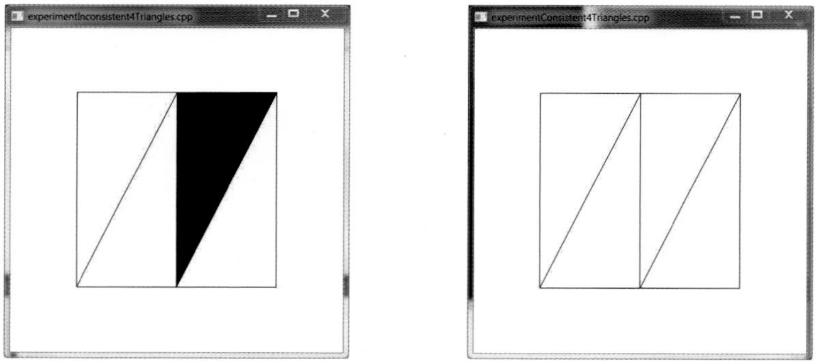

(a) (b)

Figure 9.11: Screenshots for (a) Experiment 9.2 and (b) Experiment 9.3.

Experiment 9.3. Continuing the previous experiment, next replace the polygon declaration part of `square.cpp` with (Block 6):

```
glPolygonMode(GL_FRONT, GL_LINE);
glPolygonMode(GL_BACK, GL_FILL);
glBegin(GL_TRIANGLE_STRIP);
```

```
    glVertex3f(20.0, 80.0, 0.0);
    glVertex3f(20.0, 20.0, 0.0);
    glVertex3f(50.0, 80.0, 0.0);
    glVertex3f(50.0, 20.0, 0.0);
    glVertex3f(80.0, 80.0, 0.0);
    glVertex3f(80.0, 20.0, 0.0);
glEnd();
```

The resulting triangulation is the same as before, but, as it's consistently oriented, we see only outlined front faces. (Figure 9.11(b)). **End**

In the next experiment we see an example of a consistently oriented object, both sides of which are visible.

Experiment 9.4. Run `squareOfWalls.cpp`, which shows four rectangular walls enclosing a square space. The front faces (the outside of the walls) are filled, while the back faces (the inside) are outlined. Figure 9.12(a) is a screenshot.

The triangle strip of `squareOfWalls.cpp` consists of eight triangles which are consistently oriented, because triangles in a strip are *always* consistently oriented. **End**

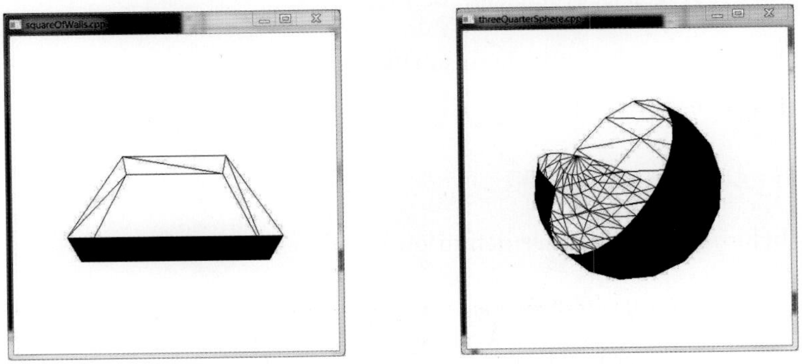

(a) (b)

Figure 9.12: Screenshots of (a) `squareOfWalls.cpp` and (b) `threeQuarterSphere.cpp`.

Experiment 9.5. Run `threeQuarterSphere.cpp`, which adds one half of a hemisphere to the bottom of the hemisphere of `hemisphere.cpp`. The two polygon mode calls ask the front faces to be drawn filled and back ones outlined. Turn the object about the axes by pressing 'x', 'X', 'y', 'Y', 'z' and 'Z'.

Unfortunately, the ordering of the vertices is such that the outside of the hemisphere appears filled, while that of the half-hemisphere outlined. Figure 9.12(b) is a screenshot. Likely, this would not be intended in a real design application where one would, typically, expect a consistent look throughout one side.

Such mixing up of orientation is not an uncommon error when assembling an object out of multiple pieces. Fix the problem in the case of threeQuarterSphere.cpp in four different ways:

(a) Replace the loop statement

```
for(i = 0; i <= p/2; i++)
```

of the half-hemisphere with

```
for(i = p/2; i >= 0; i--)
```

to reverse its orientation.

(b) Interchange the two glVertex3f() statements of the half-hemisphere, again reversing its orientation.

(c) Place the additional polygon mode calls

```
glPolygonMode(GL_FRONT, GL_LINE);
glPolygonMode(GL_BACK, GL_FILL);
```

before the half-hemisphere so that its back faces are drawn filled.

(d) Call

```
glFrontFace(GL_CCW)
```

before the hemisphere definition and

```
glFrontFace(GL_CW)
```

before the half-hemisphere to change the front-face default to be CW-facing for the latter.

Of the four, either (a) or (b) is to be preferred because they go to the source of the problem and repair the object, rather than hide it with the help of state variables, as do (c) and (d). **End**

It is not hard to orient consistently when creating objects in OpenGL because the primitives themselves tend to help. Verify from the definition of the drawing primitives in Section 2.6 that the set of triangles created by a call to GL_TRIANGLE_STRIP or GL_TRIANGLE_FAN is, in fact, consistently oriented. Therefore, a GL_TRIANGLE_STRIP or a GL_TRIANGLE_FAN call guarantees consistent orientation, at least for that particular set of triangles, so it's a good idea to use as many such as possible.

Non-Orientable Surfaces

Before concluding this section, mention must be made of non-orientability. There do exist surfaces which can be triangulated but *never* consistently oriented. The most famous two, the Möbius band and Klein bottle, are depicted in Figure 9.13. Such surfaces are said to be *non-orientable*. Surfaces for which consistently oriented triangulations do exist are *orientable*.

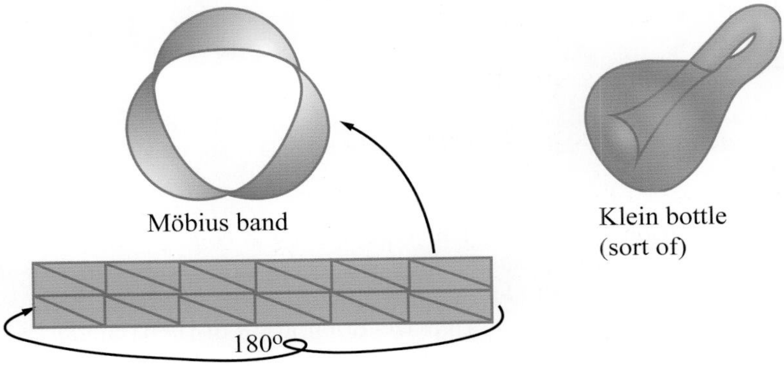

Figure 9.13: Non-orientable surfaces.

Experiment 9.6. Make a Möbius band as follows.

Take a long and thin strip of paper and draw two equal rows of triangles on one side to make a triangulation of the strip as in the bottom of Figure 9.13. Turn the strip into a Möbius band by pasting the two end edges together after twisting one 180°. The triangles you drew on the strip now make a triangulation of the Möbius band.

Try next to orient the triangles by simply drawing a curved arrow in each, in a manner such that the entire triangulation is consistently oriented. Were you able to?! **End**

We have less to worry about with the Klein bottle, at least as far as real-world applications are concerned, because it cannot be created in 3-space. It needs at least 4D space to hold it properly.

Further formalization of the notion of orientability requires knowledge of topology, but what we have discussed so far is ample from the point of view of first-level computer graphics. By the way, in case non-orientability looks like a potential can of worms, rest assured you will almost never encounter a non-orientable surface in practical applications.

9.4 Culling Obscured Faces

Consider a *closed* surface such as a sphere, cube or torus, i.e., a surface that bounds a solid. See Figure 9.14. If the surface is opaque, then a viewer outside of it sees only one side, no matter where she is located, while a viewer inside sees the other. Closedness is essential here, otherwise, a viewer may be able to see both sides, e.g., as the reader can see for the hemisphere or cylinder below.

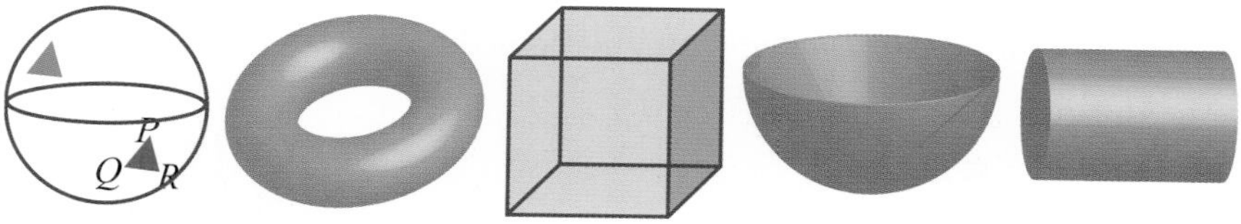

Figure 9.14: First three closed surfaces, next two non-closed. The closed sphere is not shaded to reveal the inside; the green back face of a triangle is not visible from outside the sphere.

Such a situation is replicated in OpenGL by a consistently oriented triangulation of the given closed surface. For example, suppose the outside of the sphere of Figure 9.14 is painted red, and the inside green. Suppose, too, it's consistently oriented so that the orientation of the triangle PQR appears CCW, as shown in the figure, to a viewer outside the sphere (e.g., the reader). Then *any* viewer outside the sphere sees only front-facing (assuming the default of CCW = front-facing) triangles and never any back-facing ones (e.g., the green back face of the other triangle in the figure) because, for such a viewer, all back-facing triangles are hidden behind front-facing ones. The precise opposite is true for viewers inside the sphere who see only back-facing triangles.

Now, OpenGL cannot know if a surface is closed or not because this is a global decision to be made *after* the *entire* surface has been drawn (e.g., if even one triangle were missing from the sphere then it would no longer be closed). Closedness cannot, therefore, be determined by an API which simply draws one triangle after another. As a result, what happens, for example, in the case of the sphere above with the viewer outside, is that OpenGL processes *every* triangle, and then ends up discarding back-facing ones at the time of hidden surface removal because it's only then that OpenGL discovers back-facing triangles to be obscured by front-facing ones.

Therefore, for example, knowing that a viewer located outside the closed sphere can see only front-facing triangles, the programmer can help OpenGL be more efficient by directing it not to process any further a triangle once it's been determined to be back-facing. This is called *back-face culling* or *polygon culling*.

Experiment 9.7. Run `sphereInBox1.cpp`, which draws a green ball inside a red box. Press up or down arrow keys to open or close the box. Figure 9.15(a) is a screenshot of the box partly open.

Ignore the statements to do with lighting and material properties for now. The command `glCullFace(`*face*`)` where *face* can be `GL_FRONT`,

GL_BACK or GL_FRONT_AND_BACK, is used to specify if front-facing or back-facing or all polygons are to be culled. Culling is enabled with a call to glEnable(GL_CULL_FACE) and disabled with glDisable(GL_CULL_FACE).

You can see at the bottom of the drawing routine that back-facing triangles of the sphere are indeed culled, which makes the program more efficient because these triangles are hidden in any case behind the front-facing ones.

Comment out the glDisable(GL_CULL_FACE) call and open the box. Oops! Some sides of the box have disappeared as you can see in Figure 9.15(b). The reason, of course, is that the state variable GL_CULL_FACE is set when the drawing routine is called the first time so that all back-facing triangles, including those belonging to the box, are eliminated on subsequent calls.

End

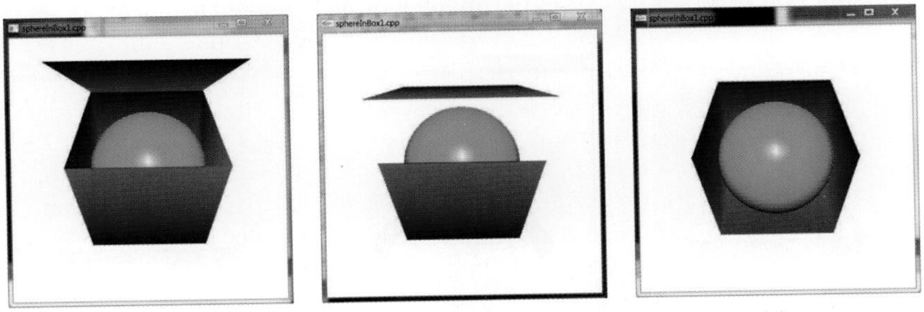

(a) (b) (c)

Figure 9.15: Screenshots for (a) Experiment 9.7 (b) Experiment 9.7 (disable culling commented out) (c) Experiment 9.8.

Exercise 9.9. So, all back-facing triangles of a closed surface are obscured to a viewer outside. Is the converse true, in particular, are all obscured triangles necessarily back-facing?

Experiment 9.8. Here's a trick often used in 3D design environments like Maya and Studio Max to open up a closed space. Suppose you've finished designing a box-like room and now want to work on objects inside it. A good way to do this is to remove only the walls obscuring your view of the inside and leave the rest; that the obscuring walls are either *all* front-facing or *all* back-facing means a cull will do the trick. Let's see this in action.

Insert the pair of statements

```
glEnable(GL_CULL_FACE);
glCullFace(GL_FRONT);
```

in the drawing routine of sphereInBox1.cpp just before the glDrawElements() calls. The top and front sides of the box are not drawn, leaving its interior visible. Figure 9.15(c) is a screenshot.

End

9.5 Transformations and the Orientation of Geometric Primitives

We know now how OpenGL uses the vertex order to determine the orientation of a primitive perceived by a viewer and, accordingly, the face seen, front or back. A reader, recollecting the theory of transformations, particularly Section 5.4.7 about orientation-preserving Euclidean transformations (i.e., rigid transformations) and orientation-reversing ones, may have already thought about and guessed the answer to the following question: how do these transformations affect the perceived orientation of a geometric primitive?

Answer: An orientation-preserving Euclidean transformation preserves the viewer's perceived orientation of the primitive, while an orientation-reversing one reverses it. An experiment will help make this clear.

(a) (b) (c)

Figure 9.16: Screenshots from Experiment 9.9: (a) Original (b) Wrongly reflected (c) Correctly reflected.

Experiment 9.9. Run `squareOfWallsReflected.cpp`, which is `squareOfWalls.cpp` with the following additional block of code, including a `glScalef(-1.0, 1.0, 1.0)` call, to reflect the scene about the yz-plane.

```
// Block to reflect the scene about the yz-plane.
if (isReflected)
{
   ...
   glScalef(-1.0, 1.0, 1.0);
   // glFrontFace(GL_CW);
}
else
{
   ...
   // glFrontFace(GL_CCW);
}
```

The original walls are as in Figure 9.16(a). Press space to reflect. Keeping in mind that front faces are filled and back faces outlined, it seems that `glScalef(-1.0, 1.0, 1.0)` not only reflects, but turns the square of walls inside out as well, as you can see in Figure 9.16(b). **End**

Well, of course! The viewer's (default) agreement with OpenGL is that if she perceives a primitive's vertex order as CCW, then she is shown the front, if not the back. Reflection about the yz-plane, an orientation-reversing Euclidean transformation, flips all perceived orientations, so those primitives whose front the viewer used to see now have their back to her, and vice versa.

We likely want the reflection to transform the primitives but not simultaneously change their orientation. This is easily done by revising the viewer's agreement with OpenGL with a call to `glFrontFace(GL_CW)`. Accordingly, uncomment the two `glFrontFace()` statements in the reflection block. Now the reflection looks right, as shown in Figure 9.16(c). The primitives are clearly still being reflected about the yz-plane, but front and back stay the same.

9.6 Summary, Notes and More Reading

In this chapter we learned how OpenGL applies orientation to determine that side of a 2D primitive which is visible. We saw as well the importance of consistently orienting a triangulation. The technique of back-face culling to improve efficiency in rendering a closed surface was a useful addition to our repertoire. We learned as well how orientation-preserving and orientation-reversing transformations impact the orientation of a primitive.

Although our discussion of orientation at the elementary level is ample for the practical programmer, a fairly sophisticated mathematical setting is required to formalize the concept of the orientability of a surface. The interested reader is urged to look up an introductory topology text. The two by Munkres [98, 99], as well as the one by Singer & Thorpe [134], are classics. Incidentally, the mathematically-inclined student of CG will find many things of use in topology. One has only to scan the latest ACM SIGGRAPH papers [131] to see the heavy application of topological ideas in cutting-edge CG. An excellent introduction to the emerging area of computational topology is by Edelsbrunner and Harer [42].

Part V

Making Things Up

CHAPTER 10

Modeling in 3D Space

The goal for this chapter is to systematically study the modeling of objects in 3D space in order to be able to populate the movies, games and other scenes that we create.

As OpenGL has only straight and flat drawing primitives, curved objects must necessarily be approximated. We'll develop general strategies to manufacture approximations of both curves and surfaces. We'll examine in depth certain special classes of curves and surfaces especially important in applications. Particular attention will be paid to Bézier primitives because of their utility, as well as the easy-to-use OpenGL syntax available to code them. Another popular class we'll study is that of fractals. Although we'll delve into some of the mathematics underlying curves and surfaces, we'll never be far from practical code: throughout this chapter are numerous illustrative programming examples and exercises.

We begin in Section 10.1 with the modeling of curves. The first two subsections, 10.1.1 and 10.1.2, describe how a curve is specified by equations, either implicitly or parametrically. A strategy to draw a curve as a polyline approximation is the topic of 10.1.3. We discuss polynomial and rational parametrizations of curves in 10.1.4, as they are computationally more efficient than other kinds. The conic sections, including parabolas, ellipses and hyperbolas, comprise a very important and commonly-occurring class of curves that we investigate briefly in 10.1.5. Section 10.1.6 is a short introduction to the mathematics of curves, particularly giving a rigorous definition of what it means to be a curve, and discussing continuity and regularity. This section can and probably should be skipped on a first reading.

We move on to surfaces in Section 10.2. We present the following 2D primitives in an informal order of increasing drawing complexity: polygons, meshes, planar surfaces and general surfaces. Subsections 10.2.1-10.2.3

describe the first three, which are straightforward to draw. The next two subsections, 10.2.4 and 10.2.5, discuss the specification of a general surface and how to model one as a mesh approximation.

The powerful technique of making a surface by sweeping a curve is the topic of 10.2.6. In 10.2.7 we pause to apply our newly-acquired skills in a bunch of modeling projects. We continue our study of surfaces in 10.2.8, discussing a special class of swept surfaces, called ruled surfaces. This class includes bilinear patches and generalized cones and cylinders. The generalization of conic sections to 3D, the quadric surfaces, is described in 10.2.9. Objects of the GLU library which are somewhat inappropriately called the GLU quadrics are introduced in 10.2.10. The beautifully symmetric regular polyhedra, or Platonic solids as they are often called, are presented in 10.2.11. Section 10.2.12 parallels 10.1.6 in a formal discussion of surfaces and the properties of continuity and regularity – and the same recommendation applies that it be skipped on a first reading.

Although we'll be discussing Bézier theory in depth in a later chapter, it turns out that a fair amount of design with Bézier curves and surfaces can be accomplished even with limited theoretical understanding. Therefore, in keeping with our aim in this chapter of equipping the reader with as many practical modeling techniques as possible, Section 10.3 introduces Bézier design – curves in 10.3.1 and surfaces in 10.3.2.

Fractal curves, ubiquitous in nature, and so often used to create surreal shapes by designers, are the topic of Section 10.4.

10.1 Curves

One-dimensional objects are unions of straight and curved segments. See Figure 10.1. Parts composed of straight segments can be drawn exactly – in an OpenGL environment one would invoke the `GL_LINES`, `GL_LINE_STRIP` and `GL_LINE_LOOP` primitives. Curved segments, on the other hand, have to be approximated.

Terminology: The term "curve" can mean any segment, curved or straight.

We'll formalize the process of approximating a curve with a polygonal line. However, let's first see how to mathematically specify a curve.

10.1.1 Specifying Plane Curves

We begin with *plane curves*, which are those that lie on a plane (or 2-space, mathematically). There are two ways to specify such a curve, implicit and parametric.

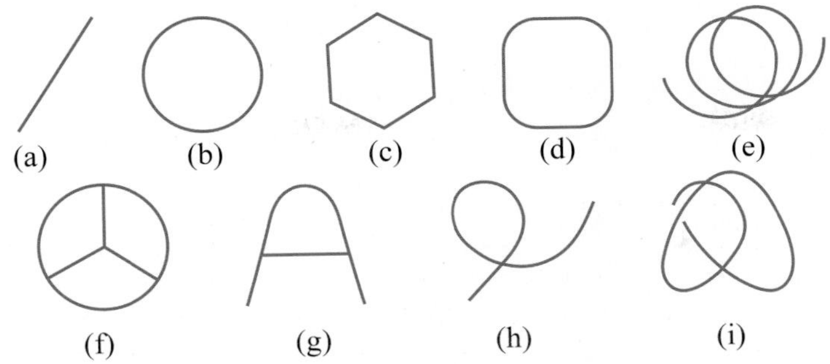

Figure 10.1: One-dimensional objects.

Implicit

A plane curve c is specified *implicitly* by the equation

$$F(x, y) = 0 \tag{10.1}$$

if the points of c are those whose coordinates (x, y) satisfy this equation. $F(x, y) = 0$ is said to be the *implicit equation* of c, and the curve c the *graph* of this equation. An implicit equation, therefore, gives a *Boolean condition* for points on the curve to satisfy: a point (a, b) lies on the curve $F(x, y) = 0$ if $F(a, b) = 0$; it doesn't if $F(a, b) \neq 0$.

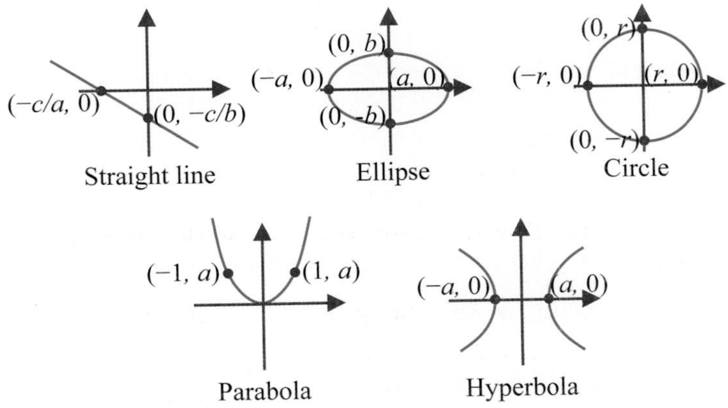

Figure 10.2: Graphs of familiar plane curves (curves are fairly accurate sketches but not exact plots).

Example 10.1. Here are examples of implicit equations of curves. Five familiar ones first (see Figure 10.2, which shows a few points on the graph of each as well):

(a) Straight line: $ax + by + c = 0$

(b) Ellipse: $\frac{x^2}{a^2} + \frac{y^2}{b^2} = 1$

(c) Circle (special case of ellipse): $x^2 + y^2 = r^2$

(d) Parabola: $y = ax^2$

(e) Hyperbola: $\frac{x^2}{a^2} - \frac{y^2}{b^2} = 1$

Remark 10.1. An implicit equation is often written in the form $F(x, y) = G(x, y)$ with the RHS not necessarily equal to 0, but, of course, it can be rearranged as $F(x, y) - G(x, y) = 0$.

The following two exotic curves (Figure 10.3) may not be as familiar:

(f) Witch of Agnesi: $y(x^2 + 4) = 8$

(g) Lemniscate of Bernoulli: $(x^2 + y^2)^2 = x^2 - y^2$

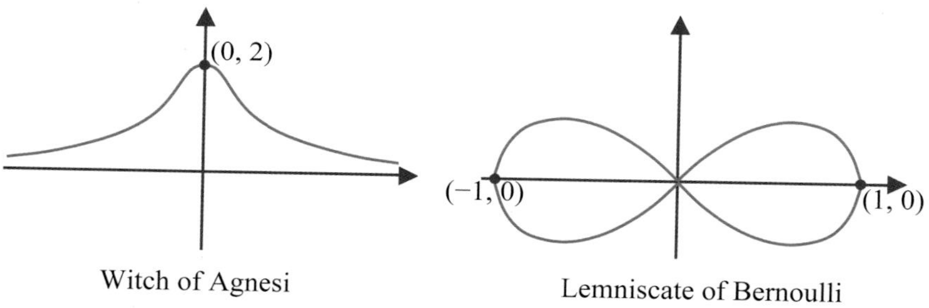

Figure 10.3: A couple of exotic curves.

Parametric

A plane curve c is specified *parametrically*, or *explicitly*, by the two equations

$$x = f(t), \ y = g(t), \text{ where } t \in T \tag{10.2}$$

if the points of c are those whose coordinates (x, y) satisfy $x = f(t)$ and $y = g(t)$, for some value of $t \in T$. Mathematically, $c = \{(f(t), g(t)) : t \in T\}$.

The functions f and g are called *parameter functions*, t the *parameter variable* and T the *parameter space* (also *parameter domain*). Typically, T is an interval of the real line $\mathbb{R}$, bounded or unbounded.

Example 10.2. Here are parametrizations of the curves earlier given implicitly in Example 10.1:

(a) Straight line: $x = t, \ y = -\frac{a}{b}t - \frac{c}{b}, \ t \in (-\infty, \infty)$, assuming $b \neq 0$.

(b) Ellipse: $x = a\cos t,\ y = b\sin t,\ t \in [-\pi, \pi]$.

Observe that the two different parameter values $t = -\pi$ and $t = \pi$ map to the same point $(-a, 0)$ on the ellipse, which is by no means illegal. A larger parameter space, e.g., $(-\infty, \infty)$, would cause even more overlap of parameter images. The half-open parameter interval $[-\pi, \pi)$ causes no overlap, but is a bit ungainly.

There's nothing special about $[-\pi, \pi]$ other than that it's symmetric about the origin. Any other closed interval of size 2π would do as well, e.g., $[0, 2\pi]$.

(c) Circle: $x = r\cos t,\ y = r\sin t,\ t \in [-\pi, \pi]$.

(d) Parabola: $x = t,\ y = at^2,\ t \in (-\infty, \infty)$.

(e) Hyperbola: $x = a\sec t,\ y = b\tan t,\ t \in [-\pi, -\pi/2) \cup (-\pi/2, \pi/2) \cup (\pi/2, \pi]$.

As you may check, the parameter space is simply $[-\pi, \pi]$ *minus* the two values $\pm\pi/2$, where sec and tan become "infinite".

(f) Witch of Agnesi: $x = 2t,\ y = \frac{2}{1+t^2},\ t \in (-\infty, \infty)$. Alternately, $x = 2\tan t,\ y = 2\cos^2 t,\ t \in (-\frac{\pi}{2}, \frac{\pi}{2})$.

(g) Lemniscate of Bernoulli: $x = \frac{\cos t}{1+\sin^2 t},\ y = \frac{\cos t \sin t}{1+\sin^2 t},\ t \in [-\pi, \pi]$.

Exercise 10.1. Prove that parametrizations of Example 10.2 are indeed those of the curves given implicitly in Example 10.1.
Hint: Plug the parametric forms for x and y into the implicit equation and verify the equality, e.g., for the circle

$$x^2 + y^2 = (r\cos t)^2 + (r\sin t)^2 = r^2(\cos^2 t + \sin^2 t) = r^2$$

Whereas an implicit equation $F(x, y) = 0$ for a curve gives a Boolean check for points "aspiring" to be on the curve, a parametric representation is more functional. If a curve c is given parametrically by the equations $x = f(t)$ and $y = g(t)$, where $t \in T$, one can think of the image $(f(t), g(t))$ of the parameter value t as traveling on c, as t travels in the parameter space. For example, as t goes from $-\pi$ to π, the point $(a\cos t, b\sin t)$ sweeps around the ellipse $\frac{x^2}{a^2} + \frac{y^2}{b^2} = 1$ from $(-a, 0)$ and back again.

In summary, while an implicit specification $F(x, y) = 0$ is ideal for *verifying* if a given point lies on a curve c, it's not as useful for the purpose of *generating* points on c; it is exactly the opposite in the case of a parametric specification such as $x = f(t)$, $y = g(t)$, $t \in T$.

Example 10.3.

(a) Verify if the points $(1, 0)$ and $(1, -1)$ lie on the Lemniscate of Bernoulli.

(b) Generate three distinct points on the Lemniscate of Bernoulli.

Answer: (a) Plugging $(1, 0)$ and $(1, -1)$ successively into the implicit equation $(x^2 + y^2)^2 = x^2 - y^2$, one sees that the first point lies on the curve, while the second doesn't.

(b) Plugging $t = 0$, $\pi/4$ and $\pi/2$ successively into the parametric equations

$$x = \frac{\cos t}{1 + \sin^2 t} \quad y = \frac{\cos t \sin t}{1 + \sin^2 t}$$

one gets the points $(1, 0)$, $(\frac{\sqrt{2}}{3}, \frac{1}{3})$ and $(0, 0)$ on the curve.

A parametric representation is to be preferred to an implicit for drawing as it enables the programmer to efficiently generate sample points on the curve. Going from one kind of representation to another often requires a bit of mathematical dexterity. The following two exercises are not particularly difficult though.

Exercise 10.2. An astroid, a curve traced by a fixed point on a circle rolling inside another circle of four times the diameter (see Figure 10.4), is given by the implicit equation

$$x^{\frac{2}{3}} + y^{\frac{2}{3}} - 1 = 0$$

Find parametric equations.

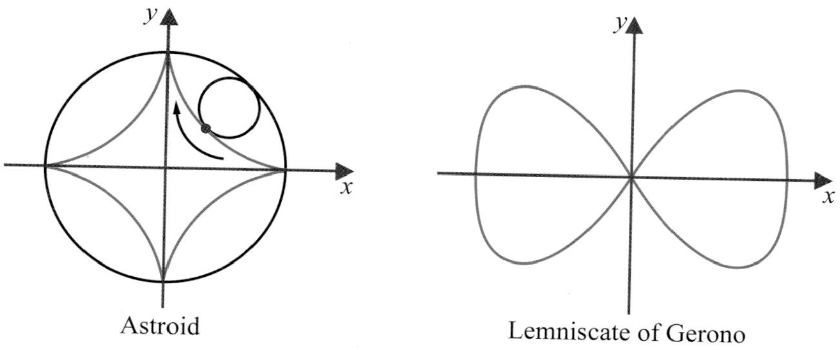

Figure 10.4: More exotic curves.

Exercise 10.3. Parametric equations for another exotic curve, the Lemniscate of Gerono (see Figure 10.4), are

$$x = \cos t, \ y = \cos t \sin t, \quad t \in [-\pi, \pi]$$

Find an implicit equation.

Remark 10.2. Neither the implicit nor parametric specification of a curve is ever unique. For example, the unit circle can be written implicitly as both $x^2 + y^2 = 1$ and $2x^2 + 2y^2 = 2$, or parametrically as $x = \cos t,\ y = \sin t,\ t \in [-\pi, \pi]$ and $x = \cos(t/2),\ y = \sin(t/2),\ t \in [-2\pi, 2\pi]$.

10.1.2 Specifying Space Curves

The extra dimension they have in which to move makes curves in 3-space – the real world – more interesting than their plane counterparts. Such curves are called *space curves*.

Implicit

The implicit specification of a space curve requires two equations:

$$\begin{aligned} F(x, y, z) &= 0\,, \\ G(x, y, z) &= 0 \end{aligned} \tag{10.3}$$

The reason for two equations rather than the one as in the case of a plane curve is as follows. $\mathbb{R}^3$ itself – unconstrained by any equations – is of dimension 3. However, each additional equation imposed reduces the resulting object's dimension by one. For example, points of $\mathbb{R}^3$ satisfying the one equation

$$x^2 + y^2 + z^2 - 1 = 0$$

make a sphere, a surface of dimension 2. Adding the equation of, say, the plane $x + y + z - 1 = 0$, one obtains the circle

$$\begin{aligned} x^2 + y^2 + z^2 - 1 &= 0, \\ x + y + z - 1 &= 0 \end{aligned}$$

which is a curve of dimension 1 at the intersection of the two (Figure 10.5). Generally, two equations imposed on $\mathbb{R}^3$ give an object of dimension $3-2 = 1$, a curve.

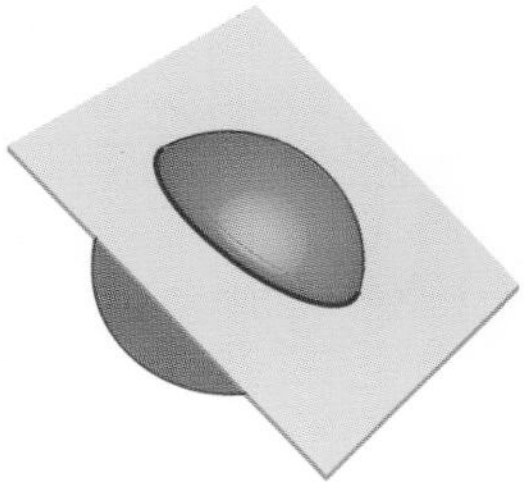

Figure 10.5: A sphere and a plane intersect in a circle.

If the implicit equation of a plane curve is already known to be $F(x, y) = 0$, then it can be written easily as a space curve by means of the two equations

$$\begin{aligned} F(x, y) &= 0, \\ z &= 0 \end{aligned}$$

Remark 10.3. We have used the term "dimension" without defining it formally. We'll not do so at this time as it would take us too far afield, but think intuitively of an object's dimension as the number of "independent directions of movement" – "degrees of freedom" would be apt as well – on it.

For example, there is only one independent direction of movement on a curve (mind that forward and backward are not independent, but merely the negative of one another). A surface allows two independent directions of movement. True, there are infinitely many directions of movement from any given point on a surface, but at most any two are independent. For example, take a point on a sphere – using only latitude and longitude one can represent any direction starting from it.

Exercise 10.4. What space curve is

$$\begin{aligned} x^2 + y^2 &= 1, \\ x + z &= 1 \end{aligned}$$

Describe or sketch the curve.

Exercise 10.5. What space curve is

$$\begin{aligned} x^2 + y^2 + z^2 &= 1, \\ x^2 + y^2 + z^2 - 2x &= 0 \end{aligned}$$

Describe or sketch the curve.

Parametric

The parametric, or explicit, specification of a space curve is similar to that of a plane one except, as one would expect, another parameter function is required to determine the z coordinate value:

$$x = f(t),\ y = g(t),\ z = h(t),\ t \in T \tag{10.4}$$

Any plane curve is a space curve as well, of course, and its parametric equations in 2-space are extended to 3-space by adding $z = 0$.

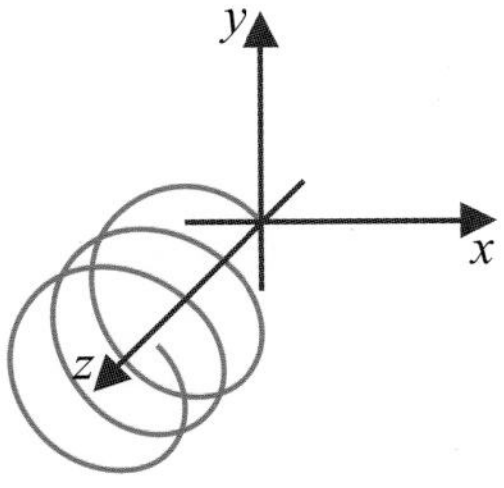

Figure 10.6: Helix.

Example 10.4. Parametric equations for a helix whose axis is along the z-axis (Figure 10.6) are

$$x = r\cos t,\ y = r\sin t,\ z = t,\ t \in \mathbb{R}$$

which, in fact, we used (slightly modified) to draw one in Section 2.8.2.

Example 10.5. Parametric equations for a general straight line in space are

$$x = a_1 t + b_1,\ y = a_2 t + b_2,\ z = a_3 t + b_3,\ t \in \mathbb{R}$$

where the $a_i, b_i, 1 \leq i \leq 3$, are constants.

Example 10.6. Give implicit equations for the helix of Example 10.4.

Answer:

$$\begin{aligned} x - r\cos z &= 0 \\ y - r\sin z &= 0 \end{aligned}$$

Exercise 10.6. Give parametric equations for the infinite straight line through (p_x, p_y, p_z) and (q_x, q_y, q_z). How is the parameter space restricted if we are interested only in the finite segment *between* the two points?

Exercise 10.7. Sketch using pencil and paper the curve, called a *conical helix*, specified parametrically by

$$x = t\cos t,\ y = t\sin t,\ z = t,\ \ t \in (-\infty, \infty)$$

10.1.3 Drawing Curves

Drawing a curve from its parametric equations is straightforward. We did this in Chapter 2 for a few particular curves, in particular, the circle, parabola and helix. We'll describe next the procedure for a general space curve c given parametrically by

$$x = f(t),\ y = g(t),\ z = h(t),\ \ t \in T \tag{10.5}$$

Assume that T is $[a, b]$, a closed interval. The point $(f(t), g(t), h(t))$ on c is denoted $c(t)$.

It's useful to imagine T as being part of the real line and to imagine c as lying in a "separate" 3-space. The parameter equations together $c(t) = (f(t), g(t), h(t))$ can then be thought to map the former to the latter or, more vividly, to lift and shape T into c. See Figure 10.7.

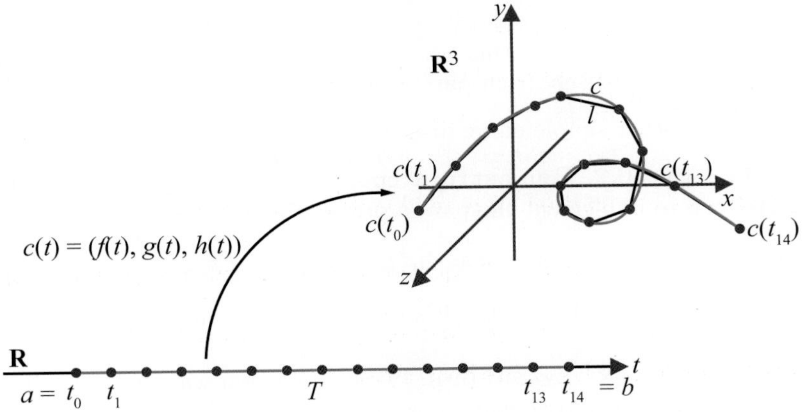

Figure 10.7: Parameter space $T = [a, b]$ mapped to a curve c. The sample grid on T, as well as its corresponding mapped sample on c, has 15 points (not all labeled). The polyline l connecting the mapped sample approximates c.

A sample

$$a = t_0 < t_1 < \ldots < t_n = b$$

of $n+1$ points from $[a, b]$ is called a *sample grid* on $[a, b]$ (note that the end points a and b are always included in the sample). It maps to a sample

$$c(t_0), c(t_1), \ldots, c(t_n)$$

of $n+1$ points of c, called the *mapped sample*.

The polyline l joining, successively, $c(t_0), c(t_1), \ldots, c(t_n)$ is an approximation of c. Each individual segment $c(t_{i-1})c(t_i)$, $1 \leq i \leq n$, of l approximates the arc of c between $c(t_{i-1})$ and $c(t_i)$. The sample grid and its corresponding mapped sample in Figure 10.7 contain 15 points each.

Keep in mind that a sample which is uniformly spaced along $[a, b]$ may *not* map to one which is uniformly spaced along c. For, the length of the arc of c between $c(t_{i-1})$ and $c(t_i)$ depends not only on the length of the interval $[t_{i-1}, t_i]$, but also on the "speed" of $c(t)$ with respect to t. We see next an example of this.

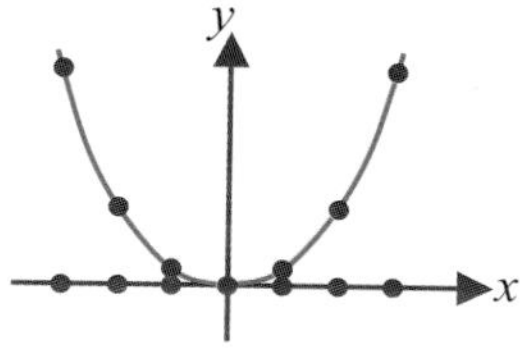

Figure 10.8: A uniformly sampled parabola $y = x^2$.

Example 10.7. Figure 10.8 shows the parabola $y = x^2$. The seven sample points on the real line are uniformly spaced, but their images on the curve are not because the rate of change of y with respect to x – equaling $\frac{dy}{dx} = 2x$ for the parabola – increases away from the origin.

Experiment 10.1. Compare the outputs of `circle.cpp`, `helix.cpp` and `parabola.cpp`, all drawn in Chapter 2.

The sample is chosen uniformly from the parameter space in all three programs. The output quality is good for both the circle – after pressing '+' a sufficient number of times for a dense enough sample – and the helix. The parabola, however, shows a difference in quality between its curved bottom and straighter sides, the sides becoming smoother more quickly than the bottom. In curves such as this, one may want to sample non-uniformly, in particular, more densely from parts of greater curvature. **End**

Here's another simple curve-drawing program.

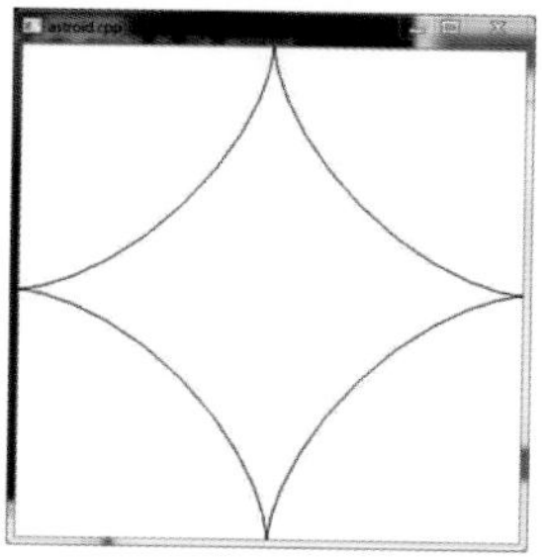

Figure 10.9: Screenshot of `astroid.cpp`.

Experiment 10.2. Run `astroid.cpp`, which was written by modifying `circle.cpp` to implement the parametric equations

$$x = \cos^3 t,\ y = \sin^3 t,\ z = 0,\ 0 \leq t \leq 2\pi$$

for the astroid of Exercise 10.2. Figure 10.9 is a screenshot. **End**

Exercise 10.8. (Programming) Draw the Lemniscate of Bernoulli with the help of the parametric equations given in Example 10.2(g).

Exercise 10.9. (Programming) Draw the conical helix of Exercise 10.7.

Exercise 10.10. (Programming) Draw a curve with a repeating pattern as close as possible to that of Figure 10.10(a). The two arcs of the "shark's fin" could be parts of circles. Your program should allow the user to specify the number of repetitions.

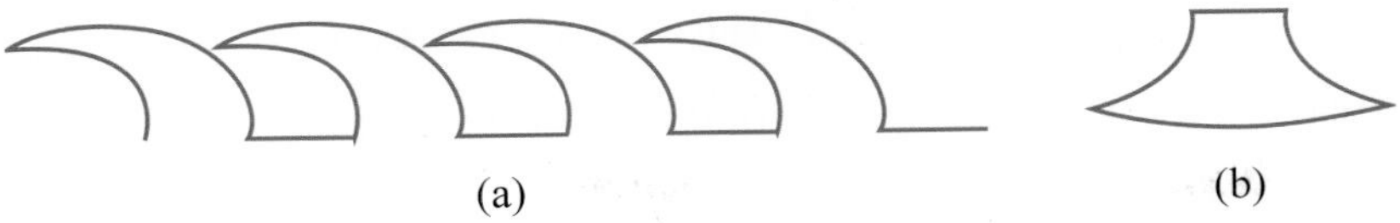

Figure 10.10: (a) Curve with a repeating pattern (b) Ax head.

Exercise 10.11. (Programming) Draw an ax head as in Figure 10.10(b).

Exercise 10.12. (Programming) The *twisted cubic* is a space curve given parametrically by the equations

$$x = t,\ y = t^2,\ z = t^3$$

Draw a part of it near the origin.

Exercise 10.13. (Programming) Animate the drawing of an astroid, as described in Exercise 10.2, as the curve traced by a point of a circle rolling inside another four times as large. The popular children's drawing toy called Spirograph can draw an astroid, among other curves.

Superellipses

A class of plane curves that generalizes both the ellipse and the astroid is that of the *superellipses*, invented by Lamé in 1818, given by the implicit equation:

$$\left|\frac{x}{a}\right|^n + \left|\frac{y}{b}\right|^n = 1$$

where a, b and n each is a positive constant.

Note: Because the exponent n can be fractional, the modulus signs are to avoid imaginaries if x or y is negative.

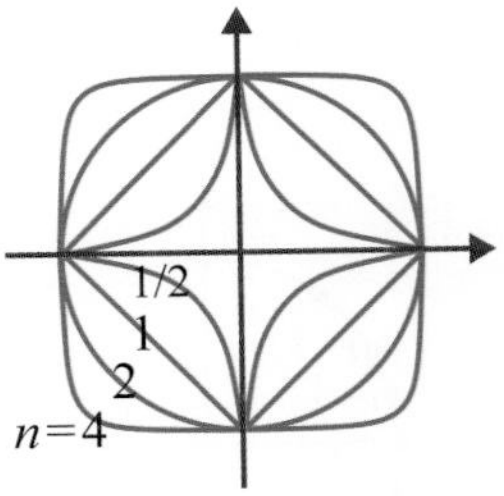

Figure 10.11: Supercircles $|x|^n + |y|^n = 1$, for $n = 1/2, 1, 2, 4$.

Figure 10.11 shows a few superellipses for $a = b = 1$ – when their equation is $|x|^n + |y|^n = 1$ – and different values of n. Generally, if $a = b$, the superellipse is called a *supercircle*. When $n = 1$ the supercircle is a square, when $n > 1$ it's convex outwards, and when $n < 1$ concave outwards.

Exercise 10.14. Justify the claim that superellipses generalize both ellipses and astroids.

Exercise 10.15. Deduce the parametric equations of a superellipse.

Exercise 10.16. (Programming) Write a program to draw a supercircle $|x|^n + |y|^n = 1$, allowing the user to choose n.

10.1.4 Polynomial and Rational Parametrizations

A *rational function* is the ratio of two polynomials. For example,

$$\frac{1+2t}{1-t+t^2-t^3} \quad \text{and} \quad \frac{x}{1+x} \tag{10.6}$$

are rational functions of t and x, respectively. A *polynomial function* is, of course, simply a special case of a rational function where the denominator is 1, e.g.,

$$1+2t-3t^2+4t^3 \quad \text{and} \quad x^3 \tag{10.7}$$

Unlike polynomial functions, rational functions may become undefined, which happens when their denominator vanishes. The first rational function of (10.6) is undefined at $t = 1$ and the second at $x = -1$.

If the parameter functions of a curve c are all rational, then it is said to be a *rational curve* and to have a *rational parametrization*; if they are, in fact, all polynomial, then c is a *polynomial curve* with a *polynomial parametrization*. E.g., Examples 10.2 (a) and (d) give polynomial parametrizations, while the first part of (f) a rational one.

The parametrization

$$x = r\cos t,\ y = r\sin t,\ t \in [-\pi, \pi]$$

of Example 10.2(c) of the circle $x^2 + y^2 = r^2$ uses trigonometric functions and is called, of course, a *trigonometric parametrization*. It turns out that there's an alternate rational parametrization of the circle, as the reader is asked to show in the following exercise.

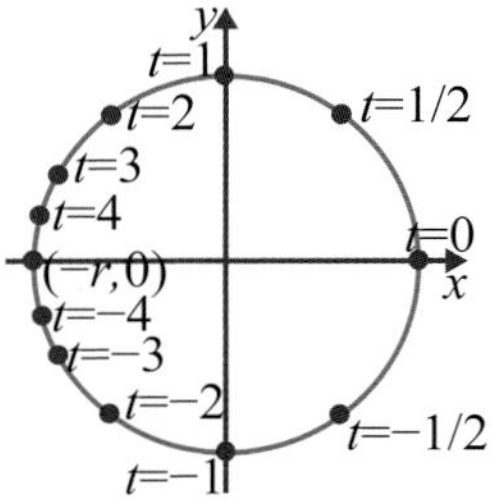

Figure 10.12: Points on a circle from a rational parametrization.

Exercise 10.17. Show, first, that

$$x = r\frac{1-t^2}{1+t^2} \quad \text{and} \quad y = r\frac{2t}{1+t^2}$$

satisfy $x^2 + y^2 = r^2$ for all values of t.

By plotting a few values of (x, y) for values of $t = 0, \pm 1, \pm 2, \ldots$ (see Figure 10.12) convince yourself that

$$x = r\frac{1-t^2}{1+t^2},\quad y = r\frac{2t}{1+t^2},\quad t \in (-\infty, \infty)$$

is a parametrization of the entire circle of radius r centered at the origin *minus* the point $(-r, 0)$. The point $(-r, 0)$, which the parametrization "cannot reach," is called a *singularity* of the curve (with respect to this particular parametrization).

Polynomial and rational parametrizations are often preferred to trigonometric ones in applications because the former can be computed exactly (up to round-off error) while a trigonometric function can at best be approximated by a series. For example, the power series $1 - x^2/2! + x^4/4! - x^6/6! + \ldots$ must be summed to some desired degree of accuracy to determine $\cos x$.

Exercise 10.18. (Programming) Draw a unit circle centered at the origin with the help of a rational parametrization. Avoid the problem of the singularity at $(-r, 0)$ in the previous exercise, as well as the infinite parameter domain there, by using the equations

$$x = \frac{1-t^2}{1+t^2}, \quad y = \frac{2t}{1+t^2}, \quad t \in [-1, 1]$$

to draw the right half semi-circle and

$$x = -\frac{1-t^2}{1+t^2}, \quad y = \frac{2t}{1+t^2}, \quad t \in [-1, 1]$$

to draw the left half.

Exercise 10.17 gave a rational parametrization of a circle. Is there a polynomial parametrization which might, in fact, be computationally better because it does not require the expensive division operation? The answer is no, as is shown in the following.

Example 10.8. Prove that the circle $x^2 + y^2 = r^2$ has no polynomial parametrization.

Answer: Suppose, if possible, that $x = f(t)$ and $y = g(t)$ is a polynomial parametrization of the circle. Then $f(t)$ and $g(t)$ are polynomial functions such that

$$f(t)^2 + g(t)^2 = r^2 \tag{10.8}$$

If $f(t)$ and $g(t)$ are both constants, i.e., neither contains a power of t, then $(f(t), g(t))$ is just one point and certainly cannot represent a circle. Therefore, either one or both of the two functions must contain a power of t. Let t^m be the highest power of t in $f(t)$ or $g(t)$. Write

$$f(t) = a_m t^m + a_{m-1} t^{m-1} + \ldots a_0 \quad \text{and} \quad g(t) = b_m t^m + b_{m-1} t^{m-1} + \ldots b_0$$

where at least one of a_m and b_m is non-zero. Then

$$f(t)^2 + g(t)^2 = (a_m^2 + b_m^2) t^{2m} + \text{ lower powers of } t$$

where the coefficient of t^{2m} is non-zero, in fact, positive, because at least one of a_m and b_m is non-zero.

We see, then, that the LHS of (10.8) contains a non-zero power of t, but, in this case, it cannot equal the RHS which is only scalar, proving that the circle $x^2 + y^2 = r^2$ indeed has no polynomial parametrization.

10.1.5 Conic Sections

The ellipse, parabola and hyperbola are well-known members of a special class of plane curves called *conic sections* or, simply, *conics*. A conic is nothing but the graph on the plane of a quadratic equation in two variables, typically written:

$$Ax^2 + Bxy + Cy^2 + Dx + Ey + F = 0 \tag{10.9}$$

(At least one of A, B and C should not be zero, or the equation is no longer that of a quadratic.)

Conditions on the coefficients determine the type of conic. If the quantity $B^2 - 4AC$, called the *discriminant* of the conic, is less than zero, then the conic is an ellipse, if it is zero then a parabola, and if it is greater than zero then a hyperbola. If $A = C$ are non-zero and $B = 0$ we get a circle, which is a special case of the ellipse. However, in all cases, there are *degenerate* instances when the equation is that of a point, straight line(s), or nothing at all, as the reader is asked to find for herself next.

Exercise 10.19. Show that the following are equations of degenerate conics by determining their graphs:

$$x^2 + 2y^2 + 1 = 0, \qquad x^2 + y^2 = 0, \qquad x^2 + 2xy + y^2 = 0, \qquad x^2 - y^2 = 0$$

In each case say as well if it is a degenerate ellipse, circle, parabola or hyperbola.

Conics arise frequently in design applications. Consequently, it's useful that they all have polynomial or rational parametrizations. In fact, any non-degenerate conic can be transformed by translation and rotation to one of the following four particular normalized forms (pictured in Figure 10.13):

Conic	Implicit	Polynomial or Rational Parametrization	Singu-larity
Ellipse	$\frac{x^2}{a^2} + \frac{y^2}{b^2} = 1$	$x = a\frac{1-t^2}{1+t^2}, y = b\frac{2t}{1+t^2}, t \in (-\infty, \infty)$	$(-a, 0)$
Circle	$x^2 + y^2 = r^2$	$x = r\frac{1-t^2}{1+t^2}, y = r\frac{2t}{1+t^2}, t \in (-\infty, \infty)$	$(-r, 0)$
Parabola	$y = ax^2$	$x = t, y = at^2, t \in (-\infty, \infty)$	None
Hyperbola	$\frac{x^2}{a^2} - \frac{y^2}{b^2} = 1$	$x = a\frac{1+t^2}{1-t^2}, y = b\frac{2t}{1-t^2}, t \in (-\infty, \infty) - \{-1, 1\}$	$(-a, 0)$

Geometric Construction

There is a rather neat geometric construction of the conics which, in fact, explains why they are called conic sections. Consider the double cone C formed from all the lines through the origin intersecting a circle c centered some distance vertically above the origin. See Figure 10.14(a).

Now, the section of C by a non-radial plane p aligned as in Figure 10.14(b) is a hyperbola (a *non-radial* line or plane is one that does not pass through the origin). A hyperbola is not the only curve that can be sectioned off a double cone, as the reader is asked to show next.

Exercise 10.20. Using paper and pencil draw three non-radial planes so that their intersections with a double cone are a circle, ellipse and parabola, respectively.

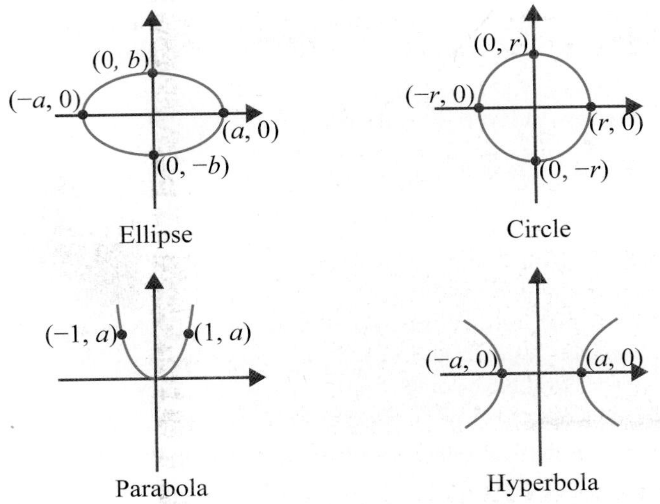

Figure 10.13: Conic sections.

Let's try to determine precisely the section of the double cone C by some given plane p. Assume that the half-angle at the vertex of C is θ and that the angle between p and the axis of C is ϕ. A typical cross-sectional view is drawn in Figure 10.14(c).

First, suppose that p is non-radial, as in Figure 10.14(c). We have then the following: *the section of C by p is an ellipse, parabola or hyperbola according as $\theta < \phi$, $\theta = \phi$ or $\theta > \phi$.* We'll leave the reader to convince herself of this fact by mentally rotating the plane p of Figure 10.14(c), where, in fact, currently $\theta < \phi$.

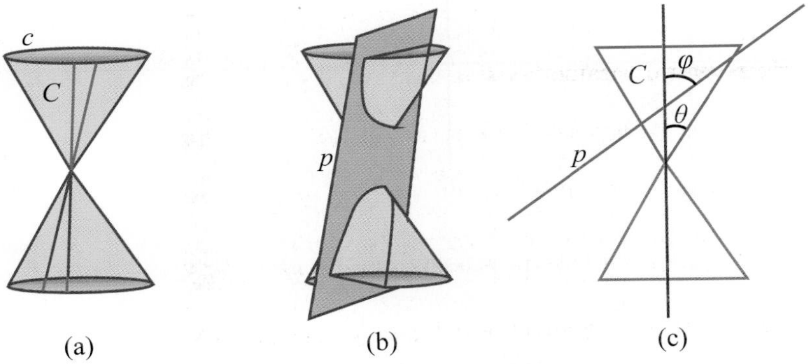

Figure 10.14: (a) A double cone C showing two of the lines through the origin lying on it (b) A hyperbolic section of C by a non-radial plane p (c) Cross-sectional view of a non-radial plane p intersecting C.

Exercise 10.21. Suppose now that p is *radial*. Determine the three different *degenerate* conic sections that arise, again according as $\theta <=> \phi$.

10.1.6 Curves More Formally

This section may be safely skipped on a first reading.

As we make our living in computer graphics drawing curves and surfaces, it's reasonable to try and understand some of their underlying mathematical formalism. We'll make a start with curves in this section. The theory of curves is a vast area within mathematics. Our objective in contrast is modest: to bring across a few definitions and results we believe most relevant to graphics applications, and in as intuitive a manner as possible.

We assume that you have some basic calculus. In other words, statements such as the function $f(x) = x^2$ is continuous and derivable (derivable, differentiable, same thing), that its derivative is the function $f'(x) = 2x$, and that its tangent at the point $(1, 1)$ has gradient 2 all make sense to you. Good!

Moving on, we'll first examine some "holes" in the rules given earlier to specify a curve. We'll then try to fix these and motivate in the process a more rigorous definition.

An implicit equation of the form $F(x, y) = 0$ on the plane may well specify an object that does not agree with our notion of what a curve should be. For example, $x^2 - y^2 = 0$ specifies two intersecting straight lines. See Figure 10.15(a). Writing $x^2 - y^2 = 0$ as $(x - y)(x + y) = 0$ explains the graph. And $x^2 + y^2 = 0$ defines just the single point $(0, 0)$, as in Figure 10.15(b)!

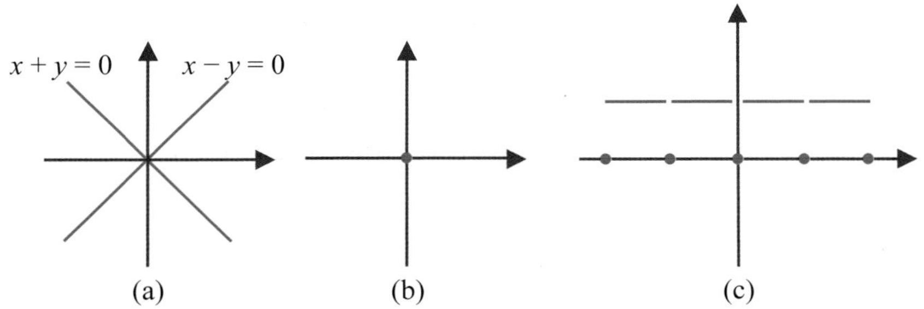

Figure 10.15: Non-curves: (a) $x^2 - y^2 = 0$ (b) $x^2 + y^2 = 0$ (c) $y = 0$, if x is an integer, 1 otherwise (gaps in the blue line indicate missing points).

Parametric equations may not fare better. The following is contrived certainly but makes the point:

$$x = t, \quad y = \begin{cases} 0, & t \text{ is an integer} \\ 1, & t \text{ is not an integer} \end{cases}, \quad t \in \mathbb{R}$$

define a disconnected union of points and straight segments (Figure 10.15(c)).

It seems, therefore, that the definition earlier of implicit and parametric curves simply by sets of equations might not be enough to agree always with our intuition at least of what a curve should be. To motivate a better definition, contemplate again the one-dimensional objects in Figure 10.1. All, except (f) and (g), seem to match the notion of a curve as being the *trajectory of a continuously-moving point*. We'll build on this simple observation.

In fact, we begin with the following definition of so-called C^0-continuity:

Definition 10.1. A real-valued function f defined on a closed interval T is said to be C^0-*continuous* or, simply, C^0, if it is continuous on T.

Remark 10.4. Yes, the definition simply re-christens what we know already as continuous. The reason to do this is that we'll soon be encountering so-called higher orders of continuity, to be called C^1, C^2, etc.

Remark 10.5. As the only functions that we consider are real-valued we won't explicitly say this any more.

Remark 10.6. A closed interval (like T in the preceding definition) is either bounded of the form $[a, b]$ or unbounded in one direction of the form $(-\infty, b]$ or $[a, \infty)$ or unbounded in both when it can only be $(-\infty, \infty)$.

The C^0-continuity of functions leads to the definition of C^0 curves:

Definition 10.2. Three C^0 functions f, g and h defined on a closed interval T give the following C^0-*parametrization* of a space curve c:

$$x = f(t),\ y = g(t),\ z = h(t),\ t \in T$$

The curve c itself is the set of all image points $\{(f(t), g(t), h(t)) : t \in T\}$. If a curve c has a C^0-parametrization, then it is said to be C^0-*continuous* or, simply, C^0.

Remark 10.7. An equivalent definition of a C^0 plane curve is obtained by dropping the "$z = h(t)$" term. Henceforth, we'll stick to 3D and give definitions only for space curves.

Example 10.9. The parametrization

$$x = t, \qquad y = \begin{cases} 0, & t \text{ is an integer} \\ 1, & t \text{ is not an integer} \end{cases}, \qquad t \in \mathbb{R}$$

given earlier is not C^0 as y is not a continuous function of t.

Example 10.10. The single point $(0, 0)$ defined implicitly by $x^2 + y^2 = 0$ is, strangely enough, a C^0 curve. For, it has the C^0-parametrization

$$x = 0,\ y = 0,\ t \in (-\infty, \infty)$$

on the plane, for the constant functions $x = 0$ and $y = 0$ are, in fact, continuous.

Example 10.11. All the parametrizations given in Example 10.2, except for the hyperbola, are C^0. The problem with the hyperbola is that its parameter space is not a (single) closed interval as we require. See the next exercise.

Exercise 10.22. Parametrize either of the two wings of the hyperbola of Example 10.2(e) so that each is a C^0 curve defined on a closed interval.

Example 10.12. Even though there are no actual parametrizations given there to go by, it's believable that, except for (f) and (g), the one-dimensional objects of Figure 10.1 are each a C^0 curve. The problem with (f) and (g) is that both seem composed of more than one trajectory.

Exercise 10.23. How about the astroid of Exercise 10.2 and the Lemniscate of Gerono of Exercise 10.3? Are they C^0?

Exercise 10.24. Is the graph of the following function C^0?

$$y = \begin{cases} 0, & x \leq 0 \\ 1, & x > 0 \end{cases}$$

Exercise 10.25. Is the graph of the function $y = |x|$ C^0?

Because of the continuity conditions, a C^0 curve is at least minimally well-behaved. However, it is not even guaranteed to possess a tangent at any given point (e.g., the one-point curve $x^2 + y^2 = 0$, or the hexagon of Figure 10.1(c) at its corners, do not seem to have meaningful tangents). In fact, it's often desirable that a curve be "smooth" in that, not only does it possess a tangent at every point, but the tangent turns continuously along the curve as well.

Note: We use the term "smooth" now as an informal descriptor. There are technical definitions of a "smooth function" and a "smooth curve" which will come up shortly.

The following two definitions are formulated to impose smoothness:

Definition 10.3. A function f defined on a closed interval T is said to be C^1-*continuous* or, simply, C^1 if its derivative f' exists and is continuous on T; equivalently, if f' exists and is C^0 on T.

Definition 10.4. Three C^1 functions f, g and h defined on a closed interval T give the following C^1-*parametrization* of a curve c:

$$x = f(t), \; y = g(t), \; z = h(t), \; t \in T$$

The curve c itself is the set of all image points $\{(f(t), g(t), h(t)) : t \in T\}$. If a curve c has a C^1-parametrization, then it is said to be C^1-*continuous* or, simply, C^1.

If, additionally, the three derivatives f', g' and h' never vanish together at any point of $[a, b]$, then the parametrization is said to be *regular*, and c is said to be a *regular curve*.

Because derivability implies continuity, C^1 curves and regular curves are C^0 as well. Regularity is "nice enough" for most CG applications. Why? Because regularity assures the smoothness of a curve c in the following sense:

(a) The tangent line to c at any point $c(t) = (f(t), g(t), h(t))$ exists. In fact, it is parallel to the vector $c'(t) = (f'(t), g'(t), h'(t))$, which is non-zero because the derivatives of the parameter functions do not vanish simultaneously (see Figure 10.16(a)). The existence of a tangent line everywhere on c means that it has a well-defined *direction* at every point.

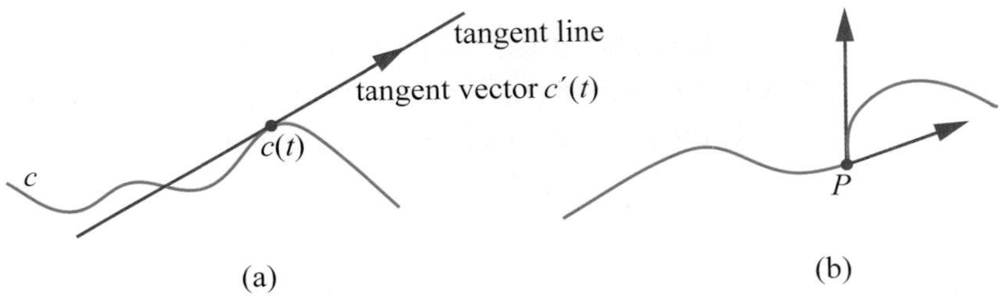

Figure 10.16: (a) A smooth curve (b) A non-smooth curve with a corner at P where the tangent changes direction abruptly.

(b) Since f', g' and h' are continuous, $c'(t) = (f'(t), g'(t), h'(t))$ is continuous along c, which means that, not only does the tangent line exist at all points of c, it turns continuously along c as well (intuitively, this means that c cannot have a corner where its direction changes abruptly as in Figure 10.16(b)).

The upshot is that a regular curve appears smooth when drawn.

It's precisely the non-vanishing property of $c'(t)$ that is not guaranteed by mere C^1-continuity, as opposed to regularity, as we see in the following example.

Example 10.13. The astroid of Exercise 10.2 has parametric equations

$$x = \cos^3 t, \ y = \sin^3 t, \ t \in [0, 2\pi]$$

As $x = \cos^3 t$ and $y = \sin^3 t$ are continuous functions of t, the astroid is C^0. Now

$$\frac{dx}{dt} = -3\cos^2 t \sin t, \quad \frac{dy}{dt} = 3\sin^2 t \cos t$$

which are continuous functions of t as well, proving that the astroid is C^1 as well. However, as both $\frac{dx}{dt}$ and $\frac{dy}{dt}$ vanish at $t = 0$, $\frac{\pi}{2}$, π and $\frac{3\pi}{2}$, the astroid is not regular. In fact, it has cusps at precisely these four parameter values. Intuitively, as well, one sees that a point traveling along the astroid has to abruptly reverse direction on reaching a cusp.

Exercise 10.26. How about the Lemniscate of Gerono of Exercise 10.3? Is it C^0, C^1, regular?

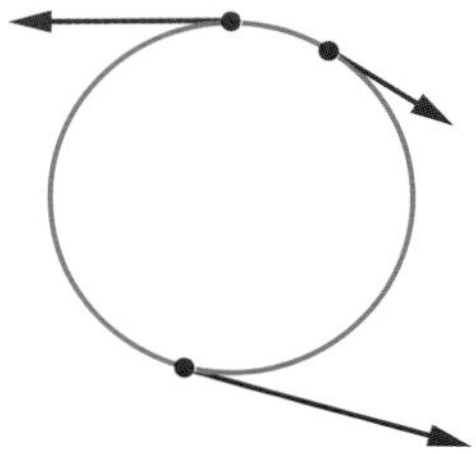

Figure 10.17: Tangent vectors to a circle.

The (non-zero) vector $c'(t) = (f'(t), g'(t), h'(t))$ is called a *tangent vector* to a regular curve $c(t) = (f(t), g(t), h(t))$. We say *a* tangent vector because any non-zero multiple of $c'(t)$, i.e., a vector collinear with it, is tangent at $c(t)$ as well. For example, Figure 10.17 shows tangent vectors, intentionally drawn of varying lengths and inconsistently oriented, at three points of a circle.

Example 10.14. Consider the helix given by the parametrization

$$x = \cos t,\ y = \sin t,\ z = t, t \in \mathbb{R}$$

A tangent vector at the point $(\cos t, \sin t, t)$ of the helix is by differentiation $(-\sin t, \cos t, 1)$, which never vanishes, so the helix is regular.

Exercise 10.27. The stationary curve

$$x = 0,\ y = 0,\ t \in (-\infty, \infty)$$

which is just the point $(0, 0)$, we saw earlier to be C^0. Is it C^1? Regular?

Exercise 10.28. What about the hexagon of Figure 10.1(c)? Is it C^0? C^1? Imagine a parametrization for it by "stringing" together parametric equations for its straight sides.

Exercise 10.29. What is a tangent vector to the graph of the function $y = f(x)$ on the xy-plane at the point $(x, f(x))$? Assume f to be differentiable.

Exercise 10.30. Is the graph of the function

$$y = \begin{cases} x^3, & x < 0 \\ x^2, & x \geq 0 \end{cases}$$

C^0? C^1? Regular? The point of interest obviously is the origin.

Exercise 10.31. (**Programming**) Animate the non-regularity of the astroid. In particular, draw the asteroid and animate its tangent vector, moving along the curve with changing t, as an arrow from $(\cos^3 t, \sin^3 t)$ to $(\cos^3 t - 3\cos^2 t \sin t,\ \sin^3 t + 3\sin^2 t \cos t)$.

The tangent vector shrinks to zero at each cusp and grows again as it leaves the cusp.

We can loosen up the definition of regularity to be a bit more inclusive.

Definition 10.5. A curve c is said to be *piecewise regular* if it can be made by sequentially joining a finite number of regular curves *end to end*.

In other words, a piecewise regular curve is regular except, possibly, for a finite number of corners inside it.

Example 10.15. Of those 1D objects depicted in Figures 10.1, the hexagon (c) is piecewise regular but not regular, while (f) and (g) are not even piecewise regular. The others are all regular curves.

Exercise 10.32. What about the curves of Figures 10.3 and 10.4? Identify those that are piecewise regular but not regular.

We define next a regular one-dimensional object as composed of pieces that are each a regular curve, except that they are not required to be joined end to end as for a piecewise regular curve.

Definition 10.6. A *regular one-dimensional object* is a finite union of regular curves.

In other words, even though composed of pieces that are regular, a regular one-dimensional object may not have the property of a curve that it can be continuously traversed end to end.

Example 10.16. Figures 10.1(f) and (g) are regular one-dimensional objects, but not piecewise regular curves. So are the letters 'K', 'Q' and 'X'.

We have the obvious proper inclusions

regular curves $\subset$ piecewise regular curves $\subset$ regular 1D objects

Remark 10.8. Curves arising in real life – strings, wires, rubber bands, edges of a car or aircraft – are almost invariably piecewise regular, if not regular. Likewise, one-dimensional objects we see around us are almost all regular one-dimensional.

It is important to keep in mind that a curve is regular (or C^0, or C^1) *if* there is *some* parametrization according to the respective definition. For example, the curve c on the plane given by the parametric equations

$$x = y = t, \quad t \in (-\infty, \infty) \tag{10.10}$$

is regular as $c'(t) = (1, 1)$ for all t. (Yes, it's the straight line $y = x$ drawn in Figure 10.18.)

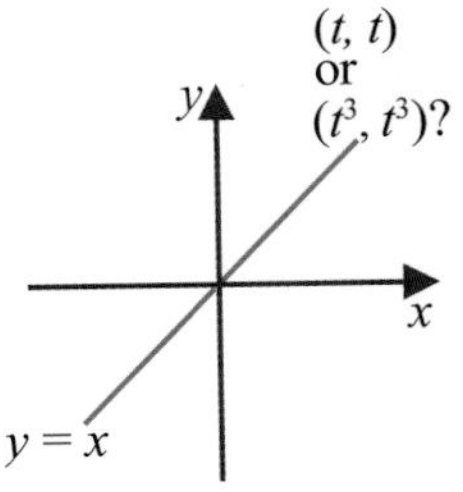

Figure 10.18: Good and bad parametrizations.

However, the same straight line is defined by the cubic equations

$$x = y = t^3, \quad t \in (-\infty, \infty) \tag{10.11}$$

but now regularity is "lost" at the origin (verify)!

In fact, to preempt the issue of finding the "best" parametrization, mathematical texts often define a curve to be the set of parameter functions itself, rather than the image, so, e.g., (10.10) and (10.11) would actually represent different curves, thus avoiding ambiguity over continuity.

If the reader is now wondering, no, there is no parametrization which will make either the hexagon or the astroid regular, though we'll not try to prove these facts.

The following intuitive proposition suggests how to join two regular curves end to end so that their union is regular.

Proposition 10.1. *If two regular curves c_1 and c_2 meet at the point P, which is an endpoint of both, and if their tangent vectors at P are collinear, then the curve $c_1 \cup c_2$ is regular as well (Figure 10.19).*

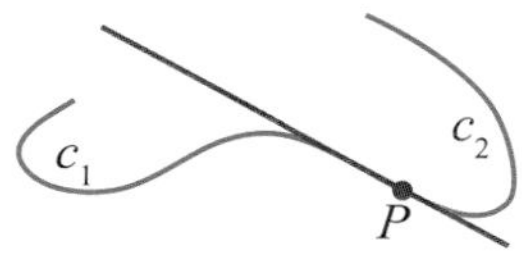

Figure 10.19: Regular curves that share a tangent line at a common endpoint join to make one regular curve.

Proof. We leave it to the reader. □

Hint: One must find a regular parametrization for the union $c_1 \cup c_2$. Change the parametrization of one of the curves, say c_1, from $t \mapsto c_1(t)$ to $t \mapsto \alpha t \mapsto c_1(\alpha t)$, choosing the constant α so that c_1's tangent vector at P becomes identical to that of c_2 (such rescaling of the parameter necessitates that the parameter interval be resized as well). Finally, "move" one parameter interval to abut the other.

Exercise 10.33. Prove that the two helixes

$$x = \cos t, \ y = \sin t, \ z = t, \ t \in [0, 10\pi]$$

and

$$x = 2 + \cos t, \ y = -\sin t, \ z = t - \pi, \ t \in [\pi, 10\pi]$$

meet at their common endpoint $(1, 0, 0)$, corresponding to $t = 0$ of the first curve and $t = \pi$ of the second, and share there a tangent vector. Can you come up with a single parametrization for the union of the two?

As expected, there are higher orders of continuity (C^0 is said to be zero-order, C^1 first-order) one can define as well, pretty much in the obvious manner:

Definition 10.7. A function f defined on a closed interval T is said to be C^m-*continuous* or, simply, C^m, where $m \geq 1$, if all of its derivatives of order m and less exist and are continuous on T.

A function f that is C^m-continuous for all m is said to be C^∞-*continuous* or, simply, C^∞. C^∞ functions are also called *smooth*.

Definition 10.8. Three C^m functions (m can be ∞ as well) f, g and h defined on a closed interval T give the following C^m-*parametrization* of a curve c:

$$x = f(t), \ y = g(t), \ z = h(t), \ t \in T$$

The curve c itself is the set of all image points $\{(f(t), g(t), h(t)) : t \in T\}$. If a curve c has a C^m-parametrization, then c is said to be C^m-*continuous* or, simply, C^m. If it has a C^∞-parametrization, then c is said to be *smooth*.

Remark 10.9. C^m-continuity implies C^n continuity for any $n < m$.

Remark 10.10. It is usual to assume regularity in addition to C^m-continuity, if $m \geq 1$.

Exercise 10.34. How continuous is a polynomial curve, in other words, what is the maximum order of continuity it possesses?

C^2 is about the highest order of continuity which can be distinguished visually. Even so, the lack of C^2-continuity is not as easy for the eye to catch as the lack of C^0 or C^1-continuity, which is, typically, obvious. The labeled continuity of all the curves of Figure 10.20, except the third, is probably easy to understand.

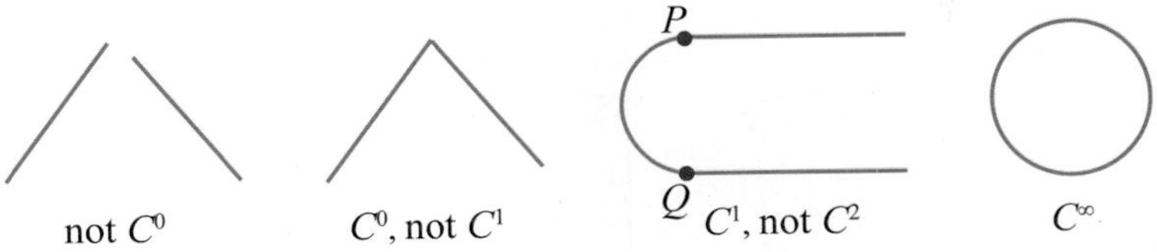

Figure 10.20: Various orders of continuity.

The third one is C^1-continuous but loses C^2-continuity at the two points P and Q where the half-circle meets straight segments, because the tangent stops turning. More explicitly, the tangent vector which is rotating at a uniform speed along the half-circle "suddenly" stops rotating altogether when it crosses into one of the straight segments, so that its rate of change drops from uniform to zero, giving rise to C^2-discontinuity.

An interesting application of C^2-continuity arises in planning the motion of a camera. We ask the reader to see this for herself in the following two exercises.

Exercise 10.35. Verify that the graph of the function (encountered earlier in Exercise 10.30)

$$y = \begin{cases} x^3, & x < 0 \\ x^2, & x \geq 0 \end{cases}$$

is not C^2 because of a second-order discontinuity at the origin.

Exercise 10.36. (Programming) Use `gluLookAt()` to simulate the view of a simple scene (populated, say, by spheres) from a camera moving along the graph of the preceding exercise and pointing always along its tangent. See Figure 10.21. Move the camera by uniformly incrementing its x-value at each time step.

The viewer perceives a jolt as the camera passes the origin. The reason is as follows. Even though the *path* of the camera is smooth, in that it is C^1, the *direction* of the camera, which is along the *tangent* to this path, does not turn smoothly past the origin because of the C^2-discontinuity there. Moreover, the location of the camera and its direction together determine the scene, so a discontinuity in either is echoed in the animation.

Bottom line: (Regular) C^1 is good enough for a curve to appear smooth, while smooth camera movement should be (regular) C^2.

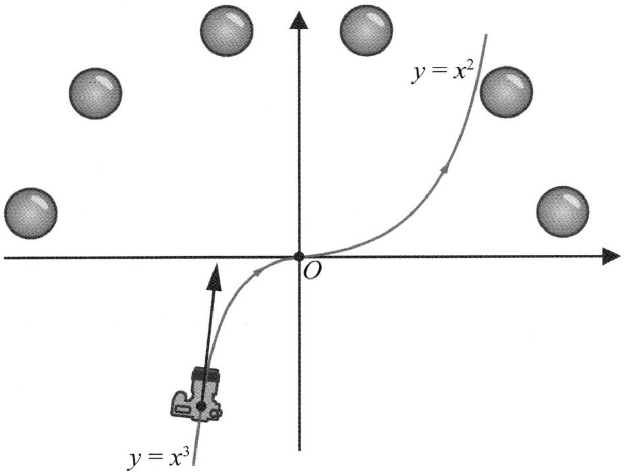

Figure 10.21: Camera moving along a path with a C^2-discontinuity at the origin O.

10.2 Surfaces

Two-dimensional objects are composed of surfaces. Analogously to the situation for one-dimensional objects, parts of a two-dimensional object which cannot be drawn exactly using triangles – OpenGL's fundamental 2D primitives as we know – must be approximated.

We'll discuss two-dimensional objects in an informal taxonomy ordered by increasing complexity from the point of view of drawing.

10.2.1 Polygons

The simplest two-dimensional object is the familiar polygon. By a polygon we shall always mean, unless stated otherwise, a *simple planar polygon*, i.e., one whose boundary lies on a plane and consists of a single component which is a non-self-intersecting line loop. All the polygons in Figure 10.22 are planar. However, (a) and (b) are non-simple polygons, while (c) and (d) are simple. A simple planar polygon can be equivalently described as a connected planar surface bounded by straight edges, and without any holes.

A convex polygon, e.g., Figure 10.22(c), can be drawn as a single `GL_POLYGON` primitive. A non-convex polygon should be drawn after decomposing it into convex pieces, otherwise (recall from Section 8.3) it may not render correctly. Figure 10.22(d) is an example of a non-convex polygon decomposed into triangles, in other words, *triangulated*. It's recommended, in fact, that *all* polygons, convex or otherwise, be first triangulated and then drawn using `GL_TRIANGLE*` primitives.

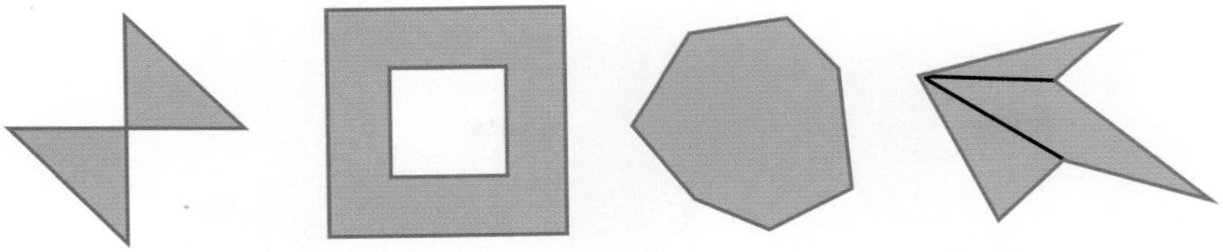

Figure 10.22: (a) and (b) Non-simple planar polygons (c) and (d) Simple planar polygons (what we call polygons).

10.2.2 Meshes

The next simplest kind of two-dimensional object is a *polygonal mesh* or, simply, *mesh*, also called a *polyhedral surface*. A mesh is a union of polygons satisfying the following two conditions:

1. Any two polygons in the union are either disjoint or intersect in a vertex of both or intersect in an edge of both.

   Note: This is a repetition of the condition for a collection of triangles to be a triangulation (see Definition 8.1) and motivated likewise to ensure deterministic rendering.

2. The "neighborhood around each vertex is sheet-like".

   We said the taxonomy would be informal. We'll not try to define what it means exactly for a neighborhood around a vertex to be sheet-like, leaving it instead to the reader's intuition with a few suggestive examples coming next.

Figure 10.23(a) is part of a hexagonal tiling of the plane with the shaded piece in the middle missing. Figure 10.23(b) is the surface of a glass with a hexagonal base and six rectangular walls. Figure 10.23(c) is the surface of an octahedron. Evident from the drawings themselves, all three objects are unions of polygons satisfying the first condition above. Moreover, for any vertex V belonging to any one of them, one can imagine fitting a small rubber sheet exactly onto the surface in an area around V, for an informal verification of the second condition above. Therefore, Figures 10.23(a)-(c) are meshes.

Both Figures 10.23(d) and (e) clearly satisfy the first condition to be a mesh as well. However, Figure 10.23(d) fails the second condition because it's crimped at U – no rubber sheet, no matter how pliable, can be squeezed to a point. Figure 10.23(e), which consists of three rectangles sharing an edge, is not a mesh either because it has multiple panels around W. One would have to tear a sheet into pieces to cover a neighborhood of W.

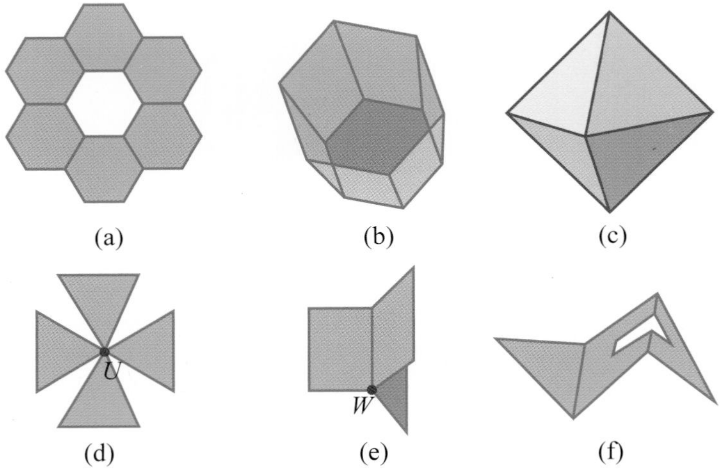

Figure 10.23: (a), (b) and (c) Meshes (d) and (e) Not meshes: parts around vertices V and W are not sheet-like (f) Your call.

Exercise 10.37. Is Figure 10.23(f) a mesh?

Remark 10.11. If you must know, in mathematical terms condition 2 above is for a mesh to be a so-called topological manifold, which guarantees a certain respectability to shapes that a mesh can take.

The polygons comprising a mesh are its *faces*. The *boundary* of a mesh consists of those edges with a polygon on only one side. The mesh of Figure 10.23(a) has two boundary components (one inside bordering the shaded missing piece and the other on the outside); that of Figure 10.23(b) has one boundary component (the rim of the glass); and that of Figure 10.23(c) has no boundary. A mesh with no boundary, a *closed* mesh as such is called, bounds a solid figure. For example, the closed mesh of Figure 10.23(c) bounds an octahedron.

It's usual to require that the faces of a mesh be convex polygons. In fact, from an OpenGL point of view it's best they all be triangles. A mesh, all of whose faces are triangles, is a *triangular mesh*. Of course, any mesh can be made triangular by triangulating every face.

Drawing a mesh is, of course, simply a matter of drawing each of its polygonal faces. Again, the recommendation is that each face be triangulated first, if it's not already a triangle, then drawn using `GL_TRIANGLE*` primitives. Because of the first condition for a mesh, the rendering is consistent whatever order the faces be drawn (for why, see the discussion in Section 8.1 of the reasons for the rules for a triangulation).

In fact, a moment's consideration of its primitives indicates that the *only* 2D objects OpenGL can draw *exactly* are meshes and unions of meshes. Others have to be approximated. So for the remaining two classes of 2D objects we will be trying to find good mesh approximations.

Example 10.17. Figures 10.23(d) and (e) are not meshes as we have seen. But are they a union of meshes?

Answer: Easily yes. For example, Figure 10.23(d) is the union of four meshes, each consisting of a single triangle.

10.2.3 Planar Surfaces

A planar surface is a generalization of a polygon. A polygon is a connected planar surface bounded by straight edges, and with no holes, while a general planar surface has no such restrictions. It may comprise multiple components and contain holes, while its boundary may be composed of both straight and curved parts; of course, the whole must lie on one plane. See Figure 10.24.

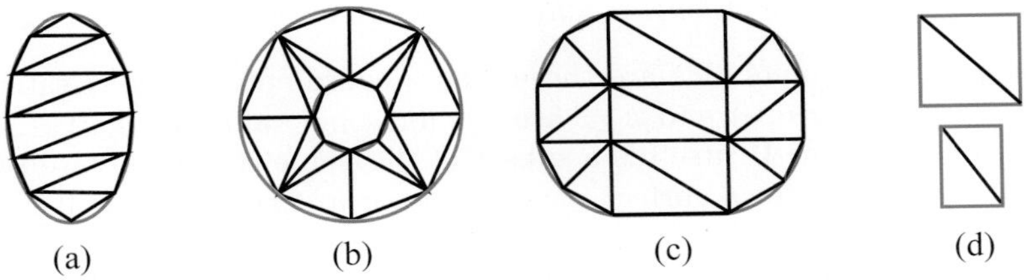

Figure 10.24: Planar surfaces with colored boundary; the last one has two components. The black edges belong to approximating meshes.

The drawing of a general planar surface can be reduced to that of drawing an approximating mesh by the following approach:

1. Apply the technique of Section 10.1.3 to make polyline approximations of the curved edges on the surface's boundary. Together with the existing straight edges, these polylines then bound a (possibly, non-simple) planar polygon, or a union of such if there are multiple components, approximating the surface.

2. Triangulate the approximating polygon(s). The result is a triangular mesh approximation of the surface.

The black edges in Figures 10.24(a)-(d) clarify the approach.

Exercise 10.38. (Programming) Draw a rounded rectangle as in Figure 10.24(c).

Figure 10.25: Wooden chair.

Exercise 10.39. (Programming) Draw a likeness in wireframe of the chair in Figure 10.25.

First of all, assume all the panels to be of zero thickness. So, for example, the seat is a flat rounded quad, while all four legs are bow-shaped planar surfaces. The only part that's evidently not planar, even at zero thickness, is the back rest, so, in fact, replace it with a rounded rectangle. Be sure to use the symmetries: the two front legs are identical, as are the two back legs, so make only one of each.

10.2.4 General Surfaces

More general surfaces, which may be neither planar nor a union of polygons, are drawn by approximation by triangular meshes as well. But, let's see how a general surface s is specified in the first place. An *implicit specification* consists of an equation

$$F(x, y, z) = 0 \tag{10.12}$$

such that points of s are those whose coordinates (x, y, z) satisfy this equation. A *parametric*, or *explicit*, *specification* consists of three equations

$$x = f(u, v),\ y = g(u, v),\ z = h(u, v),\ \text{where } (u, v) \in W \tag{10.13}$$

In this case, the points of s are those whose coordinates (x, y, z) satisfy $x = f(u, v)$, $y = g(u, v)$ and $z = h(u, v)$, for some value of $(u, v) \in W$. The parameter space W itself is a subset of the plane $\mathbb{R}^2$. There are two parameter variables for a surface, versus one for a curve, because it is of dimension two.

The point $(f(u, v), g(u, v), h(u, v))$ on the surface s is often denoted $s(u, v)$.

Example 10.18. The (infinitely long circular) cylinder with its axis along the z-axis, and a circular cross-section of radius 1, is given by the implicit equation

$$x^2 + y^2 = 1$$

It's also given by the parametric equations

$$x = \cos u,\ y = \sin u,\ z = v,\quad (u, v) \in [-\pi, \pi] \times (-\infty, \infty)$$

where the parameter space is an infinitely long rectangular subset of the plane.

Figure 10.26 shows a finite part of the parameter space, namely, the rectangle bounded by the lines $u = \pm\pi$ and $v = \pm 1$, as well as the corresponding part of the cylinder.

If the parameter variables are u and v, then the image on the surface s, of a straight line $v = \beta$ in the parameter space, is a curve, denoted $s(v = \beta)$, and called a *u-parameter curve* of s; in other words, a u-parameter curve is traced on s by fixing the parameter v and varying u. Similarly, the image $s(u = \alpha)$ on s, of the line $u = \alpha$, is called a *v-parameter curve*.

The u-parameter curves of the cylinder of the preceding example are circles, while the v-parameter curves are vertical straight lines. One curve of either class is shown on the right of Figure 10.26.

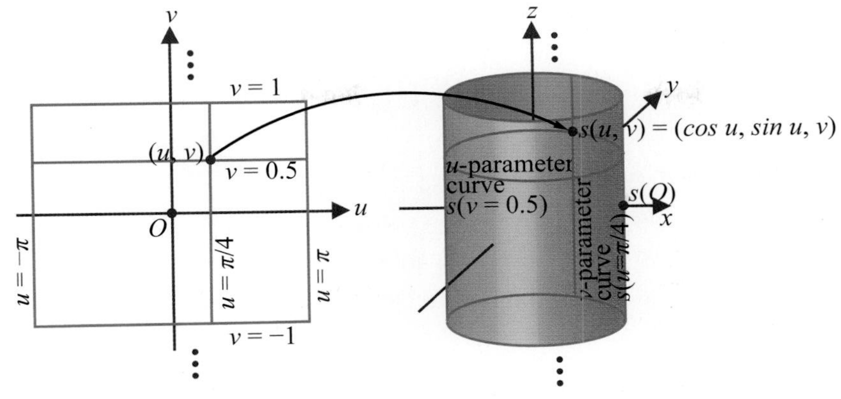

Figure 10.26: Parametric mapping of a circular cylinder s.

One can imagine the surface of s either as the union of its u-parameter curves $s(v = \beta)$ as β varies, or the union of its v-parameter curves $s(u = \alpha)$ as α varies, each over its respective range. For example, the infinite cylinder of the preceding example is the union of its u-parameter circles as the corresponding parameter β varies in $(-\infty, \infty)$, while the finite part depicted in Figure 10.26 as β varies in $[-1, 1]$. More graphically, the infinite cylinder is *swept* by its u-parameter circle as β changes from $-\infty$ to ∞. Complementarily, it's swept by its v-parameter straight lines as α varies from $-\pi$ and π.

Exercise 10.40. The implicit equation $x^2 + y^2 = 1$ of the cylinder of the preceding example did not involve z at all. In fact, it's the equation of a circle on the xy-plane simply applied to 3-space. Generally, if the implicit equation of a plane curve is $F(x, y) = 0$, then what surface is represented by the same equation in 3-space?

10.2.5 Drawing General Surfaces

The strategy to approximate a surface is similar to that for a curve. However, instead of straight line segments to approximate sub-arcs of a curve, triangles are used to approximate small patches of the surface and a triangular mesh the entire surface. We make the assumption that the parametric specification of a surface s is given as

$$x = f(u, v), \ y = g(u, v), \ z = h(u, v), \quad (u, v) \in W \qquad (10.14)$$

where the parameter space W is the plane rectangle $[a, b] \times [c, d]$. Think then of the parametric equations as lifting and shaping the rectangle W from uv-space into s in xyz-space, e.g., Figure 10.26 shows the shaping of the rectangle $[-\pi, \pi] \times [-1, 1]$ into a circular cylinder.

Sample W at the $(p+1)(q+1)$ points

$$(u_i, v_j),\ 0 \leq i \leq p,\ 0 \leq j \leq q$$

of a *rectangular sample grid* where

$$a = u_0 < u_1 < \ldots < u_p = b$$
$$c = v_0 < v_1 < \ldots < v_q = d$$

The *mapped sample*

$$s(u_i, v_j),\ 0 \leq i \leq p,\ 0 \leq j \leq q$$

of $(p+1)(q+1)$ points of s are used as vertices of a triangular mesh approximation of s. This mesh consists of the following $2pq$ triangular faces:

$$\triangle\, s(u_i, v_j)\, s(u_{i+1}, v_j)\, s(u_i, v_{j+1}) \text{ and } \triangle\, s(u_i, v_{j+1})\, s(u_{i+1}, v_j)\, s(u_{i+1}, v_{j+1}),$$

for $0 \leq i \leq p-1,\ 0 \leq j \leq q-1$. (Note that $\triangle\, ABC$ denotes the triangle with vertices at A, B and C.)

Each face approximates a patch of the surface. In particular, the face with corners at the images of the vertices of a grid triangle approximates the image of that triangle on the surface. Spelling this out for the two triangles listed above: the face $\triangle\, s(u_i, v_j)\, s(u_{i+1}, v_j)\, s(u_i, v_{j+1})$ approximates the patch $s(\triangle\,(u_i, v_j)\,(u_{i+1}, v_j)\,(u_i, v_{j+1}))$ which is the image on s of the grid triangle $\triangle\,(u_i, v_j)\,(u_{i+1}, v_j)\,(u_i, v_{j+1})$; likewise, the face $\triangle\, s(u_i, v_{j+1})\, s(u_{i+1}, v_j)$ $s(u_{i+1}, v_{j+1})$ approximates the patch $s(\triangle\,(u_i, v_{j+1})\,(u_{i+1}, v_j)\,(u_{i+1}, v_{j+1}))$.

It's easiest to understand this visually. Let's use again the circular cylinder

$$x = \cos u,\ y = \sin u,\ z = v, \quad (u, v) \in [-\pi, \pi] \times [-1, 1]$$

from the previous example. Refer to Figure 10.27. Sample points in the parameter space are seen at the upper left and the corresponding mapped sample points on the cylinder at the upper right (here, $p = 6$ and $q = 4$). The triangles of the mesh along a band of the cylinder are shown at the upper right as well.

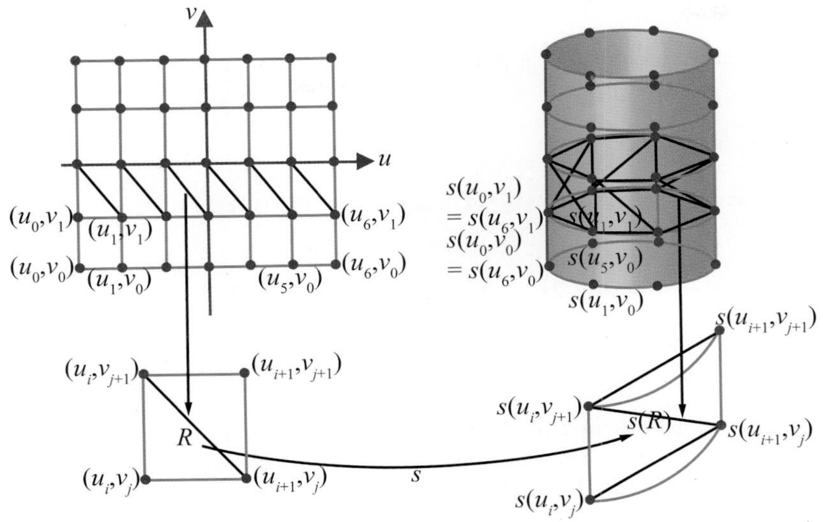

Figure 10.27: Triangular mesh approximation of a circular cylinder. *Upper*: A uniform sample grid on the parameter rectangle and its corresponding mapped sample on the cylinder. Only a few points are labeled. Vertices of a triangle strip on the rectangle maps to those of a strip approximating a band of the cylinder. *Lower*: A map from a grid rectangle to a patch of the cylinder.

The lower left and right diagrams are blow-ups, respectively, of a grid rectangle R, with corners at (u_i, v_j), (u_{i+1}, v_j), (u_{i+1}, v_{j+1}) and (u_i, v_{j+1}), and the patch $s(R)$ of the cylinder which is its image. The two triangles $s(u_i, v_j)s(u_{i+1}, v_j)s(u_i, v_{j+1})$ and $s(u_i, v_{j+1})s(u_{i+1}, v_j)s(u_{i+1}, v_{j+1})$ of the approximating mesh together approximate $s(R)$.

Cylinder

Experiment 10.3. Run `cylinder.cpp`, which shows a triangular mesh approximation of a circular cylinder, given by the parametric equations

$$x = f(u, v) = \cos u, \qquad y = g(u, v) = \sin u, \qquad z = h(u, v) = v,$$

for $(u, v) \in [-\pi, \pi] \times [-1, 1]$. Pressing arrow keys changes the fineness of the mesh. Press 'x/X', 'y/Y' or 'z/Z' to turn the cylinder itself. Figure 10.28 is a screenshot.

End

Figure 10.28: Screenshot of `cylinder.cpp`.

The approximating mesh of `cylinder.cpp` is constructed according to the method above. However, a minor technicality is that the parameter space of the program is taken to be the square $[0, 1] \times [0, 1]$, rather than the parameter rectangle $[-\pi, \pi] \times [-1, 1]$ of the definition, so the former has first to be scaled to the latter. In fact, see the definitions of the functions `f`, `g` and `h` in the program:

```
float f(int i, int j)
{
   return ( cos( (-1 + 2*(float)i/p) * PI ) );
}

float g(int i, int j)
{
   return ( sin( (-1 + 2*(float)i/p) * PI ) );
}

float h(int i, int j)
{
   return ( -1 + 2*(float)j/q );
}
```

The expression returned by f first applies the mapping $u \mapsto (-1 + 2u)\pi$ to scale $[0, 1]$ to $[-\pi, \pi]$, then applies cos; likewise, the expression returned by g applies $u \mapsto (-1 + 2u)\pi$, then sin; the expression returned by h applies $v \mapsto -1 + 2v$ to scale $[0, 1]$ to $[-1, 1]$. Figure 10.29 indicates the scheme.

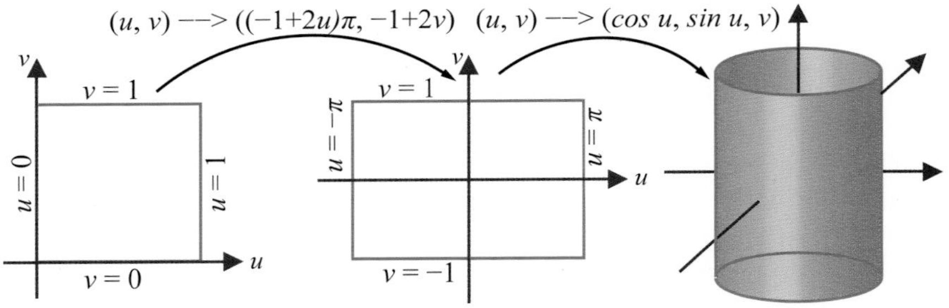

Figure 10.29: The composed mapping implemented in `cylinder.cpp`: first the parameter space is scaled, then mapped to the cylinder.

The most important part of `cylinder.cpp`'s implementing the drawing strategy of Section 10.2.5 is that the coordinate values $(i/p,\ j/q)$ run over a uniformly-spaced $(p + 1) \times (q + 1)$ grid of sample points in $[0, 1] \times [0, 1]$, as the integer argument i runs from 0 to p, and j from 0 to q. Correspondingly, $(\,f(i, j),\ g(i, j),\ h(i, j)\,)$ run over the mapped sample points on the cylinder itself. These mapped sample coordinate values are written into a vertex array by the `fillVertexArray()` routine. Triangles corresponding to each row of the grid in parameter space are drawn as a single triangle strip in the drawing routine, so that there are q triangle strips consisting of $2p$ triangles each.

The right and left arrow keys are programmed to increase and decrease p, respectively; the up and down arrow keys increase and decrease q, respectively.

Let's revisit next a surface we had first drawn in Experiment 2.24 of Chapter 2 to put it into the perspective of our general drawing strategy.

Exercise 10.41. (**Programming**) Run `hemisphere.cpp` from Chapter 2, which draws a triangulated hemisphere. Figure 10.30 is a screenshot. Press 'p/P' and 'q/Q' to coarsen or refine the triangulation. The parametric equations of the hemisphere implemented in the program are

$$x = R\cos\phi\cos\theta, \qquad y = R\sin\phi, \qquad z = R\cos\phi\sin\theta,$$

for $0 \le \theta \le 2\pi$ and $0 \le \phi \le \pi/2$.

Does the program use the drawing strategy of `cylinder.cpp` above? *Yes, it does.* Accordingly, alter only the `f`, `g` and `h` function definitions of `cylinder.cpp` to obtain a program equivalent to `hemisphere.cpp`.

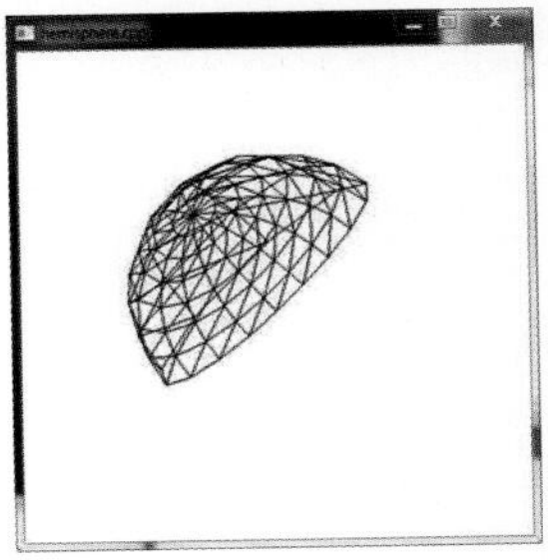

Figure 10.30: Screenshot of `hemisphere.cpp`.

You can write down *any* set of parametric equations you like and implement the corresponding surface with the help of the template of `cylinder.cpp`. All that changes in the program are the function definitions `f`, `g` and `h`. If you have a shape in mind then, of course, first deduce appropriate functions.

Helical Pipe

Experiment 10.4. Without really knowing what to expect (honestly!) we tweaked the parametric equations of the cylinder to the following:

$$x = \cos u + \sin v,\; y = \sin u + \cos v,\; z = u, \qquad (u, v) \in [-\pi, \pi] \times [-\pi, \pi]$$

It turns out the resulting shape looks like a helical pipe – run `helicalPipe.cpp`. Figure 10.31 is a screenshot.

Functionality is the same as for `cylinder.cpp`: press the arrow keys to coarsen or refine the triangulation and 'x/X', 'y/Y' or 'z/Z' to turn the pipe.

Looking at the equations again, it wasn't too hard to figure out how this particular surface came into being. See the next exercise. **End**

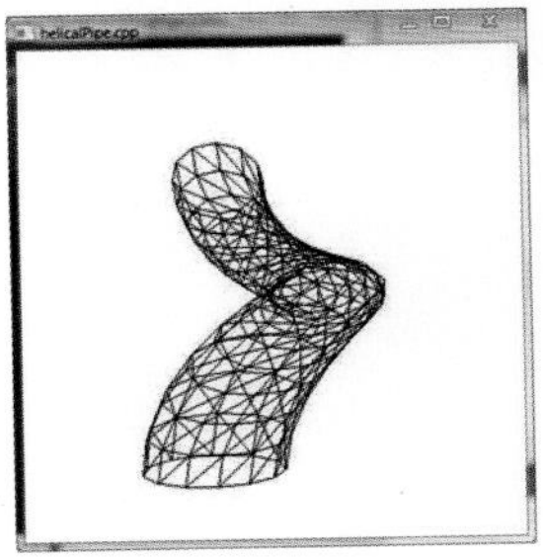

Figure 10.31: Screenshot of `helicalPipe.cpp`.

Exercise 10.42. Why do the parametric equations of the preceding experiment create a helical pipe?
Hint: The equation of the surface is

$$s(u, v) = (\cos u + \sin v,\ \sin u + \cos v,\ u) = (\cos u,\ \sin u,\ u) + (\sin v,\ \cos v,\ 0)$$

Note now that $u \mapsto (\cos u,\ \sin u,\ u)$ gives a helix, while $v \mapsto (\sin v,\ \cos v,\ 0)$ a circle.

Exercise 10.43. (**Programming**) Changing only the functions `f`, `g` and `h` of `cylinder.cpp`, draw wireframe surfaces resembling those in Figure 10.32.

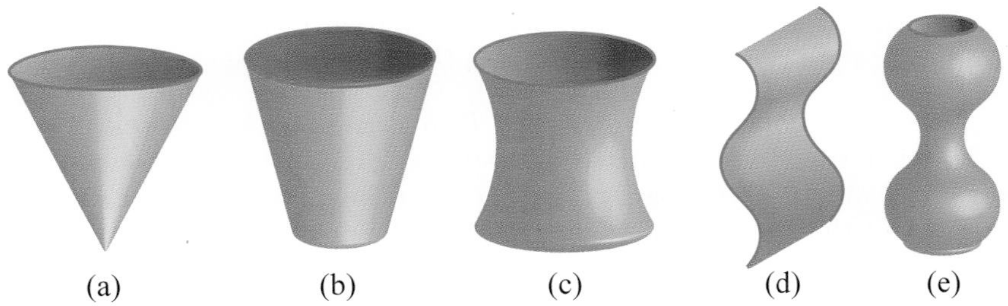

Figure 10.32: Draw these by modifying `cylinder.cpp`.

Exercise 10.44. (Programming) Nothing to do with drawing as such but practice in preparation for the newest OpenGL we'll be covering from Chapter 20 which no more has `glBegin()-glEnd()` type commands: replace the `glBegin(GL_TRIANGLE_STRIP)-glEnd()` loop in `cylinder.cpp` with *one* (see Section 3.1) `glMultiDrawElements(GL_TRIANGLE_STRIP, ...)` call.

10.2.6 Swept Surfaces

A powerful design method to create surfaces is by *sweeping* a curve. For example, consider the circular cylinder of Figure 10.33(a). One can think of it as the surface swept by a circle moving up, its center traveling along a vertical line. The curve that sweeps the surface is called its *profile curve* or, simply, *profile*. The path followed by the profile is the *trajectory*. The trajectory is actually the path of a point on the profile, or some point fixed with respect to it, such as the center of the circle sweeping the cylinder. The surface itself is the *swept surface*.

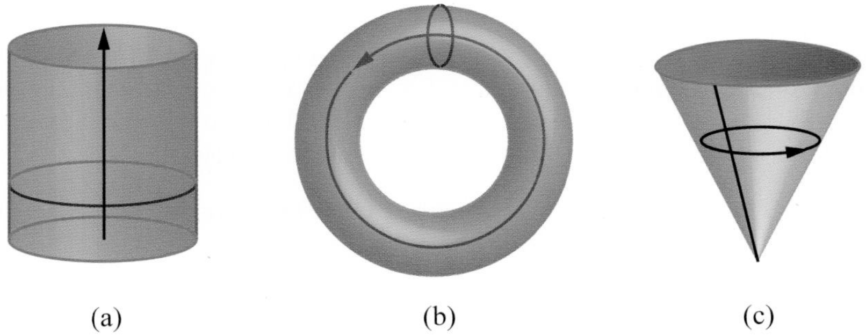

Figure 10.33: Swept surfaces: trajectories dashed arrows, profiles solid black.

A torus (Figure 10.33(b)) is swept by a circular profile itself traveling along a circular trajectory. When the trajectory is a circle, the swept surface is called a *surface of revolution*. A cone (Figure 10.33(c)) is a surface of revolution swept by a straight segment profile in a circular trajectory about the cone's axis.

When the trajectory is a straight segment, as in the case of the cylinder swept by a circle, the resulting surface is said to be an *extrusion*, or *extruded surface*, obtained by *extruding* the profile curve. In this case the profile is often called the *base curve*.

Exercise 10.45. Our description of a cylinder was as an extrusion of a circle. Can it be conceived of as a surface of revolution as well? If so, what are the profile and trajectory curves?

Exercise 10.46. How about a sphere, a hemisphere, an ellipsoid (egg shape) and (the surface of) a cube? Are they surfaces of revolution, extrusions, ...?

The advantage of being able to describe a surface as a swept surface is that its parametric equations are often, then, easy to deduce from those of its profile and trajectory curves. We see a few examples of this next.

Torus

Example 10.19. Let's compute the parametric equations of a torus. The profile is a circle c of radius r, whose center revolves along a circular trajectory C. C itself is of radius R, centered at the origin O and lying on the xy-plane. Each configuration of c, as it revolves, lies on a plane containing the z-axis and a radius of C. See Figure 10.34, where both the torus and a section through it are drawn.

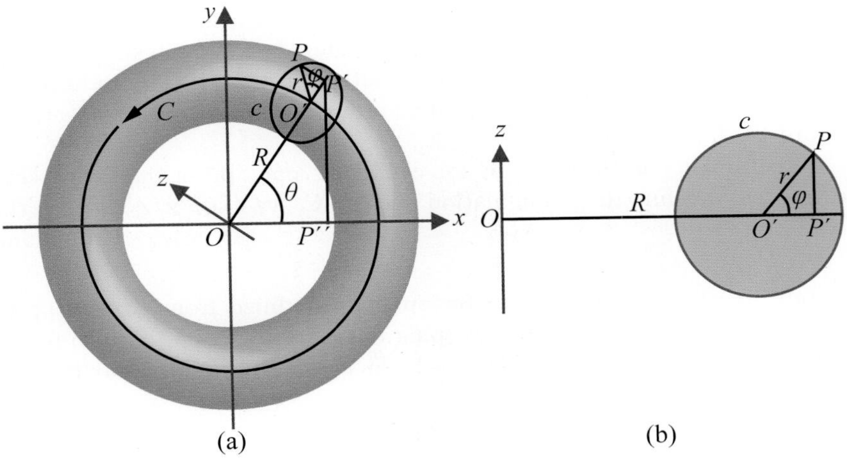

Figure 10.34: Computing parametric equations of a torus: (a) Profile circle c revolves along trajectory circle C (b) Sectional view of the left diagram along the plane containing the z-axis and OO'.

A point P on the torus is specified by two angles θ and ϕ as follows:

(a) θ is the angle made with the x-axis by the radius OO' of C from its center (the origin) to the center of the configuration of c containing P.

(b) ϕ is the angle made by $O'P$, a radius of the configuration of c containing P, with the extension of OO'.

Let P' be the projection of P on the xy-plane and P'' be the projection of P' on the x-axis.

The x coordinate of P is

$$\begin{aligned} OP'' &= OP'\cos\theta = (OO' + O'P')\cos\theta = (OO' + O'P\cos\phi)\cos\theta \\ &= (R + r\cos\phi)\cos\theta \end{aligned}$$

The y coordinate of P is

$$\begin{aligned} P'P'' &= OP'\sin\theta = (OO' + O'P')\sin\theta = (OO' + O'P\cos\phi)\sin\theta \\ &= (R + r\cos\phi)\sin\theta \end{aligned}$$

The z coordinate of P is

$$P'P = O'P\sin\phi = r\sin\phi$$

These give the parametric equations of the torus as

$$x = (R{+}r\cos\phi)\cos\theta, \quad y = (R{+}r\cos\phi)\sin\theta, \quad z = r\sin\phi, \quad -\pi \le \theta, \phi \le \pi \tag{10.15}$$

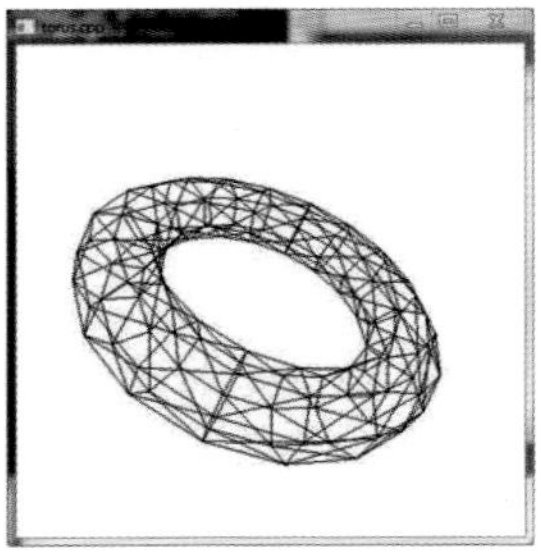

Figure 10.35: Screenshot of `torus.cpp`.

Experiment 10.5. Run `torus.cpp`, which applies the parametric equations deduced above in the template of `cylinder.cpp` (simply swapping new `f`, `g` and `h` function definitions into the latter program). The radii of the circular trajectory and the profile circle are set to 2.0 and 0.5, respectively. Figure 10.35 is a screenshot.

Functionality is the same as for `cylinder.cpp`: press the arrow keys to coarsen or refine the triangulation and 'x/X', 'y/Y' or 'z/Z' to turn the torus. **End**

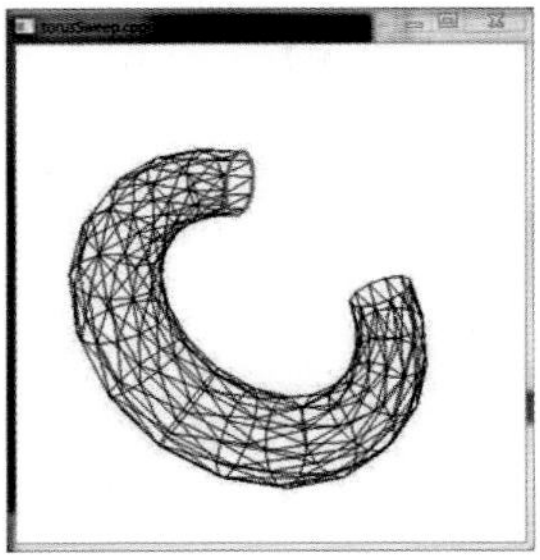

Figure 10.36: Screenshot of `torusSweep.cpp`.

Experiment 10.6. Run `torusSweep.cpp`, modified from `torus.cpp` to show the animation of a circle sweeping out a torus. Press space to toggle between animation on and off. Figure 10.36 is a screenshot part way through the animation. **End**

Exercise 10.47. (Programming) Plump a *toroidal helix* – which is a helix coiling around a torus (Figure 10.37(a)) – into a pipe (Figure 10.37(b)). Allow the user to choose the number of times the pipe coils before closing. No need to draw the torus itself.

Suggested approach: Begin by using the parametric equations of the torus itself as determined in Example 10.19, namely,

$$x = (R{+}r\cos\phi)\cos\theta, \quad y = (R{+}r\cos\phi)\sin\theta, \quad z = r\sin\phi, \quad -\pi \le \theta, \phi \le \pi$$

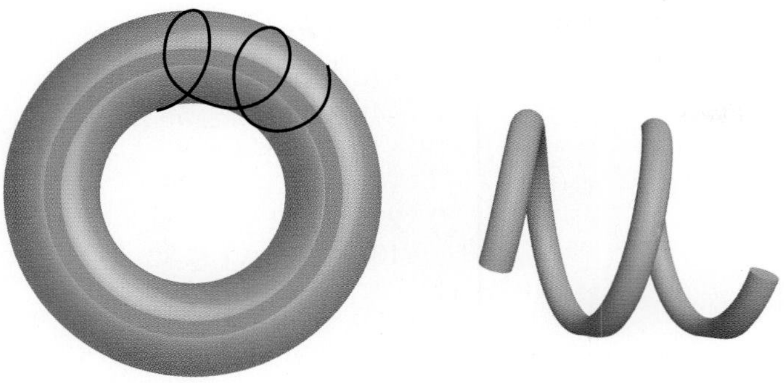

(a) (b)

Figure 10.37: (a) Part of a toroidal helix (b) Part of a toroidal helix pipe.

to find those for the toroidal helix.

The torus, being a surface, has the two degrees of freedom represented by θ and ϕ. For a curve lying on it, ϕ should depend on θ, leaving only a single degree of freedom. The function $\phi = n\theta$ yields a helix coiling n times around the torus before closing. Substituting, then, for ϕ in the equation above, one gets the parametric equations of a toroidal helix:

$$x = (R+r\cos(n\theta))\cos\theta,\ y = (R+r\cos(n\theta))\sin\theta,\ z = r\sin(n\theta), -\pi \leq \theta \leq \pi$$

To plump the toroidal helix into a pipe, observe that the pipe is swept by a circle c traveling along the toroidal helix.

Exercise 10.48. (Programming) For the mathematically inclined, a fun programming exercise is to draw a (p, q)-*torus knot*, where the user specifies p and q. You may have to look up torus knots – they are not hard at all and aesthetically pleasing.

Table

We'll draw next a table as the surface of revolution swept by revolving a profile curve c about the y-axis. The profile c is a polygonal line composed of seven segments lying on the xy-plane, starting at the point A and ending at B, as shown in Figure 10.38(a) (where the 0 z-coordinates are not written).

We'll parametrize c first by using the length t along c measured from A to P as the parameter value for a point P on c. Then the x coordinate $x_c(t)$ of a point with parameter value t is given below, as can be verified from a straightforward reading of Figure 10.38(a).

$$x_c(t) = \begin{cases} t, & 0 \le t \le 4 \\ 4, & 4 \le t \le 5 \\ 9 - t, & 5 \le t \le 8 \\ 1, & 8 \le t \le 22 \\ t - 21, & 22 \le t \le 31 \\ 10, & 31 \le t \le 32 \\ 42 - t, & 32 \le t \le 42 \end{cases}$$

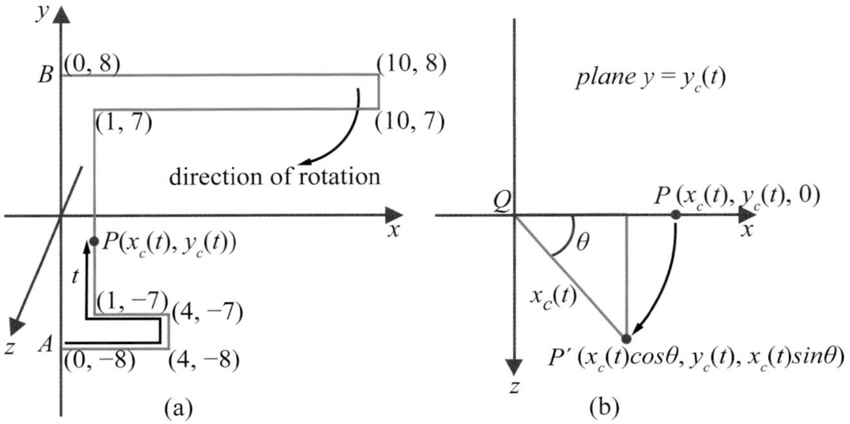

Figure 10.38: (a) The profile curve for a table on the xy-plane, with the z-coordinates, all 0, not written (b) A point on the profile curve after a rotation θ CW about the y-axis.

Likewise, the y coordinate $y_c(t)$ is given by:

$$y_c(t) = \begin{cases} -8, & 0 \le t \le 4 \\ t - 12, & 4 \le t \le 5 \\ -7, & 5 \le t \le 8 \\ t - 15, & 8 \le t \le 22 \\ 7, & 22 \le t \le 31 \\ t - 24, & 31 \le t \le 32 \\ 8, & 32 \le t \le 42 \end{cases}$$

When c revolves an angle of θ clockwise about the y-axis from its start configuration on the xy-plane, then a point P on c with coordinates $(x_c(t),\ y_c(t),\ 0)$ rotates an angle of θ on the plane $y = y_c(t)$ about the point $Q = (0, y_c(t), 0)$ to a new point P'. See Figure 10.38(b). Since $|QP'| = |QP| = x_c(t)$, the coordinates of P' are $(x_c(t)\cos\theta,\ y_c(t),\ x_c(t)\sin\theta)$.

Therefore, parametric equations for the table are

$$x = x_c(t)\cos\theta, \quad y = y_c(t), \quad z = x_c(t)\sin\theta, \qquad 0 \le t \le 42 \text{ and } -\pi \le \theta \le \pi \tag{10.16}$$

Experiment 10.7. These equations are implemented in `table.cpp`, again using the template of `cylinder.cpp`. Press the arrow keys to coarsen or refine the triangulation and 'x/X', 'y/Y' or 'z/Z' to turn the table. See Figure 10.39 for a screenshot of the table.

Note that the artifacts at the edges of the table arise because sample points may not map exactly to corners $(0, -8), (4, -8), \ldots, (0, 8)$ of the profile drawn in Figure 10.38(a) – which can be avoided by including always t values 0, 4, 5, 8, 22, 31, 32 and 42 in the sample grid. **End**

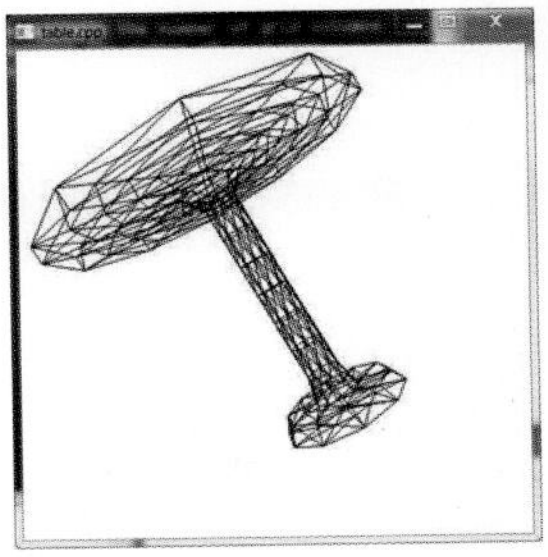

Figure 10.39: Screenshot of `table.cpp`.

Exercise 10.49. (Programming) Modify `table.cpp` to eliminate the artifacts at the edges in the manner suggested above.

Doubly-Curled Cone

Next is an experiment where the alignment of the profile curve varies as it travels along its trajectory. We want to make a "doubly-curled" cone, much like a cone made by curling a sheet of paper so that the edges don't meet, but that one wraps inside the other.

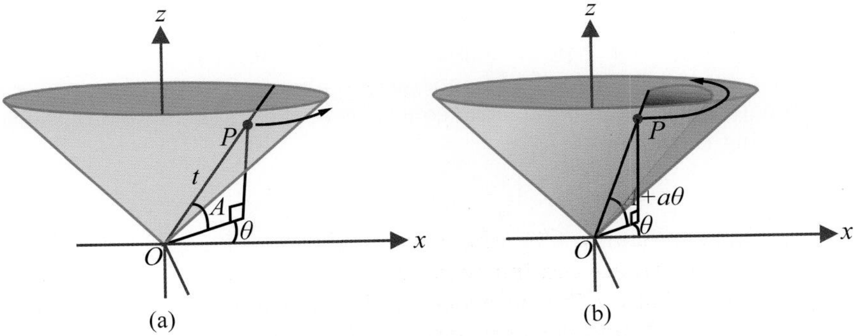

Figure 10.40: (a) A cone and (b) a doubly-curled cone as swept surfaces.

Let's write first the parametric equations for a plain-vanilla cone obtained by revolving a straight segment profile c of length 1 about the z-axis, with one end of c fixed at the origin, and with c making an angle of A with the xy-plane. We'll leave the reader to use Figure 10.40(a) to verify that the coordinates of the point P at a distance t from the origin along c, after the latter has revolved an angle of θ CCW from an original configuration on the xz-plane, are given by:

$$x = t \cos A \cos \theta, \; y = t \cos A \sin \theta, \; z = t \sin A, \quad 0 \le t \le 1 \text{ and } 0 \le \theta \le 2\pi$$

To make the cone doubly-curled we'll bring the profile c in toward the z-axis as it rotates, by increasing its angle with the xy-plane. Moreover, we'll rotate c twice about the z-axis to make a double curl.

A simple way to bring c in uniformly is to increment A, the angle that it makes with the xy-plane, by a multiple $a\theta$ of the amount of c's rotation. Figure 10.40(b) indicates the plan and it's straightforward to modify the parametric equations of the plain cone to write those of the doubly-curled:

$$x = t\cos(A + a\theta)\cos\theta,\ y = t\cos(A + a\theta)\sin\theta,\ z = t\sin(A + a\theta),$$

for $0 \le t \le 1$ and $0 \le \theta \le 4\pi$.

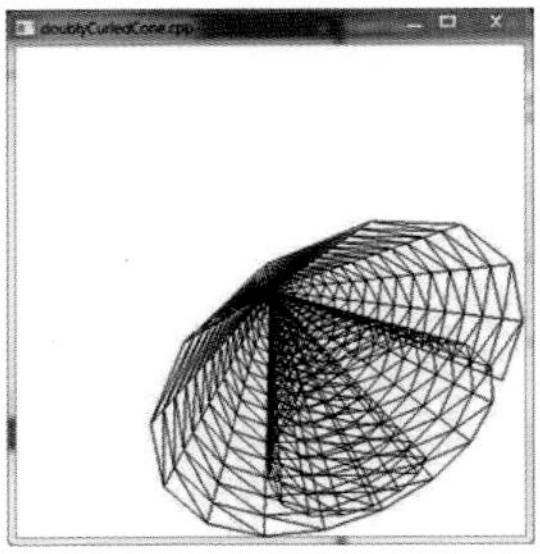

Figure 10.41: Screenshot of `doublyCurledCone.cpp`.

Experiment 10.8. The plan above is implemented in `doublyCurledCone.cpp`, again using the template of `cylinder.cpp`, with the value of A set to $\pi/4$ and a to 0.05. Press the arrow keys to coarsen or refine the triangulation and 'x/X', 'y/Y' or 'z/Z' to turn the cone. Figure 10.41 is a screenshot. **End**

Exercise 10.50. (Programming) Modify `doublyCurledCone.cpp` to change as well the length of the revolving segment c as it sweeps the cone.

Exercise 10.51. A *superellipsoid* is given generally by the implicit equation

$$\left|\frac{x}{a}\right|^n + \left|\frac{y}{b}\right|^n + \left|\frac{z}{c}\right|^n = 1 \qquad (10.17)$$

where a, b, c and n are each a positive constant. It's an extension to 3D of the superellipse of Section 10.1.3.

In fact, a special type of superellipsoid is obtained as a surface of revolution by simply revolving a superellipse about either the x- or y-axis. Accordingly, deduce the parametric equations of the superellipsoid obtained by revolving the superellipse

$$\left|\frac{x}{a}\right|^n + \left|\frac{y}{b}\right|^n = 1$$

about the y-axis.

Exercise 10.52. (Programming) Draw the general superellipsoid given by (10.17) above, allowing the user to choose parameters.

Extruded Helix

For the record, here's a simple example of extrusion.

Figure 10.42: Screenshot of `extrudedHelix.cpp`.

Experiment 10.9. Run `extrudedHelix.cpp`, which extrudes a helix, using yet again the template of `cylinder.cpp`. The parametric equations of the extrusion are

$$x = 4cos(10\pi u),\ y = 4sin(10\pi u),\ z = 10\pi u + 4v, \quad 0 \le u, v \le 1$$

the constants being chosen to size the object suitably. As the equation for z indicates, the base helix is extruded parallel to the z-axis. Figure 10.42 is a screenshot. **End**

Exercise 10.53. (Programming) Can you extrude the panels of the chair of Exercise 10.39, all of which you were then asked to draw flat, to make them now truly solid? Make the back rest curved too.

10.2.7 Drawing Projects

Here are real-life 3D projects for your drawing pleasure.

Exercise 10.54. (Programming) Draw the objects depicted in Figure 10.43: wine glass, vase, helmet with visor, extruded 'A', arch. Draw in wireframe.

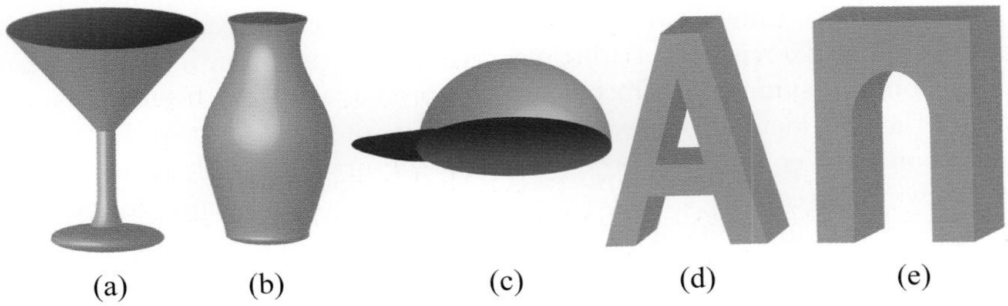

Figure 10.43: Stuff to draw.

Exercise 10.55. (Programming) Draw your name in 3D text.

Exercise 10.56. (Programming) Draw the six different chess pieces.

Exercise 10.57. (Programming) You have a chair from Exercise 10.39 (or Exercise 10.53) and a table from Experiment 10.7. Can you place these into `animateMan1.cpp` of Experiment 4.33, and get the man to walk up to the chair, sit and lean forward with his elbows on the table?

Figure 10.44: Model these? You gotta be kidding!

Exercise 10.58. (Programming) Make a likeness of an interesting structure at the place where you live, e.g., building, bridge, multi-level highway crossing, train station, or a famous one, e.g., the Eiffel Tower or Taj Mahal, of which images are available (Figure 10.44). Ignore details. Try to be as faithful as possible to the large-scale geometry.
Hint: The Eiffel Tower and Taj Mahal may seem daunting at first, but there are multiple symmetries in each which can be exploited to simplify the design process.

Consider the Eiffel Tower. The base has four identical panels reminiscent of the arch of Exercise 10.54 a little earlier. As for the tower above the base, it has fourfold symmetry. Once a suitable profile curve has been chosen for one of the four identical edges of the tower – its designer Gustave Eiffel used an exponential equation in order for the structure to be able to withstand severe wind forces – it's a matter of placing this curve four times, rotated 90° each time, and filling identical wireframes between successive pairs. Don't forget display lists from Section 3.4 in your design process.

The Taj Mahal has arches as well and numerous surfaces of revolution.

You can skip the rest of Section 10.2 on a first reading and go directly now to Section 10.3 on Bézier curves and surfaces. The reason is that Sections 10.2.8-10.2.11 deal with special classes of surfaces which you can explore later at leisure, while 10.2.12 is about the theory of surfaces which can be deferred as well.

10.2.8 Ruled Surfaces

A *ruled surface* is a swept surface whose profile curve is a straight line. In other words, a ruled surface is traced by a straight line traveling through space. Each instance of the sweeping line is called a *ruling*. See Figure 10.45.

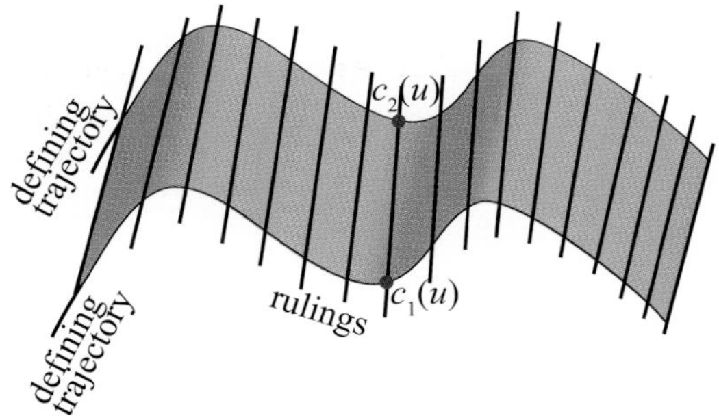

Figure 10.45: A ruled surface showing several rulings and two defining trajectories.

The parametrization of a ruled surface is particularly simple. Say that the paths of two distinct points on the profile line are $c_1(u)$ and $c_2(u)$,

$u \in [a, b]$, respectively, each called a *defining trajectory* of the surface. A parametrization then is

$$s(u, v) = (1 - v)c_1(u) + vc_2(u), \quad u \in [a, b],\ v \in (-\infty, \infty) \tag{10.18}$$

where u varies over the defining trajectories, and v over the (infinite) straight line through a pair of corresponding points on the two. If we want only the part of the surface *between* the defining trajectories, then we have to restrict the parameter space as follows:

$$s(u, v) = (1 - v)c_1(u) + vc_2(u), \quad u \in [a, b],\ v \in [0, 1] \tag{10.19}$$

Various surfaces arise as ruled surfaces. Here are three interesting ones.

Bilinear Patches

A *bilinear patch* is a ruled surface whose defining trajectories c_1 and c_2 are straight line segments both. The bilinear patch itself lies between the two trajectories. See Figure 10.46. Suppose the endpoints of c_1 are p_1 and q_1, so that it can be parametrized $c_1(u) = (1 - u)p_1 + uq_1$, $0 \le u \le 1$, while the endpoints of c_2 are p_2 and q_2, and it is parametrized $c_2(u) = (1 - u)p_2 + uq_2$, $0 \le u \le 1$. Plugging these equations into (10.19) we get the equation of the bilinear patch:

$$s(u, v) = (1 - u)(1 - v)\, p_1 + u(1 - v)\, q_1 + (1 - u)v\, p_2 + uv\, q_2, \quad u, v \in [0, 1] \tag{10.20}$$

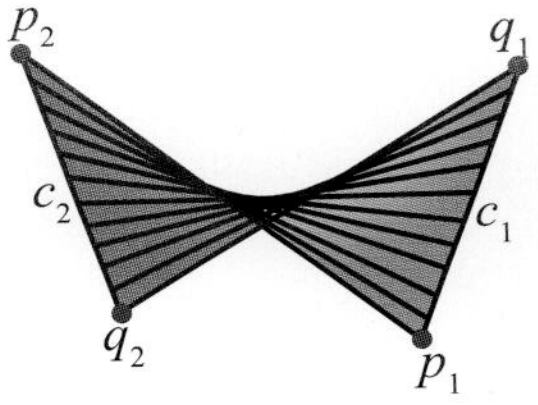

Figure 10.46: Bilinear patch.

Counter-intuitively, even though a bilinear patch is made of a family of straight segments joining points again on two straight segments c_1 and c_2, it need not be flat! In fact, it's flat only when c_1 and c_2 are coplanar; otherwise, it is a curved surface. Interestingly, it turns out, as we'll see in upcoming Section 10.2.9, that, generally, a bilinear patch is nothing but the familiar saddle surface.

Experiment 10.10. Run `bilinearPatch.cpp`, which implements precisely Equation (10.20). Press the arrow keys to refine or coarsen the wireframe and 'x/X', 'y/Y' or 'z/Z' to turn the patch. Figure 10.47 is a screenshot. **End**

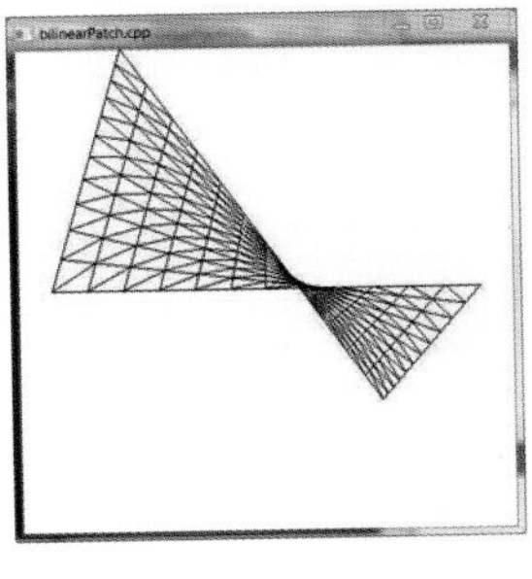

Figure 10.47: Screenshot of `bilinearPatch.cpp`.

Generalized Cones

A *generalized cone* is a ruled surface, one of whose defining trajectories is an arbitrary curve c, while the other is stationary, in other words, a single point p, which should not belong to c. The cone is said to be *over* c with *apex* at p. See Figure 10.48 for three examples. Unless the qualifier "generalized" is used, the curve c is typically presumed closed. Often, informally meant by the term cone is a *right circular cone*, which is a cone over a circle c whose

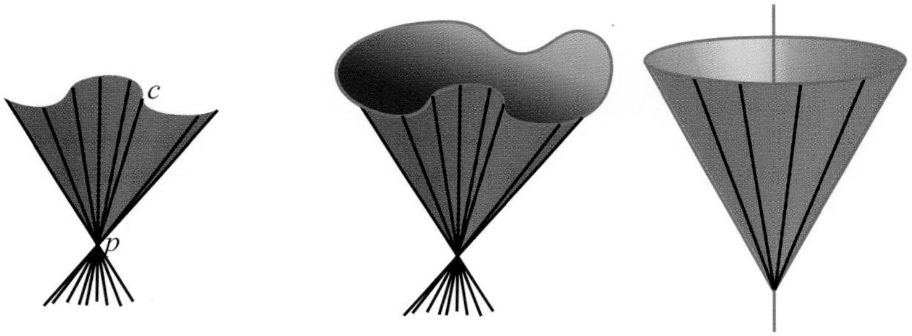

Figure 10.48: Generalized cones: (a) over a non-closed curve (b) over a closed curve (c) Right circular cone. Only the part between the two trajectories is drawn.

apex is located on the line which is through the center of c and perpendicular to its plane.

The equation of the part of the generalized cone between its trajectories – a part like those depicted in Figure 10.48 – is obtained by plugging $c_1(u) = p$ for the trajectory stationary at the apex p, and $c_2(u) = c(u)$ for the other, into (10.19):

$$s(u, v) = (1 - v)p + vc(u), \qquad u \in [a, b],\ v \in [0, 1] \tag{10.21}$$

Evidently, all rulings pass through the apex p, at $v = 0$.

Exercise 10.59. **(Programming)** Draw a cone over the astroid of `astroid.cpp`.

Generalized Cylinders

A *generalized cylinder* is a ruled surface whose defining trajectories c_1 and c_2 are translates of one another. See Figure 10.49. Colloquially, a cylinder typically means the familiar *right circular cylinder*, where c_1 and c_2 are circles whose centers are joined by a line perpendicular to the plane of both.

The equation for the generalized cylinder is obtained by writing the equation of one trajectory as $c(u)$ and the other as $c(u) + d$, where $u \in [a, b]$, and d is the vector translating the first trajectory to the second. Plugging these equations into (10.19) we get the generalized cylinder as

$$s(u, v) = (1-v)c(u)+v(c(u)+d) = c(u)+vd, \quad u \in [a, b],\ v \in [0, 1] \tag{10.22}$$

The rulings are evidently all parallel to the vector d.

Exercise 10.60. Are generalized cylinders the same as extrusions?

Exercise 10.61. **(Programming)** Draw a generalized cylinder using an astroid as a trajectory.

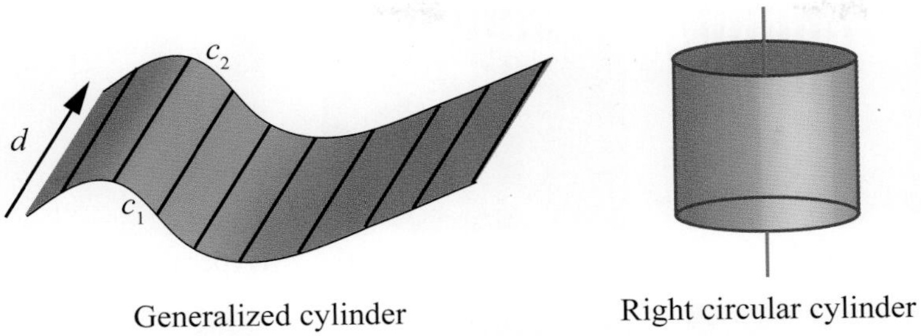

Figure 10.49: A generalized cylinder and a special case.

10.2.9 Quadric Surfaces

As we saw in 10.1.5, conics are curves on a plane given by a quadratic equation in two variables. *Quadric surfaces* or, simply, *quadrics*, are their generalization to one dimension higher. They are surfaces in 3D space given by a quadratic equation in *three* variables:

$$Ax^2 + By^2 + Cz^2 + Dyz + Ezx + Fxy + Px + Qy + Rz + H = 0$$

Excluding degenerate instances (e.g., $x^2 + y^2 + z^2 = 0$, which gives a single point) a quadric is of one of the nine kinds shown in Figure 10.50. In fact, any non-degenerate quadric can be transformed by translation and rotation to one of the normalized forms in the following table, corresponding each to one of those pictured in Figure 10.50.

Quadric	Implicit Equation
Ellipsoid	$\frac{x^2}{a^2} + \frac{y^2}{b^2} + \frac{z^2}{c^2} = 1$
Elliptic Paraboloid	$\frac{x^2}{a^2} + \frac{y^2}{b^2} - z = 0$
Hyperbolic Paraboloid	$\frac{x^2}{a^2} - \frac{y^2}{b^2} - z = 0$
Hyperboloid (1 sheet)	$\frac{x^2}{a^2} + \frac{y^2}{b^2} - \frac{z^2}{c^2} = 1$
Hyperboloid (2 sheets)	$\frac{x^2}{a^2} - \frac{y^2}{b^2} - \frac{z^2}{c^2} = 1$
Elliptic Cone	$\frac{x^2}{a^2} + \frac{y^2}{b^2} - \frac{z^2}{c^2} = 0$
Elliptic Cylinder	$\frac{x^2}{a^2} + \frac{y^2}{b^2} = 1$
Parabolic Cylinder	$y = ax^2$
Hyperbolic Cylinder	$\frac{x^2}{a^2} - \frac{y^2}{b^2} = 1$

A sphere is, of course, a special case of an ellipsoid. The hyperbolic paraboloid, for an obvious reason, is often called a *saddle surface*. The three cylindrical quadrics along the bottom row are probably the least interesting, as they are merely extrusions of plane conics. Parametrization, both trigonometric and rational, of the quadrics are not hard to derive.

Example 10.20. Find both trigonometric and rational parametrizations of the ellipsoid.

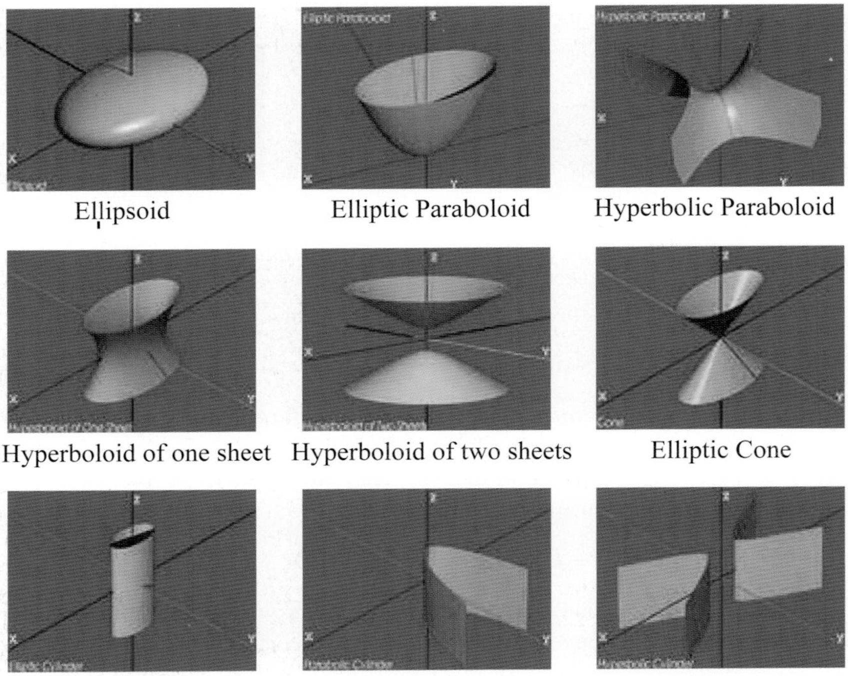

Figure 10.50: The nine non-degenerate quadric surfaces (from Wikimedia).

Answer: We begin with the ellipsoid's implicit equation

$$\frac{x^2}{a^2} + \frac{y^2}{b^2} + \frac{z^2}{c^2} = 1$$

A trigonometric parametrization is

$$x = a\cos\theta\cos\phi,\ y = b\sin\theta\cos\phi,\ z = c\sin\phi, \quad \theta \in [-\pi, \pi],\ \phi \in [-\pi/2, \pi/2]$$

while a rational one is

$$x = a\frac{1 - u^2 + v^2}{1 + u^2 + v^2},\quad y = b\frac{2uv}{1 + u^2 + v^2},\quad z = c\frac{2u}{1 + u^2 + v^2},\quad u, v \in (-\infty, \infty)$$

with a singularity at $(-a, 0, 0)$.

Exercise 10.62. Find trigonometric and rational parametrizations of the elliptic paraboloid.

Remark 10.12. If $a = b$ in the equation of the elliptic paraboloid – the equation becomes $\frac{x^2}{a^2} + \frac{y^2}{a^2} - z = 0$ in this case – then it's actually a surface of revolution obtained from revolving a parabola around its axis. This special case of an elliptic paraboloid is called a *circular paraboloid*. It is the shape used in mirrors behind headlamps because light from a bulb placed at its focal point reflects in a parallel beam.

Drawing the quadrics is simple. We'll use a trigonometric parametrization to draw next the hyperboloid of one sheet.

Experiment 10.11. Run `hyperboloid1sheet.cpp`, which draws a triangular mesh approximation of a single-sheeted hyperboloid with the help of the parametrization

$$x = \cos u \sec v, \ y = \sin u \sec v, \ z = \tan v, \quad u \in [-\pi, \pi], \ v \in (-\pi/2, \pi/2)$$

Figure 10.51(a) is a screenshot. In the implementation we restrict v to $[-0.4\pi, 0.4\pi]$ to avoid $\pm\pi/2$ where sec is undefined. **End**

(a) (b) (c)

Figure 10.51: (a) Screenshot of `hyperboloid1sheet.cpp` (b) Edible hyperbolic paraboloids (c) Hyperboloid footbridge over Corporation Street in Manchester in England supported by its rulings (courtesy of Patrick Litherland).

It's interesting that a few of the non-degenerate quadrics are, in fact, ruled surfaces and, therefore, traced by a straight line traveling through space. The ones on the bottom row of Figure 10.50 are evidently so. We'll prove a less obvious case.

Example 10.21. Show that the hyperbolic paraboloid is a ruled surface.

Answer: We'll work with the instance s given by the implicit equation

$$x^2 - y^2 = z$$

as setting the coefficients all equal to 1 simplifies calculations (without costing in generality). Write the equation as

$$(x + y)(x - y) = z$$

Setting $u = x + y$ and $v = x - y$ then leads to the following parametrization of s:

$$x = \frac{u+v}{2}, \quad y = \frac{u-v}{2}, \quad z = uv, \quad u, v \in (-\infty, \infty) \tag{10.23}$$

Now, a u-parameter curve of s is obtained by fixing $v = \beta$:

$$x = \frac{u+\beta}{2}, \quad y = \frac{u-\beta}{2}, \quad z = \beta u, \quad u \in (-\infty, \infty)$$

which is a straight line. Therefore, s is swept by a straight line profile, particularly the u-parameter curve for $v = \beta$, as β varies, proving it is indeed ruled. Evidently, it's *doubly-ruled*, the u-parameter curves and v-parameter curves defining distinct symmetric families of rulings.

In fact, it gets even more interesting! Two defining trajectories for s can be obtained as the v-parameter curves corresponding to a couple of distinct values of u, because they intersect each u-parameter curve in a distinct pair of points. Accordingly, set u equal to 0 and 1 in Equation (10.23) to get, respectively, the equations

$$x = \frac{v}{2}, \quad y = -\frac{v}{2}, \quad z = 0, \quad v \in (-\infty, \infty)$$

and

$$x = \frac{1+v}{2}, \quad y = \frac{1-v}{2}, \quad z = v, \quad v \in (-\infty, \infty)$$

which are both straight lines. So a hyperbolic paraboloid is a ruled surface with straight-line defining trajectories, which means it's a bilinear patch. This justifies an earlier remark that bilinear patches are saddle surfaces in general.

Remark 10.13. People often snack on hyperbolic paraboloids! See Figure 10.51(b).

Exercise 10.63. Prove that the single-sheeted hyperboloid is doubly-ruled. Figure 10.51(c) illustrates how this fact is applied to build a bridge – note the two sets of steel rulings and how they intersect in a grid. You likely have seen baskets woven in the shape of single-sheeted hyperboloids as well.

Exercise 10.64. (Programming) Animate a straight line segment sweeping out a single-sheeted hyperboloid.

10.2.10 GLU Quadric Objects

We are already familiar with several FreeGLUT library objects such as spheres, cubes and cones which are ready-to-use for 3D drawing. The OpenGL Utility Library GLU provides additional routines to create four kinds of so-called quadric objects: sphere, tapered cylinder, annular disc and partial annular disc. See Figure 10.52.

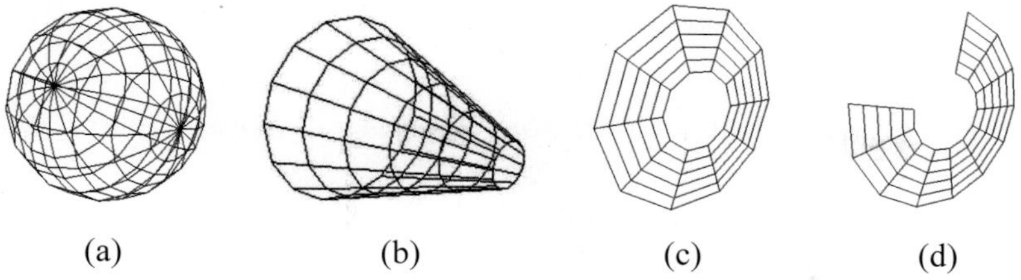

Figure 10.52: GLU quadrics: (a) Sphere (b) Tapered cylinder (c) Annular disc (d) Partial annular disc.

Experiment 10.12. Run `gluQuadrics.cpp` to see all four GLU quadrics. Press the left and right arrow keys to cycle through the quadrics and 'x/X', 'y,Y' and 'z/Z' to turn them. The images in Figure 10.52 were, in fact, generated by this program. **End**

Remark 10.14. It's a bit unfortunate that OpenGL chooses to render the quadrics quadrilateralized, rather than triangulated.

Here's how the syntax works (refer to Figure 10.53):

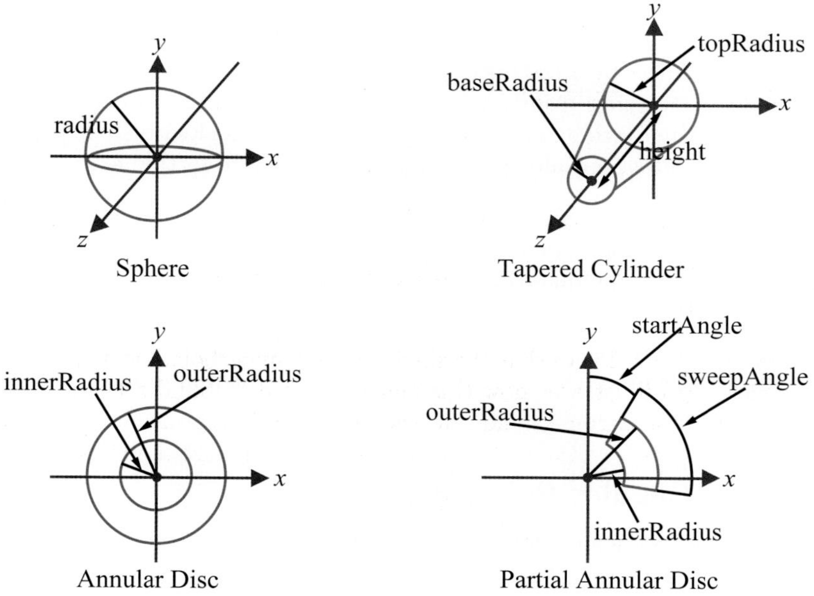

Figure 10.53: Defining the GLU quadrics.

1. `gluSphere(`**qobj*`,` *radius*`,` *slices*`,` *stacks*`)`

   Draws a *sphere* of radius *radius* centered at the origin. The parameters *slices* and *stacks* determine the fineness of the quadrilateralization.

Note: The parameter *qobj* in this case, and in the following, points to a quadric object.

2. `gluCylinder(`**qobj*, *baseRadius*, *topRadius*, *height*, *slices*, *stacks*`)`

   Draws a *tapered cylinder* with its axis along the z-axis, whose base is a circle of radius *baseRadius* lying on the $z = 0$ plane and whose top a circle of radius *topRadius* lying on the $z = height$ plane. If either *baseRadius* or *topRadius* is zero then the object is a cone. The parameters *slices* and *stacks* determine the fineness of the quadrilateralization.

3. `gluDisk(`**qobj*, *innerRadius*, *outerRadius*, *slices*, *rings*`)`

   Draws an *annular disc* centered at the origin and lying on the $z = 0$ plane, whose inner boundary is of radius *innerRadius* and outer boundary of radius *outerRadius*. The parameters *slices* and *rings* determine the fineness of the quadrilateralization.

4. `gluPartialDisk(`**qobj*, *innerRadius*, *outerRadius*, *slices*, *rings*, *startAngle*, *sweepAngle*`)`

   Draws a *partial annular disc*: precisely, the sector of the annular disc defined by `gluDisk(`**qobj*, *innerRadius*, *outerRadius*, *slices*, *rings*`)`, starting from angle *startAngle* and ending at *startAngle* + *sweepAngle*, where either angle is measured clockwise (looking from the $+z$-direction) along the xy-plane starting from the y-axis.

GLU calls of the form `gluQuadric*(`**qobj*, `*)` determine various properties of the quadric. For example, the call `gluQuadricDrawStyle(qobj, GLU_LINE)` causes the quadric to be rendered in wireframe.

Remark 10.15. The GLU quadrics are somewhat ambitiously named. Although they are each a part of one of the mathematical quadric surfaces described in the preceding section, they will hardly help in drawing the more complex ones.

10.2.11 Regular Polyhedra

To begin with here's a definition of a particular kind of polygon which should be familiar:

Definition 10.9. A *regular polygon* is a simple planar polygon whose sides are of equal length and which has equal interior angles at its vertices.

A regular polygon with n sides is convex and its vertices are spaced equally along a circle, called its *circumscribed circle*, at an angle of $2\pi/n$ apart (Figure 10.54). The larger the n, the more closely the polygon approximates its circumscribed circle.

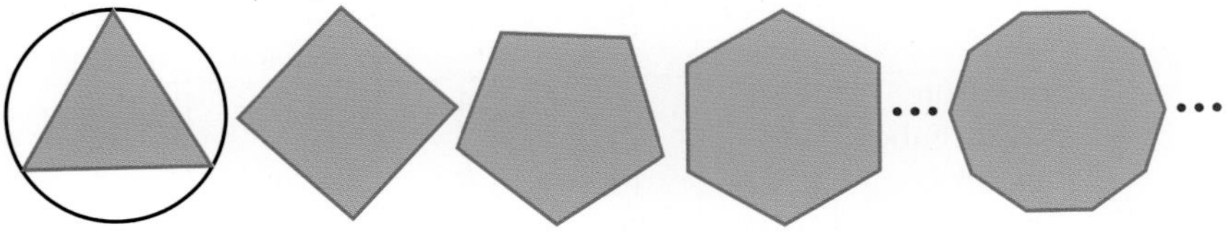

Figure 10.54: Regular polygons with number of sides indicated. The triangle shows its circumscribed circle.

Exercise 10.65. What is the interior angle at a vertex of an n-sided regular polygon?

Exercise 10.66. Show that if the condition "and which has equal interior angles at its vertices" is dropped from the definition of a regular polygon, then we could make one not belonging to the family of Figure 10.54.

Regular polyhedra are a generalization of regular polygons to three dimensions. Here first is the definition of a polyhedron based on that of a polygonal mesh (described in Section 10.2.2).

Definition 10.10. A *polyhedron* is a solid object whose boundary is a polygonal mesh (in other words, a solid whose faces are all polygons).

Definition 10.11. A *regular polyhedron* is a polyhedron all of whose faces are identical regular polygons.

Because of the symmetry constraints on their faces, there exist only five different regular polyhedra, ignoring difference in size. They are the *tetrahedron*, *hexahedron* (familiarly, *cube*), *octahedron*, *dodecahedron* and *icosahedron* in order of increasing number of faces. See Figure 10.55.

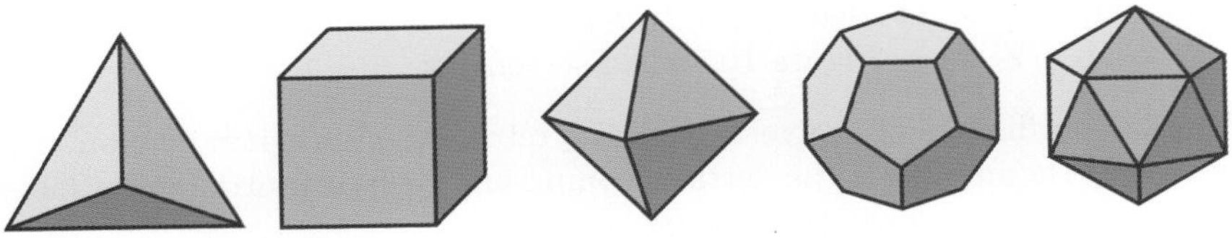

Figure 10.55: The five regular polyhedra with the number of faces indicated.

Geometric data for the five in numerical form is collected in the following table.

	Faces	Edges	Vertices	Edges of a face	Faces at a vertex
Tetrahedron	4	6	4	3	3
Hexahedron (cube)	6	12	8	4	3
Octahedron	8	12	6	3	4
Dodecahedron	12	30	20	5	3
Icosahedron	20	30	12	3	5

The hexahedron is bounded by squares, the dodecahedron by regular pentagons and the remaining three by equilateral triangles. The value (m, n) for a regular polyhedron, where m is the number of edges of a face and n the number of faces meeting at a vertex – the items in the last two columns of the table – is called its *Schläfli symbol*. The five different Schläfli symbols are $(3, 3)$, $(4, 3)$, $(3, 4)$, $(5, 3)$ and $(3, 5)$, and each uniquely identifies a regular polyhedron. It's not an accident, as we'll see, that the reverse of each Schläfli symbol is another.

Remark 10.16. Regular polyhedra are also called *Platonic solids* because they were known to Plato, as recorded in his Timaeus dialogues. In fact, archeological finds suggest that these beautiful shapes were familiar to even earlier people.

Modeling Regular Polyhedra

There's an easy way to draw regular polyhedra using OpenGL: call them from the FreeGLUT library! All five are available as FreeGLUT objects.

Experiment 10.13. Run `glutObjects.cpp`, a program we originally saw in Chapter 3. Press the left and right arrow keys to cycle through the various FreeGLUT objects and 'x/X', 'y/Y' and 'z/Z' to turn them. Among other objects you see all five regular polyhedra, both in solid and wireframe. **End**

However, in case you are the hardy do-it-yourself type, here's what you need to know.

Experiment 10.14. Run `tetrahedron.cpp`. The program draws a wireframe tetrahedron of edge length $2\sqrt{2}$ which can be turned using the 'x/X', 'y/Y' and 'z/Z' keys. Figure 10.56 is a screenshot. **End**

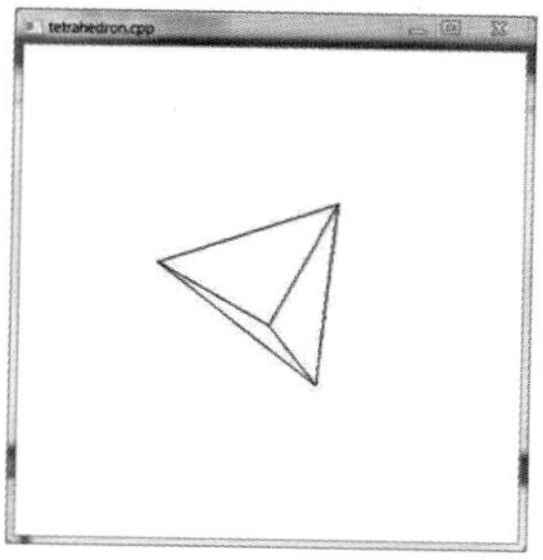

Figure 10.56: Screenshot of `tetrahedron.cpp`.

The coordinates of the vertices of the tetrahedron of `tetrahedron.cpp`, as well as the indices of the vertices comprising each of its triangular faces, are listed in the following two global arrays:

```
// Vertex coordinate vectors for the tetrahedron.
   static float vertices[] =
   {
        1.0,  1.0,  1.0, // V0
       -1.0,  1.0, -1.0, // V1
        1.0, -1.0, -1.0, // V2
```

```
    -1.0, -1.0,  1.0  // V3
};

// Vertex indices for the four triangular faces.
static int triangleIndices[4][3] =
{
    {1, 2, 3}, // F0
    {0, 3, 2}, // F1
    {0, 1, 3}, // F2
    {0, 2, 1}  // F3
};
```

For example, the face $F0$ is a triangle with corners at the vertices $V1$, $V2$ and $V3$.

Here's similar data for a cube of edge length 2:

Cube

Vertex	Coordinates
$V0$	$(1, 1, 1)$
$V1$	$(1, 1, -1)$
$V2$	$(1, -1, -1)$
$V3$	$(1, -1, 1)$
$V4$	$(-1, 1, 1)$
$V5$	$(-1, 1, -1)$
$V6$	$(-1, -1, -1)$
$V7$	$(-1, -1, 1)$

Face	Vertices
$F0$	$(V3,\ V0,\ V1,\ V2)$
$F1$	$(V2,\ V1,\ V5,\ V6)$
$F2$	$(V6,\ V5,\ V4,\ V7)$
$F3$	$(V7,\ V4,\ V0,\ V3)$
$F4$	$(V1,\ V0,\ V5,\ V4)$
$F5$	$(V3,\ V2,\ V6,\ V7)$

You may be wondering why we bothered at all with the totally trivial cube. The reason is that a cube sets up modeling an octahedron by way of the beautiful relationship of *duality* between regular polyhedra.

Duality

The *dual* of a regular polyhedron P is the polyhedron P' *inscribed* in P as follows:

(a) For each face f of P there is a vertex of P', called f's dual, located at the center of f.

(b) For each edge e of P there is an edge of P', called e's dual, joining the dual of the two faces of P adjacent to e.

(c) For each vertex v of P there is a face of P', called v's dual, with vertices at the duals of the faces of P that meet at v.

See Figure 10.57. Fascinatingly enough, it turns out that the dual of a regular polyhedron is another regular polyhedron. Cubes and octahedrons are duals of one another, as are dodecahedrons and icosahedrons, while tetrahedrons are self-dual.

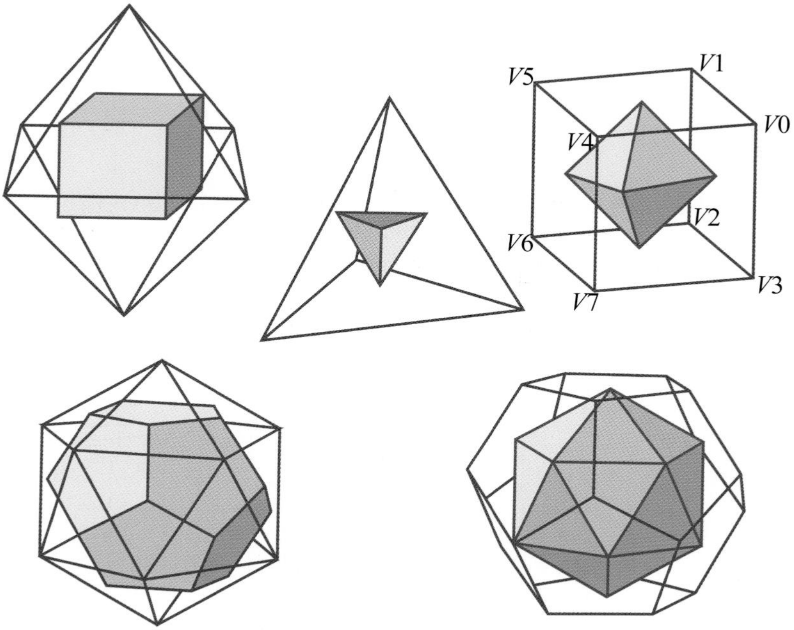

Figure 10.57: The five regular polyhedra each containing its inscribed dual (the cube is labeled to help with Exercise 10.67).

It's clear from the construction that a regular polyhedron and its dual have the same number of edges, while the number of vertices of one equals the number of faces of the other. Moreover, their Schläfli symbols are flips one of the other.

Returning to the drawing of regular polyhedra, it's easy now to compute the data for the octahedron dual to the cube whose data we listed earlier. In fact, we leave the reader to verify data for the dual octahedron in the next exercise.

Exercise 10.67. Verify the data for the octahedron, dual to the cube whose data was listed earlier, as given in the two tables just below. Note that the vertex $V'i$ of the octahedron is the dual of the face Fi of the cube, while the face $F'j$ of the octahedron dual of the face Vj of the cube. Moreover, the edge length of this particular octahedron is $\sqrt{2}$.

Octahedron

Vertex	Coordinates
$V'0$	$(1, 0, 0)$
$V'1$	$(0, 0, -1)$
$V'2$	$(-1, 0, 0)$
$V'3$	$(0, 0, 1)$
$V'4$	$(0, 1, 0)$
$V'5$	$(0, -1, 0)$

Face	Vertices
$F'0$	$(V'0,\ V'4,\ V'3)$
$F'1$	$(V'4,\ V'0,\ V'1)$
$F'2$	$(V'5,\ V'1,\ V'0)$
$F'3$	$(V'5,\ V'0,\ V'3)$
$F'4$	$(V'2,\ V'3,\ V'4)$
$F'5$	$(V'1,\ V'2,\ V'4)$
$F'6$	$(V'2,\ V'1,\ V'5)$
$F'7$	$(V'5,\ V'3,\ V'2)$

Icosahedron

Vertex	Coordinates
$V0$	$(0, 1, X)$
$V1$	$(0, 1, -X)$
$V2$	$(1, X, 0)$
$V3$	$(1, -X, 0)$
$V4$	$(0, -1, -X)$
$V5$	$(0, -1, X)$
$V6$	$(X, 0, 1)$
$V7$	$(-X, 0, 1)$
$V8$	$(X, 0, -1)$
$V9$	$(-X, 0, -1)$
$V10$	$(-1, X, 0)$
$V11$	$(-1, -X, 0)$

Note: The constant $X = (\sqrt{5} - 1)/2$ is the reciprocal of the golden ratio. Its value is approximately 0.618.

Face	Vertices
$F0$	$(V6,\ V2,\ V0)$
$F1$	$(V2,\ V6,\ V3)$
$F2$	$(V3,\ V6,\ V5)$
$F3$	$(V6,\ V7,\ V5)$
$F4$	$(V0,\ V7,\ V6)$
$F5$	$(V8,\ V2,\ V3)$
$F6$	$(V1,\ V2,\ V8)$
$F7$	$(V2,\ V1,\ V0)$
$F8$	$(V10,\ V0,\ V1)$
$F9$	$(V9,\ V10,\ V1)$
$F10$	$(V8,\ V9,\ V1)$
$F11$	$(V4,\ V8,\ V3)$
$F12$	$(V3,\ V5,\ V4)$
$F13$	$(V11,\ V4,\ V5)$
$F14$	$(V10,\ V11,\ V7)$
$F15$	$(V0,\ V10,\ V7)$
$F16$	$(V4,\ V11,\ V9)$
$F17$	$(V8,\ V4,\ V9)$
$F18$	$(V11,\ V5,\ V7)$
$F19$	$(V10,\ V9,\ V11)$

Part answer: In addition to the data for the cube, it's helpful as well to refer to the diagram of the octahedron inscribed in the labeled cube in Figure 10.57.

$V'0$, dual of the face $F0$, is located at the center of $F0$. Its coordinates, therefore, are

$$\begin{aligned}\frac{1}{4}(V3 + V0 + V1 + V2) &= \frac{1}{4}(\ (1, -1, 1) + (1, 1, 1) + (1, 1, -1) + (1, -1, -1)\ )\\ &= (1, 0, 0)\end{aligned}$$

$F'0$, dual of the vertex $V'0$, has vertices that are the duals of the faces of the cube that contain $V0$. From the cube's table the faces containing $V0$

are $F0$, $F3$ and $F4$. Therefore, $F'0$ has vertices $V'0$, $V'3$ and $V'4$.

Use the data for an icosahedron in the two tables above to solve the following two problems.

Exercise 10.68. (**Programming**) Draw an icosahedron.

Exercise 10.69. (**Programming**) Use duality to compute the data for a dodecahedron from that of an icosahedron. In fact, write a short program for this purpose which takes as input the icosahedron data.

10.2.12 Surfaces More Formally

The material in this section is fairly theoretical though we do our best to motivate it practically. We suggest skipping it on a first reading of the book and returning later. If it proves not to your taste at all then you can skip it altogether without affecting your CG skills.

In addition to calculus 101, a basic understanding of partial derivatives is required in order to formalize the notion of surfaces, particularly that of the regular surfaces. If you're not familiar with partial derivatives then a math class or calculus book, e.g., Stewart [137] or Schaum's Outlines [5, 152], is the place to pick the stuff up. We have a handy primer ourselves in Section 11.10 (independent of the rest of that chapter).

Recall that a C^0 curve was defined as the continuous image of a closed interval. Defining a surface as the continuous image of, say, a rectangle seems then a reasonable thing to do.

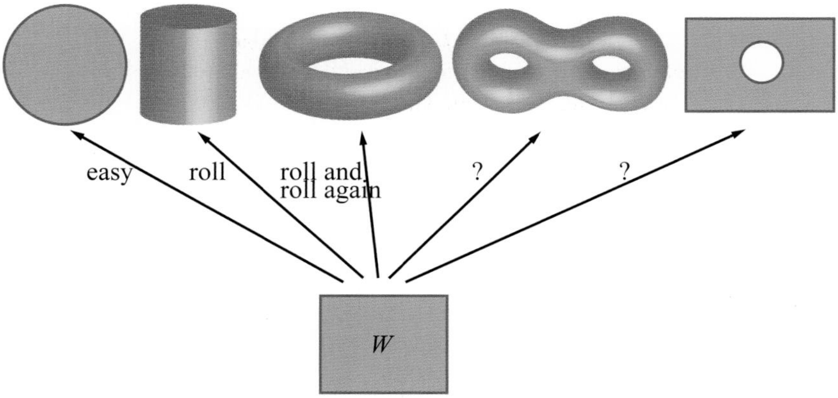

Figure 10.58: Mapping a rectangle onto surfaces.

This does indeed pass muster for simple surfaces. See the objects of Figure 10.58. It's straightforward to map the rectangle W continuously onto the disc. The cylinder requires the parametric functions to roll W – mapping an edge onto the opposite one. The torus can be made in an additional step from the cylinder, by mapping the cylinder's opposite ends onto each other. How about the double torus though, which is certainly a surface? Is it apparent how to map W onto a double torus? Or, consider something as simple as the punctured rectangle, also a surface. How can one map the (unpunctured) rectangle W onto a punctured one in a continuous manner?

It's not quite clear, then, if the view of a surface as simply the image of a rectangle can be successful. Well, even if it might not succeed *globally*, it does *locally*. Huh?

Here's a thought experiment to explain what we mean. Straighten and bring the fingers of your right hand together so that it looks like an ellipse. Actually, let's pretend that it's the rectangle W of Figure 10.58. A question now: if you coated your palm and fingers with gray ink could you color each of the surfaces of Figure 10.58 gray by patting repeatedly? Pats are allowed to overlap. Ignore size constraints as well – think of your palm or any of the surfaces to be as small or large as you like. After a couple of minutes, then, the double torus may look like Figure 10.59.

Figure 10.59: Patting gray a double torus.

Well, ...? We're hoping your answer is yes, that you could pat each of the surfaces fully gray. What would that mean then? Exactly that the surfaces can each be covered by patches, each of which is a continuous image of a rectangle – continuous in the sense that you'll probably have to bend and squeeze your palm a lot, but not do anything drastic like poke a hole through it! In other words, per patch (locally!) the surface is indeed the continuous image of a rectangle.

We're close to a definition of a surface. First, though, we have to formalize the notion of a so-called C^0 coordinate patch.

Definition 10.12. A C^0 *coordinate patch* in $\mathbb{R}^3$ is specified by three real-valued C^0 functions f, g and h, all defined on a closed rectangle $W = [a, b] \times [c, d]$ on the plane, such that the function

$$(u, v) \mapsto (f(u, v),\ g(u, v),\ h(u, v))$$

from W to its image B is one-to-one. The image set

$$B = \{(f(u, v), g(u, v), h(u, v)) : (u, v) \in W\}$$

itself is called a C^0 coordinate patch in $\mathbb{R}^3$.

Remark 10.17. The one-to-one condition ensures that B is topologically equivalent to W, which is stronger than if B were merely a continuous image of W. The examples next clarify this.

Example 10.22. Assume W to be the rectangle $[-1, 1] \times [-1, 1]$ on the plane. Refer to Figure 10.60 for diagrams of the following functions from W.

(a)

$$(u, v) \mapsto (u, v, u^2 + v^2)$$

specifies a coordinate patch that covers the bottom part of a paraboloid.

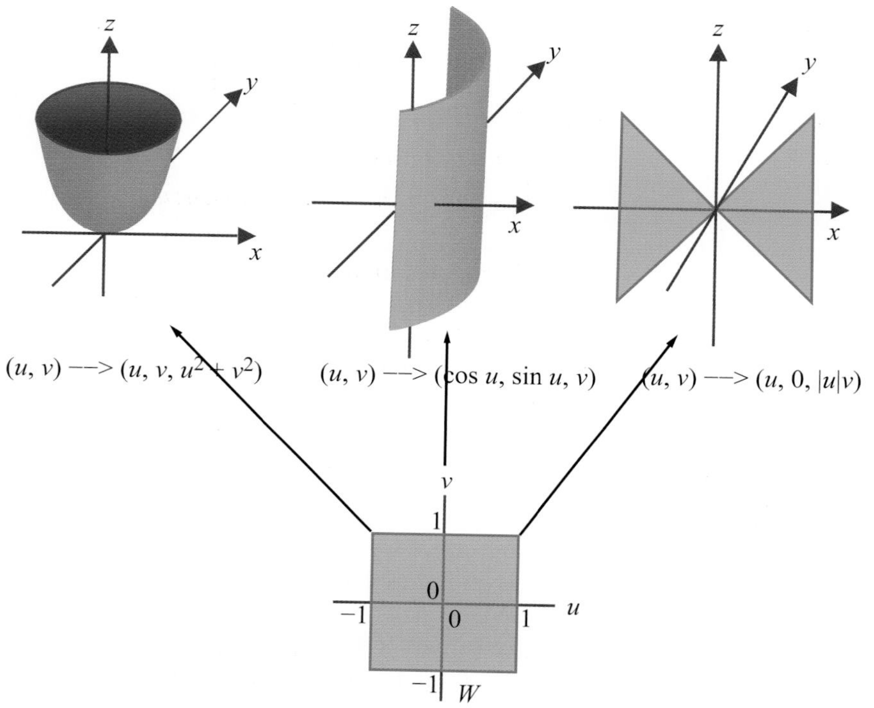

Figure 10.60: Functions $(u, v) \mapsto (f(u, v),\ g(u, v),\ h(u, v))$ and their images.

(b)

$$(u, v) \mapsto (\cos u, \sin u, v)$$

specifies a coordinate patch that covers part of a cylinder.

(c)

$$(u, v) \mapsto (u, 0, |u|v)$$

continuously maps the rectangle onto the union of two triangles *but* is not one-to-one – the entire segment $\{0\} \times [-1, 1]$ of W is mapped to the single point $(0, 0, 0)$ – so does not specify a coordinate patch.

The union cannot be patted gray either, as you can squeeze your palm as thin as you like, but never to a point.

We employ coordinate patches next to define surfaces.

Definition 10.13. A subset s of $\mathbb{R}^3$ is a C^0 *surface* if there is a collection $\mathcal{B}$ of C^0 coordinate patches in $\mathbb{R}^3$ such that

(a) the union of the coordinates patches in $\mathcal{B}$ equals s, and

(b) for each point $P \in s$, a sufficiently small neighborhood of P – i.e., consisting of points within a distance δ of P for some small positive δ – lies inside a single coordinate patch belonging to $\mathcal{B}$.

Remark 10.18. Patches in $\mathcal{B}$ can overlap.

Remark 10.19. $\mathcal{B}$ can be an infinite collection.

Remark 10.20. A C^0 surface is said to be a *two-dimensional topological manifold* or *topological surface*.

The first condition of the definition formalizes the intuition of a surface being covered by "rectangle-like" patches. The second one is to eliminate the sort of pathology shown in Figure 10.61. The object s consisting of two intersecting planes should not qualify as a surface – even though it can evidently be covered by patches – as it's not of a single sheet. In fact, a neighborhood of P, no matter how small, consists of two intersecting fragments, which can never lie in a single coordinate patch.

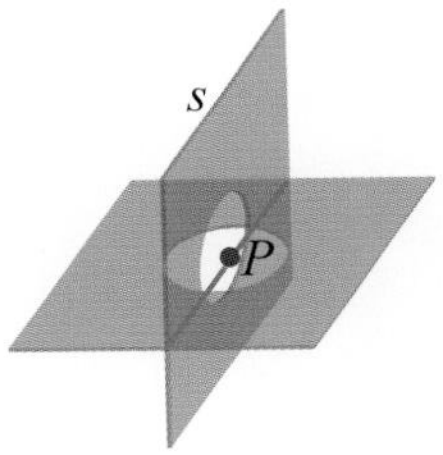

Figure 10.61: Any neighborhood of P will consist of two intersecting fragments, which cannot lie in one coordinate patch.

Example 10.23. What is the minimum number of coordinate patches required to cover the cylinder

$$x = \cos u,\ y = \sin u,\ z = v, \quad (u, v) \in [-\pi, \pi] \times [-1, 1]$$

to prove that it's a C^0 surface according to Definition 10.13? You don't have to write equations for the coordinate patches. Just sketch them on the cylinder.

Answer: Two coordinate patches are sufficient. Figure 10.62(a) indicates one patch covering a sector of more than 180°. Another patch that's a mirror image of this one would cover the rest of the cylinder. The two would overlap, of course.

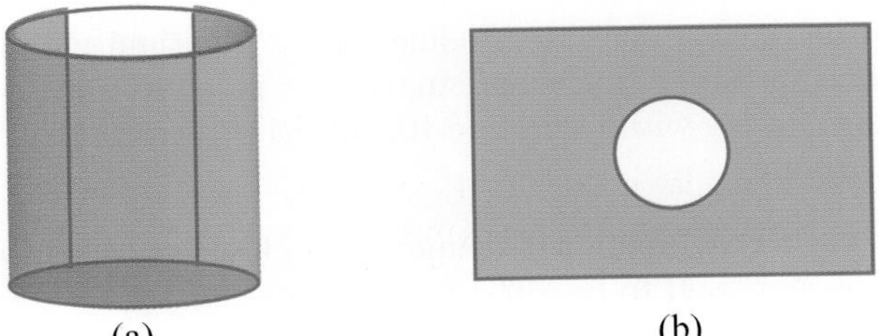

Figure 10.62: (a) One coordinate patch wrapping almost all the way around a cylinder (b) A punctured square.

One coordinate patch by itself will never do. Although this is fairly evident given the shape of the cylinder, a proof requires topology (in particular, the fact that the cylinder is not "homeomorphic" to a rectangle). Thus, two patches is the minimum.

Exercise 10.70. The two coordinate patches of the preceding example overlapped significantly. To avoid "waste" suppose we had two patches that each spanned 180°, covering exactly one half of the cylinder, one a mirror image of the other. Their intersection would then consist of two line segments.

Would these two patches do to prove a cylinder to be a C^0 surface?

Exercise 10.71. How about the following punctured square lying on the xy-plane?

$$[-1, 1] \times [-1, 1] \times \{0\} - \{(x, y, 0) : x^2 + y^2 < 0.5\}$$

See Figure 10.62(b). Note that it's closed with two boundary components – an outside one bounding the square and an inside one bounding the missing disc. How many coordinate patches does one need to prove that it's a C^0 surface? Answer with a sketch.

Exercise 10.72. Consider the following *open* disc (i.e., missing its boundary) lying on the xy-plane:

$$\{(x, y, 0) : x^2 + y^2 < 1\}$$

How does one cover it with coordinate patches to prove that it's a C^0 surface?

Suggested approach: A finite number of coordinate patches will not do. Find, first, a continuous mapping of a closed rectangle W to the closed disc

$$D_r = \{(x, y, 0) : x^2 + y^2 \leq 1 - 1/r\}$$

for any $r \geq 2$, to make D_r a coordinate patch. Consider, then, the union $\cup_{r=2}^{\infty} D_r$.

Defining C^m surfaces for values of m greater than zero is the next natural step and not hard. One must first define the C^m continuity of a function of more than one variable. Not surprisingly, this involves partial derivatives. Compare the following with Definition 10.7 of the C^m-continuity of a function of a single variable.

Definition 10.14. A function f defined on a closed rectangle $W = [a, b] \times [c, d]$ on the plane is said to be *C^m-continuous* or, simply, C^m, where $m \geq 1$, if all of its partial derivatives of order m and less exist and are continuous on W.

C^m coordinate patches will obviously invoke C^m functions of two variables:

Definition 10.15. A C^m *coordinate patch* in $\mathbb{R}^3$, where $m \geq 1$, is specified by three real-valued C^m functions f, g and h, all defined on a closed rectangle $W = [a, b] \times [c, d]$ on the plane, such that the function

$$(u, v) \mapsto (f(u, v),\ g(u, v),\ h(u, v))$$

from W to its image B is one-to-one. The image set

$$B = \{(f(u, v), g(u, v), h(u, v)) : (u, v) \in W\}$$

itself is called a C^m coordinate patch.

If, additionally, the two vectors

$$\begin{bmatrix} \frac{\partial f}{\partial u} & \frac{\partial g}{\partial u} & \frac{\partial h}{\partial u} \end{bmatrix}^T \quad \text{and} \quad \begin{bmatrix} \frac{\partial f}{\partial v} & \frac{\partial g}{\partial v} & \frac{\partial h}{\partial v} \end{bmatrix}^T$$

are linearly independent – i.e., if they are both non-zero and are not collinear – for every $(u, v) \in W$, then the coordinate patch is said to be *regular*.

It's usual to consider regular C^m coordinate patches, when $m \geq 1$, rather than just C^m. Accordingly, here's the definition of a surface covered by such patches:

Definition 10.16. A subset s of $\mathbb{R}^3$ is a *regular* C^m *surface*, where $m \geq 1$, if there is a collection $\mathcal{B}$ of regular C^m coordinate patches in $\mathbb{R}^3$ such that

(a) the union of the coordinate patches in $\mathcal{B}$ equals s, and

(b) for each point $P \in s$, all points of s sufficiently close to P lie in a single coordinate patch belonging to $\mathcal{B}$.

A regular C^m surface, $m \geq 1$, is often simply called a *regular surface*. A regular surface that is C^m for any m is regular C^∞, also called *smooth*.

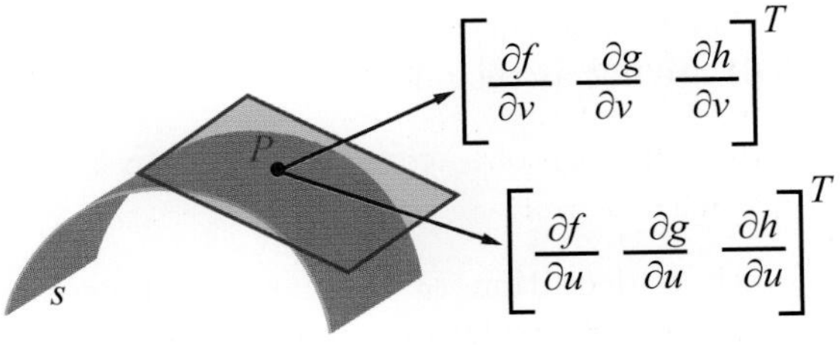

Figure 10.63: The non-zero linearly independent tangent vectors $\begin{bmatrix} \frac{\partial f}{\partial u} & \frac{\partial g}{\partial u} & \frac{\partial h}{\partial u} \end{bmatrix}^T$ and $\begin{bmatrix} \frac{\partial f}{\partial v} & \frac{\partial g}{\partial v} & \frac{\partial h}{\partial v} \end{bmatrix}^T$ span the tangent plane p at P.

Remark 10.21. Recall that the regularity condition in the case of a curve – that the tangent vector never vanishes – ensures a meaningful tangent direction at each of its points. The regularity condition for a surface ensures likewise that a meaningful *tangent plane* exists at each point. See Figure 10.63.

Figure 10.64 shows four surfaces of various orders of continuity. All labels should be clear except maybe for the second one, which is a cylinder capped by a hemisphere. This surface is regular C^1 but not regular C^2, for precisely the same reason that the third curve of Figure 10.20 is C^1 and not C^2 (we'll leave the reader to revisit the explanation there if need be).

The table in Figure 10.64 is not regular, though it is composed of regular pieces. It would be nice to have a definition of piecewise regularity to apply to such surfaces. However, formulating an analogue of Definition 10.5 of piecewise regular curves is not straightforward, as it isn't clear what it means to join a number of surfaces end to end. The following definition finesses the problem.

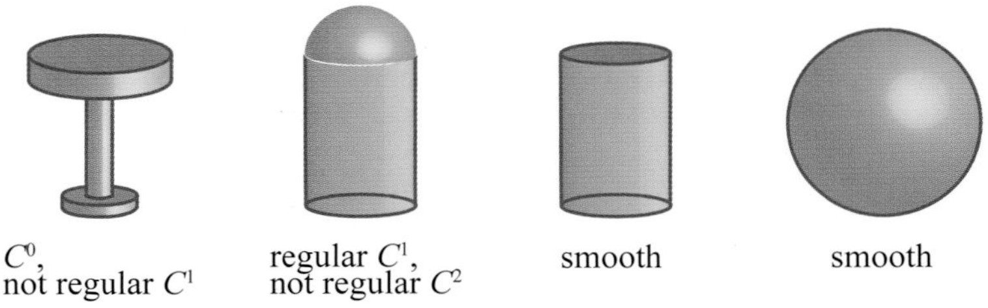

Figure 10.64: Various orders of surface continuity.

Definition 10.17. A *piecewise regular surface* in $\mathbb{R}^3$ is a C^0 surface that is the union of finitely many regular surfaces.

The requirement of C^0-continuity in the definition assures the sheet-like nature of the union. The table is then piecewise regular, but the intersecting planes of Figure 10.61 are not as they don't form a C^0 surface.

Exercise 10.73. Is the surface of a regular polyhedron C^0, C^1, piecewise regular, ...?

We finish up with a definition of regular two-dimensional objects analogous to Definition 10.6 of regular one-dimensional objects, which does apply to the intersecting planes of Figure 10.61, and pretty much everything else one is likely to run into in 3D graphics.

Definition 10.18. A *regular two-dimensional object* is a finite union of regular surfaces.

And for the record we have the proper inclusions

regular surfaces $\subset$ piecewise regular surfaces $\subset$ regular 2D objects

10.3 Bézier Phrase Book

Bézier and NURBS (Non-Uniform Rational B-Spline) curves and surfaces are two special classes of curves and surfaces widely used in 3D design. Their utility to the applications programmer lies in the ability to sculpt a primitive in an intuitive manner by manipulating so-called control points, rather than by devising equations (the equations do exist but are created and managed transparently by the API). Several 3D modeling systems, in fact, allow the user to *interactively* design Bézier and NURBS primitives in a WYSIWYG environment. OpenGL, however, offers both in a sparser code-it-yourself manner.

There is a fair amount of theory underlying both, which is the reason for their effectiveness in the first place, and it's important that designers have a reasonable understanding in order to use them effectively. We'll study polynomial Bézier and NURBS theory in depth in Chapters 15-16. The rational version of both classes of primitives is developed in Chapter 18.

Although it is most convenient to design Bézier and NURBS primitives in a WYSIWYG environment, nevertheless, with a little effort fairly complex designs can be coded in straight OpenGL. Polynomial Bézier curves and surfaces, in particular, are quite intuitive and it's perfectly possible to learn their OpenGL syntax and begin design even before fully grasping the theory. Unfortunately, such is not the case with the B-splines, polynomial or rational or, for that matter, rational Bézier primitives, as it's difficult to make sense of their OpenGL syntax without some theoretical understanding.

In keeping, therefore, with the goal of this chapter to acquaint the reader with as many 3D design techniques as possible, we'll discuss polynomial OpenGL Bézier primitives – both curves and surfaces – without, for now, much of the theory.

10.3.1 Curves

A *Bézier curve* c is specified by a sequence $P_0, P_1, \ldots, P_n$ of *control points* in 3-space, whose number $n + 1$ is called the *order* of c. The curve starts at the first control point P_0, ends at the last P_n and approaches, but does not necessarily pass through, the intermediate ones. Think of the intermediate control points as "attractors" which mold the shape of c. See Figure 10.65, a screenshot of the program `bezierCurves.cpp`, which we'll be discussing momentarily, showing six control points and their Bézier curve (the curvy line).

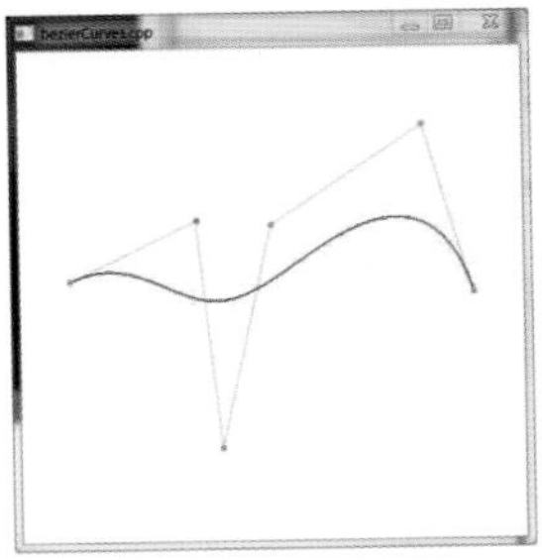

Figure 10.65: Screenshot of `bezierCurves.cpp` with six control points, showing both the Bézier curve and its control polygon.

Specifying the control points defines c in a parametric form

$$x = f(t),\ y = g(t),\ z = h(t)$$

where t belongs to a user-specified interval $[t1, t2]$. We'll leave discussing how the functions f, g and h are obtained to Chapter 15.

Experiment 10.15. Run `bezierCurves.cpp`. Press the up and down arrow keys to select an order between 2 and 6 on the first screen. Press enter to proceed to the next screen where the control points initially lie on a straight line. Press space to select a control point and then the arrow keys to move it. Press delete to start over. Figure 10.65 is a screenshot for order 6.

In addition to the black Bézier curve, drawn in light gray is its *control polygon*, the polyline through successive control points. Note how the Bézier curve tries to mimic the shape of its control polygon. **End**

We'll explain the syntax of the OpenGL Bézier curve with the help of the following two simpler programs.

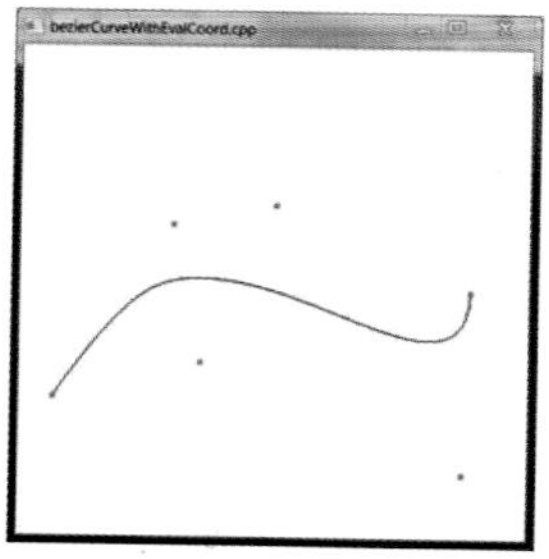

Figure 10.66: Screenshot of `bezierCurveWithEvalCoord.cpp`.

Experiment 10.16. Run `bezierCurveWithEvalCoord.cpp`, which draws a fixed Bézier curve of order 6. See Figure 10.66 for a screenshot. **End**

The pair of statements

```
glMap1f(GL_MAP1_VERTEX_3, 0.0, 1.0, 3, 6, controlPoints[0]);
glEnable(GL_MAP1_VERTEX_3);
```

in the initialization routine of `bezierCurveWithEvalCoord.cpp` specify and enable the Bézier curve. The command

`glMap1f(`*target*`, `*t1*`, `*t2*`, `*stride*`, `*order*`, `**controlPoints*`)`

defines what OpenGL calls a *one-dimensional Bézier evaluator*. Depending on how the parameter *target* is specified, the evaluator can be used to generate data for position, color, texture or normal direction. For now, we'll use it only to generate positional data and, accordingly, set *target* to `GL_MAP1_VERTEX_3`: the '1' is the dimension of the evaluator, while '3' calls for x, y and z coordinate values.

The parameters $t1$ and $t2$ specify the endpoints of the parameter interval of the curve; *order* specifies the number of control points, the coordinate values of which are to be found in the array pointed by *controlPoints*; *stride* is the number of floating point values between the start of the data set for one control point and that of the next in the array.

An evaluator of the form `glMap1f(`*target*`, ...)` must be enabled with a corresponding `glEnable(`*target*`)` command.

The Bézier curve itself of `bezierCurveWithEvalCoord.cpp` is drawn as a line strip joining vertices returned by calls to `glEvalCoord1f-((GLfloat)i/50.0)`. Generally, `glEvalCoord1f(`t`)` evaluates the coordinates of the point on the Bézier curve corresponding to the value t in the parameter interval.

Exercise 10.74. (Programming) Guess what will be displayed if the line strip definition in the drawing routine of `bezierCurveWithEvalCoord.cpp` is changed to either of the two below:

(a)
```
glBegin(GL_LINE_STRIP);
   for (i = 0; i <= 25; i++) glEvalCoord1f( (float)i/50.0 );
glEnd();
```

(b)
```
glBegin(GL_LINE_STRIP);
   for (i = 0; i <= 4; i++) glEvalCoord1f( (float)i/4.0 );
glEnd();
```

As Bézier curves are most often sampled evenly through the parameter interval, OpenGL provides a convenient way to do so, as we see in the next experiment.

Experiment 10.17. Run `bezierCurveWithEvalMesh.cpp`. This program is the same as `bezierCurveWithEval.cpp` except that, instead of calls to `glEvalCoord1f()`, the pair of statements

```
glMapGrid1f(50, 0.0, 1.0);
glEvalMesh1(GL_LINE, 0, 50);
```

are used to draw the approximating polyline.

The call `glMapGrid1f(`*n*`,` *t1*`,` *t2*`)` specifies an *evenly-spaced* grid of $n+1$ sample points in the parameter interval, starting at $t1$ and ending at $t2$. The call `glEvalMesh1(`*mode*`,` *p1*`,` *p2*`)` works in tandem with the `glMapGrid1f(`*n*`,` *t1*`,` *t2*`)` call. For example, if *mode* is `GL_LINE`, then it draws a line strip through the mapped sample points, starting with the image of the $p1$th sample point and ending at the image of the $p2$th one, which is a polyline approximation of part of the Bézier curve. **End**

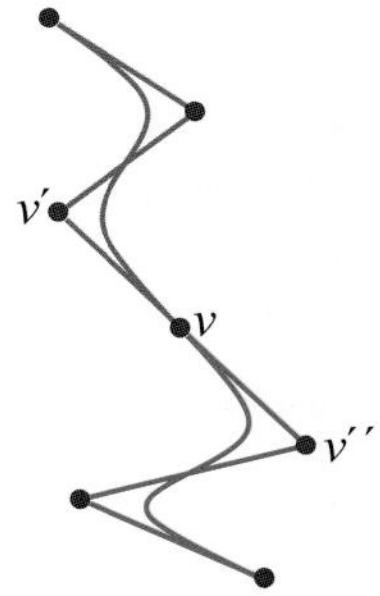

Figure 10.67: Two Bézier curves, one blue and one red, meet smoothly at an endpoint, as their control polygons meet smoothly (because v', v and v'' are collinear).

End Tangents

Not only does a Bézier curve pass through its first and last control points, the tangent at the first control point is along the straight line joining the first two control points. In other words, it lies along the first segment of the control polygon. Likewise, the tangent at the other end lies along the last control polygon segment. This makes it possible to smoothly join two Bézier curves which meet at a common end control point v, by arranging v and its adjacent control points in either curve so that all three are on one straight line. See Figure 10.67.

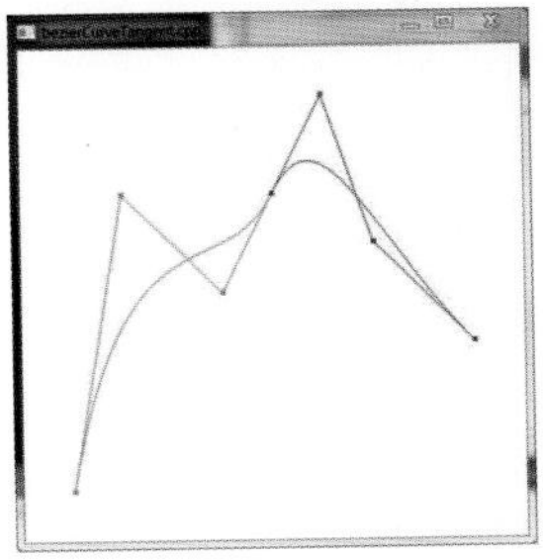

Figure 10.68: Screenshot of `bezierCurveTangent.cpp`.

Experiment 10.18. Run `bezierCurveTangent.cpp`. The blue curve may be shaped by selecting a control point with the space key and moving it with the arrow keys. Visually verify that the two curves meet smoothly when their control polygons meet smoothly. Figure 10.68 is a screenshot of such a configuration. **End**

10.3.2 Surfaces

From Bézier curves to Bézier surfaces is straightforward. A *Bézier surface* (also called *Bézier patch*) s is specified by an $(n+1) \times (m+1)$ array of control points P_{ij}, $0 \le i \le n$, $0 \le j \le m$. The surface passes through the four "corner" control points $P_{00}, P_{n0}, P_{0m}, P_{nm}$, but not necessarily the others which, nevertheless, act as attractors. Let's continue the discussion with live code in front.

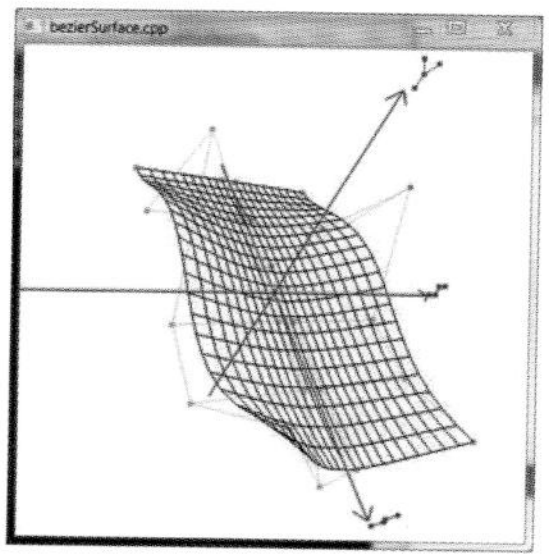

Figure 10.69: Screenshot of `bezierSurface.cpp`, showing both the surface mesh and its control polyhedron.

Experiment 10.19. Run `bezierSurface.cpp`, which allows the user herself to shape a Bézier surface by selecting and moving control points originally in a 6×4 grid. Drawn in black actually is a 20×20 quad mesh approximation of the Bézier surface. Also drawn in light gray is the *control polyhedron*, which is the polyhedral surface with vertices at control points.

Press the space and tab keys to select a control point. Use the left/right arrow keys to move the selected control point parallel to the x-axis, the up/down arrow keys to move it parallel to the y-axis, and the page up/down keys to move it parallel to the z-axis. Press 'x/X', 'y/Y' and 'z/Z' to turn the surface. Figure 10.69 is a screenshot. **End**

Specifying the control points array causes a Bézier surface s to be defined in a parametric form

$$x = f(u,v), \; y = g(u,v), \; z = h(u,v)$$

where u belongs to the interval $[u1, u2]$ and v to the interval $[v1, v2]$. We'll see in Chapter 15 how the functions f, g and h are obtained.

The statement

```
glMap2f(GL_MAP2_VERTEX_3, 0, 1, 3, 4, 0, 1, 12, 6,
        controlPoints[0][0]);
```

in the drawing routine of `bezierSurface.cpp` specifies the Bézier surface, while

```
glEnable(GL_MAP2_VERTEX_3);
```

enables it.

The syntax of the command

```
glMap2f(target, u1, u2, ustride, uorder, v1, v2, vstride,
        vorder, *controlPoints)
```

defining a *two-dimensional Bézier evaluator*, or Bézier surface, is an extension of that for a one-dimensional evaluator, taking into account the extra dimension. Like its one-dimensional counterpart, a two-dimensional evaluator can be used to generate data for position, color, texture or normal direction. We'll restrict ourselves to positional data for the present, setting *target* to `GL_MAP2_VERTEX_3` (indicating a 2D surface in 3D space).

The values $u1$ and $u2$ specify the endpoints of the u-parameter interval and $v1$ and $v2$ those of the v-parameter interval. The parameter *uorder* is $m+1$, the number of columns of the control points array P_{ij}; *vorder* is $n+1$, the number of rows.

The coordinate values of the control points are located in the array pointed by *controlPoints*. The parameter *ustride* is the number of floating point values between the starts of the data sets for control points P_{ij} and $P_{i,j+1}$; *vstride* is the number of floating point values between the starts of the data sets for control points P_{ij} and $P_{i+1,j}$.

The pair of statements

```
glMapGrid2f(20, 0.0, 1.0, 20, 0.0, 1.0);
glEvalMesh2(GL_LINE, 0, 20, 0, 20);
```

are analogous in functionality to `glMapGrid1f()` and `glEvalMesh1()` discussed in the context of drawing Bézier curves. In particular,

```
glMapGrid2f(numberU, u1, u2, numberV, v1, v2)
```

specifies a $(numberU + 1) \times (numberV + 1)$ grid of sample points in the parameter rectangle, evenly spaced along both rows and columns, each row starting with u-value $u1$ and ending with u-value $u2$, and each column starting with v-value $v1$ and ending with v-value $v2$. The call

```
glEvalMesh2(mode, i1, i2, j1, j2)
```

works in tandem with the `glMapGrid1f(nu, u1, u2, nv, v1, v2)` call. For example, if *mode* is `GL_LINE`, then a stack of outlined quad strips is drawn with vertices at the mapped sample points, making a quadrilateral mesh approximation of the Bézier surface; if *mode* is `GL_FILL` then the strips are drawn filled.

Remark 10.22. It's a minor design flaw of OpenGL that a Bézier surface is approximated with a stack of quad strips, rather than triangle strips.

Remark 10.23. The u-parameter curves and v-parameter curves of a Bézier surface are (no surprise) Bézier curves. Think of the parameter u as being associated with (i.e., varying along) the columns of the control points array, and v with the rows. Accordingly, for a fixed i, the points $P_{i0}, P_{i1}, \ldots, P_{im}$ on the ith row are the control points of a u-parameter curve. The order of a u-parameter curve, therefore, is $m+1$, which is the parameter *uorder* of `glMap2f()`.

Likewise, for a fixed j, $P_{0j}, P_{1j}, \ldots, P_{nj}$ are the control points of a v-parameter curve, whose order is $n+1$, the parameter *vorder* of `glMap2f()`.

Remark 10.24. There is a `glEvalCoord2f()` call available as well, analogous to `glEvalCoord1f()`, to evaluate the coordinates of a point on the surface corresponding to parameter point (u, v).

Next is an example of how to make a target shape by manipulating the control points of a Bézier surface.

Experiment 10.20. Run `bezierCanoe.cpp`. Repeatedly press the right arrow key for a design process that starts with a rectangular Bézier patch, and then edits the control points in each of three successive steps until a canoe is formed. The left arrow reverses the process. Press 'x/X', 'y/Y' and 'z/Z' to turn the surface.

The initial configuration is a 6×4 array of control points placed in a rectangular grid on the xz-plane, making a rectangular Bézier patch.

The successive steps are:

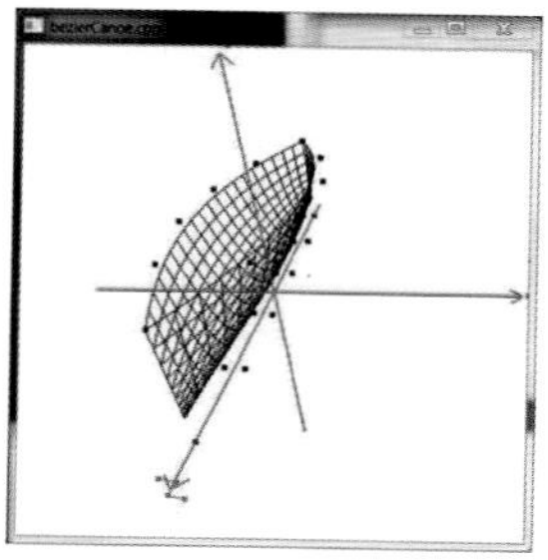

Figure 10.70: Screenshot of `bezierCanoe.cpp`.

(1) Lift the two end columns of control points up in the y-direction and bring them in along the x-direction to fold the rectangle into a deep pocket.

(2) Push the middle control points of the end columns outwards along the x-direction to plump the pocket into a "canoe" with its front and back open.

(3) Bring together the two halves of each of the two end rows of control points to stitch closed the erstwhile open front and back. Figure 10.70 is a screenshot after this step.

End

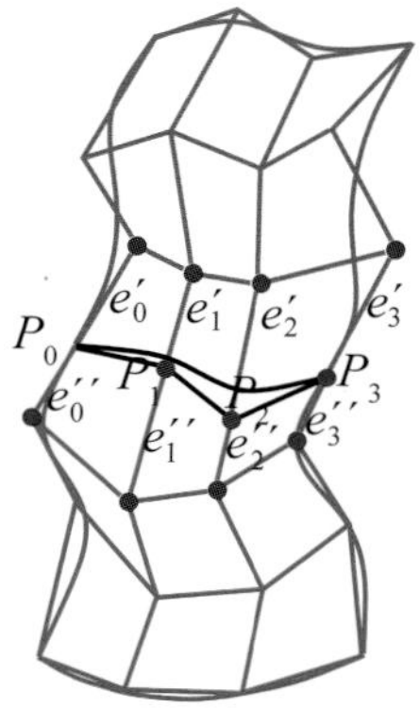

Figure 10.71: Two bicubic Bézier patches and their control polyhedrons, one pair blue and one red. The patches meet smoothly along a shared boundary curve which, together with its control polygon, is black.

Bicubic Bézier Patches and How to Join Them

Most often invoked in design are Bézier surfaces specified by a 4×4 array of control points, called *bicubic Bézier patches*. Complex shapes can be made by connecting multiple such patches. A similar principle applies to joining two bicubic patches – in fact, arbitrary Bézier patches – smoothly, as applies to joining two Bézier curves smoothly. First, two patches are contiguous if they share a common end row or end column of control points, in which case their control polyhedrons abut along that row or column. For the two patches to join smoothly, one further requires every pair of edges, one from either control polyhedron, meeting at a vertex of the shared border but not lying on the border itself, to be collinear.

For example, in Figure 10.71 two bicubic patches meet along a common boundary curve specified by their shared control points P_0, P_1, P_2 and P_3, which also specify the shared border of their control polyhedrons. The edges of the control polyhedrons on either side that meet at the border, viz. the pairs e'_i and e''_i, for $0 \leq i \leq 3$, are collinear, so the patches join smoothly.

Utah Teapot

Probably the most famous object ever made from bicubic Bézier patches is the *Utah Teapot*, created originally by Martin Newell, then at the University of Utah, in 1975. It rapidly became an iconic benchmark model within the

CG community for the testing of rendering algorithms. The FreeGLUT library's wireframe version is shown on the left of Figure 10.72.

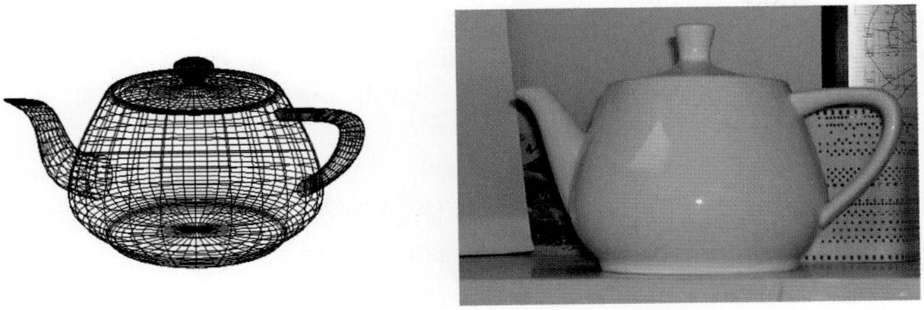

Figure 10.72: FreeGLUT library's version of the Utah teapot and Martin Newell's original porcelain Melitta model (from Wikimedia).

Newell's original design consisted of 28 patches and had neither a bottom nor a rim for the lid to rest on. The current incarnation, available from the FreeGLUT library, has both, and consists of 32 patches and a total of 306 different control points – 12 patches specify the body of the teapot, 4 the handle, 4 the spout, 8 the lid and 4 the bottom. Patches that meet obviously share control points and those composing the same part of the teapot join smoothly based on the principle described above.

Newell modeled an actual porcelain teapot that he owned, manufactured by the Melitta company. It's now on exhibit at the Computer History Museum in California. A picture is on the right of Figure 10.72. The rendered version, in fact, is squatter than the original. For the reason why and an entertaining account of the teapot's evolution read Crow's article [32]. Crow gives the patch control points data as well.

Torpedo

Experiment 10.21. Run `torpedo.cpp`, which shows a torpedo composed of a few different pieces, including bicubic Bézier patch propeller blades:

(i) Body: GLU cylinder.

(ii) Nose: hemisphere.

(iii) Three fins: identical GLU partial discs.

(iv) Backside: GLU disc.

(v) Propeller stem: GLU cylinder.

(vi) Three propeller blades: identical bicubic Bézier patches (control points arranged by trial-and-error).

Press space to start the propellers turning. Press 'x/X', 'y/Y' and 'z/Z' to turn the torpedo. Figure 10.73 is a screenshot. **End**

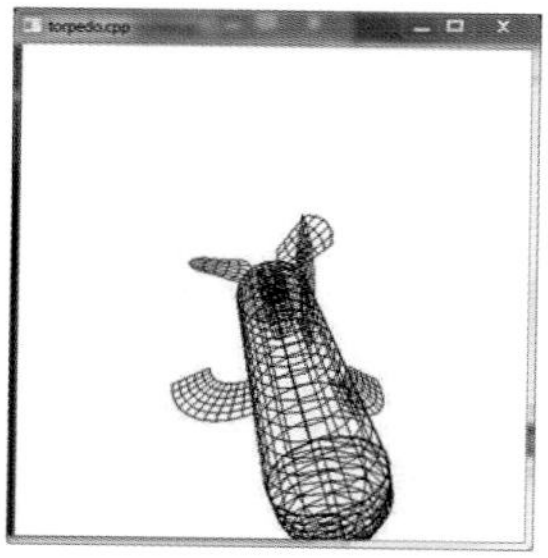

Figure 10.73: Screenshot of `torpedo.cpp`.

Exercise 10.75. (Programming) Emulate Newell by modeling an everyday object using multiple bicubic Bézier patches. First, modify `bezierSurface.cpp` for an editable bicubic patch over a 4×4 array of control points.

A shoe is about as everyday as it gets. Figure 10.74 is a screenshot of a shoe designed using three different Bézier patches by a CG student.

Exercise 10.76. (Programming) Animate a river scene with many boats. Modify the canoe from `bezierCanoe.cpp` for a couple of different kinds of boats and put them in display lists and place (scaled and colored) copies on the river. Give a split screen view, one global from the bank and one from a particular boat.

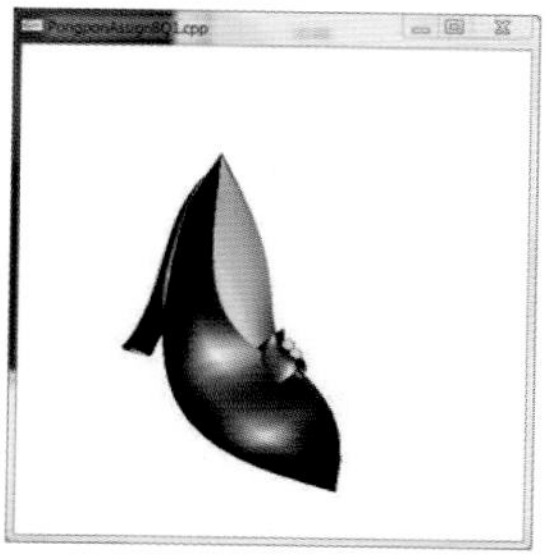

Figure 10.74: Bézier lady's shoe (courtesy of Pongpon Nilaphruek).

Figure 10.75: Aircraft and express train.

Exercise 10.77. (Programming) Model the aircraft (Figure 10.75) and make it fly. Use bicubic patches. Parts like the wings, tail and fins can be panels of zero thickness, as can be the jet engine cases. Ignore logos and details. These can be textured in later. Focus on large-scale geometry.

Exercise 10.78. (Programming) Model a running version of the express train (Figure 10.75) using bicubic patches.

Exercise 10.79. (Programming) Make a recognizable replica of some familiar automobile using bicubic patches.

Remark 10.25. We reiterate our opinion that the new programmer gains most creating objects almost by hand as it were in the minimalist OpenGL environment. However, it is no doubt true, too, that designing complicated objects and busy scenes in this manner can be hard. So, once the programmer "knows what she is doing" we recommend, in fact, she learn a sophisticated WYSIWYG modeling software like Maya, Studio Max or Blender in order to take on larger projects. Note that there are plug-ins for most 3D modelers to export their triangles and geometry to OpenGL.

10.4 Fractals

Fractals are fun, but not essential to design. You can safely skip this section if you are in a hurry to progress through the book.

A *fractal* shape, or just fractal for short, is one that possesses the characteristic of *self-similarity*. A canonical example of a fractal in nature is a coastline. The sketch on the left of Figure 10.76 is that of a hypothetical stretch of a hundred miles of coastline, as it may appear from an aircraft, outlining the shape of a country's border with the sea. To its right is a view zoomed in on a part of maybe about ten miles, seen from a low-flying aircraft, showing bays and inlets. Rightmost is an even closer zoom-in on a stretch of one mile of beach showing its own features. One notices the similarity across different levels of resolution – self-similarity – in the undulations of the coastline. Clouds, trees and neural systems of animals are among a multitude of other naturally occurring fractals.

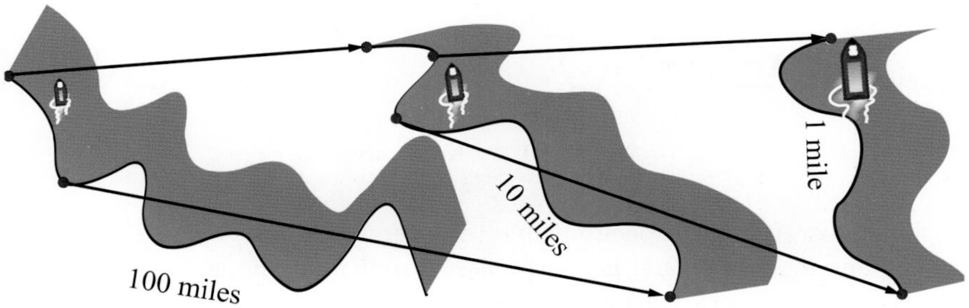

Figure 10.76: A coastline at increasing degrees of resolution: pairs of arrows indicate a blow-up.

Self-similarity lends itself immediately to programming by recursion. We'll give a semi-formal definition of fractal curves which makes self-similarity explicit. Our goal is not mathematical rigor, but a reasonable framework within which to write recursive code. The running example we'll use is a classic fractal curve – the *Koch curve* – invented by the Swedish mathematician Helge von Koch.

The first step in defining a fractal curve is to specify a *source* curve s. The location and orientation of s are not specified: it can be placed freely anywhere on the plane if we are drawing in 2D or space if in 3D. In the case of the Koch curve, s is a straight line segment on a fixed plane. See the top diagram in Figure 10.77.

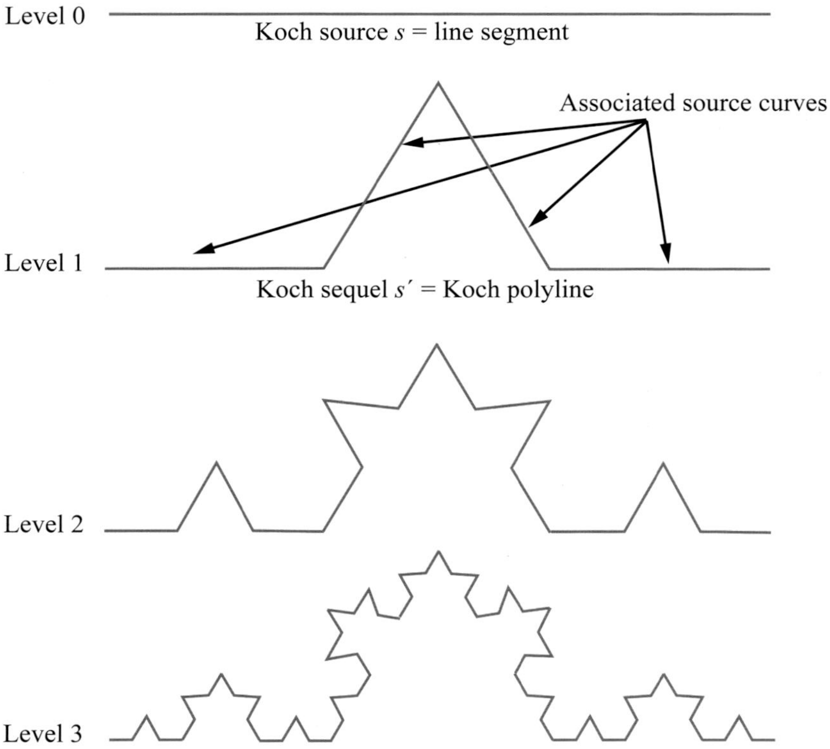

Figure 10.77: Koch curves.

The next step is to specify a rule to generate a *sequel* curve s' from the source curve. The sequel is rigidly associated with the source in that, if the location and orientation of s are fixed, then so are those of s'. The sequel for a Koch curve, in particular, is obtained by deleting the middle one-third segment of the source s and replacing it with two edges of an equilateral triangle such that the third edge would occupy the position of the now-deleted middle third. See the second diagram from the top in Figure 10.77. This particular 4-segment polyline sequel is called the *Koch polyline*.

The third and final step is to specify a rule to allow iterative reproduction of the sequel curves. In particular, for a sequel s' this rule specifies a finite set $\{s_1, s_2, \ldots, s_n\}$ of curves that are each a "transformed" copy of the source s. We'll not try to be precise as to how the transformations must be specified. It's best to imagine them being such that the s_i each are similar in shape to the source s, differing only in scale and location, which is the most common situation. Moreover, each s_i, $0 \leq i \leq n$, is rigidly associated with s' in that, if the location and orientation of s' are fixed, then so are those of s_i. The s_i,

$0 \leq i \leq n$, are said to be the source curves *associated* with the sequel. In the case of the Koch curve, the associated source curves are simply the four segments of the Koch polyline, as indicated in Figure 10.77, each obviously a scaled version of the Koch source.

We are now ready to recursively produce the fractal curve to any desired level. Level 0 is a fixed copy s of the source. Level 1 is s replaced with its sequel s'. Level 2 is s' replaced with the union of the sequels of its associated sources $s_1, s_2, \ldots, s_n$, with the proviso that transformed copies of the source each generate equally transformed copies of the sequel. Level 3 and higher are obtained by repeating the procedure. Figure 10.77 shows the Koch curves till level 3.

The following program demonstrates the flexibility afforded by the framework just described.

Experiment 10.22. Run `fractals.cpp`, which draws three different fractal curves – a Koch snowflake, a variant of the Koch snowflake and a tree – all within the framework above, by simply switching source-sequel specs! Press the left/right arrow keys to cycle through the fractals and the up/down arrow keys to change the level of recursion. Figure 10.78 shows all three at level 4. **End**

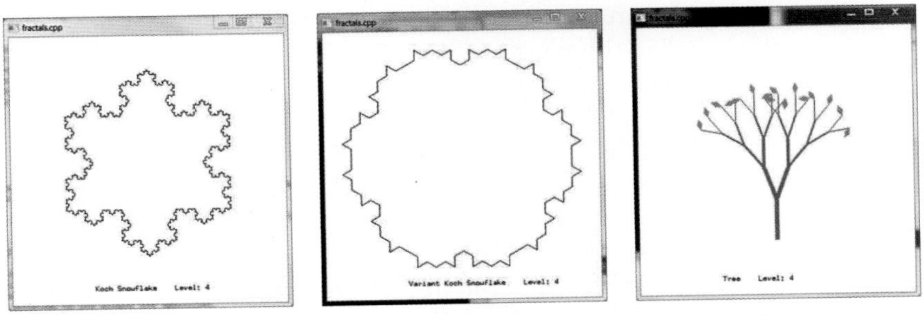

(a) (b) (c)

Figure 10.78: Screenshots from `fractals.cpp`: (a) Koch snowflake (b) Variant Koch snowflake (c) Tree.

The first curve of `fractals.cpp` drawn is the so-called *Koch snowflake* which consists of three Koch curves, as described earlier, each starting with one edge of an equilateral triangle as its level 0 source.

The variant Koch snowflake is produced with a single change from the definition of the Koch snowflake: instead of specifying the four segments of a sequel Koch polyline as its associated sources, the two segments joining the end vertices of the polyline to its middle are specified. See the upper two diagrams in Figure 10.79. Observe that it's not necessary that the associated sources be parts of the sequel.

The source for the fractal tree is a straight line segment as well, the initial copy being vertical. The sequel is a V-shaped two-segment polyline located

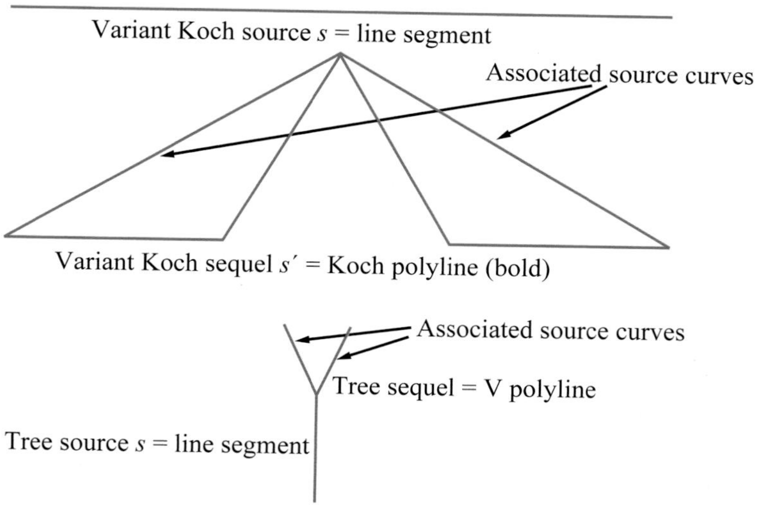

Figure 10.79: The variant Koch curve and fractal tree.

atop the source, the length of each segment being a specified fraction (the constant `RATIO` in the program) of the length of the corresponding source, with a specified angle (the constant `ANGLE`) between them. The sources associated with the sequel are its two segments. See the bottom diagram of Figure 10.79.

The tree is produced by drawing the original vertical source line segment, as also the succeeding sequels at *all* levels, till the highest level of recursion. Note the difference here with the Koch curve and its variant, *only* the highest-level sequels being drawn in the case of the latter two. Moreover, sequels at successive levels of the fractal tree are drawn thinner – not part of the fractal definition. The final drawing is also embellished with leaves – not part of the fractal structure either – which are quadrilaterals at random angles at the ends of the top-level V's.

The program combines all three fractals by providing different member functions for each in the classes `Source` and `Sequel`.

Exercise 10.80. (Programming) Create a non-uniform tree by adding randomization so that not all sequels are drawn.

Exercise 10.81. (Programming) The variant of the Koch snowflake drawn in `fractals.cpp` self-intersects at high enough levels. Draw an "interesting" variation of the snowflake which doesn't self-intersect.

Exercise 10.82. (Programming) Draw a fractal cloud.

Exercise 10.83. (Programming) Draw a fractal flower.

10.5 Summary, Notes and More Reading

In this chapter we learned how to create a range of objects in 3D space. Of course, we had already been creating objects earlier, but in this chapter we studied 3D drawing techniques systematically. These included polygonal line approximation of curves and mesh approximation of surfaces. Special classes of curves and surfaces particularly useful in drawing, including conics and quadrics, swept and ruled surfaces, regular polyhedra, Bézier curves and surfaces, and fractals were discussed. We were introduced at an elementary level to the mathematical theory underlying curves and surfaces.

The dictum that practice makes perfect applies particularly to drawing. The more drawing projects one completes the better one will know how to proceed on the next one, not to mention the reusable object parts and code one will begin to accumulate.

A great thing about drawing scenes for movies and games is that if it walks like a duck and talks like a duck, then it *is* a duck. We have already taken advantage of this notion in faking curved objects with the help of straight and flat primitives. There's more to it though as the following example shows.

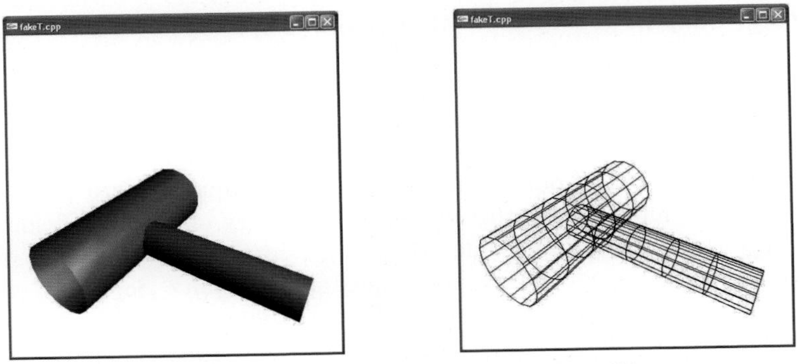

(a) (b)

Figure 10.80: A T-pipe is simulated by sticking one GLU cylinder into another.

On the left of Figure 10.80 is a picture of what seems a perfectly respectable T-pipe of a sort one might find at a plumbing goods store. The picture on the right though reveals how it's drawn – by pushing one GLU cylinder into another so that an end of it protrudes inside (run `fakeT.cpp`). This is unlikely to be an acceptable industrial design of a T-pipe, but for the purposes of 3D graphics it is. Consider the saving in complexity. Authentic industrial design would require a hole whose boundary is a non-planar loop to be excised in one cylinder and an end of the other pared to match – neither trivial operations at all!

One is reminded of the amusing story of a farmer and a mathematics

Figure 10.81: Sheared? But, what about the other side?

professor traveling together on a train through the countryside. The farmer looking out of the window remarks to the professor, "The sheep look like they've just been sheared." The ever-precise professor replies, "Well, we can say only that the side that we see has been sheared. Nothing can be concluded about the other side!" In CG one can indeed get away with shearing only the sides that the viewer sees. This liberation from real-world rectitude (or even rationality!) throws the doors wide open to creativity. If you are working on a game or a movie, now is the time to enhance it with various objects. And, if you are getting a bit tired now of wireframe, it won't be long – less than a page as a matter of fact – before we begin to color and illuminate!

All introductory texts on 3D graphics have parts devoted to drawing curves and surfaces in 3-space. Books containing a more specialized treatment of modeling include Farin [45], the two by Mortenson [94, 96] and Rogers & Adams [118]. They all include discussions of the Bézier primitives as well. In Simmons [133]) the reader will find more about conics and quadrics. The author's own paper [62] is a mathematical investigation of a rather curious irregularity of regular polyhedra.

Keep in mind that we are not at all done with modeling ourselves. This chapter laid the groundwork. More of the theory of Bézier curves and surfaces comes in Chapter 15. Chapter 16 is about B-spline curves and surfaces, which are staples in modern design. That chapter discusses the polynomial version of B-splines, while NURBS – non-uniform rational B-splines, the most general version – are a topic of the later Chapter 18. Chapter 17 is about Hermite curves and patches, which interpolate – i.e., actually pass through – their control points, rather than merely approximate.

The reader interested in the mathematical theory of curves and surfaces, especially those wishing at some point to get into the research end of 3D graphics, should refer to math books such as Lipschutz [88] and Pressley [115] for a fairly soft introduction to differential geometry, while the books by Do Carmo [37], Kreyszig [81], O'Neill [106] and Singer & Thorpe [134] are written at a higher level.

Our account of fractals, though basic, has probably enough for the person who primarily wants to draw them. There are several excellent books to learn more about this popular topic. In addition to Mandelbrot's classic [91], which had seminal influence in formalizing fractals and attracting popular interest, a couple of more recent ones are by Barnsley [9] and Falconer [44].

Part VI

Lights, Camera, Equation

CHAPTER 11

Color and Light

Our objects so far have mostly looked as if they plan to spend the afternoon home watching the game. It's time now to dress them up and go party. The goal for this chapter is to learn how to use light sources to illuminate a scene and complementarily define material properties of objects to determine how they appear when lit.

We begin with a brief discussion of the theory of vision and color models in Section 11.1, learning particularly about the RGB color model so important in CG, as well as a few other models which pop up occasionally, such as CMY, CMYK and HSV. In Section 11.2 we study Phong's lighting model and how it conceives of light coming off an object as comprised of three components – ambient, diffuse and specular – based on the nature of their reflectance. This section concludes with a formula to derive the RGB intensities of the light reflected at a vertex based on Phong's model.

We move on to OpenGL in Section 11.3 and see how faithfully it implements Phong's model. And, we begin extensively to experiment and code. Section 11.4 describes OpenGL's so-called lighting model – not to be confused with Phong's lighting model – which sets certain environmental parameters. Directional light sources, located far from the scene, and positional lights, located in or near it, are discussed in Section 11.5, as is the related notion of attenuation of light over distance. Spotlights are the topic of Section 11.6. At this point we have all the parts needed to formulate in Section 11.7 the famous lighting equation that OpenGL actually implements to calculate color intensities at a vertex.

We discuss the two so-called shading models OpenGL offers, smooth and flat, in Section 11.8. The former familiarly interpolates the vertex colors through a primitive, while the latter is a somewhat idiosyncratic discrete coloring scheme. Animation of light sources is the topic of Section 11.9.

Specifying appropriate surface normals is critical to good lighting.

OpenGL can sometimes help with automatic normals, but often the user is on her own and the task can require a fair amount of calculation. Before we begin with normal computation proper, we have an introduction in Section 11.10 to the calculus of partial derivatives, to the extent required to calculate tangent planes and normals to the kinds of surfaces typical in CG.

The long Section 11.11 is devoted to computing and applying surface normals to lighting. It begins by following the informal taxonomy of 2D objects introduced in Section 10.2, and moves on to Bézier and quadric surfaces for which automatic normals are available.

Section 11.12 contains a discussion of an alternate shading model proposed by Phong, which is more sophisticated (and more computation-intensive) than OpenGL's smooth shading.

Section 11.13 is a whole bunch of exercises. In fact, the reader will find few programming exercises before that section as we decided to collect them in one place after getting most of the theory out of the way. However, this does not mean that the interested reader should not attempt them earlier. She should keep an eye on Section 11.13 as she reads to see what comes within reach. We conclude with Section 11.14.

11.1 Vision and Color Models

We begin with a bit of the physics and biology underlying color and its perception.

Electromagnetic (EM) radiation consists of oscillating electric and magnetic fields moving through space. It is produced by the motion of electrically charged particles. From a physics point of view, EM radiation can be treated dually as waves or a stream of massless particles called photons traveling through a vacuum at the speed of light. EM radiation is characterized by its frequency or, equivalently, wavelength, the inverse of frequency. The EM *spectrum* consists of EM radiation of all possible frequencies. Visible light is a (very small) part of the EM spectrum. See Figure 11.1.

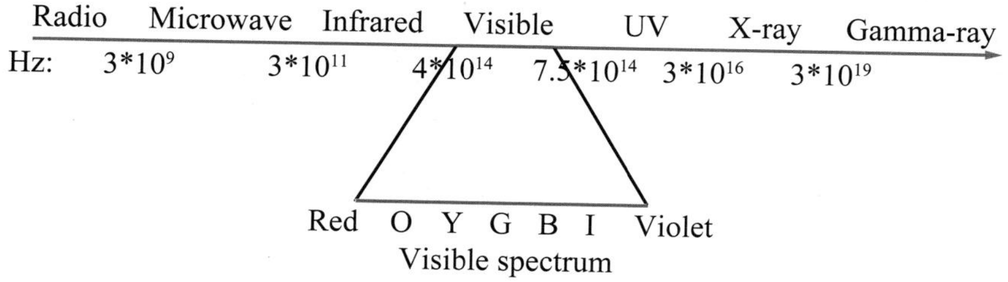

Figure 11.1: EM spectrum indicating approximate frequency ranges in Hz.

Visible light emitted from a source is rarely pure, i.e., of one particular frequency. Rather, there is an intensity distribution across the entire visible spectrum, and the perceived color depends on the particular distribution. Light from a source with an intensity distribution, for example, as in Figure 11.2(a), would be perceived as blue, as this color's intensity dominates, while one with the distribution of Figure 11.2(b) would appear white, because white is a mix of all colors in the visible spectrum.

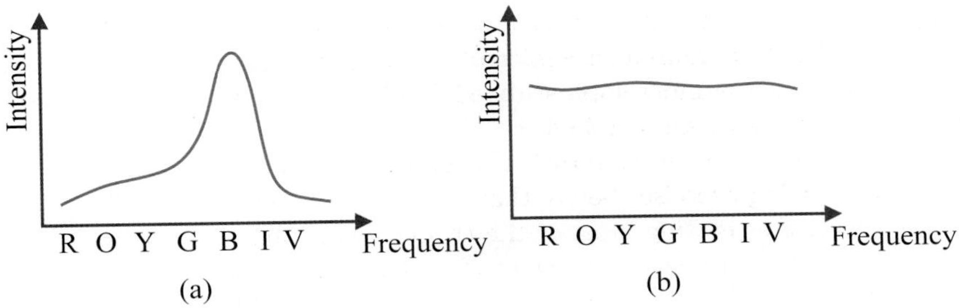

Figure 11.2: Intensity distributions across the visible spectrum: (a) appears blue (b) appears white.

We humans can see because of millions of light-sensitive cells embedded in the retinas of our eyes (see Figure 11.3 for a simplified anatomy). These cells are of two kinds, rod and cone. Rod cells are sensitive to low-intensity light, but not its frequency, which accounts for our night vision, as well as the fact that we have particular difficulty distinguishing colors in the dark. Cone cells, on the other hand, are stimulated only by fairly bright light, but can efficiently distinguish frequencies in the visible light spectrum, enabling us to perceive color. In fact, there are three kinds of cones – red, green and blue – according to the color of the light that most stimulates them. This is the basis of the *tristimulus* theory of human vision that perceived color is the net effect of the stimulation of these three kinds of cells.

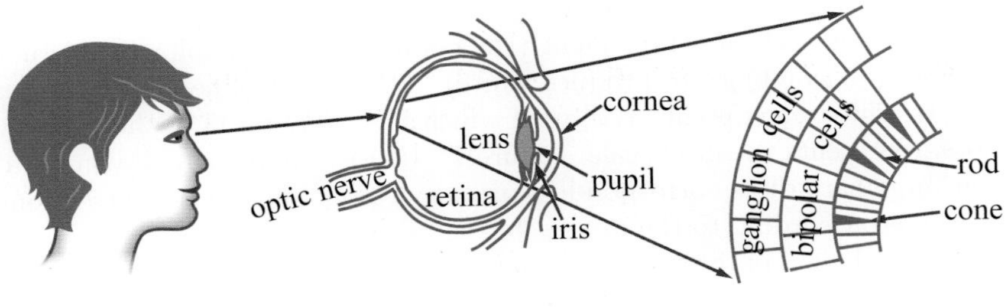

Figure 11.3: The eye.

11.1.1 RGB Color Model

A consequence of the tristimulus theory is the ubiquitous *RGB color model*: each color is represented as a sum of the three *primary colors*, red, green and blue, and each with a certain intensity, typically a value between 0 and 1 (for this reason RGB is called an *additive color model*). A color is denoted by a *color tuple* (r, g, b), where each component is the respective primary color's intensity.

Note: An intensity distribution curve, as those in Figure 11.2, one corresponding to each primary color, has been standardized by the International Commission on Illumination (CIE, from its French name Commission Internationale de L'Éclairage), as also a standard to convert intensity distributions across the visible spectrum to RGB triplets.

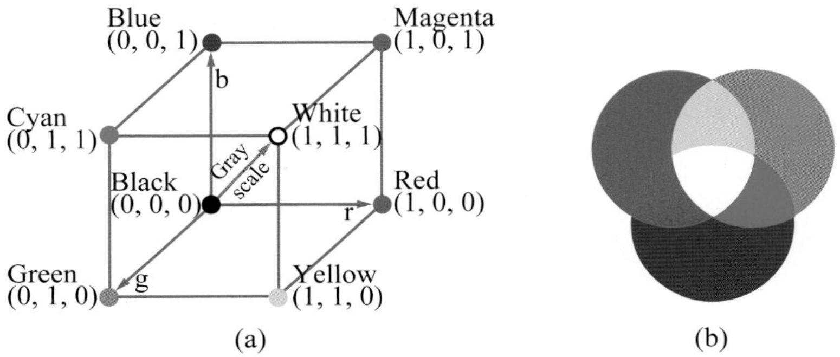

Figure 11.4: (a) The RGB color cube (b) Venn diagram combining colors.

The RGB color space can be depicted as a cube, called a *color cube*, with axes corresponding to R, G and B values (see Figure 11.4(a)). The origin $(0, 0, 0)$ of the cube corresponds to black, while its diagonally opposite vertex $(1, 1, 1)$ to white, which, of course, is the maximal equal mix of red, green and blue. The other three diagonally opposite pairs each corresponds to a primary color and its complement (the complement of a color being that which with it combines to produce white). The straight line segment from black to white, each point (x, x, x) of which has equal parts of the primary colors, represents the gray scale. Figure 11.4(b) is a popularly drawn Venn diagram, where discs correspond to primary colors and their intersections are colored according to the mixing of the primaries.

The mechanics of the "addition" of colors in the RGB model is interesting. The color cube, for instance, indicates that an equal mix of red (which is $(1, 0, 0)$) and green ($(0, 1, 0)$) produces yellow ($(1, 1, 0)$). The reason for this is that the sensation produced in the human eye by a mix of two lights, one whose red frequency dominates and another whose green dominates, is similar to that produced by a single light dominant in the yellow frequency.

This is a consequence of how our optic nerves react to the stimulation of particular combinations of cone cells, and *not* because the frequencies of red and green combine according to some law of physics to produce that of yellow! The RGB model, therefore, rests more on the biology of human vision than the physics of light.

RGB Color Model and Computer Graphics

The RGB model is implemented in millions of color display units around us, including computer monitors, both CRT and LCD. A CRT (cathode-ray tube) monitor has phosphors of the three primary colors located at each one of a rectangular array of pixels, and three electron guns that each fires a beam at phosphors of one color. A mechanism to aim and control their intensities causes the beams to travel together row by row, striking successive pixels in order to excite the RGB phosphors at each to values specified for it in the color buffer. See Figure 11.5(a).

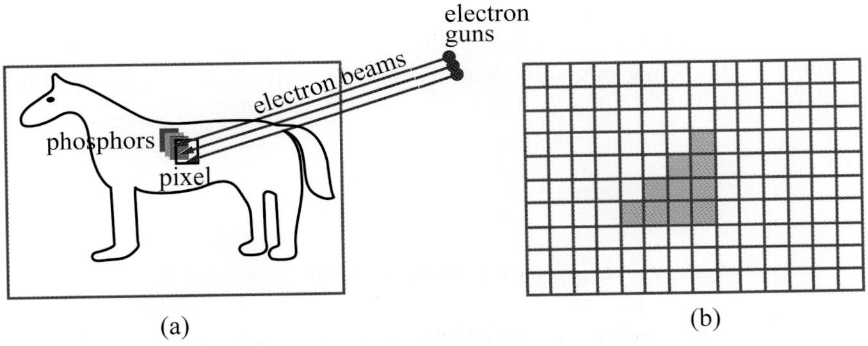

Figure 11.5: (a) Color CRT monitor with electron beams aimed at a pixel with phosphors of the 3 primaries (b) A raster of pixels showing a rasterized triangle.

Pixels in an LCD (liquid crystal display) monitor, on the other hand, each consist of three subpixels made of liquid crystal molecules, which separately filter through light of only one primary color. The amount of primary color emerging through each subpixel is controlled by an electric charge, whose intensity in turn is determined by the corresponding value in the color buffer.

From the point of view of OpenGL and, indeed, most CG theory, what matters most is that the pixels in a monitor, CRT or LCD, are, in fact, arranged in a rectangular array, called a *raster* (as depicted in Figure 11.5(b)). The number of rows and columns in the raster determines the monitor's resolution. This rectangular layout is the basis of the lowest-level CG algorithms, the so-called raster algorithms, which actually select and color the pixels to represent user-specified shapes such as lines and triangles on the monitor. Figure 11.5(b), for example, shows the rasterization of a right-angled triangle (with terrible jaggies because of the low resolution). We'll be studying raster algorithms in fair depth ourselves in Chapter 14.

Furthermore, a memory location called the *color buffer*, either in the CPU or graphics card, contains, typically, 32 bits of data per raster pixel – 8 bits for each of RGB and 8 for the alpha value (used for blending). It is the RGB values in the color buffer which determine the corresponding raster pixel's color intensities. The values in the color buffer are read by the raster – at which time the raster is said to be refreshed – at a rate called the monitor's refresh rate. Beyond this, the technology underlying a particular display device matters little practically in computer graphics.

11.1.2 CMY and CMYK Color Models

The *CMY color model*, whose augmentation CMYK is typically used in color printing, is a *subtractive color model*. CMY stands for cyan, magenta and yellow, and they are, respectively, the complements of red, green and blue. For example, cyan reflects blue and green but absorbs (or subtracts) red. Likewise, magenta and yellow subtract green and blue, respectively. Accordingly, cyan, magenta and yellow are referred to as the *subtractive primaries*. The color cube and Venn diagram for the CMY color model are depicted in Figure 11.6.

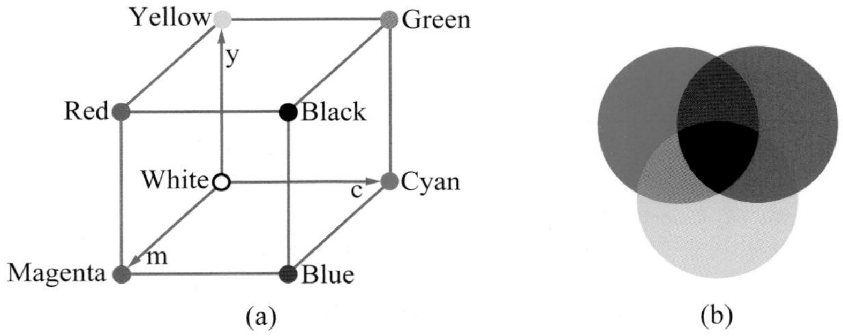

Figure 11.6: (a) The CMY color cube (b) Venn diagram of the CMY model.

Going between the RGB and CMY color spaces is simple:

$$\begin{bmatrix} c \\ m \\ y \end{bmatrix} = \begin{bmatrix} 1 \\ 1 \\ 1 \end{bmatrix} - \begin{bmatrix} r \\ g \\ b \end{bmatrix} \quad \text{and} \quad \begin{bmatrix} r \\ g \\ b \end{bmatrix} = \begin{bmatrix} 1 \\ 1 \\ 1 \end{bmatrix} - \begin{bmatrix} c \\ m \\ y \end{bmatrix} \qquad (11.1)$$

Ink of the color of each of the three subtractive primaries is coated as a grid of dots (called a *screen*) on a sheet of paper during printing. The relative proportions of CMY ink at each dot determine the amounts of light of various frequencies subtracted there; the remaining light emerges through the ink layers and imparts to the dot its perceived color.

However, in practice, the combination of CMY ink to produce RGB color does not work as well as Equations (11.1) might suggest. The CMY pigments in printer toner cartridges are never pure enough that an equal mix produces 100% black or even proper shades of gray. In addition to this technical problem there is an economic one too: making blacks and grays, by far the most common colors in printing, by mixing colored inks is expensive. Modern color printers, accordingly, supplement their CMY inks with a black ink to directly produce both blacks and the gray scale, the resulting process being called *four-color printing*.

The CMY model augmented with black is called the *CMYK color model* (for reasons to do with printing terminology black is denoted by K rather than B). Conversion formulae between the CMYK color space and the RGB and CMY color spaces are more complicated than those between the latter two and we'll not discuss them here. However, drawing and image editing programs, such as GNU's GIMP (freeware [53]), which offer both RGB and CMYK models will automatically convert between the two.

A practical point to keep in mind is that mapping from RGB to CMYK is often device-dependent and rarely 100% accurate, which is why CMYK print-outs are frequently significantly different from the original RGB display. Moreover, the space of colors representable in the RGB and CMYK color models – their *gamuts* – are not identical either, so some colors simply cannot be transferred exactly from monitor to paper (or vice versa).

11.1.3 HSV (or HSB) Color Model

The RGB color model, though pretty much ingrained into applications around us, is not particularly intuitive for the mixing of colors. For example, what RGB values should an artist combine for a jungle green, sunset orange, ocean blue, ...? The *HSV color model* was created by Alvy Ray Smith (one of the co-founders of Pixar Corporation) in 1978 as a more user-friendly alternative for designers. HSV is the abbreviation for hue, saturation and value. The model is also called *HSB*, where B stands for brightness.

The HSV model gets past the problem of having to numerically mix primaries by allowing the designer to choose a color's "coloredness" (that which we perceive as jungle green, sunset orange, ocean blue, etc.) directly with a *single* parameter, the *hue*. The hue parameter space is circular and often called the *color wheel*.

Here's how the color wheel is derived. See Figure 11.7. Begin with a triangle with corners representing the red, green and blue hues. Double the number of vertices to make a hexagon and fill in the middle hues, yellow, cyan and magenta (e.g., yellow is an equal mix of red and green, so midway between them). Again double to a dodecagon and add new hues by interpolating between previous ones. Continuing the process leads to a

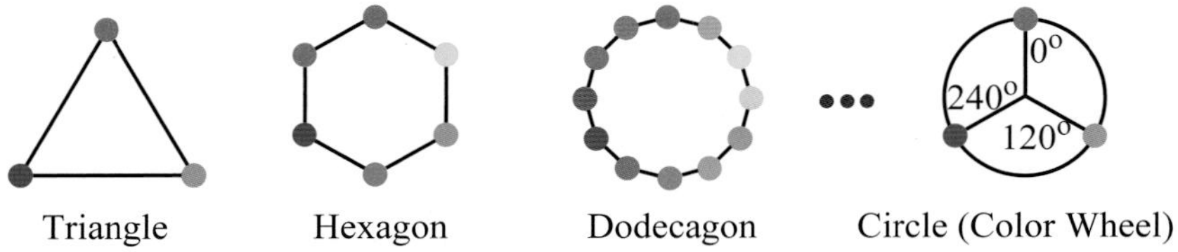

Figure 11.7: Hues on a triangle, hexagon, dodecagon and circle (color wheel).

continuum of hues in a circle. A position on this circle – or, the color wheel as it's called – thus represents a particular hue. Typically, red, green and blue are located at 0°, 120° and 240°, respectively.

The hue parameter, though, by itself is insufficient to specify a color. Two other parameters are required as well. The *saturation* of a color, typically given as a percentage, represents its purity. The higher the saturation the richer and more vibrant the color appears; conversely, less saturated colors (called *desaturated*) appear faded and grayish. The final parameter is *value*, given as a percentage as well, representing a color's intensity or brightness.

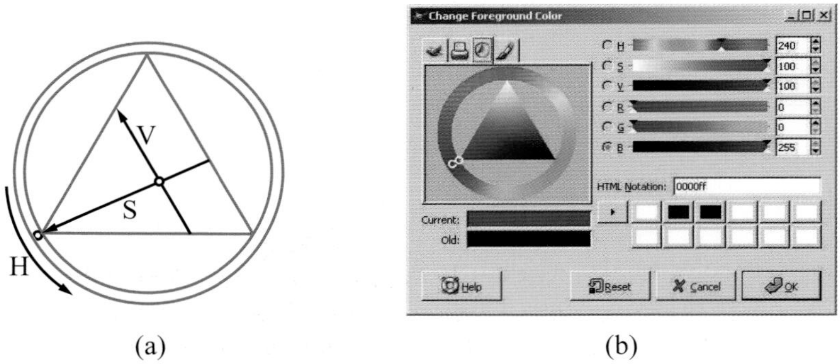

Figure 11.8: (a) The outer small hollow circle is positioned on the colored wheel to set the H value and the inner one inside the triangle to set S and V values. (b) The Gimp color dialog box making a 100% blue using the HSV color wheel option.

The saturation and value amounts of a color are often specified by positioning a point on an equilateral triangle inscribed in the color wheel, with a vertex of that triangle located at and turning with the color's hue position on the wheel. As indicated in Figure 11.8(a), value changes parallel to the edge opposite the hue vertex, while saturation increases with distance from the opposite edge. Figure 11.8(b) shows GIMP's color dialog box setting a 100% blue in HSV mode.

11.1.4 Summary of the Models

In drawing applications the predominant model is RGB and we'll really not have use for any other through the rest of this book. It's useful, though, to have a nodding acquaintance with models that may occasionally pop up elsewhere. With CMY, CMYK and HSV we have covered the ones the user is most likely to encounter. CMYK does, in fact, become particularly important when one goes from drawing to printing (e.g., a CG book!). There are other color models not used as much, such as Lab (an option in Adobe's Photoshop) and HLS (for hue, lightness, saturation).

The gold standard among color models was established by the CIE in 1931. It's called the *CIE XYZ model* (also the *CIE 1931 model*) where the X, Y and Z parameters represent, respectively, three theoretical primaries, each corresponding to a particular intensity distribution standardized by the CIE. Although not seen in practical interfaces, the CIE XYZ color model is used to calibrate implementations of the other ones.

11.2 Phong's Lighting Model

A model of interaction between light sources and objects is called a *lighting model* (or *reflection model*, or *illumination model*). In 1975 Vietnamese computer scientist Phong Bui Tuong [109] invented a particular lighting model, now known by his name, which is currently the one most widely used in practice. Despite the subsequent development of more authentic lighting models, e.g., Cook-Torrance [29], ray tracing, etc., Phong's has endured in popularity, especially because it delivers realistic lighting at moderate computational cost. OpenGL implements Phong's model. But, before we start coding up light let's first get an understanding of the model.

11.2.1 Phong Basics

In Phong's model the light reflected off an object O is the sum of three components – *ambient*, *diffuse* and *specular* – based on the *reflectance* properties of its surface. We'll describe each component next before explaining how to specify and calculate them.

Ambient: Ambient reflectance models O's reflection of background light that strikes it from multiple directions. For this reason ambient light is scattered equally in all directions from the surface of O as well. See Figure 11.9.

Of the light sources in the environment – e.g., lamps and the sun – a part of the light from each is presumed ambient in that it's scattered in the environment, e.g., by minute particles such as dust, effectively becoming part of background light before striking O. The direction of the light's source, therefore, is lost in that part of it which is ambient. Neither does

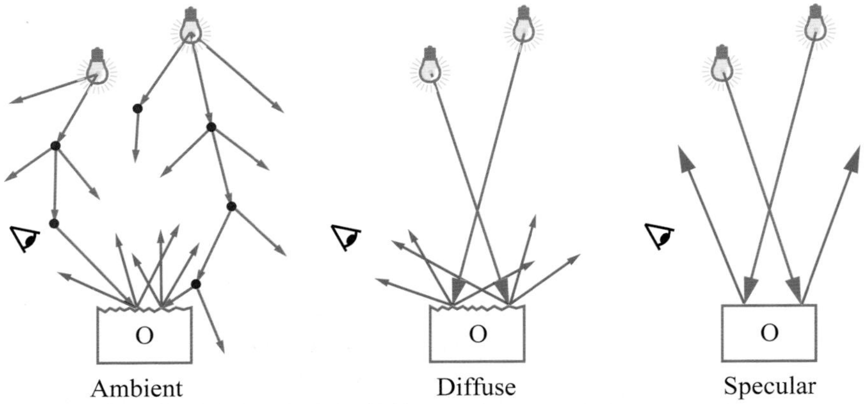

Figure 11.9: Ambient, diffuse and specular reflectance: incident light drawn blue, reflected red.

it matter where the viewer is located because of the scattered reflection from the surface of O, presumed equal in all directions. In practical terms, the ambient component models that part of light which supplies constant illumination throughout a scene. An example of a familiar light source which is mostly ambient is a tube lamp recessed behind a frosted panel (the material of the panel serving to scatter light rather than let through a focused beam).

In addition to the ambient parts of each light source, there is presumed to be a *global ambient light* as well, from no identifiable source (i.e., "true" background light). For example, when modeling a scene inside a building, we can adjust the global ambient to account for light coming in from outside through doors and windows, without trying to model every possible light source such as the sun and lights on the street, which would be very complex indeed.

The total ambient component of the light reflected from an object O is the sum of what it reflects of the ambient parts from each source, plus what it reflects of the global ambient. Informally (we'll be getting to more precise equations soon):

$$\text{ambient reflectance from } O = \sum(\text{reflectance of ambient part from each light source}) + \text{ reflectance of global ambient}$$

Diffuse: Diffuse reflectance specifically models the fine-scale graininess of the surface: the diffuse part of light from a particular source travels in a coherent beam toward O and then is scattered equally in all directions by diffuse reflectance from the surface of O. See Figure 11.9. Therefore, the direction of the light source does matter in the case of diffuse reflectance, but not that of the viewer. Practically, then, the diffuse component models the "soft" part of the light with little focus, e.g., that reflected off polished wood or silky fabric.

The total diffuse component of light reflected from O is the sum of the reflectances of the diffuse parts from each source:

$$\text{diffuse reflectance from } O = \sum(\text{reflectance of diffuse part from each light source})$$

Specular: Specular reflectance models the shininess of the surface: the specular part of light from a particular source travels in a coherent beam toward O and then is reflected in mirror-like manner, again in a coherent beam, by specular reflectance from the surface of O. Both the direction of the light source and the viewer matter in the case of specular reflectance. Specular light is "hard" light with a focus, e.g., that from a beam bouncing off a polished metal surface.

The total specular component of the light reflected from O is the sum of the reflectances of the specular parts from each source:

$$\text{specular reflectance from } O = \sum(\text{reflectance of specular part from each light source})$$

Remark 11.1. Because specular reflection is mirror-like, while the ambient and diffuse reflections are due to scattering from the surface of the object, the color of specularly reflected light depends primarily on that of the source itself, while those of the ambient and diffusely reflected on the native color of the object, as well as the light source.

Remark 11.2. The split of light into the three components of ambient, diffuse and specular according to reflectance is *independent* of the split into the primary color components of red, green and blue, in that each of the ambient, diffuse and specular components has RGB subcomponents and all nine subcomponents can be independently set. Or, one can equivalently say that each of RGB has ambient, diffuse and specular subcomponents which can all be independently set. In other words, you can think of light as being split as in Figure 11.10(a) or Figure 11.10(b) – it does not matter. Yet another way this is often phrased is by saying that color and reflectance splits are *orthogonal*.

A final component of light emerging from O is not reflected.

Emissive: The emissive component of light from an object O is that which is "manufactured" at O and unrelated to external light sources or the global ambient light. An example of an emissive object would be a lamp or the headlight of an automobile.

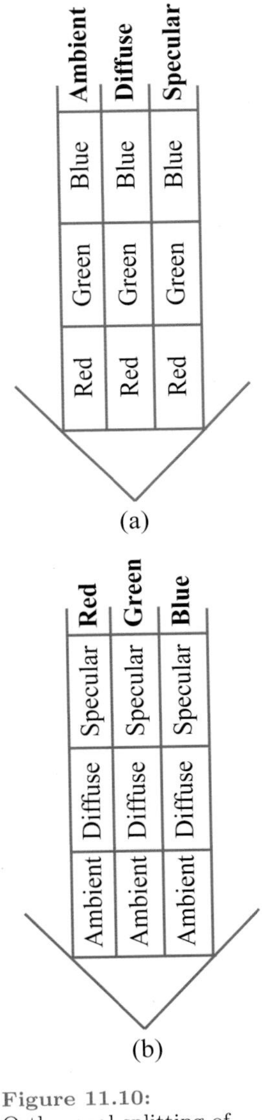

Figure 11.10: Orthogonal splitting of light: (a) Reflectance followed by color (b) Color followed by reflectance.

It is *extremely important* to keep in mind that, in OpenGL implementations, emissive light is perceived *only* by the viewer and does *not* illuminate other objects – it does *not* make O a light source for the rest of the environment.

11.2.2 Specifying Light and Material Values

OpenGL allows several light sources to be specified – the exact number depending on the implementation. For each of the, say, N specified light sources L^i, $0 \le i \le N-1$, the RGB intensities of each of its ambient, diffuse and specular components can be set to between 0 and 1, for nine values altogether per light source. These are written, typically, in a 3×3 *light properties matrix*:

$$\begin{bmatrix} L^i_{amb,\,R} & L^i_{amb,\,G} & L^i_{amb,\,B} \\ L^i_{diff,\,R} & L^i_{diff,\,G} & L^i_{diff,\,B} \\ L^i_{spec,\,R} & L^i_{spec,\,G} & L^i_{spec,\,B} \end{bmatrix} \tag{11.2}$$

Similarly, for each object O or, more precisely, each vertex V of O, one can set *scaling factors* between 0 and 1 to determine how much of each component of the incident light is reflected, for again nine values, contained in a 3×3 *material properties matrix*:

$$\begin{bmatrix} V_{amb,\,R} & V_{amb,\,G} & V_{amb,\,B} \\ V_{diff,\,R} & V_{diff,\,G} & V_{diff,\,B} \\ V_{spec,\,R} & V_{spec,\,G} & V_{spec,\,B} \end{bmatrix} \tag{11.3}$$

These so-called reflectance values represent the object's color: the higher one is, the more of the corresponding incoming light is reflected, and the more of that color the object appears to be.

The RGB values of the global ambient light are contained in a 3-vector called the *global ambient light vector*:

$$[globAmb_R \quad globAmb_G \quad globAmb_B] \tag{11.4}$$

The RGB values of the emissive light from a vertex V is a 3-vector called the *emissive light vector*:

$$[V_{emit,\,R} \quad V_{emit,\,G} \quad V_{emit,\,B}] \tag{11.5}$$

11.2.3 Calculating the Reflected Light

We come now to calculating each component of the reflected light.

Ambient

Calculating the ambient component emerging from a vertex V owing to a particular light source consists simply of scaling the light's ambient intensity by V's ambient reflectance. If the original intensity of the ambient light of some primary color from a source L (or the global ambient) is I, then that of its reflection from the surface at V is

$$I * \textit{material ambient scaling factor} \tag{11.6}$$

The *material ambient scaling factor* is the fraction of the incident ambient light that the material reflects. It is nothing but the ambient reflectance value $V_{amb,\,X}$, where X may be R, G or B, in the first row of the material properties matrix. An example will clarify use of the equation.

Example 11.1. Say the intensities of the ambient light from source L are given by

$$L_{amb,\,R} = 0.4, \quad L_{amb,\,G} = 0.9, \quad L_{amb,\,B} = 0.2$$

and the ambient reflectances of V by

$$V_{amb,\,R} = 0.9, \quad V_{amb,\,G} = 0.9, \quad V_{amb,\,B} = 0.1$$

Then the part of the red light emanating from V owing to the L ambient is

$$L_{amb,\,R} * V_{amb,\,R} = 0.36$$

and the part of the green light emanating from V owing to the L ambient is

$$L_{amb,\,G} * V_{amb,\,G} = 0.81$$

and the part of the blue light emanating from V owing to the L ambient is

$$L_{amb,\,B} * V_{amb,\,B} = 0.02$$

Exercise 11.1. If the global ambient light vector is

$$[0.2 \; 0.2 \; 0.2]$$

and all the ambient reflectances of a vertex V are as in the preceding example, calculate the parts of the RGB light emanating from V owing to the global ambient.

Diffuse

Calculation of the diffuse component of the light reflected from V is more complex than that of the ambient as, not only must the incident light be scaled by the reflectance at V, but its direction taken into account as well. The latter is done by measuring the angle between the direction of the light source and the normal to the surface at V.

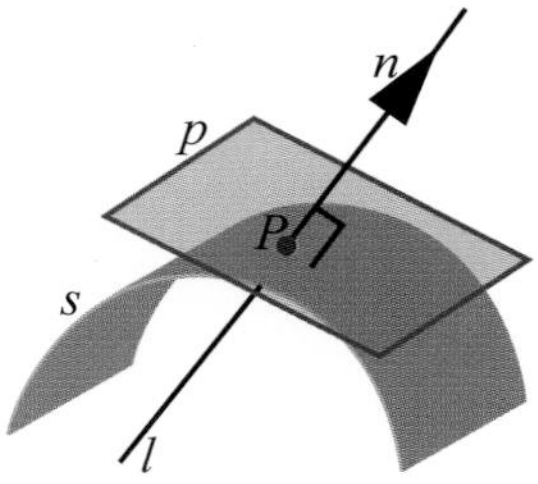

Figure 11.11: A normal vector n to a surface s at P lies along the normal line l there and is perpendicular to the tangent plane p at P.

Remark 11.3. A line l is *normal* to a surface s at the point P if it is perpendicular to the tangent plane p of s at P. Any non-zero vector n parallel to l is a *normal vector* to s at P. See Figure 11.11. (Think intuitively of the tangent plane as a hard board pressed to touch s at P.)

The light source L is modeled as a point. Further, the surface of the object O around the illuminated vertex V is assumed flat; in fact, it's taken to coincide with its own tangent plane at V. See Figure 11.12(a). Diffuse light is reflected in all directions from V.

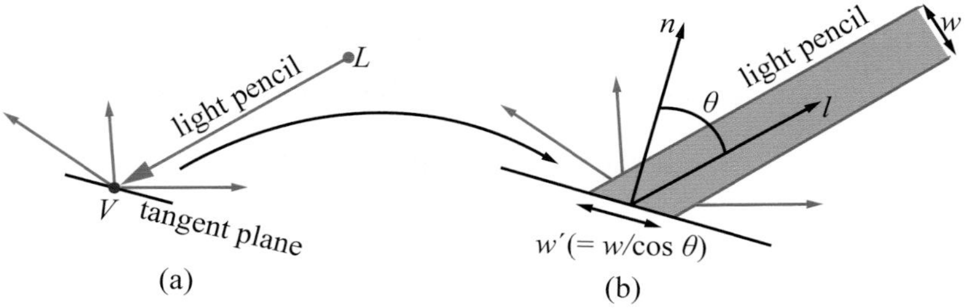

Figure 11.12: Calculating the diffuse component: (a) A light pencil from a point source L hits the surface, represented by its tangent plane at V (b) A blow-up of the pencil showing the normal vector n and the light direction vector l.

One observes from the blow-up in Figure 11.12(b) that a pencil of light of cross-sectional width w from L illuminates an area of width w', which is, typically, greater than w.

In this figure, θ is the angle between a direction vector l from V toward the light source L, called the *light direction vector*, and an outward normal vector n at V. The angle θ is called the *angle of incidence* of the light. We ask the reader to show next, by elementary trigonometry in Figure 11.12(b), that the width $w' = w/\cos\theta$.

Exercise 11.2. Verify the claim just made about the width of the area illuminated by a light of width w being $w' = w/\cos\theta$.

Since the area illuminated is greater by a factor of $1/\cos\theta$ than the cross-sectional area of the light pencil, the intensity of the light is diminished by a factor of $1/\cos\theta$ from I to $1/(1/\cos\theta) * I = \cos\theta * I$. Accordingly, if the original intensity of the diffuse light of some primary color emanating from the light source L is I, then that of its reflection from the surface of O at V is

$$\cos\theta * I * \textit{material diffuse scaling factor} \tag{11.7}$$

The *material diffuse scaling factor*, given by the values $V_{diff,\,X}$, where X is R, G or B, in the second row of material properties matrix, determines the fraction of the incident diffuse light the material reflects.

Example 11.2. Say the intensities of the diffuse light from source L are given by

$$L_{diff,\,R} = 0.3, \quad L_{diff,\,G} = 1.0, \quad L_{diff,\,B} = 1.0$$

and the diffuse reflectances of a vertex V by

$$V_{diff,\,R} = 0.8, \quad V_{diff,\,G} = 1.0, \quad V_{diff,\,B} = 0.8$$

and that the angle θ of incidence at V is $60°$.

Then the part of the red light emanating from V owing to the L diffuse is

$$\cos\theta * L_{diff,\,R} * V_{diff,\,R} = 0.5 * 0.3 * 0.8 = 0.12$$

Likewise, the part of the green light emanating from V owing to the L diffuse is

$$\cos\theta * L_{diff,\,G} * V_{diff,\,G} = 0.5$$

and the part of the blue light emanating from V owing to the L diffuse is

$$\cos\theta * L_{diff,\,B} * V_{diff,\,B} = 0.4$$

Remark 11.4. The relationship that the intensity of the reflected light varies as the cosine of the angle of incidence is known as *Lambert's law*. It is Lambert's law which explains why early mornings and late evenings, when the sun is lower in the sky, are cooler and darker than mid-day.

Specular

For specular light, as in the case of diffuse light, the light source L is modeled as a point and the surface of O identified with its tangent plane at the illuminated vertex V. An outward normal vector to the surface of O at V is n. In case of specular light, though, unlike for diffuse light, the eye comes into the equation. It is modeled as the point E.

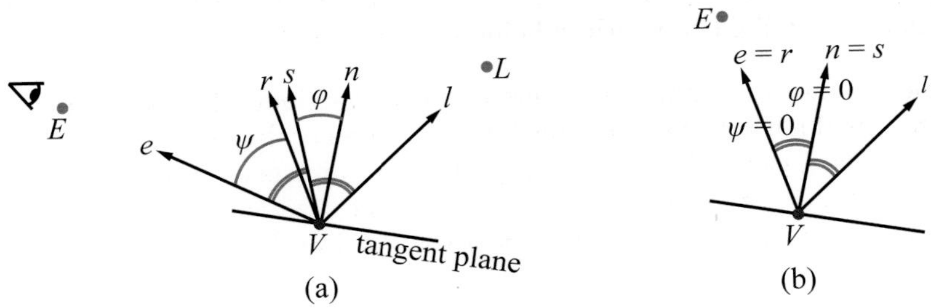

Figure 11.13: Calculating the specular component: (a) The light direction vector l, eye direction vector e, halfway vector s, normal vector n and reflection vector r (b) The special case when reflection is in the direction of the eye. (Double arcs indicate equal angles.)

The direction vector from V toward the light source L is l and the direction vector from V toward the eye E is e, Further, let s be a vector, called a *halfway vector*, which bisects the angle between l and e. See Figure 11.13(a) (ignore r for now).

We'll next state a relationship between the intensity of the reflected specular light and that of the incident which may seem unintuitive at first, but motivation will soon be apparent:

If the intensity of the specular light of some primary color emanating from the light source L is I, then that of its reflection from the surface of O at V is

$$\cos^f \phi * I * \textit{material specular scaling factor} \tag{11.8}$$

where ϕ is the angle between halfway vector s and the normal vector n, $f \geq 0$ is a scalar, called the *shininess exponent* and the *material specular scaling factor*, a value read from the material properties matrix, determines the fraction of the incident specular light the material reflects.

Here's what's happening. If the surface of O is *perfectly* mirror-like, then a ray of light from L to V reflects according to the laws of reflection, which say that the normal to O at V, the incident ray and the reflected ray all lie on the same plane and, moreover, that the incident ray and the reflected ray make the same angle with the normal. In this case of a perfect mirror, if the eye E is located in the direction of reflection, given by the *reflection vector* r, then it perceives all of the incident light from L, if not no light at all.

Note: Figure 11.13(a) should have double arcs between the pair r and n and the pair l and n, as well, to indicate equal angles, but that would have made it a bit too busy.

Say, ψ is the angle the reflection vector makes with the eye direction vector e, as in Figure 11.13(a). Figure 11.13(b) shows a particular case of the general Figure 11.13(a), where the eye is actually situated in the direction of perfect reflection, so that $\psi = 0$. Observe, in this case, that the halfway vector is aligned with the normal, in other words, $\phi = 0$ as well.

Most real surfaces, however, are not perfectly mirror-like and do not reflect light only along the direction of reflection, but, rather, spread it *about* that direction with an intensity which diminishes with increasing angle. In other word, maximum intensity is perceived by the viewer in a configuration as in Figure 11.13(b); nevertheless, even in a general configuration as in Figure 11.13(a), the eye receives light, though, with intensity inversely related to ψ.

This suggests that the intensity of specular reflection be modeled by the formula

$$\textit{angular attenuation factor} * I * \textit{material specular scaling factor} \tag{11.9}$$

where the *angular attenuation factor*, in fact, is in inverse relationship with ψ.

Phong suggested the angular attenuation factor $\cos^f \psi$, where the exponent f is larger the more mirror-like (i.e., shiny) the surface is. His considerations were empirical rather than based on actual physics. In particular, $\cos\psi$ is a function of ψ which is at its maximum of 1 when $\psi = 0$ and drops off as ψ increases, behavior expected of the angular attenuation factor. The function $\cos^f \psi$ also shows the same behavior, but more markedly, as f increases. In particular, the larger the value of f the more rapid the drop from the value of 1 as ψ increases from 0. See Figure 11.14. Practically, the shinier the surface the more rapidly the light diminishes away from the direction of reflection.

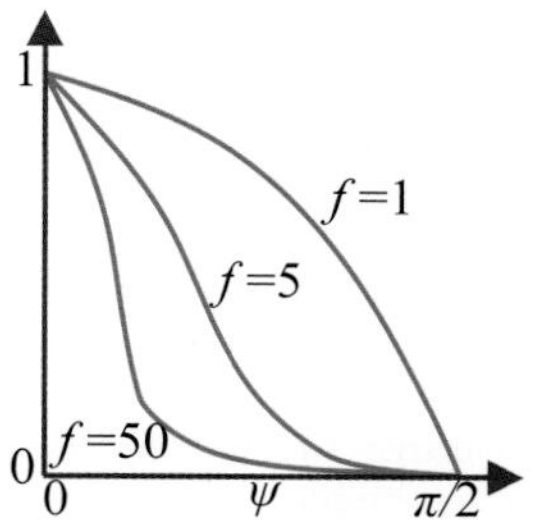

Figure 11.14: Graphs of $\cos^f \psi$ for different values of f (not exact plots).

The angle ψ is often replaced by ϕ, the angle between the halfway vector s and the normal vector n, because ϕ is easier to compute, and because it is legitimate to do so given the following linear relation between the two.

Example 11.3. Show that $\psi = 2\phi$.

Answer: Label the angles from the tangent plane to the light direction, the reflection and eye direction vectors θ_1, θ_2 and θ_3, respectively, as in Figure 11.15.

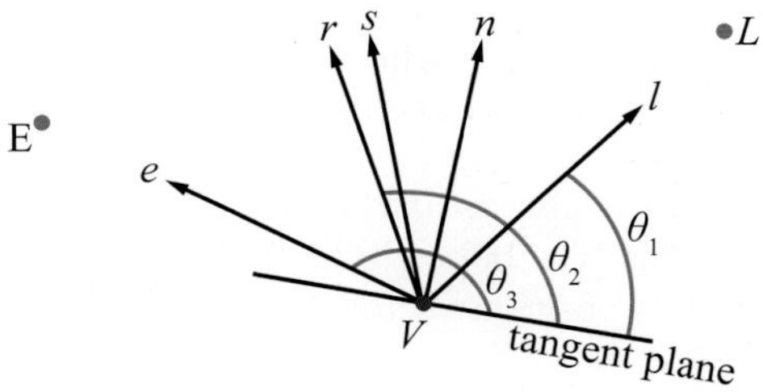

Figure 11.15: Proving that $\psi = 2\phi$.

The angle to the halfway vector, then, is $(\theta_1 + \theta_3)/2$, implying that the angle between the halfway vector and the normal is $\phi = (\theta_1 + \theta_3)/2 - \pi/2$.

The angle between the light vector and the normal, which is $\pi/2 - \theta_1$, is the same as the angle between the normal and the reflection vector, by laws of reflection. Therefore, $\theta_2 = \pi/2 + (\pi/2 - \theta_1) = \pi - \theta_1$. This, then, implies that the angle between the eye direction vector and the reflection vector is $\psi = \theta_3 - \theta_2 = \theta_3 - (\pi - \theta_1) = \theta_1 + \theta_3 - \pi$. That $\psi = 2\phi$ now follows.

Given the relationship between ϕ and ψ contained in the preceding example, substituting $\cos^f \phi$ for $\cos^f \psi$ as the angular attenuation factor in Equation (11.9) makes no qualitative difference. The result of the substitution, in fact, is Equation (11.8), which is now fully justified.

Example 11.4. Give a formula for the halfway vector s in terms of the light direction vector l and the eye direction vector e from V, which are, of course, the two vectors that s bisects. Assume that both l and e are of unit length. Give s as a unit vector as well.

Answer: See Figure 11.16, where $l = \overrightarrow{OA}$, $e = \overrightarrow{OB}$, and where $l + e$ is drawn with the help of the parallelogram law of addition of vectors. Since $|l| = |e| = 1$, all four sides of the parallelogram $OACB$ are of unit length as well. A consequence is that corresponding sides of the triangles OAC and OBC are of equal lengths. The two triangles are, therefore, congruent, so $\angle AOC = \angle BOC$. One concludes that the vector $l + e$ bisects l and e.

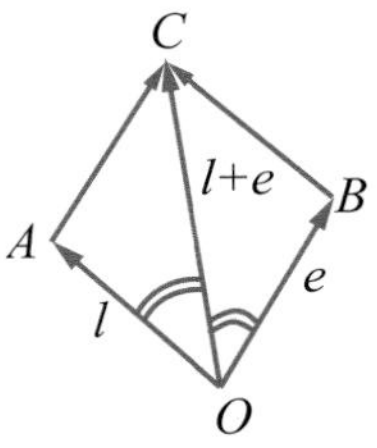

Figure 11.16: The vector $l + e$ bisects l and e.

Accordingly, the unit halfway vector

$$s = \frac{l+e}{|l+e|} \quad \text{(provided that } l + e \text{ is not the zero vector)}$$

Remark 11.5. When a vector u is used to represent a direction, so that its magnitude is not of importance, it is often convenient to scale it to unit length, a step called *normalizing* u. Normalization of a non-zero vector u (note that a vector representing a direction cannot be zero) consists simply of dividing it by its length, in other words, replacing u by $u/|u|$.

Exercise 11.3. Give a simple formula, in an OpenGL setting, for the eye direction vector e from a vertex V whose position vector is v. Accordingly, rewrite the formula for the halfway vector of the preceding example in terms of l and v.
Hint: The OpenGL eye is always at the origin, so e is the vector from V to the origin, which is just the opposite of the vector from the origin to V, this, of course, being V's position vector v.

Example 11.5. Say the intensities of the specular light from source L are given by

$$L_{spec,\,R} = 1.0, \quad L_{spec,\,G} = 1.0, \quad L_{spec,\,B} = 1.0$$

and the specular reflectances of a vertex V by

$$V_{spec,\,R} = 0.0, \quad V_{spec,\,G} = 1.0, \quad V_{spec,\,B} = 0.6$$

and that the angle ϕ between the halfway vector and the outward normal vector at V is $60°$ and that the shininess exponent is 2.0.

Then the part of the red light emanating from V owing to the L specular is

$$\cos^f \phi * L_{spec,\,R} * V_{spec,\,R} = 0.5^2 * 1.0 * 0.0 = 0.0$$

Likewise, the part of the green light emanating from V owing to the L specular is

$$\cos^f \phi * L_{spec,\,G} * V_{spec,\,G} = 0.25$$

and the part of the blue light emanating from V owing to the L specular is

$$\cos^f \phi * L_{spec,\,B} * V_{spec,\,B} = 0.15$$

It's interesting that calculation of the reflected light never actually required determination of the reflection vector r itself. It's not hard though to find r, as we see next.

Exercise 11.4. Suppose that n is the unit (outward) normal vector and l the unit light direction vector at a vertex V. Prove that the unit vector r in the direction of reflection is given by the equation

$$r = 2(n \cdot l)n - l$$

Part answer: According to the laws of reflection we have to verify that r lies on the plane of l and n and makes the same angle with n as l. We must also prove that r is of unit length.

That r lies on the plane of l and n follows from its formula above, because of the linear dependence of r on l and n. Now

$$|r|^2 = r \cdot r = (2(n \cdot l)n - l) \cdot (2(n \cdot l)n - l) = 4(n \cdot l)^2 - 4(n \cdot l)^2 + l \cdot l = |l|^2 = 1$$

proving that r indeed is a unit vector.

We'll leave the reader to prove that r makes the same angle with n as l by computing its dot product with n.

11.2.4 First Lighting Equation

At the end of the day we need a color vector for each vertex V in the scene, in other words, values for R, G and B. This is obtained by adding – for each of R, G and B – contributions of the ambient, diffuse and specular reflected light *plus* the emitted light. The corresponding equation, the so-called *lighting equation*, is straightforwardly obtained from the formulae of the last section.

Assume that we are given the values of the lighting properties matrix (11.2) for each light source L^i, $0 \le i \le N-1$, the material properties matrix (11.3) for the vertex V, the global ambient light vector (11.4), as well as the emissive light vector (11.5) at V. Further, denote the normalized light direction and halfway vectors corresponding to light source L^i at vertex V by l^i and s^i, respectively. Denote the normalized outward surface normal vector at V by n and its shininess exponent by f.

Here then is the lighting equation giving the color intensity V_X at V, where X may be any of RGB:

$$\begin{aligned} V_X \;=\;& V_{emit,\,X} \;+ \\ & globAmb_X * V_{amb,\,X} \;+ \\ & \sum_{i=0}^{N-1} \Big(L^i_{amb,\,X} * V_{amb,\,X} \;+ \\ & \quad \max\{l^i \cdot n\,,\, 0\} * L^i_{diff,\,X} * V_{diff,\,X} \;+ \\ & \quad (\max\{s^i \cdot n\,,\, 0\})^f * L^i_{spec,\,X} * V_{spec,\,X} \Big) \end{aligned} \qquad (11.10)$$

Note: The dot product of two unit vectors gives the cosine of the angle between them.

Note: If the RHS sums to more than 1, for any of X equal to R, G or B, then it is clamped to 1.

The lighting equation simply collects the components we have already discussed separately. The first summand on the RHS is the emissive component, while the second the global ambient scaled by the ambient reflectance. The third summand is a summation over the n light sources of

(a) The incident ambient component scaled by the ambient reflectance (Equation (11.6)).

(b) The incident diffuse component scaled by the diffuse reflectance and the cosine of the incident angle (Equation (11.7)).

(c) The incident specular component scaled by the specular reflectance and the angular attenuation factor (Equation (11.8)).

The reason for the $\max\{*, 0\}$ terms is not to allow a negative multiplier, which would imply the physically impossible phenomenon of light being subtracted. For example, $l^i \cdot n$ is negative when the angle between l^i and n is greater than $\pi/2$, which means that the light source L^i is behind the surface on which V is located, contributing zero light, rather than negative light.

Equation (11.10) is actually a first draft. The final lighting equation of OpenGL, which we'll see soon, enhances it by taking into account the attenuation of light over distance, as well as the spotlight effect, where light from a source emerges as a cone, rather than in all directions.

Exercise 11.5. There are two light sources, L^0 and L^1, the respective values of whose lighting properties matrices are

$$\begin{bmatrix} 0.0 & 0.0 & 0.0 \\ 0.7 & 0.1 & 0.1 \\ 0.7 & 0.1 & 0.1 \end{bmatrix} \quad \text{and} \quad \begin{bmatrix} 0.0 & 0.0 & 0.0 \\ 0.1 & 0.7 & 0.1 \\ 0.1 & 0.7 & 0.1 \end{bmatrix}$$

The material properties matrix at a vertex V is

$$\begin{bmatrix} 0.1 & 0.8 & 0.9 \\ 0.1 & 0.8 & 0.9 \\ 1.0 & 1.0 & 1.0 \end{bmatrix}$$

Furthermore, the shininess exponent of the surface at V is 2.0, there is no emission from V, and the unit outward normal vector at V is

$$[0.0\ 1.0\ 0.0]^T$$

The position vectors of L^0, L^1, V and the eye are, respectively,

$$[0.0\ 5.0\ 5.0]^T, \quad [5.0\ 5.0\ 5.0]^T, \quad [0.0\ 0.0\ 5.0]^T \text{ and } [0.0\ 0.0\ 0.0]^T$$

The global ambient light vector is

$$[0.1\ 0.1\ 0.1]^T$$

Fill out color values at V in the table below.

	R	G	B
Emission			
Global Ambient			
Ambient			
Diffuse			
Specular			
Total (color vector at V)			

Remark 11.6. It is important to realize that Phong's lighting model is *local*: the color at each vertex V depends only on the interaction between the external light properties and the material properties at V *itself*. No account is taken of whether V is obscured from a light source by another object (shadows), or of light that strikes V not directly from a light source but having bounced off other objects (reflection and secondary lighting).

For example, in Figure 11.17, the color at vertex V is determined according to Equation 11.10, where the light direction vector l is the unit vector in the direction from V to L, calculated by OpenGL to be $(L - V)/|L - V|$; *without* checking that the ray from V along l is blocked by the surface s_2 before reaching L, meaning that *really* V is in s_2's shadow. So, effectively, s_1 is drawn and lit as if s_2 does not exist!

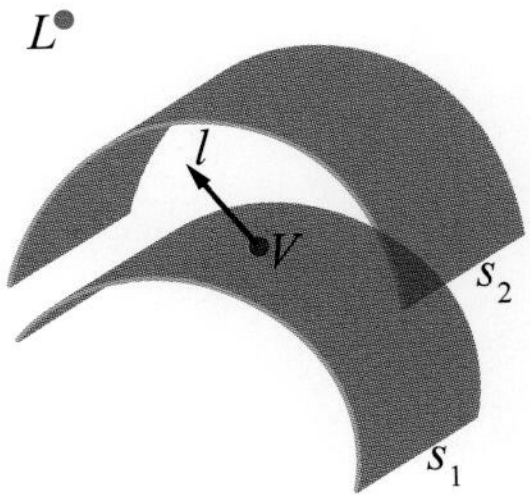

Figure 11.17: Light ray from V toward L, along l, is blocked by s_2.

Colloquially, object-object light interaction is not considered, only light-object. We will discuss two global lighting models, ray tracing and radiosity, where shadows, reflections and other secondary effects are captured, in a later chapter.

11.3 OpenGL Light and Material Properties

Time for code!

The mapping from Phong's lighting model to OpenGL syntax is pretty much one-to-one. For each light source the user defines the values in the lighting properties matrix (11.2), as also the values in the material properties matrix (11.3) for each vertex. The global ambient vector (11.4) is user-defined as well. The user, too, defines the shininess exponent f, the emission color vector (11.5) and, very importantly, the normal vector at each vertex.

If you are beginning to worry that that's a lot of values to specify to light a scene, don't! Remember that OpenGL is a state machine, so material properties – which are state variables – persist in their current setting until explicitly changed, making it convenient for the programmer to apply the same properties to all vertices of a single object. Moreover, OpenGL has sensible defaults for values the programmer doesn't care to define.

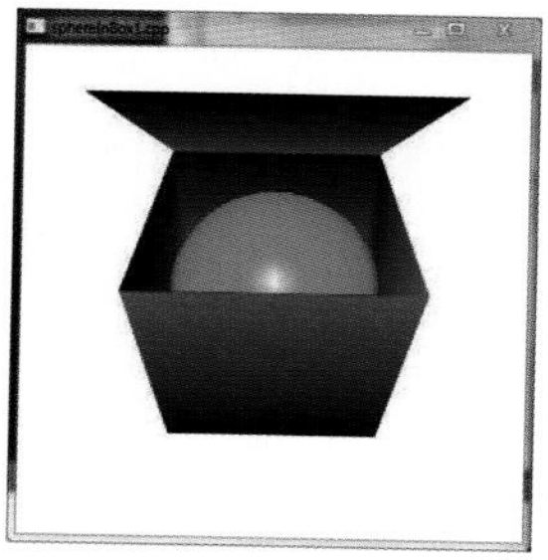

Figure 11.18: Screenshot of sphereInBox1.cpp.

Experiment 11.1. Run again `sphereInBox1.cpp`, which we ran the first time in Section 9.4. Press the up-down arrow keys to open or close the box. Figure 11.18 is a screenshot of the box partly open. We'll use this program as a running example to explain much of the OpenGL lighting and material color syntax.
End

11.3.1 Light Properties

Properties of light sources are set by statements of the form:

`glLight*(`*light*`, `*parameter*`, `*value*`)`

where *light* is the label of the light source (viz. `GL_LIGHT0`, `GL_LIGHT1`, ...) and its particular *parameter* set to *value*.

The properties of the single light source of `sphereInBox1.cpp` are specified by the following statements in the `setup()` routine:

```
glLightfv(GL_LIGHT0, GL_AMBIENT, lightAmb);
glLightfv(GL_LIGHT0, GL_DIFFUSE, lightDifAndSpec);
glLightfv(GL_LIGHT0, GL_SPECULAR, lightDifAndSpec);
glLightfv(GL_LIGHT0, GL_POSITION, lightPos);
```

The values of `GL_AMBIENT`, `GL_DIFFUSE` and `GL_SPECULAR` – `lightAmb`, `lightDifAndSpec` and `lightDifAndSpec`, respectively, above – are 4-vectors representing RGBA components. The fourth component, the alpha value, should always be 1.0 for a light source.

Typically, the diffuse and specular color vectors, i.e., the values of the `GL_DIFFUSE` and `GL_SPECULAR` parameters, respectively, are set identically to values perceived as the actual color of the light source. So, that of `sphereInBox1.cpp`, being {`1.0, 1.0, 1.0, 1.0`}, is a bright white.

It's simplifying, as well, to consolidate all light source ambients – their `GL_AMBIENT` values – into the global ambient; in other words, set light source ambient colors all to 0.0 and adjust the one global ambient light vector. We follow this approach in `sphereInBox1.cpp`, as in all our lit programs.

The value $\{x, y, z, w\}$ of `GL_POSITION` specifies the location $[x\ y\ z\ w]^T$ of the light source in homogeneous coordinates. If $w \neq 0$ then the light source is said to be positional and is located at world coordinates

$$[x/w \quad y/w \quad z/w]^T$$

The value of `lightPos` being {`0.0, 1.5, 3.0, 1.0`}, the single positional light source of `sphereInBox1.cpp` is at $[0.0\ 1.5\ 3.0]^T$, which is just above and some ways in front of the box. We'll discuss what happens if $w = 0$ in Section 11.5 when we discuss directional light sources.

Note that *no visible object* is created by OpenGL at the location of a light source! This location is simply a point used for the purpose of lighting calculation. If you want the light to appear to be from a lamp or car headlight or such object you'll have to model the object and position it yourself.

Global ambient light in `sphereInBox1.cpp` is set with the statement

```
glLightModelfv(GL_LIGHT_MODEL_AMBIENT, globAmb);
```

in the `setup()` routine, where the second parameter `globAmb` points to the global ambient vector, whose value is the mild white {`0.2, 0.2, 0.2, 1.0`}.

Finally, mind that lighting calculation is enabled with the call `glEnable(GL_LIGHTING)` and individual lights with calls to `glEnable(GL_LIGHT`*i*`)`.

Exercise 11.6. Show that nothing, in fact, is lost according to the first lighting equation (11.10) by setting all light source ambient colors to 0.0. In particular, prove that, however the light source ambients are initially set, they can all be reset to 0.0 and the global ambient adjusted accordingly so that the color computed at each vertex by (11.10) remains unchanged.

11.3.2 Material Properties

Material properties at a vertex are set by statements of the form:

```
glMaterial*(face, parameter, value)
```

where the *parameter* of *face* is set to *value*. The value of face can be `GL_FRONT`, `GL_BACK` or `GL_FRONT_AND_BACK` for both faces.

Material properties of (each vertex of) the box of `sphereInBox1.cpp` are specified by the following statements in the `drawScene()` routine:

```
glMaterialfv(GL_FRONT_AND_BACK, GL_AMBIENT_AND_DIFFUSE, matAmbAndDif1);
glMaterialfv(GL_FRONT_AND_BACK, GL_SPECULAR, matSpec);
glMaterialfv(GL_FRONT_AND_BACK, GL_SHININESS, matShine);
```

As for a light source, the values of `GL_AMBIENT`, `GL_DIFFUSE` and `GL_SPECULAR` for a material are 4-vectors representing RGBA components. The fourth, or alpha, components are currently all set to the default value of 1.0 – the alpha value pertains to blending, which is discussed in a later chapter.

Typically, the ambient and diffuse color vectors are set identically to values perceived as an object's native color. OpenGL makes it convenient to do so via the `GL_AMBIENT_AND_DIFFUSE` parameter, which, in fact, is how the ambient and diffuse values of the box are both set to the red $\{0.9, 0.0, 0.0, 1.0\}$. However, the ambient and diffuse values may be set separately as well using `GL_AMBIENT` and `GL_DIFFUSE`.

As specular light is obtained from reflection from the light source, it's reasonable to set an object's `GL_SPECULAR` value either to white $\{1.0, 1.0, 1.0, 1.0\}$, as for the box, fully reflecting the incident specular light, or a shade of gray $\{\gamma, \gamma, \gamma, 1.0\}$, *equally* diminishing each color component.

The value of `GL_SHININESS` – 50.0 for the box – is, of course, the shininess exponent f of the first lighting equation (11.10). Its value must be in the range $[0.0, 128.0]$. The default is 0.0, which causes no angular attenuation of specular reflectance.

The emissive color at a vertex can be set using the `GL_EMISSION` parameter, but we choose to go with the default of $\{0.0, 0.0, 0.0, 1.0\}$, in other words, no emission at any vertex in `sphereInBox1.cpp`.

Exercise 11.7. (Programming) What are the material ambient, diffuse and specular values and the shininess exponent of the sphere of `sphereInBox1.cpp`?

Remark 11.7. Note that, with our conventions, specifying the nine components each of the light properties and material properties matrices reduces, essentially, to specifying an RGB triple for each. Particularly, for each light source, the diffuse and specular values are an identical RGB, while ambient is black (only the global ambient being adjusted); for each vertex, the ambient and diffuse values are set to the same RGB triple, while the specular is white, reflecting all specular light (though this might be adjusted to a a shade $\{\gamma, \gamma, \gamma, 1.0\}$ of gray, equally scaling each specular component).

Remark 11.8. We'll be discussing the choice of the particular normal values in `sphereInBox1.cpp` when we begin a systematic discussion of normals in Section 11.11.

Remark 11.9. As stated at the start of the chapter, programming exercises are mostly collected in Section 11.13. However, this should not deter the reader from visiting that section as she reads and attempting exercises she can.

11.3.3 Experimenting with Properties

The two programs `lightAndMaterial1.cpp` and `lightAndMaterial2.cpp` allow the user to experiment with various material and light properties. Both show a blue ball lit by two lights, one white and one green, whose positions are indicated by small wire spheres. Figure 11.19 shows screenshots of both the programs.

Using the first program one can change material properties of the blue ball, as well as move it. The second program, on the other hand, allows properties of the white light to be controlled, as also those of the global ambient, and enables the user to rotate the white light. Text messages show property values. Let's take a quick tour of the two before experimenting with properties.

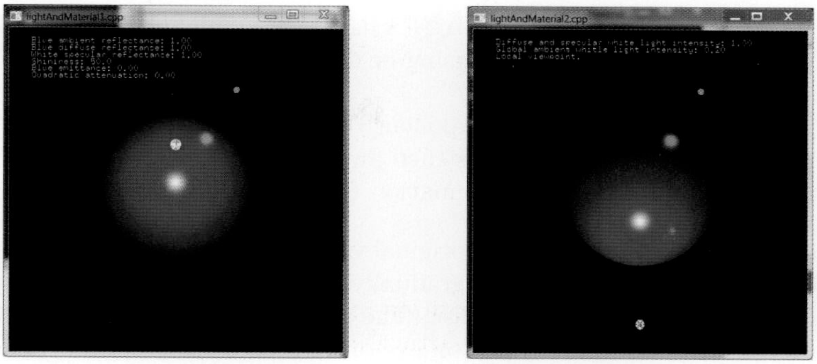

(a) (b)

Figure 11.19: Screenshots of (a) `lightAndMaterial1.cpp` (b) `lightAndMaterial2.cpp`.

Experiment 11.2. Run `lightAndMaterial1.cpp`.

The ball's current ambient and diffuse reflectances are identically set to a maximum blue of $\{0.0, 0.0, 1.0, 1.0\}$, its specular reflectance to the highest gray level $\{1.0, 1.0, 1.0, 1.0\}$ (i.e., white), shininess to 50.0 and emission to zero $\{0.0, 0.0, 0.0, 1.0\}$.

Press 'a/A' to decrease/increase the ball's blue **A**mbient and diffuse reflectance. Pressing 's/S' decreases/increases the gray level of its **S**pecular reflectance. Pressing 'h/H' decreases/increases its s**H**ininess, while pressing 'e/E' decreases/increases the blue component of the ball's **E**mission.

The program has further functionalities which we'll explain as they become relevant. **End**

Experiment 11.3. Run `lightAndMaterial2.cpp`.

The white light's current diffuse and specular are identically set to a maximum of $\{1.0, 1.0, 1.0, 1.0\}$ and it gives off zero ambient light. The green light's attributes are fixed at a maximum diffuse and specular of $\{0.0, 1.0, 0.0, 1.0\}$, again with zero ambient. The global ambient is a low intensity gray at $\{0.2, 0.2, 0.2, 1.0\}$.

Press 'w' or 'W' to toggle the **W**hite light off and on. Pressing 'g' or 'G' toggles the **G**reen light off and on. Press 'd/D' to decrease/increase the gray level of the white light's **D**iffuse and specular intensity (the ambient intensity never changes from zero). Pressing 'm/M' decreases/increases the gray intensity of the global a**M**bient. Rotate the white light about the ball by pressing the arrow keys.

This program, too, has added functionality which we'll need later. **End**

Experiment 11.4. Run `lightAndMaterial1.cpp`.

Reduce the specular reflectance of the ball. Both the white and green highlights begin to disappear, as it's the specular components of the reflected lights which appear as specular highlights. **End**

Exercise 11.8. **(Programming)** The specular highlight is sharpened or blunted, respectively, by increasing or decreasing the shininess exponent. Why?
Hint: The higher the shininess exponent the more rapidly the specular light diminishes as the vertex normals turn away from the eye direction (recall the definition of the angular attenuation factor in Section 11.2.3).

Experiment 11.5. Restore the original values of `lightAndMaterial1.cpp`.

Reduce the diffuse reflectance gradually to zero. The ball starts to lose its roundness until it looks flat as a disc. The reason for this is that the ambient intensity, which does not depend on eye or light direction, is uniform across vertices of the ball and cannot, therefore, provide the sense of depth that obtains from a contrast in color values across the surface. Diffuse light, on the other hand, which varies across the surface depending on light direction, can provide an illusion of depth.

Even though there is a specular highlight, sensitive to both eye and light direction, it's too localized to provide much depth contrast. Reducing the shininess does spread the highlight but the effect is not a realistic perception of depth.
Moral: Diffusive reflectance lends three-dimensionality.

Figure 11.20 shows the ball starting with only ambient reflectance, then adding in diffuse and specular.

End

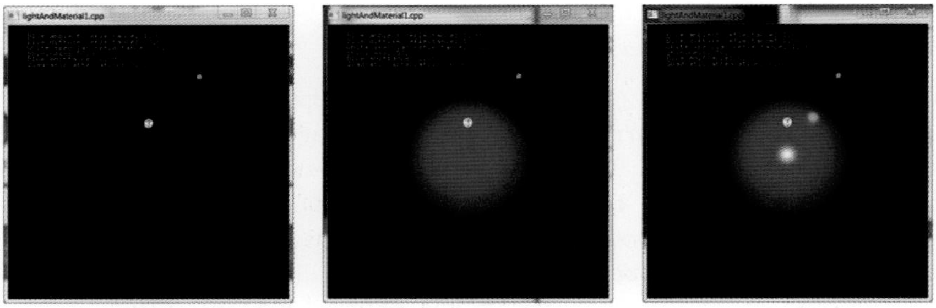

(a) (b) (c)

Figure 11.20: Screenshots of `lightAndMaterial1.cpp`: (a) Only ambient reflectance (b) Ambient and diffuse (c)Ambient, diffuse and specular.

Experiment 11.6. Restore the original values of `lightAndMaterial1.cpp`.

Now reduce the ambient reflectance gradually to zero. The ball seems to shrink! This is because the vertex normals turn away from the viewer at the now hidden ends of the ball, scaling down the diffuse reflectance there (recall the $\cos\theta$ term in the diffusive reflectance equation (11.7)). The result is that, with no ambient reflectance to offset the reduction in diffuse, the ends of the ball are dark.

Moral: Ambient reflectance provides a level of uniform lighting over a surface. **End**

Experiment 11.7. Restore the original values of `lightAndMaterial1.cpp`.

Reduce both the ambient and diffuse reflectances to nearly zero. It's like the cat disappearing, leaving only its grin! Specular light is clearly for highlights and not much else. **End**

Exercise 11.9. (Programming) Restore the original values of `lightAndMaterial1.cpp`.

Reduce all three of the ball's diffuse, ambient and specular reflectances and raise its emissive light intensity. It does appear to glow but also appears flat. Why?

Experiment 11.8. Run `lightAndMaterial1.cpp` with its original values.

With its current high ambient, diffuse and specular reflectances the ball looks a shiny plastic. Reducing the ambient and diffuse reflectances makes for a heavier and less plastic appearance. Restoring the ambient and diffuse to higher values, but reducing the specular reflectance makes it a less shiny plastic. Low values for all three of ambient, diffuse and specular reflectances give the ball a somewhat wooden appearance. **End**

Experiment 11.9. Run `lightAndMaterial2.cpp`.

Reduce the white light's diffuse and specular intensity to 0. The ball becomes a flat dull blue disc with a green highlight. This is because the ball's ambient (and diffuse) is blue and cannot reflect the green light's diffuse component, losing thereby three-dimensionality.

Raising the white global ambient brightens the ball, but it still looks flat in the absence of diffusive light. **End**

Exercise 11.10. (Programming) When the white light is switched off in `lightAndMaterial2.cpp`, the only evidence of green on the ball is the specular highlight; moreover, if the ambient is tamped down as well then the ball begins to disappear altogether.

However, this is not so in the opposite situation, when the white light is switched on and the green off – a sector of the ball is clearly visible no matter how low the ambient. Why?

Experiment 11.10. Nate Robins has a bunch of great tutorial programs at the site [100]. This is a good time to run his lightmaterial tutorial, which allows the user to control a set of parameters as well. **End**

11.3.4 Color Material Mode

Remember `glColor*()` which we used to set color in the dark days before there were light sources? Now that we do have light and `glMaterial*()` allows us to set all sorts of material properties, it seems there's no use any more for `glColor*()`. Well, it turns out that the good folk who designed OpenGL found a way to keep it on the payroll.

Here's how. Suppose you're in the not uncommon situation coloring a scene where only a particular color attribute, say the ambient and diffuse reflectances of the front faces, changes from one object to the next, other attributes remaining constant. What you can do in this case, instead of repeatedly calling `glMaterialfv(GL_FRONT, GL_AMBIENT_AND_DIFFUSE,` *value*`)`, is to:

1. Enable the so-called *color material mode* with a call to `glEnable(GL_-COLOR_MATERIAL)`.

2. Call `glColorMaterial(GL_FRONT, GL_AMBIENT_AND_DIFFUSE)`, which tells OpenGL to use the current color, set by `glColor*()`, to determine the front-face ambient and diffuse color values.

   Generally, the `glColorMaterial()` call can be of the form `glColor-Material(`*face*`,` *parameter*`)` where *face* can be `GL_FRONT`, `GL_BACK` or `GL_FRONT_AND_BACK`, and *parameter* one of `GL_AMBIENT`, `GL_DIFFUSE`, `GL_AMBIENT_AND_DIFFUSE`, `GL_SPECULAR` or `GL_EMISSION`.

3. Make a call to `glColor*()` to set the front-face ambient and diffuse color from one object to the next.

This method may, in fact, be more efficient with certain implementations of OpenGL, not to mention the convenience of not having to change a programming habit if one is used to coloring with `glColor*()`.

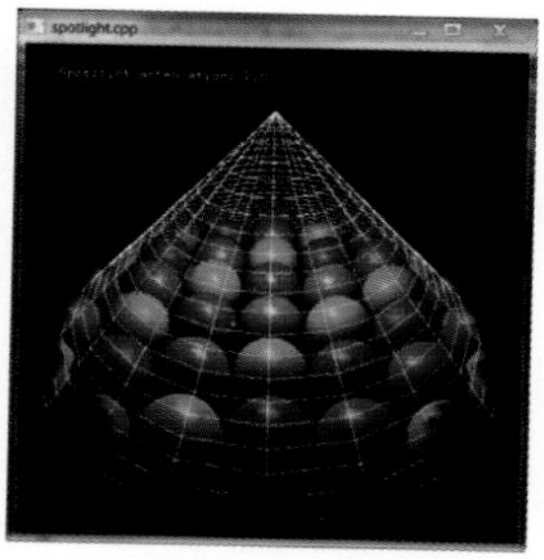

Figure 11.21: Screenshot of `spotLight.cpp`.

Experiment 11.11. Run `spotlight.cpp`. The program is primarily to demonstrate spotlighting, the topic of a forthcoming section. Nevertheless, press the page-up key to see a multi-colored array of spheres. Figure 11.21 is a screenshot.

Currently, the point of interest in the program is the invocation of the color material mode for the front-face ambient and diffuse reflectances by means of the last two statements in the initialization routine, viz.

```
glEnable(GL_COLOR_MATERIAL);
glColorMaterial(GL_FRONT, GL_AMBIENT_AND_DIFFUSE);
```

and subsequent coloring of the spheres in the drawing routine by `glColor4f()` statements.

End

11.4 OpenGL Lighting Model

The so-called OpenGL *lighting model* sets certain environmental parameters. The terminology, even though used in the red book, is somewhat unfortunate as it may suggest laws of interaction between light and objects, or a relation with Phong's model – neither of which is true. The four parameters the OpenGL lighting model sets are the following:

1. The global ambient light with the statement

   glLightModel*(GL_LIGHT_MODEL_AMBIENT, *globAmb*)

   where *globAmb* is the global ambient light vector. This we've seen already.

2. Whether to use a local or infinite viewpoint for lighting calculation.

   See again the lighting equation (11.10). The halfway vector s^i at a vertex, one for each light source, is the unit vector bisecting the angle between the direction vector l^i to the light source L^i and the direction vector e to the eye.

   The OpenGL eye being fixed at the origin $[0\ 0\ 0]^T$, evidently $e = -V$, where V is the vertex's position vector, which changes from one to another. However, it simplifies lighting computation to keep e constant, particularly $e = [0\ 0\ 1]^T$, equivalent to *assuming* an eye that is infinitely far up the z-axis and so, effectively, in the same direction from every vertex. See Figure 11.22. This simplification, often, still gives adequately authentic lighting.

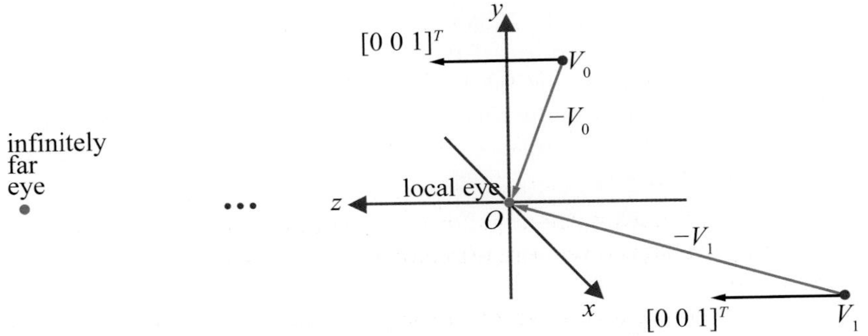

Figure 11.22: Local versus infinite viewpoint: the direction vector from each vertex toward the infinite viewpoint is black, while that toward the local viewpoint – i.e., the eye vector – is blue. V_i denotes both a vertex and its position vector.

Remark 11.10. The direction vector l^i to the light source, too, changes from vertex to vertex if the source is a positional one, i.e., if $w \neq 0$ in the value $[x\ y\ z\ w]^T$ of the source's GL_POSITION parameter. Moreover, a simplification exactly similar to that of assuming an infinite viewpoint can be achieved, not by tweaking the OpenGL lighting model, but by making the light directional by setting $w = 0$. We'll discuss this in the next section.

The OpenGL default viewpoint, in fact, is infinite. For lighting calculation to be done using a local viewpoint instead – i.e., with the eye at the origin – call

```
glLightModel*(GL_LIGHT_MODEL_LOCAL_VIEWER, GL_TRUE)
```

which is what we do in the `setup()` routines of both `sphereIn-Box1.cpp` and `lightAndMaterial1.cpp`, while `lightAndMaterial2.-cpp` provides an option. The local viewpoint is more realistic at the expense of greater computation.

Remark 11.11. The chosen light model viewpoint is used *only for lighting calculations*. The viewing frustum or box stays unchanged – therefore, in the case of a frustum, for example, we still *see the scene* from the eye at the origin.

Exercise 11.11. (Programming) Press 'l' or 'L' to toggle between the **L**ocal and the infinite viewpoint in `lightAndMaterial2.cpp`. The change seems to be only in the highlights, in other words, only the specular reflectances. Why?

3. Whether to enable two-sided lighting.

   The OpenGL default is to perform lighting calculations for each polygon based on its specified `GL_FRONT` face parameter values and its specified vertex normals, *regardless* if it is front or back facing. As the user likely sets material properties and normal values with the front faces of polygons in mind, results tend to be unrealistic for those whose back faces happen to be visible. So, when back faces might be visible, the command to use is

   ```
   glLightModel*(GL_LIGHT_MODEL_TWO_SIDE, GL_TRUE)
   ```

   which causes OpenGL to

   (a) use the `GL_BACK` (or `GL_FRONT_AND_BACK`) parameter values to color back-facing polygons, and

   (b) *reverse* the specified vertex normal for back-facing polygons.

Experiment 11.12. Run `litTriangle.cpp`, which draws a single triangle, whose front is coded red and back blue, initially front-facing and lit two-sided. Press the left and right arrow keys to turn the triangle and space to toggle two-sided lighting on and off. See Figure 11.23 for screenshots.

Notice how the back face is dark when two-sided lighting is disabled – this is because the normals are pointing oppositely of the way they should be. **End**

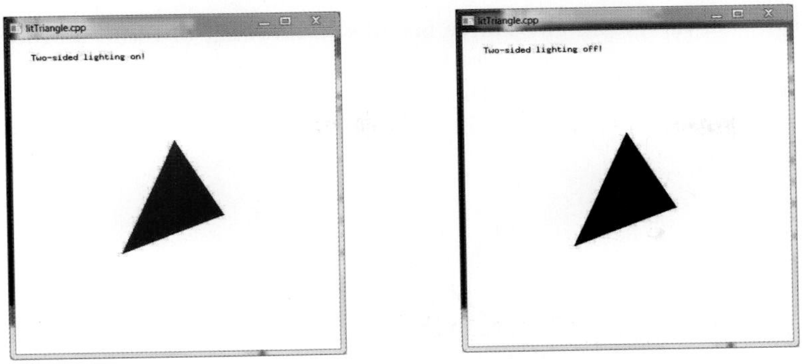

(a) (b)

Figure 11.23: Screenshots of `litTriangle.cpp` showing the back face with (a) two-sided lighting on (b) two-sided lighting off.

4. Whether to apply specular light before or after texturing.

   The following remarks will be more meaningful after the discussion of textures in the next chapter.

   The OpenGL default is to apply textures after all lighting calculations, which can cause specular highlights to be smothered. However, the command

   ```
   glLightModel*(GL_LIGHT_MODEL_COLOR_CONTROL,
                 GL_SEPARATE_SPECULAR_COLOR)
   ```

   makes OpenGL

   (a) separately produce two colors at each vertex: a primary color calculated from all incoming non-specular components and a secondary color from all incoming specular components,

   (b) combine only the primary color with texture color at the time of texture mapping and, finally,

   (c) add in the secondary color to the result of the previous step, which assures the specular highlights.

11.5 Directional Lights, Positional Lights and Attenuation of Intensity

Directional and Positional Light Sources

We know that the value of the `GL_POSITION` parameter of a light source L specifies its location $[x\ y\ z\ w]^T$ in homogeneous coordinates.

If $w \neq 0$, then the light source is called *positional*, or *local*, and located at world coordinates $[x/w \;\; y/w \;\; z/w]^T$. This is the kind of source we have used so far. If $w = 0$, then the light source is *directional* and assumed located at an infinite distance in the direction of $[x \; y \; z]^T$ from the origin. See Figure 11.24.

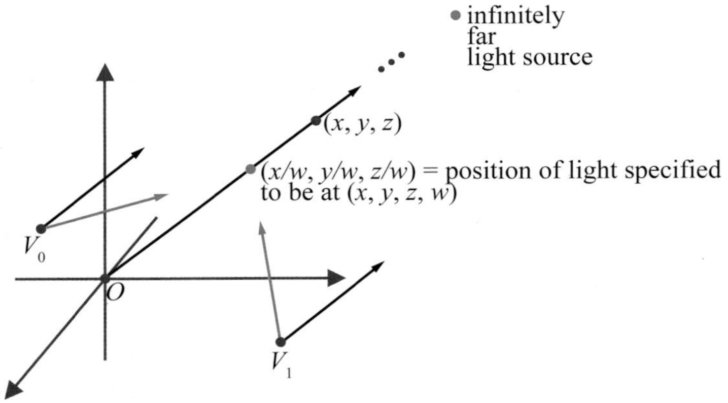

Figure 11.24: Directional versus positional light: the direction vector from each vertex toward the directional light is black and parallel to the direction of the directional light, while that toward the positional light is blue.

A positional light is located within the environment, e.g., a car headlight, while a directional light is far removed, e.g., the sun. From the point of view of lighting calculation, the difference is that the light direction vector l from a vertex V to the light source L depends on the coordinates of V if L is positional, while it is constant for all vertices if L is directional. Evidently, lighting calculation is cheaper for directional sources.

The default value for `GL_POSITION` is $[0\;0\;1\;0]^T$, which defines a directional light shining down from high up the z-axis.

Experiment 11.13. Press 'p' or 'P' to toggle between **P**ositional and directional light in `lightAndMaterial2.cpp`.

The white wire sphere indicates the positional light, while the white arrow the incoming directional light. **End**

Attenuation of Light

In the real world, the intensity of light from a source diminishes with distance from the source following an inverse square law. This phenomenon, called *distance attenuation*, can be modeled in OpenGL as well by a multiplicative *distance attenuation factor*

$$\frac{1}{k_c + k_l d + k_q d^2}$$

where d is the distance from the light source and k_c, k_l and k_d are the values of the light parameters GL_CONSTANT_ATTENUATION, GL_LINEAR_ATTENUATION and GL_QUADRATIC_ATTENUATION, respectively. These values are set by statements of the form

```
glLightf(GL_LIGHTi, GL_CONSTANT_ATTENUATION, kc);
glLightf(GL_LIGHTi, GL_LINEAR_ATTENUATION, kl);
glLightf(GL_LIGHTi, GL_QUADRATIC_ATTENUATION, kq);
```

The default values are $k_c = 1$ and $k_l = k_q = 0$, which imply no attenuation over distance at all. Attenuating the intensity of a directional light over distance is not meaningful as it's already infinitely far from every vertex; therefore, default values for the attenuation parameters cannot be changed for such a source.

Experiment 11.14. Run `lightAndMaterial1.cpp`. The current values of the constant, linear and quadratic attenuation parameters are 1, 0 and 0, respectively, so there's no attenuation. Press 't/T' to decrease/increase the quadratic a**T**tenuation parameter. Move the ball by pressing the up/down arrow keys to observe the effect of attenuation. **End**

11.6 Spotlights

The default for a light source is that it's *regular*, emitting light in all directions (a fancy word for a regular light source would be *omnidirectional*). This can be altered by turning it into a *spotlight*, in which case the emitted light is in the shape of a cone, the purpose being, of course, to simulate a real-life spotlight illuminating a limited area.

Figures 11.25(a) and (b) show, respectively, plane sections of the light from both a regular and a spotlight.

Experiment 11.15. Run `spotlight.cpp`, which shows a bright white spotlight illuminating a multi-colored array of spheres. A screenshot was shown earlier in Figure 11.21.

Press the page up/down arrows to increase/decrease the angle of the light cone. Press the arrow keys to move the spotlight. Press 't/T' to change the spotlight attenuation factor. A white wire mesh is drawn along the light cone boundary. **End**

The following block of statements in the `drawScene()` routine specify the position and all the spotlight properties of the single light source of `spotlight.cpp`:

```
glLightfv(GL_LIGHT0, GL_POSITION, lightPos);
glLightf(GL_LIGHT0, GL_SPOT_CUTOFF, spotAngle);
glLightfv(GL_LIGHT0, GL_SPOT_DIRECTION, spotDirection);
glLightf(GL_LIGHT0, GL_SPOT_EXPONENT, spotExponent);
```

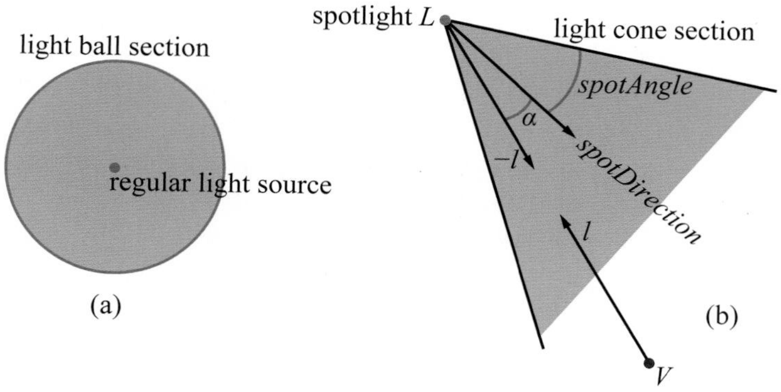

Figure 11.25: Sections of (a) regular light (b) spotlight. Shown for the spotlight are the light direction vector l from vertex V toward light source L, the inverse vector $-l$ from the light source toward the vertex, the direction vector *spotDirection* of the cone's axis and the half-angle *spotAngle* at its apex.

The first step to turning a light source L into a spotlight is to specify the half-angle at the apex of the light cone, called the *spotlight cone angle*, with the command

```
glLightf(GL_LIGHT0, GL_SPOT_CUTOFF, spotAngle)
```

which, in fact, sets the spotlight cone angle to the value *spotAngle*. This should be between 0.0 and 90.0. The default is the special value of 180.0, meaning that L is not a spotlight, but a regular source emitting in all directions. Its initial value in `spotlight.cpp`, in fact, is $10°$.

The next step is to specify the *spotlight direction* or, more specifically, that of the axis of its cone with a command

```
glLightfv(GL_LIGHT0, GL_SPOT_DIRECTION, spotDirection)
```

which sets the axis in a direction parallel to the vector

$$spotDirection = [x\ y\ z]^T$$

The default value of `GL_SPOT_DIRECTION` is $[0\ 0\ -1]^T$, aiming the spotlight down the negative z-axis. It is set to $\{0.0, -1.0, 0.0\}$ pointing in the negative y-direction in `spotlight.cpp`.

A final spotlight parameter is `GL_SPOT_EXPONENT`, whose value is called the *spotlight attenuation factor* and which controls the distribution of intensity through the light cone. If the value of `GL_SPOT_EXPONENT` is h and the angle between the axis of the light cone and the direction from source L toward vertex V is α, then the intensity of light at V is attenuated by the multiplicative factor $\cos^h \alpha$. This, of course, presumes that V lies within the light cone in the first place; if not, no light reaches V from L at all. Note that, as depicted in Figure 11.25, a vector from L toward V is, simply, $-l$, the negative of a light direction vector.

The spotlight attenuation factor is set to 2.0, initially, in `spotlight.cpp`; it can be changed by pressing 't/T'.

The motivation behind the spotlight attenuation factor is similar to that for the angular attenuation factor in the calculation of specular reflection in Equation (11.8) – so that the greater the value of h, the more rapidly the intensity of the spotlight attenuates away from the cone's axis. Equivalently, the greater h the more "concentrated" the spotlight. The default value of `GL_SPOT_EXPONENT` is 0, implying no attenuation at all.

Experiment 11.16. Run again `spotlight.cpp`. Note the change in visibility of the balls near the cone boundary as the attenuation changes.
End

Exercise 11.12. A spotlight should always be positional. Why?

For use in the upcoming final OpenGL light equation, let's write a single complete formula for a *spotlight attenuation factor* or, briefly, *saf*, at a vertex V, for a given light source L. Denote the unit vector along the spotlight axis – the normalized value of `GL_SPOT_DIRECTION` – by d and assume that l, the light direction vector from V, is normalized as well (see Figure 11.25). Then:

$$saf = \begin{cases} 1 & , \text{ if } spotAngle = 180^\circ \\ 0 & , \text{ if } \; -l \cdot d < \cos(spotAngle) \\ (-l \cdot d)^h & , \text{ otherwise} \end{cases} \tag{11.11}$$

Here's how to parse the formula.

The first line is the case when L is not a spotlight, so there's no attenuation.

For the second line, recall that $-l$ is the unit vector from L toward V. Therefore, $-l \cdot d = \cos\alpha$, where α is the angle between the axis of the light cone and the direction of V from L. Now, if $\cos\alpha < \cos(spotAngle)$, then $\alpha > spotAngle$, which means that V lies outside the light cone and gets zero light. This explains the second line.

The third line, of course, gives the angular attenuation factor.

Exercise 11.13. Why isn't it necessary to write $(\max\{-l \cdot d\, ,\, 0\})^h$, instead of $(-l \cdot d)^h$, in Equation (11.11) in a manner similar to the first lighting equation (11.10)?

11.7 OpenGL Lighting Equation

We now have the two additional pieces needed to enhance the first lighting equation (11.10) to the form that is, in fact, used by OpenGL to calculate RGB color intensities at a vertex V, namely, distance attenuation and spotlight attenuation. The enhancement is straightforward.

All symbols from the first lighting equation retain the same meaning. Additionally, d^i denotes the distance of V from the ith light source; k_c^i, k_l^i and k_q^i denote, respectively, the constant, linear and quadratic attenuation parameters for the ith light source; and saf^i is the spotlight attenuation factor for the ith light source at the vertex V, as given by Equation (11.11).

So, finally, here it is, the grand ole lighting equation of OpenGL:

$$\begin{aligned} V_X \;=\;& V_{emit,\,X} \;+ \\ & globAmb_X * V_{amb,\,X} \;+ \\ & \sum_{i=0}^{N-1} \frac{1}{k_c^i + k_l^i d^i + k_q^i (d^i)^2} * saf^i * \\ & \quad \Big(L^i_{amb,\,X} * V_{amb,\,X} \;+ \\ & \quad \max\{l^i \cdot n\,,\, 0\} * L^i_{diff,\,X} * V_{diff,\,X} \;+ \\ & \quad (\max\{s^i \cdot n\,,\, 0\})^f * L^i_{spec,\,X} * V_{spec,\,X} \Big) \end{aligned} \tag{11.12}$$

where V_X is the color intensity at V, X being any of RGB.

The additions to the first lighting equation (11.10) are exactly the two multiplicative terms on the third line of the current equation, representing distance attenuation and spotlight attenuation, respectively.

Remark 11.12. It's really Phong's lighting equation, but, given the context, we'll more often than not call it the OpenGL lighting equation.

Remark 11.13. We must revisit Exercise 11.6 at this time. Its implication that all individual light source ambients can be consolidated into the global ambient is not true any more if one uses Equation (11.12) instead of Equation (11.10), because the same light source can contribute different amounts of ambient light to different vertices owing to distance and spotlight attenuation.

Nevertheless, the simplification of setting all individual light source ambients to zero, and adjusting only the global, is probably still authentic enough for most applications.

Exercise 11.14. If there is a single directional light source in an OpenGL program, which is not distance attenuated, which of the three – ambient, diffuse and specular – reflectance components at its vertices is changed by *translating* an object?

Exercise 11.15. If there is a single positional light source in an OpenGL program, which is not a spotlight and not distance attenuated, which of the three – ambient, diffuse and specular – reflectance components at an object's vertex can change by moving the light source? By translating the object?

Exercise 11.16. Which of the three components – ambient, diffuse and specular – of light reflected from a vertex V are affected if the normal at V is altered?

11.8 OpenGL Shading Models

A *shading model* is a method to shade, or color, the interiors of primitives. Keep in mind that Phong's lighting model, as implemented through the OpenGL lighting equation, determines colors *only* at the vertices of primitives, but says nothing about how to spread them inside. OpenGL's default shading model, called *smooth shading* or *Gouraud* shading, is to interpolate color values computed at its vertices through a primitive's interior. We discussed in Section 7.2 the mechanics of interpolation by computing barycentric coordinates of interior points.

An alternate shading model, called *flat shading*, is available, as well, in OpenGL. It is specified by a call to

```
glShadeModel(GL_FLAT)
```

The default of smooth shading is restored by calling

```
glShadeModel(GL_SMOOTH)
```

When flat shading, even if the color values differ across the vertices of a primitive, OpenGL chooses *one* of them, called the *provoking vertex*, and applies its color to the entire primitive. For example, the provoking vertex of a triangle is its first (according to the order of the vertices in the code). In a triangle strip, the provoking vertex of the i th triangle is the $i+2$ th vertex. The reader is referred to the red book for a full description of provoking vertices for each primitive type.

Flat shading can be a reasonable alternative in the absence of lighting. Computationally it's, of course, far less expensive than smooth shading. One interesting application of flat shading is in applying "discrete" color schemes, which, often, is difficult with smooth shading. The following experiment is an illustration.

Experiment 11.17. Run `checkeredFloor.cpp`, which creates a checkered floor drawn as an array of flat shaded triangle strips. See Figure 11.26. Flat shading causes each triangle in the strip to be painted with the color of the last of its three vertices, according to the order of the strip's vertex list. **End**

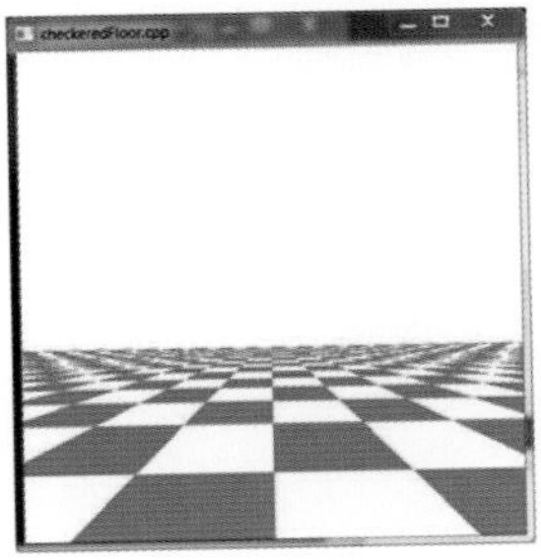

Figure 11.26: Screenshot of `checkeredFloor.cpp`.

Exercise 11.17. (**Programming**) Try and replicate the checkered floor of the preceding experiment using smooth shading instead of flat.

11.9 Animating Light

There are three ways that a light source can be animated by changing spatial properties:

1. By moving its position.

2. By changing its direction if it's a spotlight.

3. By changing the light cone angle if it's a spotlight.

We've already seen light animation in two programs in this chapter: `lightAndMaterial2.cpp` and `spotlight.cpp`.

The things to keep in mind are:

(a) A light source's position vector, specified by a `glLightfv(`*light*`, GL_POSITION, `*lightPos*`)` statement, is transformed by the value of the current modelview matrix by multiplication from the left. (See Section 4.2 if you need to review modelview matrices.)

Effectively, then, modelview transformations in the code prior to the `glLightfv(`*light*`, GL_POSITION, `*lightPos*`)` statement apply to a light's position, exactly as those prior to a `glVertex3f()` statement, defining a vertex, apply to that vertex.

(b) Likewise, a spotlight source's direction vector, specified by the `glLightfv(`*light*`, GL_SPOT_DIRECTION, `*spotDirection*`)` statement, is transformed by the value of the current modelview matrix by multiplication from the left.

For example, as the light source of `sphereInBox1.cpp` is positioned by the `glLightfv(GL_LIGHT0, GL_POSITION, lightPos)` statement in the initialization routine `setup()`, it is unaffected by any modelview transformations in `drawScene()`.

However, both lights of `lightAndMaterial1.cpp` are positioned in the display routine following the viewing command `gluLookAt()`, so their positions are, in fact, transformed by `gluLookAt()`, which effectively means that the lights stay static relative to the scene, no matter if the viewpoint is changed. The light positions of `lightAndMaterial2.cpp` are similarly transformed by its own `gluLookAt()`.

Note: The push-pop pairs surrounding the code to position the lights in both programs are to isolate the transformations applied to the spheres that depict the light sources.

The spotlight of `spotlight.cpp` is positioned in the display routine after the viewing transformation *and* a user-specified translation; moreover, its cone angle can be changed by the user too.

Remark 11.14. We've discussed only animating the spatial attributes of a light source. Obviously, color values can be animated as well.

11.10 Partial Derivatives, Tangent Planes and Normal Vectors 101

This section is an introduction to the calculus sometimes required to calculate normals to surfaces. It is not mandatory reading. We suggest you skip this section initially and consult it later if need be.

Actually, if you know how to compute derivatives of a function of a single variable, e.g., $f(x) = x^2$ or $f(x) = \sin x$, as we'll assume you do, you already know how to compute partial derivatives. Because ...

Definition 11.1. Suppose that f is a function of more than one variable $x, y, \ldots$. The *partial derivative* of f with respect to one of these variables, say x, is the derivative of f as a function *only* of x, assuming the other variables all fixed. The partial derivative of f with respect to x is denoted $\frac{\partial f}{\partial x}$.

Example 11.6. Evaluate the partial derivatives of

$$f(x, y) = x^2 + y^2$$

at the point $(1, 2)$.

Answer: We have

$$\frac{\partial f}{\partial x}(x, y) = 2x, \qquad \frac{\partial f}{\partial y}(x, y) = 2y$$

Therefore,

$$\frac{\partial f}{\partial x}(1, 2) = 2, \qquad \frac{\partial f}{\partial y}(1, 2) = 4$$

Remark 11.15. Often $\frac{\partial f}{\partial x}(x, y)$ is simply written $\frac{\partial f}{\partial x}$, e.g., the first two equations of the preceding answer could be written

$$\frac{\partial f}{\partial x} = 2x, \qquad \frac{\partial f}{\partial y} = 2y$$

Example 11.7. Evaluate the partial derivatives of

$$f(x, y) = x^2 \sin y$$

at the point $(1, \pi/2)$.

Answer: We have

$$\frac{\partial f}{\partial x} = 2x \sin y, \qquad \frac{\partial f}{\partial y} = x^2 \cos y$$

Therefore,

$$\frac{\partial f}{\partial x}(1,\ \pi/2) = 2, \qquad \frac{\partial f}{\partial y}(1,\ \pi/2) = 0$$

Example 11.8. Evaluate the partial derivatives of

$$f(x, y, z) = xz + \sin x \cos y \cos z + y$$

at the point $(\pi/2, \pi, 0)$.

Answer: We have

$$\frac{\partial f}{\partial x} = z + \cos x \cos y \cos z, \quad \frac{\partial f}{\partial y} = 1 - \sin x \sin y \cos z, \quad \frac{\partial f}{\partial z} = x - \sin x \cos y \sin z$$

Therefore,

$$\frac{\partial f}{\partial x}(\pi/2,\ \pi,\ 0) = 0, \quad \frac{\partial f}{\partial y}(\pi/2,\ \pi,\ 0) = 1, \quad \frac{\partial f}{\partial y}(\pi/2,\ \pi,\ 0) = \pi/2$$

Exercise 11.18. Evaluate the partial derivatives of

$$f(x, y) = xy$$

at the point $(2, 3)$.

Exercise 11.19. Evaluate the partial derivatives of

$$f(x, y, z) = x \cos y + y \cos z + z \cos x$$

at the point $(\pi/2, 0, \pi/2)$.

The reader may wonder that if the partial derivative $\frac{\partial f}{\partial x}$, for example, is obtained by differentiating f with respect to the single variable x, assuming the others fixed, then why do those other variables pop up again in the expression for $\frac{\partial f}{\partial x}$? Here's the reason.

Consider the function $f(x, y) = x^2 \sin y$ of Example 11.7 above. Fixing y at, say, the value $\pi/6$ gives the function $f(x, \pi/6) = x^2/2$, while fixing y at $\pi/2$ gives the function $f(x, \pi/2) = x^2$. Both $f(x, \pi/6)$ and $f(x, \pi/2)$ are functions of the one variable x, but they are *different* functions because y's been fixed at two *different* values.

Moreover,

$$\frac{\partial f}{\partial x}(x, \pi/6) = \frac{\mathrm{d}}{\mathrm{d}x}(x^2/2) = x \quad \text{and}$$
$$\frac{\partial f}{\partial x}(x, \pi/2) = \frac{\mathrm{d}}{\mathrm{d}x}(x^2) = 2x$$

are different as well, as they are derivatives of different functions. This is why $\frac{\partial f}{\partial x}$ depends on y, as well as on x.

So far so good. At least calculating partial derivatives is no different from calculating ordinary derivatives. But what do partial derivatives mean geometrically (in "real life", that is)?

For example, an ordinary derivative on a curve specifies its tangent. Let's see how first for both implicit and parametric declarations:

(a) *Implicit*: Suppose a curve is given by the equation

$$y = f(x)$$

Then the value of

$$\frac{\mathrm{d}f}{\mathrm{d}x}$$

at $x = a$ is the gradient of the tangent line to the curve at the point $(a, f(a))$.

For example, the gradient of the tangent line l at the point $(1, 1)$ of the parabola

$$y = x^2$$

is 2 as

$$\frac{\mathrm{d}}{\mathrm{d}x}(x^2) = 2x$$

which equals 2 when x is 1. See Figure 11.27(a).

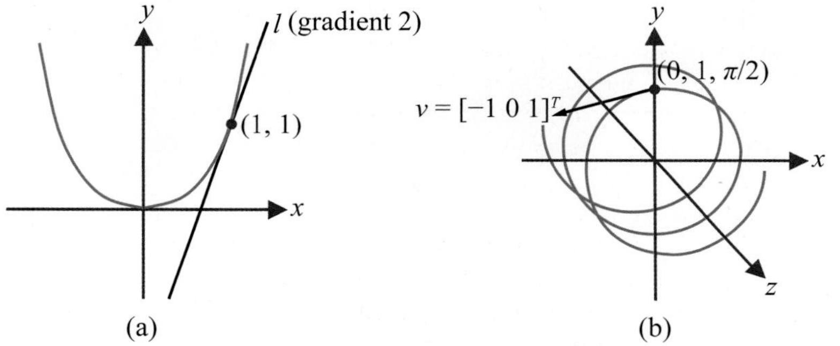

Figure 11.27: Tangents: (a) Tangent line l to the parabola $y = x^2$ at $(1, 1)$ (b) Tangent vector v to the helix $c(t) = (\cos t,\ \sin t,\ t)$ at $(0,\ 1,\ \pi/2)$.

(b) *Parametric*: Suppose a curve is given by

$$c(t) = (f(t),\ g(t),\ h(t))$$

Then the value of the vector

$$c'(t) = \begin{bmatrix} \frac{\mathrm{d}f}{\mathrm{d}t} & \frac{\mathrm{d}g}{\mathrm{d}t} & \frac{\mathrm{d}h}{\mathrm{d}t} \end{bmatrix}^T$$

at $t = a$ is a tangent vector (provided it's non-zero) to the curve at the point $(f(a),\ g(a),\ h(a))$.

For example, a tangent vector v to the helix

$$c(t) = (\cos t,\ \sin t,\ t)$$

at the point $(0,\ 1,\ \pi/2)$, corresponding to $t = \pi/2$, is $[-1\ 0\ 1]^T$, as

$$\left[\frac{\mathrm{d}}{\mathrm{d}t}(\cos t)\ \ \frac{\mathrm{d}}{\mathrm{d}t}(\sin t)\ \ \frac{\mathrm{d}}{\mathrm{d}t}(t)\right]^T = [-\sin t\ \ \cos t\ \ 1]^T$$

which equals $[-1\ 0\ 1]^T$ when $t = \pi/2$. See Figure 11.27(b).

It turns out that, just as the computation of partial derivatives is based on computing ordinary derivatives, their geometric significance obtains from that of ordinary derivatives too. Here's how:

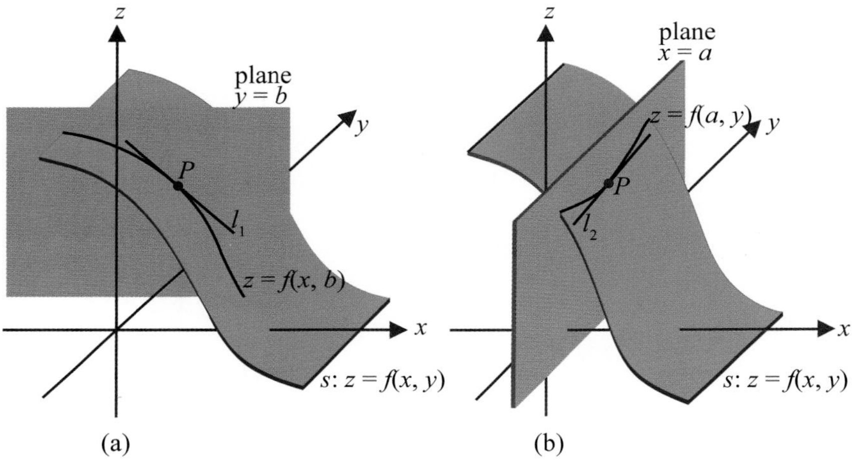

Figure 11.28: Section of the graph s of $z = f(x, y)$ by the (a) plane $y = b$, giving the tangent line l_1 at $P = (a,\ b,\ f(a, b))$, (b) plane $x = a$, giving the tangent line l_2 at P.

(a) *Implicit*: Consider $z = f(x, y)$, a function of two variables. It defines a surface s, called the graph of f. See Figure 11.28(a).

Now, if we fix y at, say, the value b, then $z = f(x, b)$ gives a curve s. In fact, this curve is the section of s by the plane $y = b$.

We know that the value of $\frac{\partial f}{\partial x}$ at (a, b) is the value at a of the ordinary derivative $\frac{\mathrm{d}}{\mathrm{d}x} f(x, b)$. This helps find geometric meaning for the partial derivative as follows.

The value of

$$\frac{\partial f}{\partial x}$$

at (a, b) is the gradient of the tangent line l_1 to the sectional curve $z = f(x, b)$ at the point $P = (a,\ b,\ f(a, b))$.

Likewise, the value of

$$\frac{\partial f}{\partial y}$$

at (a, b) is the gradient of the tangent line l_2, at the point $P = (a,\ b,\ f(a, b))$, to the curve $z = f(a, y)$, which is the section of s by the plane $x = a$ (Figure 11.28(b)).

(b) *Parametric:*

Consider next the surface s specified by the parametric equations

$$x = f(u, v),\ y = g(u, v),\ z = h(u, v),\quad (u, v) \in W$$

where $W = [u_1, u_2] \times [v_1, v_2]$ is a rectangle in uv parameter space. The function $(u, v) \mapsto s(u, v) = (f(u, v),\ g(u, v),\ h(u, v))$ maps W to the surface s in 3-space. See Figure 11.29.

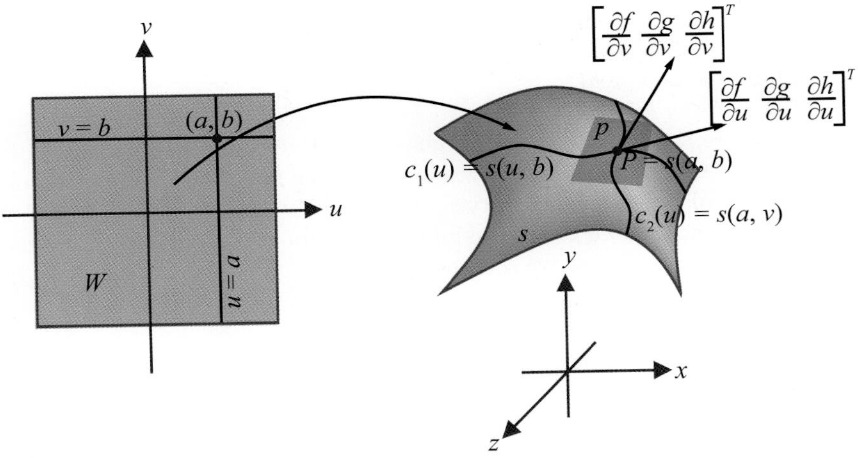

Figure 11.29: The surface s is the image of a parameter rectangle W by the map $(u, v) \mapsto s(u, v) = (f(u, v),\ g(u, v),\ h(u, v))$. Tangents to the parameter curves on s at the point $P = s(a, b)$ span the tangent plane p at P.

Fix a point $(a, b) \in W$. The image of the line $v = b$ by s is the u-parameter curve c_1 with equation

$$c_1(u) = (f(u, b),\ g(u, b),\ h(u, b)),\quad u \in [u_1, u_2]$$

The tangent vector to this curve is

$$\begin{aligned} c_1'(u) &= \left[\frac{\mathrm{d}}{\mathrm{d}u} f(u, b)\quad \frac{\mathrm{d}}{\mathrm{d}u} g(u, b)\quad \frac{\mathrm{d}}{\mathrm{d}u} h(u, b)\right]^T \\ &= \left[\frac{\partial f}{\partial u}(u, b)\quad \frac{\partial g}{\partial u}(u, b)\quad \frac{\partial h}{\partial u}(u, b)\right]^T \end{aligned}$$

at $u \in [u_1, u_2]$. Therefore, the value of the vector

$$\left[\frac{\partial f}{\partial u}\quad \frac{\partial g}{\partial u}\quad \frac{\partial h}{\partial u}\right]^T$$

at the point (a, b) is a tangent vector (provided it's non-zero) to the u-parameter curve

$$c_1(u) = s(u, b)$$

at the point $s(a, b)$.

Likewise, the value of the vector

$$\left[\frac{\partial f}{\partial v} \quad \frac{\partial g}{\partial v} \quad \frac{\partial h}{\partial v}\right]^T$$

at the point (a, b) is a tangent vector (provided it's non-zero) to the v-parameter curve

$$c_2(v) = s(a, v)$$

at the point $s(a, b)$.

So, we see that, just as an ordinary derivative on a curve specifies its tangent, the two partial derivatives on a surface specify tangents as well, particularly, of two sectional curves. In fact, these two tangents together can supply us even more geometric information about the surface – they can give us a tangent plane as we see next.

Definition 11.2. If the tangent vectors

$$\left[\frac{\partial f}{\partial u} \quad \frac{\partial g}{\partial u} \quad \frac{\partial h}{\partial u}\right]^T$$

and

$$\left[\frac{\partial f}{\partial v} \quad \frac{\partial g}{\partial v} \quad \frac{\partial h}{\partial v}\right]^T$$

to the two parameter curves through the point $P = s(a, b)$ are linearly independent – in other words, they are not collinear – then they span a plane p, called the *tangent plane* to the surface s at P. This is the case in Figure 11.29.

Any line l on p through P is said to be a *tangent line* to s at P and any non-zero vector v lying on p is said to be a *tangent vector* to s at P (v is usually drawn emanating from P). See Figure 11.30. The line perpendicular to p through P is said to be the *normal line* to s at P and any non-zero vector lying on this line a *normal vector* to s at P.

A tangent plane to a surface is precisely the geometric analogue of a tangent line to a curve. A thin straight stick pressed to a plane wire curve aligns itself along the tangent line at the point of contact; similarly, a thin flat board pressed to a surface in 3-space aligns itself along the tangent plane at the point of contact.

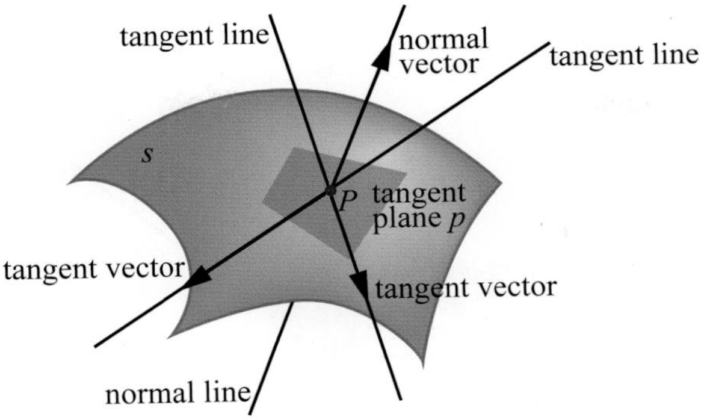

Figure 11.30: Two tangent lines and vectors on them, normal line and a normal vector to the surface s at P.

Example 11.9. Determine the tangent plane and a normal vector to the paraboloid

$$z = x^2 + y^2$$

at the point $(1, 2, 5)$.

Answer: It's easy first to write the given implicit equation in the parametric form

$$x = u, \quad y = v, \quad z = u^2 + v^2 \tag{11.13}$$

Differentiating,

$$\begin{aligned} \begin{bmatrix} \frac{\partial x}{\partial u} & \frac{\partial y}{\partial u} & \frac{\partial z}{\partial u} \end{bmatrix}^T &= [1 \ 0 \ 2u]^T \\ \begin{bmatrix} \frac{\partial x}{\partial v} & \frac{\partial y}{\partial v} & \frac{\partial z}{\partial v} \end{bmatrix}^T &= [0 \ 1 \ 2v]^T \end{aligned} \tag{11.14}$$

The point $(1, 2, 5)$ corresponds to the parameter values $u = 1$ and $v = 2$ in (11.13). Therefore, two tangent vectors to the paraboloid at $(1, 2, 5)$ are obtained by substituting these particular parameter values into the general expressions (11.14) above for tangent vectors at arbitrary points. Specifically, these two vectors are $[1 \ 0 \ 2]^T$ and $[0 \ 1 \ 4]^T$, which are evidently linearly independent. Therefore, the tangent plane to the paraboloid at $(1, 2, 5)$ is spanned by $[1 \ 0 \ 2]^T$ and $[0 \ 1 \ 4]^T$. See Figure 11.31.

A normal vector to the paraboloid at the point $(1, 2, 5)$ is perpendicular to its tangent plane there and, therefore, to both spanning vectors $[1 \ 0 \ 2]^T$ and $[0 \ 1 \ 4]^T$. It is obtained, then, as the cross-product of the latter (cross-products of vectors were reviewed in Section 5.4.3), viz.

$$[1 \ 0 \ 2]^T \times [0 \ 1 \ 4]^T = [-2 \ -4 \ 1]^T$$

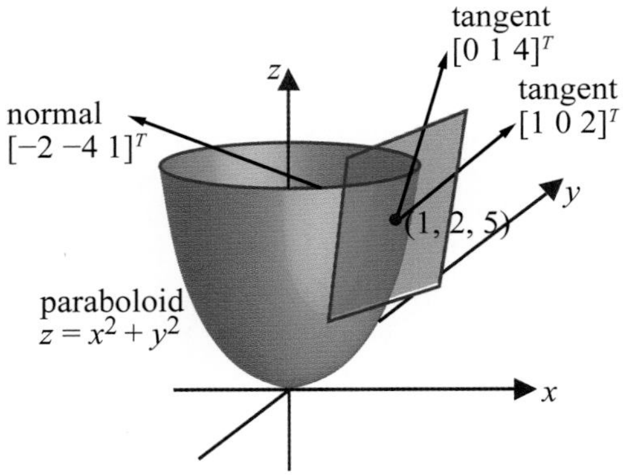

Figure 11.31: Tangent vectors, tangent plane and normal vector at the point $(1, 2, 5)$ to the paraboloid $z = x^2 + y^2$.

Remark 11.16. Computing the tangent plane at a point of a surface and computing a normal vector there are equivalent.

Exercise 11.20. Determine the tangent plane and a normal vector to the circular cylinder

$$x = \cos u, \ y = \sin u, \ z = v$$

at the point corresponding to the parameter values $(u, v) = (\pi/4, 3)$.

Exercise 11.21. Determine the tangent plane and a normal vector to the saddle-shaped surface (hyperbolic paraboloid is the mathematical name)

$$z = xy$$

at the point $(2, 3, 6)$.

Exercise 11.22. **(Programming)** Draw the paraboloid of Example 11.9 and its tangent plane at some point. The paraboloid should be wireframe and the tangent plane a finely meshed rectangle. Allow the user to press the arrow keys to slide the tangent plane over the paraboloid.

Normals from Function Gradients

Definition 11.2 of a tangent plane assumes a parametric representation of the surface. There's, however, a neat way to compute directly a normal vector at a point of a surface given implicitly.

If a surface s is specified implicitly by an equation of the form

$$F(x, y, z) = 0$$

then a normal vector to the surface at the point (a, b, c) is given by the value of the so-called *gradient* of F, denoted $grad(F)$, at that point, provided this value is not the zero vector. The gradient is defined by

$$grad(F) = \left[\frac{\partial F}{\partial x} \quad \frac{\partial F}{\partial y} \quad \frac{\partial F}{\partial z}\right]^T$$

We'll not try to prove that $grad(F)$ is indeed normal to the surface $F(x, y, z) = 0$, but simply assume so for the purpose of computation. For the actual proof and more about the gradient, as well as its related functions *divergence* and *curl*, the reader is referred to books on vector calculus, e.g., Schey [123] and Spiegel [136].

Example 11.10. Determine a normal vector to the paraboloid

$$z = x^2 + y^2$$

at the point $(1, 2, 5)$.

Answer: Write the implicit equation in the form

$$F(x, y, z) = z - x^2 - y^2 = 0$$

Then

$$grad(F) = \left[\frac{\partial F}{\partial x} \quad \frac{\partial F}{\partial y} \quad \frac{\partial F}{\partial z}\right]^T = [-2x \quad -2y \quad 1]^T$$

Therefore, a normal vector at the point $(1, 2, 5)$ is $[-2 \quad -4 \quad 1]^T$, which is obtained from putting $x = 1$ and $y = 2$ in the preceding equation. This result checks with Example 11.9.

Exercise 11.23. Verify your answer to Exercise 11.20 by finding a normal vector to the cylinder using the *grad* function. You must write an implicit equation for the cylinder first.

11.11 Computing Normals and Lighting Surfaces

Look carefully at the OpenGL lighting equation (11.12) once more. Outside of a bunch of user-specified color properties, the only data needed to compute the color intensities at a vertex V of an object O consists of the position of V, the positions of the light sources *and* the normal vector n at V.

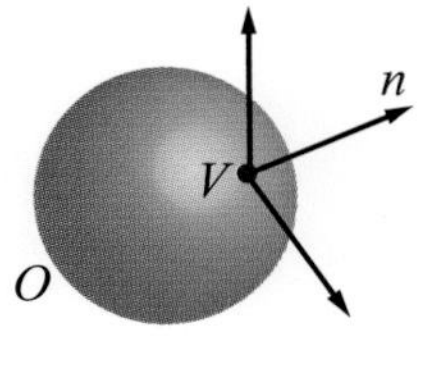

Figure 11.32: Three vectors at a vertex on a sphere, one of which has been chosen as the normal.

The position of V is, of course, part of O's design. As for the light sources, they are usually few, and the user is free to locate them as he pleases. Remaining is the normal vector n, which the user is free to set as well. However, for authentic lighting it should actually be perpendicular to the surface of O at V or at least nearly so. For example, the choice of the normal vector n at the vertex V of the sphere in Figure 11.32 seems good, though either of the other two vectors drawn there could conceivably have been picked as well.

We'll discuss computing surface normals following the informal taxonomy of 2D objects in Section 10.2 before moving on to Bézier and quadric surfaces for which OpenGL provides automatic normals.

11.11.1 Polygons and Planar Surfaces

Polygons in particular, and planar surfaces in general, are the simplest. The normal at each vertex is simply normal to the plane itself containing the surface. In particular, unit vertex normals are all identical across a given side of the surface.

So how does one determine the normal direction to a plane p? If two non-collinear vectors u and v are known to lie on p, then the cross-product $u \times v$ is normal to p (cross-products were reviewed in Section 5.4.3). For example, any two adjacent edges of a polygon determine non-collinear vectors u and v spanning the plane p containing the polygon; therefore, $u \times v$ is normal to p. In Figure 11.33, $n = (P_1 - P_0) \times (P_4 - P_0)$ is normal to p.

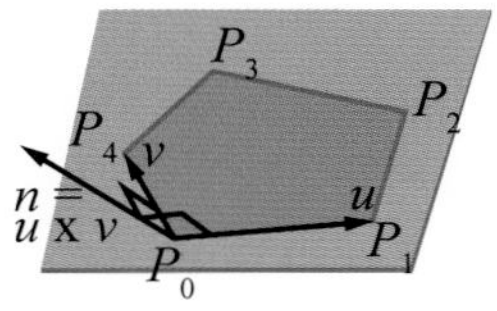

Figure 11.33: Vector n is normal to the plane p.

Exercise 11.24. Determine a normal to the plane p of the triangle with vertices at

$$P_0 = [0\ 3\ 5]^T, \quad P_1 = [1\ -2\ 0]^T, \quad P_2 = [3\ 3\ 3]^T$$

11.11.2 Meshes

Polygonal meshes are of interest next. Let's work with real examples.

Experiment 11.18. Run again `sphereInBox1.cpp`. The normal vector values at the eight box vertices of `sphereInBox1.cpp`, placed in the array `normals[]`, are

$$[\pm 1/\sqrt{3} \quad \pm 1/\sqrt{3} \quad \pm 1/\sqrt{3}]^T$$

each corresponding to one of the eight possible combinations of signs. **End**

The choice of the normals in `sphereInBox1.cpp` is easily motivated. The box being situated symmetrically about the origin, the normal values are chosen as unit vectors along the lines from the origin to each of the eight vertices, which indeed give the values above. The box is depicted in Figure 11.34(a), where only the normal vector at the lower-right vertex V of the front face is shown: it is the arrow n drawn by extending OV a unit distance from V.

In fact, probably a better rationale for this particular choice of normals – which would still hold if the same box happened to be drawn not centered at the origin, but elsewhere – is that the one at each vertex is the normalized *average* of the unit outward normals to the three faces meeting at that vertex. For example, in Figure 11.34(a) the unit outward normals to f_1, f_2

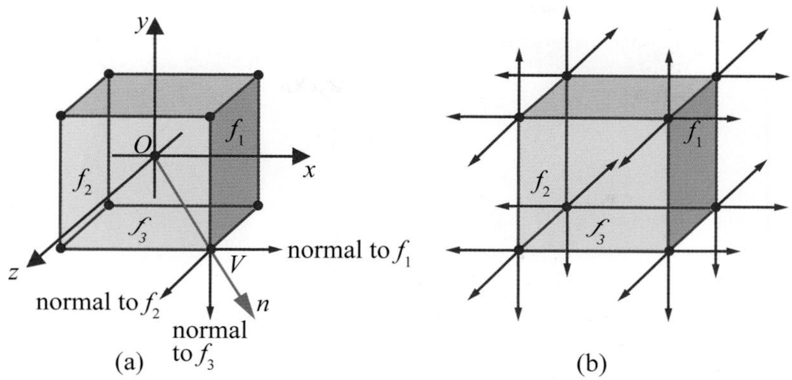

Figure 11.34: (a) The box of `sphereInBox1.cpp` with the averaged normal vector n at vertex V, together with the normals to the three faces that meet at V (f_1 right face, f_2 front face, f_3 bottom face) (b) The unaveraged normals of `sphereInBox2.cpp`.

and f_3 are $[1\ 0\ 0]^T$, $[0\ 0\ 1]^T$ and $[0\ -1\ 0]^T$, respectively, whose average is $[1/3\ \ -1/3\ \ 1/3]^T$, which normalizes to $[1/\sqrt{3}\ \ -1/\sqrt{3}\ \ 1/\sqrt{3}]^T$, which one can verify from the code is indeed the value of the normal at V in `sphereInBox1.cpp`.

Although they possess the virtue of symmetry, it's clear, nevertheless, the box normals of `sphereInBox1.cpp` are not nearly actually perpendicular to the surface of the box, in particular, not to any of its faces. This consideration leads to another approach – to set the normal at each vertex of a face as a normal to that face itself. This is implemented as an option in `sphereInBox2.cpp`.

Experiment 11.19. Run `sphereInBox2.cpp`, which modifies `sphereInBox1.cpp`. Press the arrow keys to open or close the box and space to toggle between two methods of drawing normals.

The first method is the same as that of `sphereInBox1.cpp`, specifying the normal at each vertex as an average of incident face normals. The second creates the box by first drawing one side as a square with the normal at each of its four vertices specified to be the unit vector perpendicular to the square, then placing that square in a display list and, finally, drawing it six times appropriately rotated. Figure 11.34(b) shows the vertex normals to three faces. Figure 11.35 shows screenshots of the box created with and without averaged normals. **End**

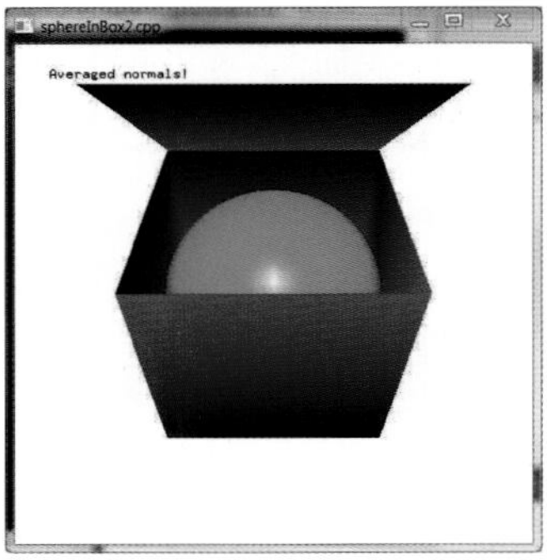

(a)

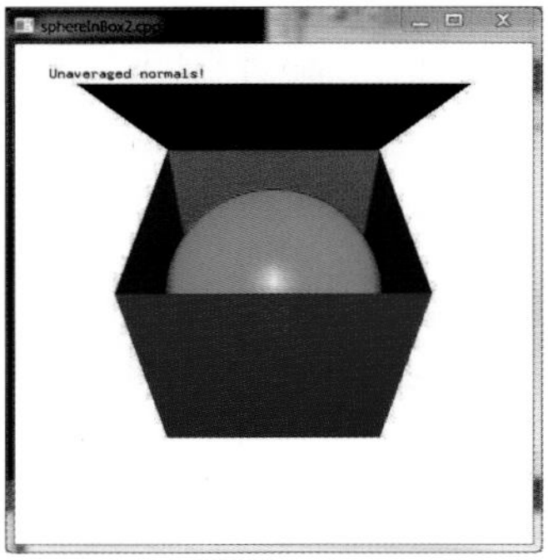

(b)

Figure 11.35: Screenshot of `sphereInBox2.cpp`: (a) Averaged box normals (b) Unaveraged box normals.

The contrast in output between the two ways of defining box normals in `sphereInBox2.cpp` is clear and the reason not hard to understand. The first method softens the edges because the averaged normal at each vertex is shared by all its three adjacent faces. Consequently, the interpolation of color values in each face's interior continues smoothly across its boundary.

The second method is significantly different. As each face is drawn separately with the normals at all its four vertices equal and perpendicular to the face itself, interpolation in the interior results in the entire face being colorized as if with that one normal value throughout. Moreover, this normal value turns abruptly by 90° from one face to the next. The upshot is that there is a significant difference in color intensities, as well, from one face to the next, throwing the edges between them into sharp relief. Which approach to choose depends on the effect desired.

Remark 11.17. Using the second method, colors at pixels along an edge are defined differently by its two adjacent faces, while pixel colors at a vertex are defined, in fact, by its three adjacent faces. At these pixels, therefore, code order determines which color prevails. This is not desirable, but it is not a serious issue because such "ambiguous" pixels lie only along edges and not in the interior of faces which constitute the bulk of the figure.

Versions of the averaging approach implemented sometimes to achieve greater realism use a *weighted* average rather than a straight one. Two possibilities are:

(a) Weight each adjacent face normal with the angle of that face at the vertex. In Figure 11.36, five faces meet at the vertex V subtending angles $\theta_1, \theta_2, \ldots, \theta_5$, respectively. The angle-weighted average value of the normal at V is:

$$n = \frac{\theta_1 n_1 + \theta_2 n_2 + \theta_3 n_3 + \theta_4 n_4 + \theta_5 n_5}{\theta_1 + \theta_2 + \theta_3 + \theta_4 + \theta_5}$$

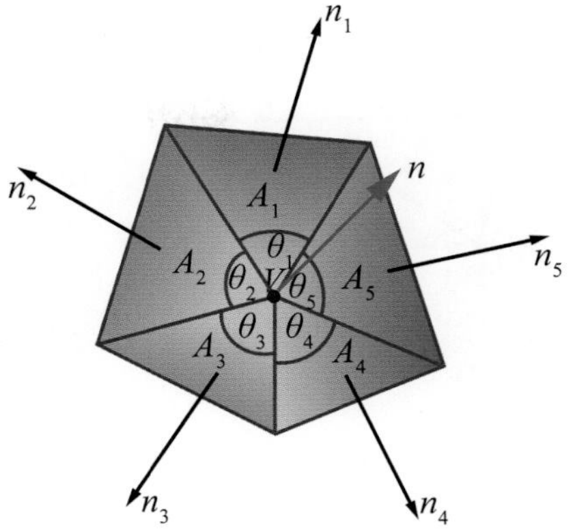

Figure 11.36: Weighted average of normals: θ_i are angles, A_i area, n_i face normals and n a weighted average normal at V.

(b) Weight each adjacent face normal with the area of that face. The areas of the five faces in Figure 11.36 meeting at V are $A_1, A_2, \ldots, A_5$, respectively. The area-weighted average value of the normal at V is then:

$$n = \frac{A_1 n_1 + A_2 n_2 + A_3 n_3 + A_4 n_4 + A_5 n_5}{A_1 + A_2 + A_3 + A_4 + A_5}$$

Important: Whatever approach you adopt to compute normals, make sure, as a last step, to normalize each to unit length (easy enough – just divide each by its length). The reason is that OpenGL uses the dot product to compute the cosine of the angle between two vectors (see Equation (11.12)), which is correct *only if* both are of unit length.

Example 11.11. For the trash can mesh whose vertices are given in Figure 11.37, compute the unit normals to the three faces adjacent to the vertex V. Then compute the (unweighted) average of these three normals and normalize to unit length.

Answer: The three edge vectors emanating from V are:

$$\begin{aligned} u_1 &= [1\ -1\ -1]^T - [1\ -1\ 1]^T = -2\mathbf{k} \\ u_2 &= [1.2\ 1\ 1.2]^T - [1\ -1\ 1]^T = 0.2\mathbf{i} + 2\mathbf{j} + 0.2\mathbf{k} \\ u_3 &= [-1\ -1\ 1]^T - [1\ -1\ 1]^T = -2\mathbf{i} \end{aligned}$$

Therefore, the outward unit normal to the face with edges u_1 and u_2 is

$$n_{12} = (u_1 \times u_2)\ /\ |u_1 \times u_2| = (4\mathbf{i} - 0.4\mathbf{j})\ /\ \sqrt{4^2 + 0.4^2} \simeq 0.995\mathbf{i} - 0.0995\mathbf{j}$$

and that to the face with edges u_2 and u_3 is

$$n_{23} = (u_2 \times u_3)\ /\ |u_2 \times u_2| = (-0.4\mathbf{j} + 4\mathbf{k})\ /\ \sqrt{4^2 + 0.4^2} \simeq -0.0995\mathbf{j} + 0.995\mathbf{k}$$

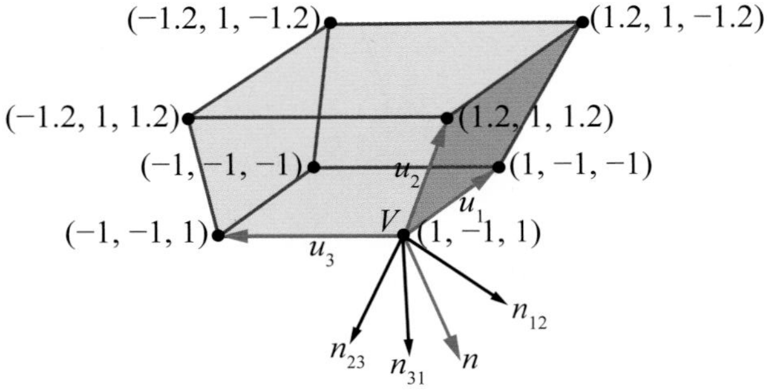

Figure 11.37: Trash can of five quadrilateral sides. The vectors n_{12}, n_{23} and n_{31} from V are normals to V's adjacent faces, while n is the averaged normal.

while the outward unit normal to the face with edges u_3 and u_1, the bottom face, is easily seen to be

$$n_{31} = -\mathbf{j}$$

The normalize average of these normals is

$$\begin{aligned} n &= (n_{12} + n_{23} + n_{31}) \;/\; |n_{12} + n_{23} + n_{31}| \\ &\simeq (0.995\mathbf{i} - 1.199\mathbf{j} + 0.995\mathbf{k}) \;/\; \sqrt{0.995^2 + 1.199^2 + 0.995^2} \\ &\simeq 0.538\mathbf{i} - 0.649\mathbf{j} + 0.538\mathbf{k} \end{aligned}$$

Exercise 11.25. (Programming) Use data from the preceding example to replace the box of `sphereInBox2.cpp` with a trash can. Omit the sphere. Let the user choose between averaged and unaveraged normals. Allow the can to be rotated keeping the light source fixed.

11.11.3 General Surfaces

As a general surface is drawn by approximating it with a polygonal mesh, the thought comes to mind to simply use the methods of the preceding section to find normals. Precisely, (a) formulate a mesh approximation of the surface and (b) specify the normal at each vertex as an average of those of its adjacent faces (we really want to use an average here, especially if the original surface is smooth, to avoid color discontinuities between adjacent mesh faces).

This approach is perfectly reasonable if the surface is known to the user only by its mesh approximation. However, if one knows, say, a parametric representation of the surface, why not get the normals from the "horse's mouth" – that being the parametrization itself? In other words, use the parametrization to *analytically* compute the normals at the mesh vertices. This makes for stable normals independent of the vagaries of the particular mesh approximation, not to mention those of the averaging process (possibly, angle-weighted or area-weighted). For example, working from the mesh approximation of the surface s in Figure 11.38, normals to the six faces adjacent to vertex V must be averaged to determine the normal n at V. However, knowledge of s itself could enable a direct computation.

So let's see how to compute normals analytically. We're going to assume in the following that you know that a tangent plane at the point $s(u, v)$ to a surface s given parametrically by the equations

$$x = f(u, v),\ y = g(u, v),\ z = h(u, v)$$

is spanned by the two vectors

$$\begin{bmatrix} \frac{\partial f}{\partial u} & \frac{\partial g}{\partial u} & \frac{\partial h}{\partial u} \end{bmatrix}^T \quad \text{and} \quad \begin{bmatrix} \frac{\partial f}{\partial v} & \frac{\partial g}{\partial v} & \frac{\partial h}{\partial v} \end{bmatrix}^T$$

evaluated at (u, v) (provided they are not collinear). Moreover, a normal vector to s at $s(u, v)$ is the cross-product

$$\begin{bmatrix} \frac{\partial f}{\partial u} & \frac{\partial g}{\partial u} & \frac{\partial h}{\partial u} \end{bmatrix}^T \times \begin{bmatrix} \frac{\partial f}{\partial v} & \frac{\partial g}{\partial v} & \frac{\partial h}{\partial v} \end{bmatrix}^T \tag{11.15}$$

evaluated at (u, v). If you need to brush up, Section 11.10 is a review of the needed calculus.

Denote the normalized value of the vector (11.15) – obtained by dividing it by its magnitude – by

$$[f_n(u, v) \quad g_n(u, v) \quad h_n(u, v)]^T \tag{11.16}$$

which, therefore, is a unit normal to s at $s(u, v)$.

Finally, we'll specify either $[f_n(u, v) \quad g_n(u, v) \quad h_n(u, v)]^T$ or its reverse, $[-f_n(u, v) \quad -g_n(u, v) \quad -h_n(u, v)]^T$, as the unit normal at $s(u, v)$ depending on which direction is appropriate for front-facing triangles. There's not much to worry about making a wrong choice, as it'll be plenty clear from the viewable output! Let's get to work on a benign surface first.

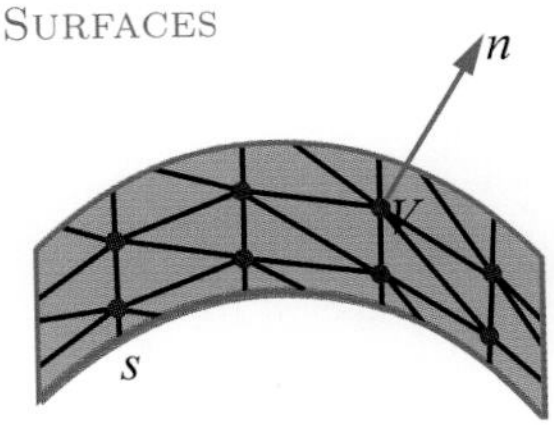

Figure 11.38: Normal vector n to the surface s at a vertex V of its mesh approximation.

Cylinder

Example 11.12. Consider the circular cylinder $s(u, v)$ with parametric equations

$$x = \cos u,\ y = \sin u,\ z = v, \quad \text{where} \quad (u, v) \in [-\pi, \pi] \times [-1, 1]$$

We drew it using these equations in `cylinder.cpp` of Experiment 10.3. To color and light, let's do normal calculations. The vectors spanning the tangent plane at $s(u, v)$ are

$$\begin{bmatrix} \frac{\partial x}{\partial u} & \frac{\partial y}{\partial u} & \frac{\partial z}{\partial u} \end{bmatrix}^T = \begin{bmatrix} \frac{\partial(\cos u)}{\partial u} & \frac{\partial(\sin u)}{\partial u} & \frac{\partial v}{\partial u} \end{bmatrix}^T = [-\sin u \quad \cos u \quad 0]^T$$

and

$$\left[\frac{\partial x}{\partial v} \;\; \frac{\partial y}{\partial v} \;\; \frac{\partial z}{\partial v} \right]^T = \left[\frac{\partial(\cos u)}{\partial v} \;\; \frac{\partial(\sin u)}{\partial v} \;\; \frac{\partial v}{\partial v} \right]^T = [0\ 0\ 1]^T$$

so a normal vector is

$$[-\sin u \quad \cos u \quad 0\,]^T \times [\,0\ 0\ 1\,]^T = [\,\cos u \quad \sin u \quad 0]^T$$

which happens to be normalized already. So, in the terminology of (11.16), for the cylinder,

$$f_n(u, v) = \cos u, \quad g_n(u, v) = \sin u, \quad h_n(u, v) = 0$$

We'll add this normal data to `cylinder.cpp` next.

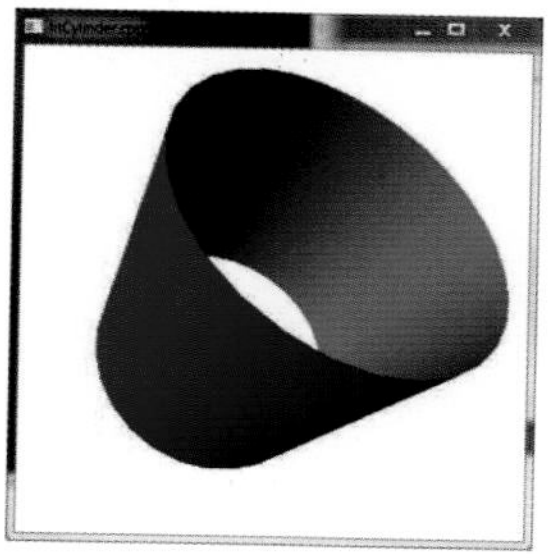

Figure 11.39: Screenshot of `litCylinder.cpp`.

Experiment 11.20. Run `litCylinder.cpp`, which builds upon `cylinder.cpp` using the normal data calculated above, together with color and a single directional light source. Press 'x/X', 'y/Y' and 'z/Z' to turn the cylinder. The functionality of being able to change the fineness of the mesh approximation has been dropped. Figure 11.39 is a screenshot. **End**

Compare the two programs `cylinder.cpp` and `litCylinder.cpp` – it's not really a lot of code from the first to the second. Essentially, the additions are (a) the `fn()`, `gn()` and `hn()` normal component functions as calculated above, (b) the `fillNormalArray()` function to fill the array `normals[]`, and (c) a bunch of routine code specifying light and material properties, which can be kept similar across most programs with lighting.

So the extra code arising from analytic normal computation is really in (a) and (b), about 20 lines all told. Not too bad, huh? And it gets better. As we used the template of `cylinder.cpp` to draw various surfaces, simply swapping in new `f()`, `g()` and `h()` functions according to the given parametrization, so we can use `litCylinder.cpp` for lit applications, additionally swapping in new `fn()`, `gn()` and `hn()` functions.

Exercise 11.26. (Programming) Reverse the normals of `litCylinder.cpp` by changing their specification in the `fillNormalArray()` routine as follows:

```
normals[k++] = -fn(i,j);
normals[k++] = -gn(i,j);
normals[k++] = -hn(i,j);
```

Not good! As we remarked earlier, wrongly-oriented normals are easy to spot. Can you fix the problem caused by the normal values above by a minimal amount of code change *only* in the drawing routine?

Hint: Think orientation, in particular, reversing the orientation of the strip triangles.

Exercise 11.27. It's a bit late now, but do we really need partial derivatives, as in Example 11.12, to determine the normal to the cylinder at the point $V = (\cos u, \sin u, v)$?

The outward normal to the cylinder at V evidently lies along a radius of the circle C which is the section of the cylinder through V by a plane perpendicular to its axis. See Figure 11.40. Use this to compute the parametric equation for a unit normal vector to the cylinder without any calculus.

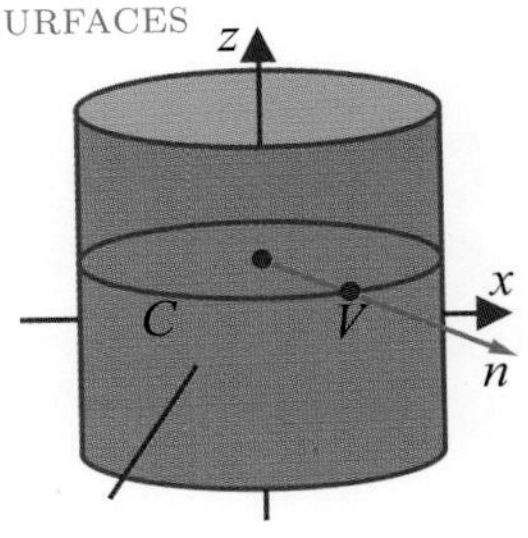

Figure 11.40: Normal n to a cylinder.

Often, as in the preceding exercise, normals to a surface can be determined by elementary geometric considerations. Unfortunately, this does not seem to be the case with the doubly-curled cone of Experiment 10.8.

Doubly-curled Cone

Next, we light the doubly-curled cone of `doublyCurledCone.cpp`. Its parametric equations are

$$x = t\cos(A + a\theta)\cos\theta,\ y = t\cos(A + a\theta)\sin\theta,\ z = t\sin(A + a\theta),$$

where $0 \le t \le 1$ and $0 \le \theta \le 4\pi$. A somewhat tedious calculation gives a normal to the cone as

$$\begin{aligned}\left[\frac{\partial x}{\partial \theta}\ \frac{\partial y}{\partial \theta}\ \frac{\partial z}{\partial \theta}\right]^T \times \left[\frac{\partial x}{\partial t}\ \frac{\partial y}{\partial t}\ \frac{\partial z}{\partial t}\right]^T & \\ = & \ [-at\sin\theta + t\sin(A + a\theta)\cos(A + a\theta)\cos\theta, \\ & \ at\cos\theta + t\sin(A + a\theta)\cos(A + a\theta)\sin\theta, \\ & \ -t\cos^2(A + a\theta)]^T \qquad (11.17)\end{aligned}$$

Moreover, the length of this normal is

$$t\sqrt{a^2 + \cos^2(A + a\theta)} \qquad (11.18)$$

Dividing the normal (11.17) by its length (11.18) gives a unit normal to the cone.

Experiment 11.21. The program `litDoublyCurledCone.cpp`, in fact, applies the preceding equations for the normal and its length. Press 'x/X', 'y/Y', 'z/Z' to turn the cone. See Figure 11.41 for a screenshot.

As promised, `litDoublyCurledCone.cpp` is pretty much a copy of `litCylinder.cpp`, except for the different `f()`, `g()`, `h()`, `fn()`, `gn()` and `hn()` functions, as also the new `normn()` to compute the normal's length. **End**

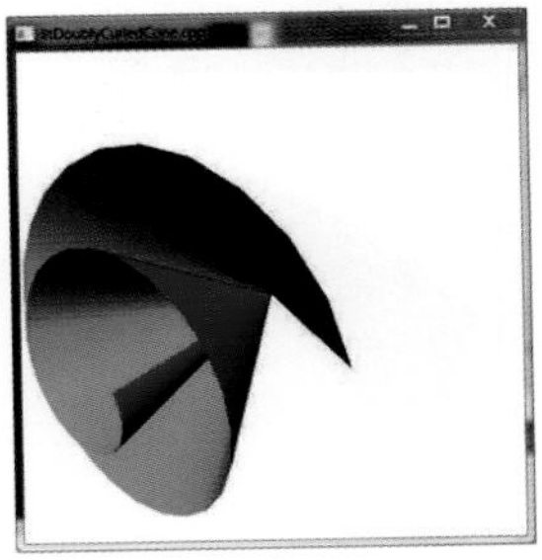

Figure 11.41: Screenshot of `litDoublyCurled-Cone.cpp`.

Exercise 11.28. Verify Equations (11.17) and (11.18) for the normal and its magnitude of the doubly-curled cone.

Programmed Normal Calculation

Exact normals deduced from the equation of a surface – as in the preceding examples of the cylinder and doubly-curled cone – are the most honest as we have observed. Nevertheless, if one wishes to avoid admittedly often tedious calculations, then the next program, a reworking of `litCylinder.cpp`, shows a simple *programmable* way to find approximate normals to a surface mesh.

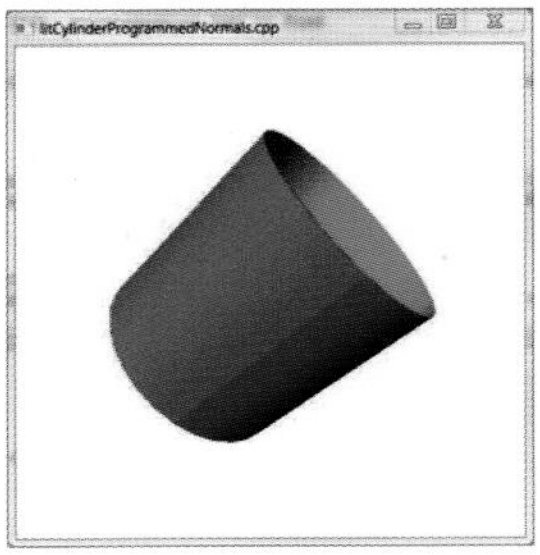

Figure 11.42: Screenshot of `litCylinder-ProgrammedNormals.cpp`.

Experiment 11.22. Run `litCylinderProgrammedNormals.cpp`. Press 'x/X', 'y/Y', 'z/Z' to turn the cylinder. Figure 11.42 is a screenshot. **End**

Let's understand now the normal calculations in `litCylinderProgrammed-Normals.cpp`.

First, for the vertex with parameters (i, j), the chord, call it $t1(i, j)$, joining the vertices with parameters $(i-1, j)$ and $(i+1, j)$, is taken to be an approximation to the tangent $\begin{bmatrix} \frac{\partial x}{\partial u} & \frac{\partial y}{\partial u} & \frac{\partial z}{\partial u} \end{bmatrix}^T$; likewise, the chord $t2(i, j)$ joining the vertices with parameters $(i, j-1)$ and $(i, j+1)$ is taken as an approximation to the tangent $\begin{bmatrix} \frac{\partial x}{\partial v} & \frac{\partial y}{\partial v} & \frac{\partial z}{\partial v} \end{bmatrix}^T$ (care taken at the boundaries of the parameter domain, where $i = 0$ or p, and $j = 0$ or q as $i \pm 1$ or $j \pm 1$ may not exist). See Figure 11.43. Evidently, the closer the mesh vertices, i.e., the finer the mesh, the better the approximations. The approximate (unnormalized) normal $un(i, j)$ is then the cross-product $t1(i, j) \times t2(i, j)$. Dividing $un(i, j)$ by its length $nl(i, j)$ gives a normalized approximate normal at the mesh vertex with parameters (i, j).

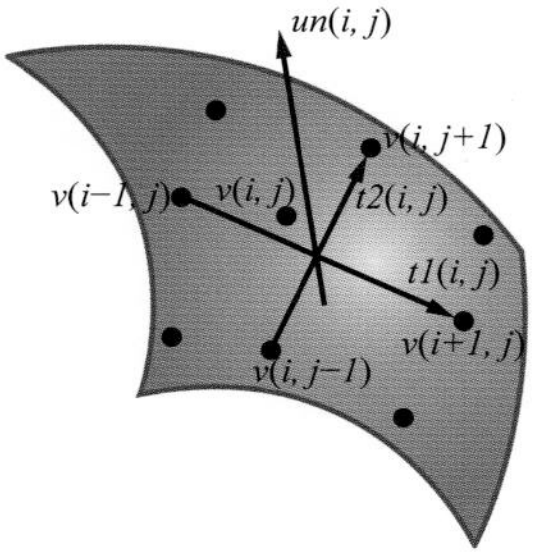

Figure 11.43: Approximate tangents and normal.

The neat thing about `litCylinderProgrammedNormals.cpp`, of course, is that the approximate normals are calculated automatically from the base functions $f()$, $g()$ and $h()$ defining the surface. Therefore, one can cut and paste in any set of $f()$, $g()$ and $h()$ into the `litCylinderProgrammed-Normals.cpp` template to instantly illuminate the corresponding surface, normals done and dusted transparently!

11.11.4 Bézier and Quadric Surfaces

Good news! All one has to do is type in the command `glEnable(GL_AUTO_-NORMAL)` for OpenGL to automatically calculate unit normals at the vertices of a Bézier surface which has been created using `glMap2f(GL_MAP2_VERTEX_3, ... )` and `glEnable(GL_MAP2_VERTEX_3)`.

Canoe

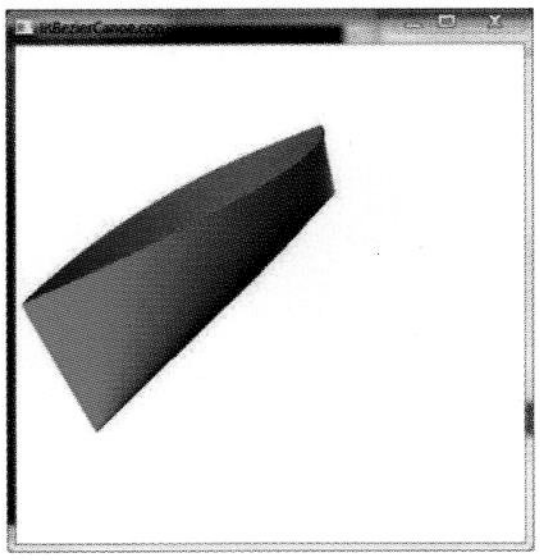

Figure 11.44: Screenshot of `litBezierCanoe.cpp`.

Experiment 11.23. Run `litBezierCanoe.cpp`. Press 'x/X', 'y/Y', 'z/Z' to turn the canoe. You can see a screenshot in Figure 11.44.

This program illuminates the final shape of `bezierCanoe.cpp` of Experiment 10.20 with a single directional light source. Other than the expected command `glEnable(GL_AUTO_NORMAL)` in the initialization routine, an important point to notice about `litBezierCanoe.cpp` is the reversal of the sample grid along the u-direction. In particular, compare the statement

```
glMapGrid2f(20, 1.0, 0.0, 20, 0.0, 1.0)
```

of `litBezierCanoe.cpp` with

```
glMapGrid2f(20, 0.0, 1.0, 20, 0.0, 1.0)
```

of `bezierCanoe.cpp`. This change reverses the directions of one of the tangent vectors evaluated at each vertex by OpenGL and, correspondingly, that of the normal (which is the cross-product of the two tangent vectors).

Modify `litBezierCanoe.cpp` by changing

```
glMapGrid2f(20, 1.0, 0.0, 20, 0.0, 1.0);
```

back to `bezierCanoe.cpp`'s

```
glMapGrid2f(20, 0.0, 1.0, 20, 0.0, 1.0);
```

Wrong normal directions! The change from `bezierCanoe.cpp` is necessary. Another solution is to leave `glMapGrid2f()` as it is in `bezierCanoe.cpp`, instead making a call to `glFrontFace(GL_CW)`. **End**

The lesson to take from this is that if you obtain normals automatically from OpenGL, then you might have to subsequently alter their orientation for authenticity, which is not unreasonable because OpenGL cannot know which you intend to be the front face of a primitive.

Remark 11.18. If the user wishes to define her own normals for a Bézier surface, she can do so with a `glMap2f(GL_MAP2_NORMAL, ...)` call. We'll not have occasion to do this ourselves.

Quadrics are even simpler. The call

```
gluQuadricNormals(qobj, GLU_SMOOTH)
```

automatically generates a normal at each vertex of the quadric pointed by `qobj`.

The next program we'll look at is a fairly substantial animation which invokes both `glEnable(GL_AUTO_NORMAL)` for Bézier surface normals and `gluQuadricNormals(qobj, GLU_SMOOTH)` for quadric surfaces.

Movie with a Ship and Torpedo

Experiment 11.24. Run `shipMovie.cpp`. Pressing space start an animation sequence which begins with a torpedo traveling toward a moving ship and which ends on its own after a few seconds. Figure 11.45 is a screenshot as the torpedo nears the ship. **End**

Figure 11.45: Screenshot of `shipMovie.cpp`.

There are a few different objects in `shipMovie.cpp`. The hull of the ship is obviously inspired by the Bézier canoe of the previous experiment. The deck is a flat Bézier surface – all its control point y-values are identical – which is designed to fit the hull. Each of the ship's three stories is a cylindrical quadric, as is its chimney.

The torpedo should be familiar from the program `torpedo.cpp` of Experiment 10.21. Each of the four grayish boats in the background is a couple of quads, while the sea itself is a solid blue cube.

The smoke from the chimney is a simple-minded particle system. In particular, we render a sequence of quadric discs in point mode and hack a coloring and animation scheme.

11.11.5 Transforming Normals

Normals are transformed by modelview transformations, but not as straightforwardly as vertices are by multiplication from the left by the transformation matrix. Let's see first how they are transformed by each of the fundamental transformations – translation, rotation and scaling.

1. Translation:

   A translation leaves a normal vector at a vertex unchanged because the normal simply translates parallely (see Figure 11.46(a)).

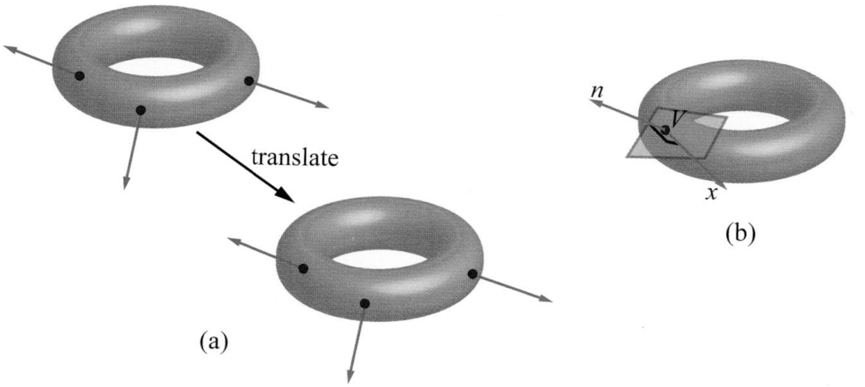

Figure 11.46: (a) Vertex normals translate parallely as the torus is translated (b) The normal n at V is perpendicular to any vector x which lies on the tangent plane at V.

2. Rotation and Scaling:

   These cases are not as simple and require a bit of calculation.

   A rotation or non-degenerate scaling, say t, corresponds to a non-singular 3×3 defining matrix, say N. Suppose that n is a normal vector at a vertex V of an object O. Therefore, n is perpendicular to an arbitrary vector, say x, tangent to the surface of O at V (see Figure 11.46(b)).

Now, if we apply t, it will transform all the vertices of O, as well as vectors tangent to O's surface, by multiplication on the left by N.

Note: To convince yourself that tangent vectors are transformed identically with vertices, think of a tangent vector as connecting two vertices infinitesimally close together on the surface of O. Therefore, these two vertices "carry" the tangent vector with them.

So x is transformed to Nx. We would, therefore, like to transform n to a vector perpendicular to Nx. Since n is perpendicular to x, we already have $n \cdot x = 0$, which is equivalent to $n^T x = 0$, the latter being a matrix equation. It follows that

$$n^T(N^{-1}N)x = n^T x = 0$$

Therefore,

$$0 = n^T(N^{-1}N)x = (n^T N^{-1})(Nx) = ((N^{-1})^T n)^T(Nx)$$

(invoking rules of matrix algebra).

One sees that $((N^{-1})^T n) \cdot Nx = 0$, so $(N^{-1})^T n$ is indeed perpendicular to Nx.

The conclusion, then, is that the appropriate transformation to apply to the normal vector n, under a rotation or non-degenerate scaling corresponding to the matrix N, is left multiplication by $(N^{-1})^T$, i.e., $n \mapsto (N^{-1})^T n$.

OpenGL actually transforms normals as just described. If the current modelview matrix is

$$M = \begin{bmatrix} a_{11} & a_{12} & a_{13} & a_{14} \\ a_{21} & a_{22} & a_{23} & a_{24} \\ a_{31} & a_{32} & a_{33} & a_{24} \\ a_{41} & a_{42} & a_{43} & a_{44} \end{bmatrix}$$

then "erasing" the translational part, which, as we know, has no impact on the normal, leaves its upper-left 3×3 submatrix

$$N = \begin{bmatrix} a_{11} & a_{12} & a_{13} \\ a_{21} & a_{22} & a_{23} \\ a_{31} & a_{32} & a_{33} \end{bmatrix}$$

and, in fact, the matrix $(N^{-1})^T$, called the *normal matrix*, is used to transform normals. It should be noted that the OpenGL normal is a 3D, not 4D, vector; in fact, recall that it was only to accommodate translations into the matrix multiplication scheme that we homogenized 3D vectors to 4D.

Exercise 11.29. We gave above a general formula for how a normal vector is transformed by a rotation or non-degenerate scaling in terms of its defining matrix. Ignoring the formula for a moment, can you deduce from elementary considerations what should happen in the particular case of a rotation? Then relate your answer to the formula.

11.11.6 Normalizing Normals

Normalizing a (non-zero) vector means dividing it by its magnitude to obtain a vector with the same direction, but of unit length. We've already seen that it's important to specify normalized normals because OpenGL uses the dot product to compute the cosine of the angle between two vectors, which is correct only if they are both of unit length.

Here's a simple modification of `litTriangle.cpp` to show what can happen if one is careless.

Figure 11.47: Screenshot of `sizeNormal.cpp`.

Experiment 11.25. Run `sizeNormal.cpp` based on `litTriangle.cpp`.

The ambient and diffuse colors of the three triangle vertices are set to red, green and blue, respectively. The normals are specified separately as well, initially each of unit length perpendicular to the plane of the triangle.

However, pressing the up/down arrow keys changes (as you can see) the size, but not the direction, of the normal at the red vertex. Observe the corresponding change in color of the triangle. Figure 11.47 is a screenshot. **End**

There are, typically, two reasons why normals turn out not normalized:

(a) The user does not specify them of unit length in the first place.

(b) Even if they are specified of unit length, a subsequent application of a scaling transformation changes the length.

If the user is not inclined to write code to ensure normals of unit length, there's a way to ask OpenGL's help. Calling `glEnable(GL_NORMALIZE)` causes OpenGL to normalize all normal vectors before lighting calculation. Beware, though, it's not a particularly efficient call and should be avoided if possible.

Experiment 11.26. Run `sizeNormal.cpp` after placing the statement `glEnable(GL_NORMALIZE)` at the end of the initialization routine. Press the up/down arrow keys. The triangle no longer changes color (though the white arrow still changes in length, of course, because its size is that of the program-specified normal). **End**

There's a cheaper renormalization call, `glEnable(GL_RESCALE_NORMAL)`, which can be used if you originally did provide unit normals that were subsequently all changed by the *same* scaling transformation.

11.12 Phong's Shading Model

Recall from Section 11.8 that a shading model is a method to color the interior of a primitive. A shading model, first proposed by Phong, different from either that OpenGL offers – Gouraud or smooth shading and flat shading – though computationally intensive, significantly improves the realism of a rendered image.

Note: Phong's shading model should not be confused with his lighting model, which we know already that OpenGL implements.

Instead of computing light values only at a primitive's vertices and then interpolating through its interior as in Gouraud shading, Phong suggested to (a) interpolate the vertex normal values through the primitive, and then (b) compute light values at each pixel using the interpolated normals.

Figure 11.48 illustrates the idea. Unit normals n_0, n_1 and n_2 are specified by the programmer at the vertices V_0, V_1 and V_2, respectively, of triangle t. These normals are then interpolated, and normalized, throughout t. For example, if the barycentric coordinates of the point V are given by

$$V = c_0 V_0 + c_1 V_1 + c_2 V_2$$

then the normal value n at V is computed to be

$$n = (c_0 n_0 + c_1 n_1 + c_2 n_2) \,/\, |c_0 n_0 + c_1 n_1 + c_2 n_2| \tag{11.19}$$

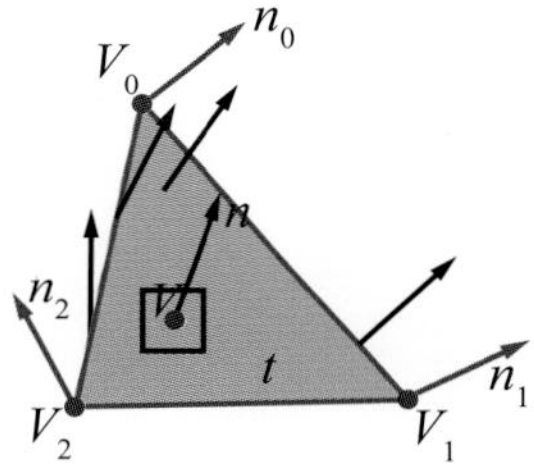

Figure 11.48: Normals n_0, n_1 and n_2 at the vertices of the triangle are programmer-specified. Shown also are (black) normalized interpolated normals at a few points and a pixel centered at V.

(provided the denominator is not zero).

The color values of a pixel which happens to be centered at V are then computed in Phong's model using the lighting equation (11.12), where, now, the normal value n applied is from (11.19) above, the color values $V_{*,X}$ are interpolated from the vertices as well, while the light direction and halfway vectors l^i and s^i are either determined from the coordinates of V itself or interpolated again from the vertices.

Remark 11.19. Phong lighting calculation at each vertex followed by Gouraud shading, OpenGL's default process, is often called *per-vertex* lighting to contrast it with the *per-pixel* lighting of Phong's shading model.

OpenGL itself, as we know, offers only flat and Gouraud shading as automatic shading options. However, the OpenGL Shading Language, or GLSL, allows individual pixels to be programmed, which means the programmer herself can code in Phong shading. We'll be doing precisely this as an application when we get to fourth generation OpenGL and the GLSL ourselves in Chapter 20.

11.13 Lighting Exercises

Here's a bunch of exercises to light up your life. Feel free to pick and choose.

Exercise 11.30. **(Programming)** As Figure 11.42 shows, there is a distinct seam on the cylinder drawn by `litCylinderProgrammedNormals.cpp` of Experiment 11.22, not present in `litCylinder.cpp`. Why is this so? Can you fix the problem?

Exercise 11.31. **(Programming)** The doubly-curled cone of Experiment 11.21 would probably benefit from at least one more light source, particularly to brighten the inside. Code this in.

Exercise 11.32. **(Programming)** Program distance attenuation into `spotlight.cpp`. Add vertical motion capability to the light source to bring out the effect of distance attenuation.

And while you're at it, why not make the light emerge from a well at the bottom of a flying saucer?!

Exercise 11.33. **(Programming)** Extending the previous exercise: add capability to aim the spotlight of `spotlight.cpp`.

Exercise 11.34. **(Programming)** Our first experiment in this chapter ran `sphereInBox1.cpp`, which, though a useful workhorse, was rather bland. Jazz it up by replacing the single large ball with a small ball (or balls) bouncing around inside the box.

Exercise 11.35. **(Programming)** Program `ballAndTorusShadowed.cpp` of Experiment 4.35, way back in Chapter 4, was actually prettified with lighting. Examine this program again. Add a satellite revolving around the ball *a la* Experiment 4.21.

Exercise 11.36. **(Programming)** Continuing the previous exercise: make the ball carry a spotlight, which is aimed always at the torus, and whose cone angle and color change as the ball travels.

Exercise 11.37. **(Programming)** It's alway fun merging projects! So, merge `ballAndTorusShadowed.cpp` and `sphereInBox1.cpp` by "placing" the former in the box of the latter; in other words, opening the box reveals a torus with a ball flying around it.

Exercise 11.38. **(Programming)** Color and light the helical pipe of Experiment 10.4 in two different ways:

(a) Using exact normals found with calculus.

(b) Using approximate normals with help of the template of `litCylinderApproximateNormals.cpp`.

Exercise 11.39. **(Programming)** Color and light the table of Experiment 10.7. You don't really need any calculus in order to compute the normals to the various component surfaces – which happen each to be either cylindrical or flat. Make sure to choose normals so that edges appear sharp.

Exercise 11.40. (Programming) Color and light the single-sheeted hyperboloid of Experiment 10.11.

Exercise 11.41. (Programming) Revisit your drawing projects from Chapter 10 and color and light the objects.

Exercise 11.42. (Programming) In Exercise 10.76 you animated a river scene using the canoe of `bezierCanoe.cpp` to make boats. Now, that we know from Experiment 11.23 how to light the canoe as well, illuminate the river scene.

Exercise 11.43. (Programming) Animate a night street scene. Buildings and cars can be boxy. Make sure to use emission to create authentic street lights.

Exercise 11.44. (Programming) The program `shipMovie.cpp` of Experiment 11.24 bears improvement. Try at least the following:

(a) Add detail to the ship.

(b) Make the water more realistic, possibly by adding movement, variation in color, etc.

(c) Put stars and a moon in the sky.

(d) Improve the smoke particle system.

(e) Make a particle system to simulate water spray from the torpedo's propeller.

Exercise 11.45. (Programming) Fill, paint and light the character of `animateMan*.cpp` of Section 4.7.1 in surroundings less bland than a plane with a ball. Make an animation sequence.

11.14 Summary, Notes and More Reading

In Chapters 4, 5 and 6 we learned to animate objects, in Chapter 10 to draw them, and now we have begun to "dress them up" with color and light. In this chapter we learned the underlying color and lighting models which OpenGL implements, the related syntax, and how to use them to specify light sources and material properties, as well as related environmental parameters. The technical issue of normal computation was an important part of our program too. We'll continue this theme in the next chapter when we learn of yet another technique to decorate an object, texturing.

For a further reference on coloring models, the somewhat encyclopedic Wyszecki and Stiles [153] is frequently called the bible of color science. The books by Berns [12] and Jackson et al. [75] are probably easier to read though.

Since the publication of Phong's model in 1975 [109] several other lighting models, both local and global, have been proposed. Local models like Phong's do not consider object-object light interaction, while global ones do, thereby displaying secondary effects such as shadows and reflections. Lighting models are often used in an application-specific manner, certain models being more realistic in rendering particular material properties and finishes.

A few of the local models which appeared after Phong's are Blinn [15], Cook-Torrance [29], He et al. [67, 68], Nayar-Oren [101], Poulin-Fournier [112] and Schlick [125]. However, the only local model that we discuss or use in this book is Phong's.

The two most commonly implemented global models are ray tracing [4, 148] and radiosity [58], which as a matter of fact complement each other. Global models, though much more realistic than local ones, are notoriously computation-intensive, so rarely apt for interactive applications. However, they are often used when frames can be created off-line, as in movies. We discuss both ray tracing and radiosity in Chapter 19.

The theory of lighting models necessarily involves a fair amount of physics and mathematics. The reader interested in learning more is best advised to start with advanced books such as those by Akenine-Möller, Haines & Hoffman [1], Buss [22] and Watt [147] and then proceed to original research papers, as the area is particularly active. The canonical source for the latest in CG research in general is the annual ACM SIGGRAPH conference [131].

CHAPTER 12

Texture

We continue to explore methods to attire objects and enhance realism, which we began in the last chapter with color and light. The topic of this chapter is texturing. Textures are a vital part of the wardrobe available to designers. Texturing makes it possible to create lifelike scenes at acceptable costs. It's an enormously important technique in modern-day CG. We'll examine how texturing is implemented in OpenGL and various aspects of texturing in practice. Textures can be combined, as well, with color and light to good effect, as we'll see.

We cover the basics of loading and applying textures in Section 12.1, as well as the so-called texture map that specifies how a texture is painted onto an object. Sections 12.2-12.3 discuss setting various texturing parameters including the very important ones to control filtering. Specifying texture coordinates to determine the texture map is the topic of Section 12.4. Section 12.5 explains how to combine texture with light and color. We learn multitexturing and how to combine multiple textures in Section 12.6 and, finally, conclude with Section 12.7.

Figure 12.1: Beer can label.

Figure 12.2: Screenshot of can with textured label and top.

12.1 Texture Basics and the Texture Map

Texturing a surface consists essentially of painting a picture onto it, a process which can be used to two great advantages when programming graphics:

(a) *Authenticity*: Realistically depicting an object which happens to be painted in real life, requires painting the surface that models the object as well.

For example, the beer can label of Figure 12.1 has been textured onto the surface of a cylinder to make a realistic-looking beer can in Figure 12.2.

(b) *Illusion of geometric detail*: Instead of faithfully recreating an object with geometric primitives, painting a picture of it in a scene can achieve a realistic result at a fraction of the cost in the number of polygons.

For example, instead of modeling the individual blades of a grass in a field, paint a picture of a field (Figure 12.3) onto a single rectangle; instead of modeling individual trees in the backdrop of a scene, paint pictures of trees (Figure 12.4) onto appropriately located polygons. The top of the can of Figure 12.2, including the fairly complex pop tab, is a texture too.

Figure 12.3: Field.

Figure 12.4: Trees on white background.

There are various ways to go about applying a texture in OpenGL. We'll discuss the one which is most common: where the texture is a rectangular array of pixels and applied to a polygonal surface. The pixels in a texture are called *texels*, each texel storing color values, such as 24-bit RGB or 32-bit RGBA, just as their counterpart pixels in the frame buffer.

The texture itself can be an external image which is imported into an OpenGL program or one created in the program itself. The former is called an *external* texture while the latter a *procedural*, or *synthetic*, texture. Once loaded though there is no difference between the two. Let's run a program using both kinds.

Figure 12.5: The two textures of `loadTextures.cpp`: shuttle launch (external, from NASA) and chessboard (synthetic).

Experiment 12.1. Run `loadTextures.cpp`, which loads an external image of a shuttle launch as one texture and generates internally a chessboard image as another.

The program paints both the external and the procedural texture onto a square. Figure 12.5 shows the two. Press space to toggle between them, the left and right arrow keys to turn the square and delete to reset it.

Important: Our own texture images are in the folder `ExperimenterSource/-Textures`.

Our programs all use the particular routine `getbmp()` to read in external image files, which is in the associated source file `getbmp.cpp`, and included in the application program via the header `getbmp.h`. Because `getbmp()` is written to accept input image files in *uncompressed 24-bit color RGB bmp* format, image files in other formats must first be converted, which can be done using image-editing software like Windows Paint, GIMP and Adobe Photoshop.

Note, though, that the input file is written *internally* into a 32-bit RGB**A** format, the A (alpha) field allowing for use in blending applications. **End**

Let's start to understand the texture-related OpenGL commands in `loadTextures.cpp`. First, the call

```
glGenTextures(2, texture)
```

in the `setup()` routine returns two texture IDs in the array `texture`. Generally, a call of the form `glGenTextures(`*n*`, `*texture*`)` returns n such IDs. Next, `setup()` calls

```
loadExternalTextures()
```

which first creates storage and loads the texture image `launch.bmp` with the statements

```
BitMapFile *image[1];
image[0] = getBMPData("../../Textures/launch.bmp");
```

Next, `loadExternalTextures()` creates a new 2D *texture object* with id `texture[0]` by the statement

```
glBindTexture(GL_TEXTURE_2D, texture[0]);
```

Then

```
glTexImage2D(GL_TEXTURE_2D, 0, GL_RGBA, image[0]->sizeX,
   image[0]->sizeY, 0, GL_RGBA, GL_UNSIGNED_BYTE, image[0]->data);
```

specifies the texture image for the currently bound texture object, this being `texture[0]`. Generally,

```
glTexImage2D(target, level, internalFormat, width, height, border, format,
             type, *pointer);
```

specifies the texture image pointed to by *pointer*, which is of (external) format *format* and type *type*; the kind of texture is *target*, the mipmap level (to be discussed later) is *level*, the size of the texture image is $width \times height$, while *internalFormat* tells OpenGL how to store the image data; *border* is a legacy parameter which must be 0.

Ignore, for now, the final four parameter-setting commands of the form `glTexParameteri()` in the `loadExternalTextures()` routine – they are, in fact, the topics of the two sections after this.

Returning to `setup()`, the

```
createChessboard();
```

routine next creates the chessboard image in the $64 \times 64 \times 4$ array `chessboard` of RGBA values (each square of the board is represented by an $8 \times 8 \times 4$ subarray of `chessboard`, consisting either of all black or all white color values). The routine

```
loadProceduralTextures();
```

binds the chessboard image to `texture[1]` similarly to how `loadExternal-Textures()` bound the launch image to `texture[0]`. Finally

```
glTexEnvf(GL_TEXTURE_ENV, GL_TEXTURE_ENV_MODE, GL_REPLACE);
glEnable(GL_TEXTURE_2D);
```

specify the texture environment parameters – in this case, asking that the texture replace the surface's current color values – and enable texturing.

On to the `drawScene()` next. The call

```
glBindTexture(GL_TEXTURE_2D, texture[id]);
```

activates texture object `texture[id]` (if a previously created texture object is bound, then it becomes active). The polygon drawing command of `drawScene()` which specifies the so-called texture coordinates of each vertex, in fact, is what we'll discuss in detail next, but first a couple of light exercises.

Exercise 12.1. (**Programming**) Replace the image of the launch with others downloaded from the web.

Figure 12.6: A striped board.

Exercise 12.2. (**Programming**) Write a routine `createStriped-Board()` that generates the image of a striped board, depicted in Figure 12.6, in a 64×64 RGB array.

Texture Coordinates

A texture, once loaded, occupies the unit square with corners at (0,0), (1,0), (1,1) and (0,1) of an imaginary plane called *texture space*. This is regardless of whether the original rectangle of texels itself is equal-sided or not. If it is not, then it is scaled to fit the square, as illustrated in Figure 12.7. The axes of texture space are usually denoted s and t.

Each of the four statements within the `glBegin(GL_POLYGON)`–`glEnd()` pair of the following piece of code, from the drawing routine of `load-Textures.cpp`, maps the vertex of a (square) polygon to a point in texture space.

```
glBegin(GL_POLYGON);
   glTexCoord2f(0.0, 0.0); glVertex3f(-10.0, -10.0, 0.0);
   glTexCoord2f(1.0, 0.0); glVertex3f(10.0, -10.0, 0.0);
   glTexCoord2f(1.0, 1.0); glVertex3f(10.0, 10.0, 0.0);
   glTexCoord2f(0.0, 1.0); glVertex3f(-10.0, 10.0, 0.0);
glEnd();
```

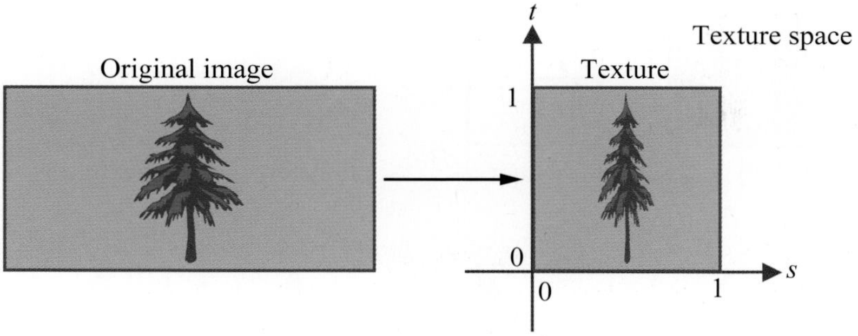

Figure 12.7: An image stored as a texture in a unit square of texture space.

The first statement, for example, maps the vertex at $(-10.0, -10.0, 0.0)$ of world space to the point $(0.0, 0.0)$ of texture space. The coordinates of the mapped point in texture space are called the *texture coordinates* of the vertex. The mapping of the polygon vertices to texture space is interpolated throughout the polygon to obtain the so-called *texture map*, which, therefore, is a map from a part of world space (that occupied by the polygon) to texture space. The texture, finally, is painted onto the polygon by applying to each point of it the RGB color values of its image by the texture map.

Exercise 12.3. In `loadTextures.cpp`, what are the texture coordinates of the following points of the world-space square?

(a) $(0.0, 0.0, 0.0)$

(b) $(5.0, 5.0, 0.0)$

(c) $(10.0, 0.0, 0.0)$

Part answer: (a) $(0.5, 0.5)$, as the midpoint of the world-space square maps to the midpoint of the texture square by linearity.

Experiment 12.2. Replace every 1.0 in each `glTexCoord2f()` command of `loadTextures.cpp` with 0.5 so that the polygon specification is (Block 1*):

```
glBegin(GL_POLYGON);
   glTexCoord2f(0.0, 0.0); glVertex3f(-10.0, -10.0, 0.0);
   glTexCoord2f(0.5, 0.0); glVertex3f(10.0, -10.0, 0.0);
   glTexCoord2f(0.5, 0.5); glVertex3f(10.0, 10.0, 0.0);
   glTexCoord2f(0.0, 0.5); glVertex3f(-10.0, 10.0, 0.0);
glEnd();
```

The lower left quarter of the texture is interpolated over the square (Figure 12.8(a)). Make sure to see both the launch and chessboard textures! **End**

*To cut-and-paste you can find the block in text format in the file `chap12codeModifications.txt` in the directory `ExperimenterSource/CodeModifications`.

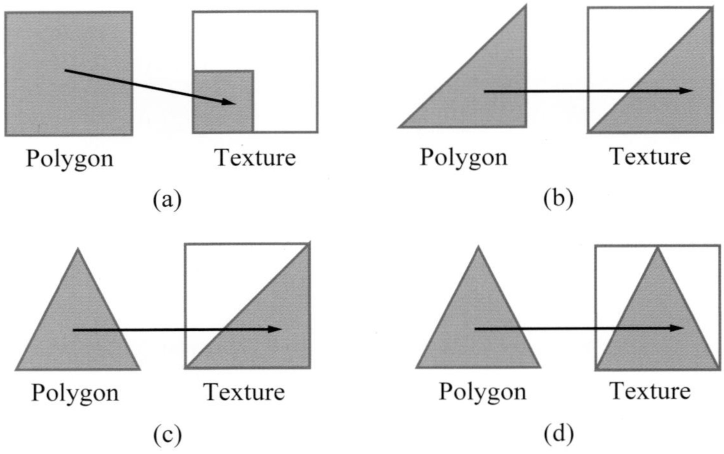

Figure 12.8: Texture maps.

Experiment 12.3. Restore the original `loadTextures.cpp` and delete the last vertex from the polygon so that the specification is that of a triangle (Block 2):

```
glBegin(GL_POLYGON);
   glTexCoord2f(0.0, 0.0); glVertex3f(-10.0, -10.0, 0.0);
   glTexCoord2f(1.0, 0.0); glVertex3f(10.0, -10.0, 0.0);
   glTexCoord2f(1.0, 1.0); glVertex3f(10.0, 10.0, 0.0);
glEnd();
```

Exactly as expected, the lower-right triangular half of the texture is interpolated over the world-space triangle (Figure 12.8(b)).

Change the coordinates of the last vertex of the world-space triangle (Block 3):

```
glBegin(GL_POLYGON);
   glTexCoord2f(0.0, 0.0); glVertex3f(-10.0, -10.0, 0.0);
   glTexCoord2f(1.0, 0.0); glVertex3f(10.0, -10.0, 0.0);
   glTexCoord2f(1.0, 1.0); glVertex3f(0.0, 10.0, 0.0);
glEnd();
```

Interpolation is clearly evident now. Parts of both launch and chessboard are skewed by texturing, as the triangle specified by texture coordinates is not similar to its world-space counterpart (Figure 12.8(c)).

Continuing, change the texture coordinates of the last vertex (Block 4):

```
glBegin(GL_POLYGON);
   glTexCoord2f(0.0, 0.0); glVertex3f(-10.0, -10.0, 0.0);
   glTexCoord2f(1.0, 0.0); glVertex3f(10.0, -10.0, 0.0);
   glTexCoord2f(0.5, 1.0); glVertex3f(0.0, 10.0, 0.0);
glEnd();
```

The textures are no longer skewed as the triangle in texture space is similar to the one being textured (Figure 12.8(d)). End

Experiment 12.4. Restore the original `loadTextures.cpp` and replace `launch.bmp` with `cray2.bmp`, an image of a Cray 2 supercomputer. View the original images in the `Textures` folder and note their sizes: the launch is 512 × 512 pixels while the Cray 2 is 512 × 256. As you can see, the Cray 2 is scaled by half width-wise to fit the square polygon. End

Exercise 12.4. (Programming) Change the polygon specs so that the Cray 2 is not distorted.

(a) (b)

Figure 12.9: Screenshots from Experiment 12.5.

Experiment 12.5. Restore the original `loadTextures.cpp` and then change the coordinates of only the third world-space vertex of the textured polygon (Block 5):

```
glBegin(GL_POLYGON);
   glTexCoord2f(0.0, 0.0); glVertex3f(-10.0, -10.0, 0.0);
   glTexCoord2f(1.0, 0.0); glVertex3f(10.0, -10.0, 0.0);
   glTexCoord2f(1.0, 1.0); glVertex3f(20.0, 0.0, 0.0);
   glTexCoord2f(0.0, 1.0); glVertex3f(-10.0, 10.0, 0.0);
glEnd();
```

The launch looks odd. The rocket rises vertically, but the flames underneath are shooting sideways! Toggle to the chessboard and it's instantly clear what's going on. Figure 12.9 shows both textures. End

The polygon and the texture have evidently been triangulated *equivalently* – in particular, triangles in the triangulation of one correspond to those in the other via the texture map. Corresponding triangles in this case, though, evidently differ in shape. Subsequently, each triangle of the texture has been *separately* interpolated over the corresponding triangle of the polygon, causing the perceived distortion. See Figure 12.10.

When we had said a little earlier that the texture map is obtained by interpolating over the entire polygon the mapping from its vertices to points in texture space, we had not taken into account the fact that there is no unambiguous way to do this if the polygon has more than three sides (recall discussions to this effect in Section 7.4). We know now how OpenGL gets past the problem: *It determines the texture map by interpolating the vertex texture coordinates not over the polygon but, after triangulation, over each triangle separately.*

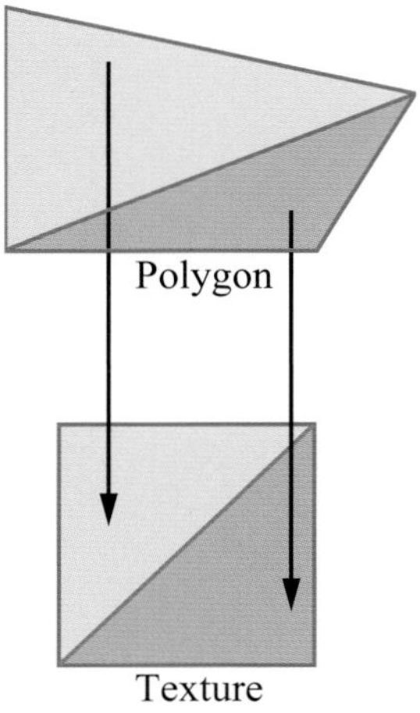

Figure 12.10: Each of the two texture triangles is interpolated over the corresponding polygon triangle.

Exercise 12.5. (Programming) Change the polygon specification in `loadTextures.cpp` to map a five-sided polygon in world space to a five-sided polygon in texture space (Block 6):

```
glBegin(GL_POLYGON);
   glTexCoord2f(0.0, 0.0); glVertex3f(-10.0, -10.0, 0.0);
   glTexCoord2f(1.0, 0.0); glVertex3f(10.0, -10.0, 0.0);
   glTexCoord2f(1.0, 0.5); glVertex3f(20.0, 0.0, 0.0);
   glTexCoord2f(0.5, 1.0); glVertex3f(0.0, 10.0, 0.0);
   glTexCoord2f(0.0, 1.0); glVertex3f(-10.0, 0.0, 0.0);
glEnd();
```

Can you make out the triangulations in world and texture space, as well as the correspondence between triangles?

12.2 Repeating and Clamping Textures

So far we've been careful to keep texture coordinates in the range $[0, 1]$, along both the s- and t-axes. What happens if they slip outside? Let's find out.

Figure 12.11: Screenshot of Experiment 12.6.

Experiment 12.6. Restore the original `loadTextures.cpp` and change the texture coordinates of the polygon as follows (Block 7):

```
glBegin(GL_POLYGON);
   glTexCoord2f(-1.0, 0.0); glVertex3f(-10.0, -10.0, 0.0);
   glTexCoord2f(2.0, 0.0); glVertex3f(10.0, -10.0, 0.0);
   glTexCoord2f(2.0, 2.0); glVertex3f(10.0, 10.0, 0.0);
   glTexCoord2f(-1.0, 2.0); glVertex3f(-10.0, 10.0, 0.0);
glEnd();
```

It seems that the texture space is *tiled* using the texture. See Figure 12.11.

In particular, the texture seems repeated in every unit square of texture space with integer vertex coordinates. As the world-space polygon is mapped to a 3×2 rectangle in texture space, it is painted with six copies of the texture, each scaled to an aspect ratio of 2:3. The scheme itself is indicated Figure 12.12. **End**

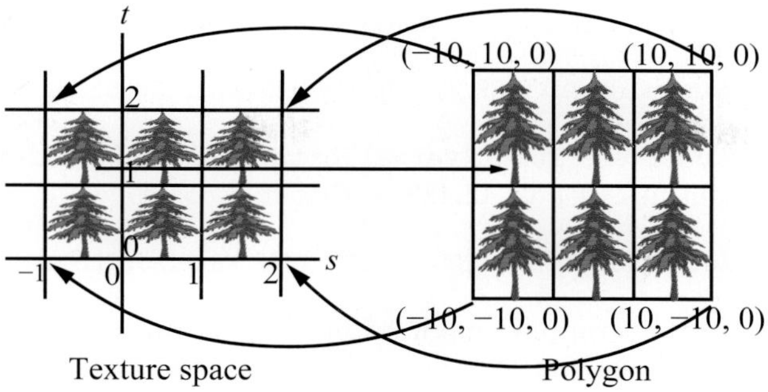

Figure 12.12: Tiling of texture space. The curved bold black arrows indicate the texture map. The straight arrow indicates the painting of one tile onto a sub-rectangle of the polygon; other tiles similarly paint corresponding sub-rectangles.

Experiment 12.7. Change the texture coordinates again by replacing each -1.0 with -0.5 (Block 8):

```
glBegin(GL_POLYGON);
   glTexCoord2f(-0.5, 0.0); glVertex3f(-10.0, -10.0, 0.0);
   glTexCoord2f(2.0, 0.0); glVertex3f(10.0, -10.0, 0.0);
   glTexCoord2f(2.0, 2.0); glVertex3f(10.0, 10.0, 0.0);
   glTexCoord2f(-0.5, 2.0); glVertex3f(-10.0, 10.0, 0.0);
glEnd();
```

Again it's apparent that the texture space is tiled with the specified texture and that the world-space polygon is painted over with its rectangular image in texture space. **End**

That the texture space is tiled with the texture is because of the following two statements in both the `loadExternalTextures()` and `loadProcedural-Textures()` routines of `loadTextures.cpp`, specifying the *wrapping mode* to be *repeated* along both s- and t-directions:

```
glTexParameteri(GL_TEXTURE_2D, GL_TEXTURE_WRAP_S, GL_REPEAT);
glTexParameteri(GL_TEXTURE_2D, GL_TEXTURE_WRAP_T, GL_REPEAT);
```

There is another option for the wrapping mode, instead of repeated, namely, *clamped*.

Experiment 12.8. Restore the original `loadTextures.cpp` and then change the texture coordinates as below, which is the same as in Experiment 12.6 (Block 7):

```
glBegin(GL_POLYGON);
   glTexCoord2f(-1.0, 0.0); glVertex3f(-10.0, -10.0, 0.0);
   glTexCoord2f(2.0, 0.0); glVertex3f(10.0, -10.0, 0.0);
   glTexCoord2f(2.0, 2.0); glVertex3f(10.0, 10.0, 0.0);
   glTexCoord2f(-1.0, 2.0); glVertex3f(-10.0, 10.0, 0.0);
glEnd();
```

Next, replace the GL_REPEAT parameter in the

```
glTexParameteri(GL_TEXTURE_2D, GL_TEXTURE_WRAP_S, GL_REPEAT);
```

statement of both the loadExternalTextures() and loadProcedural-Textures() routines with GL_CLAMP so that it becomes

```
glTexParameteri(GL_TEXTURE_2D, GL_TEXTURE_WRAP_S, GL_CLAMP);
```

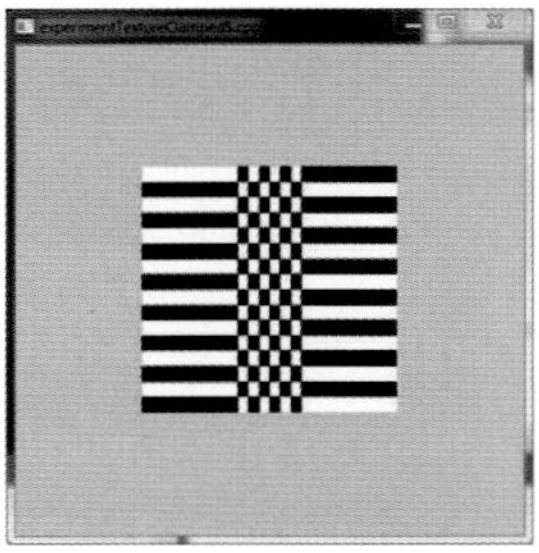

Figure 12.13: Screenshot from Experiment 12.8.

This causes the wrapping mode to be clamped in the s-direction. It's probably easiest to understand what happens in this mode by observing in particular the chessboard texture: see Figure 12.13. Texture s coordinates greater than 1 are clamped to 1, those less than 0 to 0. Precisely, instead of the texture space being tiled with the texture, points with coordinates (s, t), where $s > 1$, obtain their color values from the point $(1, t)$, while those with coordinates (s, t), where $s < 0$, obtain them from $(0, t)$. **End**

Experiment 12.9. Continue the previous experiment by clamping the texture along the t-direction as well. In particular, replace the GL_REPEAT parameter in the

```
glTexParameteri(GL_TEXTURE_2D, GL_TEXTURE_WRAP_T, GL_REPEAT);
```

statement with GL_CLAMP. We leave the reader to parse the output. **End**

The repeating option is appropriate to tile the surface of an object with a particular pattern, e.g., a wall with a brick pattern, a table with a wood grain pattern, the ground with a grass pattern and so on, while the clamping option is appropriate to paint on a single copy of the texture, e.g., the facade of a building onto a rectangle, with the texture boundary, typically, aligned with the rectangle boundary.

12.3 Filtering

Figure 12.14: Screenshot of fieldAndSky.cpp.

Experiment 12.10. Run fieldAndSky.cpp, where a grass texture is tiled over a horizontal rectangle and a sky texture clamped to a vertical rectangle. There is the added functionality of being able to transport the camera over the field by pressing the up and down arrow keys. Figure 12.14 shows a screenshot.

As the camera travels, the grass seems to *shimmer* – *flash* and *scintillate* are terms also used to describe this phenomenon. This is our first encounter with the *aliasing* problem in texturing. Any visual artifact which arises owing to the finite resolution of the display device and the correspondingly "large" size of individual pixels – at least to the extent that they are discernible to the human eye – is said to be caused by aliasing. **End**

Let's try and understand why shimmer is caused by aliasing. Recall that the texture map is obtained by interpolating through each triangle the texture coordinates specified at its vertices. Subsequently, each point of the triangle is colored with the values of its mapped texture point. However, a technicality arises at this stage we did not consider earlier. Color values in the computer are *not* associated per *point*, either in texture space or the polygon. In reality, they are associated one set (RGB or RGBA) per *pixel* in the display, as also one set per *texel* in the texture.

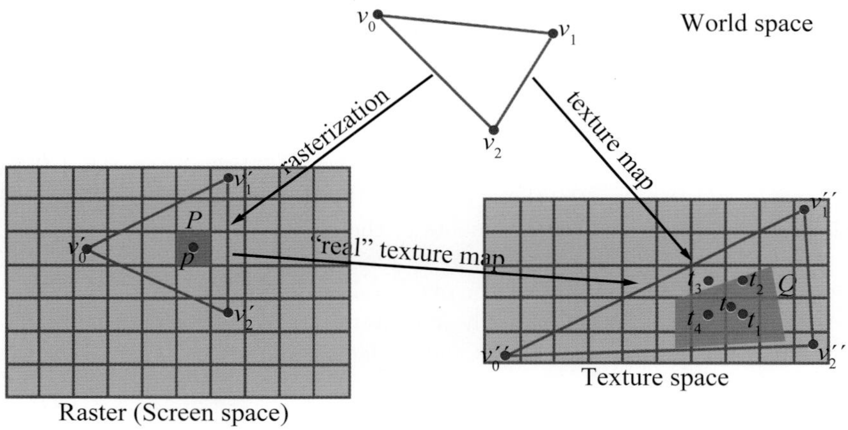

Figure 12.15: The aliasing problem in texture mapping. A single pixel P is mapped to a quadrilateral Q covering many texels (minification).

Now, once the polygon has been rasterized – i.e., its set of corresponding pixels determined – the texture map is unlikely to map pixels to texels in a one-to-one manner. The situation, more typically, is as depicted in Figure 12.15, where the triangle $v_1v_2v_3$ in world space is mapped to the raster triangle $v_1'v_2'v_3'$ – or screen space triangle, if you like, as it is the one to be displayed – via the rasterization process and to the texture space triangle $v_1''v_2''v_3''$ via the texture map. These two maps (downwards in the figure) induce the "real" texture map from raster to texture space (left to right) which takes pixels to texels (precisely, this map is rasterization reversed followed by the texture map).

The dark pixel P in the raster maps to the dark quadrilateral Q in texture space (mind that the texture map need not preserve shape). As Q intersects multiple texels, how should OpenGL choose color values for P? Particularly, which texel should OpenGL pick to apply its particular color values to P?

Here's a reasonable solution: if the texture map takes the center p of P to the point t in texture space, then choose the texel whose center is nearest to t. In Figure 12.15, then, the chosen texel is centered at t_1, so this texel's colors would be applied to P. In fact, this is precisely the so-called *filtering* option specified by the `GL_NEAREST` value in the following two parameter-setting commands in both the statement blocks that bind the grass and sky textures of the `loadExternalTextures()` routine of `fieldAndSky.cpp`:

```
glTexParameteri(GL_TEXTURE_2D, GL_TEXTURE_MIN_FILTER, GL_NEAREST);
glTexParameteri(GL_TEXTURE_2D, GL_TEXTURE_MAG_FILTER, GL_NEAREST);
```

(We'll discuss the difference between `MIN` and `MAG` filters momentarily.)

The reason for the shimmer observed in `fieldAndSky.cpp` is now not hard to grasp. See again Figure 12.15. Suppose that the object in world space moves a small distance so that its rasterization changes a small amount as well, causing the map from raster to texture space to map p to a new point just to the left of t, closer to the pixel center t_4 than t_1. Correspondingly, according to the `GL_NEAREST` filtering principle, there is a switch in the color values at pixel P, obtained now from the texel centered at t_4, rather than the one at t_1. It's exactly these relatively large discrete changes in pixel colors arising from minute movements of the object which cause shimmer.

Figure 12.15 itself suggests a way to ameliorate the problem: instead of obtaining color values from just the one texel centered at t_1, take an average of the values at the four texels whose centers (t_1, t_2, t_3 and t_4) surround t. This smooths the color transitions and, in fact, is offered by OpenGL as the `GL_LINEAR` filtering option.

Remark 12.1. The basis for linear filtering is exactly the same as for moving averages in statistics. For example, as part of analyzing the stock market, one may chart average values over a sliding window of size one week or month, instead of daily values, in order to smooth out near-term fluctuations.

Experiment 12.11. Change to linear filtering in `fieldAndSky.cpp` by replacing every `GL_NEAREST` with `GL_LINEAR`. The grass still shimmers though less severely. The sky seems okay with either `GL_NEAREST` or `GL_LINEAR`.
End

The process of selecting color values for pixels based on the texture map is called filtering. OpenGL offers a few different filtering options, in addition to `GL_NEAREST` and `GL_LINEAR`, allowing the user to trade between speed and output quality. OpenGL makes a distinction, as well, between *minification* and *magnification* when filtering. Minification occurs when a pixel is mapped onto multiple texels as in Figure 12.15, while magnification is when many pixels map onto a single texel as in Figure 12.16.

Remark 12.2. The term minification arises because the phenomenon of a pixel being mapped to many texels occurs when a painted surface moves into the distance to occupy a smaller part of the screen or, equivalently, when the viewer zooms out. For example, as an aircraft flies away from the camera, the texels comprising its logo occupy an increasingly smaller region of the screen, until, when the craft is far enough, multiple texels occupy

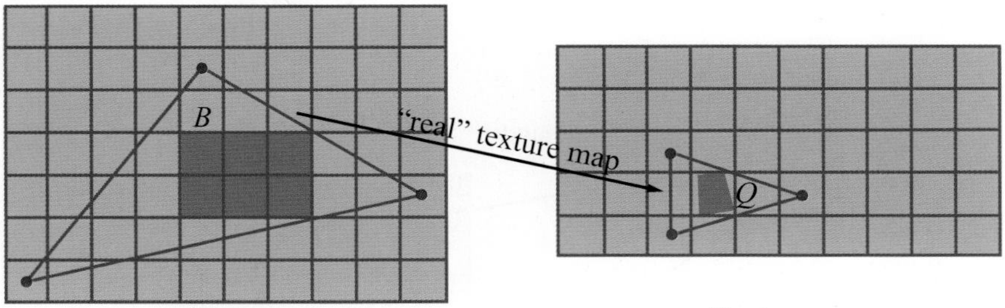

Figure 12.16: A block B of many pixels is mapped to a quadrilateral Q inside a single texel (magnification).

single pixels (which is equivalent to the texture map taking a single pixel to multiple texels). Magnification, related to zooming in, is, of course, the inverse phenomenon.

The statement

```
glTexParameteri(GL_TEXTURE_2D, case, filter)
```

causes the filtering option *filter* to be applied in the case of minification or magnification, according as the value of *case* is `GL_TEXTURE_MIN_FILTER` or `GL_TEXTURE_MAG_FILTER`. With the commands

```
glTexParameteri(GL_TEXTURE_2D, GL_TEXTURE_MIN_FILTER, GL_NEAREST);
glTexParameteri(GL_TEXTURE_2D, GL_TEXTURE_MAG_FILTER, GL_NEAREST);
```

`fieldAndSky.cpp` asks for the `GL_NEAREST` filter in the case of either minification or magnification.

In the case of minification, particularly, OpenGL offers an assortment of efficient filtering options based on *pre-assigning* a set of textures to be used at different levels of minification. The idea, conceived by Lance Williams [150], is clever yet simple. Starting with the original texture, the *base texture*, a set of textures of progressively lower resolution, called *mipmaps*, is prepared. These mipmaps are either computed by OpenGL by an averaging process – we'll see momentarily how – or supplied by the programmer.

Subsequently, during run-time, OpenGL maps a geometric primitive, based upon the size it occupies in the raster, to that particular mipmap which affords a nearly one-to-one correspondence between pixels and texels, rather than the one-to-many which would occur if the base texture were used. This (a) saves on run-time filtering computation, and (b) assures quality (provided mipmaps are initially well-chosen).

Figure 12.17 illustrates the idea with an idealized example. The base texture is of resolution 8×4 with a single scalar color value at each texel. Mipmaps of successively lower resolution till 2×1 are computed by averaging the color values in 2×2 squares of texels; finally, the 1×1 mipmap is computed by averaging the two color values in the 2×1 mipmap. For example, the

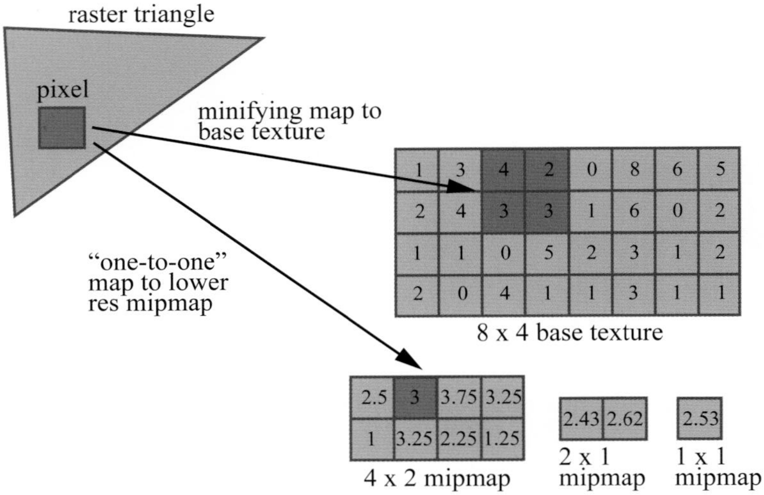

Figure 12.17: Mipmapping.

value 3 in the dark texel in the 4×2 mipmap is the average of the four values in the dark square of texels in the 8×4 mipmap. A set of mipmaps is often called a *pyramid* of mipmaps – think of them stacked one on top of the other, highest resolution at the bottom to the lowest at the top.

Now, when the triangle primitive shown is mapped to the base texture, minification causes the shaded pixel to map to the shaded square of texels, which requires run-time linear filtering to return the color value 3. On the other hand, if it is mapped to the 4×2 mipmap, then there is no minification and the color value of 3 is returned *without* run-time filtering. For this reason mipmaps are often called *pre-filtered* textures.

Generally, if a base texture of resolution $2^m \times 2^n$ is to be mipmapped, then OpenGL requires mipmaps of resolution $2^{m-1} \times 2^{n-1}$, $2^{m-2} \times 2^{n-2}, \ldots$, obtained by halving both width and height, until one of the dimensions becomes 1; subsequently, if the other dimension is still greater than 1, then it must be repeatedly halved and mipmaps provided for each resolution down to 1×1.

Example 12.1. What is the total space required to store all the mipmaps for a base texture of resolution $2^m \times 2^n$? What is the ratio of this space to that required for only the base texture?

Answer: Suppose without loss of generality that $m \geq n$. The total number of texels in all the mipmaps is

$$2^m \times 2^n \ + \ 2^{m-1} \times 2^{n-1} \ + \ldots + \ 2^{m-n+1} \times 2 \ + \ 2^{m-n} \times 1$$
$$+ \ 2^{m-n-1} \ + \ 2^{m-n-2} \ + \ldots + \ 1$$

(the first line above bringing one dimension down to 1, the second the next)

$$= 2^{m-n}(1 + 2^2 + \ldots + 2^{2n}) + (1 + 2 + \ldots + 2^{m-n-1})$$
$$= 2^{m-n}(2^{2n+2} - 1)/3 + 2^{m-n} - 1$$
$$= \frac{4}{3}\,2^{m+n} + \frac{2}{3}\,2^{m-n} - 1$$

The space required is, therefore, the above quantity multiplied by the number of bits per texel.

Now, the number of texels in the base texture is $2^m \times 2^n = 2^{m+n}$ and

$$\frac{4}{3}\,2^{m+n} + \frac{2}{3}\,2^{m-n} - 1 < \frac{4}{3}\,2^{m+n} + \frac{2}{3}\,2^{m+n} = 2 \times 2^{m+n}$$

so the ratio of the space required for all the mipmaps to that for the base texture is less than 2.

Exercise 12.6. If the base texture is 4×8 with color values at the texels as follows

1	0	4	2
3	2	1	5
0	1	2	6
8	2	7	7
2	3	1	2
6	4	3	8
7	3	6	1
3	5	0	2

then find all the mipmaps down to the one of lowest resolution.

Example 12.1 says that mipmapping offers efficiency and quality at a cost of only twice the amount of space. Once mipmaps have been set, OpenGL has four filtering options, in addition to `GL_NEAREST` and `GL_LINEAR`, available for use in case of minification. In order of increasing quality *and* computational cost they are:

(1) `GL_NEAREST_MIPMAP_NEAREST`: Applies the mipmap that's a closest fit resolution-wise to the rasterized primitive and then uses the `GL_NEAREST` filtering option within that mipmap.

(2) `GL_LINEAR_MIPMAP_NEAREST`: Applies the mipmap that's a closest fit resolution-wise to the rasterized primitive and then the `GL_LINEAR` filtering option within that mipmap.

(3) `GL_NEAREST_MIPMAP_LINEAR`: Finds the two mipmaps that are closest resolution-wise to the rasterized primitive, then uses the `GL_NEAREST` filtering option within either mipmap to produce two sets of color values and, finally, takes a weighted average of the two sets.

(4) GL_LINEAR_MIPMAP_LINEAR: Finds the two mipmaps that are closest resolution-wise to the rasterized primitive, then uses the GL_LINEAR filtering option within either mipmap to produce two sets of color values and, finally, takes a weighted average of the two sets.

Mipmaps are *not* used in the case of magnification because, with the viewer zooming in, one wants only the highest resolution, namely, the base texture. Accordingly, the only two filters available in the case of magnification are GL_NEAREST and GL_LINEAR.

Terminology: In OpenGL speak a texture object comprises a texture together with its associated parameters, specified by glTexParameter[i/f]() commands, such as the filter option and whether clamped or repeated.

It's time to see mipmapping in action. Generating mipmaps automatically is simple. The command glGenerateMipmap(*target*) – implemented in OpenGL 3.0 and on – generates a full set of mipmaps for the texture associated with *target* as in the following program.

(a) (b)

Figure 12.18: Screenshots of fieldAndSkyFiltered.cpp: (a) Weakest filter (b) Strongest filter.

Experiment 12.12. Run fieldAndSkyFiltered.cpp, identical to field-AndSky.cpp except for additional filtering options. Press the up/down arrow keys to move the camera and the left/right ones to cycle through filters for the grass texture. Messages at the top identify the current filters. **End**

The call

```
glGenerateMipmap(GL_TEXTURE_2D)
```

just after glTexImage2D(), when binding textures with min filter GL_-NEAREST_MIPMAP_NEAREST and higher, generates mipmaps.

The loadExternalTextures() routine loads the same grass image as six different textures with the min filter ranging from GL_NEAREST to

GL_LINEAR_MIPMAP_LINEAR. The mag filter used is GL_NEAREST when the min filter is GL_NEAREST as well; otherwise, it's GL_LINEAR. The sky texture is not mipmapped.

As one sees, the more expensive filters do nearly eliminate shimmering, but at the same time tamp down possibly desirable sharpness. For example, blades of grass can be distinguished easily in Figure 12.18(a), where the weakest filter is applied, but this is not so in Figure 12.18(b), which applies the strongest.

We have a couple more programs for you to experiment with mipmaps and filters.

Experiment 12.13. Run `compareFilters.cpp`, where one sees side-by-side identical images of a shuttle launch bound to a square. Press the up and down arrow keys to move the squares. Press the left arrow key to cycle through filters for the image on the left and the right arrow key to do likewise for the one on the right. Messages at the top say which filters are currently applied. Figure 12.19 is a screenshot of the initial configuration.

Compare, as the squares move, the quality of the textures delivered by the various min filters. Of course, if one of the four mipmap-based min filters – GL_NEAREST_MIPMAP_NEAREST through GL_LINEAR_MIPMAP_LINEAR – is applied, then the particular mipmap actually chosen by OpenGL depends on the screen space occupied by the square. **End**

Figure 12.19: Screenshot of `compareFilters.cpp` initially.

An unavoidable artifact of filtering using mipmaps is that of *popping* when one mipmap is replaced with another. The next program illustrates this in a purposely dramatic manner.

Experiment 12.14. Run `mipmapLevels.cpp`, where the mipmaps are supplied by the program, rather than generated automatically by `glGenerate-Mipmap()`. The mipmaps are very simple: just differently colored square images, created by the routine `createMipmaps()`, starting with the blue 64×64 `mipmapRes64` down to the black 1×1 `mipmapRes1`. Commands of the form

```
glTexImage2D(GL_TEXTURE_2D, level, GL_RGBA, width, height,
             0, GL_RGBA, GL_UNSIGNED_BYTE, image);
```

each binds a *width* × *height* mipmap image to the current texture index, starting with the highest resolution image with *level* parameter 0, and with each successive image of lower resolution having one higher *level* all the way up to 6.

Move the square using the up and down arrow keys. As it grows smaller a change in color indicates a change in the currently applied mipmap. Figure 12.20 is screenshot after the first change. As the min filter setting is GL_NEAREST_MIPMAP_NEAREST, a unique color, that of the closest mipmap, is applied to the square at any given time. **End**

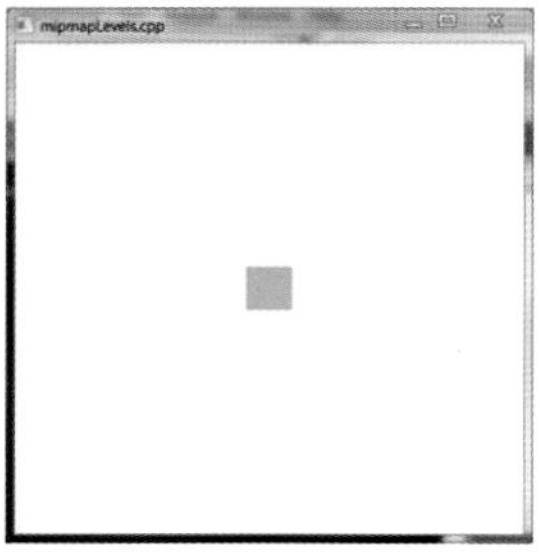

Figure 12.20: Screenshot of mipmapLevels.cpp.

Remark 12.3. OpenGL offers options in addition to those that we have discussed to fine-tune mipmapping. The interested reader is referred to the chapter on textures in the red book.

Remark 12.4. Mipmapping is one of a class of LOD (*level-of-detail*) methods, or *multiresolution* methods as they are also called, which are important in graphics from the point of view of run-time efficiency.

Representing objects by polygonal meshes of varying levels of refinement is another practically important LOD application and one related to the drawing methods that we studied in Chapter 10. For example, if the camera is close to a spacecraft, then one may want a "base mesh" of thousands, or even millions, of triangles to be rendered. However, after the ship has flown some distance off to occupy a smaller portion of the screen, a mesh with fewer triangles may not only be visually adequate, but, in fact, desirable both for quicker rendering and to avoid aliasing artifacts. Accordingly, it's often advantageous to pre-compute a set of meshes for a moving object, exactly as mipmaps for a texture, with varying numbers of triangles.

Instead of a spacecraft, we have a cow at three different levels of resolution in Figure 12.21, starting from the highest at left, and then simplified twice with the help of mesh simplification software [24] co-developed by the author.

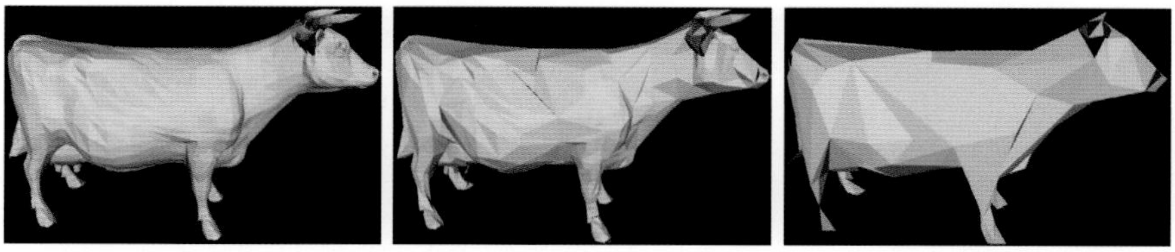

(a) (b) (c)

Figure 12.21: Cow at 3 different resolutions: (a) 5804 (b) 1772 (c) 328 triangles.

There's an extensive literature on LOD methods, of which a good starting point would be the book by Luebke et al. [89].

12.4 Specifying Texture Coordinates

Our programs so far have been simple from the point of view of specifying the texture map. The surfaces textured were all polygons, so that we simply had to specify texture coordinates at the corners. How about more complicated surfaces? It's actually surprisingly straightforward if the surface is parametrized – we can leverage the parametrization to derive texture coordinates.

12.4.1 Parametrized Surfaces

Experiment 12.15. Run `texturedTorus.cpp`, which shows the shuttle launch texture mapped onto a torus. Figure 12.22 is a screenshot. Press 'x'-'Z' to turn the torus. **End**

Figure 12.22: Screenshot of `texturedTorus.cpp`.

The program `texturedTorus.cpp` is based on `torus.cpp` of Experiment 10.5. As i runs from 0 to p and j from 0 to q, $(i/p,\ j/q)$ runs over sample points in $[0,1] \times [0,1]$, and $(\, f(i,j),\ g(i,j),\ h(i,j)\,)$ – see the corresponding function definitions `f()`, `g()` and `h()` in the program – over mapped sample points on the torus. Since the (i,j)th entry in the vertex array is the image of the point $(i/p,\ j/q)$ of the parameter rectangle $[0,1] \times [0,1]$, an obvious texture map is to associate this very same image with the point $(i/p,\ j/q)$ of texture space, effectively identifying the parameter rectangle with the texture! See Figure 12.23. This is exactly what's done in `texturedTorus.cpp` by the routine `fillTextureCoordArray()`, which fills values into a *texture coordinates array* holding texture coordinates per vertex (just like vertex coordinates arrays hold coordinates per vertex and color arrays hold colors per vertex).

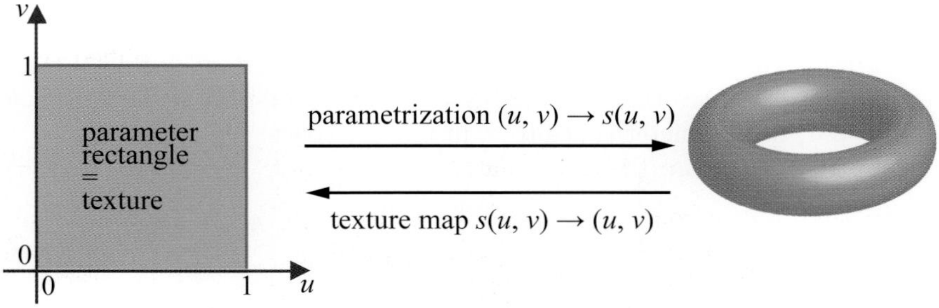

Figure 12.23: Texturing a torus by identifying the parameter rectangle with the texture.

Exercise 12.7. (Programming) If you did Exercise 12.2 to create the synthetic striped board texture of Figure 12.6, apply it now to the torus of `texturedTorus.cpp`.

Exercise 12.8. (**Programming**) Texture the helical pipe of Experiment 10.4 to give it an appearance of rusted metal.

Exercise 12.9. (**Programming**) Texture the table of Experiment 10.7 with a wood grain texture.

12.4.2 Bézier and Quadric Surfaces

It's fairly simple to use OpenGL to automatically generate texture coordinates for Bézier and quadric surfaces, as is demonstrated in the next experiment.

Figure 12.24: Screenshot of `texturedTorpedo.cpp`: propeller blades textured with the chessboard, body with stripes.

Experiment 12.16. Run `texturedTorpedo.cpp`, which textures parts of the torpedo of `torpedo.cpp` – from Experiment 10.21 – as you can see in the screenshot in Figure 12.24. Press space to start the propeller turning. **End**

The texturing of the propeller blade Bézier surface of `textured-Torpedo.cpp` is most important. We want to do this in the same manner as the torus of Experiment 12.15 – by identifying parameter rectangle and texture. However, because of the way OpenGL is set up for texture coordinate generation we are forced to a slightly roundabout approach described next.

First, we have to create a Bézier surface s' in *texture space*. The statement

```
glMap2f(GL_MAP2_TEXTURE_COORD_2, 0, 1, 2, 2, 0, 1, 4, 2,
        texturePoints[0][0])
```

in the display list for the propeller blade does just this. The syntax of this statement is similar to that of the `glMap2f(GL_MAP2_VERTEX_3, ...)` we are already familiar with to create a world-space Bézier surface. Control points of s' stored in the `texturePoints` array are $(0,0)$, $(0,1)$, $(1,0)$ and $(1,1)$, making s' the rectangle $[0,1] \times [0,1]$. See Figure 12.25. Moreover, the bilinearity of the parametric mapping of s' (it's of order 2, or degree 1, along both u and v) implies that it is simply the identity map from the rectangle $[0,1] \times [0,1]$ in parameter space to s' in texture space.

Observe, next, that the statement

```
glMap2f(GL_MAP2_VERTEX_3, 0, 1, 3, 3, 0, 1, 9, 4,
        controlPointsPropellerBlade[0][0])
```

defines the parameter space for the propeller blade Bézier surface to be $[0,1] \times [0,1]$ as well. Given, then, that the parameter spaces for the blade Bézier surface s in world space and the Bézier surface s' in texture space are identical, OpenGL defines the texture map between the two in the following simple manner: for each parameter point (u,v), the image of (u,v) on the blade surface s is mapped to the image of (u,v) on s' (see Figure 12.25). Given how s' itself is defined, evidently texture coordinates (u,v) are assigned to the image of the parameter point (u,v) on the blade

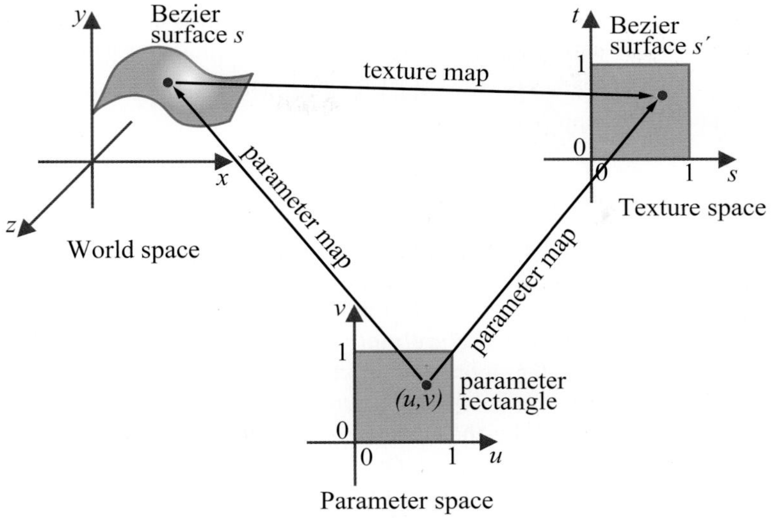

Figure 12.25: Texture mapping a Bézier surface via a Bézier surface in texture space. The parameter map on the right from parameter to texture space is the identity in the case of `texturedTorpedo.cpp`.

Bézier surface, effectively identifying the texture with the parameter rectangle as we set out to do.

The two statements

```
glEnable(GL_MAP2_TEXTURE_COORD_2);
glMapGrid2f(5, 0.0, 1.0, 5, 0.0, 1.0);
```

in the propeller blade's display list actually instigate texture coordinate generation at the image on the world-space Bézier surface – the propeller blade in this case – of an evenly-spaced 5×5 grid of sample points in the parameter domain.

OpenGL's particular mechanism to generate texture coordinates for a Bézier surface offers flexibility. For example, by changing the control points of the texture-space surface and so its extent, one can paint the real-world surface with a different region of texture space. The next exercise asks you to try this.

Exercise 12.10. (Programming) Make the red and blue squares of the chessboard pattern on the propeller blades of `texturedTorpedo.cpp` twice as big are they are currently, without changing the original texture as generated by the program.

Getting OpenGL to generate texture coordinates for a quadric is even simpler. The statement

```
gluQuadricTexture(qobj, GL_TRUE)
```

in the drawScene() routine of texturedTorpedo.cpp is all that it takes to automatically generate texture coordinates for the quadric torpedo body.

Exercise 12.11. (Programming) Texture the parts of the torpedo of texturedTorpedo.cpp which are still wire mesh.

12.4.3 Texture Matrix and Animating Textures

Recall from Section 4.4 that the current modelview matrix, say M, the topmost one in the modelview matrix stack, transforms a vertex V by multiplication from the left, in particular, $V \rightarrow MV$. Likewise, the *texture matrix stack* contains 4×4 *texture matrices*, the topmost one of which is the *current texture matrix*; moreover, texture coordinates are transformed by multiplication from the left by the current texture matrix.

When being multiplied by the current texture matrix, texture coordinates are all written as column vectors with four components, typically denoted $[s\ t\ r\ q]^T$. For example, texture coordinates represent points (s, t) in plane texture space for the 2D textures we have been using; however, extended to 4D, (s, t) would be written $[s\ t\ 0\ 1]^T$. As expected, by default, the current texture matrix is the 4×4 identity.

The texture matrix stack and the current texture matrix can be manipulated with exactly the same commands as their modelview counterparts. Transforming texture coordinates with help of the current texture matrix, one can animate a texture in various ways. Let's see a simple example.

Experiment 12.17. Run texturedTorusAnimated.cpp, which animates the launch texture of texturedTorus.cpp by applying a translation to the current texture matrix. Press space to toggle between animation on and off, the up and down arrow keys to change its speed and 'x'-'Z' to turn the torus.

The modification of texturedTorus.cpp is simple. In particular, the following block of code in the drawing routine does the trick by entering texture matrix mode and applying a translation in the t-direction, effectively translating the texture coordinates in the t-direction:

```
glMatrixMode(GL_TEXTURE);
  glLoadIdentity();
  glTranslatef(0.0, shift, 0.0);
glMatrixMode(GL_MODELVIEW);
```

The rest of the modification of texturedTorus.cpp is in managing the animation. **End**

Exercise 12.12. (Programming) Apply the rotation glRotatef(60.0, 0.0, 0.0, 1.0) first to the current texture matrix of texturedTorus-Animated.cpp followed by the existing translation.

Exercise 12.13. (Programming) Animate the sky texture of field-AndSky.cpp to make the clouds move.

12.5 Lighting Textures

By now the reader may be wondering if color and light can coexist with texture or if the two are mutually exclusive ways of adorning an object. The answer is that OpenGL, in fact, offers more than one option to combine them. An option is selected using the following texture function statement, most often located in the initialization routine:

```
glTexEnvf(GL_TEXTURE_ENV, GL_TEXTURE_ENV_MODE, parameter)
```

If *parameter* is `GL_REPLACE`, as it has been in our programs thus far, then the texture colors *overwrite* the current primitive pixel colors (i.e, the primitive's own material colors as well as light sources in the environment are ignored).

The most common way, though, to combine color and light with texture is by setting *parameter* to `GL_MODULATE`, in which case OpenGL does the following:

(a) Computes RGB values at a primitive's vertices using OpenGL's lighting equation (11.12) and interpolates these through its interior – assuming that smooth shading is on – to determine the RGB values at each of its pixels.

(b) Uses the texture map to obtain RGB values from the texture at each of the primitive's pixels.

(c) Determines the final RGB values at each pixel as the *product* of the corresponding values from the preceding two steps.

In short, OpenGL *separately* computes RGB values for color and light as if there were no texture and RGB values for texture as if there were no color and light and, finally, scales one with the other.

Example 12.2. If the RGB tuple at a pixel P is $(0.5, 0.75, 0.1)$ as obtained by interpolation from vertex RGB values computed after lighting, while that determined at P from the texture via the texture map is $(0.4, 0.5, 1.0)$, then the final color applied to P using the `GL_MODULATE` option is $(0.5 \times 0.4,\ 0.75 \times 0.5,\ 0.1 \times 1.0) = (0.2, 0.375, 0.1)$.

Experiment 12.18. Run `fieldAndSkyLit.cpp`, which applies lighting to the scene of `fieldAndSky.cpp` with help of the `GL_MODULATE` option. The light source is directional – imagine the sun – and its direction controlled using the left and right arrow keys, while its intensity can be changed using the up and down arrow keys. A white line indicates the direction and intensity of the sun. Figure 12.26(a) is a mid-morning screenshot.

The material colors are all white, as is the light. The normal to the horizontal grassy plane is vertically upwards. Strangely, we use the same normal for the sky's vertical plane, because using its "true" value toward the positive z-direction has the unpleasant, but expected, consequence of a sky that doesn't darken together with land. End

(a)

(b)

Figure 12.26: Screenshots of (a) `fieldAndSkyLit.cpp` and (b) `litTextured-Cylinder.cpp`.

Experiment 12.19. Run `litTexturedCylinder.cpp`, which adds a label texture and a can top texture to `litCylinder.cpp`. Press 'x'-'Z' to turn the can. Figure 12.26(b) is a screenshot.

Most of the program is routine – the texture coordinate generation is, in fact, a near copy of that in `texturedTorus.cpp` – except for the following lighting model statement in `setup()` which we're using for the first time:

```
glLightModeli(GL_LIGHT_MODEL_COLOR_CONTROL, GL_SEPARATE_SPECULAR_COLOR)
```

We had briefly encountered this statement as an OpenGL lighting model option in Section 11.4. It causes a modification of OpenGL's `GL_MODULATE` procedure: the specular color components are separated and not multiplied with the corresponding texture color components, as are the ambient and diffuse, but added in after. The result is that specular highlights are preserved rather than blended with the texture. **End**

Exercise 12.14. (**Programming**) Close off the bottom of the cylinder of `litTexturedCylinder.cpp` with a metal-textured disc.

Exercise 12.15. (**Programming**) Animate a lit textured flag fluttering in the wind.
Hint: A surface whose section is of the form $y = \sin(x + t)$, where t depends on time, in other words a "moving" sine curve, simulates fluttering.

Exercise 12.16. (**Programming**) Continue improvement of `shipMovie.cpp`, which you began in Exercise 11.44, now with the help of textures. There are numerous possibilities of which a few are:

(a) Texture the black back plane with the image of a night-time city skyline.

(b) Paint the surface of the sea with a water texture, possibly animated.

(c) Add detail to the ship by texturing it with images of parts of real ships.

(d) Texture the background boats. You may want to strategically place additional light sources.

Exercise 12.17. (**Programming**) Continue with Exercise 11.45 where you enhanced `animateMan*.cpp` with color and light, now using texture.

12.6 Multitexturing and Texture Combining

OpenGL allows more than one texture to be applied to a polygon in a pipelined process, each texture being combined with its predecessor in programmer-specified manner. This so-called multitexturing makes possible myriad visual effects depending upon the textures and how they are combined. Let's get to work on a particular one – transforming a night sky into day – by interpolating between night and day sky textures, illustrating in the process the key steps in multitexturing.

(a) (b) (c)

Figure 12.27: Screenshots of `multitexture.cpp`: (a) Mid-day (b) Late evening (c) Night.

Experiment 12.20. Run `multitexture.cpp`, which interpolates between night and day sky texture. Press the left/right arrow keys to transition between night and day. Figure 12.27 shows stages in the transition. **End**

Multitexturing requires more than one *texture unit* – a binding point for a texture to the OpenGL context – each with id of the form `GL_TEXTUREi`. Here's the block in the initialization routine of `multitexture.cpp` which initializes `GL_TEXTURE0`:

```
glActiveTexture(GL_TEXTURE0);
glEnable(GL_TEXTURE_2D);
glBindTexture(GL_TEXTURE_2D, texture[0]);
glTexEnvi(GL_TEXTURE_ENV, GL_TEXTURE_ENV_MODE, GL_REPLACE);
```

The first statement selects GL_TEXTURE0 as the currently active texture unit, the second enables 2D texturing, the third binds texture object texture[0] to the active texture unit, while the last specifies that surface colors come from the active texture unit.

The block initializing GL_TEXTURE1 is identical except for the last statement

```
glTexEnvi(GL_TEXTURE_ENV, GL_TEXTURE_ENV_MODE, GL_COMBINE)
```

where the third parameter GL_COMBINE, instead of GL_REPLACE, indicates that the first texture unit combines with the zeroth by application of a *texture combiner function*, which, in fact, is *interpolation* according to

```
glTexEnvi(GL_TEXTURE_ENV, GL_COMBINE_RGB, GL_INTERPOLATE);
```

the next statement in the initialization routine. This particular interpolation combiner function is

$$Arg0 * Arg2 + Arg1 * (1 - Arg2)$$

The following block of statements specify the combiner function's arguments.

```
glTexEnvi(GL_TEXTURE_ENV, GL_SRC0_RGB, GL_PREVIOUS);
glTexEnvi(GL_TEXTURE_ENV, GL_SRC1_RGB, GL_TEXTURE);
glTexEnvi(GL_TEXTURE_ENV, GL_SRC2_ALPHA, GL_CONSTANT);
glTexEnvi(GL_TEXTURE_ENV, GL_OPERAND0_RGB, GL_SRC_COLOR);
glTexEnvi(GL_TEXTURE_ENV, GL_OPERAND1_RGB, GL_SRC_COLOR);
glTexEnvi(GL_TEXTURE_ENV, GL_OPERAND2_ALPHA, GL_SRC_ALPHA);
```

The first two statements specify that the zeroth and first source's RGB value are from, respectively, GL_TEXTURE0 and GL_TEXTURE1; the third statement specifies that the second source's alpha value is from the constant environment color (specified in the display routine); the fourth statement says that $Arg0$'s RGB values are from the zeroth color source, i.e., GL_TEXTURE0; likewise, the fifth statement says that $Arg1$'s RGB values are from the first color source GL_TEXTURE1; the last statement says that $Arg2$, the interpolation parameter, is the environment color alpha value.

The two statements

```
glTexEnvfv(GL_TEXTURE_ENV, GL_TEXTURE_ENV_COLOR, constColor);
constColor[3] = alpha;
```

in the display routine specify that the texture environment color values are to be read from the global array constColor, and set its alpha value, the interpolation parameter; note that its RGB values are never defined because they are never used.

The final piece is to specify independently the texture coordinates for the two texture units, which is done within the polygon definition via statements of the form glMultiTexCoord2f(GL_TEXTUREi, *, *), e.g., the block

```
glMultiTexCoord2f(GL_TEXTURE0, 0.0, 0.0);
glMultiTexCoord2f(GL_TEXTURE1, 0.0, 0.0);
glVertex3f(-20.0, -20.0, 0.0);
```

says that the polygon vertex $(-20.0, -20.0, 0.0)$ has texture coordinates $(0.0, 0.0)$ for both texture units.

In addition to interpolation, other texture combiners, e.g., modulation, addition and subtraction, can be used to create various effects. We ask you to apply modulation, which has the combiner function $Arg0 * Arg1$, next.

Exercise 12.18. (Programming) Modulate an image (texture) of an object like a ball or table with that of stained metal, as in Figure 12.28, to give the former a tarnished finish. You may need to refer to the red book or OpenGL docs for `glTexEnv*()` specs.

Figure 12.28: Stained metal.

Remark 12.5. "*Textures may not be only images!*", a seemingly odd thing to say at the end of a chapter of applying exclusively images as textures. But it may not appear so strange upon noting that texels are bit arrays which, thus far, we have *interpreted* as RGBA values.

We don't have to, e.g., we could very well interpret a 32-bit texel as a 32-bit floating point value. Moreover, the hardware accelerated access to textures in the GPU – in order, in the case of images, for the texture map to be calculated rapidly – may be put to good use in accessing other kinds of data, as well, effectively using the texture as a giant look-up table.

In fact, there are various applications of such non-image data textures. We'll be seeing one ourselves in Section 18.2 on shadow mapping, where a so-called depth texture is used to store z-buffer values.

12.7 Summary, Notes and More Reading

Texturing, the process of painting an image onto the surface of an object, is of great practical importance in computer graphics. In this chapter we learned the basics of how to apply a texture, underlying principles of the texturing process and a fair number of techniques to effectively manage textures, amongst them being combining textures with lighting as well as multitexturing to combine multiple textures.

The seminal reference for textures in CG is Heckbert [70]. The article by Haeberli and Segal [65] is easily-readable and informative as well. More advanced CG books, e.g., the ones by Akenine-Möller, Haines & Hoffman [1] and Watt [147], all have sections on texturing that will take the reader beyond what she has learned in this chapter. The book by Reynolds and Blythe [92] is a good reference for texturing in OpenGL in particular. Texturing techniques are a field of active research and, as for CG research in general, the place to visit for the latest developments is the annual ACM SIGGRAPH conference [131].

CHAPTER 13

Special Visual Techniques

And now for special effects! Special visual techniques are the topic of the next nine sections of this chapter. These include blending, fog, billboarding, antialiasing and multisampling, point sprites, environment mapping, stencil buffer techniques, image manipulation and bump mapping.

The goal of blending objects by combining their color values, the topic of Section 13.1, is primarily to engender translucency; however, blending can also be used in effects such as reflection and morphing. Section 13.2 shows how to use fog to cue the viewer to the distance of an object. This imparts realism to a scene, as does the technique of billboarding introduced in Section 13.3, a cost-effective way to create the illusion of a 3D object by means of its 2D image.

We'll learn in Section 13.4 how to antialias points and straight lines, as well as how to multisample polygons, in order to remove jaggedness in their rendering. Section 13.5 shows how to turn plain points into spectacular point sprites, multitudes of which can then combine to form particle systems. Environment mapping, the topic of Section 13.6, enables a shiny object to reflect its surroundings, again making more realistic the rendering of things such as gleaming teapots and rocket nose cones.

The stencil buffer, described in Section 13.7, which allows drawing only to selected regions of the display, is useful in creating certain effects. Section 13.8 describes OpenGL's image manipulation commands, allowing the user to write pixel data directly to the frame buffer, and retrieve them too. We show how to use such commands to move an image around the screen. Pixel buffer objects (PBOs) for server-side storage are discussed as well.

Bump mapping, which we'll study in Section 13.9, is a technique to add the illusion of detail to an object by altering its normals, but without changing its geometry. Section 13.10 concludes the chapter.

13.1 Blending

We spend a little time first assimilating the theory of blending before putting it into practice.

13.1.1 Theory

Our plan is to understand first how OpenGL operates without blending, and then with.

No Blending

Consider the following piece of pseudo-code and how OpenGL would process it without blending:

```
glClear(GL_COLOR_BUFFER_BIT | GL_DEPTH_BUFFER_BIT);
draw triangle T;
draw quad Q;
```

See Figure 13.1. The z-values in the depth buffer are each initialized to a very large value by the `glClear()` command, as are values in the color buffer all to the clearing color white. This initial configuration is depicted in the upper left grid, where pixels only in a region of interest are labeled with their z-values, unlabeled pixels all having ∞ z-value as well.

The triangle T is then *rasterized*, i.e., the set of pixels corresponding to T determined, and color and z-values computed for each – the upper middle grid using hypothetical z-values. A raster pixel, together with its color values and z-value, is called a *fragment*.

OpenGL next *renders* T to the color buffer, which will be viewed when flushed to the screen, according to sets of rules depending on whether depth testing is on or not.

If *depth testing is enabled*, then the process is two-step:

1. The z-value of each of T's fragments, called a *source fragment*, is compared with that in the corresponding pixel, called the *destination pixel*, in the color buffer.

2. If the source fragment's z-value is less than that of the destination pixel, then it passes the depth test and its color values *overwrite* the current color values of the destination pixel and its z-value overwrites the current z-value, as well; if the source fragment's z-value is not less than that of the destination pixel, then it fails the depth test and its color values and z-value are discarded, leaving all destination values unchanged.

   Note: That its z-value be less than that in the destination pixel is the default test that a source fragment has to pass in order to overwrite the

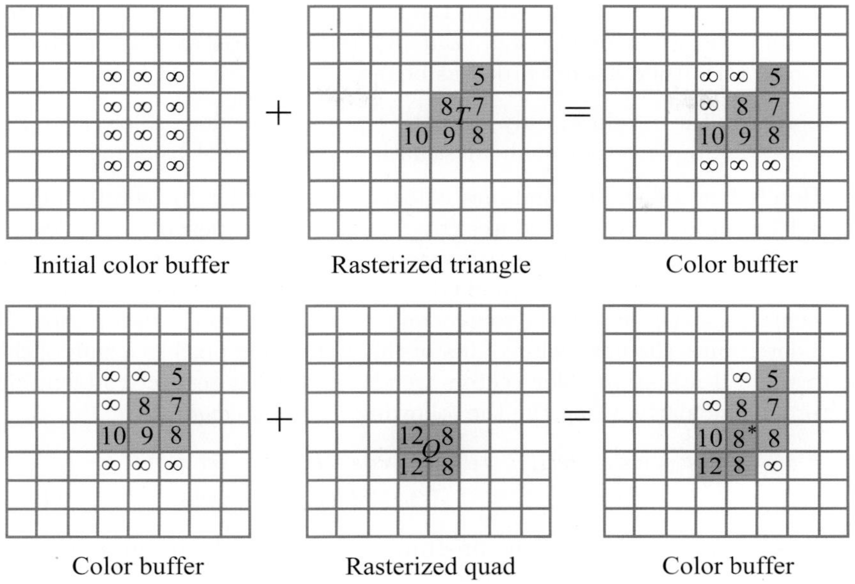

Figure 13.1: Assuming depth testing on: T rasterized and rendered (upper row), followed by Q rasterized and rendered (lower row). The starred pixel is considered in an example below.

destination. Other tests can be invoked with a call to `glDepthFunc()` (see the red book for a listing of possible tests).

If *depth testing is not enabled*, then the process is a simple single step:

1. Each one of T's fragments unconditionally overwrites the color values of its destination pixel. The z-values are not relevant and never change.

Given depth testing as enabled, the result of rendering T is depicted in the upper right grid in Figure 13.1, and also in the lower left. Following T, Q is rasterized (lower middle) and rendered (lower right) in an identical manner, Q replacing T in the two-step procedure above. One sees that three of Q's fragments pass the depth test and overwrite their destination pixels.

The Difference with Blending

Next, let's understand what happens *with* blending. The differences are precisely the following:

1. In all the cases above – without blending, that is – where a source fragment's color values are supposed to overwrite those of its destination pixel, OpenGL instead *combines*, or *blends*, the two sets and applies the result to the destination pixel. We'll discuss momentarily how color values are, in fact, combined.

2. In all the cases above, where the source fragment does not overwrite its destination pixel – e.g., if depth testing is on and its z-value is greater – then it's discarded as before.

3. As for the z-values, only if depth testing is on does a lower source z-value replace that of the destination, *regardless* of blending.

Here's how OpenGL combines color values in order to blend them in item 1 just above. Say the source fragment's color values are (src_R, src_G, src_B), while those of the destination are (dst_R, dst_G, dst_B). Based upon the programmer's specifications, OpenGL assigns so-called *blending factors*, in particular, a scalar src_b to the source, and a scalar dst_b to the destination. It then determines the final color values of the destination pixel as a sum of the source and destination color vectors weighted by their respective blending factors, particularly, using the the following *blending equation*:

$$\begin{aligned}(dst_R, dst_G, dst_B) &= src_b * (src_R, src_G, src_B) + \\ &\quad dst_b * (dst_R, dst_G, dst_B) \qquad (13.1)\end{aligned}$$

We'll say soon how, in fact, the programmer specifies the blending factors, but first an example and an exercise.

Example 13.1. Consider the starred pixel in the lower right grid of Figure 13.1. Say its RGB color values prior to the rendering of Q, in particular, those of the corresponding pixel in the triangle T in the lower left grid, are $(0.6, 0.4, 0.2)$, while those of the corresponding pixel of the quad in the lower middle are $(0.5, 0.5, 0.5)$. Suppose as well that $src_b = 0.3$ and $dst_b = 0.7$ and that depth testing is on. What are the final color and depth values of the starred pixel if blending is not enabled? If it is?

Answer: Suppose, first, that blending is off. The source fragment's z of 8 being less than the destination's 9, it overwrites the destination's colors. Final color values of the starred pixel are, therefore, $(0.5, 0.5, 0.5)$.

Next, suppose blending is enabled. Since the source would overwrite the destination colors if blending were off, given that blending is, in fact, on, their color values are combined instead. The blending equation (13.1) gives the resulting values as

$$0.3 * (0.5, 0.5, 0.5) + 0.7 * (0.6, 0.4, 0.2) = (0.57, 0.43, 0.29)$$

The resulting z of 8 for the starred pixel is the same, though, both in blended and unblended applications, as it is determined only by competition in the depth buffer.

Exercise 13.1. How about the pixel just to the left of the starred one? What are its color and depth values with blending enabled, assuming that the color values in the corresponding pixels in the left and middle grids are the same as for the starred one in the preceding example and that blending factors are identical as well? Consider when depth testing is both on and off.

Specifying Blending Factors: Alpha

It's in assigning the blending factors that the programmer can make use of the *alpha* values, the A in RGBA. For example, the *blend function* command

```
glBlendFunc(GL_SRC_ALPHA, GL_ONE_MINUS_SRC_ALPHA)
```

causes OpenGL to set

$$src_b = src_A \qquad \text{and} \qquad dst_b = 1 - src_A$$

where src_A denotes the source fragment's alpha value. The consequence of this particular blend function, as one sees from plugging the blending factors into (13.1) to get

$$\begin{aligned}(dst_R, dst_G, dst_B) \quad = \quad & src_A * (src_R, src_G, src_B) + \\ & (1 - src_A) * (dst_R, dst_G, dst_B)\end{aligned}$$

is that the greater the source's alpha the more its contribution to the final color or, intuitively, the more *opaque* it is.

In general, blending factors are set by calling

glBlendFunc(*srcFactor*, *dstFactor*)

where the values of the parameters *srcFactor* and *dstFactor* tell OpenGL how to determine the source and destination blending factors, respectively. We've already seen that the value `GL_SRC_ALPHA` chooses the source alpha value src_A, while `GL_ONE_MINUS_SRC_ALPHA` chooses $1 - src_A$. Likewise, `GL_DST_ALPHA` and `GL_ONE_MINUS_DST_ALPHA` choose dst_A and $1 - dst_A$, respectively.

The reader is referred to the red book for a full list of possible values for *srcFactor* and *dstFactor*, though `GL_SRC_ALPHA` and `GL_ONE_MINUS_SRC_ALPHA` are, in fact, most commonly used. Moreover, any of these values may be used for either parameter, e.g., even `GL_SRC_ALPHA` for *dstFactor*, which would be an odd choice indeed.

Finally, the alpha value dst_A of the destination pixel changes, too, being combined from those of the source and destination by a blending equation exactly similar to (13.1), precisely:

$$dst_A = src_b * src_A + dst_b * dst_A \qquad (13.2)$$

Remark 13.1. The blending equation (13.1), though the default and most commonly used, isn't the only option in OpenGL to combine source and destination color values. Other formulae are available as well, which can be chosen by a call to the command `glBlendEquation()`. Moreover, certain parameter choices for `glBlendFunc()` allow (13.1) to be refined to apply different blending factors to each of R, G, and B of the source, and to each of R, G and B of the destination. Refer to the red book for details.

13.1.2 Experiments

Experiment 13.1. Run `blendRectangles1.cpp`, which draws two translucent rectangles with their alpha values equal to 0.5, the red one being closer to the viewer than the blue one. The *code* order in which the rectangles are drawn can be toggled by pressing space. Figure 13.2 shows screenshots of either order. **End**

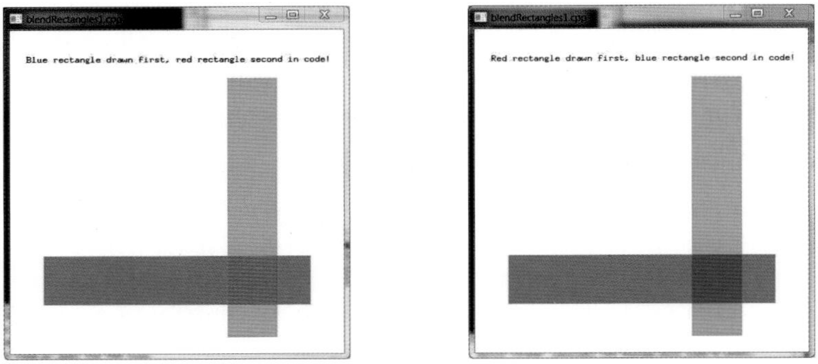

Figure 13.2: Screenshot of `blendRectangles1.cpp` with (a) the blue rectangle first in code (b) the red rectangle first in code.

Blending is enabled in `blendRectangles1.cpp` with the call

```
glEnable(GL_BLEND)
```

and the blend function used is

```
glBlendFunc(GL_SRC_ALPHA, GL_ONE_MINUS_SRC_ALPHA)
```

As depth testing is currently disabled, if there were no blending, then the rectangle second in code would overwrite the one drawn first. Therefore, from our understanding of theory now, with blending on, their colors are combined.

As $src_A = 1 - src_A = 0.5$, no matter which rectangle is drawn first, the blending equations (13.1) and (13.2) together give simply

$$\begin{aligned}(dst_R, dst_G, dst_B, dst_A) &= 0.5*(src_R, src_G, src_B, src_A) + \\ &\quad 0.5*(dst_R, dst_G, dst_B, dst_A) \qquad (13.3)\end{aligned}$$

which is symmetric in source and destination. So one may wonder why the one drawn second seems to dominate where the images intersect. The reason is that the rectangle drawn first is blended with the background white, diluting its color, while the second-drawn rectangle comes in at "full strength". Let's verify that this is actually so in the next example.

Example 13.2. For `blendRectangles1.cpp`, in the case when the blue rectangle is drawn first, use the blending equations to compute the colors of each region of the composite output shape. Verify the color values experimentally.

Answer: For the two parts of the blue rectangle which are not intersected, (13.3) gives the RGBA vector as

$$0.5 * (0.0, 0.0, 1.0, 0.5) + 0.5 * (1.0, 1.0, 1.0, 0.0) = (0.5, 0.5, 1.0, 0.25)$$

because the source fragment vector is $(0.0, 0.0, 1.0, 0.5)$ and the destination (background) vector is $(1.0, 1.0, 1.0, 0.0)$.

Where the blue and red rectangles intersect, the RGBA vector is

$$0.5 * (1.0, 0.0, 0.0, 0.5) + 0.5 * (0.5, 0.5, 1.0, 0.25) = (0.75, 0.25, 0.5, 0.375)$$

because the source fragment vector is $(1.0, 0.0, 0.0, 0.5)$ and the destination vector is $(0.5, 0.5, 1.0, 0.25)$ as calculated above.

Note: Even though the destination alpha changes after the first blend, this does not matter subsequently as the program's particular blend function `glBlendFunc(GL_SRC_ALPHA, GL_ONE_MINUS_SRC_ALPHA)` uses the source alpha to define both source and destination blending factors.

We leave it to the reader to calculate the RGBA vector for the non-intersected parts of the red rectangle.

For experimental verification, draw a point inside each region with its calculated RGB values – the points should be invisible!

Exercise 13.2. (Programming) Change the alpha values of both rectangles in `blendRectangles1.cpp` successively to 0.0, 0.25, 0.75 and 1.0. Verify the color values in each region both theoretically and experimentally, as in the preceding example, for the case when alpha is 0.75.

Exercise 13.3. (Programming) Although there is a depth buffer in `blendRectangles1.cpp`, depth testing has been disabled. Enable it by replacing the call `glDisable(GL_DEPTH_TEST)` in `setup()` with `glEnable(GL_DEPTH_TEST)`. Explain what you observe.
Hint: The z-value of the red rectangle is 0.5 (i.e., its distance from the front face of the viewing box), which is less than that of the blue, which is 0.9. Therefore, here is the situation with *no* blending: if the blue rectangle is first in code, then the red rectangle values following will overwrite the blue's in their intersection; on the other hand, if the red rectangle is first in code, then the blue rectangle values are discarded. Now, refer to the theoretical discussion earlier to see what changes, given blending is *enabled*.

Exercise 13.4. (Programming) Continuing the preceding exercise, Keeping depth testing enabled, disable blending by replacing the call `glEnable(GL_BLEND)` in `setup()` with `glDisable(GL_BLEND)`. Again, explain what you observe.

13.1.3 Opaque and Translucent Objects Together

Experiment 13.2. Run `blendRectangles2.cpp`, which draws three rectangles at different distances from the eye. The closest one at depth 0.5 is vertical and a translucent red ($\alpha = 0.5$), the next one at depth 0.7 is angled and opaque green ($\alpha = 1$), while the farthest at depth 0.9 is horizontal and a translucent blue ($\alpha = 0.5$). Figure 13.3(a) is a screenshot of the output.

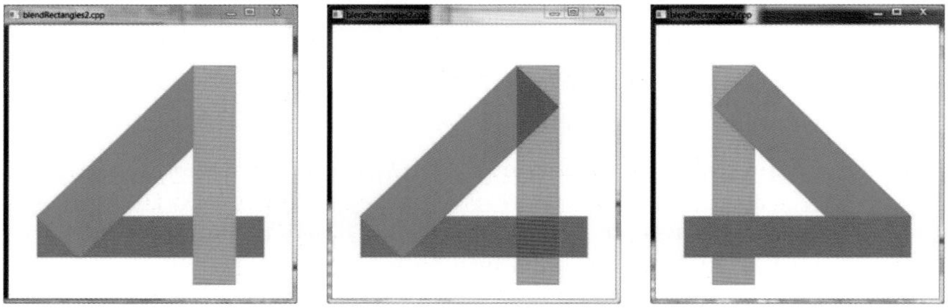

(a) (b) (c)

Figure 13.3: Screenshots of `blendRectangles2.cpp`: (a) Original (b) With rectangles re-ordered to blue, green, red in the code (c) New ordering seen from the $-z$-direction.

The scene is clearly not authentic as no translucency is evident in either of the two areas where the green and blue rectangles are behind the red. The fault is not OpenGL's as it is rendering as it's supposed to with depth testing, as we see next. **End**

Example 13.3. Verify the claim just made that the program is doing as it's supposed to, using our understanding of the rendering process.

Answer: The program's drawing order is:

```
drawRedRectangle(); // Red rectangle closest to viewer, translucent.
drawGreenRectangle(); // Green rectangle second closest, opaque.
drawBlueRectangle(); // Blue rectangle farthest, translucent.
```

Once the red rectangle, the closest, is drawn, the green and blue both fail the depth test where they intersect the red, so there's no translucency apparent in those two regions (keep in mind that only if the source fragment passes the depth test does blending kick in).

Figure 13.4: Screenshot of `blendRectangles2.cpp` with depth testing disabled.

Returning to the experiment, disabling depth testing doesn't help either, as the green blocks out the red (which it shouldn't as it's farther away), while it doesn't block out the blue (which it should as it's closer). See Figure 13.4.

In fact, trying all 6 (=3!) possible orders to draw the rectangles, it's seen that the only one producing an authentic rendering is:

```
drawBlueRectangle(); // Blue rectangle farthest, translucent.
```

```
drawGreenRectangle(); // Green rectangle second closest, opaque.
drawRedRectangle(); // Red rectangle closest to viewer, translucent.
```

See Figure 13.3(b). As one would expect, both green and blue rectangles can be seen through the red, while the green blocks out the blue. The order that the primitives are drawn in code happens to be according to their distance from the viewer, starting with the farthest; moreover, with this order it doesn't matter if depth testing is on or off.

Exercise 13.5. Explain why the (farthest-to-nearest) blue-green-red drawing order is successful, regardless of whether depth testing is enabled or not.

Unfortunately, this method is not a particularly robust way to produce an authentic scene, as the farthest-to-nearest order depends on the viewpoint. For example, keeping the same blue-green-red drawing order, replace the viewing transformation

```
gluLookAt(0.0, 0.0, 0.0, 0.0, 0.0, -1.0, 0.0, 1.0, 0.0);
```

with

```
gluLookAt(0.0, 0.0, 0.0, 0.0, 0.0, 1.0, 0.0, 1.0, 0.0);
```

to view the rectangles from the $-z$-direction, which means now the blue is closest, the green next and the red farthest away. Oops! See Figure 13.3(c). The result is no longer authentic as the closest translucent blue rectangle blocks out both red and green, which, of course, it shouldn't. The reason for the breakdown is clearly that what used to be farthest-to-nearest rendering is now no longer.

So, what is one to do? Precisely, the following:

1. Enable depth testing.

2. Draw first the opaque objects. Because of depth testing they block one another out according to distance from viewpoint, as one would want.

3. Make the depth buffer read-only with a call to `glDepthMask(GL_FALSE)`.

4. Draw next the translucent objects. As depth testing is still on, translucent objects farther than the nearest opaque one – which per-pixel has written the z-buffer – are discarded, again as one wants.

   However, as they can no longer update the z-buffer to their own z-values, closer translucent objects don't block out farther ones – which is what would happen if their z-values were recorded – but blend instead. In fact, all translucent objects, which are closer than the closest opaque one, blend successively into the latter. Exactly as the doctor ordered.

5. Restore the depth buffer's writability by calling glDepthMask(GL_-TRUE).

   Note: The depth buffer is writable by default.

Try it.

Experiment 13.3. Rearrange the rectangles and insert two glDepth-Mask() calls in the drawing routine of blendRectangles2.cpp as follows:

```
// Draw opaque objects, only one.
drawGreenRectangle();

glDepthMask(GL_FALSE); // Make depth buffer read-only.

// Draw translucent objects.
drawBlueRectangle();
drawRedRectangle();

glDepthMask(GL_TRUE); // Make depth buffer writable.
```

Try both gluLookAt(..., 0.0, 0.0, -1.0, ...) and gluLookAt(..., 0.0, 0.0, 1.0, ...) to see the rectangles from the front and back. Interchange the drawing order of the two translucent rectangles as well. The scene is authentic in every instance. **End**

Experiment 13.4. Run sphereInGlassBox.cpp, which makes the sides of the box of sphereInBox2.cpp glass-like by rendering them translucently. Only the unaveraged normals option of sphereInBox2.cpp is implemented. Press the up and down arrow keys to open or close the box and 'x/X', 'y/Y' and 'z/Z' to turn it.

The opaque sphere is drawn first and then the translucent box sides, after making the depth buffer read-only. A screenshot is Figure 13.5(a). **End**

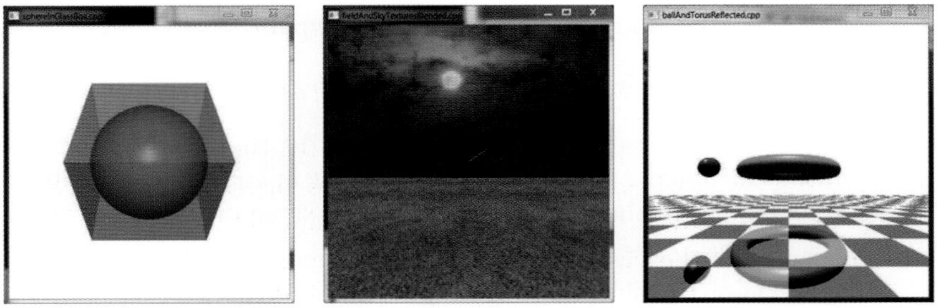

(a) (b) (c)

Figure 13.5: Screenshots of blended effects: (a) sphereInGlassBox.cpp (b) fieldAnd-SkyTexturesBlended.cpp (c) ballAndTorusReflected.cpp.

Exercise 13.6. (**Programming**) Inscribe a glass dodecahedron inside a glass icosahedron.

Exercise 13.7. (**Programming**) Make a solid ball travel through a glass helical pipe (see Experiment 10.4 for the helical pipe).

13.1.4 Blending Textures

There's no reason why textures cannot be blended – it's simply a matter of either one or both of the source fragment and destination pixel obtaining color values from a texture map. Here's a fun program doing just that.

Experiment 13.5. Run `fieldAndSkyTexturesBlended.cpp`, which is based on `fieldAndSkyLit.cpp`. Press the arrow keys to move the sun. As the sun rises the night sky morphs into a day sky (yes, we saw this very same morph as an application of multitexturing, using the interpolation combiner, in Section 12.6). Figure 13.5(b) shows late evening. **End**

The program's a fairly straightforward application of alpha blending. We point out a few interesting features:

(a) The sky rectangle is no longer lit as in `fieldAndSkyLit.cpp` because the night texture itself causes the sky to darken.

(b) Source blending factors all 1 (`GL_ONE`) and destination blending factors all 0 (`GL_ZERO` enable the grass and night sky textures to initially paint their respective rectangles without dilution.

(c) The statements

```
if (theta <= 90.0) alpha = theta/90.0;
else alpha = (180.0 - theta)/90.0;
glColor4f(1.0, 1.0, 1.0, alpha);
```

in the drawing routine link the `alpha` value to the angle `theta` of the sun in the sky, so that the former increases from 0 to 1 as the sun rises from the horizon to vertically above.

(d) The day sky is blended into the night sky because both textures paint the same rectangle and because the prior disabling of depth testing allows an incoming fragment to write to a destination pixel, even if its z-value is equal to the current one (with depth testing on it has to be less in order to do so). The call `glBlendFunc(GL_SRC_ALPHA, GL_ONE_MINUS_SRC_ALPHA)` in the drawing routine sets the source blending factor equal to `alpha` and the destination blending factor to `1 - alpha`.

Remark 13.2. The simple-minded alpha-morph just described should work fairly well if the transition required is mainly in the colors of a scene and not the geometry, because it is a straight linear interpolation between corresponding source and destination color values. For the same reason, one should not expect much from it by way of morphing *shapes*, as in the Terminator movies.

Exercise 13.8. (**Programming**) Morph a day image of a static scene (e.g., city skyline, mountainous landscape, etc.) to a night image.

13.1.5 Creating Reflections

Another neat application of blending is to simulate reflection.

Experiment 13.6. Run `ballAndTorusReflected.cpp`, which builds on `ballAndTorusShadowed.cpp`. Press space to start the ball traveling around the torus and the up and down arrow keys to change its speed.

The reflected ball and torus are obtained by drawing them scaled by a factor of -1 in the y-direction, which creates their reflections in the xz-plane, and then blending the floor into the reflection. Figure 13.5(c) shows a screenshot. **End**

Exercise 13.9. (**Programming**) Draw the reflection of the ship of `shipMovie.cpp` in the sea.

Exercise 13.10. (**Programming**) Make the character of `animate-Man1.cpp` walk along a shiny reflective floor to a window. The camera should then move to the character's point of view as he looks down at a city scene which is really a single textured image.

Figure 13.6: Screenshot of `fieldAndSkyFogged.cpp` with exponential fogging.

13.2 Fog

Fog is an *atmospheric effect* that OpenGL offers ready to use. What fog does is blend objects with a programmer-specified fog color so that the farther away an object is from the viewer the more the fog color dominates, the effect being of objects fading into the distance.

We'll explain how fog is implemented in OpenGL using the program `fieldAndSkyFogged.cpp` as a running example.

Experiment 13.7. Run `fieldAndSkyFogged.cpp`, which is based on our favorite workhorse program `fieldAndSky.cpp`, adding to it a movable black ball and controllable fog. Figure 13.6 is a screenshot.

Press the up/down arrow keys to move the ball. When the program starts there is no fog. We'll describe the fog controls after we discuss their implementation in the following. **End**

The following block of fog-related calls are near the top of the drawing routine:

```
if (isFog) glEnable(GL_FOG); else glDisable(GL_FOG);
glFogfv(GL_FOG_COLOR, fogColor);
glFogi(GL_FOG_MODE, fogMode);
glFogf(GL_FOG_START, fogStart);
glFogf(GL_FOG_END, fogEnd);
glFogf(GL_FOG_DENSITY, fogDensity);
glHint(GL_FOG_HINT, GL_NICEST);
```

Fog is enabled or disabled by a call to `glEnable(GL_FOG)` or `glDisable(GL_FOG)`, respectively. The *fog color* is specified by the statement

`glFogfv(GL_FOG_COLOR,` *fogColor*`)`

where *fogColor* points to the fog color values (a medium gray in the program).

The user can set the *fog mode* to be one of three – `GL_LINEAR`, `GL_EXP` and `GL_EXP2` – by assigning that value to the parameter *fogMode* in

`glFogi(GL_FOG_MODE,` *fogMode*`)`

It is the fog mode which, together with a few associated parameters, determines the "thickness" of the fog. Here's how.

OpenGL invokes the fog mode and the z-distance of an incoming fragment from the eye to compute a number f, called the *fog factor*, which is used to blend the fragment with the fog color. The equation which determines the fog factor f depends on the fog mode as follows:

$$\begin{aligned} \text{GL\_LINEAR:} \quad f &= \frac{fogEnd - z}{fogEnd - fogStart} \\ \text{GL\_EXP:} \quad f &= e^{-(fogDensity\, *\, z)} \\ \text{GL\_EXP2:} \quad f &= e^{-(fogDensity\, *\, z)^2} \end{aligned} \tag{13.4}$$

The values of the parameters *fogStart*, *fogEnd* and *fogDensity* are user-specified as well by the statements:

`glFogf(GL_FOG_START,` *fogStart*`);`
`glFogf(GL_FOG_END,` *fogEnd*`);`
`glFogf(GL_FOG_DENSITY,` *fogDensity*`);`

Their default values are 0, 1 and 1, respectively. One sees from Equations (13.4) that if fog mode is `GL_LINEAR`, then *fogStart* and *fogEnd* give the two endpoints of a linear ramp along the z-axis along which f decreases from 1 to 0; moreover, if fog mode is `GL_EXP` or `GL_EXP2` then *fogDensity* controls the (exponential or doubly exponential) rate of diminishment of f with increasing z – the greater *fogDensity* the more rapidly f diminishes. See Figure 13.7 for sketches of how f changes with z.

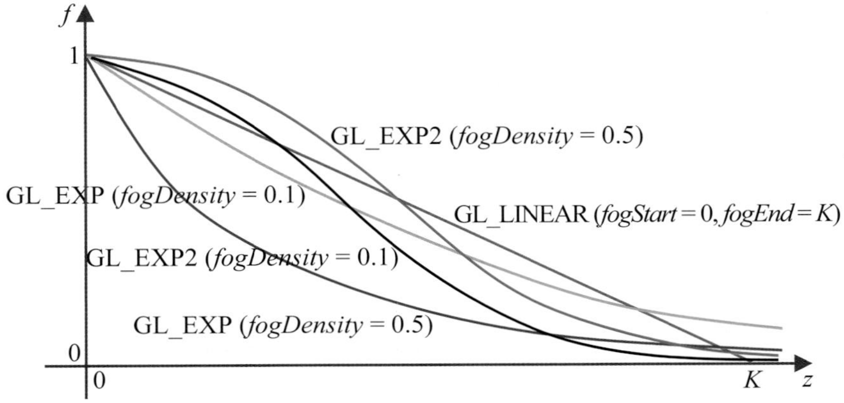

Figure 13.7: f versus z for various parameter values – the graphs are *not* mathematically exact. Values of *fogStart* and *fogEnd* are 0 and K, respectively.

After it's computed from (13.4), the fog factor f is clamped in the range $[0, 1]$. OpenGL then blends the fog color tuple *fogColor* with the incoming fragment's color tuple C_{in}, using the equation

$$C_{dest} = (1 - f)\, fogColor + f\, C_{in}$$

to determine the color tuple C_{dest} of the destination pixel. The smaller the fog factor, therefore, the more the fog dominates and the fragment fades. One sees, as well, from the equations in (13.4) that fog modes `GL_LINEAR`, `GL_EXP` and GL_EXP2 in that order create increasingly thicker fog in general, though the constants in the equations have to be taken into account as well.

Finally,

```
glHint(GL_FOG_HINT, GL_NICEST)
```

is a run-time advisory to OpenGL to use the highest-quality (and computationally most expensive) option available. Instead of `GL_NICEST` one could also pass as parameter values `GL_FASTEST` (computationally least expensive) or `GL_DONT_CARE` (no particular preference).

Interaction: Let's return now to `fieldAndSkyFogged.cpp` to describe how to control the fog. Press the space bar to cycle through the different fog modes. If the fog mode is linear, then the left and right arrow keys change the *fogEnd* parameter, while if it's exponential or doubly exponential, they change the *fogDensity* parameter. The *fogStart* parameter is fixed at 0. Messages on the display indicate the fog mode and parameter values. Press delete at any time to reset the ball. Observe how even a mild fog *depth cues* the sky and the ball when the latter travels away.

Note: The rendition of the ball suffers from the lack of lighting – loss of three-dimensionality in particular – but we wanted to keep the program simple.

Note: If you move the ball far enough it suddenly disappears altogether. That's not because of the fog, but because it's gone behind the sky rectangle!

Exercise 13.11. (Programming) Exercise 12.16 was about enhancing `shipMovie.cpp`. Fog can be used to good effect in the scene as well to depth cue the traveling torpedo and ship.

13.3 Billboarding

The technique of *billboarding* is to simulate a 3D object in a scene by placing an image of it as a texture on a rectangle, the *billboard*. The latter is then continuously rotated to keep it normal always to the direction of the viewer, giving the latter an illusion of the real object. See Figure 13.8.

As long as the object is a peripheral one, e.g., trees, road sign or furniture in the background, the device of holding up a 2D image is often authentic enough, thereby saving on the geometry required to make a "real" 3D version.

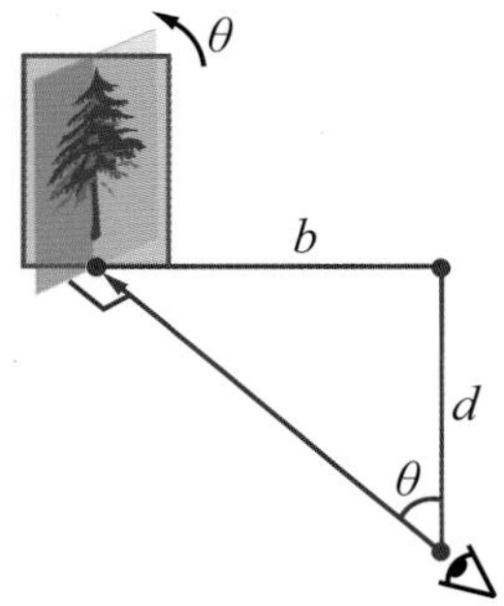

Figure 13.8: Billboarding: the original placement of the billboard (bold border) is rotated so its plane is normal to the direction of the viewer.

Experiment 13.8. Run `billboard.cpp`, where an image of two trees is textured onto a rectangle. Press the up and down arrow keys to move the viewpoint and the space bar to turn billboarding on and off. See Figure 13.9 for screenshots. **End**

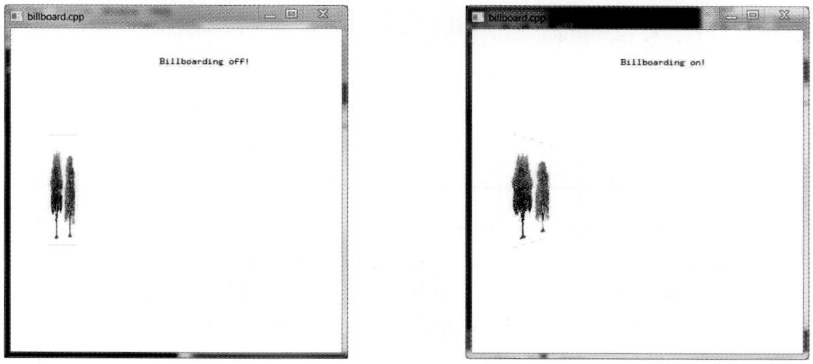

(a) (b)

Figure 13.9: Screenshots of `billboard.cpp`: (a) Billboarding off (b) Billboarding on.

The billboard rectangle of `billboard.cpp` is located in the scene by drawing it first on the xy-plane centered about the z-axis and then translating it d units down the z-direction and b units left (Figure 13.8). Therefore, the angle θ that the billboard must be rotated about the vertical line through its center to keep it normal to the viewer's direction is given by

$$\theta = \tan^{-1}(b/d)$$

This particular rotation is implemented if billboarding is on; otherwise, the billboard remains parallel to the xy-plane (the position indicated by the bordered rectangle in Figure 13.8). The effect of billboarding is marked as the viewer travels "into" the scene by pressing the up arrow key.

Remark 13.3. One way to seamlessly fit a billboard into a scene is to paint its background texels a color matching the scene's backdrop (e.g., both white in the case of `billboard.cpp`). Another is to make the billboard's background texels transparent by setting their alpha values to 0 and then blend the billboard onto the backdrop.

Exercise 13.12. (Programming) Hopefully, somewhere along the line you created a car. Make it now travel down a road against a billboard backdrop of houses and trees.

13.4 Antialiasing Points and Lines, Multisampling Polygons

A straight line segment s specified, say, to be one pixel wide can be rasterized by selecting a set of fragments that best approximates it and setting each to the color specified for s, while unselected fragments remain of the background color. See Figure 13.10(a) for a particularly low res example. Such discrete on/off rasterization protocols are computationally inexpensive, but tend to give poor visual quality at certain alignments of the segment owing to the jaggedness of the rasterization, so-called *jaggies*.

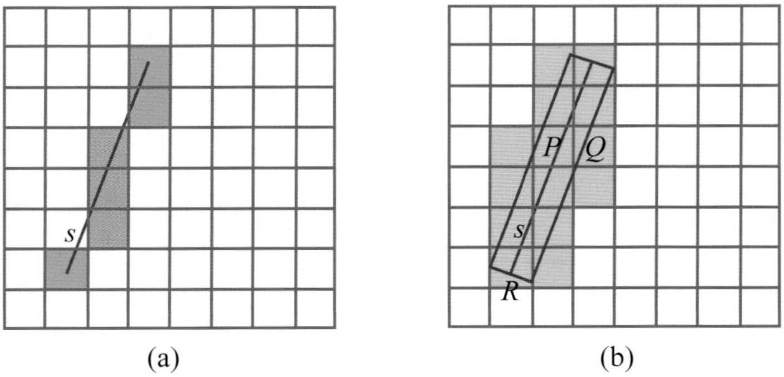

Figure 13.10: (a) Dark fragments represent a rasterization of a line segment s, specified to be one pixel wide (b) Shaded fragments are those that are intersected by the one-pixel wide rectangle R centered on s: the area that R covers of individual fragments, e.g., P and Q, varies.

Jaggies are another example of *aliasing*, a visual artifact that arises because of the limited resolution of the display device (an earlier example of aliasing that we saw was texture shimmering in the previous chapter). Since a line segment not parallel to one of the axes can at best be approximated in a raster, one cannot hope to eliminate jaggies altogether. However, there are techniques to attenuate their visual impact, not surprisingly at the cost of extra computation.

13.4.1 Antialiasing

OpenGL, in particular, offers an *antialiasing* method for line segments based on so-called *coverage* values. Consider again the line segment s. A one-pixel wide rectangle R centered on s intersects the 14 fragments shaded in Figure 13.10(b). However, the area that R covers of each, called its coverage of that fragment, varies. For example, the coverage value of fragment P is more than twice that of Q. When antialiasing is enabled, OpenGL draws the line by setting all of the 14 intersected fragments' color to that of s, then multiplying the alpha value of each with its coverage and, finally, using the resulting weighted alpha to blend the fragment with the corresponding pixel already in the color buffer. The amount of the segment's color, therefore, blended into a destination pixel is proportional to the area of its source fragment covered by R. This has the effect of smoothing out the jaggies.

Points can be antialiased too. The mechanism is simpler – the point is rounded into a disc whose edge is smoothed by blending. The experiment next illustrates antialiasing of both a line and a point. In the process we'll learn an interesting fact about how points are rendered!

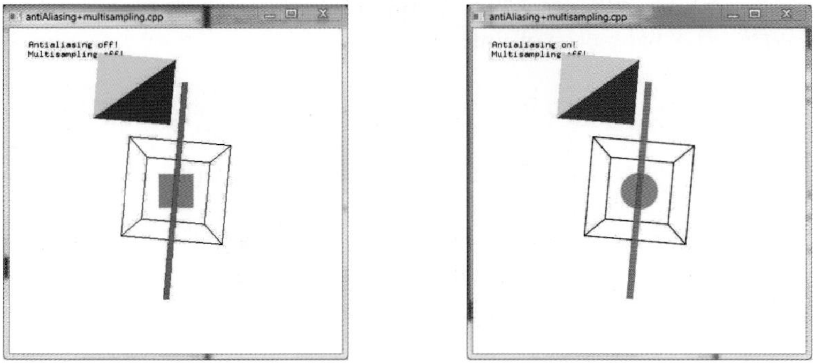

(a) (b)

Figure 13.11: Screenshots of `antiAliasing+multisampling.cpp`: (a) Antialiasing off (b) Antialiasing on. Multisampling off both cases.

Experiment 13.9. Run `antiAliasing+multisampling.cpp`. Ignore the multisampling controls, as well as the blue-yellow rectangle, for now. Focus on the red line segment and the green point, which are both either antialiased or not, 'a' or 'A' toggling between the two modes. The width of the line is changed by pressing 'l/L', while the size of the point with 'p/P'. The scene can be turned by the 'x'-'Z' keys and translated by pressing the arrow and page up/down keys. A wire cube is drawn as frame of reference to view the motion.

Figure 13.11 shows screenshots of antialiasing both off and on. The effect of antialiasing is especially marked when the line is just shy of horizontal or vertical. **End**

Antialiasing is simple to implement in OpenGL. One has to enable blending, of course. The blending factors `GL_SRC_ALPHA` and `GL_ONE_MINUS_-SRC_ALPHA`, used in `antiAliasing+multisampling.cpp`, are the best choices. A high alpha value – we use 0.8 for both line and point – enhances the antialiased effect. Line and point antialiasing are enabled, respectively, with `glEnable(GL_LINE_SMOOTH)` and `glEnable(GL_POINT_SMOOTH)`; they are disabled, of course, by the corresponding `glDisable()` calls. Note that we ask for best possible antialiasing for lines and points with the calls `glHint(GL_LINE_SMOOTH_HINT, GL_NICEST)` and `glHint(GL_POINT_-SMOOTH_HINT, GL_NICEST)`.

Rendering points

Antialiasing, of course, rounds the green point, but observe how its size and shape are never changed by translation or rotation. The reason is that a point is created zero-dimensional, so of zero size, in world space, then projected to the viewing face – equivalently, screen space – and *only there* given size as an $n \times n$ square of pixels (assuming the point is unaliased) centered at the projected point, where n, of course, may be specified by `glPointSize(`n`)`.

Therefore, its position in world space, in particular, how near or far it may be from the viewing space, has no bearing on a point's size or shape on the screen; and, of course, a point's "alignment" by rotation is a non-factor, too.

So, we see that, even though points are grouped with lines and triangles as OpenGL's fundamental drawing primitives, they are of a very different DNA at least when comes the projection step of the rendering process.

13.4.2 Multisampling

One can indeed call `glEnable()` with the parameter `GL_POLYGON_SMOOTH` to antialias polygons, particularly, their edges. However, the problem with this method is that, being based upon blending, polygons need to be properly sorted in the drawing routine as discussed in Section 13.1.3. OpenGL offers another antialiasing method, based on so-called *multisampling*, to avoid this constraint.

In multisampling, color, depth and stencil values are computed at a sample of points in a pixel's area, stored in a dedicated sample buffer, to be subsequently *resolved* into one color to apply to that pixel. Figure 13.12

shows the idea using a simple 2×2 sampling scheme: the 12 shaded fragments generated by the triangle have their color, depth and stencil values sampled at four points each.

Though we'll not go into details of the process of resolving multiple sample values into a pixel value, one can see even from Figure 13.12 how multisampling might be able to diminish jaggies at the edges of a polygon by taking into account sample values both just inside and outside. Let's observe multisampling in practice.

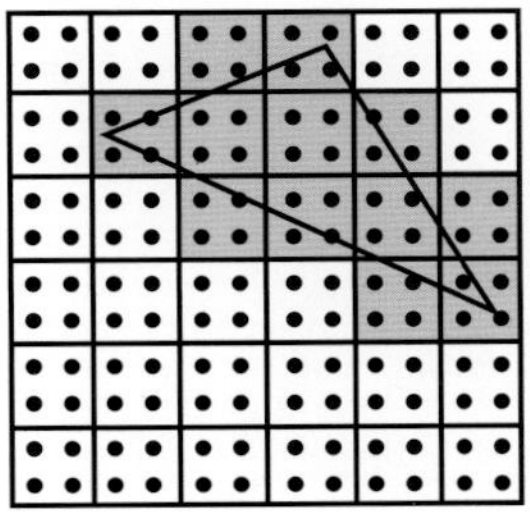

Figure 13.12: Multisampling using a 2×2 sampling scheme.

Experiment 13.10. Fire up again `antiAliasing+multisampling.cpp`. Multisampling is toggled on/off, independently of antialiasing, by pressing 'm' or 'M'.

Multisampling antialiases polygons particularly effectively and its effect in our program is best observed on the boundary of the blue-yellow rectangle, as well as the edge between its two colored halves, particularly, when they are nearly horizontal or vertical.

When multisampling is enabled, lines and points are antialiased regardless if `GL_POINT_SMOOTH` or `GL_LINE_SMOOTH` have been enabled, which you can see as well. **End**

Multisampling is simple to get going. First, one has to create an OpenGL window which supports multisampling: passing `GLUT_MULTISAMPLE` as a parameter to `glutInitDisplayMode()` does the trick for us. Next, multisampling is enabled and disabled, respectively, with `glEnable()` and `glDisable()` given `GL_MULTISAMPLE` as the parameter. That's it!

Note: The number of sample buffers, shown at the top of program's window when multisampling is enabled, should be at least one, or there is effectively no multisampling and you may need to check the settings of your graphics card.

Keep in mind that multisampling comes at a cost because of the additional processing per fragment. So, if you are particularly performance-conscious it might make sense to enable multisampling only for polygons and disabling it for lines and points, drawing the latter with only smoothing enabled.

Exercise 13.13. The unaided human eye can resolve to a minimum size of approximately 0.1 mm. (about 0.004 inch). So, what resolution levels must a 22 inch desktop monitor reach in terms of number of pixels by number of pixels for aliasing problems (and antialiasing algorithms) to go the way of the floppy disc? What's your best guess as to how long it will take for technology to get there? (By the way, do you even know what a floppy disc is, young reader?! If not, look it up in Wikipedia.)

13.5 Point Sprites

So, from the last section we know that points are created dimensionless in world space and only bestowed shape as a square of pixels after arrival by projection into screen space. Not particularly interesting is this? However, one can bring points to life by turning them into so-called *point sprites*! A point sprite is simply a point, usually of a large size, with a texture mapped on to it.

Point sprites are particularly useful in manufacturing *particle systems*, where a swarm of particles on the screen create a certain visual effect, e.g., smoke, sparks, starry sky, and so on. Because they are always flat on the display facing the viewer, point sprites can be invoked as a sort of poor man's billboard as well. Point sprites are easy to code. Let's have a look at the real thing.

Figure 13.13: Screenshot of `pointSprite.cpp`.

Experiment 13.11. Fire up `pointSprite.cpp`. The space bar toggles animation on and off. The particle system of six sprites of fluctuating size, imprinted with the same star texture, spinning in a circle is simple-minded, though, hopefully, indicative of the possibilities. Figure 13.13 is a screenshot. **End**

In addition to the usual commands to prepare a texturing environment, there are three new ones in the setup routine of `pointSprite.cpp`. The call

```
glTexEnvi(GL_POINT_SPRITE, GL_COORD_REPLACE, GL_TRUE)
```

enables texture interpolation across point sprites, while the parameter `GL_LOWER_LEFT` in

```
glPointParameteri(GL_POINT_SPRITE_COORD_ORIGIN, GL_LOWER_LEFT)
```

causes the texture t-coordinate to increase from 0 to 1 from bottom to top of the sprite. The texture s-coordinate always increases 0 to 1 from left to right.

Finally, as one would expect, `glEnable(GL_POINT_SPRITE)` enables point sprites.

If you were wondering how textures can be painted on round antialiased point sprites, the answer is that point sprites *cannot* be antialiased.

Exercise 13.14. (Programming) The only other option for the second parameter of `glPointParameteri(GL_POINT_SPRITE_COORD_ORIGIN, GL_LOWER_LEFT)` is `GL_UPPER_LEFT`, causing the texture t-coordinate to increase from 0 to 1 from top to bottom. The s-coordinate always increases from 0 to 1 from left to right of the sprite.

Try `GL_UPPER_LEFT` instead of `GL_LOWER_LEFT`. It will be hard to spot the difference with the rather symmetric star texture, but using `launch.bmp` instead should make it clear.

Exercise 13.15. (Programming) Create the effect of sparks flying by coding up a particle system of sprites.

13.6 Environment Mapping

The goal of *environment mapping* is to simulate an object reflecting its surrounding, e.g., a shiny kettle reflecting the kitchen or a well-polished car reflecting the street. As the environment can be seen by reflection off the object, it is said to be mapped onto the object. An approach to environment mapping originally invented by Blinn and Newell [19] in the seventies is still popular today because of its ease of implementation.

The Blinn-Newell method makes clever use of textures, but the basic idea is not hard. An image of the environment (presumed static) is captured in a texture or multiple textures. Subsequently, the particular texture and texture coordinates used to paint a point on the environment-mapped object are determined from the position of the viewer relative to the object.

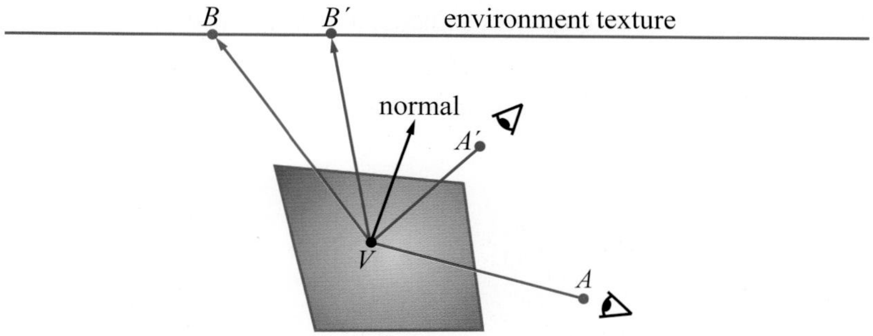

Figure 13.14: Blinn-Newell environment mapping principle: texture coordinates for a vertex V on an environment-mapped surface are obtained from the point on the texture image struck by the reflected ray originating from the eye.

Figure 13.14 illustrates the principle. The texture coordinates at the vertex V of an environment-mapped quad are determined by the point of the environment – more precisely, the corresponding point of the environment texture – seen by the viewer by reflection off the object. For example, when the viewer is at A, V is painted with the color values at B (red in the figure); when she moves to A', those of B' are used (green). The crux of the Blinn-Newell approach then is to *dynamically* compute texture coordinates, based on the laws of reflection, as the viewpoint changes.

OpenGL provides support for two methods of environment mapping: *sphere mapping* and *cube mapping*. Both are based on the Blinn-Newell approach, the difference being in the way that the environment is captured on texture and that texture coordinates are computed. OpenGL provides *automatic texture coordinate generation* for either method. We'll discuss sphere mapping in fair detail.

We'll, however, split our presentation into implementation and theory, as the former is straightforward and what the practitioner needs most to grasp, while the latter is rather more theoretical and demanding.

13.6.1 Sphere Mapping

Getting It to Work

Implementing sphere mapping using OpenGL is simple, as the following program shows.

Figure 13.15: Screenshot of `sphereMapping.cpp`.

Experiment 13.12. Run `sphereMapping.cpp`, which shows the scene of a shuttle launch with a reflective rocket cone initially stationary in the sky in front of the rocket. Press the up and down arrow keys to move the cone. As the cone flies down, the reflection on its surface of the launch image changes. Figure 13.15 is a screenshot as it's about to crash to the ground. **End**

The two commands

```
glTexGeni(GL_S, GL_TEXTURE_GEN_MODE, GL_SPHERE_MAP);
glTexGeni(GL_T, GL_TEXTURE_GEN_MODE, GL_SPHERE_MAP);
```

in the initialization routine of `sphereMapping.cpp` ask OpenGL to use functions from its library to generate the s and t texture coordinates for sphere mapping.

The pair of commands

```
glEnable(GL_TEXTURE_GEN_S);
glEnable(GL_TEXTURE_GEN_T);
```

and its inverse

```
glDisable(GL_TEXTURE_GEN_S);
glDisable(GL_TEXTURE_GEN_T);
```

in the drawing routine, bracketing the drawing of the cone, enable and disable the use of these functions. That's pretty much all there is to implementing a sphere map using OpenGL! Note that at the time sphere mapping is activated the currently bound texture is the launch image, which, of course, is why it is reflected in the cone.

Now, a reader watching the cone as it zooms down may be wondering how authentic actually is the reflection. Good question, and it leads us to investigate how OpenGL computes sphere-mapped texture coordinates.

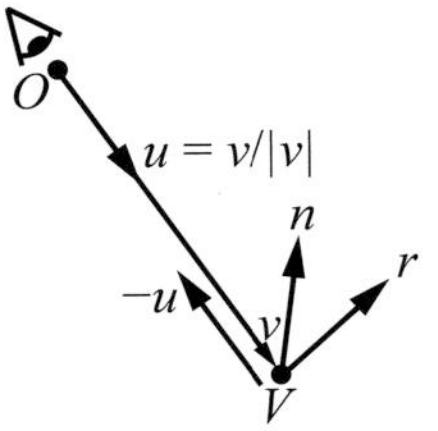

Figure 13.16: The vectors involved in generating texture coordinates.

How It Works

This part is fairly mathematical. If your interest is practical and limited to using the technique, you can safely skip it and jump to the part on preparing the environment texture.

Here's how sphere-mapped texture coordinates are generated at a vertex V. See Figure 13.16. The unit vector u from the eye (the origin O in OpenGL) toward V is $v/|v|$, where v is the position vector of V, assuming, of course, that $v \neq 0$. The unit eye direction vector from V then is $-u$. The unit normal n at V is user-provided.

OpenGL computes the reflection vector r, the unit vector in the direction that a hypothetical ray from the eye is reflected at V, with the help of the following equation (obtained by replacing light direction vector l with eye direction vector $-u$ in the formula of Exercise 11.4):

$$r = u - 2(n \cdot u)n$$

Suppose, then, OpenGL finds that $r = (r_x, r_y, r_z)$. Computed next is the quantity

$$m = 2\sqrt{r_x^2 + r_y^2 + (r_z + 1)^2}$$

Finally, the texture coordinates at V are calculated as

$$s = \frac{r_x}{m} + \frac{1}{2} \quad \text{and} \quad t = \frac{r_y}{m} + \frac{1}{2}$$

Whew!

If we parse the expressions for s and t carefully, though, it'll not be hard to understand the game plan. Using the expression for m above, write

$$s = \frac{1}{2}\frac{r_x}{\sqrt{r_x^2 + r_y^2 + (r_z + 1)^2}} + \frac{1}{2} \quad \text{and} \quad t = \frac{1}{2}\frac{r_y}{\sqrt{r_x^2 + r_y^2 + (r_z + 1)^2}} + \frac{1}{2}$$

or

$$s = \frac{1}{2}R_x + \frac{1}{2} \quad \text{and} \quad t = \frac{1}{2}R_y + \frac{1}{2} \tag{13.5}$$

where the variables

$$R_x = \frac{r_x}{\sqrt{r_x^2 + r_y^2 + (r_z + 1)^2}} \quad \text{and} \quad R_y = \frac{r_y}{\sqrt{r_x^2 + r_y^2 + (r_z + 1)^2}} \tag{13.6}$$

Once we understand what the mapping

$$(r_x, r_y, r_z) \mapsto (R_x, R_y)$$

does geometrically the rest will be straightforward.

The reflection vector $r = (r_x, r_y, r_z)$ is the position vector of some point, say P, on the unit sphere S centered at the origin. See Figure 13.17(a). Now, the position vector of P with respect to the *south pole* $(0, 0, -1)$ of S is $r' = (r_x, r_y, r_z + 1)$. And r' normalized is the vector

$$r'' = \frac{1}{\sqrt{r_x^2 + r_y^2 + (r_z + 1)^2}}(r_x, r_y, r_z{+}1) = \left(R_x, R_y, \frac{r_z + 1}{\sqrt{r_x^2 + r_y^2 + (r_z + 1)^2}}\right) \tag{13.7}$$

In fact, r'' itself is the position vector, with respect to the south pole, of the point Q of intersection of the line from the south pole to P with the

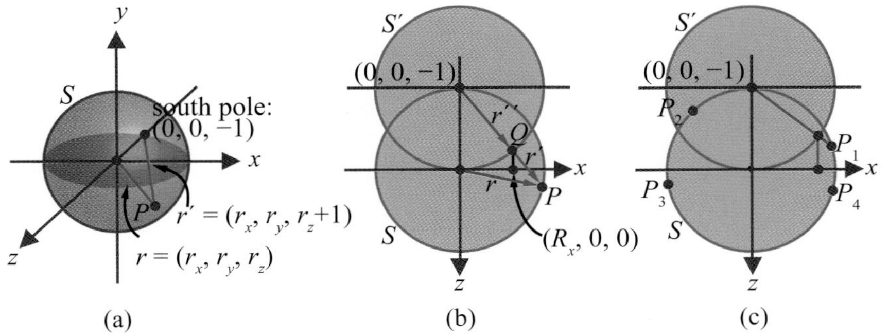

Figure 13.17: Determining R_x from r.

unit sphere S' centered at the pole. S' is not drawn in Figure 13.17(a), but Figure 13.17(b) shows both S and S' in section along the xz-plane (for this particular drawing we assume that P lies on this section). R_x being the x-value of Q by (13.7), the projection of Q on the x-axis in Figure 13.17(b) is $(R_x, 0, 0)$ (Q's y-value R_y is 0, of course, as it's on the xz-plane). Here's an exercise to reinforce your understanding of the preceding construction to find R_x from r.

Exercise 13.16. For each point $P_i, 1 \leq i \leq 4$, in Figure 13.17(c), use a ruler and pencil to draw the corresponding point $(R_x, 0, 0)$ on the x-axis.

Part answer: Red lines indicate the construction for P_1.

The reader may now agree that, at least as P varies over the xz-section of S, $(R_x, 0, 0)$ varies between $(-1, 0, 0)$ and $(1, 0, 0)$ and, correspondingly, R_x between -1 and 1. Moreover, the closer P gets to the south pole the closer is R_x to -1 or 1, depending on which side of the pole P is. However, P should never be *at* the south pole, for, otherwise, the construction to determine R_x breaks down. It follows that R_x itself reaches neither value -1 nor 1. In fact, considering now all of the sphere S, not just its xz-section, it's not hard to see that R_x varies over the open interval $(-1, 1)$ as P varies over S minus its south pole.

The mapping from P to R_y is similar. Therefore, as P moves over S minus its south pole, (R_x, R_y) moves within the interior of the square $[-1, 1] \times [-1, 1]$. For an even better understanding, let's determine analytically the dependence of (R_x, R_y) on P.

Choose a Z in $-1 < Z \leq 1$. The plane $z = Z$ intersects S in a latitudinal circle

$$x^2 + y^2 + Z^2 = 1 \quad \text{or} \quad x^2 + y^2 = 1 - Z^2$$

Now, from (13.6) we have that

$$R_x^2 + R_y^2 = \frac{r_x^2 + r_y^2}{r_x^2 + r_y^2 + (r_z + 1)^2}$$

Therefore, if P lies on the latitudinal circle $x^2 + y^2 = 1 - Z^2$, so that $r_x^2 + r_y^2 = 1 - Z^2$ and $r_z = Z$, then the preceding equation says that (R_x, R_y) lies on the circle

$$R_x^2 + R_y^2 = \frac{1 - Z^2}{1 - Z^2 + (1+Z)^2} = \frac{1 - Z^2}{2 + 2Z} \tag{13.8}$$

Now, we can see how (R_x, R_y) varies with $r = (r_x, r_y, r_z)$ as we had set out to. In fact, we'll draw a picture. See the two diagrams on the left of Figure 13.18.

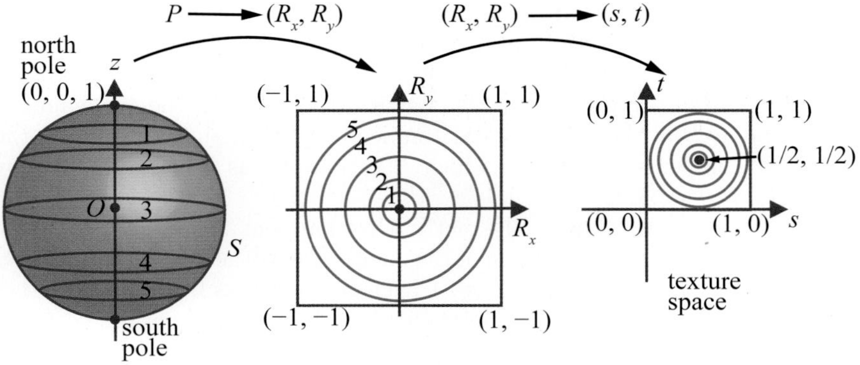

Figure 13.18: The maps $P \mapsto (R_x, R_y)$ and $(R_x, R_y) \mapsto (s, t)$.

Keep in mind that r is P's position vector, the latter varying over S. Each latitudinal circle on S (now drawn upright at left with the north pole at the top to better see these circles) maps to a circle centered at the origin and in the square $[-1, 1] \times [-1, 1]$ in $R_x R_y$-space (drawn in the middle). In particular, the north pole maps to the origin, and latitudinal circles from the north pole downward map to increasingly larger circles inside $[-1, 1] \times [-1, 1]$. Five pairs of corresponding circles have been drawn and labeled similarly in the two diagrams. As the latitudinal circles approach the south pole, the mapped circles draw nearer and nearer to the containing square.

Exercise 13.17. What is the radius of the circle to which the equator maps? What are the radii of the images of the latitudinal circles 45°N and 60°S?
Hint: Equation (13.8) gives the radius of the circle in $R_x R_y$-space, which is the image of the latitudinal circle at $z = Z$. For example, the latitudinal circle 45°N has z-value $\sin 45° = 1/\sqrt{2}$, so plug $Z = 1/\sqrt{2}$ into (13.8) to find the radius of its mapped circle in $R_x R_y$-space.

Exercise 13.18. How do longitudinal great circles on S map?
Hint: Straight lines through the origin in $R_x R_y$-space ....

The final transformation from R_xR_y-space to st-space (texture space) is simple. See the rightmost two diagrams of Figure 13.18. (R_x, R_y) is mapped to $(\frac{1}{2}R_x + \frac{1}{2}, \frac{1}{2}R_y + \frac{1}{2})$ via Equations (13.5), which linearly transform the square $[-1, 1] \times [-1, 1]$ in R_xR_y-space to the unit square $[0, 1] \times [0, 1]$ in texture space. The images in texture space of the five circles in R_xR_y-space are shown as well in the rightmost diagram.

Bottom Line

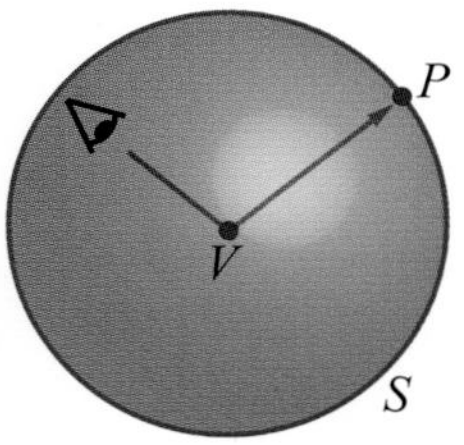

Figure 13.19: The environment S around a perfect mirror vertex V.

Time for a wrap-up in plain English. If vertex V were on a perfect mirror and the environment around it arranged along the unit sphere S centered at V, then the eye would see the point P where S is intersected by the reflection of the line of sight at V (see Figure 13.19). However, OpenGL's only knowledge of the environment is from a user-provided texture occupying a unit square in texture space. So what it does is this: if the eye wants to see the point P in the spherical environment, OpenGL shows it instead the point (s, t) in texture space to which P is mapped as described above by $P \mapsto (R_x, R_y) \mapsto (s, t)$.

The calculations above tell exactly what happens in physical terms. If the eye asks to see the north pole of the environment, then it's shown instead the center of the texture. As the eye travels to see points farther and farther from the north pole, it's shown points farther and farther from the center of the texture. Precisely, latitudinal circles in the environment are replaced for viewing by circles in the texture centered at its middle.

Preparing the Environment Texture

Given this sphere-mapped scheme to present the environment to the viewer via a texture, what is the right way to prepare the texture? Practically speaking, how then should one photograph the environment in order to create the texture image? Comparing the left and right diagrams of Figure 13.18 suggests an answer. The camera should be located at the origin O pointing up the z-axis toward the north pole and have a *very* wide-angle lens; in fact, it would be helpful if the field of view were nearly 360°! Of course, this is impossible, but a fairly wide-angle picture taken with a camera located in the vicinity of the object to be environment mapped, focused up the z-axis of world space, should be good.

Remark 13.4. Since the texel used depends only on the value of the reflection vector at a vertex, and not the vertex's location, reflections in parallel directions appear the same at all vertices. Practically, this means that the environment-mapped object should be small compared to its surroundings for authenticity.

Remark 13.5. Some practitioners advocate the application of filters to the texture prior to sphere mapping. For example, NeHe [102] suggests using the *spherizing* filter (available, e.g., in Adobe's Photoshop software).

Exercise 13.19. (**Programming**) Sphere map the torpedo in `shipMovie.cpp`.

It's fun experimenting with sphere mapping – as with any special effect – as long as you remember that in CG authenticity lies in the eyes of the beholder. If it looks real then it *is* real. Which still leaves unanswered the question that started our mathematical investigation in the first place: how authentic is the reflection in the cone of `sphereMapping.cpp`? But perhaps the reader at this point has an opinion already!

13.6.2 Cube Mapping

Cube mapping is even simpler than sphere mapping and often more authentic. The only difficulty with implementing it is that the environment has to be captured on multiple textures. In fact, the environment is imagined to be a cube and the user asked to provide an image of each of the six faces. Figure 13.20 gives the idea.

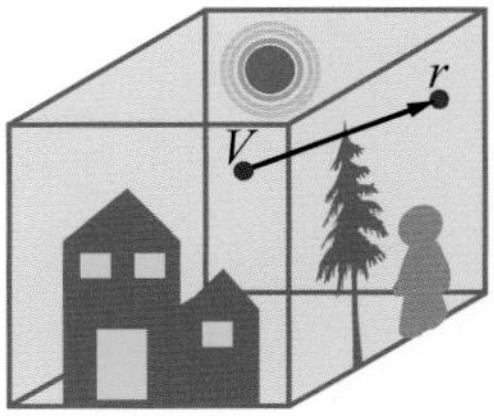

Figure 13.20: Cube mapping.

In the case of cube mapping, given the reflection vector r, first must be computed which of the six faces is struck by the reflected ray, and then the actual point, to extract color values from the texel at that location. As for sphere mapping, there is support in OpenGL for cube mapping. OpenGL automatically generates cube-mapped texture coordinates once the user loads the six environment textures. We'll leave cube mapping at this, not pursuing the topic any further, referring instead the interested reader to the red book.

13.7 Stencil Buffer Techniques

A space in memory reserved for pixel-related data is called a *buffer*. A computer system can have multiple buffers of different sizes for various purposes, though a given buffer will have an equal amount of space - in particular, the same number of bits - assigned to each pixel. We'll briefly introduce all the system buffers OpenGL supports and their uses before focusing in particular on the stencil buffer.

13.7.1 OpenGL Buffers

An OpenGL system can support the following types of buffers:

- Color
- Depth
- Stencil
- Accumulation

We are already familiar with the first two kinds and they are, in fact, the most commonly used of the buffers.

A *color buffer* stores primary RGB color values and the alpha A value, typically to a precision of 8 bits each, for a total of 32 bits per pixel. Color buffers are the only ones of the buffers the user can directly draw into, and it is the final RGB values in some particular color buffer which are *flushed* to the screen for viewing.

In fact, there may be more than one color buffer. A double-buffered system has at least two – front (viewable) and back (drawable) – critical to smooth animation, as we learned in Section 4.5.1. If stereoscopic viewing is supported, there will be left and right color buffers, possibly even front-left, back-left, front-right and back-right, if combined with double buffering. The left buffers are shown to the left eye and the right ones to the right eye, typically with the user wearing special glasses. If the images in the left and right buffers are of the same scene projected toward two points slightly offset one from the other, as are human eyes, a perception of 3D is created.

Additionally, a system may have so-called auxiliary color buffers, usually in hardware, to store intermediate steps of a complicated rendering process.

We are also familiar with the *depth buffer* (or *z-buffer* as it's popularly called) which stores depth information, usually a 24-bit integer, per pixel. When depth testing is enabled, the depth buffer helps sort out objects according to their depth from the viewer along lines of sight, permitting nearer objects to obscure further ones in the rendering phase.

Less familiar to the reader might be the *stencil buffer*. This is a buffer used to tag pixels in the color buffer. The stencil buffer most often contains 8 bits, called tags, per pixel. The typical way to use the stencil buffer is to set the tags in a first phase when nothing is drawn to the screen and then employ these tags as controls in a second phase when actual drawing takes place. The tag values allow the user to constrain drawing to limited portions of the screen, making possible various creative applications. We'll be studying the stencil buffer in fair detail shortly.

The *accumulation buffer* is yet another buffer used for special effects. Think of the accumulation buffer as a giant color buffer. It contains RGBA values per pixel as well, but often to a much higher precision. Typical accumulation buffers dedicate 16 bits to each of RGBA, for a total of 64 bits per pixel.

The primary use of the accumulation buffer is to composite several drawings into one. The accumulation buffer cannot be directly drawn into, nor can it be directly displayed. Rather, drawings are made to color buffers and combined one by one into the accumulation buffer. Successive incoming drawings can be combined with the one currently resident in the accumulation buffer in various ways, e.g., added or multiplied, the latter explaining the need for higher precision. The final composited drawing is returned from the accumulation buffer to a color buffer for display.

The collection of all buffers in a system is called its *frame buffer*. Figure 13.21(a) shows a frame buffer containing all four kinds of buffers supported by OpenGL. The size of each constituent buffer – which is its *width* × *height* as an array of bit strings – matches that of the display device. This size is also called the *resolution* of the frame buffer.

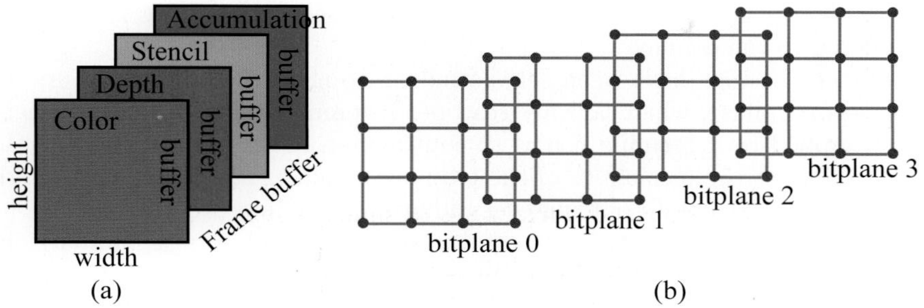

Figure 13.21: (a) A complete frame buffer (b) A 4 × 4 buffer with 4-bit precision as a stack of four bitplanes: points represent bits.

The number of bits per pixel in a buffer determines the buffer's *precision*. Bits in the buffer in some particular bit position, between 0 to *precision* − 1, form an array called a *bitplane*. So, of course, a buffer's precision is identical to its number of bitplanes. Figure 13.21(b) depicts a 4 × 4 buffer of 4-bit precision as a stack of bitplanes.

Exercise 13.20. Suppose we want a graphics card which has four 32-bit precision color buffers for stereoscopic viewing with double-buffering, an auxiliary color buffer of similar precision, a 24-bit depth buffer, an 8-bit stencil buffer and a 64-bit accumulation buffer, all to support a 1024 × 768 resolution display. How much on-card memory are we asking for?

Remark 13.6. A given OpenGL implementation may not support all the possible buffers. One can determine which are supported, as well as the number of bitplanes in each, with the help of `glGet*v()` calls, e.g., `glGetIntegerv(GL_DEPTH_BITS, *pointer)` returns the number of bitplanes in the depth buffer. Check the blue book for the specs for such calls.

13.7.2 Using the Stencil Buffer

Applications using the stencil buffer usually set the *stencil bits*, or *tags* as they are called – typically there being 8 for each pixel – in a first phase, by means of a *stencil test* applied to each incoming fragment. Stencil testing is enabled by calling `glEnable(GL_STENCIL_TEST)`. The particular test applied depends on a `glStencilFunc()` call. The stencil test is applied in the graphics pipeline just before the depth test and only if a fragment passes the stencil test does it proceed to the depth test. If a fragment fails either test then it is discarded from the drawing pipeline.

How the incoming fragments set tags depends on a `glStencilOp()` call paired with the `glStencilFunc()` call. Incoming fragments that fail the stencil test, those that pass the stencil test but fail the depth test and those that pass both tests can set tags differently. The color buffer, typically, is disabled during the tag-setting first phase so that nothing is actually rendered to the display.

Once the tags have been set, actual drawing to the display occurs in the second phase, when tags are read per incoming fragment and a stencil test applied to determine if it is to continue on down the pipeline. We'll describe next the mechanics of the twin commands `glStencilFunc()` and `glStencilOp()` before putting everything together in a program.

The `glStencilFunc(`*func*`, `*ref*`, `*mask*`)` command sets a comparison function *func* to use in the stencil test, as well as a reference value *ref* to compare the stencil tag with. For example, if *func* is `GL_LESS`, then the test passes if *ref* is less than the value of the stencil tag. The reference value is clamped to the range $[0, 2^k - 1]$ if the stencil buffer contains k bitplanes and the stencil tag, too, interpreted as an integer in the same range. However, prior to comparison, the *mask* is bitwise ANDed with both the reference value and the stencil tag. Effectively, therefore, comparison is between the two integers made from bits in the reference value and stencil tag, respectively, at positions corresponding to the 1-bits in the mask.

Example 13.4. If the call is `glStencilFunc(GL_EQUAL, 0xFF, 0x3F)` and the stencil tag corresponding to a fragment is `0xBF`, then the fragment passes the stencil test because only the lower six bits of the mask are 1, and the reference value and the given stencil tag, in fact, agree in each of these positions.

We henceforth will use always a mask value of 1 (=00000001, assuming an 8-bit stencil buffer), which means that the lowest bit of the reference value is compared with the lowest bit of the stencil tag, other bits of no matter.

Example 13.5. The call `glStencilFunc(GL_EQUAL, 1, 1)` allows a fragment to pass the stencil test only if the lowest bit in its corresponding stencil tag equals 1. Suppose this, in fact, is the call prior to drawing the square R consisting of four fragments, as shown in Figure 13.22(a), and that the contents of the stencil buffer are as in Figure 13.22(b) (only the lowest bit of each tag is drawn). Then the left two fragments of R pass the stencil test and proceed on to the depth test, while the right two fail and are ejected from the pipeline. Ignore the right grid for now.

The call `glStencilOp(`*fail*`, `*zfail*`, `*zpass*`)`, paired with `glStencil-Func()`, specifies how a stencil tag is updated following a stencil test. The values of the three parameters *fail*, *zfail* and *zpass* determine the update action

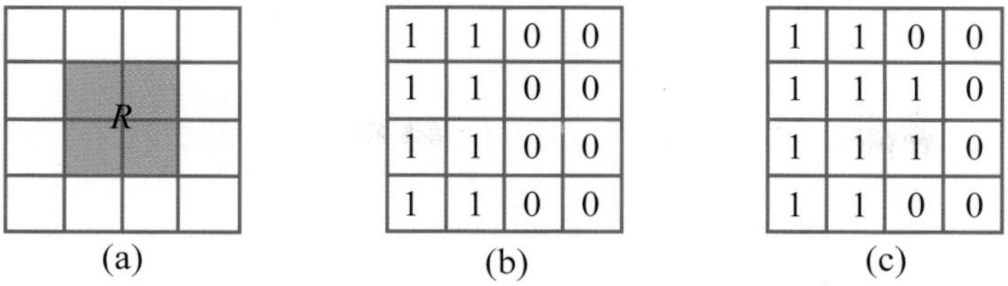

Figure 13.22: (a) Square R, which is drawn after calls to `glStencilFunc(GL_EQUAL, 1, 1)` and `glStencilOp(GL_REPLACE, GL_REPLACE, GL_REPLACE)` (b) Stencil buffer configuration before R is drawn (c) Stencil buffer configuration before R is drawn. Only each lowest bit in the stencil buffer is shown.

in case the fragment fails the stencil test, passes the stencil test but fails the ensuing depth test, and passes both tests, respectively. Figure 13.23 indicates the scheme. Values that we'll use are `GL_KEEP`, `GL_REPLACE` and `GL_INVERT` which, respectively, keep the current stencil tag unchanged, replace it with the reference value, and invert it bitwise (for the full list of possible parameter values consult the red book). Note that *fail*ed and *zfail*ed fragments are ejected from the pipeline.

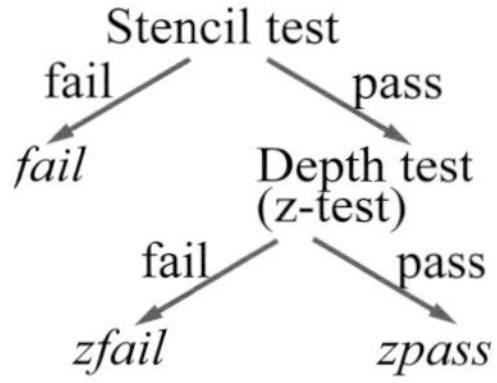

Figure 13.23: Potential outcomes for a fragment through the stencil and depth tests.

Example 13.6. The call `glStencilOp(GL_REPLACE, GL_REPLACE, GL_REPLACE)` causes the stencil tag to be replaced with the reference value in all cases. If this call were indeed paired with `glStencilFunc(GL_EQUAL, 1, 1)` prior to drawing the square R of Figure 13.22(a), then the stencil buffer would be updated as in Figure 13.22(c).

Exercise 13.21. Determine how the stencil buffer would be updated in the situation of Figure 13.22 if the call were `glStencilOp(GL_INVERT, GL_REPLACE, GL_KEEP)` instead of `glStencilOp(GL_REPLACE, GL_REPLACE, GL_REPLACE)`, and it was known that rectangle fragments all fail the depth test if they come to it. Assume that the call `glStencilFunc(GL_EQUAL, 1, 1)` remains.

Drawing reflections in a constrained area is a canonical application of the stencil buffer which we illustrate next.

Figure 13.24: Screenshot of `ballAndTorusStenciled.cpp`.

Experiment 13.13. Run `ballAndTorusStenciled.cpp`, based on `ballAndTorusReflected.cpp`. The difference is that in the earlier program the entire checkered floor was reflective, while in the current one the red floor is non-reflective except for a mirror-like disc lying on it. Pressing the arrow keys moves the disc and pressing the space key starts and stops the ball moving. As you can see in the screenshot of Figure 13.24, the ball and torus are reflected only in the disc and nowhere else. **End**

The effect of restricted reflection in `ballAndTorusStenciled.cpp` is obtained using four successive pairs of `glStencilFunc()` and `glStencilOp()` calls in the drawing routine. The first pair

```
glStencilFunc(GL_ALWAYS, 1, 1);
glStencilOp(GL_REPLACE, GL_REPLACE, GL_REPLACE);
```

causes the next drawing statement `drawReflectiveDisc(xVal, zVal)` to create a "mask" of the disc (nothing to do with the mask parameter of `glStencilFunc()`!) in the stencil buffer with 1's at positions corresponding to a disc fragment and 0's elsewhere. The reason for this is that the `GL_ALWAYS` parameter value of `glStencilFunc()` ensures that every fragment of the disc passes the stencil test, while the second two `GL_REPLACE` values of `glStencilOp()`, for *zfail* and *zpass*, respectively, ensure that the stencil tag corresponding to a disc fragment is always replaced with the reference value 1, regardless the result of the depth test. The remaining stencil tags remain at their clearing value 0 set by the call `glClearStencil(0)` in the initialization routine.

Note, as well, that both color and depth buffers are disabled prior to the drawing statement with appropriate `glColorMask()` and `glDepthMask()` commands, so that only the stencil buffer is updated and nothing is drawn to the screen by the first `drawReflectiveDisc(xVal, zVal)`.

Remark 13.7. It's because the stencil buffer itself cannot be directly drawn into, that the mask of the disc has to be created indirectly with the help of the buffer-manipulating command `glStencilOp()`.

Once the mask of the disc in the stencil buffer has been created, actual drawing to the window commences. The color and depth buffers are accordingly enabled next. The second pair of stencil-buffer manipulating calls

```
glStencilFunc(GL_EQUAL, 1, 1);
glStencilOp(GL_KEEP, GL_KEEP, GL_KEEP);
```

causes the subsequent drawing statement `drawFlyingBallAndTorus()` to draw the reflected ball and torus in the mask area of the disc, because only fragments corresponding to this area pass the stencil test and proceed on down the pipeline. The contents of the stencil buffer are kept unchanged.

The next drawing statement, `drawReflectiveDisc(xVal, zVal)`, actually draws (or, rather, blends) the disc onto the reflected ball and torus.

The third pair of stencil-buffer manipulating calls

```
glStencilFunc(GL_NOTEQUAL, 1, 1);
glStencilOp(GL_KEEP, GL_KEEP, GL_KEEP);
```

prevents the red quad drawn next from erasing the disc by allowing drawing only outside the area corresponding to the disc.

The final pair of stencil-buffer manipulating calls

```
glStencilFunc(GL_ALWAYS, 1, 1);
glStencilOp(GL_KEEP, GL_KEEP, GL_KEEP);
```

allows the real ball and torus to be drawn by a `drawFlyingBallAndTorus()` statement regardless of the stencil tags.

Exercise 13.22. (**Programming**) Reverse the roles of the floor and disc in `ballAndTorusStenciled.cpp`. In particular, make the floor reflective, while a movable red disc blocks reflection.

Exercise 13.23. (**Programming**) Draw the glass door of a roadside building reflecting a passing vehicle – all shapes in your scene being simple and boxy.

For more about drawing shadows and reflections with the help of the stencil buffer read Mark Kilgard's tutorial [78].

Scissor Test

The *scissor test* is simply a stencil test applied to a rectangular region of the display window. The command

glScissor(x, y, *width*, *height*)

specifies the lower left corner (x, y) of a *scissor rectangle* (also called *scissor box*), as well as its *width* and *height*, all in windows coordinates, such that only fragments inside the rectangle pass the scissor test. Scissoring is enabled and disabled, respectively, by `glEnable(GL_SCISSOR_TEST)` and `glDisable(GL_SCISSOR_TEST)`. The reason for singling out this special case of stencilling as a separate test is that it can be highly optimized in the GPU.

Exercise 13.24. (**Programming**) Add in `ballAndTorusStenciled.cpp` a vertical rectangular mirror on the wall directly behind the torus with help of a scissor test.

13.8 Image and Pixel Data Manipulation

OpenGL has powerful commands to manipulate images on the screen. Let's get to the first one `glDrawPixels()`, which draws a 2D array of pixels from client memory onto the frame buffer at the current raster position, in a program which applies this command to move a texture image around the OpenGL window.

Figure 13.25: Screenshot of `imageManipulation.cpp`.

Experiment 13.14. Run `imageManipulation.cpp`. An image of the numeral 1 appears at the bottom left of the OpenGL window. Clicking the mouse left button anywhere on the window will move the image to that location, while you can, as well, drag the image with the left button pressed. Figure 13.25 is a screenshot of the initial configuration. **End**

The code is fairly simple to understand. Following the loading of the image of the "1" into `image[0]` in the initialization routine, `memcpy()` is invoked to copy the image data to `textureImage`, a global 2D array of RGBA data.

In the drawing routine, the command

```
glDrawPixels(TextureWidth, TextureHeight, GL_RGBA,
             GL_UNSIGNED_BYTE, textureImage)
```

draws the pixel data contained in `textureImage` onto the OpenGL window, the lower-left of the image being positioned at the current raster position, which itself is specified by the statement

```
glRasterPos2i(rasterX, rasterY)
```

Generally, `glDrawPixels(`*width*`,` *height*`,` *format*`,` *type*`,` **pixelData*`)` reads a *width* × *height* arrays of pixel data from the memory location pointed by **pixelData* and draws it onto the OpenGL window with lower-left corner at the current raster position. The values of *format* and *type* specify, respectively, the kind of pixel data (RGBA in our case) and its data type (unsigned 8-bit integer in our case).

The two mouse function routines simply update the current raster position to track the mouse.

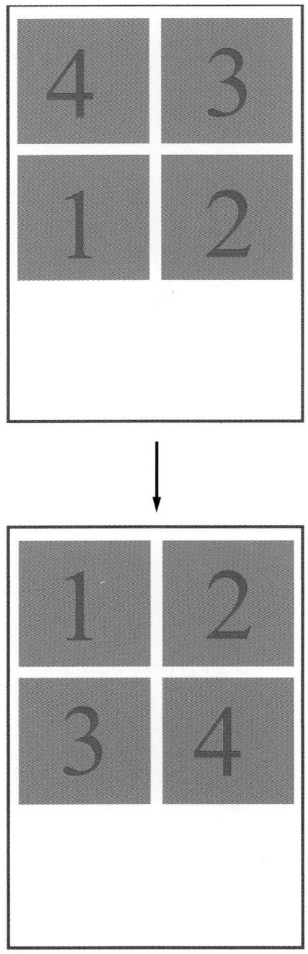

Figure 13.26: Rearrange the tiles from the order on the top to that on the bottom.

Exercise 13.25. (Programming) Program a tile-moving game. See Figure 13.26. The object is to rearrange the four tiles from the order on the top to that on the bottom by sliding them around in the big rectangle – tiles may not be picked up – the white space at the bottom being available for intermediate moves. The user should be able to select and drag a tile with the mouse. Makes sure to implement collision detection so that one tile cannot climb over another.

There are two other basic image manipulation commands available in OpenGL, in addition to `glDrawPixels()`. The command `glReadPixels()` is simply the inverse of `glDrawPixels()`, reading a rectangular array of pixel data from the frame buffer to client memory, while `glCopyPixels()` copies a rectangular array of pixels from one location of the frame buffer to another. We ask the interested reader to see the red book for the straightforward specs and attempt the following exercise.

Exercise 13.26. (Programming) Code a simple image-editing program, where the user can load an image, tweak it and save. (You might want to make use of the command `glPixelZoom()` to scale the image in connection with an invocation of a `glDrawPixels()` or `glCopyPixels()`.)

Pixel Buffer Objects

We saw in Section 3.2 how vertex-related data may be stored in the graphics server in so-called VBOs (vertex buffer objects) for reasons of efficiency.

Likewise, pixel data may be stored in the graphics server in PBOs (pixel buffer objects), the process being similar.

Experiment 13.15. Fire up `imageManipulationPBO.cpp`, which modifies `imageManipulation.cpp` to store the pixel data in a PBO. Otherwise, the functionality of both programs is identical, the image of the numeral being moved by mouse clicks and drags. **End**

The reader who has reviewed `vbo.cpp`, showing how to use VBOs, will easily understand the parallel commands of `imageManipulationPBO.cpp` to create, bind and initialize a PBO. The block

```
glGenBuffers(1, &pixelBuffer);
glBindBuffer(GL_PIXEL_UNPACK_BUFFER, pixelBuffer);
glBufferData(GL_PIXEL_UNPACK_BUFFER, TextureWidth*TextureHeight*4,
             textureImage, GL_STATIC_DRAW);
```

in the initialization routine generates a PBO buffer id, binds it to the target `GL_PIXEL_UNPACK_BUFFER` (effectively activating it for drawing) and initializes it with data from the 2D array `texureImage` of RGB pixel values. Note that `textureImage` itself has been moved from global storage in `imageManipulation.cpp` to local storage in the current program as the latter's drawing routine will read data from the PBO, not `texureImage`, so this array need no longer be kept as a permanent global.

The only other change of note from `imageManipulation.cpp` is in the drawing statement

```
glDrawPixels(TextureWidth, TextureHeight, GL_RGBA,
             GL_UNSIGNED_BYTE, BUFFER_OFFSET(0))
```

where now the fourth parameter is a byte offset value in the PBO.

13.9 Bump Mapping

Blinn [16] developed an ingenious method, called *bump mapping*, to give the illusion of geometric detail on a surface, e.g., making it appear ridged or dimpled, by means of perturbing the surface normals, but *without* actually changing any geometry. The idea is to re-align the normals to the original surface so that light reflects from it *as if* it were detailed. A one-dimensional example will make matters clear.

Consider the straight line c of Figure 13.27(a). The unit normals $n(u)$ at points $c(u)$ of c are identical vectors perpendicular to c (drawn in the figure as a discrete sequence of points).

Next, suppose that one wants to wrinkle c to make it look like the blue curve c' of Figure 13.27(b). Actually wrinkling c entails replacing it with a multi-segment polyline approximation of c', an object substantially more complex than c. The bump mapping approach is to leave c as it is, but,

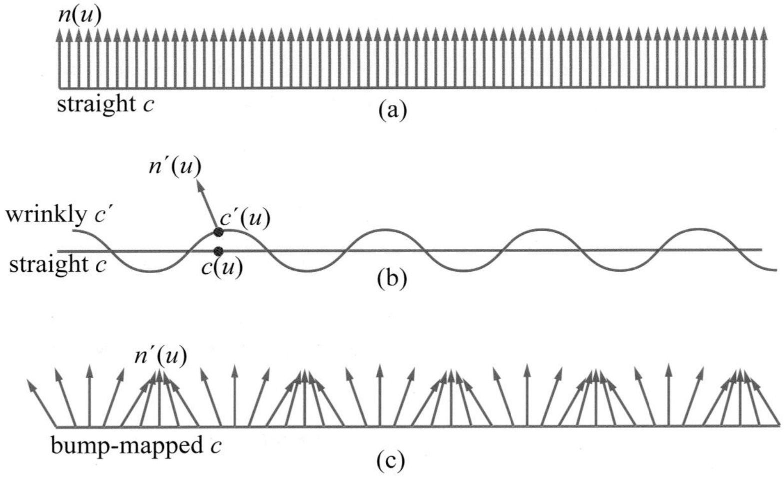

Figure 13.27: Bump mapping: (a) The original curve c and its true unit normals $n(u)$ (b) The wrinkled curve c' and its unit normal $n'(u)$ at a single point $c'(u)$ (c) Bump mapped c with redefined normals $n'(u)$.

instead, to redefine the normal at each point $c(u)$ so that it equals $n'(u)$, the normal at the corresponding point $c'(u)$ of c'. Figure 13.27(b) shows $n'(u)$ at one point $c'(u)$ of c', while Figure 13.27(c) shows the so-called *bump mapped* c with its perturbed normals.

The premise of bump mapping is that c with normals redefined to match those of c' will resemble c' when lit, because the reflection of light from a surface depends on the normals there. We describe next Blinn's method to compute the perturbed normals.

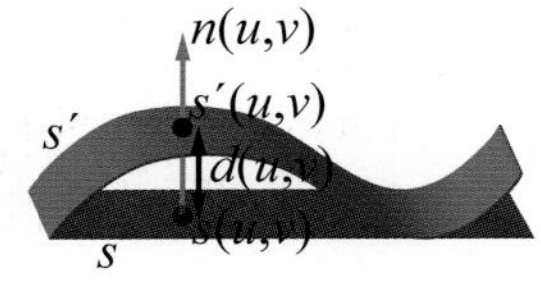

Figure 13.28: The bumped surface s' is obtained from s by displacing each point $s(u.v)$ a distance $d(u,v)$ along the normal $n(u,v)$ at $s(u,v)$.

Suppose that s is a surface in 3-space defined parametrically on some domain W by

$$s(u,v) = (f(u,v), g(u,v), h(u,v))$$

and that $n(u,v)$ is a unit normal vector to s at $s(u,v)$.

Suppose, as well, that the desired (hypothetical) detailed surface s' is obtained from s by displacing each point $s(u,v)$ a distance $d(u,v)$ along $n(u,v)$. See Figure 13.28. The scalar-valued function $d(u,v)$ giving this displacement is called the *bump map*.

Accordingly,

$$s'(u,v) = s(u,v) + d(u,v)n(u,v)$$

which we write more simply by dropping the arguments as

$$s' = s + dn \tag{13.9}$$

A normal n' to s' is given by

$$n' = \frac{\partial s'}{\partial u} \times \frac{\partial s'}{\partial v} \tag{13.10}$$

because the partial derivatives of s' with respect to u and v span the tangent plane to that surface at $s'(u, v)$, so that their cross-product is normal (see Section 11.10 for more about partial derivatives and their application to finding tangents and normals to a surface). We evaluate n' next.

First, from (13.9),

$$\begin{aligned} \frac{\partial s'}{\partial u} &= \frac{\partial s}{\partial u} + \frac{\partial d}{\partial u} n + d \frac{\partial n}{\partial u} \\ &\simeq \frac{\partial s}{\partial u} + \frac{\partial d}{\partial u} n \end{aligned}$$

where the approximation in the second line is made by dropping the term $d\frac{\partial n}{\partial u}$, which is negligibly small on the assumptions that

(a) the displacement d is small, and

(b) $\frac{\partial n}{\partial u}$ is small as well, from the reasonable premise that the original surface s lacked detail and was fairly smooth, meaning that its normal direction changed only slowly with u.

Likewise, one writes

$$\frac{\partial s'}{\partial v} \simeq \frac{\partial s}{\partial v} + \frac{\partial d}{\partial v} n$$

Plugging the preceding two approximations into (13.10) one gets

$$\begin{aligned} n' &\simeq \left(\frac{\partial s}{\partial u} + \frac{\partial d}{\partial u} n\right) \times \left(\frac{\partial s}{\partial v} + \frac{\partial d}{\partial v} n\right) \\ &= \left(\frac{\partial s}{\partial u} \times \frac{\partial s}{\partial v}\right) + \frac{\partial d}{\partial u}\left(n \times \frac{\partial s}{\partial v}\right) - \frac{\partial d}{\partial v}\left(n \times \frac{\partial s}{\partial u}\right) \\ &= n + \frac{\partial d}{\partial u}\left(n \times \frac{\partial s}{\partial v}\right) - \frac{\partial d}{\partial v}\left(n \times \frac{\partial s}{\partial u}\right) \qquad (13.11) \end{aligned}$$

which expresses the perturbed normal n' in terms of the original surface s, the original normal n and the bump map d. Finally, the new normal function for the bump mapped s is obtained by normalizing n' to unit length.

Bump mapping comes into its own in the per-pixel lighting of Phong's shading model, as the programmer can then apply Equation (13.11) to compute normal values at *each pixel* for subsequent use in the lighting equation. However, as we noted in Section 11.12, Phong's shading model is not an option in first-generation OpenGL, but can be implemented using the GLSL of the second generation and on.

In fact, when we get to fourth generation OpenGL (version 4.3) later in the book, we shall implement the per-pixel lighting of Phong's model and bump mapping itself will be a case study. Without per-pixel lighting bump mapping is at best awkward but, nevertheless, we do have a simple proof-of-concept program.

Experiment 13.16. Run `bumpMapping.cpp`, where a plane is bump mapped to make it appear corrugated. Press space to toggle between bump mapping turned on and off. Figure 13.29 shows screenshots. **End**

(a) (b)

Figure 13.29: Screenshots of `bumpMapping.cpp`: (a) Bump mapping off (b) Bump mapping on.

The equation of the plane in `bumpMapping.cpp` is

$$s(u, v) = (u, 0, -v)$$

(the minus sign in front of v is so that the normal $\frac{\partial s}{\partial u} \times \frac{\partial s}{\partial v}$ to the plane points in the upward y-direction) and that of the bump map

$$d(u, v) = \sin(2u)$$

We leave it to the reader to verify that with these equations for s and d, (13.11) gives

$$n'(u, v) = (2\cos(2u),\ 1,\ 0)$$

This formula for the perturbed normals to the plane is, in fact, implemented in `bumpMapping.cpp` when bump mapping is turned on (together with a call to `glEnable(GL_NORMALIZE)` to normalize the normals).

13.10 Summary, Notes and More Reading

In this chapter we learned a few different visual techniques to help embellish our scenes, games and movies. It's worth emphasizing that visual techniques are as much an art as a science. Experience counts a lot in knowing how to get the "right effect", a subjective notion in the first place. Collect code to save re-inventing the wheel and for inspiration. A lot of people out there are doing amazingly creative stuff. Although much of it is commercial, still there's plenty of free stuff to be found on the net. The OpenGL site [103] has numerous pointers. Check out Nehe [102] as well.

It's worth noting that we've just crossed a milestone in the progression of this book. With only the significant exceptions of NURBS, to come in a later chapter, *we've mostly covered all in this book to do with coding pre-shader OpenGL.* Pre-shader OpenGL itself is a perfectly serviceable 3D API. Moreover, as we said at the start, a grasp of pre-shader OpenGL makes the modern shader-based versions easier to learn. In particular, the reader is now well set to take on our own coverage of fourth generation OpenGL (version 4.3) which begins in Chapter 20, if she so desires.

There's *much more* to CG than OpenGL, of course, and topics we've yet to see such as rasterization, Bézier, B-spline and NURBS theory, projective spaces, ray tracing and radiosity, among others, are extremely important for a solid understanding of the field. Nevertheless, it's heartening to realize how far we have come since the first chapter, particularly from the point of view of practical programming.

Part VII

Pixels, Pixels, Everywhere

CHAPTER 14

Raster Algorithms

In this chapter we are going to be traveling almost all the way from one end of the graphics pipeline to the other – from world space to just behind screen space – to understand some of the low-level processes which take place at the time primitives are transformed to pixels in the raster. In particular, the goal for this chapter is to learn algorithms to clip and rasterize lines and polygons. These are operations a user cannot herself call directly or interact with via a high-level API such as OpenGL, which is not a bad thing as she can devote herself then to modeling and animation. Nevertheless, it's useful to have a grasp overall of the functioning of the pipeline. We'll not attempt here a comprehensive coverage of raster algorithms, but focus instead on four which are commonly implemented and fairly representative.

First comes clipping, which is the process of determining the part of a primitive within some restricted area. We're already familiar with the functionality of OpenGL's clipping to a viewing volume and in this chapter we'll learn how the operation is implemented in two 2D cases, namely, those of a straight line segment and a convex polygon, both clipped to a rectangle. In particular, the Cohen-Sutherland line clipper is the topic of Section 14.1 and the Sutherland-Hodgeman polygon clipper that of Section 14.2. Both clippers can be straightforwardly extended to 3D to clip a line segment or convex polygon against a box, the version actually implemented in the rendering pipeline.

Next, we'll investigate rasterization, the process of selecting and coloring pixels from the raster to represent a given primitive. We'll again limit ourselves to the two cases of straight segments and polygons. Moreover, we shall only be choosing pixels to comprise a primitive, leaving the problem of coloring them to a later chapter. Section 14.3 presents Bresenham's line rasterizer, actually as an improvement over the DDA (Digital Differential

Analyzer) line rasterizing algorithm. Section 14.4 next discusses scan-based polygon rasterization. Section 14.5 concludes the chapter.

14.1 Cohen-Sutherland Line Clipper

Straight line segments traveling down the OpenGL graphics pipeline are clipped to within an axis-aligned 3D viewing box prior to rasterization. Cohen-Sutherland is the classic algorithm for this purpose. Our exposition of Cohen-Sutherland, though, will be in 2D – clipping a segment to a rectangle – for the sake of simplicity. However, extension of the 2D version to 3D, where a segment is clipped to a box, is fairly straightforward.

Remark 14.1. How about clipping to a viewing volume which is not a box, but a frustum, as when the projection statement is `glFrustum()` or `gluPerspective()`? It turns out that in a stage in the pipeline, prior to clipping, the frustum is "straightened" into a box by a so-called projective transformation, so one need clip only to a box.

The inputs to (2D) Cohen-Sutherland are, then, the endpoints p_i and p_j of a straight line segment S on a plane, and an axis-aligned rectangle R on the same plane bounded by the lines $x = a$, $x = b$, $y = c$ and $y = d$. See Figure 14.1(a) for a diagram with multiple input segments. R is called the *clipping rectangle*.

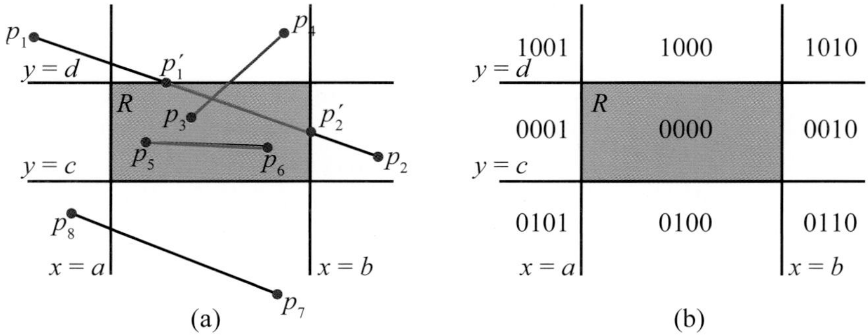

Figure 14.1: (a) A clipping rectangle R and four straight line segments with their parts clipped to R colored (b) Nine regions of the plane by outcode.

The output consists of the endpoints p_i' and p_j' of the intersection $S \cap R$ of S with R if it is non-empty, and *empty* otherwise. The output is said to be the segment S clipped to R. The parts of the segments p_1p_2, p_3p_4 and p_5p_6 clipped to R are indicated with color in Figure 14.1(a), while the output for p_7p_8 is *empty* (the end points of the colored segments are not all labeled to avoid clutter).

Outcodes and Trivial Termination

Critical to Cohen-Sutherland is a classification of points of the plane, according to their disposition with respect to the input rectangle R, by means of so-called outcodes. A point $p = (x, y)$ is said to have *outcode* the 4-bit string $k_3k_2k_1k_0$, whose value is determined by comparing the x- and y-values of p with those of the edges of R as follows:

- $k_0 = 0$, if $x \geq a$; $k_0 = 1$, if $x < a$
- $k_1 = 0$, if $x \leq b$; $k_1 = 1$, if $x > b$
- $k_2 = 0$, if $y \geq c$; $k_2 = 1$, if $y < c$
- $k_3 = 0$, if $y \leq d$; $k_3 = 1$, if $y > d$

It's easily seen that the four infinite straight lines, viz. $x = a$, $x = b$, $y = c$ and $y = d$, bounding R divide the plane into nine regions by outcode, as indicated in Figure 14.1(b). All points in each region have the outcode with which the region is labeled. For each of the lines $x = a$, $x = b$, $y = c$ and $y = d$, call the side of it containing R the *inside*, the other the *outside*. The line itself is included on its inside (so, e.g., the inside of $x = a$ is the closed half-plane $x \geq a$, while its outside is the open half-plane $x < a$). The rules above say, then, that each of the four lines determines an outcode bit, which is 0 if the point is on its inside, 1 if it's outside.

Cohen-Sutherland begins with the first step of determining the outcodes o_i and o_j of the endpoints p_i and p_j, respectively, of the input segment S, using the rules above. The next step is to determine if one of the following two cases applies, when the algorithm can immediately return the answer and terminate:

(a) S lies entirely inside R. In this case, both outcodes o_i and o_j are 0000, which can be verified by performing the logical bitwise operation $o_i \vee o_j$ and checking that the result is false, i.e., 0000.

(b) S lies entirely outside one of the four straight lines bounding R. In this case, both outcodes must have a 1-bit in the same position, which can be verified by performing the operation $o_i \wedge o_j$ and checking that the result is true, i.e., *not* 0000.

If (a) holds, the algorithm *trivially accepts*, returning the endpoints of S itself; if (b) holds, it *trivially rejects*, returning *empty*.

Recursion

If the outcome of the second step is neither a trivial accept nor a trivial reject, then Cohen-Sutherland proceeds recursively as follows.

There exists a bit position where o_i and o_j differ for, otherwise, one of the cases (a) and (b) above would have occurred. Scan the bits of o_i and

o_j from right to left to find the first (i.e., lowest) bit position, say the rth, where they differ. It follows that p_i and p_j lie on opposite sides of the infinite straight line bounding R, call it L, corresponding to the rth bit. Therefore, S intersects L and the algorithm calculates the point of intersection q. As they lie on opposite sides of L, exactly one of p_i and p_j lies inside L. The algorithm, then, calls itself recursively on the segment p_iq or p_jq, according as p_i or p_j is inside L.

The next example illustrates the working of Cohen-Sutherland.

Example 14.1. Apply the Cohen-Sutherland algorithm to the segment p_1p_2 in Figure 14.1(a).

Answer: The rectangle R and segment p_1p_2 of Figure 14.1(a) are drawn separately in Figure 14.2. The outcodes of p_1 and p_2 are 1001 and 0010, respectively, and neither of the two conditions that terminate Cohen-Sutherland trivially holds.

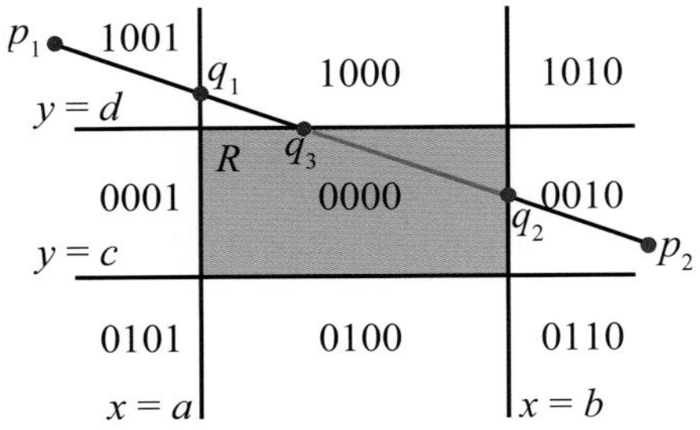

Figure 14.2: Cohen-Sutherland, called on segment p_1p_2, recursively calls itself on p_2q_1, q_1q_2 and q_2q_3, successively.

Scanning the bits of the two outcodes from right to left, they are seen to differ in the rightmost bit k_0, corresponding to the line $x = a$. Accordingly, the intersection q_1 of p_1p_2 with $x = a$ is determined. As p_2 lies inside $x = a$, a recursive call is made on the segment p_2q_1.

The outcodes of p_2 and q_1 are, respectively, 0010 and 1000. Again, there is no trivial termination. The rightmost bit at which the outcodes differ is k_1, so the intersection q_2 of p_2q_1 with $x = b$ is computed. As q_1 lies inside $x = b$, a recursive call is made next on the segment q_1q_2.

The outcodes of q_1 and q_2 are, respectively, 1000 and 0000. There is no trivial termination. The rightmost bit the outcodes differ at is k_3, so the intersection q_3 of q_1q_2 with $y = d$ is computed. As q_2 lies inside $y = d$, a recursive call is made on the segment q_2q_3.

The outcodes of q_2 and q_3 are both 0000 and the call terminates trivially, returning q_2 and q_3.

Exercise 14.1. Apply Cohen-Sutherland to the other three segments in Figure 14.1(a).

Exercise 14.2. Show that the maximum number of times Cohen-Sutherland can recursively call itself on a single input segment is four. Give an example of a segment where, in fact, four calls are made.

Exercise 14.3. (Programming) Animate Cohen-Sutherland using OpenGL. Draw a fixed clipping rectangle R, but allow the user to specify the endpoints of an arbitrary segment. Subsequently, highlight its subsegments as they are recursively processed.

The following two exercises extend Cohen-Sutherland.

Exercise 14.4. Extend Cohen-Sutherland to handle *semi-infinite* segments. A semi-infinite segment S is specified by one finite endpoint p_1 and another point p_2 in the *direction* of which it is infinite. See Figure 14.3.

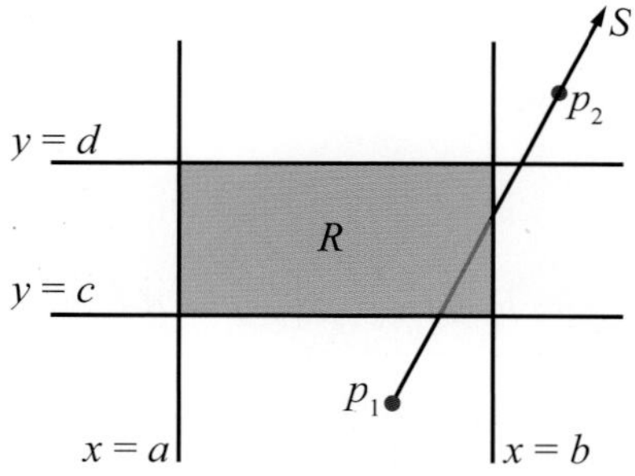

Figure 14.3: Clipping a semi-infinite segment to a rectangle.

Exercise 14.5. Extend Cohen-Sutherland to 3D: pseudo-code a 3D version to clip a straight-line segment S in 3-space against an axis-aligned box. See Figure 14.4. This is the actual clipper invoked in a 3D synthetic-camera rendering pipeline, like OpengGL's.

Questions to ponder: How does one define outcodes in 3D? How many bits are there in an outcode? Into how many regions is 3-space divided by outcode? Beyond these, need there be a significant difference between the 2D and 3D algorithms?

Extend your 3D clipper to handle semi-infinite segments as well.

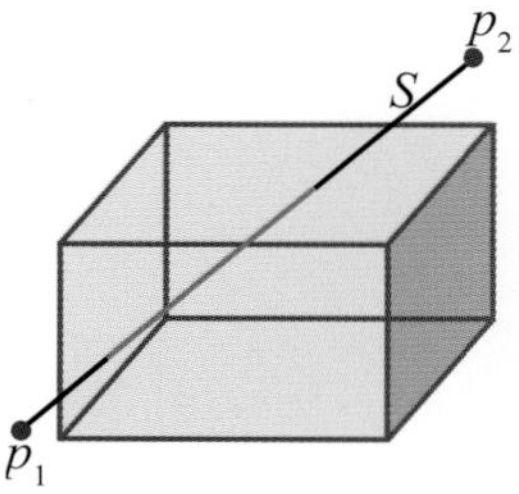

Figure 14.4: Clipping a line segment to an axis-aligned box.

Complexity

The complexity of Cohen-Sutherland lies mainly in the intersection computation resulting if the logical operation on the outcodes in the second step fails to terminate the algorithm trivially. Suppose the endpoints of the input segment S are $p_1 = (x_1, y_1)$ and $p_2 = (x_2, y_2)$ as in Example 14.1, drawn again in Figure 14.5. The first intersection to be calculated is $q_1 = (x_3, y_3)$, where S and $x = a$ meet.

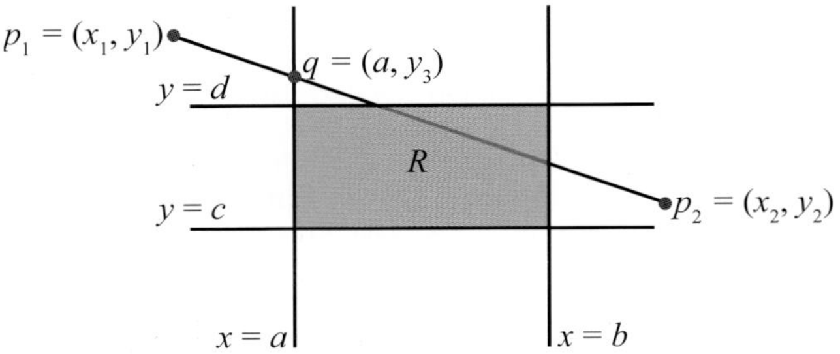

Figure 14.5: Cohen-Sutherland intersection computation.

The slope-intercept form of the equation of the straight line on which S lies is

$$y = \frac{y_2 - y_1}{x_2 - x_1} x + x_1$$

Obviously, $x_3 = a$, so y_3 can be found by plugging $x = a$ into this equation to get

$$y_3 = \frac{y_2 - y_1}{x_2 - x_1} a + x_1$$

Intersections with $x = b$, $y = c$ and $y = d$ can be found similarly as required.

If the input segments are pre-processed to determine their slope-intercept form, e.g., in the case above this entails pre-computing $\frac{y_2 - y_1}{x_2 - x_1}$, then, evidently, one floating point multiplication and one addition is performed per intersection finding.

Note: When pre-processing the slope-intercept form, one has to be careful to check for vertical line segments for which this form is not defined. However, once detected, vertical segments are obviously easy to clip to any rectangle.

We know from Exercise 14.2 that a call to Cohen-Sutherland to clip an input segment spawns at most four more recursive calls. The conclusion then is that a call to Cohen-Sutherland may require four floating point multiplications and four additions per input segment beyond pre-processing, in the worst case, which is fairly expensive. Refinements of Cohen-Sutherland, e.g., Liang-Barsky [86], invest in greater pre-processing in order to reduce subsequent intersection computation.

14.2 Sutherland-Hodgeman Polygon Clipper

The Sutherland-Hodgeman strategy for clipping a convex polygon P to a rectangle R is to successively clip off parts of P lying outside the four straight lines bounding R – the outside being the side of the straight line not containing R. The process can be conceived of as a pipeline of four clippers, as in Figure 14.6. The clipping rectangle is drawn only in outline in this section.

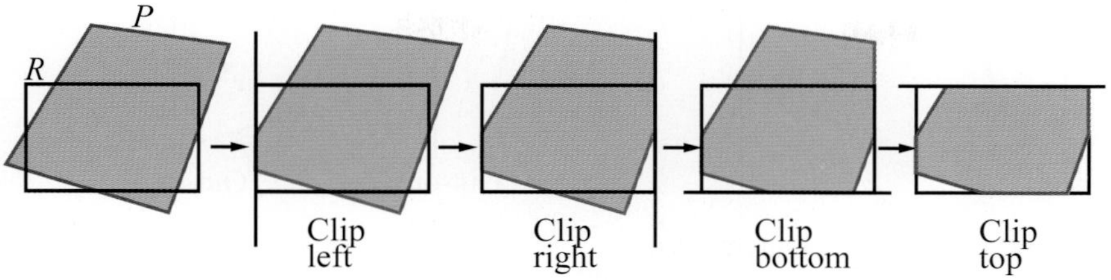

Figure 14.6: A pipeline of clippers.

Remark 14.2. There is no particular merit to the left-right-bottom-top ordering in Figure 14.6 – any other ordering could have been chosen.

The implementation of the four clippers is similar and we explain in detail only the leftmost one, which clips off outside the line bordering the left of R. In general, the input to this clipper is an ordered list $\{v_0, v_1, \ldots, v_n\}$ of the vertices of a convex polygon P and a vertical straight line L. See Figure 14.7. The output is an ordered list of vertices of the polygon P' resulting from clipping P off to the left of L.

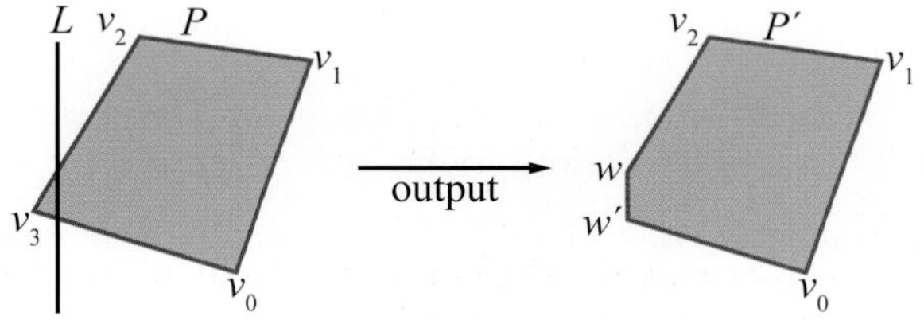

Figure 14.7: The left clipper in action: input $= \{v_0, v_1, v_2, v_3\}$, output $= \{v_0, v_1, v_2, w, w'\}$.

Output Rules for Left Clipper

The input list $\{v_0, v_1, \ldots, v_n\}$ of vertices is processed, in fact, in successive pairs, plus a final pair containing the first and last vertices, in particular, $v_0v_1, v_1v_2, \ldots, v_nv_0$. Equivalently, processing is edge by edge around the polygon P. Each edge outputs zero, one or two vertices to the output list. The output of an edge depends on the respective disposition of its end vertices with respect to L, in particular, if either is inside or outside L.

There is, therefore, a total of four possible dispositions. The four output rules, one corresponding to each disposition, are listed below. Refer to Figure 14.8 as you read (take v_{i+1} to be v_0, if v_i is v_n).

(a) In-in: Both v_i and v_{i+1} are inside L. Output is v_{i+1}.

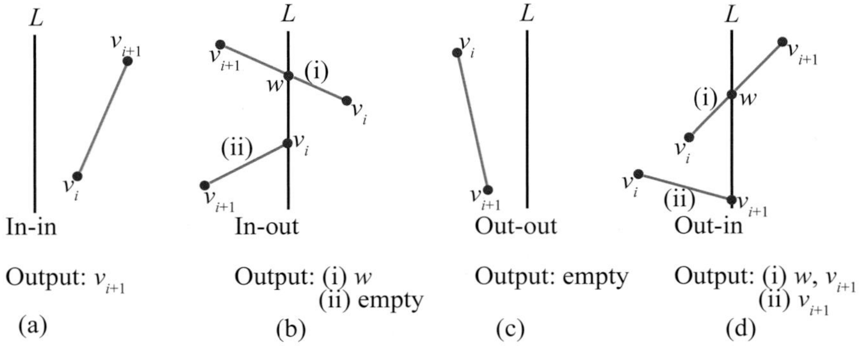

Figure 14.8: Output of a pair of successive vertices (equivalently, edge) entering the left clipper. Note both in-out and out-in dispositions have a special case, labeled (ii), where the vertex on the inside actually lies *on* L.

(b) In-out: v_i is inside, while v_{i+1} outside L. There are two subcases, according as v_i lies strictly right of L or on it:

(i) v_i is strictly right of L. Output is w, the point where the segment $v_i v_{i+1}$ intersects L.

(ii) v_i lies on L. Output is *empty*.

(c) Out-out: Both v_i and v_{i+1} are outside L. Output is *empty*.

(d) Out-in: v_i is outside, while v_{i+1} inside L. There are two subcases, according as v_{i+1} lies strictly right of L or on it:

(i) v_{i+1} is strictly right of L. Output is w, v_{i+1}, where w is the point where the segment $v_i v_{i+1}$ intersects L (this is the only case when two vertices are output).

(ii) v_{i+1} lies on L. Output is v_{i+1}.

Observe that, if L is $x = a$, it is easy to determine if a point is inside or outside L by simply comparing its x-value with a. Therefore, it's easy as well to decide which of the four rules above to apply to each successive pair of vertices. We ask the reader next to determine the new vertex w, in case it arises, in the following exercise.

Exercise 14.6. Suppose $v_i = (x_i, y_i)$, $0 \le i \le n$, and L is the line $x = a$. Observe, from the rules above, that if a new vertex – one not belonging to the original input sequence – is at all output, then there is only one such. Give a formula to determine the new vertex.

Example 14.2. Let's apply the rules above to the initial polygon P, at the leftmost of Figure 14.6, and the vertical straight line L along the left

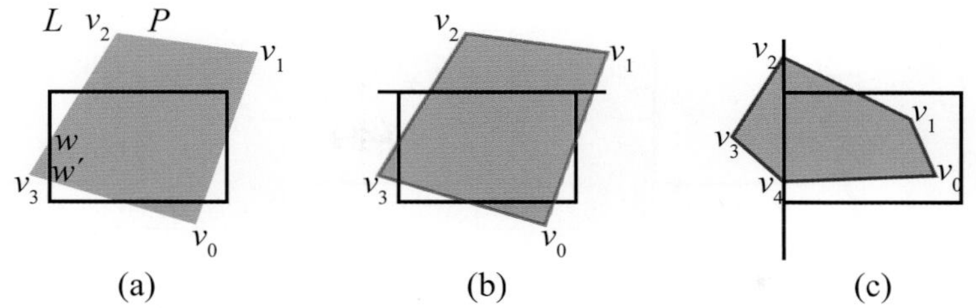

Figure 14.9: Applying the clipping rules.

edge of the rectangle R. See Figure 14.9(a). The output of successive vertex pairs is as follows:

$$\begin{array}{lcll} v_0v_1 & \Rightarrow & v_1 & \text{(in-in)} \\ v_1v_2 & \Rightarrow & v_2 & \text{(in-in)} \\ v_2v_3 & \Rightarrow & w & \text{(in-out (i))} \\ v_3v_0 & \Rightarrow & w', v_0 & \text{(out-in (i))} \end{array}$$

Accordingly, the vertex list returned by the left clipper is $\{v_1, v_2, w, w', v_0\}$, which indeed is the sequence of vertices around the polygon resulting from clipping off the part of P outside L.

Exercise 14.7. Clip the initial polygon P of Figure 14.6 to the top, in other words, apply the top clipper to it first. See Figure 14.9(b). Rules, exactly as for those given already for the left clipper, apply to the top.

Exercise 14.8. Clip the polygon P of Figure 14.9(c) to the left.

Pipelining

Now that we understand the implementation of its individual clippers, we come to the beauty of the Sutherland-Hodgeman algorithm: that it can be *pipelined* with all four clippers running in *parallel*, each one after the first using as input the output of its predecessor. This follows from observing that each clipper *incrementally* produces its output list as it *incrementally* consumes its input vertices. Therefore, the next clipper in the sequence does not have to wait till its predecessor completes processing – it can begin to operate *as soon* as it receives the first two vertices output by its predecessor.

For example, we saw in Example 14.2 that the successive vertices output by the left clipper operating on P are $\{v_1, v_2, w, w', v_0\}$. Observe, now, that as soon as v_1 and v_2 enter the next clipper in the sequence – the right one according to the scheme in Figure 14.6 – the latter can process them to output w'', v_2. See Figure 14.10, where the disposition of v_1 and v_2 with respect to L' is case (i) of out-in.

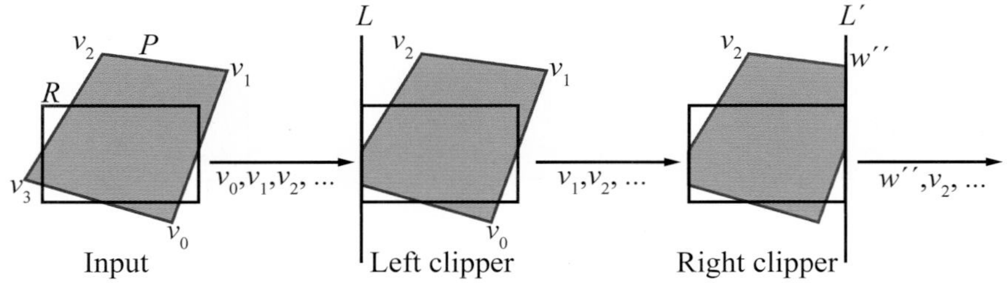

Figure 14.10: The right clipper consumes the first two vertices v_1 and v_2 entering into it from the left clipper to output w'' and v_2.

The Sutherland-Hodgeman pipeline of four clippers is often implemented in hardware, the clippers being identical but separate modules.

Exercise 14.9. Assuming that each clipper takes unit time to perform the operation of applying one of the rules of Figure 14.8 to a pair of successive vertices and that vertices move from one clipper to the next in zero time, how long does it take for the clipping pipeline of Figure 14.6 to process the particular polygon P in that figure?

Exercise 14.10. (Programming) Code and creatively animate the Sutherland-Hodgeman clipping pipeline.

Exercise 14.11. Extend Sutherland-Hodgeman to 3D to clip a convex polygon against an axis-aligned box. This, in fact, is simpler than generalizing the Cohen-Sutherland line clipper to three dimensions, which we considered in Exercise 14.5, because the Sutherland-Hodgeman output rules go through pretty much verbatim in one higher dimension, except that the part of the polygon to one side of a plane, rather than a line, is clipped off.

14.3 DDA and Bresenham's Line Rasterizers

Rasterization of a straight line segment consists of picking and coloring the pixels to comprise that segment, given its start and end vertices and their color attributes. Both the DDA and Bresenham's algorithm actually only pick pixels. Coloring them is a straightforward application of linear interpolation which we learned in Chapter 7, except for the added twist of "perspective correction" needed in case of perspective projection. We'll fill out details of the coloring process when we examine the synthetic-camera pipeline in Chapter 19.

Let's take the raster to be a rectangular $m \times n$ grid of square pixels, each of side length one, located axis-aligned on a plane, so that the pixel centers have integer coordinates (i, j), $0 \leq i \leq m-1$ and $0 \leq j \leq n-1$. See Figure 14.11(a).

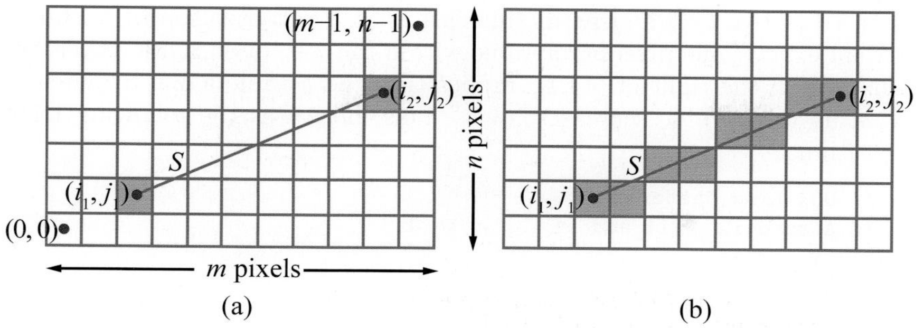

Figure 14.11: An $m \times n$ raster of pixels and a rasterized straight line segment.

We'll assume that both end vertices of an input segment S have already been projected and scaled onto the raster – shot on film and printed according to the analogy in the second chapter – so that they lie at the pixels centered at (i_1, j_1) and (i_2, j_2), respectively, as shown in Figure 14.11(a).

Note: Even though S is actually a segment in world space, we'll call its image on the raster S as well – which of the two we mean should be clear from the context.

The rasterization task then is to choose the pixels between its two ends to represent S, as, say, in Figure 14.11(b). We can assume that one end of S lies strictly to the right of the other because, otherwise, either both ends are at the same pixel or one is vertically above the other, both cases being trivial to rasterize. Suppose, without loss of generality then, that (i_2, j_2) is to the right of (i_1, j_1), meaning $i_2 > i_1$. The line equation of S is:

$$\frac{y - j_1}{x - i_1} = \frac{j_2 - j_1}{i_2 - i_1} \qquad \text{or} \qquad y = m(x - i_1) + j_1 \qquad (14.1)$$

where $m = \frac{j_2 - j_1}{i_2 - i_1}$ is the slope of S.

DDA Algorithm – Floating Point Heavy

We'll warm up with the very simple DDA (Digital Differential Analyzer) rasterization algorithm which Bresenham subsequently improves by eliminating floating point computation.

Remark 14.3. The rather fancy name Digital Differential Analyzer comes first from Differential Analyzer, a class of mechanical machines invented in the 1800s to solve differential equations. A Digital Differential Analyzer, or DDA, is a digital version of the Differential Analyzer. The DDA line rasterizer implements an incrementing loop borrowed from the DDA, hence the name.

Suppose, first, that its slope m lies between -1 and 1, so that the input segment S makes an angle of at most $45°$ with the x-axis (as in

Figure 14.11(a)). It's clear in this case that a rasterization of S should contain exactly one pixel per x-value within the x-span of S. Moreover, the equation on the right of (14.1) implies that the y-value along S increases by m as the x-value increases by 1. This motivates the following DDA algorithm:

```
// DDA Line Rasterizer
// Assume i2 > i1 and -1 <= m <= 1
float y = j1;
float m = (j2 - j1)/(i2 - i1);
for (int x = i1; x <= i2; x++)
{
   pickPixel(x, round(y));
   y += m;
}
```

Note: Due to obvious typesetting constraints we write variables such as i_1 and j_1 as `i1` and `j1` in the code snippet.

The algorithm starts with the pixel centered at (i_1, j_1), then increases the x-value by 1 and the y-value by m at each step, until x equals i_2. The pixel chosen at each step by the `pickPixel()` function follows rounding of the y-value to an integer.

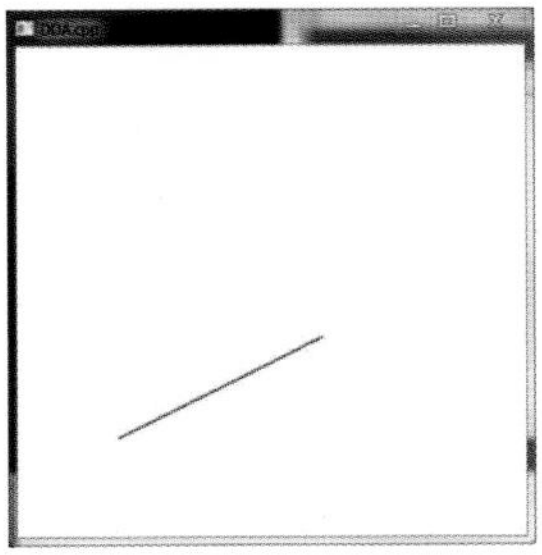

Figure 14.12: Screenshot of `DDA.cpp`.

Experiment 14.1. Run `DDA.cpp`, which is pretty much a word for word implementation of the DDA algorithm above. A point of note is the *simulation* of the raster by the OpenGL window: the statement `gluOrtho2D(0.0, 500.0, 0.0, 500.0)` identifies "pixel-to-pixel" the viewing face with the 500×500 OpenGL window (a "pixel" of the viewing face being a 1×1 square with corners at integer coordinates).

There's no interaction and the endpoints of the line are fixed in the code at $(100, 100)$ and $(300, 200)$. Figure 14.12 is a screenshot. **End**

Exercise 14.12. (Programming) The restriction on the slope m in the DDA rasterizer can be removed by noting that a rasterization of S should contain exactly one pixel per y-value in its y-span, if it makes an angle of more than $45°$ with the x-axis.

Accordingly, rewrite `DDA.cpp` so that there are no restrictions at all and that it interactively accepts arbitrary input end vertices (i_1, j_1) and (i_2, j_2).

The problem with the DDA algorithm is that it invokes two floating point operations per loop iteration – an addition and a rounding – floating point operations always being computationally expensive. We'll see next how Bresenham manages to dispense with floating point calculations.

Eliminating Floating Points – Bresenham's Algorithm

Bresenham's algorithm avoids floating point operations altogether. Here's the idea in a nutshell. The algorithm incrementally picks one pixel after another starting with the left end vertex of S, the smarts being in using the location of the current pixel to cost-effectively deduce that of the next.

Assume first that the slope of S satisfies $0 < m < 1$; we'll handle other cases later. Given this constraint, there is exactly one pixel of S for each x-value within its x-span.

Suppose that pixel (i, j) has just been picked – pixel (i, j) means, of course, the pixel centered at (i, j). See Figure 14.13(a), where only pixel centers are shown in a grid. One can assume that the point $p = (p_x, p_y)$ of the intersection of S with the line $x = i$ is at least as close to (i, j) as it is to the centers of either pixel above or below it, for, otherwise, pixel (i, j) would not have been chosen. In other words, $j - \frac{1}{2} \leq p_y \leq j + \frac{1}{2}$. These inequalities, together with the condition $0 < m < 1$ on the gradient of S, imply that S intersects $x = i + 1$ at the point $q = (q_x, q_y)$, where $j - \frac{1}{2} < q_y < j + \frac{3}{2}$, as we'll show next.

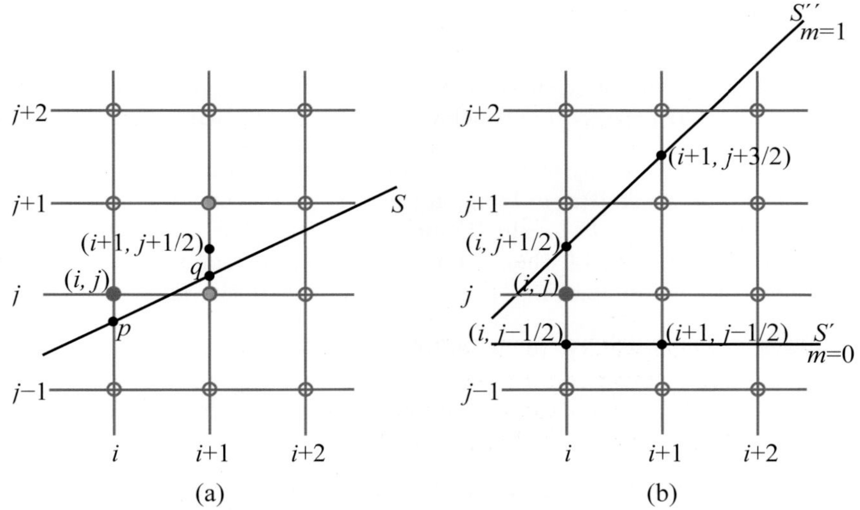

Figure 14.13: The larger circles are pixel centers. (a) Pixel (i, j) shown filled blue has just been chosen, the two candidate pixels for the next step are filled orange (b) Diagram for Example 14.3.

Example 14.3. Prove the inequalities claimed on q_y in the preceding statement.

Answer: Let the line S' go through $(i, j - \frac{1}{2})$ with gradient 0 and the line S'' go through $(i, j + \frac{1}{2})$ with gradient 1. See Figure 14.13(b), where S' and S'' are shown but not S. Since $j - \frac{1}{2} \leq p_y \leq j + \frac{1}{2}$ and $0 < m < 1$, S intersects $x = i + 1$ at a point $q = (q_x, q_y)$ strictly between the points where S' and S'' intersect $x = i + 1$. It's straightforward geometry to see that S' intersects $x = i + 1$ at $(i + 1, j - \frac{1}{2})$ and S'' intersects $x = i + 1$ at $(i + 1, j + \frac{3}{2})$. It follows that, indeed, $j - \frac{1}{2} < q_y < j + \frac{3}{2}$.

As $j - \frac{1}{2} < q_y < j + \frac{3}{2}$, the pixel chosen when x is incremented to $i+1$ is either $(i+1, j)$ or $(i+1, j+1)$, depending on which of the two is closer to q. Now, q is closer to $(i+1, j)$ if the midpoint $(i+1, j+\frac{1}{2})$ between the two candidate pixels is above S (the situation depicted in Figure 14.13(a)); it's closer to $(i+1, j+1)$ if the midpoint is below S.

Evidently, we need to determine the disposition of q with respect to S. So, how does one generally determine the disposition of some given point (x, y) with respect to S? Rewrite the line equation of S, on the left of Equation (14.1), as

$$(i_2 - i_1)y - (j_2 - j_1)x + i_1 j_2 - i_2 j_1 = 0$$

Denote the LHS of the preceding equation by $D'(x, y)$, i.e.,

$$D'(x, y) = (i_2 - i_1)y - (j_2 - j_1)x + i_1 j_2 - i_2 j_1 \qquad (14.2)$$

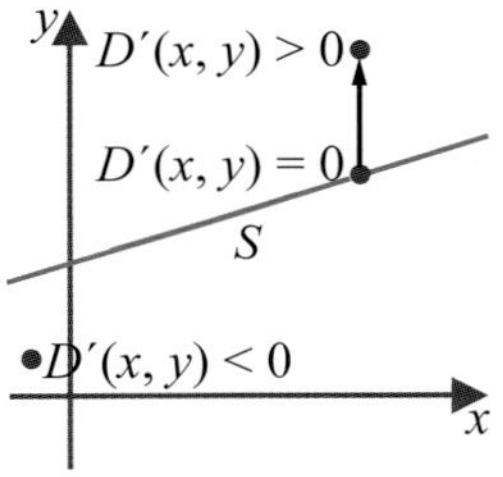

Figure 14.14: Discriminating the position of a point with respect to a segment: S is bold, while the straight line through it is thin.

Therefore, if (x, y) satisfies $D'(x, y) = 0$, then the point lies on the straight line containing S. In fact, it lies on S itself if, additionally, its x-value lies within the x-span of S. Furthermore, (x, y) lies above that straight line if $D'(x, y) > 0$ and below if $D'(x, y) < 0$. See Figure 14.14.

Exercise 14.13. Verify the claim made in the last statement.
Hint: Consider a point (x, y) lying on the straight line containing S, so that $D'(x, y) = 0$. If only its y-value is increased to raise it above the line, as indicated in Figure 14.14, then the value of $D'(x, y)$ increases because the coefficient of y in the formula (14.2) for $D'(x, y)$ is positive.

For a reason which will soon be apparent, we'll use

$$D(x, y) = 2D'(x, y) = 2(i_2 - i_1)y - 2(j_2 - j_1)x + 2(i_1 j_2 - i_2 j_1) \qquad (14.3)$$

whose sign is always the same as that of $D'(x, y)$, as the *discriminant* to determine the disposition of (x, y) with respect to the straight line on S.

Returning to the question of which of the candidate pixels $(i+1, j)$ or $(i+1, j+1)$ to choose when $x = i+1$, we see the answer now as:

- Choose $(i+1, j)$ if $D(i+1, j+\frac{1}{2}) > 0$ (when the midpoint $(i+1, j+\frac{1}{2})$ between the candidate pixels lies above S, implying that S is closer to the center of the lower one).

- Choose $(i+1, j+1)$ if $D(i+1, j+\frac{1}{2}) < 0$ (complementary to the first case).

- Choose $(i+1, j+1)$ if $D(i+1, j+\frac{1}{2}) = 0$ (this choice being made arbitrarily as S is equidistant from the centers of both candidates).

We near the crux of Bresenham's algorithm, which is to use the old value of D to compute its new value after x is incremented once again to $i+2$. There are two cases:

(a) $D(i+1, j+\frac{1}{2}) > 0$, which implies, according to the pixel-choosing process just described, that $(i+1, j)$ was chosen when $x = i+1$.

Repeating the reasoning that took us from $x = i$ to $x = i+1$, we see that the two candidate pixels when $x = i+2$ are $(i+2, j)$ and $(i+2, j+1)$; moreover, $(i+2, j)$ is chosen if $D(i+2, j+\frac{1}{2}) > 0$ and $(i+2, j+1)$ if $D(i+2, j+\frac{1}{2}) \leq 0$. As for the new value of the discriminant, we ask the reader to verify from (14.3) that it can be calculated from its previous value by:

$$D(i+2, j+\frac{1}{2}) = D(i+1, j+\frac{1}{2}) - 2(j_2 - j_1) \qquad (14.4)$$

(b) $D(i+1, j+\frac{1}{2}) \leq 0$, which implies that $(i+1, j+1)$ was chosen when $x = i+1$.

The two candidate pixels when $x = i+2$ are now $(i+2, j+1)$ and $(i+2, j+2)$; moreover, $(i+2, j+1)$ is chosen if $D(i+2, j+\frac{3}{2}) > 0$ and $(i+2, j+2)$ if $D(i+2, j+\frac{3}{2}) \leq 0$. Again, we ask the reader to verify that:

$$D(i+2, j+\frac{3}{2}) = D(i+1, j+\frac{1}{2}) + 2(i_2 - i_1 - j_2 + j_1) \qquad (14.5)$$

We have in hand now the crux of Bresenham's algorithm: Equations (14.4) and (14.5) together say how D changes as x is incremented by 1. It only remains to get the algorithm started by initializing D. The first pixel picked is at the left endpoint (i_1, j_1) of S. Accordingly, the candidate pixels for the next value of x, namely, $x = i_1 + 1$, are (i_1+1, j_1) and (i_1+1, j_1+1), whose midpoint is $(i_1, j_1 + \frac{1}{2})$. Therefore, the first value of D is

$$D(i_1+1, j_1+\frac{1}{2}) = i_2 - i_1 - 2(j_2 - j_1) \qquad (14.6)$$

It's here that taking $D(x, y) = 2D'(x, y)$, rather than $D'(x, y)$ itself, pays off. For, $D'(i_1+1, j_1+\frac{1}{2}) = \frac{1}{2}(i_2 - i_1) - (j_2 - j_1)$, which may be fractional (remember, we want to stay away from floating points).

With (14.4)-(14.6) all pieces now are in place to implement Bresenham's strategy: initialize D using (14.6) and then successively increment it with the help of (14.4)-(14.5), picking at each step the next pixel based on the sign of D. Here's code:

```
// Bresenham's Line Rasterizer
// Assume i2 > i1 and 0 < m < 1
int y = j1;
int diff1 = -2*(j2 - j1);
int diff2 =  2*(i2 - i1 - j2 + j1);
int D = i2 - i1 - 2*(j2 - j1);
for (int x = i1; x <= i2; x++)
```

```
{
   pickPixel(x, y);
   if (D > 0) D += diff1;
   else {y++; D += diff2;}
}
```

Compare this with the DDA algorithm earlier which required floating point rounding and addition – there's nary a floating point value in sight in Bresenham's procedure! Bresenham is extremely efficient and, often, implemented in the graphics card itself for even higher rendering speeds.

Exercise 14.14. Use Bresenham's Line Rasterizer as coded above to pick the pixels on the straight line segment joining pixel centers $(9, 17)$ and $(25, 25)$.

Part answer: We'll get the reader started. Set $(i1, j1) = (9, 17)$ and $(i2, j2) = (25, 25)$ in the code above. Accordingly,

$$\begin{aligned} diff1 &= -2 * (j2 - j1) = -16 \\ diff2 &= 2 * (i2 - i1 - j2 + j1) = 16 \\ D &= i2 - i1 - 2 * (j2 - j1) = 0 \quad \text{(initially)} \end{aligned}$$

The first pixel picked is the left endpoint $(9, 17)$. As $D = 0$, the next pixel picked is $(10, 18)$ and D changes to $D + diff2 = 16 > 0$. Therefore, the next pixel picked is $(11, 18)$ and D changes to $D + diff1 = 0, \ldots$.

The reader should sketch the rasterization on graph paper and verify that it indeed starts and ends at the given endpoints.

Exercise 14.15. (Programming) We ask the reader now to implement Bresenham's algorithm, simulating the raster using the OpenGL window, as in `DDA.cpp`. Remove, as well, the assumptions in our formulation of Bresenham's algorithm. In particular, allow the two input end vertices (i_1, j_1) and (i_2, j_2) to be arbitrary.

Moreover, ensure that your program is not sensitive to the order that the end vertices are input. In other words, the rasterization should be identical if (i_1, j_1) and (i_2, j_2) are swapped in the routine. This is as a user would expect: it should not matter in what order the end vertices happen to appear in the segment definition in code.

14.4 Scan-Based Polygon Rasterizer

We'll describe a commonly-implemented scan-based rasterization algorithm to draw a polygon, or *fill* it, as is commonly said, because pixels comprising the interior of the polygon are selected. The notion of scanning arises from CRT technology where an electron gun repeatedly scans the screen, pixel row by pixel row. One row of pixels is called a *scan line*. Although rasterization algorithms are implemented when filling pixel values in the frame buffer and have really nothing to do with the front-end display technology, this particular algorithm for polygon rasterization happens to mimic the scan process, hence the CRT reference.

A central part of the scan-based algorithm is to use the so-called *parity test*, also called the *inside-outside test*, to determine if a point lies inside or outside a polygon P.

Parity Test

Suppose P is a simple planar polygon, i.e., one which lies on a plane and whose boundary is a single line loop which doesn't self-intersect. Let q be a point known to lie outside P (possibly, by choosing q's x- or y-value to be either very large or small). Now, suppose we wish to determine if another given point p lies inside or outside P. Consider the ray R from q to p. The ray, obviously, starts from outside P. Upon its first intersection (if any) with the boundary of P, R enters P; upon its second intersection with the boundary of P, it exits P; upon its third intersection, it re-enters P; and so on. See Figure 14.15.

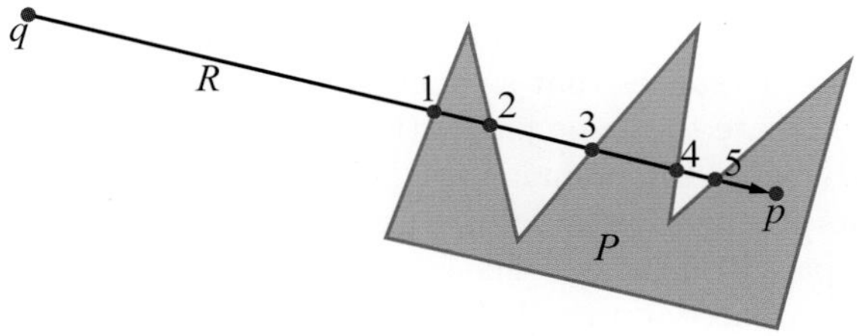

Figure 14.15: Intersections of R with the boundary of P are labeled with their respective ranks.

This leads to the following test to decide if p lies inside P or outside.

Parity Test: If the ray R from a point q, known to lie outside P, to some given point p, intersects the boundary of P an odd number of times, then p lies inside P, while if it intersects the boundary of P an even number of times, then P lies outside P. We assume, when applying the parity test, that p itself does not lie *on* the boundary of P, i.e., it is either inside or outside.

The parity of the number of intersections of R with the boundary of P, odd or even, is often called the *parity* of p. So the parity test may be rephrased to say that points of odd parity lie inside P, while those of even parity outside.

Note: If R does not intersect the boundary of P at all, then P's parity is that of zero, which is even, and, of course, p lies outside P.

Exercise 14.16. For the five points p_1, p_2, p_3, p_4, p_5 in Figure 14.16, use the parity test to verify where they lie with respect to P. The rays from an outside point q are already drawn. Consider how you would handle the particular singularities seen in the case of p_3 and p_4 without revising the statement of the parity test itself (we'll be discussing this matter in some detail momentarily).

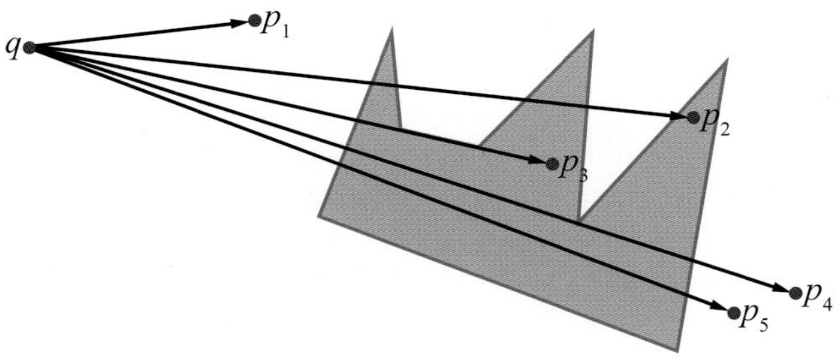

Figure 14.16: Testing parity: the points p_3 and p_4 require handling a singularity.

See Figure 14.17. The integer label beneath each segment of the semi-infinite ray R from q indicates the number of intersections of R with the boundary of P before it reaches that segment. Therefore, all points in the *interior* of a segment share the same parity (the restriction to the interior is to avoid endpoints lying on the boundary of P, whose parity is not defined). For example, points in the interior of qp_1 have even parity, those in the interior of p_1p_2 odd parity and so on. The segments w and w' are not labeled because they each lie entirely on the boundary.

One has to be careful, evidently, when counting intersections, to take into account edges such as w and w', as well as points such as p_6, where the ray touches the polygon without properly intersecting it. Here are the conventions to follow, and which have been followed in Figure 14.17, to ensure that these singularities are handled correctly.

Conventions for Singularities:

(a) When the ray R passes through a vertex v, but neither of the edges adjacent to v lie along R, there are two cases:

  (i) if these two edges lie on opposite sides of R, then v is a proper intersection point and counted once (p_4 in Figure 14.17);

  (ii) otherwise, R *touches* P at v, and v is counted as two intersections (p_6, explaining the difference in count on either side of it).

(b) When the ray R passes through a vertex v and one of the edges, say e, adjacent to v lies along R (we'll assume that successive edges of P are never collinear, so that both edges adjacent to v cannot lie on R):

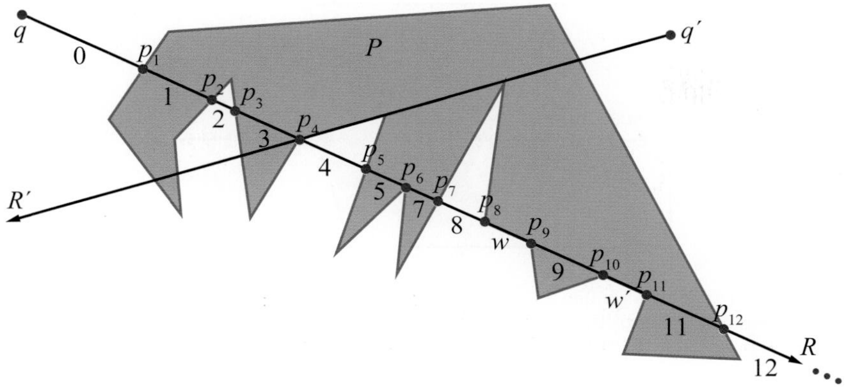

Figure 14.17: Applying the parity test: the integer label beneath a segment of R indicates the number of intersections of R with the boundary of P prior to reaching that segment. The reason for the change from 5 to 7 after p_6 and from 9 to 11 after edge w' is explained in the text. The edges w and w' are not labeled because they lie on the boundary.

(i) if the two edges adjacent to e lie on opposite sides of R, then count the entire edge e as one intersection (w in Figure 14.17);

(ii) otherwise, count the entire edge e as two intersections (edge w').

In other words, case (b) is the same as (a) if one imagines e as one giant vertex.

Exercise 14.17. Label the segments of the ray R' emanating from the point q' in Figure 14.17 in a manner similar to that of segments along R.

Exercise 14.18. (Programming) Implement the parity test. Allow the user to select a polygon P, as well as a test point p, by clicking on the OpenGL window. Either ask the user to, or automatically, choose a point q outside P. Then display the working of the test with some creative animation. For example, you might show the ray from q approaching p, highlighting and labeling intersections and segments on the way.

It's easy now to explain the strategy of the scan-based polygon-filling algorithm. Treating each scan line as a horizontal ray starting from some point far to the left of the screen, pixels along it are selected to fill a polygon P according to the parity test. Figure 14.18 shows one such scan line. In particular, each scan line is split into segments, each with endpoints at intersections with the boundary of P. Moreover, pixels in one segment all have the same parity. The result is alternate runs of pixels filling and not filling P.

Before we proceed to specify exactly a scan-based polygon filling algorithm, we need first to resolve how to deal with pixels which happen to

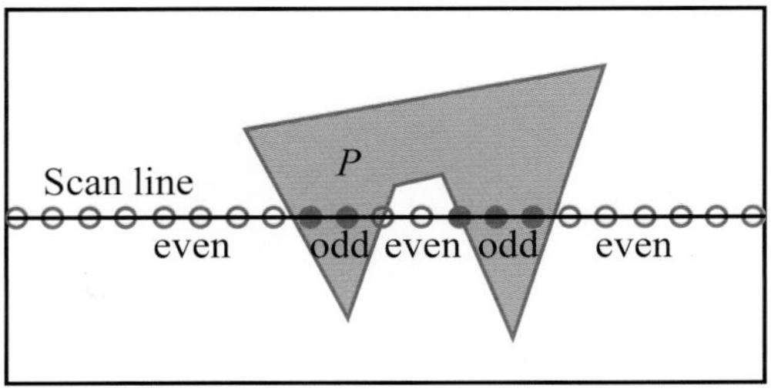

Figure 14.18: The scan line as a ray with parities along segments indicated. Pixels fill the polygon according to the parity test: pixels in the polygon are drawn solid, others are hollow.

lie on the boundary of a polygon, whose parity, therefore, is indeterminate. We do this next.

Ownership of Boundary Pixels

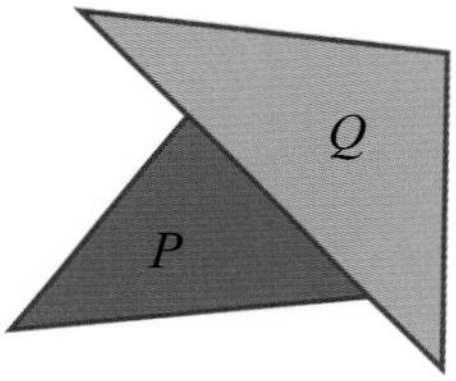

Figure 14.19: Abutting polygons.

One must be cautious in scan-based filling about pixels lying *on* the boundary of a polygon. For example, given a couple of *abutting* polygons – ones whose edges overlap, e.g., P and Q in Figure 14.19 – one has to decide ownership of the pixels along the overlapping part, which, in turn, decides the coloring of these pixels. Clearly, it is desirable to do this in a consistent manner, independent of the order in which the polygons happen to appear in the code.

For example, if each polygon is awarded ownership of all its boundary pixels, then the color of shared boundary pixels depends, in fact, on the order in which the abutting polygons appear in the code, the color of the one appearing last prevailing. At the other extreme, if boundary pixels are excluded altogether from a polygon, then there will arise gaps in the rendering between abutting polygons.

Remark 14.4. If one follows the rules of triangulation as described in Chapter 8, then there will be no anomalies, as we saw then, if every triangle simply owns its boundary. However, the intention here is to set up rules robust enough to hold *even* given an invalid triangulation, as in Figure 14.19.

A simple and oft-implemented method to resolve the ownership of a polygon's boundary and, accordingly, boundary pixels is by means of the following:

Edge Ownership Rule: Among its non-horizontal edges, a polygon owns only the *left* edges, while, among its horizontal edges, it owns only the *bottom* ones. A vertex shared by a left and right edge is owned only if both incident edges extend above the vertex; it is not owned otherwise.

Note: A left edge of a polygon is one such that its interior lies to the right of the edge; the edge does not have to be physically located at a left extreme of the polygon. Similar remarks apply to right, bottom and top edges. Figure 14.20 labels the edges of the polygon P accordingly.

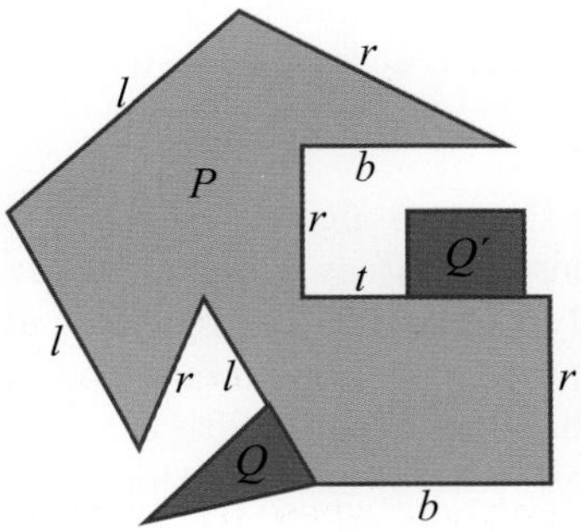

Figure 14.20: The non-horizontal edges of P are labeled left or right, while the horizontal ones top or bottom. P owns the part of the boundary it shares with Q; P does not own the part of the boundary it shares with Q'.

Remark 14.5. The rule above does not resolve the ownership question in every case of a vertex being incident to one edge which is owned by the polygon, and another which is not. Generally, this depends on the particular filling algorithm implemented.

In Figure 14.20, from the edge ownership rule, P owns the pixels on the part of the edge it shares with Q, while those on the part of the edge it shares with Q' are owned by the latter. The two pixels in Figure 14.18 on the boundary of the polygon P have been processed according to this rule as well.

Remark 14.6. The edge ownership rule explains why the expected ambiguity in edge color did not arise in Exercise 8.3.

14.4.1 Algorithms

Assume that P is a simple polygon input as the list of its edges, each edge specified by the coordinates of its end vertices. Here's a first cut at an algorithm to rasterize P:

Scan-based Polygon-filling Algorithm (Version 1)

for each scan line s
{
   1. for each non-horizontal edge e of P intersecting s
         determine the intersection point between s and e;
   2. sort the points from the preceding step from left to right along s in a list $p_1, \ldots, p_k$;
   3. fill, as belonging to P, pixels *strictly* between each of the pairs

p_1 and p_2, p_3 and p_4, . . .;

4. if a pixel coincides with a p_i, which means it's a boundary pixel, fill it or not according as i is odd or even;

5. for each horizontal edge e intersecting s
    include/exclude e according as it is a bottom/top edge;

}

The reader will agree that this algorithm is at least a forthright attempt to implement the scan-based strategy: the first three statements inside the bracketed `for` loop apply the parity test to select alternate runs of pixels to fill, while the fourth statement implements the disambiguation rule for ownership of left and right edges and the fifth that for bottom and top edges.

However, upon a more careful examination, Version 1 is seen to have three significant flaws:

(1) When a scan line passes through a vertex whose adjacent edges are non-horizontal and lie on either side of the scan line, that vertex is listed twice as an intersection point, in violation of clause (i) of convention (a) for singularities listed earlier.

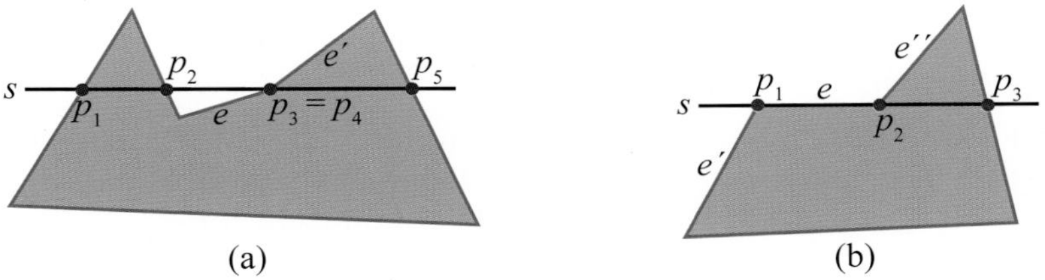

Figure 14.21: Problems with Version 1.

For example, in Figure 14.21(a), the sorted list of intersection points for scan line s is $\{p_1, p_2, p_3, p_4, p_5\}$, where p_3 $(= p_4)$ appears twice on the list, once as an intersection with the edge e and once as an intersection with edge e', leading to pixels between p_4 and p_5 being wrongfully left unfilled by statement 3 of Version 1.

(2) If a pixel coincides with more than one p_i – in particular, if it's at a vertex adjacent to two non-horizontal edges – how do we decide to include/exclude it when executing statement 4?

For example, in Figure 14.21(a), do we treat the pixel at the vertex shared by e and e' as a p_i with an odd subscript (i.e., p_3), or even subscript (p_4), at the time of processing statement 4?

(3) There may be ambiguity in rendering pixels along a horizontal edge because its vertices are those as well of its non-horizontal neighbors and will be processed as such for the scan line upon which it lies.

For example, in Figure 14.21(b), pixels in the interior of the horizontal edge e (strictly between p_1 and p_2) are filled when processing scan line s at statement 3, because s's sorted list of intersections is $\{p_1, p_2, p_3\}$. However, when processing e itself at statement 5, we find that these pixels should be excluded, as e is a top edge.

It looks like rescuing Version 1 will be tricky, but it turns out, in fact, that a fix is not hard at all. Just add in the following two rules:

(i) For a non-horizontal edge, don't list the intersection of a scan line with the upper endpoint of that edge.

(ii) Moreover, don't process horizontal edges at all.

That's it! Here's the amended algorithm with the few changes from Version 1 highlighted with comments:

Scan-based Polygon-filling Algorithm (Version 2)

for each scan line s
{
  1. for each non-horizontal edge e of P intersecting s
  {
    determine the intersection point between s and e;
    ignore this intersection point if it is the upper endpoint of e; // New line.
  }
  2. sort the points from the preceding step from left to right along s in a list $p_1, \ldots, p_k$;
  3. fill, as belonging to P, pixels *strictly* between each of the pairs p_1 and p_2, p_3 and p_4, . . .;
  4. if a pixel coincides with a *single* p_i, which means it's a boundary pixel, fill it or not according as i is odd or even; // Modified.
  5. if a pixel coincides with *more than one* p_i, which also means it's a boundary pixel, fill it; // New line. Processing of horizontal edges // is now omitted.
}

We'll soon see why this version is correct, but let's take it for a spin first. Figure 14.22 shows how many times each intersection point along the boundary of a polygon P appears in the sorted list for the scan line containing it according to statements 1 and 2 of Version 2. Based on these labels is an exercise next to run the new version (with a part answer).

Exercise 14.19. Describe how pixels are filled along each of the eight scan lines drawn in Figure 14.22 by Version 2.

Part answer: We'll assume first that all the vertices of P are located exactly at pixels.

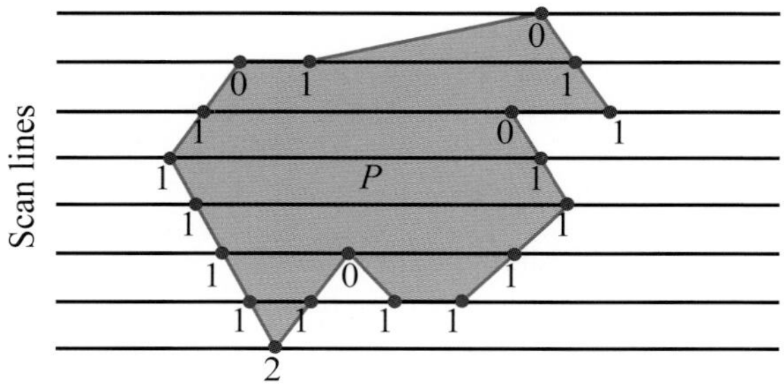

Figure 14.22: Counting intersections according to Version 2: shown are eight scan lines, their intersections with the polygon boundary and the number of times each intersection point appears in the sorted list for that scan line, according to statements 1 and 2 of algorithm Version 2. Polygon vertices and intersection points are not named to avoid clutter.

Bottom scan line: Denote the single vertex on this scan line by p_1 (note that the labels p_i are not drawn in the figure to avoid clutter). Then the sorted list on this scan line, after statements 1 and 2, is $\{p_1, p_1\}$. Only the pixel at p_1 is filled (by statement 5); all others on the scan line are not.

Second scan line: Denote the four intersection points along this scan line from left to right by p_1, p_2, p_3, p_4. Then the sorted list (statements 1 and 2) is $\{p_1, p_2, p_3, p_4\}$. All pixels strictly between p_1 and p_2 and strictly between p_3 and p_4 are filled (statement 3). If there is a pixel at p_1, it is filled (statement 4). The pixel at p_3 is filled (statement 4 again). All other pixels on the scan line are not filled.

Third scan line: Denote the three intersection points along this scan line from left to right by p_1, p_2, p_3. Then the sorted list (statements 1 and 2) is $\{p_1, p_3\}$. All pixels strictly between p_1 and p_3 are filled (statement 3). Note that the pixel at p_2 is filled. If there is a pixel at p_1, it is filled (statement 4). All other pixels on the scan line are not filled.

As the reader may have surmised from completing the exercise, the new version gets past the three problems of the earlier one as follows:

(1) Ignoring upper endpoints eliminates the first flaw.

(2) Noting that a pixel coincides with more than one p_i only when it is the bottom endpoint of two adjacent edges – upper endpoints are ignored, remember – and, therefore, must be included, removes the second flaw.

(3) Noting that the pixels on a horizontal edge lie, as well, between the endpoints of the two non-horizontal edges adjacent to it suggests that we can leave off separately processing horizontal edges altogether. Version 2 does so, resolving the last flaw.

Exercise 14.20. How are the problems with Version 1, as indicated particularly in Figures 14.21(a) and (b), resolved in Version 2?

14.4.2 Optimizing Using Edge Coherence – Active Edge List

Version 2 is pretty much ready to go as long as a couple of statements requiring efficient implementation in code are kept in mind: in particular, statements 1 and 2 to determine and sort the intersection points of a scan line with polygon edges. These two tasks need not be done from scratch per scan line – which wouldn't be efficient at all – if one exploits so-called *edge coherence*.

This simple but very useful concept is illustrated in Figure 14.23, which shows an edge e straddling a run of scan lines. Now, if scan lines are processed in order, say from bottom to top, then e first appears in the intersection list for scan line s, remains in the intersection list for each scan line until s' and then disappears forever. Moreover, the intersection of e with successive scan lines clearly travels uniformly along the x-direction.

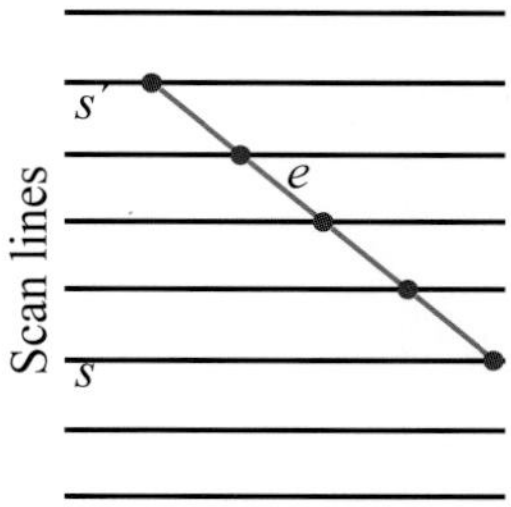

Figure 14.23: Edge coherence.

Consideration of edge coherence leads to the creation and maintenance of a particular dynamic data structure, called the *active edge list* (AEL). The AEL is a linked list of records, one for each non-horizontal edge e intersecting the *current* scan line s (hence the qualifier "active"). The record for an edge e contains three data items:

1. The y-value of the upper endpoint of e.

2. The reciprocal $1/m$ of e's gradient (as e can't be horizontal, $m \neq 0$).

3. The x-value of the intersection of e with s.

The reason for choosing these particular items will become apparent as we learn to process the AEL. The AEL, additionally, is sorted according to the left to right order of the intersections of its member edges with s, a left edge (of the polygon) preceding a right one if their intersections with s coincide. Observe that the first two items of each record are static, depending only on the particular edge e, while the value of the third item varies as s sequences through successive scan lines.

See Figure 14.24, which shows a polygon P in a 19×11 raster, as well as the AEL values for scan lines 0, 4 and 5. For example, the data values in the first record of the bottom AEL arise because it corresponds to the edge joining vertices at $(12, 7)$ and $(13, 0)$, the reciprocal of whose gradient, therefore, is $-1/7 \simeq -0.143$, and which intersects scan line 0 at x-value 13.

Exercise 14.21. Verify the data values in each record of each of the three AEL values shown in Figure 14.24.

Given the value of the AEL for a scan line s, it's straightforward to apply Version 2 to compute the runs of filled pixels along s by traversing the AEL from left to right, reading off the third data value from successive records.

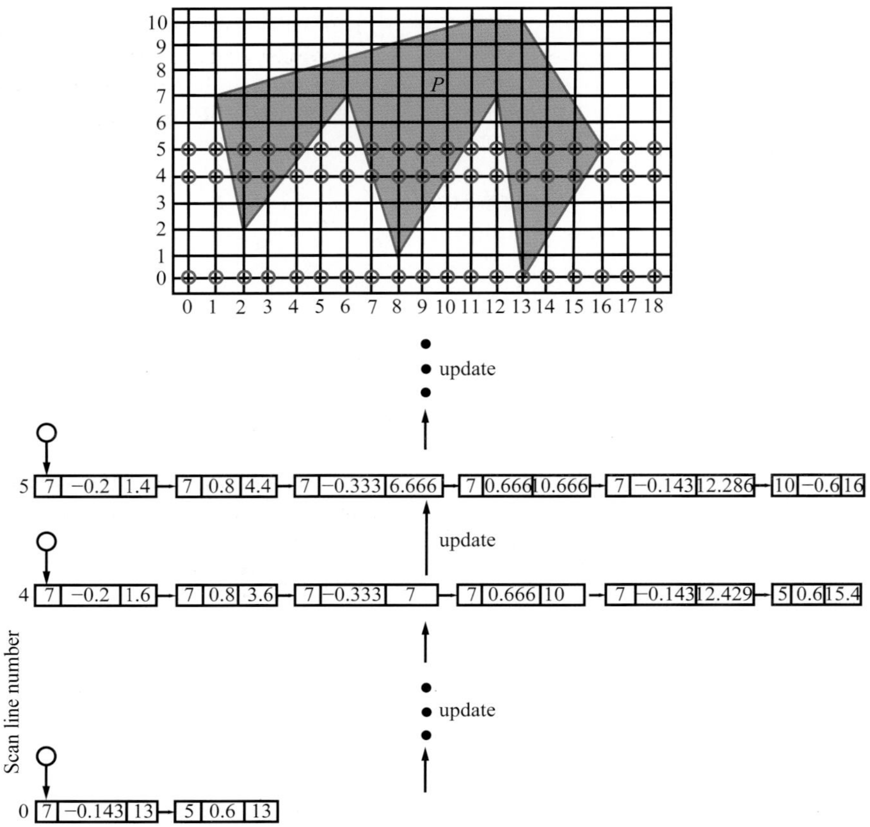

Figure 14.24: The AEL values for scan lines 0, 4 and 5 for a polygon P in a 19×11 raster. Pixels are drawn as hollow circles only for these three scan lines.

Initialization – Edge Table

All that's left now to do is initialize the AEL data structure and say how to update it from scan line to scan line. To this end, a bit of pre-processing first is needed to put the non-horizontal edges of P into a data structure called the *edge table* (ET).

The ET is an array of linked lists, one for each scan line. The linked list for a scan line consists of one record for each non-horizontal edge which has a lower endpoint on that scan line. Each record is similar in structure to an AEL record and, in fact, the first two data items are identical, while the third contains the x-value of the lower endpoint of the edge. Moreover, each list is sorted according to the left to right order of its lower endpoints, with a left polygon edge again preceding a right one if their lower endpoints coincide.

Figure 14.25 shows the ET for the polygon P of Figure 14.24. The ET is evidently a static data structure, consisting, essentially, of a bucket sort of the non-horizontal edges keyed on their lower y-value.

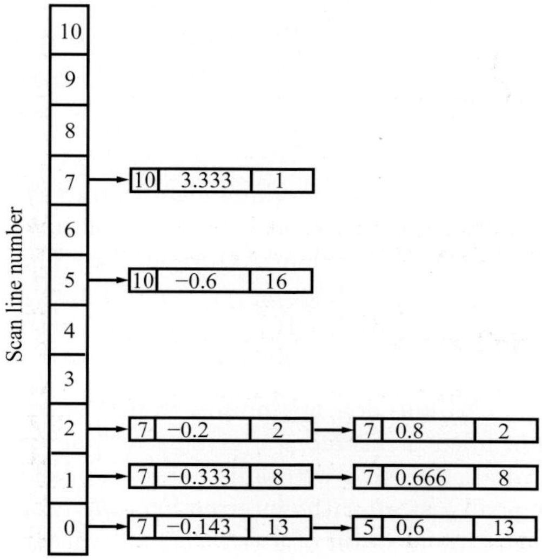

Figure 14.25: Edge table for the polygon P of Figure 14.24.

Processing the AEL

With the ET in hand, processing the AEL is straightforward. Set $j = -1$, the number of an imaginary scan line just below the drawing area and initialize the AEL to empty. To update the AEL from scan line j to scan line $j+1$, do the following:

<u>Updating the AEL</u>
for each record in the AEL
{
> **1**. if the first data value (i.e., the y-value of the upper endpoint of the corresponding edge) equals $j+1$, delete the record;
> **2**. else, update the third data value (i.e., the x-value of the intersection of the edge with the scan line) by adding to it the second data value (i.e., the reciprocal of the edge's gradient);

}
if the ET list for scan line $j+1$ is not empty, merge that list with the AEL using the third data value as the key;

The reader is asked next to prove the one fact required to assure the validity of the AEL update procedure.

Exercise 14.22. Show that the third data value is correctly computed for each record in the new AEL by the update procedure.

Exercise 14.23. Refer to Figure 14.24. Apply the update procedure to compute the AEL values for scan lines 1, 2, 3 and 6.

Exercise 14.24. Incorporate into the Scan-based Polygon-filling Algorithm (Version 2) both initialization of the edge table and subsequent AEL-driven processing.

If you're coming to this polygon-filling algorithm fresh from Bresenham's line rasterizer, then alarm bells are probably going off in your head right now regarding floating point computation. And rightly so. The update procedure as just described does require a floating point addition in its step 2 to update a record. This, in fact, can be avoided by optimizing the implementation. We'll let you figure out how in the next exercise.

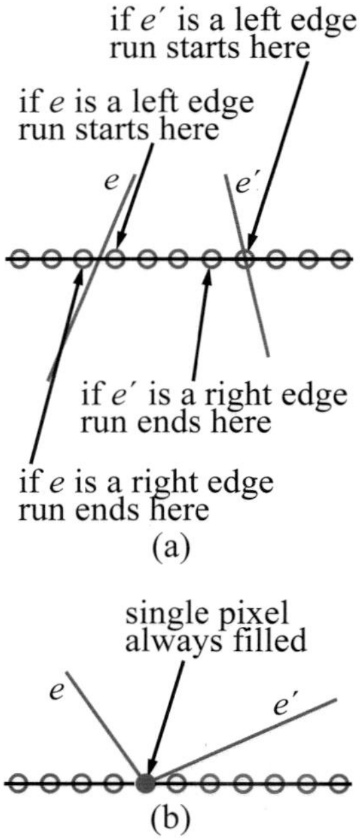

Figure 14.26: (a) Typical dispositions of a left and right edge with respect to a run of pixels (b) The one exceptional case where a left and a right edge meet at a common lower endpoint.

Exercise 14.25. To cut floating points from Version 2, observe that, generally, a run of filled pixels along a given scan line s will (1) start at at the intersection of a left edge e with s if there is a pixel at the intersection or, if not, at the pixel just after the intersection, *and* (2) the run will end at the pixel before the intersection of a right edge e with s, even if there is a pixel at the intersection. See Figure 14.26(a). The one exceptional case – easily detected – is shown in Figure 14.26(b).

Therefore, instead of the floating point x-value, call it X, of the intersection of a segment e with a scan line s, one can store in AEL records the smallest integer greater than or equal to X if e is a left edge, or the largest integer strictly less than X if it is a right edge.

Accordingly, suggest a new form of the AEL with no floating point data items and a method to update it.

Remark 14.7. Polygon rasterization in OpenGL is particularly simple, as it consists only of rasterizing triangles in the case of polygon drawing calls – recall that OpenGL polygons are always automatically fan-triangulated – or axis-aligned rectangles in the case of `glRectf()` calls.

Flood-fill

The scan-based polygon rasterization algorithm does not explicitly draw a polygon P's boundary given its vertices. Rather, its output is directly a set of pixels filling an area corresponding to P. Another approach to rasterizing P is, in fact, to draw first its boundary by, say, repeatedly applying Bresenham's line rasterizer for each edge, and then fill its interior.

Once the boundary of P has been drawn, a particularly intuitive algorithm to fill its interior is the so-called *flood-fill* algorithm. Here's how flood-fill works:

Start with a pixel p known to be in P's interior, found, possibly, by following a ray through the raster and applying the parity test. Fill p. Then examine four of p's neighboring pixels, in particular, the ones to its north, south, east and west. Of these pixels, which are said to be *4-adjacent* to p, fill those that don't belong to the boundary and have not yet been filled. Next, examine the pixels 4-adjacent to the ones just filled and, again, fill those that don't belong to the boundary and have not yet been filled. Continue in this manner until no more pixels can be filled. See Figure 14.27 for the initial configuration and the next two steps of a flood-fill.

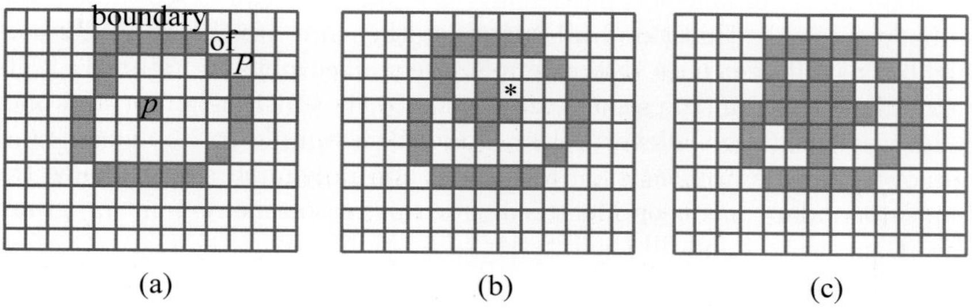

Figure 14.27: Flood-fill: (a) Initially (b) Fill pixels 4-adjacent to p (c) Fill pixels 4-adjacent to the ones filled in the previous step. The starred pixel of (b) is examined by both its south and west neighbors at this step.

Exercise 14.26. In how many more steps after Figure 14.27(c) does the flood-fill algorithm terminate?

Coding flood-fill is simple and we ask the reader to do this next.

Exercise 14.27. Assume that initially all pixels in the raster are of `background_color`, that a line rasterizer has subsequently been invoked on the edges of an input polygon P to set its boundary pixels to `foreground_color` and that the pixel at location (x, y) is known to be in the interior of P.

Pseudo-code a recursive flood-fill procedure to set pixels in the interior of P to `foreground_color` as well.

Now, the flood-fill algorithm will fill the interior of a polygon P provided that it is *4-connected*, in that any pair of pixels in the interior can be joined by a path of pixels, each consecutive pair of which is 4-adjacent. The interior of the polygon P of Figure 14.27 is 4-connected and flood-fill indeed terminates successfully after filling the whole interior. On the other hand, the boundary of a polygon Q where flood-fill will not succeed is shown in Figure 14.28. Evidently, Q's interior is not 4-connected and, whichever of its two pixels is chosen to start with, flood-fill will terminate after filling only that one.

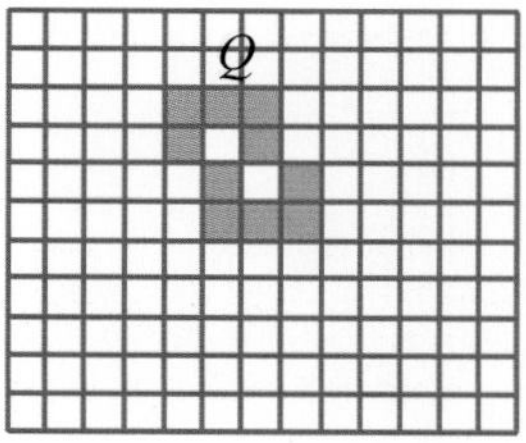

Figure 14.28: Flood-fill fails on the polygon Q whose boundary is drawn.

It might seem, then, that one need only enhance flood-fill to examine all eight neighbors of the current pixel, which are said to be 8-adjacent to it, instead of only the 4-adjacent ones. However, care is needed. For, consider the two interior pixels of Q in Figure 14.28. Both are 8-adjacent to pixels belonging neither to Q's boundary nor its interior. Therefore, a simple-minded enhancement of flood-fill could "leak" and fill the whole raster. We'll leave the reader to ponder an appropriate enhancement.

Exercise 14.28. Enhance flood-fill to examine all 8-adjacent neighbors of the current pixel, instead of only the 4-adjacent ones. Make sure your procedure does not leak.

Flood-fill is not efficient, particularly compared to scan-based filling. The primary reason is that flood-fill examines the same pixel repeatedly as a neighbor of different ones. For example, the starred pixel in Figure 14.27(b) is examined twice in the second step, once by its south neighbor and once by its west neighbor. Nevertheless, flood-fill is simple to implement and, moreover, can be applied even to curved non-polygonal shapes, once the shape's boundary has been identified. For this reason many paint programs allow the user to flood-fill the interior of a "blob".

14.5 Summary, Notes and More Reading

In this chapter we learned a few important algorithms from deep in the graphics pipeline. These included line and polygon clipping, as well as the raster-conversion of these primitives. This knowledge will hardly impact our programming of graphics, particularly because high-level API's like OpenGL allow no access to these algorithms. Still, it's good to have some understanding of what goes on at the far end of the pipeline. And, in Section 19.1, when we assemble the entire synthetic-camera rendering pipeline, clipping and rasterization algorithms will be incorporated into its final stage.

The algorithms in this chapter are all from the foundational period of modern CG, particularly the sixties and early seventies. The Cohen-Sutherland line clipper is from Sketchpad [139], the pioneering interactive CG system Sutherland developed in 1963. The Sutherland-Hodgeman polygon clipper [140] is from that period as well, as is Bresenham's line rasterizer [21], the latter originally being proposed as a method to plot lines on real paper using a computer-controlled pen. The scan-based filling techniques migrated from pen-plotters to raster displays as well.

The classic books by Foley et al. [47] and Rogers [117] contain extensive discussions of various raster algorithms, while the two by Akenine-Möller, Haines & Hoffman [1] and Watt [147] describe modern-day implementations.

Part VIII

Anatomy of Curves and Surfaces

CHAPTER 15

Bézier

The goal for this chapter is an understanding of the theory underlying Bézier primitives. We are already familiar with many of their practical aspects. In Section 10.3 of the chapter on drawing we saw how to specify Bézier curves and surfaces and incorporate them into our designs. This was possible then – even before theory – as an intuitive understanding of control points and their role as attractors in shaping Bézier primitives is sufficient to grasp the OpenGL syntax. We went even further in Sections 11.11.4 and 12.4.2, learning how to light and texture Bézier surfaces.

We'll restrict ourselves in this chapter to *polynomial* Bézier primitives. The more general form is *rational* – a rational function being the ratio of two polynomials. We'll postpone the discussion of the rational primitives to Chapter 18, as an application of projective spaces, which are the natural setting for these primitives.

Several 3D modeling systems support rational Bézier primitives – in fact, often, the even more general class of NURBS primitives – in a WYSIWYG environment where users create primitives interactively by manipulating control points. Of course, a system supporting rational primitives supports as well its polynomial subclass. OpenGL is often the front-end of such modelers, itself offering both rational Bézier and NURBS primitives in a low-level "code-it-yourself" manner.

There is a bit of math in the development of Bézier theory, but behind it always is the fairly intuitive "mechanics" of Bézier primitives, which we'll try to make as apparent as possible. Illustrative code is interspersed throughout this chapter as well.

We begin with Bézier curves in Section 15.1. First, de Casteljau's procedural approach to defining linear and quadratic Bézier curves is explained in Sections 15.1.1-15.1.2. The reader is asked to do most of the work for cubic Bézier curves in 15.1.3. Section 15.1.4 generalizes the development to Bézier curves of arbitrary order and we see as well a host of properties that make Bézier curves so useful in design. From Bézier curves to Bézier surfaces in Section 15.2 is a fairly intuitive progression. Section 15.3 concludes the chapter.

15.1 Bézier Curves

Suppose a programmer specifies a sequence $P_0, P_1, \ldots, P_n$ of $n+1$ *control points*, asking for a curve not necessarily passing through them, but, rather, whose shape is molded by the control points. In other words, the control points are expected to act as "attractors", each exerting a pull on the curve. The generated curve is said to *approximate* the control points. Figure 15.1 is illustrative of the situation.

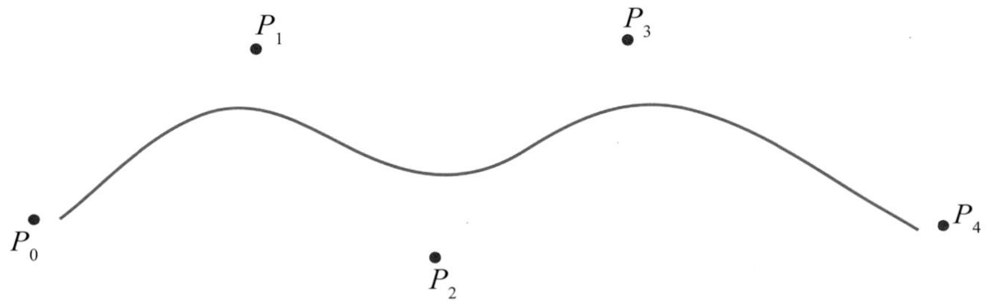

Figure 15.1: A curve approximating five control points.

Circa 1960, two French automotive designers, Bézier [13, 14] and de Casteljau [35, 36], independently invented a particular method to approximate a sequence of control points. It bears the name of Bézier because his publications had earlier circulation in the design community. However, de Casteljau's approach to Bézier curves is actually the more intuitive and it is what we'll first describe.

15.1.1 Linear Bézier Curves

Let's start with the simplest case, where there are only two control points P_0 and P_1. The Bézier curve c approximating P_0 and P_1 is, simply, the straight line segment joining the two. We write the parametric equation of c as follows:

$$c(u) = (1-u)P_0 + uP_1 \qquad (0 \le u \le 1) \tag{15.1}$$

See Figure 15.2(a). This Bézier curve is said to be *linear*, or of *degree one*, or of *second order*, order being the number of control points.

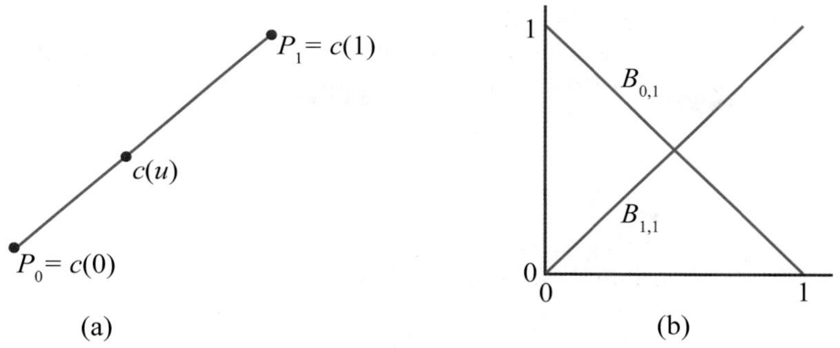

Figure 15.2: (a) Bézier curve of degree 1 (b) Bernstein polynomials of degree 1: $B_{0,1}(u) = 1 - u,\ \ B_{1,1}(u) = u$.

If the ambient space is $\mathbb{R}^2$, and $P_0 = [x_0\ y_0]^T$ and $P_1 = [x_1\ y_1]^T$, we can write (15.1) as

$$c(u) = [(1-u)\,x_0 + u\,x_1 \quad (1-u)\,y_0 + u\,y_1]^T \qquad (0 \le u \le 1) \qquad (15.2)$$

Exercise 15.1. Write an equation analogous to (15.2) if the ambient space is $\mathbb{R}^3$.

Remark 15.1. As far as the theory goes, the control points can belong to a real space of arbitrary dimension, but practical applications are in $\mathbb{R}^2$ or $\mathbb{R}^3$.

Remark 15.2. It is evident from either (15.1) or (15.2) that a linear Bézier curve linearly interpolates between its two control points (recall linear interpolation from Section 7.2).

Example 15.1. What is the equation of the linear Bézier curve c with control points $[5\ 1]^T$ and $[-1\ 0]^T$? What are the points on c corresponding to the values 0, 0.3 and 1 of the parameter u?

Answer: The equation of c is

$$c(u) = (1-u)[5\ 1]^T + u[-1\ 0]^T = [5 - 6u \quad 1 - u]^T \qquad (0 \le u \le 1)$$

The point corresponding to $u = 0$ is $[5\ 1]^T$, the first control point.
The point corresponding to $u = 0.3$ is $[3.2\ 0.7]^T$.
The point corresponding to $u = 1$ is $[-1\ 0]^T$, the second control point.

Exercise 15.2. What is the equation of the linear Bézier curve c with control points $[0\ \ -2\ \ -4]^T$ and $[3\ 8\ 0]^T$? What are the points on c corresponding to the values 0, 0.3 and 1 of the parameter u?

We make the following observations, all straightforward, for a linear Bézier curve c:

1. The parametric equation (15.1) for c is linear in u, which, of course, is why it's called a linear Bézier curve.

2. The *point* $c(u)$ of the curve c is a weighted sum of the control points P_0 and P_1, the weights of P_0 and P_1 being the *values* of $1 - u$ and u, respectively. Accordingly, the *curve* c can be thought of as a weighted sum of P_0 and P_1, where the weights of P_0 and P_1 are the *functions* $1 - u$ and u, respectively.

   These functions are called the *blending functions* of the respective control points (the term *basis function* is used as well).

   The blending functions $1 - u$ (of the first control point) and u (of the second control point) are known as the *Bernstein polynomials* of degree 1. They are denoted $B_{0,1}(u)$ and $B_{1,1}(u)$, respectively. Figure 15.2(b) shows their graphs. Accordingly, Equation (15.1) can be written as

$$c(u) = B_{0,1}(u)P_0 + B_{1,1}(u)P_1 \qquad (0 \le u \le 1) \qquad (15.3)$$

   The blending function $B_{0,1}(u)$ of the first control point decreases from 1 to 0 as u goes from 0 to 1, while exactly the opposite is true of that of the second control point.

3. Because

   (a) $B_{0,1}(u)$ and $B_{1,1}(u)$ both lie between 0 and 1, and

   (b) $B_{0,1}(u) + B_{1,1}(u) = 1$,

   for each u in $0 \le u \le 1$, every point of c is a convex combination (recall Definition 7.3) of the control points P_0 and P_1 and, therefore, lies in their convex hull, which is actually pretty obvious in this simple case.

4. c starts at the first control point P_0, when $u = 0$, and ends at the second one P_1, when $u = 1$.

   If an approximating curve passes through a control point, then it is said to *interpolate* it. So a linear Bézier curve interpolates both its control points.

15.1.2 Quadratic Bézier Curves

Consider next three control points P_0, P_1 and P_2. We want to construct the Bézier curve approximating these control points by means of a process of linear interpolation. Is there, though, an evident way to linearly interpolate a curve between three control points "simultaneously"? What does this even mean?

A possibility, of course, is to linearly interpolate between P_0 and P_1 and then between P_1 and P_2, to get the two-segment polyline $P_0P_1P_2$, which is not particularly attractive because of the corner at P_1 (see Figure 15.3(a)). De Casteljau's method, however, succeeds by adding a third interpolation

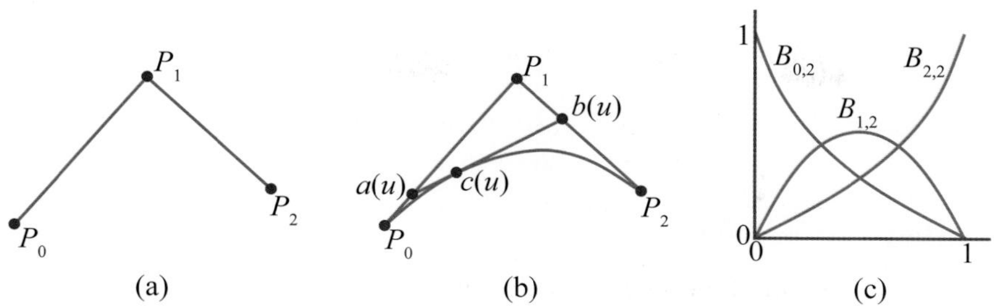

Figure 15.3: (a) An "unhappy" way of approximating three control points (b) $c(u)$ describes a Bézier curve of degree 2 interpolating P_0, P_1 and P_2 after a "triple" interpolation (c) Bernstein polynomials of degree 2:
$B_{0,2}(u) = (1-u)^2$, $B_{1,2}(u) = 2(1-u)u$, $B_{2,2}(u) = u^2$.

step to "amalgamate" the two segments P_0P_1 and P_1P_2, smoothening thereby the corner. Here's how it works.

Given a u, $0 \le u \le 1$ (see Figure 15.3(b)):

1. First interpolate between P_0 and P_1 to find the point
$$a(u) = (1-u)P_0 + uP_1$$
2. Next interpolate between P_1 and P_2 to find the point
$$b(u) = (1-u)P_1 + uP_2$$
3. Finally, interpolate between $a(u)$ and $b(u)$ to determine the point
$$c(u) = (1-u)a(u) + ub(u)$$

Substituting the expressions for $a(u)$ and $b(u)$ into that for $c(u)$, one obtains the parametric equation for a curve c:

$$c(u) = (1-u)^2P_0 + 2(1-u)uP_1 + u^2P_2 \qquad (0 \le u \le 1) \tag{15.4}$$

As u varies from 0 to 1, $c(u)$ describes the *quadratic*, or *degree two*, or *third-order*, Bézier curve approximating three control points P_0, P_1 and P_2, which is indeed smooth.

Note: Curves drawn in this chapter are fairly accurate sketches, but not necessarily exact plots of their equations.

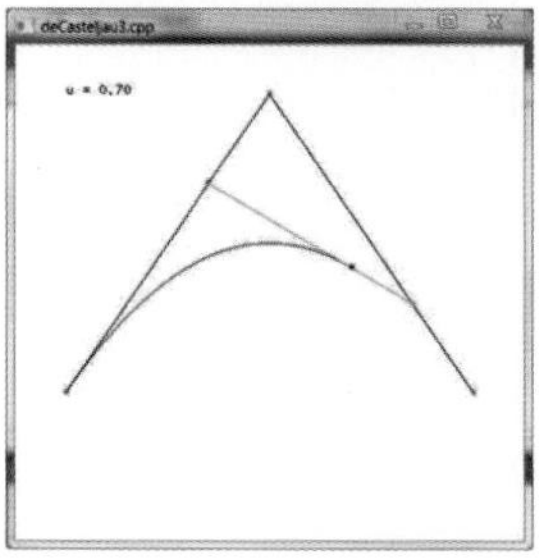

Figure 15.4: Screenshot of `deCasteljau3.cpp`.

Experiment 15.1. Run `deCasteljau3.cpp`, which shows an animation of de Casteljau's method for three control points. Press the left or right arrow keys to decrease or increase the curve parameter u. The interpolating points $a(u)$, $b(u)$ and $c(u)$ are colored red, green and blue, respectively. Figure 15.4 is a screenshot. **End**

If the ambient space is $\mathbb{R}^2$, and $P_0 = [x_0\ y_0]^T$, $P_1 = [x_1\ y_1]^T$ and $P_2 = [x_2\ y_2]^T$, one can write (15.4) as

$$\begin{aligned} c(u) &= [(1-u)^2\, x_0 + 2(1-u)u\, x_1 + u^2\, x_2 \\ &\qquad (1-u)^2\, y_0 + 2(1-u)u\, y_1 + u^2\, y_2]^T \quad (0 \le u \le 1) \end{aligned} \quad (15.5)$$

Example 15.2. What is the equation of the third-order Bézier curve c with control points $[0\ -1]^T$, $[1\ 2]^T$ and $[5\ -1]^T$?

Answer: The equation of c is

$$\begin{aligned} c(u) &= (1-u)^2[0\ -1]^T + 2(1-u)u[1\ 2]^T + u^2[5\ -1]^T \\ &= [2u + 3u^2 \quad -1 + 6u - 6u^2]^T \qquad (0 \le u \le 1) \end{aligned}$$

Exercise 15.3. What is the equation of the third-order Bézier curve c with control points $[-2\ 2\ 2]^T$, $[0\ 3\ 5]^T$ and $[-6\ 0\ 2]^T$? What are the points on c corresponding to the values 0, 0.5 and 1 of the parameter u?

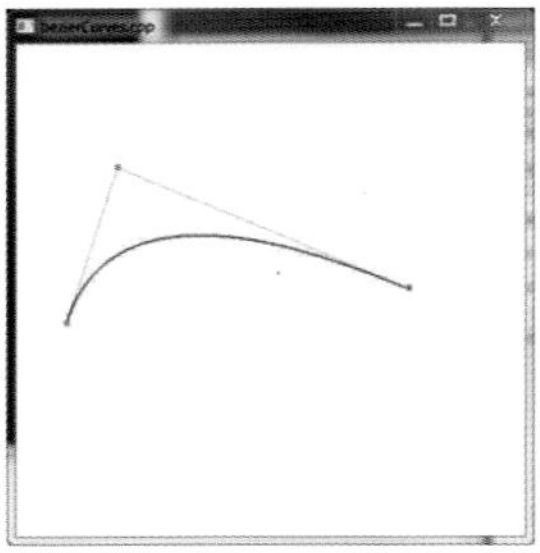

Figure 15.5: Screenshot of `bezierCurves.cpp` with three control points, showing both the Bézier curve and its control polygon.

Experiment 15.2. Run `bezierCurves.cpp`, which allows the user to choose a Bézier curve of order 2-6 and move each control point.

You can choose an order in the first screen by pressing the up and down arrow keys. Select 3. Press enter to go to the next screen to find the control points initially on a straight line. Press space to select a control point – the selected one is red – and then arrow keys to move it. Delete resets to the first screen. Figure 15.5 is a screenshot.

The polygonal line joining the control points, called the *control polygon* of the curve, is drawn in light gray. Evidently, the Bézier curve "mimics" its control polygon, but smoothly, avoiding a corner. **End**

Compare the following observations for a third-order Bézier curve c with the corresponding ones, made earlier, for a second-order curve:

1. c is quadratic in u.

2. c is a weighted sum of the control points P_0, P_1 and P_2, where the weights of P_0, P_1 and P_2 are the blending functions $(1-u)^2$, $2(1-u)u$ and u^2, respectively. These blending functions are called the Bernstein polynomials of degree 2, and denoted $B_{0,2}(u)$, $B_{1,2}(u)$ and $B_{2,2}(u)$, respectively. Figure 15.3(c) shows their graphs. Accordingly, Equation (15.4) can be written as

$$c(u) = B_{0,2}(u)P_0 + B_{1,2}(u)P_1 + B_{2,2}(u)P_2 \qquad (0 \le u \le 1) \qquad (15.6)$$

It's useful to think of the value of a control point's blending function at $c(u)$ as the amount of its "attraction" (or "pull", or "weight") on that point of the Bézier curve and of u as a dial propelling $c(u)$ along the curve by altering these attractions.

The blending function $B_{0,2}(u)$ of the first control point P_0 decreases from 1 to 0 as u goes from 0 to 1; the blending function $B_{1,2}(u)$ of the middle control point P_0 starts and ends at 0, reaching a maximum value of $\frac{1}{2}$ at $u = \frac{1}{2}$; finally, the blending function $B_{2,2}(u)$ of the last control point P_0 increases from 0 to 1. So, e.g., the attraction of the middle control point is greatest on the point of the curve corresponding to $u = \frac{1}{2}$.

3. Every point of c is a convex combination of the control points P_0, P_1 and P_2, because

   (a) $B_{0,2}(u)$, $B_{1,2}(u)$ and $B_{2,2}(u)$ all lie between 0 and 1, and

   (b) $B_{0,2}(u) + B_{1,2}(u) + B_{2,2}(u) = (1-u)^2 + 2(1-u)u + u^2 = 1$,

   for each u in $0 \le u \le 1$. It follows that the entire curve c is contained in the convex hull of P_0, P_1 and P_2.

   Colloquially, (a) and (b) say that the attraction of each control point is between 0 and 1 and that the total attraction of all three is always 1.

4. c interpolates the first and last control points, but not necessarily the middle one.

Exercise 15.4. What is the attraction of each of the control points P_0, P_1 and P_2 on $c(0.2)$?

Exercise 15.5. Verify that Equation (15.4) can be written in the matrix form:

$$c(u) = [P_0\ P_1\ P_2] \begin{bmatrix} 1 & -2 & 1 \\ -2 & 2 & 0 \\ 1 & 0 & 0 \end{bmatrix} [u^2\ u\ 1]^T \tag{15.7}$$

15.1.3 Cubic Bézier Curves

Consider, now, four control points P_0, P_1, P_2 and P_3. We're going to ask you, dear reader, to do most of the work. Perform step-by-step interpolation, similarly to the case of three control points, as follows.

Given a u in $0 \le u \le 1$ (see Figure 15.6(a)):

1. $a(u)$ interpolates between P_0 and P_1.
2. $b(u)$ between P_1 and P_2.
3. $d(u)$ between P_2 and P_3.
4. $e(u)$ between $a(u)$ and $b(u)$.

5. $f(u)$ between $b(u)$ and $d(u)$.

6. Finally, $c(u)$ between $e(u)$ and $f(u)$.

As u varies from 0 to 1, $c(u)$ describes the *cubic*, or *degree three*, or *fourth-order*, Bézier curve approximating the four control points P_0, P_1, P_2 and P_3.

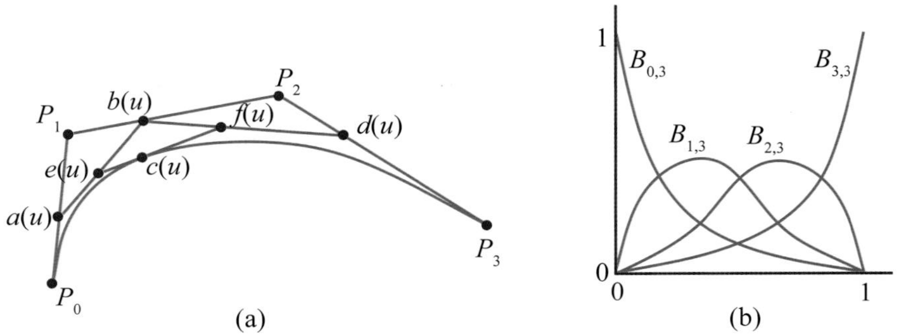

Figure 15.6: (a) Bézier curve of degree 3 (b) Bernstein polynomials of degree 3: $B_{0,3}(u) = (1-u)^3$, $B_{1,3}(u) = 3(1-u)^2u$, $B_{2,3}(u) = 3(1-u)u^2$, $B_{3,3}(u) = u^3$.

Exercise 15.6. Prove that the parametric equation of the Bézier curve c approximating the four control points P_0, P_1, P_2 and P_3 is

$$c(u) = B_{0,3}(u)P_0 + B_{1,3}(u)P_1 + B_{2,3}(u)P_2 + B_{3,3}(u)P_3 \qquad (0 \le u \le 1) \tag{15.8}$$

where the Bernstein polynomials are

$$B_{0,3}(u) = (1-u)^3, \ B_{1,3}(u) = 3(1-u)^2u, \ B_{2,3}(u) = 3(1-u)u^2, \ B_{3,3}(u) = u^3$$

Figure 15.6(b) shows their graphs.

Exercise 15.7. If the ambient space is $\mathbb{R}^2$ write an equation analogous to (15.5) for the cubic Bézier curve.

Exercise 15.8. What is the equation of the cubic Bézier curve c with control points $[-2\ 2]^T$, $[0\ -3]^T$, $[3\ 4]^T$ and $[7\ 0]^T$? What are the points on c corresponding to the values 0, 0.5 and 1 of the parameter u?

Experiment 15.3. Run `bezierCurves.cpp` and choose order 4 to get a feel for cubic Bézier curves. Note again how the curve mimics its control polygon. **End**

Exercise 15.9. Make four observations for a fourth-order Bézier curve c, corresponding to the four made for Bézier curves of orders 2 and 3.

Exercise 15.10. What is the attraction of each of the control points P_0, P_1, P_2 and P_3 at $c(0.2)$, where c is their approximating Bézier curve?

Exercise 15.11. Write Equation (15.8) in a matrix form similar to (15.7).

Exercise 15.12. **(Programming)** Write a program `deCasteljau4.cpp`, in the style of `deCasteljau3.cpp`, to illustrate de Casteljau's method for four control points.

Remark 15.3. Cubic Bézier curves are the ones most commonly used in design applications as three is a sort of "Goldilocks" degree, high enough to allow the curve good flexibility, yet not too high as to be computationally cumbersome.

Here's an exercise to get you warmed up for the general case coming next.

Exercise 15.13. From only the cases $n = 1$, 2 and 3, that we have seen, it's clear how the *variable part* of the Bernstein polynomial changes from $B_{0,n}(u)$ to $B_{1,n}(u)$, ..., finally, to $B_{n,n}(u)$. In fact, the variable part of $B_{0,n}(u)$ is $(1-u)^n u^0$. (Of course, $u^0 = 1$.) Next, for $B_{1,n}(u)$, the power of $1-u$ decreases by one and that of u increases by one, so its variable part is $(1-u)^{n-1}u^1$. And so it continues, until the variable part of $B_{n,n}(u)$ is $(1-u)^0 u^n$.

How about the *constant coefficients* though? Let's see what they are.

For Bernstein polynomials of degree 1: 1 1

For Bernstein polynomials of degree 2: 1 2 1

For Bernstein polynomials of degree 3: 1 3 3 1

Do you see a pattern? (*Hints*: Pascal's triangle, binomial coefficients.) Can you write down now the parametric equation for a fifth-order Bézier curve, without going through a de Casteljau process?

15.1.4 General Bézier Curves

It should now be fairly clear how to generalize de Casteljau's method to construct the Bézier curve approximating an arbitrary number of control points. We'll show that the parametric equation for the Bézier curve c approximating $n+1$ control points $P_0, P_1, \ldots, P_n$ is

$$c(u) = \sum_{i=0}^{n} B_{i,n}(u) P_i \qquad (0 \le u \le 1) \tag{15.9}$$

where $B_{i,n}(u)$, $0 \le i \le n$, called the ith Bernstein polynomial of degree n, is given by

$$B_{i,n}(u) = \binom{n}{i} (1-u)^{n-i} u^i \tag{15.10}$$

where $\binom{n}{i} = \frac{n!}{(n-i)!i!}$ is a binomial coefficient. The curve c is called a Bézier curve of *degree* n, or *order* $n+1$.

We'll verify Equation (15.9) by induction based on the following recursive specification of de Casteljau's method.

Recursive de Casteljau

We'll start the recursive definition by specifying (again) that the Bézier curve approximating two control points P_0 and P_1 is the straight segment joining them, given by (repeating (15.3)):

$$c(u) = B_{0,1}(u)P_0 + B_{1,1}(u)P_1 \qquad (0 \leq u \leq 1) \tag{15.11}$$

Assume, then, that we can specify the Bézier curve approximating any n control points, for some given $n \geq 2$, and that, next, we are given $n+1$ control points $P_0, P_1, \ldots P_n$. Say the Bézier curve $c_0(u)$, $0 \leq u \leq 1$, approximates the first n of these $P_0, P_1, \ldots P_{n-1}$, and that $c_1(u)$, $0 \leq u \leq 1$, approximates the last n $P_1, P_2, \ldots P_n$.

Recursive de Casteljau says, then, that the Bézier curve approximating all $n+1$ control points $P_0, P_1, \ldots P_n$ is

$$c(u) = (1-u)c_0(u) + uc_1(u) \qquad (0 \leq u \leq 1) \tag{15.12}$$

which is an "interpolation" between $c_0(u)$ and $c_1(u)$. The scheme is indicated in Figure 15.7.

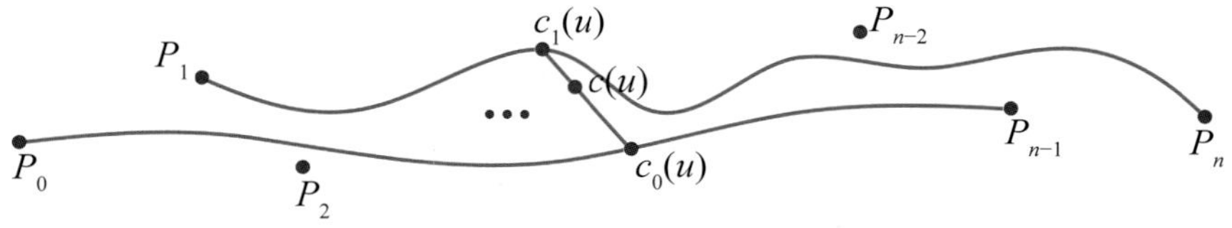

Figure 15.7: Recursive de Casteljau scheme: $c_0(u)$ approximates $P_0, P_1, \ldots, P_{n-1}$; $c_1(u)$ approximates $P_1, P_2, \ldots, P_n$; $c(u)$ interpolates between $c_0(u)$ and $c_1(u)$ to approximate $P_0, P_1, \ldots, P_n$.

Exercise 15.14. Verify that the recursive specification above yields the same equations as de Casteljau's method for three and four control points described earlier.

Note that in the case of three control points the earlier description matches exactly the recursive one above. However, in the case of four, we earlier did repeated linear interpolation between pairs of points on separate straight segments, while the recursive specification above would have us linearly interpolate just once between points on two separate quadratic Bézier curves.

Let's turn now to proving the general formula (15.9) by induction. Starting the induction is simply a matter of noting that (15.9) is identical to (15.11) when $n = 1$.

Suppose, inductively, that (15.9) is true with $n-1$ in place of n, i.e., it is true for n control points. We'll prove it next for $n+1$ control points $P_0, P_1, \dots P_n$. By the inductive hypothesis, the Bézier curve approximating the first n of these, $P_0, P_1, \dots P_{n-1}$, is given by

$$c_0(u) = \sum_{i=0}^{n-1} B_{i,n-1}(u) P_i \qquad (0 \le u \le 1)$$

and that approximating the last n points, $P_1, P_2, \dots P_n$, by

$$c_1(u) = \sum_{i=0}^{n-1} B_{i,n-1}(u) P_{i+1} \qquad (0 \le u \le 1)$$

The Bézier curve approximating all $n+1$ control points $P_0, P_1, \dots P_n$ by the recursive formula (15.12), therefore, is

$$\begin{aligned}
c(u) &= (1-u)c_0(u) + uc_1(u) \\
&= (1-u)\sum_{i=0}^{n-1} B_{i,n-1}(u)\, P_i + u\sum_{i=0}^{n-1} B_{i,n-1}(u)\, P_{i+1} \\
&= (1-u)\sum_{i=0}^{n-1} \binom{n-1}{i} (1-u)^{n-i-1}\, u^i\, P_i + \\
&\qquad u\sum_{i=0}^{n-1} \binom{n-1}{i} (1-u)^{n-i-1}\, u^i\, P_{i+1}
\end{aligned}$$

(applying formula (15.10) for Bernstein polynomials of degree $n-1$)

$$\begin{aligned}
&= (1-u)\sum_{i=0}^{n-1} \binom{n-1}{i} (1-u)^{n-i-1}\, u^i\, P_i + \\
&\qquad u\sum_{i=1}^{n} \binom{n-1}{i-1} (1-u)^{n-i}\, u^{i-1}\, P_i
\end{aligned}$$

(changing the limits on the second summation by replacing i by $i-1$)

$$\begin{aligned}
&= (1-u)^n\, P_0 + \sum_{i=1}^{n-1} \left(\binom{n-1}{i} (1-u)^{n-i}\, u^i + \binom{n-1}{i-1} (1-u)^{n-i}\, u^i \right) P_i + \\
&\qquad u^n\, P_n
\end{aligned}$$

(bringing together terms for $i = 1, \dots, n-1$ from the two summations)

$$= (1-u)^n\, P_0 + \sum_{i=1}^{n-1} \left(\binom{n}{i} (1-u)^{n-i}\, u^i \right) P_i + u^n\, P_n$$

(using the property of binomial coefficients that $\binom{n-1}{i} + \binom{n-1}{i-1} = \binom{n}{i}$)

$$= \sum_{i=0}^{n} B_{i,n}(u)\, P_i$$

This completes the inductive verification of Equation (15.9).

If the control points are $P_i = [x_i\ y_i\ z_i]^T$, $0 \le i \le n$, in real-world 3-space, then (15.9) can be written as

$$c(u) = \left[\sum_{i=0}^{n} B_{i,n}(u)x_i \quad \sum_{i=0}^{n} B_{i,n}(u)y_i \quad \sum_{i=0}^{n} B_{i,n}(u)z_i \right]^T \quad (0 \le u \le 1) \tag{15.13}$$

We collect facts about general Bézier curves in the following:

Proposition 15.1. *If c is the Bézier curve approximating the sequence of $n+1$ control points $P_0, P_1, \ldots, P_n$ – called a Bézier curve of order $n+1$, or degree n – then the following hold:*

(a) c is polynomial of degree n in the parameter u. In particular, each coordinate value of c is polynomial of degree n in u.

(b) c is a weighted sum of the control points $P_0, P_1, \ldots, P_n$, where the weight of P_i, for $0 \le i \le n$, is its blending function $B_{i,n}(u)$.

Definition 15.1. A set of functions is said to a form a *partition of unity* over some domain if they are each non-negative and add up to 1 everywhere in that domain.

(c) The blending functions $B_{i,n}(u)$, $0 \le i \le n$, form a partition of unity over the parameter space $[0, 1]$.

(d) Every point of c is a convex combination of the control points $P_0, P_1, \ldots, P_n$; therefore, c lies inside the convex hull of $P_0, P_1, \ldots, P_n$.

(e) c interpolates the first and last control points, but not necessarily intermediate ones.

(f) (Affine Invariance) If the control points $P_0, P_1, \ldots, P_n$ belong to $\mathbb{R}^3$ and $g : \mathbb{R}^3 \to \mathbb{R}^3$ is an affine transformation, then the image curve $g(c)$ is the Bézier curve approximating the images $g(P_0), g(P_1), \ldots, g(P_n)$ of the control points.

In other words, the transformed curve approximates the transformed control points.

(g) (End Tangents) The tangent to c at P_0 lies along the straight line from P_0 to P_1 and the tangent to c at P_n lies along the straight line from P_{n-1} to P_n.

Note: Further discussions of affine invariance and end tangents follow the proof.

Proof. Items (a) and (b) follow straightforwardly from Equation (15.9).

It's easily seen that Bernstein polynomials $B_{i,n}(u) = \binom{n}{i}(1-u)^{n-i}u^i$ all lie between 0 and 1, for each u in $0 \le u \le 1$. Further, for any such u,

$$\sum_{i=0}^{n} B_{i,n}(u) = \sum_{i=0}^{n} \binom{n}{i}(1-u)^{n-i}u^i = (\,(1-u)+u\,)^n = 1 \qquad (15.14)$$

by the Binomial Theorem. This proves that the blending functions form a partition of unity over the parameter space $[0, 1]$, establishing (c). It follows, as well, that the point $c(u) = \sum_{i=0}^{n} B_{i,n}(u)P_i$, for $0 \le u \le 1$, is indeed a convex combination of the points $P_0, P_1, \ldots, P_n$, proving (d).

Item (e) is verified by checking that $c(0) = P_0$ and $c(1) = P_n$.

The proof of (f) exploits the fact that the blending functions form a partition of unity over the parameter space. Let the affine transformation $g : \mathbb{R}^3 \to \mathbb{R}^3$ be given by $g(P) = MP + D$, where M is a non-singular 3×3 matrix and D a 3-vector (the translational component). For any u in $0 \le u \le 1$, we then have

$$\begin{aligned}
g(c(u)) &= Mc(u) + D \\
&= M(\sum_{i=0}^{n} B_{i,n}(u)P_i\,) + D \\
&= \sum_{i=0}^{n} B_{i,n}(u)(MP_i) + (\sum_{i=0}^{n} B_{i,n}(u))D \\
&\qquad \text{(invoking the partition-of-unity property } \textstyle\sum_{i=0}^{n} B_{i,n}(u) = 1) \\
&= \sum_{i=0}^{n} B_{i,n}(u)(MP_i + D) \\
&= \sum_{i=0}^{n} B_{i,n}(u)g(P_i)
\end{aligned}$$

proving (e).

For (g), observe that the derivative

$$\begin{aligned}
c'(u) &= \frac{\mathrm{d}}{\mathrm{d}u}(\,(1-u)^n P_0 + n(1-u)^{n-1}uP_1 + \ldots) \\
&= -n(1-u)^{n-1}P_0 + n(1-u)^{n-1}P_1 + \textit{ terms that} \\
&\quad \textit{each contain the factor } u \qquad (15.15)
\end{aligned}$$

Therefore, the tangent vector at P_0, when $u = 0$, is

$$c'(0) = -nP_0 + nP_1 = n(P_1 - P_0)$$

as terms after the second of (15.15) all vanish. Evidently, then, the tangent at P_0 is in the direction toward P_1, proving the first part of (g). The second part follows symmetrically. □

Remark 15.4. From (a) it follows that a Bézier curve is C^∞, or smooth (recall definitions from Section 10.1.6).

Experiment 15.4. Run `bezierCurves.cpp` and choose the higher orders. It's straightforward to enhance the code for orders even greater than 6. **End**

Affine Invariance of Bézier Curves

The affine invariance of Bézier curves given by Proposition 15.1(f) is extremely useful. Here's a simple example of what it means practically. Suppose a cubic Bézier curve c approximating the four control points

$$P_0 = [5\ 5]^T \qquad P_1 = [7\ 8]^T \qquad P_2 = [8\ 4]^T \qquad P_3 = [11\ 5]^T$$

is drawn as a 10-segment polyline l (Figure 15.8(a), where l is drawn at an offset to c).

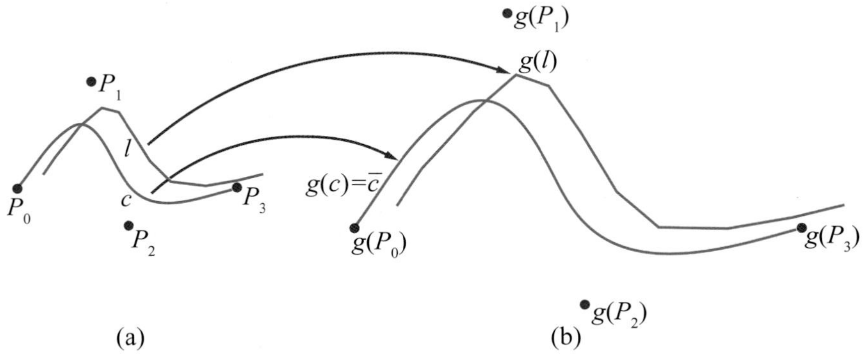

Figure 15.8: (a) A 10-segment fairly smooth-looking polyline l approximation (drawn at an offset) of the Bézier curve c with control points P_0, P_1, P_2 and P_3 (b) The magnification $g(l)$ (also at an offset) is not a good approximation of $g(c)$.

Next, suppose we want to magnify c twofold by scaling it a factor of 2 in each coordinate direction. For the purpose of drawing, if the scaling transformation, call it g, is applied simply to the polyline l, then the result $g(l)$ (Figure 15.8(b)) is likely too coarse an approximation of the magnified curve $g(c)$.

Affine invariance, however, says that $g(c)$ is the same as the cubic Bézier curve $\bar{c}$ approximating the four control points

$$g(P_0) = [10\ 10]^T \quad g(P_1) = [14\ 16]^T \quad g(P_2) = [16\ 8]^T \quad g(P_3) = [22\ 10]^T$$

Therefore, one can "forget" the original polyline and, instead, approximate $\bar{c}$ at a resolution of one's choosing (e.g., with a 20-segment polyline).

Bottom Line: Affine invariance means that if a Bézier curve is affinely transformed to a new one, then the original control points transform to the new. Therefore, the only "data" required to generate the transformed Bézier curve are the transformed control points.

Exercise 15.15. The transformations

```
glScalef(1.0, 2.0, 2.0);
glTranslatef(2.0, 3.0, 0.0);
```

are applied to the cubic Bézier curve with control points

$$[2\ 1\ 1]^T \quad [3\ 3\ 2]^T \quad [-2\ 7\ -1]^T \quad [0\ 0\ 4]^T$$

Describe the resulting curve.

End Tangents and Joining Bézier Curves

Proposition 15.1(g) enables the user to smoothly join two Bézier curves c_0 (approximating control points $P_0, P_1, \ldots, P_n$) and c_1 (approximating control points $Q_0, Q_1, \ldots, Q_m$) by, for instance, making P_n coincide with Q_0 and arranging the three points $P_{n-1}, P_n(= Q_0)$ and Q_1 in that order on one straight line. See Figure 15.9. Another way to say this is that two Bézier curves meet smoothly if their control polygons meet smoothly.

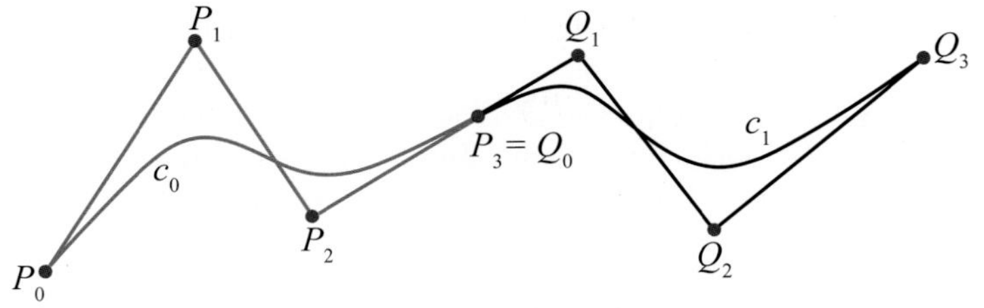

Figure 15.9: Two cubic Bézier curves (one blue, other black) meet smoothly at an endpoint.

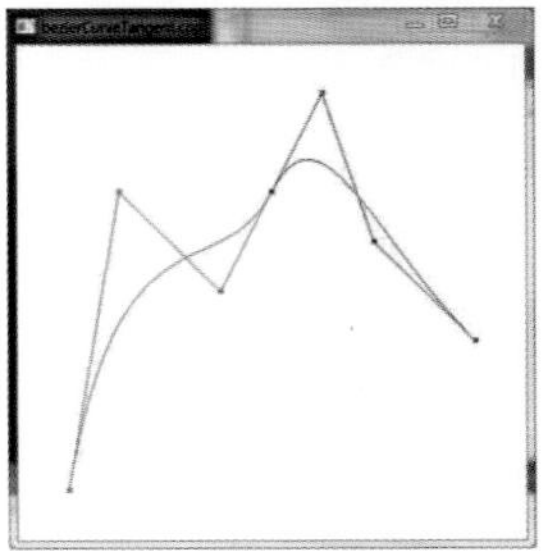

Figure 15.10: Screenshot of `bezierCurveTangent.cpp`.

Experiment 15.5. Run `bezierCurveTangent.cpp` which shows two cubic Bézier curves. The second curve may be shaped by selecting a control point with the space bar and moving it with the arrow keys. See Figure 15.10. Visually verify Proposition 15.1(g). **End**

Exercise 15.16. **(Programming)** Show how to use Proposition 15.1(g) to arrange a sequence of control points, so that the approximating Bézier curve is a *smooth closed* loop. Illustrate with the help of `bezierCurves.cpp`.

Exercise 15.17. A Bézier loop is drawn approximating the six control points

$$[0\ 0\ 0]^T \quad [4\ 0\ 0]^T \quad [8\ -5\ 3]^T \quad [-1\ -5\ -3]^T \quad [x\ y\ z]^T \quad [0\ 0\ 0]^T$$

Suggest values for x, y and z for the second to last control point to make the loop smooth at $[0\ 0\ 0]^T$.

A curve made by joining Bézier curves end to end, but not necessarily smoothly, is called *piecewise Bézier*. See Figure 15.11.

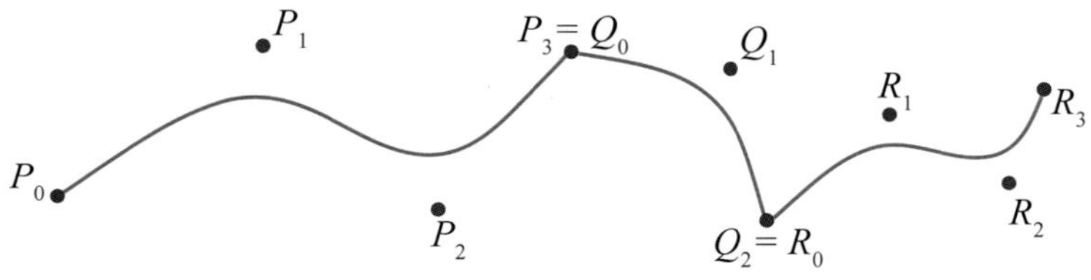

Figure 15.11: A piecewise Bézier curve consisting of three Bézier arcs.

Exercise 15.18. Can a piecewise Bézier curve be Bézier? For example, in Figure 15.11 one may ask if the union of the three Bézier curves is merely the ninth-order Bézier curve with control points $P_0, P_1, P_2, P_3 = Q_0, Q_1, Q_2 = R_0, R_1, R_2, R_3$. Consider, in particular, the case when the Bézier pieces happen all to join smoothly (which is not the case in Figure 15.11).

Exercise 15.19. The sequence $P_0, P_1, \ldots, P_n$ of the control points of a Bézier curve is obviously important. Jumbling them up will not give the same curve. How about if the sequence is reversed to $P_n, P_{n-1}, \ldots, P_0$?

Exercise 15.20. There is nothing special about the parameter space $[0, 1]$. Show how to change the parameter space of the Bézier curve, given by Equation (15.9), to $[u_1, u_2]$, where $u_1 < u_2$ may be arbitrary, without changing the curve's shape.

Exercise 15.21. Show that the blending function $B_{i,n}(u)$ of the ith control point P_i reaches its maximum at $u = \frac{i}{n}$, and at this point the value of $B_{i,n}(u)$ exceeds that of all the other blending functions. This means that the attraction of P_i is greatest on the point $c(\frac{i}{n})$ of the curve.

Polynomial Curves and Bézier Curves

As noted in Proposition 15.1(a), each coordinate value of a degree n Bézier curve is a polynomial of degree n in the parameter u. Recall from Section 10.1.4 that a polynomial curve (in $\mathbb{R}^3$) is of the form

$$b(u) = [f(u)\ g(u)\ h(u)]^T$$

where each coordinate value $f(u)$, $g(u)$ and $h(u)$ is polynomial in u. Bézier curves are, therefore, polynomial. How about the other way around? Are polynomial curves Bézier? It's nice to know, in fact, that all polynomial curves are Bézier. Precisely:

Proposition 15.2. *If*

$$b(u) = [f(u)\ g(u)\ h(u)]^T \qquad (0 \le u \le 1)$$

is a polynomial curve, each coordinate value being a polynomial of degree at most n, then one can find $n+1$ control points $P_0, P_1, \ldots P_n$, such that $b(u) = c(u)$, where c is the Bézier approximation of the P_i, $0 \le i \le n$. In other words, the Bézier approximation of these control points is the given polynomial curve.

Proof. The proof is beyond our scope here and the interested reader is referred to the text by Buss [22]. □

Remark 15.5. It's certainly gratifying that, despite their arising from the very special de Casteljau construction, the proposition assures us that the class of Bézier curves is just as general as the class of polynomial curves.

At this point let's pause a moment to appreciate the power and utility of Bézier curves, particularly in light of the preceding proposition. Suppose that a developers' group set out to design 1D primitives for a modeler. They might quite reasonably decide to support, in addition to straight lines and polylines, cubic polynomial curves, namely, those of the form

$$p(u) = [f_0{+}f_1u{+}f_2u^2{+}f_3u^3 \quad g_0{+}g_1u{+}g_2u^2{+}g_3u^3 \quad h_0{+}h_1u{+}h_2u^2{+}h_3u^3]^T$$

for $0 \le u \le 1$. That's 12 scalar coefficients $f_0, f_2, \ldots, h_3$ required to specify such a curve and a (very simple-minded) design decision might be to allow the user to edit the curve by changing each.

Contrast this with representing a cubic polynomial curve as a cubic Bézier curve specified by four control points. The size of the representation is still 12 scalars – three coordinates per control point – and the preceding proposition says that we still get all cubic polynomial curves. Consider, though, how much more convenient it is to mold the curve by manipulating control points rather than coefficients!

In fact, is there at all an easy-to-understand relationship between the coefficients $f_0, f_2, \ldots, h_3$ and the shape of $p(u)$ as given above? Even for the simple plane paper graph of a curve, say, $y = 3x^3 - x^2 + 5x + 7$, do the four coefficients 3, −1, 5 and 7 themselves convey anything immediately meaningful about its shape?

15.2 Bézier Surfaces

From an understanding of Bézier curves it's a fairly intuitive next step to defining Bézier surfaces. Suppose we have an $(n+1) \times (m+1)$ *array* of control points

$$P_{i,j}\ , \quad \text{for} \quad 0 \le i \le n,\ 0 \le j \le m$$

and wish to approximate these with a surface s. A construction of s via Bézier curves is as follows:

Think of the $(n+1) \times (m+1)$ array $P_{i,j}$ as $n+1$ different *sequences*, each of $m+1$ control points. In particular, the ith sequence, for $0 \le i \le n$, consists of $P_{i,0}, P_{i,0}, \ldots, P_{i,m}$, these being the points along the ith row of the control points array. Construct the Bézier curve approximating each of these $n+1$ sequences to obtain $n+1$ different Bézier curves, each of order $m+1$. Say the Bézier curve approximating the ith sequence is c_i, $0 \le i \le n$. See Figure 15.12, where both n and m are 3.

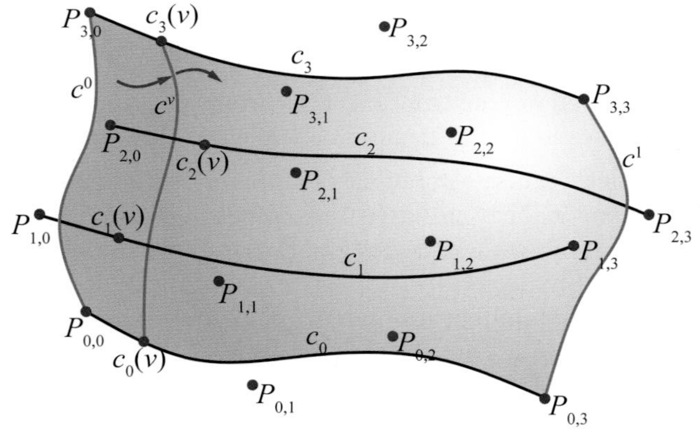

Figure 15.12: Constructing the Bézier surface approximating a 4 × 4 array of control points by sweeping a Bézier curve of order 4.

For each v in $0 \le v \le 1$, there are $n+1$ points, one on each curve c_i, corresponding to the parameter value v, namely, the sequence $c_0(v), c_1(v), \ldots, c_n(v)$. Say the Bézier curve c^v of order $n+1$ approximates these points. One such c^v is shown in the figure.

The union of all the Bézier curves c^v, for $0 \le v \le 1$, is the Bézier surface s approximating the control points array $P_{i,j}$, $0 \le i \le n$, $0 \le j \le m$. One can, as well, think of s as being *swept* by c^v, as v changes from 0 to 1.

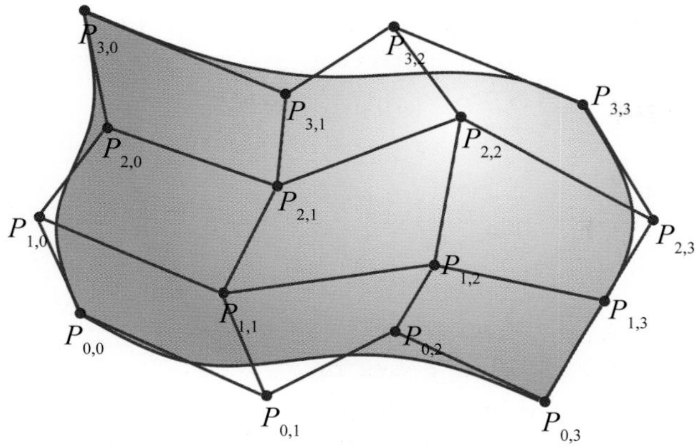

Figure 15.13: The Bézier surface approximating a 4×4 array of control points and its control polyhedron (dashed).

The polyhedral surface composed of the quadrilateral faces $P_{i,j}P_{i+1,j}P_{i+1,j+1}P_{i,j+1}$, $0 \le i \le n-1$, $0 \le j \le m-1$, is called the *control polyhedron* of the Bézier surface specified by the control points $P_{i,j}$, $0 \le i \le n$, $0 \le j \le m$. As a Bézier curve mimics its control polygon, so a Bézier surface mimics its control polyhedron. See Figure 15.13.

Experiment 15.6. Run `sweepBezierSurface.cpp` to see an animation of the procedure. Press the left/right (or up/down) arrow keys to move the sweeping curve and the space bar to toggle between the two possible sweep directions. Figure 15.14 is a screenshot.

The 4×4 array of the Bézier surface's control points (drawn as small squares) consists of a blue, red, green and yellow row of four control points each. The four fixed Bézier curves of order 4 are drawn blue, red, green and yellow, respectively (the curves are in 3-space, which is a bit hard to make out because of the projection). The sweeping Bézier curve is black and its (moving) control points are drawn as larger squares. The currently swept part of the Bézier surface is the dark mesh. The current parameter value is shown at the top left. **End**

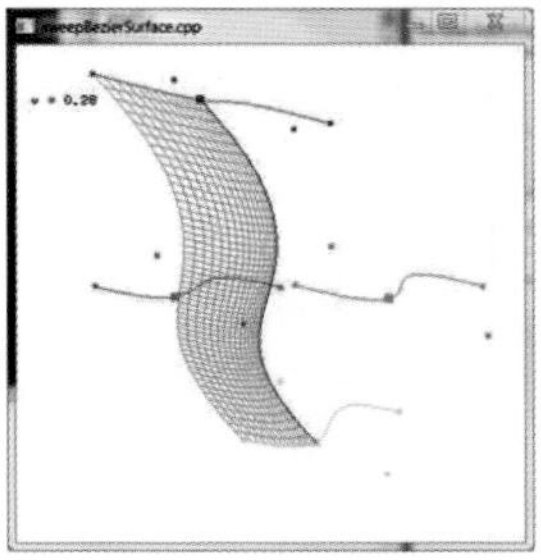

Figure 15.14: Screenshot of `sweepBezierSurface.cpp`.

Determining the parametric equation of the Bézier surface s constructed as above is not difficult. The equation of c_i, the Bézier curve along the ith row of control points, is

$$c_i(v) = \sum_{j=0}^{m} B_{j,m}(v) P_{i,j} \qquad (0 \le v \le 1)$$

for $0 \le i \le n$. Therefore, the equation of the Bézier curve c^v approximating the "column" control sequence $c_0(v), c_1(v), \ldots, c_n(v)$ is

$$
\begin{aligned}
c^v(u) &= \sum_{i=0}^{n} B_{i,n}(u) c_i(v) \\
&= \sum_{i=0}^{n} B_{i,n}(u) \left(\sum_{j=0}^{m} B_{j,m}(v) P_{i,j} \right) \\
&= \sum_{i=0}^{n} \sum_{j=0}^{m} B_{i,n}(u) B_{j,m}(v) P_{i,j} \qquad (0 \le u \le 1)
\end{aligned}
$$

Letting both u and v vary one obtains the following parametric equation for the Bézier surface s approximating the control points array $P_{i,j}$, $0 \le i \le n$, $0 \le j \le m$:

$$
s(u,v) = \sum_{i=0}^{n} \sum_{j=0}^{m} B_{i,n}(u) B_{j,m}(v) P_{i,j} \qquad (0 \le u \le 1,\ 0 \le v \le 1) \qquad (15.16)
$$

Exercise 15.22. If the control points of a Bézier surface are $P_{i,j} = [x_{i,j} \ \ y_{i,j} \ \ z_{i,j}]^T$, $0 \le i \le n$, $0 \le j \le m$, write a parametric equation for its x-, y- and z-values, analogous to (15.13) for Bézier curves. There will now, of course, be two parameter variables instead of the one for curves.

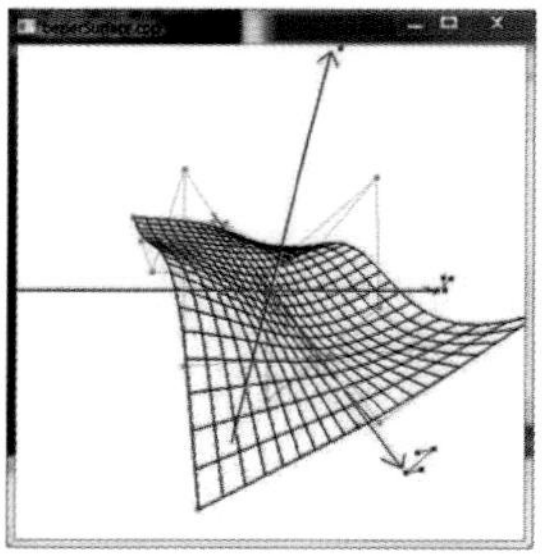

Figure 15.15: Screenshot of `bezierSurace.cpp`.

Experiment 15.7. Run `bezierSurface.cpp`, which allows the user to shape a Bézier surface by selecting and moving control points. Press the space and tab keys to select a control point. Use the left/right arrow keys to move the control point parallel to the x-axis, the up/down arrow keys to move it parallel to the y-axis and the page up/down keys to move it parallel to the z-axis.

Press 'x/X', 'y/Y' and 'z/Z' to turn the viewpoint. Figure 15.15 is a screenshot. **End**

Exercise 15.23. If the procedure to construct the Bézier surface s via Bézier curves is "inverted" to first (a) construct $m+1$ different Bézier curves, each of order $n+1$, approximating a *column* of control points, and then (b) sweep the Bézier curve approximating the points corresponding to the same parameter value on each of these $m+1$ curves, prove that the same surface s is obtained. In particular, derive the parametric form of the surface resulting from the inverted process and show it to be identical to Equation (15.16).

The program `sweepBezierSurface.cpp` of Experiment 15.6 allows the user to toggle between either process by pressing the space bar.

The following proposition is similar to Proposition 15.1 for Bézier curves:

Proposition 15.3. *If s is the Bézier surface approximating an $(n+1) \times (m+1)$ array of control points $P_{i,j}$, $0 \le i \le n$, $0 \le j \le m$, then the following hold:*

(a) s is polynomial of degree n in one parameter variable u and polynomial of degree m in the other parameter variable v.

(b) s is a weighted sum of the control points $P_{i,j}$, $0 \le i \le n$, $0 \le j \le m$, where the weight of $P_{i,j}$ is the blending function $B_{i,n}(u)B_{j,m}(v)$ (a product of Bézier curve blending functions).

(c) The blending functions $B_{i,n}(u)B_{j,m}(v)$, $0 \le i \le n$, $0 \le j \le m$, form a partition of unity over the parameter space $[0,1] \times [0,1]$.

(d) Every point of s is a convex combination of the control points $P_{i,j}$, $0 \le i \le n$, $0 \le j \le m$, and, therefore, c lies inside the convex hull of the $P_{i,j}$.

(e) s passes through the four corner control points $P_{0,0}$, $P_{n,0}$, $P_{0,m}$ and $P_{n,m}$, but not necessarily the others.

(f) (Affine Invariance) If the control points $P_{i,j}$, $0 \le i \le n$, $0 \le j \le m$, belong to $\mathbb{R}^3$ and $g : \mathbb{R}^3 \to \mathbb{R}^3$ is an affine transformation, then the image surface $g(s)$ is the Bézier surface approximating the images $g(P_{i,j})$, $0 \le i \le n$, $0 \le j \le m$, of the control points.

In other words, the transformed surface approximates the transformed control points.

Proof. We begin by observing that the blending functions $B_{i,n}(u)B_{j,m}(v)$, $0 \le i \le n$, $0 \le j \le m$, form a partition of unity over the parameter space $[0,1] \times [0,1]$ because

$$\sum_{i=0}^{n}\sum_{j=0}^{m} B_{i,n}(u)B_{j,m}(v) = \sum_{i=0}^{n} B_{i,n}(u) \sum_{j=0}^{m} B_{j,m}(v) = 1 * 1 = 1$$

proving (c). We leave the rest of the proof, which is similar to that of Proposition 15.1, to the reader. □

Exercise 15.24. What kinds of curves are the u- and v-parameter curves – recall these from Section 10.2.4 – on the following Bézier surface?

$$s(u,v) = \sum_{i=0}^{n}\sum_{j=0}^{m} B_{i,n}(u)B_{j,m}(v)P_{i,j}$$

Exercise 15.25. Recall Equation (10.20)

$$s(u,v) = (1-u)(1-v)\,p_1 + u(1-v)\,q_1 + (1-u)v\,p_2 + uv\,q_2, \qquad u,v \in [0,1]$$

of a bilinear patch from Section 10.2.8. It's again a weighted sum of the "control" points p_1, p_2, q_1 and q_2. Do the blending functions form a partition of unity? Is a bilinear patch a Bézier surface?

Exercise 15.26. What condition would you impose on the control polyhedrons of two abutting Bézier surfaces, say, s defined by an $(n+1) \times (m+1)$ array of control points and s' defined by an $(n+1) \times (m'+1)$ array of control points – the number of rows is the same – so that they join smoothly?
Hint: A similar discussion for bicubic patches was in Section 10.3.2.

15.3 Summary, Notes and More Reading

This chapter was a fairly thorough introduction to the theory of the Bézier primitives. Our exploration was restricted, however, to the polynomial version, which itself is popularly used in design and, moreover, sets the stage for the rational primitives in a forthcoming chapter. Theory too has now caught up with practice: we learned to code polynomial Bézier curves and surfaces much earlier in Chapter 10.

There are a number of excellent books – Farin [45], Mortenson [94, 96], Rogers & Adams [118] and Vince [146] to name a few – which both complement the material here and take the reader beyond it. It is interesting to read in the first chapter of Farin's book an account by Bézier himself of the invention of the UNISURF CAD system that uses his primitives. In addition to those just mentioned, which are mostly math and modeling books, any CG book itself will likely have a section or two on Bézier theory and practice. The reader should have no trouble now in following discussions of Bézier primitives in even advanced CG texts, such as Akenine-Möller, Haines & Hoffman [1], Buss [22], Slater et al. [135] and Watt [147].

CHAPTER 16

B-Spline

Our aim in this chapter is to master the theory underpinning B-spline primitives, the dominant class of primitives used in freeform design nowadays. As in the preceding chapter on Bézier theory, we'll restrict ourselves here to the polynomial version, reserving the more general rational class of NURBS (Non-Uniform Rational B-Spline) primitives for Chapter 18, as an application of projective spaces, which are the natural setting for these primitives.

Almost all 3D modelers support NURBS primitives – and so, of course, their polynomial subclass as well – in a WYSIWYG design environment. In such a setting, the user can get by merely pushing control points around, with little understanding of theory. OpenGL, on the other hand, provides an interface at a much lower level. In fact, there is almost a one-to-one correspondence between NURBS theory and OpenGL syntax. Consequently, some knowledge at least of the former is required in order to use the latter.

Unfortunately, as NURBS theory is more complex than Bézier, there really is no use-now-learn-later approach. This is the reason we did not introduce NURBS, or even its polynomial subclass, in the earlier chapter on drawing, as we did polynomial Bézier primitives. True, the lack of shortcuts and a fancy interface will be seen as drawbacks by those who care only about design and not so much about what is under the hood. On the other hand, OpenGL's minimalist setting is ideal for the purpose of grasping the underlying theory.

Our account of B-splines begins in Section 16.1 with an analysis of the weakness of Bézier primitives, motivating the progression to B-splines as a search, in fact, for better blending functions. The investigation of the B-spline primitives themselves begins with curves in Section 16.2, setting the stage with so-called knot vectors in anticipation of new blending functions that are polynomial in knot intervals. In subsections 16.2.1-16.2.3, we go from (uniform) first-order to quadratic B-spline curves, applying an intuitive "break-and-make" procedure to repeatedly increase the degree of the spline functions. The reader is asked to apply this procedure herself in 16.2.4 to fill in the details for cubic B-splines. A significant generalization is made in 16.2.5, not only by extending the theory to B-splines of arbitrary order, but by allowing the knot vector to be non-uniform as well. We'll see the utility of non-uniform knot vectors, particularly of repeated knots which empower the designer with the best of both worlds, Bézier and B-spline.

From B-spline curves to surfaces in Section 16.3 is exactly the same process as from Bézier curves to surfaces. The topic of Section 16.4 is the OpenGL NURBS drawing primitives, though we use them in this chapter only to the extent of their polynomial functionality. Subsections 16.4.1 and 16.4.2 discuss drawing B-spline curves and surfaces, respectively. We describe how to light and texture a B-spline surface in 16.4.3. The useful technique of trimming a B-spline surface is described in 16.4.4. Section 16.5, with notes and suggestions for future reading, concludes the chapter.

16.1 Problems with Bézier Primitives: Motivating B-Splines

Bézier curves and surfaces, the topics of the previous chapter, are easy to use, especially in an interactive environment, and powerful enough to create complex designs. However, they suffer from two weaknesses:

1. *Lack of local control.*

   Observe that the blending function of each control point of a Bézier curve is non-zero over the entire open parameter interval $(0, 1)$; in other words, each has non-zero weight (attraction, pull, ...) at every point of the curve, except, possibly, the endpoints. For example, Figure 16.1(a) shows the blending functions of a cubic Bézier curve, which are, of course, the Bernstein polynomials of degree three.

   This makes modifying a Bézier curve difficult: moving any one control point alters the *entire* curve, not just near the control point. Albeit points on the curve far from the relocated control point move little because its weight is small at distant points, nevertheless, there is change. Moving control point P_1 in Figure 16.1(b), for example, from a reading of its blending function $B_{1,3}(u)$ in Figure 16.1(a) maximally affects the curve in the vicinity of $c(0.33)$, but all points on the curve, except for the endpoints, are altered to some extent.

   The situation for Bézier surfaces is similar, as each control point has non-zero weight at every point of the surface, except, possibly, the corners.

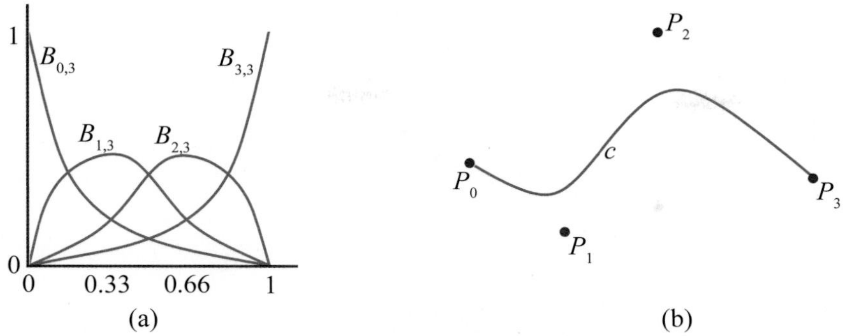

Figure 16.1: (a) Bernstein polynomials of degree 3: $B_{0,3}(u) = (1-u)^3$, $B_{1,3}(u) = 3(1-u)^2u$, $B_{2,3}(u) = 3(1-u)u^2$, $B_{3,3}(u) = u^3$ (b) A cubic Bézier curve.

Typically, in designing a complex object with numerous control points a designer would prefer to be able to modify parts of the object independently, in other words, have local control, which in turn would necessitate restricting each control point to its own limited "region of influence". For example, in arranging Boris's smirk – see Figure 16.2 – the designer may want to leave his nose and eyes exactly as they are.

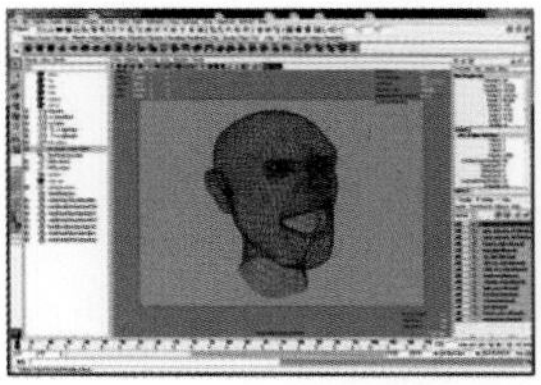

Figure 16.2: Mesh of Boris's head (courtesy of Sateesh Malla at `www.sateeshmalla.com`).

2. *The degree increases with the number of control points.*

   The Bézier curve $c(u)$ approximating $n+1$ control points is polynomial of degree n in u. Evaluating a high-degree polynomial is expensive and repeated products lead to numerical instability. Complex curves, therefore, with multiple control points present a computational problem. And ditto for surfaces.

What to do about these problems? First, let's step back a bit to take the following abstract view of Bézier curves: a Bézier curve is the sum

$$c(u) = f_0(u)P_0 + f_1(u)P_1 + \ldots + f_n(u)P_n \qquad (0 \leq u \leq 1) \qquad (16.1)$$

of its control points P_i weighted by blending functions f_i which *happen to be* Bernstein polynomials. There's no reason they *have to be* Bernstein polynomials, provided that the resulting curve c – maybe no longer Bézier – does a satisfactory job of approximating the control points. The plan then is to try and find new blending functions which, hopefully, alleviate the Bézier difficulties.

Before proceeding, here's a bit of useful terminology: if a function f, defined on the interval domain $[a, b]$, is non-zero everywhere inside the subinterval $[a', b']$, excepting possibly its endpoints a' and b', and zero on the rest of $[a, b]$, then it is said to have *support* in $[a', b']$. Figure 16.3(a) depicts a function $f_i(u)$ defined on $[0, 1]$ with support in the subinterval $[a', b']$.

Exercise 16.1. If the blending function f_i of control point P_i in expression (16.1) has support in the proper subinterval $[a', b']$ of the parameter interval

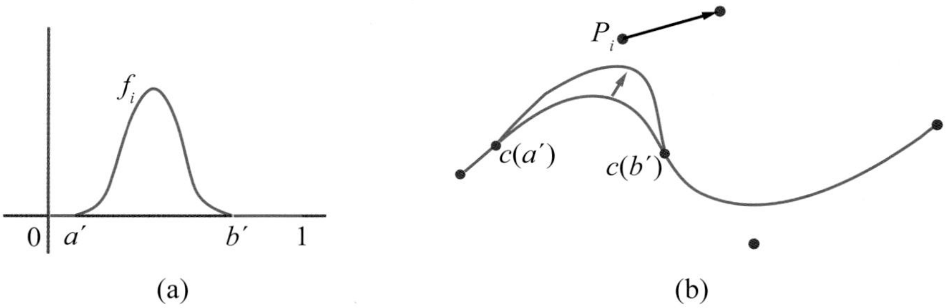

Figure 16.3: (a) Function f_i defined on $[0, 1]$ has support in $[a', b']$ (b) Moving P_i, with associated blending function f_i, changes c only between $c(a')$ and $c(b')$.

$[0, 1]$, then show that moving P_i changes the arc of the approximating curve c only between $c(a')$ and $c(b')$. See Figure 16.3(b).

Exercise 16.2. Prove that the ith Bernstein polynomial of degree n for every i, $0 \leq i \leq n$, has support in the entire parameter interval $[0, 1]$ (keep in mind that the behavior of the polynomial outside of $[0, 1]$ is of no interest).

From the preceding two exercises, it seems, then, that the first problem with Bézier curves mentioned above arises because the blending function of every control point has support in the entire parameter interval $[0, 1]$. A solution, therefore, would be to find blending functions each having support in only part of that interval.

Moreover, the second problem would be solved if the degree of the blending functions could be *decoupled* from the number of control points, so that increasing the latter did not necessarily raise the former. So now we have an idea of what we want, let's see what we can find.

Suppose, to begin with, that we ask for blending functions all quadratic, *no matter* the number of control points. The first thing to do then is find quadratics with limited support – whose graphs resemble that of f_i in Figure 16.3(a). Unfortunately, this is a hopeless task because there are none such! For, a quadratic is zero only at its at most two roots, not on any interval stretch like that between 0 and a', or b' and 1. But, look again at f_i. Except for the two straight zero parts at either end, the graph of f_i does resemble somewhat an upside-down parabola – see the graph of the parabola $f(u) = u^2$ in Figure 16.4(a).

Note: Curves drawn in this chapter are fairly accurate sketches, but not necessarily exact plots of their equations.

Here, then, is a drastic solution. Let's make a blending function f like f_i by assembling it from three parts – one quadratic (an upside-down parabola) and two straight zero – as follows:

$$f(u) = \begin{cases} 0, & -2 \leq u \leq -1 \\ -u^2 + 1, & -1 \leq u \leq 1 \\ 0, & 1 \leq u \leq 2 \end{cases}$$

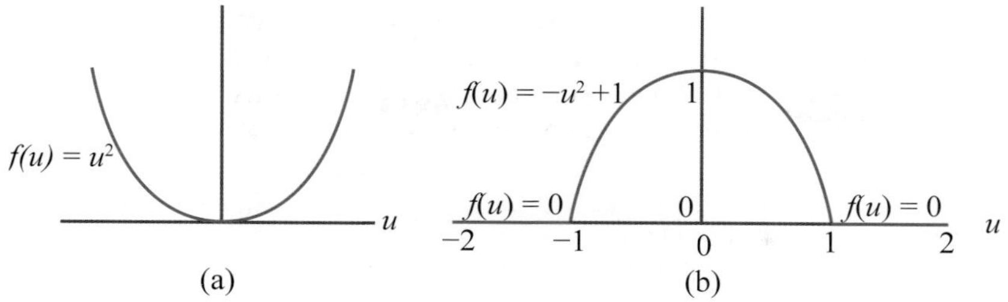

Figure 16.4: (a) Parabola (b) Three-part function: one upside-down parabola and two straight.

There's no law that says that a formula has to be one line! So the specification of f is fine. Figure 16.4(b) shows its graph. As the two end parts are constant, they are actually "cheaper" than a quadratic!

We seem to be headed in the right direction. We have a blending function which is at most quadratic and which has support in $[-1, 1]$, just half of its whole domain $[-2, 2]$.

Note: If the reader is wondering about the new parameter interval $[-2, 2]$, keep in mind that there's nothing special about the parameter interval $[0, 1]$ we use most often, other than that it's convenient to write. Parameter intervals can be any $[a, b]$, with $a < b$. In the case above, $[-2, 2]$ helps avoid fractions.

The corners (C^1-discontinuities, to be precise) at $u = \pm 1$, where the straight parts of f meet the parabolic, are undesirable though, because discontinuities in the blending function will carry over to discontinuities in the approximating curve employing such a function. It'll be nice to be rid of them. How do we get a parabolic part to join a straight part without making a corner? Oddly enough, Figure 16.4(a) suggests a solution. Consider the part of the parabola $f(u) = u^2$ to the *right* of the y-axis and the (straight) part of the x-axis to the *left* of the y-axis: they meet smoothly at the origin! See Figure 16.5.

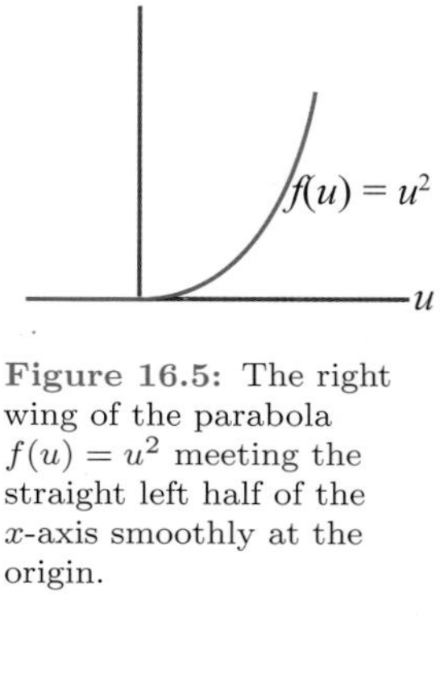

Figure 16.5: The right wing of the parabola $f(u) = u^2$ meeting the straight left half of the x-axis smoothly at the origin.

So here's the next draft. For $u \leq 0$ and $u \geq 4$, define $f(u)$ to be 0, giving two long straight parts; define $f(u) = u^2$ between 0 and 1; and, $f(u) = (u - 4)^2$ between 3 and 4. See the blue curves in Figure 16.6. Particularly, $f(u) = u^2$ in $[0, 1]$ is part of the right wing of the parabola of Figure 16.4(a), while $f(u) = (u - 4)^2$ in $[3, 4]$ from the left wing of the same parabola (but shifted 4 units to the right). The two quadratics meet the straight parts smoothly, so that's taken care of, but there's a piece missing in between (pretend you don't see the black curve!). Now, if we could only find a quadratic to sit smoothly atop the two side quadratics and cap the gap.

It turns out that a fairly intuitive choice works: drag $f(u) = u^2$ two units to the right, flip it upside down and then raise it two units. The equation is $f(u) = -(u - 2)^2 + 2$, giving the black curve in Figure 16.6. We leave verification to the reader in the next two exercises.

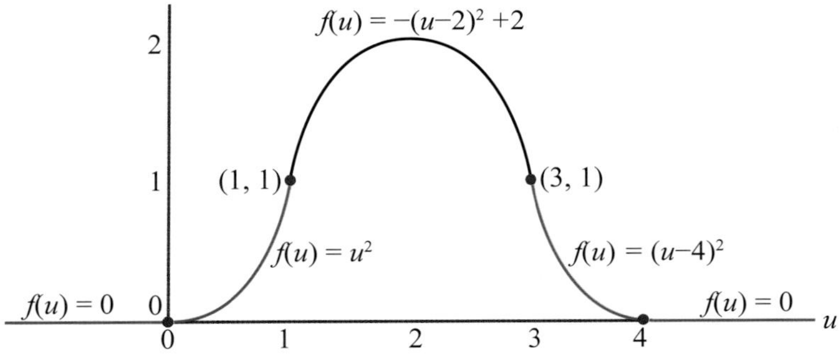

Figure 16.6: Five-part function: three parabolic and two straight parts. Joints are black points.

Exercise 16.3. Show that the curve $f(u) = -(u-2)^2 + 2$ indeed meets $f(u) = u^2$ at $(1, 1)$ and $f(u) = (u-4)^2$ at $(3, 1)$.

Exercise 16.4. Show that at each of the four *joints* $(0, 0)$, $(1, 1)$, $(3, 1)$ and $(4, 0)$ of the five-part function depicted in Figure 16.6 the tangent lines of the curves on either side are equal. Therefore, there is no C^1-discontinuity at a joint and the function is C^1-continuous everywhere.

Part answer: At $(1, 1)$, where $u = 1$, the tangent on the left is from $f(u) = u^2$ and on the right from $f(u) = -(u-2)^2 + 2$. Now, $\frac{\mathrm{d}}{\mathrm{d}u} u^2 = 2u$, which is 2 at $u = 1$, and $\frac{\mathrm{d}}{\mathrm{d}u}(-(u-2)^2 + 2) = -2(u-2)$, which is also 2 at $u = 1$, so, indeed, the tangent lines on either side of the joint $(1, 1)$ are equal.

For the record, here's the 5-line formula specifying f:

$$f(u) = \begin{cases} 0, & u \le 0 \\ u^2, & 0 \le u \le 1 \\ -(u-2)^2 + 2, & 1 \le u \le 3 \\ (u-4)^2, & 3 \le u \le 4 \\ 0, & 4 \le u \end{cases} \tag{16.2}$$

f has support in $[0, 4]$ and, from the preceding exercise, is C^1-continuous throughout. Moreover, if its parameter interval is chosen to be an interval larger than $[0, 4]$, e.g., $[-2, 6]$, then we have indeed a C^1-continuous blending function with limited support.

The moral then is to look for blending functions among the class of piecewise polynomial functions: a function is *piecewise polynomial* if its domain can be split into subintervals in each of which it's polynomial. For example, f above is composed of five polynomial pieces. From a computational point of view, evaluating a piecewise polynomial is not much harder than evaluating a polynomial. If one thinks in terms of C or C++ code, then there is simply an extra `if/else` ladder to determine the appropriate subinterval and corresponding polynomial.

The piecewise polynomials to be used as blending functions must be chosen carefully though. For example, looking back at Propositions 15.1 and 15.3 of the last chapter, it's desirable for the set of blending functions to form a partition of unity over the parameter space. Good things happen then: (a) points on the curve (or surface) are convex combinations of its control points, so the whole lies in the convex hull of its control points and (b) affine invariance.

Writing down all the properties we want, then, we put together a Wish List for blending functions. We ask that they

(a) be at least a C^1-continuous piecewise polynomial,

(b) be of a low degree independent of the number of control points,

(c) each have support in only part of the parameter space, and,

(d) together form a partition of unity over the parameter space.

We're indeed led to B-splines.

16.2 B-Spline Curves

Let's set the stage for the *B-spline blending functions* (or, as they are also called, *B-spline functions*, or *B-splines*, or *spline functions*) that we are going to define. Each will be piecewise polynomial, in other words, polynomial on subintervals. In anticipation, then, let's fix a particular parameter space and chop it up into subintervals. For convenience now, we choose $[0, r]$, where r is some positive integer, and its r subintervals to be the equally sized

$$[0, 1], \quad [1, 2], \quad \ldots, \quad [r-1, r]$$

See Figure 16.7. The sequence

$$\{0, 1, \ldots, r\}$$

Figure 16.7: Parameter space $[0, r]$ with uniformly-spaced knots.

of successive interval endpoints is called the *knot vector* and the endpoints $0, 1, \ldots, r$ themselves, *knots*. Each subinterval $[i, i+1]$, for $0 \le i \le r-1$, is a *knot interval*. We expect to define blending functions which are polynomial in each knot interval.

Remark 16.1. A knot vector as above with equally spaced knots is called a *uniform* knot vector. Later in this chapter we'll see non-uniform knot vectors as well.

Remark 16.2. The "B" in B-splines, the name given these functions by Schoenberg [127], a pioneer in their use, comes from "basis".

16.2.1 First-Order B-Splines

We'll start at the lowest level possible and define the *B-splines of degree 0* by means of constant functions. There are r B-splines of degree 0, each equal to 1 on one knot interval and 0 outside it. Precisely, the ith B-spline of degree 0, for $0 \le i \le r-1$, denoted $N_{i,1}$, is defined as follows.

When $i = 0$:

$$N_{0,1}(u) = \begin{cases} 1, & 0 \le u \le 1 \\ 0, & \text{otherwise} \end{cases} \tag{16.3}$$

When $1 \le i \le r-1$:

$$N_{i,1}(u) = \begin{cases} 1, & i < u \le i+1 \\ 0, & \text{otherwise} \end{cases} \tag{16.4}$$

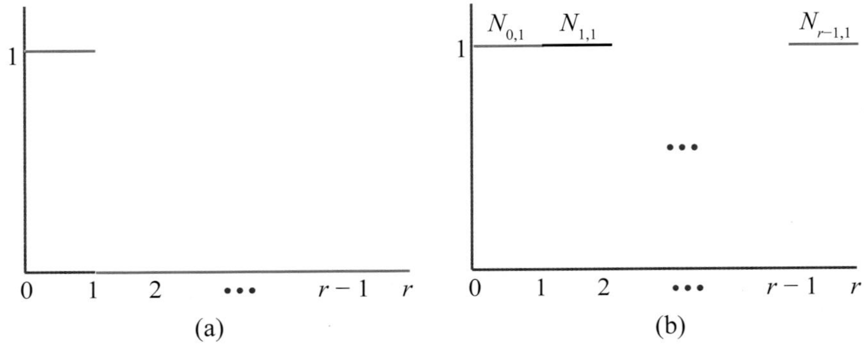

Figure 16.8: First-order B-splines: (a) $N_{0,1}$ (b) Non-zero parts of $N_{i,1}$, $0 \le i \le r-1$, distinguished by alternate blue and black colors.

In other words, each $N_{i,1}$ is 1 on the knot interval $[i, i+1]$, except, possibly, at the endpoints, and 0 outside it. Figure 16.8(a) shows the graph of $N_{0,1}$ over the entire parameter space $[0, r]$, while Figure 16.8(b) only the non-zero parts of the graphs of $N_{i,1}$, for $0 \le i \le r-1$. The niggling technicality – see the first line of the two equations above – of having to define $N_{0,1}$ to be 1 on a closed interval, while the other B-splines of degree 0 are equal to 1 on a half-open interval, is unavoidable. For, we want the r B-splines of degree 0 to form together a partition of unity over $[0, r]$, so no two are allowed to be 1 at the same point.

Experiment 16.1. Run `bSplines.cpp`, which shows the non-zero parts of the spline functions from first order to cubic over the uniformly spaced knot vector

$$[0, 1, 2, 3, 4, 5, 6, 7, 8]$$

Press the up/down arrow keys to choose the order. Figure 16.9 is a screenshot of the first order. The knot values can be changed as well, but there's no need to now.
End

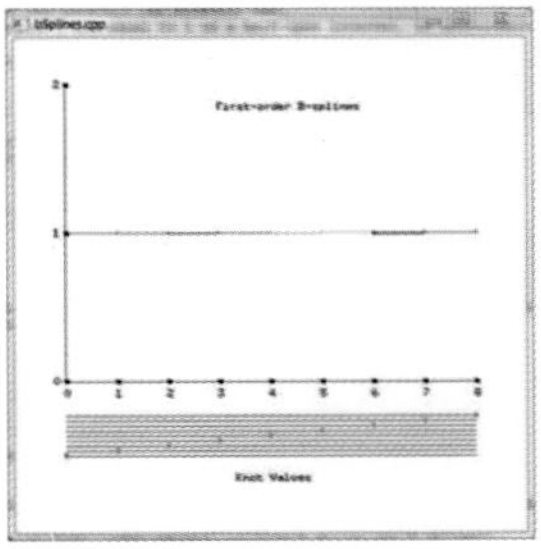

Figure 16.9: Screenshot of `bSplines.cpp` at first order.

B-splines of degree 0 are commonly called *first-order B-splines*. If the knot vector is uniform, as above, they are called *uniform first-order B-splines*.

Interestingly, as the reader may easily verify, all items on the Wish List at the end of Section 16.1 are fulfilled by the first-order B-splines, except for C^1-continuity, where, in fact, they fail badly because the $N_{i,1}$ are not even continuous (i.e., not even C^0-continuous). As we see next, unfortunately, this deficiency carries over to approximating curves made from first-order B-splines as well.

First-Order B-Spline Curves

A *first-order B-spline approximation* of r control points $P_0, P_1, \ldots, P_{r-1}$ is called a first order B-spline curve. This is the curve c obtained from applying the first-order B-splines as blending functions to these control points, namely,

$$c(u) = \sum_{i=0}^{r-1} N_{i,1}(u)P_i \qquad (0 \leq u \leq r) \tag{16.5}$$

What sort of a curve is c? Well, one would be hard pressed to call c a curve in the first place! Applying the definitions of $N_{i,1}$ from Equations (16.3)-(16.4) to Equation (16.5) above, one sees that c is *stationary* at P_0 for u from 0 to 1. When u crosses 1, c *jumps* to P_1, staying stationary again till u crosses 2, when c jumps to P_2 and so on. The graph of c is then just the collection of its own control points! See Figure 16.10. Obviously, if there are even two distinct control points then c is not C^0. Clearly, we'll have to move to higher orders of B-splines for satisfaction.

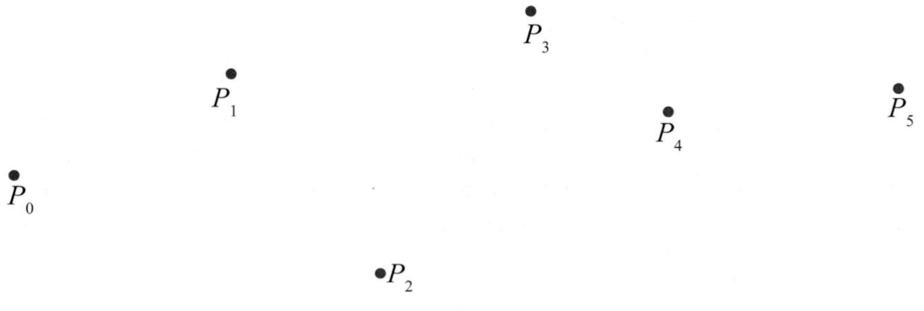

Figure 16.10: First-order B-spline approximation – the "curve" consists of its control points.

First-order B-Spline Properties

However, before leaving the first order, here are a few of their properties for future reference:

1. Each $N_{i,1}$ is piecewise polynomial, consisting of at most three pieces, each of which is constant.
2. $N_{i,1}$ has support in the single knot interval $[i, i+1]$.
3. Each $N_{i,1}$ is not C^0 only at the endpoints of its supporting interval; elsewhere, it's C^∞. In other words, it's smooth – remember from Definition 10.7 that C^∞ is also called smooth – apart from its joints.
4. Together, the $N_{i,1}$ form a partition of unity over the parameter space $[0, r]$.
5. Except for $N_{0,1}$, the $N_{i,1}$ are translates of one another, i.e., the graph of one is a translate of that of another. This is a consequence of the knots being uniformly spaced.
6. A first-order B-spline approximation is, generally, not even C^0.

16.2.2 Linear B-Splines

The clear problem with first-order B-splines is that their polynomial degree 0 is too low, allowing them little flexibility in shape. Straight and horizontal is all they can be. Let's go one higher to degree 1. We'll do this in a particular way which will be easy to generalize down the road.

The trivial formula that

$$1 = u + (-u + 1) \tag{16.6}$$

allows one to "break" each B-spline $N_{i,1}$, of degree 0, into two functions $N^0_{i,1}$ and $N^1_{i,1}$ of degree 1. For example, $N_{0,1}$ breaks into $N^0_{0,1}$ and $N^1_{0,1}$, where

$$N^0_{0,1}(u) = \begin{cases} u, & 0 \le u \le 1 \\ 0, & \text{otherwise} \end{cases} \tag{16.7}$$

and $$N^1_{0,1}(u) = \begin{cases} -u + 1, & 0 \le u \le 1 \\ 0, & \text{otherwise} \end{cases} \tag{16.8}$$

The two obviously add up to give back $N_{0,1}$, viz.

$$N_{0,1}(u) = N^0_{0,1}(u) + N^1_{0,1}(u) \tag{16.9}$$

$N_{i,1}$, when $i > 0$, can likewise be broken into $N^0_{i,1}$ and $N^1_{i,1}$, where

$$N^0_{i,1}(u) = \begin{cases} u - i, & i < u \le i + 1 \\ 0, & \text{otherwise} \end{cases} \tag{16.10}$$

$$\text{and} \qquad N_{i,1}^1(u) = \begin{cases} -u+i+1, & i < u \leq i+1 \\ 0, & \text{otherwise} \end{cases} \tag{16.11}$$

which actually applies the i-shift of (16.6), in particular, $1 = (u-i) + (-u + i + 1)$. And, again

$$N_{i,1}(u) = N_{i,1}^0(u) + N_{i,1}^1(u) \tag{16.12}$$

Figure 16.11 shows the two parts of each first-order B-spline. For obvious reasons, we call the $N_{i,1}^0$'s "up" and the $N_{i,1}^1$'s "down". The up parts are all left or right translates of one another, as are the down parts, except that the technicality that their values at the left end of the knot interval $[0, 1]$ are different from those at the left end of other knot intervals persists from first-order.

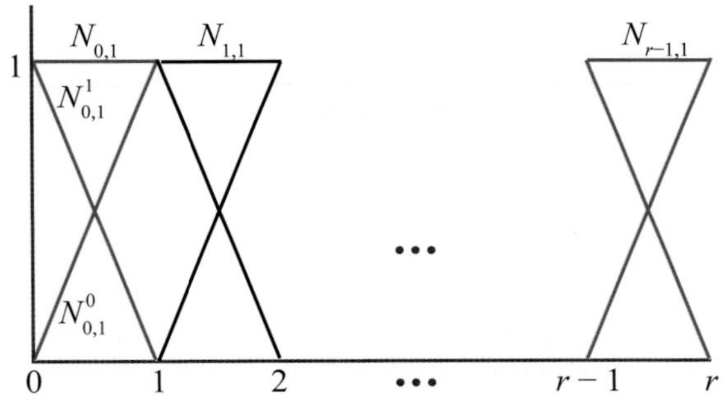

Figure 16.11: First-order B-splines each broken into an up part (dashed) $N_{i,1}^0$ and a down part (dotted) $N_{i,1}^1$. Successive $N_{i,1}$'s are distinguished by color.

Remark 16.3. This is important! For future reference, think of what we have just done as the following: each $N_{i,1}$ is broken into two functions over its support, one obtained from multiplying $N_{i,1}$ by a *straight-line function increasing from 0 to 1* from the left end of its support to the right (namely, $u - i$), while the other from multiplying it by a *straight-line function decreasing from 1 to 0* over the same interval (namely, $-u + i + 1$).

Equations (16.9) and (16.12) evidently guarantee that $N_{i,1}^0$ and $N_{i,1}^1$, for $0 \leq i \leq r-1$, together form a partition of unity because the $N_{i,1}$, $0 \leq i \leq r-1$, do. But there are $2r$ of the former, which is twice as many as we need to blend r control points. Figure 16.11, in fact, suggests a way to pair them up nicely – join each up part to the following down part! Accordingly, define the *second-order B-splines* (or *linear B-splines*), for $0 \leq i \leq r-2$, as follows:

$$N_{i,2}(u) = N_{i,1}^0(u) + N_{i+1,1}^1(u) = \begin{cases} 0, & u \leq i \\ u-i, & i \leq u \leq i+1 \\ -u+i+2, & i+1 \leq u \leq i+2 \\ 0, & i+2 \leq u \end{cases} \tag{16.13}$$

Figure 16.12 shows the non-zero parts of the linear B-splines $N_{i,2}$, $0 \leq i \leq r-2$, on the domain $[0, r]$. *See the magic*: pairing has removed all C^0-discontinuities! The linear B-splines are each continuous everywhere.

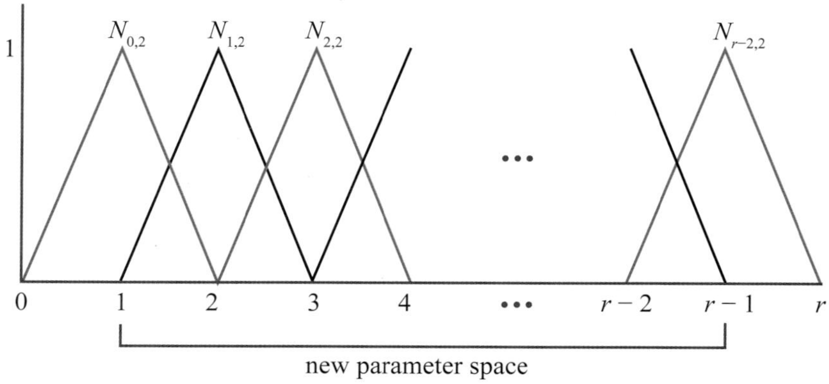

Figure 16.12: Non-zero parts of linear B-splines. Each is an inverted V. Successive ones are distinguished by color. The down part in the first knot interval and the up part in the last are discarded. The new (truncated) parameter space is $[1, r-1]$.

Experiment 16.2. Run again `bSplines.cpp` and select the linear B-splines over the knot vector

$$[0, 1, 2, 3, 4, 5, 6, 7, 8]$$

Figure 16.13 is a screenshot. **End**

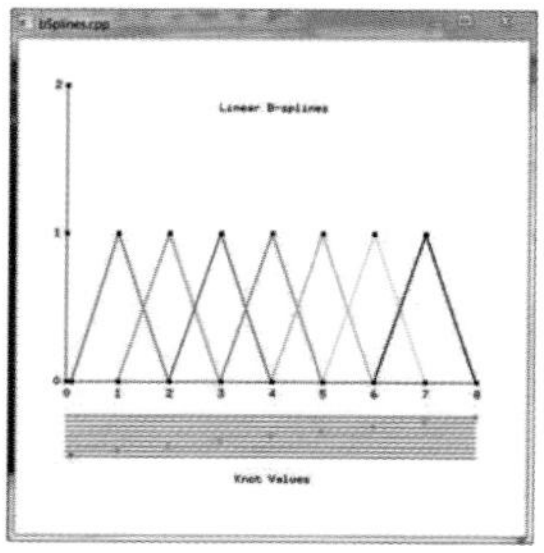

Figure 16.13: Screenshot of `bSplines.cpp` at second-order.

Exercise 16.5. Verify that the multi-part formula above for $N_{i,2}(u)$ indeed follows from joining up and down parts (using the equations for $N_{i,1}^0$ and $N_{i,1}^1$ given earlier).

Remark 16.4. The technicality at the left endpoint of a supporting interval is now gone. The definition of $N_{i,2}$ is the same for all i in $0 \leq i \leq r-2$.

Remark 16.5. Second-order B-splines as defined above are often called *uniform linear B-splines* to emphasize the use of a uniform knot vector.

Note that the down part $N_{0,1}^1$ of $N_{0,1}$ and the up part $N_{r-1,1}^0$ of $N_{r-1,1}$ have no partners, so are discarded, which is why we have $r-1$ linear B-splines $N_{i,2}$, for $i = 0$ to $r-2$, versus the r first-order B-splines we started with. It's clear from Figure 16.12 that the parameter space must be truncated from $[0, r]$ to $[1, r-1]$ as well, for, otherwise, there's a problem with the partition-of-unity property in the two end knot intervals $[0, 1]$ and $[r-1, r]$. Once this is done, though, we're in good shape, or at least in significantly better shape than the first-order B-splines. All items in the Wish List at the end of Section 16.1 are now fulfilled except for C^1-continuity, but now the functions are at least C^0, if not quite C^1 (because of corners at the joints).

Linear B-Spline Curves

What sort of curve is the linear B-spline approximation c of $r-1$ control points $P_0, P_1, \ldots, P_{r-2}$, which uses the linear B-splines as blending functions? It's defined as follows:

$$c(u) = \sum_{i=0}^{r-2} N_{i,2}(u) P_i \qquad (1 \le u \le r-1) \tag{16.14}$$

Exercise 16.6. Verify that the linear B-spline approximation c given by Equation (16.14) is the polygonal line through the control points in the sequence they are given. See Figure 16.14, where $r = 7$. This is certainly more respectable a curve than the first-order approximation.

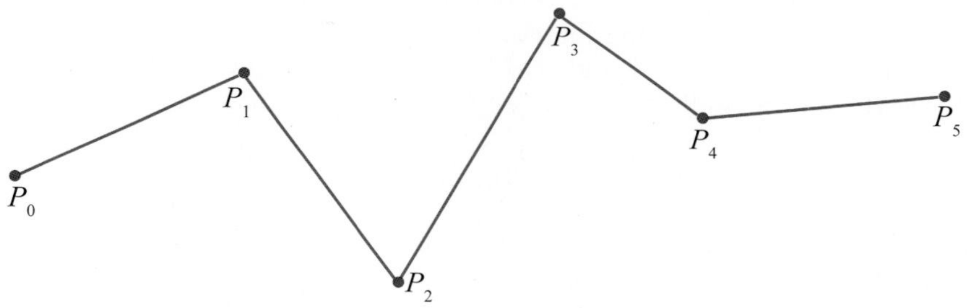

Figure 16.14: Linear B-spline approximation.

Terminology: A B-spline approximation of a sequence of control points is often called a *B-spline curve*, a *spline curve* or, simply, a *spline*. There is ambiguity sometimes, therefore, with the terminology for B-spline blending functions, but it'll be clear from the context if the term refers to a blending function or an approximating curve.

Linear B-Spline Properties

Here's a list of properties of linear B-splines similar to the one made earlier for first-order B-splines:

1. Each $N_{i,2}$ is piecewise polynomial, consisting of at most four pieces, each of which is linear, except for zero end pieces.

2. $N_{i,2}$ has support in $[i, i+2]$, the union of two consecutive knot intervals.

3. Each $N_{i,2}$ is C^0, but not C^1, at its joints. Apart from its joints it's smooth everywhere.

4. Together, the $N_{i,2}$ form a partition of unity over the parameter space $[1, r-1]$.

5. The $N_{i,2}$ are translates of one another.

6. A linear B-spline approximation is C^0, but, generally, not C^1.

16.2.3 Quadratic B-Splines

Linear B-splines are certainly preferable to first-order ones, but we're still shy of C^1-continuity. If we could raise the degree of the polynomial pieces yet again, from 1 to 2, we might do better continuity-wise.

It turns out that the approach introduced in the last section of breaking first-order B-splines into up and down parts of one higher degree, and then pairing them up to make linear ones, generalizes. Consider first $N_{0,2}$, graphed in Figure 16.15(a). Recall Remark 16.3: to break $N_{0,2}$ into two, multiply it by a straight-line function increasing from 0 at the left end of its support to 1 at the right, as well as by the complementary function decreasing from 1 to 0. Since the supporting interval of $N_{0,2}$ is $[0, 2]$, the two straight-line functions called for are $u/2$ and $-u/2 + 1$, respectively, which are shown in Figure 16.15(a) as well.

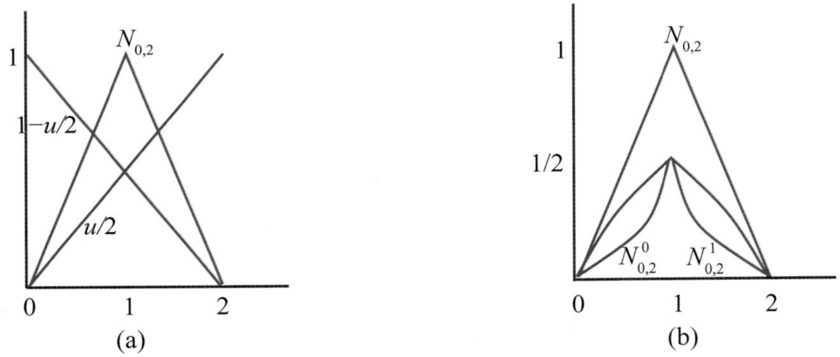

Figure 16.15: (a) The graphs of the two straight-line multiplying functions for $N_{0,2}$, one dashed and one dotted (b) The result of the multiplication: the up part $N^0_{0,2}$ (dashed) and the down part $N^1_{0,2}$ (dotted).

Accordingly, break $N_{0,2}$ as follows:

$$N_{0,2} = \frac{u}{2} N_{0,2} + (-\frac{u}{2} + 1)\, N_{0,2} \tag{16.15}$$

where the up part – it's not really increasing any more but we'll stick with the term – is

$$N^0_{0,2}(u) = \frac{u}{2} N_{0,2} = \begin{cases} 0, & u \le 0 \\ \frac{1}{2}u^2, & 0 \le u \le 1 \\ -\frac{1}{2}u^2 + u, & 1 \le u \le 2 \\ 0, & 2 \le u \end{cases} \tag{16.16}$$

and the down part is

$$N_{0,2}^1(u) = (-\frac{u}{2} + 1)\, N_{0,2} = \begin{cases} 0, & u \le 0 \\ -\frac{1}{2}u^2 + u, & 0 \le u \le 1 \\ \frac{1}{2}u^2 - 2u + 2, & 1 \le u \le 2 \\ 0, & 2 \le u \end{cases} \tag{16.17}$$

The graphs of the two parts, resembling opposing shark fins, are shown in Figure 16.15(b).

Exercise 16.7. Verify the formulae for $N_{0,2}^0$ and $N_{0,2}^1$ by multiplying that for $N_{0,2}$ by $u/2$ and $-u/2+1$, respectively.

The other linear B-splines $N_{i,2}$, for $1 \le i \le r-2$, can similarly be broken. Figure 16.16 shows the graphs of the up and down parts.

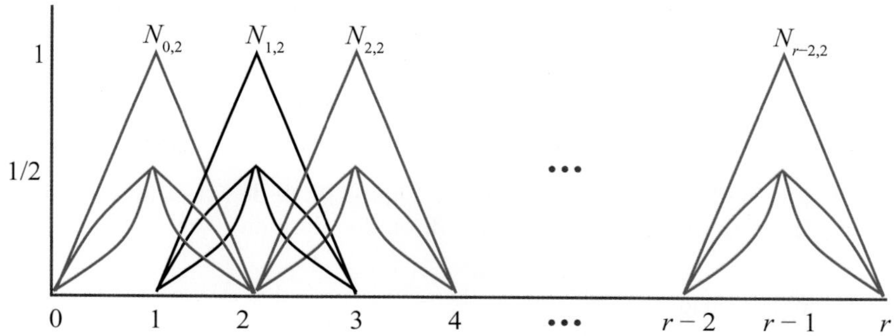

Figure 16.16: Linear B-splines each broken into an up (dashed) part $N_{i,2}^0$ and down (dotted) part $N_{i,2}^1$. Successive ones are distinguished by color.

Next, as in the first-order case, pair them up, adding each up part to the down part of the following linear B-spline. Non-zero pieces of the up and down parts did not overlap in the first-order case, so adding meant simply splicing graphs end to end. Now we do actually have to add on the overlaps.

And again magic! Two adjacent and opposing shark fins, one dashed and the other dotted, both with a sharp corner in the middle, add up to a smooth-looking floppy hat! See Figure 16.17. Precisely, the up part of one linear B-spline adds to the down part of the following one to make a *quadratic B-spline* (or, *third order B-spline*).

Figure 16.17 explains exactly what's happening. The graph of $N_{0,2}^0$ is blue dashed, while that of $N_{1,2}^1$ black dotted. The graph $N_{0,3}$ of their sum consists of the outer blue dashed arc on $[0,1]$, the outer black dotted arc on $[2,3]$ and the unbroken red arc on $[1,2]$ in the middle, the latter being the sum of the inner blue dashed and the inner black dotted. So it's in the middle interval $[1,2]$ that actual summing takes place. We'll see the summed equation itself momentarily.

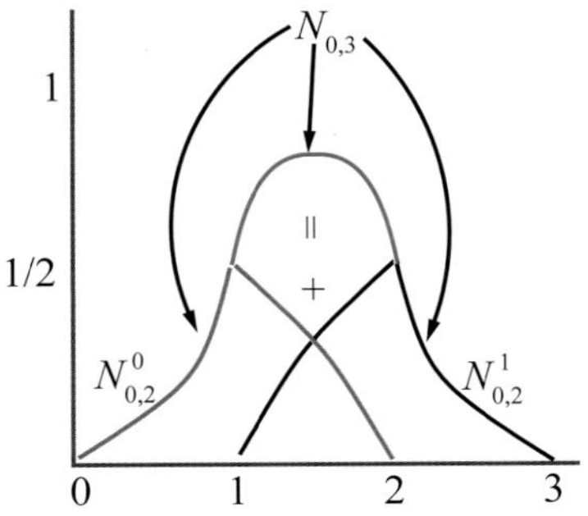

Figure 16.17: Adding $N^0_{0,2}$ and $N^1_{1,2}$ to make $N_{0,3}$. $N_{0,3}$ consists of three parts: on $[0,1]$ it's just $N^0_{0,2}$, on $[2,3]$ it's $N^1_{1,2}$, while in the middle, on $[1,2]$ it is the sum of $N^0_{0,2}$ and $N^1_{1,2}$.

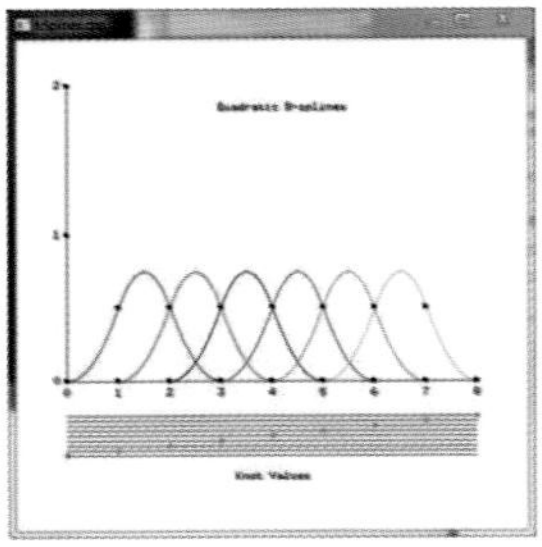

Figure 16.18: Screenshot of `bSplines.cpp` at third order.

Experiment 16.3. Run again `bSplines.cpp` and select the quadratic B-splines over the knot vector

$$[0, 1, 2, 3, 4, 5, 6, 7, 8]$$

Figure 16.18 is a screenshot. Note the joints indicated as black points. **End**

$N_{0,3}$ is the first quadratic B-spline. Figure 16.19 depicts the sequence of quadratic B-splines $N_{i,3}$, $0 \le i \le r-3$, on the domain $[0, r]$.

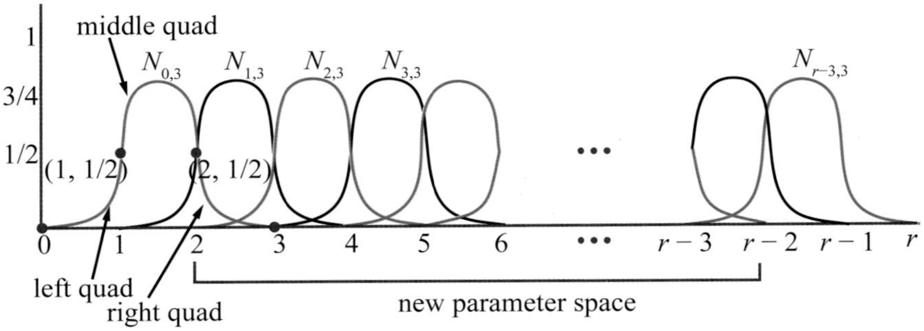

Figure 16.19: Non-zero parts of the quadratic B-splines; the four joints of the first one are indicated as points as well. Successive splines are distinguished by color.

Now for the equations of the quadratic B-splines. As they are evidently translates of one another, it's sufficient to write only that of the first one:

$$N_{0,3}(u) = N^0_{0,2}(u) + N^1_{1,2}(u) = \begin{cases} 0, & u \le 0 \\ \frac{1}{2}u^2, & 0 \le u \le 1 \\ \frac{3}{4} - (u - \frac{3}{2})^2, & 1 \le u \le 2 \\ \frac{1}{2}(-u+3)^2, & 2 \le u \le 3 \\ 0, & 3 \le u \end{cases} \quad (16.18)$$

Exercise 16.8. Verify the preceding formula with the help of (16.16) and (16.17). Don't forget to shift the second equation one unit to the right for the formula for $N^1_{1,2}$.

Exercise 16.9. Use Equation (16.18) to determine the equation of $N_{1,3}(u)$ and, generally, $N_{i,3}(u)$.

Exercise 16.10. Verify that the first quadratic B-spline $N_{0,3}$ is C^1 *everywhere* by differentiating the functions on the RHS of (16.18) and comparing the tangents on either side at each joint (which is only where discontinuity might occur). The four joints of $N_{0,3}$, with x-values 0, 1, 2 and 3, are indicated in Figure 16.19.

Differentiating again, verify that $N_{0,3}$ is *not* C^2 at its joints.

As the quadratic B-splines are translates one of one another, it follows from the preceding exercise that they are all C^1 everywhere, though not C^2 at their joints.

Remark 16.6. Compare the 5-line formulas (16.2) and (16.18) to see that we've come now full circle back to almost the same piecewise quadratic blending function which we used to motivate B-splines in the first place!

As in the linear case, the parameter space must be truncated, this time to $[2, r-2]$, to ensure that the partition-of-unity property holds. The key to keep in mind is that partition-of-unity holds in those knot intervals on which there is defined a left, a middle and a right quadratic arc – from successive quadratic B-splines.

Pop the champagne bottle: we now officially have every item on the Wish List!

Quadratic B-Spline Curves

So what sort of curve is the quadratic B-spline approximation c of $r-2$ control points $P_0, P_1, \ldots, P_{r-3}$, defined by

$$c(u) = \sum_{i=0}^{r-3} N_{i,3}(u) P_i \qquad (2 \le u \le r-2) \tag{16.19}$$

where the quadratic B-splines are used as blending functions?

First, and importantly, since the quadratic B-splines are all C^1, so is a quadratic B-spline approximation. We've gained at least respectable continuity then. However, as we ask the reader to show next, the property of interpolating the first and last control points has been lost (though not on our Wish List, this, nevertheless, is desirable).

Exercise 16.11. Prove that the quadratic spline curve c defined by Equation (16.19) begins at the midpoint of P_0 and P_1, ends at the midpoint of P_{r-4} and P_{r-3} and doesn't necessarily interpolate *any* of the control points. See Figure 16.20.

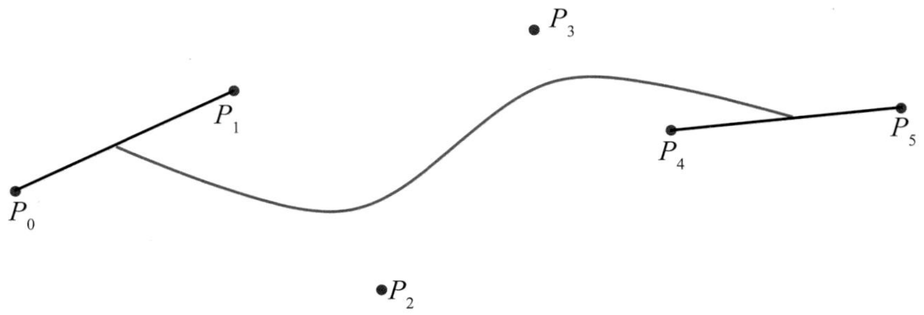

Figure 16.20: Quadratic B-spline approximation.

Darn, just when we thought things were going our way, a potentially nasty bug rears its ugly head. Not to worry, as soon as we are able to loosen up the knot vector from being uniform, we'll be happily interpolating first and last control points again.

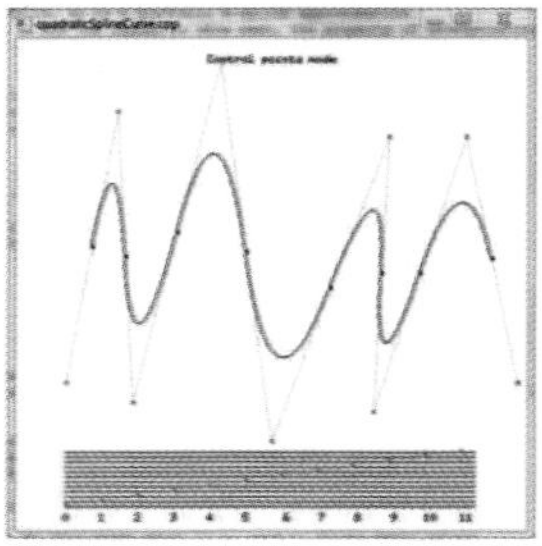

Figure 16.21: Screenshot of `quadraticSpline-Curve.cpp`.

Experiment 16.4. Run `quadraticSplineCurve.cpp`, which shows the quadratic spline approximation of nine control points over a uniformly spaced vector of 12 knots. Figure 16.21 is a screenshot.

The control points are green. Press the space bar to select a control point – the selected one turns red – and the arrow keys to move it. The knots are the green points on the black bars at the bottom. At this stage there is no need to change their values. The blue points are the joints of the curve, i.e., images of the knots. Also drawn in light gray is the control polygon. **End**

Exercise 16.12. What part of the quadratic spline curve c approximating the control points $P_0, P_1, \ldots, P_{r-3}$ is altered by moving only P_i? Your answer should be in terms of an arc of c between a particular pair of its joints. Verify using `quadraticSplineCurve.cpp`.

Quadratic B-Spline Properties

A list of properties for quadratic B-splines:

1. Each $N_{i,3}$ is piecewise polynomial, consisting of at most five pieces, each of which is quadratic, except for zero end pieces.

2. $N_{i,3}$ has support in $[i, i+3]$, the union of three consecutive knot intervals.

3. Each $N_{i,3}$ is C^1, but not C^2, at its joints. Apart from its joints it's smooth everywhere.

4. Together, the $N_{i,3}$ form a partition of unity over the parameter space $[2, r-2]$.

5. The $N_{i,3}$ are translates of one another.

6. A quadratic B-spline approximation is C^1, but, generally, not C^2.

When placing it on our Wish List, we expected to be rewarded for the partition-of-unity property by felicitous behavior of the B-spline approximating curves. The reader is asked to show next that indeed we are.

Exercise 16.13.

(a) Prove that the quadratic spline curve approximating a sequence of control points lies in the convex hull of the latter.

(b) Affine invariance: prove that an affine transformation of a quadratic spline curve is the same as the quadratic spline curve approximating the transformed control points.

16.2.4 Cubic B-Splines

We're going to ask you to do most of the lifting in this section.

To start with, break the first quadratic B-spline $N_{0,3}$ into two parts: an "up" part obtained from multiplying it by a straight-line function increasing from 0 at the left end of its support to 1 at the right end and a "down" part from multiplying it by the complementary function decreasing from 1 to 0 over its support. Here's the equation showing the split:

$$N_{0,3} = \frac{u}{3}\,N_{0,3} + (-\frac{u}{3} + 1)\,N_{0,3} \tag{16.20}$$

Exercise 16.14. Write equations for the up part

$$N^0_{0,3}(u) = \frac{u}{3}\,N_{0,3}$$

and the down part

$$N^1_{0,3}(u) = (-\frac{u}{3} + 1)\,N_{0,3}$$

in a manner analogous to Equations (16.16) and (16.17) for the quadratic B-splines. Both up and down parts are piecewise cubic.

Exercise 16.15. Verify by adding $N^0_{0,3}(u)$ and $N^1_{1,3}(u)$ that the equation of the first cubic B-spline is:

$$N_{0,4}(u) = \begin{cases} 0, & u \le 0 \\ p(2-u), & 0 \le u \le 1 \\ q(2-u), & 1 \le u \le 2 \\ q(u-2), & 2 \le u \le 3 \\ p(u-2), & 3 \le u \le 4 \\ 0, & 4 \le u \end{cases} \tag{16.21}$$

where the functions p and q are given by:

$$p(u) = \frac{1}{6}(2-u)^3$$

and

$$q(u) = \frac{1}{6}(3u^3 - 6u^2 + 4)$$

See Figure 16.22.

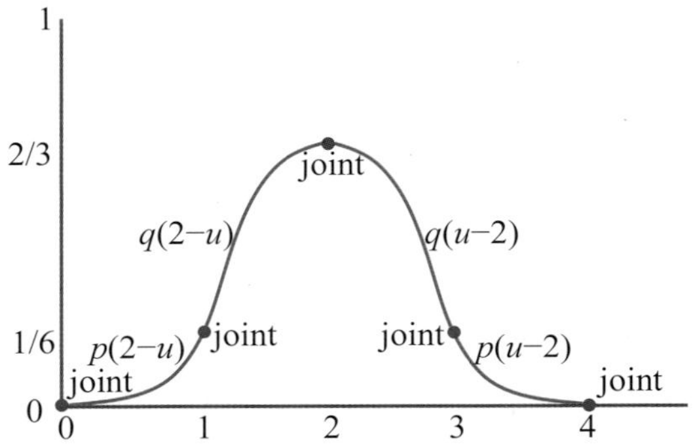

Figure 16.22: The first cubic B-spline function $N_{0,4}$.

Exercise 16.16. Verify that cubic B-splines are C^2, but not C^3, at their joints.

Exercise 16.17. Sketch the sequence of cubic B-splines $N_{i,4}$, for $0 \leq i \leq r - 4$, over $[0, r]$ similarly to Figure 16.19 for quadratic B-splines. What should be the new parameter range?

Experiment 16.5. Run `bSplines.cpp` and change the order to see a sequence of cubic B-splines. **End**

Cubic B-Spline Curves

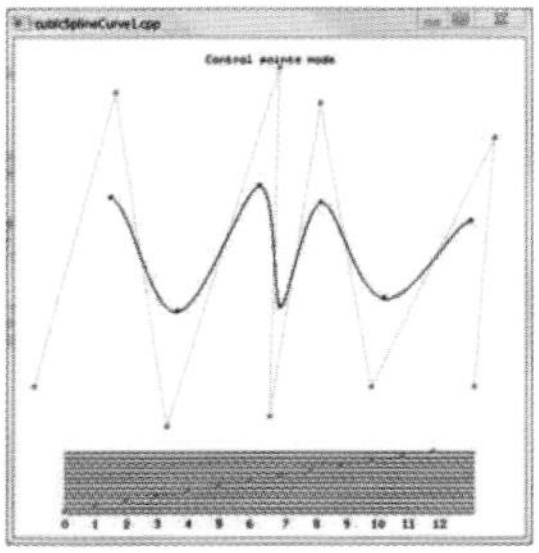

Figure 16.23: Screenshot of `cubicSplineCurve1.cpp`.

The cubic spline curve c approximating $r - 3$ control points $P_0, P_1, \ldots, P_{r-4}$ is obtained as

$$c(u) = \sum_{i=0}^{r-4} N_{i,4}(u)P_i \qquad (3 \leq u \leq r - 3) \tag{16.22}$$

Experiment 16.6. Run `cubicSplineCurve1.cpp`, which shows the cubic spline approximation of nine control points over a uniformly-spaced vector of 13 knots. The program is similar to `quadraticSplineCurve.cpp`. See Figure 16.23 for a screenshot.

The control points are green. Press the space bar to select a control point – the selected one is colored red – then the arrow keys to move it. The knots are the green points on the black bars at the bottom. The blue points are the joints of the curve. The control polygon is a light gray. **End**

Cubic B-Spline Properties

A list of properties for cubic B-splines:

1. Each $N_{i,4}$ is piecewise polynomial, consisting of at most six pieces, each of which is cubic, except for zero end pieces.
2. $N_{i,4}$ has support in $[i, i+4]$, the union of four consecutive knot intervals.
3. Each $N_{i,4}$ is C^2, but not C^3, at its joints. Apart from its joints it's smooth everywhere.
4. Together, the $N_{i,4}$ form a partition of unity over the parameter space $[3, r-3]$.
5. The $N_{i,4}$ are translates of one another.
6. A cubic B-spline approximation is C^2, but, generally, not C^3.

Remark 16.7. Cubic B-splines are the most commonly used in design applications, because they offer the best trade-off between continuity and computational efficiency.

16.2.5 General B-Splines and Non-uniform Knot Vectors

It's probably evident now how to manufacture B-splines of arbitrary order over the uniform knot vector $\{0, 1, \ldots, r\}$. One would apply the *break-and-make* procedure to B-splines of each order to derive ones of one higher order. We formalize the derivation of B-splines of arbitrary order over $\{0, 1, \ldots, r\}$ recursively as follows:

Definition 16.1. The first-order B-splines $N_{i,1}$, $0 \le i \le r-1$, are as defined in Section 16.2.1.

Suppose, recursively, that the B-splines $N_{i,m-1}$, for $0 \le i \le r-m+1$, have been defined for some order $m-1 \ge 1$. Then define the ith B-spline $N_{i,m}$ of order m, for $0 \le i \le r-m$, by the equation:

$$N_{i,m}(u) = \left(\frac{u-i}{m-1}\right) N_{i,m-1}(u) \; + \; \left(\frac{i+m-u}{m-1}\right) N_{i+1,m-1}(u) \quad (16.23)$$

Equation (16.23) comes from a straightforward application of break-and-make. The summand

$$\left(\frac{u-i}{m-1}\right) N_{i,m-1}(u)$$

is the up part of $N_{i,m-1}(u)$ obtained from multiplying it by the straight-line function $(u-i)/(m-1)$ increasing from 0 at i, the left end of its support, to 1 at $i+m-1$, the right end.

Likewise, the summand

$$\left(\frac{i+m-u}{m-1}\right) N_{i+1,m-1}(u)$$

is the down part of $N_{i+1,m-1}(u)$ obtained from multiplying it by the straight-line function $(i+m-u)/(m-1)$ decreasing from 1 to 0 from the left end $i+1$ to the right $i+m$ of its support.

Terminology: The *degree* of a B-spline is that of its polynomial pieces, while its *order* is its degree plus one.

Exercise 16.18. Make a six-point list of properties for uniform B-splines of the mth order like the ones earlier for uniform lower-order splines.

Before proceeding further, though, we are going to loosen restrictions on the knot vector, which till now had been the uniform sequence

$$\{0, 1, \ldots, r\}$$

Keep in mind that the operative word is *uniform*, in particular, that knots are equally spaced; it does not matter that they are integers. For instance, if the knot vector were of the form

$$\{a, a+\delta, a+2\delta, \ldots, a+r\delta\}$$

for some a, and some $\delta > 0$, e.g.,

$$\{1.3, 2.8, 4.3, \ldots, 1.3+1.5r\}$$

all calculations made so far would clearly go through again, though with different (and awkward) number values, and all properties of B-splines deduced previously would hold, too.

The restriction of uniformity is removed by allowing the knot vector to be any sequence of knots of the form

$$T = \{t_0, t_1, \ldots, t_r\}$$

where the t_i are *non-decreasing*, i.e.,

$$t_0 \leq t_1 \leq \ldots \leq t_r \tag{16.24}$$

Such knot vectors are called *non-uniform*. Yes, successive knots can even be equal and such so-called multiple knots have important applications, as we'll see.

Remark 16.8. The term non-uniform knot vector is a little unfortunate in that it actually means *not necessarily* uniform, because a uniform knot vector evidently satisfies (16.24) as well!

Hmm, do we start afresh working our way up from first-order splines, this time around over non-uniform knot vectors? Not at all. Pretty much all our earlier discussions go through again, including break-and-make. Without further ado then, here's the recursive definition of B-splines over non-uniform knot vectors.

Definition 16.2. Let

$$T = \{t_0, t_1, \ldots, t_r\} \tag{16.25}$$

be a non-uniform knot vector, where $r \geq 1$.

The (non-uniform) first-order B-spline functions $N_{i,1}$, for $0 \leq i \leq r-1$, are defined as follows.

When $i = 0$:

$$N_{0,1}(u) = \begin{cases} 1, & t_0 \leq u \leq t_1 \\ 0, & \text{otherwise} \end{cases} \tag{16.26}$$

When $1 \leq i \leq r-1$:

$$N_{i,1}(u) = \begin{cases} 1, & t_i < u \leq t_{i+1} \\ 0, & \text{otherwise} \end{cases} \tag{16.27}$$

The (non-uniform) mth order B-spline functions $N_{i,m}$, where the order m lies within $1 < m \leq r$, and the index i in $0 \leq i \leq r-m$, are recursively defined by:

$$N_{i,m}(u) = \left(\frac{u - t_i}{t_{i+m-1} - t_i}\right) N_{i,m-1}(u) + \left(\frac{t_{i+m} - u}{t_{i+m} - t_{i+1}}\right) N_{i+1,m-1}(u) \tag{16.28}$$

Note: The convention to follow in case the denominator of either of the two fractional terms is 0 – which may occur if there are equal knots – is the following: if the term is of the form $\frac{0}{0}$, then declare its value to be 1; if it is of the form $\frac{a}{0}$, where a is not 0, then declare its value to be 0.

This recursive formula (16.28), discovered by Cox, de Boor and Mansfield independently in 1972, known accordingly as the Cox-de Boor-Mansfield (CdM) formula or recurrence, was an important milestone in B-spline theory. However, it's really straightforward for us to understand now, given our development of the topic so far:

Equations (16.26) and (16.27), respectively, replicate, with obvious changes, (16.3) and (16.4) for first-order B-splines over a uniform knot vector. Equation (16.28) follows (16.23). It formalizes break-and-make – the summands are the up and down parts, respectively, of two successive spline functions of one lower order. Figure 16.24 shows graphs of all four functions on the RHS of Equation (16.28).

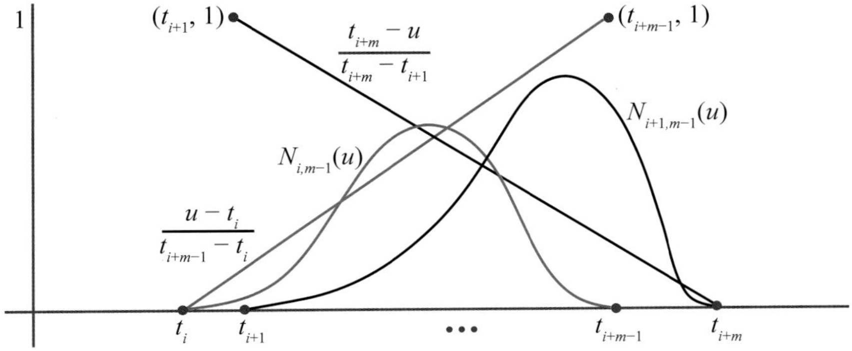

Figure 16.24: Graphs of the functions on the RHS of Equation (16.28): $N_{i,m-1}$ and $N_{i+1,m-1}$ and their respective linear multipliers $\frac{u-t_i}{t_{i+m-1}-t_i}$ and $\frac{t_{i+m}-u}{t_{i+m}-t_{i+1}}$.

Figure 16.25 shows the graphs of the first-order B-splines over a non-uniform knot vector, while Figure 16.26 those of linear B-splines over the same knot vector.

The equations of the spline functions themselves are a little more complicated than in the case of integer knots for the simple reason that they now involve variables for knot values. For example, here's the equation, analogous to (16.18), for the first quadratic B-spline over a non-uniform knot vector:

$$N_{0,3}(u) = \begin{cases} 0, & u \le t_0 \\ \frac{u-t_0}{t_2-t_0}\frac{u-t_0}{t_1-t_0}, & t_0 \le u \le t_1 \\ \frac{u-t_0}{t_2-t_0}\frac{t_2-u}{t_2-t_1} + \frac{t_3-u}{t_3-t_1}\frac{u-t_1}{t_2-t_1}, & t_1 \le u \le t_2 \\ \frac{t_3-u}{t_3-t_1}\frac{t_3-u}{t_3-t_2}, & t_2 \le u \le t_3 \\ 0, & t_3 \le u \end{cases} \tag{16.29}$$

Not pretty, but B-spline computations are invariably done recursively, so a formula like this rarely needs to be written explicitly.

Experiment 16.7. Run again `bSplines.cpp`. Change the knot values by selecting one with the space bar and then pressing the left/right arrow keys. Press delete to reset knot values. Note that the routine `Bspline()` implements the CdM formula (and its convention for 0 denominators).

In particular, observe the quadratic and cubic spline functions. Note how they lose their symmetry about a vertical axis through the center, and that no longer are they translates of one another.

Play around with making knot values equal – we'll soon be discussing the utility of multiple knots.

Figures 16.27(a) and (b) are screenshots of the quadratic and cubic functions, respectively, both over the same non-uniform knot vector with a triple knot at the right end. **End**

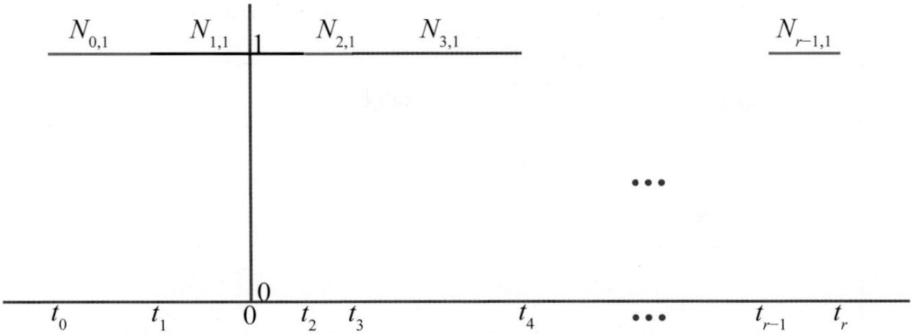

Figure 16.25: Non-zero parts of the first-order B-splines over a non-uniform knot vector.

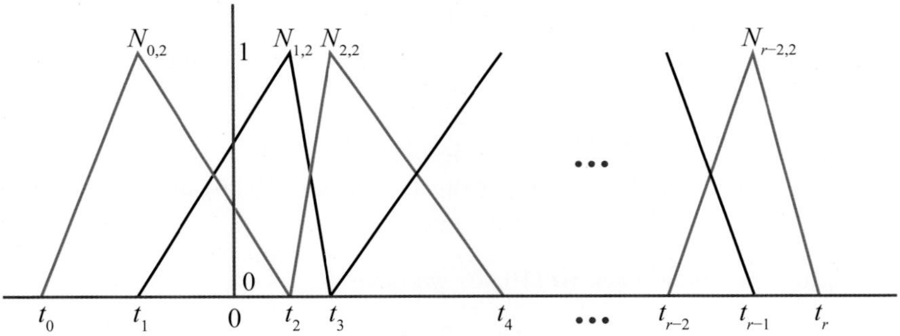

Figure 16.26: Non-zero parts of the linear B-splines over a non-uniform knot vector.

Example 16.1. Find the values of (a) $N_{3,3}(5)$ and (b) $N_{4,3}(5)$, if the knot vector is $\{0, 1, 2, 3, 4, 5, 5, 5, 6, 7, \ldots\}$, the non-negative integers, except that 5 has multiplicity three.

Answer: The successive knot values are

$$t_0=0,\ t_1=1,\ t_2=2,\ t_3=3,\ t_4=4,\ t_5=5,\ t_6=5,\ t_7=5,\ t_8=6,\ t_9=7,\ \ldots$$

(a) Instantiating the CdM formula (16.28):

$$N_{3,3}(u) \quad = \quad \frac{u-t_3}{t_5-t_3}\, N_{3,2}(u) \;+\; \frac{t_6-u}{t_6-t_4}\, N_{4,2}(u)$$

Plugging in $u = 5$ and the given knot values:

$$N_{3,3}(5) \quad = \quad \frac{5-3}{5-3}\, N_{3,2}(5) \;+\; \frac{5-5}{5-4}\, N_{4,2}(5) \quad = \quad N_{3,2}(5) \qquad (16.30)$$

Using CdM again,

$$N_{3,2}(u) \quad = \quad \frac{u-t_3}{t_4-t_3}\, N_{3,1}(u) \;+\; \frac{t_5-u}{t_5-t_4}\, N_{4,1}(u)$$

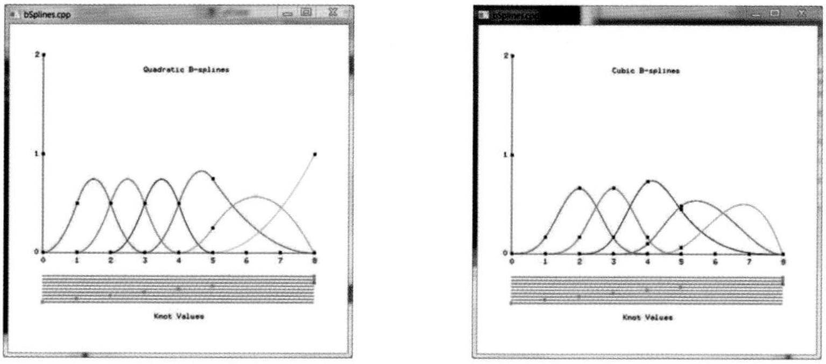

(a) (b)

Figure 16.27: Screenshots of `bSplines.cpp` over a non-uniform knot vector with a triple knot at the right end: (a) Quadratic (b) Cubic.

so that

$$\begin{aligned} N_{3,2}(5) &= \frac{5-3}{4-3}\, N_{3,1}(5) \;+\; \frac{5-5}{5-4}\, N_{4,1}(5) \\ &= 2*0 \;+\; 0*1 \quad \text{(from Equations (16.26) and (16.27))} \\ &= 0 \end{aligned}$$

Taking the above back to (16.30) we have

$$N_{3,3}(5) = 0$$

(b)

$$N_{4,3}(u) \;=\; \frac{u-t_4}{t_6-t_4}\, N_{4,2}(u) \;+\; \frac{t_7-u}{t_7-t_5}\, N_{5,2}(u)$$

giving

$$\begin{aligned} N_{4,3}(5) &= \frac{5-4}{5-4}\, N_{4,2}(5) \;+\; \frac{5-5}{5-5}\, N_{5,2}(5) \\ &= N_{4,2}(5) \;+\; \frac{0}{0} N_{5,2}(5) \\ &= N_{4,2}(5) \;+\; N_{5,2}(5) \quad \text{(using convention } \tfrac{0}{0} = 1\text{)} \end{aligned} \qquad (16.31)$$

Using CdM again,

$$N_{4,2}(u) \;=\; \frac{u-t_4}{t_5-t_4}\, N_{4,1}(u) \;+\; \frac{t_6-u}{t_6-t_5}\, N_{5,1}(u)$$

so that

$$\begin{aligned} N_{4,2}(5) &= \frac{5-4}{5-4}\, N_{4,1}(5) \;+\; \frac{5-5}{5-5}\, N_{5,1}(5) \\ &= 1*1 \;+\; 1*0 \quad \text{(note by (16.27) that } N_{5,1} \text{ is zero everywhere)} \\ &= 1 \end{aligned} \qquad (16.32)$$

CdM again gives

$$N_{5,2}(u) \;=\; \frac{u - t_5}{t_6 - t_5}\, N_{5,1}(u) \;+\; \frac{t_7 - u}{t_7 - t_6}\, N_{6,1}(u)$$

implying

$$\begin{aligned} N_{5,2}(5) &= \frac{5-5}{5-5}\, N_{5,1}(5) \;+\; \frac{5-5}{5-5}\, N_{6,1}(5) \\ &= 1 * 0 \;+\; 1 * 0 \\ &= 0 \end{aligned} \qquad (16.33)$$

Using (16.32) and (16.33) in (16.31) we have

$$N_{4,3}(5) = 1$$

Exercise 16.19. Find the values of $N_{5,3}(5)$ and $N_{6,3}(5)$ for the same knot vector as in the preceding example.

Exercise 16.20. Compute $N_{4,3}(7)$ again over the knot vector of the preceding example. You will have to invoke the convention that $\frac{a}{0} = 0$, if a is not 0.

General B-Spline Curves

The mth order B-spline approximation c of $r - m + 1$ control points $P_0, P_1, \ldots, P_{r-m}$ is the curve obtained by applying the mth order B-splines as blending functions. Its equation is:

$$c(u) = \sum_{i=0}^{r-m} N_{i,m}(u) P_i \qquad (t_{m-1} \le u \le t_{r-m+1}) \qquad (16.34)$$

Experiment 16.8. Run again `quadraticSplineCurve.cpp`. Press 'k' to enter knots mode and alter knot values using the left/right arrow keys and 'c' to return to control points mode. Press delete in either mode to reset.

Try to understand what happens if knots are repeated. Do you notice a loss of C^1-continuity when knots in the interior of the knot vector coincide? What if knots at the ends coincide? Figure 16.28 is a screenshot of `quadraticSplineCurve.cpp` with a double knot at 5 and a triple at the end at 11. **End**

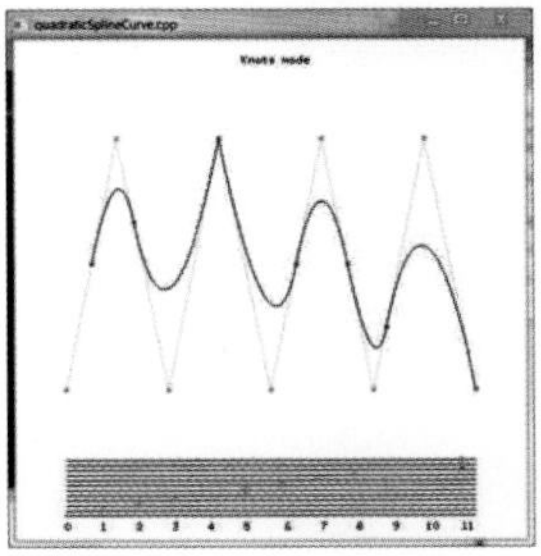

Figure 16.28: Screenshot of `quadraticSpline-Curve.cpp` with one double knot and one triple knot.

Exercise 16.21. Can you find an arrangement of the knots for the quadratic spline curve to interpolate its first and last control points?

Exercise 16.22. Why does changing the value of only the first, or only the last knot, not affect the quadratic spline curve?

Experiment 16.9. Run again `cubicSplineCurve1.cpp`. Press 'k' to enter knots mode and alter knot values using the left/right arrow keys and 'c' to return to control points mode. Press delete in either mode to reset. **End**

Exercise 16.23. Can you find an arrangement of the knots so that the cubic spline curve interpolates its first and last control points?

Exercise 16.24. What part of the mth order spline curve c approximating the control points $P_0, P_1, \ldots, P_{r-m}$ is altered by moving only P_i? Your answer should be in terms of an arc of c between a particular pair of its joints.

We collect information about mth order B-spline functions and their corresponding approximating spline curves in the following proposition.

Proposition 16.1. *Let*

$$T = \{t_0, t_1, \ldots, t_r\}$$

be a non-uniform knot vector, where $r \geq 1$.

The mth order B-spline functions $N_{i,m}$, for some order m lying within $1 \leq m \leq r$, and, where $0 \leq i \leq r - m$, satisfy the following properties:

(a) Each $N_{i,m}$ is piecewise polynomial, consisting of at most $m + 2$ pieces, each of which is a degree $m - 1$ polynomial, except possibly for zero end pieces.

(b) $N_{i,m}$ has support in $[t_i, t_{i+m}]$, the union of m consecutive knot intervals.

(c) If the knots in T are distinct, each $N_{i,m}$ is C^{m-2}, but not C^{m-1}, at its joints. In this case, apart from its joints, each $N_{i,m}$ is smooth everywhere.

(d) The $N_{i,m}$ together form a partition of unity over the parameter space $[t_{m-1}, t_{r-m+1}]$.

(e) Every point of the mth order B-spline approximation c of $r - m + 1$ control points $P_0, P_1, \ldots, P_{r-m}$, defined by Equation (16.34), over the parameter space $[t_{m-1}, t_{r-m+1}]$, is a convex combination of the control points and lies inside their convex hull.

(f) (Affine Invariance) If $g : \mathbb{R}^3 \to \mathbb{R}^3$ is an affine transformation, and c is the mth order B-spline approximation of $r - m + 1$ control points $P_0, P_1, \ldots, P_{r-m}$ in $\mathbb{R}^3$, then the image curve $g(c)$ is the mth order B-spline approximation of the images $g(P_0), g(P_1), \ldots, g(P_{r-m})$ of the control points.

(g) If the knots in T are distinct, the mth order B-spline approximation c of $r-m+1$ control points $P_0, P_1, \ldots, P_{r-m}$ defined by Equation (16.34) is C^{m-2}, but, generally, not C^{m-1}.

Proof. The proofs are a straightforward technical slog and we'll not write them out. □

The following relation for a B-spline curve is useful to remember:

$$\textit{number of knots} = \textit{number of control points} + \textit{order} \tag{16.35}$$

Exercise 16.25. Deduce (16.35).
Hint: Count the number of knots and control points in (16.34).

Non-uniform Knot Vectors

So, of what use are non-uniform knot vectors?

One is to be able to control the influence that a control point has over an approximating curve. For example, consider the cubic spline curve c approximating control points $P_0, P_1, \ldots$ over the knot vector $\{t_0, t_1, \ldots\}$, as in Figure 16.29(a), which shows a few intermediate control points. Moving control point, say, P_5 alters only the arc of c between $a = c(t_5)$ and $b = c(t_9)$, as $N_{5,4}$ has support in $[t_5, t_9]$. Consequently, the closer or farther apart are the knots from t_5 to t_9, the more concentrated or diffuse the influence of P_5. This generalizes, of course, to all P_i, allowing the designer to vary the domain of influence of control points by rearranging knots.

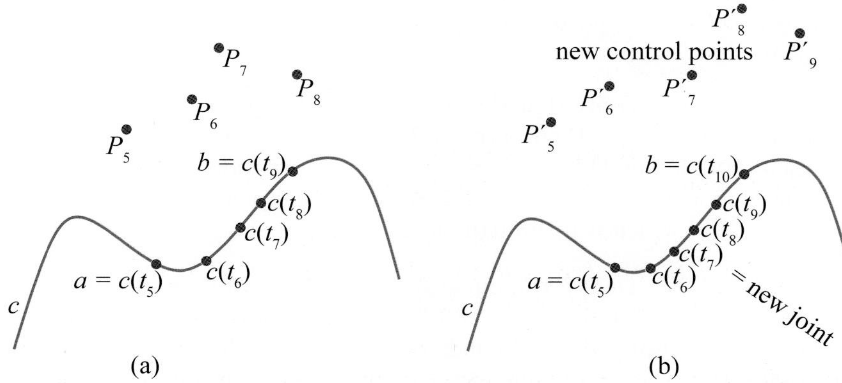

Figure 16.29: (a) Part of a cubic spline curve (b) With a new knot inserted.

Another practical consequence of non-uniform knot vectors is the technique of *knot insertion*, implemented in many commercial modelers, to allow the designer increasingly fine control over part of a spline curve. Clearly, the more knot images (joints, that is) there are in an arc of a curve, the more control points have influence over it and, therefore, the more finely it can be edited. Refer again to Figure 16.29(a). Currently, the shape of the arc between a and b is determined by the four control points P_5, P_6, P_7 and P_8. If one could insert a new knot, say, between t_6 and t_7 *without* changing the shape of the curve, there would then be five control points, instead of four, acting upon the same arc, affording the designer an added level of control.

Knots can, in fact, be inserted without changing either the shape of a spline curve or its degree, though, with a newly computed set of control points. See Figure 16.29(b), where a new knot has been inserted between t_6 and t_7, giving rise to a corresponding new joint. The joints have been re-labeled in sequence and a (hypothetical) new set of control points shown. We'll not go into the theory of knot insertion ourselves, referring the reader instead to more mathematical texts such as Buss [22], Farin [45] and Piegl & Tiller [111].

Multiple Knots

Coincident knots – *multiple knots* and *repeated knots* are the terms most commonly used – have a special application.

We'll motivate our discussion with a running example using the knot vector

$$T = \{t_0\!=\!0,\ t_1\!=\!1,\ t_2\!=\!2,\ t_3\!=\!3,\ t_4\!=\!3,\ t_5\!=\!4,\ t_6\!=\!5,\ t_7\!=\!6,\ \ldots\}$$

which has a double knot at $t_3 = t_4 = 3$. Generally, the *multiplicity* of a knot is the number of times it repeats.

The graphs of some of the B-spline functions over T are shown in Figure 16.30.

Exercise 16.26. Verify that the graphs of the first-order B-splines over T are correctly depicted in the top row of Figure 16.30 by applying the defining Equations (16.26) and (16.27). In particular, the first-order B-splines are all 1 on their supporting intervals, excluding possibly endpoints, and 0 elsewhere, *except* for $N_{3,1}$, which is 0 throughout.

Exercise 16.27. Derive the equations of the linear B-splines from the first-order ones – by plugging $m = 2$ into the recursive Equation (16.28) – to verify their graphs in the second row of Figure 16.30, as well as at the leftmost in the third. In particular, the linear B-splines over T are all C^0 and translates of one another, *except* for $N_{2,2}$ and $N_{3,2}$, neither of which is C^0.

Unfortunately, the artifact of vertical edges in the display when knots coincide makes it tricky to use `bSplines.cpp` to visually verify the linear B-spline graphs in Figure 16.30. However, there is no such issue with quadratic B-splines, so we ask the reader to do the following.

Exercise 16.28. (Programming) Arrange the knots of `bSplines.cpp` to make their nine successive values 0, 1, 2, 3, 3, 4, 5, 6 and 7, which are the first few knots of T. Then verify visually the graphs of the five quadratic

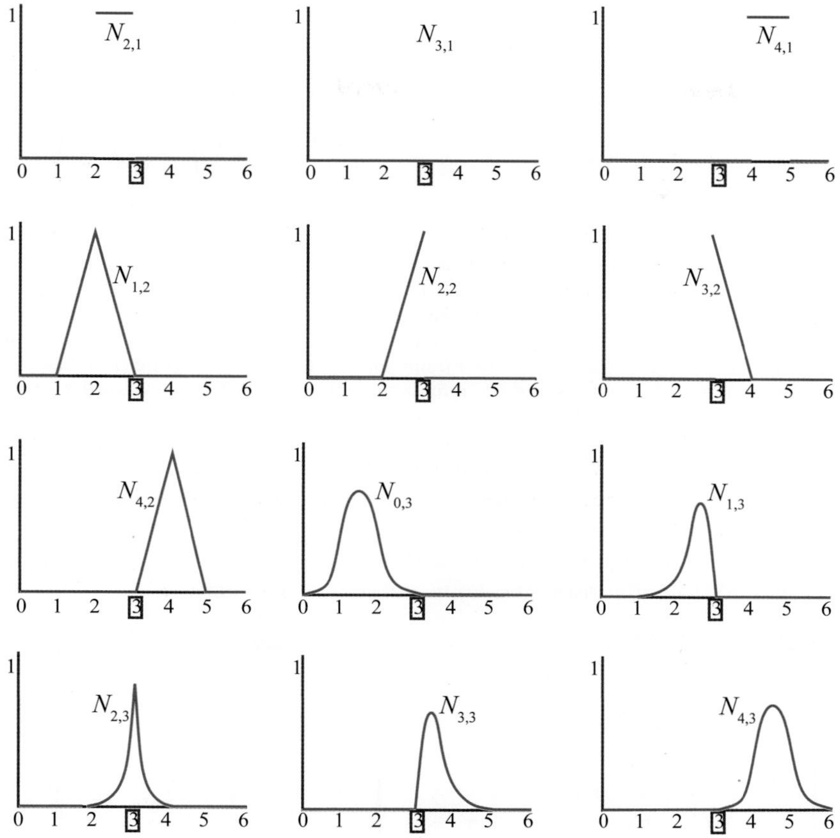

Figure 16.30: B-spline functions over the knot vector $T = \{0, 1, 2, 3, 3, 4, 5, \ldots\}$ with a double knot at 3 (distinguished inside a box).

B-splines in Figure 16.30. In fact, all the quadratic B-splines over T are C^1 and translates of one another, *except* for $N_{1,3}$, $N_{2,3}$ and $N_{3,3}$, which are C^0 but not C^1.

Next, we investigate the behavior of the approximating B-spline curve in the presence of repeated knot values.

Exercise 16.29. Use Equation (16.34) and the graphs already drawn of the first-order and linear spline functions over T to verify that the first-order and linear spline curves approximating nine control points – arranged, alternately, in two horizontal rows – are correctly drawn in Figures 16.31(a) and (b), respectively.

In particular, the first-order approximation loses the control point P_3 (drawn hollow) altogether, while the linear approximation loses the segment P_2P_3 and, therefore, is no longer C^0.

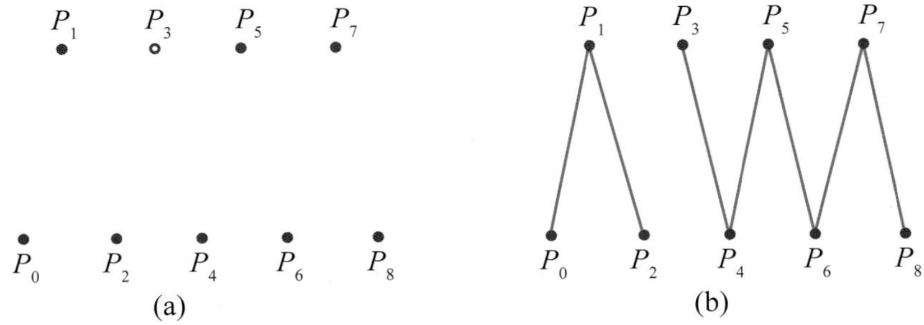

Figure 16.31: (a) First-order and (b) linear spline curves over the knot vector $T = \{0, 1, 2, 3, 3, 4, 5, 6, 7, \ldots\}$, approximating nine control points arranged alternately in two horizontal rows. The (hollow) control point P_3 is the only one missing from the first-order "curve", which consists of the remaining eight points. The second-order curve is the polyline $P_0P_1 \ldots P_8$ *minus* P_2P_3.

Experiment 16.10. Use the programs `quadraticSplineCurve.cpp` and `cubicSplineCurve1.cpp` to make the quadratic and cubic B-spline approximations over the knot vector $T = \{0, 1, 2, 3, 3, 4, 5, 6, 7, \ldots\}$ of nine control points placed as in Figure 16.31(a) (or (b)). See Figure 16.32(a) and (b) for screenshots of the quadratic and cubic curves, respectively.

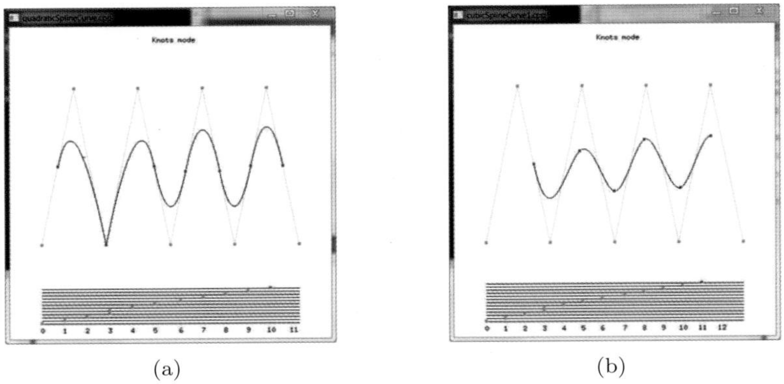

Figure 16.32: Screenshots of (a) `quadraticSplineCurve.cpp` and (b) `cubicSplineCurve1.cpp` over the knot vector $T = \{0, 1, 2, 3, 3, 4, 5, 6, 7, \ldots\}$ and approximating nine control points arranged in two horizontal rows.

The quadratic approximation loses C^1-continuity precisely at the control point P_2, which it now *interpolates* as the curve point $c(3)$. It's still C^0 everywhere.

It's not easy to discern visually, but the cubic spline drops from C^2 to C^1-continuous at $c(3)$. **End**

Let's see next what happens with even higher multiplicity.

Experiment 16.11. Continuing with `cubicSplineCurve1.cpp` with control points as in the preceding experiment, press delete to reset and then make equal t_4, t_5 and t_6, creating a triple knot. Figure 16.33 is a screenshot of this configuration. Evidently, the control point P_3 is now interpolated at the cost of a drop in continuity there to mere C^0. Elsewhere, the curve is still C^2. **End**

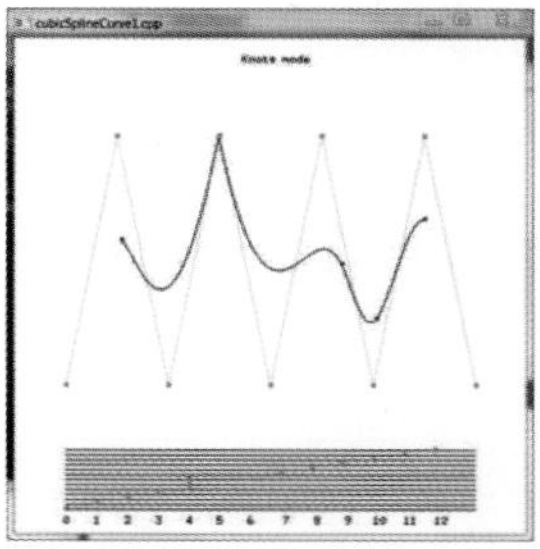

Figure 16.33: Screenshot of `cubicSplineCurve1.cpp` with a triple knot.

It seems, generally, that repeating a knot increases the influence of a particular control point, to the extent that if the repetition is sufficient then that control point itself is interpolated, though at the cost of continuity at the control point itself. This does not appear to be a particularly appealing trade-off unless a low-continuity artifact, e.g., a corner, is itself a design goal.

Let's examine more closely how the loss arises – evidently, because of the difference in the value of the derivative (of some order) of c *on either side* of a control point P. For example, the tangents to the arcs on either side of the interpolated control point P_2 of the quadratic spline curve in Figure 16.32(a) are different.

Consider now if P were an *endpoint* of c. Then continuity cannot be lost by derivatives differing on the two sides of P, for the simple reason that the curve is only to one side! And, yet, there is no reason why the influence of P cannot still be increased by repeating knots. We are on our way to recovering the property of interpolating end control points that was lost at first by quadratic spline curves.

Experiment 16.12. Make the first three and last three knots separately equal in `quadraticSplineCurve.cpp` (Figure 16.34(a)). Make the first four and last four knots separately equal in `cubicSplineCurve1.cpp` (Figure 16.34(b)). The first and last control points are interpolated in both. Do you notice any impairment in continuity? *No*! **End**

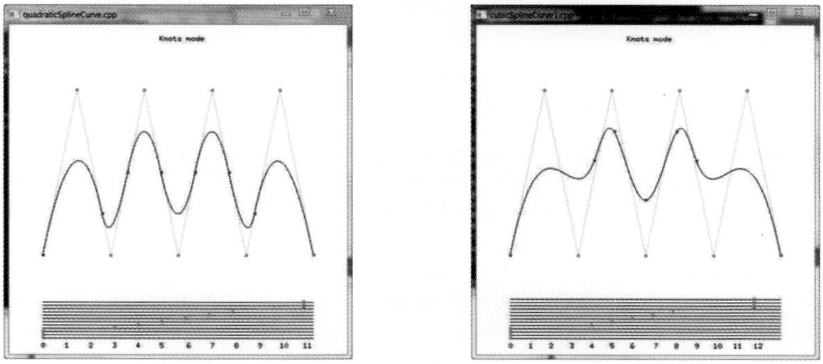

(a) (b)

Figure 16.34: Screenshots of (a) `quadraticSplineCurve.cpp` and (b) `cubicSplineCurve1.cpp`, both with knots repeated at the end to interpolate the first and last control points.

Generally, if the first m and last m knots of an mth order spline curve are coincident, and there are no other multiple knots, then the curve interpolates its first and last control points without losing C^{m-2}-continuity anywhere. In fact, a knot vector which starts and ends with a multiplicity of m and whose intermediate knots are uniformly spaced is called a *standard knot vector*.

A standard knot vector for a quadratic spline with nine control points is

$$\{0, 0, 0, 1, 2, 3, 4, 5, 6, 7, 7, 7\}$$

The above is the canonical (and simplest) way to write standard knot vectors, though, for example

$$\{2.7,\ 2.7,\ 2.7,\ 3.5,\ 4.3,\ 5.1,\ 5.9,\ 6.7,\ 7.5,\ 8.3,\ 8.3,\ 8.3\}$$

would be equivalent.

The size of the standard knot vector is calculated from formula (16.35), viz.

$$\textit{number of knots} = \textit{number of control points} + \textit{order}$$

when given the two quantities on its RHS.

Exercise 16.30. Jot down a standard knot vector for a quadratic spline over 10 control points and for a cubic spline over 9 control points.

Exercise 16.31. Use the CdM formula to show that $N_{0,3}(t_2) = 1$ over the standard knot vector

$$T = \{0,\ 0,\ 0,\ 1,\ 2,\ \ldots,\ r-6,\ r-5,\ r-5,\ r-5\}$$

of size r for a quadratic spline. Use this to prove that the quadratic spline

$$c(u) = \sum_{i=0}^{r-3} N_{i,3}(u) P_i \qquad (t_2 = 0 \le u \le r-5 = t_{r-2})$$

approximating the $r-m+1$ control points P_i, $0 \le i \le r-m$, over T indeed interpolates the first one, in particular, $c(t_2) = P_0$.

For the record here's a proposition:

Proposition 16.2. *A spline curve over a standard knot vector interpolates its first and last control points.*

Proof. The proof is a generalization of the preceding exercise to establish that the first control point is always interpolated. We'll leave the reader to do this by an induction. That the last control point is interpolated as well follows by symmetry. □

The use of a standard knot vector for splines bequeaths yet another Bézier-like property – recall Proposition 15.1(f) – in addition to the interpolation of the end control points:

Proposition 16.3. *The tangent lines at the endpoints of a spline curve over a standard knot vector each pass through the adjacent control point.*

Proof. We'll not prove this in full generality, but only for quadratic splines in the next example. The general proof is not difficult, but tedious. □

Example 16.2. Prove that the tangent lines at the endpoints of a quadratic spline curve over a standard knot vector each pass through the adjacent control point.

Answer: We'll show that the tangent vector at the first control point passes through the second. The result at the other end follows by symmetry.

For quadratic splines, the standard knot vector is

$$T = \{0,\ 0,\ 0,\ 1,\ 2,\ \ldots\}$$

The quadratic spline curve approximating the control points $P_0, P_1, P_2, \ldots$ is

$$c(u) = N_{0,3}(u)P_0 + N_{1,3}(u)P_1 + N_{2,3}(u)P_2 + N_{3,3}(u)P_3 + \ldots$$

Now, the blending functions $N_{i,3}$, for $i \geq 3$, all vanish in $[t_2, t_3] = [0, 1]$. Consequently, in $[0, 1]$:

$$c(u) = N_{0,3}(u)P_0 + N_{1,3}(u)P_1 + N_{2,3}(u)P_2$$

Plugging the standard knot vector values into formula (16.29) for $N_{0,3}$ we get

$$N_{0,3}(u) = 1 - 2u + u^2, \quad u \in [0, 1]$$

One can use (16.29) to determine $N_{1,3}(u)$ as well by incrementing the subscripts on its RHS by 1. This gives

$$N_{1,3}(u) = 2u - \frac{3}{2}u^2, \quad u \in [0, 1]$$

Likewise, it's found that

$$N_{2,3}(u) = \frac{1}{2}u^2, \quad u \in [0, 1]$$

Therefore,

$$c(u) = (1 - 2u + u^2)\, P_0 + (2u - \frac{3}{2}u^2)\, P_1 + (\frac{1}{2}u^2)\, P_2, \quad u \in [0, 1]$$

Differentiating,

$$c'(u) = (-2 + 2u)\, P_0 + (2 - 3u)\, P_1 + u\, P_2, \quad u \in [0, 1]$$

Plugging in $u = 0$, one sees that

$$c'(0) = 2(P_1 - P_0)$$

which is indeed in the direction from P_0 to P_1.

We see it's for good reason, therefore, that standard knot vectors are most often used in B-spline design.

Exercise 16.32. Proposition 16.1(e) says that a spline curve is contained in the convex hull of (all) its control points. Prove the stronger statement that a spline curve of order m can be divided into successive stretches that each lie in the convex hull of only some m of its control points.

Bézier Curves and Spline Curves

It turns out that Bézier curves are special cases of spline curves:

Proposition 16.4. *The $(n+1)$th order Bézier curve approximating the $n+1$ control points*

$$P_0, P_1, \ldots, P_n$$

coincides with the $(n+1)$th order spline curve approximating the same control points over the particular standard knot vector

$$\{0, 0, \ldots, 0, 1, 1, \ldots, 1\}$$

consisting of $n+1$ 0's followed by $n+1$ 1's.

Proof. Again, in the following example, we'll restrict ourselves to establishing the quadratic case, leaving the general proof by induction to the mathematically inclined reader. □

Example 16.3. Show that the quadratic Bézier curve approximating the three control points P_0, P_1 and P_2 coincides with the quadratic spline curve approximating the same control points over the particular standard knot vector $\{0, 0, 0, 1, 1, 1\}$.

Answer: Recall from the previous chapter that the Bézier curve approximating P_0, P_1 and P_2 is

$$c_B(u) = (1-u)^2 P_0 + 2(1-u)uP_1 + u^2 P_2, \quad u \in [0, 1]$$

The quadratic spline approximating the same three points over the knot vector $T = \{t_0=0,\ t_1=0,\ t_2=0,\ t_3=1,\ t_4=1,\ t_5=1\}$ is

$$c_S(u) = N_{0,3}(u)P_0 + N_{1,3}(u)P_1 + N_{2,3}(u)P_2, \quad u \in [t_2, t_3] = [0, 1]$$

Therefore, we must show that spline blending functions of the preceding equation match the Bernstein polynomial blending functions of the one before it, over the knot interval $[0, 1]$. Refer to formula (16.29) for $N_{0,3}$. The fourth line on the RHS gives

$$\begin{aligned} N_{0,3}(u) &= \frac{t_3 - u}{t_3 - t_1} \frac{t_3 - u}{t_3 - t_2} \\ &= (1-u)^2 \\ &\quad \text{(after plugging in the knot values } t_0 = t_1 = t_2 = 0 \text{ and } t_3 = 1) \end{aligned}$$

in $t_2 = 0 \leq u \leq 1 = t_3$, confirming a match with the first Bernstein polynomial.

We can use (16.29) for $N_{1,3}$ as well, making sure to increment the subscripts on the RHS by 1. This gives

$$\begin{aligned} N_{1,3}(u) &= \frac{u - t_1}{t_3 - t_1} \frac{t_3 - u}{t_3 - t_2} + \frac{t_4 - u}{t_4 - t_2} \frac{u - t_2}{t_3 - t_2} \\ &= 2(1-u)u \\ &\quad \text{(after plugging in the knot values } t_0 = t_1 = t_2 = 0 \text{ and } t_3 = 1) \end{aligned}$$

in $0 \leq u \leq 1$, matching the second Bernstein polynomial. We'll leave the reader to verify that $N_{2,3}(u) = u^2$, $u \in [0,1]$, completing the answer.

In the opposite direction, the following is true because spline curves are piecewise polynomial (from the way they are constructed) and polynomial curves are Bézier (from Proposition 15.2).

Proposition 16.5. *A spline curve is piecewise Bézier.* □

Exercise 16.33. Why is it not possible that the preceding proposition can somehow be strengthened to say that spline curves are, in fact, Bézier, not just piecewise?
Hint: Bézier curves are smooth throughout.

16.3 B-Spline Surfaces

The construction of B-spline surfaces as a continuum of B-spline curves parallels exactly the construction of Bézier surfaces from Bézier curves described in Section 15.2. See Figure 16.35 for the following.

Suppose that we are given an $(n+1) \times (n'+1)$ array of control points

$$P_{i,j}, \quad \text{for} \quad 0 \leq i \leq n, \ 0 \leq j \leq n'$$

and two spline orders m and m', and a knot vector

$$T = \{t_0, t_1, \ldots, t_r\}, \quad \text{whose size satisfies} \quad |T| = r + 1 = n + 1 + m$$

(to ensure that *number of knots = number of control points + order*) and another knot vector

$$T' = \{t_0, t_1, \ldots, t'_r\}, \quad \text{whose size satisfies} \quad |T'| = r' + 1 = n' + 1 + m'$$

Think of the control points array as $n+1$ different sequences, each of $n'+1$ control points. In particular, the ith sequence, for $0 \leq i \leq n$, consists of $P_{i,0}, P_{i,1}, \ldots, P_{i,n'}$, lying along the ith row of the control points array. Construct the m'th order B-spline curve c_i, for $0 \leq i \leq n$, approximating the ith sequence, each using the knot vector T' over the parameter space $[t_{m'-1}, t_{r'-m'+1}]$.

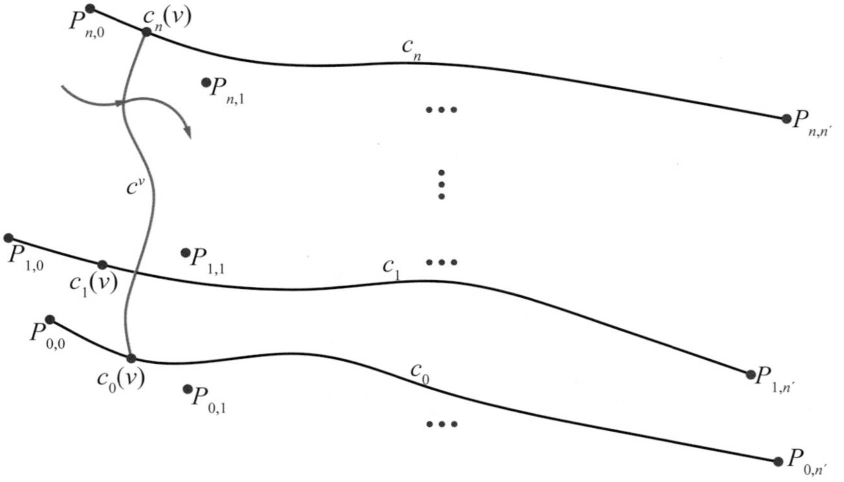

Figure 16.35: Constructing the B-spline surface approximating an array of control points by sweeping a B-spline curve. The B-spline curves depicted all interpolate both end control points, which need not always be the case in practice.

For each v in $t_{m'-1} \leq v \leq t_{r'-m'+1}$, generate the mth order B-spline curve c^v approximating the control points sequence $c_0(v), c_1(v), \ldots, c_n(v)$, using the knot vector T over the parameter space $[t_{m-1}, t_{r-m+1}]$. The union of all these B-spline curves c^v, for $t_{m'-1} \leq v \leq t_{r'-m'+1}$, then, is the B-spline surface s approximating the control points array $P_{i,j}$, $0 \leq i \leq n$, $0 \leq j \leq m$. One may imagine s as being swept by c^v, as v varies from $t_{m'-1}$ to $t_{r'-m'+1}$.

Exercise 16.34. Prove that the parametric equation of the B-spline surface s constructed as above is

$$s(u,v) = \sum_{i=0}^{n} \sum_{j=0}^{n'} N_{i,m}^{T}(u) N_{j,m'}^{T'}(v) P_{i,j} \tag{16.36}$$

for $t_{m-1} \leq u \leq t_{r-m+1}$ and $t_{m'-1} \leq v \leq t_{r'-m'+1}$, and where $N_{i,m}^{T}$ (respectively, $N_{j,m'}^{T'}$) denotes the B-spline function $N_{i,m}$ over the knot vector T (respectively, $N_{j,m'}$ over T').

In other words, the surface is obtained from applying the blending function $N_{i,m}^{T}(u) N_{j,m'}^{T'}(v)$ to the control point $P_{i,j}$, over the parameter domain $t_{m-1} \leq u \leq t_{r-m+1}$, $t_{m'-1} \leq v \leq t_{r'-m'+1}$.
Hint: Mimic the proof of (15.16) for a Bézier surface.

Exercise 16.35. Formulate an analogue for B-spline surfaces of Proposition 16.1 for curves.

We've given thus far an account of NURBS (non-uniform rational B-spline) theory, *except* for the 'R', or rational, part. Instead of generally rational, our functions have all been polynomial. You could say that we have covered NUPBS, or simply NUBS, as the default for B-splines is polynomial! We'll put the 'R' into NURBS in Chapter 18 with the help of projective spaces.

16.4 Drawing B-Spline Curves and Surfaces

NURBS – the full-blown rational version of B-splines – curves and surfaces are implemented in the GLU library of OpenGL. Now that we have a fair amount of the theory, the GLU NURBS interface will turn out to be fairly simple to use, as the mapping between theory and syntax is almost one-to-one. We'll, of course, restrict ourselves to polynomial B-spline primitives for now, leaving the rational ones to a later chapter.

16.4.1 B-Spline Curves

We had already used OpenGL to draw polynomial B-spline curves in the programs `quadraticSplineCurve.cpp` and `cubicSplineCurve1.cpp` earlier this chapter, without caring then about the drawing syntax itself. Let's look at this now.

The command

`gluNurbsCurve(`**nurbsObject*, *knotCount*, **knots*, *stride*, **controlPoints*, *order*, *type*`)`

defines a B-spline curve which is pointed by *nurbsObject*. The parameter *knotCount* is the number of knots in the knot vector – a one-dimensional array – pointed by *knots*. The parameter *order* is the order of the spline curve, *controlPoints* points to the one-dimensional array of control points, and *stride* is the number of floating point values between the start of the data set for one control point and that of the next in the control points array. The number of control points is not explicitly specified, but computed by OpenGL with the help of (16.35):

$$\textit{number of control points} = \textit{number of knots} - \textit{order}$$

The parameter *type* is `GL_MAP1_VERTEX_3` or `GL_MAP1_VERTEX_4`, according as the spline curve is polynomial or rational.

A `gluNurbsCurve()` command must be bracketed between a `gluBeginCurve()-gluEndCurve()` pair of statements. The following statements from the drawing routine of `quadraticSplineCurve.cpp`, defining a quadratic B-spline curve approximating nine control points, should now be clear:

```
gluBeginCurve(nurbsObject);
gluNurbsCurve(nurbsObject, 12, knots, 3, ctrlpoints[0], 3,
              GL_MAP1_VERTEX_3);
gluEndCurve(nurbsObject);
```

Exercise 16.36. Refer to Section 10.3.1 for the syntax of the command `glMap1f()` defining a Bézier curve and compare it with that of `gluNurbsCurve()`.

There are certain initialization steps to be completed prior to a `gluNurbsCurve()` call. First, `gluNewNurbsRenderer()` returns the pointer to a NURBS object, which is passed to the subsequent `gluNurbsCurve()` call. Then optional `gluNurbsProperty()` calls control the quality of the rendering. They can activate as well a callback interface. There are several possible attributes for `gluNurbsProperty()` and we refer to the red book for details. Our own usage is kept to a simple minimum – the relevant statements from the `setup()` routine of `quadraticSplineCurve.cpp` are the following:

```
nurbsObject = gluNewNurbsRenderer();
gluNurbsProperty(nurbsObject, GLU_SAMPLING_METHOD, GLU_PATH_LENGTH);
gluNurbsProperty(nurbsObject, GLU_SAMPLING_TOLERANCE, 10.0);
```

The last two statements specify that the longest length of a line segment in a strip approximating a NURBS curve (or that of a quad edge, in the case of a mesh approximating a NURBS surface) is at most 10.0 pixels.

Experiment 16.13. Change the last parameter of the statement

```
gluNurbsProperty(nurbsObject, GLU_SAMPLING_TOLERANCE, 10.0);
```

in the initialization routine of `quadraticSplineCurve.cpp` from 10.0 to 100.0. The fall in resolution is noticeable. **End**

If you are wondering whether a B-spline curve can be drawn in a manner similar to that using `glMapGrid1f()` followed by `glEvalMesh1()` for a Bézier curve – sampling the curve uniformly through the parameter domain – the answer is yes. We don't use them ourselves but the needed calls are `gluNurbsProperty(`**nurbsObject*`, GLU_SAMPLING_METHOD, GLU_DOMAIN_DISTANCE)` and `gluNurbsProperty(`**nurbsObject*`, GLU_U_STEP,` *value*`)`. The reader is referred to the red book for implementation details.

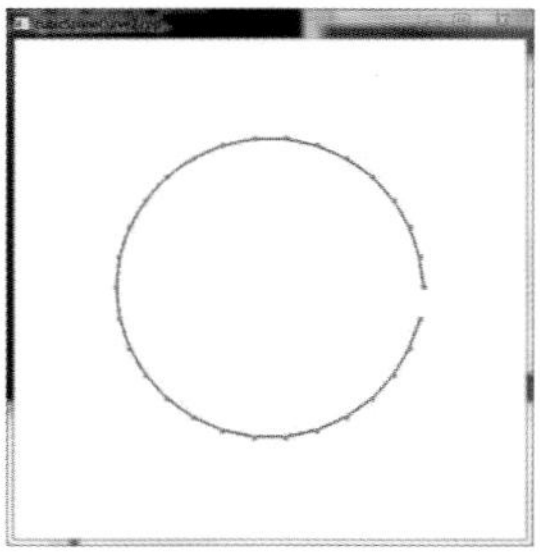

Figure 16.36: Screenshot of `cubicSplineCurve2-.cpp`.

Experiment 16.14. Run `cubicSplineCurve2.cpp`, which draws the cubic spline approximation of 30 movable control points, initially laid out on a circle, over a fixed standard knot vector. Press space and backspace to cycle through the control points and the arrow keys to move the selected control point. The delete key resets the control points. Figure 16.36 is a screenshot of the initial configuration.

The number of control points being much larger than the order, the user has good local control. **End**

Exercise 16.37. (Programming) Use `cubicSplineCurve2.cpp` to draw two smooth closed loops like those in Figure 16.37.

Figure 16.37: Use cubicSplineCurve2.cpp to draw a man and his cat.

16.4.2 B-Spline Surfaces

The OpenGL syntax for a B-spline surface is a straightforward extension of that for a B-spline curve. The `gluNurbsSurface()` command, which must be bracketed between a `gluBeginSurface()-gluEndSurface()` pair of statements, has the following form:

gluNurbsSurface(**nurbsObject, uknotCount, *uknots, vknotCount, *vknots, ustride, vstride, *controlPoints, uorder, vorder, type*)

**vknots* points to the knot vector used with the control point row, in other words, to make the parameter curves c_i in the discussion in Section 16.3 of a B-spline curve sweeping a surface; **uknots* points to the knot vector used with the control point columns, i.e., to make the curves c^v in that discussion.

The parameter *vknotCount* is the number of knots in the vector pointed by **vknots*, *vorder* is the order of the B-spline curves c_i and *vstride* is the number of floating point values between the data set for one control point and the next in a row of the control points array. The parameters *uknotCount*, *uorder* and *ustride* represent similar values for the control point columns.

The parameter *type* is GL_MAP2_VERTEX_3 or GL_MAP2_VERTEX_4 for polynomial or rational surfaces, respectively; it can have other values as well to specify surface normals and texture coordinates.

Experiment 16.15. Run bicubicSplineSurface.cpp, which draws a spline surface approximation to a 15 × 10 array of control points, each movable in 3-space. The spline is cubic in both parameter directions and a standard knot vector is specified in each as well.

Press the space, backspace, tab and enter keys to select a control point. Move the selected control point using the arrow and page up and down keys. The delete key resets the control points. Press 'x/X', 'y/Y' and 'z/Z' to turn the surface. Figure 16.38 is a screenshot. **End**

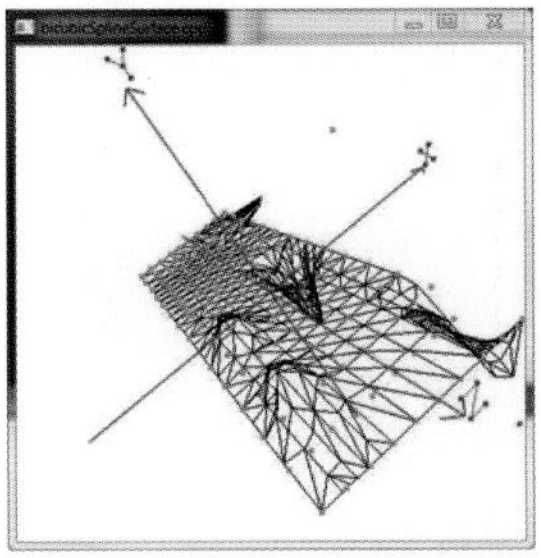

Figure 16.38: Screenshot of bicubicSpline-Surface.cpp.

Exercise 16.38. (Programming) Use bicubicSplineSurface.cpp to draw a hilly terrain and a boat.

16.4.3 Lighting and Texturing a B-Spline Surface

Lighting and texturing a B-spline surface is similar to doing likewise for a Bézier surface. Normals are required for lighting and the quickest way to create normals for a B-spline surface is to generate them automatically with a call, as for Bézier surfaces, to `glEnable(GL_AUTO_NORMAL)`.

And, again as for Bézier surfaces, determining texture coordinates for a B-spline surface requires, first, the creation of a "fake" B-spline surface in texture space on the same parameter rectangle as the real one – the reader should review if need be the discussion in Section 12.4 on specifying texture coordinates for a Bézier surfaces. OpenGL, subsequently, assigns as texture coordinates to the image on the real surface of a particular parameter point the image of that same point on the fake surface in texture space. Code will clarify.

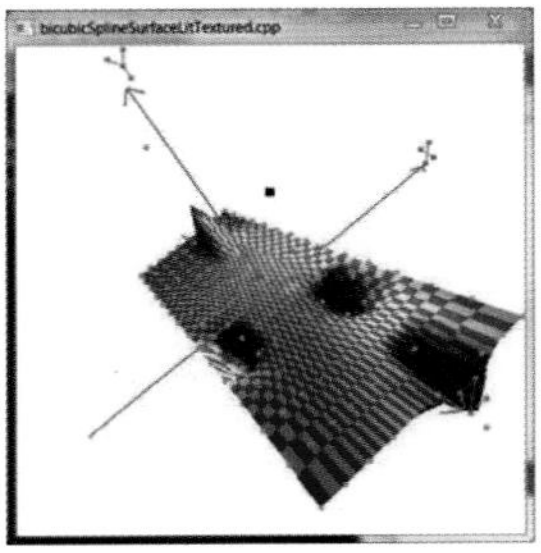

Figure 16.39: Screenshot of `bicubicSplineSurface-LitTextured.cpp`.

Experiment 16.16. Run `bicubicSplineSurfaceLitTextured.cpp`, which textures the spline surface of `bicubicSplineSurface.cpp` with a red-white chessboard texture. Figure 16.39 is a screenshot. The surface is illuminated by a single positional light source whose location is indicated by a large black point. User interaction remains as in `bicubicSplineSurface.cpp`. Note that pressing the 'x'-'Z' keys turns only the surface, not the light source.

The bicubic B-spline surface, as well as the fake bilinear one in texture space, are created by the following statements in the drawing routine:

```
gluBeginSurface(nurbsObject);
gluNurbsSurface(nurbsObject, 19, uknots, 14, vknots,
       30, 3, controlPoints[0][0], 4, 4, GL_MAP2_VERTEX_3);
gluNurbsSurface(nurbsObject, 4, uTextureknots, 4, vTextureknots,
       4, 2, texturePoints[0][0], 2, 2, GL_MAP2_TEXTURE_COORD_2);
gluEndSurface(nurbsObject);
```

We'll leave the reader to parse in particular the third statement and verify that it creates a "pseudo-surface" – a 10×10 rectangle – in texture space on the same parameter domain $[0, 12] \times [0, 7]$ as the real one. **End**

Exercise 16.39. (**Programming**) Light and texture the B-spline surfaces you created for Exercise 16.38.

16.4.4 Trimmed B-Spline Surface

A powerful design tool is to *trim* (i.e., excise or remove) part of a B-spline surface. Here, first, is what happens theoretically.

Say the parametric specification of a surface s is given to be

$$x = f(u, v),\ y = g(u, v),\ z = h(u, v), \quad \text{where} \quad (u, v) \in W = [u_1, u_2] \times [v_1, v_2]$$

The parametric equations map the rectangle W from uv-space onto the surface s in xyz-space. Moreover, a loop (closed curve) c on W maps to a loop c' on s. See Figure 16.40.

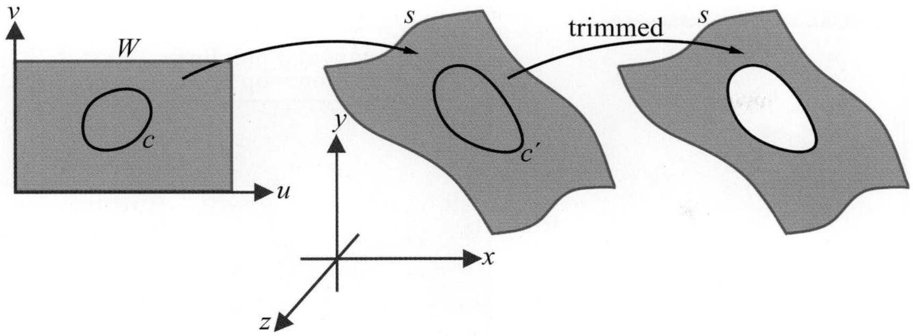

Figure 16.40: The loop c on the parameter space W is mapped to the loop c' on the surface s by the parametric equations for s. Then s is trimmed by c.

If the part of s inside, or outside, the loop c' is excised, then s is said to be trimmed by the loop c (probably, more accurate would be to say that it is trimmed by c', but the given usage is common). Figure 16.40 shows the inside trimmed. Loop c itself is called the *trimming loop*.

OpenGL allows B-spline surfaces to be trimmed. We use the program `trimmedBicubicBsplineSurface.cpp`, as a running example to explain OpenGL syntax for trimming.

Experiment 16.17. Run `trimmedBicubicBsplineSurface.cpp`, which shows the surface of `cubicBsplineSurface.cpp` trimmed by multiple loops. The code is modified from `bicubicBsplineSurface.cpp`, functionality remaining same. Figure 16.41(a) is a screenshot. **End**

All the code relevant to trimming is in the drawing routine:

```
gluBeginSurface(nurbsObject);
gluNurbsSurface(nurbsObject, 19, uknots, 14, vknots,
        30, 3, controlPoints[0][0], 4, 4, GL_MAP2_VERTEX_3);

gluBeginTrim(nurbsObject);
   gluPwlCurve(nurbsObject, 5, boundaryPoints[0], 2,
               GLU_MAP1_TRIM_2);
gluEndTrim(nurbsObject);

gluBeginTrim(nurbsObject);
   gluPwlCurve(nurbsObject, 11, circlePoints[0], 2,
               GLU_MAP1_TRIM_2);
gluEndTrim(nurbsObject);

gluBeginTrim(nurbsObject);
   gluNurbsCurve(nurbsObject, 10, curveKnots, 2, curvePoints[0], 4,
                 GLU_MAP1_TRIM_2);
gluEndTrim(nurbsObject);

gluEndSurface(nurbsObject);
```

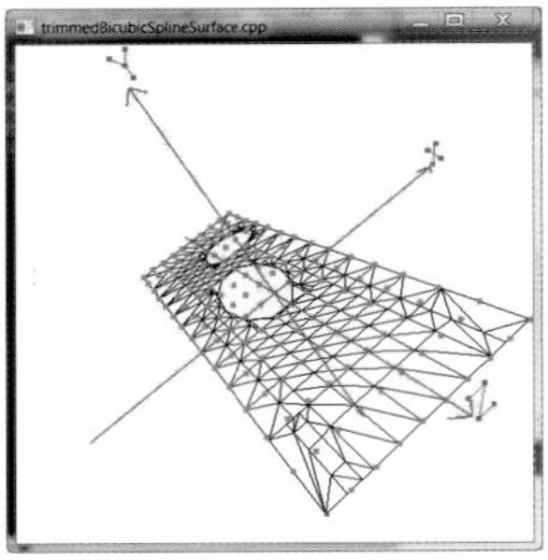

(a)

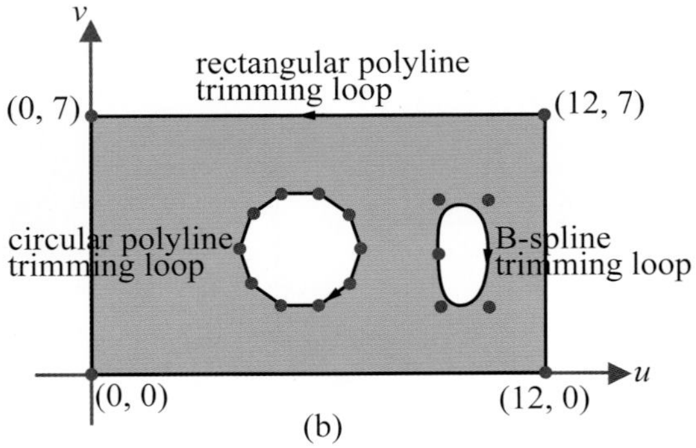

(b)

Figure 16.41: (a) Screenshot of `trimmedBicubicBsplineSurface.cpp` (b) The three trimming loops – two polygonal and one B-spline.

Points to note:

1. Each trimming loop is defined within a `gluBeginTrim()-gluEndTrim()` pair of statements, which itself must lie within the `gluBeginSurface()-gluEndSurface()` pair. The trimming loop definitions are located after the `gluNurbsSurface()` definition.

2. Each trimming loop must be a closed curve in the parameter space.

3. There are two ways to define a trimming loop:

   (a) As a polygonal line loop defined by a

   `glPwlCurve(`**nurbsObject*`,` *pointsCount*`,` **pointsArray*`,`
   stride`,` *type*`)`

   statement, where *pointsCount* is the number of vertices in an array of the form $\{v_0, v_1, \ldots, v_n\}$ pointed by *pointsArray* (it is required that $v_0 = v_n$).

   There are two such polyline trimming loops in the program (see Figure 16.41(b)):

   (i) The five vertices (first and last equal) of one are in the array `boundaryPoints`, describing the rectangular boundary of the parameter space itself, oriented *counter-clockwise*. We'll soon see why this particular bounding trimming loop is required.

   (ii) The eleven vertices (again, first and last equal) of the other are in the array `circlePoints`, equally spaced along a circle, oriented *clockwise*.

   (b) As a B-spline loop defined by

   `gluNurbsCurve(`*nurbsObject*`,` *knotCount*`,` **knots*`,` *stride*`,`
   **controlPoints*`,` *order*`,` *type*`)`

In the program there is a single such B-spline trimming loop, whose six control points (first and last equal) are in the array `curvePoints` oriented *clockwise* (Figure 16.41(b)).

4. The part outside a trimming loop oriented counter-clockwise is trimmed, while that inside a trimming loop oriented clockwise is trimmed.

   Accordingly, the first trimming polyline loop of the program, which bounds the parameter space going counter-clockwise, trims off the *exterior* of the drawn surface, not trimming the surface itself per se. The other two trimming loops actually create holes in the surface.

Exercise 16.40. **(Programming)** Draw a terrain with a few extinct volcanoes (or smoking ones if you like particle systems).

16.5 Summary, Notes and More Reading

We have learned a fair amount of the theory underlying the widely-used class of 3D design primitives – B-splines, both curves and surfaces. Emphasis was on motivating each new concept. We did *not* want to pull stuff out of a hat. A test if we were successful is for the reader to deduce some formula, e.g., (16.18) for the first quadratic B-spline $N_{0,3}$ over a uniform knot vector or the Cox-de Boor-Mansfield recurrence (16.28), using just pencil and paper, and not referring again to the text! This chapter prepares the reader, as well, for the rational version of the theory – NURBS – coming up in Chapter 18.

As for OpenGL, we learned not only how to draw B-spline curves and surfaces, but to illuminate, texture and trim the latter as well.

While B-spline theory is extensive, material we covered in this chapter of the polynomial B-spline primitives, together with what is covered in Chapter 18 of NURBS, is ample for an applications programmer to function knowledgeably. However, the reader is well-advised to expand her knowledge, particularly, of such practical topics as "knot insertion", "degree elevation", etc. It's easy enough given the number of excellent books available – Bartels et al. [10], Farin [45], Mortenson [94], Piegl & Tiller [111] and Rogers & Adams [118] are a few that come to mind. The mathematically inclined reader, in particular, will find much to fascinate her in the more specialized nooks and crannies. Advanced 3D CG books, e.g., Akenine-Möller, Haines & Hoffman [1], Buss [22], Slater et al. [135] and Watt [147], should each have a thorough presentation of B-spline theory as well.

B-spline functions were first studied in the 1800s by the Russian mathematician Nicolai Lobachevsky. However, the modern theory began with Schoenberg's [126] application of spline functions to data smoothing and received particular impetus with the discovery in 1972 of the recursive formula (16.28) for B-spline functions by Cox [30], de Boor [34] and Mansfield. It has since seen explosive growth and B-spline (and NURBS) primitives are *de rigueur* in modern-day CG design.

CHAPTER 17

Hermite

Our objective in this chapter is to learn a method of interpolating a set of control points, in other words, finding a curve (or surface) that passes through each. Bézier curves, as we know, mandatorily interpolate only their first and last control points, while Bézier surfaces only the four corner control points. B-spline curves and surfaces of quadratic and higher degree do not necessarily interpolate any of their control points. Nevertheless, we learned in Section 16.2.5 how to force a B-spline curve to interpolate a control point by raising the multiplicity of a knot. In fact, the so-called standard knot vector, with repeated end knots, is often used to ensure the interpolation of end control points.

However, if a designer wishes to draw a curve or surface interpolating *all* its control points, then it's best to apply an intrinsically interpolating technique, rather than try to coax an approximating one like Bézier or B-spline into interpolating. A popular class of interpolating curves is that of the Hermite splines and this short chapter introduces this class, together with two special subclasses, that of the natural cubic splines and the cardinal splines. We discuss Hermite surface patches, as well, to interpolate 2D arrays of control points.

We begin with a discussion of general Hermite splines in Section 17.1. These curves, unfortunately, are guaranteed only to be piecewise smooth – they can have corners at control points. Moreover, the user is required to specify tangent vectors at all the control points. The subclass of natural cubic splines, the topic of Section 17.2, automatically determines these tangent vectors by imposing an additional C^2-continuity requirement. Cardinal splines, in Section 17.3, are based upon yet another scheme to automatically specify tangent vectors at control points.

We make a brief presentation of Hermite surfaces in Section 17.4 and conclude in Section 17.5.

17.1 Hermite Splines

A *Hermite spline*, also called a *cubic spline*, interpolating a sequence $P_0, P_1, \ldots, P_n$ of $n+1$ control points, is a *piecewise cubic* curve c passing through the control points. Each cubic arc of c joins successive pairs of control points, so that the entire spline comprises n cubic arcs joined end to end. Figure 17.1 shows a Hermite spline through four control points on a plane. There are corners at the middle two because the tangents of the cubics on either side don't agree.

Terminology: A *cubic arc* is a part of a cubic curve; e.g., an arc of the graph of $y = x^3$ is a cubic arc on the plane. Sometimes we'll loosen cubic to mean a polynomial of degree at most three, rather than exactly three.

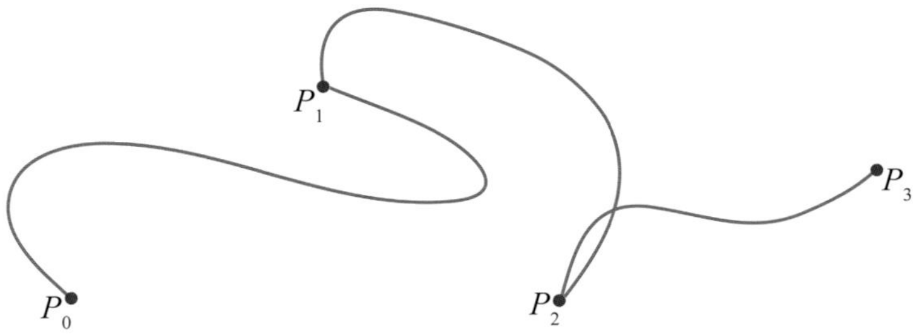

Figure 17.1: A (non-smooth) Hermite spline through four control points, composed of three cubic arcs.

Remark 17.1. Hermite splines are named after the nineteenth-century French mathematician Charles Hermite.

Remark 17.2. Curves of degree higher than three could be used to interpolate, or even lower, e.g., quadratic. However, three is a "Goldilocks" degree: a happy medium value, high enough to assure flexibility, and yet low enough to be computationally efficient.

Hermite interpolation for evident reasons is often called *cubic interpolation*.

We'll soon find a way to eliminate the corners in the interior and create a smooth Hermite spline through a given sequence of control points, but let's see first how to make a single cubic arc joining two arbitrary points P and Q.

Write the parametric equation of a general cubic curve c as

$$c(u) = A_3u^3 + A_2u^2 + A_1u + A_0 \qquad (0 \leq u \leq 1) \tag{17.1}$$

where each A_i, $0 \leq i \leq 3$, is a point – precisely, its vector of coordinates – in the ambient space. If you are wondering about polynomial coefficients which are vectors rather than scalars, then consider the following example.

Example 17.1. Suppose that we are interested in Hermite splines in the real world so that our ambient space is $\mathbb{R}^3$. Then the equation of a cubic curve is

$$c(u) = A_3u^3 + A_2u^2 + A_1u + A_0 \qquad (0 \le u \le 1)$$

where each A_i, $0 \le i \le 3$, is a point in 3-space.

To illustrate, say,

$$A_3 = [-1\ 2\ 0]^T, \quad A_2 = [3\ 0\ -2]^T, \quad A_1 = [4\ 3\ 4]^T, \quad \text{and} \quad A_0 = [0\ 8\ 7]^T$$

Then,

$$\begin{aligned} c(u) &= [-1\ 2\ 0]^T u^3 + [3\ 0\ -2]^T u^2 + [4\ 3\ 4]^T u + [0\ 8\ 7]^T \\ &= [-u^3 + 3u^2 + 4u \quad 2u^3 + 3u + 8 \quad -2u^2 + 4u + 7]^T \end{aligned}$$

over the interval $[0, 1]$. As one would expect, the cubic c in $\mathbb{R}^3$ is simply a scalar cubic in *each* of its three coordinates.

Example 17.2. Express in the form (17.1) the twisted cubic given parametrically by

$$x = t,\ y = t^2,\ z = t^3$$

Answer:

$$c(t) = [t\ t^2\ t^3]^T = [0\ 0\ 1]^T t^3 + [0\ 1\ 0]^T t^2 + [1\ 0\ 0]^T t$$

Returning to the general form (17.1) of the cubic, rewrite it as a matrix equation:

$$c(u) = [u^3\ u^2\ u\ 1] \begin{bmatrix} A_3 \\ A_2 \\ A_1 \\ A_0 \end{bmatrix} \qquad (0 \le u \le 1) \qquad (17.2)$$

Note: The RHS is a product of a 1×4 matrix of scalars with a 4×1 matrix of vectors, but this is not a problem if we appropriately multiply a vector by a scalar while following the usual rules of matrix multiplication.

Differentiating (17.2) one obtains the derivative of c as

$$c'(u) = [3u^2\ \ 2u\ \ 1\ \ 0] \begin{bmatrix} A_3 \\ A_2 \\ A_1 \\ A_0 \end{bmatrix} \qquad (0 \le u \le 1) \qquad (17.3)$$

Substitute 0 and 1 for u in Equations (17.2) and (17.3) to find that

$$c(0) = A_0, \quad c(1) = A_3 + A_2 + A_1 + A_0, \quad c'(0) = A_1, \quad c'(1) = 3A_3 + 2A_2 + A_1 \qquad (17.4)$$

It seems that if one could specify $c(0)$, $c(1)$, $c'(0)$ and $c'(1)$, then one would have four equations in the four unknowns A_0, A_1, A_2 and A_3, which should solve to find these coefficients and specify c (*alert*: that's four equations in four *vector* unknowns, so, e.g., if we are in 3-space, we'll have actually twelve equations in twelve *scalar* unknowns!). Since c goes from P to Q we know at least that $c(0) = P$ and $c(1) = Q$; as for the tangent vectors $c'(0)$ and $c'(1)$, we have freedom to specify them as we please. Let's choose them to be two vectors denoted P' and Q', respectively. See Figure 17.2.

Figure 17.2: Four boundary constraints on a cubic curve c.

Accordingly, write (17.4) as

$$P = A_0, \qquad Q = A_3 + A_2 + A_1 + A_0, \qquad P' = A_1, \qquad Q' = 3A_3 + 2A_2 + A_1 \tag{17.5}$$

which in matrix form is the equation

$$\begin{bmatrix} P \\ Q \\ P' \\ Q' \end{bmatrix} = \begin{bmatrix} 0 & 0 & 0 & 1 \\ 1 & 1 & 1 & 1 \\ 0 & 0 & 1 & 0 \\ 3 & 2 & 1 & 0 \end{bmatrix} \begin{bmatrix} A_3 \\ A_2 \\ A_1 \\ A_0 \end{bmatrix} \tag{17.6}$$

Solve this equation by inverting the coefficient matrix as follows

$$\begin{aligned} \begin{bmatrix} A_3 \\ A_2 \\ A_1 \\ A_0 \end{bmatrix} &= \begin{bmatrix} 0 & 0 & 0 & 1 \\ 1 & 1 & 1 & 1 \\ 0 & 0 & 1 & 0 \\ 3 & 2 & 1 & 0 \end{bmatrix}^{-1} \begin{bmatrix} P \\ Q \\ P' \\ Q' \end{bmatrix} \\ &= \begin{bmatrix} 2 & -2 & 1 & 1 \\ -3 & 3 & -2 & -1 \\ 0 & 0 & 1 & 0 \\ 1 & 0 & 0 & 0 \end{bmatrix} \begin{bmatrix} P \\ Q \\ P' \\ Q' \end{bmatrix} \end{aligned} \tag{17.7}$$

to see that the four coefficients A_0, A_1, A_2 and A_3 can indeed be derived from the four boundary constraints P, Q, P' and Q'. The 4×4 matrix in the second line of the equation is called the *Hermite matrix* and denoted M_H, so (17.7) is written concisely as

$$[A_3\ A_2\ A_1\ A_0]^T = M_H\,[P\ Q\ P'\ Q']^T \tag{17.8}$$

Finally, let's use (17.2) to write c's equation in terms of its boundary constraints:

$$\begin{aligned} c(u) &= [u^3\ u^2\ u\ 1]\,[A_3\ A_2\ A_1\ A_0]^T \\ &= [u^3\ u^2\ u\ 1]\,M_H\,[P\ Q\ P'\ Q']^T \\ &= (2u^3 - 3u^2 + 1)\,P + (-2u^3 + 3u^2)\,Q + \\ &\qquad (u^3 - 2u^2 + u)\,P' + (u^3 - u^2)\,Q' \end{aligned} \tag{17.9}$$

in $0 \le u \le 1$, after performing the matrix multiplications in the second line.

Therefore,

$$c(u) = H_0(u)\,P + H_1(u)\,Q + H_2(u)\,P' + H_3(u)\,Q' \qquad (0 \le u \le 1) \tag{17.10}$$

where the polynomials

$$\begin{aligned} H_0(u) &= 2u^3 - 3u^2 + 1, & H_1(u) &= -2u^3 + 3u^2, \\ H_2(u) &= u^3 - 2u^2 + u, & H_3(u) &= u^3 - u^2 \end{aligned}$$

are called *Hermite blending polynomials*, which, of course, are blending functions, but obviously different from those used earlier in Bézier and B-spline theory (moreover, not just control points, but tangent vectors, too, enter the mix!). Their graphs are sketched in Figure 17.3. Certain symmetries are evident. Observe, as well, that $H_3(u)$ is non-positive in $0 \le u \le 1$, reaching a minimum value of nearly -0.15.

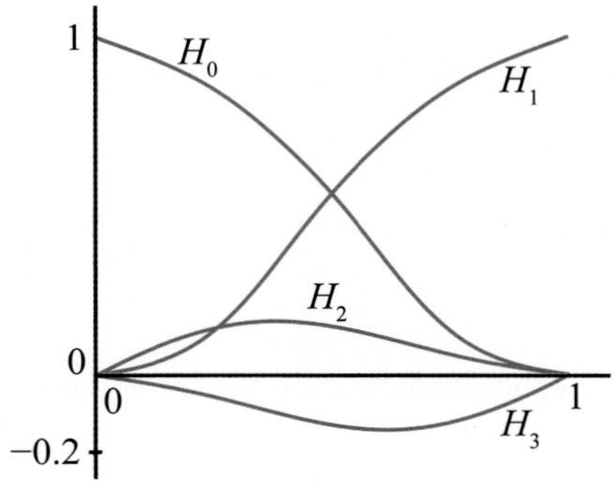

Figure 17.3: Hermite blending polynomials (not exact plots).

The curve $c(u)$ itself is called a *Hermite cubic*. Equation (17.10) is called the *geometric* form of the cubic because its expression is in terms of c's boundary constraints, while (17.1) is its *algebraic* form.

Remark 17.3. Readers familiar with programs such as Adobe Photoshop or Illustrator will recognize that the pen tool can be used to draw Hermite cubics by specifying endpoints and the tangents at the endpoints.

Exercise 17.1. Use calculus to determine the maximum value of $H_2(u)$ and the minimum value of $H_3(u)$ in the interval $[0, 1]$.

Exercise 17.2. Determine the symmetries among the Hermite blending polynomials. For example, that $H_0(u)$ and $H_1(u)$ are mirror images across the vertical line $u = \frac{1}{2}$ down the middle of the parameter interval $[0, 1]$ can be seen by substituting $(1 - u)$ for u in the equation of one to obtain that of the other.

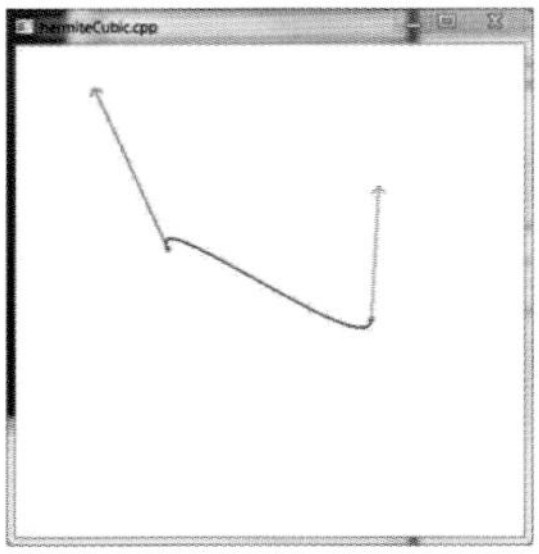

Figure 17.4: Screenshot of `hermiteCubic.cpp`.

Experiment 17.1. Run `hermiteCubic.cpp`, which implements Equation (17.10) to draw a Hermite cubic on a plane. Press space to select either a control point or tangent vector and the arrow keys to change it. Figure 17.4 is a screenshot. The actual cubic is simple to draw, but as you can see in the program we invested many lines of code to get the arrow heads right! **End**

Exercise 17.3. What sort of curve is c if the two boundary constraints P' and Q' are both zero (i.e., if the two end velocities vanish)? Determine this from the geometric form of the Hermite cubic and verify in the program.

It's interesting to contrast (17.10) with the equation of the cubic Bézier curve (Equation (15.8)):

$$c(u) = B_{0,3}(u)P_0 + B_{1,3}(u)P_1 + B_{2,3}(u)P_2 + B_{3,3}(u)P_3 \qquad (0 \le u \le 1)$$

In the case of the Bézier curve, the control points are blended with weights equal to the Bernstein polynomials of degree 3; in the case of the Hermite cubic, the two end control points and their respective tangents are blended with weights equal to the Hermite blending polynomials, which are of degree 3 as well.

Remark 17.4. Since a Hermite cubic interpolates not only its two specified control points, but also the specified tangents there, it's said to make a first-order interpolation (versus a zeroth-order one which would interpolate merely control points).

Exercise 17.4. Prove the affine invariance of the cubic curve c given by Equation (17.10).

Note: Keeping in mind that an affine transformation is a linear transformation followed by a translation, we'll want its linear transformation part applied to all four boundary constraints P, Q, P' and Q', while the translation should apply only to P and Q.

Let's return to the original problem of joining successive pairs of the $n+1$ control points $P_0, P_1, \dots, P_n$ by means of cubic arcs so that the resulting Hermite spline is smooth. A strategy that comes to mind from the discussion above is to ask the designer to specify, in addition to the $n+1$ control points, the tangent vectors $P'_0, P'_1, \dots, P'_n$ at each, as indicated in Figure 17.5.

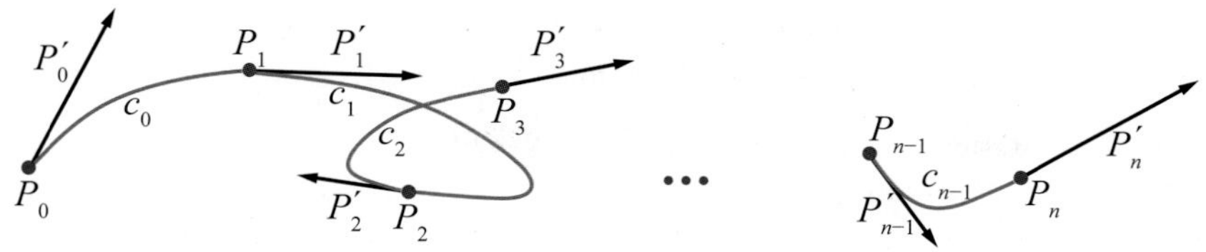

Figure 17.5: Specifying a Hermite spline by specifying the tangent vector at each control point.

Then, using (17.10) to manufacture each of the n successive Hermite cubic arcs c_i, $0 \leq i \leq n$, subject to the respective boundary constraints P_i, P_{i+1}, P'_i and P'_{i+1} yields a C^1-continuous Hermite spline, as the derivatives on either side of each internal control point agree by design.

However, asking the designer for $n + 1$ tangent values, in addition to the control points themselves, may be a bit much. It would be nice to have an *automatic* way to deduce these tangent values from other constraints, *transparently* to the user. In fact, there is and we'll discuss next two popular types of Hermite splines arising from particular sets of constraints. These are the natural cubic and cardinal splines.

17.2 Natural Cubic Splines

A *natural cubic spline* is a Hermite spline which is C^2-continuous (i.e., its second derivative is continuous) *and* whose second derivative vanishes at its two end control points. It turns out, as we'll see, that these constraints are enough to uniquely determine the spline.

Assume that the $n + 1$ control points through which a natural cubic spline passes are $P_0, P_1, \ldots, P_n$. Because of C^1-continuity (remember that C^2-continuity implies C^1-continuity) one assumes that the tangents at the control points are uniquely defined as well – say they are $P'_0, P'_1, \ldots, P'_n$, respectively. The values P'_i, $0 \leq i \leq n$, are *not* user-specified; rather, we'll compute them.

Rewrite (17.9) as the equation of the cubic arc c_i from P_i to P_{i+1}:

$$\begin{aligned} c_i(u) &= (2u^3 - 3u^2 + 1)\, P_i + (-2u^3 + 3u^2)\, P_{i+1} + \\ &\quad (u^3 - 2u^2 + u)\, P'_i + (u^3 - u^2)\, P'_{i+1} \quad (0 \leq u \leq 1) \end{aligned} \tag{17.11}$$

Differentiating twice one finds the second derivative

$$c''_i(u) = (12u-6)\, P_i + (-12u+6)\, P_{i+1} + (6u-4)\, P'_i + (6u-2)\, P'_{i+1} \quad (0 \leq u \leq 1) \tag{17.12}$$

Observe now that the second-order constraints on a natural cubic spline through $P_0, P_1, \ldots, P_n$ can be written as the $n + 1$ equations:

$$c''_0(0) = 0, \qquad c''_{i-1}(1) = c''_i(0) \text{ for } 1 \leq i \leq n-1, \qquad c''_{n-1}(1) = 0 \tag{17.13}$$

(the middle equations say that the values of the second derivative on either side of each internal control point are equal, assuring C^2-continuity). Expand the constraint equations using (17.12):

$$\begin{aligned}
-6P_0 + 6P_1 - 4P_0' - 2P_1' &= 0 \\
6P_{i-1} - 6P_i + 2P_{i-1}' + 4P_i' &= -6P_i + 6P_{i+1} - 4P_i' - 2P_{i+1}', 1 \le i \le n-1 \\
6P_{n-1} - 6P_n + 2P_{n-1}' + 4P_n' &= 0
\end{aligned}$$

Simplifying and rearranging, we have the system

$$\begin{aligned}
2P_0' + P_1' &= -3P_0 + 3P_1 \\
P_{i-1}' + 4P_i' + P_{i+1}' &= -3P_{i-1} + 3P_{i+1}, \qquad 1 \le i \le n-1 \\
P_{n-1}' + 2P_n' &= -3P_{n-1} + 3P_n \qquad (17.14)
\end{aligned}$$

of $n+1$ equations in $n+1$ unknowns, which can be solved for the P_i' in terms of the P_i. In fact, writing out the system (17.14) in matrix form one obtains

$$\begin{bmatrix}
2 & 1 & 0 & 0 & 0 & 0 & \dots & 0 & 0 & 0 & 0 \\
1 & 4 & 1 & 0 & 0 & 0 & \dots & 0 & 0 & 0 & 0 \\
0 & 1 & 4 & 1 & 0 & 0 & \dots & 0 & 0 & 0 & 0 \\
0 & 0 & 1 & 4 & 1 & 0 & \dots & 0 & 0 & 0 & 0 \\
\dots & & & \dots & & & \dots & & & & \dots \\
0 & 0 & 0 & 0 & 0 & 0 & \dots & 0 & 1 & 4 & 1 \\
0 & 0 & 0 & 0 & 0 & 0 & \dots & 0 & 0 & 1 & 2
\end{bmatrix}
\begin{bmatrix} P_0' \\ P_1' \\ P_2' \\ P_3' \\ \dots \\ P_{n-1}' \\ P_n' \end{bmatrix}
=
\begin{bmatrix} -3P_0 + 3P_1 \\ -3P_0 + 3P_2 \\ -3P_1 + 3P_3 \\ -3P_2 + 3P_4 \\ \dots \\ -3P_{n-2} + 3P_n \\ -3P_{n-1} + 3P_n \end{bmatrix} \qquad (17.15)$$

where the coefficient matrix is *tridiagonal* because it has non-zero entries only along the principal diagonal and its two neighbors. Tridiagonal matrices are particularly efficient to invert and, accordingly, equation systems with a tridiagonal coefficient matrix are efficiently solvable [114].

Finally, using the solved tangent values $P_0', P_1', \dots, P_n'$ and the geometric form (17.10) of the Hermite cubic, one determines the n Hermite cubic arcs between successive pairs from $P_0, P_1, \dots, P_n$. These arcs then join end to end to give the natural cubic spline through these $n+1$ control points.

Exercise 17.5. (Programming) Solve (17.15) by hand for only three control points P_0, P_1 and P_2. Write a program to draw a natural cubic spline through three control points, each of which can be moved on a plane.

Exercise 17.6. Investigate the local control (or lack thereof) of natural cubic splines. In particular, which of the cubic arcs of a natural cubic spline are affected by moving only one control point?
Hint: Playing with a natural cubic spline applet (there are many on the web) should suggest an answer.

17.3 Cardinal Splines

A *cardinal spline* is a C^1 Hermite spline whose tangent vector at each internal control point is determined by the location of its two adjacent control points in the following simple manner. Say the control points through which a cardinal spline passes are $P_0, P_1, \dots, P_n$. The tangent vector at P_i, $1 \le i \le n-1$, is specified to be *parallel* to the vector from P_{i-1} to P_{i+1} by the equation

$$P_i' = \frac{1}{2}(1 - t_i)(P_{i+1} - P_{i-1}) \quad (17.16)$$

See Figure 17.6.

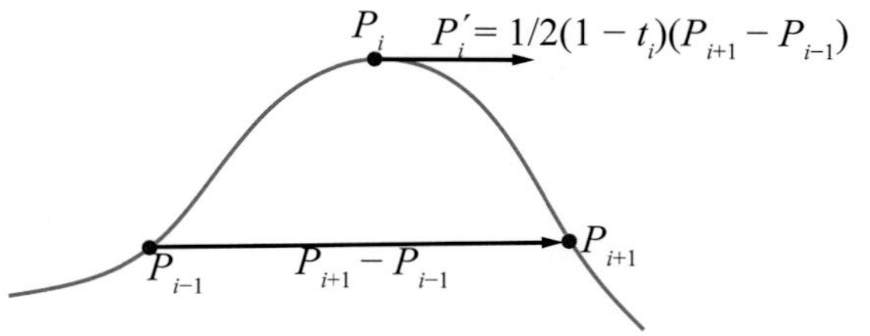

Figure 17.6: The tangent vector at an internal control point of a cardinal spline is parallel to the vector joining the adjacent control points – the tension parameter t_i is user-specified.

The constant of proportionality $\frac{1}{2}(1 - t_i)$ in (17.16) involves a designer-specified parameter t_i, called the *tension parameter*. The tension parameter is usually set between -1 and 1 at each internal control point. If it is set to 0 at *every* internal control point, one gets a popularly used special kind of cardinal spline called a *Catmull-Rom* spline. Specifically, the tangent vector at the internal control point P_i of a Catmull-Rom spline is

$$P_i' = \frac{1}{2}(P_{i+1} - P_{i-1}) \quad (17.17)$$

Now, from (17.16), $1 \le i \le n-1$, one has only $n-1$ equations in the $n+1$ unknowns P_i', $0 \le i \le n$. Therefore, two more are required to uniquely solve for these unknowns and determine the cardinal spline through P_i, $0 \le i \le n$. Typically, as in the case of a natural cubic spline, these are obtained from requiring the second derivatives to vanish at the two end control points.

Exercise 17.7. Write a matrix equation analogous to (17.15) relating P_i' to P_i for a cardinal spline, assuming the additional constraints that the second derivatives vanish at the terminal control points. Is the coefficient matrix tridiagonal?

Exercise 17.8. What can you say of local control in cardinal splines? In other words, which of the cubic arcs of a cardinal spline are affected by moving a specific control point?

Exercise 17.9. Natural cubic splines are C^2 by definition. How about cardinal splines – are they C^2?
Hint: The answer is no in general and we ask the reader to try and come up with a counter-example. A Catmull-Rom spline through three control points which loses C^2-continuity in the middle is probably easiest.

17.4 Hermite Surface Patches

We'll give a brief introduction to the 2D version of Hermite curves, namely, Hermite surfaces. Analogously to (17.1), one can write the parametric equation of a *Hermite surface patch* (or *bicubic surface patch*) in algebraic form as

$$\begin{aligned} s(u,v) &= \sum_{i=0}^{3}\sum_{j=0}^{3} A_{i,j}u^i v^j \\ &= A_{3,3}\,u^3v^3 + A_{3,2}\,u^3v^2 + A_{3,1}\,u^3v + A_{3,0}\,u^3 \\ &\quad + A_{2,3}\,u^2v^3 + A_{2,2}\,u^2v^2 + A_{2,1}\,u^2v + A_{2,0}\,u^2 \\ &\quad + A_{1,3}\,uv^3 + A_{1,2}\,uv^2 + A_{1,1}\,uv + A_{1,0}\,u \\ &\quad + A_{0,3}\,v^3 + A_{0,2}\,v^2 + A_{0,1}\,v + A_{0,0} \end{aligned} \tag{17.18}$$

for $0 \leq u, v \leq 1$. The expression after the second equality consists of 16 monomial summands, where A_{ij}, $0 \leq i \leq 3$, $0 \leq j \leq 3$, are points in the ambient space.

Going back to curves for a moment, observe that the geometric form (17.10), viz.

$$c(u) = H_0(u)\,P + H_1(u)\,Q + H_2(u)\,P' + H_3(u)\,Q'$$

of the equation of a Hermite cubic is more useful than the algebraic (17.1), viz.

$$c(u) = A_3u^3 + A_2u^2 + A_1u + A_0$$

because it gives an equation in terms of *perceptible* boundary constraints, in particular, the endpoints P and Q and the tangent vectors P' and Q' there. Moreover, we were able to derive the algebraic form from the geometric because these four boundary constraints were sufficient to uniquely recover the four coefficients A_i, $0 \leq i \leq 3$, of the algebraic form.

So what would be a suitable set of boundary constraints for a geometric form of the equation of a Hermite patch? Clearly, one would want sixteen constraints leading to a unique determination of the sixteen coefficients A_{ij}, $0 \leq i \leq 3$, $0 \leq j \leq 3$, on the RHS of (17.18).

Twelve choices are fairly clear. See Figure 17.7. They are the four corners $s(0,0), s(1,0), s(1,1), s(0,1)$ of the patch s (or, more precisely, the position vectors of these corners) and values of the partial derivatives with respect

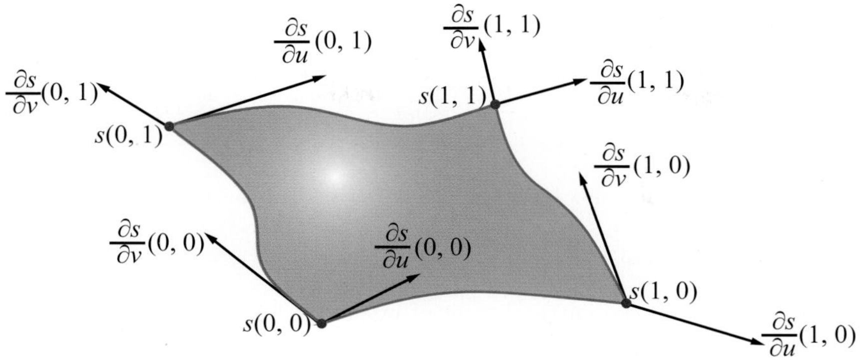

Figure 17.7: Twelve boundary constraints on a bicubic patch.

to u and v at each corner. The two partial derivatives at each corner are nothing but the tangent vectors to the two boundary curves meeting there. The four remaining boundary constraints are usually taken to be values of the second-order mixed partial derivatives at the corners, namely,

$$\frac{\partial^2 s}{\partial u \partial v}(0,0), \quad \frac{\partial^2 s}{\partial u \partial v}(1,0), \quad \frac{\partial^2 s}{\partial u \partial v}(0,1), \quad \frac{\partial^2 s}{\partial u \partial v}(1,1)$$

These four are called *twist vectors* and have geometric significance too – though not as straightforwardly as the first twelve – which we'll not go into here.

We'll conclude our discussion by saying that it turns out that, indeed, the four corner position vectors, the eight tangent vectors at the corners and the four twist vectors together provide sixteen boundary constraints which are sufficient to uniquely specify a Hermite patch. We'll not go further into the derivation ourselves, but refer the interested reader to the chapter on Hermite surfaces in the book by Mortenson [94].

Lagrange Interpolation

At the conclusion of this chapter, we'll briefly describe a method of polynomial (in fact, *entirely* polynomial, not piecewise like Hermite) interpolation, called *Lagrange interpolation*, actually of more theoretical interest than practical value in design.

The *Lagrange polynomial* $f_{i,n}$, where n is a positive integer and i is an integer between 0 and n, is defined by the equation

$$f_{i,n}(u) = \prod_{0 \le j \le n, \; j \ne i} \frac{u-j}{i-j}$$

For example,

$$\begin{aligned} f_{2,4}(u) &= \frac{(u-0)(u-1)(u-3)(u-4)}{(2-0)(2-1)(2-3)(2-4)} \\ &= \frac{1}{4}u(u-1)(u-3)(u-4) \end{aligned}$$

Lagrange polynomials have the easily verified property that

$$f_{i,n}(u) = \begin{cases} 1, & u = i \\ 0, & u \in \{0, 1, \ldots, n\},\ u \neq i \end{cases}$$

In other words, on the particular set of integers $\{0, 1, \ldots, n\}$, the Lagrange polynomial $f_{i,n}$ is 1 at exactly one point, namely i, and 0, elsewhere.

Exercise 17.10. Write the formula for $f_{0,4}(u)$ and check it for the above-mentioned property.

If, now, one uses the Lagrange polynomials as blending functions for $n+1$ control points P_i, $0 \leq i \leq n$, obtaining the curve

$$c(u) = f_{0,n}(u)P_0 + f_{1,n}(u)P_1 + \ldots + f_{n,n}(u)P_n \qquad (0 \leq u \leq n)$$

then c, called a *Lagrange curve*, is a polynomial curve of degree n. It's seen easily from its definition that c interpolates all its control points; in particular, c is equal to P_i at the point i of the parameter domain $[0, n]$, for $0 \leq i \leq n$.

Exercise 17.11. Write the formula for the Lagrange curve interpolating the four control points

$$[0\ -1\ 3]^T \qquad [1\ 2\ -3]^T \qquad [5\ -1\ 4]^T \qquad [2\ 0\ 8]^T$$

Lagrange interpolation is rarely used in practice because it suffers from the Bézier-like problem that the degree of the interpolating curve grows with its number of control points. It lacks local control as well.

17.5 Summary, Notes and More Reading

After a couple of chapters on Bézier and B-spline approximation of control points, we learned in this chapter practical methods to interpolate. These will come in handy in design applications that do require interpolation and most 3D modelers, in fact, offer at least a flavor or two of Hermite interpolation, such as natural cubic and Catmull-Rom splines. It's true, though, in the majority of real-life applications that the only known constraints on a curve or surface are at its boundary, e.g., by the way a surface patch meets its neighbors, so the designer typically prefers using internal control points as attractors *a la* Bézier or B-spline, rather than having them tightly latched to an interpolating curve or surface.

For more about Hermite interpolation the reader should consult Farin [45] and Mortenson [94].

Part IX

Well Projected

CHAPTER 18

Applications of Projective Spaces

Projective spaces and transformations play an important role in computer graphics and our goal in this chapter is to study three powerful applications. The first is in the "shoot" part of shoot-and-print in the OpenGL pipeline, which comes down to devising a so-called projection transformation. The next is in shadow mapping, a technique to authentically cast shadows of a local light source on curved surfaces, which is based on an understanding of the projection transformation. The final application is developing rational versions of Bézier and B-spline theory.

It's best to come to this chapter with some familiarity with projective spaces. If you have this already, maybe from a college math course or from books on projective geometry such as Henle [71], Jennings [76] and Pedoe [108], you are set; if not, Appendix A, which is an introduction to projective spaces and transformations, has all you need. Appendix A has been written particularly for a CG audience, with connections constantly drawn to familiar CG settings. In fact, you are strongly urged to flip through this appendix even if already acquainted with projective geometry.

However, we do realize there might be a significant readership as yet unfamiliar with projective spaces who, nevertheless, would like a view of their applications without necessarily going through all the math first. This chapter has been arranged to be accessible to such persons as far as possible. Before each part that invokes projective theory, the reader is alerted with a note containing the minimum information needed to make sense of it. Of course, understanding will not be 100%, but, hopefully, good enough for a first light on the applications. Familiarity at least with Section 5.2 on affine transformations, though, particularly the use of homogeneous coordinates, is assumed on everyone's part.

The first application of projective transformations in Section 18.1 is to accomplish the so-called projection transformation step in the synthetic-camera graphics pipeline – mapping the viewing volume to a box. This leads to a derivation of OpenGL's 4×4 projection matrices, as well as an understanding of how these matrices compose in the graphics pipeline with modelview matrices. An immediate practical bonus of this insight into matrix computations in the pipeline is the ubiquitous technique of shadow mapping used to draw projective shadows in real-time.

An understanding of the projection transformation is needed in Section 18.2 which explains the shadow mapping technique.

The final application is in Section 18.3 where we learn the rational versions of both Bézier and B-spline theory. This lengthy section begins with an extensive discussion of rational Bézier curves which, once assimilated, lends itself to fairly straightforward generalization, first to rational Bézier surfaces and then rational B-spline, or NURBS, primitives. Section 18.4 concludes the chapter.

18.1 OpenGL Projection Transformations

Way back in Section 2.2 we described OpenGL's rendering as conceptually a two-step process, shoot-and-print. Shooting consists of projecting – parallely in the case of a viewing box and perspectively in that of a viewing frustum – the scene onto the viewing face. Printing consists of scaling the viewing face to fit the OpenGL window. This account, though simplified, is not far from the actual implementation in the OpenGL graphics pipeline.

The second step of scaling is evidently straightforward, but the first of projection is more difficult. Projection itself is performed in two stages.

In the first stage, OpenGL transforms the viewing volume – a box defined by `glOrtho()` or a frustum by `glFrustum()` and `gluPerspective()` – into a *canonical viewing box*. The canonical viewing box is an axis-aligned cubical box centered at the origin with side lengths two. Figure 18.1 shows the canonical viewing box, as well as a generic viewing box and a generic viewing frustum. The transformation from the given viewing volume to the canonical viewing box is called the *projection transformation* of OpenGL.

We'll deduce equations for the projection transformation soon, but the crux of what it does geometrically is to take lines of sight to lines of sight, "straightening" them out in the process in the case of a frustum. See Figure 18.2 for a sectional view along the xz-plane. For example, the lines of sight l_1 and l_2, both in the box and frustum, are mapped by the projection transformation to the corresponding lines of sight l_1' and l_2' in the canonical viewing box. Note the little quirk that orientation of the lines of sight seem reversed by the transformation. The points p, q and r on both lines of sight l_1 are mapped to p', q' and r', respectively, on the line of sight l_1'. Rectangle X in the box and rectangle Y in the frustum are transformed to rectangle X' and the trapezoid Y', respectively, bold edge going to bold edge. The distortion from Y to Y' is precisely the foreshortening one would

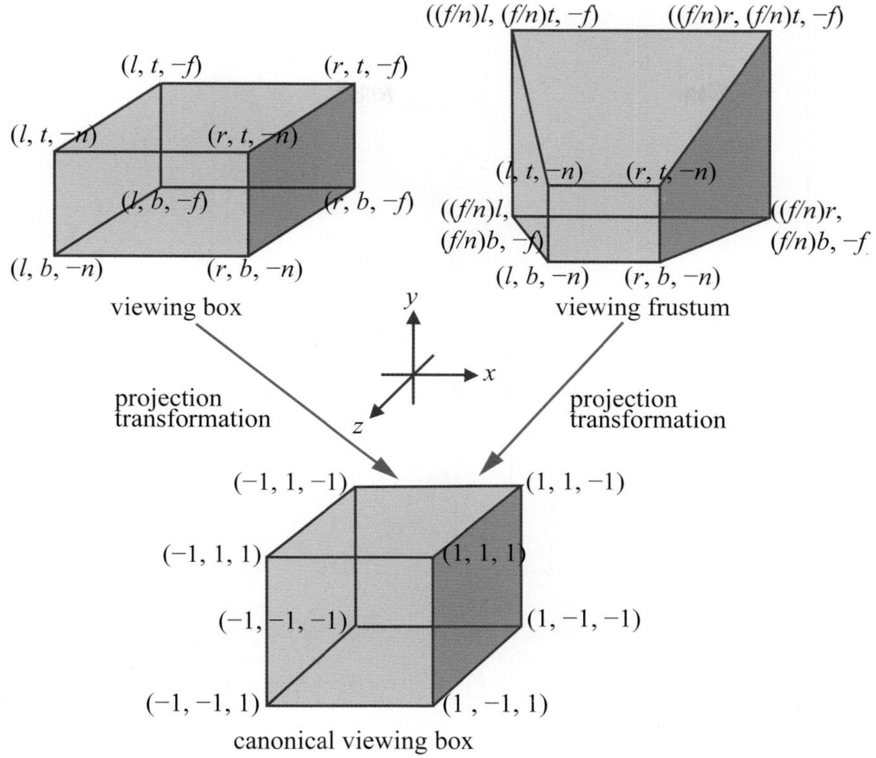

Figure 18.1: As part of the OpenGL rendering pipeline a `glOrtho(`l, r, b, t, n, f`)`-defined viewing box or `glFrustum(`l, r, b, t, n, f`)`-defined viewing frustum is transformed into the canonical viewing box by a projection transformation.

expect from a perspective view.

In the second stage of the two-stage projection process, OpenGL projects primitives in the canonical viewing box parallely onto its back face, the one lying on $z = -1$. It's because of this reversal of the direction of projection – a quirk of OpenGL as a projection to the front face would have worked just as well – that the orientation of the lines of sight is reversed.

Observe, now, that projection in the canonical viewing box is *exactly equivalent* to that in the original viewing volume, precisely because the projection transformation preserves lines of sight. In Figure 18.2, for example, the point p, in both box and frustum, projects to the point r on the respective viewing face, while p' (the image of p by the projection transformation) projects to r' (the image of r by the projection transformation) in the canonical viewing box.

Of the two stages of the projection process, the second one of back-face projection is certainly computationally simpler, as it's a matter simply of tossing the z-values *after* they've been used in depth testing if need be.

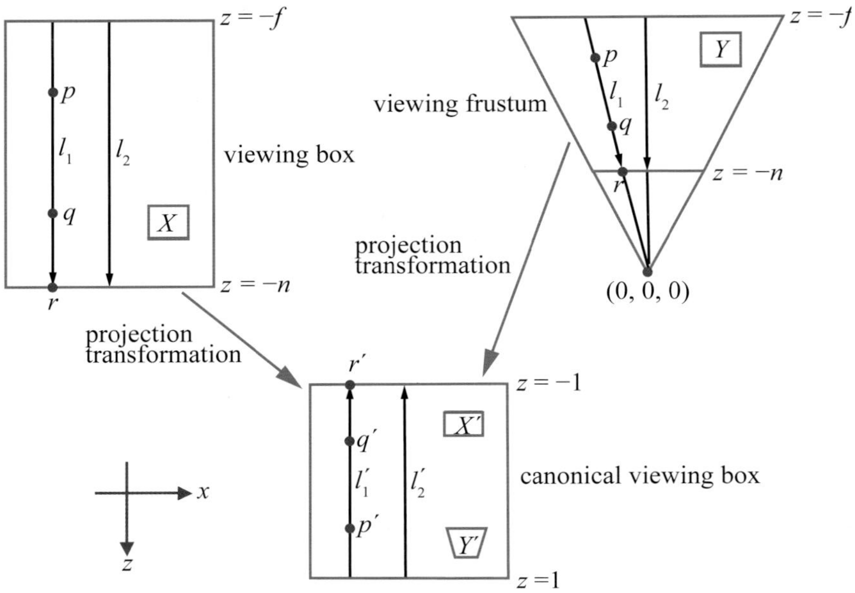

Figure 18.2: Sectional view along the xz-plane of the viewing volumes from Figure 18.1. Black arrows inside the viewing volumes are lines of sight: they are directed toward the $+z$ direction in the viewing box and frustum and toward the $-z$ direction in the canonical viewing box.

If depth testing is enabled, then z-values in the canonical box are used for this purpose, rather than the original ones from world space. It's valid to do so because if, say, point q obscures point p in the viewing volume prior to transformation, as in Figure 18.2, then transformed point q' obscures transformed point p' as well, again because lines of sight are preserved. Of course, given the direction of the lines of sight in the canonical box, lower z-values win the depth competition.

OpenGL accomplishes the projection transformation, from programmer-specified viewing volume to canonical viewing box, by means of a 4×4 *projection matrix* whose nature depends on whether it is a box or frustum to be transformed into the canonical box. Our next objective is to derive the projection matrix in both cases.

Remark 18.1. To be fastidious we should now rephrase our earlier description of the print part of shoot-and-print to say that it scales the back face of the canonical box, rather than the front face of the viewing volume, to fit the OpenGL window.

18.1.1 Viewing Box to Canonical Viewing Box

The strategy to transform a `glOrtho()`-defined viewing box into the canonical box is straightforward: translate the viewing box so that its center coincides with that of the canonical one, then scale its sides so that they match those of the canonical box.

The center of the viewing box defined by a call to `glOrtho(`l`, `r`, `b`, `t`, `n`, `f`)` is at $[(r+l)/2 \quad (t+b)/2 \quad -(f+n)/2]^T$, while the center of the canonical box is at the origin $[0\ 0\ 0]^T$. Therefore, the displacement vector translating the first to the second is $[-(r+l)/2 \quad -(t+b)/2 \quad (f+n)/2]^T$. The corresponding 4×4 translation matrix is

$$T(\ -(r+l)/2,\ -(t+b)/2,\ (f+n)/2\ )$$

(see Section 5.4 for a listing of affine transformation matrices in homogeneous form).

Since the viewing box is of size $(r-l) \times (t-b) \times (f-n)$, while the canonical box is of size $2 \times 2 \times 2$, the scaling transformation matching the sides of the former with those of the latter has the matrix

$$S(\ 2/(r-l),\ 2/(t-b),\ 2/(f-n)\ )$$

Finally, to account for the reversal in direction of the lines of sight, the needed transformation is $(x, y, z) \mapsto (x, y, -z)$, whose matrix is

$$S(1, 1, -1)$$

Composing the preceding three transformations, one obtains the projection transformation, denoted $P($`glOrtho(`l`, `r`, `b`, `t`, `n`, `f`)`$)$, mapping the viewing box of the call `glOrtho(`l`, `r`, `b`, `t`, `n`, `f`)` to the canonical viewing box. The projection matrix corresponding to $P($`glOrtho(`l`, `r`, `b`, `t`, `n`, `f`)`$)$, using eponymous notation, is

$$\begin{aligned}
& P(\texttt{glOrtho}(l, r, b, t, n, f)) \\
= \ & S(1,1,-1)\ \ S(\ 2/(r-l),\ 2/(t-b),\ 2/(f-n)\ ) \\
& T(\ -(l+r)/2,\ -(b+t)/2,\ (n+f)/2\ ) \\
= \ & \begin{bmatrix} 1 & 0 & 0 & 0 \\ 0 & 1 & 0 & 0 \\ 0 & 0 & -1 & 0 \\ 0 & 0 & 0 & 1 \end{bmatrix} \begin{bmatrix} \frac{2}{r-l} & 0 & 0 & 0 \\ 0 & \frac{2}{t-b} & 0 & 0 \\ 0 & 0 & \frac{2}{f-n} & 0 \\ 0 & 0 & 0 & 1 \end{bmatrix} \begin{bmatrix} 1 & 0 & 0 & -\frac{r+l}{2} \\ 0 & 1 & 0 & -\frac{t+b}{2} \\ 0 & 0 & 1 & \frac{f+n}{2} \\ 0 & 0 & 0 & 1 \end{bmatrix} \\
= \ & \begin{bmatrix} \frac{2}{r-l} & 0 & 0 & -\frac{r+l}{r-l} \\ 0 & \frac{2}{t-b} & 0 & -\frac{t+b}{t-b} \\ 0 & 0 & -\frac{2}{f-n} & -\frac{f+n}{f-n} \\ 0 & 0 & 0 & 1 \end{bmatrix} \qquad (18.1)
\end{aligned}$$

As it is a composition of a translation and scalings, $P($`glOrtho(`l`, `r`, `b`, `t`, `n`, `f`)`$)$ is an affine transformation of $\mathbb{R}^3$.

Example 18.1. Determine how the point $[20\ 80\ 0]^T$ is transformed by the projection transformation corresponding to `glOrtho(0, 100, 0, 100, -1, 1)`.

Answer: Now

$$P(\texttt{glOrtho}(0,100,0,100,-1,1)) = \begin{bmatrix} \frac{2}{100} & 0 & 0 & -\frac{100}{100} \\ 0 & \frac{2}{100} & 0 & -\frac{100}{100} \\ 0 & 0 & -\frac{2}{2} & \frac{0}{2} \\ 0 & 0 & 0 & 1 \end{bmatrix}$$

$$= \begin{bmatrix} 0.02 & 0 & 0 & -1 \\ 0 & 0.02 & 0 & -1 \\ 0 & 0 & -1 & 0 \\ 0 & 0 & 0 & 1 \end{bmatrix} \tag{18.2}$$

Writing $[20\ 80\ 0]^T$ in homogeneous coordinates as $[20\ 80\ 0\ 1]^T$, one sees that it's transformed to the point

$$\begin{bmatrix} 0.02 & 0 & 0 & -1 \\ 0 & 0.02 & 0 & -1 \\ 0 & 0 & -1 & 0 \\ 0 & 0 & 0 & 1 \end{bmatrix} \begin{bmatrix} 20 \\ 80 \\ 0 \\ 1 \end{bmatrix} = \begin{bmatrix} -0.6 \\ 0.6 \\ 0 \\ 1 \end{bmatrix}$$

which is $[-0.6\ 0.6\ 0]^T$ in Cartesian coordinates.

Exercise 18.1. Determine how the following points are transformed by the projection transformation corresponding to `glOrtho(-20, 20, 0, 50, -1, 1)`: (a) $[0\ 40\ 0]^T$ (b) $[50\ 20\ 0.5]^T$ (see the following remark)

Remark 18.2. Just as points inside the viewing box are transformed to points inside the canonical box, those outside, e.g., (b) of the preceding exercise, are transformed to points outside the canonical box. The latter are clipped subsequently in the pipeline prior to rendering; in other words, operationally, clipping is done against the canonical box.

18.1.2 Viewing Frustum to Canonical Viewing Box

No affine transformation can map the viewing frustum defined by the call `glFrustum(`l`, `r`, `b`, `t`, `n`, `f`)` to the canonical viewing box, for the simple reason that this requires mapping intersecting lines (along edges of the frustum) to parallel ones (along edges of the box), while we know (see Proposition 5.1) that an affine transformation of $\mathbb{R}^3$ takes parallel straight lines to parallel straight lines and intersecting ones again to intersecting ones. We've run into a brick wall as far as affine transformations go. It's time to appeal to the projective.

Note to Readers Unfamiliar with Projective Geometry: Here's what you need to know for the rest of this particular section. Projective 3-space $\mathbb{P}^3$

consists of 4-tuples of the form $[x\ y\ z\ w]^T$, where these so-called homogeneous coordinates cannot all be zero. Two tuples represent the same point if one's a scalar multiple of the other, e.g., $[2\ 4\ 1\ -3]^T$ and $[4\ 8\ 2\ -6]^T$. Real 3-space $\mathbb{R}^3$ is embedded in $\mathbb{P}^3$ by mapping the point $[x\ y\ z]^T$ of $\mathbb{R}^3$ to $[x\ y\ z\ 1]^T$ of $\mathbb{P}^3$, e.g., $[2\ 4\ 1]^T$ maps to $[2\ 4\ 1\ 1]^T$.

Yet another thing to keep in mind is that, in addition to the points of $\mathbb{R}^3$ embedded into it as above, $\mathbb{P}^3$ has points corresponding to "directions" in $\mathbb{R}^3$. These points, which are called points at infinity, have a w-value of 0; e.g., $[0\ 0\ 1\ 0]^T$ is the point at infinity corresponding to the direction along the z-axis (both up and down directions along any line are regarded equal). Points with non-zero w-values, embedded from $\mathbb{R}^3$, are called regular points.

Finally, a projective transformation of $\mathbb{P}^3$ is defined by a non-singular 4×4 matrix and acts on tuples of $\mathbb{P}^3$ by multiplication from the left (similarly to how linear transformations of $\mathbb{R}^3$ act on 3-tuples).

You should jump now to the paragraph below containing Equation (18.3).

(Resuming from just before the note ....) However, our experience with projective transformations – Example A.17 which illustrates a projective transformation of $\mathbb{P}^3$ mapping a trapezoid to a rectangle is particularly motivating – suggests applying one.

Projectively transforming $\mathbb{R}^3$ is analogous to projectively transforming $\mathbb{R}^2$. For the latter, we identified $\mathbb{R}^2$ with a "film" in $\mathbb{R}^3$, almost always the plane $z = 1$, to capture the transformation of 2D objects lifted to $\mathbb{P}^2$ (note that the allusion to films is developed in Appendix A). Likewise, to projectively transform $\mathbb{R}^3$, we'll identify it with the hyperplane $w = 1$ in four-dimensional $xyzw$-space $\mathbb{R}^4$ in order to capture the transformation of 3D objects lifted to $\mathbb{P}^3$.

In particular, for the current application, we seek a projective transformation h^M of $\mathbb{P}^3$ which is captured on $\mathbb{R}^3$ as taking the viewing frustum specified by the call `glFrustum(`l`, `r`, `b`, `t`, `n`, `f`)` to the canonical box – as indicated by the thick blue arrow on the right of both Figures 18.1 and 18.2. Suppose its defining matrix is

$$M = \begin{bmatrix} a_{11} & a_{12} & a_{13} & a_{14} \\ a_{21} & a_{22} & a_{23} & a_{24} \\ a_{31} & a_{32} & a_{33} & a_{34} \\ a_{41} & a_{42} & a_{43} & a_{44} \end{bmatrix} \qquad (18.3)$$

so that it maps the point $[x\ y\ z\ w]^T$ of $\mathbb{P}^3$ to $M[x\ y\ z\ w]^T$.

The four lines along the four sides of the frustum that meet at its apex, which is the regular point $[0\ 0\ 0\ 1]^T$, are mapped, respectively, to four lines along edges of the canonical box all parallel to the z-axis, meeting, therefore, at the point at infinity $[0\ 0\ 1\ 0]^T$. Accordingly, we ask that

$$h^M([0\ 0\ 0\ 1]^T) = [0\ 0\ 1\ 0]^T$$

giving the matrix equation

$$\begin{bmatrix} a_{11} & a_{12} & a_{13} & a_{14} \\ a_{21} & a_{22} & a_{23} & a_{24} \\ a_{31} & a_{32} & a_{33} & a_{34} \\ a_{41} & a_{42} & a_{43} & a_{44} \end{bmatrix} \begin{bmatrix} 0 \\ 0 \\ 0 \\ 1 \end{bmatrix} = \begin{bmatrix} 0 \\ 0 \\ d \\ 0 \end{bmatrix}$$

where d can be any non-zero scalar because homogeneous coordinates $[0\ 0\ 1\ 0]^T$ and $[0\ 0\ d\ 0]^T$ represent the same point, implying that

$$a_{14} = 0, \quad a_{24} = 0, \quad a_{34} = d, \quad a_{44} = 0$$

It turns out that choosing $d = -\frac{2fn}{f-n}$ simplifies manipulations down the road so we'll write

$$M = \begin{bmatrix} a_{11} & a_{12} & a_{13} & 0 \\ a_{21} & a_{22} & a_{23} & 0 \\ a_{31} & a_{32} & a_{33} & -\frac{2fn}{f-n} \\ a_{41} & a_{42} & a_{43} & 0 \end{bmatrix}$$

The mappings

$$\begin{aligned} h^M([l\ b\ -n\ 1]^T) &= [-1\ -1\ -1\ 1]^T \\ h^M([r\ b\ -n\ 1]^T) &= [1\ -1\ -1\ 1]^T \\ h^M([l\ t\ -n\ 1]^T) &= [-1\ 1\ -1\ 1]^T \\ h^M([r\ t\ -n\ 1]^T) &= [1\ 1\ -1\ 1]^T \end{aligned}$$

from the mapping of the four vertices at the front of the frustum to the corresponding ones at the back of the canonical box give the four matrix equations

$$\begin{bmatrix} a_{11} & a_{12} & a_{13} & 0 \\ a_{21} & a_{22} & a_{23} & 0 \\ a_{31} & a_{32} & a_{33} & -\frac{2fn}{f-n} \\ a_{41} & a_{42} & a_{43} & 0 \end{bmatrix} \begin{bmatrix} l \\ b \\ -n \\ 1 \end{bmatrix} = \begin{bmatrix} -c_1 \\ -c_1 \\ -c_1 \\ c_1 \end{bmatrix}$$

$$\begin{bmatrix} a_{11} & a_{12} & a_{13} & 0 \\ a_{21} & a_{22} & a_{23} & 0 \\ a_{31} & a_{32} & a_{33} & -\frac{2fn}{f-n} \\ a_{41} & a_{42} & a_{43} & 0 \end{bmatrix} \begin{bmatrix} r \\ b \\ -n \\ 1 \end{bmatrix} = \begin{bmatrix} c_2 \\ -c_2 \\ -c_2 \\ c_2 \end{bmatrix}$$

$$\begin{bmatrix} a_{11} & a_{12} & a_{13} & 0 \\ a_{21} & a_{22} & a_{23} & 0 \\ a_{31} & a_{32} & a_{33} & -\frac{2fn}{f-n} \\ a_{41} & a_{42} & a_{43} & 0 \end{bmatrix} \begin{bmatrix} l \\ t \\ -n \\ 1 \end{bmatrix} = \begin{bmatrix} -c_3 \\ c_3 \\ -c_3 \\ c_3 \end{bmatrix}$$

$$\begin{bmatrix} a_{11} & a_{12} & a_{13} & 0 \\ a_{21} & a_{22} & a_{23} & 0 \\ a_{31} & a_{32} & a_{33} & -\frac{2fn}{f-n} \\ a_{41} & a_{42} & a_{43} & 0 \end{bmatrix} \begin{bmatrix} r \\ t \\ -n \\ 1 \end{bmatrix} = \begin{bmatrix} c_4 \\ c_4 \\ -c_4 \\ c_4 \end{bmatrix}$$

(c_i, $1 \le i \le 4$, are non-zero scalars) leading to 16 equations simultaneously in 16 unknowns:

$$\begin{aligned} l\,a_{11} + b\,a_{12} - n\,a_{13} &= -c_1 \\ l\,a_{21} + b\,a_{22} - n\,a_{23} &= -c_1 \\ l\,a_{31} + b\,a_{32} - n\,a_{33} - \frac{2fn}{f-n} &= -c_1 \\ l\,a_{41} + b\,a_{42} - n\,a_{23} &= c_1 \\ \ldots &= \ldots \\ r\,a_{41} + t\,a_{42} - n\,a_{43} &= c_4 \end{aligned}$$

These can be solved – not difficult, but tedious – to find

$$a_{11} = \frac{2n}{r-l}, \quad a_{12} = 0, \quad a_{13} = \frac{r+l}{r-l}, \quad a_{21} = 0, \quad a_{22} = \frac{2n}{t-b},$$
$$a_{23} = \frac{t+b}{t-b}, \quad a_{31} = 0, \quad a_{32} = 0, \quad a_{33} = -\frac{f+n}{f-n}, \quad a_{41} = 0,$$
$$a_{42} = 0, \quad a_{43} = -1, \quad c_1 = n, \quad c_2 = n, \quad c_3 = n, \quad c_4 = n$$

It follows that the projection transformation mapping the viewing frustum of the call `glFrustum(`l`, `r`, `b`, `t`, `n`, `f`)` to the canonical viewing box is given by the matrix

$$P(\texttt{glFrustum}(l, r, b, t, n, f)) = \begin{bmatrix} \frac{2n}{r-l} & 0 & \frac{r+l}{r-l} & 0 \\ 0 & \frac{2n}{t-b} & \frac{t+b}{t-b} & 0 \\ 0 & 0 & -\frac{f+n}{f-n} & -\frac{2fn}{f-n} \\ 0 & 0 & -1 & 0 \end{bmatrix} \qquad (18.4)$$

That $a_{43} \neq 0$ confirms that P(`glFrustum(`l`, `r`, `b`, `t`, `n`, `f`)`) is not affine, as a 4×4 projective transformation matrix is affine if and only if its last row is all 0 except for the last element.

Exercise 18.2. Characterize those regular points that P(`glFrustum(`l`, `r`, `b`, `t`, `n`, `f`)`) maps to points at infinity.

At this time we ask the reader to open the red book to Appendix F, where OpenGL's 4×4 projection matrices are given and compare their values to those in Equations (18.1) and (18.4) above. Seeing these together with its 4×4 matrices for translation, rotation and scaling, listed in Appendix F as well, and derived by us in Section 5.4, the reader may tend to agree that OpenGL "lives" in projective 3-space.

Example 18.2. Determine how the point $[0\ 0\ -10]^T$ is transformed by the projection transformation corresponding to `glFrustum(-5, 5, -5, 5, 5, 100)`, a command used frequently in earlier chapters.

Answer:

$$P(\texttt{glFrustum}(-5,5,-5,5,5,25)) = \begin{bmatrix} \frac{10}{10} & 0 & \frac{0}{10} & 0 \\ 0 & \frac{10}{10} & \frac{0}{10} & 0 \\ 0 & 0 & -\frac{105}{95} & -\frac{1000}{95} \\ 0 & 0 & -1 & 0 \end{bmatrix} = \begin{bmatrix} 1 & 0 & 0 & 0 \\ 0 & 1 & 0 & 0 \\ 0 & 0 & -1.105 & -10.526 \\ 0 & 0 & -1 & 0 \end{bmatrix}$$

Writing $[0\ 0\ -10]^T$ in homogeneous coordinates as $[0\ 0\ -10\ 1]^T$, one sees that it's transformed to the point

$$\begin{bmatrix} 1 & 0 & 0 & 0 \\ 0 & 1 & 0 & 0 \\ 0 & 0 & -1.105 & -10.526 \\ 0 & 0 & -1 & 0 \end{bmatrix} \begin{bmatrix} 0 \\ 0 \\ -10 \\ 1 \end{bmatrix} = \begin{bmatrix} 0 \\ 0 \\ 0.524 \\ 10 \end{bmatrix}$$

which is $[0\ 0\ 0.0524]^T$ in Cartesian coordinates.

The last step, where the homogeneous coordinates are divided by the w-value – in this case $[0\ 0\ 0.524\ 10]^T$ by 10 – to project the transformed point back into xyz-space ($w = 1$) is often called *perspective division*, especially when performed as a part of the graphics pipeline.

Exercise 18.3. Determine how the following points are transformed by the projection transformation corresponding to `glFrustum(-10, 10, -10, 10, 1, 10)`: (a) $[1\ 1\ -2]^T$ (b) $[10\ 20\ -1]^T$ (c) $[5\ 5\ 0]^T$.

If any of them is mapped to a point at infinity – whose w-value is 0 – simply identify it as such. Obviously, you'll not be able to complete the projection transformation for such points, as they will not pass perspective division. We'll discuss in Section 19.1 of the next chapter how they are, in fact, handled in the pipeline.

Remark 18.3. Do keep in mind the terminological distinction that a *projective transformation* is one of projective space, while a *projection transformation* is a particular transformation in the graphics pipeline, which is implemented *by means of* a projective transformation, if the viewing volume is a frustum.

Projection Matrix of gluPerspective()

We are going to leave the reader to solve the following:

Exercise 18.4. Write an equation similar to (18.4) for the projection matrix corresponding to the GLU call `gluPerspective(`*fovy*`, `*aspect*`, `*n*`, `*f*`)`.

18.1.3 Projection Matrix in the Pipeline

We know now how the projection matrix corresponding to a programmer-specified projection command – `glOrtho()`, `glFrustum()` or `gluPerspective()` – is computed by OpenGL. How, then, is this matrix stored? And how is it applied in the graphics pipeline?

The answer to the first question is in a manner exactly similar to modelview matrices. As the current modelview matrix is at the top of the modelview matrix stack, so the *current projection matrix* is the topmost of the *projection matrix stack*. Again, as for modelview statements, a projection statement is applied by multiplying the current projection matrix on the right by the matrix corresponding to that statement. Moreover, the projection matrix stack can be pushed and popped, and the current projection matrix accessed and manipulated, just as the modelview matrix stack. Refer to Section 5.4.6 for commands to access the current modelview matrix.

The reader can probably guess the answer to the second question: if the current modelview and projection matrices are M and P, respectively, then the vertex V at $[x\ y\ z\ 1]^T$ in world space is transformed to the vertex V' at

$$P\,M\,[x\ y\ z\ 1]^T \tag{18.5}$$

In fact, Figure 18.3 illustrates what is actually the first part of the graphics pipeline. We'll be digging deeper into the pipeline in the next chapter.

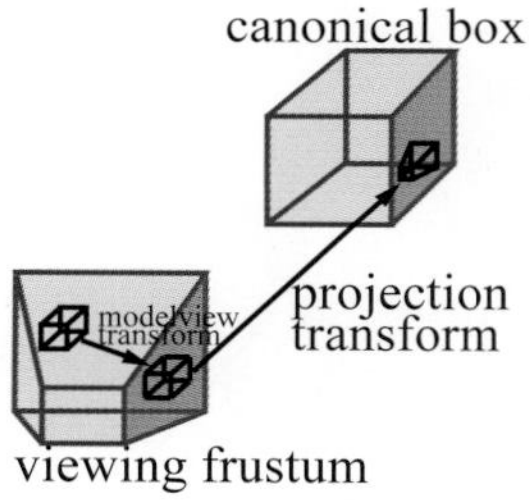

Figure 18.3: Combining modelview and projection transformations.

Remark 18.4. The transformation above is one of world space which takes the viewing frustum into the canonical box. Moreover, points inside the frustum map inside the box, while those outside the frustum map outside the box.

Experiment 18.1. Run `manipulateProjectionMatrix.cpp`, a simple modification of `manipulateModelviewMatrix.cpp` of Chapter 5. Figure 18.4 is a screenshot, though the output to the OpenGL window is of little interest. Of interest, though, are the new statements in the `resize()` routine that output the current projection matrix just before and after the call `glFrustum(-5.0, 5.0, -5.0, 5.0, 5.0, 100.0)`.

Compare the second matrix output to the command window with P(`glFrustum(-5.0, 5.0, -5.0, 5.0, 5.0, 100.0)`) computed with the help of Equation (18.4). **End**

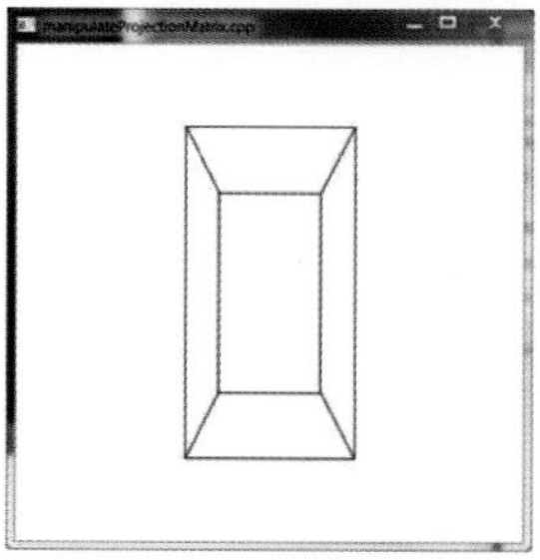

Figure 18.4: Screenshot of `manipulateProjectionMatrix.cpp`.

Exercise 18.5. (Programming) Continue with the preceding experiment by replacing the projection statement

```
glFrustum(-5.0, 5.0, -5.0, 5.0, 5.0, 100.0)
```

with

```
glOrtho(-10.0, 10.0, -10.0, 10.0, 0.0, 20.0)
```

Compare the second matrix output to the command window with

P(`glOrtho(-10.0, 10.0, -10.0, 10.0, 0.0, 20.0)`)

as computed using Equation (18.4).

Remark 18.5. It's unlikely that you'll ever need to access the projection matrix stack in an ordinary program, other than in the mandatory definition of the viewing volume, which would be exactly one of `glOrtho()` or `glFrustum()` or `gluPerspective()`.

18.2 Shadow Mapping

We'll now apply our insight into the matrix computations in the pipeline, learned from the preceding subsection, to understand the very popular technique of shadow mapping, formulated by Lance Williams in his 1978 paper "Casting curved shadows on curved surfaces" [149], to authentically draw shadows cast by a local light source – such shadows are called *perspective* or *projective*.

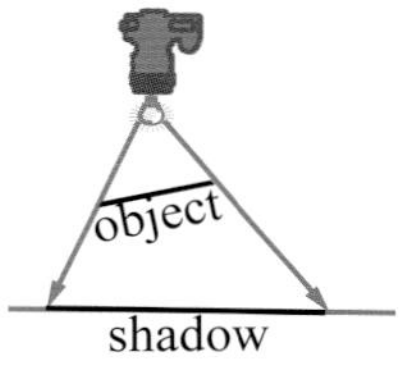

Figure 18.5: A camera and light source at the "same" location.

Consider a scene with a single point light source L. The first key insight of shadow mapping is that the lit region is the one to every point of which there is an unobstructed straight-line ray from L, in other words, it consists precisely of those points which would be visible from L's viewpoint, equivalently, from a camera located at L (Figure 18.5). Now, if we had access to the z-buffer for such a camera, then we could determine which scene fragments are visible – lit by L – by comparing their z-values to corresponding ones in the z-buffer. Fragments whose z-values are greater than the corresponding ones in the z-buffer are hidden from L and not lit. Shadow mapping takes exactly this approach: it starts off by locating a camera at L and saving its z-buffer values in a texture, such a texture being called a *depth texture* for obvious reasons.

But, given a scene viewed through a programmer-defined camera, not necessarily located at the light, how do we determine z-values from the light's viewpoint? Here is where the second key insight of shadow mapping comes in. Suppose, the programmer's camera, call it C, sees a vertex at $[x\ y\ z\ 1]^T$. Now, the vertex arrived at these (world space) coordinates after modeling transformations in the program *followed by* a viewing transformation, namely, C's `gluLookAt()`, which itself, as we know from Section 4.6, is simulated by a sequence of modeling transformations. However, from the light's viewpoint C's viewing transformation is "fake" – equivalent to a sequence of modeling transformations applied to the scene simply to keep the camera at its fixed disposition at the origin.

For example, say the only modeling transformation applied to a scene is `glTranslatef(10.0, 0.0, 0.0)`, while the camera's viewing transformation is `gluLookAt(0.0, 0.0, 10.0, 0.0, 0.0, 0.0, 0.0, 1.0, 0.0)`. Then, the combined transformation applied is `glTranslatef(10.0, 0.0,`

-10.0). However, from the viewpoint of another stationary observer, the scene is transformed by glTranslatef(10.0, 0.0, 0.0) *only*, and *not* glTranslatef(10.0, 0.0, -10.0). In other words, one needs to undo the camera's viewing transformation in order to perceive the scene from the second observer's viewpoint.

Denote, then, C's viewing transformation, defined by its gluLookAt(), by V_{camera}; suppose, as well, there is a second camera located at the light, call it L too to keep notation simple, with its projection transformation (derived from its own frustum) being P_{light} and its viewing transformation (derived from its gluLookAt()) V_{light}. So, the transformation

$$P_{light}\, V_{light}\, V_{camera}^{-1}\, [x\ y\ z\ 1]^T \tag{18.6}$$

applied to a vertex at $[x\ y\ z\ 1]^T$ undoes C's viewing transformation *and*, by Equation (18.5), takes L's viewing frustum to the canonical box, whence the vertex's z-value from L's viewpoint may simply be read off! Keep in mind, though, Remark 18.4 of the previous section, noting that vertices mapping outside the canonical box are clipped, so safely ignored.

There is one last technicality to navigate before we can put shadow mapping into production. It arises from the fact that a $[0,1] \times [0,1]$ depth texture is used to store z-values, these values themselves ranging in $[0,1]$, while the canonical box is $[-1,1] \times [-1,1] \times [-1,1]$. So, we need a final transformation of the box $[-1,1] \times [-1,1] \times [-1,1]$ to the box $[0,1] \times [0,1] \times [0,1]$ in order to make correct z-value comparisons. This is not hard: the linear 2-transformation sequence

```
glTranslatef(0.5, 0.5, 0.5);
glScalef(0.5, 0.5, 0.5);
```

does the needful, its corresponding matrix, often called the *bias matrix*, being

$$B = \begin{bmatrix} 0.5 & 0 & 0 & 0.5 \\ 0 & 0.5 & 0 & 0.5 \\ 0 & 0 & 0.5 & 0.5 \\ 0 & 0 & 0 & 1 \end{bmatrix}$$

Tacking B onto (18.6), we have finally the transformation to use in shadow mapping to read z-values from the light's viewpoint:

$$B\, P_{light}\, V_{light}\, V_{camera}^{-1}\, [x\ y\ z\ 1]^T \tag{18.7}$$

Time to turn words into deeds!

Experiment 18.2. Before running ballAndTorusShadowMapped.cpp you may want to run again ballAndTorusShadowed.cpp from Section 4.7.2, which implements a simple-minded blacken-and-flatten strategy, with help of a scaling degenerate along the y-direction, to draw shadows on the floor.

The scenes of the two programs are almost identical, except, of course, that `ballAndTorusShadowMapped.cpp` shadow maps a local light source, whose position is indicated by a red sphere. Controls are identical too: press space to start the ball traveling around the torus and the up and down arrow keys to change its speed. Figure 18.6 is a screenshot. **End**

Figure 18.6: Screenshot of `ballAndTorusShadowMapped.cpp`.

We'll narrate `ballAndTorusShadowMapped.cpp` starting from the top. First comes a bunch of globals whose purpose will be clear as we go along. The `drawFlyingBallAndTorus()`, `drawCheckeredFloor()` and timer routines are copied from `ballAndTorusShadowed.cpp`, except now there is no option in the first one to blacken the ball and torus as it's not needed in the current program.

There are three points of note in the initialization routine:

1. With `glDepthFunc(GL_LEQUAL)` the comparison in the z-buffer is set to "$\leq$", rather than the default "$<$", so that brightly lit regions, drawn in the third drawing pass, can overwrite their dark versions drawn earlier in the second pass.

2. The shadow map texture is created in a manner similar to image (RGB) textures, except now the `GL_DEPTH_COMPONENT` parameter indicates that it is a depth texture intended to store z-values.

3. The camera and the light's projection and viewing transformation matrices are computed and stored in globals. By the light's transformations, of course, we mean those of a camera located there. The parameters for the respective `gluPerspective()` and `gluLookAt()` commands are set as globals.

   The computations are done all in the modelview matrix stack, which we use through the program as our personal matrix calculator! (Indeed, it may seem odd to apply a `gluPerspective()` to the modelview stack, but it's not illegal and we care only about the corresponding matrix.)

The drawing routine next has four passes which we discuss one after another.

FIRST PASS: We set the shadow map texture values in this pass. First, we load the light's projection and viewing transformation matrices to draw the scene from its viewpoint. The viewport dimensions are set to match those of the shadow map texture so that there is one-to-one correspondence between pixels and texels. Rendering of the scene to the color buffer is disabled by passing `GL_FALSE` parameters to `glColorMask()`, as we don't want to actually see the scene, but only fill the depth buffer.

Front faces are culled prior to drawing so that only back faces go into the depth buffer, ensuring that in ensuing z-value competitions front faces with their lower z-values will prevail over back faces (as they should from the light's viewpoint). Finally, the scene is drawn and z-values captured in the shadow map texture using a `glCopyTexImage2D()` command.

SECOND PASS: In this pass the whole scene is drawn as if shadowed. First, the camera's projection and viewing transformation matrices are loaded. The viewport is set to the OpenGL window size and back face culling is enabled as is usual for efficient drawing. Rendering to the color buffer is enabled as well. Lighting is enabled prior to drawing the scene, but the local light source `GL_LIGHT0` is *not* enabled, permitting only the dim global ambient light, which, of course, is the only light which should illuminate shadowed regions.

THIRD PASS: Now, we'll draw with the local light source turned on, eliminating first, however, with help of the shadow map texture, fragments which are shadowed. Thus, the lit parts only, which were drawn shadowed in the previous pass, are redrawn illuminated by the light source.

First, the light source `GL_LIGHT0` is enabled. Next, we want to calculate and apply the transformation matrix $B\,P_{light}\,V_{light}\,V_{camera}^{-1}$ of (18.7). More than one step will be needed to do this. To begin with, the matrix product $B\,P_{light}\,V_{light}$, which we call the texture matrix, is calculated in the modelview matrix stack and saved in the global `texMat`. Post-multiplying the texture matrix by V_{camera}^{-1} and then using it to transform vertices is up next. Unfortunately, though, there is no inversion operation available in any OpenGL matrix stack. Still, there is another resource which solves our problem exactly. The automatic texture coordinate generation sequence

```
glEnable(GL_TEXTURE_GEN_X);
glTexGeni(GL_X, GL_TEXTURE_GEN_MODE, GL_EYE_LINEAR);
glTexGenfv(GL_X, GL_EYE_PLANE, eyePlaneParams);
```

X being any one of the four texture coordinates S, T, R and Q and *eyePlaneParams* pointing to a 4-vector $[p_x\ p_y\ p_z\ p_w]$, generates the coordinate value

$$p'_x x + p'_y y + p'_z z + p'_w w$$

for X, where $[x\ y\ z\ w]^T$ are the homogeneous coordinates of the current vertex and where

$$[p'_x\ p'_y\ p'_z\ p'_w] = [p_x\ p_y\ p_z\ p_w]\,M^{-1}$$

M being the current modelview matrix. Now, consider the program. The current modelview matrix, in fact, is V_{camera}; moreover, the code sequence above is repeated once for each of the texture coordinates S, T, R and Q, with *eyePlaneParams* pointing to successive rows of the texture matrix $B\,P_{light}\,V_{light}$. It's not hard to check then that

$$[S\ T\ R\ Q]^T = B\,P_{light}\,V_{light}\,V_{camera}^{-1}\,[x\ y\ z\ w]^T$$

which is precisely the transformation the doctor ordered if you see again Equation (18.7). So, R is the z-value from the light's viewpoint, which we must compare with the corresponding value in the shadow map texture. OpenGL has specialized support for exactly this. The three statements

```
glTexParameteri(GL_TEXTURE_2D, GL_TEXTURE_COMPARE_MODE,
                GL_COMPARE_R_TO_TEXTURE);
glTexParameteri(GL_TEXTURE_2D, GL_TEXTURE_COMPARE_FUNC, GL_LEQUAL);
glTexParameteri(GL_TEXTURE_2D, GL_DEPTH_TEXTURE_MODE, GL_INTENSITY);
```

cause a fragment's R-value to be compared with the corresponding value in the currently bound depth texture, the comparison passing if the former is less than or equal to the latter, failed comparisons generating an intensity value of 0, successful ones an intensity value of 1 (an intensity value of ρ corresponds to a gray scale signal of $R = G = B = A = \rho$). At the end of these statements, then, shadowed fragments, which failed the comparison because of an R-value greater than their shadow map counterpart, will have an alpha value of 0, while lit fragments have an alpha value of 1.

Finally, the two statements

```
glEnable(GL_ALPHA_TEST);
glAlphaFunc(GL_GREATER, 0.5);
```

prior to drawing the scene set up an alpha test which will be failed by fragments with an alpha value of 0, namely, shadowed ones. The net result is that the lit region is redrawn illuminated by the light source in this pass. Combined with the dim lighting of the whole scene from the previous pass, shadow mapping is now complete.

FOURTH PASS: This is a trivial pass whose purpose is only to draw a red sphere at the light's position.

The reshape routine is interesting too in that, instead of simply declaring a viewing frustum, it goes on to compute the frustum's projection matrix and accordingly update the `cameraProjMat` global, because it is this global from which the program obtains the camera's projection matrix. The rest of the program is straightforward.

Exercise 18.6. **(Programming)** Shadow map `sphereInBox1.cpp` of Experiment 11.1, but, first, make the ball smaller, move it up a bit and tilt the box toward the viewer so that the ball's shadow can be clearly seen inside the box.

Remark 18.6. Shadow mapping is a powerful and much-implemented real-time shadowing technique. However, a couple of its drawbacks are seen even in our program `ballAndTorusShadowMapped.cpp`. Firstly, aliasing is evidently an issue with the shadows – it is, in fact, inversely correlated to the resolution of the shadow map. Secondly, multiple drawing passes may become computationally taxing, especially with more than one light source.

Refinements of shadow mapping have been developed, though, in order to circumvent these issues. A popular alternative to shadow mapping, that of *shadow volumes*, was first developed by Crow [33] in 1977. The reader is referred to more advanced texts such as those by Akenine-Möller et al. [1] and McReynolds-Blythe [92], as well as the research literature, for more on shadow mapping and alternatives such as shadow volumes.

18.3 Rational Bézier and NURBS Curves and Surfaces

Our final application of projective geometry is to set the stage for rational Bézier primitives, as well as to put the 'R' – 'R' stands for rational, of course – into NURBS. In Chapters 15 and 16 we investigated the polynomial versions of Bézier and NURBS theory, respectively.

We'll begin with rational Bézier curves, as conceptually they are the simplest and notationally least cumbersome. Once we have rational Bézier curves under our belts, extending our understanding to rational Bézier surfaces and then to NURBS curves and surfaces will not be difficult.

18.3.1 Rational Bézier Curves Basics

Recall Equation (15.13) of a Bézier curve in $\mathbb{R}^3$ specified by $n+1$ control points $P_i = [x_i\ y_i\ z_i]^T$, $0 \le i \le n$:

$$C(u) = \left[\sum_{i=0}^{n} B_{i,n}(u)x_i \quad \sum_{i=0}^{n} B_{i,n}(u)y_i \quad \sum_{i=0}^{n} B_{i,n}(u)z_i \right]^T \quad (0 \le u \le 1) \tag{18.8}$$

Note to Readers Unfamiliar with Projective Geometry: Here's what you need to know for most of this particular section. Projective 2-space $\mathbb{P}^2$ consists of 3-tuples of the form $[x\ y\ z]^T$, where these so-called homogeneous coordinates cannot all be zero. Two tuples represent the same point if one's a scalar multiple of the other, e.g., $[0\ -1\ 2]^T$ and $[0\ -4\ 8]^T$.

Real 2-space $\mathbb{R}^2$ is embedded in $\mathbb{P}^2$ by mapping the point $[x\ y]^T$ of $\mathbb{R}^2$ to $[x\ y\ 1]^T$ of $\mathbb{P}^2$, e.g., $[2\ 4]^T$ maps to $[2\ 4\ 1]^T$. Conversely, a point $[x\ y\ z]^T$ of $\mathbb{P}^2$, with $z \neq 0$, is an image by this embedding of the point $[x/z\ y/z]^T$ of $\mathbb{R}^2$.

Getting back to the equation at the top of the section, what if none of the P_i has coordinates all zero, so that one can imagine each to be a projective point with homogeneous coordinates $[x_i\ y_i\ z_i]^T$, rather than the real point $[x_i\ y_i\ z_i]^T$? Certainly, then, Equation (18.8) defines a point $C(u)$ in $\mathbb{P}^2$ for every u in $0 \le u \le 1$, as long as all its three components are not simultaneously zero either. In this case, one could call C *the* projective Bézier curve over the projective control points $P_i, 0 \le i \le n$, *provided* that it doesn't depend on the choice of the P_i's homogeneous coordinates, for, otherwise, (18.8) would give different curves $C(u)$ for different choices and not be a proper definition at all.

Let's see if C is, in fact, independent of the choice of homogeneous coordinates for its control points. Accordingly, write $P_i = [w_i x_i\ \ w_i y_i\ \ w_i z_i]^T$, where $w_i \neq 0$, for $0 \leq i \leq n$, and plug into (18.8):

$$D(u) = \left[\sum_{i=0}^{n} B_{i,n}(u) w_i x_i \quad \sum_{i=0}^{n} B_{i,n}(u) w_i y_i \quad \sum_{i=0}^{n} B_{i,n}(u) w_i z_i \right]^T \quad (0 \leq u \leq 1)$$

Is $D(u) = C(u)$? Not necessarily! Playing a bit with the equation it's clear that there's no way to "pull the w_i's out of the square brackets" and write

$$\begin{aligned} & \left[\sum_{i=0}^{n} B_{i,n}(u) w_i x_i \quad \sum_{i=0}^{n} B_{i,n}(u) w_i y_i \quad \sum_{i=0}^{n} B_{i,n}(u) w_i z_i \right]^T \\ = \quad & w \left[\sum_{i=0}^{n} B_{i,n}(u) x_i \quad \sum_{i=0}^{n} B_{i,n}(u) y_i \quad \sum_{i=0}^{n} B_{i,n}(u) z_i \right]^T \end{aligned}$$

for some one scalar w, unless all the w_i's happen to be equal to w.

Ouch! Major road block? Quit and start over? Nah, we'll just make a feature of the bug! Choosing different homogeneous coordinates for the projective control points gives different projective Bézier curves? Well, then, more choice for the designer!

Let's start with control points all on the real plane as, at the end of the day, we'll be modeling in real space, not projective. However, first, identify $\mathbb{R}^2$ with the plane $z = 1$ in $\mathbb{R}^3$, i.e., $[x \ y]^T \in \mathbb{R}^2$ with $[x \ y \ 1]^T \in \mathbb{R}^3$.

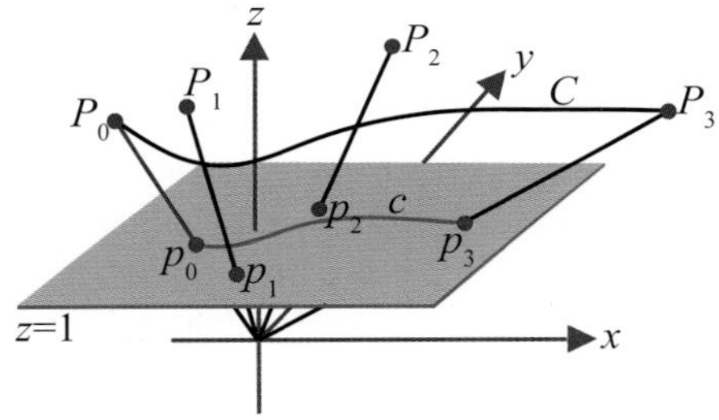

Figure 18.7: Four real control points $p_i = [x_i \ y_i \ 1]^T$, with weights w_i, are lifted to the projective control points $P_i = [w_i x_i \ \ w_i y_i \ \ w_i]^T$, $0 \leq i \leq 3$. The (black) polynomial projective Bézier curve C projects to the (blue) rational real Bézier curve c.

Choose $n + 1$ control points $p_i = [x_i \ y_i \ 1]^T$, $0 \leq i \leq n$, in $\mathbb{R}^2$, as well as $n + 1$ non-zero scalars w_i, $0 \leq i \leq n$. Lift each p_i to the projective point $P_i = [w_i x_i \ \ w_i y_i \ \ w_i]^T$ in $\mathbb{P}^2$, expressed using *these particular* homogeneous coordinates. See Figure 18.7. The scalar w_i is called the *weight* of the control point p_i. The projective *polynomial* Bézier curve C specified by the control points $P_i = [w_i x_i \ \ w_i y_i \ \ w_i]^T$ is

$$C(u) = \left[\sum_{i=0}^{n} B_{i,n}(u) w_i x_i \quad \sum_{i=0}^{n} B_{i,n}(u) w_i y_i \quad \sum_{i=0}^{n} B_{i,n}(u) w_i \right]^T \quad (0 \leq u \leq 1) \tag{18.9}$$

To return to $\mathbb{R}^2$, divide C throughout by its z-coordinate to get the plane curve

$$c(u) = \left[\frac{\sum_{i=0}^{n} B_{i,n}(u) w_i x_i}{\sum_{i=0}^{n} B_{i,n}(u) w_i} \quad \frac{\sum_{i=0}^{n} B_{i,n}(u) w_i y_i}{\sum_{i=0}^{n} B_{i,n}(u) w_i} \quad 1 \right]^T \quad (0 \le u \le 1) \tag{18.10}$$

assuming that $\sum_{i=0}^{n} B_{i,n}(u) w_i \neq 0$ in $0 \le u \le 1$, so there's never division by zero. Rewriting (18.10) as a proper equation in $\mathbb{R}^2$ by dropping the z-value 1, we have

$$c(u) = \left[\frac{\sum_{i=0}^{n} B_{i,n}(u) w_i x_i}{\sum_{i=0}^{n} B_{i,n}(u) w_i} \quad \frac{\sum_{i=0}^{n} B_{i,n}(u) w_i y_i}{\sum_{i=0}^{n} B_{i,n}(u) w_i} \right]^T \quad (0 \le u \le 1) \tag{18.11}$$

which is said to be the *rational Bézier curve* in $\mathbb{R}^2$ approximating the control points $p_i = [x_i \; y_i]^T$, with respective weights w_i, $0 \le i \le n$.

If the weights w_i, $0 \le i \le n$, are all positive, then it's guaranteed that the denominator $\sum_{i=0}^{n} B_{i,n}(u) w_i$ in (18.11) is positive, and so non-zero, in $0 \le u \le 1$. Consequently, this condition on the weights is, typically, assumed as a design constraint. We'll make a tacit assumption ourselves of positive weights henceforth.

It's a bit hard to make out from (18.11) exactly what's going on. Let's consider a particular case with few control points, say $n = 2$, for three control points. Write out the Bernstein polynomials in (18.11) to obtain the following equation for the quadratic rational Bézier curve on three control points $[x_0 \; y_0]^T$, $[x_1 \; y_1]^T$ and $[x_2 \; y_2]^T$, with weights w_0, w_1 and w_2, respectively:

$$\begin{aligned} c(u) \quad = \quad & \Bigg[\frac{w_0 x_0 (1-u)^2 + 2 w_1 x_1 (1-u) u + w_2 x_2 u^2}{w_0 (1-u)^2 + 2 w_1 (1-u) u + w_2 u^2} \\ & \quad \frac{w_0 y_0 (1-u)^2 + 2 w_1 y_1 (1-u) u + w_2 y_2 u^2}{w_0 (1-u)^2 + 2 w_1 (1-u) u + w_2 u^2} \Bigg]^T \end{aligned} \tag{18.12}$$

in $0 \le u \le 1$.

Compare with (15.5), which is

$$c(u) = [x_0 (1-u)^2 + 2 x_1 (1-u) u + x_2 u^2 \quad y_0 (1-u)^2 + 2 y_1 (1-u) u + y_2 u^2]^T$$

in $0 \le u \le 1$, the equation of the quadratic polynomial Bézier curve on the same three control points. Observe, first, that both the x- and y-values on the RHS of (18.12) are rational functions, i.e., *ratios* of two polynomials, particularly of two quadratics in this case. The values on the RHS of the equation for the polynomial curve, on the other hand, are simply quadratic polynomials. The following three exercises shed further light on the quadratic rational Bézier curve.

Exercise 18.7. Putting $u = 0$ and 1 in Equation (18.12), show that, whatever the assignment of weights, a quadratic rational Bézier curve always interpolates both its first and last control points.

Exercise 18.8. Show that if the weights of its three control points are equal, then a quadratic rational Bézier curve coincides with the quadratic polynomial Bézier curve specified by the same control points.

Evidently, then, at least for three control points, a polynomial Bézier curve is simply a special case of a rational one.

Exercise 18.9. The quadratic polynomial Bézier curve is a weighted sum of its control points. One can rewrite (15.5) above as follows to see this.

$$\begin{aligned} & c(u) \\ &= [x_0(1-u)^2 + 2x_1(1-u)u + x_2u^2 \;\; y_0(1-u)^2 + 2y_1(1-u)u + y_2u^2]^T \\ &= (1-u)^2\,[x_0\; y_0]^T \;+\; 2(1-u)u\,[x_1\; y_1]^T \;+\; u^2\,[x_2\; y_2]^T \end{aligned}$$

in $0 \le u \le 1$, where the weights in the second line, namely, $(1-u)^2$, $2(1-u)u$ and u^2, are the so-called blending functions of the control points. As we know, these particular blending functions, called degree 2 Bernstein polynomials, form a partition of unity.

How about the quadratic rational Bézier curve of (18.12)? Write it similarly as a sum

$$c(u) = (\ldots)\,[x_0\; y_0]^T + (\ldots)\,[x_1\; y_1]^T + (\ldots)\,[x_2\; y_2]^T$$

weights being rational blending functions, rather than polynomial. Do these new blending functions still form a partition of unity?

Going from quadratic rational to cubic rational with four weighted control points is straightforward, as we ask the reader to show next.

Exercise 18.10. Write an equation analogous to (18.12) for a cubic rational Bézier curve.

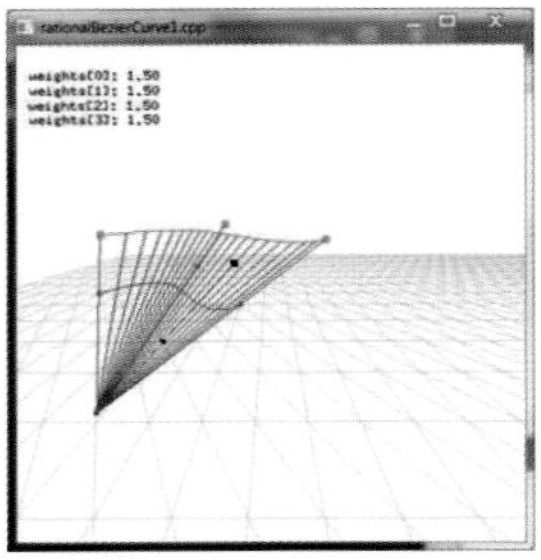

Figure 18.8: Screenshot of `rationalBezierCurve1.cpp`.

Experiment 18.3. Run `rationalBezierCurve1.cpp`, which draws the cubic rational Bézier curve specified by four control points on the plane at *fixed* locations, but with *changeable* weights.

The control points on the plane (light gray triangular mesh) are all red, except for the currently selected one, which is black. Press space to cycle through the control points. The control point weights are shown at the upper-left, that of the currently selected one being changed by pressing the up/down arrow keys. The rational Bézier curve on the plane is red as well. Figure 18.8 is a screenshot.

Drawn in green are all the lifted control points, except for that of the currently selected control point, which is black. The projective polynomial Bézier curve approximating the lifted control points is green too. The lifted control points are a larger size as well.

Note: The lifted control points and the projective Bézier curve are primitives in $\mathbb{P}^2$, of course, but represented in $\mathbb{R}^3$ using their homogeneous coordinates.

Also drawn is a cone of several gray lines through the projective Bézier curve which intersects the plane in its projection, the rational Bézier curve.

Observe that increasing the weight of a control point pulls the (red rational Bézier) curve toward it, while reducing it has the opposite effect. Moreover, the end control points are always interpolated regardless of assigned weights. It's sometimes hard to discern the very gradual change in the shape of the curve as one varies the weights. A trick is to press delete for the curve to spring back to its original configuration, at which moment the difference should be clear.

It seems, then, that the control point weights are an additional set of "dials" at the designer's disposal for use to edit the curve.

The code of `rationalBezierCurve1.cpp` is instructive as well, as we'll see in the next section on drawing. **End**

We've been studying rational Bézier curves on the plane for really no other reason than that, though we started with the 3D equation (18.8), we soon projected it down to 2D. Deriving the equation for a rational Bézier curve in 3-space is not hard and, in fact, almost a repeat of the 2D process, as we ask the reader to show next.

Exercise 18.11. Identify $\mathbb{R}^3$ with the hyperplane $w = 1$ in $xyzw$-space $\mathbb{R}^4$, just as we identify $\mathbb{R}^2$ with the plane $z = 1$ in xyz-space $\mathbb{R}^3$. Suppose $p_i = [x_i\ y_i\ z_i\ 1]^T$, $0 \le i \le n$, are $n+1$ control points in $\mathbb{R}^3$ with assigned weights w_i, $0 \le i \le n$.

Lift each p_i to the projective point $P_i = [w_i x_i \quad w_i y_i \quad w_i z_i \quad w_i]^T$ in $\mathbb{P}^3$, expressed using those particular homogeneous coordinates. Reasoning as earlier in the 2D case, show that the equation of the rational Bézier curve in $\mathbb{R}^3$ approximating the control points $p_i = [x_i\ y_i\ z_i]^T$, with respective weights w_i, $0 \le i \le n$, is

$$c(u) = \left[\frac{\sum_{i=0}^{n} B_{i,n}(u) w_i x_i}{\sum_{i=0}^{n} B_{i,n}(u) w_i} \quad \frac{\sum_{i=0}^{n} B_{i,n}(u) w_i y_i}{\sum_{i=0}^{n} B_{i,n}(u) w_i} \quad \frac{\sum_{i=0}^{n} B_{i,n}(u) w_i z_i}{\sum_{i=0}^{n} B_{i,n}(u) w_i} \right]^T \tag{18.13}$$

for $0 \le u \le 1$, which, of course, is the analogue of the 2D equation (18.11) that we have already derived.

18.3.2 Drawing Rational Bézier Curves

OpenGL can draw rational Bézier curves in 3-space. To draw the curve with control points $p_i = [x_i\ y_i\ z_i]^T$ and weights w_i, $0 \le i \le n$, the command is

`glMap1f(GL_MAP1_VERTEX_4,` *t1*, *t2*, *stride*, *order*, **controlPoints*)

where *controlPoints* points to the $(n+1) \times 4$ array

$$\{\{w_0 x_0\ w_0 y_0\ w_0 z_0\ w_0\}, \{w_1 x_1\ w_1 y_1\ w_1 z_1\ w_1\}, \ldots, \{w_n x_n\ w_n y_n\ w_n z_n\ w_n\}\} \tag{18.14}$$

and other parameters have the same meaning as for the command `glMap1f(GL_MAP1_VERTEX_3,` *t1*, *t2*, *stride*, *order*, **controlPoints*)

with which we are familiar from drawing polynomial Bézier curves in Section 10.3.1.

Returning to `rationalBezierCurve1.cpp`, let's see if OpenGL's commands to draw a rational Bézier curve in 3-space have been correctly invoked. Say the four planar control points of the program are $[x_i\ y_i]^T$, $0 \le i \le 3$, represented in homogeneous coordinates by $[x_i\ y_i\ 1]^T$, the values of the latter being stored in the array `controlPoints`. Their respective weights w_i are stored in the array `weights`.

Note: We are using variable names for convenience, of course. The actual values in the program, as you can see, are $[7.0\ 2.0]^T$ for $[x_1\ y_1]^T$, 1.5 for w_1, and so on.

The array `ControlPointsLifted` is filled by the routine `compute-ControlPointsLifted()` with the lifted coordinate values $[w_i x_i\ \ w_i y_i\ \ w_i]^T$, $0 \le i \le 3$. The green Bézier curve is the 3D polynomial Bézier approximation of the lifted points drawn using `glMap1f(GL_MAP1_VERTEX_3, ...)`.

The array `controlPointsHomogeneous` is filled by `computeControl-PointsHomogeneous()` with the values $[w_i x_i\ \ w_i y_i\ \ w_i\ \ w_i]^T$. From our understanding of the syntax of `glMap1f`(GL_MAP1_VERTEX_4, ...) – compare, in particular, array `controlPointsHomogeneous` with array (18.14) above – the red rational Bézier curve approximates the control points $[x_i\ y_i\ 1]^T$ in $\mathbb{R}^3$ with weights w_i, $0 \le i \le 3$. By (18.13) the equation of the latter is seen to be

$$c(u) = \begin{bmatrix} \frac{\sum_{i=0}^{n} B_{i,n}(u) w_i x_i}{\sum_{i=0}^{n} B_{i,n}(u) w_i} & \frac{\sum_{i=0}^{n} B_{i,n}(u) w_i y_i}{\sum_{i=0}^{n} B_{i,n}(u) w_i} & 1 \end{bmatrix}^T \quad (0 \le u \le 1)$$

which is precisely the 2D rational Bézier approximation of the control points $[x_i\ y_i]^T$, $0 \le i \le 3$, drawn on the plane $z = 1$.

We have verified, therefore, that the green and red curves of `rational-BezierCurve1.cpp` are indeed the particular Bézier approximations claimed in Experiment 18.3.

So what do rational Bézier curves have that the polynomial curves do not? Let's see ....

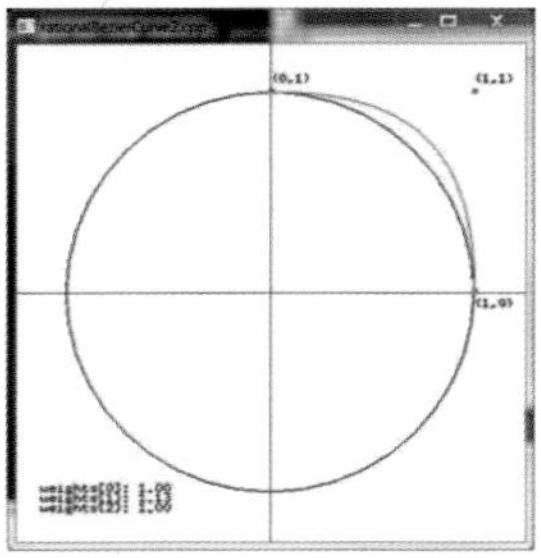

Figure 18.9: Screenshot of `rationalBezier-Curve2.cpp` with the weight of the middle control point 1.13.

18.3.3 Rational Bézier Curves and Conic Sections

Experiment 18.4. Run `rationalBezierCurve2.cpp`, which draws a red quadratic rational Bézier curve on the plane specified by the three control points $[1, 0]^T$, $[1, 1]^T$ and $[0, 1]^T$. See Figure 18.9. Also drawn is the unit circle centered at the origin. Press the up/down arrow keys to change the weight of the middle control point $[1, 1]^T$. The weights of the two end control points are fixed at 1.

Decrease the weight of the control point $[1, 1]^T$ from its initial value of 1.5. It seems that at some value between 0.70 and 0.71 the curve lies exactly along a quarter of the circle (the screenshot of Figure 18.9 is at 1.13). This is no accident, as the following exercise shows. **End**

Exercise 18.12. Plug the values

$$[x_0\ y_0]^T = [1,0]^T, \qquad [x_1\ y_1]^T = [1,1]^T, \qquad [x_2\ y_2]^T = [0,1]^T$$

of the control points of the preceding experiment, together with the weights

$$w_0 = 1, \qquad w_1 = 1/\sqrt{2}, \qquad w_2 = 1$$

into Equation (18.12). Show, then, that the rational functions $x(u)$ giving the x-value and $y(u)$ the y-value, satisfy $x(u)^2 + y(u)^2 = 1$.

One sees from the preceding exercise that the quadratic rational Bézier curve specified by the control points $[1,0]^T$ with weight 1, $[1,1]^T$ with weight $1/\sqrt{2}$ ($\simeq 0.7071$) and $[0,1]^T$ with weight 1 is indeed a quarter of a circle. It follows that any whole circle can be obtained by joining end to end at most four quadratic rational Bézier curves. In fact, this generalizes to a very close relationship between quadratic rational Bézier curves and conic sections:

Proposition 18.1. *Any bounded arc of a conic section can be obtained by joining end to end a finite number of quadratic rational Bézier curves.*

In the other direction, any quadratic rational Bézier curve is an arc of a conic section. □

The proof is beyond our scope here. We refer the interested reader to the text by Buss [22].

Remark 18.7. The qualifier "bounded" in the proposition is necessary simply because a rational Bézier curve is bounded by definition, so that no unbounded arc of a conic section (e.g., an entire parabola or wing of a hyperbola) can be assembled from a finite number of rational Bézier curves.

Remark 18.8. If the reader is wondering how a quadratic rational Bézier curve which happens to be a straight line segment, e.g., if its three control points are collinear, can be an arc of a conic section, keep in mind that straight lines are, in fact, degenerate conic sections (refer to Exercise 10.21).

Now, not even a circle, the simplest of conic sections, can be constructed from polynomial Bézier curves, because no non-trivial arc of a circle has a polynomial parametrization, as we saw in Example 10.8. This is important! Using rational Bézier curves, though not polynomial ones, one can draw conic sections, including circles, ellipses, parabolas and hyperbolas, all curves which arise naturally in diverse applications.

Score one for the rationals!

Remark 18.9. The original Utah Teapot, discussed toward the end of Section 10.3.2, composed of bicubic polynomial Bézier patches, is not – and can never be – perfectly round! To make it so, it has to be redesigned with the help of rational patches.

18.3.4 Properties of Rational Bézier Curves

We ask the reader to establish some properties of rational Bézier curves in general. Observe first that the x and y components of the rational Bézier curve c given by Equation (18.11), written below again,

$$c(u) = \begin{bmatrix} \frac{\sum_{i=0}^{n} B_{i,n}(u) w_i x_i}{\sum_{i=0}^{n} B_{i,n}(u) w_i} & \frac{\sum_{i=0}^{n} B_{i,n}(u) w_i y_i}{\sum_{i=0}^{n} B_{i,n}(u) w_i} \end{bmatrix}^T \qquad (0 \leq u \leq 1)$$

are both ratios of polynomials of degree n in u. Accordingly, c is said to be the rational Bézier curve of *degree* n, or *order* $n+1$, the latter being the number of control points.

The earlier Exercises 18.7 and 18.8 both generalize to rational Bézier curves of arbitrary order, as we see next.

Exercise 18.13. Prove that a rational Bézier curve (of arbitrary order) always interpolates both its first and last control points, no matter what the assignment of weights.

Exercise 18.14. Prove that a rational Bézier curve (of arbitrary order) whose control points have all equal weights coincides with the polynomial Bézier curve specified by the same control points.

Therefore, generally, a polynomial Bézier curve is simply a special case of a rational one.

Let's massage (18.11) into a form which will afford us a familiar way of understanding rational Bézier curves:

$$\begin{aligned} c(u) &= \begin{bmatrix} \frac{\sum_{i=0}^{n} B_{i,n}(u) w_i x_i}{\sum_{i=0}^{n} B_{i,n}(u) w_i} & \frac{\sum_{i=0}^{n} B_{i,n}(u) w_i y_i}{\sum_{i=0}^{n} B_{i,n}(u) w_i} \end{bmatrix}^T \\ &= \begin{bmatrix} \sum_{i=0}^{n} \frac{B_{i,n}(u) w_i}{\sum_{i=0}^{n} B_{i,n}(u) w_i} x_i & \sum_{i=0}^{n} \frac{B_{i,n}(u) w_i}{\sum_{i=0}^{n} B_{i,n}(u) w_i} y_i \end{bmatrix}^T \\ &= \sum_{i=0}^{n} \frac{B_{i,n}(u) w_i}{\sum_{i=0}^{n} B_{i,n}(u) w_i} p_i \end{aligned} \qquad (18.15)$$

in $0 \leq u \leq 1$, where the control point $p_i = [x_i \; y_i]^T$, $0 \leq i \leq n$.

One sees from Equation (18.15) that a rational Bézier curve is a weighted sum of its control points, as is a polynomial Bézier curve, but using a different set of blending functions as weights: instead of the Bernstein polynomial $B_{i,n}(u)$, the rational function

$$\frac{B_{i,n}(u) w_i}{\sum_{i=0}^{n} B_{i,n}(u) w_i}$$

blends control point p_i.

Exercise 18.15. Verify that the blending functions of a rational Bézier curve form a partition of unity. Therefore, a rational Bézier curve is (a) constrained to lie in the convex hull of its control points, and (b) affinely invariant.
Hint: See the proof of Proposition 15.1.

Exercise 18.16. Prove that if the weight w_i of one particular control point p_i is increased in Equation (18.15), then the value

$$\frac{B_{i,n}(u)w_i}{\sum_{i=0}^{n} B_{i,n}(u)w_i}$$

of its blending function increases everywhere in the open interval $0 < u < 1$, while that of every other control point decreases.

This explains the phenomenon observed in Experiment 18.3, that increasing a control point's weight attracts the curve to it.

Experiment 18.5. Run `rationalBezierCurve3.cpp`, which shows a rational Bézier curve on the plane specified by six control points. See Figure 18.10 for a screenshot. A control point is selected by pressing the space key, moved with the arrow keys and its weight changed by the page up/down keys. Pressing delete resets. **End**

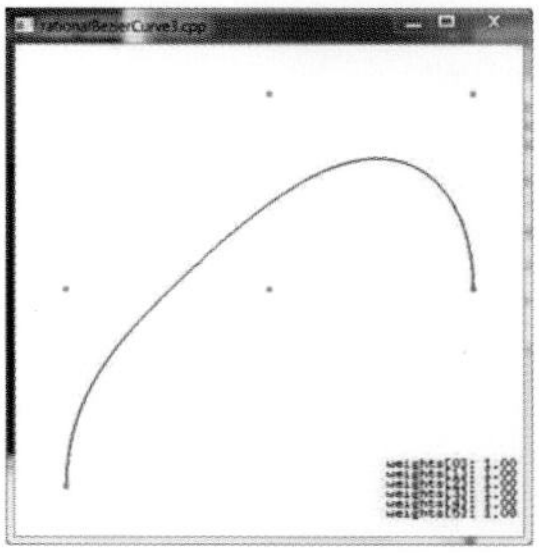

Figure 18.10: Screenshot of `rationalBezierCurve3.cpp`.

From a design point of view then a control point's weight is a dial to turn up or down its attraction on the curve. It adds a level of control to edit a rational Bézier curve beyond what is available for a polynomial one.

That's score two for the rationals!

18.3.5 Rational Bézier Curves and Projective Invariance

Note to Readers Unfamiliar with Projective Geometry: This section investigates how a projective transformation transforms a Bézier curve. It begins, though, with the effect of so-called snapshot transformations, a subclass of the projective, defined in Appendix A. Informally, a snapshot transformation is the change induced in how an object is seen by altering the alignment of a point camera. Unfortunately, just this much may not be enough to follow the entire discussion in this section, but to go farther it seems unavoidable to refer to Appendix A. Our suggestion, therefore, to the reader not inclined to peruse that appendix is to simply read once Proposition 18.2, which describes how a rational Bézier curve changes through projective transformation, and take it for granted.

What happens when a snapshot transformation (snapshot transformations, a special subclass of the projective, are introduced in Section A.5 of Appendix A) is applied to a Bézier curve, either polynomial or rational? Let's try and repeat Experiment A.1, where we ran the program `turnFilm1.cpp` to compare snapshots of parallel power lines taken with the film along the $z = 1$ and $x = 1$ plane, respectively, but, with power lines now replaced by a polynomial Bézier curve drawn on the $z = 1$ plane.

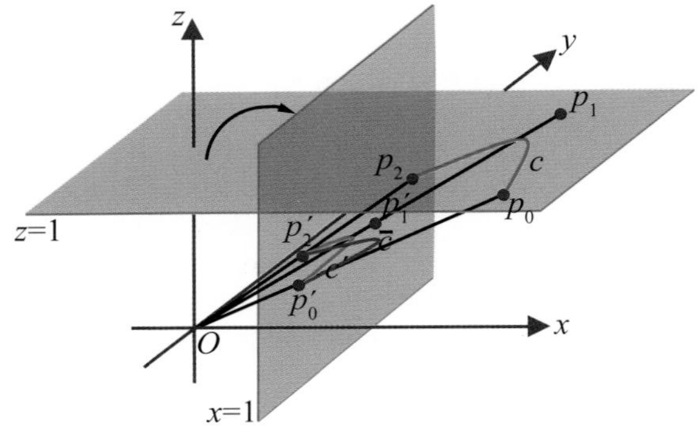

Figure 18.11: The (red solid) quadratic polynomial Bézier curve c on the $z = 1$ plane is specified by the control points p_0, p_1 and p_2. The points p_0, p_1 and p_2 and the curve c project to the points p'_0, p'_1 and p'_3 and the (red dashed) curve c', respectively, on the plane $x = 1$. The (green dashed) curve $\bar{c}$, different from c', is the polynomial Bézier approximation of p'_0, p'_1 and p'_3.

See Figure 18.11, where the control points p_0, p_1 and p_2 lie on the $z = 1$ plane, and the (red solid) quadratic polynomial Bézier curve c approximates them. The points p_0, p_1 and p_2 and the curve c are projected along lines toward the origin to the points p'_0, p'_1 and p'_3 and the (red dashed) curve c' on the plane $x = 1$. Therefore, c' is the snapshot transformation of c.

Is c' the polynomial Bézier curve approximating p'_0, p'_1 and p'_3? No! That happens to be a different (green dashed) curve $\bar{c}$. Coding is believing ....

Experiment 18.6. Run `turnFilm2.cpp`, which animates the snapshot transformation of a polynomial Bézier curve described above. Three control points and their red dashed approximating polynomial Bézier curve are initially drawn on the $z = 1$ plane. See Figure 18.12(a). The locations of the control points, and so of their approximating curve as well, are *fixed* in world space. However, they will *appear* to move as the film rotates.

Initially, the film lies along the $z = 1$ plane. Pressing the right arrow key rotates it toward the $x = 1$ plane, while pressing the left arrow key rotates it back. The film itself, of course, is never seen. As the film changes position, so do the control points and the red dashed curve, these being the *projections* (snapshot transformations, particularly) onto the current film of the control points and their approximating curve (all fixed, as said, in world space). Also drawn on the film is a green dashed curve, which is the polynomial Bézier curve approximating the current projections of the control points.

Note: The control points and their approximating curve, all fixed on the $z = 1$ plane, and corresponding to the control points p_0, p_1 and p_2 and the solid red curve in Figure 18.11, are *not* drawn by the program – only their snapshot transformations on the turning film.

Initially, when the plane of the film coincides with that on which the control points are drawn, viz. $z = 1$, the projection onto the film of the polynomial Bézier curve approximating the control points (the red dashed curve) coincides with the polynomial Bézier curve approximating the projected control points (the green dashed curve). This is to be expected because the control points coincide with their projections. However, as the film turns away from the $z = 1$ plane, the red and green dashed curves begin to separate. Their final configuration, when the film lies along $x = 1$, is shown in Figure 18.12(b).

There is more functionality to the program that we'll discuss momentarily.

End

(a) (b)

Figure 18.12: Screenshots of `turnFilm2.cpp`: (a) Initial configuration (b) Final.

So, if the snapshot transformation c' of the approximating polynomial Bézier curve (the red dashed curve of `turnFilm2.cpp`) is not the polynomial Bézier curve $\overline{c}$ approximating the transformed control points (the green dashed curve), then what is it?

It's not hard to deduce the answer by comparing the earlier Figure 18.7 with Figure 18.11. Imagine the plane $z = 1$ of the former figure replaced by $x = 1$ of the latter. Accordingly, points p_0, p_1 and p_2 of Figure 18.11 are liftings of their respective projections p'_0, p'_1 and p'_2 on $x = 1$, the weights associated with the latter three being the respective *x-values* of the first three.

Conclusion: the snapshot transformation of the polynomial Bézier curve on the control points p_0, p_1 and p_2, from the plane $z = 1$ to $x = 1$, is not the polynomial Bézier curve approximating the projected control points p'_0,

p'_1 and p'_2, but, rather, the rational Bézier curve approximating them, with the weight of p'_i equal to the x-value of p_i, $0 \le i \le 2$.

What about snapshot transforming an arbitrary rational Bézier curve, rather than a polynomial one? Exactly the same principle applies. The result is a rational Bézier curve approximating the transformed control points, with *new* weights.

Figure 18.13 explains how the new weights are calculated in the simple case of transforming from $z = 1$ to $x = 1$. If the control point p on the $z = 1$ plane is $[x\ y\ 1]^T$ with weight w, then its lifted control point P in $\mathbb{R}^3$ is $[wx\ wy\ w]^T$. The projection of p, as that of P, on $x = 1$ is the point $p' = [1\ \ y/x\ \ 1/x]^T$. Therefore, the weight of p' so that its lifting coincides with P is wx.

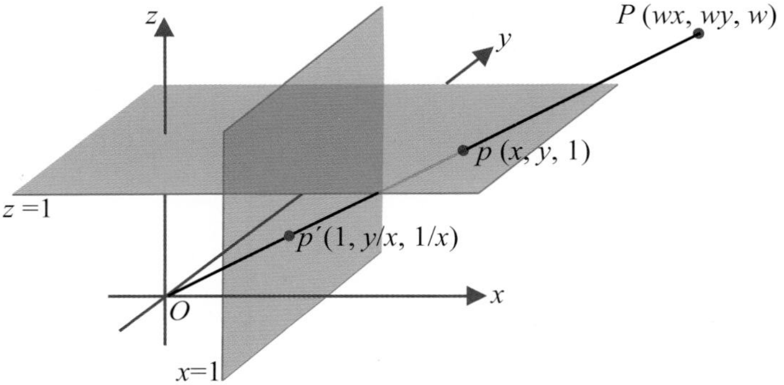

Figure 18.13: Both the control point p on $z = 1$, with weight w, and its lifting P project to the point p' on $x = 1$.

Suppose, now, that c is the rational Bézier curve approximating $n + 1$ control points $[x_i\ y_i\ 1]^T$ on the plane $z = 1$, with weights w_i, $0 \le i \le n$. It follows that the snapshot transformation of c from $z = 1$ to $x = 1$ is the rational Bézier curve on the transformed control points $p'_i = [1\ \ y_i/x_i\ \ 1/x_i]^T$, with weights $w_i x_i$, $0 \le i \le n$.

Example 18.3. Compute the snapshot transformation of the rational Bézier curve c on the $z = 1$ plane with control points $[1\ -1\ 1]^T$, $[2\ 1\ 1]^T$ and $[4\ 3\ 1]^T$, and respective weights 0.5, 2.0 and 1.0, onto the $x = 1$ plane.

Answer: From the preceding discussion the transformation of c onto the $x = 1$ plane is the rational Bézier curve with control points $[1\ -1\ 1]^T$, $[1\ 0.5\ 0.5]^T$ and $[1\ 0.75\ 0.25]^T$, and respective weights 0.5, 4 and 4.

Example 18.4. A polynomial Bézier curve c is drawn in 3-space with control points at $[2\ 2\ 5]^T$, $[3\ 1\ 4]^T$ and $[0\ 4\ 2]^T$. What is its projection on the $z = 1$ plane?

Answer: The projection of c on $z = 1$ is the rational Bézier curve with control points at $[0.4\ 0.4\ 1]^T$, $[0.75\ 0.25\ 1]^T$ and $[0\ 2\ 1]^T$, and respective weights 5, 4 and 2.

Exercise 18.17. Compute the snapshot transformation of the rational Bézier curve c on the $z = 1$ plane with control points at $[2\ 2\ 1]^T$, $[1\ 4\ 1]^T$ and $[5\ 1\ 1]^T$, and respective weights 1.0, 4.0 and 0.5, onto the $y = 1$ plane.

Exercise 18.18. A polynomial Bézier curve c is drawn in 3-space with control points at $[4\ 1\ 5]^T$, $[2\ 2\ 3]^T$ and $[1\ 2\ 2]^T$. What is its projection on the $z = 1$ plane?

Experiment 18.7. Fire up `turnFilm2.cpp` once again. Pressing space at any time draws, instead of the green dashed curve, a blue dashed *rational* Bézier curve approximating the projected control points on the current plane of the film. The control point weights of the blue dashed curve are computed according to the strategy just described. Voilà! The blue dashed rational curve and the red dashed projection are inseparable. **End**

Exercise 18.19. (**Programming**) Verify that `turnFilm2.cpp` does as just claimed. In particular, check that the weights of the projected control points used to draw the blue dashed curve are correctly calculated as the new weights following a snapshot transformation.
Hint: The code is a little tricky as the projection of the control points on the turning film are computed "by hand", via the routine `computeProjectedControlPoints()`. What this routine does, in fact, is simulate the rotation of the film clockwise about the y-axis by computing the projection of the control points on the plane $z = 1$, after rotating the control points *counter-clockwise* about the y-axis (but leaving the film fixed). For this reason, the first viewing transformation, which is used to turn the film, is not applied to the projected control points, but rather a second one keeping the camera pointed at the $z = 1$ plane.

The routine `computeWeightedProjectedControlPoints()` assigns the new weights to the projected control points that then are used to draw the blue dashed curve.

Let's pause a moment to take stock. A snapshot transformation of a polynomial Bézier curve may not even be a polynomial Bézier curve. However, that of a rational Bézier curve is not only a rational Bézier curve, but the control points of the transformed curve are transformations of the original control points. Moreover, their new weights can be computed from values of the original weights and original control points. We call this property the *snapshot invariance* of rational Bézier curves.

In fact, rational Bézier curves are *projectively invariant*:

Proposition 18.2. *Applying a projective transformation of $\mathbb{P}^2$ to a rational Bézier curve in $\mathbb{R}^2$ gives another rational Bézier curve in $\mathbb{R}^2$ whose control* *points are the transformations of the original control points, and with altered weights which can be computed from the values of the original weights and original control points.*

Proof. Again, do keep in mind that a projective transformation acts on a curve in $\mathbb{R}^2$ by transforming its lifting (which belongs to $\mathbb{P}^2$). The proof of the proposition, though not difficult, involves a fair amount of algebraic manipulation which we'll not get into. The mathematically inclined reader should try to prove it for herself. Otherwise, refer to Piegl and Tiller [111]. □

Projective invariance versus only affine: make that 3-0 in favor of the rationals then!

In Exercise 18.15 we deduced the affine invariance of rational Bézier curves as a consequence of the partition-of-unity property of their blending functions. It's also a consequence of the preceding proposition because affine transformations are a subclass of the projective.

Exercise 18.20. Prove that an affine transformation of a rational Bézier curve in $\mathbb{R}^2$ does not alter its weights.

Exercise 18.21. Find the flaw in the following argument:

Rational Bézier curves are projectively invariant. Polynomial Bézier curves are special cases of rational Bézier curves with weights all equal. Therefore, polynomial Bézier curves are projectively invariant as well.

Recall that a projective transformation can map a regular point to a point at infinity (and vice versa) with respect to a particular embedding of $\mathbb{R}^2$ in $\mathbb{P}^2$. In fact, one may even contemplate control points of a Bézier curve at infinity! Here's an interesting application to obtain a very familiar curve as a rational Bézier curve with one control point indeed at infinity:

Exercise 18.22. Embed $\mathbb{R}^2$ in $\mathbb{P}^2$ as the plane $z = 1$. Prove that the polynomial Bézier curve in $\mathbb{P}^2$ with control points $[1\ 0\ 1]^T$, $[0\ 1\ 0]^T$ and $[-1\ 0\ 1]^T$ projects to the upper-half of the unit circle centered at the origin of $\mathbb{R}^2$. Observe that the middle control point is at infinity with respect to $z = 1$.

18.3.6 Rational Bézier Curves in the Real World

Except for Exercise 18.11, our discussion thus far in this section has been exclusively of rational Bézier curves on the plane. Extension to curves in 3-space, however, is straightforward. For example, Equation (18.13)

$$c(u) = \left[\frac{\sum_{i=0}^n B_{i,n}(u) w_i x_i}{\sum_{i=0}^n B_{i,n}(u) w_i} \quad \frac{\sum_{i=0}^n B_{i,n}(u) w_i y_i}{\sum_{i=0}^n B_{i,n}(u) w_i} \quad \frac{\sum_{i=0}^n B_{i,n}(u) w_i z_i}{\sum_{i=0}^n B_{i,n}(u) w_i} \right]^T$$

$0 \le u \le 1$, of a rational Bézier curve in $\mathbb{R}^3$ approximating the control points $[x_i\ y_i\ z_i]^T$, with respective weights w_i, $0 \le i \le n$, which the reader was asked to deduce in Exercise 18.11, adds the expected z-component to its 2D counterpart (18.11)

$$c(u) = \left[\frac{\sum_{i=0}^{n} B_{i,n}(u) w_i x_i}{\sum_{i=0}^{n} B_{i,n}(u) w_i} \quad \frac{\sum_{i=0}^{n} B_{i,n}(u) w_i y_i}{\sum_{i=0}^{n} B_{i,n}(u) w_i} \right]^T \qquad (0 \le u \le 1)$$

Exercise 18.23. Show that the projection of a rational 3D Bézier curve on any plane is a rational 2D Bézier curve.

Exercise 18.24. (**Programming**) Write a 3D version of `rationalBezierCurve3.cpp` of Experiment 18.5 with control points which can be moved in 3-space, and with changeable weights. Add functionality to rotate the viewpoint.

18.3.7 Rational Bézier Surfaces

With the spadework for rationalization already mostly done, the next step up to rational Bézier surfaces is not going to be much more than a matter of jotting down equations one by one with an eye still on curves.

Recall from Section 15.2 the equation

$$s(u, v) = \sum_{i=0}^{n} \sum_{j=0}^{m} B_{i,n}(u) B_{j,m}(v) p_{i,j} \qquad (0 \le u \le 1,\ 0 \le v \le 1) \qquad (18.16)$$

of a polynomial Bézier surface in 3-space with control points $p_{i,j}$, for $0 \le i \le n$ and $0 \le j \le m$, and the process of "sweeping by a Bézier curve" by which it was derived.

Following a similar process, one can write the equation of a *rational Bézier surface* specified by control points $p_{i,j}$ with respective weights $w_{i,j}$, $0 \le i \le n$ and $0 \le j \le m$:

$$s(u, v) = \sum_{i=0}^{n} \sum_{j=0}^{m} \frac{B_{i,n}(u) B_{j,m}(v)\, w_i}{\sum_{i=0}^{n} \sum_{j=0}^{m} B_{i,n}(u) B_{j,m}(v)\, w_i} p_{i,j} \quad (0 \le u \le 1,\ 0 \le v \le 1)$$

(18.17)

From (18.16) to (18.17) the change is simply in the blending functions, now rational, rather than polynomial. We ask the reader next to determine equations for a rational Bézier surface in forms analogous to those that we have already deduced for curves.

Exercise 18.25. Find equations for rational Bézier surfaces in $\mathbb{R}^3$ analogous to Equations (18.8)-(18.12) for rational Bézier curves.

It should come as no surprise to the reader that rational Bézier surfaces are projectively invariant and, therefore, affine and snapshot invariant as well. Moreover, they can represent exactly parts of paraboloids, ellipsoids and hyperboloids, and other quadric surfaces. From a designer's perspective, a control point's weight, as in the case of a rational Bézier curve, is an additional dial to turn up or down its attractive pull on the surface.

All the advantages of rational Bézier curves over polynomial propagate, therefore, to rational Bézier surfaces.

Drawing Rational Bézier Surfaces

As expected, the main change in drawing polynomial versus rational Bézier surfaces, as we saw in Section 18.3.2 going from polynomial to rational Bézier curves, is replacing "`VERTEX_3`" with "`VERTEX_4`" to include the extra weight parameter w, in addition to x, y and z, per control point.

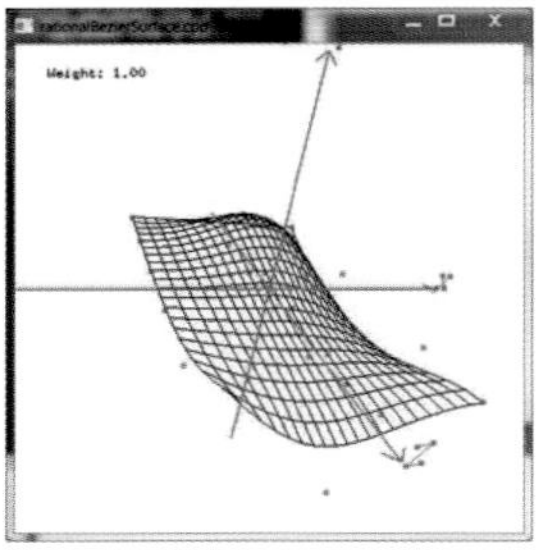

Figure 18.14: Screenshot of `rationalBezier-Surface.cpp`.

Experiment 18.8. Run `rationalBezierSurface.cpp`, based on `bezier-Surface.cpp`, which draws a rational Bézier surface with the functionality that the location and weight of each control point can be changed. Press the space and tab keys to select a control point. Use the arrow and page up/down keys to translate the selected control point. Press '</>' to change its weight. Press delete to reset. The 'x/X', 'y/Y' and 'z/Z' keys turn the viewpoint. Figure 18.14 is a screenshot.

Mark the use of `glMap2f(GL_MAP2_VERTEX_4, ...)`, as also of `glEnable-(GL_MAP2_VERTEX_4)`. The 2's in the syntax are for a surface. **End**

18.3.8 The 'R' in NURBS

With all the groundwork laid in rational Bézier theory, putting the 'R' now into NURBS (Non-Uniform Rational B-Splines) is going to be rather anti-climactic.

Recall Equation (16.34) of the mth order B-spline curve c approximating $r-m+1$ control points $p_0, p_1, \ldots, p_{r-m}$ over the knot vector $\{t_0, t_1, \ldots, t_r\}$:

$$c(u) = \sum_{i=0}^{r-m} N_{i,m}(u) p_i \qquad (t_{m-1} \le u \le t_{r-m+1}) \tag{18.18}$$

where the blending function of the ith control point is the mth order B-spline function $N_{i,m}$, $0 \le i \le r-m$.

Following a development *exactly parallel* to that for rational Bézier curves, one can write for a NURBS curve an equation analogous to (18.15), which expresses a rational Bézier curve as a weighted sum of its control points, the weights being rational blending functions. In fact, the equation for a NURBS curve approximating $r-m+1$ control points p_i, with weights w_i, $0 \le i \le r-m$, over the knot vector $\{t_0, t_1, \ldots, t_r\}$, is

$$c(u) = \sum_{i=0}^{r-m} \frac{N_{i,m}(u) w_i}{\sum_{i=0}^{r-m} N_{i,m}(u) w_i} p_i \qquad (t_{m-1} \le u \le t_{r-m+1}) \tag{18.19}$$

where, of course, the blending functions are now ratios of terms composed of B-splines.

We'll leave finding the equation of a NURBS surface to the reader in the following exercise.

Exercise 18.26. Recall Equation (16.36) of a B-spline surface approximating an $(n+1) \times (n'+1)$ array of control points over a pair of non-uniform knot vectors. Rewrite it for a NURBS surface, taking into account control point weights. What is the blending function of the control point $p_{i,j}$, $0 \leq i \leq n$, $0 \leq j \leq n'$?

Exercise 18.27. Prove that NURBS curves and surfaces are affinely invariant. (In fact, they are projectively invariant.)
Hint: Think partition of unity.

Drawing NURBS Curves and Surfaces

The reader may wish to review Section 16.4 where the GLU NURBS interface is explained and used to draw polynomial B-spline curves and surfaces. With the practicalities of the transition from drawing polynomial Bézier primitives to rational ones already learned from earlier in this chapter, those for drawing rational NURBS primitives are straightforward and left to the reader. The following two exercises ask her to apply the NURBS interface to draw a rational curve and a rational surface, respectively.

Exercise 18.28. (Programming) Modify `cubicSplineCurve1.cpp` of Experiment 16.6 to draw a cubic NURBS curve so that the weight of the selected control point can be changed, in addition to all the original functionality. You must use the call `gluNurbsCurve`(GL_MAP1_VERTEX_4, ...).

Exercise 18.29. (Programming) Modify `bicubicSplineSurface.cpp` of Experiment 16.15 to draw a NURBS surface.

18.4 Summary, Notes and More Reading

In this chapter we studied three important applications of projective spaces to CG: (a) the projection transformation to convert a viewing volume into the canonical box in the synthetic-camera pipeline, (b) shadow mapping based on applying the math from (a), and, finally, (c) rational Bézier and B-spline theory. The first demonstrates the practical importance of projective geometry in the CG rendering pipeline. The second of shadow mapping is a particularly useful addition to our repertoire of authentic rendering techniques. The final application is important for a deeper understanding of design because rational primitives, in particular NURBS, are the de facto standard in CAD.

There are several excellent sources for the reader to follow up on both rational Bézier and NURBS primitives. A few are the books by Buss [22], Farin [45, 46], Mortenson [94], Piegl & Tiller [111] and Rogers [116].

Part X

The Time is Pipe

CHAPTER 19

Fixed-Functionality Pipelines

At the end of Chapter 4 about moving and shaping objects and manipulating the OpenGL camera, we said that it was like having got our driver's license. It's time now to look at the engine under the hood to understand the whole process, from ignition to motion. So in this chapter we are going to study graphics rendering pipelines – processes that transform a user-defined scene into an image on a raster display.

The particular topic of this chapter, though, is fixed-functionality pipelines where, once the programmer has specified the scene, she has little further say in the rendering process. In this category falls the first-generation synthetic-camera pipeline, which, in fact, our OpenGL programs thus far have all invoked. The basic ray tracing pipeline, based on a global illumination model – versus a local in the case of the synthetic camera – is fixed-functionality as well, as is radiosity, another global illumination model often implemented in tandem with ray tracing.

We begin in Section 19.1 with the fixed-functionality synthetic-camera pipeline. This is the pipeline that the first generation of OpenGL (versions 1.x) implements and the one used so far in this book. Our description of this particular pipeline began, in fact, with the shoot-and-print analogy of Chapter 2. We'll put all the pieces together now to get a fairly complete idea of its implementation.

Section 19.2 introduces ray tracing, the most popular global illumination model and its rendering pipeline. As its name suggests, ray tracing is based upon following individual light rays through a scene. It is a near photorealistic way of rendering, but computationally so expensive as to be almost never used in real-time applications such as games. For off-line applications, though, e.g., movies, where computational resources and time are not major constraints, ray tracing is far more authentic an alternative to synthetic-camera-based rendering.

Radiosity, another global lighting model and the topic of Section 19.3, is frequently implemented together with ray tracing, as the two have complementary models of light transport.

We conclude in Section 19.4.

19.1 Synthetic-Camera Pipeline

We're in a position now to put together, with pieces learned so far in this book, a complete synthetic-camera rendering pipeline, though without a lot of the bells and whistles that practical implementations come with. Let's begin with a review of our understanding of how a piece of OpenGL code turns into a picture on the monitor.

First, following the programmer's definition of a scene (points, lines, triangles, ...), the modelview transformation, as described, particularly in Sections 4.1-4.2, is applied to the scene. Section 5.4 explains how to compute the matrix corresponding to any given modelview transformation – the transformation being applied to a vertex by multiplying it by the transformation's matrix from the left. Next, the scene is "captured on film" by applying the shoot-and-print paradigm of Section 2.2, where primitives are projected to the front face of the viewing box or frustum (shoot) and then scaled to fit the OpenGL window (print).

In Section 18.1 we saw that the shoot process itself is implemented in two stages. First comes a projection transformation mapping the viewing volume to the canonical viewing box, which itself consists of multiplying the vertices in homogeneous $xyzw$-coordinates by the projection matrix, and then, possibly, a perspective division step to divide out the w-value. The next stage is a parallel projection to the canonical box's back plane of the parts of primitives *inside* it, because only these are rendered to the screen.

Implicit in the second stage, then, is the clipping of primitives to within the canonical box. This can be accomplished for 1D and 2D primitives, respectively, with use of the Cohen-Sutherland line clipper from Section 14.1 (particularly its extension to 3-space suggested in Exercise 14.5) and the Sutherland-Hodgeman polygon clipper from Section 14.2 (Exercise 14.11 suggests the 3-space version). Note that clipping 0D primitives, i.e., points, is a trivial matter of tossing those whose coordinates place them outside the canonical box.

The last print step, where the back face of the canonical box is scaled to the OpenGL window, involves choosing and coloring a set of pixels in the latter for each primitive on the former, which, of course, is rasterization. Again, 0D primitives, or points, are rasterized in obvious manner, while 1D and 2D primitives can be processed with the use, respectively, of the line and polygon rasterizers from Sections 14.3 and 14.4. Just prior to rasterization, depth testing may be invoked to decide which part, if any, of a primitive is obscured by others, in which case this part is not allowed to colorize its corresponding pixels.

That's it. This is enough to give us a skeletal code-to-image pipeline. Texture, lighting and other capabilities can come in later. Let's keep it simple to start with.

19.1.1 Pipeline: Preliminary Version

Time now for specifics. Let's follow a vertex, given in homogeneous coordinates, down a simple pipeline which follows the strategy outlined above (description of each stage is just below its line to the right; see also the notes below the pipeline):

Synthetic-camera Rendering Pipeline (Preliminary Version)

1.	$[x\ y\ z\ 1]^T$	$\longrightarrow$	$[x^M\ y^M\ z^M\ 1]^T$ *Modelview transformation = multiplication by the modelview matrix.*
2.		$\longrightarrow$	$[x^{PM}\ \ y^{PM}\ \ z^{PM}\ \ w^{PM}]^T$ *Multiplication by the projection matrix.*
3.		$\longrightarrow$	$\left[\frac{x^{PM}}{w^{PM}}\ \ \frac{y^{PM}}{w^{PM}}\ \ \frac{z^{PM}}{w^{PM}}\right]^T$ *Perspective division.*
4.		$\longrightarrow$	$\left[\frac{x^{PM}}{w^{PM}}\ \ \frac{y^{PM}}{w^{PM}}\ \ \frac{z^{PM}}{w^{PM}}\right]^T$ *Clipping to the canonical box.*
5.		$\longrightarrow$	$\left[\frac{x^{PM}}{w^{PM}}\ \ \frac{y^{PM}}{w^{PM}}\right]^T$ *Projection to the back of the canonical box (z-values possibly retained for depth testing).*
6.		$\longrightarrow$	$[i\ j]^T$ *Rasterization.*

Notes:

(i) Superscripts indicate the transforming matrix, e.g., $[x^M\ y^M\ z^M\ 1]^T = M[x\ y\ z\ 1]^T$. Notation: M = modelview, P = projection and PM their product.

(ii) Multiplication by the projection matrix P (Stage 2) + perspective division (Stage 3) = projection transformation of Section 18.1, which transforms the viewing volume into the canonical viewing box.

(iii) Perspective division is a non-operation in the case of a `glOrtho()`-defined viewing box as w^{PM} are all 1.

(iv) All the x-, y-, z- and w-values, with or without superscripts, are floating points up to and including Stage 5. It's only at the final Stage 6 that the vertex "jumps" from $\mathbb{R}^2$ (where lies the back face of the canonical box) to a discrete $m \times n$ raster (in the frame buffer), in other words, from being an $[x\ y]^T$ *floating point* tuple to an $[i\ j]^T$ *integer* tuple.

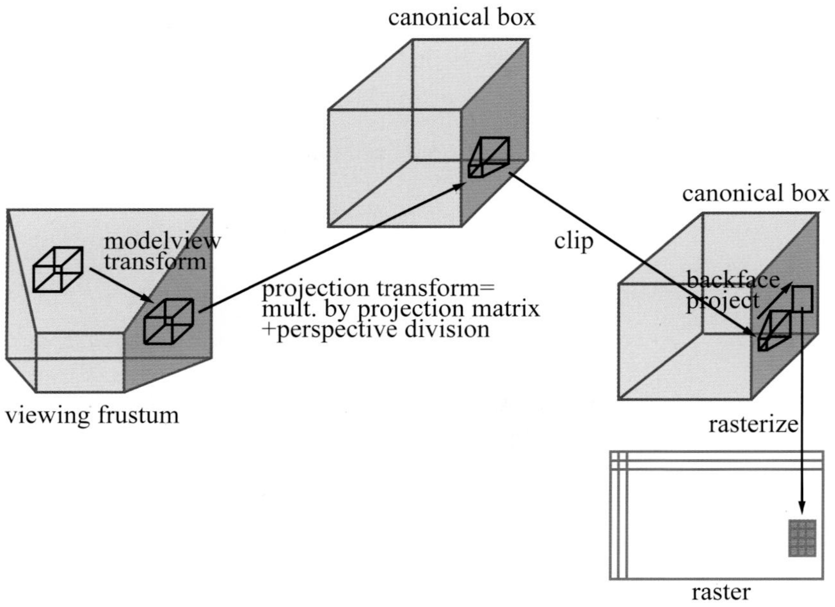

Figure 19.1: Synthetic-camera rendering pipeline (the dashed part of the small drawn box is outside the viewing frustum; the corresponding transformed part is outside the canonical box, so clipped).

Figure 19.1 is a pictorial view, implicit in which is an additional *primitive assembly* step when geometric data is used to connect points into line, triangles, etc. There are two significant technicalities, though, to deal with before the pipeline can be put into production. First is the problem of handling zero w-values in the perspective division of Stage 3; the next is the issue of so-called perspective correction needed to be taken into account when colorizing a raster primitive in Stage 6. The next two subsections discuss these two technicalities, respectively. *Both are fairly mathematical, so if you are not so inclined skip to Section 19.1.4, taking the revised pipeline there for granted.*

19.1.2 Perspective Division by Zero

You may need to review Section 18.1 as our discussion here is a follow-on

of the account in that section of the projection transformation.
Perspective division in Stage 3 of the pipeline could involve division by zero. Prima facie, however, this appears to be not much of a problem because the canonical box, into which the viewing volume is transformed by multiplication by the projection matrix, consists only of regular points (with respect to $w = 1$). Therefore, a point that's mapped to a point at infinity – with w-value 0 and, hence, outside the canonical box – never belonged to the viewing volume in the first place. So it seems all we have to do is add a filter before the perspective division stage to simply eject points with w-value equal to 0. Unfortunately, the problem is a bit more complicated, as the following experiment indicates.

Experiment 19.1. Replace the box `glutWireCube(5.0)` of `box.cpp` with the line segment

```
glBegin(GL_LINES);
   glVertex3f(1.0, 0.0, -10.0);
   glVertex3f(1.0, 0.0, 0.0);
glEnd();
```

and delete the `glTranslatef(0.0, 0.0, -15.0)` statement. You see a short segment, the clipped part of the defined line segment, whose first endpoint $[1\ 0\ -10]^T$ is inside the viewing frustum defined by the program's projection statement `glFrustum(-5.0, 5.0, -5.0, 5.0, 5.0, 100.0)`, while the second $[1\ 0\ 0]^T$ is outside (as is easily checked). Figure 19.2 is a screenshot. **End**

Figure 19.2: Screenshot of Experiment 19.1.

Here's what's interesting though about the experiment – the second endpoint of the drawn segment is mapped to a point at infinity by multiplication by OpenGL's projection matrix! This is easy to verify. Simply take the dot product of $[0\ 0\ -1\ 0]$, which is the last row of the projection matrix corresponding to `glFrustum(-5.0, 5.0, -5.0, 5.0, 5.0, 100.0)` as given by Equation (18.4), and $[1\ 0\ 0\ 1]$, the homogeneous coordinates of the second endpoint, to find that the endpoint's transformed w-value is 0 (the other coordinate values are irrelevant).

The conclusion from the experiment is that even though vertices that map to infinity don't belong in the viewing volume, they may be corners of primitives which partially do. So we just can't toss them – we'll have to make sure that the primitives they belong to are handed off correctly to the clipper. This requires a little care. It's convenient at this time to climb a dimension down to 2D to visualize the right strategy. Example A.17 of Appendix A is perfect for this purpose.

Note: If you have not yet read Appendix A on projective spaces and transformations, then simply take the following transformation for granted. However, to understand how it was derived you will need to refer to the appendix.

The projective transformation h^M of $\mathbb{P}^2$ given by

$$M = \begin{bmatrix} -1/2 & 0 & 0 \\ 0 & -3/2 & 1 \\ 0 & -1/2 & 0 \end{bmatrix}$$

transforms the trapezoid q on the plane $z = 1$, aka $\mathbb{R}^2$, with vertices at

$$[-1\ 1]^T,\ [1\ 1]^T,\ [2\ 2]^T \text{ and } [-2\ 2]^T$$

to the rectangle q' with vertices at

$$[-1\ 1]^T,\ [1\ 1]^T,\ [1\ 2]^T \text{ and } [-1\ 2]^T$$

precisely the 2D analogue of transforming a frustum to a box – instead of a frustum we now have a trapezoid and instead of a box a rectangle. See Figure 19.3.

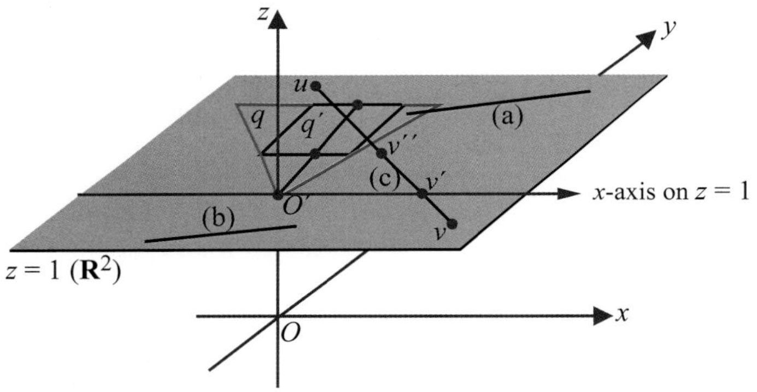

Figure 19.3: The synthetic camera in Flatland: the point camera is at O', the "viewing trapezoid" is q, the "canonical rectangle" q'.

Note: Keep in mind that h^M maps the plane point $[x\ y]^T$ to the one on the plane $z = 1$ with homogeneous coordinates $M[x\ y\ 1]^T$. For example, to determine $h^M([-1\ 1]^T)$, we compute $M[-1\ 1\ 1]^T = [1/2\ \ -1/2\ \ -1/2]^T$. Dividing the latter through by its z-value, and then dropping the z-value, we see that $h^M([-1\ 1]^T) = [-1\ 1]^T$.

Let's see how to deal with a line segment primitive, say uv, on $\mathbb{R}^2$, subject to transformation by h^M, and, subsequently, clipping to the "canonical" rectangle q'. Exercise A.28 of Appendix A tells us the nature of $h^M(uv)$. It is either a segment (if no point of uv maps to a point at infinity) or two semi-infinite segments (if an interior point maps to infinity) or one semi-infinite segment (if one endpoint maps to infinity) or empty (if both endpoints map to infinity).

It's checked easily that points of the plane mapped by h^M to points at infinity are precisely those on the x-axis. Here, then, is how to clip $h^M(uv)$ to the canonical rectangle – assumed available is a Cohen-Sutherland line clipper for this rectangle with the additional ability to clip semi-infinite segments *a la* Exercise 14.4. The three cases that may arise are listed below and for each an example segment correspondingly labeled is seen in Figure 19.3.

(a) Both u and v are above the x-axis (i.e., with positive y-values):

Pass the transformed segment $h^M(u)h^M(v)$ to the clipper. Note that the transformed segment itself is not drawn in the figure.

(b) Both u and v are below or on the x-axis:

Pre-clip uv altogether as it doesn't intersect the viewing trapezoid.

(c) One endpoint, say u, is above the x-axis and the other endpoint v is on or below:

Determine the point v' of intersection of uv with the x-axis. Pass the image $h^M(uv')$, which is a semi-infinite segment, to the clipper (even if v is below the x-axis, the image of $v'v$, another semi-infinite segment, cannot intersect the trapezoid and need not be transmitted).

Note: If the clipper has been extended to handle semi-infinite segments in the manner suggested in Exercise 14.4, then it will need as input the finite endpoint of $h^M(uv')$, as well as the direction in which it is infinite. The finite endpoint is, of course, $h^M(u)$, while the direction it is infinite is toward $h^M(v'')$, where v'' is any point between u and v', e.g., the midpoint. Mind that $h^M(v')$, being a point at infinity, cannot decide the direction itself.

Exercise 19.1. Determine what is transmitted to the extended clipper in the 2D scenario above in the following cases.

(1) $u = [0\ 2]^T$ and $v = [3\ 1]^T$

(2) $u = [0\ -2]^T$ and $v = [3\ 0]^T$

(3) $u = [0\ 2]^T$ and $v = [3\ -1]^T$

Part answer:

(3) Here, u is above and v below the x-axis, so we are in case (c) above. The point where uv intersects the x-axis is $v' = [2\ 0]^T$. Take v'' to be the midpoint $[1\ 1]^T$ of uv'.

Therefore, passed to the extended clipper is a semi-infinite segment which has a finite end at the point on $z = 1$ with homogeneous coordinates

$$h^M(u) = \begin{bmatrix} -1/2 & 0 & 0 \\ 0 & -3/2 & 1 \\ 0 & -1/2 & 0 \end{bmatrix} [0\ 2\ 1]^T = [0\ -2\ -2]^T$$

and is infinite toward the point on $z = 1$ with homogeneous coordinates

$$h^M(v'') = \begin{bmatrix} -1/2 & 0 & 0 \\ 0 & -3/2 & 1 \\ 0 & -1/2 & 0 \end{bmatrix} [1\ 1\ 1]^T = [-\frac{1}{2}\ -\frac{1}{2}\ -\frac{1}{2}]^T$$

One sees, therefore, that the segment passed to the clipper has a finite end at $[0\ 1]^T$ and is infinite toward $[1\ 1]^T$.

We're going to leave it at this, hoping the reader is convinced that the approach just described to handle vertices, otherwise leading to perspective division by zero, can be implemented, even in 3D, by appropriately enhancing Stage 3 of the six-stage synthetic-camera rendering pipeline.

19.1.3 Rasterization with Perspectively Correct Interpolation

Rasterization is more than a matter of plugging in Bresenham's rasterizer for lines and the scan-based rasterizer for polygons (Sections 14.3 and 14.4, respectively). The reason is that both these rasterizing algorithms choose the pixels comprising a primitive but say nothing about how to color them.

However, coloring the pixels of a rasterized primitive seems merely a question of linearly interpolating the values specified at its vertices through its interior. It really ought to be since we made such a fuss in Chapter 7 about how nice are points, line segments and triangles – the fundamental primitives of OpenGL – because values at their vertices can, in fact, be unambiguously interpolated through their interiors! Well, it is, pretty much, except ...

Figure 19.4: Screenshot of `perspective-Correction.cpp`.

Experiment 19.2. Run `perspectiveCorrection.cpp`. You see a thick straight line segment which starts at a red vertex at its left and ends at a green one at its right. Also seen is a big point just above the line, which can be slid along it by pressing the left/right arrow keys. The point's color can be changed, as well, between red and green by pressing the up/down arrow keys. Figure 19.4 is a screenshot.

The color-tuple of the segment's left vertex, as you can verify in the code, is $(1.0, 0.0, 0.0)$, a pure red, while that of the right is $(0.0, 1.0, 0.0)$, a pure green. As expected by interpolation, therefore, there is a color transition from red at the left end of the segment to green at its right.

The number at the topmost right of the display indicates the fraction of the way the big movable point is from the left vertex of the segment to the right. The number below it indicates the fraction of the "way" its color is from red to green – precisely, if the value is u then the color of the point is $(1 - u, u, 0)$.

Initially, the point is at the left and a pure red; in other words, it is 0 distance from the left end, and its color 0 distance from red. Change both values to 0.5 – the color of the point does *not* match that of the segment below it any more. It seems, therefore, that the midpoint of the line is not colored $(0.5, 0.5, 0.0)$, which is the color of the point. Shouldn't it be so, though, by linear interpolation, as it is half-way between two end vertices colored $(1.0, 0.0, 0.0)$ and $(0.0, 1.0, 0.0)$, respectively? **End**

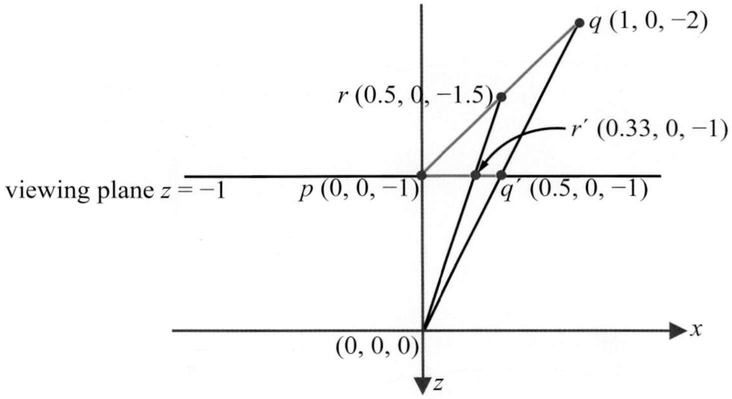

Figure 19.5: The line segment drawn in `perspectiveCorrection.cpp` is pq and its projection on the viewing face pq'.

The apparent conundrum of the preceding experiment is not hard to resolve. Figure 19.5, an xz-section of world space, shows what's happening. The line segment drawn in `perspectiveCorrection.cpp` is from $p = [0\ 0\ -1]^T$ to $q = [1\ 0\ -2]^T$, as specified in the `drawScene()` routine. The midpoint of pq is $r = [0.5\ \ 0\ \ -1.5]^T$. Moreover, the perspective projections of p, q and r on the viewing plane $z = -1$ are p itself, q' and r', respectively. The coordinates shown in the figure of q' and r' can be easily verified by properties of similar triangles.

One sees, then, that r', the projection of the midpoint of the segment pq, is *not* the midpoint of the projected segment pq', but rather approximately 0.66 $(= 0.33/0.5)$ of the way from its left end p. With this in mind, return to the program to set the color fraction of the movable point to 0.5 and its distance fraction to 0.66 – now you do see a match! If perspective projections preserved convex combinations, like linear transformations, then midpoints would map to midpoints, but, unfortunately, as we have just found out, they do not.

The conclusion, then, is that the colors at its endpoints *are* linearly interpolated along the user-specified line segment, which is a virtual object in world space; however, as perspective projection does not preserve convex combination, colors of the projected endpoints *are not* linearly interpolated through the segment drawn on screen.

We understand the issue now, so let's square it with our rasterization procedure by incorporating an additional *perspective correction* factor.

A point $p = [p_x\ p_y\ p_z]^T$ in the viewing frustum is mapped by projection transformation to the point $p' = [p'_x\ p'_y\ p'_z]^T$ in the canonical viewing box; parallel projection to the back face of the box then maps p' to $\overline{p} = [p'_x\ p'_y]^T$. See Figure 19.6(a).

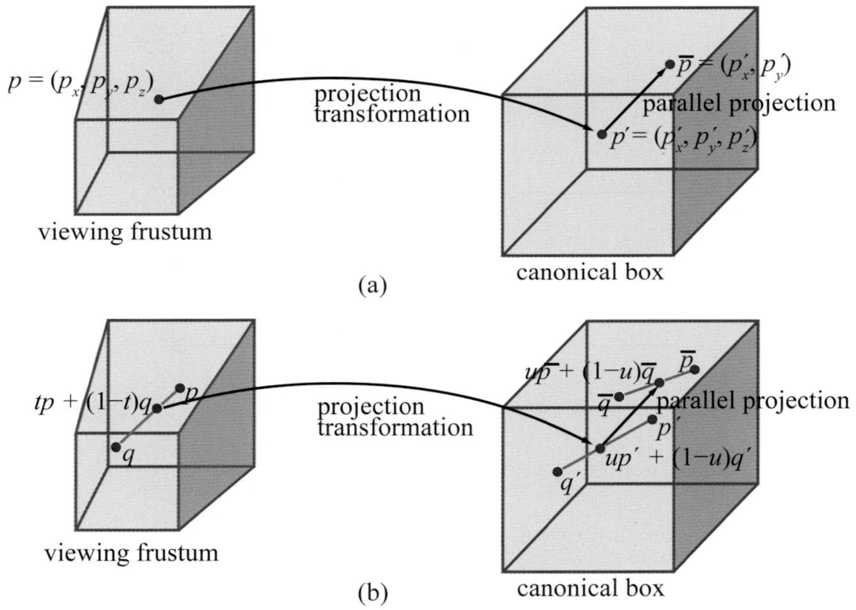

Figure 19.6: (a) A point is mapped by the projection transformation from a viewing frustum to the canonical viewing box, followed by parallel projection to the latter's back face. (b) Likewise for a line segment: the projection transformation does not preserve convex combinations, but parallel projection does.

Moreover, a point $tp + (1-t)q$ on the segment joining two points p and q in the viewing frustum maps to a point $up' + (1-u)q'$ on the segment joining their respective images p' and q', though, as we understand now, not necessarily does $u = t$. See Figure 19.6(b).

We want to find the function $u \to t$ that gives the pre-image $tp + (1-t)q$ of $up' + (1-u)q'$. This will serve our purpose of perspective correction, for we'll color the point $u\overline{p} + (1-u)\overline{q}$ – now identifying the box's back face with the raster – with the color values $tC(\overline{p}) + (1-t)C(\overline{q})$, instead of $uC(\overline{p})+(1-u)C(\overline{q})$ as in the case of uncorrected interpolation. $C(\overline{p}) = C(p)$ and $C(\overline{q}) = C(q)$ are, of course, the programmer-specified colors at p and q, respectively. Observe that correction is required only for the projection transformation, not the parallel projection to the back face of the canonical box, as the latter is a linear map preserving convex combinations.

Finding the function $u \to t$ is a matter of some calculation. Write $p = [p_x\ p_y\ p_z]^T$ in homogeneous coordinates as $[p\ 1]^T = [p_x\ p_y\ p_z\ 1]^T$ and q as $[q\ 1]^T = [q_x\ q_y\ q_z\ 1]^T$. Let P be the projection matrix. Denote the results of multiplying $[p\ 1]^T$ and $[q\ 1]^T$ by P as follows:

$$P[p\ 1]^T = [p_x^*\ p_y^*\ p_z^*\ \ -p_z]^T \qquad \text{and} \qquad P[q\ 1]^T = [q_x^*\ q_y^*\ q_z^*\ \ -q_z]^T \quad (19.1)$$

where the starred symbols are variables to be determined, while the two w-values on the RHS's follow because the last row of the projection matrix P is always $[0\ 0\ -1\ 0]$ (see Equation (18.4)). Applying perspective division, denoted D, next, we have

$$DP[p\ 1]^T = [-\frac{p_x^*}{p_z}\quad -\frac{p_y^*}{p_z}\quad -\frac{p_z^*}{p_z}\quad 1]^T = [p_x'\ p_y'\ p_z'\ 1]^T = [p'\ 1]^T$$

and

$$DP[q\ 1]^T = [-\frac{q_x^*}{q_z}\quad -\frac{q_y^*}{q_z}\quad -\frac{q_z^*}{q_z}\quad 1]^T = [q_x'\ q_y'\ q_z'\ 1]^T = [q'\ 1]^T$$

where the second equality in both equations above follows because DP, in fact, is the projection transformation mapping p to p' and q to q'. The preceding two equations imply that

$$p_x^* = -p_z p_x',\ p_y^* = -p_z p_y',\ p_z^* = -p_z p_z',\ q_x^* = -q_z q_x',\ q_y^* = -q_z q_y',\ q_z^* = -q_z q_z' \quad (19.2)$$

Consider, next, an interpolated point $tp + (1-t)q$ between p and q. Multiplying it by P:

$$\begin{aligned} &P\,[tp + (1-t)q \quad 1]^T \\ &= P\,(t\,[p\ 1]^T \ + \ (1-t)\,[q\ 1]^T\,) \\ &= t\,(P\,[p\ 1]^T) \ + \ (1-t)\,(P\,[q\ 1]^T) \\ &= t[p_x^*\ \ p_y^*\ \ p_z^*\ \ -p_z]^T \ + \ (1-t)[q_x^*\ \ q_y^*\ \ q_z^*\ \ -q_z]^T \ \text{(applying (19.1))} \\ &= [tp_x^* + (1-t)q_x^*\ \ tp_y^* + (1-t)q_y^*\ \ tp_z^* + (1-t)q_z^*\ \ -tp_z - (1-t)q_z]^T \end{aligned}$$

Applying D by dividing through by the w-value:

$$
\begin{aligned}
& DP\,[tp + (1-t)q \quad 1]^T \\
&= \left[-\frac{tp^*_x + (1-t)q^*_x}{tp_z + (1-t)q_z} \quad -\frac{tp^*_y + (1-t)q^*_y}{tp_z + (1-t)q_z} \quad -\frac{tp^*_z + (1-t)q^*_z}{tp_z + (1-t)q_z} \quad 1 \right]^T \\
&= \left[\frac{tp_z p'_x + (1-t)q_z q'_x}{tp_z + (1-t)q_z} \quad \frac{tp_z p'_y + (1-t)q_z q'_y}{tp_z + (1-t)q_z} \quad \frac{tp_z p'_z + (1-t)q_z q'_z}{tp_z + (1-t)q_z} \quad 1 \right]^T \\
&\quad \text{(using (19.2))} \\
&= \frac{tp_z}{tp_z + (1-t)q_z}\,[p'_x \; p'_y \; p'_z \; 1]^T + \frac{(1-t)q_z}{tp_z + (1-t)q_z}\,[q'_x \; q'_y \; q'_z \; 1]^T \\
&= \frac{tp_z}{tp_z + (1-t)q_z}\,[p' \; 1]^T + \frac{(1-t)q_z}{tp_z + (1-t)q_z}\,[q' \; 1]^T \\
&= u\,[p' \; 1]^T + (1-u)\,[q' \; 1]^T
\end{aligned}
$$

where

$$u = \frac{tp_z}{tp_z + (1-t)q_z}$$

Inverting the preceding relationship gives the desired function $u \to t$:

$$
\begin{aligned}
t &= \frac{uq_z}{(1-u)p_z + uq_z} \\
&= \frac{q_z}{(\frac{1}{u} - 1)p_z + q_z}, \text{ if } u > 0\,; \qquad 0, \text{ if } u = 0
\end{aligned}
\tag{19.3}
$$

Whew! But now we know exactly what to do: Referring back to Figure 19.6(b), we'll color the point $u\overline{p} + (1-u)\overline{q}$ with the color values $tC(\overline{p})+(1-t)C(\overline{q})$, instead of $uC(\overline{p})+(1-u)C(\overline{q})$ as in the case of uncorrected interpolation, where t is given by the formula 19.3. This process is called *perspectively correct interpolation* or *linear interpolation with perspective correction*.

Here, then, is how to apply perspectively correct interpolation in coloring pixels along a line segment. Say the rasterization $R(S)$ of a line segment S joining the points $p = [p_x \; p_y \; p_z]^T$ and $q = [q_x \; q_y \; q_z]^T$ in the viewing frustum consists of $N+1$ pixels in the raster, as depicted in Figure 19.7.

The end pixel of $R(S)$ corresponding to p is (i_1, j_1) and that to q is (i_2, j_2). Precisely, (i_1, j_1) is obtained from mapping p to $\overline{p}$ on the back face of the canonical box by projection transformation and parallel projection, followed by mapping $\overline{p}$ to a point on the raster by the scaling transformation matching the back face of the canonical box to the raster and, finally, followed by a rounding to integer coordinates. Likewise, (i_2, j_2) is obtained from $\overline{q}$. Suppose, as well, that $R(S)$ makes an angle of at most 45° with the positive i-axis – other dispositions of $R(S)$ can be handled by symmetry – so that $i_2 = i_1 + N$, which means that each successive pixel of $R(S)$ from left to right has one higher i-value.

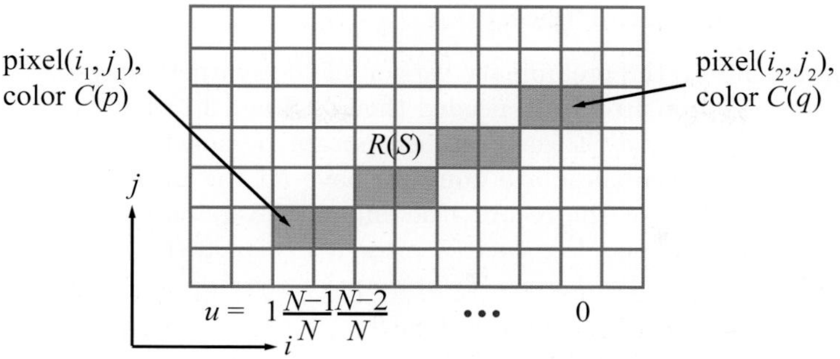

Figure 19.7: The rasterization $R(S)$ of a line segment S consists of $N+1$ pixels, each corresponding to a particular u-value (a few u-values are shown vertically below the corresponding pixel).

Each of the $N+1$ pixels of $R(S)$, counting from the left, corresponds successively to a point $u\overline{p} + (1-u)\overline{q}$ of $\overline{pq}$, where $u = 1, \frac{N-1}{N}, \frac{N-2}{N}, \ldots, 0$. The first few u-values are indicated at the bottom of a pixel's column in the figure.

The color tuples $C(p)$ and $C(q)$ of the two end pixels are, of course, the programmer-specified colors of the corresponding end vertices. It remains to color the in-between pixels. The pixel next to the leftmost corresponds to $u = \frac{N-1}{N}$, therefore, in turn, by Equation (19.3), to the perspectively correct

$$t = \frac{q_z}{(\frac{1}{\frac{N-1}{N}} - 1)p_z + q_z} = \frac{q_z}{\frac{p_z}{N-1} + q_z}$$

In other words, that pixel corresponds to the point $tp + (1-t)q$ of S, where t is given by the preceding equation. Consequently, the color to apply is $tC(p) + (1-t)C(q)$. Likewise, the color to apply to the next pixel is $tC(p) + (1-t)C(q)$, after updating t to

$$t = \frac{q_z}{\frac{2p_z}{N-2} + q_z}$$

by setting $u = \frac{N-2}{N}$ in (19.3). The procedure of decrementing u by $\frac{1}{N}$, updating t and applying the interpolated colors $tC(p) + (1-t)C(q)$ to the next pixel is repeated until the pixel just before the rightmost is colored, which, of course, completes the coloring of $R(S)$. This procedure can be integrated into Bresenham's line rasterizer: simultaneously picking the pixels along a line segment *and* coloring them with perspective correction.

We'll leave the reader to convince herself that, going from 1D to 2D, a similar perspective correction can be incorporated into triangle and polygon rasterization.

Remark 19.1. Not only color values, but other numerical data defined per vertex, e.g., normals, can be linearly interpolated with perspective correction as well.

19.1.4 Revised Pipeline

Enhancements to the preliminary version of the synthetic-camera rendering pipeline in Section 19.1.1 are needed then to Stage 3 to handle perspective division by zero and Stage 6 to incorporate perspective correction into rasterization. Once these are done we have all the pieces in place to go into production. For the record, the enhanced version (with additions in bold) is shown below. Figure 19.8 after it is a pictorial summary – note the additional primitive assembly step, prior to clipping, when points are assembled into lines, triangles, and such.

Synthetic-camera Rendering Pipeline

1. $[x\ y\ z\ 1]^T \longrightarrow [x^M\ y^M\ z^M\ 1]^T$
   Modelview transformation = multiplication by the modelview matrix.

2. $\longrightarrow [x^{PM}\ \ y^{PM}\ \ z^{PM}\ \ w^{PM}]^T$
   Multiplication by the projection matrix.

3. $\longrightarrow \left[\frac{x^{PM}}{w^{PM}}\ \ \frac{y^{PM}}{w^{PM}}\ \ \frac{z^{PM}}{w^{PM}}\right]^T$
   Perspective division **with mechanism to handle zero w-values**.

4. $\longrightarrow \left[\frac{x^{PM}}{w^{PM}}\ \ \frac{y^{PM}}{w^{PM}}\ \ \frac{z^{PM}}{w^{PM}}\right]^T$
   Clipping to the canonical box.

5. $\longrightarrow \left[\frac{x^{PM}}{w^{PM}}\ \ \frac{y^{PM}}{w^{PM}}\right]^T$
   Projection to the back of the canonical box (z-values possibly retained for depth testing).

6. $\longrightarrow [i\ j]^T$
   Rasterization **with perspective correction**.

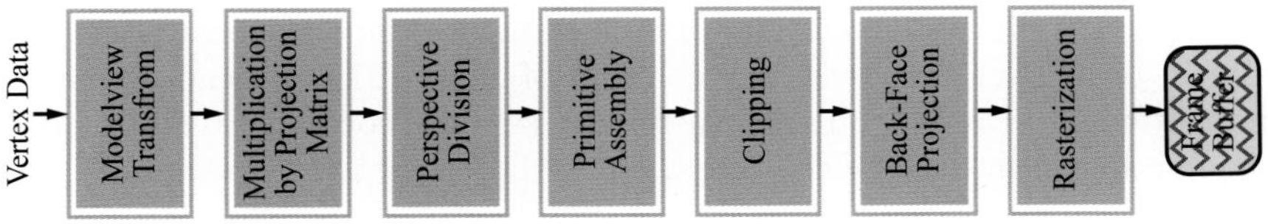

Figure 19.8: Complete minimal synthetic-camera rendering pipeline.

This is a complete synthetic-camera pipeline in that it will transform a user-specified scene correctly into a picture on the monitor. However, it is skeletal. The OpenGL pipeline, as we'll see next, adds several features.

19.1.5 OpenGL Fixed-function Pipeline

The OpenGL 1.x fixed-function rendering pipeline, while keeping the gist of the synthetic-camera pipeline as described above, enhances it with several new capabilities to make it significantly more powerful. The additions are indicated in a darker shade in Figure 19.9.

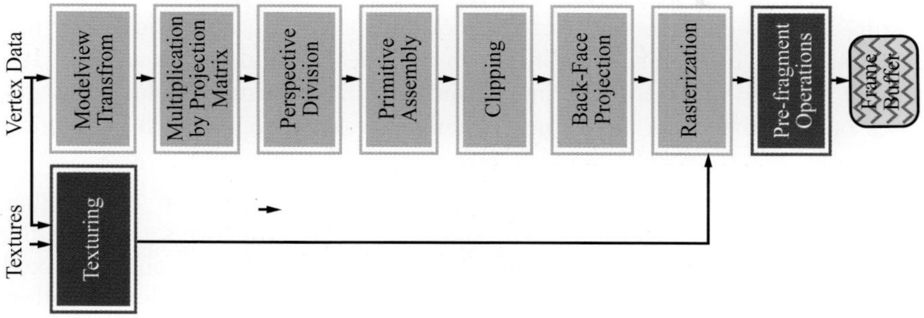

Figure 19.9: OpenGL fixed-function pipeline. Additions to the minimal synthetic-camera pipeline are darkly shaded.

The first addition is texturing, where vertex data (vertex and texture coordinates, particularly) and a set of controlling parameters (filters, environment settings, etc.) are used to combine the texture images into the raster. We learned the fundamentals of texturing ourselves in Chapter 12.

Next, instead of simply copying the raster into the frame buffer, the user can define *per-fragment* operations – a raster pixel with color data and z-value is called a fragment. The per-fragment operations allowed in OpenGL consist of four tests in the order

1. Scissor test
2. Alpha test (discarded from OpenGL 3.1 on)
3. Stencil test
4. Depth test

followed by

5. Blending

If a fragment fails an early test then it is eliminated immediately and does not proceed to subsequent tests. A fragment which survives all tests graduates into a pixel.

We are already familiar with the stencil test (Section 13.7) and depth test (from as far back as Section 2.8). The scissor test is just a special case of the stencil test, where stencil tags are used to mask a rectangular region of the OpenGL window. It has been made a separate test because it can be optimized in hardware, unlike the general stencil test. The alpha test allows the user to accept or reject a fragment depending upon its alpha value.

Note: The alpha test has been discarded since OpenGL 3.1 because its functionality can be executed in a fragment shader, part of the programmable pipeline.

If the reader is wondering why lighting is missing from the pipeline of Figure 19.9, then note that lighting computations are done along the top path, starting from vertex data, which includes normal values as well. It's simply to avoid clutter in the figure that specific lighting calculation stages have been omitted, as have some other processing stages, such as fogging and antialiasing.

Remark 19.2. The shading language in second-generation and higher OpenGL, called GLSL for GL shading language, transforms the pipeline of Figure 19.9 by making programmable major sections currently of fixed functionality. In particular, so-called vertex and fragment shaders allow the programmer to dictate to a great extent how vertices and fragments are processed. This permits greater flexibility and, therefore, creativity, in modeling scenes than with the fixed-functionality pipeline. We'll be studying the GLSL ourselves from the next chapter.

19.1.6 1D Primitive Example

Let's chase a 1D primitive down the synthetic-camera rendering pipeline of Section 19.1.4, which is *the* OpenGL pipeline minus bells and whistles.

Example 19.1. The projection statement in the reshape routine of a program is

```
glFrustum(-5, 5, -5, 5, 5, 25)
```

The only primitive definition and only modelview transformation in the drawing routine are

```
glTranslatef(0, 5, 0);
glBegin(GL_LINES);
   glColor3f(1, 0, 0); glVertex3f(0, 0, -10);
   glColor3f(0, 1, 0); glVertex3f(25, 5, -20);
glEnd();
```

All other statements in the program are routine.

Apply the synthetic-camera rendering pipeline to determine the rasterization of the line segment drawn by the program in a 100×100 raster. Determine as well the z-values corresponding to the segment's pixels.

Answer: The segment endpoints in homogeneous coordinates are

$$p = [0\ 0\ -10\ 1]^T \quad \text{and} \quad q = [25\ 5\ -20\ 1]^T$$

The matrix corresponding to the translation is (from Equation (5.28))

$$M = \begin{bmatrix} 1 & 0 & 0 & 0 \\ 0 & 1 & 0 & 5 \\ 0 & 0 & 1 & 0 \\ 0 & 0 & 0 & 1 \end{bmatrix}$$

The matrix corresponding to the projection statement is (from Equation (18.4))

$$P = \begin{bmatrix} 1 & 0 & 0 & 0 \\ 0 & 1 & 0 & 0 \\ 0 & 0 & -1.5 & -12.5 \\ 0 & 0 & -1 & 0 \end{bmatrix}$$

Apply the modelview transformation first, multiplying both endpoints by M:

$$Mp = [0\ 5\ -10\ 1]^T \quad \text{and} \quad Mq = [25\ 10\ -20\ 1]^T$$

At this point we note that both z-values are negative, so we are in a situation analogous to case (a) at the end of the discussion in Section 19.1.2 and can proceed down the pipeline without worrying about invoking enhancements to handle zero w-values in Stage 3. Accordingly, multiplying by P next:

$$PMp = P[0\ 5\ -10\ 1]^T = [0\ 5\ 2.5\ 10]^T$$

and

$$PMq = P[25\ 10\ -20\ 1]^T = [25\ 10\ 17.5\ 20]^T$$

Perspective division, then, gives the Cartesian coordinates of the transformed endpoints as follows:

$$[0/10\ 5/10\ 2.5/10]^T = [0\ 0.5\ 0.25]^T$$

and

$$[25/20\ 10/20\ 17.5/20]^T = [1.25\ 0.5\ 0.875]^T$$

As the first point lies in the canonical box, while the second outside of only the $x = 1$ plane, clipping involves a single intersection computation – that of the transformed segment with the $x = 1$ plane. We'll leave the reader to verify by means of elementary geometry that the intersection, in fact, is $[1\ 0.5\ 0.75]^T$, so that the endpoints of the clipped segment are

$$[0\ 0.5\ 0.25]^T \text{ and } [1\ 0.5\ 0.75]^T$$

We must determine the color tuple to assign the new second endpoint. It's checked that the new endpoint as a convex combination of the old ones is:

$$[1\ 0.5\ 0.75]^T = u[0\ 0.5\ 0.25]^T + (1-u)[1.25\ 0.5\ 0.875]^T$$

where $u = 0.2$. Therefore, by Equation (19.3), it corresponds to the point $tp + (1-t)q$ on pq, where

$$t = \frac{0.875}{(\frac{1}{1/0.2} - 1)0.25 + 0.875} = 0.467$$

Accordingly, the color tuple assigned the new endpoint is

$$0.467(1,0,0) + 0.533(0,1,0) = (0.467, 0.533, 0)$$

Projecting the first endpoint of the clipped segment to the back face of the canonical box gives then the point $[0\ 0.5]^T$, with color value $(1,0,0)$ and z-value 0.25 (retained possibly for depth testing). Likewise, the second endpoint projects to $[1\ 0.5]^T$ with color value $(0.467, 0.533, 0)$ and z-value 0.75.

Time to leap from world space to screen space!

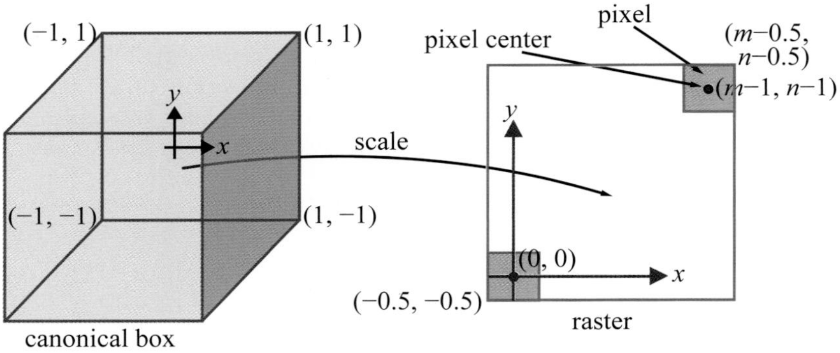

Figure 19.10: Scaling from the 2×2 back face of the canonical box, located on an xy-plane, to the $m \times n$ area of the raster.

Generally, if the raster is $m \times n$ and pixel centers have integer coordinates (i, j), where $0 \le i \le m-1$ and $0 \le j \le n-1$, then the area of the raster is an axis-aligned rectangle, whose lower-left corner is $(-0.5, -0.5)$ and upper-right $(m - 0.5, n - 0.5)$. See Figure 19.10. The back face of the canonical box, on the other hand, can be imagined as a 2×2 square with corner coordinates $x = \pm 1$ and $y = \pm 1$ on the xy-plane (now that projection is done, we can "forget" z-values, except when they are needed for depth testing). Accordingly, functions that scale the back face onto the raster – we're doing the print part of shoot-and-print now – are:

$$x \to \frac{x+1}{2} m - 0.5 \quad \text{and} \quad y \to \frac{y+1}{2} n - 0.5$$

Applying these functions to the projected endpoints on the back face of the canonical box, with $m = n = 100$, we get:

$$[0\ 0.5]^T \to [49.5\ 74.5]^T \quad \text{and} \quad [1\ 0.5]^T \to [99.5\ 74.5]^T$$

Rounding, one has the endpoint pixels on the raster as $(49, 74)$ and $(99, 74)$, respectively. Data associated with these pixels are the color values $(1, 0, 0)$ and $(0.467, 0.533, 0)$ and z-values 0.25 and 0.75, respectively.

As the rasterized segment is horizontal, choosing pixels along it is trivial – $(49, 74), (50, 74), \ldots, (99, 74)$ – obtaining a raster line segment of length $N = 50$ (containing 51 pixels).

It remains to color the in-between pixels using perspective correction, as well as assign their z-values. The u-value corresponding to pixel $(50, 74)$, second from left, is $1 - \frac{1}{50} = 0.98$, and t-value, therefore (applying (19.3)):

$$\frac{0.75}{(\frac{1}{0.98} - 1)0.25 + 0.75} = 0.993$$

Accordingly, its color tuple is

$$0.993(1, 0, 0) + 0.007(0.467, 0.533, 0) = (0.996, 0.004, 0)$$

and z-value

$$0.98 * 0.25 + 0.02 * 0.75 = 0.26$$

using u itself, rather than t, to interpolate.

Important: z-values *need not* be perspectively corrected as their values in the canonical box, following projection transformation, are valid.

We'll leave the reader to calculate the color and z-values of a few more pixels or, better still, write a routine to generate them all.

19.1.7 Exercising the Pipeline

Exercise 19.2. Redo the preceding example with only the part in the drawing routine changed to

```
glRotatef(45, 0, 0, 1);
glBegin(GL_LINES);
   glColor3f(1, 1, 1); glVertex3f(5, 0, -10);
   glColor3f(0, 0, 0); glVertex3f(10, 10, -5);
glEnd();
```

Exercise 19.3. Repeat the previous exercise with the drawing routine changed again to

```
glRotatef(90, 0, 1, 0);
glBegin(GL_LINES);
   glColor3f(1, 0, 0); glVertex3f(1, 0, -1);
   glColor3f(1, 0, 0); glVertex3f(-4, 4, 0);
glEnd();
```

and the projection statement, as well, to

```
glOrtho(-5, 5, -5, 5, 5, -5)
```

Exercise 19.4. Repeat the previous exercise with the drawing routine changed once more to

```
glTranslatef(1, 1, 1);
glBegin(GL_LINES);
   glColor3f(1, 1, 1); glVertex3f(5, 0, -10);
   glColor3f(0, 0, 0); glVertex3f(10, 10, 5);
glEnd();
```

and the projection statement back to

```
glFrustum(-5, 5, -5, 5, 5, 25)
```

Note: You can roughly check your result for each of the preceding exercises by comparing it with the output of a minimal OpenGL program containing the given statements.

Exercise 19.5. (**Programming**) This is a substantial programming project: implement the synthetic-camera pipeline to render (only) 0D and 1D primitives (drawn in 3-space, of course). Use the OpenGL window to simulate the raster as in `DDA.cpp`.

19.2 Ray Tracing Pipeline

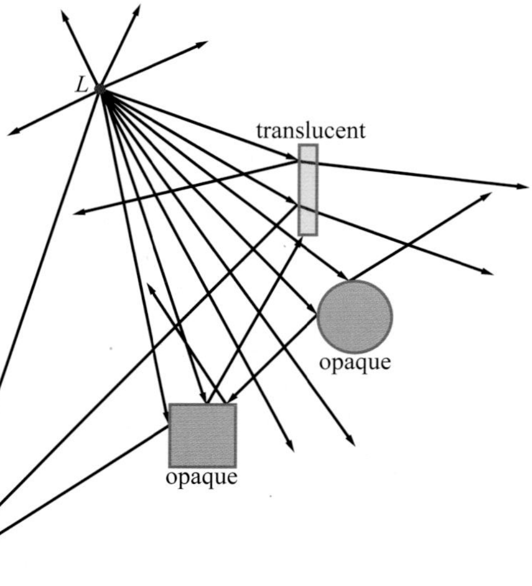

Figure 19.11: Tracing rays *from* a light source L – only few reach the eye.

The ray tracing pipeline is an "alternate" to the synthetic-camera graphics pipeline. The reason for the quotes is that the ray tracing approach is very different from that of the synthetic-camera-based approach and rarely does a programmer have the option of simply exchanging one for the other. Why this is the case will be apparent once we understand how ray tracing works.

The idea behind ray tracing is straightforward: to follow light rays from each source as they interact with the scene – reflecting off opaque objects one to another, and both reflecting off and refracting through translucent ones, in the process casting shadows and creating reflections – till they finally reach the eye. However, implementing this idea as just stated is not a particularly well-advised endeavor, as (a) there is an infinite continuum of light rays emanating from each source, and (b) even after somehow discretizing them to a finite number, only a fraction thereof reach the viewer. See Figure 19.11 for an idea of the situation.

Ray tracing, instead, implements the plan "backwards". Rays are traced *from* the eye, one through each pixel, so that no computation is expended on rays which are ultimately invisible. Each ray is followed through the pixel and into and around the scene, possibly bouncing off opaque objects and passing through translucent ones, for a *finite* amount of time, till a determination is made of its color. Of course, an implementation has to "cut off" each ray after a finite number of steps and determine the color it has picked up through interactions with objects up to that point, or the ray tracing process will continue indefinitely.

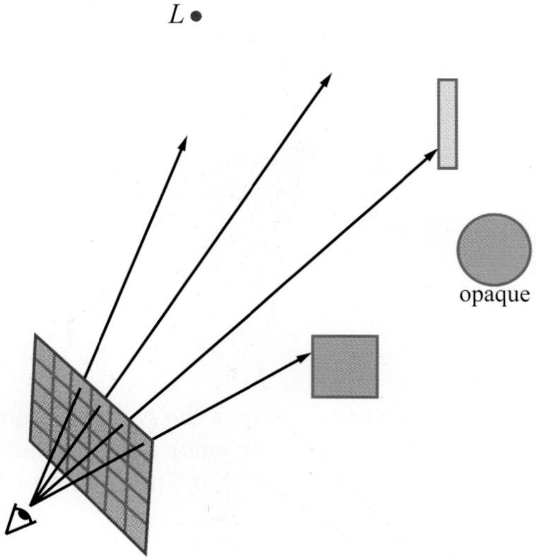

Figure 19.12: Tracing rays *from* the eye, one through each pixel. Rays are "stopped" when they strike an object.

See Figure 19.12 for a very simple scene. The screen is virtual – akin to the front face of an OpenGL viewing box or frustum. In this particular figure, rays either go off to infinity (there are two such drawn) or stop upon hitting the surface of an object (there are two such as well). We don't (as yet) follow rays beyond their first intersection with an object's surface.

This, in fact, suggests a simple first implementation of ray tracing: color pixels, rays through which go off to infinity without collision, the background color; assign every other pixel the color of the first point of intersection of the ray through it with an object's surface. This particular color is determined from Phong's lighting model – see Chapter 11, in particular the lighting equation (11.12).

Here's pseudo-code:

Ray Tracer Version 1: Non-recursive local

```
void topLevelRoutineCallsTheRayTracer()
{
   positionEye = position of eye in world space;
   for (each pixel P of the virtual screen)
   {
      d = unit vector from positionEye toward the center point of P;
      color of P = rayTracer(positionEye, d);
   }
}

Color rayTracer(p, d)
{
   if (ray from point p in the direction d does not intersect the
        surface of any object in the scene) return backgroundColor;
   else
   {
      q = first point of intersection with an object's surface;
      computedColor = color computed at q using Phong's lighting model;
      return computedColor;
   }
}
```

Notes:

1. The *base* case of Version 1, when the ray is stopped at an intersection with an object, uses Phong for color calculation. It is typical, in fact, of ray tracers to invoke a local lighting model at the base case.

2. Intersection detection, implicit in the code, is the most computationally intensive part of the ray tracer. We'll not go into intersection computation in our account of ray tracing, but focus instead on color calculations.

Interestingly, the ray tracer version above renders the same image as a synthetic-camera pipeline – *a la* OpenGL – implementing Phong's lighting model with depth testing. The only difference is that depth testing via the

z-buffer has been replaced by ray tracing to determine visible surfaces (one surface obscuring another if it blocks rays from reaching the other).

19.2.1 Going Global: Shadows

The next step up is shadow computation. This is simple to do. If a ray through a pixel intersects a surface, then send a *feeler ray* from the point of intersection toward each light source. If the feeler ray hits an object before reaching a light source, then the point of intersection is in the shadow of the struck object and not illuminated directly by the source. Recall in this connection that, according to Phong's model, only the diffuse and specular components of light reflected off a surface depend upon direct, i.e., straight-line, illumination from the light source, while the ambient does not.

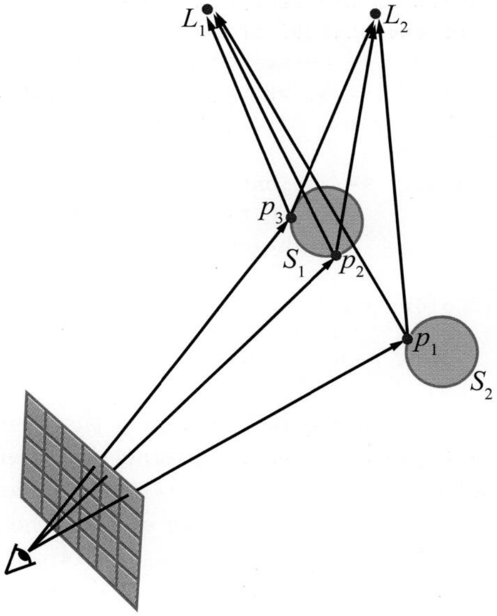

Figure 19.13: Shadow computation: feeler rays are dashed.

See Figure 19.13. Point p_1 on ball S_2 is in the shadow of the ball S_1 cast by light from L_1 – because the feeler ray from it toward L_1 is cut off by S_1 – so it reflects only the ambient component of light from that particular source. On the other hand, p_1 is directly illuminated by L_2, so reflects all components of light from that source. Point p_2 is in the shadows of S_1, of which it is a point itself, cast both by L_1 and L_2. Point p_3 is illuminated directly by both sources.

Here's pseudo-code for a shadow-computing ray tracer (a top-level routine the same as that of the first version is not repeated):

Ray Tracer Version 2: Non-recursive global, with shadows

```
void topLevelRoutineCallsTheRayTracer(); // See Version 1.

Color rayTracer(p, d)
{
   if (ray from point p in the direction d does not intersect the
       surface of any object) return backgroundColor;
   else
   {
      q = first point of intersection;
      computedColor = black; // Color values all set to zero.
      for (each light source L)
      {
         // Object not shadowed.
         if (feeler ray from q toward L does not intersect the
             surface of any object before reaching L)
             computedColor  += color computed at q due to light from
                               L, using Phong's lighting model;

         // Object shadowed.
         else computedColor += ambient component of color computed at
                               q due to light from L using Phong's
                               lighting model;
      }
      return computedColor;
   }
}
```

A hugely significant development in Version 2 is that the lighting model has now gone *global*: object-object light interaction comes into play in computing shadows. A local lighting model, like OpenGL's default Phong, on the other hand, does not take into account other objects when coloring a particular one. Recollect how we computed shadows ourselves in the programs `ballAndTorusShadowed.cpp` and `ballAndTorusShadowMapped.cpp`. Now, this global version of ray tracing not only gives us shadows automatically, but ones as authentic as those drawn by Mother Nature, in particular, the laws of light.

19.2.2 Going Even More Global: Recursive Reflection and Transmission

We are ready now for the full blast of ray tracing power. So far, we've stopped at the first intersection of a ray from the eye with a surface. In reality, rays from a light source can bounce from object to object, or even pass through them, several times before reaching the viewer, giving rise to such phenomena as reflection and translucence. To model this in keeping with ray tracing's backward approach of following rays from the eye into the scene, one must allow a ray to continue even after it hits an object. The physics of light suggests that a ray striking an object is partially reflected off its surface and partially transmitted through it, depending on the characteristics of the material, as well as the color of the light. For example, an opaque object transmits almost zero light and reflects the remainder according to its surface finish and color, while a translucent one transmits most.

Accordingly, we'll enhance Ray Tracer Version 2 such that each ray from the eye that strikes a surface spawns two additional rays: a *reflected ray* in the direction of perfect reflection and a *transmitted ray* passing through the surface, possibly with its direction altered by refraction. The two spawned rays are treated *exactly* as the incoming ray and may each spawn additional rays themselves upon subsequent intersection with a surface. If you are thinking recursion, then that's exactly where we're headed.

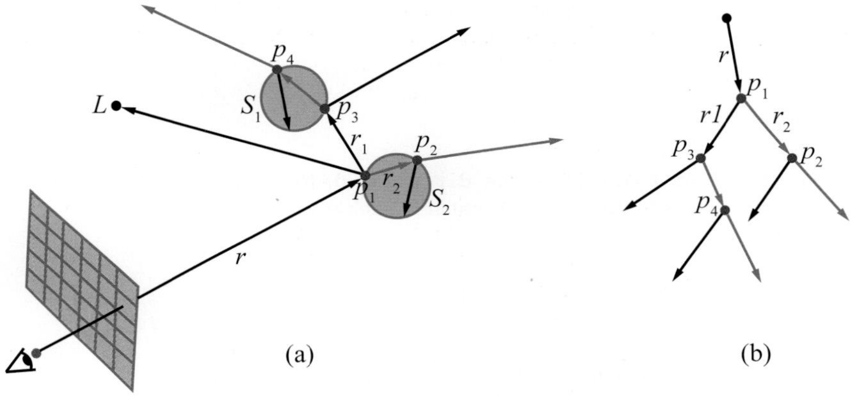

Figure 19.14: (a) Reflection and transmission: reflected rays are black, transmitted blue. One dashed feeler ray is drawn. (b) Ray tree (not all edges are labeled).

As an example, Figure 19.14(a) follows a single ray r from the eye through a few intersections with two translucent balls. The resulting binary *ray tree* data structure is shown in Figure 19.14(b). Observe that the transmitted rays are refracted by the material of the balls. The color computed at a point now has three components – one computed locally, one returned by the reflected ray, and one by the transmitted ray – as given by the following equation:

$$computedColor = color_{local} + coef_{refl}\, color_{refl} + coef_{tran}\, color_{tran}$$

For example, at point p_1 of the figure, $color_{local}$ is computed using Phong (exactly as in Version 2, with the help of feeler rays to find "visible" light sources – the feeler ray from p_1 to L_1 is shown in the figure); $color_{refl}$ is the value returned recursively by the reflected ray r_1, attenuated by a material-dependent multiplicative factor $coef_{refl}$, which specifies the fraction of the incoming ray r that is reflected; $color_{tran}$ is likewise returned recursively by the transmitted ray r_2 and attenuated by $coef_{tran}$.

Pseudo-code is below. The new top-level routine passes a non-negative integer depth parameter **maxDepth** to the ray tracer to cut off recursion after a finite number of levels.

Ray Tracer Version 3: Recursive global, with shadows, reflection and transmission

```
void topLevelRoutineCallsTheRayTracer()
{
   positionEye = position of eye in world space;
   for (each pixel P of the virtual screen)
   {
      d = unit vector from positionEye toward the center point of P;
      color of P = rayTracer(positionEye, d, maxDepth);
   }
}

Color rayTracer(p, d, depth)
{
   if (ray from point p in the direction d does not intersect the
        surface of any object) return backgroundColor;
   else
   {
      q = first point of intersection;
      computedColor = black; // Color values all set to zero.

     // Local component, copy of Version 2 calculations.
      for (each light source L)
      {
         // Object not shadowed.
         if (feeler ray from q toward L does not intersect the
              surface of any object before reaching L)
             computedColor  += color computed at q, due to light from
                                L, using Phong's lighting model;

         // Object shadowed.
         else computedColor += ambient component of color computed at
                                q, due to light from L, using Phong's
                                lighting model;
      }

      // Global component.
      if (depth > 0)
      {
         d1 = unit vector from q in direction of perfect reflection;
         d2 = unit vector from q in direction of transmission;
```

```
            // Reflected component added in recursively.
            computedColor += coefRefl * rayTracer(q, d1, depth-1)

            // Transmitted component added in recursively.
            computedColor += coefTran * rayTracer(q, d2, depth-1)
        }
        return computedColor;
    }
}
```

Notes:

1. Determining where an incident ray strikes an object and spawns a reflected and a transmitted ray obviously requires intersection computation. Subsequent calculation of the direction of the reflected and transmitted rays requires computation of the normal to the surface at the point of incidence as well:

   (a) The direction of the reflected ray is given by the laws of reflection, which say that both the incident and reflected rays make the same angle with the normal to the surface, and that all three lie on the same plane. See Figure 19.15, where the equation for reflection is $A = B$.

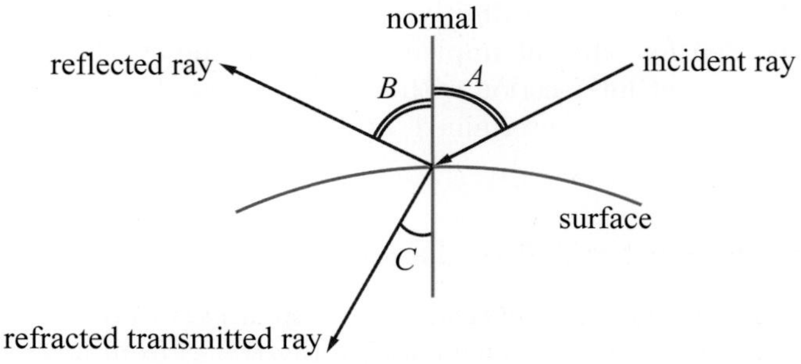

Figure 19.15: Calculating the direction of the reflected and transmitted rays: A = angle of incidence, B = angle of reflection, C = angle of refraction.

   (b) Routines to compute the direction of transmission can be simple or as fancy as the need for realism dictates.

   For instance, refraction is often taken into account with the help of Snell's law, which says that the ratio of the sine of the angle of incidence to the sine of the angle of refraction is equal to the ratio of the speed of light in the medium of the incident ray to that in the medium of the refracted ray; moreover, the incident ray, refracted ray and normal to the surface all lie on the same plane.

The ratio of the speed of light in two different media is the inverse ratio of the refractive indices of the media. Therefore, in Figure 19.15, one can write the equation for refraction as $\frac{\sin A}{\sin C} = \frac{\eta_2}{\eta_1}$, where η_1 is the refractive index of the medium on the side of the incident ray and η_2 that on the side of the refracted ray.

2. The direction of the reflected ray is taken to be that of perfect mirror-like reflection. This models well the transport of specular light but not that of diffuse. For the latter is needed *multiple* reflected rays – remember that diffuse light is scattered in all directions by the lit object – which would make the ray tracing process computationally overwhelming.

   This inability of ray tracing to realistically model diffuse illumination is a weakness often overcome by combining it with radiosity, another global lighting model which is specially designed to track the dispersion of diffuse light. We'll discuss radiosity in the next section.

Exercise 19.6. Neither version 2 nor 3 of our ray tracer seems to take into account global ambient light in their Phong base case. Revise both versions to do so.

Remark 19.3. It's interesting to observe that a ray tracer does not ask for a small set of simple drawing primitives, e.g., points, line segments and triangles, as needed for efficient implementation of the synthetic-camera model. As long as their intersection with a given ray can be computed and the normal at a given point determined, arbitrary curved surfaces may be rendered.

19.2.3 Implementing Ray Tracing

We're going to implement ray tracing with the help of POV-Ray (Persistence of Vision Ray Tracer), a freely downloadable ray tracer from `povray.org` [113]. Download and install POV-Ray. The executable is about 10 MB and there are Linux, Mac OS and Windows versions. It comes packaged with a nicely written tutorial and a reference manual.

Here's POV-Ray code to show off how realistic ray traced rendering can be.

Experiment 19.3. If you have successfully installed POV-Ray, then open `sphereInBoxPOV.pov` from that program; if not, use any editor.

If you have installed POV-Ray, then press the Run button at the top; otherwise, open the output image file `sphereInBoxPOV.jpg` in the same folder as `sphereInBoxPOV.pov`. Figure 19.16(a) is a screenshot. Impressive, is it not, especially if you compare with the output in Figure 19.16(b) of `sphereInBox1.cpp`? The inside of the box, with the interplay of light evident

in shadows and reflections, is far more realistic in the ray-traced picture.
End

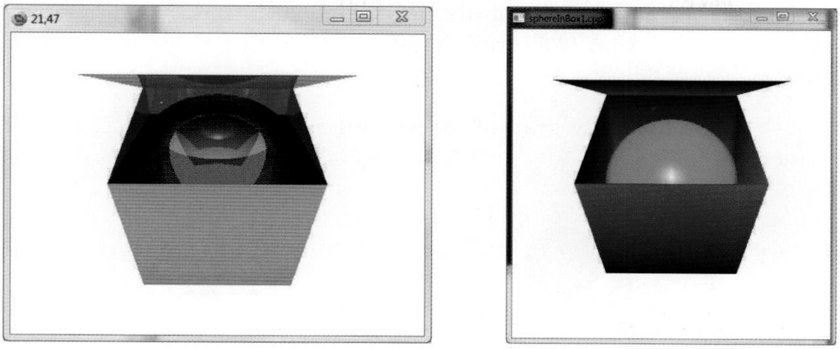

(a) (b)

Figure 19.16: Ray tracing versus OpenGL: screenshot of (a) `sphereInBoxPOV.pov` (b) `sphereInBox1.cpp`.

The code itself is fairly self-explanatory. It's written in POV-Ray's scene description language (SDL), which, unlike OpenGL, is *not* a library meant to be called from a C++ program – the SDL is stand-alone. We've obviously tried to follow the settings in our OpenGL program `sphereInBox1.cpp` as far as possible. The camera and a white light source are placed identically as in `sphereInBox1.cpp`. The red box, as in `sphereInBox1.cpp`, is an axis-aligned cube of side lengths two centered at the origin. It comprises six polygonal faces, each originally drawn as a square with vertices at $(-1,-1)$, $(1,-1)$, $(1,1)$ and $(-1,1)$ on the xy-plane, and then appropriately rotated and translated. The top face is opened to an angle of $60°$. Finally drawn is a green sphere of radius one. The material finishes are minimally complex, just enough to obtain reflection and a specular highlight on the sphere.

So what gives? If ray tracing is so much more realistic than the synthetic-camera-based OpenGL pipeline, then why bother with the latter (or, for that matter, write fat books about it)?! If you noticed how long it took to render the output of `sphereInBoxPOV.pov` – a few seconds at least on a decent desktop – then you have the answer. Ray tracing is *very very* computationally intensive. Intersection computations don't come cheap and they have to be done for every ray at every level in every ray tree generated, and there's one ray tree for each of maybe a million pixels. To even open the lid of the simple box of `sphereInBoxPOV.pov` in *real-time*, in the manner of `sphereInBox1.cpp`, is beyond the power of any modern-day desktop. Interactive animation, therefore, of remotely complex scenes (read games) is likely to remain beyond the reach of ray traced rendering for a while now.

On the other hand, still-life and movies, where either there is either no animation or it is all done off-line, are perfect applications for ray tracing.

Computer animation in Hollywood is almost exclusively ray traced, individual frames of complex and highly realistic animated sequences sometimes taking hours each to render on special-purpose hardware (often clusters of computers called *render farms*). Incidentally, POV-Ray, too, has the capability to sequence an animation from individually generated frames (refer to their tutorial).

Remark 19.4. The holy grail of ray tracing research is, in fact, real-time ray-traced rendering.

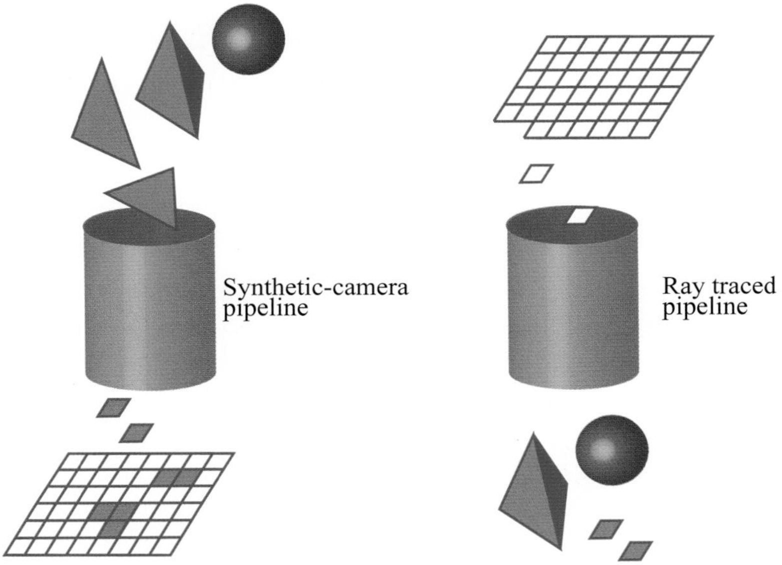

Figure 19.17: The (object-oriented) synthetic-camera pipeline versus the (screen-oriented) ray traced pipeline.

A somewhat amusing, though fairly authentic, comparison of the synthetic-camera pipeline with ray tracing is to say that the former is "object-oriented", while the latter "screen-oriented". See Figure 19.17: on the left, objects (primitives) are dropped into the synthetic-camera pipeline to emerge rasterized, while on the right, pixels are dropped into the ray tracing pipeline to emerge colorized.

In case you enjoyed the little of POV-Ray that we showed and want to try your own hand at ray tracing, here are a couple of exercises.

Exercise 19.7. (Programming) Use POV-Ray to generate a ray traced "combination" of `ballAndTorusReflected.cpp` and `ballAndTorus-PerspectivelyShadowed.cpp` with both shadows and reflections. It will be a single still shot, of course. Make sure that both the ball and torus, in addition to the floor and wall, are highly reflective, so that, in fact, they all reflect each other. You may need additional light sources to liven the scene.

Exercise 19.8. (Programming) Animate the opening of the lid of the box of `sphereInBoxPOV.pov` by generating a sequence of stills – one for every degree the lid turns would mimic `sphereInBox1.cpp`.

19.3 Radiosity

19.3.1 Introduction

Radiosity is a global lighting model which uses principles of heat transfer to track the dispersion of diffuse light around a scene.

Figure 19.18: Living room lit mostly with diffuse light (courtesy `www.freshome.com`).

It is quite often that a significant component of the light illuminating a scene is, in fact, multiply reflected diffuse light. For an example, consider a living room scene like the one depicted in Figure 19.18, populated with non-shiny furniture and lit by early morning rays. In such a setting there is little specular transport of light (i.e., by mirror-like reflection). Instead, in addition to the ambient component, which is fairly constant throughout, there tends to be mostly diffuse activity. For example, light from the floor and walls reflect diffusely onto the shelves and furniture fabric. Even parts of the environment obscured from direct lighting, such as the floor between the sofas, are not in a well-defined shadow, but mildly illuminated by light reflecting off adjacent objects.

Such multiple diffuse reflections are not modeled by ray tracing at all. In fact, if you see again the design of Ray Tracer Version 3 in the previous section, a ray upon intersecting an object spawns a single ray in the direction of transmission and another in the direction of perfect reflection. Radiosity complements ray tracing by modeling diffuse illumination. Together, they can deliver highly realistic rendering, ray tracing emphasizing the shadows and highlights, and radiosity recording the softer lights.

19.3.2 Basic Theory

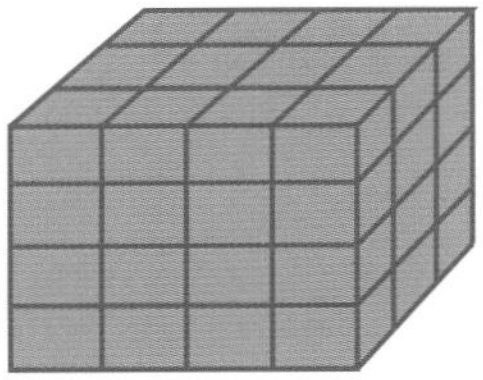

Figure 19.19: Patchified box.

The radiosity algorithm that we'll describe begins by dividing the scene into some number n of small flat, typically polygonal, *patches*, $P_i, 1 \leq i \leq n$, e.g., see Figure 19.19. A triangulated scene is, of course, automatically patchified. However, even then, one may want to refine certain triangles, or combine others to coarsen the given triangulation in response to two opposing forces in patchification: the smaller and more numerous the patches the more authentic is the lighting calculation; on the other hand, the time complexity of the radiosity algorithm, which is $O(n^2)$, increases rapidly with the number of patches. The best strategy is an adaptive one where a region over which light intensities are expected to vary rapidly is finely patchified, while one of steadier light levels more coarsely.

Remark 19.5. Patchifying to compute radiosity means that it is a *finite element* method.

The *brightness* or *radiosity* of a patch is the light energy per unit time per unit area leaving the patch, measured, typically, in a unit such as $joules/(second \times meter^2)$ (equivalent to $watts/meter^2$). The brightness varies with the frequency of the light in a manner that determines the perceived color of the patch; e.g., a red patch emits the greatest intensity at the red end of the visible spectrum. However, for simplicity's sake, we'll develop the theory assuming the brightness of a patch P_i as a single scalar value B_i, while a real implementation will have three different scalar values corresponding to the RGB brightnesses.

Our starting point is the following equation which, in fact, holds for each i, $1 \leq i \leq n$,

$$A_i B_i = A_i E_i + R_i \sum_{j=1}^{n} F_{ji} A_j B_j \tag{19.4}$$

where A_i is the area of patch P_i, B_i its brightness, E_i its emission rate, R_i the reflective scaling factor and, finally, F_{ji}, the so-called *form factor* between patches P_j and P_i.

The equation simply states that the amount of light energy leaving a patch P_i, which is the *area* $\times$ *brightness* term on the LHS, is equal to (a) the amount it emits as a source *plus* (b) the amount it reflects of incoming light, the two additive terms, respectively, on the RHS.

The value of (a) as the product of the patch's area and emission rate is clear. For (b), note first that the form factor F_{ji} denotes the fraction of the total light energy leaving patch P_j that reaches P_i. Therefore, $F_{ji}(A_j B_j)$ is, in fact, the amount of light leaving P_j for P_i; multiplied by P_i's reflective scaling factor R_i, this gives the amount of light from P_j actually reflected from P_i. Accordingly, the value of (b) is the summation of $R_i F_{ji}(A_j B_j)$

over all patches P_j, which is precisely the second term on the RHS of the equation above.

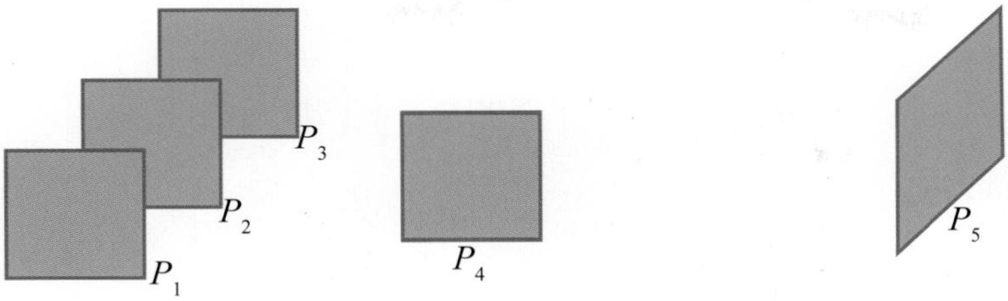

Figure 19.20: Form factor between patches depends on their respective orientation, the distance between them and if there is occlusion by other patches.

Shortly, we'll be computing form factors mathematically but it's easy to understand intuitively that F_{ji} depends on the orientation of P_j and P_i relative to each other, their distance and, further, if there is occlusion by intermediate patches between the two. For example, in Figure 19.20, the form factor between patches P_1 and P_2 and between P_2 and P_3 is high, because the two pairs are side by side and parallel, while that between P_1 and P_3 low because P_2 is between them. The form factor between P_4 and any one of P_1, P_2 and P_3 is low because of unfavorable orientation. The form factor between P_5 and any one of P_1, P_2 and P_3 is low as well because of distance.

Form factors will be seen to satisfy the *reciprocity equation*:

$$F_{ij} A_i = F_{ji} A_j \tag{19.5}$$

Assuming this reciprocity for now, rewrite Equation (19.4) as

$$A_i B_i = A_i E_i + R_i \sum_{j=1}^{n} F_{ij} A_i B_j \tag{19.6}$$

Dividing out A_i, one gets the *radiosity equations*:

$$B_i = E_i + R_i \sum_{j=1}^{n} F_{ij} B_j, \quad 1 \le i \le n \tag{19.7}$$

which is a set of simultaneous linear equations in the brightnesses B_i, the latter being the only unknowns, provided we already have at hand the emissivities E_i and reflectivities R_i from a knowledge of material properties, and provided we can compute, as well, the form factors F_{ij} from the patch geometry.

Rearranging the radiosity equations as

$$(1 - R_i F_{ii}) B_i \quad - \sum_{1 \le j \le n,\ j \ne i} R_i F_{ij} B_j = E_i, \quad 1 \le i \le n$$

one can write them in matrix form as

$$\begin{bmatrix} 1-R_1F_{11} & -R_1F_{12} & \dots & -R_1F_{1n} \\ -R_2F_{21} & 1-R_2F_{22} & \dots & -R_2F_{2n} \\ \vdots & \vdots & \dots & \vdots \\ -R_nF_{n1} & -R_nF_{n2} & \dots & 1-R_nF_{nn} \end{bmatrix} \begin{bmatrix} B_1 \\ B_2 \\ \vdots \\ B_n \end{bmatrix} = \begin{bmatrix} E_1 \\ E_2 \\ \vdots \\ E_n \end{bmatrix} \quad (19.8)$$

Denoting

$$\mathbf{B} = \begin{bmatrix} B_1 \\ B_2 \\ \vdots \\ B_n \end{bmatrix}, \ \mathbf{E} = \begin{bmatrix} E_1 \\ E_2 \\ \vdots \\ E_n \end{bmatrix}, \text{ and } Q = \begin{bmatrix} R_1F_{11} & R_1F_{12} & \dots & R_1F_{1n} \\ R_2F_{21} & R_2F_{22} & \dots & R_2F_{2n} \\ \vdots & \vdots & \dots & \vdots \\ R_nF_{n1} & R_nF_{n2} & \dots & R_nF_{nn} \end{bmatrix}$$

a succinct matrix form of the radiosity equations is obtained from (19.8):

$$(I_n - Q)\mathbf{B} = \mathbf{E} \quad (19.9)$$

where, of course, I_n is the $n \times n$ identity matrix.

Therefore, once we know how to do the following two tasks efficiently, we'll be in a position to practically implement the theory developed thus far:

(a) Compute form factors.

(b) Solve the radiosity equation (19.9) to determine patch brightnesses.

We discuss the two in the next sections.

19.3.3 Computing Form Factors

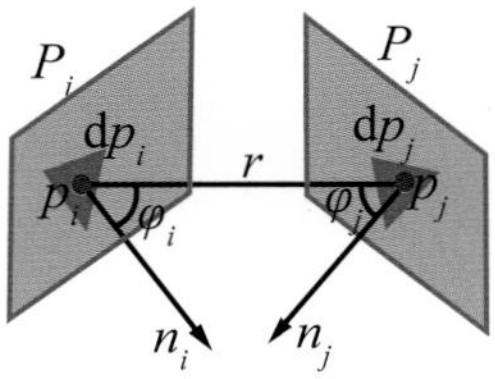

Figure 19.21: Computing form factors.

Consider two patches P_i and P_j. Even though the respective normal directions, n_i and n_j, remain constant over the patches, assumed flat, the amount of light from a point on P_i reaching a point on P_j, e.g., p_i and p_j in Figure 19.21, varies as the points vary, depending on the distance between them and the angle the line joining them makes with n_i and n_j, respectively. Therefore, one must integrate over the two patches – points being represented by infinitesimal areas – in order to determine the total light reaching P_j from P_i. In the figure, small triangles indicate the infinitesimal areas $\mathrm{d}p_i$ and $\mathrm{d}p_j$ at p_i and p_j, respectively.

In fact, if patches are presumed to be Lambertian, i.e., they reflect light uniformly in all directions from every surface point, then it can be proved that the form factor F_{ij}, the fraction of the total light emanating from F_i that reaches F_j, is given by:

$$F_{ij} = \frac{1}{A_i} \int_{p_i \in P_i} \int_{p_j \in P_j} v_{ij} \frac{\cos\phi_i \cos\phi_j}{\pi r^2} \, \mathrm{d}p_j \, \mathrm{d}p_i \quad (19.10)$$

where A_i is the area of P_i, ϕ_i and ϕ_j are the angles between the segment $p_i p_j$ and the normals n_i and n_j, respectively, r is the length of $p_i p_j$, and v_{ij} is a Boolean which is 1 if p_j is visible from p_i and 0 otherwise.

Exercise 19.9. It is easy now to deduce the reciprocity equation (19.5) from the formula for a form factor. Do so.

Except for the simplest cases, the double integral in Formula (19.10) is impossible to compute exactly. The *hemicube method*, however, is a clever approximation algorithm developed by Cohen and Greenberg [25], which takes advantage of fast hardware-based z-buffers.

Write formula (19.10) as

$$F_{ij} = \frac{1}{A_i} \int\limits_{p_i \in P_i} \left(\int\limits_{p_j \in P_j} v_{ij} \frac{\cos\phi_i \cos\phi_j}{\pi r^2}\, \mathrm{d}p_j \right) \mathrm{d}p_i \tag{19.11}$$

The inner integral can be imagined to be the form factor between p_i – or, more precisely, an infinitesimal patch $\mathrm{d}p_i$ containing p_i as in Figure 19.21 – and P_j, while F_{ij} itself is the average of these form factors over points of P_i.

The first assumption of the hemicube method is that the form factor between p_i and P_j does not vary significantly as p_i varies over P_i, which is justified if the distance between P_i and P_j is large in comparison to their respective sizes. In such a case, the average F_{ij} can be approximated by a single value, say that of the form factor at a fixed point p_i located centrally in P_i; precisely,

$$F_{ij} = \int\limits_{p_j \in P_j} v_{ij} \frac{\cos\phi_i \cos\phi_j}{\pi r^2}\, \mathrm{d}p_j \tag{19.12}$$

obtained from assuming the inner integral of (19.11) to be constant over P_i.

The next assumption is that this p_i–P_j form factor itself can be approximated by replacing P_j with its projection P'_j on an (imaginary) hemicube – half a cube – with its base lying on the plane of P_i and centered at p_i. Figure 19.22(a) shows such a hemicube, being half of a cube of side lengths 2. The justification for this assumption is as follows.

As P_i is Lambertian, light from each of its points emanates uniformly in all directions, which means that the light from p_i uniformly illuminates a hemisphere with its base along P_i and center at p_i. So the projection of P_j onto such a hemisphere would be an "ideal" replacement for P_j. However, for the sake of computational advantages, which will be perceived momentarily, the hemi*sphere* is replaced with a hemi*cube* centered at p_i and of dimensions indicated in Figure 19.22. So we get the following approximation from (19.12) by replacing the patch P_j with its projection P'_j on the replacement hemicube:

$$F_{ij} = \int\limits_{p'_j \in P'_j} v_{ij} \frac{\cos\phi_i \cos\phi_j}{\pi r^2}\, \mathrm{d}p'_j \tag{19.13}$$

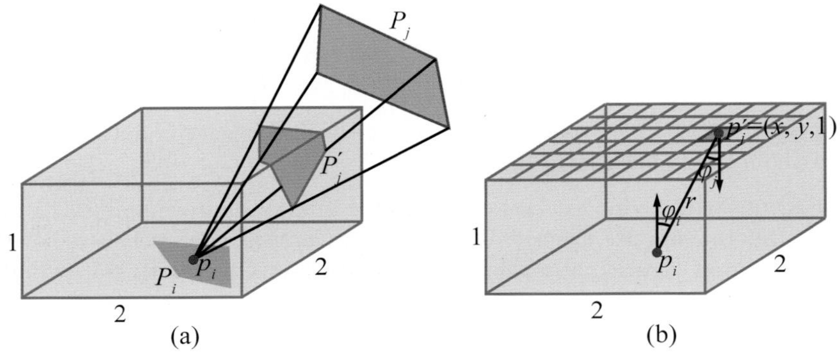

Figure 19.22: (a) Projecting a patch onto a hemicube (b) Computing the delta form factor.

The hemicube algorithm, next, discretizes the computation of the preceding integral by dividing the hemicube into a grid of squares, called (suggestively, as we shall see) pixels, and treating each as an infinitesimal area dp'_j. Figure 19.22(b) shows a division into pixels of the top face. This process effectively replaces the integral with a finite sum.

It's in the evaluation of this sum that the beauty of the hemicube method lies. Here's what it asks: for each of the five faces of the hemicube – top and four sides – render the scene with the eye at p_i and the front face of the viewing frustum coinciding with that hemicube face. Presto! Screen pixels now correspond to pixels on the hemicube face so that occlusion – the pesky v_{ij} in the integral – is automatically taken care of by means of the z-buffer.

To determine all v_{ij}, then, color code each patch – with, typically, 2^{24} colors to choose from, there should be plenty to assign a unique one to each patch – and render the scene with depth testing to find the screen pixels of a given color, which determines the projection of the patch of that color on a hemicube face. For example, the projection P'_j of patch P_j in Figure 19.22(a) has a part on the top and one on the side of the hemicube. If P_j were coded, say, red, then the red pixels, when the scene is rendered with the hemicube top as the viewing face, comprise the part of the top not occluded (in the figure there happens to be no occlusion of P_j at all).

Consider, next, a single pixel belonging to patch P'_j, lying on the top face and centered at the point $p'_j = (x, y, 1)$, e.g., the darker one in Figure 19.22(b). We have $r = \sqrt{x^2 + y^2 + 1}$, $\phi_i = \phi_j$ and $\cos\phi_i = \cos\phi_j = 1/r$. Moreover, the area of the pixel is $\frac{4}{wh}$, where the screen size is w pixels $\times$ h pixels. Therefore, the contribution of the top face of the hemicube to the integral

(19.13) is approximated by the sum

$$\sum_{\substack{\text{pixel on top face} \\ \text{of color code of } P_j}} \frac{(1/r)(1/r)}{\pi r^2}\frac{4}{wh} = \frac{4}{\pi wh}\sum_{\substack{\text{pixel on top face} \\ \text{of color code of } P_j}} \frac{1}{(x^2+y^2+1)^2} \tag{19.14}$$

Exercise 19.10. Write sums analogous to (19.14) for the contributions of each of the four side faces of the hemicube to the integral (19.13).

The implementation of the hemicube algorithm should now be clear. First, color code patches. Next, for each patch, and for each of the five faces of the hemicube centered at the middle of the patch, render the scene using that particular face as the viewing face and then process the resulting screen by tallying the contribution of each pixel according to its color. The contribution of a pixel to the form factor between p_i and the patch of the pixel's color, e.g.,

$$\frac{4}{\pi(x^2+y^2+1)^2 wh}$$

for a pixel on top of the hemicube, is called a *delta form factor*. Accordingly, the computation of each form factor is reduced to the process of incrementing it from zero, by a delta form factor at each step, as the screen is swept row by row, pixel by pixel, for each of the five renderings.

19.3.4 Solving the Radiosity Equation to Determine Patch Brightnesses

The second and final piece before we can practically implement the radiosity method is an efficient algorithm to solve the radiosity equation (copied from (19.9))

$$(I_n - Q)\mathbf{B} = \mathbf{E}$$

to determine the *patch brightness vector* $\mathbf{B}$, where the matrix

$$I_n - Q = \begin{bmatrix} 1 - R_1F_{11} & R_1F_{12} & \dots & R_1F_{1n} \\ R_2F_{21} & 1 - R_2F_{22} & \dots & R_2F_{2n} \\ \vdots & \vdots & \dots & \vdots \\ R_nF_{n1} & R_nF_{n2} & \dots & 1 - R_nF_{nn} \end{bmatrix}$$

Trying to solve the preceding equation by writing $\mathbf{B} = (I_n - Q)^{-1}\mathbf{E}$ and straightforwardly inverting $I_n - Q$ would be prohibitively expensive as the computation involved is $O(n^3)$, where the number n of patches is typically in the thousands. However, certain properties of the matrix $I_n - Q$, derived from properties of the form factors, lead to an efficient method to approximate its inverse.

First,

$$F_{ii} = 0, \quad 1 \le i \le n$$

because patches, being flat, cannot self-reflect. Moreover, we can assume as well that

$$\sum_{j=1}^{n} F_{ij} = 1, \quad 1 \le i \le n$$

which means all light leaving any given patch strikes other patches, by closing off the environment with black (i.e., non-reflective) patches. These properties of the form factors, together with that each reflectivity R_i is at most 1, imply that the principal diagonal of $I_n - Q$ consists of all 1's and that the sum of non-diagonal entries in any row of $I_n - Q$ is at most 1. One can then prove that

$$(I_n - Q)^{-1} = I_n + Q + Q^2 + \ldots$$

where the series on the right converges (which might remind the reader of the power series expansion $(1-x)^{-1} = 1 + x + x^2 + \ldots$, which converges if $|x| < 1$). Therefore,

$$\mathbf{B} = (I_n - Q)^{-1}\mathbf{E} = \mathbf{E} + Q\mathbf{E} + Q^2\mathbf{E} + \ldots$$

allowing the patch brightness vector $\mathbf{B}$ to be approximated to arbitrary accuracy by adding sufficiently many terms of the series on the right. Care needs still to be taken, as a simple-minded computation of the term $Q^k\mathbf{E}$ by repeatedly multiplying by Q is nearly $O(n^3)$ again. A simple observation, however, helps cut the cost.

Denote successive partial sums of the series $\mathbf{E} + Q\mathbf{E} + Q^2\mathbf{E} + \ldots$ by $\mathbf{B}_0, \mathbf{B}_1, \ldots$. In other words, $\mathbf{B}_0 = \mathbf{E}$, $\mathbf{B}_1 = \mathbf{E} + Q\mathbf{E}$, $\mathbf{B}_2 = \mathbf{E} + Q\mathbf{E} + Q^2\mathbf{E}$ and so on. Then we have the recurrence

$$\mathbf{B}_{k+1} = \mathbf{E} + Q\mathbf{B}_k, \quad k \ge 0$$

so that each successive term of the sequence $\mathbf{B}_0, \mathbf{B}_1, \mathbf{B}_2, \ldots$ of partial sums converging to $\mathbf{B}$ can be computed from the previous by a matrix-vector product and a vector-vector addition, the two operations together being of $O(n^2)$ complexity. In fact, if the matrix Q is sparse, likely if the form factor between patches at a distance greater than some threshold value are set to 0, then the complexity may be closer to linear.

Exercise 19.11. Consider the operator Ψ that acts on n-vectors by

$$\Psi(X) = \mathbf{E} + QX$$

Prove that the solution $\mathbf{B}$ of the radiosity equation is a *fixed point* of Ψ, i.e., a vector X such that $\Psi(X) = X$.

The preceding exercise leads to *Jacobi's iterative method* to approximate the fixed point $\mathbf{B}$ of Ψ as follows. Choose arbitrarily a start vector X'. Then repeatedly apply Ψ to X' to obtain a sequence $X', \Psi(X'), \Psi(\Psi(X')), \ldots$. Properties of the radiosity equations guarantee that this sequence converges to the unique fixed point of the operator Ψ. We'll not discuss the theory underlying Jacobi's method any further ourselves, but the interested reader is referred to Hageman & Young [66].

Exercise 19.12. Prove that Jacobi's iterative method to approximately determine the solution $\mathbf{B}$ of the radiosity equation, using a *zero* start vector, is precisely equivalent to the power series method of approximating $\mathbf{B}$.

Remark 19.6. A useful physical insight into the sequence $\mathbf{B}_0, \mathbf{B}_1, \mathbf{B}_2, \ldots$ of partial sums converging to $\mathbf{B}$ is that the first term represents only emitted light, the second emitted light together with diffuse light coming to the eye after a single reflection, the third emitted light together with diffuse light after at most two reflections and so on.

In practical terms, this means that using the sequence $\mathbf{B}_0, \mathbf{B}_1, \mathbf{B}_2, \ldots$ as brightness vectors illuminates the scene in an increasingly authentic manner, which leads to the cost-saving technique of executing each iteration to compute $\mathbf{B}_i$ only on demand, called *progressive refinement*.

19.3.5 Implementing Radiosity

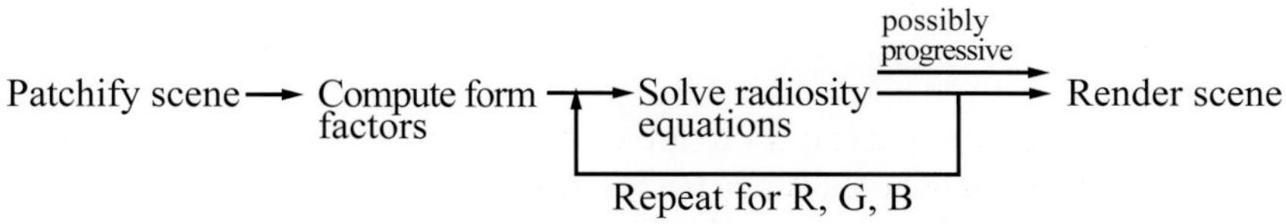

Figure 19.23: The radiosity algorithm.

Figure 19.23 shows the four steps of the radiosity algorithm. The first three are *view-independent* and may be pre-computed for a scene whose geometry does not change. The last rendering step, of course, depends on the location of the viewer. To reduce aliasing artifacts at patch borders, instead of rendering each patch with its computed brightness, each *vertex* is assigned a brightness computed from its adjacent patches, e.g., a weighted average. Subsequently, vertex colors are interpolated through each patch via Gouraud shading.

Remark 19.7. As the first three steps of the radiosity algorithm are view-independent, while the last, even though dependent on the viewer's location, is not particularly computationally intensive, radiosity can be efficiently incorporated into a real-time walk-through of a static scene, e.g., a building interior.

Experiment 19.4. Run again `sphereInBoxPOV.pov`. Then run again after uncommenting the line

```
global_settings{radiosity{}}
```

at the top to enable radiosity computation with default settings. The difference is significant, is it not?

Figure 19.24(a) is the ray-traced output without radiosity, while Figure 19.24(b) is the output with radiosity (both images are in the folder `ExperimenterSource/Chapter19/ExperimentRadiosity`). There clearly is much more light going around inside the box in the latter rendering. **End**

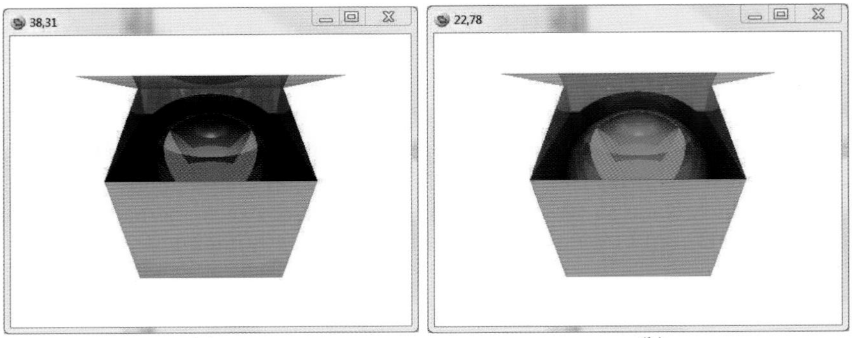

(a) (b)

Figure 19.24: Without and with radiosity: screenshot of `sphereInBoxPOV.pov` with (a) radiosity disabled (b) radiosity enabled.

Exercise 19.13. If the lighting in a scene changes, then which steps of the radiosity algorithm need to be redone? How about if the geometry changes, e.g., with a ball looping in and out of a torus?

19.4 Summary, Notes and More Reading

In this chapter we went into particularly gory detail about the synthetic-camera pipeline that OpenGL implements, the fixed-functionality variant in particular. The reader should now be in a position to even implement a barebones version of her own. The synthetic-camera pipeline is based on a local illumination model. We were introduced as well to two global models, those of ray tracing and radiosity, and saw how much more realism they afford than the synthetic camera, though at hugely more computation cost.

The book by Jim Blinn [17], a CG pioneer, has several insightful articles, written in his particularly entertaining style, on various pipeline-related topics. Segal-Akeley [128] is a must-read high-level overview of the OpenGL pipeline written by two members of the original design team and, of course, the red book itself is a canonical source.

The seminal work on ray tracing was by Appel [4] and Whitted [148], and on radiosity by Goral [58]. A classic introduction to ray tracing is by Glassner [56]. For more advanced reading about ray tracing and radiosity, some useful textbooks are Akenine-Möller, Haines & Hoffman [1], Buss [22] and Watt [147]. Cohen & Wallace [26] and Sillion & Puech [132] are especially about radiosity.

Part XI

Rendering Pipe Dreams

CHAPTER 20

OpenGL 4.3, Shaders and the Programmable Pipeline: Liftoff

Programmers mutiny! We're going to throw off our shackles and take over the engine room!

The first radical advancement of OpenGL, since its creation in 1992, was the inclusion of shaders in OpenGL 2.0 released in 2004. Shaders are ancillary programs, attached to an OpenGL program, that run on the GPU and are written by the user to supplant and enhance parts of the graphics pipeline formerly of fixed-functionality. They are written in a C-like language called the OpenGL Shading Language (abbreviated GLSL).

Historically, shaders evolved as a response to the increasing capabilities of GPUs and the need to expose these to the application programmer. Before the standardization of the GLSL for shaders a programmer had to write code in vendor-specific language to access individual GPU features – a difficult and inefficient task at best. Just as high-level programming languages like C evolved from assembly in order to hide low-level calls from the developer and give her a structured environment, so did the GLSL.

As a language the GLSL itself is based on C, so coding will not be a problem for us. In addition to much of C's functionality, the GLSL necessarily has features for shaders to interact with each other, as well as with the application OpenGL program to which they are attached.

The first version of the GLSL – that included in OpenGL 2.0 in 2004 – was GLSL 1.1. Since then both OpenGL and the GLSL have progressed in tandem through several versions to, as of this writing, OpenGL 4.3 and GLSL 4.3 (since OpenGL 3.3, released in 2010, OpenGL and GLSL version numbers have matched). In this chapter and the next we are going to cover the OpenGL 4.3 *forward-compatible core* version, the snooty 1%er which brooks no legacy or even deprecated commands, and its sister GLSL 4.3.

With this chapter and the next, we intend to take the reader from the fixed-functionality pre-shader OpenGL thus far in this book to a treatment of core 4.3 complete except maybe for a few advanced features. We are confident though that, as she continues, the reader will agree that pre-shader OpenGL was well worth the effort, even if her goal all along was mastery of the latest version of the API, because fundamental concepts remain the same and, most importantly, fixed-functionality allows for a comfortable learning curve to 4.3, which, otherwise, could well prove vexing for the novice graphics programmer.

Section 20.1 is an overview of the programmable pipeline with a brief discussion of its most significant consequences to actually writing OpenGL code. GLSL, particularly its data types, is introduced in Section 20.2. In Section 20.3 we take apart our first core 4.3 program to see what makes it tick – this program, in fact, is a rewrite of our very first OpenGL program `square.cpp`. Animation, lighting and textures, as managed in the 4.3 pipeline, are the topics, respectively, of Sections 20.4, 20.5 and 20.6. We conclude with Section 20.7.

20.1 New Pipeline for OpenGL

We discuss the role of shaders in the programmable pipeline and changes in OpenGL as it has grown to accommodate shaders.

20.1.1 Shaders in the Rendering Pipeline

There are four possible *shader stages* in the programmable pipeline: vertex, tessellation, geometry and fragment. Each stage, except tessellation, corresponds to a single shader program written in the GLSL; the tessellation stage can consist of two shader programs, namely, tessellation control and tessellation evaluation.

Now take a good hard stare again at the fixed-function OpenGL pipeline of the previous chapter, copied below in Figure 20.1. Next, look at Figure 20.2 which shows the programmable pipeline. As you see, the fixed-function sequence from perspective division to rasterization remains intact, but now there are new shading stages before this sequence, as well as between it and the per-fragment operations which also remain. The texturing path, as we'll find, has been subsumed into the fragment shader. Let's make a quick first acquaintance of the shaders.

Vertex shader: This is a mandatory shading stage which was part of the original programmable pipeline specification in OpenGL 2.0. The vertex shader runs once per input vertex, processing the data associated with it, e.g., world coordinates, color values, normal values, texture coordinates. At minimum, it must output $xyzw$-coordinates for each vertex, presumably after

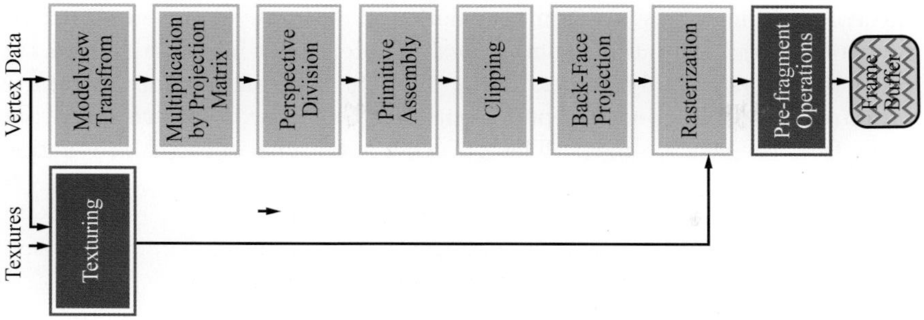

Figure 20.1: OpenGL fixed-function pipeline.

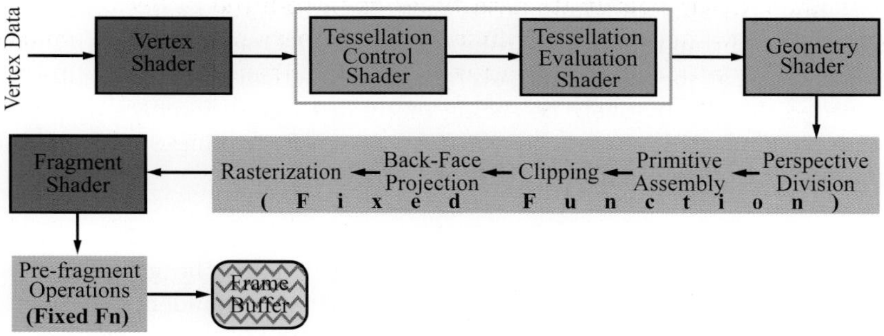

Figure 20.2: OpenGL programmable pipeline.

multiplication by the modelview matrix (i.e., after modelview transformation) and by the projection matrix. This make perfect sense if you see the first two stages of the fixed-function pipeline in Figure 20.2, which, in fact, have been supplanted by the vertex shader.

Tessellation shader: This is an optional stage comprising two shaders, the tessellation control and evaluation shaders. In fact, even the tessellation control shader is optional and tessellation shading can be done with only the tessellation evaluation shader. Tessellation shading was introduced in OpenGL 4.0 for the purpose of LOD (level-of-detail) management. It can adaptively refine or coarsen an object's mesh, e.g., adding more triangles as it comes closer to the eye.

Geometry shader: This is yet another optional stage, following tessellation in the pipeline, but, actually, introduced earlier in OpenGL 3.2. The geometry shader allows the programmer to transform the original geometry, e.g., replacing triangles with lines, or new triangles of a different size, or replacing lines with points, and such.

Fragment shader: This is a mandatory shading stage which, like the vertex shader, was part of the original programmable pipeline of OpenGL 2.0. The fragment shader runs once per output fragment either setting its color or discarding it (which means simply not drawing the fragment).

The color value set by a fragment shader may, however, be modified by the per-fragment operations coming next before finally being written to the frame buffer. Typically, for example, the fragment shader may compute the interpolated color values per fragment from color values received per vertex from the vertex shader. The texturing operations of the fixed-function pipeline are now the responsibility of the fragment shader.

20.1.2 New OpenGL

As shaders have evolved so has OpenGL, the shared aim being to (a) leverage the computational power of the GPU to the maximum, and (b) minimize traffic, particularly, transfer of vertex data, along the relatively slow CPU-GPU bus. To this end, numerous new features have been added in the progression through versions from (pre-shader) OpenGL 1.0 to OpenGL version 4.3 as, at the same time, several old ones were deprecated or altogether discarded. We'll obviously be seeing many of these differences as we code 4.3. Nevertheless, it's worth noting even now a few changes with palpably large footprints on programming:

1. *Elimination of immediate mode (viz.* `glBegin()-glEnd()`*) drawing*: 4.3 allows only retained-mode drawing calls of the `glDraw*()` and `glMultiDraw*()` type. The reason is not hard to understand from the following analogy.

   Compare issuing a stream of instructions via cell phone to a friend in a supermarket along the lines of "Got the milk? Good! Now, pick up a loaf of whole wheat from across the aisle. Great! Frosted Flakes next in cereals on aisle 9 ..." with, instead, setting him off once and for all with a shopping list which he can himself optimize together (possibly) with other stuff that he has to buy.

   The first option might make sense if your buddy happens to be a bit thick like GPUs from a decade ago, but certainly not if he is as quick as even the low-end ones nowadays.

   Fortunately, we switched (well, mostly) from immediate mode to retained way back in Section 3.1, though our motivation then was separating data out of drawing procedures more so than efficient GPU usage.

2. *Requirement that all data must be stored in buffer objects*: We met buffer objects a long time ago too, in Section 3.2, learning even then their utility in saving CPU-GPU traffic by providing server-side storage for vertex and pixel data. That section was, particularly, about VBOs (vertex buffer objects), while the next, Section 3.3, was about VAOs (vertex array objects) which help encapsulate the buffer objects related to a given geometric object.

At the end of Section 3.3 we, in fact, counseled the user against coding VBOs and VAOs then because the added complexity would detract from our focus on fundamentals at that time. Now, though, the situation is changed: VBOs and VAOs are *compulsory* in 4.3. So, it's time for a review exercise if need be.

Exercise 20.1. (Programming) Review Section 3.2 and Section 3.3 on VBOs and VAOs, respectively, and do the exercises therein.

It's important to be comfortable with their usage before starting to code 4.3.

3. *Elimination of modelview transformation commands*: `glTranslatef()`, `glRotatef()`, `glScalef()` and `gluLookAt()` are all gone from 4.3, the management of the modelview matrix stack (in fact, also the projection and texture matrix stacks) now the responsibility of the programmer who has to write, store and operate on the matrices herself (don't worry, we'll import a library to help with this!).

   The principle behind this change is that the vertex shader already allows the user to change vertex coordinates with her own matrices, so why not give her full charge, instead of leaving some coordinate processing to fixed-function.

4. *Do-it-yourself lighting*: `glLight*()` and `glMaterial*()` commands? All gone from 4.3! You have to calculate the color values at lit vertices yourself and then if, say, you want to Gouraud (i.e., smooth) shade the interiors of triangle, you're going to have to do that yourself too (no, there is no `glShadeModel()` either).

   Again, this approach makes sense if one observes that lighting is all about coloring pixels, which is exactly what the fragment shader is for.

One might conclude that 4.3 tends to puts the programmer in a master-slave relationship with herself.

Defining the OpenGL Context

With changing OpenGL versions, there is naturally demand sometimes from consumers to ensure continued usability of code written in an older version, as equally from developers to protect their products from future obsolescence. For this reason, since OpenGL 3.0, there is a fairly refined way that one can ask an *OpenGL context* from the operating system (think of an OpenGL context as the interface between an OpenGL program and the drivers and hardware which run it).

Firstly, the command

```
glutInitContextVersion(major, minor);
```

specifies the OpenGL version number *major.minor*. Next

```
glutInitContextProfile(profile);
```

where *profile* can be GLUT_CORE_PROFILE or GLUT_COMPATIBILITY_PROFILE specifies the profile. The core profile excludes all features discarded from the specification of the current or earlier versions, while the compatibility profile includes them all. This is why all our programs thus far, till Chapter 19, which declared a version 4.3 compatibility profile with the statements

```
glutInitContextVersion(4, 3);
glutInitContextProfile(GLUT_COMPATIBILITY_PROFILE);
```

in `main()` were able to use legacy non-4.3 commands. From now on, though, we'll be replacing the above block with

```
glutInitContextVersion(4, 3);
glutInitContextProfile(GLUT_CORE_PROFILE);
glutInitContextFlags(GLUT_FORWARD_COMPATIBLE);
```

which asks for a 4.3 core profile; in fact, the last statement asks for forward-compatibility, which means excluding features marked for deprecation in the current version, thus ensuring compatibility with future versions.

20.2 GLSL Basics

We'll come to grips in earnest with shaders in the next section but before that our goal in this one is to give a quick overview of their language, the GLSL.

To begin with, GLSL has the "Cish" (though, not exactly C) *basic data types*:

`float` 32-bit floating point number

`double` 64-bit floating point number

`int` signed 32-bit integer

`uint` unsigned 32-bit integer

`bool` Boolean (true/false)

GLSL has, as well, the *aggregate data types* vectors and matrices to more accurately support OpenGL functionality. Particularly, 2-, 3- and 4-component vectors are available in each of the five basic types:

float:	`vec2`	`vec3`	`vec4`
double:	`dvec2`	`dvec3`	`dvec4`
int:	`ivec2`	`ivec3`	`ivec4`
uint:	`uvec2`	`uvec3`	`uvec4`
bool:	`bvec2`	`bvec3`	`bvec4`

Moreover, GLSL implements floating point and double matrices:

float:	mat2x2	mat2x3	mat2x4
	mat3x2	mat3x3	mat3x4
	mat4x2	mat4x3	mat4x4
	mat2	mat3	mat4
double:	dmat2x2	dmat2x3	dmat2x4
	dmat3x2	dmat3x3	dmat3x4
	dmat4x2	dmat4x3	dmat4x4
	dmat2	dmat3	dmat4

Note that mat$p$xq and dmat$p$xq have each p columns and q rows, while mat$p$ and dmat$p$ are square of size p.

Construction and initialization of both the basic and aggregate types are pretty much along the lines one would expect coming from C, e.g.,

```
vec4 color;
color = vec4(1.0, 0.0, 1.0, 1.0);
vec3 rgbColor = vec3(color); // rgbColor has the first
                             // three components of color
```

A matrix, e.g.,

$$M = \begin{bmatrix} 1.0 & 3.0 & 5.0 \\ 2.0 & 4.0 & 6.0 \end{bmatrix}$$

can be initialized in multiple ways including

```
mat3x2 M = mat3x2(1.0, 2.0, 3.0, 4.0, 5.0, 6.0); // Column major!
```

and

```
vec2 column0 = vec2(1.0, 2.0);
vec2 column1 = vec2(3.0, 4.0);
vec2 column2 = vec2(5.0, 6.0);
mat3x2 M = mat3x2(column0, column1, column2);
```

There's a simple way to initialize a scalar matrix, e.g.,

```
mat2 M = mat2(3.0);
```

sets

$$M = \begin{bmatrix} 3.0 & 0.0 \\ 0.0 & 3.0 \end{bmatrix}$$

Accessing the components of an aggregate type is Cish too (e.g., v[i], M[i][j] to access elements of a vector and matrix, respectively). Additionally, one has the particularly OpenGL-friendly way of accessing the first, second, third or fourth component of a vector via the corresponding member of any one of the following three sets of accessors:

```
x, y, z, w          r, g, b, a          s, t, p, q
```

Particularly, if v is a vec4 variable then $v.x$ (or $v.r$ or $v.s$) is its first component, $v.y$ (or $v.g$ or $v.t$) its second component, and so on. The above sets of accessors, of course, are meant to be used in connection with position coordinates, color values and texture coordinates, respectively. The sets cannot be mixed, though, in the same statement. The "." in the middle is often called the *swizzle* operator. The following snippet illustrates both traditional access and *swizzling*, which is the use of the swizzle operator to both access and rearrange components.

```
vec4 pos1 = vec4(1.0, 2.0, 3.0, 4.0);
float xVal = pos1[0]; // xVal = 1.0
float xVal = pos1.x;  // xVal = 1.0
float yVal = pos1.y;  // yVal = 2.0
float yVal = pos1.g;  // yVal = 2.0

vec4 pos2 = pos1.yxzw; // Rearranging: pos2 = (2.0, 1.0, 3.0, 4.0)
vec4 pos3 = pos1.rrba; // Duplication: pos3 = (1.0, 1.0, 3.0, 4.0)
vec4 pos4 = vec4(pos1.xyz, 5.0); // pos4 = (1.0, 2.0, 3.0, 5.0).
vec2 pos5 = pos1.xy; // pos5 = (1.0, 2.0).
vec4 pos6 = pos1.xgga; // Illegal: mixing names from two sets.
```

One can swizzle on the left-hand side as well, but repeated components are disallowed, e.g.,

```
vec4 pos1 = vec4(1.0, 2.0, 3.0, 4.0);
pos1.xy = vec2(5.0, 6.0); // pos1 = (5.0, 6.0, 3.0, 4.0).
pos1.yx = vec2(5.0, 6.0); // pos1 = (6.0, 5.0, 3.0, 4.0).
pos1.xx = vec2(5.0, 6.0); // Illegal: x is repeated.
```

Here is a snippet illustrating access of elements of a matrix:

```
mat3x2 M = mat3x2(1.0, 2.0, 3.0, 4.0, 5.0, 6.0);
vec2 column2 = M[2]; // column2 = vec2(5.0, 6.0)
float xTan = M[2][1]; // xTan = 6.0
float xTan = M[2].y; // xTan = 6.0
```

Note: Keep in mind the column-major order for matrices! The usual math convention is that $M[i][j]$ is the element in row i and column j; in GLSL it's the element in column i and row j.

The fun with GLSL vectors and matrices starts with applying the operators "$*$" and "$+$" between them, when they start behaving as in linear algebra, e.g.,

```
mat2 M = mat2(1.0, 2.0, 3.0, 4.0);
mat2 N = mat2(1.0, 0.0, 0.0, 2.0);
mat2 P = M + N; // P = mat2(2.0, 2.0, 3.0, 6.0)
P = M * N; // P = mat2(1.0, 2.0, 6.0, 8.0)

vec2 V = vec2(1.0, 2.0);
vec2 W = M * V; // W = vec2(7.0, 10.0)
```

The reader can guess how this feature will, amongst other things, support modelview and projections transforms.

The *complex data types* structures (`struct`) and arrays (`[ ]`), functioning as in C, are implemented in GLSL as well. Additionally, and usefully, one can query the length of an array with help of the Java-like `length` method, e.g., if the array `A` contains 10 elements then `A.length()` returns 10.

Variable types discussed thus far are all *transparent* in that they can be read and written directly. GLSL has a class of *opaque* types, as well, which can be accessed only via built-in functions. Opaque variables are always handles to other objects. The only opaque type that we'll use is `sampler2D`, a handle to a 2D texture.

Variable declarations may be preceded by at most one of the *storage qualifiers* listed in Table 20.1.

`const`	Read-only variable whose value is fixed after initialization.
`in`	Variable whose value is input from a previous shader stage or the application program.
`out`	Variable whose value is output to a subsequent shader stage.
`uniform`	Variable whose value is supplied to the shader by the application and is constant across a primitive.
`buffer`	Variable which can be read and written by both the shader and the application.
`shared`	Variable shared within a local work group (only compute shaders).

Table 20.1: Storage Qualifiers

A conceptually simple classification of variables is as *attribute* and *uniform*: attribute variables (or, simply, attributes) are those that vary from vertex to vertex, while uniform variables (or, uniforms) vary from primitive to primitive (just like its namesake storage qualifier, of course). The coordinates of a vertex are classic examples of an attribute variable; the modelview matrix is an example of a uniform; color, e.g., could be attribute if we define a different color for each vertex, or uniform if we choose to keep colors constant across objects.

A typical example of the declaration of an attribute variable in the vertex shader is

```
layout(location=1) in vec4 coordinates;
```

which, in fact, introduces the *layout qualifier* which in/out variables, uniforms and buffers may have additionally. Generally, the syntax of a layout qualifier is of the form

`layout(`*parameter1* `or` *parameter1* = *value*`,` *parameter2* `...)` *variable definition* where the parameter values specify properties of the variable. Particularly, in the declaration of `coordinates` above the value of the parameter `location` indicates the buffer which supplies the values of `coordinates`.

An example of a uniform declaration which might be in the vertex shader, as well, is

```
uniform mat4 modelViewMatrix;
```

As another example, the declaration

```
out vec4 colors;
```

in the vertex shader might pair with the declaration

```
in vec4 colors;
```

in the fragment shader. The latter, too, might declare

```
uniform sampler2D aTexture;
```

as a handle to a texture.

We'll leave it at this for now. Don't worry if the definitions don't all make sense at this time. It should all come together when we get to live GLSL code in a bit. However, if you are interested now in a fuller specification of the language, then see the red book, or even better, refer to the horse's mouth, that being the GLSL spec sheet at `opengl.org`.

20.3 First Core GL 4.3 Program (Dissected)

For our first program we'll *shaderize* our trusty guinea pig from way back when, namely, `square.cpp` – we coin the rather ugly verb "shaderize" to indicate converting one of our existing pre-shader programs to compliance with the latest forward-compatible core profile (GL 4.3 in our case). Shaderizing the simple `square.cpp`, making sure we understand each and every step of the process, should set us on our way.

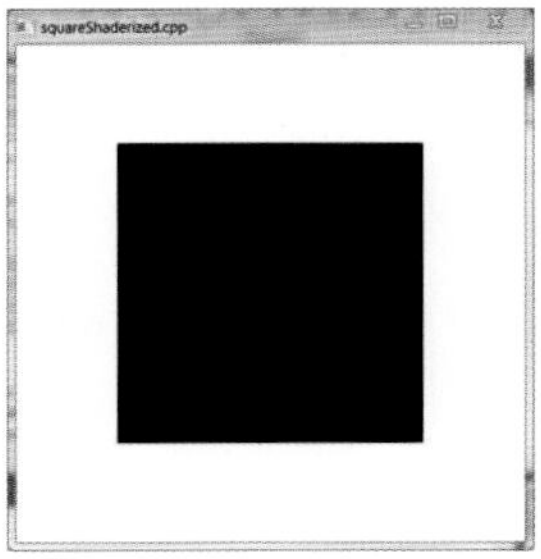

Figure 20.3: Screenshot of `squareShaderized.cpp`.

Experiment 20.1. Fire up the application program `squareShaderized.cpp`, which comes with its two sidekick shaders, the imaginatively named `vertexShader.glsl` and `fragmentShader.glsl`. Compiling and running the application program should automatically suck in the shaders provided they are all in the same folder. Not only is the functionality of `squareShaderized.cpp` – drawing a black square over white background, see Figure 20.3 – *exactly* that of `square.cpp`, but, as we'll see, so are its internals (modulo shaders). **End**

We're going to step through the code of `squareShaderized.cpp` line by line from top to bottom, but first point out the three statements

```
   glutInitContextVersion(4, 3);
   glutInitContextProfile(GLUT_CORE_PROFILE);
   glutInitContextFlags(GLUT_FORWARD_COMPATIBLE);
```

in `main()`, which assert that the program is compliant with the 4.3 *forward-compatible core* profile vs. the 4.3 *compatibility* (i.e., backward-compatibility) profile as in all earlier programs. Yes, indeed, from now on we are going to write the latest and greatest* OpenGL!

Okay, so back to the top of `squareShaderized.cpp`. First, to better manage buffer data we set up the `Vertex` structure

```
struct Vertex
{
   float coords[4];
   float colors[4];
};
```

with one array field for (x, y, z, w) coordinates and another for RGBA colors. Next, to manage matrices – 4.3 asks us to do modelview and projection matrices ourselves – we define the `Matrix4x4` structure

```
struct Matrix4x4
{
   float entries[16];
};
```

to hold the 16 entries of a 4×4 matrix. Then, we define the 4×4 identity matrix

```
static const Matrix4x4 IDENTITY_MATRIX4x4 =
{
   {
      1.0, 0.0, 0.0, 0.0,
      0.0, 1.0, 0.0, 0.0,
      0.0, 0.0, 1.0, 0.0,
      0.0, 0.0, 0.0, 1.0
   }
};
```

as a constant. Next, a couple of enums, namely,

```
static enum buffer {SQUARE_VERTICES};
static enum object {SQUARE};
```

contain names for buffer and vertex array objects, respectively.

Next, we have a block of global variables:

```
static Vertex squareVertices[] =
{
```

*As of this writing.

```
   { { 20.0, 20.0, 0.0, 1.0 }, { 0.0, 0.0, 0.0, 1.0 } },
   { { 80.0, 20.0, 0.0, 1.0 }, { 0.0, 0.0, 0.0, 1.0 } },
   { { 20.0, 80.0, 0.0, 1.0 }, { 0.0, 0.0, 0.0, 1.0 } },
   { { 80.0, 80.0, 0.0, 1.0 }, { 0.0, 0.0, 0.0, 1.0 } }
};

static Matrix4x4
   modelViewMat = IDENTITY_MATRIX4x4,
   projMat = IDENTITY_MATRIX4x4;

static unsigned int
   programId,
   vertexShaderId,
   fragmentShaderId,
   modelViewMatLoc,
   projMatLoc,
   buffer[1],
   vao[1];
```

First, above, come the coordinates (same as in `square.cpp`) and color values (black) for the four vertices of the square to be drawn. Then, we declare and initialize our very own modelview and projection matrices. We'll explain the seven `unsigned int` variables declared at the bottom when they are initialized later on.

The routine `readTextFile()` to read an external text file into a character string, obviously to be used to read the shaders, is one whose innards are not of particular interest.

On to the initialization routine `setup()` next.

Compiling and Linking Shaders

After the mandatory `glClearColor()`, the first part of `setup()` compiles and links the two shaders into a single shader program executable. Here's the first block of statements from this part:

```
char* vertexShader = readTextFile("vertexShader.glsl");
vertexShaderId = glCreateShader(GL_VERTEX_SHADER);
glShaderSource(vertexShaderId, 1, (const char**) &vertexShader,
               NULL);
glCompileShader(vertexShaderId);
```

The first statement reads in the text file `vertexShader.glsl` containing the vertex shader into the character string `vertexShader`, while the next creates an empty vertex shader object, returning the non-zero id `vertexShaderId`. The third statement sets the source code of the shader with id `vertexShaderId` to the value of the character string `vertexShader`, the second and fourth parameters of `glShaderSource()` indicating there is only one null-terminated string. The last statement compiles the source code for shader id `vertexShaderId`. So, on completion of this block we have a compiled vertex shader object.

The second block of statements processes likewise the fragment shader source file `fragmentShader.glsl` to produce a compiled fragment shader object with id `fragmentShaderId`.

The final block

```
programId = glCreateProgram();
glAttachShader(programId, vertexShaderId);
glAttachShader(programId, fragmentShaderId);
glLinkProgram(programId);
glUseProgram(programId);
```

of the first part of `setup()` is fairly self-explanatory. The first statement creates an empty shader program object, returning its non-zero id `programId`, the next two statements attach the shader objects identified by `vertexShaderId` and `fragmentShaderId` to the program object identified by `programId`, while the final two statements, respectively, link the shader objects attached to the program id `programId` to create an executable shader program and install it within the current rendering state. The entire process is diagrammed in Figure 20.4.

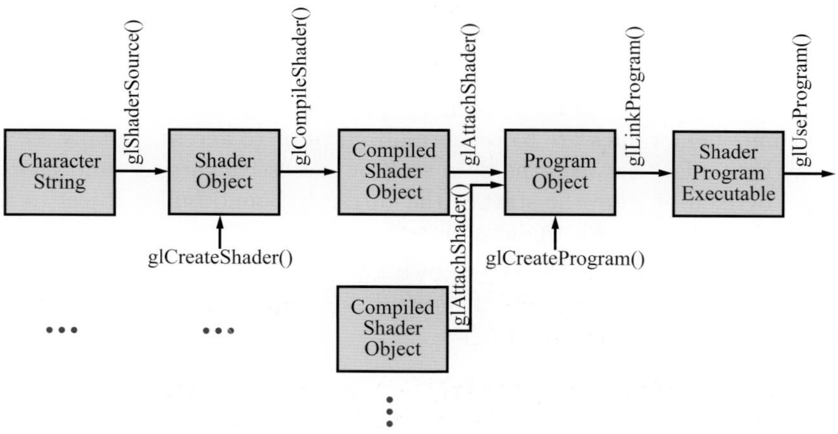

Figure 20.4: Process to create a shader program executable.

Initializing Data and Communicating with the Vertex Shader

The second part of the `setup()` routine sets up and initializes both a VAO (vertex array object) and VBO (vertex buffer object) and associates the data in the latter with variables in the vertex shader. Now, we are going to assume that, having reviewed Sections 3.2 and 3.3, the reader understands how, in fact, the first block of statements

```
glGenVertexArrays(1, vao);
glGenBuffers(1, buffer);
glBindVertexArray(vao[SQUARE]);
glBindBuffer(GL_ARRAY_BUFFER, buffer[SQUARE_VERTICES]);
glBufferData(GL_ARRAY_BUFFER, sizeof(squareVertices),
             squareVertices, GL_STATIC_DRAW);
```

creates the VAO with id `vao[SQUARE]` containing the buffer with id `buffer[SQUARE_VERTICES]`, filling the latter with the square's vertex data.

Here, then, is the next block which, critically, is the first step connecting the application program to the vertex shader:

```
glVertexAttribPointer(0, 4, GL_FLOAT, GL_FALSE,
                      sizeof(squareVertices[0]), 0);
glEnableVertexAttribArray(0);
glVertexAttribPointer(1, 4, GL_FLOAT, GL_FALSE,
                      sizeof(squareVertices[0]),
                      (GLvoid*)sizeof(squareVertices[0].coords));
glEnableVertexAttribArray(1);
```

Let's begin to understand it. Generally, the command

`glVertexAttribPointer(`*index*`,` *size*`,` *type*`,` *normalized*`,` *stride*`,` *pointer*`)`

specifies where data for the shader attribute at location *index* is to be accessed. The value of *size* is the number of components to be read per attribute, while *type* is the data type of a component. The Boolean *normalized* specifies if the component values are to be normalized prior to access. Finally, *stride* is the byte offset between the data sets for successive vertices, and *pointer* is the byte offset from the start of the currently-bound buffer object to the start of the data set for the first vertex.

Time now for our first peek at a shader, bringing it into the discussion. Open the vertex shader `vertexShader.glsl` in any text editor. The statements

```
layout(location=0) in vec4 squareCoords;
layout(location=1) in vec4 squareColors;
```

declare the `vec4` attribute variables `squareCoords` and `squareColors`, respectively, the storage qualifier `in` indicating that both get data values from the application program. The variable `squareCoords` is at location 0 and `squareColors` at location 1.

So, the `glVertexAttribPointer(0, ...)` statement above in the application program causes the vertex shader to read four floats (a `vec4`, in other words) for `squareCoords` per vertex from the GPU-side `buffer[SQUARE_VERTICES]` starting from the beginning of the buffer, with a stride of `sizeof(squareVertices[0])` (i.e., the size of data for one vertex) between data sets.

Likewise, `glVertexAttribPointer(1, ...)` means it will read four floats for `squareColors` from `buffer[SQUARE_VERTICES]` starting from byte offset `sizeof(squareVertices[0].coords)` (i.e., the size of the coordinates field) with the same stride of `sizeof(squareVertices[0])`.

The statements `glEnableVertexAttribArray(0)` and `glEnableVertexAttribArray(1)`, above in the application program as well, activate, respectively, the attribute data pointed by `glVertexAttribPointer(0, ...)` and `glVertexAttribPointer(1, ...)` (this data often being called a *vertex attribute array*, hence the name).

See Figure 20.5 for a diagram of how the application program and vertex shader link up.

Continuing with `setup()`, the next part, setting the projection matrix and connecting it to the vertex shader, is

```
Matrix4x4 projMat =
{
   {
      0.02, 0.0,  0.0, -1.0,
      0.0,  0.02, 0.0, -1.0,
      0.0,  0.0, -1.0,  0.0,
      0.0,  0.0,  0.0,  1.0
   }
};
projMatLoc = glGetUniformLocation(programId, "projMat");
glUniformMatrix4fv(projMatLoc, 1, GL_TRUE, projMat.entries);
```

Referring back to Example 18.1, particularly matrix (18.2), we see that the matrix corresponding to the projection command of `square.cpp`, viz. `glOrtho(0.0, 100.0, 0.0, 100.0, -1.0, 1.0)`, is, indeed, the value of `projMat` as defined in the first line above. At this time we swing over once more to the vertex shader to note the two uniform declarations

```
uniform mat4 projMat;
uniform mat4 modelViewMat;
```

Returning to the application program part earlier above, `glGetUniformLocation()` in the next statement returns the location of the shader uniform variable named `projMat`, within the (current) program object `programId`, into the variable `projMatLoc`. The last command `glUniformMatrix4fv()` updates the uniform at the location given by its first parameter, namely, `projMatLoc` – this uniform, indeed, being `projMat` by the previous line – with the value pointed by its fourth parameter, namely, that of the application program variable `projMat`.

Moreover, the second parameter of `glUniformMatrix4fv()` specifies that there is one matrix to update, while the third specifies to use the transpose of `projMat` in updating (because GLSL matrices are stored in column-major order).

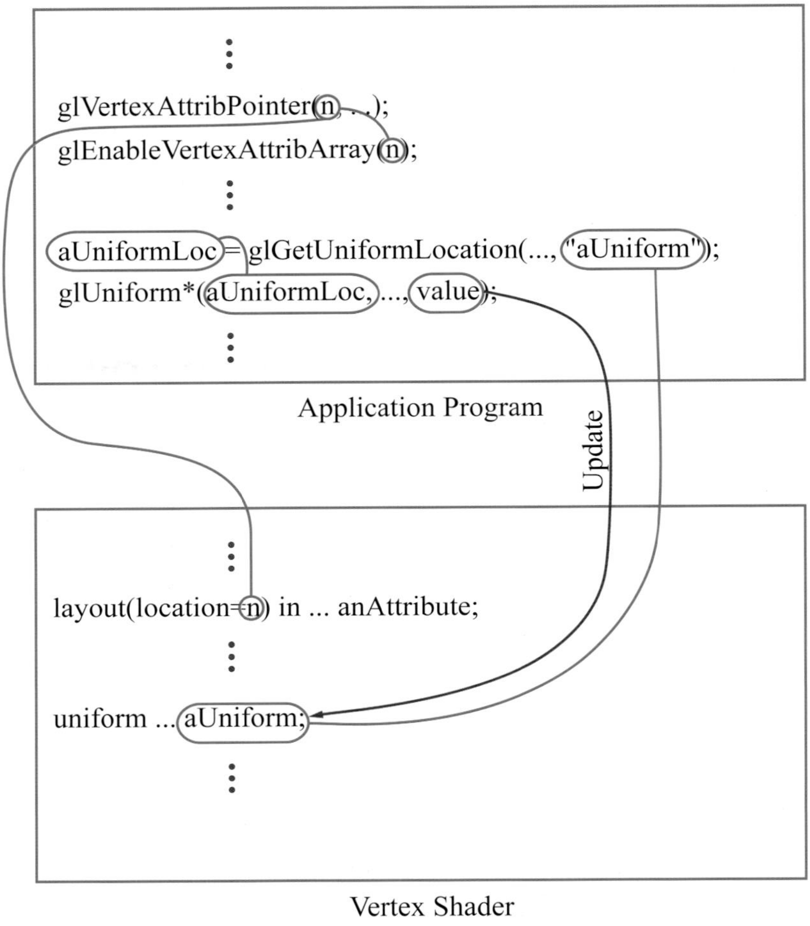

Figure 20.5: Linkages between the application program and vertex shader. During run-time, the source data for the attribute variables are read from GPU buffers, while uniform values are shipped from the CPU by the application program.

Note: As the application program and shaders have different name spaces, we can use identical names (e.g., `projMat` above) for application and shader variables, when this makes sense, without risk of ambiguity.

The final part of `setup()` setting the modelview matrix and connecting it to the vertex shader is a process exactly similar to that for the projection matrix and we'll leave the reader to parse it.

Remaining Routines

The drawing routine drawScene() is extremely simple: the one non-trivial call glDrawArrays(GL_TRIANGLE_STRIP, 0, 4) draws the square as a 4-vertex (2-triangle) triangle strip. Keep in mind that immediate mode drawing, precisely, glBegin()-glEnd() code, has been banished from 4.3 – we must use retained mode calls of the glDraw*() and glMultiDraw*() type.

The window reshape routine resize() in squareShaderized.cpp is simpler, too, than in square.cpp as the vertex shader has now charge of the projection and modelview matrices: the one statement glViewport(0, 0, w, h) sets the viewport to the entire OpenGL window.

The keyInput() routine is unchanged from square.cpp, while the only change in main(), as we noted right at the start, is from (backward-) compatibility to forward-compatible core profile.

Finally, we examine the two shaders, both of which, fortunately, are fairly minimal.

Vertex Shader

The first line of the vertex shader is the preprocessor command

```
#version 430 core
```

which declares the shader's version of GLSL to be 4.3 core. We've already discussed the four vertex shader variables declared in

```
layout(location=0) in vec4 squareCoords;
layout(location=1) in vec4 squareColors;

uniform mat4 projMat;
uniform mat4 modelViewMat;
```

in connection with the application program – these four variables, in fact, form the communication interface between the application program and the vertex shader. The remaining vertex shader variable, declared

```
out vec4 colorsExport;
```

to match with the fragment shader declaration

```
in vec4 colorsExport;
```

is used evidently to communicate color values from the vertex to the fragment shader. Moreover, as can be seen from its main routine

```
void main(void)
{
   gl_Position = projMat * modelViewMat * squareCoords;
   colorsExport = squareColors;
}
```

our vertex shader does little work. The first line applies modelview and then projection transformation on input vertex world coordinates, writing the result into the *built-in variable* `gl_Position` – built-in variables are system-defined variables which shaders use to communicate with fixed-function pipeline stages. In fact, it is `gl_Position` which continues into the fixed-function perspective division stage of the pipeline (see again Figure 20.2, noting there is neither tessellation nor geometry shader associated with `squareShaderized.cpp`). The second line of `main()` simply copies the input color values into the `colorsExport` variable for output to the fragment shader.

Fragment Shader

The first line

```
#version 430 core
```

of the fragment shader is the same preprocessor GLSL version declaration as in the vertex shader. Next,

```
in vec4 colorsExport;
```

expectedly matches the namesake `colorsExport` of type `out` in the vertex shader. Lastly,

```
out vec4 colorsOut;
```

is declared to output the color values of a fragment.

Remark 20.1. OpenGL automatically identifies the fragment shader's output variable – it must have exactly one such and of type `vec4` – as supplying the fragment's final color values.

The one-line main routine

```
void main(void)
{
   colorsOut = colorsExport;
}
```

simply copies the color values input from the vertex shader into `colorsExport` over to `colorsOut` for output. Such a shader is often called a *pass-through* shader because it simply relays incoming values on to the next stage of the pipeline.

However, there is a bit more going on in that one line in main than meets the eye. The fragment shader receives color values from the vertex shader per *vertex*, while it outputs them per *fragment* (remember, a fragment shader runs once per fragment). So, how does it propagate values from vertices to fragments? Interpolation seems the answer, and indeed is the case with this fragment shader though, evidently, the process is being done with a bit of "secret" help from fixed-function because there is no pertinent code in the fragment shader.

The GLSL actually offers options for coloring interior fragments of a triangle in the form of three *interpolation qualifiers* – namely, `smooth`, `noperspective` and `flat` – whose respective functions are described in Table 20.2. We ask the reader to compare the `smooth` and `flat` options next.

`smooth`	Perspectively correct interpolation (see Section 19.1.3). This is the default. Works exactly the same as `glShadeModel(GL_SMOOTH)`, the default shading model (see Section 11.8)) in pre-shader OpenGL.
`noperspective`	Linear interpolation without perspective correction (rarely used).
`flat`	No interpolation: all fragments given same color value, which is that of the *provoking vertex* (see discussion of flat shading in Section 11.8) of the triangle. Works like `glShadeModel(GL_FLAT)` in pre-shader OpenGL.

Table 20.2: Interpolation Qualifiers

Exercise 20.2. (Programming) Change the color values of the square's four vertices in `squareShaderized.cpp` to red, green, blue and yellow, respectively. Smooth interpolation should now be evident. Next, change the declaration of the `colorsExport` variable in the vertex shader to

```
flat out vec4 colorsExport;
```

and in the fragment shader to

```
flat in vec4 colorsExport;
```

Does what you see tally with the rule for provoking vertices of triangles in a strip described in Section 11.8?

Finally, we are done with our first forward-compatible GL 4.3 core program! A long slog it was but well worth the effort because subsequent programs are going to follow pretty much the same template. Before moving on though here are a couple for you to write.

Exercise 20.3. (Programming) Shaderize `squareAnnulus4.cpp` from Chapter 3.

Exercise 20.4. (Programming) Shaderize `hemisphereMultidraw-VBO.cpp` from Chapter 3.

20.4 Animation

We move right on to animation. Now, as we have seen, 4.3 asks us to manage our own modelview matrix; in fact, there is no `glTranslatef()`, `glRotatef()` or `glScalef()` at our disposal. We are on our own! To save doing a bunch of 4×4 matrix computation in code, therefore, we first import the GLM library.

OpenGL Mathematics (GLM)

GLM, standing for OpenGL Mathematics, is a header-only C++ library for graphics applications meant to replicate the math functionality of the GLSL. So, a programmer familiar with the GLSL will automatically be able to use GLM. However, not only does GLM simulate GLSL math, it (amongst other things) provides replacements for discarded OpenGL functions like, well, `glTranslatef()`, `glRotatef()`, `glScalef()` (yay!) and the like.

We'll be using GLM in all our programs from now on so you should install it right away. Installing GLM is easy. Download the latest version from the GLM site `http://glm.g-truc.net` and place the `glm` folder in the include path of your environment. Then, include `<glm/glm.hpp>` in your code for full GLSL math functionality. Additional functionality requires including dedicated header files as described in the GLM manual.

With GLM installed we are all set for our first 4.3 animation program which, in fact, is a shaderization of `ballAndTorus.cpp` from Chapter 4.

Figure 20.6: Screenshot of `ballAndTorus-Shaderized.cpp`.

Experiment 20.2. Run `ballAndTorusShaderized.cpp`. The program's two shaders are `vertexShader.glsl` and `fragmentShader.glsl`. If you care to run `ballAndTorus.cpp` again you see that the functionality of `ballAndTorusShaderized.cpp` is exactly same. Figure 20.6 is a screenshot. **End**

The code is not hard to understand – in fact, it's mostly made up of pieces from older programs. Firstly, though, note the GLM headers included near the top. Further, use of the `glm` namespace saves us, for example, from writing `glm::vec4` instead of, simply, `vec4`. Included as well is the header file `shader.h` which lists function declarations from the separate source `shader.cpp`, the latter containing the routines to create the shader program executable, all of which we have already seen in `squareShaderized.cpp`.

Sadly, the GLUT objects we saw in Section 3.10 are no longer available in 4.3 so we'll have to make our own ball and torus. The ball will be two hemispheres face-to-face and, fortunately, we can pretty much copy in the code from `hemisphereMultidrawVBO.cpp` of Chapter 3, which draws a hemisphere using VBOs to hold vertex and index arrays: see our separate source `hemisphere.cpp` included via the header `hemisphere.h`.

Similarly, we have a source `torus.cpp` and header `torus.h` to initialize the torus. Now, comparing `hemisphere.cpp` and `torus.cpp`, one sees that only the first functions fill*Obj*VertexArray() differ depending on the object's geometry (for the torus we refer back to the program torus.cpp of Chapter 10 for formulas for the sample point coordinates), while the remaining functions are identical save for naming. Hopefully, then, drawing our own objects in future will not be hard if we follow this template.

Both hemisphere.cpp and torus.cpp use the Vertex structure defined in the header file vertex.h.

Exercise 20.5. Why can't GLUT calls to draw objects from pre-shader OpenGL be included as such in 4.3?
Hint: Immediate mode ....

The first block of globals of ballAndTorusShaderized.cpp, from latAngle to animationPeriod, are animation-related parameters copied over from ballAndTorus.cpp. Following are blocks of storage for hemisphere and torus data, and, finally, a couple of blocks, which should be familiar from squareShaderized.cpp, containing uniform values and locations, in addition to shader, buffer and vao ids.

The first few parts of the initialization routine where we create the shader program executable, call functions to initialize data for a hemisphere and a torus, and create VAOs and VBOs to hold this data should be clear from squareShaderized.cpp and hemisphereMultidrawVBO.cpp. Let's move on then to where the projection matrix uniform is set:

```
projMatLoc = glGetUniformLocation(programId,"projMat");
projMat = frustum(-5.0, 5.0, -5.0, 5.0, 5.0, 100.0);
glUniformMatrix4fv(projMatLoc, 1, GL_FALSE, value_ptr(projMat));
```

The first line retrieves the location of the uniform projMat in the vertex shader. The second line sees our first use of GLM. As the user might guess, frustum(*left*, *right*, *bottom*, *top*, *near*, *far*) returns the matrix corresponding to glFrustum(*left*, *right*, *bottom*, *top*, *near*, *far*). The last line updates that uniform's value with the value of projMat – note the GLM call value_ptr(*variable*) which returns a pointer to *variable*'s storage.

The next part of setup() sets the hemisphere and torus color uniforms in the fragment shader, their values being obtained from hemisphere.h and torus.h, respectively. In the final part, we obtain the locations of the modelview matrix and object name uniforms for future reference.

On to the drawing routine next where the modelview transformations are, in fact, copied line for line from ballAndTorus.cpp, except, of course, that they now are implemented with the help of GLM, rather than pre-shader OpenGL calls. Here are the first couple:

```
modelViewMat = mat4(1.0);
modelViewMat = translate(modelViewMat, vec3(0.0, 0.0, -25.0));
```

The first statement, equivalent to glLoadIdentity(), sets the modelview matrix to the 4×4 identity. To understand the next, note that the

GLM call `translate(`*matrix*`, vec3(`*p*`, `*q*`, `*r*`))` returns the result of post-multiplying *matrix* by the matrix corresponding to `glTranslatef(`*p*`, `*q*`, `*r*`)`. Effectively, then, the second statement post-multiplies the current value of the modelview matrix `modelViewMat` with the matrix corresponding to the translation `glTranslatef(0.0, 0.0, -25.0)`. Keeping in mind that the GLM calls `rotate()` and `scale()` function similarly to `translate()`, the reader should have no difficulty following the rest of the modelview transformation sequence.

Prior to each `glMultiDrawElements()` draw call the `modelViewMat` uniform value is updated with a `glUniformMatrix4fv(modelViewMatLoc, ...)` statement so that the correct modelview transformation is applied. Updated, as well, with a `glUniform1ui(objectLoc, ...)` statement, is the `object` uniform value in both shaders with the name of the object to be drawn – the value of `object` determines in the vertex shader if the `vec4` coordinates variable `coords` will read its values from the attribute variable `hemCoords` or `torCoords`, while in the fragment shader it determines the output colors.

Note: Observe that the uniform `object` has been declared identically in both vertex and fragment shader. Their values will be updated simultaneously too.

Note: As `enum` isn't a part of GLSL 4.3, we are unable to replicate the `enum object {HEMISPHERE, TORUS}` declaration of the application program, but, rather, define the values of `HEMISPHERE` and `TORUS` in both shaders via preprocessor directives.

Except for `resize()` and `main()`, both of which are straightforward, the remaining routines of `ballAndTorusShaderized.cpp` are copied over from `ballAndTorus.cpp`.

Exercise 20.6. Doesn't using GLM to recreate fixed-function commands defeat the whole purpose of programmable pipelines?!

Exercise 20.7. **(Programming)** Shaderize `floweringPlant.cpp` from Chapter 4.

Exercise 20.8. **(Programming)** If you did animation projects from Section 4.5.3 then shaderize at least a couple of them now.

20.5 Lighting

As noted in Section 20.1.2 on major facets of the newer OpenGL, 4.3 asks us to do our own lighting – which is good and bad. The bad, or rather tedious, is that, even for classical lighting, we have to implement ourselves the OpenGL lighting equation, which used to be a free service of the fixed-function pipeline. To be fair though, this is just a one-time job writing the appropriate shader, which can then be reused. The good is that, now that we can write our own equations, we can actually go beyond classical lighting to fancier effects.

Our case study will be bump mapping, which we first encountered as a special effect in Section 13.9. The idea of bump mapping is to give an illusion of detail to a surface by perturbing its normals so that light reflects off it as though it were actually detailed. We applied this idea in Experiment 13.16 of Section 13.9 to make a plane appear corrugated in the program `bumpMapping.cpp`. We remarked then that bump mapping is particularly effective with the per-pixel lighting of Phong's shading model, where normal values are interpolated across primitives, rather than being fixed at vertices as in classical lighting (aka per-vertex lighting). However, the fixed-function pipeline cannot do Phong shading; on the other hand, we can program the pipeline of 4.3 to do so.

So, here's the plan: first, we'll shaderize `bumpMapping.cpp` which'll mean essentially setting up per-vertex lighting ourselves; then, to reinforce our ideas, we'll shaderize `litCylinder.cpp` of Experiment 11.20 of Section 11.11; finally, we'll move definitively beyond fixed-function by applying per-pixel lighting to the setting of `bumpMapping.cpp`.

20.5.1 Per-Vertex Lighting

Per-vertex lighting comprises Phong lighting at each vertex, governed by the OpenGL lighting equation, followed by Gouraud (smooth) shading to interpolate colors through triangles. Per-vertex lighting is the default of the fixed-function pipeline and is what we have used in all our lit programs to date, including, of course, `bumpMapping.cpp`.

Let's see then how to write a shader to implement per-vertex lighting with no assistance from fixed-function.

Experiment 20.3. Run `bumpMappingShaderized.cpp`. Interaction is the same as `bumpMapping.cpp`: press space to toggle between bump mapping on and off. Figure 20.7(b) is a screenshot with bump mapping enabled, evidently identical to that of `bumpMapping.cpp` in Figure 20.7(a). **End**

So, let's understand the code of `bumpMappingShaderized.cpp`. Note, first, that the structure `Vertex` defined in `vertex.h` now has fields for the normal and bump mapped normal, in addition to position. The two simple new header files `light.h` and `material.h` define the structures `Light` and `Material` to hold light and material properties, respectively. In `bumpMappingShaderized.cpp`, `light0` of type `Light` holds light property values, `planeFandB` of type `Material` holds material property values (identical for the front and back of the plane), and `globAmb` holds the global ambient vector, all these being copied from `bumpMapping.cpp`.

The first part of the `setup()` routine (as usual) creates the shader program executable. The data for the plane, drawn as exactly the same stack of triangle strips as in bumpMapping.cpp, is initialized next. Interesting here, if you see plane.cpp, is the definition of two kinds of normals, the"real" planeVertices[k].normal and the bump mapped planeVertices[k].bumpedNormal. The real normal is a unit vector in the y-direction at every vertex as in bumpMapping.cpp, while the bump mapped normal varies according to a formula again copied from bumpMapping.cpp.

Returning to the initialization routine of bumpMappingShaderized.cpp, the VAO and VBOs associated with the plane are created next and, then, as expected, the plane coordinate, real normal and bump mapped normal values are input via glVertexAttribPointer()-glEnableVertexAttribArray() statement pairs to attribute variables in the vertex shader – namely, planeCoords, planeNormal and planeBumpedNormal at locations 0, 1 and 2, respectively.

The projection matrix uniform is set next, while the locations of the modelview matrix and normal matrix uniforms are retrieved for future reference. The toggle isBumPMapped uniform is set as well.

In the final parts of the initialization routine, light property values, the global ambient and material property values are copied, respectively, from the variables light0, globAmb and planeFandB in the application program to corresponding fields in like-named structure uniforms in the vertex shader.

The drawing routine is mundane save, possibly, for the calculation of the normal matrix, following Section 11.11.5, as the transpose inverse of the upper-left 3×3 submatrix of the modelview matrix in the line

```
normalMat = transpose(inverse(mat3(modelViewMat)));
```

On to the vertex shader next where all of the action is. At this time the reader may want to review the short Section 11.7 on the OpenGL lighting equation as we'll be pretty much implementing the latter line by line. Here's the first part of the vertex shader's main():

```
normal = (isBumpMapped == 1)? planeBumpedNormal : planeNormal;
normal = normalize(normalMat * normal);
lightDirection = normalize(vec3(light0.coords));
eyeDirection = vec3(0.0, 0.0, 1.0);
halfway = (length(lightDirection + eyeDirection) == 0.0f) ?
               vec3(0.0) : (lightDirection + eyeDirection)/
               length(lightDirection + eyeDirection);
```

The first line reads into the variable normal the real or bumped normal values depending on the value of the toggle isBumpMapped, while the second line transforms the normal by the normal matrix and then (re-)normalizes it. The third line sets the light direction vector as the xyz-values of the light's position, ignoring its w-value of 0 because the light is directional.

Since bumpMapping.cpp has an infinite viewpoint (the pre-shader default) for lighting calculation, we accordingly set the eye direction vector pointing up the z-axis in the fourth line. The halfway vector is set in the last line as the unit vector in the direction of the sum of the light direction and eye vectors (following the formula derived in Example 11.4) with a zero-division check first.

Here's the final part of the vertex shader's `main()`, except for the routine setting of `gl_Position`:

```
fAndBEmit = planeFandB.emitCols;
fAndBGlobAmb = globAmb * planeFandB.ambRefl;
fAndBAmb = light0.ambCols * planeFandB.ambRefl;
fAndBDif = max(dot(normal, lightDirection), 0.0f) *
           light0.difCols * planeFandB.difRefl;
fAndBSpec = pow(max(dot(normal, halfway), 0.0f),
                planeFandB.shininess) *
            light0.specCols * planeFandB.specRefl;
fAndBColsExport =  vec4(vec3(min(fAndBEmit + fAndBGlobAmb +
                   fAndBAmb + fAndBDif + fAndBSpec,
                   vec4(1.0))), 1.0);
```

We ask the reader to now refer to the OpenGL lighting equation (11.12) to see that the above implements that equation term for term, except there is no distance or spotlight attenuation factor. The last statement which sums the color terms and writes them into `fAndBColsExport` for output to the fragment shader clamps, in a slightly screwy manner, the individual RGB values to a maximum of 1, and sets the A value to 1.

The fragment shader is pass-through, the default `smooth` interpolation qualifier of the input variable `fAndBColsExport` assuring Gouraud shading.

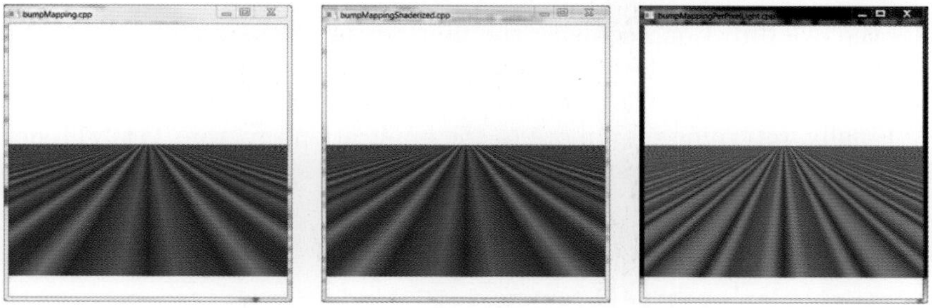

(a) (b) (c)

Figure 20.7: Screenshots of (a) `bumpMapping.cpp` (b) `bumpMappingShaderized.cpp` (c) `bumpMappingPerPixelLight.cpp`.

Let's shaderize `litCylinder.cpp` from Chapter 11 for more practice.

Experiment 20.4. Run `litCylinderShaderized.cpp`. Interaction is the same as `litCylinder.cpp`: press the 'x'-'Z' keys to turn the cylinder. As far as lighting is concerned, the main twist from `bumpMappingShaderized` is that now there is two-sided lighting, following `litCylinder.cpp` where it was activated by the statement

```
glLightModeli(GL_LIGHT_MODEL_TWO_SIDE, GL_TRUE);
```

Material properties differ, as well, between the inside and outside of the cylinder. Figure 20.8 is a screenshot. **End**

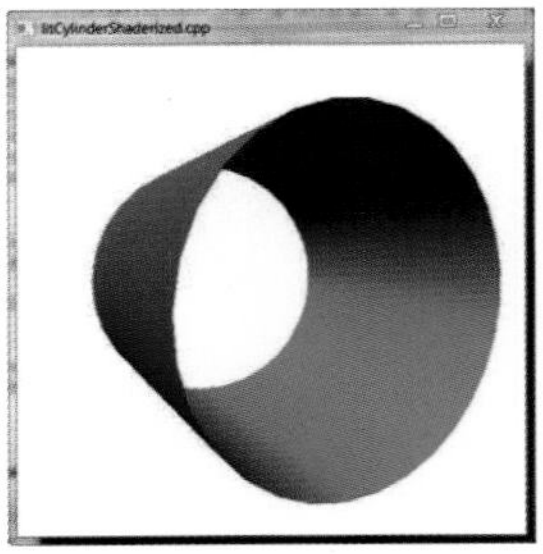

Figure 20.8: Screenshot of `litCylinder-Shaderized.cpp`.

If you followed `bumpMappingShaderized.cpp` there's not much really that needs to be explained in `litCylinderShaderized.cpp`. As expected, though, there are two sets of material properties in the latter, namely, `cylFront` for the front (actually, outside) of the cylinder and `cylBack` for the back (inside), which are input, respectively, to like-named structure uniforms in the vertex shader.

Lighting calculations in the vertex shader of `litCylinderShaderized.-cpp` are almost exactly as in that of `bumpMappingShaderized.cpp`, too, except now there are two sets of calculations, identical save for normal reversal, to determine, respectively, front and back colors, both of which are output to the fragment shader. The pass-through fragment shader outputs either the front or back color depending on the value of the Boolean built-in `gl_FrontFacing` which is true if the current fragment belongs to a front-facing triangle, false otherwise.

The one difference, though, in the lighting calculations themselves is because `litCylinder.cpp` asks for a local viewpoint for lighting calculation with the command

```
glLightModeli(GL_LIGHT_MODEL_LOCAL_VIEWER, GL_TRUE);
```

To this end, `litCylinderShaderized.cpp`'s vertex shader statement

```
eyeDirection = -1.0 * normalize(vec3(modelViewMat * cylCoords));
```

sets the eye direction vector as the unit vector in the direction from the vertex (in its current location in world space after transformation by the modelview matrix) to the origin.

Finally, returning once more to the application program `litCylinder-Shaderized.cpp`, we point out a minor change: in order to be faithful to the command

```
gluPerspective(60.0, (float)w/(float)h, 1.0, 50.0);
```

in `litCylinder`'s `resize()` routine, `litCylinderShaderized.cpp` uses the OpenGL window width and height, obtained from its own `resize()` routine via the globals `width` and `height`, respectively, to set the projection matrix in `drawScene()` by

```
projMat = perspective(60.0f, (float)width/(float)height,
                      1.0f, 50.0f);
```

As the reader might guess, the GLM command `perspective(`*fovy*, *aspect*, *near*, *far*`)` returns the projection matrix corresponding to `glu-Perspective(`*fovy*, *aspect*, *near*, *far*`)`.

Exercise 20.9. (Programming) Shaderize `spotlight.cpp` from Chapter 11.

20.5.2 Per-Pixel Lighting

Finally, we are going write a 4.3 program to do something that the fixed-function pipeline could never, namely, per-pixel lighting. You might want, though, to first review the discussion of Phong shading in Section 11.12.

Phong shading, also called per-pixel lighting, is a shading model alternate to Gouraud, where (a) vertex normal values are interpolated through each triangle, and then (b) light values computed at each pixel using the interpolated normals. Let's bump map the exact same plane of `bumpMappingShaderized.cpp`, but now applying per-pixel lighting instead.

Experiment 20.5. Run `bumpMappingPerPixelLight.cpp`. Again, press space to toggle between bump mapping on and off. Figure 20.7(c) is a screenshot. The program, its associated C++ source and header files are all *exactly* same as for `bumpMappingShaderized.cpp` – the difference is only in the shaders! **End**

A comparison of the shaders of `bumpMappingPerPixelLight.cpp` with those of `bumpMappingShaderized.cpp` reveals quickly the former's modus operandi. The star turn is now the fragment shader's: the lighting equation statements are brought over from the vertex shader to the fragment shader, which, as well, gets light and material property values directly from the application program *and imports normal values from the vertex shader*. This last is crucial. The default `smooth` interpolation qualifier of the input variable `normalExport` causes normal values to be interpolated through each triangle, following which the lighting equation gives precisely Phong shading.

Compare the identical Figures 20.7(a) and (b) with Figure 20.7(c) to see the crisper waves on the plane of the latter, a consequence of its more sophisticated shading model.

Exercise 20.10. (Programming) Apply per-pixel lighting to `litCylinder.cpp`. Compare the result with `litCylinderShaderized.cpp` which is per-vertex lit. Do you see a difference? How about the highlights?

Exercise 20.11. (Programming) Write a per-pixel lit version of `lightAndMaterial1.cpp`. Make sure to take into account distance attenuation and that both lights are positional.

Well, programming shader-based lighting wasn't too bad, was it? Mostly, a straightforward implementation of theory that we learned earlier in Chapter 11, it seemed.

20.6 Textures

Textures can not only be imported into the programmable pipeline without difficulty, but manipulated there to great effect as well. Let's start with shaderizing `fieldAndSkyFiltered.cpp` from Chapter 12 on textures, which the reader might want to quickly review at this time.

Experiment 20.6. Run `fieldAndSkyFilteredShaderized.cpp`. As in `fieldAndSkyFiltered.cpp` press the up and down arrow keys to move the viewpoint. However, unlike the earlier program, `fieldAndSkyFiltered-Shaderized.cpp` implements (to keep it simple) only one fixed filter for the grass texture and no options. Figure 20.9 is a screenshot. **End**

Figure 20.9: Screenshot of `fieldAndSkyFiltered-Shaderized.cpp`.

Let's see if texturing in 4.3 is indeed straightforward. The initialization routine of `fieldAndSkyFilteredShaderized.cpp` loads the sky and grass images with

```
image[0] = getbmp("../../Textures/grass.bmp");
image[1] = getbmp("../../Textures/sky.bmp");
```

and then generates two texture ids with

```
glGenTextures(2, texture);
```

The next block of statements, which binds the grass texture to texture unit zero, is

```
glActiveTexture(GL_TEXTURE0);
glBindTexture(GL_TEXTURE_2D, texture[0]);
glTexImage2D(GL_TEXTURE_2D, 0, GL_RGBA, image[0]->sizeX,
             image[0]->sizeY, 0, GL_RGBA,
             GL_UNSIGNED_BYTE, image[0]->data);
glTexParameteri(GL_TEXTURE_2D, GL_TEXTURE_WRAP_S, GL_REPEAT);
glTexParameteri(GL_TEXTURE_2D, GL_TEXTURE_WRAP_T, GL_REPEAT);
glTexParameteri(GL_TEXTURE_2D, GL_TEXTURE_MIN_FILTER,
                GL_LINEAR_MIPMAP_LINEAR);
glTexParameteri(GL_TEXTURE_2D, GL_TEXTURE_MAG_FILTER, GL_LINEAR);
glGenerateMipmap(GL_TEXTURE_2D);
grassTexLoc = glGetUniformLocation(programId, "grassTex");
glUniform1i(grassTexLoc, 0);
```

The first statement above selects `GL_TEXTURE0` as the active texture unit, the second binds the 2D texture object `texture[0]` to this unit, the third sets the image data to that of grass, the next four set texture parameters, while the eighth statement generates the mipmaps required for the `GL_LINEAR_MIPMAP_LINEAR` min filter. All of these statements, thus far, should be familiar from Chapter 12.

It's the last two statements above which connect the grass texture to the fragment shader. The second last statement reads the location of `grassTex`, a uniform of type `sampler2D`, in particular, a handle to a 2D texture, in the fragment shader; the last statement sets `grassTex` to texture unit `GL_TEXTURE0`.

A similar block of statements binds the sky texture to texture unit one. And that's it for the application program as far as texture-related statements are concerned. The action moves to the fragment shader next, where the relevant statements are

```
...
in vec2 texCoordsExport;

uniform sampler2D grassTex;
uniform sampler2D skyTex;
uniform uint object;

...
void main(void)
{
   fieldTexColor = texture(grassTex, texCoordsExport);
   skyTexColor = texture(skyTex, texCoordsExport);

   if (object == FIELD) colorsOut = fieldTexColor;
   if (object == SKY) colorsOut = skyTexColor;
}
```

Firstly, the `in` variable `texCoordsExport` reads vertex texture coordinates from the namesake `out` variable in the vertex shader, which in turn obtains them from the vertex attribute array `fieldTexCoords`. As we have seen already, the samplers `grassTex` and `skyTex` have been set to texture units 0 and 1, respectively, corresponding to the grass and sky images. The uniform `object` tells the shader the object currently being processed.

Let's examine the main routine next. Now, generally, `texture(`*sampler*, *texCoords*`)` is a built-in GLSL *texture lookup function* which computes the colors at the location *texCoords* of the texture space corresponding to the texture bound to *sampler*.

So, for example, the first statement in `main()` above returns in `fieldTexColor` the color values of the grass texture at location `texCoordsExport`, the latter being interpolated from the texture coordinates at the field's vertices; in other words, `fieldTexColor` are the color values at `texCoordsExport`, the latter being precisely the point in texture space corresponding to the current fragment by the texture map (which is exactly consistent with how we understood texture coordinates to work in Section 12.1).

Note: The value of `texCoordsExport` is, in fact, interpolated from the vertex texture coordinates of the field by the (default) `smooth` interpolation qualifier of the input variable `texCoordsExport`.

The rest of `main()` in the fragment shader should now be clear. Not too hard, was it, getting texturing going?

Let's shaderize a couple more earlier texture programs: `texturedTorus.cpp` from Section 12.4 specified texture coordinates of a parametrized surface, a torus, using, in fact, the surface parametrization, while `litTexturedCylinder.cpp` from Section 12.5 combined texturing with light.

Experiment 20.7. Run `texturedTorusShaderized.cpp`. As in `texturedTorus.cpp` press 'x'-'Z' to turn the torus.

The point to note is how the associated source `torus.cpp` defines texture coordinates for the torus following exactly `texturedTorus.cpp`. Beyond that, the application program and shaders should be easily understood. See Figure 20.10 for a screenshot. **End**

Figure 20.10: Screenshot of `textureTorus-Shaderized.cpp`.

Experiment 20.8. Run `litTexturedCylinderShaderized.cpp`. As in `litTexturedCylinder.cpp` press 'x'-'Z' to turn the beer can. Figure 20.11 is a screenshot. **End**

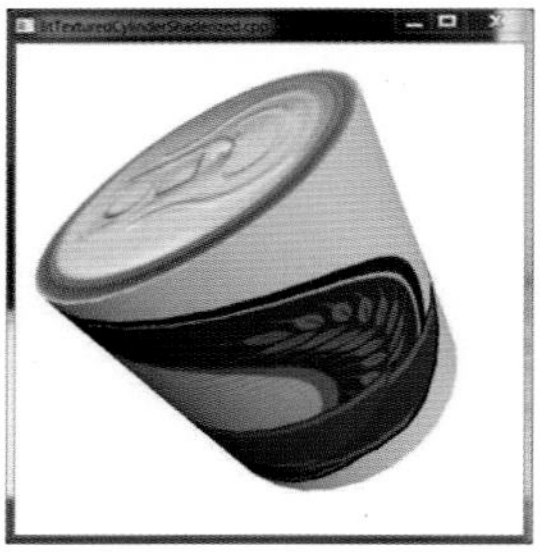

Figure 20.11: Screenshot of `litTexturedCylinder-Shaderized.cpp`.

Obviously, the application program `litTexturedCylinderShaderized.-cpp` and its shaders piggyback on `litCylinderShaderized.cpp`, from the previous section, and its shaders, respectively, as far as possible. Let's pick up the former program then where it brings in texture to combine with light. For this we need to go to its shaders.

Observe, first, how `litTexturedCylinderShaderized.cpp`'s vertex shader separates out the specular component of the computed light for front and back faces in `frontSpecExport` and `backSpecExport`, respectively, while the rest of the computed light is in `frontAmbDiffExport` and `backAmbDiffExport`, again respectively. This is as opposed to `litCylinderShaderized.cpp`'s vertex shader storing all the computed front colors in `frontColsExport` and back colors in `backColsExport`. For the reason for this, refer back to the statement

```
glLightModeli(GL_LIGHT_MODEL_COLOR_CONTROL,
              GL_SEPARATE_SPECULAR_COLOR);
```

in `litTexturedCylinder`'s initialization routine asking that specular colors be added in only after mixing the texture with the non-specular colors (in order not to dilute specular highlights: see the discussion in Section 11.4 on the OpenGL lighting model).

All four color variables `frontAmbDiffExport`, `frontSpecExport`, `back-AmbDiffExport` and `backSpecExport` are exported from `litTextured-CylinderShaderized.cpp`'s vertex shader to its fragment shader, whose main is

```
void main(void)
{
   if (object == CYLINDER) texColor =
                            texture(canLabelTex, texCoordsExport);
   if (objectLet == DISC) texColor =
                           texture(canTopTex, texCoordsExport);

   colorsOut = gl_FrontFacing?
             (frontAmbDiffExport * texColor + frontSpecExport) :
             (backAmbDiffExport * texColor + backSpecExport);
}
```

The first two lines compute the texture color values at the fragment depending on if the current object is the cylinder or the disc. The last line combines the non-specular color component with the texture color by *multiplication*, as asked by the GL_MODULATE parameter of

```
glTexEnvf(GL_TEXTURE_ENV, GL_TEXTURE_ENV_MODE, GL_MODULATE);
```

in litTexturedCylinder's initialization routine, and *afterward* adds in the specular component.

We mentioned earlier that the make-it-yourself lighting of 4.3 allows you to do your own thing. Here's an exercise asking you to implement an effect impossible in fixed-function.

Exercise 20.12. (Programming) Modify the fragment shader of litTexturedCylinderShaderized.cpp, changing its main's last line to

```
colorsOut = gl_FrontFacing? (frontAmbDiffExport * texColor +
                              2.0*frontSpecExport) :
                             (backAmbDiffExport * texColor +
                              2.0*backSpecExport);
```

In other words, separately *double* the specular light. See the difference? (Rotate about the y-axis and you certainly will.) Not particularly pretty, but one sees the possibilities.

Exercise 20.13. (Programming) Shaderize texturedTorpedo.cpp from Chapter 12. Texturing the Bézier propeller blades is a challenge.

20.7 Summary, Notes and More Reading

With this chapter began our coverage of the programmable pipeline, particularly OpenGL and GLSL versions 4.3. After learning the basics of GLSL we dissected the 4.3 version of, in fact, our very first OpenGL program from Chapter 2. Then we saw how to do animation, lighting and textures in 4.3. And, as always, we saw plenty of live code along the way, so the student by this point should be fairly comfortable with the new way of doing things. Hopefully, she will agree that the earlier pre-shader pipeline helped in a solid laying of the foundations. Moreover, she may agree as well that there are hardly any bad habits from pre-shader to shed, but, rather, new ones to acquire.

Particularly exciting for students of shader-based programming might be that, since OpenGL ES (ES for Embedded Systems) 2.0, shaders have gone mobile with a vengeance: anything that can be done in a shader has been removed from fixed-functionality and *must be done* in a shader! OpenGL ES – a "lean, mean, shadin' machine" as the OpenGL site called it – is by far the most common 3D API on mobile devices like smartphones and tablets. The mobile shading language, GLSL ES, itself is very similar to the desktop version. Moreover, OpenGL 4.3 is fully compatible with OpenGL ES 3.0 (the latest version). Therefore, the reader should now be able to begin coding OpenGL for small devices without trouble if she's interested.

There is even more good news. WebGL, the emerging standard for 3D graphics on the web, is based on OpenGL ES 2.0 and, therefore, just a stone's throw in programming methodology from 4.3 (though, not surprisingly as it's for the web, WebGL is written in JavaScript, rather than C++, using the HTML5 canvas element). Currently, WebGL is fully supported on the latest versions of major browsers including Mozilla Firefox, Google Chrome, Safari and Opera – sadly, Internet Explorer provides only partial support – and should see the next few years the exponential growth in application development that mobile 3D has already experienced.

So, the 4.3 programmer has multiple platforms to strut her stuff, with pretty much recession-proof job skills for the foreseeable future. Don't go away though! This chapter got us off the ground. There's more, much more, to come.

Chapter 21

OpenGL 4.3, Shaders and the Programmable Pipeline: Escape Velocity

We'll now pick up where we left off end of the last chapter. The last chapter set the foundations of OpenGL 4.3. In this we'll study more advanced features, as well as the two optional shader stages, particularly, tessellation and geometry.

We begin in Section 21.1 by trying to reconstitute the pre-shader toolbox of Chapter 3. In the process we'll find that even though a few of the older gadgets have been discarded from 4.3, new ones have been added and, in fact, the programmable pipeline often affords a more efficient way to do what an old gadget used to.

Section 21.2 introduces shader subroutines, which offer an elegant method of switching threads of control at run-time. In Section 21.3 we'll learn two powerful techniques related to animation. The first is for picking an object on the screen, familiar from Chapter 4, but done entirely differently via the fragment shader in 4.3. The second is that of transform feedback, which is a way to look ahead in an animation sequence.

Just as we try to make ourselves a new toolbox in the first section, in Section 21.4 we revisit Chapter 13 on special visual techniques to see how they can be done in 4.3. Finally, tessellation and geometry shaders are the topics of Sections 21.5 and 21.6, respectively, and we conclude the chapter in Section 21.7.

21.1 Toolbox

What would be a nice collection of gadgets to keep handy while coding 4.3? Let's start by opening again the pre-shader OpenGL toolbox of Chapter 3 to see if the contents still work in 4.3, or if we might not add a few new ones. Let's run through the gadgets from that earlier chapter in the order they were presented.

Vertex Arrays, VBOs and VAOs

Vertex arrays and VBOs, the topics, respectively, of Sections 3.1 and 3.2, obviously need no further comment in 4.3!

21.1.1 VAOs and Instanced Rendering Instead of Display Lists

VAOs of Section 3.3, too, need little further introduction at this time. Display lists, on the other hand, so convenient as we saw in Section 3.4 to encapsulate objects and transformations, sadly, are no more in 4.3: `glNewList()`, `glEndList()` and `glCallList()` are all gone. Now, VAOs themselves can take up some of the slack, as VAOs and display lists do share the purpose of encapsulating objects server-side, though, of course, they go about things rather differently: VAOs package storage states while display lists package a set of commands.

Interestingly, 4.3 has a new gadget, namely, *instanced rendering*, to address a weakness of both VAOs and display lists, that neither can be parametrized at run-time. Instanced rendering allows the same drawing command to be repeated multiple times with certain attributes changing per instance based on the value of a so-called instance counter. There are actually two different ways to do instanced rendering – by setting *instanced vertex attributes* and by using the *instance counter in the shader*. We'll describe both by shaderizing `helixList.cpp` of Section 3.4 in the two ways.

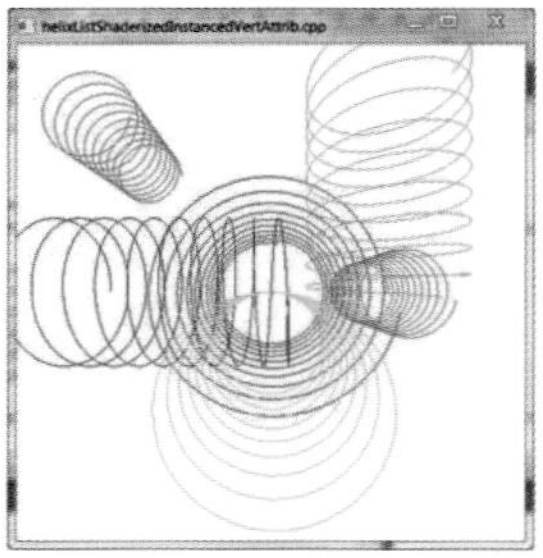

Figure 21.1: Screenshot of `helixListShaderizedInstancedVertAttrib.cpp`.

Instanced Vertex Attributes

Experiment 21.1. Run `helixListShaderizedInstancedVertAttrib.cpp`. The output of six different helixes is exactly the same as that of `helixList.cpp`. See Figure 21.1. **End**

The helix itself is created, of course, in the obligatory separate source `helix.cpp` included in the application program via `helix.h`. Now, first see the instanced drawing call in the drawing routine:

```
glDrawArraysInstanced(GL_LINE_STRIP, 0, HEL_SEGS, 6);
```

What it does is simply execute the statement `glDrawArrays(GL_LINE_STRIP, 0, HEL_SEGS)` successively six times, incrementing the *instance counter* `gl_InstanceID`, a built-in shader variable, after each instance (starting from 0). Generally,

`glDrawArrays(` *primitive*, *first*, *count*, *primitivesCount*`)`

executes `glDrawArrays(`*primitive*, *first*, *count*`)` successively *primitivesCount* times, likewise incrementing `gl_InstanceID` after each instance, starting from 0.

Next, note the code in `setup()` associating the helix color data, copied over from `helixList.cpp` in the array `helColors[6]`, with the vertex shader:

```
glBindBuffer(GL_ARRAY_BUFFER, buffer[HEL_COLORS]);
glBufferData(GL_ARRAY_BUFFER, sizeof(helColors),
             helColors, GL_STATIC_DRAW);
glVertexAttribPointer(1, 4, GL_FLOAT, GL_FALSE,
                      sizeof(helColors[0]), 0);
glEnableVertexAttribArray(1);
glVertexAttribDivisor(1, 1);
```

It's all routine except for the last statement which makes the color attribute instanced. Generally `glVertexAttribDivisor(`*location*, *divisor*`)` declares that successive values of the vertex attribute at *location* of the vertex shader are read every *divisor* instances of the instanced drawing statement. So, the statement `glVertexAttribDivisor(1, 1)` above means that color values for the attribute `helColors` of the vertex shader will be read once for each of the six instances a helix is drawn by `glDrawArraysInstanced(GL_LINE_STRIP, 0, HEL_SEGS, 6)`. If the statement `glVertexAttribDivisor(1, 1)` had not been there, then color values would have been read once per vertex, rather than once per instance.

In fact, the previous block

```
glBindVertexArray(vao[HELIX]);
...
glEnableVertexAttribArray(0);
```

which associates `buffer[HEL_VERTICES]` with the attribute `helCoords` has no `glVertexAttribDivisor()` command, meaning `helCoords` is uninstanced and will be read once per vertex as, of course, one wants.

Finally, the block

```
glBindBuffer(GL_ARRAY_BUFFER, buffer[HEL_TRANSFORM_MATS]);
glBufferData(GL_ARRAY_BUFFER, sizeof(helTransformMats),
             helTransformMats, GL_STATIC_DRAW);
for (int i = 0; i < 4; i++)
{
   glVertexAttribPointer(2 + i, 4, GL_FLOAT, GL_FALSE,
                         sizeof(mat4), (void*)(sizeof(vec4) * i));
   glEnableVertexAttribArray(2 + i);
   glVertexAttribDivisor(2 + i, 1); // Set attribute instancing.
}
```

in `setup()` instances the shader vertex attribute `helTransformMats`, the transforming matrices (again, copied over from `helixList.cpp`), so that successive ones are applied to successive instances of the helix. The thing to note is that since `helTransformMats` is of type `mat4`, the declaration

```
layout(location=2) in mat4 helTransformMats;
```

in the vertex shader causes it to actually occupy locations 2, 3, 4 and 5. The application program block above, accordingly, associates the data in `buffer[HEL_TRANSFORM_MATS]` with these locations in its `for` loop.

Finally, of course

```
gl_Position = projMat * helTransformMats * helCoords;
colorsExport = helColors;
```

of the vertex shader's main together generate one matrix transform and one color per helix instance.

Instance Counter in the Shader

The primary difference of this method from the previous is that the data to be instanced is stored in server-side buffers, rather than vertex attribute arrays.

Experiment 21.2. Run `helixListShaderizedShaderCounter.cpp`. The output, just like that of `helixListShaderizedInstancedVertAttrib.cpp`, is the same as that of `helixList.cpp`. See Figure 21.2. **End**

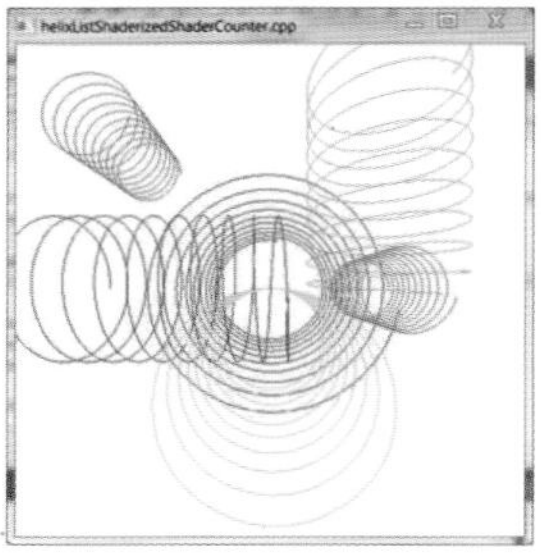

Figure 21.2: Screenshot of `helixListShaderized-ShaderCounter.cpp`.

Since we want the vertex shader itself to access helix color and transformation matrix values depending on the value of the instance counter `gl_InstanceID`, we need to store these particular colors and matrices somewhat differently. In fact, we'll use *texture buffer objects* (*TBO*s), which are randomly accessible data buffers bound to a texture unit. First, see the block

```
glActiveTexture(GL_TEXTURE0);
glBindTexture(GL_TEXTURE_BUFFER, texture[0]);
glTexBuffer(GL_TEXTURE_BUFFER, GL_RGBA32F, buffer[HEL_COLORS]);
helColorsTexLoc = glGetUniformLocation(programId, "helColorsTex");
glUniform1i(helColorsTexLoc, 0);
```

in the initialization routine binding the color data. The first statement activates texture unit `GL_TEXTURE0`, the second binds the texture buffer `texture[0]` to this unit, creating a TBO, while the third specifies `buffer[HEL_COLORS]` as the data source for the TBO kept in the 4-component floating point internal format `GL_RGBA32F`. The fourth statement reads the location of `helColorsTex`, a uniform of type `samplerBuffer`, in the vertex shader; the last statement sets `helColorsTex` to texture unit `GL_TEXTURE0`.

The next block

```
glActiveTexture(GL_TEXTURE1);
glBindTexture(GL_TEXTURE_BUFFER, texture[1]);
glTexBuffer(GL_TEXTURE_BUFFER, GL_RGBA32F,
            buffer[HEL_TRANSFORM_MATS]);
helTransformMatsTexLoc = glGetUniformLocation(programId,
                         "helTransformMatsTex");
glUniform1i(helTransformMatsTexLoc, 1);
```

creates, likewise, a TBO containing transformation matrix values, which is referenced by the samplerBuffer uniform helTransformMatsTex in the vertex shader. The rest of the application program is routine and, mostly, copied over from helixListShaderizedInstancedVertAttrib.cpp.

So, let's turn next to the vertex shader where the instance counter is actually used. The statement

```
helColors = texelFetch(helColorsTex, gl_InstanceID);
```

in the vertex shader's main returns in helColors the value at location gl_InstanceID of the TBO referenced by helColorsTex. Generally, texelFetch(*samplerBuffer*, *intCoord*) returns the value at integer location *intCoord* of the TBO referenced by *samplerBuffer*.

Note: One can, therefore, think of a TBO as a one-dimensional texture with texels located at integer coordinates.

Likewise, the block

```
col0 = texelFetch(helTransformMatsTex, gl_InstanceID * 4);
col1 = texelFetch(helTransformMatsTex, gl_InstanceID * 4 + 1);
col2 = texelFetch(helTransformMatsTex, gl_InstanceID * 4 + 2);
col3 = texelFetch(helTransformMatsTex, gl_InstanceID * 4 + 3);
helTransformMats = mat4(col0, col1, col2, col3);
```

in the vertex shader's main accesses the TBO referenced by helTransform-MatsTex to retrieve four column vectors at a time, subsequently assembling them into a mat4 transformation matrix.

Finally,

```
gl_Position = projMat * helTransformMats * helCoords;
colorsExport = helColors;
```

together generate one matrix transform and one color per helix instance.

Exercise 21.1. (**Programming**) Apply instanced rendering to create a starry night sky.

Text Drawing, Mouse Programming, Special Keys and Menus

The GLUT text drawing calls of Section 3.5, both glutBitmap*() and glutStroke*(), are gone from 4.3 because of their immediate mode of operation. We are unaware at this time of a convenient library to use, so the only recourse seems to be to create a font oneself, a not-difficult but tedious task which we have avoided. If labeling is essential, as sometimes it can be, then we recommend sneaking back into compatibility mode to invoke old text drawing calls.

Programming the mouse in 4.3 stays exactly as first discussed in Section 3.6: the commands `glutMouseFunc()`, `glutMotionFunc()`, and `glutMouseWheelFunc()` to register callbacks for mouse clicks, mouse motion and wheel rotation, respectively, are used as before.

We have already been using `glutSpecialFunc()` in 4.3 code to register the handler for non-ASCII key presses, in exactly same manner as discussed earlier in Section 3.7.

Pop-up menus, too, can be attached to a 4.3 program exactly as described in Section 3.8; specifically, by invoking `glutCreateMenu()`, `glutAddMenuEntry()`, `glutAddSubMenu()` and `glutAttachMenu()`.

21.1.2 Do-It-Yourself Line Stipples

Unfortunately, we can no longer stipple lines in 4.3 as we learned in the pre-shader Section 3.9. The command `glEnable()` no longer accepts the parameter `GL_LINE_STIPPLE` and there is no `glLineStipple()` in 4.3.

However, we can stipple ourselves with little difficulty by programming the fragment shader. For this purpose, we need help of the built-in input variable, `gl_FragCoord` of type `vec4`, accessible by the fragment shader; its (x, y)-values are the coordinates of the fragment in the windows system, z-value is the depth of the fragment and w-value is $1/W$, the perspective division factor (recall Section 18.1.2 for the latter). We ask the reader to do the rest in the next exercise.

Exercise 21.2. (Programming) Stipple a line so that groups of four pixels are successively off and on by writing a statement of the form

```
if ( mod(gl_FragCoord.x, 8.0) < 4.0 ) discard;
```

in the fragment shader.

Here's the capstone program from Chapter 3 for you to redo.

Exercise 21.3. (Programming) Shaderize `canvas.cpp` of Chapter 3.

FreeGLUT Objects not Free Anymore

We have already been doing without the FreeGLUT object calls of Section 3.10, unavailable in 4.3 because of their immediate mode of operation, instead creating objects - e.g., spheres and tori – ourselves in small ancillary source programs, which has proved not too hard.

Hopefully, someone will figure out soon a benign way to import readymade objects into shader programs.

21.1.3 Clipping Planes

Setting up user-defined clipping planes, in addition to the six automatic ones bounding the viewing volume, was discussed in Section 3.11. We can set up clipping planes in 4.3 as well, though the process, mostly conducted in the vertex shader, is different. Let's get straight to code, particularly, a program made from a quick modification of `ballAndTorusShaderized.cpp`.

Experiment 21.3. Run `ballAndTorusClipped.cpp`. The controls are exactly as for `ballAndTorusShaderized.cpp`: space to toggle animation on and off, up/down arrows to change its speed and 'x'-'Z' to rotate the scene.

However, there is now, as will be evident as soon as the animation is started, a clipping plane slicing through the torus's initial position. You can see this from Figure 21.3. Let's see how the clipping plane is set up. **End**

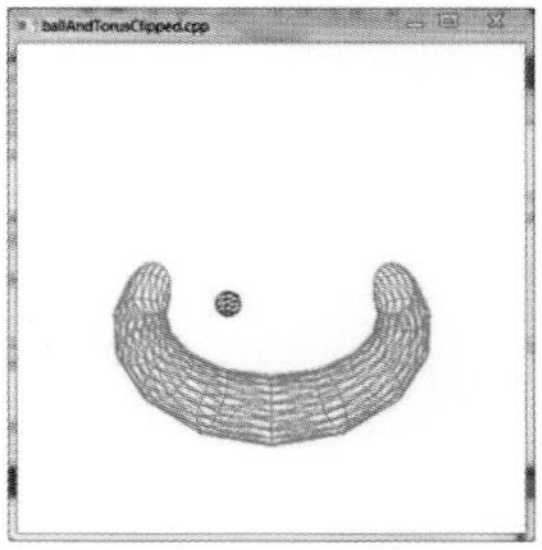

Figure 21.3: Screenshot of `ballAndTorusClipped.cpp`.

The statements

```
clipPlaneLoc = glGetUniformLocation(programId, "clipPlane");
glUniform4fv(clipPlaneLoc, 1, &clipPlane[0]);

glEnable(GL_CLIP_PLANE0);
```

at the bottom of the application program's `setup()` routine, firstly, pass the `vec4` $(0.0, 0.0, 1.0, 25.0)$, stored in the application program's `clipPlane`, to the vertex shader's corresponding uniform `clipPlane`. Of course, `vec4` (A, B, C, D) is meant to represent the clipping plane $Ax+By+Cz+D = 0$, as in Section 3.11, so $(0.0, 0.0, 1.0, 25.0)$ represents the plane $z = -25.0$, which is parallel to the xy-plane and located 25 units in the negative z-direction. Moreover, a single clip plane `GL_CLIP_PLANE0` is enabled.

To the vertex shader next where

```
float gl_ClipDistance[1];
```

initializes the built-in vertex shader array `gl_ClipDistance` in order to implement a single clip plane. The statement

```
gl_ClipDistance[0] = dot(clipPlane, modelViewMat * coords);
```

in the vertex shader's main then sets the value of `gl_ClipDistance[0]` to the dot product

$$(A, B, C, D) \cdot ((x, y, z, 1) = Ax + By + Cz + D$$

where (A, B, C, D) is the plane's coefficient vector (here, $(0.0, 0.0, 1.0, 25.0)$) and $(x, y, z, 1)$ are the point's post-transformation world coordinates. It's not hard to see that this dot product is zero on the plane $Ax + By + Cz + D = 0$, negative in one half-plane and positive in the other. The per-vertex values of `gl_ClipDistance[0]` are then interpolated across the primitive, and fragments with value less than 0 are culled.

In other words, fragments whose interpolated value of $Ax+By+Cz+D$ is less than 0 will be culled, which is exactly how clip planes in the fixed-function pipeline worked. In our case above, then, fragments with $z < -25.0$ are culled, which are those on the far side of the clipping plane $z = -25.0$.

We defined only one clip plane in `ballAndTorusClipped.cpp`, so our array `gl_ClipDistance` was initialized to size 1. Generally, it can be initialized to any size n (subject to a system-dependent maximum) by

```
float gl_ClipDistance[n];
```

in which case OpenGL culls fragments with at least one interpolated `gl_ClipDistance[i]` value, from $0 \leq i \leq n-1$, being negative.

Keep in mind that it's up to the user to set per-vertex values for each `gl_ClipDistance[i]`, just as we did above with the statement `gl_ClipDistance[0] = dot(...)`. She should also enable the corresponding clip plane `GL_CLIP_PLANE`*i* in the application program.

Most often `gl_ClipDistance[i]` is calculated so that its sign indicates on which side of some actual clipping plane a vertex lies, as in `ballAndTorusClipped.cpp`. Interestingly, though, this does not have to be the case and the programmer may set `gl_ClipDistance[i]` in any manner she finds useful, which may have nothing to do with any plane whatsoever.

Exercise 21.4. **(Programming)** Shaderize `clippingPlanes.cpp` of Chapter 3.

`gluPerspective()` and GLU in General

The GLU call `gluPerspective()`, explained in Section 3.12, is gone from 4.3 for obvious reasons: the projection matrix must be managed by the user in the shader. In fact, GLU itself is deprecated so we had best steer away from all `glu*()` calls.

However, we have already seen a workaround for `gluPerspective()`. In fact, in `litCylinderShaderized.cpp` of the last chapter we applied the GLM command `perspective(`*fovy*`, `*aspect*`, `*near*`, `*far*`)`, which returns the projection matrix corresponding to `gluPerspective(`*fovy*`, `*aspect*`, `*near*`, `*far*`)`.

Viewports and Multiple Windows

Finally, creating viewports and top-level windows, topics of Sections 3.13 and 3.14, respectively, remains identical in 4.3 to what it was pre-shader.

21.2 Shader Subroutines

To dynamically choose at run-time between alternate threads of control in a shader, what we've been doing so far is use `if`-type clauses conditioned on the value of a uniform, as exemplified in the following extract from the vertex shader of the program `ballAndTorusShaderized.cpp` of the previous chapter:

```
uniform uint object;
...
   if (object == HEMISPHERE) coords = hemCoords;
   if (object == TORUS) coords = torCoords;
...
```

Shader subroutines offer a somewhat more elegant alternative to the above, allowing the user to dynamically select a subroutine, albeit at a higher set-up cost. In short, what the user does is specify a subroutine type, then a bunch of subroutines of that type, and, finally, a subroutine uniform variable whose value decides the subroutine to execute. The conceptual forebear of shader subroutines is the C function pointer.

Let's get to work. We'll rewrite `ballAndTorusShaderized.cpp` to use shader subroutines instead of the `if` clauses in the extract above.

Experiment 21.4. Run `ballAndTorusShaderSubroutines.cpp`. The controls are exactly as for `ballAndTorusShaderized.cpp`: space to toggle animation on and off, up/down arrows to change its speed, and 'x'-'Z' to rotate the scene. Figure 21.4 is a screenshot. **End**

Figure 21.4: Screenshot of `ballAndTorusShader-Subroutines.cpp`.

The subroutines are in the vertex shader. First, see the two statements

```
subroutine void objectAction(void);
subroutine uniform objectAction object;
```

The first statement declares a *subroutine type* named `objectAction` with both `void` parameter list and return type. The general form of this declaration is

`subroutine` *returnType subroutineTypeName*`(type` *parameter*`, type` *parameter*`, ...)`

The second statement declares the *subroutine uniform* variable `object` corresponding to the just-declared subroutine `objectAction`. The general form is

`subroutine uniform` *subroutineTypeName subroutineUniformName*

Next, two simple subroutines, named `hemisphere` and `torus`, respectively, of type `objectAction` are defined by

```
subroutine (objectAction) void hemisphere(void)
{
   coords = hemCoords;
   colorsExport = hemColor;
}
```

```
subroutine (objectAction) void torus(void)
{
   coords = torCoords;
   colorsExport = torColor;
}
```

Their function is evidently to set the coordinate values and colors to export (to the fragment shader) for the hemisphere and torus, respectively. Observe that both subroutines have return type and parameter list matching the subroutine type, namely `objectAction`, to which they belong, as they must.

Now, to the initialization routine of `ballAndTorusShaderSubroutines.-cpp`, where the statements

```
hemSubroutineIndex = glGetSubroutineIndex(programId,
                     GL_VERTEX_SHADER, "hemisphere");
torSubroutineIndex = glGetSubroutineIndex(programId,
                     GL_VERTEX_SHADER, "torus");
```

obtain the indices of the subroutines `hemisphere` and `torus`, respectively.

Finally, back again to the vertex shader where the first line of

```
void main(void)
{
   object();
   gl_Position = projMat * modelViewMat * coords;
}
```

invokes the subroutine whose index is chosen in the application program's drawing routine by statements of the form

```
glUniformSubroutinesuiv(GL_VERTEX_SHADER, 1, &XSubroutineIndex);
```

where `X` is either `hem` or `tor`. Such statements replace those of the form

```
glUniform1ui(objectLoc, X);
```

of the earlier `ballAndTorusShaderized.cpp` using `if`-based shader control, where `X` was either `HEMISPHERE` or `TORUS`.

We have as a rule used `if`-based shader control for simplicity's sake. In fact, for relatively small programs like ours it's unlikely there is any significant difference in performance between `if`-based control and shader subroutines. However, for large programs the user might want to choose between the two, considering:

(a) Run-time performance: though, certainly, subroutines are more easily optimizable, there is a trade-off with the overhead in setting them up; the proof of the pudding is in the running.

(b) Clarity of presentation: shader subroutines have an advantage in that they can make it easier to follow the program logic, especially if there is a lot going on in the individual threads of control; accordingly, modifying or extending the program in future might be a less hair-raising enterprise as well.

Remark 21.1. There's actually yet another way to dynamically control shader flow, which is to make *separate* shader objects. So, for example, instead of choosing between two lines of control in, say, the one vertex shader via subroutines or `if` cases, one could create *two* different vertex shaders, dynamically attaching and detaching one or the other during run-time. We'll leave the interested reader to peruse the red book for more on this.

21.3 More Animation

We'll add next to our animation repertoire two powerful practical techniques. The first, that of picking an object on the screen, is actually familiar from Chapter 4, but we'll do it entirely differently in 4.3. The second, so-called transform feedback, was introduced in OpenGL 3.0 to allow vertices to be intercepted and their attributes recorded, right after they have been transformed by the vertex, tessellation and geometry shaders (of course, in the case of the latter two, only if they are present). These recorded values may then be used by the program, for example, to modify subsequent passes through the pipeline.

21.3.1 Picking

Essential to interaction with animated programs is for the user to be able to pick an object on the screen. Section 4.8 from our pre-shader days showed how to do this with the help of a pick matrix and entering selection mode via `glRenderMode(GL_SELECT)`.

Sadly, `glRenderMode()` is gone from 4.3, but the fragment shader rides to our rescue. In fact, the power of the fragment shader affords a much more direct approach to picking than our earlier rather roundabout method via checking for object intersection with a pretend selection volume. Let's see this by shaderizing `ballAndTorusPicking.cpp`.

Figure 21.5: Screenshot of `ballAndTorusPicking-Shaderized.cpp` moments after the ball has been clicked.

Experiment 21.5. Run `ballAndTorusPickingShaderized.cpp`. The controls are exactly as for `ballAndTorusPicking.cpp`: space to toggle animation on and off, up/down arrows to change its speed, 'x'-'Z' to rotate the scene, and, most importantly, left mouse click to pick either ball or torus. See Figure 21.5 for a screenshot. **End**

We saw when discussing stippling in the previous section that the fragment shader has access to the input built-in `gl_FragCoord` of type `vec4`, whose (x, y)-values are the coordinates of the fragment in the windows system and z-value is the depth of the fragment. So here's the plan for `ballAndTorusPickingShaderized.cpp`: obtain mouse click coordinates from the application program and, then, match these in the fragment shader with fragment coordinates; then, pick the object which has a matching fragment of least depth.

What could be simpler? There's a minor new technicality to navigate though. When running the "depth competition" to find the fragment nearest the viewer, we'll evidently need storage which can be shared amongst fragment shader invocations.

The solution is to use variables of the `buffer` storage type (one of those listed in Table 20.1). Buffer variables must be placed in so-called *interface blocks*, which are blocks of variables declared with a `struct`-like syntax. Here's the definition of the interface block, named `shaderStorage`, in `ballAndTorusPickingShaderized.cpp`'s fragment shader:

```
layout(std430, binding=0) buffer shaderStorage
{
   ivec2 clickedCoords;
   uint clickedObj;
   float minClickedDepth;
};
```

The qualifier `buffer` defines the storage type of the member variables, while the `layout` directive specifies that member variables will be stored in memory following the particular `std430` rules – which will tell us what offsets to use to access these variables from the application program – and that the binding index of the block is 0.

As for the member variables themselves, it's fairly evident what they are intended to hold: the mouse click coordinates place in `clickedCoords`, while `minClickedDepth` will hold the smallest depth of a fragment matching the clicked coordinates seen so far, and `clickedObj` the name of the object to which it belongs.

Note, as well, the redeclaration of `gl_FragCoord` in the fragment shader:

```
layout(origin_upper_left, pixel_center_integer) in  vec4 gl_FragCoord;
```

which moves its origin of pixel coordinates to the upper-left-most pixel (from the default of the lower-left-most) and makes the pixel coordinate values, both `gl_FragCoord.x` and `gl_FragCoord.y`, the integral $0.0, 1.0, \ldots$ (rather than the default of $0.5, 1.5, \ldots$). The purpose of this redeclaration is so the coordinate system of the fragment shader matches exactly that of the OpenGL window which, of course, the application's mouse control uses.

To the application program next, where the statements

```
glBindBuffer(GL_SHADER_STORAGE_BUFFER, buffer[SHADER_STORAGE]);
glBufferData(GL_SHADER_STORAGE_BUFFER, 128, NULL,
             GL_DYNAMIC_COPY);
glBindBufferBase(GL_SHADER_STORAGE_BUFFER, 0,
```

```
                buffer[SHADER_STORAGE]);

storageBufferPtrInt = (int*)glMapBuffer(GL_SHADER_STORAGE_BUFFER,
                      GL_READ_WRITE);

storageBufferPtrFloat = (float*)&storageBufferPtrInt[0];
```

in the initialization routine set up a shader storage buffer in GPU memory to hold the buffer variables declared in the interface block `shaderStorage` (in particular, note that the second parameter of `glBindBufferBase()`, being 0, specifies the binding index of the associated interface block) and map pointers to the buffer data (the integer pointer is also cast as a float pointer in the last statement because we'll need to access both kinds of data in the buffer).

Moreover, the `std430` layout rules (see the red book for specs) mean that the variables of the fragment shader's interface block `shaderStorage` are packed into 128 bits: two successive 32-bit `int`s for `clickedCoords`, followed by one 32-bit `uint` for `clickedObj`, and then one 32-bit `float` for `minClickedDepth`.

Now that we have all the data structures in place, the logic of the program is simple to understand. On a left click, the `mouseControl()` function executes the following:

```
storageBufferPtrInt[0] = x;
storageBufferPtrInt[1] = y;
storageBufferPtrInt[2] = 2;
storageBufferPtrFloat[3] = 1.0;

isSelecting = 1;
glUniform1i(isSelectingLoc, isSelecting);

glutPostRedisplay();
```

These statements first of all put the click coordinates into `clickedCoords`, the value `NONE` (defined to be 2 in the fragment shader) in `clickedObj` and 1 in `minClickedDepth` (note that the fragment depth in `gl_FragCoord.z` is normalized from 0 to 1, 0 being nearest the eye and 1 farthest away, so a value of 1 means, effectively, infinity or that no object is selected).

Next, "selection mode" (our own and nothing to do with the OpenGL environment as in using pre-shader `glRenderMode()`!) is entered by setting `isSelecting` to 1, and the drawing routine called, which means, of course, that the fragment shader will run for each fragment generated. Accordingly, see the following block in the fragment shader's main which runs the "depth competition" for fragments matching the click:

```
if (isSelecting == 1)
if (     (abs(gl_FragCoord.x - clickedCoords.x) <= 1)
     && (abs(gl_FragCoord.y - clickedCoords.y) <= 1)
```

```
        && (gl_FragCoord.z < minClickedDepth)
     )
  {
     minClickedDepth = gl_FragCoord.z;
     clickedObj = object;
  }
```

As you see, we allow a tolerance of one pixel in both the x and y directions for a fragment to match the click; if a fragment does match and it is closer to the eye than the currently closest fragment, then we accordingly update `minClickedDepth` to the depth of the current fragment and `clickedObj` to the object to which this fragment belongs. The remaining statements

```
  if (object == HEMISPHERE)
  ...
  if (object == TORUS)
  ...
```

in the fragment shader's main either highlight the picked object in red if indeed either ball or torus is picked, or draw both in their respective default colors.

Finally, returning to the application program, the statements

```
  if (isSelecting == 1)
  {
     highlightFrames = 10;
     glUniform1i(highlightFramesLoc, highlightFrames);

     isSelecting = 0;
     glUniform1i(isSelectingLoc, isSelecting);
  }
```

at the end of the drawing routine set the number of frames to highlight the picked object and restore non-selection mode.

Exercise 21.5. (**Programming**) Redo Exercise 4.70 from Section 4.8 in OpenGL 4.3.

21.3.2 Transform Feedback

Transform feedback is an operation which may be used to tremendous effect, particularly in animated programs. What it does is record into a buffer selected attributes of vertices after they have been transformed by the vertex, tessellation and geometry shaders (or, more precisely, the attributes are recorded after the last of these three shader stages, depending on which are present). The contents of this recording buffer may then serve as a look-ahead.

For example, one might run a transformation step of an animation without drawing the results to the frame buffer, but using transform feedback to capture, say, the world coordinates of objects post-transformation. These coordinates may then be used to rearrange the animation, e.g., if a collision is detected. In fact, we have a program exactly along these lines.

Experiment 21.6. Run `ballsAndTorusTransformFeedback.cpp`. The controls are exactly as for `ballAndTorusShaderized.cpp`: space to toggle animation on and off, up/down arrows to change its speed, and 'x'-'Z' to rotate the scene.

However, instead of one ball, now there are two, initially coincident, which travel in opposite directions around the torus. When the balls intersect they are red, when they are close (closer than a particular threshold distance) they turn orange, and beyond that they are blue. Figure 21.6 is a screenshot when the balls are close. **End**

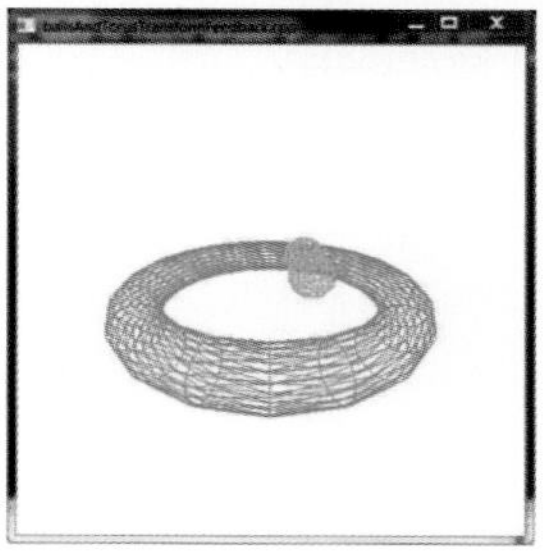

Figure 21.6: Screenshot of `ballAndTorus-TransformFeedback.cpp` when the balls are close.

Exercise 21.6. To better appreciate the power of transform feedback, contemplate for a minute how to do what the preceding program does in a pre-shader world.

Here's the plan how to use transform feedback in the program. Firstly, observe that since the balls are of radius 2 each, they intersect when the distance between their centers is at most 4 units. So, we'll use transform feedback to record the world coordinates of the two centers in order to calculate how far apart they are: if they are at most 4 apart then the balls intersect and we draw them red, if they are at most 8 (an arbitrarily chosen value) apart then we draw the balls orange.

Now, let's see how transform feedback is set up in the initialization routine. The statement

```
glTransformFeedbackVaryings(programId, 1, varyings,
                            GL_INTERLEAVED_ATTRIBS);
```

specifies the shader outputs, aka *varyings*, to be recorded through transform feedback. Specifically, the third parameter `varyings` is the array of names of the varyings – only one in our case defined in the globals by

```
static const char* varyings[] = "centerWorldCoords";
```

The second parameter is the number of names, while the first parameter identifies the program, and the last specifies the mode in which to record the varyings. The options for this last are `GL_INTERLEAVED_ATTRIBS`, when all the varyings are recorded one after another in one buffer, and `GL_SEPARATE_ATTRIBS`, when each varying is recorded in its own buffer. Since we have a single varying `centerWorldCoords`, both options are effectively identical for us.

The program must be linked *after* the transform feedback varyings are defined which explains the location of the `glTransformFeedbackVaryings()` statement in the code.

Now, if you see the vertex shader, then you find the one varying `centerWorldCoords`, in fact, is set by

```
layout(location=2) in vec4 centerCoords;
...
out vec4 centerWorldCoords;
vec4 coords;
...
   if (object == CENTER)
   {
      coords = centerCoords;
      centerWorldCoords = modelViewMat * coords;
   }
```

In other words, `centerWorldCoords` are the world coordinates of the ball center after modelview transformation.

Returning to the application program's initialization, a *transform feedback object* is created and bound by the statement pair

```
glGenTransformFeedbacks(1, transformFeedback);
glBindTransformFeedback(GL_TRANSFORM_FEEDBACK, transformFeedback[0]);
```

similarly to how VAO and VBOs are created and bound, the id of the object, obviously, stored in `transformFeedback[0]`.

Further down the initialization routine the statement block

```
glBindBuffer(GL_TRANSFORM_FEEDBACK_BUFFER,
             buffer[TRANSFORM_FEEDBACK]);
glBufferData(GL_TRANSFORM_FEEDBACK_BUFFER, 1024,
             NULL, GL_DYNAMIC_COPY);
glBindBufferBase(GL_TRANSFORM_FEEDBACK_BUFFER, 0,
                 buffer[TRANSFORM_FEEDBACK]);
```

sets up a *transform feedback buffer* in a manner exactly similar to how we set up a shader storage buffer in the previous section on picking.

Next,

```
glGenTextures(1, texture);
glActiveTexture(GL_TEXTURE1);
glBindTexture(GL_TEXTURE_BUFFER,  texture[0]);
glTexBuffer(GL_TEXTURE_BUFFER,  GL_RGBA32F,
            buffer[TRANSFORM_FEEDBACK]);
tfBufferLoc = glGetUniformLocation(programId,
                                   "transformFeedbackTex");
glUniform1i(tfBufferLoc, 1);
```

creates a TBO – see the discussion of program `helixListShaderized-InstanceCounter.cpp` in Section 21.1 where we first used texture buffer objects – whose data source is the transform feedback buffer set up earlier. The reason to bind a TBO to the transform feedback buffer is for the fragment shader to be able to access its values.

Finally, the application program's drawing routine, as clearly indicated by comments there, is two phase: first transform feedback, then actual drawing. The first two statements

```
glBeginTransformFeedback(GL_POINTS);
glEnable(GL_RASTERIZER_DISCARD);
```

of the first phase start transform feedback, specifying that the type of primitive to be recorded is points and that they be discarded before rasterization, so that recording takes place in this phase with nothing actually drawn to the screen. As expected, the second phase begins with the inverse pair

```
glDisable(GL_RASTERIZER_DISCARD);
glEndTransformFeedback();
```

to begin drawing to the screen without recording. Let's now see what exactly is drawn in the two phases. It's best to examine the second phase first, where the drawing commands are exactly as in `ballAndTorusShaderized.cpp`, except that there is an additional ball revolving in a direction opposite to the first, the command

```
modelViewMat = rotate(modelViewMat, -1.0f*longAngle,
                      vec3(0.0, 0.0, 1.0));
```

in the transformation block preceding the second ball, vs. the corresponding command

```
modelViewMat = rotate(modelViewMat, longAngle, vec3(0.0, 0.0, 1.0));
```

in the block preceding the first, doing the needful.

Now, if you compare, the first recording phase is identical in its drawing commands to the second, *except* that

(a) The torus is not drawn, as it is irrelevant to computing the distance between the two balls.

(b) Instead of drawing a ball as a hemisphere pair, only its center is drawn, because it's only the two balls' centers' respective world coordinates which we need for our calculations, as explained earlier.

The rest of the program logic now falls in place from a reading of the fragment shader. Here is the latter's main:

```
void main(void)
{
   center0 = texelFetch(transformFeedbackTex, 0).xyz;
   center1 = texelFetch(transformFeedbackTex, 1).xyz;
   distBetweenCenters = distance(center0 , center1);

   if (object == HEMISPHERE)
   {
      colorsOut = hemColor;
      if (distBetweenCenters <= 8.0) colorsOut = orangeColor;
```

```
      if (distBetweenCenters <= 4.0) colorsOut = redColor;
   }
   if (object == TORUS) colorsOut = torColor;
}
```

The two ball center xyz-coordinates, as recorded in Phase 1 of the drawing routine, are fetched from the transform feedback buffer, their distance calculated, and the ball colors set accordingly.

Exercise 21.7. (**Programming**) Exercise 4.36 from Chapter 4 was to animate a single cue ball rolling on a pool table. Add a second ball, redoing the exercise in OpenGL 4.3 to use transform feedback to detect and react to collisions (between balls and between ball and table edge).

21.4 Special Visual Techniques

We had a glimpse in Exercise 20.12 of the preceding chapter of the kind of effect one can code up in the fragment shader: we separated out the specular light and doubled it, something which the conventional pipeline, of course, can never be made to do. It's reasonable then to revisit the pre-shader Chapter 13 on special visual techniques to see how the few there might be recreated in GL 4.3, and if we can do more.

Blending

First is blending done pre-shader in Section 13.1, which it turns out becomes almost trivial in the fragment shader. Let's get straight to blending textures – we'll shaderize `fieldAndSkyTexturesBlended.cpp`.

Experiment 21.7. Run `fieldAndSkyTexturesBlendedShaderized.cpp`. As in `fieldAndSkyTexturesBlended.cpp` press the arrow keys to move the sun, the transition between day and night happening from a blending of day and night textures. See Figure 21.7 for a screenshot. **End**

Figure 21.7: Screenshot of `fieldAndSkyTextures-BlendedShaderized.cpp` late morning.

All the code of `fieldAndSkyTexturesBlendedShaderized.cpp` should be routine for the reader at this point except, maybe, for where the texture blending actually takes place: see the line

```
if (object == SKY) colorsOut = mix(nightSkyTexColor,
                                   skyTexColor, alpha);
```

in the main routine of the fragment shader. Now, the built-in GLSL function `mix(`x`, `y`, `a`)` returns the linear combination $(1 - a) * y + a * y$ of x and y, so the line above obtains the exact same blending of the night and day sky textures as in `fieldAndSkyTexturesBlended.cpp`.

So, we have total control of blending in 4.3 and it's not hard. However, blending can still be enabled in 4.3 with `glEnable(GL_BLEND)` and glBlendFunc(), as well as a host of other glBlend*() commands, remaining in the fixed-function part of the 4.3 pipeline. This may seem odd but we'll see soon the reason why when we discuss antialiasing.

Not the Foggiest

Fog is gone from 4.3. There is no support for fog in the fixed-function part of 4.3 – glFog*() commands are all now obsolete. The reason is simple: as is clear from the discussion in Section 13.2 of the mechanics of fog, to fog is to blend. Ergo, the reader can code fog herself, as we ask her to do next.

Exercise 21.8. (Programming) Shaderize fieldAndSkyFogged.cpp from Section 13.2. You don't have to implement all the fog modes exactly. Just make a realistic fog.

Billboarding

Billboarding, discussed earlier in Section 13.3, is simply the clever placement of a textured rectangle to give the illusion of a 3D object. Obviously, this principle does not change whatever OpenGL version we use.

Exercise 21.9. (Programming) Shaderize billboard.cpp.

Antialiasing Lines

We can antialias lines based on the coverage value of their fragments just as in pre-shader OpenGL – see the discussion in Section 13.4.1 – and using the very same commands. We ask the reader to verify this next.

Exercise 21.10. (Programming) Antialias a line in 4.3 with the commands

```
glEnable(GL_BLEND);
glBlendFunc(GL_SRC_ALPHA, GL_ONE_MINUS_SRC_ALPHA);
glEnable(GL_LINE_SMOOTH);
```

in the initialization routine.

Now we see a reason for keeping blending in the fixed-function part of OpenGL 4.3; it would be asking a bit much of the programmer to calculate coverage value per fragment for a line segment and blend accordingly into the color buffer.

21.4.1 Points

Points are a whole different ball game. In fact, GLSL 4.3 has a rich set of controls allowing the programmer to do pretty much as she pleases with points. You might now want to quickly review the discussion about how points are rendered at the end of Section 13.4.1 and also point sprites in Section 13.5. Good, let's get to code.

Experiment 21.8. Run `points.cpp`. Drawn are three large points. Press space to cycle between four different point renderings and the up and down arrow keys to move the points parallel to the z-axis. Figure 21.8 is a screenshot of what is seen at first, particularly, three unedited points. **End**

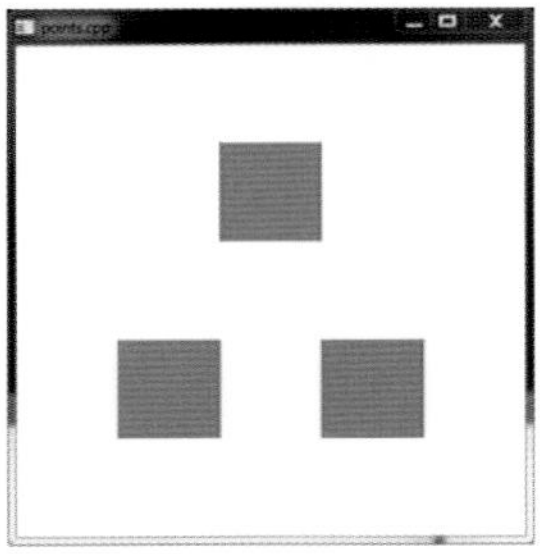

Figure 21.8: Screenshot of `points.cpp` initially.

The four different renderings correspond to the four cases, respectively, of the switch statement in the fragment shader's main. The first,

```
case 0:
   colorsOut = pointSetColor;
   break;
```

simply draws the points unedited in the fragment shader. However, as the up or down arrow keys are pressed to move the points away or near, their size, respectively, decreases or increases. This change is effected by the statement

```
gl_PointSize = 100.0 - zTrans * 10.0;
```

in the vertex shader's main which sets a simple linear ramp on the built-in variable `gl_PointSize`. Note, of course, that, if the program didn't change the point size, then all three points would be rendered a fixed size, no matter how they were moved (validate this by changing the statement above to `gl_PointSize = 100.0`; the reason is explained in the part about rendering points at the end of Section 13.4.1).

Importantly, it is

```
glEnable(GL_PROGRAM_POINT_SIZE);
```

in the application program's initialization routine which allows the vertex shader to set the point size in the first place.

The fragment shader's next switch case

```
case 1:
   coordWRTcenter = gl_PointCoord - pointCoordCenter;
   distFromCenter = sqrt(dot(coordWRTcenter, coordWRTcenter));
   if (distFromCenter > pointCoordRadius) discard;
   colorsOut = pointSetColor;
   break;
```

first computes the fragment's coordinates w.r.t. the center of the point within the latter's built-in texture coordinate system. The fragment's coordinates themselves in this system are accessible through the built-in variable `gl_PointCoord`. The point's center is `pointCoordCenter`, set to $(0.5, 0.5)$, as both texture coordinates s and t run from 0 to 1 along the point.

Next, the distance of the fragment from the point center is determined and the fragment *discarded* if it is greater than the value of `pointCoordRadius`, which is 0.5. The function `discard` is built-in as well. The output of case 1, then, is a rounded point.

One still sees jaggies along the boundary of the rounded point of case 1. We try to eliminate these in

```
case 2:
   coordWRTcenter = gl_PointCoord - pointCoordCenter;
   distFromCenter = sqrt(dot(coordWRTcenter, coordWRTcenter));
   if (distFromCenter > pointCoordRadius) discard;
   alpha = clamp( (distFromCenter - startBlend)/
                  (pointCoordRadius - startBlend), 0.0, 1.0 );
   colorsOut = mix(pointSetColor, backgroundColor, alpha);
   break;
```

which is similar to the preceding case, except that, finally, a linear color ramp blends the color of the points with the background color, starting from a distance of `startBlend`, currently 0.4, from the point's center and ending at its border. Note that the built-in GLSL function `clamp(`x`,` a`,` b`)` returns, as one would expect, x if $a \leq x \leq b$, a if $x < a$, and b if $x > b$.

Point Sprites

Finally,

```
case 3:
   colorsOut = texture(starTex, gl_PointCoord);
   break;
```

very simply implements point sprites – the topic of the pre-shader Section 13.5 – obviously based on the point's built-in texture coordinate system. Note that

```
glPointParameteri(GL_POINT_SPRITE_COORD_ORIGIN, GL_LOWER_LEFT);
glEnable(GL_POINT_SPRITE);
```

in `setup()` of the application program cause the texture t-coordinate to increase from 0 to 1 from bottom to top of the sprite (s is always from left to right) and enable point sprites.

Exercise 21.11. (Programming) Experiment with antialiasing the rounded point by changing the value of `startBlend` in case 2 of the `points.cpp`'s fragment shader's switch statement.

Try, as well, to use the built-in GLSL function `smoothstep` (look up the GLSL docs) for a smoother transition from 0 to 1 of the value of the blending parameter `alpha` (the linear ramp of case 2 is not smooth where it reaches 0 at the bottom or 1 at the top).

Multisampling

Multisampling, which we first encountered in Section 13.4.2, is enabled in 4.3 via the command `glEnable(GL_MULTISAMPLE)`, just as in pre-shader OpenGL. Remember, though, to first create an OpenGL window which supports multisampling by passing `GLUT_MULTISAMPLE` as a parameter to `glutInitDisplayMode()`.

Exercise 21.12. (**Programming**) Implement multisampling to antialias polygons in OpenGL 4.3.

Difficult Environment for Mapping

Environment mapping, sadly, can be a chore in 4.3. The reason is that neither `glTexGeni()` nor `glEnable(GL_TEXTURE_GEN_?)`, which made environment mapping such a joy in pre-shader OpenGL, as we saw in Section 13.6, is available in 4.3. If you followed the math of texture generation in Section 13.6, though, you can probably implement environment mapping on your own. However, we would recommend taking the easy way out, viz. sneak back to compatibility mode and use pre-shader OpenGL.

Stencils and Scissors

Thankfully, both stencil and scissor tests remain very much alive in the fixed-function part of the OpenGL 4.3 pipeline, and in exactly the same manner as discussed in Section 13.7, as we ask the reader to investigate next.

Exercise 21.13. (**Programming**) Shaderize `ballAndTorusStenciled.cpp` from Section 13.7.

Image Manipulation and PBOs

Image manipulation statements such as `glDrawPixels()`, `glReadPixels()` and `glCopyPixels()`, discussed in Section 13.8, are, unsurprisingly, gone from 4.3 as the fragment shader now gives direct access to fragments. All the functionality of these commands can now be user-programmed. However, PBOs (pixel buffer objects) to store pixel data can be implemented in 4.3, exactly as in pre-shader OpenGL as described in Section 13.8.

Exercise 21.14. (**Programming**) Shaderize `imageManipulation.cpp` from Section 13.8.

Bump Mapping

Finally, of course, the pre-shader bump mapping of Section 13.9 was what we fully redid in the programmable pipeline when discussing 4.3 lighting in Section 20.5 of the last chapter.

21.5 Tessellation Shaders

Vertex shaders, while they can accomplish much as we have seen, have two major limitations: (a) they can only update attributes per vertex without access to data from other, say, neighboring, vertices, and (b) cannot generate additional geometry, e.g., new vertices. We'll run into these limitations, for example, if we want to adaptively refine the representation of an object depending on its closeness to the camera. This will require the generation of new geometry, namely, more detail, as the object occupies more screen area, something the vertex shader is incapable of doing. In fact, it was to solve precisely this kind of problem *within the GPU* – one can always generate new vertices in the application program, but then comes the cost of communication over the CPU-GPU bus – that the tessellation shader was conceived (and, in fact, got its name).

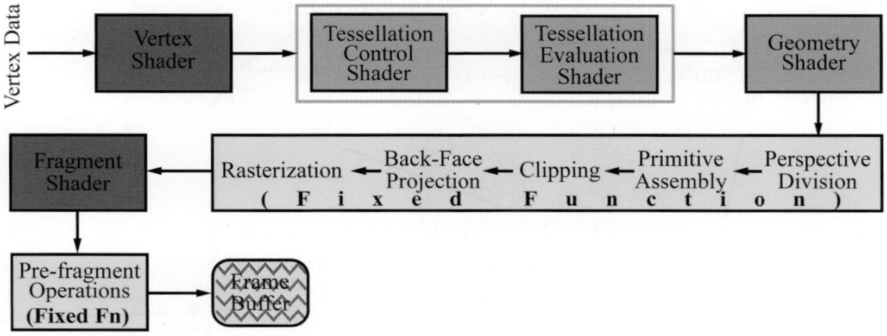

Figure 21.9: OpenGL programmable pipeline.

The tessellation shader is a fairly complex *optional* component of the programmable pipeline consisting of four modules – the tessellation control shader (TCS), tessellation primitive generator (TPG), tessellation evaluation shader (TES) and a primitive assembler (PA) – operating (pretty much but not exactly as we'll see) in that sequence. Figure 21.9 (copy of Figure 20.2) shows how the tessellation shader is situated within the OpenGL pipeline. Of the four modules, the TCS and TES are programmable, while the TPG and PA are fixed function. So, the tessellation shader is actually two-shaders-in-one, plus fixed function support.

We begin with a simple scenario, showing how the four modules combine, which should make the subsequent discussion fairly easy to follow.

Suppose that input is an ordered sequence of three vertices p_0, p_1 and p_2. Assume they are not on a straight line, which means there's a unique arc of a circle through the three – see Figure 21.10 for the elementary geometry to construct the arc. Now, suppose we have already made a design decision that all curves are to be drawn as cubic Bézier curves and all surfaces as cubic Bézier surfaces. In particular, we want to approximate our circular arc with a cubic Bézier curve drawn as a 5-segment polyline. Think of 5 segments as the currently desirable level of refinement if you will.

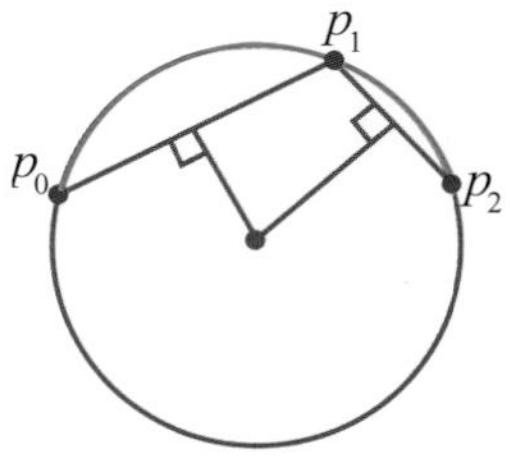

Figure 21.10: Arc of a circle defined by 3 points.

Here's how this would happen in a tessellation shader. Refer to Figure 21.11 as you read on.

The three wannabe arc vertices p_0, p_1 and p_2 are first input to the TCS – together they are the *input patch*. The TCS waves a math wand over the

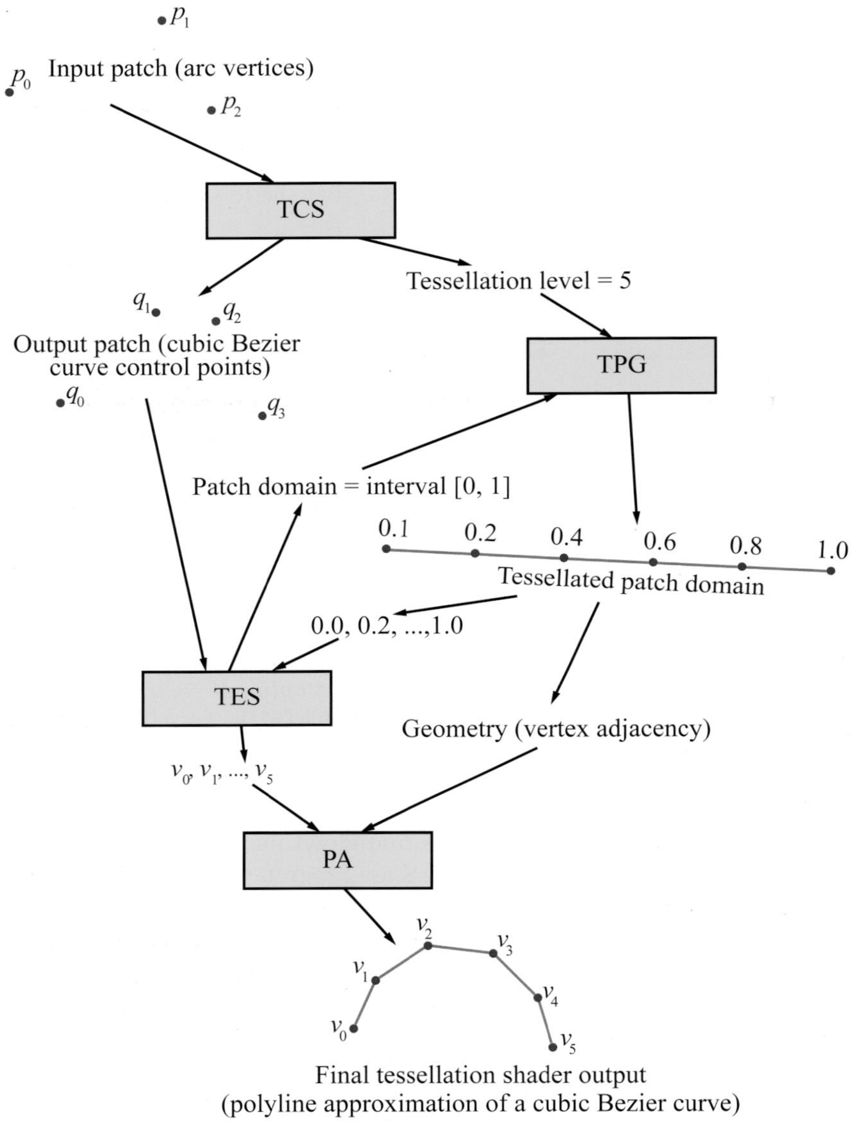

Figure 21.11: Tessellation shader scheme.

input patch to conjure up a sequence q_0, q_1, q_2 and q_3 of 4 control points of a cubic Bézier curve nicely approximating the circular arc through p_0, p_1 and p_2 – we'll trust this can be done and not sweat the details. The 4 control points together form the *output patch*. The output patch is delivered to the TES. Note, then, that a patch, input or output, is simply an ordered list of vertices with *no* intrinsic geometry.

Next, the TPG receives from the TCS the value of 5 for the *tessellation level*, and from the TES the instruction that the primitive to be tessellated, called the *patch domain*, is the interval [0, 1]. All this, of course, corresponds to asking a 5-interval grid on the [0, 1] parameter interval of the Bézier curve. The TPG computes, say, a regular grid, which, geometrically, is the 5-segment line strip with vertices at $0.0, 0.2, \ldots, 1.0$.

These line strip, or patch domain, vertices are transmitted from the TPG to the TES. Remember the TES already has at its disposal the output patch, i.e., all 4 Bézier control points q_0, q_1, q_2 and q_3. What the TES does next is, for each patch domain vertex, i.e., for each parameter value $0.0, 0.2, \ldots, 1.0$, compute the world space coordinates of the corresponding cubic Bézier curve vertex, denoted, respectively, $v_0, v_1, \ldots, v_5$. Chapter 15 on Bézier theory tells us that the formula to use is

$$c(u) = (1-u)^3 q_0 + 3(1-u)^2 u q_1 + 3(1-u)u^2 q_2 + u^3 q_3 \tag{21.1}$$

so the value of v_i is $c(0.2i)$. The v_i are sent by the TES to the PA. The latter also receives the tessellation geometry (line strip in this case) from the TPG and uses this to assemble the v_i into the final output polyline in world space.

On a first encounter with tessellations shaders, the TPG is often most confusing. First of all, it's a bit of a misfit in the OpenGL pipeline – it does nothing at all with any application data, e.g., patch vertex. The TPG is more of a refugee from a math app like MATLAB! Given a patch domain, which is an abstract mathematical domain (the [0, 1] interval in our case), and a tessellation specification (regular 5-interval grid in our case), the TPG simply computes a tessellation of the patch domain. It's the TCS and TES which actually get grease on their elbows with patch vertex values.

Summing up, roughly, the TCS has first dibs at the input patch with license to transform it into an output patch in any way it likes, which it then hands over to the TES. The TES next manufactures a real-world vertex from each patch vertex it receives from the TCS. Finally, the PA assembles a real-world object from the TES's output vertices with help of the abstract template produced by the TPG.

Moreover, it's worth repeating that the TCS's input and output patches do not carry any geometry. They are simply ordered lists of vertices.

Time now to get all OpenGLly and technical. We're going to explore, successively, in fair detail the TCS, TES and TPG, which are tessellation-specific. The PA, however, is a generic module, which assembles real-world primitives given their geometry and vertex coordinates. As it is not specific to tessellation, we won't discuss it further here.

And, of course, we're going to back things up with live code. In fact, we have programmed the example scenario above in `tessellatedCurve.cpp` and, as we describe technicalities in general terms, we'll point to their specific incarnations in this program.

Experiment 21.9. Run `tessellatedCurve.cpp`. You see the three input vertices and an initially 5-segment Bézier curve polyline. Press the up/down arrow keys to increase/decrease the number of segments. Figure 21.12 is a screenshot with the initial five segments.

The Bézier curve is evidently a poor approximation to the circular arc (not drawn) through the three input vertices, the reason being, as we'll see in the upcoming discussion, an intentionally simple-minded choice of the control points; keep in mind, though, that the actual curve itself is of far less interest here than the programming steps to generate it.

The code is explained through the following sections. **End**

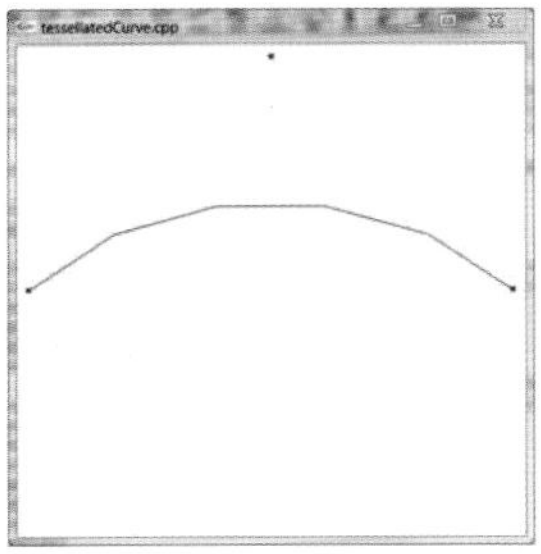

Figure 21.12: Screenshot of `tessellatedCurve.cpp` initially.

21.5.1 TCS (Tessellation Control Shader)

The input to the TCS is a new OpenGL primitive, namely *patch* – the symbol being `GL_PATCHES` – which is just an ordered list of vertices of the size *inputPatchSize* set by the statement

```
glPatchParameteri(GL_PATCH_VERTICES, inputPatchSize)
```

in the application program. The last line

```
glPatchParameteri(GL_PATCH_VERTICES, 3);
```

of the initialization routine of `tessellatedCurve.cpp` sets *inputPatchSize* to 3.

Of course, as is evident from the layout of the OpenGL pipeline (Figure 21.9), each vertex entering into a patch is first processed by the vertex shader. In `tessellatedCurve.cpp`, though, the vertex shader is particularly hands-off. It simply reads each input vertex's position coordinates into the per-vertex built-in variable `gl_Position`, transforming them no further.

Exercise 21.15. The vertex shader of `tessellatedCurve.cpp` does not apply modelview or projection transforms to the input coordinates. In fact, the program seems to dispense with modelview and projection transforms and their matrices altogether. How does it get away with this?
Hint: See discussion of the canonical viewing box in Section 18.1.

Patches themselves are specified by using `GL_PATCHES` in a drawing command – one being created every *inputPatchSize* vertices. Therefore, given that `tessellatedCurve.cpp`'s *inputPatchSize* is 3, the command

```
glDrawArrays(GL_PATCHES, 0, 3);
```

in its drawing routine means that its TCS reads exactly one patch.

Generally, the pair of statements

```
glPatchParameteri(GL_PATCH_VERTICES, inputPatchSize);
...
glDrawArrays(GL_PATCHES, first, count);
```

in the application program would cause the TCS to read *count*/*inputPatchSize* patches from the vertex shader.

The TCS has access to all the per-vertex built-in attributes, set by the vertex shader, for every vertex in each input patch. Accordingly, the data for each input patch, as read by the TCS, is a built-in variable `gl_in`, which is an array of structures – the three fields of the structure corresponding to the three built-in vertex attributes – of the size of the patch. Here is its declaration:

```
in gl_PerVertex
{
   vec4 gl_Position;
   float gl_PointSize;
   float gl_ClipDistance[];
} gl_in[gl_PatchVerticesIn];
```

The declaration is implicit as `gl_in` is built-in, so the programmer can simply access the variable without making any declarations of her own. The size of the above array is the value of the built-in variable `gl_PatchVerticeIn` accessible by the TCS – it is, of course, equal to *inputPatchSize*, the size of an input patch. The TCS may, additionally, be supplied user-defined per-vertex attribute values, beyond the three built-ins, by the vertex shader.

If you recall the example scenario described at the start of this section, the TCS has two responsibilities: (a) produce an output patch for each input patch to send the TES, and (b) set the tessellation levels per output patch to control operation of the TPG. It is time now to look at `tessellatedCurve.cpp`'s TCS, namely, `tessControlShader.glsl`.

The number of vertices per output patch is set by a statement of the form

```
layout(vertices=outputPatchSize) out;
```

in the TCS. It is accessible by the TCS in the built-in input variable `gl_PatchVerticesOut`. The TCS of `tessellatedCurve.cpp` sets *outputPatchSize* to 4 with the statement

```
layout(vertices=4) out;
```

The TCS executes once per output patch vertex. The sequence number of the currently processing output patch vertex within the current patch is the value of the built-in input variable `gl_InvocationID`. Moreover, the sequence number of the current patch in the current drawing statement is contained in the built-in input variable `gl_PrimitiveID`. The per-patch output is a built-in of the same form as the input, viz.

```
out gl_PerVertex
{
   vec4 gl_Position;
   float gl_PointSize;
```

```
   float gl_ClipDistance[];
} gl_in[gl_PatchVerticesOut];
```

the size of the array being that of an output patch.

The switch statement below in the main routine of tessControlShader.glsl creates the four vertices of the output patch (to be used as control points of a cubic Bézier curve, of course) from the three input vertices in a simple-minded manner: the first output vertex is the first input vertex, the second is midway between the first and second input vertices, the third is midway between the second and third input vertices, while the fourth is the third input vertex (the particular accuracy of the Bézier approximation not being of interest in this discussion, we chose output vertices simple to code in).

```
switch(gl_InvocationID)
{
   case 0: gl_out[ gl_InvocationID ].gl_Position =
           gl_in[0].gl_Position;
   break;
   case 1: gl_out[ gl_InvocationID ].gl_Position =
           (gl_in[0].gl_Position + gl_in[1].gl_Position)/2.0;
   break;
   case 2: gl_out[ gl_InvocationID ].gl_Position =
           (gl_in[1].gl_Position + gl_in[2].gl_Position)/2.0;
   break;
   case 3: gl_out[ gl_InvocationID ].gl_Position =
           gl_in[2].gl_Position;
   break;
   default:
   break;
}
```

The TCS tessControlShader.glsl has now discharged its first responsibility of specifying the one output patch. It discharges the second of fixing the tessellation levels for this output patch with the statement pair

```
gl_TessLevelOuter[0] = 1.0;
gl_TessLevelOuter[1] = tessLevelOuter1;
```

Now, the so-called inner and outer tessellation levels are contained in the following built-in per-patch arrays, respectively:

```
patch out float gl_TessLevelInner[2];
patch out float gl_TessLevelOuter[4];
```

So tessControlShader.glsl sets gl_TessLevelOuter[0] to 1.0 and gl_TessLevelOuter[1] to the value of the application-provided uniform tessLevelOuter1 (initially, 5.0). It does not set the two inner tessellation levels or the other two outer tessellation levels because they are never used in this particular program, as we'll see.

The TCS may additionally set user-defined per-vertex and per-patch output variables (`tessControlShader.glsl` does not).

A very important point to note is that each invocation of the TCS, for an output patch vertex, has access to the *entire* `gl_in` and `gl_out` arrays. This is very different from, say, the vertex shader which, when processing a vertex, has access to values for that vertex only, effectively "blind" to its neighbors. Not so for the TCS, which sees *all* the vertices in its current input and output patches. In fact, we see from the switch statement above of `tessellatedCurve.cpp`'s TCS how it uses the position of more than one input vertex to calculate the position of an output vertex.

In fact, this global view is precisely the source of the TCS's computational power. However, it may give rise to synchronization issues if the TCS, for example, asks a value from an invocation, for some output patch vertex, which has not yet completed (keep in mind that the GPU is free to parallelize invocations in any order for efficiency). The GLSL `barrier()` command provides a mechanism to resolve this problem: it causes all invocations to complete execution up to that command before proceeding further; therefore, it is guaranteed that all writes prior to a `barrier()` command will have completed.

Remark 21.2. Oddly enough, the TCS is not a mandatory part of the tessellation shader. In fact, if there is no need to transform the input patches, simply copying them over as output patches being enough, then one can omit the TCS altogether, leaving the application program to set the tessellation levels via calls

glPatchParameterfv(GL_PATCH_DEFAULT_INNER_LEVEL, **pointerToArray*)

and

glPatchParameterfv(GL_PATCH_DEFAULT_OUTER_LEVEL, **pointerToArray*)

where the pointers are to arrays of inner and outer tessellation levels, respectively.

21.5.2 TES (Tessellation Evaluation Shader)

The input to the TES includes all the built-in vertex attributes output for each patch by the TCS, in particular, the array

```
in gl_PerVertex
{
   vec4 gl_Position;
   float gl_PointSize;
   float gl_ClipDistance[]
} gl_in[gl_PatchVerticesIn];
```

which means, just as for the TCS, the TES sees *all* vertices in its own input patch. The size of the above array is the value of the built-in variable `gl_PatchVerticesIn` accessible by the TES – it is the size of a patch input to the TES, the same, of course, as the size of a patch output from the TCS (whose we saw was written in the TCS's built-in `gl_PatchVerticesOut`).

The TES can also read the input per-patch built-in arrays `gl_TessLevelInner[2]` and `gl_TessLevelOuter[4]` set by the TCS to control the TPG.

The TES has two functions: (a) produce a world space vertex for each patch domain vertex (recall patch domain vertices are produced by the TPG from tessellating its assigned patch domain), and (b) configure the TPG by assigning values to a particular set of parameters (details coming up).

The TES executes once for each patch domain vertex emitted by the TPG. For each patch domain vertex, the TES computes and outputs to the PA the corresponding world space vertex, in particular, setting values for the following built-in variables

```
out gl_PerVertex
{
   vec4 gl_Position;
   float gl_PointSize;
   float gl_ClipDistance[];
}
```

For this purpose, it has access to the coordinates of the current patch domain vertex in the built-in 3-vector `gl_TessCoord`. Typically, the TES uses these coordinates, as well as the input patch vertex values, to compute the current world space vertex. The sequence number of the current patch in the current rendering statement may be read, too, by the TES in the built-in `gl_PrimitiveID`.

In the case of `tessellatedCurve.cpp`, here is the main routine of its TES `tessEvaluationShader.glsl`:

```
void main( )
{
   q0 = gl_in[0].gl_Position;
   q1 = gl_in[1].gl_Position;
   q2 = gl_in[2].gl_Position;
   q3 = gl_in[3].gl_Position;

   u = gl_TessCoord.x;

   c0 = (1.0-u) * (1.0-u) * (1.0-u);
   c1 = 3.0 * u * (1.0-u) * (1.0-u);
   c2 = 3.0 * u * u * (1.0-u);
   c3 = u * u * u;

   gl_Position = c0*q0 + c1*q1 + c2*q2 + c3*q3;
}
```

Evidently, the TES is applying the cubic Bézier curve Equation (21.1): the control points are its input patch vertices, while the curve parameter is the x-value of the patch domain (which as we'll see is actually one-dimensional in this case).

The TES, additionally, configures the TPG with a command of the form

layout(*primitive*, *tessellationSpacing*, *orientation*, *pointMode*) in

where *primitive* is one of `quads`, `isolines` and `triangles`, which specifies not only the kind of primitive output by the TPG, but as well the base primitive, called *patch domain*, tessellated by the TPG to produce the output primitives; *tessellationSpacing* is one of `equal_spacing`, `fractional_even_spacing` and `fractional_odd_spacing`; *orientation*, required only if output is 2D, is one of `cw` and `ccw`; and, *pointMode* is an optional parameter which may be `points`. We'll have more to say about these in the next section on the TPG.

The command configuring the TPG in `tessEvaluationShader.glsl` is

```
layout(isolines, equal_spacing) in;
```

whose meaning will be clear once we understand the TPG.

21.5.3 TPG (Tessellation Primitive Generator)

As we have noted before, the mathy gadget, the TPG, lives in fixed-function – it has no corresponding shader to program. However it is *configured* by the TCS and TES, specifically, by

(a) the floating point tessellation level arrays `gl_TessLevelInner[2]` and `gl_TessLevelOuter[4]` set by the TCS, and

(b) the TES command layout(*primitive*, *tessellationSpacing*, *orientation*, *pointMode*) in.

The patch domain itself, the base primitive tessellated by the TPG, may be a rectangle or a triangle depending on the value of *primitive* – `quads`, `isolines` or `triangles` – the latter being the type of the TPG's output primitives. The value of *tessellationSpacing* may be one of `equal_spacing`, `fractional_even_spacing` and `fractional_odd_spacing`, which determines how the TPG subdivides an edge. The parameter *orientation*, either `cw` or `ccw`, causes the TPG to output vertices in such an order that 2D primitives produced (if any) are appropriately oriented. If the optional parameter *pointMode* is `points`, then only vertex values are output by the TPG, no adjacency data being supplied, i.e., the geometry is suppressed.

We'll understand the exact mechanism momentarily. Roughly, though, the inner and outer tessellation levels contained in the arrays `gl_TessLevelInner[2]` and `gl_TessLevelOuter[4]`, set by the TCS, determine the fineness of the subdivision of the interior and perimeter of the patch domain, respectively.

It is the values of the parameters *primitive* and *tessellationSpacing* set by the TES which, together with the tessellation levels set by the TCS, determine exactly the tessellation. We are going to make the assumptions, reasonable for most applications, that the value of *tessellationSpacing* is `equal_spacing` and that tessellation levels are all integer-valued.

Now, not all tessellation level values may actually be used: which are depends on *primitive*. Depending on *primitive* as well is the number of coordinates (either 2 or 3) produced per output vertex by the TPG. The following table summarizes these dependencies.

Primitive	**Domain**	**gl_TessLevel... Values Used**	**Tessellation Coordinates per Vertex**
`quads`	rectangle	`Inner[0]-[1], Outer[0]-[3]`	u, v
`isolines`	rectangle	`Outer[0]-[1]`	u, v
`triangles`	triangle	`Inner[0], Outer[0]-[2]`	u, v, w

Table 21.1: Dependencies on Primitive

We'll make concrete the discussion above successively for each possible *primitive*, namely, `quads`, `isolines` and `triangles`.

quads

Let's work an example, simultaneously explaining the general tessellation procedure. We choose the following values for tessellation levels.

```
gl_TessLevelInner[0] = 5.0;
gl_TessLevelInner[1] = 4.0;
gl_TessLevelOuter[0] = 1.0;
gl_TessLevelOuter[1] = 2.0;
gl_TessLevelOuter[2] = 3.0;
gl_TessLevelOuter[3] = 6.0;
```

The patch domain for a `quads` is the unit square $[0, 1] \times [0, 1]$ on the uv-plane. The tessellation steps are shown from left to right in Figure 21.13.

Figure 21.13(a): Using the inner tessellation values, first subdivide equally the patch domain into a 4×5 mesh of rectangles – `gl_TessLevelInner[0]` subdivisions in the v direction, `gl_TessLevelInner[1]` subdivisions in the u direction.

Figure 21.13(b): Next, triangulate each rectangle, except those along the square's perimeter, into two triangles each; erase all rectangles along the perimeter.

Figure 21.13(c): Finally, use the outer values to equally subdivide the four perimeter edges – `gl_TessLevelOuter[0]` subdivisions of the edge from

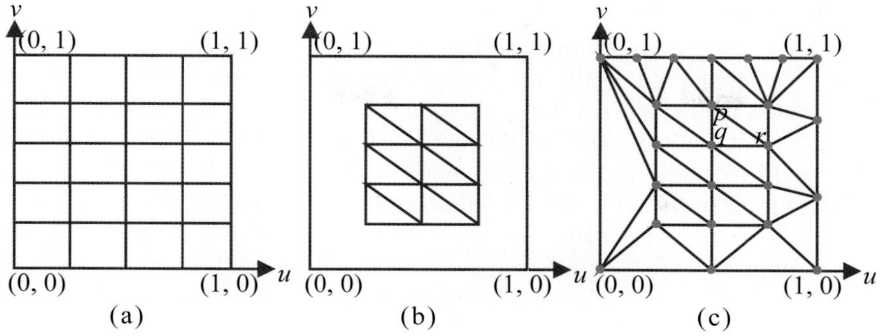

Figure 21.13: Tessellating `quads`.

$(0,1)$ to $(0,0)$, `gl_TessLevelOuter[1]` subdivisions of the edge from $(0,0)$ to $(1,0)$, and so on counterclockwise; triangulate the annular region along the perimeter using edges connecting the subdivision vertices on the perimeter with the outermost vertices of the inner mesh (the particular triangulation is implementation-dependent).

Figure 21.13(c) shows all the vertices as solid points. These vertices, in fact, are sent to the TES, which uses their (u, v) coordinates to compute for each the corresponding world space vertex. The latter are sent next to the PA, which assembles them into the final object according to the geometry of the TPG's tessellated quad, e.g., the world vertices corresponding to vertices p, q and r of Figure 21.13(c) will define the vertices of one triangle.

Note: Even though *primitive* is `quads`, the output primitive type is actually triangle (the non-perimeter triangles, arising in pairs from quadrilaterals, somewhat justifying the name).

Note: The reason for separate tessellation levels for the interior and perimeter is for the user to be able to independently refine the latter, it being the interface with adjacent patches.

Isolines

Again, let's do a running example (`tessellatedCurve.cpp` will actually provide another). Only the first two outer tessellation levels are used for isolines. We'll set them to be

```
gl__TessLevelOuter[0] = 5.0;
gl_TessLevelOuter[1] = 4.0;
```

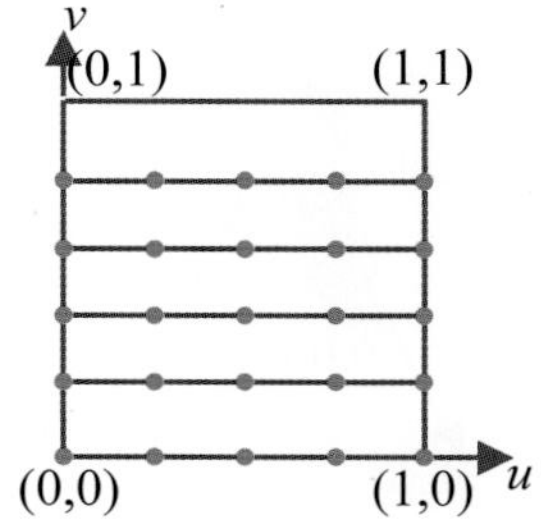

Figure 21.14: Tessellating `isolines`.

Refer to Figure 21.14 as you read on. As for `quads`, the patch domain for `isolines` is the unit square $[0, 1] \times [0, 1]$ on the uv-plane. However, instead of triangles, the primitives output are parallel line strips (isolines being the fancy name) – in particular, `gl_TessLevelOuter[0]` horizontal line strips, starting with the bottom one on the u-axis ($v = 0$), splitting the $[0, 1]$ interval of v into `gl_TessLevelOuter[0]` equal subintervals. Note that this means that there is no line strip along $v = 1$, for, otherwise, the $[0, 1]$ interval would be split into only `gl_TessLevelOuter[0]`-1 subintervals. The reason not to place a line strip along $v = 1$ is to avoid overlap between two adjacent isolines patches. Further, each horizontal line strip consists of `gl_TessLevelOuter[1]` equal segments.

The final tessellated patch domain output for our example tessellation levels above consists of the solid lines and vertices in Figure 21.14. Again, as for quads, it's the vertices which are delivered to the TES, which uses the (u, v) coordinates of each input vertex to produce the corresponding world space vertex.

Back to `tessellatedCurve.cpp` next. The statement

```
layout(isolines, equal_spacing) in;
```

in `tessellatedCurve.cpp`'s TES and the two

```
gl_TessLevelOuter[0] = 1.0;
gl_TessLevelOuter[1] = tessLevelOuter1;
```

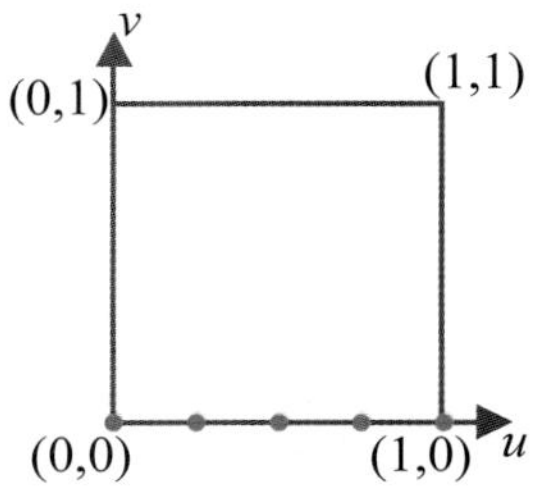

Figure 21.15: `tessellatedCurve.cpp`'s tessellated patch domain.

in its TCS mean that the tessellated patch domain output is as in Figure 21.15 initially (when `tessLevelOuter1` $= 5.0$) – a single isoline divided equally into 5 parts – which is why the TES uses only the u-coordinate of patch domain vertices to compute the corresponding world vertex.

Remark 21.3. Its interesting to note how `tessellatedCurve.cpp` goes about drawing the three input points and the Bézier curve together. Its initialization routine actually attaches only the vertex and fragment shaders. The drawing routine, then, first attaches the TCS and TES to draw the Bézier curve polyline, and, next, detaches both to draw the points (of course, one does not want any tessellation happening when drawing just points).

Triangles

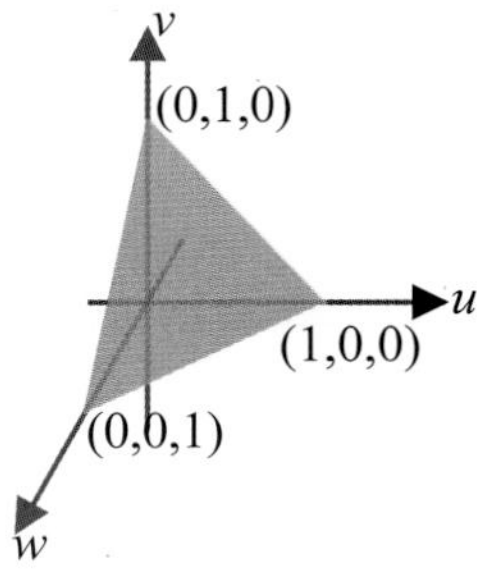

Figure 21.16: Patch domain for `triangles`.

The patch domain for `triangles` is a triangle situated in uvw 3-space with corners a unit distance along each axis, as in Figure 21.16. The reason for this, rather than a flat triangle on the uv-plane, is the convenience of barycentric coordinates (see Section 7.2). In fact, this patch domain triangle consists exactly of the points (u, v, w) such that $u + v + w = 1$ and $u, v, w \geq 0$, where (u, v, w) serve as barycentric coordinates as well. Evidently, the triangle is equilateral. Only the first inner tessellation level and the first three outer levels are used for triangles. So, let's set up an example with the following levels:

```
gl_TessLevelInner[0] = 4.0;
gl_TessLevelOuter[0] = 1.0;
gl_TessLevelOuter[1] = 2.0;
gl_TessLevelOuter[2] = 3.0;
```

First, subdivide each triangle edge into `gl_TessLevelInner[0]` equal segments. From each subdivision vertex drop the perpendicular into the triangle to make a nested triangle with each edge subdivided into `gl_TessLevelInner[0]- 2` equal segments (see Figure 21.17(a) for our example value `gl_TessLevelInner[0]` = 4.0).

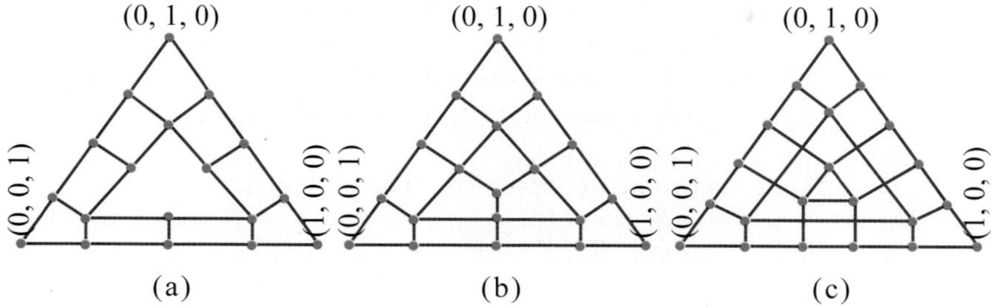

Figure 21.17: Tessellating `triangles`.

Continue the process of generating nested triangles until one reaches either a single point (which happens when `gl_TessLevelInner[0]` is even) or an innermost triangle with no subdivided edge (when `gl_TessLevelInner[0]` is odd). Figure 21.17(b) shows the end figure of this process for our example, where `gl_TessLevelInner[0]` = 4.0, while Figure 21.17(c) shows it for `gl_TessLevelInner[0]` = 5.0, an odd number.

Finally, we'll replicate very nearly the final steps for `quads`: triangulate each quad inside the triangle, except those on the perimeter, into two triangles; erase quads adjacent to the perimeter; use the outer values to equally subdivide the three perimeter edges – `gl_TessLevelOuter[0]` subdivisions of the edge from $(0, 1, 0)$ to $(0, 0, 1)$, `gl_TessLevelOuter[1]` subdivisions of the edge from $(0, 0, 1)$ to $(1, 0, 0)$, and `gl_TessLevelOuter[2]` subdivisions of the edge from $(1, 0, 0)$ to $(0, 1, 0)$; triangulate the annular region along the perimeter using edges connecting the subdivision vertices on the perimeter with the outermost vertices of the inner mesh (as for `quads`, the triangulation itself is implementation-dependent).

Figure 21.18 shows the final tessellation for our example above.

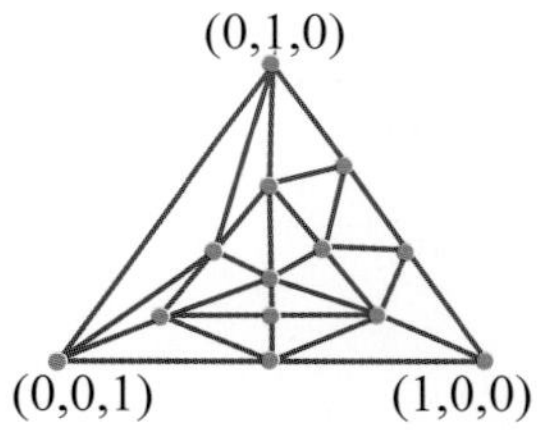

Figure 21.18: Final tessellated triangle for the example tessellation levels.

Before we leave tessellation shaders here's a program which tessellates a surface, rather than a curve.

Experiment 21.10. Run `tessellatedHemisphere.cpp`. You see a tessellated hemisphere. Press 'x'-'Z' to turn it, the up/down arrow keys to increase/decrease the inner tessellation levels (both equal) and the right/left arrow keys to increase/decrease the outer tessellation levels (all four equal). Figure 21.19 is a screenshot after raising both inner and outer tessellation levels a fair amount.

Figure 21.19: Screenshot of `tessellated-Hemisphere.cpp`.

We ask the reader to examine the program in the following exercise.
End

Exercise 21.16. The following are about `tessellatedHemisphere.cpp`.

(a) The vertex shader is, well, empty, yet we have a vertex-rich object being drawn. What gives? (*Hint*: Tessellation shaders can create and transform geometry ....)

(b) The program has no TCS. So, how are tessellation levels being set?

(c) Draw a figure like Figure 21.13(c) showing the tessellated patch domain for the initial tessellation levels.

(d) Narrate exactly how the TES draws the hemisphere, beginning with the patch domain vertices it gets from the TPG (for the formula for `hemCoords` see Section 2.10).

Exercise 21.17. (Programming) Use the tessellation shader to draw a cylinder whose tessellation levels are user-controlled.

Figure 21.20: Hemisphere on top of a cylinder.

Exercise 21.18. (Programming) Place a hemisphere on top of a cylinder as in Figure 21.20. Allow their respective tessellations to be controlled separately, making sure, though, that there is agreement on the shared circular boundary.

21.6 Geometry Shaders

Geometry shaders are an optional component of the programmable pipeline meant, just as tessellation shaders, to shift geometry processing from the application program to the GPU. However, geometry shaders are structured quite differently from their tessellation sisters and, in fact, far less complex. A vital commonality, though, is that both derive their computational power from having a global view of input primitives (vs. the one-vertex-at-a-time vertex shader).

Here's an overview of how a geometry shader works: it takes input primitives of a particular specified type (from one of the familiar points, lines and triangles, plus a couple of new so-called "adjacency" types, namely, lines_adjacency and triangles_adjacency) and produces output primitives of a particular type (precisely, one of points, line strips and triangle strips). Importantly, there is no *a priori* relationship between the input and output types: points may generate line strips, triangles points, and so on. Moreover, one input primitive may generate zero, one or more output primitives. That's it barring gory details (coming up)! Not too bad, huh?

The input to a geometry shader arrives from the vertex shader if there is no tessellation, and from the tessellation evaluation shader if there is. So, let's see what sort of input primitive it accepts. The declaration `layout(`*inputPrimitiveType*`) in;`

in the geometry shader sets *inputPrimitiveType* as that accepted, where its value is one of the five on the left of Table 21.2.

Input Primitive Type	**Drawing Command Modes**	**Atomic Input Primitive**	**Size**
points	GL_POINTS, GL_PATCHES	GL_POINTS	1
lines	GL_LINES, GL_LINE_STRIP, GL_LINE_LOOP, GL_PATCHES	GL_LINES	2
triangles	GL_TRIANGLES, GL_TRIANGLE_STRIP, GL_TRIANGLE_FAN, GL_PATCHES	GL_TRIANGLES	3
lines_adjacency	GL_LINES_ADJACENCY, GL_LINE_STRIP_ADJACENCY	GL_LINES_-ADJACENCY	4
triangles_adjacency	GL_TRIANGLES_ADJACENCY, GL_TRIANGLE_STRIP_ADJACENCY	GL_TRIANGLES_-ADJACENCY	6

Table 21.2: Dependencies on Input Primitive Type

Note: The reader may be wondering why we aren't narrating with example snippets from a live program as we did for tessellation. We certainly have an example program to show off the geometry shader but thought the technical presentation for this particular topic would go better if we deferred the program till after.

The corresponding drawing command modes which may be actually used to draw primitives in the application program are in the second column of the preceding table. Whatever drawing command is used, though, input primitives of a given type are each split into a sequence of atomic primitives of the type in the third column. For example, the declaration

```
layout(triangles) in
```

is consistent with the drawing command `glDrawElements(GL_TRIANGLE_-STRIP, ...)`, each triangle strip being split into a sequence of triangles which are input successively to the geometry shader. The last column shows the size, in number of vertices, of the atomic primitive. We'll be explaining the `GL_*_ADJACENCY` primitives momentarily.

Note: The tessellation shader, if there is one, converts `GL_PATCHES` into points, lines or triangles before shipping to the geometry shader.

Output from a geometry shader goes to the fragment shader for rendering. The output type is set to *outputPrimitiveType* by the declaration

`layout(`*outputPrimitiveType*`, maxVertices=`*num*`) out;`

in the geometry shader, where the value of *outputPrimitiveType* may be one of `points`, `line_strip` and `triangle_strip`, while `max_vertices` is the maximum number of vertices a single output primitive may have. Observe that 1D and 2D output types are both strips. So, for example, to declare single triangles as output, the appropriate statement would be

```
layout(triangle_strip, max_vertices=3) out;
```

Note, further, that output primitives are produced one per atomic input primitive. So, if the drawing command is `glDrawElements(GL_TRIANGLE_STRIP, ...)`, then the statement just above asks one triangle to be output for each one into which the input strip is split.

Adjacency Primitive Types

The purpose of the input adjacency primitive types is to provide the geometry shader with data about the *neighborhood* of a primitive in the object of which it is part. Let's examine the four of them one by one.

`GL_LINES_ADJACENCY`: Just as a `GL_LINES` primitive is a sequence of individual line segment primitives, a `GL_LINES_ADJACENCY` primitive is a sequence of individual 4-vertex base primitives, each as in Figure 21.21(a): the middle vertices v_1 and v_2 represent the start and end vertices of a line segment, respectively, while v_0 and v_3 represent v_1's predecessor and v_2's successor, respectively; particularly, v_0 and v_3 *represent* adjacency (i.e., neighborhood) information.

Note that the operative word in the preceding sentence is "represent". It is up to the programmer how she wishes to interpret each 4-vertex primitive. For, keep in mind that input to a geometry shader is not an object in world space per se; it's only its output which is intended to be rendered. So, for example, the four vertices may even be interpreted as four isolated points.

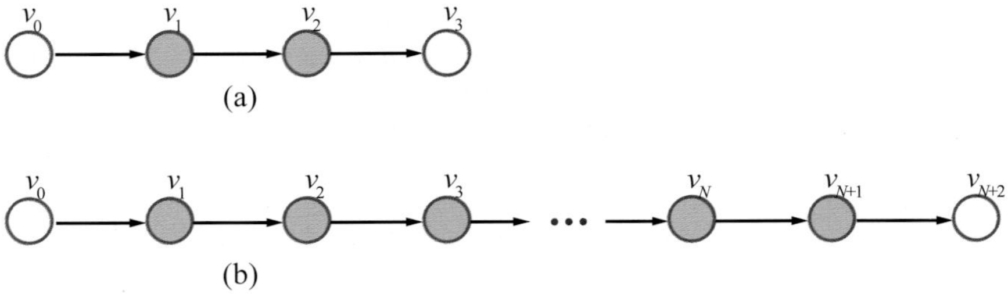

Figure 21.21: (a) `GL_LINES_ADJACENCY` primitive (b) `GL_LINE_STRIP_ADJACENCY` primitive. Arrows indicate line orientation. Unfilled points and broken arrows represent adjacency information – if there is no geometry shader present, these are discarded and only the filled points and unbroken segments rendered.

`GL_LINE_STRIP_ADJACENCY`: A `GL_LINE_STRIP_ADJACENCY` primitive representing a strip of N segments consists of a sequence of $N + 3$ vertices, as shown in Figure 21.21(b). The line strip runs from vertex v_1 to v_{N+1}, while the predecessor of v_1 is v_0 and the successor of v_{N+1} is v_{N+2}.

We see from Table 21.2 that the atomic primitive corresponding to a `GL_LINE_STRIP_ADJACENCY` primitive is `GL_LINE_ADJACENCY`. So, for example, the N-segment primitive of Figure 21.21(b) is, in fact, split into N `GL_LINE_ADJACENCY` primitives which are sent sequentially to the geometry shader: the first one consists of v_0, v_1, v_2, v_3, the next v_1, v_2, v_3, v_4, and so on, till, finally, v_{N-1}, v_N, v_{N+1}, v_{N+2}.

`GL_TRIANGLES_ADJACENCY`: Just as a `GL_TRIANGLES` primitive is a sequence of individual triangle primitives, a `GL_TRIANGLES_ADJACENCY` primitive is a sequence of single 6-vertex base primitives, each as in Figure 21.22(a): the vertices v_0, v_2 and v_4 represent the vertices of a triangle, while v_1, v_3 and v_5 each represent the third vertex of one of the three abutting triangles; specifically, v_1, v_3 and v_5 represent adjacency information.

`GL_TRIANGLE_STRIP_ADJACENCY`: A `GL_TRIANGLE_STRIP_ADJACENCY` primitive representing a strip of N triangles, $t_0, \ldots, t_{N-1}$, consists of $2N+4$ vertices, configured as in Figure 21.22(b) (N is assumed even for this particular configuration; if N is odd, then the last triangle t_{N-1} will look like t_0, rather than t_1). For each triangle of the strip, we see as well vertices defining its three abutting triangles, i.e., adjacency information.

The atomic primitive corresponding to a `GL_TRIANGLE_STRIP_ADJACENCY` primitive is `GL_TRIANGLE_ADJACENCY`. So, for example, the N-triangle primitive of Figure 21.22(b) is split into N `GL_TRIANGLE_ADJACENCY` primitives which are sent sequentially to the geometry shader: the first one consists of the triangle t_0 vertices v_0, v_2 and v_4 together with the neighboring vertices v_1, v_3 and v_6, the next consists of the triangle t_1 vertices v_2, v_4 and v_6 together with the neighboring vertices v_0, v_5 and v_8, and so on, till, finally, the triangle t_{N-1} vertices t_{2N-2}, t_{2N} and t_{2N+2} together with neighboring vertices t_{2N-4}, t_{2N+1} and t_{2N+3}.

The geometry shader executes once per input atomic primitive. The geometry shader has access to all the per-vertex built-in attributes, set by the vertex shader or tessellation evaluation shader, for every vertex in each input atomic primitive. *This global view is precisely the source of the geometry shader's computational power.* Accordingly, the data for each such input primitive, as read by the geometry shader, is a built-in variable `gl_in`, which is an array of structures – the three fields corresponding to the three built-in vertex attributes – of the size of the input primitive (see the last column of Table 21.2). Here is its definition:

```
in gl_PerVertex
{
   vec4 gl_Position;
   float gl_PointSize;
   float gl_ClipDistance[];
```

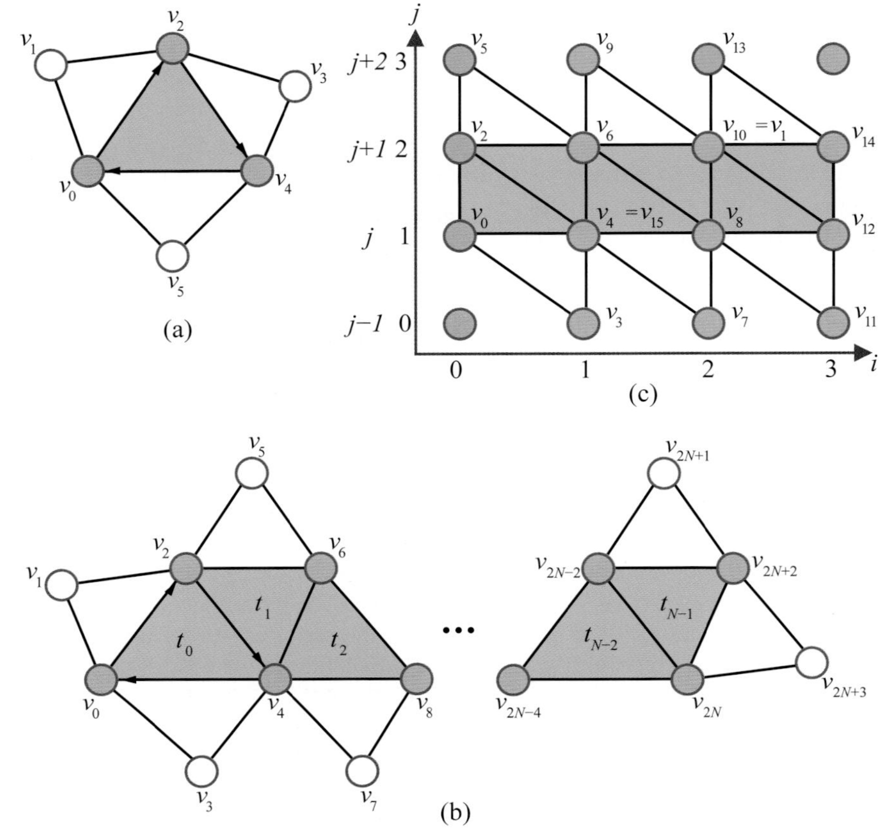

Figure 21.22: (a) `GL_TRIANGLES_ADJACENCY` primitive (b) `GL_TRIANGLE_STRIP_ADJACENCY` primitive. Arrows indicate triangle orientation (the shaded strip of triangles is consistently oriented). Unfilled points and broken arrows represent adjacency information – if there is no geometry shader present, these are discarded and only the shaded triangles rendered. (c) Torus mesh with one `GL_TRIANGLE_STRIP_ADJACENCY` primitive labeled.

```
} gl_in[];
```

(Note the similarity with the built-in `gl_in` defined for the tessellation control shader.) So, for example, if the input primitive type is `lines_adjacency` then the size of `gl_in[]` is 4. The geometry shader may additionally be supplied user-defined per-vertex attribute values, beyond the three built-ins. The sequence number of the current atomic primitive in the current drawing statement is contained in the geometry shader's built-in input variable `gl_PrimitiveIDIn`.

The output data structure of the geometry shader, which goes to the primitive assembler and then on to the fragment shader, is of the same form per vertex as the input, namely,

```
out gl_PerVertex
```

```
{
   vec4 gl_Position;
   float gl_PointSize;
   float gl_ClipDistance[];
};
```

Unlike input, though, there is no built-in array to hold `gl_PerVertex` values to be output. Rather, the geometry shader calls the special function `EmitVertex()` to produce a vertex from the current value of `gl_PerVertex`, and the special function `EndPrimitive()` to assemble all the vertices produced since the last `EndPrimitive()` call (or the beginning of the draw command) into an output primitive of the type specified by the `layout(...) out` declaration.

Time now to bring the preceding discussion to life with code. We have programmed a silhouette-extraction algorithm, a popular application for geometry shaders.

Experiment 21.11. Run `torusSilhouette.cpp`. Press the space bar to toggle between the silhouette and mesh of a torus. Press 'x'-'Z' to turn the torus. Figure 21.23 is a screenshot of the torus in silhouette. We explain the program below. **End**

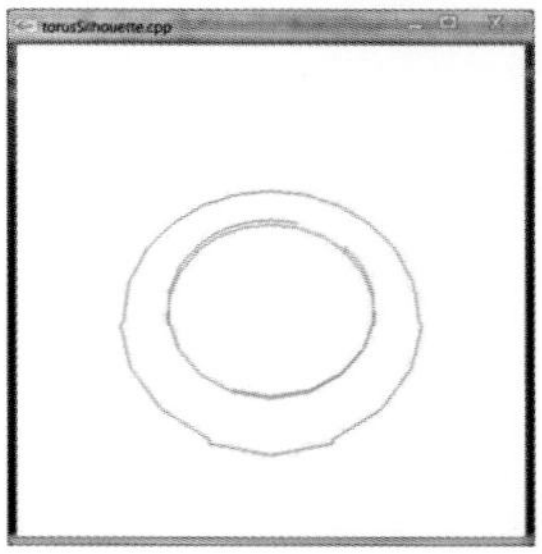

Figure 21.23: Screenshot of `torusSilhouette.cpp` in silhouette mode.

Let's first understand the simple geometric principle underlying the algorithm. Given a consistent orientation of an object's mesh, an edge is part of its silhouette if it's shared by a front-facing triangle and a back-facing triangle (see Figure 21.24(a)); an edge shared by two front-facing triangles (Figure 21.24(b)) or two back-facing ones is not on the silhouette. Accordingly, our plan is to run through the edges of the torus mesh, comparing the orientations of the triangles on either side.

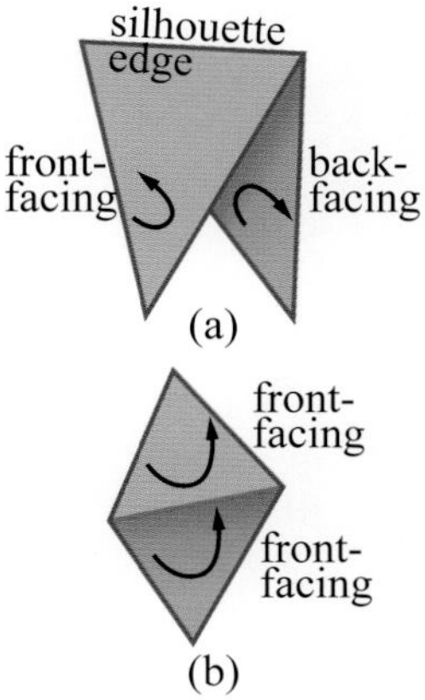

Figure 21.24: Making a silhouette edge.

Since our plan obviously calls for adjacency information to implement, we draw the torus with use of adjacency primitives, precisely, the call

```
glMultiDrawElements(GL_TRIANGLE_STRIP_ADJACENCY, torCounts,
                    GL_UNSIGNED_INT, (const void **)torOffsets,
                    TOR_LATS);
```

in the drawing routine. For this call to work correctly, the associated source `torus.cpp` is changed from versions associated earlier with programs like `ballAndTorusShaderized.cpp`. The main change is the function to create the array of index arrays:

```
void fillTorIndices(unsigned int torIndices[TOR_LATS][4*(TOR_LONGS+1)])
{
   int i, j;
   for(j = 0; j < TOR_LATS; j++)
   {
      for (i = 0; i <= TOR_LONGS; i++)
      {
```

```
            torIndices[j][4*i] = j * (TOR_LONGS+1) + i;
            torIndices[j][4*i+2] = (j+1) * (TOR_LONGS+1) + i;
        }
        for (i = 0; i < TOR_LONGS; i++)
        {
            torIndices[j][4*i+3] = (j > 0 ? j-1 : TOR_LATS) *
                                   (TOR_LONGS+1) + i + 1;
            torIndices[j][4*i+5] = ((j+2) % (TOR_LATS+1)) *
                                   (TOR_LONGS+1) + i;
        }
        torIndices[j][1] = (j+1) * (TOR_LONGS + 1) + TOR_LONGS - 1;
        torIndices[j][4*TOR_LONGS+3] = j * (TOR_LONGS + 1) + 1;
    }
}
```

For each of the TOR_LATS iterations of the outer loop this function computes the 4*(TOR_LONGS+1) indices of a GL_TRIANGLE_STRIP_ADJACENCY primitive. The best way to understand the computation is from a small concrete example. In fact, see Figure 21.22(c), which shows a torus mesh with both TOR_LATS and TOR_LONGS equal to 3 (note that since this is a torus, the four vertices along the top row are identified with the corresponding vertices along the bottom, and, likewise, for the leftmost and rightmost vertex columns).

Assume that the current value of the outer loop variable j is 1, as indicated by the leftmost column of notations. The corresponding shaded GL_TRIANGLE_STRIP_ADJACENCY primitive is a case of the general such primitive shown in Figure 21.22(b) with $N = 6$. Match Figures 21.22(b) and (c) to see that the 16 vertices $v_0, \ldots, v_{15}$ of the latter are correctly labeled.

That the four vertices v_0, v_4, v_8 and v_{12}, with indices a multiple of 4, lie successively on row j of the mesh explains the line

```
torIndices[j][4*i] = j * (TOR_LONGS+1) + i;
```

in the function above. Similarly, that the four vertices v_2, v_6, v_{10} and v_{14} lie successively on row $j+1$ of the mesh explains the line

```
torIndices[j][4*i+2] = (j+1) * (TOR_LONGS+1) + i;
```

We'll leave the reader to parse the rest of the function keeping in mind the caveat above identifying the top and bottom rows of vertices, as well as the leftmost and rightmost columns.

The geometry shader takes the stage next. It receives for each of the TOR_LATS number of GL_TRIANGLE_STRIP_ADJACENCY primitives produced by the

```
glMultiDrawElements(GL_TRIANGLE_STRIP_ADJACENCY, ...)
```

drawing command, a sequence of 2 *TOR_LONGS of its corresponding atomic GL_TRIANGLE_ADJACENCY primitive.

Exercise 21.19. Verify that the number 2 ∗TOR_LONGS of GL_TRIANGLE_-ADJACENCY primitives per GL_TRIANGLE_STRIP_ADJACENCY primitive as claimed above is correct.

Based on the geometric principle stated earlier, the geometry shader must check for each input GL_TRIANGLE_ADJACENCY primitive and for each edge (refer to Figure 21.22(a)) v_0v_2, v_2v_4 and v_4v_0, if the triangles on either side are front- or back-facing. If for such an edge a triangle on one side is front-facing and the other back-facing, then it is to be output as a silhouette edge.

But, first, see the declarations at the top:

```
layout(triangles_adjacency) in;
layout(line_strip, max_vertices=2) out;
```

The first line is consistent with the drawing mode GL_TRIANGLE_STRIP_-ADJACENCY (see Table 21.2), while the second declares that output will be individual line segments.

Next, to determine if a triangle is front- or back-facing to a viewer located at the origin we have to determine its orientation as perceived by this viewer (by default, CCW means front-facing, CW back-facing). Now, Exercise 9.5 of Chapter 9 tells us that a triangle with vertices (x_1, y_1, z_1), $(x_2, y_2 z_2)$ and (x_3, y_3, z_3), in that order, is perceived by a viewer at the origin to be oriented CW, CCW or viewed edge-on according as the determinant

$$\begin{vmatrix} x_1 & x_2 & x_3 \\ y_1 & y_2 & y_3 \\ z_1 & z_2 & z_3 \end{vmatrix}$$

is greater than, less than or equal to zero. Accordingly, the block

```
float orient024 = determinant(mat3(gl_in[0].gl_Position.xyz,
                                   gl_in[2].gl_Position.xyz,
                                   gl_in[4].gl_Position.xyz));
```

in the geometry shader's main calculates the corresponding determinant for triangle $v_0v_2v_4$. Similar blocks compute the determinant for triangles $v_1v_2v_0$, $v_2v_3v_4$ and $v_0v_4v_5$.

Subsequently, in silhouette drawing mode, edge v_0v_2 is output if the triangles on either side of it, particularly, $v_0v_2v_4$ and $v_1v_2v_0$, are oppositely oriented, in which case their determinant product is negative; the implementing code block is

```
if ( orient024 * orient120 <= 0.0 )
{
   gl_Position = gl_in[0].gl_Position;
   EmitVertex( );
   gl_Position = gl_in[2].gl_Position;
   EmitVertex( );
   EndPrimitive( );
}
```

(observe that we output the edge to silhouette, as well, if one of the triangles is seen edge-on, when its determinant is zero). Similar blocks decide if to output edges v_2v_4 and v_4v_0.

In mesh drawing mode, of course, all three edges v_0v_2, v_2v_4 and v_4v_0 are always output.

Exercise 21.20. (Programming) Comment out the line

```
glAttachShader(programId, geometryShaderId);
```

of `torusSilhouette.cpp`. What this shows is that, even if there is no geometry shader active, a drawing primitive of adjacency type is still rendered, though, not surprisingly, without use of the adjacency data.

21.7 Summary, Notes and More Reading

As they say, all good things must come to an end. This chapter concludes our coverage of the programmable pipeline, and the book as well.

Following up on the previous chapter, we dove deep into OpenGL 4.3, learning an assortment of fairly advanced features, including instanced rendering, shader subroutines and transform feedback amongst others, and learned as well of a bunch of new ways to do old things in the programmable pipeline. And, of course, we studied the two new shader stages, tessellation and geometry, thus rounding off our understanding of the programmable pipeline.

The canonical source for all things OpenGL, including the GLSL is, of course, the OpenGL site [103]. Interestingly, just as there is the red book (programming guide) and the blue book (reference manual), there, too, is the so-called orange book by Rost & Licea-Kane [120] on the OpenGL shading language. However, it is somewhat dated and, in fact, the newest edition of the red book subsumes most of the shader material. A couple of more recent textbooks devoted to the OpenGL shading language are Bailey & Cunningham [7] and Wolff [151].

What next? Well, the reader should now be ready to take on fairly complex 3D projects, and not only in the desktop environment. As we pointed out at the end of the last chapter, grasp of 4.3 means that one is ready as well to code OpenGL ES for mobile devices and WebGL for browsers. The practical-minded reader should indeed now take on a significant project. Gym work's all done, time to hit the tracks!

APPENDIX **A**

Projective Spaces and Transformations

Projective geometry is at the heart of computer graphics whichever view you take of it, practical or theoretical. The various transformations of real 3-space we learned to use for the purpose of animation in Chapter 4 and studied mathematically in Chapter 5 are, in fact, most naturally viewed as transformations of projective 3-space, following a so-called lifting of the scene from real to projective space. A consequence is that representing these transformations as projective is more efficient from a computational point of view, a fact that OpenGL takes constant advantage of in its design. Capturing the scene after a perspective projection on film – "shooting" as we imagine the OpenGL point camera to do – involves a projective transformation as well.

In fact, it's not an exaggeration to say that projective geometry is the mathematical foundation of modern-day CG, and that API's such as OpenGL "live" in projective 3-space. Unfortunately, though, because projective geometry works its magic deep inside the graphics pipeline, its importance often is not realized.

There are several books out there which discuss projective geometry – Coxeter [31], Henle [71], Jennings [76], Pedoe [108] and Samuel [122] come to mind – from mainly a geometer's point of view, as well as a few, such as Baer [6] and Kadison & Kromann [77], which take an algebraic standpoint. All these books, however, seem written primarily for a student of mathematics. There seems none yet dedicated to answering the computer scientist's (almost certainly a CG person) question of projective geometry, "What can you do for me?"

This appendix is a small attempt to fill this gap in the literature and introduce projective spaces and transformations from a CG point of view.

Projective spaces generalize real space. They are not difficult to understand, but geometric primitives, such as lines and planes, behave somewhat differently in a projective space than a real one. By applying a camera-view analogy from the outset, we try to convey a physical-based intuition for basic concepts, establishing at the same time connection with CG.

This appendix is long and the mathematics often admittedly abstract, but the payback for persevering through it comes in the form of a wealth of applications, including the projection transformation in the graphics pipeline, as well as the rational Bézier and all-important NURBS primitives, which are all topics of Chapter 18 on applications of projective spaces.

Logically, this appendix could as well have been a chapter of the book, just prior to Chapter 18. However, we decided against upsetting the fairly easy gradient of the book from the first chapter to the last with the insertion of a mathematical "hill". In fact, Chapter 18 on applications has been written so that the reader reluctant to take on the venture into projective theory can still make her way through it with minimal loss. This is not in any way to diminish the importance of the material in this appendix, but merely recognition of the reality that there are numbers of people out there who would make fine CG professionals, but care little for abstract mathematics.

We begin in Section A.1 by invoking a camera's point of view to motivate the definition of the projective plane. The geometry of this plane, including its surprising point-line duality and coordinatization by means of the homogeneous coordinate system, is the topic of Sections A.2 and A.3. In Section A.4 we study the structure of the projective plane and learn that the real plane can be embedded in the projective, which in turn yields a classification of projective points into regular ones and those at infinity.

A particularly intuitive kind of projective transformation, the so-called snapshot transformation, comes next in Section A.5. Section A.6 covers a few applications of homogeneous polynomial equations, including an algebraic insight into the projective plane's point-line duality, and an algebraic method to compute the outcome of a snapshot transformation. Following a brief discussion of projective spaces of arbitrary dimension in Section A.7, we move on to projective transformations.

Projective transformations are first defined algebraically in Section A.8 and then understood geometrically in A.9. In Section A.10 we relate projective, snapshot and affine transformations, and see that projective transformations are more powerful than either of the other two. The process of determining the projective transformation to accomplish some particular mapping – often beyond the reach of an affine transformation – is the topic of Section A.11.

A.1 Motivation and Definition of the Projective Plane

Consider a viewer taking pictures with a point camera with a film in front of it. Light rays from objects in the scene travel toward the camera and their intersections with the film render the scene. See Figure A.1. Captured on film is the (perspective) projection of the objects. In the case of OpenGL, this is precisely the situation when the user defines a viewing frustum: the point camera is at the apex of the frustum, while the film lies along its front face.

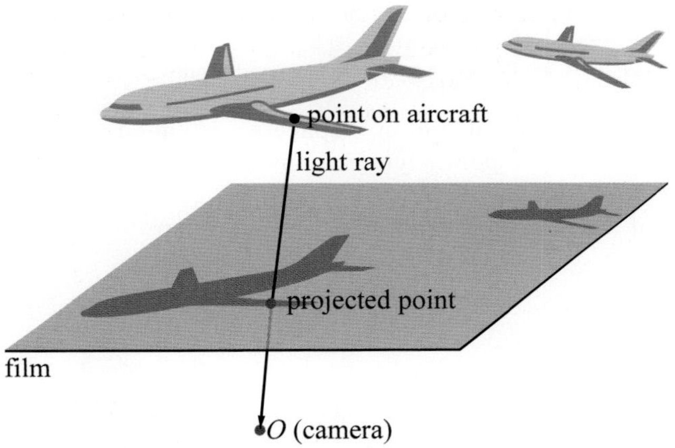

Figure A.1: Perceiving objects with a point camera and a plane film.

Clearly, points in the scene that lie on the same (straight) line through the camera cannot be distinguished by the viewer. In fact, all objects, e.g., points and line segments, lying on one line l through the camera cannot be distinguished by the viewer. They all project to and are perceived as a single point on the film. See Figure A.2(a). Assume for the moment that the film is two-sided and that objects behind project onto it as well (depicted is one such point). For now, ignore as well that lines through the camera parallel to the film, e.g., l', do not intersect the latter at all. This is owing to the alignment of the film, which can always be changed.

So, one can say that the viewer perceives *every* line through the camera as a point. What then does he perceive as a line? The likely answer is a plane. Indeed, any plane through the camera intersects the film in a line, though, again, the film may have to be re-aligned so as not to be parallel to the plane. See Figure A.2(b).

Lines are points, planes are lines, ... Let's take a moment to formalize, as a new space, the world as it is perceived through a point camera at the origin. Recall that a *radial* primitive is one which passes through the origin.

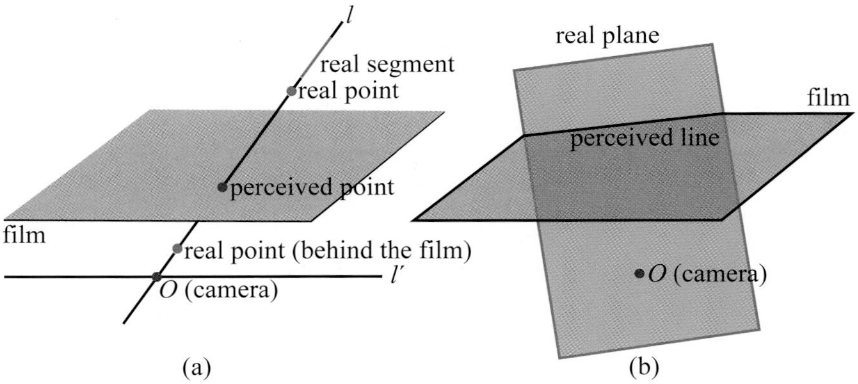

Figure A.2: Perceiving points, lines and planes by projection.

Definition A.1. A radial line in 3-space $\mathbb{R}^3$ is called a *projective point*. The set of all projective points lying on any one radial plane in $\mathbb{R}^3$ is called a *projective line*. (See Figure A.3.)

The set of all projective points is called 2-dimensional *projective space* and denoted $\mathbb{P}^2$. $\mathbb{P}^2$ is also called the *projective plane*.

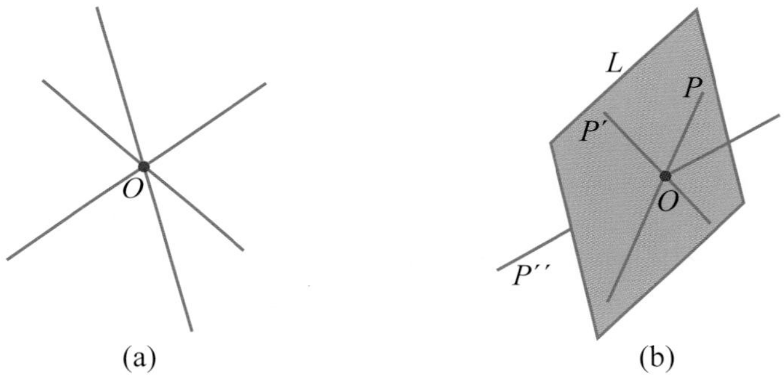

Figure A.3: (a) Projective points are radial lines (b) A projective line consists of all projective points on a radial plane: projective points P and P' belong to the projective line L, while P'' does not. *Keep the distinction in mind that, though we have labeled the plane L, the projective line L actually consists of all the projective points, e.g., P and P', that lie on this plane, and is different from the plane itself.*

Remark A.1. We are taking a significant step up in abstraction in leaving $\mathbb{R}^2$ for $\mathbb{P}^2$. The real plane $\mathbb{R}^2$ is easy to visualize as, well, a real plane, e.g., a table top or a sheet of paper. Not so the projective plane. There is no real object to which it corresponds nicely.

Things such as a line, which is a set of points in one space, being only a point of another may seem a bit strange as well. It's mostly a matter of getting used to it though – like learning a foreign language. As with a new language, some words translate literally, but some don't simply because the concept isn't familiar (what's sandstorm in Eskimoan?).

It's recommended that the reader stick close to the real-based definitions at first. A thought process like "Hmm, the projective point P belongs to the projective line L. Well, then, this means that the real line which is P sits inside the real plane which is L" may seem cumbersome at first, but projective primitives will seem less and less strange as we go along.

The term "projective" arose because objects on the projective plane are perceived by projection onto a real one, which for us is the film. Observe that in Figure A.3(b) we denote by L both a radial plane (a primitive in $\mathbb{R}^3$), as well as the projective line (a primitive in $\mathbb{P}^2$) consisting of projective points that lie on that plane. There should be no cause for ambiguity as it'll be clear from the context which we mean.

Terminology: We'll generally use lower case letters to denote primitives in $\mathbb{R}^2$ and upper case for those in $\mathbb{P}^2$.

Remark A.2. The dimension of $\mathbb{P}^2$ is two (as indicated by the superscript). This is because, while points in $\mathbb{R}^3$ have three "degrees of freedom", radial lines in $\mathbb{R}^3$ have only two. We'll elaborate on the dimension of the projective plane in Section A.7.

A.2 Geometry on the Projective Plane and Point-Line Duality

We have, then, on the projective plane $\mathbb{P}^2$ projective points and projective lines, just as on the real plane $\mathbb{R}^2$ we have real points and lines. It's interesting to compare the relationship between points and lines in the two spaces.

Recall the following two facts from Euclidean geometry (geometry in real space is called Euclidean):

(a) There is a unique line containing two distinct points in $\mathbb{R}^2$.

(b) Two distinct lines in $\mathbb{R}^2$ intersect in a unique point, *except* if they are parallel, in which case they do not intersect at all.

What is the situation in projective geometry?

Two distinct projective points P and P' correspond to two distinct radial lines in $\mathbb{R}^3$, and, in fact, there is a unique radial plane L in $\mathbb{R}^3$ containing the latter two. See Figure A.4(a).

It follows that:

(A) There is a unique projective line containing two distinct projective points in $\mathbb{P}^2$.

How about two distinct projective lines? Observe that the corresponding two distinct radial planes, say, L and L' in $\mathbb{R}^3$, intersect in a unique radial

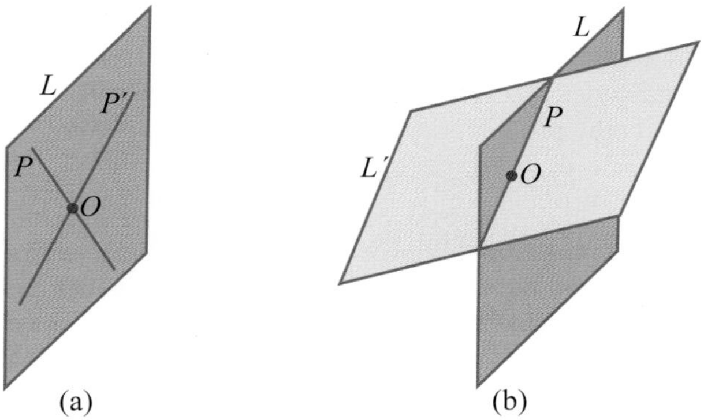

Figure A.4: (a) Radial lines corresponding to projective points P and P' are contained in a unique radial plane corresponding to the projective line L (b) Radial planes corresponding to projective lines L and L' intersect in a unique radial line corresponding to the projective point P.

line corresponding, in fact, to some projective point P. See Figure A.4(b). We have:

(B) Two distinct projective lines in $\mathbb{P}^2$ intersect in a unique projective point.

No exceptions! There's no such thing as parallelism in $\mathbb{P}^2$! Any two different lines always intersect in a point. Two points–one line, two lines–one point, *always*: $\mathbb{P}^2$ has better so-called *point-line duality* than $\mathbb{R}^2$. We'll have more to say about the point-line duality of $\mathbb{P}^2$ as we go along.

Exercise A.1. Consider three distinct projective lines L, L' and L''. We know that their pairwise intersections are three projective points, say, P, P' and P''. Give examples where (a) all three points are identical and (b) all three are distinct. Can only two of them be distinct? If all three are distinct can they be collinear, i.e., lie on one projective line?

A.3 Homogeneous Coordinates

We want to *coordinatize* $\mathbb{P}^2$, if possible, in a manner similar to that of $\mathbb{R}^2$ by Cartesian coordinates. This is important for the purpose of geometric calculations. For example, Cartesian coordinates on the real plane allow us to make a statement such as "The equation of the line through the $[-2\ -5]^T$ and $[1\ 1]^T$ is $y - 2x + 1 = 0$, which is satisfied as well by $[0\ -1]^T$, so that all three points are collinear."

So how does one coordinatize $\mathbb{P}^2$? As follows:

Definition A.2. The *homogeneous coordinates* of a projective point are the Cartesian coordinates of any real point on it, other than the origin. (No, homogeneous coordinates are not unique, a projective point having many different homogeneous coordinates. This may seem strange at first but read on ....)

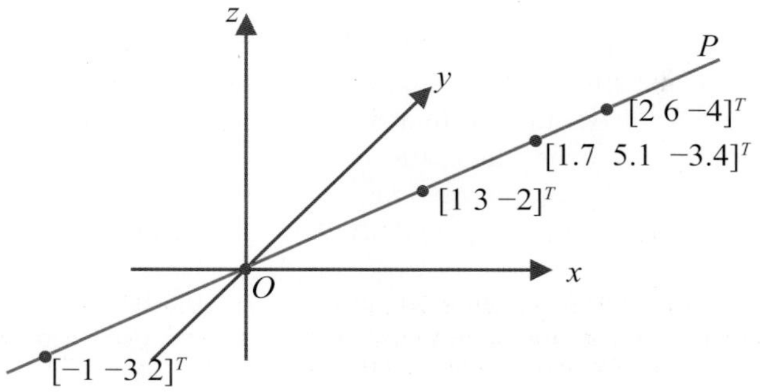

Figure A.5: The coordinates of any point on P, except the origin, can be used as its homogeneous coordinates – four possibilities are shown.

Example A.1. The projective point P corresponding to the radial line through $[1\ 3\ -2]^T$ has, as shown in Figure A.5, among others, homogeneous coordinates $[1\ 3\ -2]^T$, $[2\ 6\ -4]^T$, $[-1\ -3\ 2]^T$ and $[1.7\ 5.1\ -3.4]^T$. In fact, any tuple of the form $[c\ 3c\ -2c]^T$, where $c \neq 0$, can serve as homogeneous coordinates for P.

Terminology: To avoid clutter in diagrams, we'll often write homogeneous coordinates $[x\ y\ z]^T$ as (x, y, z).

That a projective point has infinitely many different homogeneous coordinates may seem odd, but it's not really a problem because two distinct projective points cannot share the same homogeneous coordinates. This is because two distinct radial lines do not share any point other than the origin. In other words, even though projective points have non-unique homogeneous coordinates, there is no risk of ambiguity. As an analogy, think of a roomful of people, each having multiple nicknames, but no two having a nickname in common – there is no danger of confusion then. As a non-zero tuple $[x\ y\ z]^T$ gives homogeneous coordinates of a unique projective point, we'll often refer to *the* projective point $[x\ y\ z]^T$ or write, say, the projective point $P = [x\ y\ z]^T$.

If you are wondering if $\mathbb{P}^2$ can at all be coordinatized in a unique manner, as is $\mathbb{R}^2$ by Cartesian coordinates, the answer is that there is no "natural" way to do this. Don't take our word for it, but give the question a bit of thought and you'll see the pitfalls. For example, a likely approach is to choose the coordinates of *one* real point from the radial line corresponding to each projective point. But then one has to come up with a *well-defined* way of choosing such a point; in other words, an *algorithm* that, given input a radial line, uniquely outputs a point from it. Try and devise such an algorithm! (The point on the line a unit distance from the origin? There are two such! The one in the positive direction? Be careful now: exactly which direction is this?)

Remark A.3. An important difference between the Cartesian and homogeneous coordinate systems is the lack of an origin in the latter. No matter how one sets up a Cartesian coordinate system in $\mathbb{R}^3$, i.e., no matter how one sets up the coordinate axes, the origin $(0, 0, \ldots, 0)$ is always distinguished as a special point. This is not the case for the homogeneous coordinate system in $\mathbb{P}^2$ – no projective point is special. It is truly homogeneous!

Example A.2. Find homogeneous coordinates of the projective point P of intersection of the projective lines L and L', corresponding, respectively, to the radial planes $2x + 2y - z = 0$ and $x - y + z = 0$.

Answer: Solving the simultaneous equations

$$\begin{aligned} 2x + 2y - z &= 0 \\ x - y + z &= 0 \end{aligned}$$

one finds that points on their intersecting line are of the form

$$y = -3x, \quad z = -4x$$

Therefore, homogeneous coordinates of P are (arbitrarily choosing $x = 1$)

$$[1 \ -3 \ -4]^T$$

Exercise A.2. Find homogeneous coordinates of the projective point P of intersection of the projective lines L and L' corresponding, respectively, to the radial planes $-x - y + z = 0$ and $3x + 2y = 0$.

Exercise A.3. Find the equation of the radial plane in $\mathbb{R}^3$ corresponding to the projective line L which intersects the two projective points $P = [1 \ 2 \ 3]^T$ and $P' = [2 \ -1 \ 0]^T$.

A.4 Structure of the Projective Plane

We're going to try and understand the structure of $\mathbb{P}^2$ by relating it to that of $\mathbb{R}^2$. In fact, we'll start off by using the homogeneous coordinate system of $\mathbb{P}^2$ to *embed* $\mathbb{R}^2$ inside $\mathbb{P}^2$.

A.4.1 Embedding the Real Plane in the Projective Plane

Associate a point $p = [x\ y]^T$ of $\mathbb{R}^2$ with the projective point $\phi(p) = [x\ y\ 1]^T$. The easiest way to picture this association is to first identify $\mathbb{R}^2$ with the plane $z = 1$; particularly, $[x\ y]^T$ of $\mathbb{R}^2$ is identified with $[x\ y\ 1]^T$ of $z = 1$. See Figure A.6. Following this, the association $p \mapsto \phi(p)$ is simply each real point with the radial line through it, in particular, the real point $[x\ y\ 1]^T$ (Cartesian coordinates) with the projective point $[x\ y\ 1]^T$ (homogeneous coordinates).

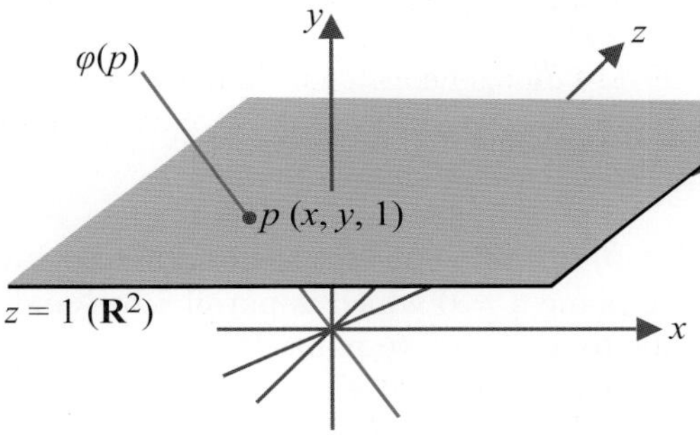

Figure A.6: Real point p on the plane $z = 1$ is associated with the projective point $\phi(p)$. Projective point Q, lying on the plane $z = 0$, is not associated with any real point.

The association $p \mapsto \phi(p)$ is clearly one-to-one as distinct points of $z = 1$ give rise to distinct radial lines through them. It's not onto as points of $\mathbb{P}^2$ that lie *on* the plane $z = 0$ or, equivalently, are parallel to $z = 1$, do not intersect $z = 1$ and, therefore, are not associated with any point of $\mathbb{R}^2$ (e.g., Q in the figure). Precisely, points of $\mathbb{P}^2$ with homogeneous coordinates of the form $[x\ y\ 0]^T$ are not associated with any point of $\mathbb{R}^2$.

$\mathbb{R}^2$, therefore, is embedded by ϕ as the *proper* subset of $\mathbb{P}^2$ consisting of radial lines intersecting $z = 1$. We're at the point now where we can try to understand how we ended up trading parallelism in $\mathbb{R}^2$ for perfect point-line duality in $\mathbb{P}^2$.

A.4.2 A Thought Experiment

Here's a thought experiment. Two parallel lines l and l' lie on $\mathbb{R}^2$, aka the plane $z = 1$ in $\mathbb{R}^3$, a distance of d apart. Points p and p' on l and l', respectively, start a distance d apart and begin to travel at the same speed and in the same direction on their individual lines. See Figure A.7. Evidently, they remain d apart no matter how far they go. Well, of course, as l and l' are parallel!

Consider next what happens to the projective points $\phi(p)$ and $\phi(p')$ associated with p and p', respectively. See again Figure A.7 to convince

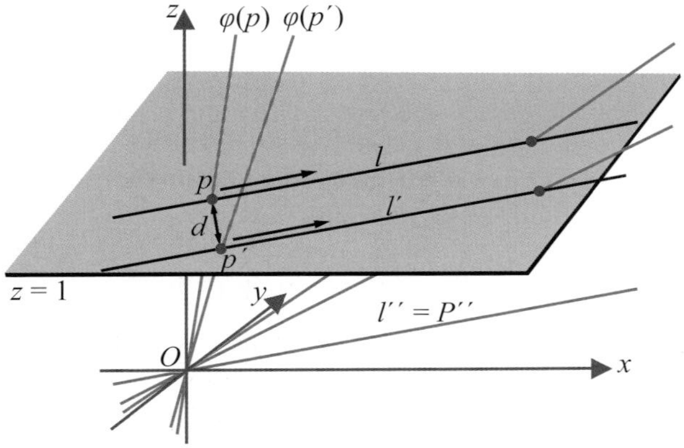

Figure A.7: The real points p and p' travel along parallel lines l and l'. Associated projective points $\phi(p)$ and $\phi(p')$ travel with p and p'.

yourself that both $\phi(p)$ and $\phi(p')$ draw closer and closer to that particular radial line l'' on the plane $z = 0$ which is parallel to l and l'. As it lies on $z = 0$, l'' corresponds to a projective point P'' not associated with any real; in fact, P'''s homogeneous coordinates are of the form $[x\ y\ 0]^T$.

Observe that the projective point $\phi(p)$ itself travels along a projective line L – the one whose radial plane contains l. We'll call L the projective line corresponding to l. Likewise, the projective point $\phi(p')$ travels along the projective line L' corresponding to l'. Moreover, L and L' intersect in P''. See Figure A.8.

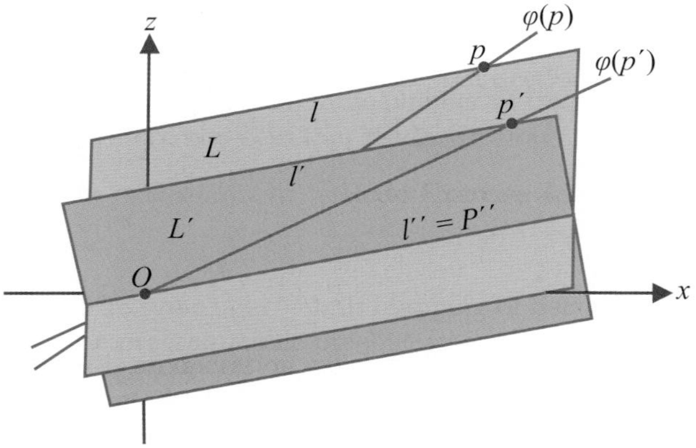

Figure A.8: $\phi(p)$ travels along L and $\phi(p')$ along L'. L and L' meet at P''.

Let's take stock of the situation so far. The parallel lines l and l' on the real plane never meet, but the projective lines L and L' corresponding to them in $\mathbb{P}^2$ meet in P''. Moreover, every point of L or L', *except* for P'', is associated by ϕ to a point of l or l', respectively. We can say then that the projective line L equals its real counterpart l *plus* the extra point P''; L', likewise, is its real counterpart l' *plus* P''. And, it's at this point P'', beyond the reals, that the two projective lines meet, while their real counterparts never do.

Example A.3. What if both points p and p', and together with them $\phi(p)$ and $\phi(p')$, travel along their respective lines in directions opposite to those indicated in Figure A.7? What if only one reversed its direction?

Answer: If both p and p' reversed directions, then again they would travel forever exactly d apart. If only one of the two reversed its direction, then, of course, the distance between them would continuously increase.

However, in either case, $\phi(p)$ and $\phi(p')$ draw closer, again both to P''. It seems that, whatever the sense of travel is of $\phi(p)$ and $\phi(p')$ along their respective projective lines L and L', they approach that one point of intersection of these two lines. Two points traveling in opposite directions along a real line ultimately grow farther and farther apart. A projective line, on the other hand, apparently behaves more like a circle.

A.4.3 Regular Points and Points at Infinity

Recall *equivalence relations* and *equivalence classes* from undergrad discrete math. In particular, recall that the lines of $\mathbb{R}^2$ can be split into equivalence classes by the equivalence relation of being parallel. Consider any equivalence class $\mathbf{l}$ of parallel lines of $\mathbb{R}^2$, the latter being identified with the plane $z = 1$ in $\mathbb{R}^3$ as before. There is a unique radial line l on the plane $z = 0$ parallel to the members of $\mathbf{l}$. See Figure A.9.

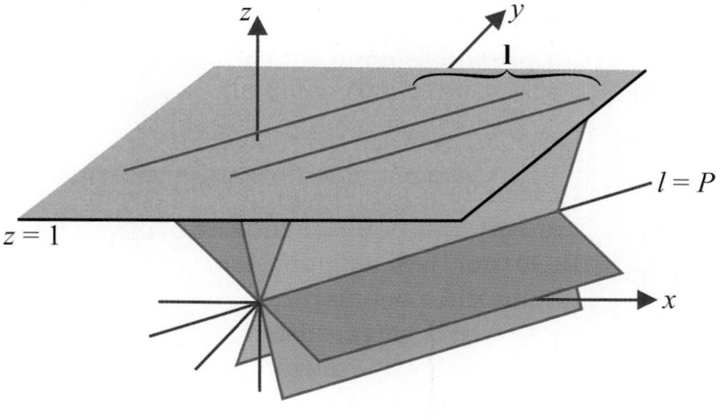

Figure A.9: The line l (= projective point P) is parallel to lines in $\mathbf{l}$. P is said to be the point at infinity along the equivalence class $\mathbf{l}$ of parallel lines.

Denote the projective point corresponding to l by P. Projective lines corresponding to lines in $\mathbf{l}$ all meet at P, because their radial planes each contain l. The point P, which is not associated with any real point by ϕ as it lies on $z = 0$, is called the *point at infinity* along **l** or, simply, the point at infinity along any one of the lines in **l**. Conversely, any radial line l on the plane $z = 0$ is the point at infinity along the equivalence class of lines in $\mathbb{R}^2$ parallel to it. In other words, the correspondences

$$\begin{aligned}\text{equivalence class of parallel lines in } \mathbb{R}^2 \quad &\leftrightarrow \quad \text{radial line on } z = 0 \\ &\leftrightarrow \quad \text{point at infinity of } \mathbb{P}^2\end{aligned}$$

are both one-to-one. Note that points at infinity of $\mathbb{P}^2$ are precisely those with homogeneous coordinates of the form $[x\ y\ 0]^T$.

Returning to the thought experiment of Section A.4.2, one can imagine points at infinity plugging the "holes" along the "border" of $\mathbb{R}^2$ through which parallel lines "run off" without meeting, which explains why every pair of lines on the projective plane meets.

Projective points which are not points at infinity are called *regular points*. Regular points have homogeneous coordinates of the form $[x\ y\ z]^T$, where z is *not* zero. Moreover, regular points intersect $z = 1$, so are associated each by ϕ^{-1} with a point of $\mathbb{R}^2$ (remember ϕ takes a real point of $\mathbb{R}^2$, represented by the plane $z = 1$, to the projective point whose corresponding radial line passes through that point). Accordingly, one can write:

$$\mathbb{P}^2 = \mathbb{R}^2 \cup \{\text{points at infinity}\} = \{\text{regular points}\} \cup \{\text{points at infinity}\}$$

The union of all points at infinity, called the *line at infinity*, is the projective line whose radial plane is $z = 0$. Therefore, one can as well write:

$$\mathbb{P}^2 = \mathbb{R}^2 \cup \text{line at infinity} = \{\text{regular points}\} \cup \text{line at infinity}$$

Our embedding ϕ of $\mathbb{R}^2$ as a subset of $\mathbb{P}^2$ depends on the plane $z = 1$, particularly because we identify $z = 1$ with $\mathbb{R}^2$ and subsequently associate each point of $\mathbb{R}^2$ with the radial line in $\mathbb{R}^3$ through it. Is there anything special about the plane $z = 1$? Not at all. It just seemed convenient. In fact, we could have used any *any* non-radial plane p.

Exercise A.4. Why does p have to be non-radial?

Example A.4. Instead of $z = 1$, identify $\mathbb{R}^2$ with the plane $x = 2$ in $\mathbb{R}^3$. Accordingly, embed $\mathbb{R}^2$ into $\mathbb{P}^2$ by associating $[x\ y]^T$ with the radial line through $[2\ x\ y]^T$. Which now are the regular points and which are the points at infinity of $\mathbb{P}^2$?

Answer: The regular points of $\mathbb{P}^2$ are the radial lines in $\mathbb{R}^3$ which intersect the plane $x = 2$. These are precisely the radial lines which do not lie on the plane $x = 0$. The points at infinity are the radial lines which do lie on the plane $x = 0$. Equivalently, regular points have homogeneous coordinates of the form $[x\ y\ z]^T$, where $x \neq 0$, while points at infinity have homogeneous coordinates of the form $[0\ y\ z]^T$.

Exercise A.5. Identify $\mathbb{R}^2$ with the plane $x + y + z = 1$ in $\mathbb{R}^3$, embedding it into $\mathbb{P}^2$ by associating $[x\ y]^T$ with the radial line through $[x\ y\ \ 1 - x - y]^T$. Which now are the regular points and which the points at infinity of $\mathbb{P}^2$?

It may seem strange at first that the separation of $\mathbb{P}^2$ into regular points and points at infinity depends on the particular embedding of $\mathbb{R}^2$ in $\mathbb{P}^2$. However, this situation becomes clearer after a bit of thought. It's related, as a matter of fact, to the discussion at the beginning of the chapter, where we motivated projective spaces by observing that lines through a point camera are perceived as points on the plane film. Even though all lines through the camera do not intersect the film, we argued this to be merely an artifact of the alignment of the film, the latter being changeable. Therefore, we concluded that all radial lines should be taken as points in projective space.

We come now full circle back to this initially motivating scenario. Embedding $\mathbb{R}^2$ in $\mathbb{P}^2$ corresponds exactly to choosing an alignment of the film – the film is a copy of $\mathbb{R}^2$ and each point on it associated with the light ray (= radial line in $\mathbb{R}^3$ = point of $\mathbb{P}^2$) through that point to the camera. Light rays toward the camera which intersect the film are regular points of $\mathbb{P}^2$ and visible, while those which do not are points at infinity and invisible. Moreover, the line at infinity corresponds to the plane through the camera parallel to the film. And, of course, we are at perfect liberty to align the film, i.e., embed $\mathbb{R}^2$ in $\mathbb{P}^2$, as we like, different choices leading to different sets of visible and invisible light rays.

A.5 Snapshot Transformations

Here's another interesting thought experiment.

Example A.5. A point camera is at the origin with two power lines passing over it, both parallel to the x-axis. One lies along the line $y = 2, z = 2$ (i.e., the intersection of the planes $y = 2$ and $z = 2$) and the other along the line $y = -2, z = 2$.

Take "snapshots" of the power lines with the film aligned along (a) the plane $z = 1$ and (b) the plane $x = 1$. Sketch and compare the two snapshots.

Answer: This is one you might want to try yourself before reading on!

See Figure A.10. Figure A.10(a) shows the snapshot (or, projection) of the power lines on the plane $z = 1$. They are two *parallel* lines $y = 1$ and $y = -1$. This is not hard to understand: by simple geometry, the line $y = 2, z = 2$ projects toward the origin (the camera) to the line $y = 1$ on the plane $z = 1$; likewise, $y = -2, z = 2$ projects to $y = -1$ on $z = 1$.

Figure A.10(b) shows the snapshot on the plane $x = 1$. It is the two *intersecting* lines $z = y$ and $z = -y$ making an X-shape. This requires explanation. The top of the X, above its center, is formed from intersections with the film of light rays through points on the power lines with x-value greater than zero, while the bottom from rays through points with x-value

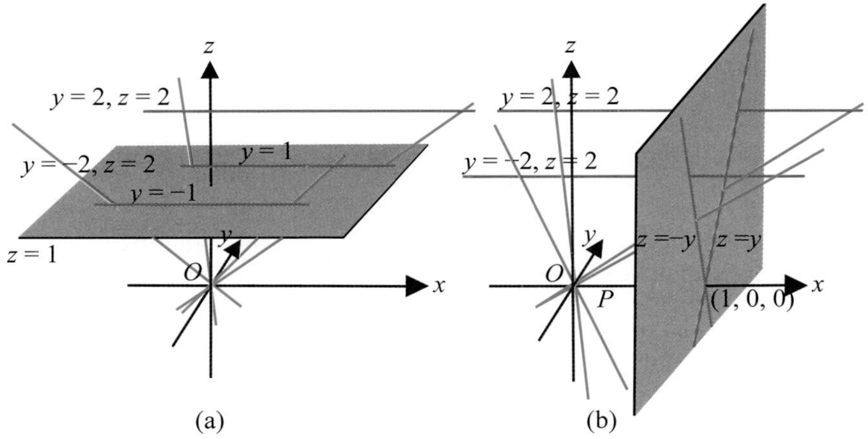

Figure A.10: Power lines $y = \pm 2, z = 2$ projected onto the planes (a) $z = 1$ and (b) $x = 1$. Red lines depict light rays. The x-axis corresponds to the projective point P.

less than zero. The rays from the points on either power line with x-value equal to zero do not strike the film.

The point $[1\ 0\ 0]^T$ at the middle of the X is included in the snapshot, though no ray from either power line passes through it, because it's the intersection with the film of the "limit" of the rays from points on either power line as they run off to infinity. It's convenient to imagine the limits of visible rays as being visible as well and we ask the reader to accept this. In geometric drawing parlance $[1\ 0\ 0]^T$ is the *vanishing point* of the power lines – it's where they *seem* to meet on the film $x = 1$.

Contemplate now the situation from the point of view of projective geometry. The projective lines corresponding to the two power lines meet at the projective point P corresponding to the x-axis, as the radial planes through the power lines intersect in the x-axis. Now, P is a point at infinity with respect to the plane $z = 1$ (because the x-axis doesn't intersect this plane), while it's a regular point with respect to the plane $x = 1$ (because the x-axis intersects this plane at $[1\ 0\ 0]^T$). In terms of shooting pictures, then, the camera with its film along $z = 1$ cannot see where the two power lines meet, so they appear parallel. However, with its film along $x = 1$ the camera sees them meet at $[1\ 0\ 0]^T$.

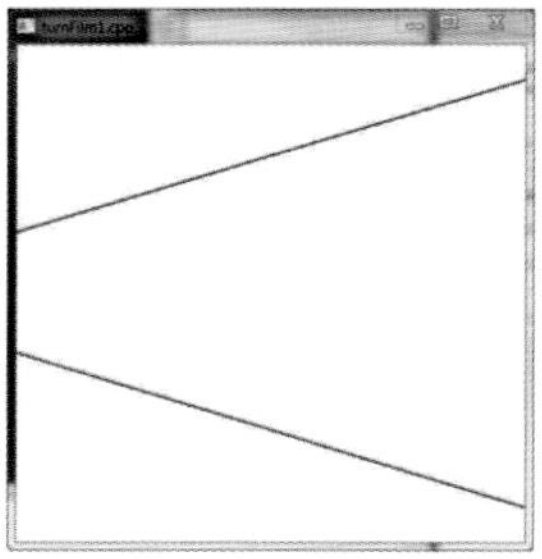

Figure A.11: Screenshot of `turnFilm1.cpp`.

Experiment A.1. Run `turnFilm1.cpp`, which animates the setting of the preceding exercise by means of a viewing transformation. Initially, the film lies along the $z = 1$ plane. Pressing the right arrow key rotates it toward the $x = 1$ plane, while pressing the left one reverses the rotation. Figure A.11 is a screenshot midway. You cannot, of course, see the film, only the view of the lines as captured on it.

The reason that the lower part of the X-shaped image of the power lines cannot be seen is that OpenGL film doesn't capture rays hitting it from behind, as the viewing plane is a clipping plane too. Moreover, if the lines seem to actually meet to make a V after the film turns a certain finite amount, that's because they are very long and your monitor has limited resolution!

This program itself is simple with the one statement of interest being `gluLookAt()`, which we ask the reader to examine next. **End**

Exercise A.6. (Programming) Verify that the `gluLookAt()` statement of `turnFilm1.cpp` indeed simulates the film's rotation as claimed between the $z = 1$ and $x = 1$ planes.

Example A.6. Refer to Figure A.10(b). Suppose two power lines *actually* lie along the two intersecting lines $z = y$ and $z = -y$ on the plane $x = 1$, which is the snapshot on the plane $x = 1$ of the power lines of the preceding example. What would *their* snapshot look like on the films $z = 1$ and $x = 1$?

Answer: Exactly as in the preceding Example A.5, as depicted in Figures A.10(a) and (b)! It's not possible to distinguish between these two pairs of power lines – the pair in Example A.5 being "really" parallel and the current one "really" intersecting – with a point camera at the origin.

A somewhat whimsical take on all this is to imagine a Matrix-like world where one can never know reality. Perception is limited to whatever is captured on film. Therefore, one agent's intersecting power lines are just as real as the other's parallel ones!

It's useful to think of one snapshot of Example A.5 or A.6 as a *transformation* of the other. Keep in mind that if a snapshot appears as the two parallel lines $y = \pm 1$ on the film $z = 1$, then it always appears as the two intersecting lines $z = \pm y$ on the film $x = 1$, *regardless* of what the "real" objects are.

Convince yourself of this by mentally tilting one of the power lines in Figure A.10(a) on the radial plane (not drawn) through it, so that its projection on the $z = 1$ plane does not change. The power line's projection on the $x = 1$ plane remains unchanged, as well, because the set of light rays from it through the camera doesn't change. For this reason, it makes sense to talk of transforming one snapshot to another, without any reference to the real scene. We'll informally call such transformations *snapshot transformations*.

Remark A.4. Snapshot transformations as described are not really transformations in the mathematical sense, as they don't map some space to itself but, rather, one plane (film) to another. A rigorous formulation is possible, though likely not worth the effort, as we'll see soon that snapshot transformations are subsumed within the class of projective transformations, which we'll be studying in depth. Nevertheless, the notion of a snapshot transformation is geometrically intuitive and useful.

Here are more for you to ponder.

Exercise A.7. In each case below you are told what the snapshot looks like on the film $z = 1$, aka $\mathbb{R}^2$, and asked what is captured on the film $x = 1$. The $z = 1$ shots are drawn in Figure A.12, each labeled the same as the item to which it corresponds. You don't have to find equations for your answer for $x = 1$. Just a sketch or a verbal description is enough.

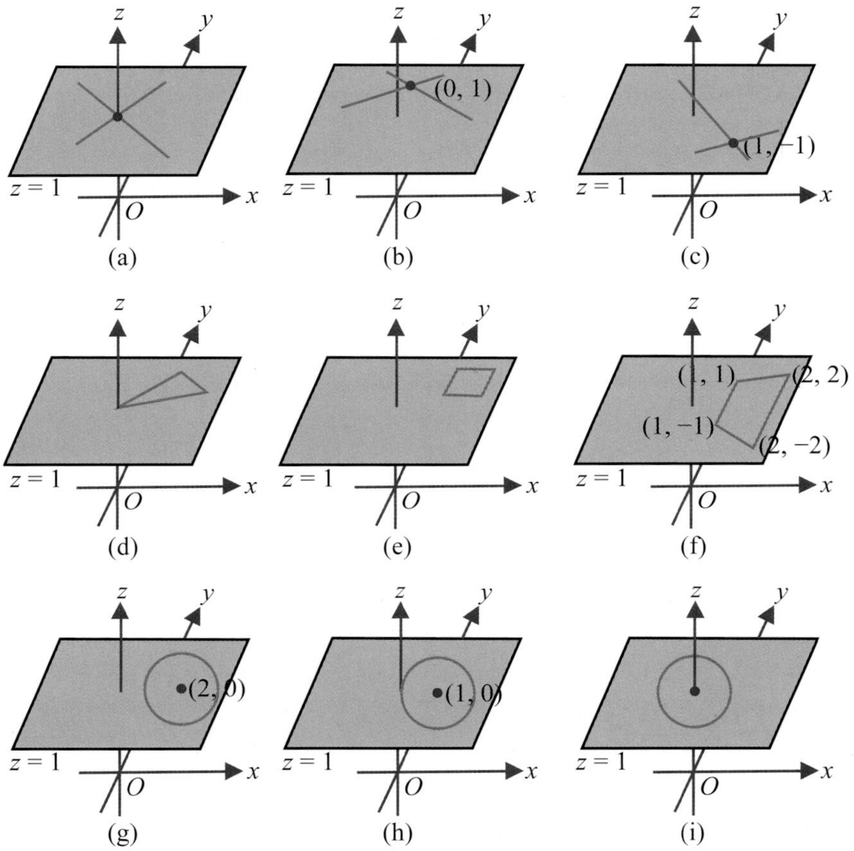

Figure A.12: Transform these snapshots on the plane $z = 1$ to the plane $x = 1$. Some points on the plane $z = 1$ are shown with their xy coordinates. Labels correspond to items of Exercise A.7.

Answers are in italics. Figure A.13 justifies the answer to (h).

(a) Two lines that intersect at the origin, neither being the y-axis (on $\mathbb{R}^2$). *Two parallel lines.* Why the caveat? What happens if one is the y-axis?

(b) Two lines that intersect at the point $[0\ 1]^T$, neither being the y-axis. *Two parallel lines.*

(c) Two lines that intersect at the point $[1 \ -1]^T$.
Two intersecting lines.

(d) A triangle in the upper-right quadrant with one vertex at the origin but, otherwise, not touching the axes.
An infinitely long U-shape with straight sides.

(e) A square in the upper-right quadrant not touching any of the axes.
A quadrilateral with no two parallel sides.

(f) A trapezoid symmetric about the x-axis with vertices at $[1 \ 1]^T$, $[1 \ -1]^T$, $[2 \ -2]^T$ and $[2 \ 2]^T$.
A rectangle.

(g) A unit radius circle centered at $[2 \ 0]^T$.
An ellipse.

(h) A unit radius circle centered at $[1 \ 0]^T$.
A parabola – see Figure A.13.

(i) A unit radius circle centered at the origin.
A hyperbola.

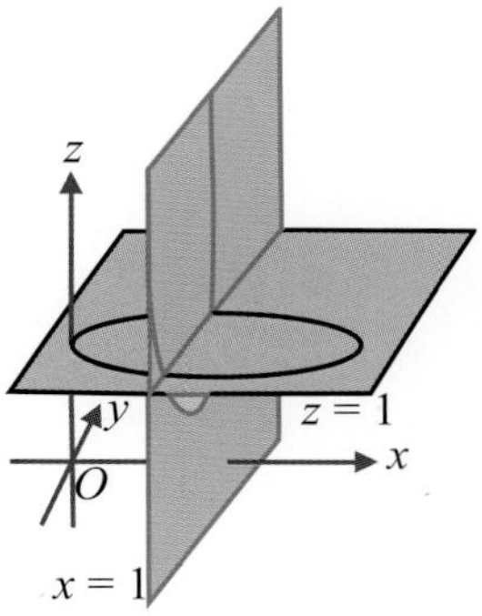

Figure A.13: Answer to Exercise A.7(h).

Remark A.5. Exercise A.7(f) seems innocuous enough, but it is very important. Its generalization to 3D will help convert viewing frustums to rectangular boxes in the graphics pipeline.

Exercise A.8. Refer to the geometric construction of conic sections in Section 10.1.5 as plane sections of a double cone, and show that any non-degenerate conic section can be snapshot transformed to another such.

Exercise A.9. (Programming) Write code similar to `turnFilm1.cpp` to animate the snapshot transformation of Exercise A.7(h). Again, you'll see only part of the parabola because OpenGL cannot see behind its film.

It's not hard to see that none of the snapshot transformations of Exercise A.7, except for (c) and (g), can be accomplished using OpenGL modeling transformations. This is because they are not affine – recall from Section 5.4.5 that OpenGL implements only affine transformations.

Remark A.6. We just said that most of the snapshot transformations of Exercise A.7 are not affine and yet seem to be suggesting with the preceding Exercise A.9 that they may be implemented by means of an OpenGL viewing transformation. We know, however, that the latter is equivalent to a sequence of modeling transformations and, therefore, affine.

The apparent conundrum is not hard to resolve. The result of the viewing transformation of, e.g., `turnFilm1.cpp`, is indeed a snapshot transformation in terms of what is *seen on the screen*. In other words, the transformation from the OpenGL window prior to applying the viewing transformation to that after is a snapshot transformation. However, the viewing transformation serves only to change the scene to one which OpenGL *projects* onto the window as the new one. A snapshot transformation, therefore, is more than a viewing transformation – it's a viewing transformation *plus* a projection.

Exercise A.10. By considering how to turn the film, i.e., viewing plane, show that implementing a snapshot transformation in OpenGL is equivalent to:

(a) setting the *centerx*, *centery*, *centerz*, *upx*, *upy* and *upz* parameters of the viewing transformation

```
gluLookAt(0, 0, 0, centerx, centery, centerz, upx, upy, upz)
```

and

(b) setting the *near* parameter of the perspective projection call

```
glFrustum(left, right, bottom, top, near, far)
```

where the other five parameters can be kept fixed at some initially chosen values.

A.6 Homogeneous Polynomial Equations

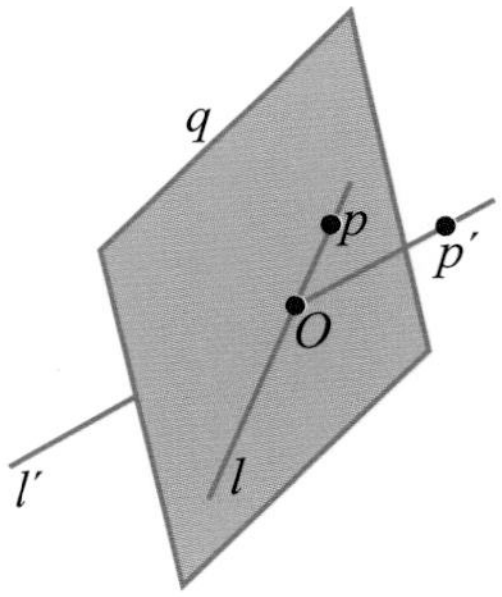

Figure A.14: Point p of radial line l lies on radial plane q, implying that l lies on q; point p' of l' doesn't lie on q, implying that no point of l', other than the origin, lies on q.

The only application we've made so far of homogeneous coordinates is to embed $\mathbb{R}^2$ in $\mathbb{P}^2$. We haven't used them yet to write equations of curves on the projective plane. Let's try now to do this.

We'll start with the simplest curve on the projective plane, in fact, a projective line. We want an equation – as for straight lines in real geometry – that will say if a projective point belongs to a projective line. For example, an equation such as $2x + y - 1 = 0$ for a straight line on the real plane gives the condition for a real point $[x\ y]^T$ to lie on that line.

Now, a projective point is a radial line and a projective line a radial plane. Moreover, a radial line lies on a radial plane if and only if any point of it, other than the origin, lies on that plane (the origin always does). See Figure A.14.

Therefore, a projective point $P = [x\ y\ z]^T$ belongs to a projective line L, whose radial plane has the equation $ax + by + cz = 0$, if and only if the real point $[x\ y\ z]^T$ lies on the real plane $ax + by + cz = 0$. It follows that the equation of L is identical to that of its radial plane:

$$ax + by + cz = 0 \tag{A.1}$$

Accordingly, a projective point $P = [x\ y\ z]^T$ belongs to L if it satisfies (A.1). Does it matter if we choose some other homogeneous coordinates for P? No, because

$$a(kx) + b(ky) + c(kz) = k(ax + by + cz) = 0$$

so any homogeneous coordinates $[kx\ ky\ kz]^T$ for P satisfy Equation (A.1).

Exercise A.11. Prove that if the projective line L is specified by the equation

$$ax + by + cz = 0$$

then it is specified by any equation of the form

$$(ma)x + (mb)y + (mc)z = 0$$

where $m \neq 0$, as well.

Exercise A.12. What is the equation of the projective line through the projective points $[2\ 1\ -1]^T$ and $[3\ 4\ 2]^T$?

Answer: Suppose that the line is L with equation

$$ax + by + cz = 0$$

Since $[2\ 1\ -1]^T$ and $[3\ 4\ 2]^T$ lie on L they must satisfy its equation, giving

$$\begin{aligned} 2a + b - c &= 0 \\ 3a + 4b + 2c &= 0 \end{aligned}$$

Any solution to these simultaneous equations, not all zero, then determines L. As there are more variables then equations, let's set one of them, say c, arbitrarily to 1, to get the equations

$$\begin{aligned} 2a + b - 1 &= 0 \\ 3a + 4b + 2 &= 0 \end{aligned}$$

These solve to give $a = 1.2$ and $b = -1.4$. The equation of the projective line L is, therefore,

$$1.2x - 1.4y + z = 0$$

(or, equivalently, $6x - 7y + 5z = 0$, from Exercise A.11.)

Exercise A.13. What is the projective point of intersection of the projective lines $3x + 2y - 4z = 0$ and $x - y + z = 0$?

Exercise A.14. When are three projective points $[x\ y\ z]^T$, $[x'\ y'\ z']^T$ and $[x''\ y''\ z'']^T$ collinear, i.e., when do they belong to the same projective line? Find a simple condition involving a determinant.

A.6.1 More About Point-Line Duality

In Section A.4.2 we tried to understand the point-line duality of the projective plane from a geometric point of view. We'll examine the phenomenon now from an algebraic standpoint.

The correspondence from the set of projective points to the set of projective lines given by

$$\text{projective point } [a\ b\ c]^T \quad \mapsto \quad \text{projective line } ax + by + cz = 0 \qquad \text{(A.2)}$$

is well-defined as, whatever homogeneous coordinates we choose for a projective point, the image is the same projective line (by Exercise A.11). Moreover, the correspondence is easily seen to be one-to-one and onto.

Definition A.3. The projective line $ax + by + cz = 0$ is said to be the *dual* of the projective point $[a\ b\ c]^T$ and vice versa.

Exercise A.15. Prove that a projective point P belongs to a projective line L if and only if the dual of L belongs to the dual of P.

The preceding exercise implies that if some statement about the incidence of projective points and lines is true, then so is the dual statement, obtained by replacing "point" with "line" and "line" with "point".

Exercise A.16. What is the dual of the following statement? "There is a unique projective line incident to two distinct projective points."

From this last exercise one sees, then, the point-line duality of the projective plane as a consequence of the one-to-one correspondence (A.2) between projective points and lines. We ask the reader to contemplate if there exists a similar correspondence between real points and lines.

A.6.2 Lifting an Algebraic Curve from the Real to the Projective Plane

Let's see next projective curves more complex than a line. Consider, then, the curve Q' in $\mathbb{P}^2$ consisting of the projective points intersecting the parabola q

$$y - x^2 = 0 \qquad \text{(A.3)}$$

on $\mathbb{R}^2$ (the plane $z = 1$). See Figure A.15.

The intersection of the projective point $P = [x\ y\ z]^T$ with the plane $z = 1$ is the real point $[x/z\ \ y/z\ \ 1]^T$, assuming $z \neq 0$, for, otherwise, there is no intersection. Now, $[x/z\ \ y/z\ \ 1]^T$ satisfies the equation of the parabola q if

$$y/z - (x/z)^2 = 0 \qquad \Longrightarrow \qquad yz - x^2 = 0$$

Accordingly, the curve consisting of projective points $[x\ y\ z]^T$ which satisfy

$$yz - x^2 = 0 \qquad \text{(A.4)}$$

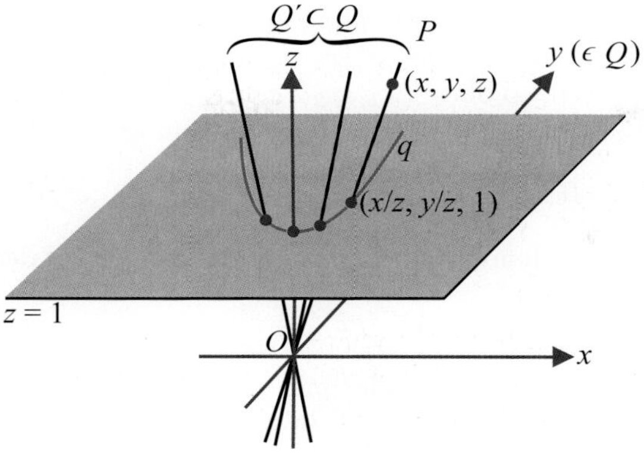

Figure A.15: Lifting a parabola drawn on the real plane $z = 1$ to the projective plane.

is called the *lifting* Q of q from the real to the projective plane. Q is sometimes simply called the lifting of q and also the *projectivization* of q. In this particular case, as a lifting of a parabola, Q is a *parabolic projective curve*.

The camera analogy is that Q is the set of rays seen, by intersection with the film $z = 1$, as q. However, Q is actually one point *bigger* than Q', the set of projective points intersecting the parabola q on $\mathbb{R}^2$, as it includes the projective point $[0\ 1\ 0]^T$, the y-axis of 3-space, which satisfies (A.4), but does not intersect q. We can justify the inclusion of this extra point, with the help of the proviso from Section A.5 that a limit of visible rays is visible, as follows.

From its equation $y - x^2 = 0$, a point of q is of the form $[x\ x^2\ 1]^T$, for any x. Therefore, the homogeneous coordinates of a projective point intersecting q are $[x\ x^2\ 1]^T$, for any x, as well. Rewriting these coordinates as $[\frac{1}{x}\ 1\ \frac{1}{x^2}]^T$ we see that its limit as $x \to \infty$ is indeed $[0\ 1\ 0]^T$. More intuitively, *a la* the thought experiment of Section A.4.2, as a point p travels off along either wing of the parabola, the projective point $\phi(p)$, corresponding to the line through p, approaches $[0\ 1\ 0]^T$, the projective point corresponding to the y-axis.

Definition A.4. A *homogeneous polynomial* is one whose terms each have the same degree, the degree of a term being the sum of the powers of the variables in the term. This common degree is called the degree of the homogeneous polynomial.

An equation with a homogeneous polynomial on the left and 0 on the right is called a homogeneous polynomial equation.

The equations $ax + by + cz = 0$ of a projective line and $yz - x^2 = 0$ of a parabolic projective curve are homogeneous polynomial equations of degree one and two, respectively. That they are both homogeneous is no accident, as we'll soon see.

Exercise A.17. Suppose that $p(x_1, x_2, \ldots, x_n)$ is a homogeneous polynomial in n variables. Then, if $[x_1\ x_2\ \ldots\ x_n]^T$ satisfies the equation $p(x_1, x_2, \ldots, x_n) = 0$, so does $[cx_1\ cx_2\ \ldots\ cx_n]^T$, for any scalar c.
Hint: Show, first, that, if $p(x_1, x_2, \ldots, x_n)$ is homogeneous of degree r, then

$$p(cx_1, cx_2, \ldots, cx_n) = c^r p(x_1, x_2, \ldots, x_n)$$

For example, for the homogeneous polynomial $yz - x^2$ of degree 2,

$$(cy)(cz) - (cx)^2 = c^2(yz - x^2)$$

So, in this case, if (x, y, z) satisfies $yz - x^2 = 0$, then so does (cx, cy, cz), because $(cy)(cz) - (cx)^2 = 0$ as well, by the equation just above.

The preceding exercise implies that a homogeneous polynomial equation of the form $p(x, y, z) = 0$ is legitimately an equation in $\mathbb{P}^2$, because a point of $\mathbb{P}^2$ can be tested if it satisfies $p(x, y, z) = 0$, independently of the homogeneous coordinates used to represent it. Here are some definitions.

Definition A.5. An *algebraic curve* on the real plane consists of points satisfying an equation of the form

$$p(x, y) = 0$$

where p is a polynomial in the two variables x and y. The degree of the curve is the highest degree of a term belonging to $p(x, y)$.

A *projective algebraic curve* on the projective plane consists of points satisfying an equation of the form

$$p(x, y, z) = 0$$

where p is a homogeneous polynomial in the three variables x, y and z. The degree of the curve is the degree of $p(x, y, z)$.

Familiar algebraic curves of degree one include straight lines, e.g., $2x + y - 3 = 0$, and of degree two conic sections, e.g., the hyperbola $xy - 1 = 0$.

At the start of this section we lifted a parabola from the real to the projective plane. Here are a couple more examples of lifting.

Example A.7. Lift the straight line

$$ax + by + c = 0$$

drawn on the plane $z = 1$, to $\mathbb{P}^2$.

Answer: The projective point $[x\ y\ z]^T$ intersects the plane $z = 1$ at the real point $[x/z\ y/z\ 1]^T$ (assuming $z \neq 0$), which belongs to the given straight line if, replacing x by x/z and y by y/z in the latter's equation:

$$ax/z + by/z + c = 0$$

Multiplying throughout by z, one gets the homogeneous polynomial equation of degree 1

$$ax + by + cz = 0$$

confirming that the lifting of a straight line on $z = 1$ is the projective line corresponding to it. A projective line is, of course, a projective algebraic curve of degree 1.

Example A.8. Lift the algebraic curve of degree 3

$$x^3 + 3x^2y + y^2 + x + 2 = 0$$

drawn on the plane $z = 1$, to $\mathbb{P}^2$.

Answer: The projective point $[x\ y\ z]^T$ intersects the plane $z = 1$ at the real point $[x/z\ y/z\ 1]^T$ (assuming $z \neq 0$). Accordingly, replace x by x/z and y by y/z in the given polynomial equation:

$$\begin{aligned} & (x/z)^3 + 3(x/z)^2(y/z) + (y/z)^2 + x/z + 2 = 0 \\ \Longrightarrow \quad & x^3/z^3 + 3x^2y/z^3 + y^2/z^2 + x/z + 2 = 0 \\ \Longrightarrow \quad & x^3 + 3x^2y + y^2z + xz^2 + 2z^3 = 0 \end{aligned}$$

defining the lifted curve, a projective algebraic curve of degree 3.

Exercise A.18. Lift the algebraic curve of degree 5

$$xy^4 - 2x^2y^2 + 3xy^2 + y^3 - xy + 2 = 0$$

drawn on the plane $z = 1$, to $\mathbb{P}^2$.

It should be fairly clear at this point that the lifting of an algebraic curve $p(x, y) = 0$ is a projective algebraic curve $\overline{p}(x, y, z) = 0$ of the same degree. We leave a formal proof to the reader in the following exercise.

Exercise A.19. Show that the lifting of an algebraic curve $p(x, y) = 0$ of degree r is a projective algebraic curve $\overline{p}(x, y, z) = 0$ of degree r.

Definition A.6. The process of going from the equation of an algebraic curve on the real plane to the homogeneous polynomial equation of its lifting is called *homogenization*.

It's worth keeping mind that the process of homogenization depends on the particular plane on which the algebraic equation holds. For example, in Examples A.7 and A.8 and Exercise A.18 the plane was $z = 1$. This need not always be the case as we see in the next example.

Example A.9. Homogenize the polynomial equation

$$y^2 + z^2 + z = 0$$

drawn on the plane $x = 2$.

Answer: The projective point $[x\ y\ z]^T$ intersects the plane $x = 2$ at the real point $[2\ 2y/x\ 2z/x]^T$ (assuming $x \neq 0$, and multiplying $[x\ y\ z]^T$ by $2/x$). Accordingly, replace y by $2y/x$ and z by $2z/x$ in the given polynomial equation:

$$(2y/x)^2 + (2z/x)^2 + 2z/x = 0 \quad \Longrightarrow \quad 4y^2/x^2 + 4z^2/x^2 + 2z/x = 0$$

Multiplying throughout by x^2 one gets the homogenized polynomial equation

$$4y^2 + 4z^2 + 2xz = 0$$

Not surprisingly, giving the algebraic equation on different real planes corresponds, simply, to specifying the algebraic curve as seen by the viewer on differently aligned films. The lifting itself, of course, is the set of rays intersecting the film in the given curve, which does not change.

Exercise A.20. Homogenize the polynomial equation

$$3x^4 + 2x^2y + 2y^3 + 2x^2 + xy + x = 0$$

drawn on the plane $z = 1$.

Exercise A.21. Homogenize the polynomial equation

$$x^3 + 2xz - z^4$$

drawn on the plane $y = 2$.

Remark A.7. It's possible to define the homogenization of a polynomial in an abstract manner independent of reference to a particular plane. See Jennings [76].

One sees, then, that the algebraic analogue of lifting an algebraic curve from the real to the projective plane is homogenization. The reverse process of projecting a (projective algebraic) curve onto a real plane consists of taking the section of the projective points composing the curve with the given plane. Algebraically, this means simultaneously solving the equation of the curve and that of the plane – a process not surprisingly called *de-homogenization*.

Example A.10. Project the curve

$$yz - x^2 = 0$$

in $\mathbb{P}^2$ onto the real plane $z = 1$.

Answer: De-homogenize the equation of the curve by simultaneously solving

$$\begin{aligned} yz - x^2 &= 0 \\ z &= 1 \end{aligned}$$

to get

$$y - x^2 = 0$$

which is the equation of a parabola.

Exercise A.22. Project the curve of the preceding example onto the real plane $x = 1$.

Exercise A.23. Project the curve

$$4y^2 + 4z^2 + 2xz = 0$$

in $\mathbb{P}^2$ onto the real plane $y = -2$.

A.6.3 Snapshot Transformations Algebraically

It should make sense now that the snapshot transformation of an algebraic curve c from one real plane p to another p' can be determined by (a) first homogenizing the equation of c to lift it to the projective plane, and, then (b) de-homogenizing to project it back onto p'.

Example A.11. Let's solve the snapshot transformation problem of Exercise A.7(h) algebraically. The equation of the unit circle, centered at $[1\ 0]^T$ on the $z = 1$ plane, is

$$x^2 + y^2 - 2x = 0$$

Homogenizing, one gets

$$x^2 + y^2 - 2xz = 0$$

To project onto the plane $x = 1$, de-homogenize by simultaneously solving

$$\begin{aligned} x^2 + y^2 - 2xz &= 0 \\ x &= 1 \end{aligned}$$

to get

$$y^2 - 2z + 1 = 0 \quad \Longrightarrow \quad z = \frac{1}{2}y^2 + \frac{1}{2}$$

which indeed agrees with the sketch of a parabola in Figure A.13.

Exercise A.24. Solve Exercises A.7(g) and (i) algebraically.

A.7 The Dimension of the Projective Plane and Its Generalization to Higher Dimensions

Note: The next few paragraphs about $\mathbb{P}^2$ as a surface require recollecting some of the material from Section 10.2.12 on surface theory. If the reader is not inclined to do so, then she can safely skip ahead to Definition A.7. It won't affect her understanding of anything that follows.

Why do we say that the projective plane is a projective space of dimension 2? Because, as we'll see momentarily, $\mathbb{P}^2$ is a surface. In fact, it's a regular C^∞ surface, *except* that it is not a subset of $\mathbb{R}^3$: it's not possible to embed $\mathbb{P}^2$ in $\mathbb{R}^3$. One must go at least one dimension higher to $\mathbb{R}^4$.

Ignoring for now the question of the space in which it's embedded, it's not hard to find a coordinate patch containing any given point $P \in \mathbb{P}^2$. Suppose, for the moment, that P intersects the point p on the plane $z = 1$ (our favorite copy of $\mathbb{R}^2$). See Figure A.16. Let W be a closed rectangle containing p and B be the set of projective points intersecting W. The function

$$\text{point} \quad \mapsto \quad \text{the radial line through it}$$

from W to B is a one-to-one correspondence that makes B a coordinate patch.

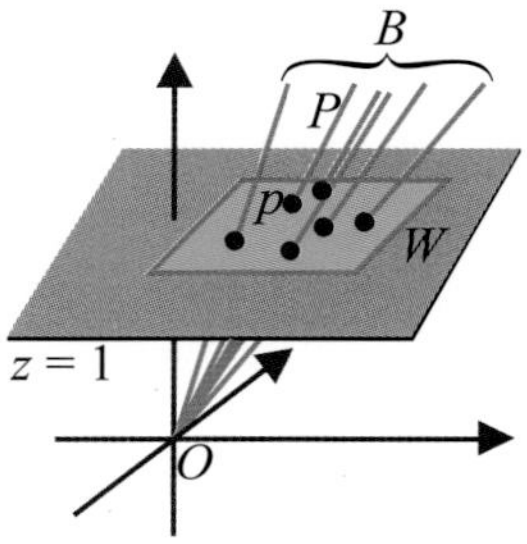

Figure A.16: The coordinate patch B containing P in $\mathbb{P}^2$ is in one-to-one correspondence with the rectangle W containing p in $\mathbb{R}^2$ (a few points in W and their corresponding projective points are shown).

And what if P doesn't intersect $z = 1$, i.e., if P is a point at infinity with respect to $z = 1$? Remember, there's nothing special about $z = 1$! Simply choose another non-radial plane with respect to which P is regular.

The reader has guessed by now that there exist projective spaces of various dimensions. True.

Definition A.7. A radial line in $\mathbb{R}^{n+1}$ is said to be an *n-dimensional projective point*. The set of all n-dimensional projective points is *n-dimensional projective space*, denoted $\mathbb{P}^n$.

$\mathbb{P}^0$, not very interestingly, is a one-point space as there is only one line, radial or otherwise, in $\mathbb{R}^1$. We'll try to convince the reader next, without being mathematically precise, that $\mathbb{P}^1$ is a circle.

Let U be the upper-half of a circle centered at the origin of $\mathbb{R}^2$. Associate with each radial line in $\mathbb{R}^2$ its intersection(s) with U. See Figure A.17, where, e.g., the radial line P is associated with the point p. Each radial line in $\mathbb{R}^2$ is then associated with a unique point of U, except for the x-axis, which we denote Q; Q intersects U in two points q_1 and q_2. And, the other way around, every point of U is associated with a unique radial line, except only for q_1 and q_2, which are associated with the same one Q. It follows, then, that the set $\mathbb{P}^1$ of all radial lines in $\mathbb{R}^2$ is in one-to-one correspondence with the space obtained by "identifying" the two endpoints q_1 and q_2 of U as one. But this latter space is clearly a circle (imagine U as a length of string whose ends are brought together).

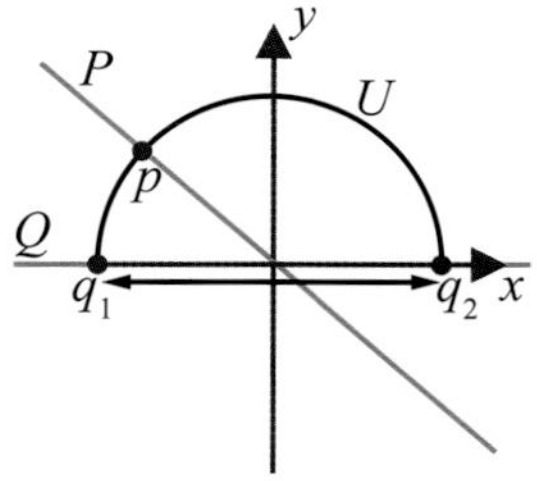

Figure A.17: Identifying $\mathbb{P}^1$ with a circle.

One can set up homogeneous coordinates for an arbitrary $\mathbb{P}^n$ in a manner similar to what we did for $\mathbb{P}^2$. For example, the homogeneous coordinates of a point $P \in \mathbb{P}^3$ are the coordinates of any point, other than the origin, on the radial line in $\mathbb{R}^4$ to which it corresponds. So the homogeneous coordinates of the point in $\mathbb{P}^3$ corresponding to the radial line through $[x\ y\ z\ w]^T$, where

x, y, z and w are not all zero, is any tuple of the form $[cx\ cy\ cz\ cw]^T$, where $c \neq 0$.

It's hard to visualize $\mathbb{P}^3$ and higher-dimensional projective spaces for the same reason that it's hard to visualize $\mathbb{R}^4$ and higher-dimensional real spaces. The trick is to develop one's intuition in $\mathbb{P}^2$, as many of its properties do generalize.

A.8 Projective Transformations Defined

That the homogeneous coordinates of a point $P \in \mathbb{P}^2$ are of the form $[x\ y\ z]^T$ suggests defining transformations of $\mathbb{P}^2$ by mimicking the definition of a linear transformation of real 3-space. In particular, if

$$M = \begin{bmatrix} a_{11} & a_{12} & a_{13} \\ a_{21} & a_{22} & a_{23} \\ a_{31} & a_{32} & a_{33} \end{bmatrix}$$

is a 3×3 matrix, then tentatively define a transformation of $\mathbb{P}^2$ by

$$[x\ y\ z]^T \mapsto M[x\ y\ z]^T \tag{A.5}$$

This definition has the virtue at least of being unambiguous because

$$[cx\ cy\ cz]^T \mapsto M[cx\ cy\ cz]^T = c(M[x\ y\ z]^T)$$

which represents the same point as $M[x\ y\ z]^T$, implying that the choice of any homogeneous coordinates for P gives the same image by the transformation.

The potential glitch to consider before putting (A.5) into production is if it maps a non-zero tuple to a zero tuple, for then it would map the homogeneous coordinates of a point $P \in \mathbb{P}^2$ to a value not even belonging to $\mathbb{P}^2$. However, we know from basic linear algebra that there is a non-zero tuple $[x\ y\ z]^T$ such that

$$M[x\ y\ z]^T = [0\ 0\ 0]^T$$

if and only if M is a singular matrix; otherwise, M maps non-zero tuples to non-zero tuples. We conclude that defining a transformation of $\mathbb{P}^2$ by (A.5) is indeed valid provided M is non-singular. Ergo:

Definition A.8. If M is a non-singular 3×3 matrix, then the transformation

$$[x\ y\ z]^T \mapsto M[x\ y\ z]^T$$

denoted h^M, is called a *projective transformation* of the projective plane. The transformation f^M of $\mathbb{R}^3$ – the linear transformation defined by M – is called a *related* linear transformation.

A simple relation between h^M and f^M is the following: if the radial line corresponding to a point P of $\mathbb{P}^2$ is l, then that corresponding to $h^M(P)$ is $f^M(l)$, the image of l by f^M.

Exercise A.25. Prove that if M is a non-singular 3×3 matrix and c is a scalar such that $c \neq 0$, then M and cM define the same projective transformation of $\mathbb{P}^2$, i.e., $h^M = h^{cM}$.

Remark A.8. The preceding exercise implies that actually there is not a unique linear transformation related to a projective transformation h^M, because f^{cM} is related to $h^{cM} = h^M$, for any non-zero c. However, when we do have a specific M that we are using to define h^M, then we'll often speak of *the* related linear transformation f^M.

Exercise A.26. Prove that a projective transformation h^M of $\mathbb{P}^2$ takes projective lines to projective lines.
Hint: The related (non-singular) linear transformation f^M takes radial planes in $\mathbb{R}^3$ to radial planes in $\mathbb{R}^3$.

Exercise A.27. Prove that the composition $h^M \circ h^N$ of two projective transformations of $\mathbb{P}^2$ is equal to the projective transformation h^{MN}.

A.9 Projective Transformations Geometrically

Our definition of projective transformations was purely algebraic. We would like to picture, if possible, how they transform primitives in $\mathbb{P}^2$. Now, projective primitives are "seen" by projection onto the real plane – by capture on a point camera's film as we've been putting it. Let's find out, then, what a projective transformation looks like through a point camera.

Here's what we plan to do. Start with a primitive s, on the plane $z = 1$, our favorite copy of $\mathbb{R}^2$, as the designated film. Suppose that the given projective transformation is h^M. Then we'll transform the lifting S of s by h^M to $h^M(S)$. Finally, we'll project $h^M(S)$ back to $z = 1$ to obtain a new primitive s'. It's precisely the change from s to s' which is seen as the transformation h^M by a point camera at the origin. For example, in Figure A.18, a boxy car is changed (fancifully) into a sleek convertible.

Back to reality, let's begin with a simple example. Consider a straight segment s joining two points p and q on $z = 1$. Given a projective transformation h^M, we want to determine s'. The lifting S of s, which is the set of all radial lines intersecting s, is not hard to visualize: it forms an "infinite double triangle" which lies on the radial plane containing s and the origin. See Figure A.19(a). The radial lines through p and q are denoted P and Q, respectively.

The related linear transformation f^M transforms s to a segment $\overline{s} = \overline{p}\,\overline{q}$, where $f^M(p) = \overline{p}$ and $f^M(q) = \overline{q}$. See Figure A.19(b). Note that $\overline{s}$ can be

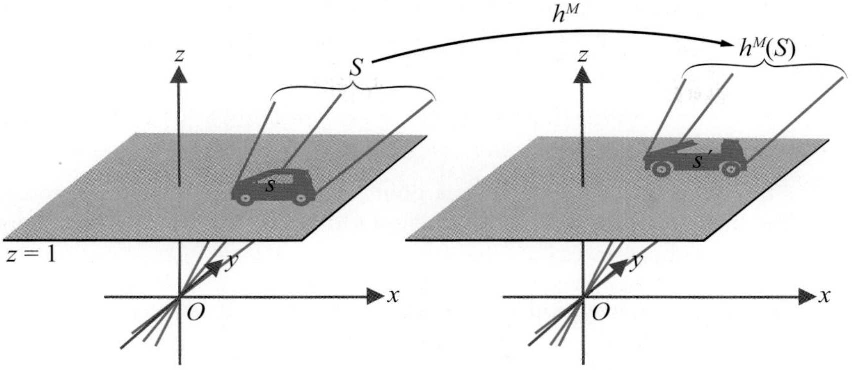

Figure A.18: Projective transformation of a car (purely conceptual!).

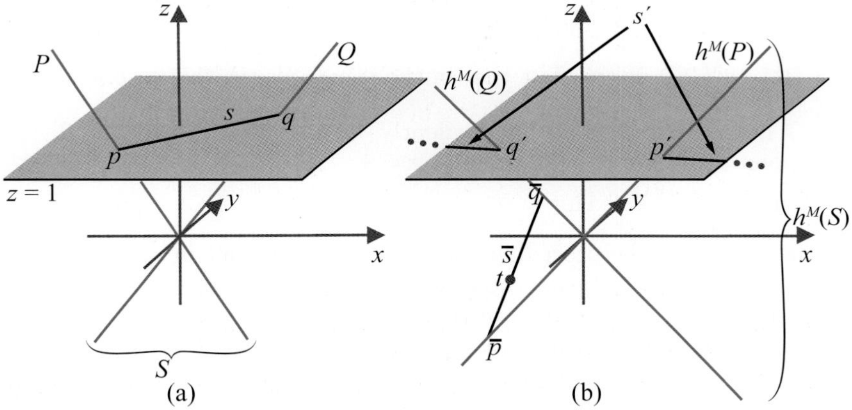

Figure A.19: (a) A segment s on $\mathbb{R}^2$ and its lifting S (b) f^M transforms s to $\overline{s}$ and S to $h^M(S)$, while s' is the intersection of $h^M(S)$ with $z = 1$.

anywhere in 3-space, depending on f^M, unlike s and s', which are both on $z = 1$.

Moreover, each radial line in S, the lifting of s, is transformed by f^M to a radial line in $h^M(S)$. Each radial line in $h^M(S)$, of course, intersects $\overline{s}$. A diagram depicting a particular disposition of $\overline{s}$, where it intersects the xy-plane in a single point t, is shown in Figure A.19(b).

The transformed primitive s' is the intersection of the radial lines in $h^M(S)$ with $z = 1$. At this time we ask the reader to complete the following exercise to find out for herself what it looks like, depending on the situation of $\overline{s}$.

Exercise A.28. Show that exactly one of (a)-(c) is true:

(a) $\overline{s}$ does not intersect the xy-plane, equivalently, every radial line in

$h^M(S)$ is a regular point with respect to $z = 1$.

In this case, s' is the segment between the points p' and q' where $h^M(P)$ and $h^M(Q)$, respectively, intersect $z = 1$ (remember that P and Q are the radial lines through p and q, the endpoints of s, respectively). Sketch this case.

(b) $\overline{s}$ intersects the xy-plane at one point, equivalently, exactly one radial line in $h^M(S)$ is a point at infinity with respect to $z = 1$. Now, there are two subcases:

(b1) If the intersection point, call it t, is in the interior of $\overline{s}$, then s' consists of the *entire* infinite straight line through p' and q', where $h^M(P)$ and $h^M(Q)$, respectively, intersect $z = 1$, *minus* the finite open segment between p' and q'. This situation is sketched in Figure A.19(b).

(b2) If the intersection is an endpoint of $\overline{s}$, say $\overline{p}$, then s' is a straight line infinite in one direction and with an endpoint at q', where $h^M(Q)$ intersects $z = 1$, in the other. Sketch this case.

(c) $\overline{s}$ lies on the xy-plane, equivalently, every radial line in $h^M(S)$ is a point at infinity with respect to $z = 1$.

In this case, s' is empty.

The answer to the preceding exercise is not tidy, but in most practical situations it will be case (a), the most benign of the three, which applies.

So we know now what we set out to find: how the projective transformation of the lifting of a segment looks like on film. Generally, for any primitive s on the plane, if s' is the "film-capture" of the transformation by h^M of the lifting of s, we'll call s' the *projective transformation* of s by h^M, and denote it $h^M(s)$ – giving thus a geometric counterpart of the algebraic definition of a projective transformation in Section A.8. Although h^M is well-defined, it is not a transformation of $\mathbb{R}^2$ in general because $h^M(p)$ may not even exist for a point $p \in \mathbb{R}^2$, particularly if p's corresponding projective point is taken by h^M to a point at infinity (which has no film-capture).

In our usage, therefore, h^M can represent either a transformation of projective space (as defined in Section A.8) or a transformation of real primitives (as just defined above). There is no danger of ambiguity as the nature of the argument in $h^M(*)$ will make clear how it's being used.

Example A.12. The segment s joins $p = [1 \ -1]^T$ and $q = [-2 \ -2]^T$ on the plane $z = 1$, the latter identified with $\mathbb{R}^2$. The projective transformation $h^M : \mathbb{P}^2 \to \mathbb{P}^2$ is specified by

$$M = \begin{bmatrix} 0 & 0 & -1 \\ 0 & 1 & 0 \\ 1 & 0 & 0 \end{bmatrix}$$

which is the matrix corresponding to a rotation f^M of $\mathbb{R}^3$ by 90° about the y-axis, clockwise when seen from the positive side of the y-axis. Determine $h^M(s)$.

Answer: f^M transforms s to the segment $\overline{s} = \overline{p}\,\overline{q}$, where $\overline{p}$ and $\overline{q}$ are the images by f^M of p and q, respectively. Multiplying p and q, written as points of $z = 1$, on the left by M we get:

$$\overline{p} = M[1 \ -1 \ 1]^T = [-1 \ -1 \ 1]^T$$

and

$$\overline{q} = M[-2 \ -2 \ 1]^T = [-1 \ -2 \ -2]^T$$

As the z-values of $\overline{p}$ and $\overline{q}$ are of different signs, an interior point of $\overline{s}$ lies on the xy-plane. Therefore, we are in case (b1) of Exercise A.28 above.

Let P and Q denote the radial lines through p and q, respectively. The radial line $h^M(P)$ through $\overline{p}$ meets $z = 1$ at $h^M(p) = [-1 \ -1 \ 1]^T$, which is $\overline{p}$ itself. The radial line $h^M(Q)$ through $\overline{q}$ meets $z = 1$ at $h^M(q) = [\frac{1}{2} \ 1 \ 1]^T$ (multiplying the coordinate tuple of $\overline{q}$ by $-\frac{1}{2}$ to make its z-value equal to 1).

Applying Exercise A.28 case (b1), $h^M(s)$ is the entire straight line through the points $[-1 \ -1]^T$ and $[\frac{1}{2} \ 1]^T$ minus the finite open segment joining $[-1 \ -1]^T$ to $[\frac{1}{2} \ 1]^T$.

Example A.13. The rectangle r lies on the plane $z = 1$, the latter identified with $\mathbb{R}^2$. Its vertices are $p_1 = [0.5 \ 1]^T$, $p_2 = [0.5 \ -1]^T$, $p_3 = [1 \ -1]^T$ and $p_4 = [1 \ 1]^T$. See Figure A.20. Determine $h^M(r)$, where h^M is the same projective transformation as in the preceding example.

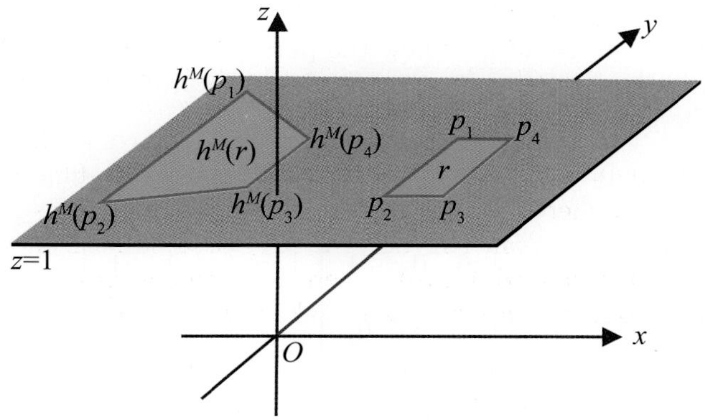

Figure A.20: Rectangle r is transformed to the trapezoid $h^M(r)$.

Answer: f^M transforms r to the rectangle $\overline{r}$ with vertices $\overline{p_i} = f^M(p_i)$, $1 \leq i \leq 4$. Multiplying each p_i, written as points of $z = 1$, on the left by M

we get:

$$\begin{aligned}
\overline{p_1} &= M[0.5\ 1\ 1]^T = [-1\ 1\ 0.5]^T \\
\overline{p_2} &= M[0.5\ -1\ 1]^T = [-1\ -1\ 0.5]^T \\
\overline{p_3} &= M[1\ -1\ 1]^T = [-1\ -1\ 1]^T \\
\overline{p_4} &= M[1\ 1\ 1]^T = [-1\ 1\ 1]^T
\end{aligned}$$

As the z-value of every $\overline{p_i}$, $1 \leq i \leq 4$, is greater than 0, none of the edges of $\overline{r}$ intersects the xy-plane. According to case (a) of Exercise A.28 then, $h^M(r)$ is the quadrilateral with vertices at the points $h^M(p_i)$, where the radial lines through $\overline{p_i}$, $1 \leq i \leq 4$, intersect $z = 1$. See Figure A.20. Multiply the coordinate tuple of each $\overline{p_i}$ by a scalar to make its z-value equal to 1, to find that

$$\begin{aligned}
h^M(p_1) &= [-2\ 2\ 1]^T \\
h^M(p_2) &= [-2\ -2\ 1]^T \\
h^M(p_3) &= [-1\ -1\ 1]^T \\
h^M(p_4) &= [-1\ 1\ 1]^T
\end{aligned}$$

One sees, therefore, that $h^M(r)$ has vertices at $[-2\ 2]^T$, $[-2\ -2]^T$, $[-1\ -1]^T$ and $[-1\ 1]^T$, which makes it a trapezoid.

It's interesting to note that no affine transformation of $\mathbb{R}^2$ can map a rectangle to a trapezoid: as affine transformations preserve parallelism (see Proposition 5.1), at most they can transform a rectangle to a parallelogram.

Clearly, with the help of Exercise A.28 we can determine the projective transformation of any shape specified by straight edges. More general shapes are curved and curves specified by equations. Let's see, for example, how a parabola is projectively transformed.

Example A.14. Determine how the parabola $y - x^2 = 0$ on $z = 1$, the latter identified with $\mathbb{R}^2$, is mapped by the same projective transformation h^M as in the previous example.

Answer: The point $[x\ y]^T$ on $z = 1$, which has coordinates $[x\ y\ 1]^T$ in $\mathbb{R}^3$, is transformed by f^M to the point $[\overline{x}\ \ \overline{y}\ \ \overline{z}]^T$, where

$$\begin{bmatrix} \overline{x} \\ \overline{y} \\ \overline{z} \end{bmatrix} = \begin{bmatrix} 0 & 0 & -1 \\ 0 & 1 & 0 \\ 1 & 0 & 0 \end{bmatrix} \begin{bmatrix} x \\ y \\ 1 \end{bmatrix} = \begin{bmatrix} -1 \\ y \\ x \end{bmatrix}$$

which gives

$$\overline{x} = -1, \qquad \overline{y} = y, \qquad \overline{z} = x$$

The image $[x'\ y']^T$ of $[x\ y]^T$ by h^M, then, is the point $[\overline{x}/\overline{z}\ \ \overline{y}/\overline{z}]^T$, where the radial line through $[\overline{x}\ \overline{y}\ \overline{z}]^T$ intersects $z = 1$. Therefore:

$$x' = \overline{x}/\overline{z} = -1/x \implies x = -1/x'$$

and

$$y' = \overline{y}/\overline{z} = y/x \quad \Longrightarrow \quad y = y'x = -y'/x' \text{ (using } x = -1/x' \text{ from above)}$$

Plugging these expressions for x and y into the equation of the parabola $y - x^2 = 0$, we have the equation

$$-y'/x' - 1/x'^2 = 0, \text{ equivalently, } \quad x'y' + 1 = 0$$

of the transformed curve, which describes a hyperbola.

Here's another rather interesting example.

Example A.15. Determine how points of $\mathbb{R}^2$, identified with $z = 1$, are transformed by the projective transformation h^M of $\mathbb{P}^2$ specified by

$$M = \begin{bmatrix} 1 & 0 & 7 \\ 0 & 1 & 0 \\ 0 & 0 & 1 \end{bmatrix}$$

Answer: The point $[x\ y]^T$ on $z = 1$, which has coordinates $[x\ y\ 1]^T$ in $\mathbb{R}^3$, is transformed by f^M to the point $[\overline{x}\ \overline{y}\ \overline{z}]^T$, where

$$\begin{bmatrix} \overline{x} \\ \overline{y} \\ \overline{z} \end{bmatrix} = \begin{bmatrix} 1 & 0 & 7 \\ 0 & 1 & 0 \\ 0 & 0 & 1 \end{bmatrix} \begin{bmatrix} x \\ y \\ 1 \end{bmatrix} = \begin{bmatrix} x+7 \\ y \\ 1 \end{bmatrix}$$

giving

$$\overline{x} = x + 7, \qquad \overline{y} = y, \qquad \overline{z} = 1$$

The image $[x'\ y']^T$ of $[x\ y]^T$ by h^M, then, is the point $[\overline{x}/\overline{z}\ \ \overline{y}/\overline{z}\ \ 1]^T$, where the radial line through $[\overline{x}\ \overline{y}\ \overline{z}]^T$ intersects $z = 1$. Therefore,

$$x' = \overline{x}/\overline{z} = x + 7 \quad \text{and} \quad y' = \overline{y}/\overline{z} = y$$

which is nothing but a *translation* by 7 units in the x-direction.

A projection transformation has just done something beyond the reach of linear transformations, for a linear transformation cannot translate. Translations, as we learned in Chapter 5, are in the domain of affine transformations. In fact, in Example A.13, we saw a projective transformation convert a rectangle into a trapezoid, something beyond even affine transformations. For transformations inspired by and defined by matrix-vector multiplication, just like linear transformations, projective transformations certainly seem to carry plenty of firepower. It turns out that this makes them particularly worthy allies in the advancement of computer graphics.

Incidentally, we did not pull the matrix M above out of a hat: it is the transformation matrix of a 3D shear whose plane is the xy-plane and line the x-axis (recall 3D shears from Section 5.4.8).

Exercise A.29. Find a projective transformation to translate points of $\mathbb{R}^2$ 3 units in the x-direction and 2 in the y-direction, i.e., whose displacement vector is $[3\ 2]^T$.
Hint: Think another shear.

Exercise A.30. Determine how the segment s on $\mathbb{R}^2$, the latter identified with the plane $z = 1$, joining $p = [2\ \ -2]^T$ and $q = [-2\ 1]^T$, is mapped by the projective transformation h^M of $\mathbb{P}^2$ specified by

$$M = \begin{bmatrix} 0 & 1 & 1 \\ 1 & 0 & 1 \\ 1 & 1 & 0 \end{bmatrix}$$

Exercise A.31. Determine how the hyperbola $xy = 1$ on $z = 1$, the latter identified with $\mathbb{R}^2$, is mapped by the same projective transformation h^M as in the previous exercise.

Part answer: The problem is not hard but there is a fair amount of manipulation.

The point $[x\ y]^T$ on $z = 1$, which has coordinates $[x\ y\ 1]^T$ in $\mathbb{R}^3$, is transformed by f^M to the point $[\overline{x}\ \overline{y}\ \overline{z}]^T$, where

$$\begin{bmatrix} \overline{x} \\ \overline{y} \\ \overline{z} \end{bmatrix} = \begin{bmatrix} 0 & 1 & 1 \\ 1 & 0 & 1 \\ 1 & 1 & 0 \end{bmatrix} \begin{bmatrix} x \\ y \\ 1 \end{bmatrix}$$

Let's flip this equation over with the help of an inverse matrix:

$$\begin{bmatrix} x \\ y \\ 1 \end{bmatrix} = \begin{bmatrix} 0 & 1 & 1 \\ 1 & 0 & 1 \\ 1 & 1 & 0 \end{bmatrix}^{-1} \begin{bmatrix} \overline{x} \\ \overline{y} \\ \overline{z} \end{bmatrix} = \frac{1}{2} \begin{bmatrix} -1 & 1 & 1 \\ 1 & -1 & 1 \\ 1 & 1 & -1 \end{bmatrix} \begin{bmatrix} \overline{x} \\ \overline{y} \\ \overline{z} \end{bmatrix}$$

which gives

$$x = \frac{1}{2}(-\overline{x} + \overline{y} + \overline{z}) \qquad y = \frac{1}{2}(\overline{x} - \overline{y} + \overline{z}) \qquad 1 = \frac{1}{2}(\overline{x} + \overline{y} - \overline{z})$$

Plugging these expressions into the equation of the hyperbola $xy = 1 = 1^2$ we get:

$$\frac{1}{4}(-\overline{x} + \overline{y} + \overline{z})(\overline{x} - \overline{y} + \overline{z}) = \frac{1}{4}(\overline{x} + \overline{y} - \overline{z})^2$$

Now, the image $[x'\ y']^T$ of $[x\ y]^T$ by h^M is the point $[\overline{x}/\overline{z}\ \ \overline{y}/\overline{z}]^T$, where the radial line through $[\overline{x}\ \overline{y}\ \overline{z}]^T$ intersects $z = 1$. We ask the reader to complete the exercise by dividing the preceding equation by $\overline{z}^2$ throughout to obtain an equation relating x' and y', and identifying the corresponding curve.

Exercise A.32. Determine how the straight line $x + y + 1 = 0$ on $z = 1$ is mapped by the same projective transformation h^M as in the previous exercise.

Exercise A.33. We saw in Example 5.4 that affine transformations preserve convex combinations and barycentric coordinates. Show that projective transformations in general do not.

Remark A.9. Projective transformations of $\mathbb{P}^2$ can be thought of as a powerful class of *pseudo-transformations* of $\mathbb{R}^2$ – pseudo because a projective transformation may map a regular point to a point at infinity, in which case the corresponding point of $\mathbb{R}^2$ has no valid image. If one is careful, however, to restrict its domain to a region of $\mathbb{R}^2$ where it *is* valid throughout, one may be able to exploit the ability of a projective transformation to do more than an affine one.

A.10 Relating Projective, Snapshot and Affine Transformations

We'll explore in this section the inter-relationships between projective, snapshot and affine transformations.

A.10.1 Snapshot Transformations via Projective Transformations

Snapshot transformations, being transformations of an object seen through a point camera as the film changes alignment, are geometrically intuitive. They are, in fact, a kind of projective transformation, as we'll now see.

Consider again Example A.13 for motivation. We saw that the rectangle r on the plane $z = 1$ (aka $\mathbb{R}^2$) with vertices at $p_1 = [0.5\ 1]^T$, $p_2 = [0.5\ -1]^T$, $p_3 = [1\ -1]^T$ and $p_4 = [1\ 1]^T$ is mapped to the trapezoid $r' = h^M(r)$ with vertices at $p'_1 = [-2\ 2]^T$, $p'_2 = [-2\ -2]^T$, $p'_3 = [-1\ -1]^T$ and $p'_4 = [-1\ 1]^T$, by the projective transformation h^M specified by

$$M = \begin{bmatrix} 0 & 0 & -1 \\ 0 & 1 & 0 \\ 1 & 0 & 0 \end{bmatrix}$$

See Figure A.21(a).

We observed, as well, that the related linear transformation f^M is a rotation of $\mathbb{R}^3$ by $90°$ about the y-axis, which is clockwise when seen from the positive side of the y-axis.

Denote the radial line through p_i by P_i, $1 \le i \le 4$, and their respective images $h^M(P_i)$ by P'_i. Now rotate the radial lines P'_i, *as well as* the plane $z = 1$, an angle of $90°$ about the y-axis, this time counter-clockwise when

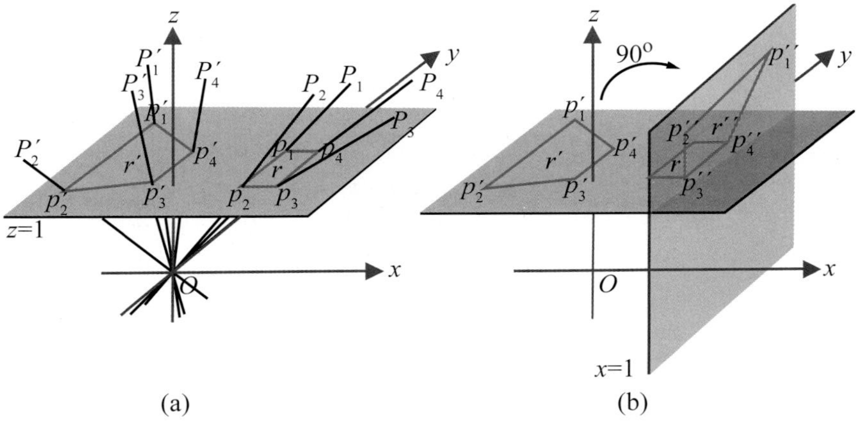

Figure A.21: (a) Projective transformation h^M maps rectangle r to trapezoid $r' = h^M(r)$ (b) r' is the "same" as r'', the picture of r captured on a film along $x = 1$.

seen from the positive side of the y-axis, in order to undo the effect of f^M; in other words, apply $f^{M^{-1}}$. We see the following:

(a) The radial line P_i', of course, rotates back onto (its pre-image) the radial line P_i, $1 \leq i \leq 4$.

(b) The plane $z = 1$ is taken by the rotation onto the plane $x = 1$.

(c) The trapezoid r', as a consequence of (a) and (b), rotates onto a trapezoid r'' with vertices at the intersections p_i'' of P_i with $x = 1$, for $1 \leq i \leq 4$. See Figure A.21(b) (note that the edge of r that happens to lie on the intersection of the planes $z = 1$ and $x = 1$ is shared with r'').

But r'' is precisely the snapshot transformation of r from the film along $z = 1$ to the one along $x = 1$! Here's what is happening. The image r' is obtained by applying the rotation f^M to the radials P_i and intersecting them with the plane $z = 1$, while r'' is obtained from r' by applying the reverse rotation $f^{M^{-1}}$, which takes the radials back to the where they were, and, at the same time, changes the intersecting plane from $z = 1$ to $x = 1$. Therefore, the transformation from r to r'' comes from a change in the plane (= film) intersecting the radials, which is precisely a snapshot transformation.

One sees, therefore, that, generally, a snapshot transformation in which the film is re-aligned by a rotation f about a radial axis is equivalent to a projective transformation whose related linear transformation is f^{-1}, in that the images are identical, though situated differently in space (precisely, the two images differ by a rigid transformation of $\mathbb{R}^3$). But, how about snapshot transformations where the new alignment of the film cannot be obtained from the original by mere rotation? To answer this question, we ask the reader, first, to prove the following, which says that an arbitrary snapshot transformation can be composed from two very simple ones.

Exercise A.34. Prove that any plane p in $\mathbb{R}^3$ can be aligned with any other p' by a translation parallel to itself followed by a rotation about a radial axis.

Therefore, any snapshot transformation is the composition of two: first, a snapshot transformation from one film to a parallely translated one and then another, where one film is obtained from the other by a rotation about a radial axis.
Hint: See Figure A.22.

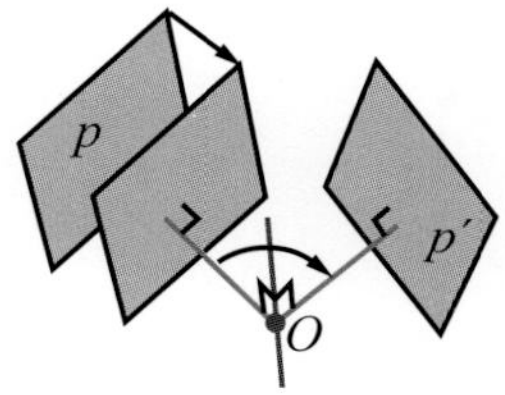

Figure A.22: Aligning plane p with p' by a parallel displacement, so that their respective distances from the origin are equal, followed by a rotation.

We have already seen how a snapshot transformation from one film to a rotated one is equivalent to a projective transformation. A snapshot transformation to a parallely translated one is also equivalent to a projective transformation, as the next exercise asks the reader to show.

Exercise A.35. Suppose that two parallel non-radial planes p and p' in $\mathbb{R}^3$ are at a distance of c and c' from the origin, respectively. Then the snapshot transformation from p to p' is equivalent to the projective transformation h^M, where

$$M = \begin{bmatrix} \frac{c'}{c} & 0 & 0 \\ 0 & \frac{c'}{c} & 0 \\ 0 & 0 & \frac{c'}{c} \end{bmatrix} = \frac{c'}{c} I$$

(i.e., a projective transformation whose related linear transformation is a *uniform* scaling of $\mathbb{R}^3$ by a factor of $\frac{c'}{c}$ in all directions).
Hint: See Figure A.23.

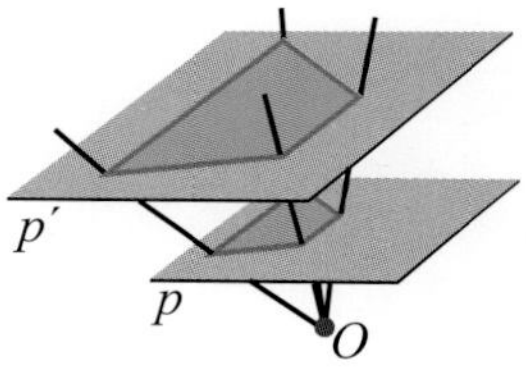

Figure A.23: A snapshot transformation to a parallel plane is equivalent to a scaling by a constant factor in all directions.

Putting the pieces together we have the following proposition:

Proposition A.1. *A snapshot transformation k from a non-radial plane p in $\mathbb{R}^3$ to another p' is equivalent to a projective transformation h^M of $\mathbb{P}^2$, in the sense that the images of primitives by k and h^M are identical modulo a rigid transformation of $\mathbb{R}^3$.*

In particular, k is equivalent to the projective transformation h^M which is the composition of a projective transformation h^{dI}, whose related linear transformation is a uniform scaling, with a projective transformation h^N, whose related linear transformation is a rotation of $\mathbb{R}^3$ about a radial axis.

In other words, k is equivalent to h^{dN}, where d is a scalar and N is the matrix of a rotation of $\mathbb{R}^3$ about a radial axis. □

Exercise A.36. Determine the projective transformation equivalent to the snapshot transformation from the plane $z = 1$ to the plane $x = 2$.

A.10.2 Affine Transformations via Projective Transformations

We begin by asking if there exist projective transformations that respect regular points, i.e., map regular points to regular points. Such a transformation could then be *entirely* captured on film because it takes no point of the film out of it, as would happen, say, if a regular point were mapped to one at infinity. Looking back at Remark A.9, one could then say that such a projective transformation is no longer "pseudo", but a true transformation of $\mathbb{R}^2$.

So suppose the film lies along the plane (surprise) $z = 1$. What condition must a projective transformation h^M, where

$$M = \begin{bmatrix} a_{11} & a_{12} & a_{13} \\ a_{21} & a_{22} & a_{23} \\ a_{31} & a_{32} & a_{33} \end{bmatrix}$$

satisfy in order to transform each point regular with respect to $z = 1$ to another such? Homogeneous coordinates of regular points are of the form $[x\ y\ 1]^T$. Now,

$$h^M([x\ y\ 1]^T) = M[x\ y\ 1]^T = \begin{bmatrix} a_{11}x + a_{12}y + a_{13} \\ a_{21}x + a_{22}y + a_{23} \\ a_{31}x + a_{32}y + a_{33} \end{bmatrix}$$

For this image point to be regular we must have

$$a_{31}x + a_{32}y + a_{33} \neq 0$$

However, if either one of a_{31} and a_{32} is non-zero, or if a_{33} is zero, then it's possible to find values of x and y such that $a_{31}x + a_{32}y + a_{33} = 0$. The conclusion then is that for h^M to transform all regular points to regular points, one must have both a_{31} and a_{32} equal to zero and a_{33} non-zero. Therefore, M must be of the form

$$\begin{bmatrix} a_{11} & a_{12} & a_{13} \\ a_{21} & a_{22} & a_{23} \\ 0 & 0 & a_{33} \end{bmatrix}$$

with $a_{33} \neq 0$. By Exercise A.25, M can be multiplied by $1/a_{33}$ to still represent the same projective transformation, so one can assume $a_{33} = 1$, implying that the form of M is

$$\begin{bmatrix} a_{11} & a_{12} & a_{13} \\ a_{21} & a_{22} & a_{23} \\ 0 & 0 & 1 \end{bmatrix}$$

In this case, h^M transforms $[x\ y\ 1]^T$ to

$$\begin{bmatrix} a_{11} & a_{12} & a_{13} \\ a_{21} & a_{22} & a_{23} \\ 0 & 0 & 1 \end{bmatrix} \begin{bmatrix} x \\ y \\ 1 \end{bmatrix} = \begin{bmatrix} a_{11}x + a_{12}y + a_{13} \\ a_{21}x + a_{22}y + a_{23} \\ 1 \end{bmatrix}$$

Tossing the last coordinate, it transforms $[x\ y]^T \in \mathbb{R}^2$ to

$$\begin{bmatrix} a_{11}x + a_{12}y + a_{13} \\ a_{21}x + a_{22}y + a_{23} \end{bmatrix} = \begin{bmatrix} a_{11} & a_{12} \\ a_{21} & a_{22} \end{bmatrix} \begin{bmatrix} x \\ y \end{bmatrix} + \begin{bmatrix} a_{13} \\ a_{23} \end{bmatrix}$$

which is precisely an affine transformation!

We conclude that a projective transformation of $\mathbb{P}^2$ that respects regular points gives nothing but an affine transformation of $\mathbb{R}^2$. Conversely, it's not hard to see that any affine transformation of $\mathbb{R}^2$ can be obtained as a projective transformation preserving regular points. We record these facts in the following proposition.

Proposition A.2. *An affine transformation of $\mathbb{R}^2$ is equivalent to a projective transformation of $\mathbb{P}^2$, in particular, one that respects regular points.*

Conversely, a projective transformation of $\mathbb{P}^2$ that respects regular points is equivalent to an affine transformation of $\mathbb{R}^2$.

Evidently, the constraint to respect regular points is a burden on projective transformations. It dumbs them down to affine and all the excitement of parallel lines turning into intersecting ones, rectangles into trapezoids, and circles into hyperbolas is lost!

However, one does see now a good reason for the use of homogeneous coordinates of real points in computing affine transformations. When first we did this in Section 5.2.3, it seemed merely a neat maneuver to obtain an affine transformation as a single matrix-vector multiplication. The bigger picture is that affine transformations are a subclass of the projective. Therefore, as the latter are obtained (by definition) from matrix-vector multiplication, so can the former, provided we relocate to projective space, in other words, use homogeneous coordinates.

A Roundup of the Three Kinds of Transformations

Snapshot and affine transformations are subclasses of the projective, as we have just seen. How about the relationship between these two subclasses themselves? Are snapshot transformations affine or affine transformations snapshot?

At the start of Section A.10.1 we saw a projective transformation, equivalent, in fact, to a snapshot transformation, map a rectangle to a trapezoid. This is not possible for an affine transformation to do, as it is obliged to preserve parallelism (Proposition 5.1). Therefore, snapshot transformations are certainly not all affine.

A shear on the plane, an affine transformation, can map a rectangle to a non-rectangular parallelogram. We leave the reader to convince herself that this is not possible for a snapshot transformation. So not all affine transformations are snapshot.

We see then that neither of the two subclasses, snapshot and affine, of projective transformations contains the other. However, what transformations, if any, do they have in common? We ask the reader herself to characterize the transformations at the intersection of affine and snapshot in the next exercise.

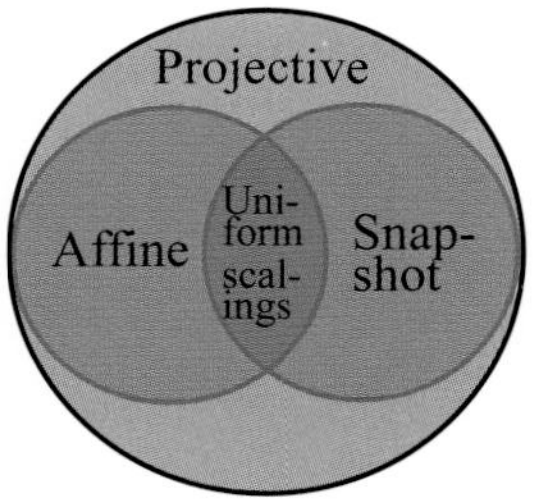

Figure A.24: Venn diagram of transformation classes of $\mathbb{R}^2$.

Exercise A.37. Prove that projective transformations which are both affine and snapshot are precisely those whose related linear transformation is a uniform scaling.

The final important question on the relationship between the three classes is if the union of snapshot and affine covers projective transformations or if the latter is strictly bigger. In Exercise A.40 in the next section we'll see an example of a projective transformation neither snapshot nor affine. Therefore, indeed, the class of projective transformations is strictly bigger than the union of snapshot and affine. Figure A.24 summarizes the relationship between the three classes.

A.11 Designer Projective Transformations

We know from elementary linear algebra that a linear transformation is uniquely specified by defining its values on a basis. Here's a like-minded claim for projective transformations of $\mathbb{P}^2$.

Proposition A.3. *If two sets $\{P_1, P_2, P_3, P_4\}$ and $\{Q_1, Q_2, Q_3, Q_4\}$ of four points each from $\mathbb{P}^2$ are such that no three in any one set are collinear, then there is a unique projective transformation of $\mathbb{P}^2$ that maps P_i to Q_i, for $1 \leq i \leq 4$.*

Proof. Choose non-zero vectors p_1, p_2, p_3 and p_4 from $\mathbb{R}^3$ lying on P_1, P_2, P_3 and P_4, respectively, and non-zero vectors q_1, q_2, q_3 and q_4 lying on Q_1, Q_2, Q_3 and Q_4, respectively.

Since P_1, P_2 and P_3 do not lie on one projective line, p_1, p_2 and p_3 do not lie on one radial plane. The latter three form, therefore, a basis of $\mathbb{R}^3$. Likewise, q_1, q_2 and q_3 form a basis of $\mathbb{R}^3$ as well.

Let c_1, c_2 and c_3 be arbitrary scalars, all three non-zero, whose values will be determined. As q_1, q_2 and q_3 form a basis of $\mathbb{R}^3$, so do c_1q_1, c_2q_2 and c_3q_3. Therefore, there is a unique non-singular linear transformation $f^M : \mathbb{R}^3 \to \mathbb{R}^3$ such that

$$f^M(p_i) = c_iq_i, \text{ for } 1 \leq i \leq 3$$

which, then, is related to a projective transformation $h^M : \mathbb{P}^2 \to \mathbb{P}^2$, such that

$$h^M(P_i) = Q_i, \text{ for } 1 \le i \le 3$$

It remains to make $h^M(P_4) = Q_4$.

As p_1, p_2 and p_3 form a basis of $\mathbb{R}^3$, there exist unique scalars α, β and γ such that

$$p_4 = \alpha p_1 + \beta p_2 + \gamma p_3$$

Now, α, β and γ are all three non-zero, for, otherwise, p_4 lies on the same radial plane as two of p_1, p_2 and p_3, which implies that P_4 lies on the same projective line as two of P_1, P_2 and P_3, contradicting an initial hypothesis. Likewise, there exist unique non-zero scalars, λ, μ and ν such that

$$q_4 = \lambda q_1 + \mu q_2 + \nu q_3$$

For

$$h^M(P_4) = Q_4$$

to hold, then, one requires a scalar $c_4 \neq 0$ such that

$$\begin{aligned} f^M(p_4) &= c_4 q_4 \\ &= c_4(\lambda q_1 + \mu q_2 + \nu q_3) \\ &= \lambda c_4 q_1 + \mu c_4 q_2 + \nu c_4 q_3 \end{aligned} \tag{A.6}$$

However,

$$\begin{aligned} f^M(p_4) &= f^M(\alpha p_1 + \beta p_2 + \gamma p_3) \\ &= \alpha f^M(p_1) + \beta f^M(p_2) + \gamma f^M(p_3) \\ &= \alpha c_1 q_1 + \beta c_2 q_2 + \gamma c_3 q_3 \end{aligned} \tag{A.7}$$

Combining (A.6) and (A.7) one has

$$\alpha c_1 q_1 + \beta c_2 q_2 + \gamma c_3 q_3 = \lambda c_4 q_1 + \mu c_4 q_2 + \nu c_4 q_3$$

As q_1, q_2 and q_3 is a basis of $\mathbb{R}^3$, it follows that

$$\alpha c_1 = \lambda c_4, \qquad \beta c_2 = \mu c_4, \qquad \gamma c_3 = \nu c_4$$

giving

$$c_1 = (\lambda/\alpha) c_4, \qquad c_2 = (\mu/\beta) c_4, \qquad c_3 = (\nu/\gamma) c_4$$

determining c_1, c_2, c_3 and c_4 uniquely, up to a constant of proportionality, so completing the proof. □

The following corollary, which is a straightforward application of the proposition, is particularly important.

Corollary A.1. *Any non-degenerate quadrilateral, i.e., one with no three collinear vertices, in* $\mathbb{R}^2$ *can be projectively transformed to any other such.*
□

More than just theoretically, the proposition is important in that it suggests how to go about finding projective transformations specified at only a few points.

Example A.16. Determine the projective transformation h^M of $\mathbb{P}^2$ mapping the projective points

$$P_1 = [1\ 0\ 0]^T,\ P_2 = [0\ 1\ 0]^T,\ P_3 = [0\ 0\ 1]^T \text{ and } P_4 = [1\ 1\ 1]^T$$

to the respective images

$$Q_1 = [2\ 1\ 3]^T,\ Q_2 = [-1\ -1\ 1]^T,\ Q_3 = [0\ 1\ 1]^T \text{ and } Q_4 = [0\ 0\ 6]^T$$

Answer: Choose (not particularly imaginatively)

$$p_1 = [1\ 0\ 0]^T,\ p_2 = [0\ 1\ 0]^T,\ p_3 = [0\ 0\ 1]^T \text{ and } p_4 = [1\ 1\ 1]^T$$

in $\mathbb{R}^3$ lying on P_i, $1 \le i \le 4$, and

$$q_1 = [2\ 1\ 3]^T,\ q_2 = [-1\ -1\ 1]^T,\ q_3 = [0\ 1\ 1]^T \text{ and } q_4 = [0\ 0\ 6]^T$$

lying on Q_i, $1 \le i \le 4$.

The linear transformation $f^M : \mathbb{R}^3 \to \mathbb{R}^3$ such that $f^M(p_i) = c_i q_i$, for $1 \le i \le 3$, where c_1, c_2 and c_3 are non-zero scalars, is easily verified to be given by

$$M = \begin{bmatrix} 2c_1 & -c_2 & 0 \\ c_1 & -c_2 & c_3 \\ 3c_1 & c_2 & c_3 \end{bmatrix}$$

One can verify as well that

$$p_4 = p_1 + p_2 + p_3$$

and

$$q_4 = q_1 + 2q_2 + q_3$$

Therefore,

$$f^M(p_4) = f^M(p_1+p_2+p_3) = f^M(p_1)+f^M(p_2)+f^M(p_3) = c_1q_1+c_2q_2+c_3q_3$$

Accordingly, if $f^M(p_4) = c_4 q_4$, for some $c_4 \neq 0$, then

$$c_1q_1 + c_2q_2 + c_3q_3 = c_4(q_1 + 2q_2 + q_3) = c_4q_1 + 2c_4q_2 + c_4q_3$$

which implies that

$$c_1 = c_4, \quad c_2 = 2c_4, \quad c_3 = c_4$$

Setting $c_4 = 1$, one has: $c_1 = 1,\ c_2 = 2,\ c_3 = 1,\ c_4 = 1$. One concludes that the required projective transformation h^M is given by

$$M = \begin{bmatrix} 2 & -2 & 0 \\ 1 & -2 & 1 \\ 3 & 2 & 1 \end{bmatrix}$$

The following example will help in an application of projective transformations in the graphics pipeline.

Example A.17. Determine the projective transformation h^M of $\mathbb{P}^2$ that transforms the trapezoid q on the plane $z = 1$ (aka $\mathbb{R}^2$) with vertices at

$$p_1 = [-1\ 1]^T,\ p_2 = [1\ 1]^T,\ p_3 = [2\ 2]^T \text{ and } p_4 = [-2\ 2]^T$$

to the rectangle q' on the same plane with vertices at

$$p_1' = [-1\ 1]^T,\ p_2' = [1\ 1]^T,\ p_3' = [1\ 2]^T \text{ and } p_4' = [-1\ 2]^T$$

See Figure A.25.

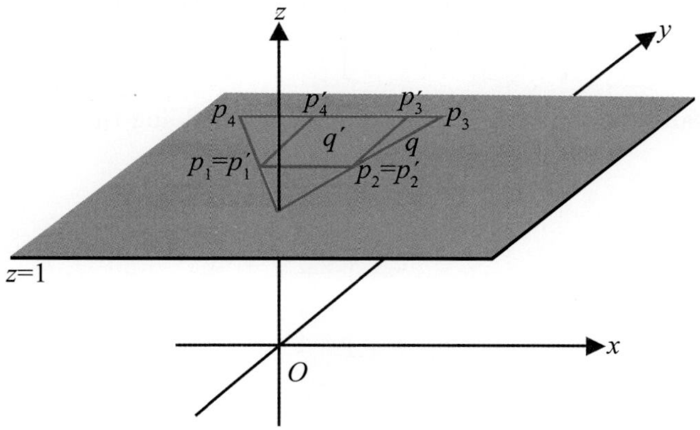

Figure A.25: Transforming the trapezoid q on $z = 1$ to the rectangle (bold) q'.

Answer: Suppose that the required projective transformation h^M is defined by the matrix

$$M = \begin{bmatrix} a_{11} & a_{12} & a_{13} \\ a_{21} & a_{22} & a_{23} \\ a_{31} & a_{32} & a_{33} \end{bmatrix}$$

We have to determine the a_{ij} up to a non-zero multiplicative constant.

The two sides p_1p_4 and p_2p_3 of the trapezoid q meet at the regular point (with respect to $z = 1$) $[0\ 0\ 1]^T$, while the corresponding sides $p_1'p_4'$ and $p_2'p_3'$

of the rectangle q' are parallel and meet at the point at infinity $[0\ 1\ 0]^T$. The transformation must, therefore, map $[0\ 0\ 1]^T$ to $[0\ 1\ 0]^T$, yielding our first equation

$$h^M([0\ 0\ 1]^T) = [0\ 1\ 0]^T$$

(the RHS could be $c[0\ 1\ 0]^T$ for any non-zero scalar c, but there's no loss in assuming that $c = 1$) which translates to the matrix equation

$$\begin{bmatrix} a_{11} & a_{12} & a_{13} \\ a_{21} & a_{22} & a_{23} \\ a_{31} & a_{32} & a_{33} \end{bmatrix} \begin{bmatrix} 0 \\ 0 \\ 1 \end{bmatrix} = \begin{bmatrix} 0 \\ 1 \\ 0 \end{bmatrix}$$

giving

$$a_{13} = 0, \qquad a_{23} = 1, \qquad a_{33} = 0$$

So we write

$$M = \begin{bmatrix} a_{11} & a_{12} & 0 \\ a_{21} & a_{22} & 1 \\ a_{31} & a_{32} & 0 \end{bmatrix}$$

That we have $h^M(p_1) = p_1'$ and $h^M(p_2) = p_2'$ gives two more matrix equations

$$\begin{bmatrix} a_{11} & a_{12} & 0 \\ a_{21} & a_{22} & 1 \\ a_{31} & a_{32} & 0 \end{bmatrix} \begin{bmatrix} -1 \\ 1 \\ 1 \end{bmatrix} = \begin{bmatrix} -c \\ c \\ c \end{bmatrix} \quad \text{and} \quad \begin{bmatrix} a_{11} & a_{12} & 0 \\ a_{21} & a_{22} & 1 \\ a_{31} & a_{32} & 0 \end{bmatrix} \begin{bmatrix} 1 \\ 1 \\ 1 \end{bmatrix} = \begin{bmatrix} d \\ d \\ d \end{bmatrix}$$

where c and d are arbitrary non-zero scalars, yielding the six equations

$$\begin{aligned} -a_{11} + a_{12} &= -c \\ -a_{21} + a_{22} + 1 &= c \\ -a_{31} + a_{32} &= c \\ a_{11} + a_{12} &= d \\ a_{21} + a_{22} + 1 &= d \\ a_{31} + a_{32} &= d \end{aligned} \tag{A.8}$$

Subtracting the first equation from the fourth, adding the second and fifth, and adding the third and sixth, one gets

$$a_{11} = \frac{c+d}{2}, \qquad a_{22} = \frac{c+d}{2} - 1, \qquad a_{32} = \frac{c+d}{2}$$

implying that

$$a_{22} = a_{11} - 1 \quad \text{and} \quad a_{32} = a_{11}$$

Likewise, adding the first and fourth equations, subtracting the second from the fifth, and subtracting the third from the sixth, one gets

$$a_{12} = a_{21} = a_{31}$$

We can now write

$$M = \begin{bmatrix} a_{11} & a_{12} & 0 \\ a_{12} & a_{11}-1 & 1 \\ a_{12} & a_{11} & 0 \end{bmatrix}$$

That $h^M(p_3) = p_3'$ and $h_M(p_4) = p_4'$ give another two matrix equations

$$\begin{bmatrix} a_{11} & a_{12} & 0 \\ a_{12} & a_{11}-1 & 1 \\ a_{12} & a_{11} & 0 \end{bmatrix} \begin{bmatrix} 2 \\ 2 \\ 1 \end{bmatrix} = \begin{bmatrix} e \\ 2e \\ e \end{bmatrix} \quad \text{and}$$

$$\begin{bmatrix} a_{11} & a_{12} & 0 \\ a_{12} & a_{11}-1 & 1 \\ a_{12} & a_{11} & 0 \end{bmatrix} \begin{bmatrix} -2 \\ 2 \\ 1 \end{bmatrix} = \begin{bmatrix} -f \\ 2f \\ f \end{bmatrix}$$

where e and f are arbitrary non-zero scalars. Again one obtains six equations, as in (A.8), which can be solved to find that

$$a_{11} = -1/2 \quad \text{and} \quad a_{12} = 0$$

We have, finally, that

$$M = \begin{bmatrix} -1/2 & 0 & 0 \\ 0 & -3/2 & 1 \\ 0 & -1/2 & 0 \end{bmatrix}$$

(or, a non-zero scalar multiple of the matrix on the RHS).

Exercise A.38. The projective transformation h^M of the preceding example mapped, by design, the regular point $[0\ 0\ 1]^T$ to the point at infinity $[0\ 1\ 0]^T$. What other regular points, if any, does it map to a point at infinity?

Exercise A.39. Determine the projective transformation h^M of $\mathbb{P}^2$ that transforms the rectangle q on the plane $z = 1$ with vertices at

$$p_1 = [0.5\ 1]^T,\ p_2 = [0.5\ -1]^T,\ p_3 = [1\ -1]^T \text{ and } p_4 = [1\ 1]^T$$

to the trapezoid q' on $z = 1$ with vertices at

$$p_1' = [-2\ 2]^T,\ p_2' = [-2\ -2]^T,\ p_3' = [-1\ -1]^T \text{ and } p_4' = [-1\ 1]^T$$

(see Example A.13 earlier for the solution).

Exercise A.40. Prove that there exist projective transformations that are neither affine nor snapshot.

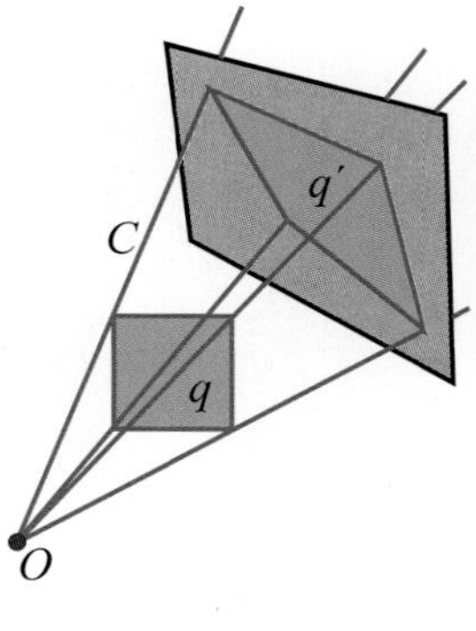

Figure A.26: The square q is mapped to the quadrilateral q' by a snapshot transformation.

Suggested approach: Corollary A.1 implies that a square can be projectively transformed to any non-degenerate quadrilateral. Non-degenerate quadrilaterals q' that can be obtained from a square q by a snapshot transformation are the intersections of a non-radial plane with the "cone" C through q (see Figure A.26). Those that can be obtained by an affine transformation, on the other hand, are parallelograms.

Therefore, if one can find a non-degenerate quadrilateral q'' which is neither a parallelogram nor the intersection of C with a plane, then one shows that there exists a projective transformation neither affine nor snapshot.

APPENDIX **B**

Math Self-Test

This self-test is designed to help you assess your math readiness. The essentials in order to study computer graphics are coordinate geometry, trigonometry, linear algebra and calculus, all at elementary level.

Try to answer all 30 questions. Time is not an issue. And feel free to dust off old math books you may have stashed away in some corner, stroll over to the school or public library, or, even, check into the internet. The principle, of course, is that each and every one of these activities will be allowed throughout your career as a student and practitioner of CG (except, maybe, when you are actually in an exam). Having just the right formula or solution method pop off the top of your head is fantastic, but it's fine as well, given a problem, that you know how to *go about* solving it.

Give yourself 4 points for each correct answer (solutions follow in the next section). If you score at least 100, come on in, the water's fine.* If you're between 80 and 100 then the questions you missed tell where the rust is and, as long as you are willing to put in the extra work, you should be okay. If less than 80 then you need to sit down with yourself and be perfectly honest: is it simply rust that will come off or things that I've just never had in school but trust myself to be able to pick up or is this the kind of stuff that makes me want to curl up into a fetal position?

A word about math and CG, especially to those who did not fare well in the test. If you are motivated to study CG then picking up the math on the way isn't just possible, it can be a lot of fun. Its application to CG will bring to life stuff that caused your eyes to roll in high school. "The middle of the spacecraft is light because of the interpolated color values from the ends of the long triangle" or "This matrix will skew the evil character's head" are a

*There's more math you'll learn while studying CG (some from this book itself) than is covered in the test. Doing well here simply means you're unlikely to have serious problems.

lot different from "Groan, that's 12 different theorems and a chapter-load of trig formulas I have to cram for the mid-term."

If you are interested, there are several books out there dedicated to teaching the math needed for CG. A few that come to mind are Dunn [38], Lengyel [85], Mortenson [96] and Vince [146].

Use the following if you need to (some are approximations): $\sin 30° = \cos 60° = 0.5$, $\sin 45° = \cos 45° = 0.707$, $\sin 60° = \cos 30° = 0.866$, $\pi = 3.141$, $\sqrt{2} = 1.414$, $\sqrt{3} = 1.732$.

The first seven questions refer to Figure B.1.

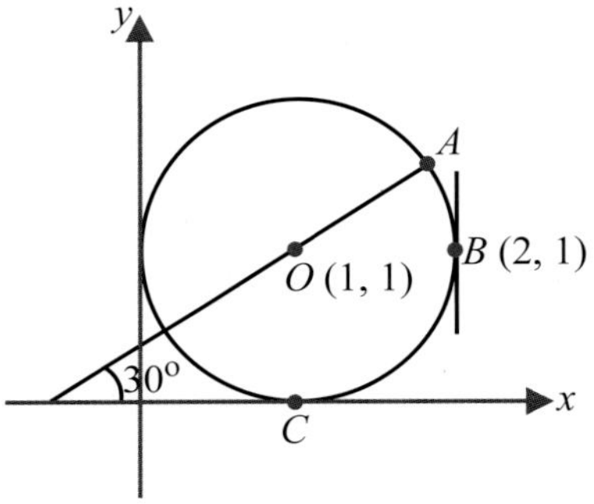

Figure B.1: Circle of unit radius.

1. What is the equation of the circle?

2. What are the coordinates of point C?

3. What is the equation of the tangent to the circle at B?

4. What is the length of the short arc of the circle from A to B?

5. What are the coordinates of point A?

6. If the circle is moved (without turning) so that its center lies at $(-3, -4)$, where then does point B lie?

7. Suppose another circle is drawn with center at A and passing through O. The two circles intersect in two points. What angles do their tangents make at these points?

8. If a straight line on a plane passes through the points $(3, 1)$ and $(5, 2)$, which, if any, of the following two points does it pass through as well: $(9, 4)$ and $(12, 6)$?

9. What are the coordinates of the midpoint of the straight line segment joining the points $(3, 5)$ and $(4, 7)$?

10. At what point do the straight lines $3x + 4y - 6 = 0$ and $4x + 7y - 8 = 0$ intersect?

11. What is the equation of the straight line through the point $(3, 0)$ that is parallel to the straight line $3x - 4y - 6 = 0$?

12. What is the equation of the straight line through the point $(3, 0)$ that is perpendicular to the straight line $3x - 4y - 6 = 0$?

13. What are the coordinates of the point that is the reflection across the line $y = x$ of the point $(3, 1)$?

14. What is the length of the straight line segment on the plane joining the origin $(0, 0)$ to the point $(3, 4)$? In 3-space (xyz-space) what is the length of the straight line segment joining the points $(1, 2, 3)$ and $(4, 6, 8)$?

15. Determine the value of $\sin 75°$ using only the trigonometric values given at the top of the test (in other words, don't use your calculator to do anything other than arithmetic operations).

16. What is the dot product (or, scalar product, same thing) of the two vectors u and v in 3-space, where u starts at the origin and ends at $(\frac{1}{\sqrt{2}}, \frac{1}{\sqrt{2}}, 0)$ and v starts at the origin and ends at $(\frac{1}{\sqrt{2}}, \frac{1}{\sqrt{2}}, \sqrt{3})$, i.e., $u = \frac{1}{\sqrt{2}}\mathbf{i} + \frac{1}{\sqrt{2}}\mathbf{j}$ and $v = \frac{1}{\sqrt{2}}\mathbf{i} + \frac{1}{\sqrt{2}}\mathbf{j} + \sqrt{3}\mathbf{k}$.

    Use the dot product to calculate the angle between u and v.

17. Determine a vector that is perpendicular to *both* the vectors u and v of the preceding question.

18. For the block in Figure B.2, what are the coordinates of the corner point F?

19. For the block again, what is the angle CDE?

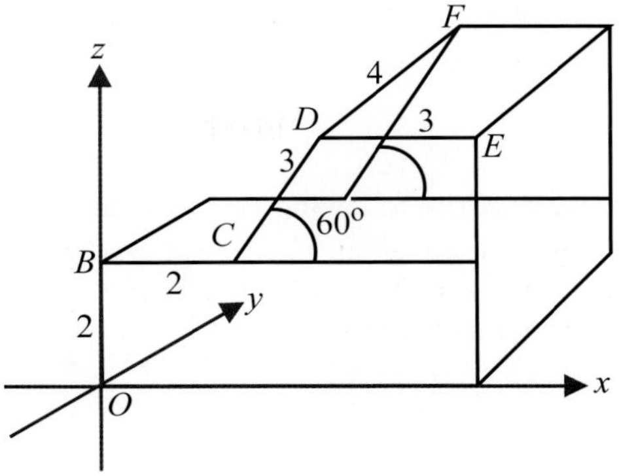

Figure B.2: Solid block (some edges are labeled with their length).

20. For the unit sphere (i.e., of radius 1) centered at the origin, depicted in Figure B.3, the equator (0° latitude) is the great circle cut by the xy-plane, while 0° longitude is that half of the great circle cut by the xz-plane where x-values are non-negative.

    What are the xyz coordinates of the point P whose latitude and longitude are both 45°?

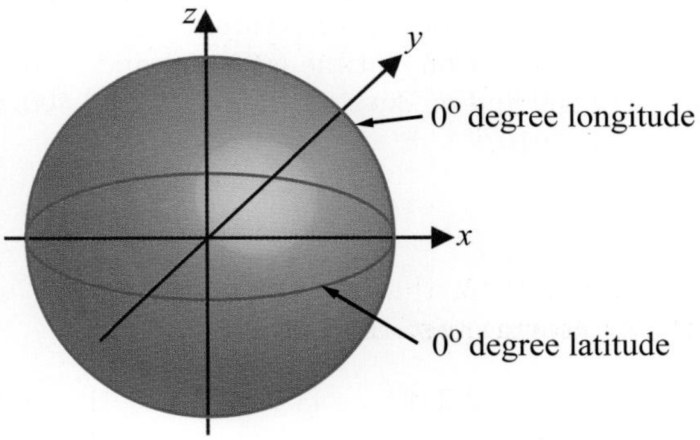

Figure B.3: Unit sphere.

21. Multiply two matrices:

$$\begin{bmatrix} 2 & 4 \\ 3 & 1 \end{bmatrix} \times \begin{bmatrix} 1 & 1 & 0 \\ 2 & 1 & -1 \end{bmatrix}$$

22. Calculate the value of the following two determinants:

$$\begin{vmatrix} 2 & 4 \\ 3 & 1 \end{vmatrix} \quad \text{and} \quad \begin{vmatrix} -1 & 2 & -3 \\ 0 & 5 & -2 \\ 0 & 3 & 3 \end{vmatrix}$$

23. Calculate the inverse of the following matrix:

$$\begin{bmatrix} 4 & 7 \\ 2 & 4 \end{bmatrix}$$

24. If the Dow Jones Industrial Average were a straight-line (or, linear, same thing) function of time (it isn't) and if its value on January 1, 2007 is 12,000 and on January 1, 2009 it's 13,500, what is the value on January 1, 2010?

25. Are the following vectors linearly independent?

$$[2\ 3\ 0]^T \qquad [3\ 7\ -1]^T \qquad [1\ -6\ 3]^T$$

26. Determine the linear transformation of $\mathbb{R}^3$ that maps the standard basis vectors

$$[1\ 0\ 0]^T \qquad [0\ 1\ 0]^T \qquad [0\ 0\ 1]^T$$

to the respective vectors

$$[-1\ -1\ 1]^T \qquad [-2\ 3\ 2]^T \qquad [-3\ 1\ -2]^T$$

27. What is the equation of the normal to the parabola

$$y = 2x^2 + 3$$

at the point $(2, 11)$?

28. If x and y are related by the equation

$$xy + x + y = 1$$

find a formula for $\frac{\mathrm{d}y}{\mathrm{d}x}$.

29. The formula for the height at time t of a projectile shot vertically upward from the ground with initial velocity u is

$$h = ut - \frac{1}{2}gt^2$$

assuming only the action of gravitational acceleration g (ignoring wind resistance and other factors).

What is the velocity of the projectile at time t? What is the maximum height reached by the projectile?

30. At what points do the curves $y = \sin x$ and $y = \cos x$ meet for values of x between 0 and 2π? What angles do they make at these points?

Appendix C

Math Self-Test Solutions

Score 4 points for each correct answer. For an assessment of your total, read the part before the start of the test.

Use the following if you need to (some are approximations): $\sin 30° = \cos 60° = 0.5$, $\sin 45° = \cos 45° = 0.707$, $\sin 60° = \cos 30° = 0.866$, $\pi = 3.141$, $\sqrt{2} = 1.414$, $\sqrt{3} = 1.732$.

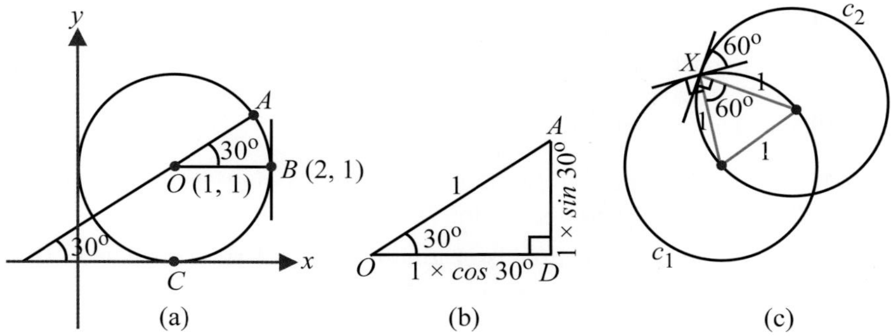

Figure C.1: (a) Circle of unit radius (b) Right-angled triangle (c) Two circles of unit radius.

The first seven questions refer to Figure C.1(a).

1. What is the equation of the circle?

   Answer: Generally, the equation of a circle on the xy-plane centered at (a, b) and of radius r is

$$(x - a)^2 + (y - b)^2 = r^2$$

So the equation of the drawn circle is

$$(x-1)^2+(y-1)^2 = 1^2 \quad \text{which evaluates to} \quad x^2+y^2-2x-2y+1=0$$

2. What are the coordinates of point C?

   Answer: $(1, 0)$.

3. What is the equation of the tangent to the circle at B?

   Answer: $x = 2$.

4. What is the length of the short arc of the circle from A to B?

   Answer: The angle subtended at the center by this arc is $\angle AOB = 30°$ (Figure C.1(a)).

   Therefore, the length of the arc is $\frac{30}{360}$ of the circumference $= \frac{30}{360} \times 2\pi \times 1 = \frac{\pi}{6} = 3.141/6 = 0.5235$.

5. What are the coordinates of point A?

   Answer: Suppose the horizontal line through O and the vertical line through A intersect at D (Figure C.1(b)). The hypotenuse OA of the right-angled triangle ODA is of length 1 (= radius of the circle). Therefore, the length of $OD = 1 \times \cos 30° = 0.866$ and the length of $AD = 1 \times \sin 30° = 0.5$.

   These lengths are the displacements of A in the x- and y-direction, respectively, from O. Therefore, the coordinates of A are $(1, 1) + (0.866, 0.5) = (1.866, 1.5)$.

6. If the circle is moved (without turning) so that its center lies at $(-3, -4)$, where then does point B lie?

   Answer: Since the center is originally at $(1, 1)$ the translation that moves it to $(-3, -4)$ consists of a displacement of -4 in the x-direction and -5 in the y direction. The same translation applies to B, so B moves $(2, 1) + (-4, -5) = (-2, -4)$.

7. Suppose another circle is drawn with center at A and passing through O. The two circles intersect in two points. What angles do their tangents make at these points?

   Answer: The two circles c_1 and c_2 intersect at points X and Y (Figure C.1(c)). The triangle OAX is equilateral as all its sides are of length 1, the radius of either circle. All its angles, therefore, are $60°$.

   Now, the tangents to c_1 and c_2 at X are perpendicular, respectively, to OX and AX. Therefore, the angle between them is the same as that between OX and AX, which is $60°$.

Symmetrically, the tangents to the two circles at Y intersect at $60°$ as well.

8. If a straight line on a plane passes through the points $(3, 1)$ and $(5, 2)$, which, if any, of the following two points does it pass through as well: $(9, 4)$ and $(12, 6)$?

   Answer: Suppose the equation of the straight line is $y = mx + c$ (any non-vertical straight line has an equation of this form, called the slope-intercept form). Since it passes through $(3, 1)$ and $(5, 2)$, we have the two equations

$$\begin{aligned} 1 &= 3m + c \\ 2 &= 5m + c \end{aligned}$$

   Solving simultaneously, we get $m = \frac{1}{2}$ and $c = -\frac{1}{2}$, yielding the equation $y = \frac{1}{2}x - \frac{1}{2}$ for the line. Of the two points $(9, 4)$ and $(12, 6)$, only $(9, 4)$ satisfies this equation and so lies on it.

9. What are the coordinates of the midpoint of the straight line segment joining the points $(3, 5)$ and $(4, 7)$?

   Answer: The coordinates of the midpoint are $(\frac{3+4}{2}, \frac{5+7}{2}) = (3.5, 6)$.

10. At what point do the straight lines $3x + 4y - 6 = 0$ and $4x + 7y - 8 = 0$ intersect?

    Answer: Simultaneously solving the two equations we get the intersection as $(2, 0)$.

11. What is the equation of the straight line through the point $(3, 0)$ that is parallel to the straight line $3x - 4y - 6 = 0$?

    Answer: Any line parallel to $3x - 4y - 6 = 0$ may be written as $3x - 4y - c = 0$, where c can be an arbitrary number. If such a line passes through $(3, 0)$ then this point's coordinates must satisfy $3x - 4y - c = 0$.

    In other words, $3 \times 3 - 4 \times 0 - c = 0 \implies c = 9$.

    Therefore, the required equation is $3x - 4y - 9 = 0$.

12. What is the equation of the straight line through the point $(3, 0)$ that is perpendicular to the straight line $3x - 4y - 6 = 0$?

    Answer: Rewrite the given straight line's equation in slope-intercept form: $y = \frac{3}{4}x - \frac{3}{2}$. Its gradient, therefore, is $\frac{3}{4}$. Now the gradient of a straight line that is perpendicular to one of gradient m is $-\frac{1}{m}$.

    Therefore, the gradient of the straight line perpendicular to the given one is $-\frac{4}{3}$ and its equation is of the form $y = -\frac{4}{3}x + c$. Since it passes through $(3, 0)$ we have $0 = -\frac{4}{3} \times 3 + c \implies c = 4$.

Therefore, the required equation is $y = -\frac{4}{3}x + 4$ or $3y + 4x - 12 = 0$.

13. What are the coordinates of the point that is the reflection across the line $y = x$ of the point $(3, 1)$?

    Answer: Reflecting a point across the line $y = x$ interchanges its x and y coordinates. So the reflection of $(3, 1)$ is $(1, 3)$.

14. What is the length of the straight line segment on the plane joining the origin $(0, 0)$ to the point $(3, 4)$? In 3-space (xyz-space) what is the length of the straight line segment joining the points $(1, 2, 3)$ and $(4, 6, 8)$?

    Answer: The (Euclidean) distance between two points (x_1, y_1) and (x_2, y_2) on the plane is $\sqrt{(x_2 - x_1)^2 + (y_2 - y_1)^2}$, while that between two points (x_1, y_1, z_1) and (x_2, y_2, z_2) in 3-space is (similarly) $\sqrt{(x_2 - x_1)^2 + (y_2 - y_1)^2 + (z_2 - z_1)^2}$.

    Therefore, the length of the straight line segment joining $(0, 0)$ and $(3, 4)$ is $\sqrt{(3-0)^2 + (4-0)^2} = \sqrt{25} = 5$ and that joining $(1, 2, 3)$ and $(4, 6, 8)$ is

$$\begin{aligned} & \sqrt{(4-1)^2 + (6-2)^2 + (8-3)^2} \\ = \; & \sqrt{50} = \sqrt{25 \times 2} = \sqrt{25} \times \sqrt{2} = 5 \times 1.414 = 7.07 \end{aligned}$$

    using only the value of $\sqrt{2}$ given above.

15. Determine the value of $\sin 75°$ using only the trigonometric values given at the top of the test (in other words, don't use your calculator to do anything other than arithmetic operations).

    Answer: Use the formula

$$\sin(A + B) = \sin A \cos B + \cos A \sin B$$

    to write

$$\begin{aligned} \sin 75° &= \sin(45° + 30°) \\ &= \sin 45° \cos 30° + \cos 45° \sin 30° \\ &= 0.707 \times 0.866 + 0.707 \times 0.5 \\ &= 0.966 \end{aligned}$$

16. What is the dot product (or, scalar product, same thing) of the two vectors u and v in 3-space, where u starts at the origin and ends at $(\frac{1}{\sqrt{2}}, \frac{1}{\sqrt{2}}, 0)$ and v starts at the origin and ends at $(\frac{1}{\sqrt{2}}, \frac{1}{\sqrt{2}}, \sqrt{3})$, i.e., $u = \frac{1}{\sqrt{2}}\mathbf{i} + \frac{1}{\sqrt{2}}\mathbf{j}$ and $v = \frac{1}{\sqrt{2}}\mathbf{i} + \frac{1}{\sqrt{2}}\mathbf{j} + \sqrt{3}\mathbf{k}$.

    Use the dot product to calculate the angle between u and v.

Answer:

$u \cdot v = (\frac{1}{\sqrt{2}}, \frac{1}{\sqrt{2}}, 0) \cdot (\frac{1}{\sqrt{2}}, \frac{1}{\sqrt{2}}, \sqrt{3}) = \frac{1}{\sqrt{2}} \times \frac{1}{\sqrt{2}} + \frac{1}{\sqrt{2}} \times \frac{1}{\sqrt{2}} + 0 \times \frac{1}{\sqrt{3}} = 1$

Moreover,

$|u| = \sqrt{(\frac{1}{\sqrt{2}})^2 + (\frac{1}{\sqrt{2}})^2 + 0^2} = 1$ and $|v| = \sqrt{(\frac{1}{\sqrt{2}})^2 + (\frac{1}{\sqrt{2}})^2 + (\sqrt{3})^2} = 2$.

Now, $u \cdot v = |u||v| \cos\theta$, where θ is the angle between u and v. Therefore, $\cos\theta = \frac{u \cdot v}{|u||v|} = \frac{1}{1 \times 2} = \frac{1}{2}$, which means $\theta = 60°$.

17. Determine a vector that is perpendicular to *both* the vectors u and v of the preceding question.

Answer: The cross-product of two (non-zero and non-collinear) vectors is perpendicular to both of them. Now, $u \times v = (\frac{1}{\sqrt{2}}\mathbf{i} + \frac{1}{\sqrt{2}}\mathbf{j}) \times (\frac{1}{\sqrt{2}}\mathbf{i} + \frac{1}{\sqrt{2}}\mathbf{j} + \sqrt{3}\mathbf{k}) =$

$$\begin{vmatrix} \mathbf{i} & \frac{1}{\sqrt{2}} & \frac{1}{\sqrt{2}} \\ \mathbf{j} & \frac{1}{\sqrt{2}} & \frac{1}{\sqrt{2}} \\ \mathbf{k} & 0 & \sqrt{3} \end{vmatrix} = \frac{\sqrt{3}}{\sqrt{2}}\mathbf{i} - \frac{\sqrt{3}}{\sqrt{2}}\mathbf{j}$$

Therefore, the vector that starts at the origin and ends at $(\frac{\sqrt{3}}{\sqrt{2}}, -\frac{\sqrt{3}}{\sqrt{2}}, 0)$ is perpendicular to both u and v (the answer is not unique).

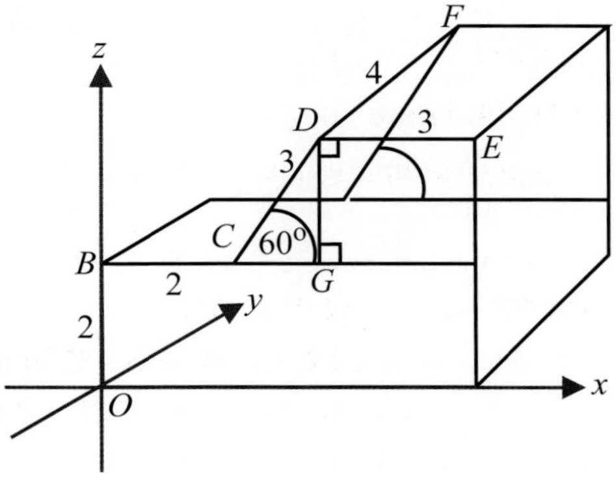

Figure C.2: Solid block (some edges are labeled with their length).

18. For the block in Figure C.2, what are the coordinates of the corner point F?

Answer: Drop the perpendicular DG from D to the straight line through B and C (Figure C.2).

The x-coordinate of F is $|BC|+|CG| = 2+3\cos 60° = 2+3\times 0.5 = 3.5$.

The y-coordinate of F is $|DF| = 4$.

The z-coordinate of F is $|AB|+|GD| = 2+3\sin 60° = 2+3\times 0.866 = 4.598$.

Therefore, $F = (3.5, 4, 4.598)$.

19. For the block again, what is the angle CDE?

Answer: $\angle CDE = \angle CDG + \angle GDE = 30° + 90° = 120°$.

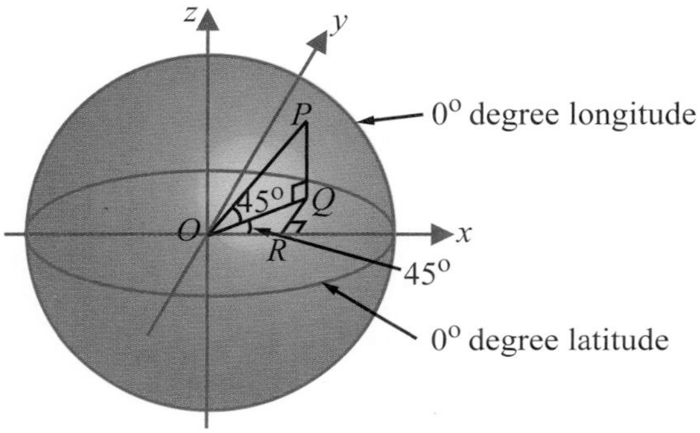

Figure C.3: Unit sphere.

20. For the unit sphere (i.e., of radius 1) centered at the origin, depicted in Figure C.3, the equator (0° latitude) is the great circle cut by the xy-plane, while 0° longitude is that half of the great circle cut by the xz-plane where x-values are non-negative.

What are the xyz coordinates of the point P whose latitude and longitude are both 45°?

Answer: Drop the perpendicular from P to Q on the xy-plane and then the perpendicular from Q to R on the x-axis (Figure C.3).

Now, $|OP| = 1$, so that $|PQ| = 1 \times \sin 45° = \frac{1}{\sqrt{2}}$ and $|OQ| = 1 \times \cos 45° = \frac{1}{\sqrt{2}}$.

Moreover, $QR = |OQ| \sin 45° = \frac{1}{\sqrt{2}} \times \frac{1}{\sqrt{2}} = \frac{1}{2}$ and $OR = |OQ| \cos 45° = \frac{1}{\sqrt{2}} \times \frac{1}{\sqrt{2}} = \frac{1}{2}$.

Now, the x, y and z coordinates of P are $|OR|$, $|QR|$ and $|PQ|$, respectively. Therefore, $P = (\frac{1}{2}, \frac{1}{2}, \frac{1}{\sqrt{2}})$.

21. Multiply two matrices:

$$\begin{bmatrix} 2 & 4 \\ 3 & 1 \end{bmatrix} \times \begin{bmatrix} 1 & 1 & 0 \\ 2 & 1 & -1 \end{bmatrix}$$

Answer:

$$\begin{bmatrix} 2 & 4 \\ 3 & 1 \end{bmatrix} \times \begin{bmatrix} 1 & 1 & 0 \\ 2 & 1 & -1 \end{bmatrix} = \begin{bmatrix} 10 & 6 & -4 \\ 5 & 4 & -1 \end{bmatrix}.$$

22. Calculate the value of the following two determinants:

$$\begin{vmatrix} 2 & 4 \\ 3 & 1 \end{vmatrix} \quad \text{and} \quad \begin{vmatrix} -1 & 2 & -3 \\ 0 & 5 & -2 \\ 0 & 3 & 3 \end{vmatrix}$$

Answer:

$$\begin{vmatrix} 2 & 4 \\ 3 & 1 \end{vmatrix} = 2 \times 1 - 4 \times 3 = -10$$

$$\begin{aligned} \begin{vmatrix} -1 & 2 & -3 \\ 0 & 5 & -2 \\ 0 & 3 & 3 \end{vmatrix} &= -1 \times \begin{vmatrix} 5 & -2 \\ 3 & 3 \end{vmatrix} - 0 \times \begin{vmatrix} 2 & -3 \\ 3 & 3 \end{vmatrix} + 0 \times \begin{vmatrix} 2 & -3 \\ 5 & -2 \end{vmatrix} \\ &= -(5 \times 3 - (-2) \times 3) = -21 \end{aligned}$$

23. Calculate the inverse of the following matrix:

$$\begin{bmatrix} 4 & 7 \\ 2 & 4 \end{bmatrix}$$

Answer: To obtain the inverse we have to replace each element by its cofactor, take the transpose and, finally, divide by the determinant of the original matrix.

Replacing each element by its cofactor we get the matrix

$$\begin{bmatrix} 4 & -2 \\ -7 & 4 \end{bmatrix}$$

Taking the transpose next gives

$$\begin{bmatrix} 4 & -7 \\ -2 & 4 \end{bmatrix}$$

Finally, dividing by the determinant $4 \times 4 - 2 \times 7 = 2$ of the original matrix, we have its inverse

$$\begin{bmatrix} 2 & -3.5 \\ -1 & 2 \end{bmatrix}$$

24. If the Dow Jones Industrial Average were a straight-line (or, linear, same thing) function of time (it isn't) and if its value on January 1, 2007 is 12,000 and on January 1, 2009 it's 13,500, what is the value on January 1, 2010?

    Answer: As a linear function of time then the DJIA grows 1500 points in two years, or 750 per year, which takes it to 14,250 on January 1, 2010.

25. Are the following vectors linearly independent?

$$[2\ 3\ 0]^T \qquad [3\ 7\ -1]^T \qquad [1\ -6\ 3]^T$$

    Answer: The vectors are linearly independent if the only solution to the equation

$$c_1[2\ 3\ 0]^T + c_2[3\ 7\ -1]^T + c_3[1\ -6\ 3]^T = [0\ 0\ 0]^T$$

    is $c_1 = c_2 = c_3 = 0$.

    The vector equation above is equivalent to the following set of three simultaneous equations – one from each position in the vectors – in c_1, c_2 and c_3:

$$\begin{aligned} 2c_1 + 3c_2 + c_3 &= 0 \\ 3c_1 + 7c_2 - 6c_3 &= 0 \\ -c_2 + 3c_3 &= 0 \end{aligned}$$

    Solving we find solutions not all 0, e.g., $c_1 = 5$, $c_2 = -3$ and $c_3 = -1$, proving that the given set of vectors is not linearly independent.

26. Determine the linear transformation of $\mathbb{R}^3$ that maps the standard basis vectors

$$[1\ 0\ 0]^T \qquad [0\ 1\ 0]^T \qquad [0\ 0\ 1]^T$$

    to the respective vectors

$$[-1\ -1\ 1]^T \qquad [-2\ 3\ 2]^T \qquad [-3\ 1\ -2]^T$$

    Answer: The required linear transformation is defined by the matrix whose columns are, respectively, the images of the successive basis vectors. In particular, then, its matrix is

$$\begin{bmatrix} -1 & -2 & -3 \\ -1 & 3 & 1 \\ 1 & 2 & -2 \end{bmatrix}$$

27. What is the equation of the normal to the parabola

$$y = 2x^2 + 3$$

at the point $(2, 11)$?

Answer: $\frac{\mathrm{d}y}{\mathrm{d}x} = 4x$, so at the point $(2, 11)$ the gradient of the tangent is $4 \times 2 = 8$. The gradient of the normal, therefore, is $-\frac{1}{8}$. Its equation, accordingly, is

$$\frac{y - 11}{x - 2} = -\frac{1}{8} \quad \text{or} \quad x + 8y - 90 = 0$$

28. If x and y are related by the equation

$$xy + x + y = 1$$

find a formula for $\frac{\mathrm{d}y}{\mathrm{d}x}$.

Answer: Differentiating the equation throughout with respect to x:

$$(\frac{\mathrm{d}}{\mathrm{d}x}x)y + x(\frac{\mathrm{d}}{\mathrm{d}x}y) + \frac{\mathrm{d}}{\mathrm{d}x}x + \frac{\mathrm{d}}{\mathrm{d}x}y = \frac{\mathrm{d}}{\mathrm{d}x}1$$

which simplifies to

$$y + x\frac{\mathrm{d}y}{\mathrm{d}x} + 1 + \frac{\mathrm{d}y}{\mathrm{d}x} = 0$$

giving

$$\frac{\mathrm{d}y}{\mathrm{d}x} = -\frac{y + 1}{x + 1}$$

29. The formula for the height at time t of a projectile shot vertically upward from the ground with initial velocity u is

$$h = ut - \frac{1}{2}gt^2$$

assuming only the action of gravitational acceleration g (ignoring wind resistance and other factors).

What is the velocity of the projectile at time t? What is the maximum height reached by the projectile?

Answer: Its velocity at time t is

$$\frac{\mathrm{d}h}{\mathrm{d}t} = u - gt$$

When the projectile attains maximum height, its velocity is 0. Therefore, $u - gt = 0$, implying that $t = \frac{u}{g}$. Therefore, the maximum height reached by the projectile is obtained by substituting $t = \frac{u}{g}$ in the formula for its height, which gives the maximum height as

$$u\frac{u}{g} - \frac{1}{2}g\left(\frac{u}{g}\right)^2 = \frac{u^2}{g} - \frac{u^2}{2g} = \frac{u^2}{2g}$$

30. At what points do the curves $y = \sin x$ and $y = \cos x$ meet for values of x between 0 and 2π? What angles do they make at these points?

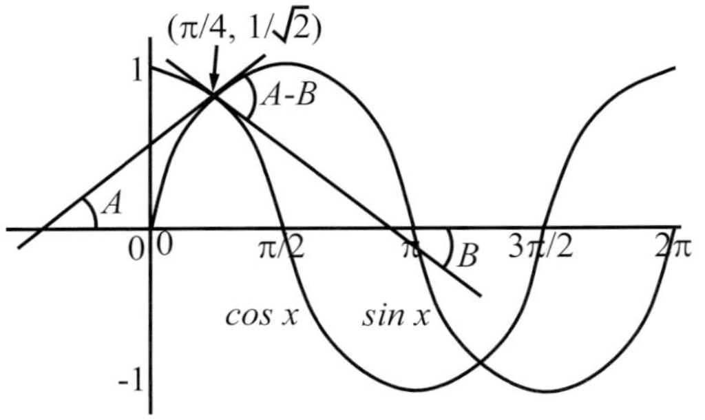

Figure C.4: Graphs of $\sin x$ and $\cos x$ (*sketch, not exact*).

Answer: When the curves (see Figure C.4) meet, their y-values are equal, so $\sin x = \cos x$, giving $\tan x = \frac{\sin x}{\cos x} = 1$. The two values in the interval $[0, 2\pi]$ where $\tan x = 1$ are $\frac{\pi}{4}$ and $\frac{5\pi}{4}$. Therefore, the two points at which the curves meet are $(\frac{\pi}{4}, \frac{1}{\sqrt{2}})$ and $(\frac{5\pi}{4}, -\frac{1}{\sqrt{2}})$ ($\cos\frac{\pi}{4} = \sin\frac{\pi}{4} = \frac{1}{\sqrt{2}}$ and $\cos\frac{5\pi}{4} = \sin\frac{5\pi}{4} = -\frac{1}{\sqrt{2}}$).

For the first curve $\frac{dy}{dx} = \cos x$, while for the second $\frac{dy}{dx} = -\sin x$. At the point of intersection $(\frac{\pi}{4}, \frac{1}{\sqrt{2}})$, therefore, the gradients of the two curves are $\cos\frac{\pi}{4} = \frac{1}{\sqrt{2}}$ and $-\sin\frac{\pi}{4} = -\frac{1}{\sqrt{2}}$, respectively.

If the tangent lines at $(\frac{\pi}{4}, \frac{1}{\sqrt{2}})$ make angles A and B, respectively, with the x-axis, then the gradients are precisely the tan of these angles. Therefore, $\tan A = \frac{1}{\sqrt{2}}$ and $\tan B = -\frac{1}{\sqrt{2}}$ (B is a negative angle). The angle between the curves – which by definition is the angle between their tangents – at $(\frac{\pi}{4}, \frac{1}{\sqrt{2}})$ is $A - B$. Now,

$$\tan(A-B) = \frac{\tan A - \tan B}{1 + \tan A \tan B} = \frac{\frac{1}{\sqrt{2}} + \frac{1}{\sqrt{2}}}{1 - \frac{1}{\sqrt{2}}\frac{1}{\sqrt{2}}} = 2\sqrt{2}$$

which means $A - B = \tan^{-1} 2\sqrt{2} = \tan^{-1} 2.828 = 1.231$ radians or $70.526°$ (approximately). The angle between the curves at the other point of intersection is the same by symmetry.

Bibliography

[1] T. Akenine-Möller, E. Haines, N. Hoffman, *Real-Time Rendering*, 3rd Edn., A K Peters, 2008.

[2] E. Angel, *Interactive Computer Graphics: A Top-Down Approach with Shader-Based OpenGL*, 6th Edn., Addison-Wesley, 2011.

[3] E. Angel, *OpenGL; A Primer*, 3rd Edn., Addison-Wesley, 2007.

[4] A. Appel, Some techniques for the machine rendering of solids, *Proceedings of the Spring Joint Computer Conference*, 1968, 37-45.

[5] F. Ayres, E. Mendelson, *Schaum's Outlines: Calculus*, 6th Edn., McGraw-Hill, 2012.

[6] R. Baer, *Linear Algebra and Projective Geometry*, Kindle Edn., Dover Publications, 2012.

[7] M. Bailey, S. Cunningham, *Graphics Shaders: Theory and Practice*, 2nd Edn., A K Peters, 2012.

[8] T. Banchoff, J. Wermer, *Linear Algebra through Geometry*, 2nd Edn., Springer-Verlag, 1993.

[9] M. F. Barnsley, *Fractals Everywhere*, Kindle Edn., Dover Publications, 2013.

[10] R. H. Bartels, J. C. Beatty, B. A. Barsky, *An Introduction to Splines for Use in Computer Graphics and Geometric Modeling*, Morgan Kaufmann, 1987.

[11] M. de Berg, O. Cheong, M. van Kreveld, M. Overmars, *Computational Geometry: Algorithms and Applications*, 3rd Edn., Springer-Verlag, 2010.

[12] R. S. Berns, *Billmeyer and Saltzman's Principles of Color Technology*, 3rd Edn., Wiley-Interscience, 2000.

[13] P. E. Bézier, How Renault uses numerical control for car body design and tooling, *Society of Automotive Engineers' Congress*, SAE paper 680010, Detroit, 1968.

[14] P. E. Bézier, Mathematical and practical possibilities of UNISURF, in *Computer Aided Geometric Design: Proceedings of a Conference Held at the University of Utah*, R. Barnhill and R. Riesenfeld, editors, Academic Press, 1974, 127-152.

[15] J. F. Blinn, Models of light reflection for computer synthesized pictures, *Computer Graphics (Proceedings SIGGRAPH 1977)* **11** (1977), 192-198.

[16] J. F. Blinn, Simulation of wrinkled surfaces, *Computer Graphics (Proceedings SIGGRAPH 1978)* **12** (1978), 286-292.

[17] J. F. Blinn, *Jim Blinn's Corner: A Trip Down the Graphics Pipeline*, Morgan Kaufmann, 1996.

[18] J. F. Blinn, *Jim Blinn's Corner: Dirty Pixels*, Morgan Kaufmann, 1998.

[19] J. Blinn, M. Newell, Texture and reflection in computer generated images, *Communications of the ACM*, **19** (1976), 456-547.

[20] D. Bourg, B. Bywalec, *Physics for Game Developers*, 2nd Edn., O'Reilly, 2013.

[21] J. E. Bresenham, Algorithm for computer control of digital plotter, *IBM Systems Journal* **4** (1965), 25-30.

[22] S. R. Buss, *3-D Computer Graphics: A Mathematical Introduction with OpenGL*, Cambridge University Press, 2003.

[23] E. Catmull, R. Rom, A class of local interpolating splines, in *Computer Aided Geometric Design: Proceedings of a Conference Held at the University of Utah*, R. Barnhill and R. Riesenfeld, editors, Academic Press, 1974, 317-326.

[24] C. Chuon, S. Guha, Volume Cost Based Mesh Simplification, *Proceedings 6th International Conference on Computer Graphics, Imaging and Visualization (CGIV 09)* (2009), 164-169.

[25] M. F. Cohen, D. P. Greenberg, The hemi-cube: a radiosity solution for complex environments, *Computer Graphics (Proceedings SIGGRAPH 1985)* **19** (1985), 31-40.

[26] M. F. Cohen, J. R. Wallace, *Radiosity and Realistic Image Synthesis*, Morgan Kaufmann, 1993.

[27] Comparison of OpenGL and Direct3D (in Wikipedia), `http://en.wikipedia.org/wiki/Direct3D_vs._OpenGL`.

[28] Computational Geometry Algorithms Library, `http://www.cgal.org`.

[29] R. L. Cook, K. E. Torrance, A reflectance model for computer graphics, *ACM Transaction on Graphics* **1** (1982), 7-24.

[30] M. G. Cox, The numerical evaluation of B-splines, *Journal of the Institute of Mathematics and Its Applications* **10** (1972), 134-149.

[31] H. S. M. Coxeter, *Projective Geometry*, Springer, 2nd Edn., 2013.

[32] F. C. Crow, The origins of the teapot, *IEEE Computer Graphics and Applications* **7** (1987), 8-19.

[33] F. C. Crow, Shadow algorithms for computer graphics, *Computer Graphics (Proceedings SIGGRAPH 1977)* **11** (1977), 242-248.

[34] C. de Boor, On calculating with B-splines, *Journal of Approximation Theory* **6** (1972), 50-62.

[35] P. de Casteljau, Outillages méthodes calcul, Technical report, A. Citroen, Paris, 1959.

[36] P. de Casteljau, Courbes et surfaces à poles, Technical report, A. Citroen, Paris, 1963.

[37] M. P. Do Carmo, *Differential Geometry of Curves and Surfaces*, Prentice Hall, 1976.

[38] F. Dunn, I. Parberry, *3D Math Primer for Graphics and Game Development*, 2nd Edn., A K Peters, 2012.

[39] D. H. Eberly, *3D Game Engine Design: A Practical Approach to Real-Time Computer Graphics*, 2nd Edn., Morgan Kaufmann, 2006.

[40] D. H. Eberly, *Game Physics*, 2nd Edn., Morgan Kaufmann, 2010.

[41] H. Edelsbrunner, *Geometry and Topology for Mesh Generation*, Cambridge University Press, 2006.

[42] H. Edelsbrunner, J. L. Harer, *Computational Topology*, American Mathematical Society, 2009.

[43] C. Ericson, *Real-Time Collision Detection*, Morgan Kaufmann, 2005.

[44] K. Falconer, *Fractal Geometry: Mathematical Foundations and Applications*, 3rd Edn., John Wiley & Sons, 2013.

[45] G. Farin, *Curves and Surfaces for CAGD: A Practical Guide*, 5th Edn., Morgan Kaufmann, 2001.

[46] G. Farin, *NURBS: from Projective Geometry to Practical Use*, 2nd Edn., A K Peters, 1999.

[47] J. D. Foley, A. van Dam, S. K. Feiner, J. F. Hughes, *Computer Graphics: Principles and Practice*, 2nd Edn., Addison-Wesley, 1995.

[48] J. D. Foley, A. van Dam, S. K. Feiner, J. F. Hughes, R. L. Phillips, *Introduction to Computer Graphics*, Addison-Wesley, 1993.

[49] FreeGLUT, `http://freeglut.sourceforge.net`.

[50] S. H. Friedberg, A. J. Insel, L. E. Spence, *Linear Algebra*, 4th Edn., Prentice Hall, 2002.

[51] H. Fuchs, Z. M. Kedem, B. F. Naylor, On visible surface generation by a priori tree structures, *Computer Graphics (Proceedings SIGGRAPH 1980)* **14** (1980), 124-133.

[52] D. C. Giancoli, *Physics for Scientists and Engineers with Modern Physics*, 4th Edn., Addison-Wesley, 2008.

[53] GIMP, `http://www.gimp.org`.

[54] A. S. Glassner, *Andrew Glassner's Notebook: Recreational Computer Graphics*, Morgan Kaufmann, 1999.

[55] A. S. Glassner, *Andrew Glassner's Other Notebook: Further Recreations in Computer Graphics*, A K Peters, 2002.

[56] A. S. Glassner, *An Introduction to Ray Tracing*, Academic Press, 1989.

[57] GLUI, `http://www.cs.unc.edu/~rademach/glui`.

[58] C. M. Goral, K. E. Torrance, D. P. Greenberg, B. Battaile, Modeling the interaction of light between diffuse surfaces, *Computer Graphics (Proceedings SIGGRAPH 1984)* **18** (1984), 213-222.

[59] H. Gouraud, Continuous shading of curved surfaces, *IEEE Transactions on Computers* **20** (1971), 623-629.

[60] S. Govil-Pai, *Principles of Computer Graphics: Theory and Practice Using OpenGL and Maya*, Springer, 2005.

[61] W. H. Greub, *Linear Algebra*, 4th Edn., Springer, 1981.

[62] S. Guha, Joint separation of geometric clusters and the extreme irregularities of regular polyhedra, *International Journal of Computational Geometry and Applications* **15** (2005), 491-510.

[63] A. J. Hanson, *Visualizing Quaternions*, Morgan Kaufmann, 2006.

[64] P. Haeberli, K. Akeley, The accumulation buffer: hardware support for high-quality rendering, *Computer Graphics (Proceedings 17th Annual Conference on Computer Graphics and Interactive Techniques)* **24** (1990), 309-318.

[65] P. Haeberli, M. Segal, Texture mapping as a fundamental drawing primitive, *Proceedings Fourth Eurographics Workshop on Rendering* (1993), 259-266 (on-line version at `http://www.sgi.com/misc/grafica/texmap`).

[66] L. A. Hageman, D. M. Young, *Applied Iterative Methods*, Kindle Edn., Dover Publications, 2012.

[67] X. D. He, K. E. Torrance, F. X. Sillion, D. P. Greenberg, A comprehensive physical model for light reflection, *Computer Graphics (Proceedings SIGGRAPH 1991)* **25** (1991), 175-186.

[68] X. D. He, P. O. Heynen, R. L. Phillips, K. E. Torrance, D. H. Salesin, D. P. Greenberg, A fast and accurate light reflection model, *Computer Graphics (Proceedings SIGGRAPH 1992)* **26** (1992), 253-254.

[69] D. Hearn, M. P. Baker, *Computer Graphics with OpenGL*, 4th Edn., Prentice Hall, 2010.

[70] P. S. Heckbert, Survey of texture mapping, *IEEE Computer Graphics and Applications* **6** (1986), 56-67.

[71] M. Henle, *Modern Geometries: Non-Euclidean, Projective, and Discrete Geometry*, 2nd Edn., Prentice Hall, 2001.

[72] F. S. Hill, Jr., S. M. Kelley, *Computer Graphics Using OpenGL*, 3rd Edn., Prentice Hall, 2006.

[73] K. M. Hoffman, R. Kunze, *Linear Algebra*, 2nd Edn., Prentice Hall, 1971.

[74] International Meshing Roundtable, `http://www.imr.sandia.gov`.

[75] R. Jackson, L. MacDonald, K. Freeman, *Computer Generated Colour: A Practical Guide to Presentation and Display*, John Wiley & Sons, 1994.

[76] G. A. Jennings, *Modern Geometry with Applications*, Springer, 1994.

[77] L. Kadison, M. T. Kromann, *Projective Geometry and Modern Algebra*, Birkhäuser, 1996.

[78] M. J. Kilgard, Improving shadows and reflections via the stencil buffer, `https://developer.nvidia.com/sites/default/files/akamai/gamedev/docs/stencil.pdf`.

[79] Khronos Group OpenGL ES, `http://www.opengles.org`.

[80] B. Kolman, D. R. Hill, *Introductory Linear Algebra: An Applied First Course*, 8th Edn., Prentice Hall, 2004.

[81] E. Kreyszig, *Differential Geometry*, Kindle Edn., Dover Publications, 2013.

[82] J. B. Kuipers, *Quaternions and Rotation Sequences: A Primer with Applications to Orbits, Aerospace and Virtual Reality*, Princeton University Press, 2002.

[83] A. Kumar, V. Kwatra, B. Singh, S. Kapoor, Dynamic Binary Space Partitioning for Hidden Surface Removal, *Proceedings Indian Conference on Computer Vision, Graphics and Image Processing (ICVGIP 1998)*, 1998.

[84] D. C. Lay, *Linear Algebra and Its Applications*, 4th Edn., Addison-Wesley, 2011.

[85] E. Lengyel, *Mathematics for 3D Game Programming & Computer Graphics*, 3rd Edn., Cengage Learning, 2011.

[86] Y. Liang, B. Barsky, A new concept and method for line clipping, *ACM Transactions on Graphics* **3** (1984), 1-22.

[87] Lighthouse 3D, `http://www.lighthouse3d.com/opengl`.

[88] M. M. Lipschutz, *Schaum's Outlines: Differential Geometry*, McGraw-Hill, 1969.

[89] D. Luebke, M. Reddy, J. D. Cohen, A. Varshney, B. Watson, R. Huebner, *Level of Detail for 3D Graphics*, Morgan Kaufmann, 2002.

[90] F. D. Luna, *Introduction to 3D Game Programming with DirectX 11*, Kindle Edn., Mercury Learning and Information, 2012.

[91] B. Mandelbrot, *The Fractal Geometry of Nature*, Kindle Edn., W. H. Freeman, 2010.

[92] T. McReynolds, D. Blythe, *Advanced Graphics Programming Using OpenGL*, Morgan Kaufmann, 2005.

[93] Mesa 3D, `http://www.mesa3d.org`.

[94] M. E. Mortenson, *Geometric Modeling*, 3rd Edn., Industrial Press, 2006.

[95] M. E. Mortenson, *Geometric Transformations for 3D Modeling*, 2nd Edn., Industrial Press, 2007.

[96] M. E. Mortenson, *Mathematics for Computer Graphics Applications*, 2nd Edn., Industrial Press, 1999.

[97] T. K. Mukherjee, *private communication*, 2008.

[98] J. Munkres, *Elements of Algebraic Topology*, Westview Press, 1996.

[99] J. Munkres, *Topology*, 2nd Edn., Prentice Hall, 2000.

[100] Nate Robins, `http://www.xmission.com/~nate/tutors.html`.

[101] S. K. Nayar, M. Oren, Generalization of the Lambertian model and implications for machine vision, *International Journal of Computer Vision* **14** (1995), 227-251.

[102] NeHe Productions, `http://nehe.gamedev.net`.

[103] OpenGL, `http://www.opengl.org`.

[104] OpenGL Architecture Review Board, *OpenGL Programming Guide*, 8th Edn., Addison-Wesley, 2013.

[105] OpenGL Architecture Review Board, *OpenGL Reference Manual*, 4th Edn., Addison-Wesley, 2004.

[106] B. O'Neill, *Elementary Differential Geometry*, 2nd Edn., Academic Press, 2006.

[107] J. O'Rourke, *Computational Geometry in C*, 2nd Edn., Cambridge University Press, 2013.

[108] D. Pedoe, *Geometry: A Comprehensive Course*, Kindle Edn., Dover Publications, 2013.

[109] B. T. Phong, Illumination for computer generated pictures, *Communications of the ACM* **18** (1975), 311-317.

[110] Physics in Graphics, `http://physicsingraphics.endofinternet.org`.

[111] L. Piegl, W. Tiller, *The NURBS Book*, 2nd Edn., Springer, 1996.

[112] P. Poulin, A. Fournier, A model for anisotropic reflection, *Computer Graphics (Proceedings SIGGRAPH 1990)* **24** (1990), 273-282.

[113] POV-Ray, `http://www.povray.org`.

[114] W. H. Press, S. A. Teukolsky, W. T. Vetterling, B. P. Flannery, *Numerical Recipes: The Art of Scientific Computing*, 3rd Edn., Cambridge University Press, 2007.

[115] A. Pressley, *Elementary Differential Geometry*, 2nd Edn., Springer, 2010.

[116] D. F. Rogers, *An Introduction to NURBS: With Historical Perspective*, Morgan Kaufmann, 2000.

[117] D. F. Rogers, *Procedural Elements for Computer Graphics*, 2nd Edn., McGraw-Hill, 1997.

[118] D. F. Rogers, J. A. Adams, *Mathematical Elements for Computer Graphics*, 2nd Edn., McGraw-Hill, 1989.

[119] S. Roman, *Advanced Linear Algebra*, 3rd Edn., Springer, 2007.

[120] R. J. Rost, B. Licea-Kane, *OpenGL Shading Language*, 3rd Edn., Addison-Wesley, 2009.

[121] H. Samet, *Foundations of Multidimensional and Metric Data Structures*, Morgan Kaufmann, 2006.

[122] P. Samuel, *Projective Geometry*, Springer-Verlag, 1988.

[123] H. M. Schey, *Div, Grad, Curl, and All That: An Informal Text on Vector Calculus*, 4th Edn., W. W. Norton & Co., 2004.

[124] H. Schildt, *STL Programming from the Ground Up*, Osborne/McGraw-Hill, 1998.

[125] C. Schlick, An inexpensive BRDF model for physically-based rendering, *Computer Graphics Forum* **13** (1994), 233-246.

[126] I. Schoenberg, Contributions to the problem of approximation of equidistant data by analytic functions, *Quarterly of Applied Mathematics* **4** (1946), 45-99.

[127] I. Schoenberg, On spline functions, in *Inequalities*, O. Sisha, editor, Academic Press, 1967, 249-274.

[128] M. Segal, K. Akeley, The design of the OpenGL graphics interface, 1994 (on-line version available from `http://citeseerx.ist.psu.edu/viewdoc/summary?doi=10.1.1.126.8268`).

[129] G. Sellers, R. S. Wright, Jr., N. Haemel, *OpenGL Superbible*, 6th Edn., Addison-Wesley, 2013.

[130] P. Shirley, S. Marschner, *Fundamentals of Computer Graphics*, 3rd Edn., A K Peters, 2009.

[131] ACM SIGGRAPH, `http://www.siggraph.org`.

[132] F. X. Sillion, C. Puech, *Radiosity and Global Illumination*, Morgan Kaufmann, 1994.

[133] G. F. Simmons, *Calculus With Analytic Geometry*, 2nd Edn., McGraw-Hill, 1996.

[134] I. M. Singer, J. A. Thorpe, *Lecture Notes on Elementary Topology and Geometry*, Springer-Verlag, 1976.

[135] M. Slater, A. Steed, Y. Chrysanthou, *Computer Graphics and Virtual Environments: From Realism to Real-Time*, Addison-Wesley, 2001.

[136] M. R. Spiegel, *Schaum's Outlines: Vector Analysis and an Introduction to Tensor Analysis*, McGraw-Hill, 1968.

[137] J. Stewart, *Calculus*, 7th Edn., Cengage Learning, 2012.

[138] G. Strang, *Introduction to Linear Algebra*, 4th Edn., Wellesley-Cambridge Press, 2009.

[139] I. E. Sutherland, *Sketchpad: A Man-Machine Graphical Communication System*, MIT Thesis, 1963.

[140] I. E. Sutherland, G. W. Hodgeman, Reentrant polygon clipping, *Communications of the ACM* **17** (1974), 32-42.

[141] A. Thorn, *DirectX 9 Graphics: The Definitive Guide to Direct3D*, Jones & Bartlett Publishers, 2005.

[142] Trolltech, `http://www.trolltech.com`.

[143] UNC GAMMA Research Group, `http://www.cs.unc.edu/~geom`.

[144] G. van den Bergen, *Collision Detection in Interactive 3D Environments*, CRC Press, 2003.

[145] J. M. van Verth, L. M. Bishop, *Essential Mathematics for Games and Interactive Applications: A Programmer's Guide*, 2nd Edn., Morgan Kaufmann, 2008.

[146] J. Vince, *Mathematics for Computer Graphics*, 4th Edn., Springer, 2013.

[147] A. Watt, *3D Computer Graphics*, 3rd Edn., Addison-Wesley, 1999.

[148] J. T. Whitted, An improved illumination model for shaded display, *Communications of the ACM* **23** (1980), 343-349.

[149] L. Williams, Casting curved shadows on curved surfaces, *Computer Graphics (Proceedings SIGGRAPH 1978)* **12** (1978), 270-274.

[150] L. Williams, Pyramidal parametrics, *Computer Graphics (Proceedings SIGGRAPH 1983)* **17** (1983), 1-11.

[151] D, Wolff, *OpenGL Shading Language Cookbook*, Packt Publishing, 2011.

[152] R. C. Wrede, M. Spiegel, *Schaum's Outlines: Advanced Calculus*, 3rd Edn., McGraw-Hill, 2010.

[153] G. Wyszecki, W. S. Stiles, *Color Science: Concepts and Methods, Quantitative Data and Formulae*, 2nd Edn., Wiley-Interscience, 2000.

[154] Z. Xiang, R. Plastock, *Schaum's Outlines: Computer Graphics*, 2nd Edn., McGraw-Hill, 2000.

[155] I. M. Yaglom, *Geometric Transformations I*, Mathematical Association of America, 1962.

[156] I. M. Yaglom, *Geometric Transformations II*, Mathematical Association of America, 1968.

[157] I. M. Yaglom, *Geometric Transformations III*, Mathematical Association of America, 1973.

Subject Index

D

S

X

Y

Z

Program Index